RECIPROCALS OF BASIC FUNCTIONS

18. $\displaystyle\int \frac{1}{1 \pm \sin u}\, du = \tan u \mp \sec u + C$

19. $\displaystyle\int \frac{1}{1 \pm \cos u}\, du = -\cot u \pm \csc u + C$

20. $\displaystyle\int \frac{1}{1 \pm \tan u}\, du = \tfrac{1}{2}(u \pm \ln|\cos u \pm \sin u|) + C$

21. $\displaystyle\int \frac{1}{\sin u \cos u}\, du = \ln|\tan u| + C$

22. $\displaystyle\int \frac{1}{1 \pm \cot u}\, du = \tfrac{1}{2}(u \mp \ln|\sin u \pm \cos u|) + C$

23. $\displaystyle\int \frac{1}{1 \pm \sec u}\, du = u + \cot u \mp \csc u + C$

24. $\displaystyle\int \frac{1}{1 \pm \csc u}\, du = u - \tan u \pm \sec u + C$

25. $\displaystyle\int \frac{1}{1 \pm e^u}\, du = u - \ln(1 \pm e^u) + C$

POWERS OF TRIGONOMETRIC FUNCTIONS

26. $\displaystyle\int \sin^2 u\, du = \tfrac{1}{2}u - \tfrac{1}{4}\sin 2u + C$

27. $\displaystyle\int \cos^2 u\, du = \tfrac{1}{2}u + \tfrac{1}{4}\sin 2u + C$

28. $\displaystyle\int \tan^2 u\, du = \tan u - u + C$

29. $\displaystyle\int \sin^n u\, du = -\frac{1}{n}\sin^{n-1} u \cos u + \frac{n-1}{n}\int \sin^{n-2} u\, du$

30. $\displaystyle\int \cos^n u\, du = \frac{1}{n}\cos^{n-1} u \sin u + \frac{n-1}{n}\int \cos^{n-2} u\, du$

31. $\displaystyle\int \tan^n u\, du = \frac{1}{n-1}\tan^{n-1} u - \int \tan^{n-2} u\, du$

32. $\displaystyle\int \cot^2 u\, du = -\cot u - u + C$

33. $\displaystyle\int \sec^2 u\, du = \tan u + C$

34. $\displaystyle\int \csc^2 u\, du = -\cot u + C$

35. $\displaystyle\int \cot^n u\, du = -\frac{1}{n-1}\cot^{n-1} u - \int \cot^{n-2} u\, du$

36. $\displaystyle\int \sec^n u\, du = \frac{1}{n-1}\sec^{n-2} u \tan u + \frac{n-2}{n-1}\int \sec^{n-2} u\, du$

37. $\displaystyle\int \csc^n u\, du = -\frac{1}{n-1}\csc^{n-2} u \cot u + \frac{n-2}{n-1}\int \csc^{n-2} u\, du$

PRODUCTS OF TRIGONOMETRIC FUNCTIONS

38. $\displaystyle\int \sin mu \sin nu\, du = -\frac{\sin(m+n)u}{2(m+n)} + \frac{\sin(m-n)u}{2(m-n)} + C$

39. $\displaystyle\int \cos mu \cos nu\, du = \frac{\sin(m+n)u}{2(m+n)} + \frac{\sin(m-n)u}{2(m-n)} + C$

40. $\displaystyle\int \sin mu \cos nu\, du = -\frac{\cos(m+n)u}{2(m+n)} - \frac{\cos(m-n)u}{2(m-n)} + C$

41. $\displaystyle\int \sin^m u \cos^n u\, du = -\frac{\sin^{m-1} u \cos^{n+1} u}{m+n} + \frac{m-1}{m+n}\int \sin^{m-2} u \cos^n u\, du$

$\displaystyle\qquad = \frac{\sin^{m+1} u \cos^{n-1} u}{m+n} + \frac{n-1}{m+n}\int \sin^m u \cos^{n-2} u\, du$

PRODUCTS OF TRIGONOMETRIC AND EXPONENTIAL FUNCTIONS

42. $\displaystyle\int e^{au} \sin bu\, du = \frac{e^{au}}{a^2 + b^2}(a \sin bu - b \cos bu) + C$

43. $\displaystyle\int e^{au} \cos bu\, du = \frac{e^{au}}{a^2 + b^2}(a \cos bu + b \sin bu) + C$

POWERS OF *u* MULTIPLYING OR DIVIDING BASIC FUNCTIONS

44. $\displaystyle\int u \sin u\, du = \sin u - u \cos u + C$

45. $\displaystyle\int u \cos u\, du = \cos u + u \sin u + C$

46. $\displaystyle\int u^2 \sin u\, du = 2u \sin u + (2 - u^2) \cos u + C$

47. $\displaystyle\int u^2 \cos u\, du = 2u \cos u + (u^2 - 2) \sin u + C$

48. $\displaystyle\int u^n \sin u\, du = -u^n \cos u + n \int u^{n-1} \cos u\, du$

49. $\displaystyle\int u^n \cos u\, du = u^n \sin u - n \int u^{n-1} \sin u\, du$

50. $\displaystyle\int u^n \ln u\, du = \frac{u^{n+1}}{(n+1)^2}[(n+1)\ln u - 1] + C$

51. $\displaystyle\int u e^u\, du = e^u(u - 1) + C$

52. $\displaystyle\int u^n e^u\, du = u^n e^u - n \int u^{n-1} e^u\, du$

53. $\displaystyle\int u^n a^u\, du = \frac{u^n a^u}{\ln a} - \frac{n}{\ln a}\int u^{n-1} a^u\, du + C$

54. $\displaystyle\int \frac{e^u\, du}{u^n} = -\frac{e^u}{(n-1)u^{n-1}} + \frac{1}{n-1}\int \frac{e^u\, du}{u^{n-1}}$

55. $\displaystyle\int \frac{a^u\, du}{u^n} = -\frac{a^u}{(n-1)u^{n-1}} + \frac{\ln a}{n-1}\int \frac{a^u\, du}{u^{n-1}}$

56. $\displaystyle\int \frac{du}{u \ln u} = \ln|\ln u| + C$

POLYNOMIALS MULTIPLYING BASIC FUNCTIONS

57. $\displaystyle\int p(u) e^{au}\, du = \frac{1}{a}p(u)e^{au} - \frac{1}{a^2}p'(u)e^{au} + \frac{1}{a^3}p''(u)e^{au} - \cdots$ [signs alternate: $+ - + - \cdots$]

58. $\displaystyle\int p(u) \sin au\, du = -\frac{1}{a}p(u)\cos au + \frac{1}{a^2}p'(u)\sin au + \frac{1}{a^3}p''(u)\cos au - \cdots$ [signs alternate in pairs after first term: $+ + - - + + - - \cdots$]

59. $\displaystyle\int p(u) \cos au\, du = \frac{1}{a}p(u)\sin au + \frac{1}{a^2}p'(u)\cos au - \frac{1}{a^3}p''(u)\sin au - \cdots$ [signs alternate in pairs: $+ + - - + + - - \cdots$]

Calculus provides a way of viewing and analyzing the physical world. As with all mathematics courses, calculus involves equations and formulas. However, if you successfully learn to use all the formulas and solve all of the problems in the text but do not master the underlying *ideas*, you will have missed the most important part of calculus. If you master these ideas, you will have a widely applicable tool that goes far beyond textbook exercises.

Before starting your studies, you may find it helpful to leaf through this text to get a general feeling for its different parts:

- The opening page of each chapter gives you an overview of what that chapter is about, and the opening page of each section within a chapter gives you an overview of what that section is about. To help you locate specific information, sections are subdivided into topics that are marked with a box like this ■.

- Each section ends with a set of exercises. The answers to most odd-numbered exercises appear in the back of the book. If you find that your answer to an exercise does not match that in the back of the book, do not assume immediately that yours is incorrect—there may be more than one way to express the answer. For example, if your answer is $\sqrt{2}/2$ and the text answer is $1/\sqrt{2}$, then both are correct since your answer can be obtained by "rationalizing" the text answer. In general, if your answer does not match that in the text, then your best first step is to look for an algebraic manipulation or a trigonometric identity that might help you determine if the two answers are equivalent. If the answer is in the form of a decimal approximation, then your answer might differ from that in the text because of a difference in the number of decimal places used in the computations.

- The section exercises include regular exercises and four special categories: *Quick Check*, *Focus on Concepts*, *True/False*, and *Writing*.

  - The *Quick Check* exercises are intended to give you quick feedback on whether you understand the key ideas in the section; they involve relatively little computation, and have answers provided at the end of the exercise set.

  - The *Focus on Concepts* exercises, as their name suggests, key in on the main ideas in the section.

  - *True/False* exercises focus on key ideas in a different way. You must decide whether the statement is true in *all possible circumstances*, in which case you would declare it to be "true," or whether there are some circumstances in which it is not true, in which case you would declare it to be "false." In each such exercise you are asked to "Explain your answer." You might do this by noting a

theorem in the text that shows the statement to be true or by finding a particular example in which the statement is not true.

  - *Writing* exercises are intended to test your ability to explain mathematical ideas in words rather than relying solely on numbers and symbols. All exercises requiring writing should be answered in complete, correctly punctuated logical sentences—not with fragmented phrases and formulas.

- Each chapter ends with two additional sets of exercises: *Chapter Review Exercises*, which, as the name suggests, is a select set of exercises that provide a review of the main concepts and techniques in the chapter, and *Making Connections*, in which exercises require you to draw on and combine various ideas developed throughout the chapter.

- Your instructor may choose to incorporate technology in your calculus course. Exercises whose solution involves the use of some kind of technology are tagged with icons to alert you and your instructor. Those exercises tagged with the icon ∼ require graphing technology—either a graphing calculator or a computer program that can graph equations. Those exercises tagged with the icon C require a computer algebra system (CAS) such as *Mathematica*, *Maple*, or available on some graphing calculators.

- At the end of the text you will find a set of four appendices covering various topics such as a detailed review of trigonometry and graphing techniques using technology. Inside the front and back covers of the text you will find endpapers that contain useful formulas.

- The ideas in this text were created by real people with interesting personalities and backgrounds. Pictures and biographical sketches of many of these people appear throughout the book.

- Notes in the margin are intended to clarify or comment on important points in the text.

A Word of Encouragement

As you work your way through this text you will find some ideas that you understand immediately, some that you don't understand until you have read them several times, and others that you do not seem to understand, even after several readings. Do not become discouraged—some ideas are intrinsically difficult and take time to "percolate." You may well find that a hard idea becomes clear later when you least expect it.

Wiley Web Site for this Text

www.wiley.com/college/anton

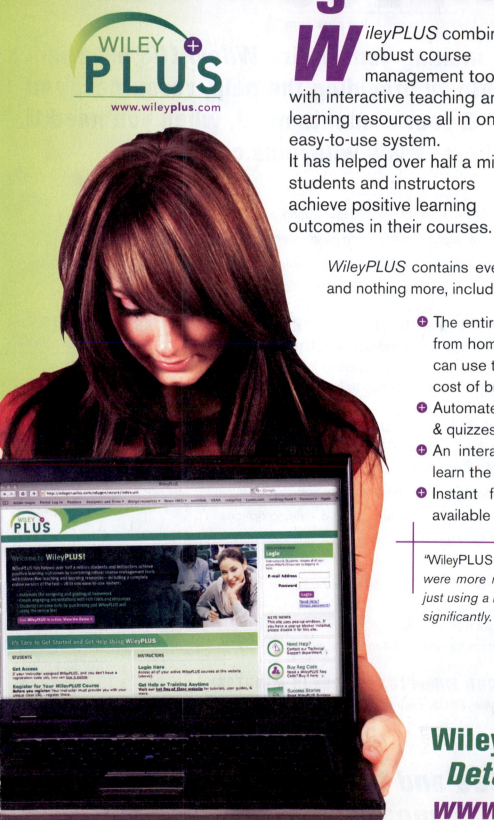
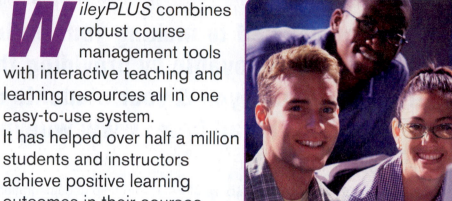

WILEY PLUS

www.wileyplus.com

Wiley is committed to making your entire *WileyPLUS* experience productive & enjoyable by providing the help, resources, and personal support you & your students need, when you need it. It's all here: www.wileyplus.com –

TECHNICAL SUPPORT:

- A fully searchable knowledge base of FAQs and help documentation, available 24/7
- Live chat with a trained member of our support staff during business hours
- A form to fill out and submit online to ask any question and get a quick response
- **Instructor-only** phone line during business hours: 1.877.586.0192

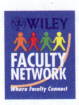

FACULTY-LED TRAINING THROUGH THE WILEY FACULTY NETWORK:
Register online: www.wherefacultyconnect.com

Connect with your colleagues in a complimentary virtual seminar, with a personal mentor in your field, or at a live workshop to share best practices for teaching with technology.

1ST DAY OF CLASS...AND BEYOND!
Resources You & Your Students Need to Get Started & Use *WileyPLUS* from the first day forward.

- 2-Minute Tutorials on how to set up & maintain your *WileyPLUS* course
- User guides, links to technical support & training options
- ***WileyPLUS for Dummies***: Instructors' quick reference guide to using *WileyPLUS*
- Student tutorials & instruction on how to register, buy, and use *WileyPLUS*

YOUR *WileyPLUS* ACCOUNT MANAGER:

Your personal *WileyPLUS* connection for any assistance you need!

SET UP YOUR *WileyPLUS* COURSE IN MINUTES!

Selected *WileyPLUS* courses with QuickStart contain pre-loaded assignments & presentations created by subject matter experts who are also experienced *WileyPLUS* users.

Interested? See and try WileyPLUS *in action!*
Details and Demo: *www.wileyplus.com*

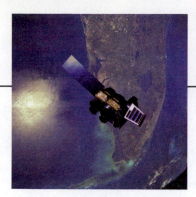

9th
EDITION

CALCULUS

EARLY TRANSCENDENTALS
SINGLE VARIABLE

- **HOWARD ANTON** *Drexel University*
- **IRL BIVENS** *Davidson College*
- **STEPHEN DAVIS** *Davidson College*

with contributions by

Thomas Polaski *Winthrop University*

JOHN WILEY & SONS, INC.

Publisher: Laurie Rosatone
Acquisitions Editor: David Dietz
Freelance Developmental Editor: Anne Scanlan-Rohrer
Marketing Manager: Jaclyn Elkins
Associate Editor: Michael Shroff/Will Art
Editorial Assistant: Pamela Lashbrook
Full Service Production Management: Carol Sawyer/The Perfect Proof
Senior Production Editor: Ken Santor
Senior Designer: Madelyn Lesure
Associate Photo Editor: Sheena Goldstein
Freelance Illustration: Karen Heyt
Cover Photo: © Eric Simonsen/Getty Images

This book was set in LaTeX by Techsetters, Inc., and printed and bound by R.R. Donnelley/Jefferson City. The cover was printed by R.R. Donnelley.

This book is printed on acid-free paper.

The paper in this book was manufactured by a mill whose forest management programs include sustained yield harvesting of its timberlands. Sustained yield harvesting principles ensure that the numbers of trees cut each year does not exceed the amount of new growth.

ISBN 978-0470-18204-8

Printed in the United States of America

10 9 8 7 6 5 4 3 2 1

About HOWARD ANTON

Howard Anton wrote the original version of this text and was the author of the first six editions. He obtained his B.A. from Lehigh University, his M.A. from the University of Illinois, and his Ph.D. from the Polytechnic University of Brooklyn, all in mathematics. In the early 1960s he worked for Burroughs Corporation and Avco Corporation at Cape Canaveral, Florida, where he was involved with the manned space program. In 1968 he joined the Mathematics Department at Drexel University, where he taught full time until 1983. Since that time he has been an adjunct professor at Drexel and has devoted the majority of his time to textbook writing and activities for mathematical associations. Dr. Anton was president of the EPADEL Section of the Mathematical Association of America (MAA), served on the Board of Governors of that organization, and guided the creation of the Student Chapters of the MAA. He has published numerous research papers in functional analysis, approximation theory, and topology, as well as pedagogical papers. He is best known for his textbooks in mathematics, which are among the most widely used in the world. There are currently more than one hundred versions of his books, including translations into Spanish, Arabic, Portuguese, Italian, Indonesian, French, Japanese, Chinese, Hebrew, and German. For relaxation, Dr. Anton enjoys traveling and photography.

About IRL BIVENS

Irl C. Bivens, recipient of the George Polya Award and the Merten M. Hasse Prize for Expository Writing in Mathematics, received his A.B. from Pfeiffer College and his Ph.D. from the University of North Carolina at Chapel Hill, both in mathematics. Since 1982, he has taught at Davidson College, where he currently holds the position of professor of mathematics. A typical academic year sees him teaching courses in calculus, topology, and geometry. Dr. Bivens also enjoys mathematical history, and his annual History of Mathematics seminar is a perennial favorite with Davidson mathematics majors. He has published numerous articles on undergraduate mathematics, as well as research papers in his specialty, differential geometry. He has served on the editorial boards of the MAA Problem Book series and *The College Mathematics Journal* and is a reviewer for *Mathematical Reviews*. When he is not pursuing mathematics, Professor Bivens enjoys juggling, swimming, walking, and spending time with his son Robert.

About STEPHEN DAVIS

Stephen L. Davis received his B.A. from Lindenwood College and his Ph.D. from Rutgers University in mathematics. Having previously taught at Rutgers University and Ohio State University, Dr. Davis came to Davidson College in 1981, where he is currently a professor of mathematics. He regularly teaches calculus, linear algebra, abstract algebra, and computer science. A sabbatical in 1995–1996 took him to Swarthmore College as a visiting associate professor. Professor Davis has published numerous articles on calculus reform and testing, as well as research papers on finite group theory, his specialty. Professor Davis has held several offices in the Southeastern section of the MAA, including chair and secretary-treasurer. He is currently a faculty consultant for the Educational Testing Service Advanced Placement Calculus Test, a board member of the North Carolina Association of Advanced Placement Mathematics Teachers, and is actively involved in nurturing mathematically talented high school students through leadership in the Charlotte Mathematics Club. He was formerly North Carolina state director for the MAA. For relaxation, he plays basketball, juggles, and travels. Professor Davis and his wife Elisabeth have three children, Laura, Anne, and James, all former calculus students.

About THOMAS POLASKI, contributor to the ninth edition

Thomas W. Polaski received his B.S. from Furman University and his Ph.D. in mathematics from Duke University. He is currently a professor at Winthrop University, where he has taught since 1991. He was named Outstanding Junior Professor at Winthrop in 1996. He has published articles on mathematics pedagogy and stochastic processes and has authored a chapter in a forthcoming linear algebra textbook. Professor Polaski is a frequent presenter at mathematics meetings, giving talks on topics ranging from mathematical biology to mathematical models for baseball. He has been an MAA Visiting Lecturer and is a reviewer for *Mathematical Reviews*. Professor Polaski has been a reader for the Advanced Placement Calculus Tests for many years. In addition to calculus, he enjoys travel and hiking. Professor Polaski and his wife, LeDayne, have a daughter, Kate, and live in Charlotte, North Carolina.

To
my wife Pat and my children: Brian, David, and Lauren

In Memory of
my mother Shirley
my father Benjamin
my thesis advisor and inspiration, George Bachman
my benefactor in my time of need, Stephen Girard (1750–1831)
—HA

To
my son Robert
—IB

To
my wife Elisabeth
my children: Laura, Anne, and James
—SD

PREFACE

This ninth edition of *Calculus* maintains those aspects of previous editions that have led to the series' success—we continue to strive for student comprehension without sacrificing mathematical accuracy, and the exercise sets are carefully constructed to avoid unhappy surprises that can derail a calculus class. However, this edition also has many new features that we hope will attract new users and also motivate past users to take a fresh look at our work. We had two main goals for this edition:

- To make those adjustments to the order and content that would align the text more precisely with the most widely followed calculus outlines.
- To add new elements to the text that would provide a wider range of teaching and learning tools.

All of the changes were carefully reviewed by an advisory committee of outstanding teachers comprised of both users and nonusers of the previous edition. The charge of this committee was to ensure that all changes did not alter those aspects of the text that attracted users of the eighth edition and at the same time provide freshness to the new edition that would attract new users. Some of the more substantive changes are described below.

NEW FEATURES IN THIS EDITION

New Elements in the Exercises   We added new true/false exercises, new writing exercises, and new exercise types that were requested by reviewers of the eighth edition.

Making Connections   We added this new element to the end of each chapter. A Making Connections exercise synthesizes concepts drawn across multiple sections of its chapter rather than using ideas from a single section as is expected of a regular or review exercise.

Reorganization of Review Material   The precalculus review material that was in Chapter 1 of the eighth edition forms Chapter 0 of the ninth edition. The body of material in Chapter 1 of the eighth edition that is not generally regarded as precalculus review was moved to appropriate sections of the text in this edition. Thus, Chapter 0 focuses exclusively on those preliminary topics that students need to start the calculus course.

Parametric Equations Reorganized   In the eighth edition, parametric equations were introduced in the first chapter and picked up again later in the text. Many instructors asked that we return to the traditional organization, and we have done so; the material on parametric equations is now first introduced and then discussed in detail in Section 10.1 (*Parametric Curves*). However, to support those instructors who want to continue the eighth edition path of giving an early exposure to parametric curves, we have provided Web materials (Web Appendix I) as well as self-contained exercise sets on the topic in Section 6.4 (*Length of a Plane Curve*) and Section 6.5 (*Area of a Surface of Revolution*).

Also, Section 14.4 (*Surface Area; Parametric Surfaces*) has been reorganized so surfaces of the form $z = f(x, y)$ are discussed before surfaces defined parametrically.

Differential Equations Reorganized   We reordered and revised the chapter on differential equations so that instructors who cover only separable equations can do so without a forced diversion into general first-order equations and other unrelated topics. This chapter can be skipped entirely by those who do not cover differential equations at all in calculus.

New 2D Discussion of Centroids and Center of Gravity   In the eighth edition and earlier, centroids and center of gravity were covered only in three dimensions. In this edition we added a new section on that topic in Chapter 6 (*Applications of the Definite Integral*), so centroids and center of gravity can now be studied in two dimensions, as is common in many calculus courses.

Related Rates and Local Linearity Reorganized   The sections on related rates and local linearity were moved to follow the sections on implicit differentiation and logarithmic, exponential, and inverse trigonometric functions, thereby making a richer variety of techniques and functions available to study related rates and local linearity.

Rectilinear Motion Reorganized   The more technical aspects of rectilinear motion that were discussed in the introductory discussion of derivatives in the eighth edition have been deferred so as not to distract from the primary task of developing the notion of the derivative. This also provides a less fragmented development of rectilinear motion.

Other Reorganization   The section *Graphing Functions Using Calculators and Computer Algebra Systems*, which appeared in the text body of the eighth edition, is now a text appendix (Appendix A), and the sections *Mathematical Models* and *Second-Order Linear Homogeneous Differential Equations* are now posted on the Web site that supports the text.

OTHER FEATURES

Flexibility   This edition has a built-in flexibility that is designed to serve a broad spectrum of calculus philosophies—from traditional to "reform." Technology can be emphasized or not, and the order of many topics can be permuted freely to accommodate each instructor's specific needs.

Rigor   The challenge of writing a good calculus book is to strike the right balance between rigor and clarity. Our goal is to present precise mathematics to the fullest extent possible in an introductory treatment. Where clarity and rigor conflict, we choose clarity; however, we believe it to be important that the student understand the difference between a careful proof and an informal argument, so we have informed the reader when the arguments being presented are informal or motivational. Theory involving ϵ-δ arguments appears in a separate section so that it can be covered or not, as preferred by the instructor.

Rule of Four   The "rule of four" refers to presenting concepts from the verbal, algebraic, visual, and numerical points of view. In keeping with current pedagogical philosophy, we used this approach whenever appropriate.

Visualization   This edition makes extensive use of modern computer graphics to clarify concepts and to develop the student's ability to visualize mathematical objects, particularly those in 3-space. For those students who are working with graphing technology, there are many exercises that are designed to develop the student's ability to generate and analyze mathematical curves and surfaces.

Quick Check Exercises   Each exercise set begins with approximately five exercises (answers included) that are designed to provide students with an immediate assessment of whether they have mastered key ideas from the section. They require a minimum of computation and are answered by filling in the blanks.

Focus on Concepts Exercises   Each exercise set contains a clearly identified group of problems that focus on the main ideas of the section.

Technology Exercises   Most sections include exercises that are designed to be solved using either a graphing calculator or a computer algebra system such as *Mathematica*, *Maple*, or the open source program *Sage*. These exercises are marked with an icon for easy identification.

Applicability of Calculus   One of the primary goals of this text is to link calculus to the real world and the student's own experience. This theme is carried through in the examples and exercises.

Career Preparation   This text is written at a mathematical level that will prepare students for a wide variety of careers that require a sound mathematics background, including engineering, the various sciences, and business.

Trigonometry Review   Deficiencies in trigonometry plague many students, so we have included a substantial trigonometry review in Appendix B.

Appendix on Polynomial Equations   Because many calculus students are weak in solving polynomial equations, we have included an appendix (Appendix C) that reviews the Factor Theorem, the Remainder Theorem, and procedures for finding rational roots.

Principles of Integral Evaluation   The traditional Techniques of Integration is entitled "Principles of Integral Evaluation" to reflect its more modern approach to the material. The chapter emphasizes general methods and the role of technology rather than specific tricks for evaluating complicated or obscure integrals.

Historical Notes   The biographies and historical notes have been a hallmark of this text from its first edition and have been maintained. All of the biographical materials have been distilled from standard sources with the goal of capturing and bringing to life for the student the personalities of history's greatest mathematicians.

Margin Notes and Warnings   These appear in the margins throughout the text to clarify or expand on the text exposition or to alert the reader to some pitfall.

SUPPLEMENTS

SUPPLEMENTS FOR THE STUDENT

Print Supplements

The Student Solutions Manual (978-0470-37958-5) provides students with detailed solutions to odd-numbered exercises from the text. The structure of solutions in the manual matches those of worked examples in the textbook.

Student Companion Site

The Student Companion Site provides access to the following student supplements:

- Web Quizzes, which are short, fill-in-the-blank quizzes that are arranged by chapter and section.
- Additional textbook content, including answers to odd-numbered exercises and appendices.

WileyPLUS

WileyPLUS, Wiley's digital-learning environment, is loaded with all of the supplements above, and also features the following:

- The E-book, which is an exact version of the print text, but also features hyperlinks to questions, definitions, and supplements for quicker and easier support.
- The Student Study Guide provides concise summaries for quick review, checklists, common mistakes/pitfalls, and sample tests for each section and chapter of the text.
- The Graphing Calculator Manual helps students to get the most out of their graphing calculator and shows how they can apply the numerical and graphing functions of their calculators to their study of calculus.
- Guided Online (GO) Exercises prompt students to build solutions step by step. Rather than simply grading an exercise answer as wrong, GO problems show students precisely where they are making a mistake.
- Are You Ready? quizzes gauge student mastery of chapter concepts and techniques and provide feedback on areas that require further attention.
- Algebra and Trigonometry Refresher quizzes provide students with an opportunity to brush up on material necessary to master calculus, as well as to determine areas that require further review.

SUPPLEMENTS FOR THE INSTRUCTOR

Print Supplements

The Instructor's Solutions Manual (978-0470-37957-8) contains detailed solutions to all exercises in the text.

The Instructor's Manual (978-0470-37956-1) suggests time allocations and teaching plans for each section in the text. Most of the teaching plans contain a bulleted list of key points to emphasize. The discussion of each section concludes with a sample homework assignment.

The Test Bank (978-0470-40856-8) features nearly 7000 questions and answers for every section in the text.

Instructor Companion Site

The Instructor Companion Site provides detailed information on the textbook's features, contents, and coverage and provides access to the following instructor supplements:

- The Computerized Test Bank features nearly 7000 questions—mostly algorithmically generated—that allow for varied questions and numerical inputs.
- PowerPoint slides cover the major concepts and themes of each section in a chapter.
- Personal-Response System questions ("Clicker Questions") appear at the end of each PowerPoint presentation and provide an easy way to gauge classroom understanding.
- Additional textbook content, such as Calculus Horizons and Explorations, back-of-the-book appendices, and selected biographies.

WileyPLUS

WileyPLUS, Wiley's digital-learning environment, is loaded with all of the supplements above, and also features the following:

- Homework management tools, which easily allow you to assign and grade questions, as well as gauge student comprehension.
- QuickStart features predesigned reading and homework assignments. Use them as-is or customize them to fit the needs of your classroom.
- The E-book, which is an exact version of the print text but also features hyperlinks to questions, definitions, and supplements for quicker and easier support.
- Animated applets, which can be used in class to present and explore key ideas graphically and dynamically—especially useful for display of three-dimensional graphs in multivariable calculus.

ACKNOWLEDGMENTS

It has been our good fortune to have the advice and guidance of many talented people whose knowledge and skills have enhanced this book in many ways. For their valuable help we thank the following people.

Reviewers and Contributors to the Ninth Edition of Early Transcendentals Calculus

Frederick Adkins, *Indiana University of Pennsylvania*
Bill Allen, *Reedley College–Clovis Center*
Jerry Allison, *Black Hawk College*
Seth Armstrong, *Southern Utah University*
Przemyslaw Bogacki, *Old Dominion University*
Wayne P. Britt, *Louisiana State University*
Kristin Chatas, *Washtenaw Community College*
Michele Clement, *Louisiana State University*
Ray Collings, *Georgia Perimeter College*
David E. Dobbs, *University of Tennessee, Knoxville*
H. Edward Donley, *Indiana University of Pennsylvania*
Jim Edmondson, *Santa Barbara City College*
Michael Filaseta, *University of South Carolina*
Jose Flores, *University of South Dakota*
Mitch Francis, *Horace Mann*
Jerome Heaven, *Indiana Tech*
Patricia Henry, *Drexel University*
Danrun Huang, *St. Cloud State University*
Alvaro Islas, *University of Central Florida*
Bin Jiang, *Portland State University*
Ronald Jorgensen, *Milwaukee School of Engineering*
Raja Khoury, *Collin County Community College*

Carole King Krueger, *The University of Texas at Arlington*
Thomas Leness, *Florida International University*
Kathryn Lesh, *Union College*
Behailu Mammo, *Hofstra University*
John McCuan, *Georgia Tech*
Daryl McGinnis, *Columbus State Community College*
Michael Mears, *Manatee Community College*
John G. Michaels, *SUNY Brockport*
Jason Miner, *Santa Barbara City College*
Darrell Minor, *Columbus State Community College*
Kathleen Miranda, *SUNY Old Westbury*
Carla Monticelli, *Camden County College*
Bryan Mosher, *University of Minnesota*
Ferdinand O. Orock, *Hudson County Community College*
Altay Ozgener, *Manatee Community College*
Chuang Peng, *Morehouse College*
Joni B. Pirnot, *Manatee Community College*
Elise Price, *Tarrant County College*
Holly Puterbaugh, *University of Vermont*
Hah Suey Quan, *Golden West College*
Joseph W. Rody, *Arizona State University*
Constance Schober, *University of Central Florida*

Kurt Sebastian, *United States Coast Guard*
Paul Seeburger, *Monroe Community College*
Bradley Stetson, *Schoolcraft College*
Walter E. Stone, Jr., *North Shore Community College*
Eleanor Storey, *Front Range Community College, Westminster Campus*
Stefania Tracogna, *Arizona State University*
Francis J. Vasko, *Kutztown University*
Jim Voss, *Front Range Community College*
Anke Walz, *Kutztown Community College*
Xian Wu, *University of South Carolina*
Yvonne Yaz, *Milwaukee School of Engineering*
Richard A. Zang, *University of New Hampshire*

The following people read the ninth edition at various stages for mathematical and pedagogical accuracy and/or assisted with the critically important job of preparing answers to exercises:

Dean Hickerson, *University of California, Davis*
Ron Jorgensen, *Milwaukee School of Engineering*
Roger Lipsett
Georgia Mederer
David Ryeburn, *Simon Fraser University*
Neil Wigley

Reviewers and Contributors to the Ninth Edition of Late Transcendentals and Multivariable Calculus

David Bradley, *University of Maine*
Dean Burbank, *Gulf Coast Community College*
Jason Cantarella, *University of Georgia*
Yanzhao Cao, *Florida A&M University*
T.J. Duda, *Columbus State Community College*
Nancy Eschen, *Florida Community College, Jacksonville*
Reuben Farley, *Virginia Commonwealth University*
Zhuang-dan Guan, *University of California, Riverside*
Greg Henderson, *Hillsborough Community College*

Micah James, *University of Illinois*
Mohammad Kazemi, *University of North Carolina, Charlotte*
Przemo Kranz, *University of Mississippi*
Steffen Lempp, *University of Wisconsin, Madison*
Wen-Xiu Ma, *University of South Florida*
Vania Mascioni, *Ball State University*
David Price, *Tarrant County College*
Jan Rychtar, *University of North Carolina, Greensboro*
John T. Saccoman, *Seton Hall University*

Charlotte Simmons, *University of Central Oklahoma*
Don Soash, *Hillsborough Community College*
Bryan Stewart, *Tarrant County College*
Helene Tyler, *Manhattan College*
Pavlos Tzermias, *University of Tennessee, Knoxville*
Raja Varatharajah, *North Carolina A&T*
David Voss, *Western Illinois University*
Richard Watkins, *Tidewater Community College*
Xiao-Dong Zhang, *Florida Atlantic University*
Diane Zych, *Erie Community College*

Reviewers and Contributors to the Eighth Edition of Calculus

Gregory Adams, *Bucknell University*

Bill Allen, *Reedley College–Clovis Center*

Jerry Allison, *Black Hawk College*

Stella Ashford, *Southern University and A&M College*

Mary Lane Baggett, *University of Mississippi*

Christopher Barker, *San Joaquin Delta College*

Kbenesh Blayneh, *Florida A&M University*

David Bradley, *University of Maine*

Paul Britt, *Louisiana State University*

Judith Broadwin, *Jericho High School*

Andrew Bulleri, *Howard Community College*

Christopher Butler, *Case Western Reserve University*

Cheryl Cantwell, *Seminole Community College*

Judith Carter, *North Shore Community College*

Miriam Castroconde, *Irvine Valley College*

Neena Chopra, *The Pennsylvania State University*

Gaemus Collins, *University of California, San Diego*

Fielden Cox, *Centennial College*

Danielle Cross, *Northern Essex Community College*

Gary Crown, *Wichita State University*

Larry Cusick, *California State University–Fresno*

Stephan DeLong, *Tidewater Community College–Virginia Beach Campus*

Debbie A. Desrochers, *Napa Valley College*

Ryness Doherty, *Community College of Denver*

T.J. Duda, *Columbus State Community College*

Peter Embalabala, *Lincoln Land Community College*

Phillip Farmer, *Diablo Valley College*

Laurene Fausett, *Georgia Southern University*

Sally E. Fishbeck, *Rochester Institute of Technology*

Bob Grant, *Mesa Community College*

Richard Hall, *Cochise College*

Noal Harbertson, *California State University, Fresno*

Donald Hartig, *California Polytechnic State University*

Karl Havlak, *Angelo State University*

J. Derrick Head, *University of Minnesota–Morris*

Konrad Heuvers, *Michigan Technological University*

Tommie Ann Hill-Natter, *Prairie View A&M University*

Holly Hirst, *Appalachian State University*

Joe Howe, *St. Charles County Community College*

Shirley Huffman, *Southwest Missouri State University*

Gary S. Itzkowitz, *Rowan University*

John Johnson, *George Fox University*

Kenneth Kalmanson, *Montclair State University*

Grant Karamyan, *University of California, Los Angeles*

David Keller, *Kirkwood Community College*

Dan Kemp, *South Dakota State University*

Vesna Kilibarda, *Indiana University Northwest*

Cecilia Knoll, *Florida Institute of Technology*

Carole King Krueger, *The University of Texas at Arlington*

Holly A. Kresch, *Diablo Valley College*

John Kubicek, *Southwest Missouri State University*

Theodore Lai, *Hudson County Community College*

Richard Lane, *University of Montana*

Jeuel LaTorre, *Clemson University*

Marshall Leitman, *Case Western Reserve University*

Phoebe Lutz, *Delta College*

Ernest Manfred, *U.S. Coast Guard Academy*

James Martin, *Wake Technical Community College*

Vania Mascioni, *Ball State University*

Tamra Mason, *Albuquerque TVI Community College*

Thomas W. Mason, *Florida A&M University*

Roy Mathia, *The College of William and Mary*

John Michaels, *SUNY Brockport*

Darrell Minor, *Columbus State Community College*

Darren Narayan, *Rochester Institute of Technology*

Doug Nelson, *Central Oregon Community College*

Lawrence J. Newberry, *Glendale College*

Judith Palagallo, *The University of Akron*

Efton Park, *Texas Christian University*

Joanne Peeples, *El Paso Community College*

Gary L. Peterson, *James Madison University*

Lefkios Petevis, *Kirkwood Community College*

Thomas W. Polaski, *Winthrop University*

Richard Ponticelli, *North Shore Community College*

Holly Puterbaugh, *University of Vermont*

Douglas Quinney, *University of Keele*

B. David Redman, Jr., *Delta College*

William H. Richardson, *Wichita State University*

Lila F. Roberts, *Georgia Southern University*

Robert Rock, *Daniel Webster College*

John Saccoman, *Seton Hall University*

Avinash Sathaye, *University of Kentucky*

George W. Schultz, *St. Petersburg Junior College*

Paul Seeburger, *Monroe Community College*

Richard B. Shad, *Florida Community College–Jacksonville*

Mary Margaret Shoaf-Grubbs, *College of New Rochelle*

Charlotte Simmons, *University of Central Oklahoma*

Ann Sitomer, *Portland Community College*

Jeanne Smith, *Saddleback Community College*

Rajalakshmi Sriram, *Okaloosa-Walton Community College*

Mark Stevenson, *Oakland Community College*

Bryan Stewart, *Tarrant County College*

Bradley Stoll, *The Harker School*

Eleanor Storey, *Front Range Community College*

John A. Suvak, *Memorial University of Newfoundland*

Richard Swanson, *Montana State University*

Skip Thompson, *Radford University*

Helene Tyler, *Manhattan College*

Paramanathan Varatharajah, *North Carolina A&T State University*

David Voss, *Western Illinois University*

Jim Voss, *Front Range Community College*

Richard Watkins, *Tidewater Community College*

Bruce R. Wenner, *University of Missouri–Kansas City*

Jane West, *Trident Technical College*

Ted Wilcox, *Rochester Institute of Technology*

Janine Wittwer, *Williams College*

Diane Zych, *Erie Community College–North Campus*

The following people read the eighth edition at various stages for mathematical and pedagogical accuracy and/or assisted with the critically important job of preparing answers to exercises:

Elka Block, *Twin Prime Editorial*

Dean Hickerson, *University of California, Davis*

Thomas Polaski, *Winthrop University*

Frank Purcell, *Twin Prime Editorial*

David Ryeburn, *Simon Fraser University*

CONTENTS

WEB APPENDICES

WEB PROJECTS (Expanding the Calculus Horizon)

THE ROOTS OF CALCULUS

Today's exciting applications of calculus have roots that can be traced to the work of the Greek mathematician Archimedes, but the actual discovery of the fundamental principles of calculus was made independently by Isaac Newton (English) and Gottfried Leibniz (German) in the late seventeenth century. The work of Newton and Leibniz was motivated by four major classes of scientific and mathematical problems of the time:

- Find the tangent line to a general curve at a given point.

- Find the area of a general region, the length of a general curve, and the volume of a general solid.

- Find the maximum or minimum value of a quantity—for example, the maximum and minimum distances of a planet from the Sun, or the maximum range attainable for a projectile by varying its angle of fire.

- Given a formula for the distance traveled by a body in any specified amount of time, find the velocity and acceleration of the body at any instant. Conversely, given a formula that specifies the acceleration of velocity at any instant, find the distance traveled by the body in a specified period of time.

Newton and Leibniz found a fundamental relationship between the problem of finding a tangent line to a curve and the problem of determining the area of a region. Their realization of this connection is considered to be the "discovery of calculus." Though Newton saw how these two problems are related ten years before Leibniz did, Leibniz published his work twenty years before Newton. This situation led to a stormy debate over who was the rightful discoverer of calculus. The debate engulfed Europe for half a century, with the scientists of the European continent supporting Leibniz and those from England supporting Newton. The conflict was extremely unfortunate because Newton's inferior notation badly hampered scientific development in England, and the Continent in turn lost the benefit of Newton's discoveries in astronomy and physics for nearly fifty years. In spite of it all, Newton and Leibniz were sincere admirers of each other's work.

ISAAC NEWTON (1642–1727)

Newton was born in the village of Woolsthorpe, England. His father died before he was born and his mother raised him on the family farm. As a youth he showed little evidence of his later brilliance, except for an unusual talent with mechanical devices—he apparently built a working water clock and a toy flour mill powered by a mouse. In 1661 he entered Trinity College in Cambridge with a deficiency in geometry. Fortunately, Newton caught the eye of Isaac Barrow, a gifted mathematician and teacher. Under Barrow's guidance Newton immersed himself in mathematics and science, but he graduated without any special distinction. Because the bubonic plague was spreading rapidly through London, Newton returned to his home in Woolsthorpe and stayed there during the years of 1665 and 1666. In those two momentous years the entire framework of modern science was miraculously created in Newton's mind. He discovered calculus, recognized the underlying principles of planetary motion and gravity, and determined that "white" sunlight was composed of all colors, red to violet. For whatever reasons he kept his discoveries to himself. In 1667 he returned to Cambridge to obtain his Master's degree and upon graduation became a teacher at Trinity. Then in 1669 Newton succeeded his teacher, Isaac Barrow, to the Lucasian chair of mathematics at Trinity, one of the most honored chairs of mathematics in the world.

Thereafter, brilliant discoveries flowed from Newton steadily. He formulated the law of gravitation and used it to explain the motion of the moon, the planets, and the tides; he formulated basic theories of light, thermodynamics, and hydrodynamics; and he devised and constructed the first modern reflecting telescope. Throughout his life Newton was hesitant to publish his major discoveries, revealing them only to a select circle of friends, perhaps because of a fear of criticism or controversy. In 1687, only after intense coaxing by the astronomer, Edmond Halley (disoverer of Halley's comet), did Newton publish his masterpiece, *Philosophiae Naturalis Principia Mathematica* (The Mathematical Principles of Natural Philosophy). This work is generally considered to be the most important and influential scientific book ever written. In it Newton explained the workings of the solar system and formulated the basic laws of motion, which to this day are fundamental in engineering and physics. However, not even the pleas of his friends could convince Newton to publish his discovery of calculus. Only after Leibniz published his results did Newton relent and publish his own work on calculus.

After twenty-five years as a professor, Newton suffered depression and a nervous breakdown. He gave up research in 1695 to accept a position as warden and later master of the London mint. During the twenty-five years that he worked at the mint, he did virtually no scientific or mathematical work. He was knighted in 1705 and on his death was buried in Westminster Abbey with all the honors his country could bestow. It is interesting to note that Newton was a learned theologian who viewed the primary value of his work to be its support of the existence of God. Throughout his life he worked passionately to date biblical events by relating them to astronomical phenomena. He was so consumed with this passion that he spent years searching the Book of Daniel for clues to the end of the world and the geography of hell.

Newton described his brilliant accomplishments as follows: "I seem to have been only like a boy playing on the seashore and diverting myself in now and then finding a smoother pebble or prettier shell than ordinary, whilst the great ocean of truth lay all undiscovered before me."

GOTTFRIED WILHELM LEIBNIZ (1646–1716)

This gifted genius was one of the last people to have mastered most major fields of knowledge—an impossible accomplishment in our own era of specialization. He was an expert in law, religion, philosophy, literature, politics, geology, metaphysics, alchemy, history, and mathematics.

Leibniz was born in Leipzig, Germany. His father, a professor of moral philosophy at the University of Leipzig, died when Leibniz was six years old. The precocious boy then gained access to his father's library and began reading voraciously on a wide range of subjects, a habit that he maintained throughout his life. At age fifteen he entered the University of Leipzig as a law student and by the age of twenty received a doctorate from the University of Altdorf. Subsequently, Leibniz followed a career in law and international politics, serving as counsel to kings and princes. During his numerous foreign missions, Leibniz came in contact with outstanding mathematicians and scientists who stimulated his interest in mathematics—most notably, the physicist Christian Huygens. In mathematics Leibniz was self-taught, learning the subject by reading papers and journals. As a result of this fragmented mathematical education, Leibniz often rediscovered the results of others, and this helped to fuel the debate over the discovery of calculus.

Leibniz never married. He was moderate in his habits, quick-tempered but easily appeased, and charitable in his judgment of other people's work. In spite of his great achievements, Leibniz never received the honors showered on Newton, and he spent his final years as a lonely embittered man. At his funeral there was one mourner, his secretary. An eyewitness stated, "He was buried more like a robber than what he really was—an ornament of his country."

© Arco Images/Alamy

0

BEFORE CALCULUS

The development of calculus in the seventeenth and eighteenth centuries was motivated by the need to understand physical phenomena such as the tides, the phases of the moon, the nature of light, and gravity.

One of the important themes in calculus is the analysis of relationships between physical or mathematical quantities. Such relationships can be described in terms of graphs, formulas, numerical data, or words. In this chapter we will develop the concept of a "function," which is the basic idea that underlies almost all mathematical and physical relationships, regardless of the form in which they are expressed. We will study properties of some of the most basic functions that occur in calculus, including polynomials, trigonometric functions, inverse trigonometric functions, exponential functions, and logarithmic functions.

0.1 FUNCTIONS

In this section we will define and develop the concept of a "function," which is the basic mathematical object that scientists and mathematicians use to describe relationships between variable quantities. Functions play a central role in calculus and its applications.

■ DEFINITION OF A FUNCTION

Many scientific laws and engineering principles describe how one quantity depends on another. This idea was formalized in 1673 by Gottfried Wilhelm Leibniz (see p. xx) who coined the term *function* to indicate the dependence of one quantity on another, as described in the following definition.

> **0.1.1 DEFINITION** If a variable y depends on a variable x in such a way that each value of x determines exactly one value of y, then we say that ***y is a function of x***.

Four common methods for representing functions are:

- Numerically by tables
- Algebraically by formulas
- Geometrically by graphs
- Verbally

1

Table 0.1.1

INDIANAPOLIS 500
QUALIFYING SPEEDS

YEAR t	SPEED S (mi/h)
1989	223.885
1990	225.301
1991	224.113
1992	232.482
1993	223.967
1994	228.011
1995	231.604
1996	233.100
1997	218.263
1998	223.503
1999	225.179
2000	223.471
2001	226.037
2002	231.342
2003	231.725
2004	222.024
2005	227.598
2006	228.985

The method of representation often depends on how the function arises. For example:

- Table 0.1.1 shows the top qualifying speed S for the Indianapolis 500 auto race as a function of the year t. There is exactly one value of S for each value of t.

- Figure 0.1.1 is a graphical record of an earthquake recorded on a seismograph. The graph describes the deflection D of the seismograph needle as a function of the time T elapsed since the wave left the earthquake's epicenter. There is exactly one value of D for each value of T.

- Some of the most familiar functions arise from formulas; for example, the formula $C = 2\pi r$ expresses the circumference C of a circle as a function of its radius r. There is exactly one value of C for each value of r.

- Sometimes functions are described in words. For example, Isaac Newton's Law of Universal Gravitation is often stated as follows: The gravitational force of attraction between two bodies in the Universe is directly proportional to the product of their masses and inversely proportional to the square of the distance between them. This is the verbal description of the formula

$$F = G\frac{m_1 m_2}{r^2}$$

in which F is the force of attraction, m_1 and m_2 are the masses, r is the distance between them, and G is a constant. If the masses are constant, then the verbal description defines F as a function of r. There is exactly one value of F for each value of r.

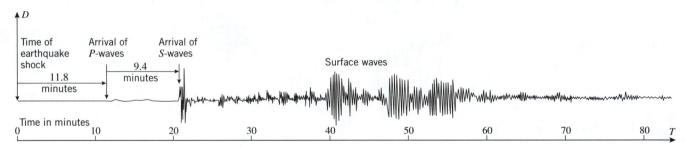

▲ **Figure 0.1.1**

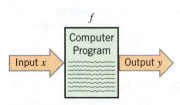

▲ **Figure 0.1.2**

In the mid-eighteenth century the Swiss mathematician Leonhard Euler (pronounced "oiler") conceived the idea of denoting functions by letters of the alphabet, thereby making it possible to refer to functions without stating specific formulas, graphs, or tables. To understand Euler's idea, think of a function as a computer program that takes an *input x*, operates on it in some way, and produces exactly one *output y*. The computer program is an object in its own right, so we can give it a name, say f. Thus, the function f (the computer program) associates a unique output y with each input x (Figure 0.1.2). This suggests the following definition.

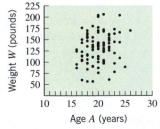

▲ **Figure 0.1.3**

0.1.2 DEFINITION A *function* f is a rule that associates a unique output with each input. If the input is denoted by x, then the output is denoted by $f(x)$ (read "f of x").

In this definition the term *unique* means "exactly one." Thus, a function cannot assign two different outputs to the same input. For example, Figure 0.1.3 shows a plot of weight versus age for a random sample of 100 college students. This plot does *not* describe W as a function of A because there are some values of A with more than one corresponding value of W. This is to be expected, since two people with the same age can have different weights.

■ INDEPENDENT AND DEPENDENT VARIABLES

For a given input x, the output of a function f is called the *value* of f at x or the *image* of x under f. Sometimes we will want to denote the output by a single letter, say y, and write

$$y = f(x)$$

This equation expresses y as a function of x; the variable x is called the *independent variable* (or *argument*) of f, and the variable y is called the *dependent variable* of f. This terminology is intended to suggest that x is free to vary, but that once x has a specific value a corresponding value of y is determined. For now we will only consider functions in which the independent and dependent variables are real numbers, in which case we say that f is a *real-valued function of a real variable*. Later, we will consider other kinds of functions.

Table 0.1.2

x	0	1	2	3
y	3	4	−1	6

▶ **Example 1**   Table 0.1.2 describes a functional relationship $y = f(x)$ for which

$f(0) = 3$ | f associates $y = 3$ with $x = 0$. |

$f(1) = 4$ | f associates $y = 4$ with $x = 1$. |

$f(2) = -1$ | f associates $y = -1$ with $x = 2$. |

$f(3) = 6$ | f associates $y = 6$ with $x = 3$. | ◀

▶ **Example 2**   The equation

$$y = 3x^2 - 4x + 2$$

has the form $y = f(x)$ in which the function f is given by the formula

$$f(x) = 3x^2 - 4x + 2$$

Leonhard Euler (1707–1783)   Euler was probably the most prolific mathematician who ever lived. It has been said that "Euler wrote mathematics as effortlessly as most men breathe." He was born in Basel, Switzerland, and was the son of a Protestant minister who had himself studied mathematics. Euler's genius developed early. He attended the University of Basel, where by age 16 he obtained both a Bachelor of Arts degree and a Master's degree in philosophy. While at Basel, Euler had the good fortune to be tutored one day a week in mathematics by a distinguished mathematician, Johann Bernoulli. At the urging of his father, Euler then began to study theology. The lure of mathematics was too great, however, and by age 18 Euler had begun to do mathematical research. Nevertheless, the influence of his father and his theological studies remained, and throughout his life Euler was a deeply religious, unaffected person. At various times Euler taught at St. Petersburg Academy of Sciences (in Russia), the University of Basel, and the Berlin Academy of Sciences. Euler's energy and capacity for work were virtually boundless. His collected works form more than 100 quarto-sized volumes and it is believed that much of his work has been lost. What is particularly astonishing is that Euler was blind for the last 17 years of his life, and this was one of his most productive periods! Euler's flawless memory was phenomenal. Early in his life he memorized the entire *Aeneid* by Virgil, and at age 70 he could not only recite the entire work but could also state the first and last sentence on each page of the book from which he memorized the work. His ability to solve problems in his head was beyond belief. He worked out in his head major problems of lunar motion that baffled Isaac Newton and once did a complicated calculation in his head to settle an argument between two students whose computations differed in the fiftieth decimal place.

Following the development of calculus by Leibniz and Newton, results in mathematics developed rapidly in a disorganized way. Euler's genius gave coherence to the mathematical landscape. He was the first mathematician to bring the full power of calculus to bear on problems from physics. He made major contributions to virtually every branch of mathematics as well as to the theory of optics, planetary motion, electricity, magnetism, and general mechanics.

For each input x, the corresponding output y is obtained by substituting x in this formula. For example,

$$f(0) = 3(0)^2 - 4(0) + 2 = 2$$

f associates $y = 2$ with $x = 0$.

$$f(-1.7) = 3(-1.7)^2 - 4(-1.7) + 2 = 17.47$$

f associates $y = 17.47$ with $x = -1.7$.

$$f(\sqrt{2}) = 3(\sqrt{2})^2 - 4\sqrt{2} + 2 = 8 - 4\sqrt{2}$$

f associates $y = 8 - 4\sqrt{2}$ with $x = \sqrt{2}$. ◄

■ GRAPHS OF FUNCTIONS

Figure 0.1.4 shows only portions of the graphs. Where appropriate, and unless indicated otherwise, it is understood that graphs shown in this text extend indefinitely beyond the boundaries of the displayed figure.

If f is a real-valued function of a real variable, then the **graph** of f in the xy-plane is defined to be the graph of the equation $y = f(x)$. For example, the graph of the function $f(x) = x$ is the graph of the equation $y = x$, shown in Figure 0.1.4. That figure also shows the graphs of some other basic functions that may already be familiar to you. In Appendix A we discuss techniques for graphing functions using graphing technology.

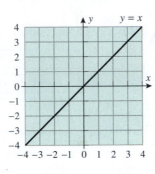
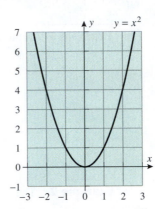
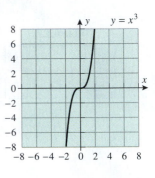

Since $\sqrt{x}$ is imaginary for negative values of x, there are no points on the graph of $y = \sqrt{x}$ in the region where $x < 0$.

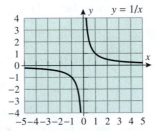
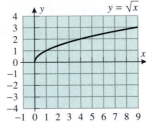
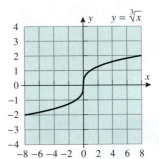

▲ **Figure 0.1.4**

Graphs can provide valuable visual information about a function. For example, since the graph of a function f in the xy-plane is the graph of the equation $y = f(x)$, the points on the graph of f are of the form $(x, f(x))$; that is, *the y-coordinate of a point on the graph of f is the value of f at the corresponding x-coordinate* (Figure 0.1.5). The values of x for which $f(x) = 0$ are the x-coordinates of the points where the graph of f intersects the x-axis (Figure 0.1.6). These values are called the **zeros** of f, the **roots** of $f(x) = 0$, or the **x-intercepts** of the graph of $y = f(x)$.

▲ **Figure 0.1.5** The y-coordinate of a point on the graph of $y = f(x)$ is the value of f at the corresponding x-coordinate.

■ THE VERTICAL LINE TEST

Not every curve in the xy-plane is the graph of a function. For example, consider the curve in Figure 0.1.7, which is cut at two distinct points, (a, b) and (a, c), by a vertical line. This curve cannot be the graph of $y = f(x)$ for any function f; otherwise, we would have

$$f(a) = b \quad \text{and} \quad f(a) = c$$

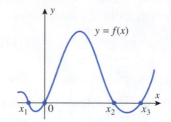

▲ Figure 0.1.6 f has zeros at $x_1, 0, x_2,$ and x_3.

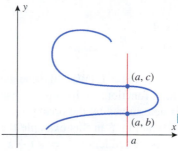

▲ Figure 0.1.7 This curve cannot be the graph of a function.

> Symbols such as $+x$ and $-x$ are deceptive, since it is tempting to conclude that $+x$ is positive and $-x$ is negative. However, this need not be so, since x itself can be positive or negative. For example, if x is negative, say $x = -3$, then $-x = 3$ is positive and $+x = -3$ is negative.

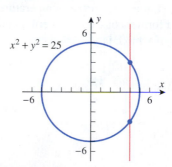

▲ Figure 0.1.8

WARNING

> To denote the negative square root you must write $-\sqrt{x}$. For example, the positive square root of 9 is $\sqrt{9} = 3$, whereas the negative square root of 9 is $-\sqrt{9} = -3$. (Do not make the mistake of writing $\sqrt{9} = \pm 3$.)

which is impossible, since f cannot assign two different values to a. Thus, there is no function f whose graph is the given curve. This illustrates the following general result, which we will call the ***vertical line test***.

> **0.1.3 THE VERTICAL LINE TEST**   *A curve in the xy-plane is the graph of some function f if and only if no vertical line intersects the curve more than once.*

▶ **Example 3**   The graph of the equation

$$x^2 + y^2 = 25$$

is a circle of radius 5 centered at the origin and hence there are vertical lines that cut the graph more than once (Figure 0.1.8). Thus this equation does not define y as a function of x. ◀

■ **THE ABSOLUTE VALUE FUNCTION**

Recall that the ***absolute value*** or ***magnitude*** of a real number x is defined by

$$|x| = \begin{cases} x, & x \geq 0 \\ -x, & x < 0 \end{cases}$$

The effect of taking the absolute value of a number is to strip away the minus sign if the number is negative and to leave the number unchanged if it is nonnegative. Thus,

$$|5| = 5, \quad \left|-\tfrac{4}{7}\right| = \tfrac{4}{7}, \quad |0| = 0$$

A more detailed discussion of the properties of absolute value is given in Web Appendix F. However, for convenience we provide the following summary of its algebraic properties.

> **0.1.4 PROPERTIES OF ABSOLUTE VALUE**   *If a and b are real numbers, then*
>
> (a) $|-a| = |a|$ A number and its negative have the same absolute value.
>
> (b) $|ab| = |a|\,|b|$ The absolute value of a product is the product of the absolute values.
>
> (c) $|a/b| = |a|/|b|, b \neq 0$ The absolute value of a ratio is the ratio of the absolute values.
>
> (d) $|a + b| \leq |a| + |b|$ The ***triangle inequality***

The graph of the function $f(x) = |x|$ can be obtained by graphing the two parts of the equation

$$y = \begin{cases} x, & x \geq 0 \\ -x, & x < 0 \end{cases}$$

separately. Combining the two parts produces the V-shaped graph in Figure 0.1.9.

Absolute values have important relationships to square roots. To see why this is so, recall from algebra that every positive real number x has two square roots, one positive and one negative. By definition, the symbol $\sqrt{x}$ denotes the *positive* square root of x.

Care must be exercised in simplifying expressions of the form $\sqrt{x^2}$, since it is *not* always true that $\sqrt{x^2} = x$. This equation is correct if x is nonnegative, but it is false if x is negative. For example, if $x = -4$, then

$$\sqrt{x^2} = \sqrt{(-4)^2} = \sqrt{16} = 4 \neq x$$

Verify (1) by using a graphing utility to show that the equations $y = \sqrt{x^2}$ and $y = |x|$ have the same graph.

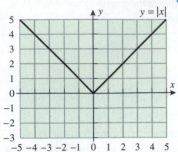

▲ **Figure 0.1.9**

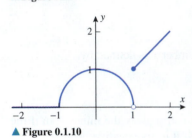

▲ **Figure 0.1.10**

A statement that is correct for all real values of x is

$$\sqrt{x^2} = |x| \tag{1}$$

■ **PIECEWISE-DEFINED FUNCTIONS**

The absolute value function $f(x) = |x|$ is an example of a function that is defined *piecewise* in the sense that the formula for f changes, depending on the value of x.

▶ **Example 4**   Sketch the graph of the function defined piecewise by the formula

$$f(x) = \begin{cases} 0, & x \le -1 \\ \sqrt{1 - x^2}, & -1 < x < 1 \\ x, & x \ge 1 \end{cases}$$

Solution.   The formula for f changes at the points $x = -1$ and $x = 1$. (We call these the *breakpoints* for the formula.) A good procedure for graphing functions defined piecewise is to graph the function separately over the open intervals determined by the breakpoints, and then graph f at the breakpoints themselves. For the function f in this example the graph is the horizontal ray $y = 0$ on the interval $(-\infty, -1]$, it is the semicircle $y = \sqrt{1 - x^2}$ on the interval $(-1, 1)$, and it is the ray $y = x$ on the interval $[1, +\infty)$. The formula for f specifies that the equation $y = 0$ applies at the breakpoint -1 [so $y = f(-1) = 0$], and it specifies that the equation $y = x$ applies at the breakpoint 1 [so $y = f(1) = 1$]. The graph of f is shown in Figure 0.1.10. ◀

REMARK   In Figure 0.1.10 the solid dot and open circle at the breakpoint $x = 1$ serve to emphasize that the point on the graph lies on the ray and not the semicircle. There is no ambiguity at the breakpoint $x = -1$ because the two parts of the graph join together continuously there.

© Brian Horisk/Alamy
The wind chill index measures the sensation of coldness that we feel from the combined effect of temperature and wind speed.

▶ **Example 5**   Increasing the speed at which air moves over a person's skin increases the rate of moisture evaporation and makes the person feel cooler. (This is why we fan ourselves in hot weather.) The *wind chill index* is the temperature at a wind speed of 4 mi/h that would produce the same sensation on exposed skin as the current temperature and wind speed combination. An empirical formula (i.e., a formula based on experimental data) for the wind chill index W at 32°F for a wind speed of v mi/h is

$$W = \begin{cases} 32, & 0 \le v \le 3 \\ 55.628 - 22.07v^{0.16}, & 3 < v \end{cases}$$

A computer-generated graph of $W(v)$ is shown in Figure 0.1.11. ◀

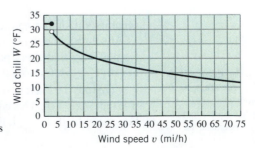

▶ **Figure 0.1.11** Wind chill versus wind speed at 32°F

DOMAIN AND RANGE

If x and y are related by the equation $y = f(x)$, then the set of all allowable inputs (x-values) is called the ***domain*** of f, and the set of outputs (y-values) that result when x varies over the domain is called the ***range*** of f. For example, if f is the function defined by the table in Example 1, then the domain is the set $\{0, 1, 2, 3\}$ and the range is the set $\{-1, 3, 4, 6\}$.

Sometimes physical or geometric considerations impose restrictions on the allowable inputs of a function. For example, if y denotes the area of a square of side x, then these variables are related by the equation $y = x^2$. Although this equation produces a unique value of y for every real number x, the fact that lengths must be nonnegative imposes the requirement that $x \geq 0$.

When a function is defined by a mathematical formula, the formula itself may impose restrictions on the allowable inputs. For example, if $y = 1/x$, then $x = 0$ is not an allowable input since division by zero is undefined, and if $y = \sqrt{x}$, then negative values of x are not allowable inputs because they produce imaginary values for y and we have agreed to consider only real-valued functions of a real variable. In general, we make the following definition.

One might argue that a physical square cannot have a side of length zero. However, it is often convenient mathematically to allow zero lengths, and we will do so throughout this text where appropriate.

> **0.1.5   DEFINITION**   If a real-valued function of a real variable is defined by a formula, and if no domain is stated explicitly, then it is to be understood that the domain consists of all real numbers for which the formula yields a real value. This is called the ***natural domain*** of the function.

The domain and range of a function f can be pictured by projecting the graph of $y = f(x)$ onto the coordinate axes as shown in Figure 0.1.12.

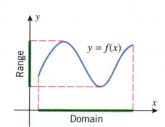

▲ **Figure 0.1.12**   The projection of $y = f(x)$ on the x-axis is the set of allowable x-values for f, and the projection on the y-axis is the set of corresponding y-values.

For a review of trigonometry see Appendix B.

▶ **Example 6**   Find the natural domain of

(a) $f(x) = x^3$ (b) $f(x) = 1/[(x-1)(x-3)]$
(c) $f(x) = \tan x$ (d) $f(x) = \sqrt{x^2 - 5x + 6}$

Solution (a).   The function f has real values for all real x, so its natural domain is the interval $(-\infty, +\infty)$.

Solution (b).   The function f has real values for all real x, except $x = 1$ and $x = 3$, where divisions by zero occur. Thus, the natural domain is

$$\{x : x \neq 1 \text{ and } x \neq 3\} = (-\infty, 1) \cup (1, 3) \cup (3, +\infty)$$

Solution (c).   Since $f(x) = \tan x = \sin x / \cos x$, the function f has real values except where $\cos x = 0$, and this occurs when x is an odd integer multiple of $\pi/2$. Thus, the natural domain consists of all real numbers except

$$x = \pm\frac{\pi}{2}, \pm\frac{3\pi}{2}, \pm\frac{5\pi}{2}, \ldots$$

Solution (d).   The function f has real values, except when the expression inside the radical is negative. Thus the natural domain consists of all real numbers x such that

$$x^2 - 5x + 6 = (x-3)(x-2) \geq 0$$

This inequality is satisfied if $x \leq 2$ or $x \geq 3$ (verify), so the natural domain of f is

$$(-\infty, 2] \cup [3, +\infty) \blacktriangleleft$$

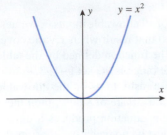

In some cases we will state the domain explicitly when defining a function. For example, if $f(x) = x^2$ is the area of a square of side x, then we can write

$$f(x) = x^2, \quad x \geq 0$$

to indicate that we take the domain of f to be the set of nonnegative real numbers (Figure 0.1.13).

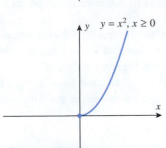

■ THE EFFECT OF ALGEBRAIC OPERATIONS ON THE DOMAIN

Algebraic expressions are frequently simplified by canceling common factors in the numerator and denominator. However, care must be exercised when simplifying formulas for functions in this way, since this process can alter the domain.

▲ **Figure 0.1.13**

▶ **Example 7** The natural domain of the function

$$f(x) = \frac{x^2 - 4}{x - 2} \qquad (2)$$

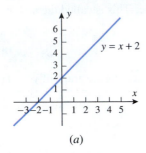

(a)

consists of all real x except $x = 2$. However, if we factor the numerator and then cancel the common factor in the numerator and denominator, we obtain

$$f(x) = \frac{(x - 2)(x + 2)}{x - 2} = x + 2 \qquad (3)$$

Since the right side of (3) has a value of $f(2) = 4$ and $f(2)$ was undefined in (2), the algebraic simplification has changed the function. Geometrically, the graph of (3) is the line in Figure 0.1.14a, whereas the graph of (2) is the same line but with a hole at $x = 2$, since the function is undefined there (Figure 0.1.14b). In short, the geometric effect of the algebraic cancellation is to eliminate the hole in the original graph. ◀

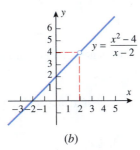

(b)

Sometimes alterations to the domain of a function that result from algebraic simplification are irrelevant to the problem at hand and can be ignored. However, if the domain must be preserved, then one must impose the restrictions on the simplified function explicitly. For example, if we wanted to preserve the domain of the function in Example 7, then we would have to express the simplified form of the function as

$$f(x) = x + 2, \quad x \neq 2$$

▲ **Figure 0.1.14**

▶ **Example 8** Find the domain and range of

(a) $f(x) = 2 + \sqrt{x - 1}$ (b) $f(x) = (x + 1)/(x - 1)$

Solution (a). Since no domain is stated explicitly, the domain of f is its natural domain, $[1, +\infty)$. As x varies over the interval $[1, +\infty)$, the value of $\sqrt{x - 1}$ varies over the interval $[0, +\infty)$, so the value of $f(x) = 2 + \sqrt{x - 1}$ varies over the interval $[2, +\infty)$, which is the range of f. The domain and range are highlighted in green on the x- and y-axes in Figure 0.1.15.

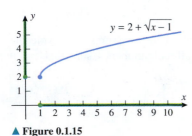

▲ **Figure 0.1.15**

Solution (b). The given function f is defined for all real x, except $x = 1$, so the natural domain of f is

$$\{x : x \neq 1\} = (-\infty, 1) \cup (1, +\infty)$$

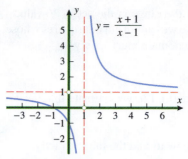

▲ **Figure 0.1.16**

To determine the range it will be convenient to introduce a dependent variable

$$y = \frac{x+1}{x-1} \tag{4}$$

Although the set of possible y-values is not immediately evident from this equation, the graph of (4), which is shown in Figure 0.1.16, suggests that the range of f consists of all y, except $y = 1$. To see that this is so, we solve (4) for x in terms of y:

$$(x-1)y = x+1$$
$$xy - y = x+1$$
$$xy - x = y+1$$
$$x(y-1) = y+1$$
$$x = \frac{y+1}{y-1}$$

It is now evident from the right side of this equation that $y = 1$ is not in the range; otherwise we would have a division by zero. No other values of y are excluded by this equation, so the range of the function f is $\{y : y \neq 1\} = (-\infty, 1) \cup (1, +\infty)$, which agrees with the result obtained graphically. ◄

■ DOMAIN AND RANGE IN APPLIED PROBLEMS

In applications, physical considerations often impose restrictions on the domain and range of a function.

► **Example 9** An open box is to be made from a 16-inch by 30-inch piece of cardboard by cutting out squares of equal size from the four corners and bending up the sides (Figure 0.1.17a).

(a) Let V be the volume of the box that results when the squares have sides of length x. Find a formula for V as a function of x.

(b) Find the domain of V.

(c) Use the graph of V given in Figure 0.1.17c to estimate the range of V.

(d) Describe in words what the graph tells you about the volume.

Solution (a). As shown in Figure 0.1.17b, the resulting box has dimensions $16 - 2x$ by $30 - 2x$ by x, so the volume $V(x)$ is given by

$$V(x) = (16 - 2x)(30 - 2x)x = 480x - 92x^2 + 4x^3$$

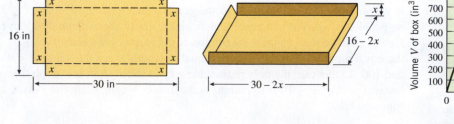
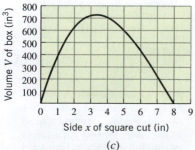

(a)   (b)   (c)

▲ **Figure 0.1.17**

Solution (b).   The domain is the set of x-values and the range is the set of V-values. Because x is a length, it must be nonnegative, and because we cannot cut out squares whose sides are more than 8 in long (why?), the x-values in the domain must satisfy

$$0 \leq x \leq 8$$

Solution (c).   From the graph of V versus x in Figure 0.1.17c we estimate that the V-values in the range satisfy
$$0 \leq V \leq 725$$

Note that this is an approximation. Later we will show how to find the range exactly.

Solution (d).   The graph tells us that the box of maximum volume occurs for a value of x that is between 3 and 4 and that the maximum volume is approximately 725 in^3. The graph also shows that the volume decreases toward zero as x gets closer to 0 or 8, which should make sense to you intuitively. ◄

In applications involving time, formulas for functions are often expressed in terms of a variable t whose starting value is taken to be $t = 0$.

► **Example 10**   At 8:05 A.M. a car is clocked at 100 ft/s by a radar detector that is positioned at the edge of a straight highway. Assuming that the car maintains a constant speed between 8:05 A.M. and 8:06 A.M., find a function $D(t)$ that expresses the distance traveled by the car during that time interval as a function of the time t.

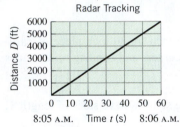

Radar Tracking

▲ Figure 0.1.18

Solution.   It would be clumsy to use the actual clock time for the variable t, so let us agree to use the *elapsed* time in seconds, starting with $t = 0$ at 8:05 A.M. and ending with $t = 60$ at 8:06 A.M. At each instant, the distance traveled (in ft) is equal to the speed of the car (in ft/s) multiplied by the elapsed time (in s). Thus,

$$D(t) = 100t, \quad 0 \leq t \leq 60$$

The graph of D versus t is shown in Figure 0.1.18. ◄

■ **ISSUES OF SCALE AND UNITS**

In geometric problems where you want to preserve the "true" shape of a graph, you must use units of equal length on both axes. For example, if you graph a circle in a coordinate system in which 1 unit in the y-direction is smaller than 1 unit in the x-direction, then the circle will be squashed vertically into an elliptical shape (Figure 0.1.19).

However, sometimes it is inconvenient or impossible to display a graph using units of equal length. For example, consider the equation

$$y = x^2$$

If we want to show the portion of the graph over the interval $-3 \leq x \leq 3$, then there is no problem using units of equal length, since y only varies from 0 to 9 over that interval. However, if we want to show the portion of the graph over the interval $-10 \leq x \leq 10$, then there is a problem keeping the units equal in length, since the value of y varies between 0 and 100. In this case the only reasonable way to show all of the graph that occurs over the interval $-10 \leq x \leq 10$ is to compress the unit of length along the y-axis, as illustrated in Figure 0.1.20.

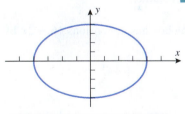

The circle is squashed because 1 unit on the y-axis has a smaller length than 1 unit on the x-axis.

▲ Figure 0.1.19

In applications where the variables on the two axes have unrelated units (say, centimeters on the y-axis and seconds on the x-axis), then nothing is gained by requiring the units to have equal lengths; choose the lengths to make the graph as clear as possible.

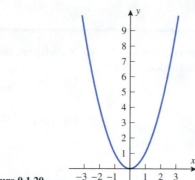
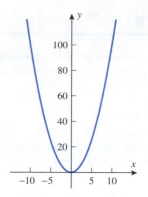

▶ **Figure 0.1.20**

✔ **QUICK CHECK EXERCISES 0.1**   *(See page 15 for answers.)*

---

1. Let $f(x) = \sqrt{x + 1} + 4$.
   (a) The natural domain of f is _____.
   (b) $f(3) = $_____
   (c) $f(t^2 - 1) = $_____
   (d) $f(x) = 7$ if $x = $_____
   (e) The range of f is _____.

2. Line segments in an xy-plane form "letters" as depicted.

   (a) If the y-axis is parallel to the letter I, which of the letters represent the graph of $y = f(x)$ for some function f?
   (b) If the y-axis is perpendicular to the letter I, which of the letters represent the graph of $y = f(x)$ for some function f?

3. The accompanying figure shows the complete graph of $y = f(x)$.
   (a) The domain of f is _____.
   (b) The range of f is _____.
   (c) $f(-3) = $_____
   (d) $f\left(\frac{1}{2}\right) = $_____
   (e) The solutions to $f(x) = -\frac{3}{2}$ are $x = $_____ and $x = $_____.

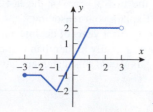

◀ **Figure Ex-3**

4. The accompanying table gives a 5-day forecast of high and low temperatures in degrees Fahrenheit (°F).
   (a) Suppose that x and y denote the respective high and low temperature predictions for each of the 5 days. Is y a function of x? If so, give the domain and range of this function.
   (b) Suppose that x and y denote the respective low and high temperature predictions for each of the 5 days. Is y a function of x? If so, give the domain and range of this function.

	MON	TUE	WED	THURS	FRI
HIGH	75	71	65	70	73
LOW	52	56	48	50	52

▲ **Table Ex-3**

5. Let l, w, and A denote the length, width, and area of a rectangle, respectively, and suppose that the width of the rectangle is half the length.
   (a) If l is expressed as a function of w, then $l = $_____.
   (b) If A is expressed as a function of l, then $A = $_____.
   (c) If w is expressed as a function of A, then $w = $_____.

EXERCISE SET 0.1   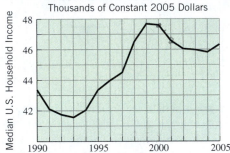 Graphing Utility

1. Use the accompanying graph to answer the following questions, making reasonable approximations where needed.
   (a) For what values of x is $y = 1$?
   (b) For what values of x is $y = 3$?
   (c) For what values of y is $x = 3$?
   (d) For what values of x is $y \leq 0$?
   (e) What are the maximum and minimum values of y and for what values of x do they occur?

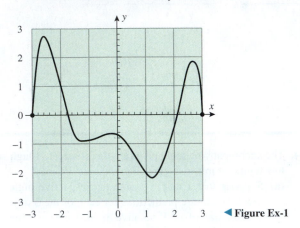

◀ **Figure Ex-1**

2. Use the accompanying table to answer the questions posed in Exercise 1.

x	-2	-1	0	2	3	4	5	6
y	5	1	-2	7	-1	1	0	9

▲ **Table Ex-2**

3. In each part of the accompanying figure, determine whether the graph defines y as a function of x.

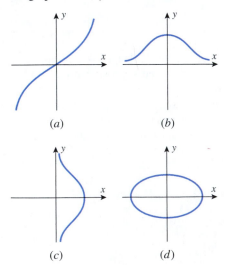

▲ **Figure Ex-3**

4. In each part, compare the natural domains of f and g.

   (a) $f(x) = \dfrac{x^2 + x}{x + 1}$; $g(x) = x$

   (b) $f(x) = \dfrac{x\sqrt{x} + \sqrt{x}}{x + 1}$; $g(x) = \sqrt{x}$

FOCUS ON CONCEPTS

5. The accompanying graph shows the median income in U.S. households (adjusted for inflation) between 1990 and 2005. Use the graph to answer the following questions, making reasonable approximations where needed.
   (a) When was the median income at its maximum value, and what was the median income when that occurred?
   (b) When was the median income at its minimum value, and what was the median income when that occurred?
   (c) The median income was declining during the 2-year period between 2000 and 2002. Was it declining more rapidly during the first year or the second year of that period? Explain your reasoning.

Median U.S. Household Income in Thousands of Constant 2005 Dollars

Source: U.S. Census Bureau, August 2006.

▲ **Figure Ex-5**

6. Use the median income graph in Exercise 5 to answer the following questions, making reasonable approximations where needed.
   (a) What was the average yearly growth of median income between 1993 and 1999?
   (b) The median income was increasing during the 6-year period between 1993 and 1999. Was it increasing more rapidly during the first 3 years or the last 3 years of that period? Explain your reasoning.
   (c) Consider the statement: "After years of decline, median income this year was finally higher than that of last year." In what years would this statement have been correct?

7. Find $f(0)$, $f(2)$, $f(-2)$, $f(3)$, $f(\sqrt{2})$, and $f(3t)$.

(a) $f(x) = 3x^2 - 2$

(b) $f(x) = \begin{cases} \dfrac{1}{x}, & x > 3 \\ 2x, & x \leq 3 \end{cases}$

8. Find $g(3)$, $g(-1)$, $g(\pi)$, $g(-1.1)$, and $g(t^2 - 1)$.

(a) $g(x) = \dfrac{x+1}{x-1}$

(b) $g(x) = \begin{cases} \sqrt{x+1}, & x \geq 1 \\ 3, & x < 1 \end{cases}$

9–10 Find the natural domain and determine the range of each function. If you have a graphing utility, use it to confirm that your result is consistent with the graph produced by your graphing utility. [*Note:* Set your graphing utility in radian mode when graphing trigonometric functions.] ■

9. (a) $f(x) = \dfrac{1}{x-3}$

(b) $F(x) = \dfrac{x}{|x|}$

(c) $g(x) = \sqrt{x^2 - 3}$

(d) $G(x) = \sqrt{x^2 - 2x + 5}$

(e) $h(x) = \dfrac{1}{1 - \sin x}$

(f) $H(x) = \sqrt{\dfrac{x^2 - 4}{x - 2}}$

10. (a) $f(x) = \sqrt{3 - x}$

(b) $F(x) = \sqrt{4 - x^2}$

(c) $g(x) = 3 + \sqrt{x}$

(d) $G(x) = x^3 + 2$

(e) $h(x) = 3\sin x$

(f) $H(x) = (\sin\sqrt{x})^{-2}$

FOCUS ON CONCEPTS

11. (a) If you had a device that could record the Earth's population continuously, would you expect the graph of population versus time to be a continuous (unbroken) curve? Explain what might cause breaks in the curve.

(b) Suppose that a hospital patient receives an injection of an antibiotic every 8 hours and that between injections the concentration C of the antibiotic in the bloodstream decreases as the antibiotic is absorbed by the tissues. What might the graph of C versus the elapsed time t look like?

12. (a) If you had a device that could record the temperature of a room continuously over a 24-hour period, would you expect the graph of temperature versus time to be a continuous (unbroken) curve? Explain your reasoning.

(b) If you had a computer that could track the number of boxes of cereal on the shelf of a market continuously over a 1-week period, would you expect the graph of the number of boxes on the shelf versus time to be a continuous (unbroken) curve? Explain your reasoning.

13. A boat is bobbing up and down on some gentle waves. Suddenly it gets hit by a large wave and sinks. Sketch a rough graph of the height of the boat above the ocean floor as a function of time.

14. A cup of hot coffee sits on a table. You pour in some cool milk and let it sit for an hour. Sketch a rough graph of the temperature of the coffee as a function of time.

15–18 As seen in Example 3, the equation $x^2 + y^2 = 25$ does not define y as a function of x. Each graph in these exercises is a portion of the circle $x^2 + y^2 = 25$. In each case, determine whether the graph defines y as a function of x, and if so, give a formula for y in terms of x. ■

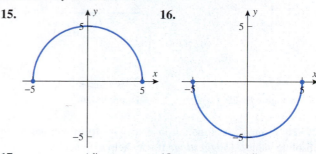

15.   **16.**

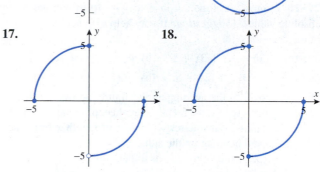

17.   **18.**

19–22 True–False Determine whether the statement is true or false. Explain your answer. ■

19. A curve that crosses the x-axis at two different points cannot be the graph of a function.

20. The natural domain of a real-valued function defined by a formula consists of all those real numbers for which the formula yields a real value.

21. The range of the absolute value function is all positive real numbers.

22. If $g(x) = 1/\sqrt{f(x)}$, then the domain of g consists of all those real numbers x for which $f(x) \neq 0$.

23. Use the equation $y = x^2 - 6x + 8$ to answer the following questions.

(a) For what values of x is $y = 0$?

(b) For what values of x is $y = -10$?

(c) For what values of x is $y \geq 0$?

(d) Does y have a minimum value? A maximum value? If so, find them.

24. Use the equation $y = 1 + \sqrt{x}$ to answer the following questions.

(a) For what values of x is $y = 4$?

(b) For what values of x is $y = 0$?

(c) For what values of x is $y \geq 6$?

(cont.)

(d) Does y have a minimum value? A maximum value? If so, find them.

25. As shown in the accompanying figure, a pendulum of constant length L makes an angle θ with its vertical position. Express the height h as a function of the angle θ.

26. Express the length L of a chord of a circle with radius 10 cm as a function of the central angle θ (see the accompanying figure).

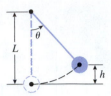

▲ **Figure Ex-25**   ▲ **Figure Ex-26**

27–28 Express the function in piecewise form without using absolute values. [*Suggestion:* It may help to generate the graph of the function.] ■

27. (a) $f(x) = |x| + 3x + 1$ (b) $g(x) = |x| + |x - 1|$

28. (a) $f(x) = 3 + |2x - 5|$ (b) $g(x) = 3|x - 2| - |x + 1|$

29. As shown in the accompanying figure, an open box is to be constructed from a rectangular sheet of metal, 8 in by 15 in, by cutting out squares with sides of length x from each corner and bending up the sides.
    (a) Express the volume V as a function of x.
    (b) Find the domain of V.
    (c) Plot the graph of the function V obtained in part (a) and estimate the range of this function.
    (d) In words, describe how the volume V varies with x, and discuss how one might construct boxes of maximum volume.

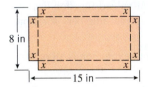
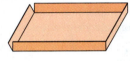

▲ **Figure Ex-29**

30. Repeat Exercise 29 assuming the box is constructed in the same fashion from a 6-inch-square sheet of metal.

31. A construction company has adjoined a 1000 ft² rectangular enclosure to its office building. Three sides of the enclosure are fenced in. The side of the building adjacent to the enclosure is 100 ft long and a portion of this side is used as the fourth side of the enclosure. Let x and y be the dimensions of the enclosure, where x is measured parallel to the building, and let L be the length of fencing required for those dimensions.
    (a) Find a formula for L in terms of x and y.
    (b) Find a formula that expresses L as a function of x alone.
    (c) What is the domain of the function in part (b)?

(d) Plot the function in part (b) and estimate the dimensions of the enclosure that minimize the amount of fencing required.

32. As shown in the accompanying figure, a camera is mounted at a point 3000 ft from the base of a rocket launching pad. The rocket rises vertically when launched, and the camera's elevation angle is continually adjusted to follow the bottom of the rocket.
    (a) Express the height x as a function of the elevation angle θ.
    (b) Find the domain of the function in part (a).
    (c) Plot the graph of the function in part (a) and use it to estimate the height of the rocket when the elevation angle is $\pi/4 \approx 0.7854$ radian. Compare this estimate to the exact height.

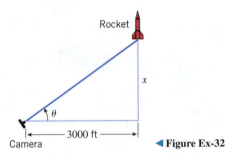

◀ **Figure Ex-32**

33. A soup company wants to manufacture a can in the shape of a right circular cylinder that will hold 500 cm³ of liquid. The material for the top and bottom costs 0.02 cent/cm², and the material for the sides costs 0.01 cent/cm².
    (a) Estimate the radius r and the height h of the can that costs the least to manufacture. [*Suggestion:* Express the cost C in terms of r.]
    (b) Suppose that the tops and bottoms of radius r are punched out from square sheets with sides of length $2r$ and the scraps are waste. If you allow for the cost of the waste, would you expect the can of least cost to be taller or shorter than the one in part (a)? Explain.
    (c) Estimate the radius, height, and cost of the can in part (b), and determine whether your conjecture was correct.

34. The designer of a sports facility wants to put a quarter-mile (1320 ft) running track around a football field, oriented as in the accompanying figure on the next page. The football field is 360 ft long (including the end zones) and 160 ft wide. The track consists of two straightaways and two semicircles, with the straightaways extending at least the length of the football field.
    (a) Show that it is possible to construct a quarter-mile track around the football field. [*Suggestion:* Find the shortest track that can be constructed around the field.]
    (b) Let L be the length of a straightaway (in feet), and let x be the distance (in feet) between a sideline of the football field and a straightaway. Make a graph of L versus x. *(cont.)*

(c) Use the graph to estimate the value of x that produces the shortest straightaways, and then find this value of x exactly.

(d) Use the graph to estimate the length of the longest possible straightaways, and then find that length exactly.

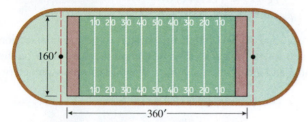

▲ **Figure Ex-34**

35–36 (i) Explain why the function f has one or more holes in its graph, and state the x-values at which those holes occur. (ii) Find a function g whose graph is identical to that of f, but without the holes. ■

35. $f(x) = \dfrac{(x+2)(x^2-1)}{(x+2)(x-1)}$ **36.** $f(x) = \dfrac{x^2+|x|}{|x|}$

37. In 2001 the National Weather Service introduced a new wind chill temperature (WCT) index. For a given outside temper-ature T and wind speed v, the wind chill temperature index is the equivalent temperature that exposed skin would feel with a wind speed of v mi/h. Based on a more accurate model of cooling due to wind, the new formula is

$$\text{WCT} = \begin{cases} T, & 0 \le v \le 3 \\ 35.74 + 0.6215T - 35.75v^{0.16} + 0.4275Tv^{0.16}, & 3 < v \end{cases}$$

where T is the temperature in °F, v is the wind speed in mi/h, and WCT is the equivalent temperature in °F. Find the WCT to the nearest degree if $T = 25°$F and
(a) $v = 3$ mi/h   (b) $v = 15$ mi/h   (c) $v = 46$ mi/h.

Source: Adapted from UMAP Module 658, *Windchill*, W. Bosch and L. Cobb, COMAP, Arlington, MA.

38–40 Use the formula for the wind chill temperature index described in Exercise 37. ■

38. Find the air temperature to the nearest degree if the WCT is reported as $-60°$F with a wind speed of 48 mi/h.

39. Find the air temperature to the nearest degree if the WCT is reported as $-10°$F with a wind speed of 48 mi/h.

40. Find the wind speed to the nearest mile per hour if the WCT is reported as $5°$F with an air temperature of $20°$F.

✔ **QUICK CHECK ANSWERS 0.1**

1. (a) $[-1, +\infty)$ (b) 6 (c) $|t| + 4$ (d) 8 (e) $[4, +\infty)$ **2.** (a) M (b) I   **3.** (a) $[-3, 3)$ (b) $[-2, 2]$ (c) -1 (d) 1 (e) $-\frac{3}{4}$; $-\frac{3}{2}$ **4.** (a) yes; domain: $\{65, 70, 71, 73, 75\}$; range: $\{48, 50, 52, 56\}$ (b) no   **5.** (a) $l = 2w$ (b) $A = l^2/2$ (c) $w = \sqrt{A/2}$

0.2 NEW FUNCTIONS FROM OLD

Just as numbers can be added, subtracted, multiplied, and divided to produce other numbers, so functions can be added, subtracted, multiplied, and divided to produce other functions. In this section we will discuss these operations and some others that have no analogs in ordinary arithmetic.

■ **ARITHMETIC OPERATIONS ON FUNCTIONS**

Two functions, f and g, can be added, subtracted, multiplied, and divided in a natural way to form new functions $f + g$, $f - g$, fg, and f/g. For example, $f + g$ is defined by the formula

$$(f + g)(x) = f(x) + g(x) \tag{1}$$

which states that for each input the value of $f + g$ is obtained by adding the values of f and g. Equation (1) provides a formula for $f + g$ but does not say anything about the domain of $f + g$. However, for the right side of this equation to be defined, x must lie in the domains of both f and g, so we define the domain of $f + g$ to be the intersection of these two domains. More generally, we make the following definition.

0.2.1 **DEFINITION** Given functions f and g, we define

$$(f + g)(x) = f(x) + g(x)$$
$$(f - g)(x) = f(x) - g(x)$$
$$(fg)(x) = f(x)g(x)$$
$$(f/g)(x) = f(x)/g(x)$$

For the functions $f + g$, $f - g$, and fg we define the domain to be the intersection of the domains of f and g, and for the function f/g we define the domain to be the intersection of the domains of f and g but with the points where $g(x) = 0$ excluded (to avoid division by zero).

If f is a constant function, that is, $f(x) = c$ for all x, then the product of f and g is cg, so multiplying a function by a constant is a special case of multiplying two functions.

▶ **Example 1** Let

$$f(x) = 1 + \sqrt{x - 2} \quad \text{and} \quad g(x) = x - 3$$

Find the domains and formulas for the functions $f + g$, $f - g$, fg, f/g, and $7f$.

Solution. First, we will find the formulas and then the domains. The formulas are

$$(f + g)(x) = f(x) + g(x) = (1 + \sqrt{x - 2}) + (x - 3) = x - 2 + \sqrt{x - 2} \quad (2)$$
$$(f - g)(x) = f(x) - g(x) = (1 + \sqrt{x - 2}) - (x - 3) = 4 - x + \sqrt{x - 2} \quad (3)$$
$$(fg)(x) = f(x)g(x) = (1 + \sqrt{x - 2})(x - 3) \quad (4)$$
$$(f/g)(x) = f(x)/g(x) = \frac{1 + \sqrt{x - 2}}{x - 3} \quad (5)$$
$$(7f)(x) = 7f(x) = 7 + 7\sqrt{x - 2} \quad (6)$$

The domains of f and g are $[2, +\infty)$ and $(-\infty, +\infty)$, respectively (their natural domains). Thus, it follows from Definition 0.2.1 that the domains of $f + g$, $f - g$, and fg are the intersection of these two domains, namely,

$$[2, +\infty) \cap (-\infty, +\infty) = [2, +\infty) \quad (7)$$

Moreover, since $g(x) = 0$ if $x = 3$, the domain of f/g is (7) with $x = 3$ removed, namely,

$$[2, 3) \cup (3, +\infty)$$

Finally, the domain of $7f$ is the same as the domain of f. ◀

We saw in the last example that the domains of the functions $f + g$, $f - g$, fg, and f/g were the natural domains resulting from the formulas obtained for these functions. The following example shows that this will not always be the case.

▶ **Example 2** Show that if $f(x) = \sqrt{x}$, $g(x) = \sqrt{x}$, and $h(x) = x$, then the domain of fg is not the same as the natural domain of h.

Solution. The natural domain of $h(x) = x$ is $(-\infty, +\infty)$. Note that

$$(fg)(x) = \sqrt{x}\sqrt{x} = x = h(x)$$

on the domain of fg. The domains of both f and g are $[0, +\infty)$, so the domain of fg is

$$[0, +\infty) \cap [0, +\infty) = [0, +\infty)$$

by Definition 0.2.1. Since the domains of fg and h are different, it would be misleading to write $(fg)(x) = x$ without including the restriction that this formula holds only for $x \geq 0$. ◄

■ COMPOSITION OF FUNCTIONS

We now consider an operation on functions, called *composition*, which has no direct analog in ordinary arithmetic. Informally stated, the operation of composition is performed by substituting some function for the independent variable of another function. For example, suppose that

$$f(x) = x^2 \quad \text{and} \quad g(x) = x + 1$$

If we substitute $g(x)$ for x in the formula for f, we obtain a new function

$$f(g(x)) = (g(x))^2 = (x + 1)^2$$

which we denote by $f \circ g$. Thus,

$$(f \circ g)(x) = f(g(x)) = (g(x))^2 = (x + 1)^2$$

In general, we make the following definition.

Although the domain of $f \circ g$ may seem complicated at first glance, it makes sense intuitively: To compute $f(g(x))$ one needs x in the domain of g to compute $g(x)$, and one needs $g(x)$ in the domain of f to compute $f(g(x))$.

0.2.2 DEFINITION Given functions f and g, the *composition* of f with g, denoted by $f \circ g$, is the function defined by

$$(f \circ g)(x) = f(g(x))$$

The domain of $f \circ g$ is defined to consist of all x in the domain of g for which $g(x)$ is in the domain of f.

▶ **Example 3** Let $f(x) = x^2 + 3$ and $g(x) = \sqrt{x}$. Find

$$\text{(a)} \ (f \circ g)(x) \qquad \text{(b)} \ (g \circ f)(x)$$

Solution (a). The formula for $f(g(x))$ is

$$f(g(x)) = [g(x)]^2 + 3 = (\sqrt{x})^2 + 3 = x + 3$$

Since the domain of g is $[0, +\infty)$ and the domain of f is $(-\infty, +\infty)$, the domain of $f \circ g$ consists of all x in $[0, +\infty)$ such that $g(x) = \sqrt{x}$ lies in $(-\infty, +\infty)$; thus, the domain of $f \circ g$ is $[0, +\infty)$. Therefore,

$$(f \circ g)(x) = x + 3, \quad x \geq 0$$

Solution (b). The formula for $g(f(x))$ is

$$g(f(x)) = \sqrt{f(x)} = \sqrt{x^2 + 3}$$

Since the domain of f is $(-\infty, +\infty)$ and the domain of g is $[0, +\infty)$, the domain of $g \circ f$ consists of all x in $(-\infty, +\infty)$ such that $f(x) = x^2 + 3$ lies in $[0, +\infty)$. Thus, the domain of $g \circ f$ is $(-\infty, +\infty)$. Therefore,

Note that the functions $f \circ g$ and $g \circ f$ in Example 3 are not the same. Thus, the order in which functions are composed can (and usually will) make a difference in the end result.

$$(g \circ f)(x) = \sqrt{x^2 + 3}$$

There is no need to indicate that the domain is $(-\infty, +\infty)$, since this is the natural domain of $\sqrt{x^2 + 3}$. ◄

Compositions can also be defined for three or more functions; for example, $(f \circ g \circ h)(x)$ is computed as

$$(f \circ g \circ h)(x) = f(g(h(x)))$$

In other words, first find $h(x)$, then find $g(h(x))$, and then find $f(g(h(x)))$.

▶ **Example 4** Find $(f \circ g \circ h)(x)$ if

$$f(x) = \sqrt{x}, \quad g(x) = 1/x, \quad h(x) = x^3$$

Solution.

$$(f \circ g \circ h)(x) = f(g(h(x))) = f(g(x^3)) = f(1/x^3) = \sqrt{1/x^3} = 1/x^{3/2} \quad ◀$$

■ EXPRESSING A FUNCTION AS A COMPOSITION

Many problems in mathematics are solved by "decomposing" functions into compositions of simpler functions. For example, consider the function h given by

$$h(x) = (x + 1)^2$$

To evaluate $h(x)$ for a given value of x, we would first compute $x + 1$ and then square the result. These two operations are performed by the functions

$$g(x) = x + 1 \quad \text{and} \quad f(x) = x^2$$

We can express h in terms of f and g by writing

$$h(x) = (x + 1)^2 = [g(x)]^2 = f(g(x))$$

so we have succeeded in expressing h as the composition $h = f \circ g$.

The thought process in this example suggests a general procedure for decomposing a function h into a composition $h = f \circ g$:

- Think about how you would evaluate $h(x)$ for a specific value of x, trying to break the evaluation into two steps performed in succession.
- The first operation in the evaluation will determine a function g and the second a function f.
- The formula for h can then be written as $h(x) = f(g(x))$.

For descriptive purposes, we will refer to g as the "inside function" and f as the "outside function" in the expression $f(g(x))$. The inside function performs the first operation and the outside function performs the second.

▶ **Example 5** Express $\sin(x^3)$ as a composition of two functions.

Solution. To evaluate $\sin(x^3)$, we would first compute x^3 and then take the sine, so $g(x) = x^3$ is the inside function and $f(x) = \sin x$ the outside function. Therefore,

$$\sin(x^3) = f(g(x)) \qquad \boxed{g(x) = x^3 \text{ and } f(x) = \sin x} \quad ◀$$

Table 0.2.1 gives some more examples of decomposing functions into compositions.

Table 0.2.1
COMPOSING FUNCTIONS

FUNCTION	$g(x)$ INSIDE	$f(x)$ OUTSIDE	COMPOSITION
$(x^2 + 1)^{10}$	$x^2 + 1$	x^{10}	$(x^2 + 1)^{10} = f(g(x))$
$\sin^3 x$	$\sin x$	x^3	$\sin^3 x = f(g(x))$
$\tan(x^5)$	x^5	$\tan x$	$\tan(x^5) = f(g(x))$
$\sqrt{4 - 3x}$	$4 - 3x$	$\sqrt{x}$	$\sqrt{4 - 3x} = f(g(x))$
$8 + \sqrt{x}$	$\sqrt{x}$	$8 + x$	$8 + \sqrt{x} = f(g(x))$
$\dfrac{1}{x + 1}$	$x + 1$	$\dfrac{1}{x}$	$\dfrac{1}{x + 1} = f(g(x))$

REMARK   There is always more than one way to express a function as a composition. For example, here are two ways to express $(x^2 + 1)^{10}$ as a composition that differ from that in Table 0.2.1:

$$(x^2 + 1)^{10} = [(x^2 + 1)^2]^5 = f(g(x)) \qquad \boxed{g(x) = (x^2 + 1)^2 \text{ and } f(x) = x^5}$$

$$(x^2 + 1)^{10} = [(x^2 + 1)^3]^{10/3} = f(g(x)) \qquad \boxed{g(x) = (x^2 + 1)^3 \text{ and } f(x) = x^{10/3}}$$

■ **NEW FUNCTIONS FROM OLD**

The remainder of this section will be devoted to considering the geometric effect of performing basic operations on functions. This will enable us to use known graphs of functions to visualize or sketch graphs of related functions. For example, Figure 0.2.1 shows the graphs of yearly new car sales $N(t)$ and used car sales $U(t)$ over a certain time period. Those graphs can be used to construct the graph of the total car sales

$$T(t) = N(t) + U(t)$$

by adding the values of $N(t)$ and $U(t)$ for each value of t. In general, the graph of $y = f(x) + g(x)$ can be constructed from the graphs of $y = f(x)$ and $y = g(x)$ by adding corresponding y-values for each x.

▶ **Example 6**   Referring to Figure 0.1.4 for the graphs of $y = \sqrt{x}$ and $y = 1/x$, make a sketch that shows the general shape of the graph of $y = \sqrt{x} + 1/x$ for $x \geq 0$.

Solution.   To add the corresponding y-values of $y = \sqrt{x}$ and $y = 1/x$ graphically, just imagine them to be "stacked" on top of one another. This yields the sketch in Figure 0.2.2.   ◀

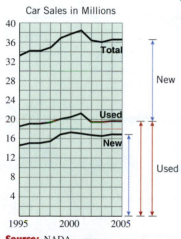

Car Sales in Millions

Source: NADA.

▲ **Figure 0.2.1**

Use the technique in Example 6 to sketch the graph of the function

$$\sqrt{x} - \frac{1}{x}$$

▶ **Figure 0.2.2**

Add the y-coordinates of $\sqrt{x}$ and $1/x$ to obtain the y-coordinate of $\sqrt{x} + 1/x$.

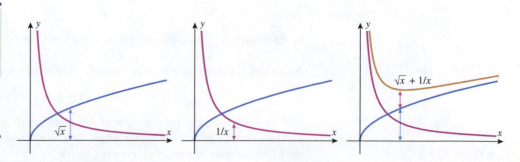

■ TRANSLATIONS

Table 0.2.2 illustrates the geometric effect on the graph of $y = f(x)$ of adding or subtracting a *positive* constant c to f or to its independent variable x. For example, the first result in the table illustrates that adding a positive constant c to a function f adds c to each y-coordinate of its graph, thereby shifting the graph of f up by c units. Similarly, subtracting c from f shifts the graph down by c units. On the other hand, if a positive constant c is added to x, then the value of $y = f(x + c)$ at $x - c$ is $f(x)$; and since the point $x - c$ is c units to the left of x on the x-axis, the graph of $y = f(x + c)$ must be the graph of $y = f(x)$ shifted left by c units. Similarly, subtracting c from x shifts the graph of $y = f(x)$ right by c units.

Table 0.2.2
TRANSLATION PRINCIPLES

OPERATION ON $y = f(x)$	Add a positive constant c to $f(x)$	Subtract a positive constant c from $f(x)$	Add a positive constant c to x	Subtract a positive constant c from x
NEW EQUATION	$y = f(x) + c$	$y = f(x) - c$	$y = f(x + c)$	$y = f(x - c)$
GEOMETRIC EFFECT	Translates the graph of $y = f(x)$ up c units	Translates the graph of $y = f(x)$ down c units	Translates the graph of $y = f(x)$ left c units	Translates the graph of $y = f(x)$ right c units
EXAMPLE	$y = x^2 + 2$, $y = x^2$	$y = x^2$, $y = x^2 - 2$	$y = (x + 2)^2$, $y = x^2$	$y = x^2$, $y = (x - 2)^2$

Before proceeding to the next examples, it will be helpful to review the graphs in Figures 0.1.4 and 0.1.9.

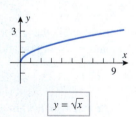

$y = \sqrt{x}$

▶ **Example 7**   Sketch the graph of

$$\text{(a) } y = \sqrt{x - 3} \qquad \text{(b) } y = \sqrt{x + 3}$$

Solution.   Using the translation principles given in Table 0.2.2, the graph of the equation $y = \sqrt{x - 3}$ can be obtained by translating the graph of $y = \sqrt{x}$ right 3 units. The graph of $y = \sqrt{x + 3}$ can be obtained by translating the graph of $y = \sqrt{x}$ left 3 units (Figure 0.2.3).   ◀

$y = \sqrt{x - 3}$

▶ **Example 8**   Sketch the graph of $y = x^2 - 4x + 5$.

Solution.   Completing the square on the first two terms yields

$$y = (x^2 - 4x + 4) - 4 + 5 = (x - 2)^2 + 1$$

(see Web Appendix H for a review of this technique). In this form we see that the graph can be obtained by translating the graph of $y = x^2$ right 2 units because of the $x - 2$, and up 1 unit because of the $+1$ (Figure 0.2.4).   ◀

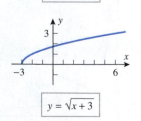

$y = \sqrt{x + 3}$

▲ **Figure 0.2.3**

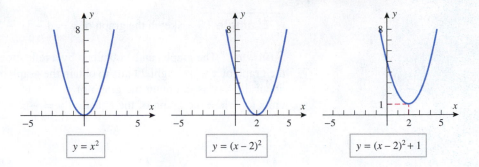

► **Figure 0.2.4**

■ REFLECTIONS

The graph of $y = f(-x)$ is the reflection of the graph of $y = f(x)$ about the y-axis because the point (x, y) on the graph of $f(x)$ is replaced by $(-x, y)$. Similarly, the graph of $y = -f(x)$ is the reflection of the graph of $y = f(x)$ about the x-axis because the point (x, y) on the graph of $f(x)$ is replaced by $(x, -y)$ [the equation $y = -f(x)$ is equivalent to $-y = f(x)$]. This is summarized in Table 0.2.3.

Table 0.2.3
REFLECTION PRINCIPLES

OPERATION ON $y = f(x)$	Replace x by $-x$	Multiply $f(x)$ by -1
NEW EQUATION	$y = f(-x)$	$y = -f(x)$
GEOMETRIC EFFECT	Reflects the graph of $y = f(x)$ about the y-axis	Reflects the graph of $y = f(x)$ about the x-axis
EXAMPLE	$y = \sqrt{-x}$ $y = \sqrt{x}$	$y = \sqrt{x}$ $y = -\sqrt{x}$

► **Example 9**   Sketch the graph of $y = \sqrt[3]{2 - x}$.

Solution.   Using the translation and reflection principles in Tables 0.2.2 and 0.2.3, we can obtain the graph by a reflection followed by a translation as follows: First reflect the graph of $y = \sqrt[3]{x}$ about the y-axis to obtain the graph of $y = \sqrt[3]{-x}$, then translate this graph right 2 units to obtain the graph of the equation $y = \sqrt[3]{-(x - 2)} = \sqrt[3]{2 - x}$ (Figure 0.2.5). ◄

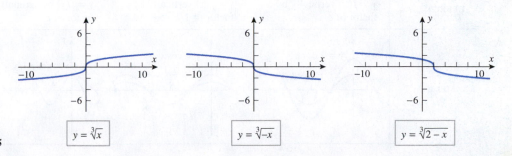

► **Figure 0.2.5**

▶ **Example 10** Sketch the graph of $y = 4 - |x - 2|$.

Solution. The graph can be obtained by a reflection and two translations: First translate the graph of $y = |x|$ right 2 units to obtain the graph of $y = |x - 2|$; then reflect this graph about the x-axis to obtain the graph of $y = -|x - 2|$; and then translate this graph up 4 units to obtain the graph of the equation $y = -|x - 2| + 4 = 4 - |x - 2|$ (Figure 0.2.6). ◀

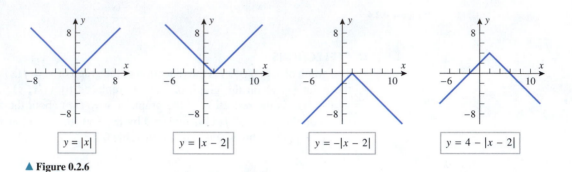

| $y = |x|$ | $y = |x - 2|$ | $y = -|x - 2|$ | $y = 4 - |x - 2|$ |

▲ **Figure 0.2.6**

■ **STRETCHES AND COMPRESSIONS**

Multiplying $f(x)$ by a *positive* constant c has the geometric effect of stretching the graph of $y = f(x)$ in the y-direction by a factor of c if $c > 1$ and compressing it in the y-direction by a factor of $1/c$ if $0 < c < 1$. For example, multiplying $f(x)$ by 2 doubles each y-coordinate, thereby stretching the graph vertically by a factor of 2, and multiplying by $\frac{1}{2}$ cuts each y-coordinate in half, thereby compressing the graph vertically by a factor of 2. Similarly, multiplying x by a *positive* constant c has the geometric effect of compressing the graph of $y = f(x)$ by a factor of c in the x-direction if $c > 1$ and stretching it by a factor of $1/c$ if $0 < c < 1$. [If this seems backwards to you, then think of it this way: The value of $2x$ changes twice as fast as x, so a point moving along the x-axis from the origin will only have to move half as far for $y = f(2x)$ to have the same value as $y = f(x)$, thereby creating a horizontal compression of the graph.] All of this is summarized in Table 0.2.4.

> Describe the geometric effect of multiplying a function f by a *negative* constant in terms of reflection and stretching or compressing. What is the geometric effect of multiplying the independent variable of a function f by a *negative* constant?

Table 0.2.4
STRETCHING AND COMPRESSING PRINCIPLES

OPERATION ON $y = f(x)$	Multiply $f(x)$ by c $(c > 1)$	Multiply $f(x)$ by c $(0 < c < 1)$	Multiply x by c $(c > 1)$	Multiply x by c $(0 < c < 1)$
NEW EQUATION	$y = cf(x)$	$y = cf(x)$	$y = f(cx)$	$y = f(cx)$
GEOMETRIC EFFECT	Stretches the graph of $y = f(x)$ vertically by a factor of c	Compresses the graph of $y = f(x)$ vertically by a factor of $1/c$	Compresses the graph of $y = f(x)$ horizontally by a factor of c	Stretches the graph of $y = f(x)$ horizontally by a factor of $1/c$
EXAMPLE	$y = 2\cos x$ $y = \cos x$	$y = \cos x$ $y = \frac{1}{2}\cos x$	$y = \cos x$ $y = \cos 2x$	$y = \cos \frac{1}{2}x$ $y = \cos x$

■ SYMMETRY

Figure 0.2.7 illustrates three types of symmetries: ***symmetry about the x-axis***, ***symmetry about the y-axis***, and ***symmetry about the origin***. As illustrated in the figure, a curve is symmetric about the x-axis if for each point (x, y) on the graph the point $(x, -y)$ is also on the graph, and it is symmetric about the y-axis if for each point (x, y) on the graph the point $(-x, y)$ is also on the graph. A curve is symmetric about the origin if for each point (x, y) on the graph, the point $(-x, -y)$ is also on the graph. (Equivalently, a graph is symmetric about the origin if rotating the graph $180°$ about the origin leaves it unchanged.) This suggests the following symmetry tests.

Explain why the graph of a nonzero function cannot be symmetric about the x-axis.

Symmetric about
the x-axis

Symmetric about
the y-axis

Symmetric about
the origin

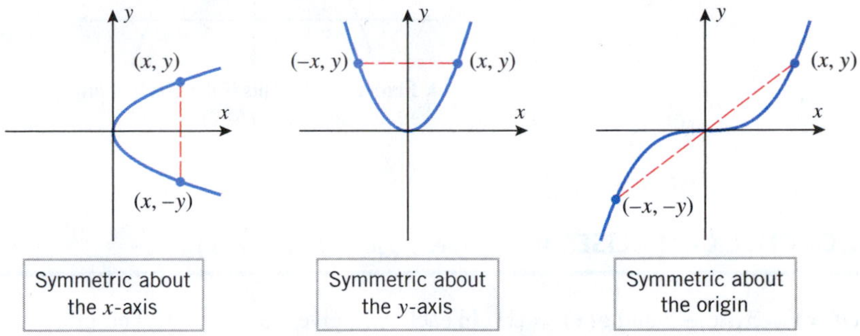

▶ **Figure 0.2.7**

0.2.3   THEOREM (*Symmetry Tests*)

(*a*)   *A plane curve is symmetric about the y-axis if and only if replacing x by −x in its equation produces an equivalent equation.*

(*b*)   *A plane curve is symmetric about the x-axis if and only if replacing y by −y in its equation produces an equivalent equation.*

(*c*)   *A plane curve is symmetric about the origin if and only if replacing both x by −x and y by −y in its equation produces an equivalent equation.*

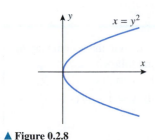

▲ **Figure 0.2.8**

▶ **Example 11**   Use Theorem 0.2.3 to identify symmetries in the graph of $x = y^2$.

Solution.   Replacing y by $-y$ yields $x = (-y)^2$, which simplifies to the original equation $x = y^2$. Thus, the graph is symmetric about the x-axis. The graph is not symmetric about the y-axis because replacing x by $-x$ yields $-x = y^2$, which is not equivalent to the original equation $x = y^2$. Similarly, the graph is not symmetric about the origin because replacing x by $-x$ and y by $-y$ yields $-x = (-y)^2$, which simplifies to $-x = y^2$, and this is again not equivalent to the original equation. These results are consistent with the graph of $x = y^2$ shown in Figure 0.2.8. ◀

■ EVEN AND ODD FUNCTIONS

A function f is said to be an ***even function*** if

$$f(-x) = f(x) \tag{8}$$

and is said to be an ***odd function*** if

$$f(-x) = -f(x) \tag{9}$$

Geometrically, the graphs of even functions are symmetric about the y-axis because replacing x by $-x$ in the equation $y = f(x)$ yields $y = f(-x)$, which is equivalent to the original equation $y = f(x)$ by (8) (see Figure 0.2.9). Similarly, it follows from (9) that graphs of odd functions are symmetric about the origin (see Figure 0.2.10). Some examples of even functions are x^2, x^4, x^6, and $\cos x$; and some examples of odd functions are x^3, x^5, x^7, and $\sin x$.

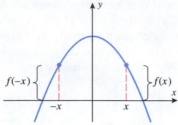

▲ **Figure 0.2.9**   This is the graph of an even function since $f(-x) = f(x)$.

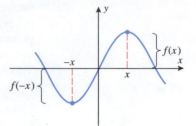

▲ **Figure 0.2.10**   This is the graph of an odd function since $f(-x) = -f(x)$.

✔ **QUICK CHECK EXERCISES 0.2**   *(See page 27 for answers.)*

1. Let $f(x) = 3\sqrt{x} - 2$ and $g(x) = |x|$. In each part, give the formula for the function and state the corresponding domain.
   (a) $f + g$: _____   Domain: _____
   (b) $f - g$: _____   Domain: _____
   (c) fg: _____   Domain: _____
   (d) f/g: _____   Domain: _____

2. Let $f(x) = 2 - x^2$ and $g(x) = \sqrt{x}$. In each part, give the formula for the composition and state the corresponding domain.
   (a) $f \circ g$: _____   Domain: _____
   (b) $g \circ f$: _____   Domain: _____

3. The graph of $y = 1 + (x - 2)^2$ may be obtained by shifting the graph of $y = x^2$ _____ (left/right) by _____ unit(s) and then shifting this new graph _____ (up/down) by _____ unit(s).

4. Let
$$f(x) = \begin{cases} |x + 1|, & -2 \le x \le 0 \\ |x - 1|, & 0 < x \le 2 \end{cases}$$
   (a) The letter of the alphabet that most resembles the graph of f is _____.
   (b) Is f an even function?

EXERCISE SET 0.2   Graphing Utility

FOCUS ON CONCEPTS

1. The graph of a function f is shown in the accompanying figure. Sketch the graphs of the following equations.
   (a) $y = f(x) - 1$ (b) $y = f(x - 1)$
   (c) $y = \frac{1}{2}f(x)$ (d) $y = f\left(-\frac{1}{2}x\right)$

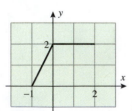

◀ **Figure Ex-1**

2. Use the graph in Exercise 1 to sketch the graphs of the following equations.
   (a) $y = -f(-x)$ (b) $y = f(2 - x)$
   (c) $y = 1 - f(2 - x)$ (d) $y = \frac{1}{2}f(2x)$

3. The graph of a function f is shown in the accompanying figure. Sketch the graphs of the following equations.
   (a) $y = f(x + 1)$ (b) $y = f(2x)$
   (c) $y = |f(x)|$ (d) $y = 1 - |f(x)|$

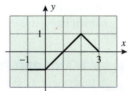

◀ **Figure Ex-3**

4. Use the graph in Exercise 3 to sketch the graph of the equation $y = f(|x|)$.

 5–24 Sketch the graph of the equation by translating, reflecting, compressing, and stretching the graph of $y = x^2$, $y = \sqrt{x}$, $y = 1/x$, $y = |x|$, or $y = \sqrt[3]{x}$ appropriately. Then use a graphing utility to confirm that your sketch is correct. ■

5. $y = -2(x+1)^2 - 3$ **6.** $y = \frac{1}{2}(x-3)^2 + 2$

7. $y = 1 + 2x - x^2$ **8.** $y = \frac{1}{2}(x^2 - 2x + 3)$

9. $y = 3 - \sqrt{x+1}$ **10.** $y = 1 + \sqrt{x-4}$

11. $y = \frac{1}{2}\sqrt{x} + 1$ **12.** $y = -\sqrt{3x}$

13. $y = \dfrac{1}{x-3}$ **14.** $y = \dfrac{1}{1-x}$

15. $y = 2 - \dfrac{1}{x+1}$ **16.** $y = \dfrac{x-1}{x}$

17. $y = |x+2| - 2$ **18.** $y = 1 - |x-3|$

19. $y = |2x-1| + 1$ **20.** $y = \sqrt{x^2 - 4x + 4}$

21. $y = 1 - 2\sqrt[3]{x}$ **22.** $y = \sqrt[3]{x-2} - 3$

23. $y = 2 + \sqrt[3]{x+1}$ **24.** $y + \sqrt[3]{x-2} = 0$

25. (a) Sketch the graph of $y = x + |x|$ by adding the corresponding y-coordinates on the graphs of $y = x$ and $y = |x|$.

    (b) Express the equation $y = x + |x|$ in piecewise form with no absolute values, and confirm that the graph you obtained in part (a) is consistent with this equation.

26. Sketch the graph of $y = x + (1/x)$ by adding corresponding y-coordinates on the graphs of $y = x$ and $y = 1/x$. Use a graphing utility to confirm that your sketch is correct.

27–28 Find formulas for $f + g$, $f - g$, fg, and f/g, and state the domains of the functions. ■

27. $f(x) = 2\sqrt{x-1}$, $g(x) = \sqrt{x-1}$

28. $f(x) = \dfrac{x}{1+x^2}$, $g(x) = \dfrac{1}{x}$

29. Let $f(x) = \sqrt{x}$ and $g(x) = x^3 + 1$. Find

    (a) $f(g(2))$ (b) $g(f(4))$ (c) $f(f(16))$

    (d) $g(g(0))$ (e) $f(2+h)$ (f) $g(3+h)$.

30. Let $g(x) = \sqrt{x}$. Find

    (a) $g(5s+2)$ (b) $g(\sqrt{x}+2)$ (c) $3g(5x)$

    (d) $\dfrac{1}{g(x)}$ (e) $g(g(x))$ (f) $(g(x))^2 - g(x^2)$

    (g) $g(1/\sqrt{x})$ (h) $g((x-1)^2)$ (i) $g(x+h)$.

31–34 Find formulas for $f \circ g$ and $g \circ f$, and state the domains of the compositions. ■

31. $f(x) = x^2$, $g(x) = \sqrt{1-x}$

32. $f(x) = \sqrt{x-3}$, $g(x) = \sqrt{x^2 + 3}$

33. $f(x) = \dfrac{1+x}{1-x}$, $g(x) = \dfrac{x}{1-x}$

34. $f(x) = \dfrac{x}{1+x^2}$, $g(x) = \dfrac{1}{x}$

35–40 Express f as a composition of two functions; that is, find g and h such that $f = g \circ h$. [*Note:* Each exercise has more than one solution.] ■

35. (a) $f(x) = \sqrt{x+2}$ (b) $f(x) = |x^2 - 3x + 5|$

36. (a) $f(x) = x^2 + 1$ (b) $f(x) = \dfrac{1}{x-3}$

37. (a) $f(x) = \sin^2 x$ (b) $f(x) = \dfrac{1}{5 + \cos x}$

38. (a) $f(x) = 3\sin(x^2)$ (b) $f(x) = 3\sin^2 x + 4\sin x$

39. (a) $f(x) = (1 + \sin(x^2))^3$ (b) $f(x) = \sqrt{1 - \sqrt[3]{x}}$

40. (a) $f(x) = \dfrac{1}{1 - x^2}$ (b) $f(x) = |5 + 2x|$

41–44 True–False Determine whether the statement is true or false. Explain your answer. ■

41. The domain of $f + g$ is the intersection of the domains of f and g.

42. The domain of $f \circ g$ consists of all values of x in the domain of g for which $g(x) \neq 0$.

43. The graph of an even function is symmetric about the y-axis.

44. The graph of $y = f(x+2) + 3$ is obtained by translating the graph of $y = f(x)$ right 2 units and up 3 units.

FOCUS ON CONCEPTS

45. Use the data in the accompanying table to make a plot of $y = f(g(x))$.

x	-3	-2	-1	0	1	2	3
$f(x)$	-4	-3	-2	-1	0	1	2
$g(x)$	-1	0	1	2	3	-2	-3

▲ **Table Ex-45**

46. Find the domain of $g \circ f$ for the functions f and g in Exercise 45.

47. Sketch the graph of $y = f(g(x))$ for the functions graphed in the accompanying figure.

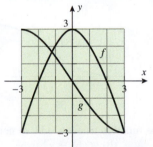

◀ **Figure Ex-47**

48. Sketch the graph of $y = g(f(x))$ for the functions graphed in Exercise 47.

49. Use the graphs of f and g in Exercise 47 to estimate the solutions of the equations $f(g(x)) = 0$ and $g(f(x)) = 0$.

50. Use the table given in Exercise 45 to solve the equations $f(g(x)) = 0$ and $g(f(x)) = 0$.

51–54 Find

$$\frac{f(x+h)-f(x)}{h} \quad \text{and} \quad \frac{f(w)-f(x)}{w-x}$$

Simplify as much as possible. ∎

51. $f(x) = 3x^2 - 5$ **52.** $f(x) = x^2 + 6x$

53. $f(x) = 1/x$ **54.** $f(x) = 1/x^2$

55. Classify the functions whose values are given in the accompanying table as even, odd, or neither.

x	−3	−2	−1	0	1	2	3
$f(x)$	5	3	2	3	1	−3	5
$g(x)$	4	1	−2	0	2	−1	−4
$h(x)$	2	−5	8	−2	8	−5	2

▲ Table Ex-55

56. Complete the accompanying table so that the graph of $y = f(x)$ is symmetric about
(a) the y-axis   (b) the origin.

x	−3	−2	−1	0	1	2	3
$f(x)$	1		−1	0		−5	

▲ Table Ex-56

57. The accompanying figure shows a portion of a graph. Complete the graph so that the entire graph is symmetric about
(a) the x-axis   (b) the y-axis   (c) the origin.

58. The accompanying figure shows a portion of the graph of a function f. Complete the graph assuming that
(a) f is an even function   (b) f is an odd function.

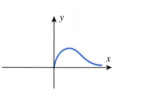

▲ Figure Ex-57

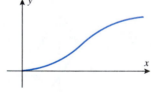

▲ Figure Ex-58

59. In each part, classify the function as even, odd, or neither.
(a) $f(x) = x^2$ (b) $f(x) = x^3$
(c) $f(x) = |x|$ (d) $f(x) = x + 1$
(e) $f(x) = \dfrac{x^5 - x}{1 + x^2}$ (f) $f(x) = 2$

60. Suppose that the function f has domain all real numbers. Determine whether each function can be classified as even or odd. Explain.
(a) $g(x) = \dfrac{f(x) + f(-x)}{2}$ (b) $h(x) = \dfrac{f(x) - f(-x)}{2}$

61. Suppose that the function f has domain all real numbers. Show that f can be written as the sum of an even function and an odd function. [*Hint:* See Exercise 60.]

62–63 Use Theorem 0.2.3 to determine whether the graph has symmetries about the x-axis, the y-axis, or the origin. ∎

62. (a) $x = 5y^2 + 9$ (b) $x^2 - 2y^2 = 3$
(c) $xy = 5$

63. (a) $x^4 = 2y^3 + y$ (b) $y = \dfrac{x}{3 + x^2}$
(c) $y^2 = |x| - 5$

64–65 (i) Use a graphing utility to graph the equation in the first quadrant. [*Note:* To do this you will have to solve the equation for y in terms of x.] (ii) Use symmetry to make a hand-drawn sketch of the entire graph. (iii) Confirm your work by generating the graph of the equation in the remaining three quadrants. ∎

64. $9x^2 + 4y^2 = 36$ **65.** $4x^2 + 16y^2 = 16$

66. The graph of the equation $x^{2/3} + y^{2/3} = 1$, which is shown in the accompanying figure, is called a ***four-cusped hypocycloid***.
(a) Use Theorem 0.2.3 to confirm that this graph is symmetric about the x-axis, the y-axis, and the origin.
(b) Find a function f whose graph in the first quadrant coincides with the four-cusped hypocycloid, and use a graphing utility to confirm your work.
(c) Repeat part (b) for the remaining three quadrants.

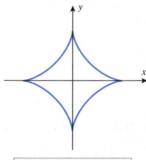

Four-cusped hypocycloid

◀ Figure Ex-66

67. The equation $y = |f(x)|$ can be written as

$$y = \begin{cases} f(x), & f(x) \geq 0 \\ -f(x), & f(x) < 0 \end{cases}$$

which shows that the graph of $y = |f(x)|$ can be obtained from the graph of $y = f(x)$ by retaining the portion that lies on or above the x-axis and reflecting about the x-axis the portion that lies below the x-axis. Use this method to obtain the graph of $y = |2x - 3|$ from the graph of $y = 2x - 3$.

68–69 Use the method described in Exercise 67. ∎

68. Sketch the graph of $y = |1 - x^2|$.

69. Sketch the graph of
(a) $f(x) = |\cos x|$
(b) $f(x) = \cos x + |\cos x|$.

70. The **greatest integer function**, $\lfloor x \rfloor$, is defined to be the greatest integer that is less than or equal to x. For example, $\lfloor 2.7 \rfloor = 2$, $\lfloor -2.3 \rfloor = -3$, and $\lfloor 4 \rfloor = 4$. In each part, sketch the graph of $y = f(x)$.
(a) $f(x) = \lfloor x \rfloor$
(b) $f(x) = \lfloor x^2 \rfloor$
(c) $f(x) = \lfloor x \rfloor^2$
(d) $f(x) = \lfloor \sin x \rfloor$

71. Is it ever true that $f \circ g = g \circ f$ if f and g are nonconstant functions? If not, prove it; if so, give some examples for which it is true.

✔ **QUICK CHECK ANSWERS 0.2**

1. (a) $(f + g)(x) = 3\sqrt{x} - 2 + x; \ x \geq 0$ (b) $(f - g)(x) = 3\sqrt{x} - 2 - x; \ x \geq 0$ (c) $(fg)(x) = 3x^{3/2} - 2x; \ x \geq 0$
(d) $(f/g)(x) = \dfrac{3\sqrt{x} - 2}{x}; \ x > 0$ **2.** (a) $(f \circ g)(x) = 2 - x; \ x \geq 0$ (b) $(g \circ f)(x) = \sqrt{2 - x^2}; \ -\sqrt{2} \leq x \leq \sqrt{2}$
3. right; 2; up; 1   **4.** (a) W (b) yes

0.3 **FAMILIES OF FUNCTIONS**

Functions are often grouped into families according to the form of their defining formulas or other common characteristics. In this section we will discuss some of the most basic families of functions.

■ **FAMILIES OF CURVES**

The graph of a constant function $f(x) = c$ is the graph of the equation $y = c$, which is the horizontal line shown in Figure 0.3.1a. If we vary c, then we obtain a set or *family* of horizontal lines such as those in Figure 0.3.1b.

Constants that are varied to produce families of curves are called *parameters*. For example, recall that an equation of the form $y = mx + b$ represents a line of slope m and y-intercept b. If we keep b fixed and treat m as a parameter, then we obtain a family of lines whose members all have y-intercept b (Figure 0.3.2a), and if we keep m fixed and treat b as a parameter, we obtain a family of parallel lines whose members all have slope m (Figure 0.3.2b).

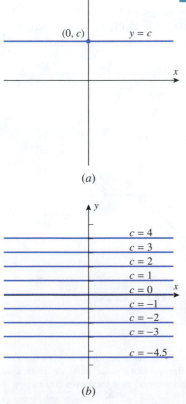

(a)

(b)

▲ **Figure 0.3.1**

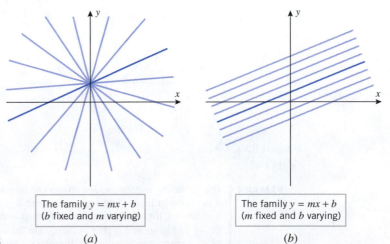

The family $y = mx + b$
(b fixed and m varying)

(a)

The family $y = mx + b$
(m fixed and b varying)

(b)

▶ **Figure 0.3.2**

■ POWER FUNCTIONS; THE FAMILY $y = x^n$

A function of the form $f(x) = x^p$, where p is constant, is called a **power function**. For the moment, let us consider the case where p is a positive integer, say $p = n$. The graphs of the curves $y = x^n$ for $n = 1, 2, 3, 4,$ and 5 are shown in Figure 0.3.3. The first graph is the line with slope 1 that passes through the origin, and the second is a parabola that opens up and has its vertex at the origin (see Web Appendix H).

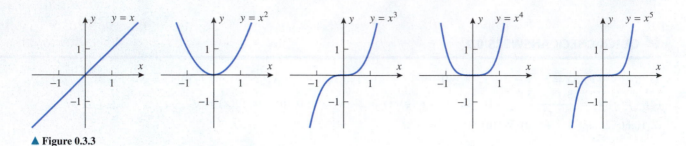

▲ **Figure 0.3.3**

For $n \geq 2$ the shape of the curve $y = x^n$ depends on whether n is even or odd (Figure 0.3.4):

- For even values of n, the functions $f(x) = x^n$ are even, so their graphs are symmetric about the y-axis. The graphs all have the general shape of the graph of $y = x^2$, and each graph passes through the points $(-1, 1)$, $(0, 0)$, and $(1, 1)$. As n increases, the graphs become flatter over the interval $-1 < x < 1$ and steeper over the intervals $x > 1$ and $x < -1$.

- For odd values of n, the functions $f(x) = x^n$ are odd, so their graphs are symmetric about the origin. The graphs all have the general shape of the curve $y = x^3$, and each graph passes through the points $(-1, -1)$, $(0, 0)$, and $(1, 1)$. As n increases, the graphs become flatter over the interval $-1 < x < 1$ and steeper over the intervals $x > 1$ and $x < -1$.

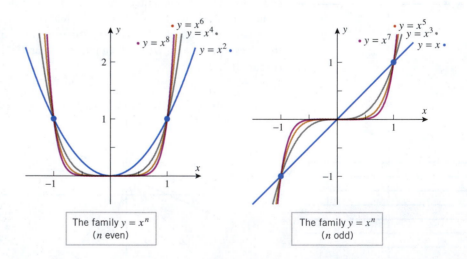

▶ **Figure 0.3.4**

REMARK   The flattening and steepening effects can be understood by considering what happens when a number x is raised to higher and higher powers: If $-1 < x < 1$, then the absolute value of x^n *decreases* as n increases, thereby causing the graphs to become flatter on this interval as n increases (try raising $\frac{1}{2}$ or $-\frac{1}{2}$ to higher and higher powers). On the other hand, if $x > 1$ or $x < -1$, then the absolute value of x^n *increases* as n increases, thereby causing the graphs to become steeper on these intervals as n increases (try raising 2 or -2 to higher and higher powers).

THE FAMILY $y = x^{-n}$

If p is a negative integer, say $p = -n$, then the power functions $f(x) = x^p$ have the form $f(x) = x^{-n} = 1/x^n$. Figure 0.3.5 shows the graphs of $y = 1/x$ and $y = 1/x^2$. The graph of $y = 1/x$ is called an ***equilateral hyperbola*** (for reasons to be discussed later).

As illustrated in Figure 0.3.5, the shape of the curve $y = 1/x^n$ depends on whether n is even or odd:

- For even values of n, the functions $f(x) = 1/x^n$ are even, so their graphs are symmetric about the y-axis. The graphs all have the general shape of the curve $y = 1/x^2$, and each graph passes through the points $(-1, 1)$ and $(1, 1)$. As n increases, the graphs become steeper over the intervals $-1 < x < 0$ and $0 < x < 1$ and become flatter over the intervals $x > 1$ and $x < -1$.

- For odd values of n, the functions $f(x) = 1/x^n$ are odd, so their graphs are symmetric about the origin. The graphs all have the general shape of the curve $y = 1/x$, and each graph passes through the points $(1, -1)$ and $(-1, -1)$. As n increases, the graphs become steeper over the intervals $-1 < x < 0$ and $0 < x < 1$ and become flatter over the intervals $x > 1$ and $x < -1$.

- For both even and odd values of n the graph $y = 1/x^n$ has a break at the origin (called a ***discontinuity***), which occurs because division by zero is undefined.

> By considering the value of $1/x^n$ for a fixed x as n increases, explain why the graphs become flatter or steeper as described here for increasing values of n.

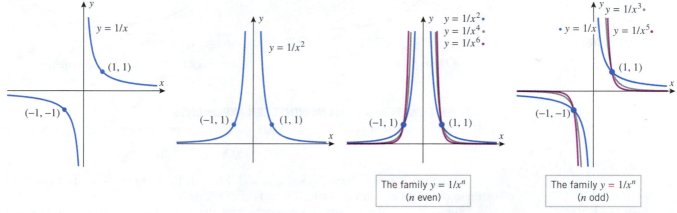

▲ Figure 0.3.5

INVERSE PROPORTIONS

Recall that a variable y is said to be ***inversely proportional to a variable*** x if there is a positive constant k, called the ***constant of proportionality***, such that

$$y = \frac{k}{x} \tag{1}$$

Since k is assumed to be positive, the graph of (1) has the same shape as $y = 1/x$ but is compressed or stretched in the y-direction. Also, it should be evident from (1) that doubling x multiplies y by $\frac{1}{2}$, tripling x multiplies y by $\frac{1}{3}$, and so forth.

Equation (1) can be expressed as $xy = k$, which tells us that the product of inversely proportional variables is a positive constant. This is a useful form for identifying inverse proportionality in experimental data.

Table 0.3.1

x	0.8	1	2.5	4	6.25	10
y	6.25	5	2	1.25	0.8	0.5

▶ **Example 1** Table 0.3.1 shows some experimental data.

(a) Explain why the data suggest that y is inversely proportional to x.

(b) Express y as a function of x.

(c) Graph your function and the data together for $x > 0$.

Solution. For every data point we have $xy = 5$, so y is inversely proportional to x and $y = 5/x$. The graph of this equation with the data points is shown in Figure 0.3.6. ◄

Inverse proportions arise in various laws of physics. For example, **Boyle's law** in physics states that *if a fixed amount of an ideal gas is held at a constant temperature, then the product of the pressure P exerted by the gas and the volume V that it occupies is constant*; that is,

$$PV = k$$

This implies that the variables P and V are inversely proportional to one another. Figure 0.3.7 shows a typical graph of volume versus pressure under the conditions of Boyle's law. Note how doubling the pressure corresponds to halving the volume, as expected.

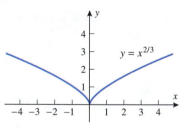

▲ **Figure 0.3.8**

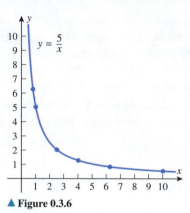

▲ **Figure 0.3.6**

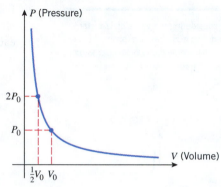

▲ **Figure 0.3.7** Doubling pressure corresponds to halving volume

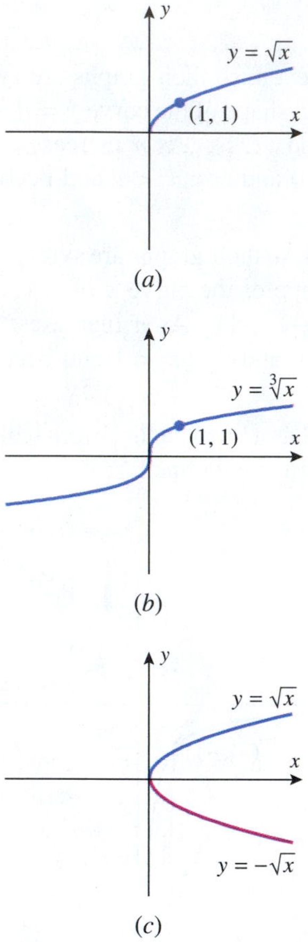

■ POWER FUNCTIONS WITH NONINTEGER EXPONENTS

If $p = 1/n$, where n is a positive integer, then the power functions $f(x) = x^p$ have the form

$$f(x) = x^{1/n} = \sqrt[n]{x}$$

In particular, if $n = 2$, then $f(x) = \sqrt{x}$, and if $n = 3$, then $f(x) = \sqrt[3]{x}$. The graphs of these functions are shown in parts (a) and (b) of Figure 0.3.8.

Since every real number has a real cube root, the domain of the function $f(x) = \sqrt[3]{x}$ is $(-\infty, +\infty)$, and hence the graph of $y = \sqrt[3]{x}$ extends over the entire x-axis. In contrast, the graph of $y = \sqrt{x}$ extends only over the interval $[0, +\infty)$ because $\sqrt{x}$ is imaginary for negative x. As illustrated in Figure 0.3.8c, the graphs of $y = \sqrt{x}$ and $y = -\sqrt{x}$ form the upper and lower halves of the parabola $x = y^2$. In general, the graph of $y = \sqrt[n]{x}$ extends over the entire x-axis if n is odd, but extends only over the interval $[0, +\infty)$ if n is even.

Power functions can have other fractional exponents. Some examples are

$$f(x) = x^{2/3}, \quad f(x) = \sqrt[5]{x^3}, \quad f(x) = x^{-7/8} \tag{2}$$

The graph of $f(x) = x^{2/3}$ is shown in Figure 0.3.9. We will discuss expressions involving irrational exponents later.

▲ **Figure 0.3.9**

TECHNOLOGY MASTERY

Graphing utililties sometimes omit portions of the graph of a function involving fractional exponents (or radicals). If $f(x) = x^{p/q}$, where p/q is a positive fraction in *lowest terms*, then you can circumvent this problem as follows:

- If p is even and q is odd, then graph $g(x) = |x|^{p/q}$ instead of $f(x)$.
- If p is odd and q is odd, then graph $g(x) = (|x|/x)|x|^{p/q}$ instead of $f(x)$.

Use a graphing utility to generate graphs of $f(x) = \sqrt[5]{x^3}$ and $f(x) = x^{-7/8}$ that show all of their significant features.

POLYNOMIALS

A *polynomial in x* is a function that is expressible as a sum of finitely many terms of the form cx^n, where c is a constant and n is a nonnegative integer. Some examples of polynomials are

$$2x + 1, \quad 3x^2 + 5x - \sqrt{2}, \quad x^3, \quad 4(= 4x^0), \quad 5x^7 - x^4 + 3$$

The function $(x^2 - 4)^3$ is also a polynomial because it can be expanded by the binomial formula (see the inside front cover) and expressed as a sum of terms of the form cx^n:

$$(x^2 - 4)^3 = (x^2)^3 - 3(x^2)^2(4) + 3(x^2)(4^2) - (4^3) = x^6 - 12x^4 + 48x^2 - 64 \quad (3)$$

A general polynomial can be written in either of the following forms, depending on whether one wants the powers of x in ascending or descending order:

$$c_0 + c_1 x + c_2 x^2 + \cdots + c_n x^n$$
$$c_n x^n + c_{n-1} x^{n-1} + \cdots + c_1 x + c_0$$

> A more detailed review of polynomials appears in Appendix C.

The constants $c_0, c_1, \ldots, c_n$ are called the *coefficients* of the polynomial. When a polynomial is expressed in one of these forms, the highest power of x that occurs with a nonzero coefficient is called the *degree* of the polynomial. Nonzero constant polynomials are considered to have degree 0, since we can write $c = cx^0$. Polynomials of degree 1, 2, 3, 4, and 5 are described as *linear*, *quadratic*, *cubic*, *quartic*, and *quintic*, respectively. For example,

> The constant 0 is a polynomial called the *zero polynomial*. In this text we will take the degree of the zero polynomial to be undefined. Other texts may use different conventions for the degree of the zero polynomial.

$$3 + 5x \qquad x^2 - 3x + 1 \qquad 2x^3 - 7$$

| Has degree 1 (linear) | Has degree 2 (quadratic) | Has degree 3 (cubic) |

$$8x^4 - 9x^3 + 5x - 3 \qquad \sqrt{3} + x^3 + x^5 \qquad (x^2 - 4)^3$$

| Has degree 4 (quartic) | Has degree 5 (quintic) | Has degree 6 [see (3)] |

The natural domain of a polynomial in x is $(-\infty, +\infty)$, since the only operations involved are multiplication and addition; the range depends on the particular polynomial. We already know that the graphs of polynomials of degree 0 and 1 are lines and that the graphs of polynomials of degree 2 are parabolas. Figure 0.3.10 shows the graphs of some typical polynomials of higher degree. Later, we will discuss polynomial graphs in detail, but for now it suffices to observe that graphs of polynomials are very well behaved in the sense that they have no discontinuities or sharp corners. As illustrated in Figure 0.3.10, the graphs of polynomials wander up and down for awhile in a roller-coaster fashion, but eventually that behavior stops and the graphs steadily rise or fall indefinitely as one travels along the curve in either the positive or negative direction. We will see later that the number of peaks and valleys is less than the degree of the polynomial.

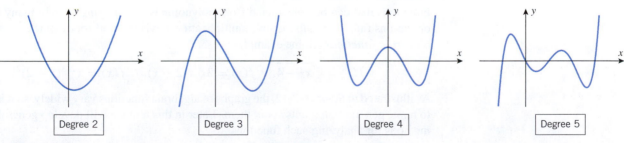

Degree 2   Degree 3   Degree 4   Degree 5

▲ **Figure 0.3.10**

RATIONAL FUNCTIONS

A function that can be expressed as a ratio of two polynomials is called a *rational function*. If $P(x)$ and $Q(x)$ are polynomials, then the domain of the rational function

$$f(x) = \frac{P(x)}{Q(x)}$$

consists of all values of x such that $Q(x) \neq 0$. For example, the domain of the rational function

$$f(x) = \frac{x^2 + 2x}{x^2 - 1}$$

consists of all values of x, except $x = 1$ and $x = -1$. Its graph is shown in Figure 0.3.11 along with the graphs of two other typical rational functions.

The graphs of rational functions with nonconstant denominators differ from the graphs of polynomials in some essential ways:

- Unlike polynomials whose graphs are continuous (unbroken) curves, the graphs of rational functions have discontinuities at the points where the denominator is zero.

- Unlike polynomials, rational functions may have numbers at which they are not defined. Near such points, many rational functions have graphs that closely approximate a vertical line, called a ***vertical asymptote***. These are represented by the dashed vertical lines in Figure 0.3.11.

- Unlike the graphs of nonconstant polynomials, which eventually rise or fall indefinitely, the graphs of many rational functions eventually get closer and closer to some horizontal line, called a ***horizontal asymptote***, as one traverses the curve in either the positive or negative direction. The horizontal asymptotes are represented by the dashed horizontal lines in the first two parts of Figure 0.3.11. In the third part of the figure the x-axis is a horizontal asymptote.

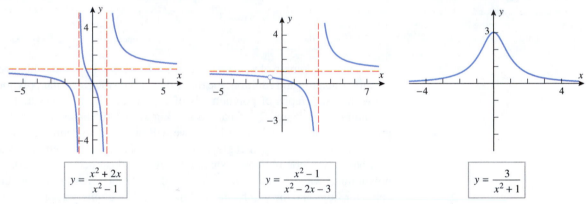

$$y = \frac{x^2 + 2x}{x^2 - 1}$$

$$y = \frac{x^2 - 1}{x^2 - 2x - 3}$$

$$y = \frac{3}{x^2 + 1}$$

▲ **Figure 0.3.11**

■ ALGEBRAIC FUNCTIONS

Functions that can be constructed from polynomials by applying finitely many algebraic operations (addition, subtraction, multiplication, division, and root extraction) are called ***algebraic functions***. Some examples are

$$f(x) = \sqrt{x^2 - 4}, \quad f(x) = 3\sqrt[3]{x}(2 + x), \quad f(x) = x^{2/3}(x + 2)^2$$

As illustrated in Figure 0.3.12, the graphs of algebraic functions vary widely, so it is difficult to make general statements about them. Later in this text we will develop general calculus methods for analyzing such functions.

■ THE FAMILIES $y = A \sin Bx$ AND $y = A \cos Bx$

Many important applications lead to trigonometric functions of the form

$$f(x) = A \sin(Bx - C) \quad \text{and} \quad g(x) = A \cos(Bx - C) \tag{4}$$

where A, B, and C are nonzero constants. The graphs of such functions can be obtained by stretching, compressing, translating, and reflecting the graphs of $y = \sin x$ and $y = \cos x$

In this text we will assume that the independent variable of a trigonometric function is in radians unless otherwise stated. A review of trigonometric functions can be found in Appendix B.

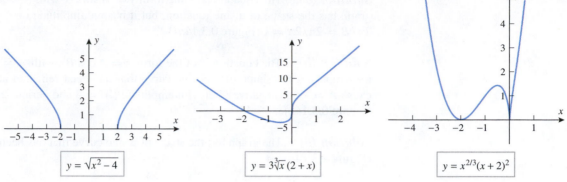

$$y = \sqrt{x^2 - 4} \qquad y = 3\sqrt[3]{x}\,(2 + x) \qquad y = x^{2/3}(x + 2)^2$$

▲ Figure 0.3.12

appropriately. To see why this is so, let us start with the case where $C = 0$ and consider how the graphs of the equations

$$y = A \sin Bx \quad \text{and} \quad y = A \cos Bx$$

relate to the graphs of $y = \sin x$ and $y = \cos x$. If A and B are positive, then the effect of the constant A is to stretch or compress the graphs of $y = \sin x$ and $y = \cos x$ vertically and the effect of the constant B is to compress or stretch the graphs of $\sin x$ and $\cos x$ horizontally. For example, the graph of $y = 2 \sin 4x$ can be obtained by stretching the graph of $y = \sin x$ vertically by a factor of 2 and compressing it horizontally by a factor of 4. (Recall from Section 0.2 that the multiplier of x *stretches* when it is less than 1 and *compresses* when it is greater than 1.) Thus, as shown in Figure 0.3.13, the graph of $y = 2 \sin 4x$ varies between -2 and 2, and repeats every $2\pi/4 = \pi/2$ units.

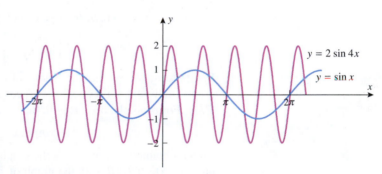

$y = 2 \sin 4x$

$y = \sin x$

► Figure 0.3.13

In general, if A and B are positive numbers, then the graphs of

$$y = A \sin Bx \quad \text{and} \quad y = A \cos Bx$$

oscillate between $-A$ and A and repeat every $2\pi/B$ units, so we say that these functions have **amplitude** A and **period** $2\pi/B$. In addition, we define the **frequency** of these functions to be the reciprocal of the period, that is, the frequency is $B/2\pi$. If A or B is negative, then these constants cause reflections of the graphs about the axes as well as compressing or stretching them; and in this case the amplitude, period, and frequency are given by

$$\text{amplitude} = |A|, \quad \text{period} = \frac{2\pi}{|B|}, \quad \text{frequency} = \frac{|B|}{2\pi}$$

► **Example 2**   Make sketches of the following graphs that show the period and amplitude.

(a) $y = 3 \sin 2\pi x$ (b) $y = -3 \cos 0.5x$ (c) $y = 1 + \sin x$

Solution (a). The equation is of the form $y = A \sin Bx$ with $A = 3$ and $B = 2\pi$, so the graph has the shape of a sine function, but it has an amplitude of $A = 3$ and a period of $2\pi/B = 2\pi/2\pi = 1$ (Figure 0.3.14a).

Solution (b). The equation is of the form $y = A \cos Bx$ with $A = -3$ and $B = 0.5$, so the graph has the shape of a cosine curve that has been reflected about the x-axis (because $A = -3$ is negative), but with amplitude $|A| = 3$ and period $2\pi/B = 2\pi/0.5 = 4\pi$ (Figure 0.3.14b).

Solution (c). The graph has the shape of a sine curve that has been translated up 1 unit (Figure 0.3.14c). ◄

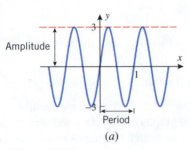 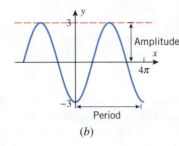 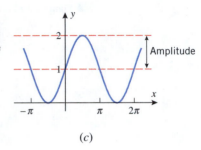

(a)   (b)   (c)

▲ **Figure 0.3.14**

▪ THE FAMILIES $y = A \sin(Bx - C)$ AND $y = A \cos(Bx - C)$

To investigate the graphs of the more general families

$$y = A \sin(Bx - C) \quad \text{and} \quad y = A \cos(Bx - C)$$

it will be helpful to rewrite these equations as

$$y = A \sin\left[B\left(x - \frac{C}{B}\right)\right] \quad \text{and} \quad y = A \cos\left[B\left(x - \frac{C}{B}\right)\right]$$

In this form we see that the graphs of these equations can be obtained by translating the graphs of $y = A \sin Bx$ and $y = A \cos Bx$ to the left or right, depending on the sign of C/B. For example, if $C/B > 0$, then the graph of

$$y = A \sin[B(x - C/B)] = A \sin(Bx - C)$$

can be obtained by translating the graph of $y = A \sin Bx$ to the right by C/B units (Figure 0.3.15). If $C/B < 0$, the graph of $y = A \sin(Bx - C)$ is obtained by translating the graph of $y = A \sin Bx$ to the left by $|C/B|$ units.

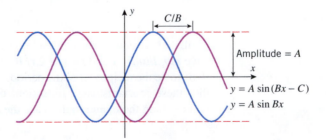

▶ **Figure 0.3.15**

▶ **Example 3** Find the amplitude and period of

$$y = 3 \cos\left(2x + \frac{\pi}{2}\right)$$

and determine how the graph of $y = 3\cos 2x$ should be translated to produce the graph of this equation. Confirm your results by graphing the equation on a calculator or computer.

Solution. The equation can be rewritten as

$$y = 3\cos\left[2x - \left(-\frac{\pi}{2}\right)\right] = 3\cos\left[2\left(x - \left(-\frac{\pi}{4}\right)\right)\right]$$

which is of the form

$$y = A\cos\left[B\left(x - \frac{C}{B}\right)\right]$$

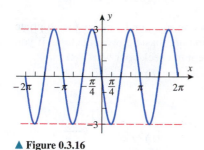

▲ **Figure 0.3.16**

with $A = 3$, $B = 2$, and $C/B = -\pi/4$. It follows that the amplitude is $A = 3$, the period is $2\pi/B = \pi$, and the graph is obtained by translating the graph of $y = 3\cos 2x$ left by $|C/B| = \pi/4$ units (Figure 0.3.16). ◄

✔ QUICK CHECK EXERCISES 0.3   *(See page 38 for answers.)*

1. Consider the family of functions $y = x^n$, where n is an integer. The graphs of $y = x^n$ are symmetric with respect to the y-axis if n is _____. These graphs are symmetric with respect to the origin if n is _____. The y-axis is a vertical asymptote for these graphs if n is _____.

2. What is the natural domain of a polynomial?

3. Consider the family of functions $y = x^{1/n}$, where n is a nonzero integer. Find the natural domain of these functions if n is
   (a) positive and even   (b) positive and odd
   (c) negative and even   (d) negative and odd.

4. Classify each equation as a polynomial, rational, algebraic, or not an algebraic function.
   (a) $y = \sqrt{x} + 2$ (b) $y = \sqrt{3}x^4 - x + 1$
   (c) $y = 5x^3 + \cos 4x$ (d) $y = \dfrac{x^2 + 5}{2x - 7}$
   (e) $y = 3x^2 + 4x^{-2}$

5. The graph of $y = A\sin Bx$ has amplitude _____ and is periodic with period _____.

EXERCISE SET 0.3   Graphing Utility

1. (a) Find an equation for the family of lines whose members have slope $m = 3$.
   (b) Find an equation for the member of the family that passes through $(-1, 3)$.
   (c) Sketch some members of the family, and label them with their equations. Include the line in part (b).

2. Find an equation for the family of lines whose members are perpendicular to those in Exercise 1.

3. (a) Find an equation for the family of lines with y-intercept $b = 2$.
   (b) Find an equation for the member of the family whose angle of inclination is $135°$.
   (c) Sketch some members of the family, and label them with their equations. Include the line in part (b).

4. Find an equation for
   (a) the family of lines that pass through the origin
   (b) the family of lines with x-intercept $a = 1$
   (c) the family of lines that pass through the point $(1, -2)$
   (d) the family of lines parallel to $2x + 4y = 1$.

5. Find an equation for the family of lines tangent to the circle with center at the origin and radius 3.

6. Find an equation for the family of lines that pass through the intersection of $5x - 3y + 11 = 0$ and $2x - 9y + 7 = 0$.

7. The U.S. Internal Revenue Service uses a 10-year linear depreciation schedule to determine the value of various business items. This means that an item is assumed to have a value of zero at the end of the tenth year and that at intermediate times the value is a linear function of the elapsed time. Sketch some typical depreciation lines, and explain the practical significance of the y-intercepts.

8. Find all lines through $(6, -1)$ for which the product of the x- and y-intercepts is 3.

FOCUS ON CONCEPTS

9–10 State a geometric property common to all lines in the family, and sketch five of the lines. ■

9. (a) The family $y = -x + b$
   (b) The family $y = mx - 1$
   (c) The family $y = m(x + 4) + 2$
   (d) The family $x - ky = 1$

10. (a) The family $y = b$
    (b) The family $Ax + 2y + 1 = 0$
    (c) The family $2x + By + 1 = 0$
    (d) The family $y - 1 = m(x + 1)$

11. In each part, match the equation with one of the accompanying graphs.
    (a) $y = \sqrt[5]{x}$ (b) $y = 2x^5$
    (c) $y = -1/x^8$ (d) $y = \sqrt{x^2 - 1}$
    (e) $y = \sqrt[4]{x - 2}$ (f) $y = -\sqrt[5]{x^2}$

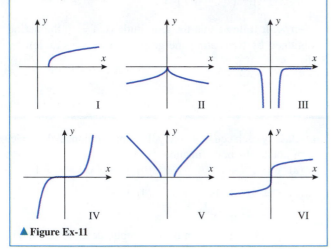

▲ **Figure Ex-11**

12. The accompanying table gives approximate values of three functions: one of the form kx^2, one of the form kx^{-3}, and one of the form $kx^{3/2}$. Identify which is which, and estimate k in each case.

x	0.25	0.37	2.1	4.0	5.8	6.2	7.9	9.3
$f(x)$	640	197	1.08	0.156	0.0513	0.0420	0.0203	0.0124
$g(x)$	0.0312	0.0684	2.20	8.00	16.8	19.2	31.2	43.2
$h(x)$	0.250	0.450	6.09	16.0	27.9	30.9	44.4	56.7

▲ **Table Ex-12**

13–14 Sketch the graph of the equation for $n = 1, 3$, and 5 in one coordinate system and for $n = 2, 4$, and 6 in another coordinate system. If you have a graphing utility, use it to check your work. ■

13. (a) $y = -x^n$ (b) $y = 2x^{-n}$ (c) $y = (x - 1)^{1/n}$

14. (a) $y = 2x^n$ (b) $y = -x^{-n}$
    (c) $y = -3(x + 2)^{1/n}$

15. (a) Sketch the graph of $y = ax^2$ for $a = \pm 1, \pm 2$, and ± 3 in a single coordinate system.
    (b) Sketch the graph of $y = x^2 + b$ for $b = \pm 1, \pm 2$, and ± 3 in a single coordinate system.
    (c) Sketch some typical members of the family of curves $y = ax^2 + b$.

16. (a) Sketch the graph of $y = a\sqrt{x}$ for $a = \pm 1, \pm 2$, and ± 3 in a single coordinate system.

    (b) Sketch the graph of $y = \sqrt{x} + b$ for $b = \pm 1, \pm 2$, and ± 3 in a single coordinate system.
    (c) Sketch some typical members of the family of curves $y = a\sqrt{x} + b$.

17–18 Sketch the graph of the equation by making appropriate transformations to the graph of a basic power function. If you have a graphing utility, use it to check your work. ■

17. (a) $y = 2(x + 1)^2$ (b) $y = -3(x - 2)^3$

    (c) $y = \dfrac{-3}{(x + 1)^2}$ (d) $y = \dfrac{1}{(x - 3)^5}$

18. (a) $y = 1 - \sqrt{x + 2}$ (b) $y = 1 - \sqrt[3]{x + 2}$

    (c) $y = \dfrac{5}{(1 - x)^3}$ (d) $y = \dfrac{2}{(4 + x)^4}$

19. Use the graph of $y = \sqrt{x}$ to help sketch the graph of $y = \sqrt{|x|}$.

20. Use the graph of $y = \sqrt[3]{x}$ to help sketch the graph of $y = \sqrt[3]{|x|}$.

21. As discussed in this section, Boyle's law states that at a constant temperature the pressure P exerted by a gas is related to the volume V by the equation $PV = k$.
    (a) Find the appropriate units for the constant k if pressure (which is force per unit area) is in newtons per square meter (N/m^2) and volume is in cubic meters (m^3).
    (b) Find k if the gas exerts a pressure of 20,000 N/m^2 when the volume is 1 liter (0.001 m^3).
    (c) Make a table that shows the pressures for volumes of 0.25, 0.5, 1.0, 1.5, and 2.0 liters.
    (d) Make a graph of P versus V.

22. A manufacturer of cardboard drink containers wants to construct a closed rectangular container that has a square base and will hold $\frac{1}{10}$ liter (100 cm^3). Estimate the dimension of the container that will require the least amount of material for its manufacture.

23–24 A variable y is said to be *inversely proportional to the square of a variable* x if y is related to x by an equation of the form $y = k/x^2$, where k is a nonzero constant, called the *constant of proportionality*. This terminology is used in these exercises. ■

23. According to *Coulomb's law*, the force F of attraction between positive and negative point charges is inversely proportional to the square of the distance x between them.
    (a) Assuming that the force of attraction between two point charges is 0.0005 newton when the distance between them is 0.3 meter, find the constant of proportionality (with proper units).
    (b) Find the force of attraction between the point charges when they are 3 meters apart.
    (c) Make a graph of force versus distance for the two charges.   *(cont.)*

(d) What happens to the force as the particles get closer and closer together? What happens as they get farther and farther apart?

24. It follows from Newton's Law of Universal Gravitation that the weight W of an object (relative to the Earth) is inversely proportional to the square of the distance x between the object and the center of the Earth, that is, $W = C/x^2$.
    (a) Assuming that a weather satellite weighs 2000 pounds on the surface of the Earth and that the Earth is a sphere of radius 4000 miles, find the constant C.
    (b) Find the weight of the satellite when it is 1000 miles above the surface of the Earth.
    (c) Make a graph of the satellite's weight versus its distance from the center of the Earth.
    (d) Is there any distance from the center of the Earth at which the weight of the satellite is zero? Explain your reasoning.

25–28 True–False Determine whether the statement is true or false. Explain your answer. ■

25. Each curve in the family $y = 2x + b$ is parallel to the line $y = 2x$.

26. Each curve in the family $y = x^2 + bx + c$ is a translation of the graph of $y = x^2$.

27. If a curve passes through the point $(2, 6)$ and y is inversely proportional to x, then the constant of proportionality is 3.

28. Curves in the family $y = -5\sin(A\pi x)$ have amplitude 5 and period $2/|A|$.

FOCUS ON CONCEPTS

29. In each part, match the equation with one of the accompanying graphs, and give the equations for the horizontal and vertical asymptotes.
    (a) $y = \dfrac{x^2}{x^2 - x - 2}$ (b) $y = \dfrac{x - 1}{x^2 - x - 6}$
    (c) $y = \dfrac{2x^4}{x^4 + 1}$ (d) $y = \dfrac{4}{(x + 2)^2}$

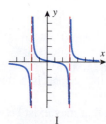

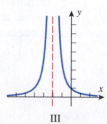

▲ Figure Ex-29

30. Find an equation of the form $y = k/(x^2 + bx + c)$ whose graph is a reasonable match to that in the accompanying figure. If you have a graphing utility, use it to check your work.

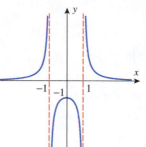

◀ Figure Ex-30

31–32 Find an equation of the form $y = D + A\sin Bx$ or $y = D + A\cos Bx$ for each graph. ■

31.

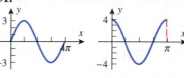

Not drawn to scale *Not drawn to scale* *Not drawn to scale*
(a) (b) (c)

▲ Figure Ex-31

32.

Not drawn to scale *Not drawn to scale* *Not drawn to scale*
(a) (b) (c)

▲ Figure Ex-32

33. In each part, find an equation for the graph that has the form $y = y_0 + A\sin(Bx - C)$.

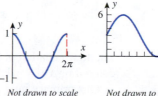

Not drawn to scale *Not drawn to scale* *Not drawn to scale*
(a) (b) (c)

▲ Figure Ex-33

34. In the United States, a standard electrical outlet supplies sinusoidal electrical current with a maximum voltage of $V = 120\sqrt{2}$ volts (V) at a frequency of 60 hertz (Hz). Write an equation that expresses V as a function of the time t, assuming that $V = 0$ if $t = 0$. [*Note:* 1 Hz = 1 cycle per second.]

35–36 Find the amplitude and period, and sketch at least two periods of the graph by hand. If you have a graphing utility, use it to check your work. ■

35. (a) $y = 3 \sin 4x$ (b) $y = -2 \cos \pi x$
(c) $y = 2 + \cos\left(\dfrac{x}{2}\right)$

36. (a) $y = -1 - 4 \sin 2x$ (b) $y = \frac{1}{2} \cos(3x - \pi)$
(c) $y = -4 \sin\left(\dfrac{x}{3} + 2\pi\right)$

37. Equations of the form

$$x = A_1 \sin \omega t + A_2 \cos \omega t$$

arise in the study of vibrations and other periodic motion. Express the equation

$$x = 5\sqrt{3} \sin 2\pi t + \tfrac{5}{2} \cos 2\pi t$$

in the form $x = A \sin(\omega t + \theta)$, and use a graphing utility to confirm that both equations have the same graph.

38. Determine the number of solutions of $x = 2 \sin x$, and use a graphing or calculating utility to estimate them.

✔ **QUICK CHECK ANSWERS 0.3**

---

1. even; odd; negative   **2.** $(-\infty, +\infty)$ **3.** (a) $[0, +\infty)$ (b) $(-\infty, +\infty)$ (c) $(0, +\infty)$ (d) $(-\infty, 0) \cup (0, +\infty)$ **4.** (a) algebraic (b) polynomial (c) not algebraic (d) rational (e) rational   **5.** $|A|$; $2\pi/|B|$

0.4   INVERSE FUNCTIONS; INVERSE TRIGONOMETRIC FUNCTIONS

*In everyday language the term "inversion" conveys the idea of a reversal. For example, in meteorology a temperature inversion is a reversal in the usual temperature properties of air layers, and in music a melodic inversion reverses an ascending interval to the corresponding descending interval. In mathematics the term **inverse** is used to describe functions that reverse one another in the sense that each undoes the effect of the other. In this section we discuss this fundamental mathematical idea. In particular, we introduce inverse trigonometric functions to address the problem of recovering an angle that could produce a given trigonometric function value.*

■ **INVERSE FUNCTIONS**

The idea of solving an equation $y = f(x)$ for x as a function of y, say $x = g(y)$, is one of the most important ideas in mathematics. Sometimes, solving an equation is a simple process; for example, using basic algebra the equation

$$y = x^3 + 1 \qquad \boxed{y = f(x)}$$

can be solved for x as a function of y:

$$x = \sqrt[3]{y - 1} \qquad \boxed{x = g(y)}$$

The first equation is better for computing y if x is known, and the second is better for computing x if y is known (Figure 0.4.1).

Our primary interest in this section is to identify relationships that may exist between the functions f and g when an equation $y = f(x)$ is expressed as $x = g(y)$, or conversely. For example, consider the functions $f(x) = x^3 + 1$ and $g(y) = \sqrt[3]{y - 1}$ discussed above. When these functions are composed in either order, they cancel out the effect of one another in the sense that

$$g(f(x)) = \sqrt[3]{f(x) - 1} = \sqrt[3]{(x^3 + 1) - 1} = x$$

$$f(g(y)) = [g(y)]^3 + 1 = (\sqrt[3]{y - 1})^3 + 1 = y$$

(1)

Pairs of functions with these two properties are so important that there is special terminology for them.

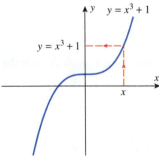

▲ **Figure 0.4.1**

> **0.4.1 DEFINITION** If the functions f and g satisfy the two conditions
>
> $$g(f(x)) = x \text{ for every } x \text{ in the domain of } f$$
> $$f(g(y)) = y \text{ for every } y \text{ in the domain of } g$$
>
> then we say that *f is an inverse of g* and *g is an inverse of f* or that *f and g are inverse functions*.

It can be shown (Exercise 60) that if a function f has an inverse, then that inverse is unique. Thus, if a function f has an inverse, then we are entitled to talk about "the" inverse of f, in which case we denote it by the symbol f^{-1}.

▶ **Example 1** The computations in (1) show that $g(y) = \sqrt[3]{y-1}$ is the inverse of $f(x) = x^3 + 1$. Thus, we can express g in inverse notation as

$$f^{-1}(y) = \sqrt[3]{y-1}$$

and we can express the equations in Definition 0.4.1 as

$$f^{-1}(f(x)) = x \quad \text{for every } x \text{ in the domain of } f$$
$$f(f^{-1}(y)) = y \quad \text{for every } y \text{ in the domain of } f^{-1} \tag{2}$$

We will call these the *cancellation equations* for f and f^{-1}. ◀

CHANGING THE INDEPENDENT VARIABLE

The formulas in (2) use x as the independent variable for f and y as the independent variable for f^{-1}. Although it is often convenient to use different independent variables for f and f^{-1}, there will be occasions on which it is desirable to use the same independent variable for both. For example, if we want to graph the functions f and f^{-1} together in the same xy-coordinate system, then we would want to use x as the independent variable and y as the dependent variable for both functions. Thus, to graph the functions $f(x) = x^3 + 1$ and $f^{-1}(y) = \sqrt[3]{y-1}$ of Example 1 in the same xy-coordinate system, we would change the independent variable y to x, use y as the dependent variable for both functions, and graph the equations

$$y = x^3 + 1 \quad \text{and} \quad y = \sqrt[3]{x-1}$$

We will talk more about graphs of inverse functions later in this section, but for reference we give the following reformulation of the cancellation equations in (2) using x as the independent variable for both f and f^{-1}:

$$f^{-1}(f(x)) = x \quad \text{for every } x \text{ in the domain of } f$$
$$f(f^{-1}(x)) = x \quad \text{for every } x \text{ in the domain of } f^{-1} \tag{3}$$

▶ **Example 2** Confirm each of the following.

(a) The inverse of $f(x) = 2x$ is $f^{-1}(x) = \frac{1}{2}x$.

(b) The inverse of $f(x) = x^3$ is $f^{-1}(x) = x^{1/3}$.

Solution (a).

$$f^{-1}(f(x)) = f^{-1}(2x) = \tfrac{1}{2}(2x) = x$$
$$f(f^{-1}(x)) = f\left(\tfrac{1}{2}x\right) = 2\left(\tfrac{1}{2}x\right) = x$$

The results in Example 2 should make sense to you intuitively, since the operations of multiplying by 2 and multiplying by $\frac{1}{2}$ in either order cancel the effect of one another, as do the operations of cubing and taking a cube root.

Solution (b).

$$f^{-1}(f(x)) = f^{-1}(x^3) = \left(x^3\right)^{1/3} = x$$

$$f(f^{-1}(x)) = f(x^{1/3}) = \left(x^{1/3}\right)^3 = x \quad \blacktriangleleft$$

In general, if a function f has an inverse and $f(a) = b$, then the procedure in Example 3 shows that $a = f^{-1}(b)$; that is, f^{-1} maps each output of f back into the corresponding input (Figure 0.4.2).

► **Example 3**   Given that the function f has an inverse and that $f(3) = 5$, find $f^{-1}(5)$.

Solution.   Apply f^{-1} to both sides of the equation $f(3) = 5$ to obtain

$$f^{-1}(f(3)) = f^{-1}(5)$$

and now apply the first equation in (3) to conclude that $f^{-1}(5) = 3$. ◄

▲ **Figure 0.4.2**   If f maps a to b, then f^{-1} maps b back to a.

■ DOMAIN AND RANGE OF INVERSE FUNCTIONS

The equations in (3) imply the following relationships between the domains and ranges of f and f^{-1}:

$$\begin{aligned} \text{domain of } f^{-1} &= \text{range of } f \\ \text{range of } f^{-1} &= \text{domain of } f \end{aligned} \tag{4}$$

One way to show that two sets are the same is to show that each is a subset of the other. Thus we can establish the first equality in (4) by showing that the domain of f^{-1} is a subset of the range of f and that the range of f is a subset of the domain of f^{-1}. We do this as follows: The first equation in (3) implies that f^{-1} is defined at $f(x)$ for all values of x in the domain of f, and this implies that the range of f is a subset of the domain of f^{-1}. Conversely, if x is in the domain of f^{-1}, then the second equation in (3) implies that x is in the range of f because it is the image of $f^{-1}(x)$. Thus, the domain of f^{-1} is a subset of the range of f. We leave the proof of the second equation in (4) as an exercise.

■ A METHOD FOR FINDING INVERSE FUNCTIONS

At the beginning of this section we observed that solving $y = f(x) = x^3 + 1$ for x as a function of y produces $x = f^{-1}(y) = \sqrt[3]{y - 1}$. The following theorem shows that this is not accidental.

> **0.4.2   THEOREM**   *If an equation $y = f(x)$ can be solved for x as a function of y, say $x = g(y)$, then f has an inverse and that inverse is $g(y) = f^{-1}(y)$.*

PROOF   Substituting $y = f(x)$ into $x = g(y)$ yields $x = g(f(x))$, which confirms the first equation in Definition 0.4.1, and substituting $x = g(y)$ into $y = f(x)$ yields $y = f(g(y))$, which confirms the second equation in Definition 0.4.1. ■

Theorem 0.4.2 provides us with the following procedure for finding the inverse of a function.

A Procedure for Finding the Inverse of a Function f

Step 1.   Write down the equation $y = f(x)$.

Step 2.   If possible, solve this equation for x as a function of y.

Step 3.   The resulting equation will be $x = f^{-1}(y)$, which provides a formula for f^{-1} with y as the independent variable.

An alternative way to obtain a formula for $f^{-1}(x)$ with x as the independent variable is to reverse the roles of x and y at the outset and solve the equation $x = f(y)$ for y as a function of x.

Step 4.   If y is acceptable as the independent variable for the inverse function, then you are done, but if you want to have x as the independent variable, then you need to interchange x and y in the equation $x = f^{-1}(y)$ to obtain $y = f^{-1}(x)$.

▶ **Example 4** Find a formula for the inverse of $f(x) = \sqrt{3x - 2}$ with x as the independent variable, and state the domain of f^{-1}.

Solution. Following the procedure stated above, we first write

$$y = \sqrt{3x - 2}$$

Then we solve this equation for x as a function of y:

$$y^2 = 3x - 2$$
$$x = \tfrac{1}{3}(y^2 + 2)$$

which tells us that

$$f^{-1}(y) = \tfrac{1}{3}(y^2 + 2) \tag{5}$$

Since we want x to be the independent variable, we reverse x and y in (5) to produce the formula

$$f^{-1}(x) = \tfrac{1}{3}(x^2 + 2) \tag{6}$$

We know from (4) that the domain of f^{-1} is the range of f. In general, this need not be the same as the natural domain of the formula for f^{-1}. Indeed, in this example the natural domain of (6) is $(-\infty, +\infty)$, whereas the range of $f(x) = \sqrt{3x - 2}$ is $[0, +\infty)$. Thus, if we want to make the domain of f^{-1} clear, we must express it explicitly by rewriting (6) as

$$f^{-1}(x) = \tfrac{1}{3}(x^2 + 2), \quad x \geq 0 \blacktriangleleft$$

■ EXISTENCE OF INVERSE FUNCTIONS

The procedure we gave above for finding the inverse of a function f was based on solving the equation $y = f(x)$ for x as a function of y. This procedure can fail for two reasons—the function f may not have an inverse, or it may have an inverse but the equation $y = f(x)$ cannot be solved explicitly for x as a function of y. Thus, it is important to establish conditions that ensure the existence of an inverse, even if it cannot be found explicitly.

If a function f has an inverse, then it must assign distinct outputs to distinct inputs. For example, the function $f(x) = x^2$ cannot have an inverse because it assigns the same value to $x = 2$ and $x = -2$, namely,

$$f(2) = f(-2) = 4$$

Thus, if $f(x) = x^2$ were to have an inverse, then the equation $f(2) = 4$ would imply that $f^{-1}(4) = 2$, and the equation $f(-2) = 4$ would imply that $f^{-1}(4) = -2$. But this is impossible because $f^{-1}(4)$ cannot have two different values. Another way to see that $f(x) = x^2$ has no inverse is to attempt to find the inverse by solving the equation $y = x^2$ for x as a function of y. We run into trouble immediately because the resulting equation $x = \pm\sqrt{y}$ does not express x as a *single* function of y.

A function that assigns distinct outputs to distinct inputs is said to be ***one-to-one*** or ***invertible***, so we know from the preceding discussion that if a function f has an inverse, then it must be one-to-one. The converse is also true, thereby establishing the following theorem.

0.4.3 **THEOREM** *A function has an inverse if and only if it is one-to-one.*

Stated algebraically, a function f is one-to-one if and only if $f(x_1) \neq f(x_2)$ whenever $x_1 \neq x_2$; stated geometrically, a function f is one-to-one if and only if the graph of $y = f(x)$ is cut at most once by any horizontal line (Figure 0.4.3). The latter statement together with Theorem 0.4.3 provides the following geometric test for determining whether a function has an inverse.

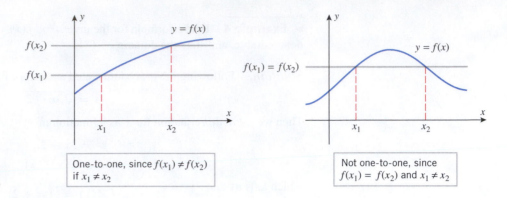

One-to-one, since $f(x_1) \neq f(x_2)$ if $x_1 \neq x_2$

Not one-to-one, since $f(x_1) = f(x_2)$ and $x_1 \neq x_2$

▶ **Figure 0.4.3**

0.4.4   **THEOREM** (*The Horizontal Line Test*)   *A function has an inverse function if and only if its graph is cut at most once by any horizontal line.*

▶ **Example 5**   Use the horizontal line test to show that $f(x) = x^2$ has no inverse but that $f(x) = x^3$ does.

Solution.   Figure 0.4.4 shows a horizontal line that cuts the graph of $y = x^2$ more than once, so $f(x) = x^2$ is not invertible. Figure 0.4.5 shows that the graph of $y = x^3$ is cut at most once by any horizontal line, so $f(x) = x^3$ is invertible. [Recall from Example 2 that the inverse of $f(x) = x^3$ is $f^{-1}(x) = x^{1/3}$.] ◀

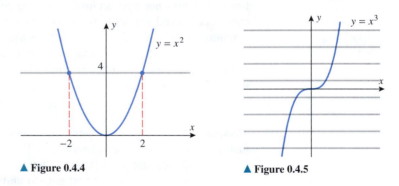

▲ **Figure 0.4.4**   ▲ **Figure 0.4.5**

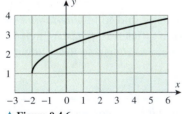

▲ **Figure 0.4.6**

The function $f(x) = x^3$ in Figure 0.4.5 is an example of an increasing function. Give an example of a decreasing function and compute its inverse.

▶ **Example 6**   Explain why the function f that is graphed in Figure 0.4.6 has an inverse, and find $f^{-1}(3)$.

Solution.   The function f has an inverse since its graph passes the horizontal line test. To evaluate $f^{-1}(3)$, we view $f^{-1}(3)$ as that number x for which $f(x) = 3$. From the graph we see that $f(2) = 3$, so $f^{-1}(3) = 2$. ◀

■ **INCREASING OR DECREASING FUNCTIONS ARE INVERTIBLE**

A function whose graph is always rising as it is traversed from left to right is said to be an *increasing function*, and a function whose graph is always falling as it is traversed from left to right is said to be a *decreasing function*. If x_1 and x_2 are points in the domain of a function f, then f is increasing if

$$f(x_1) < f(x_2) \quad \text{whenever } x_1 < x_2$$

and f is *decreasing* if

$$f(x_1) > f(x_2) \quad \text{whenever } x_1 < x_2$$

(Figure 0.4.7). It is evident geometrically that increasing and decreasing functions pass the horizontal line test and hence are invertible.

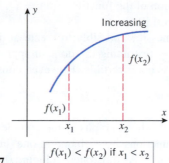
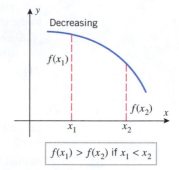

▶ **Figure 0.4.7**

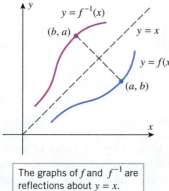

The points (a, b) and (b, a) are reflections about $y = x$.

▲ **Figure 0.4.8**

■ GRAPHS OF INVERSE FUNCTIONS

Our next objective is to explore the relationship between the graphs of f and f^{-1}. For this purpose, it will be desirable to use x as the independent variable for both functions so we can compare the graphs of $y = f(x)$ and $y = f^{-1}(x)$.

If (a, b) is a point on the graph $y = f(x)$, then $b = f(a)$. This is equivalent to the statement that $a = f^{-1}(b)$, which means that (b, a) is a point on the graph of $y = f^{-1}(x)$. In short, reversing the coordinates of a point on the graph of f produces a point on the graph of f^{-1}. Similarly, reversing the coordinates of a point on the graph of f^{-1} produces a point on the graph of f (verify). However, the geometric effect of reversing the coordinates of a point is to reflect that point about the line $y = x$ (Figure 0.4.8), and hence the graphs of $y = f(x)$ and $y = f^{-1}(x)$ are reflections of one another about this line (Figure 0.4.9). In summary, we have the following result.

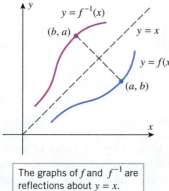

0.4.5   THEOREM   *If f has an inverse, then the graphs of $y = f(x)$ and $y = f^{-1}(x)$ are reflections of one another about the line $y = x$; that is, each graph is the mirror image of the other with respect to that line.*

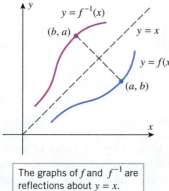

The graphs of f and f^{-1} are reflections about $y = x$.

▲ **Figure 0.4.9**

▶ **Example 7**   Figure 0.4.10 shows the graphs of the inverse functions discussed in Examples 2 and 4. ◀

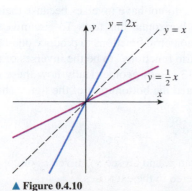
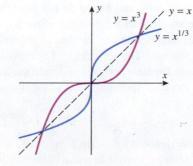
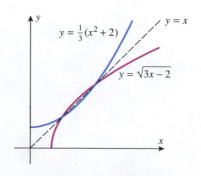

▲ **Figure 0.4.10**

■ RESTRICTING DOMAINS FOR INVERTIBILITY

If a function g is obtained from a function f by placing restrictions on the domain of f, then g is called a *restriction* of f. Thus, for example, the function

$$g(x) = x^3, \quad x \geq 0$$

is a restriction of the function $f(x) = x^3$. More precisely, it is called the restriction of x^3 to the interval $[0, +\infty)$.

Sometimes it is possible to create an invertible function from a function that is not invertible by restricting the domain appropriately. For example, we showed earlier that $f(x) = x^2$ is not invertible. However, consider the restricted functions

$$f_1(x) = x^2, \quad x \geq 0 \quad \text{and} \quad f_2(x) = x^2, \quad x \leq 0$$

the union of whose graphs is the complete graph of $f(x) = x^2$ (Figure 0.4.11). These restricted functions are each one-to-one (hence invertible), since their graphs pass the horizontal line test. As illustrated in Figure 0.4.12, their inverses are

$$f_1^{-1}(x) = \sqrt{x} \quad \text{and} \quad f_2^{-1}(x) = -\sqrt{x}$$

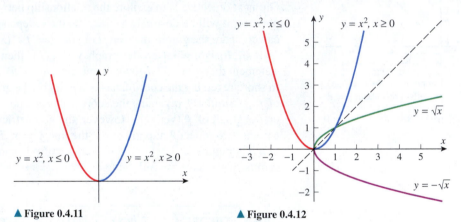

▲ Figure 0.4.11   ▲ Figure 0.4.12

■ INVERSE TRIGONOMETRIC FUNCTIONS

A common problem in trigonometry is to find an angle x using a known value of $\sin x$, $\cos x$, or some other trigonometric function. Recall that problems of this type involve the computation of "arc functions" such as $\arcsin x$, $\arccos x$, and so forth. We will conclude this section by studying these arc functions from the viewpoint of general inverse functions.

The six basic trigonometric functions do not have inverses because their graphs repeat periodically and hence do not pass the horizontal line test. To circumvent this problem we will restrict the domains of the trigonometric functions to produce one-to-one functions and then define the "inverse trigonometric functions" to be the inverses of these restricted functions. The top part of Figure 0.4.13 shows geometrically how these restrictions are made for $\sin x$, $\cos x$, $\tan x$, and $\sec x$, and the bottom part of the figure shows the graphs of the corresponding inverse functions

$$\sin^{-1} x, \quad \cos^{-1} x, \quad \tan^{-1} x, \quad \sec^{-1} x$$

(also denoted by $\arcsin x$, $\arccos x$, $\arctan x$, and $\operatorname{arcsec} x$). Inverses of $\cot x$ and $\csc x$ are of lesser importance and will be considered in the exercises.

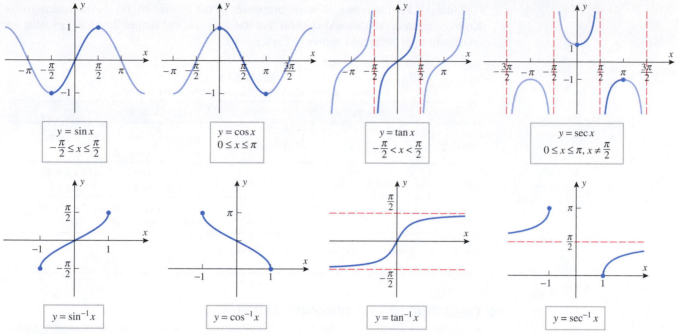

▲ **Figure 0.4.13**

If you have trouble visualizing the correspondence between the top and bottom parts of Figure 0.4.13, keep in mind that a reflection about $y = x$ converts vertical lines into horizontal lines, and vice versa; and it converts x-intercepts into y-intercepts, and vice versa.

The following formal definitions summarize the preceding discussion.

0.4.6   DEFINITION   The *inverse sine function*, denoted by $\sin^{-1}$, is defined to be the inverse of the restricted sine function

$$\sin x, \quad -\pi/2 \leq x \leq \pi/2$$

0.4.7   DEFINITION   The *inverse cosine function*, denoted by $\cos^{-1}$, is defined to be the inverse of the restricted cosine function

$$\cos x, \quad 0 \leq x \leq \pi$$

0.4.8   DEFINITION   The *inverse tangent function*, denoted by $\tan^{-1}$, is defined to be the inverse of the restricted tangent function

$$\tan x, \quad -\pi/2 < x < \pi/2$$

WARNING

The notations $\sin^{-1} x, \cos^{-1} x, \ldots$ are reserved exclusively for the inverse trigonometric functions and are not used for reciprocals of the trigonometric functions. If we want to express the reciprocal $1/\sin x$ using an exponent, we would write $(\sin x)^{-1}$ and *never* $\sin^{-1} x$.

0.4.9   DEFINITION*   The *inverse secant function*, denoted by $\sec^{-1}$, is defined to be the inverse of the restricted secant function

$$\sec x, \quad 0 \leq x \leq \pi \text{ with } x \neq \pi/2$$

---

*There is no universal agreement on the definition of $\sec^{-1} x$, and some mathematicians prefer to restrict the domain of $\sec x$ so that $0 \leq x < \pi/2$ or $\pi \leq x < 3\pi/2$, which was the definition used in some earlier editions of this text. Each definition has advantages and disadvantages, but we will use the current definition to conform with the conventions used by the CAS programs *Mathematica*, *Maple*, and *Sage*.

Table 0.4.1 summarizes the basic properties of the inverse trigonometric functions we have considered. You should confirm that the domains and ranges listed in this table are consistent with the graphs shown in Figure 0.4.13.

Table 0.4.1
PROPERTIES OF INVERSE TRIGONOMETRIC FUNCTIONS

FUNCTION	DOMAIN	RANGE	BASIC RELATIONSHIPS
$\sin^{-1}$	$[-1, 1]$	$[-\pi/2, \pi/2]$	$\sin^{-1}(\sin x) = x$ if $-\pi/2 \le x \le \pi/2$ $\sin(\sin^{-1} x) = x$ if $-1 \le x \le 1$
$\cos^{-1}$	$[-1, 1]$	$[0, \pi]$	$\cos^{-1}(\cos x) = x$ if $0 \le x \le \pi$ $\cos(\cos^{-1} x) = x$ if $-1 \le x \le 1$
$\tan^{-1}$	$(-\infty, +\infty)$	$(-\pi/2, \pi/2)$	$\tan^{-1}(\tan x) = x$ if $-\pi/2 < x < \pi/2$ $\tan(\tan^{-1} x) = x$ if $-\infty < x < +\infty$
$\sec^{-1}$	$(-\infty, -1] \cup [1, +\infty)$	$[0, \pi/2) \cup (\pi/2, \pi]$	$\sec^{-1}(\sec x) = x$ if $0 \le x \le \pi, x \ne \pi/2$ $\sec(\sec^{-1} x) = x$ if $|x| \ge 1$

■ EVALUATING INVERSE TRIGONOMETRIC FUNCTIONS

A common problem in trigonometry is to find an angle whose sine is known. For example, you might want to find an angle x in radian measure such that

$$\sin x = \tfrac{1}{2} \qquad (7)$$

and, more generally, for a given value of y in the interval $-1 \le y \le 1$ you might want to solve the equation

$$\sin x = y \qquad (8)$$

Because $\sin x$ repeats periodically, this equation has infinitely many solutions for x; however, if we solve this equation as

$$x = \sin^{-1} y$$

then we isolate the specific solution that lies in the interval $[-\pi/2, \pi/2]$, since this is the range of the inverse sine. For example, Figure 0.4.14 shows four solutions of Equation (7), namely, $-11\pi/6$, $-7\pi/6$, $\pi/6$, and $5\pi/6$. Of these, $\pi/6$ is the solution in the interval $[-\pi/2, \pi/2]$, so

$$\sin^{-1}\left(\tfrac{1}{2}\right) = \pi/6 \qquad (9)$$

In general, if we view $x = \sin^{-1} y$ as an angle in radian measure whose sine is y, then the restriction $-\pi/2 \le x \le \pi/2$ imposes the geometric requirement that the angle x in standard position terminate in either the first or fourth quadrant or on an axis adjacent to those quadrants.

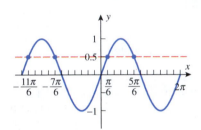

▲ Figure 0.4.14

TECHNOLOGY MASTERY

Refer to the documentation for your calculating utility to determine how to calculate inverse sines, inverse cosines, and inverse tangents; and then confirm Equation (9) numerically by showing that

$$\sin^{-1}(0.5) \approx 0.523598775598\ldots$$
$$\approx \pi/6$$

If $x = \cos^{-1} y$ is viewed as an angle in radian measure whose cosine is y, in what possible quadrants can x lie? Answer the same question for

$$x = \tan^{-1} y \quad \text{and} \quad x = \sec^{-1} y$$

▶ **Example 8** Find exact values of

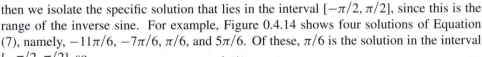

by inspection, and confirm your results numerically using a calculating utility.

Solution (a). Because $\sin^{-1}(1/\sqrt{2}) > 0$, we can view $x = \sin^{-1}(1/\sqrt{2})$ as that angle in the first quadrant such that $\sin\theta = 1/\sqrt{2}$. Thus, $\sin^{-1}(1/\sqrt{2}) = \pi/4$. You can confirm this with your calculating utility by showing that $\sin^{-1}(1/\sqrt{2}) \approx 0.785 \approx \pi/4$.

Solution (b). Because $\sin^{-1}(-1) < 0$, we can view $x = \sin^{-1}(-1)$ as an angle in the fourth quadrant (or an adjacent axis) such that $\sin x = -1$. Thus, $\sin^{-1}(-1) = -\pi/2$. You can confirm this with your calculating utility by showing that $\sin^{-1}(-1) \approx -1.57 \approx -\pi/2$. ◀

TECHNOLOGY
MASTERY

Most calculators do not provide a direct method for calculating inverse secants. In such situations the identity

$$\sec^{-1} x = \cos^{-1}(1/x) \tag{10}$$

is useful (Exercise 48). Use this formula to show that

$$\sec^{-1}(2.25) \approx 1.11 \quad \text{and} \quad \sec^{-1}(-2.25) \approx 2.03$$

If you have a calculating utility (such as a CAS) that can find $\sec^{-1} x$ directly, use it to check these values.

◾ IDENTITIES FOR INVERSE TRIGONOMETRIC FUNCTIONS

If we interpret $\sin^{-1} x$ as an angle in radian measure whose sine is x, and if that angle is *nonnegative*, then we can represent $\sin^{-1} x$ geometrically as an angle in a right triangle in which the hypotenuse has length 1 and the side opposite to the angle $\sin^{-1} x$ has length x (Figure 0.4.15a). Moreover, the unlabeled acute angle in Figure 0.4.15a is $\cos^{-1} x$, since the cosine of that angle is x, and the unlabeled side in that figure has length $\sqrt{1-x^2}$ by the Theorem of Pythagoras (Figure 0.4.15b). This triangle motivates a number of useful identities involving inverse trigonometric functions that are valid for $-1 \leq x \leq 1$; for example,

$$\sin^{-1} x + \cos^{-1} x = \frac{\pi}{2} \tag{11}$$

$$\cos(\sin^{-1} x) = \sqrt{1-x^2} \tag{12}$$

$$\sin(\cos^{-1} x) = \sqrt{1-x^2} \tag{13}$$

$$\tan(\sin^{-1} x) = \frac{x}{\sqrt{1-x^2}} \tag{14}$$

There is little to be gained by memorizing these identities. What is important is the mastery of the *method* used to obtain them.

In a similar manner, $\tan^{-1} x$ and $\sec^{-1} x$ can be represented as angles in the right triangles shown in Figures 0.4.15c and 0.4.15d (verify). Those triangles reveal additional useful identities; for example,

$$\sec(\tan^{-1} x) = \sqrt{1+x^2} \tag{15}$$

$$\sin(\sec^{-1} x) = \frac{\sqrt{x^2-1}}{x} \quad (x \geq 1) \tag{16}$$

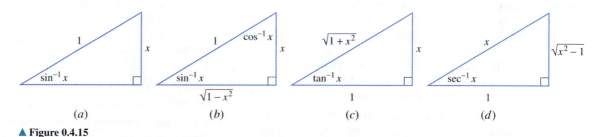

(a)   (b)   (c)   (d)

▲ **Figure 0.4.15**

REMARK

The triangle technique does not always produce the most general form of an identity. For example, in Exercise 59 we will ask you to derive the following extension of Formula (16) that is valid for $x \leq -1$ as well as $x \geq 1$:

$$\sin(\sec^{-1} x) = \frac{\sqrt{x^2-1}}{|x|} \quad (|x| \geq 1) \tag{17}$$

Referring to Figure 0.4.13, observe that the inverse sine and inverse tangent are odd functions; that is,

$$\sin^{-1}(-x) = -\sin^{-1}(x) \quad \text{and} \quad \tan^{-1}(-x) = -\tan^{-1}(x) \tag{18–19}$$

▶ **Example 9** Figure 0.4.16 shows a computer-generated graph of $y = \sin^{-1}(\sin x)$. One might think that this graph should be the line $y = x$, since $\sin^{-1}(\sin x) = x$. Why isn't it?

Solution. The relationship $\sin^{-1}(\sin x) = x$ is valid on the interval $-\pi/2 \leq x \leq \pi/2$, so we can say with certainty that the graphs of $y = \sin^{-1}(\sin x)$ and $y = x$ coincide on this interval (which is confirmed by Figure 0.4.16). However, outside of this interval the relationship $\sin^{-1}(\sin x) = x$ does not hold. For example, if the quantity x lies in the interval $\pi/2 \leq x \leq 3\pi/2$, then the quantity $x - \pi$ lies in the interval $-\pi/2 \leq x \leq \pi/2$, so

$$\sin^{-1}[\sin(x - \pi)] = x - \pi$$

Thus, by using the identity $\sin(x - \pi) = -\sin x$ and the fact that $\sin^{-1}$ is an odd function, we can express $\sin^{-1}(\sin x)$ as

$$\sin^{-1}(\sin x) = \sin^{-1}[-\sin(x - \pi)] = -\sin^{-1}[\sin(x - \pi)] = -(x - \pi)$$

This shows that on the interval $\pi/2 \leq x \leq 3\pi/2$ the graph of $y = \sin^{-1}(\sin x)$ coincides with the line $y = -(x - \pi)$, which has slope -1 and an x-intercept at $x = \pi$. This agrees with Figure 0.4.16. ◀

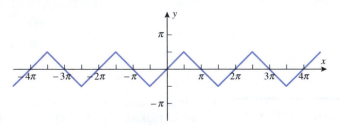

▶ **Figure 0.4.16**

✔ **QUICK CHECK EXERCISES 0.4**   *(See page 52 for answers.)*

1. In each part, determine whether the function f is one-to-one.
   (a) $f(t)$ is the number of people in line at a movie theater at time t.
   (b) $f(x)$ is the measured high temperature (rounded to the nearest °F) in a city on the xth day of the year.
   (c) $f(v)$ is the weight of v cubic inches of lead.

2. A student enters a number on a calculator, doubles it, adds 8 to the result, divides the sum by 2, subtracts 3 from the quotient, and then cubes the difference. If the resulting number is x, then _____ was the student's original number.

3. If $(3, -2)$ is a point on the graph of an odd invertible function f, then _____ and _____ are points on the graph of f^{-1}.

4. In each part, determine the exact value without using a calculating utility.
   (a) $\sin^{-1}(-1) =$ _____
   (b) $\tan^{-1}(1) =$ _____
   (c) $\sin^{-1}\left(\frac{1}{2}\sqrt{3}\right) =$ _____
   (d) $\cos^{-1}\left(\frac{1}{2}\right) =$ _____
   (e) $\sec^{-1}(-2) =$ _____

5. In each part, determine the exact value without using a calculating utility.
   (a) $\sin^{-1}(\sin \pi/7) =$ _____
   (b) $\sin^{-1}(\sin 5\pi/7) =$ _____
   (c) $\tan^{-1}(\tan 13\pi/6) =$ _____
   (d) $\cos^{-1}(\cos 12\pi/7) =$ _____

EXERCISE SET 0.4   ⌁ Graphing Utility

---

1. In (a)–(d), determine whether f and g are inverse functions.
   (a) $f(x) = 4x$, $g(x) = \frac{1}{4}x$
   (b) $f(x) = 3x + 1$, $g(x) = 3x - 1$
   (c) $f(x) = \sqrt[3]{x - 2}$, $g(x) = x^3 + 2$
   (d) $f(x) = x^4$, $g(x) = \sqrt[4]{x}$

2. Check your answers to Exercise 1 with a graphing utility by determining whether the graphs of f and g are reflections of one another about the line $y = x$.

3. In each part, use the horizontal line test to determine whether the function f is one-to-one.
(a) $f(x) = 3x + 2$
(b) $f(x) = \sqrt{x-1}$
(c) $f(x) = |x|$
(d) $f(x) = x^3$
(e) $f(x) = x^2 - 2x + 2$
(f) $f(x) = \sin x$

4. In each part, generate the graph of the function f with a graphing utility, and determine whether f is one-to-one.
(a) $f(x) = x^3 - 3x + 2$ (b) $f(x) = x^3 - 3x^2 + 3x - 1$

FOCUS ON CONCEPTS

5. In each part, determine whether the function f defined by the table is one-to-one.

(a)

x	1	2	3	4	5	6
$f(x)$	−2	−1	0	1	2	3

(b)

x	1	2	3	4	5	6
$f(x)$	4	−7	6	−3	1	4

6. A face of a broken clock lies in the xy-plane with the center of the clock at the origin and 3:00 in the direction of the positive x-axis. When the clock broke, the tip of the hour hand stopped on the graph of $y = f(x)$, where f is a function that satisfies $f(0) = 0$.
(a) Are there any times of the day that cannot appear in such a configuration? Explain.
(b) How does your answer to part (a) change if f must be an invertible function?
(c) How do your answers to parts (a) and (b) change if it was the tip of the minute hand that stopped on the graph of f?

7. (a) The accompanying figure shows the graph of a function f over its domain $-8 \le x \le 8$. Explain why f has an inverse, and use the graph to find $f^{-1}(2)$, $f^{-1}(-1)$, and $f^{-1}(0)$.
(b) Find the domain and range of f^{-1}.
(c) Sketch the graph of f^{-1}.

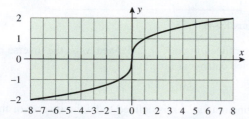

▲ Figure Ex-7

8. (a) Explain why the function f graphed in the accompanying figure has no inverse function on its domain $-3 \le x \le 4$.

(b) Subdivide the domain into three adjacent intervals on each of which the function f has an inverse.

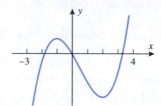

◀ Figure Ex-8

9–16 Find a formula for $f^{-1}(x)$. ■

9. $f(x) = 7x - 6$

10. $f(x) = \dfrac{x+1}{x-1}$

11. $f(x) = 3x^3 - 5$

12. $f(x) = \sqrt[5]{4x + 2}$

13. $f(x) = 3/x^2, \quad x < 0$

14. $f(x) = 5/(x^2 + 1), \quad x \ge 0$

15. $f(x) = \begin{cases} 5/2 - x, & x < 2 \\ 1/x, & x \ge 2 \end{cases}$

16. $f(x) = \begin{cases} 2x, & x \le 0 \\ x^2, & x > 0 \end{cases}$

17–20 Find a formula for $f^{-1}(x)$, and state the domain of the function f^{-1}. ■

17. $f(x) = (x + 2)^4, \quad x \ge 0$

18. $f(x) = \sqrt{x + 3}$ **19.** $f(x) = -\sqrt{3 - 2x}$

20. $f(x) = 3x^2 + 5x - 2, \quad x \ge 0$

21. Let $f(x) = ax^2 + bx + c, a > 0$. Find f^{-1} if the domain of f is restricted to
(a) $x \ge -b/(2a)$
(b) $x \le -b/(2a)$.

FOCUS ON CONCEPTS

22. The formula $F = \frac{9}{5}C + 32$, where $C \ge -273.15$ expresses the Fahrenheit temperature F as a function of the Celsius temperature C.
(a) Find a formula for the inverse function.
(b) In words, what does the inverse function tell you?
(c) Find the domain and range of the inverse function.

23. (a) One meter is about 6.214×10^{-4} miles. Find a formula $y = f(x)$ that expresses a length y in meters as a function of the same length x in miles.
(b) Find a formula for the inverse of f.
(c) Describe what the formula $x = f^{-1}(y)$ tells you in practical terms.

24. Let $f(x) = x^2, x > 1$, and $g(x) = \sqrt{x}$.
(a) Show that $f(g(x)) = x, x > 1$, and $g(f(x)) = x$, $x > 1$.
(b) Show that f and g are *not* inverses by showing that the graphs of $y = f(x)$ and $y = g(x)$ are not reflections of one another about $y = x$.
(c) Do parts (a) and (b) contradict one another? Explain.

25. (a) Show that $f(x) = (3 - x)/(1 - x)$ is its own inverse.

(b) What does the result in part (a) tell you about the graph of f?

26. Sketch the graph of a function that is one-to-one on $(-\infty, +\infty)$, yet not increasing on $(-\infty, +\infty)$ and not decreasing on $(-\infty, +\infty)$.

27. Let $f(x) = 2x^3 + 5x + 3$. Find x if $f^{-1}(x) = 1$.

28. Let $f(x) = \dfrac{x^3}{x^2 + 1}$. Find x if $f^{-1}(x) = 2$.

29. Prove that if $a^2 + bc \neq 0$, then the graph of

$$f(x) = \frac{ax + b}{cx - a}$$

is symmetric about the line $y = x$.

30. (a) Prove: If f and g are one-to-one, then so is the composition $f \circ g$.

(b) Prove: If f and g are one-to-one, then

$$(f \circ g)^{-1} = g^{-1} \circ f^{-1}$$

31–34 True–False Determine whether the statement is true or false. Explain your answer. ◼

31. If f is an invertible function such that $f(2) = 2$, then $f^{-1}(2) = \frac{1}{2}$.

32. If f and g are inverse functions, then f and g have the same domain.

33. A one-to-one function is invertible.

34. The range of the inverse tangent function is the interval $-\pi/2 \leq y \leq \pi/2$.

35. Given that $\theta = \tan^{-1}\left(\frac{4}{3}\right)$, find the exact values of $\sin\theta$, $\cos\theta$, $\cot\theta$, $\sec\theta$, and $\csc\theta$.

36. Given that $\theta = \sec^{-1} 2.6$, find the exact values of $\sin\theta$, $\cos\theta$, $\tan\theta$, $\cot\theta$, and $\csc\theta$.

37. For which values of x is it true that

(a) $\cos^{-1}(\cos x) = x$ (b) $\cos(\cos^{-1} x) = x$

(c) $\tan^{-1}(\tan x) = x$ (d) $\tan(\tan^{-1} x) = x$?

38–39 Find the exact value of the given quantity. ◼

38. $\sec\left[\sin^{-1}\left(-\frac{3}{4}\right)\right]$ **39.** $\sin\left[2\cos^{-1}\left(\frac{3}{5}\right)\right]$

40–41 Complete the identities using the triangle method (Figure 0.4.15). ◼

40. (a) $\sin(\cos^{-1} x) = ?$ (b) $\tan(\cos^{-1} x) = ?$

(c) $\csc(\tan^{-1} x) = ?$ (d) $\sin(\tan^{-1} x) = ?$

41. (a) $\cos(\tan^{-1} x) = ?$ (b) $\tan(\cos^{-1} x) = ?$

(c) $\sin(\sec^{-1} x) = ?$ (d) $\cot(\sec^{-1} x) = ?$

42. (a) Use a calculating utility set to radian measure to make tables of values of $y = \sin^{-1} x$ and $y = \cos^{-1} x$ for $x = -1, -0.8, -0.6, \ldots, 0, 0.2, \ldots, 1$. Round your answers to two decimal places.

(b) Plot the points obtained in part (a), and use the points to sketch the graphs of $y = \sin^{-1} x$ and $y = \cos^{-1} x$. Confirm that your sketches agree with those in Figure 0.4.13.

(c) Use your graphing utility to graph $y = \sin^{-1} x$ and $y = \cos^{-1} x$; confirm that the graphs agree with those in Figure 0.4.13.

43. In each part, sketch the graph and check your work with a graphing utility.

(a) $y = \sin^{-1} 2x$ (b) $y = \tan^{-1} \frac{1}{2} x$

44. The *law of cosines* states that

$$c^2 = a^2 + b^2 - 2ab\cos\theta$$

where a, b, and c are the lengths of the sides of a triangle and θ is the angle formed by sides a and b. Find θ, to the nearest degree, for the triangle with $a = 2$, $b = 3$, and $c = 4$.

FOCUS ON CONCEPTS

45. (a) Use a calculating utility to evaluate the expressions $\sin^{-1}(\sin^{-1} 0.25)$ and $\sin^{-1}(\sin^{-1} 0.9)$, and explain what you think is happening in the second calculation.

(b) For what values of x in the interval $-1 \leq x \leq 1$ will your calculating utility produce a real value for the function $\sin^{-1}(\sin^{-1} x)$?

46. A soccer player kicks a ball with an initial speed of 14 m/s at an angle θ with the horizontal (see the accompanying figure). The ball lands 18 m down the field. If air resistance is neglected, then the ball will have a parabolic trajectory and the horizontal range R will be given by

$$R = \frac{v^2}{g} \sin 2\theta$$

where v is the initial speed of the ball and g is the acceleration due to gravity. Using $g = 9.8$ m/s^2, approximate two values of θ, to the nearest degree, at which the ball could have been kicked. Which angle results in the shorter time of flight? Why?

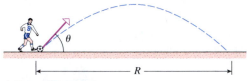

▲ **Figure Ex-46**

47–48 The function $\cot^{-1} x$ is defined to be the inverse of the restricted cotangent function

$$\cot x, \quad 0 < x < \pi$$

and the function $\csc^{-1} x$ is defined to be the inverse of the restricted cosecant function

$$\csc x, \quad -\pi/2 < x < \pi/2, \quad x \neq 0$$

Use these definitions in these and in all subsequent exercises that involve these functions. ◼

47. (a) Sketch the graphs of $\cot^{-1} x$ and $\csc^{-1} x$.

(b) Find the domain and range of $\cot^{-1} x$ and $\csc^{-1} x$.

48. Show that

(a) $\cot^{-1} x = \begin{cases} \tan^{-1}(1/x), & \text{if } x > 0 \\ \pi + \tan^{-1}(1/x), & \text{if } x < 0 \end{cases}$

(b) $\sec^{-1} x = \cos^{-1} \dfrac{1}{x}$, if $|x| \ge 1$

(c) $\csc^{-1} x = \sin^{-1} \dfrac{1}{x}$, if $|x| \ge 1$.

49. Most scientific calculators have keys for the values of only $\sin^{-1} x$, $\cos^{-1} x$, and $\tan^{-1} x$. The formulas in Exercise 48 show how a calculator can be used to obtain values of $\cot^{-1} x$, $\sec^{-1} x$, and $\csc^{-1} x$ for positive values of x. Use these formulas and a calculator to find numerical values for each of the following inverse trigonometric functions. Express your answers in degrees, rounded to the nearest tenth of a degree.

(a) $\cot^{-1} 0.7$ (b) $\sec^{-1} 1.2$ (c) $\csc^{-1} 2.3$

50. An Earth-observing satellite has horizon sensors that can measure the angle θ shown in the accompanying figure. Let R be the radius of the Earth (assumed spherical) and h the distance between the satellite and the Earth's surface.

(a) Show that $\sin \theta = \dfrac{R}{R + h}$.

(b) Find θ, to the nearest degree, for a satellite that is 10,000 km from the Earth's surface (use $R = 6378$ km).

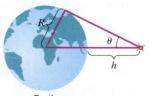

Earth   ◀ **Figure Ex-50**

51. The number of hours of daylight on a given day at a given point on the Earth's surface depends on the latitude λ of the point, the angle γ through which the Earth has moved in its orbital plane during the time period from the vernal equinox (March 21), and the angle of inclination ϕ of the Earth's axis of rotation measured from ecliptic north ($\phi \approx 23.45°$). The number of hours of daylight h can be approximated by the formula

$$h = \begin{cases} 24, & D \ge 1 \\ 12 + \frac{2}{15} \sin^{-1} D, & |D| < 1 \\ 0, & D \le -1 \end{cases}$$

where

$$D = \frac{\sin \phi \sin \gamma \tan \lambda}{\sqrt{1 - \sin^2 \phi \sin^2 \gamma}}$$

and $\sin^{-1} D$ is in degree measure. Given that Fairbanks, Alaska, is located at a latitude of $\lambda = 65°$ N and also that $\gamma = 90°$ on June 20 and $\gamma = 270°$ on December 20, approximate

(a) the maximum number of daylight hours at Fairbanks to one decimal place

(b) the minimum number of daylight hours at Fairbanks to one decimal place.

Source: This problem was adapted from *TEAM, A Path to Applied Mathematics*, The Mathematical Association of America, Washington, D.C., 1985.

52. A camera is positioned x feet from the base of a missile launching pad (see the accompanying figure). If a missile of length a feet is launched vertically, show that when the base of the missile is b feet above the camera lens, the angle θ subtended at the lens by the missile is

$$\theta = \cot^{-1} \frac{x}{a + b} - \cot^{-1} \frac{x}{b}$$

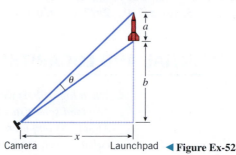

Camera   Launchpad ◀ **Figure Ex-52**

53. An airplane is flying at a constant height of 3000 ft above water at a speed of 400 ft/s. The pilot is to release a survival package so that it lands in the water at a sighted point P. If air resistance is neglected, then the package will follow a parabolic trajectory whose equation relative to the coordinate system in the accompanying figure is

$$y = 3000 - \frac{g}{2v^2} x^2$$

where g is the acceleration due to gravity and v is the speed of the airplane. Using $g = 32$ ft/s^2, find the "line of sight" angle θ, to the nearest degree, that will result in the package hitting the target point.

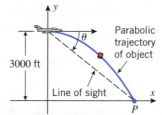

◀ **Figure Ex-53**

54. Prove:

(a) $\sin^{-1}(-x) = -\sin^{-1} x$

(b) $\tan^{-1}(-x) = -\tan^{-1} x$.

55. Prove:

(a) $\cos^{-1}(-x) = \pi - \cos^{-1} x$

(b) $\sec^{-1}(-x) = \pi - \sec^{-1} x$.

56. Prove:

(a) $\sin^{-1} x = \tan^{-1} \dfrac{x}{\sqrt{1 - x^2}}$ $(|x| < 1)$

(b) $\cos^{-1} x = \dfrac{\pi}{2} - \tan^{-1} \dfrac{x}{\sqrt{1 - x^2}}$ $(|x| < 1)$.

57. Prove:

$$\tan^{-1} x + \tan^{-1} y = \tan^{-1} \left(\frac{x+y}{1-xy} \right)$$

provided $-\pi/2 < \tan^{-1} x + \tan^{-1} y < \pi/2$. [*Hint:* Use an identity for $\tan(\alpha + \beta)$.]

58. Use the result in Exercise 57 to show that
  (a) $\tan^{-1} \frac{1}{2} + \tan^{-1} \frac{1}{3} = \pi/4$
  (b) $2\tan^{-1} \frac{1}{3} + \tan^{-1} \frac{1}{7} = \pi/4$.

59. Use identities (10) and (13) to obtain identity (17).

60. Prove: A one-to-one function f cannot have two different inverses.

✔ QUICK CHECK ANSWERS 0.4

1. (a) not one-to-one (b) not one-to-one (c) one-to-one   **2.** $\sqrt[3]{x} - 1$ **3.** $(-2, 3);\ (2, -3)$ **4.** (a) $-\pi/2$ (b) $\pi/4$ (c) $\pi/3$
(d) $\pi/3$ (e) $2\pi/3$ **5.** (a) $\pi/7$ (b) $2\pi/7$ (c) $\pi/6$ (d) $2\pi/7$

0.5 EXPONENTIAL AND LOGARITHMIC FUNCTIONS

When logarithms were introduced in the seventeenth century as a computational tool, they provided scientists of that period computing power that was previously unimaginable. Although computers and calculators have replaced logarithm tables for numerical calculations, the logarithmic functions have wide-ranging applications in mathematics and science. In this section we will review some properties of exponents and logarithms and then use our work on inverse functions to develop results about exponential and logarithmic functions.

■ IRRATIONAL EXPONENTS

Recall from algebra that if b is a nonzero real number, then nonzero *integer* powers of b are defined by

$$b^n = \underbrace{b \times b \times \cdots \times b}_{n \text{ factors}} \quad \text{and} \quad b^{-n} = \frac{1}{b^n}$$

and if $n = 0$, then $b^0 = 1$. Also, if p/q is a positive *rational* number expressed in lowest terms, then

$$b^{p/q} = \sqrt[q]{b^p} = (\sqrt[q]{b})^p \quad \text{and} \quad b^{-p/q} = \frac{1}{b^{p/q}}$$

If b is negative, then some fractional powers of b will have imaginary values—the quantity $(-2)^{1/2} = \sqrt{-2}$, for example. To avoid this complication, we will assume throughout this section that $b > 0$, even if it is not stated explicitly.

There are various methods for defining *irrational* powers such as

$$2^\pi, \quad 3^{\sqrt{2}}, \quad \pi^{-\sqrt{7}}$$

One approach is to define irrational powers of b via successive approximations using rational powers of b. For example, to define 2^π consider the decimal representation of π:

$$3.1415926\ldots$$

From this decimal we can form a sequence of rational numbers that gets closer and closer to π, namely,

$$3.1, \quad 3.14, \quad 3.141, \quad 3.1415, \quad 3.14159$$

and from these we can form a sequence of *rational* powers of 2:

$$2^{3.1}, \quad 2^{3.14}, \quad 2^{3.141}, \quad 2^{3.1415}, \quad 2^{3.14159}$$

Since the exponents of the terms in this sequence get successively closer to π, it seems plausible that the terms themselves will get successively closer to some number. It is that number that we *define* to be 2^π. This is illustrated in Table 0.5.1, which we generated using

Table 0.5.1

x	2^x
3	8.000000
3.1	8.574188
3.14	8.815241
3.141	8.821353
3.1415	8.824411
3.14159	8.824962
3.141592	8.824974
3.1415926	8.824977

a calculator. The table suggests that to four decimal places the value of 2^π is

$$2^\pi \approx 8.8250 \qquad (1)$$

TECHNOLOGY MASTERY

Use a calculating utility to verify the results in Table 0.5.1, and then verify (1) by using the utility to compute 2^π directly.

With this notion for irrational powers, we remark without proof that the following familiar laws of exponents hold for all real values of p and q:

$$b^p b^q = b^{p+q}, \qquad \frac{b^p}{b^q} = b^{p-q}, \qquad \left(b^p\right)^q = b^{pq}$$

■ THE FAMILY OF EXPONENTIAL FUNCTIONS

A function of the form $f(x) = b^x$, where $b > 0$, is called an *exponential function with base b*. Some examples are

$$f(x) = 2^x, \qquad f(x) = \left(\tfrac{1}{2}\right)^x, \qquad f(x) = \pi^x$$

Note that an exponential function has a constant base and variable exponent. Thus, functions such as $f(x) = x^2$ and $f(x) = x^\pi$ would *not* be classified as exponential functions, since they have a variable base and a constant exponent.

Figure 0.5.1 illustrates that the graph of $y = b^x$ has one of three general forms, depending on the value of b. The graph of $y = b^x$ has the following properties:

- The graph passes through $(0, 1)$ because $b^0 = 1$.
- If $b > 1$, the value of b^x increases as x increases. As you traverse the graph of $y = b^x$ from left to right, the values of b^x increase indefinitely. If you traverse the graph from right to left, the values of b^x decrease toward zero but never reach zero. Thus, the x-axis is a horizontal asymptote of the graph of b^x.
- If $0 < b < 1$, the value of b^x decreases as x increases. As you traverse the graph of $y = b^x$ from left to right, the values of b^x decrease toward zero but never reach zero. Thus, the x-axis is a horizontal asymptote of the graph of b^x. If you traverse the graph from right to left, the values of b^x increase indefinitely.
- If $b = 1$, then the value of b^x is constant.

Some typical members of the family of exponential functions are graphed in Figure 0.5.2. This figure illustrates that the graph of $y = (1/b)^x$ is the reflection of the graph of $y = b^x$ about the y-axis. This is because replacing x by $-x$ in the equation $y = b^x$ yields

$$y = b^{-x} = (1/b)^x$$

The figure also conveys that for $b > 1$, the larger the base b, the more rapidly the function $f(x) = b^x$ increases for $x > 0$.

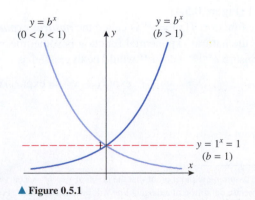

▲ **Figure 0.5.1**

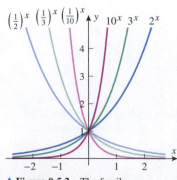

▲ **Figure 0.5.2** The family $y = b^x \, (b > 0)$

The domain and range of the exponential function $f(x) = b^x$ can also be found by examining Figure 0.5.1:

- If $b > 0$, then $f(x) = b^x$ is defined and has a real value for every real value of x, so the natural domain of every exponential function is $(-\infty, +\infty)$.

- If $b > 0$ and $b \neq 1$, then as noted earlier the graph of $y = b^x$ increases indefinitely as it is traversed in one direction and decreases toward zero but never reaches zero as it is traversed in the other direction. This implies that the range of $f(x) = b^x$ is $(0, +\infty)$.[*]

▶ **Example 1**   Sketch the graph of the function $f(x) = 1 - 2^x$ and find its domain and range.

Solution.   Start with a graph of $y = 2^x$. Reflect this graph across the x-axis to obtain the graph of $y = -2^x$, then translate that graph upward by 1 unit to obtain the graph of $y = 1 - 2^x$ (Figure 0.5.3). The dashed line in the third part of Figure 0.5.3 is a horizontal asymptote for the graph. You should be able to see from the graph that the domain of f is $(-\infty, +\infty)$ and the range is $(-\infty, 1)$. ◀

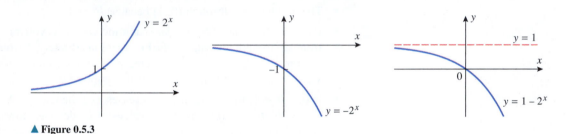

▲ **Figure 0.5.3**

■ THE NATURAL EXPONENTIAL FUNCTION

Among all possible bases for exponential functions there is one particular base that plays a special role in calculus. That base, denoted by the letter e, is a certain irrational number whose value to six decimal places is

$$e \approx 2.718282 \tag{2}$$

This base is important in calculus because, as we will prove later, $b = e$ is the only base for which the slope of the tangent line[**] to the curve $y = b^x$ at any point P on the curve is equal to the y-coordinate at P. Thus, for example, the tangent line to $y = e^x$ at $(0, 1)$ has slope 1 (Figure 0.5.4).

The function $f(x) = e^x$ is called the ***natural exponential function***. To simplify typography, the natural exponential function is sometimes written as $\exp(x)$, in which case the relationship $e^{x_1+x_2} = e^{x_1}e^{x_2}$ would be expressed as

$$\exp(x_1 + x_2) = \exp(x_1)\exp(x_2)$$

The use of the letter e is in honor of the Swiss mathematician Leonhard Euler (biography on p. 3) who is credited with recognizing the mathematical importance of this constant.

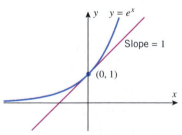

▲ **Figure 0.5.4**   The tangent line to the graph of $y = e^x$ at $(0, 1)$ has slope 1.

---

[*]We are assuming without proof that the graph of $y = b^x$ is a curve without breaks, gaps, or holes.
[**]The precise definition of a tangent line will be discussed later. For now your intuition will suffice.

TECHNOLOGY
MASTERY

Your technology utility should have keys or commands for approximating e and for graphing the natural exponential function. Read your documentation on how to do this and use your utility to confirm (2) and to generate the graphs in Figures 0.5.2 and 0.5.4.

The constant e also arises in the context of the graph of the equation

$$y = \left(1 + \frac{1}{x}\right)^x \tag{3}$$

As shown in Figure 0.5.5, $y = e$ is a horizontal asymptote of this graph. As a result, the value of e can be approximated to any degree of accuracy by evaluating (3) for x sufficiently large in absolute value (Table 0.5.2).

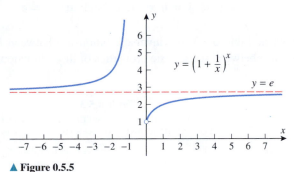

▲ **Figure 0.5.5**

Table 0.5.2

APPROXIMATIONS OF e BY $(1 + 1/x)^x$
FOR INCREASING VALUES OF x

x	$1 + \frac{1}{x}$	$\left(1 + \frac{1}{x}\right)^x$
1	2	≈ 2.000000
10	1.1	2.593742
100	1.01	2.704814
1000	1.001	2.716924
10,000	1.0001	2.718146
100,000	1.00001	2.718268
1,000,000	1.000001	2.718280

■ **LOGARITHMIC FUNCTIONS**

Recall from algebra that a logarithm is an exponent. More precisely, if $b > 0$ and $b \neq 1$, then for a positive value of x the expression

$$\log_b x$$

(read "the logarithm to the base b of x") denotes that exponent to which b must be raised to produce x. Thus, for example,

$$\log_{10} 100 = 2, \quad \log_{10}(1/1000) = -3, \quad \log_2 16 = 4, \quad \log_b 1 = 0, \quad \log_b b = 1$$

$$\boxed{10^2 = 100} \qquad \boxed{10^{-3} = 1/1000} \qquad \boxed{2^4 = 16} \qquad \boxed{b^0 = 1} \qquad \boxed{b^1 = b}$$

Logarithms with base 10 are called *common logarithms* and are often written without explicit reference to the base. Thus, the symbol $\log x$ generally denotes $\log_{10} x$.

We call the function $f(x) = \log_b x$ the ***logarithmic function with base b***.

Logarithmic functions can also be viewed as inverses of exponential functions. To see why this is so, observe from Figure 0.5.1 that if $b > 0$ and $b \neq 1$, then the graph of $f(x) = b^x$ passes the horizontal line test, so b^x has an inverse. We can find a formula for this inverse with x as the independent variable by solving the equation

$$x = b^y$$

for y as a function of x. But this equation states that y is the logarithm to the base b of x, so it can be rewritten as

$$y = \log_b x$$

Thus, we have established the following result.

0.5.1 THEOREM *If $b > 0$ and $b \neq 1$, then b^x and $\log_b x$ are inverse functions.*

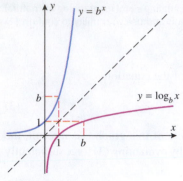

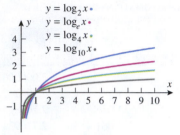

▲ **Figure 0.5.6**

▲ **Figure 0.5.7** The family
$y = \log_b x \ (b > 1)$

TECHNOLOGY MASTERY

Use your graphing utility to generate
the graphs of $y = \ln x$ and $y = \log x$.

It follows from this theorem that the graphs of $y = b^x$ and $y = \log_b x$ are reflections of one another about the line $y = x$ (see Figure 0.5.6 for the case where $b > 1$). Figure 0.5.7 shows the graphs of $y = \log_b x$ for various values of b. Observe that they all pass through the point $(1, 0)$.

The most important logarithms in applications are those with base e. These are called *natural logarithms* because the function $\log_e x$ is the inverse of the natural exponential function e^x. It is standard to denote the natural logarithm of x by $\ln x$ (read "ell en of x"), rather than $\log_e x$. For example,

$$\ln 1 = 0, \qquad \ln e = 1, \qquad \ln 1/e = -1, \qquad \ln(e^2) = 2$$

| Since $e^0 = 1$ | Since $e^1 = e$ | Since $e^{-1} = 1/e$ | Since $e^2 = e^2$ |

In general,

$$y = \ln x \quad \text{if and only if} \quad x = e^y$$

As shown in Table 0.5.3, the inverse relationship between b^x and $\log_b x$ produces a correspondence between some basic properties of those functions.

Table 0.5.3

CORRESPONDENCE BETWEEN PROPERTIES OF
LOGARITHMIC AND EXPONENTIAL FUNCTIONS

PROPERTY OF b^x	PROPERTY OF $\log_b x$
$b^0 = 1$	$\log_b 1 = 0$
$b^1 = b$	$\log_b b = 1$
Range is $(0, +\infty)$	Domain is $(0, +\infty)$
Domain is $(-\infty, +\infty)$	Range is $(-\infty, +\infty)$
x-axis is a horizontal asymptote	y-axis is a vertical asymptote

It also follows from the cancellation properties of inverse functions [see (3) in Section 0.4] that

$$\begin{aligned} \log_b(b^x) &= x \quad \text{for all real values of } x \\ b^{\log_b x} &= x \quad \text{for } x > 0 \end{aligned} \tag{4}$$

In the special case where $b = e$, these equations become

$$\begin{aligned} \ln(e^x) &= x \quad \text{for all real values of } x \\ e^{\ln x} &= x \quad \text{for } x > 0 \end{aligned} \tag{5}$$

In words, the functions b^x and $\log_b x$ cancel out the effect of one another when composed in either order; for example,

$$\log 10^x = x, \quad 10^{\log x} = x, \quad \ln e^x = x, \quad e^{\ln x} = x, \quad \ln e^5 = 5, \quad e^{\ln \pi} = \pi$$

■ **SOLVING EQUATIONS INVOLVING EXPONENTIALS AND LOGARITHMS**
You should be familiar with the following properties of logarithms from your earlier studies.

0.5.2   THEOREM (*Algebraic Properties of Logarithms*)   *If $b > 0, b \neq 1, a > 0, c > 0$, and r is any real number, then:*

(*a*) $\log_b(ac) = \log_b a + \log_b c$ Product property

(*b*) $\log_b(a/c) = \log_b a - \log_b c$ Quotient property

(*c*) $\log_b(a^r) = r \log_b a$ Power property

(*d*) $\log_b(1/c) = -\log_b c$ Reciprocal property

WARNING

Expressions of the form $\log_b(u + v)$ and $\log_b(u - v)$ have no useful simplifications. In particular,

$\log_b(u + v) \neq \log_b(u) + \log_b(v)$

$\log_b(u - v) \neq \log_b(u) - \log_b(v)$

These properties are often used to expand a single logarithm into sums, differences, and multiples of other logarithms and, conversely, to condense sums, differences, and multiples of logarithms into a single logarithm. For example,

$$\log \frac{xy^5}{\sqrt{z}} = \log xy^5 - \log \sqrt{z} = \log x + \log y^5 - \log z^{1/2} = \log x + 5 \log y - \tfrac{1}{2} \log z$$

$$5 \log 2 + \log 3 - \log 8 = \log 32 + \log 3 - \log 8 = \log \frac{32 \cdot 3}{8} = \log 12$$

$$\tfrac{1}{3} \ln x - \ln(x^2 - 1) + 2 \ln(x + 3) = \ln x^{1/3} - \ln(x^2 - 1) + \ln(x + 3)^2 = \ln \frac{\sqrt[3]{x}(x+3)^2}{x^2 - 1}$$

An equation of the form $\log_b x = k$ can be solved for x by rewriting it in the exponential form $x = b^k$, and an equation of the form $b^x = k$ can be solved by rewriting it in the logarithm form $x = \log_b k$. Alternatively, the equation $b^x = k$ can be solved by taking *any* logarithm of both sides (but usually log or ln) and applying part (*c*) of Theorem 0.5.2. These ideas are illustrated in the following example.

▶ **Example 2**   Find x such that

(a) $\log x = \sqrt{2}$ (b) $\ln(x + 1) = 5$ (c) $5^x = 7$

Solution (a).   Converting the equation to exponential form yields

$$x = 10^{\sqrt{2}} \approx 25.95$$

Solution (b).   Converting the equation to exponential form yields

$$x + 1 = e^5 \quad \text{or} \quad x = e^5 - 1 \approx 147.41$$

Solution (c).   Converting the equation to logarithmic form yields

$$x = \log_5 7 \approx 1.21$$

Alternatively, taking the natural logarithm of both sides and using the power property of logarithms yields

$$x \ln 5 = \ln 7 \quad \text{or} \quad x = \frac{\ln 7}{\ln 5} \approx 1.21 \blacktriangleleft$$

Erik Simonsen/Getty Images

Power to satellites can be supplied by batteries, fuel cells, solar cells, or radio-isotope devices.

▶ **Example 3** A satellite that requires 7 watts of power to operate at full capacity is equipped with a radioisotope power supply whose power output P in watts is given by the equation

$$P = 75e^{-t/125}$$

where t is the time in days that the supply is used. How long can the satellite operate at full capacity?

Solution. The power P will fall to 7 watts when

$$7 = 75e^{-t/125}$$

The solution for t is as follows:

$$7/75 = e^{-t/125}$$
$$\ln(7/75) = \ln(e^{-t/125})$$
$$\ln(7/75) = -t/125$$
$$t = -125\ln(7/75) \approx 296.4$$

so the satellite can operate at full capacity for about 296 days. ◀

Here is a more complicated example.

▶ **Example 4** Solve $\dfrac{e^x - e^{-x}}{2} = 1$ for x.

Solution. Multiplying both sides of the given equation by 2 yields

$$e^x - e^{-x} = 2$$

or equivalently,

$$e^x - \frac{1}{e^x} = 2$$

Multiplying through by e^x yields

$$e^{2x} - 1 = 2e^x \quad \text{or} \quad e^{2x} - 2e^x - 1 = 0$$

This is really a quadratic equation in disguise, as can be seen by rewriting it in the form

$$\left(e^x\right)^2 - 2e^x - 1 = 0$$

and letting $u = e^x$ to obtain

$$u^2 - 2u - 1 = 0$$

Solving for u by the quadratic formula yields

$$u = \frac{2 \pm \sqrt{4+4}}{2} = \frac{2 \pm \sqrt{8}}{2} = 1 \pm \sqrt{2}$$

or, since $u = e^x$,

$$e^x = 1 \pm \sqrt{2}$$

But e^x cannot be negative, so we discard the negative value $1 - \sqrt{2}$; thus,

$$e^x = 1 + \sqrt{2}$$
$$\ln e^x = \ln(1 + \sqrt{2})$$
$$x = \ln(1 + \sqrt{2}) \approx 0.881 \quad ◀$$

■ CHANGE OF BASE FORMULA FOR LOGARITHMS

Scientific calculators generally have no keys for evaluating logarithms with bases other than 10 or e. However, this is not a serious deficiency because it is possible to express a logarithm with any base in terms of logarithms with any other base (see Exercise 42). For example, the following formula expresses a logarithm with base b in terms of natural logarithms:

$$\log_b x = \frac{\ln x}{\ln b} \tag{6}$$

We can derive this result by letting $y = \log_b x$, from which it follows that $b^y = x$. Taking the natural logarithm of both sides of this equation we obtain $y \ln b = \ln x$, from which (6) follows.

▶ **Example 5** Use a calculating utility to evaluate $\log_2 5$ by expressing this logarithm in terms of natural logarithms.

Solution. From (6) we obtain

$$\log_2 5 = \frac{\ln 5}{\ln 2} \approx 2.321928 \quad ◀$$

■ LOGARITHMIC SCALES IN SCIENCE AND ENGINEERING

Logarithms are used in science and engineering to deal with quantities whose units vary over an excessively wide range of values. For example, the "loudness" of a sound can be measured by its *intensity* I (in watts per square meter), which is related to the energy transmitted by the sound wave—the greater the intensity, the greater the transmitted energy, and the louder the sound is perceived by the human ear. However, intensity units are unwieldy because they vary over an enormous range. For example, a sound at the threshold of human hearing has an intensity of about 10^{-12} W/m^2, a close whisper has an intensity that is about 100 times the hearing threshold, and a jet engine at 50 meters has an intensity that is about $10,000,000,000,000 = 10^{13}$ times the hearing threshold. To see how logarithms can be used to reduce this wide spread, observe that if

$$y = \log x$$

then increasing x by a *factor* of 10 *adds* 1 unit to y since

$$\log 10x = \log 10 + \log x = 1 + y$$

Physicists and engineers take advantage of this property by measuring loudness in terms of the *sound level* β, which is defined by

$$\beta = 10 \log(I/I_0)$$

where $I_0 = 10^{-12}$ W/m^2 is a reference intensity close to the threshold of human hearing. The units of β are *decibels* (dB), named in honor of the telephone inventor Alexander Graham Bell. With this scale of measurement, *multiplying* the intensity I by a factor of 10 *adds* 10 dB to the sound level β (verify). This results in a more tractable scale than intensity for measuring sound loudness (Table 0.5.4).   Some other familiar logarithmic scales are the *Richter scale* used to measure earthquake intensity and the **pH** *scale* used to measure acidity in chemistry, both of which are discussed in the exercises.

Table 0.5.4

β (dB)	I/I_0
0	$10^0 = 1$
10	$10^1 = 10$
20	$10^2 = 100$
30	$10^3 = 1000$
40	$10^4 = 10,000$
50	$10^5 = 100,000$
⋮	⋮
120	$10^{12} = 1,000,000,000,000$

Regina Mitchell-Ryall, Tony Gray/NASA/Getty Images

The roar of a space shuttle near the launch pad would damage your hearing without ear protection.

▶ **Example 6** A space shuttle taking off generates a sound level of 150 dB near the launch pad. A person exposed to this level of sound would experience severe physical injury. By comparison, a car horn at one meter has a sound level of 110 dB, near the threshold of pain for many people. What is the ratio of sound intensity of a space shuttle takeoff to that of a car horn?

Solution. Let I_1 and $\beta_1 \ (= 150 \text{ dB})$ denote the sound intensity and sound level of the space shuttle taking off, and let I_2 and $\beta_2 \ (= 110 \text{ dB})$ denote the sound intensity and sound level of a car horn. Then

$$I_1/I_2 = (I_1/I_0)/(I_2/I_0)$$
$$\log(I_1/I_2) = \log(I_1/I_0) - \log(I_2/I_0)$$
$$10\log(I_1/I_2) = 10\log(I_1/I_0) - 10\log(I_2/I_0) = \beta_1 - \beta_2$$
$$10\log(I_1/I_2) = 150 - 100 = 40$$
$$\log(I_1/I_2) = 4$$

Thus, $I_1/I_2 = 10^4$, which tells us that the sound intensity of the space shuttle taking off is 10,000 times greater than a car horn! ◀

■ EXPONENTIAL AND LOGARITHMIC GROWTH

The growth patterns of e^x and $\ln x$ illustrated in Table 0.5.5 are worth noting. Both functions increase as x increases, but they increase in dramatically different ways—the value of e^x increases extremely rapidly and that of $\ln x$ increases extremely slowly. For example, the value of e^x at $x = 10$ is over 22,000, but at $x = 1000$ the value of $\ln x$ has not even reached 7.

A function f is said to *increase without bound* as x increases if the values of $f(x)$ eventually exceed any specified positive number M (no matter how large) as x increases indefinitely. Table 0.5.5 strongly suggests that $f(x) = e^x$ increases without bound, which is consistent with the fact that the range of this function is $(0, +\infty)$. Indeed, if we choose any positive number M, then we will have $e^x = M$ when $x = \ln M$, and since the values of e^x increase as x increases, we will have

$$e^x > M \quad \text{if} \quad x > \ln M$$

(Figure 0.5.8). It is not clear from Table 0.5.5 whether $\ln x$ increases without bound as x increases because the values grow so slowly, but we know this to be so since the range of this function is $(-\infty, +\infty)$. To see this algebraically, let M be any positive number. We will have $\ln x = M$ when $x = e^M$, and since the values of $\ln x$ increase as x increases, we will have

$$\ln x > M \quad \text{if} \quad x > e^M$$

(Figure 0.5.9).

Table 0.5.5

x	e^x	$\ln x$
1	2.72	0.00
2	7.39	0.69
3	20.09	1.10
4	54.60	1.39
5	148.41	1.61
6	403.43	1.79
7	1096.63	1.95
8	2980.96	2.08
9	8103.08	2.20
10	22026.47	2.30
100	2.69×10^{43}	4.61
1000	1.97×10^{434}	6.91

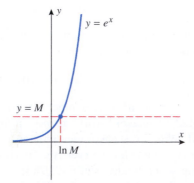

▲ **Figure 0.5.8** The value of $y = e^x$ will exceed an arbitrary positive value of M when $x > \ln M$.

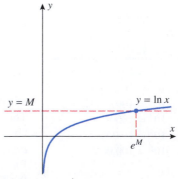

▲ **Figure 0.5.9** The value of $y = \ln x$ will exceed an arbitrary positive value of M when $x > e^M$.

✔ **QUICK CHECK EXERCISES 0.5**   *(See page 63 for answers.)*

1. The function $y = \left(\frac{1}{2}\right)^x$ has domain _____ and range _____.

2. The function $y = \ln(1 - x)$ has domain _____ and range _____.

3. Express as a power of 4:
   (a) 1   (b) 2   (c) $\frac{1}{16}$ (d) $\sqrt{8}$ (e) 5.

4. Solve each equation for x.
   (a) $e^x = \frac{1}{2}$ (b) $10^{3x} = 1{,}000{,}000$
   (c) $7e^{3x} = 56$

5. Solve each equation for x.
   (a) $\ln x = 3$ (b) $\log(x - 1) = 2$
   (c) $2\log x - \log(x + 1) = \log 4 - \log 3$

EXERCISE SET 0.5   ◠ Graphing Utility

1–2 Simplify the expression without using a calculating utility. ■

1. (a) $-8^{2/3}$ (b) $(-8)^{2/3}$ (c) $8^{-2/3}$

2. (a) 2^{-4} (b) $4^{1.5}$ (c) $9^{-0.5}$

3–4 Use a calculating utility to approximate the expression. Round your answer to four decimal places. ■

3. (a) $2^{1.57}$ (b) $5^{-2.1}$

4. (a) $\sqrt[5]{24}$ (b) $\sqrt[8]{0.6}$

5–6 Find the exact value of the expression without using a calculating utility. ■

5. (a) $\log_2 16$ (b) $\log_2\left(\frac{1}{32}\right)$
   (c) $\log_4 4$ (d) $\log_9 3$

6. (a) $\log_{10}(0.001)$ (b) $\log_{10}(10^4)$
   (c) $\ln(e^3)$ (d) $\ln(\sqrt{e})$

7–8 Use a calculating utility to approximate the expression. Round your answer to four decimal places. ■

7. (a) $\log 23.2$ (b) $\ln 0.74$

8. (a) $\log 0.3$ (b) $\ln \pi$

9–10 Use the logarithm properties in Theorem 0.5.2 to rewrite the expression in terms of r, s, and t, where $r = \ln a$, $s = \ln b$, and $t = \ln c$. ■

9. (a) $\ln a^2 \sqrt{bc}$ (b) $\ln \dfrac{b}{a^3 c}$

10. (a) $\ln \dfrac{\sqrt[3]{c}}{ab}$ (b) $\ln \sqrt{\dfrac{ab^3}{c^2}}$

11–12 Expand the logarithm in terms of sums, differences, and multiples of simpler logarithms. ■

11. (a) $\log(10x\sqrt{x - 3})$ (b) $\ln \dfrac{x^2 \sin^3 x}{\sqrt{x^2 + 1}}$

12. (a) $\log \dfrac{\sqrt[3]{x + 2}}{\cos 5x}$ (b) $\ln \sqrt{\dfrac{x^2 + 1}{x^3 + 5}}$

13–15 Rewrite the expression as a single logarithm. ■

13. $4\log 2 - \log 3 + \log 16$

14. $\frac{1}{2}\log x - 3\log(\sin 2x) + 2$

15. $2\ln(x + 1) + \frac{1}{3}\ln x - \ln(\cos x)$

16–23 Solve for x without using a calculating utility. ■

16. $\log_{10}(1 + x) = 3$ 17. $\log_{10}(\sqrt{x}) = -1$

18. $\ln(x^2) = 4$ 19. $\ln(1/x) = -2$

20. $\log_3(3^x) = 7$ 21. $\log_5(5^{2x}) = 8$

22. $\ln 4x - 3\ln(x^2) = \ln 2$

23. $\ln(1/x) + \ln(2x^3) = \ln 3$

24–29 Solve for x without using a calculating utility. Use the natural logarithm anywhere that logarithms are needed. ■

24. $3^x = 2$ 25. $5^{-2x} = 3$

26. $3e^{-2x} = 5$ 27. $2e^{3x} = 7$

28. $e^x - 2xe^x = 0$ 29. $xe^{-x} + 2e^{-x} = 0$

30. Solve $e^{-2x} - 3e^{-x} = -2$ for x without using a calculating utility. [*Hint:* Rewrite the equation as a quadratic equation in $u = e^{-x}$.]

FOCUS ON CONCEPTS

31–34 In each part, identify the domain and range of the function, and then sketch the graph of the function without using a graphing utility. ■

31. (a) $f(x) = \left(\frac{1}{2}\right)^{x-1} - 1$ (b) $g(x) = \ln|x|$

32. (a) $f(x) = 1 + \ln(x - 2)$ (b) $g(x) = 3 + e^{x-2}$

33. (a) $f(x) = \ln(x^2)$ (b) $g(x) = e^{-x^2}$

34. (a) $f(x) = 1 - e^{-x+1}$ (b) $g(x) = 3\ln\sqrt[3]{x - 1}$

35–38 True–False Determine whether the statement is true or false. Explain your answer. ■

35. The function $y = x^3$ is an exponential function.

36. The graph of the exponential function with base b passes through the point $(0, 1)$.

37. The natural logarithm function is the logarithmic function with base e.

38. The domain of a logarithmic function is the interval $x > 1$.

39. Use a calculating utility and the change of base formula (6) to find the values of $\log_2 7.35$ and $\log_5 0.6$, rounded to four decimal places.

 40–41 Graph the functions on the same screen of a graphing utility. [Use the change of base formula (6), where needed.] ■

40. $\ln x$, e^x, $\log x$, 10^x

41. $\log_2 x$, $\ln x$, $\log_5 x$, $\log x$

42. (a) Derive the general change of base formula

$$\log_b x = \frac{\log_a x}{\log_a b}$$

  (b) Use the result in part (a) to find the exact value of $(\log_2 81)(\log_3 32)$ without using a calculating utility.

FOCUS ON CONCEPTS

 43. (a) Is the curve in the accompanying figure the graph of an exponential function? Explain your reasoning.
  (b) Find the equation of an exponential function that passes through the point $(4, 2)$.
  (c) Find the equation of an exponential function that passes through the point $\left(2, \frac{1}{4}\right)$.
  (d) Use a graphing utility to generate the graph of an exponential function that passes through the point $(2, 5)$.

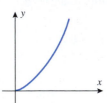

◄ **Figure Ex-43**

 44. (a) Make a conjecture about the general shape of the graph of $y = \log(\log x)$, and sketch the graph of this equation and $y = \log x$ in the same coordinate system.
  (b) Check your work in part (a) with a graphing utility.

45. Find the fallacy in the following "proof" that $\frac{1}{8} > \frac{1}{4}$. Multiply both sides of the inequality $3 > 2$ by $\log \frac{1}{2}$ to get

$$3 \log \tfrac{1}{2} > 2 \log \tfrac{1}{2}$$

$$\log \left(\tfrac{1}{2}\right)^3 > \log \left(\tfrac{1}{2}\right)^2$$

$$\log \tfrac{1}{8} > \log \tfrac{1}{4}$$

$$\tfrac{1}{8} > \tfrac{1}{4}$$

46. Prove the four algebraic properties of logarithms in Theorem 0.5.2.

47. If equipment in the satellite of Example 3 requires 15 watts to operate correctly, what is the operational lifetime of the power supply?

48. The equation $Q = 12e^{-0.055t}$ gives the mass Q in grams of radioactive potassium-42 that will remain from some initial quantity after t hours of radioactive decay.
  (a) How many grams were there initially?
  (b) How many grams remain after 4 hours?
  (c) How long will it take to reduce the amount of radioactive potassium-42 to half of the initial amount?

49. The acidity of a substance is measured by its pH value, which is defined by the formula

$$\text{pH} = -\log[H^+]$$

where the symbol $[H^+]$ denotes the concentration of hydrogen ions measured in moles per liter. Distilled water has a pH of 7; a substance is called *acidic* if it has pH < 7 and *basic* if it has pH > 7. Find the pH of each of the following substances and state whether it is acidic or basic.

	SUBSTANCE	$[H^+]$
(a)	Arterial blood	3.9×10^{-8} mol/L
(b)	Tomatoes	6.3×10^{-5} mol/L
(c)	Milk	4.0×10^{-7} mol/L
(d)	Coffee	1.2×10^{-6} mol/L

50. Use the definition of pH in Exercise 49 to find $[H^+]$ in a solution having a pH equal to
  (a) 2.44   (b) 8.06.

51. The perceived loudness β of a sound in decibels (dB) is related to its intensity I in watts per square meter (W/m^2) by the equation

$$\beta = 10 \log(I/I_0)$$

where $I_0 = 10^{-12}$ W/m^2. Damage to the average ear occurs at 90 dB or greater. Find the decibel level of each of the following sounds and state whether it will cause ear damage.

	SOUND	I
(a)	Jet aircraft (from 50 ft)	1.0×10^2 W/m^2
(b)	Amplified rock music	1.0 W/m^2
(c)	Garbage disposal	1.0×10^{-4} W/m^2
(d)	TV (mid volume from 10 ft)	3.2×10^{-5} W/m^2

52–54 Use the definition of the decibel level of a sound (see Exercise 51). ■

52. If one sound is three times as intense as another, how much greater is its decibel level?

53. According to one source, the noise inside a moving automobile is about 70 dB, whereas an electric blender generates 93 dB. Find the ratio of the intensity of the noise of the blender to that of the automobile.

54. Suppose that the intensity level of an echo is $\frac{2}{3}$ the intensity level of the original sound. If each echo results in another echo, how many echoes will be heard from a 120 dB sound given that the average human ear can hear a sound as low as 10 dB?

55. On the **Richter scale**, the magnitude M of an earthquake is related to the released energy E in joules (J) by the equation
$$\log E = 4.4 + 1.5M$$
(a) Find the energy E of the 1906 San Francisco earthquake that registered $M = 8.2$ on the Richter scale.

(b) If the released energy of one earthquake is 10 times that of another, how much greater is its magnitude on the Richter scale?

56. Suppose that the magnitudes of two earthquakes differ by 1 on the Richter scale. Find the ratio of the released energy of the larger earthquake to that of the smaller earthquake. [*Note:* See Exercise 55 for terminology.]

✔ QUICK CHECK ANSWERS 0.5

1. $(-\infty, +\infty)$; $(0, +\infty)$ **2.** $(-\infty, 1)$; $(-\infty, +\infty)$ **3.** (a) 4^0 (b) $4^{1/2}$ (c) 4^{-2} (d) $4^{3/4}$ (e) $4^{\log_4 5}$ **4.** (a) $\ln \frac{1}{2} = -\ln 2$ (b) 2 (c) $\ln 2$ **5.** (a) e^3 (b) 101 (c) 2

CHAPTER 0 REVIEW EXERCISES   ∿ Graphing Utility

1. Sketch the graph of the function

$$f(x) = \begin{cases} -1, & x \leq -5 \\ \sqrt{25 - x^2}, & -5 < x < 5 \\ x - 5, & x \geq 5 \end{cases}$$

2. Use the graphs of the functions f and g in the accompanying figure to solve the following problems.
(a) Find the values of $f(-2)$ and $g(3)$.
(b) For what values of x is $f(x) = g(x)$?
(c) For what values of x is $f(x) < 2$?
(d) What are the domain and range of f?
(e) What are the domain and range of g?
(f) Find the zeros of f and g.

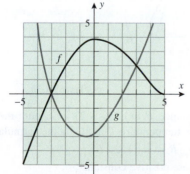

◀ **Figure Ex-2**

3. A glass filled with water that has a temperature of 40°F is placed in a room in which the temperature is a constant 70°F. Sketch a rough graph that reasonably describes the temperature of the water in the glass as a function of the elapsed time.

4. You want to paint the top of a circular table. Find a formula that expresses the amount of paint required as a function of the radius, and discuss all of the assumptions you have made in finding the formula.

5. A rectangular storage container with an open top and a square base has a volume of 8 cubic meters. Material for the base costs $5 per square meter and material for the sides $2 per square meter.
(a) Find a formula that expresses the total cost of materials as a function of the length of a side of the base.
(b) What is the domain of the cost function obtained in part (a)?

6. A ball of radius 3 inches is coated uniformly with plastic.
(a) Express the volume of the plastic as a function of its thickness.
(b) What is the domain of the volume function obtained in part (a)?

∿ **7.** A box with a closed top is to be made from a 6 ft by 10 ft piece of cardboard by cutting out four squares of equal size (see the accompanying figure), folding along the dashed lines, and tucking the two extra flaps inside.
(a) Find a formula that expresses the volume of the box as a function of the length of the sides of the cut-out squares.
(b) Find an inequality that specifies the domain of the function in part (a).
(c) Use the graph of the volume function to estimate the dimensions of the box of largest volume.

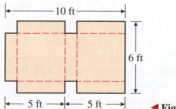

◀ **Figure Ex-7**

∿ **8.** Let C denote the graph of $y = 1/x$, $x > 0$.
(a) Express the distance between the point $P(1, 0)$ and a point Q on C as a function of the x-coordinate of Q.
(b) What is the domain of the distance function obtained in part (a)?

(cont.)

(c) Use the graph of the distance function obtained in part (a) to estimate the point Q on C that is closest to the point P.

9. Sketch the graph of the equation $x^2 - 4y^2 = 0$.

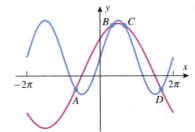 10. Generate the graph of $f(x) = x^4 - 24x^3 - 25x^2$ in two different viewing windows, each of which illustrates a different property of f. Identify each viewing window and a characteristic of the graph of f that is illustrated well in the window.

11. Complete the following table.

x	-4	-3	-2	-1	0	1	2	3	4
$f(x)$	0	-1	2	1	3	-2	-3	4	-4
$g(x)$	3	2	1	-3	-1	-4	4	-2	0
$(f \circ g)(x)$									
$(g \circ f)(x)$									

▲ **Table Ex-11**

12. Let $f(x) = -x^2$ and $g(x) = 1/\sqrt{x}$. Find formulas for $f \circ g$ and $g \circ f$ and state the domain of each composition.

13. Given that $f(x) = x^2 + 1$ and $g(x) = 3x + 2$, find all values of x such that $f(g(x)) = g(f(x))$.

14. Let $f(x) = (2x - 1)/(x + 1)$ and $g(x) = 1/(x - 1)$.
    (a) Find $f(g(x))$.
    (b) Is the natural domain of the function $h(x) = (3 - x)/x$ the same as the domain of $f \circ g$? Explain.

15. Given that

$$f(x) = \frac{x}{x - 1}, \quad g(x) = \frac{1}{x}, \quad h(x) = x^2 - 1$$

find a formula for $f \circ g \circ h$ and state the domain of this composition.

16. Given that $f(x) = 2x + 1$ and $h(x) = 2x^2 + 4x + 1$, find a function g such that $f(g(x)) = h(x)$.

17. In each part, classify the function as even, odd, or neither.
    (a) $x^2 \sin x$ (b) $\sin^2 x$ (c) $x + x^2$ (d) $\sin x \tan x$

18. (a) Write an equation for the graph that is obtained by reflecting the graph of $y = |x - 1|$ about the y-axis, then stretching that graph vertically by a factor of 2, then translating that graph down 3 units, and then reflecting that graph about the x-axis.
    (b) Sketch the original graph and the final graph.

19. In each part, describe the family of curves.
    (a) $(x - a)^2 + (y - a^2)^2 = 1$
    (b) $y = a + (x - 2a)^2$

20. Find an equation for a parabola that passes through the points $(2, 0)$, $(8, 18)$, and $(-8, 18)$.

21. Suppose that the expected low temperature in Anchorage, Alaska (in °F), is modeled by the equation

$$T = 50 \sin \frac{2\pi}{365}(t - 101) + 25$$

where t is in days and $t = 0$ corresponds to January 1.
    (a) Sketch the graph of T versus t for $0 \le t \le 365$.
    (b) Use the model to predict when the coldest day of the year will occur.
    (c) Based on this model, how many days during the year would you expect the temperature to be below 0°F?

22. The accompanying figure shows a model for the tide variation in an inlet to San Francisco Bay during a 24-hour period. Find an equation of the form $y = y_0 + y_1 \sin(at + b)$ for the model, assuming that $t = 0$ corresponds to midnight.

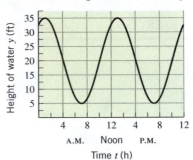

◄ **Figure Ex-22**

23. The accompanying figure shows the graphs of the equations $y = 1 + 2 \sin x$ and $y = 2 \sin(x/2) + 2 \cos(x/2)$ for $-2\pi \le x \le 2\pi$. Without the aid of a calculator, label each curve by its equation, and find the coordinates of the points A, B, C, and D. Explain your reasoning.

◄ **Figure Ex-23**

24. The electrical resistance R in ohms (Ω) for a pure metal wire is related to its temperature T in °C by the formula

$$R = R_0(1 + kT)$$

in which R_0 and k are positive constants.
    (a) Make a hand-drawn sketch of the graph of R versus T, and explain the geometric significance of R_0 and k for your graph.
    (b) In theory, the resistance R of a pure metal wire drops to zero when the temperature reaches absolute zero ($T = -273°C$). What information does this give you about k?
    (c) A tungsten bulb filament has a resistance of 1.1 Ω at a temperature of 20°C. What information does this give you about R_0 for the filament?   *(cont.)*

(d) At what temperature will the tungsten filament have a resistance of 1.5 Ω?

25. (a) State conditions under which two functions, f and g, will be inverses, and give several examples of such functions.
    (b) In words, what is the relationship between the graphs of $y = f(x)$ and $y = g(x)$ when f and g are inverse functions?
    (c) What is the relationship between the domains and ranges of inverse functions f and g?
    (d) What condition must be satisfied for a function f to have an inverse? Give some examples of functions that do not have inverses.

26. (a) State the restrictions on the domains of $\sin x$, $\cos x$, $\tan x$, and $\sec x$ that are imposed to make those functions one-to-one in the definitions of $\sin^{-1} x$, $\cos^{-1} x$, $\tan^{-1} x$, and $\sec^{-1} x$.
    (b) Sketch the graphs of the restricted trigonometric functions in part (a) and their inverses.

27. In each part, find $f^{-1}(x)$ if the inverse exists.
    (a) $f(x) = 8x^3 - 1$
    (b) $f(x) = x^2 - 2x + 1$
    (c) $f(x) = (e^x)^2 + 1$
    (d) $f(x) = (x + 2)/(x - 1)$
    (e) $f(x) = \sin\left(\dfrac{1 - 2x}{x}\right)$, $\dfrac{2}{4 + \pi} \le x \le \dfrac{2}{4 - \pi}$
    (f) $f(x) = \dfrac{1}{1 + 3\tan^{-1} x}$

28. Let $f(x) = (ax + b)/(cx + d)$. What conditions on a, b, c, and d guarantee that f^{-1} exists? Find $f^{-1}(x)$.

29. In each part, find the exact numerical value of the given expression.
    (a) $\cos[\cos^{-1}(4/5) + \sin^{-1}(5/13)]$
    (b) $\sin[\sin^{-1}(4/5) + \cos^{-1}(5/13)]$

30. In each part, sketch the graph, and check your work with a graphing utility.
    (a) $f(x) = 3\sin^{-1}(x/2)$
    (b) $f(x) = \cos^{-1} x - \pi/2$
    (c) $f(x) = 2\tan^{-1}(-3x)$
    (d) $f(x) = \cos^{-1} x + \sin^{-1} x$

31. Suppose that the graph of $y = \log x$ is drawn with equal scales of 1 inch per unit in both the x- and y-directions. If a bug wants to walk along the graph until it reaches a height of 5 ft above the x-axis, how many miles to the right of the origin will it have to travel?

32. Suppose that the graph of $y = 10^x$ is drawn with equal scales of 1 inch per unit in both the x- and y-directions. If a bug wants to walk along the graph until it reaches a height of 100 mi above the x-axis, how many feet to the right of the origin will it have to travel?

33. Express the following function as a rational function of x:
$$3\ln\left(e^{2x}(e^x)^3\right) + 2\exp(\ln 1)$$

34. Suppose that $y = Ce^{kt}$, where C and k are constants, and let $Y = \ln y$. Show that the graph of Y versus t is a line, and state its slope and Y-intercept.

35. (a) Sketch the curves $y = \pm e^{-x/2}$ and $y = e^{-x/2}\sin 2x$ for $-\pi/2 \le x \le 3\pi/2$ in the same coordinate system, and check your work using a graphing utility.
    (b) Find all x-intercepts of the curve $y = e^{-x/2}\sin 2x$ in the stated interval, and find the x-coordinates of all points where this curve intersects the curves $y = \pm e^{-x/2}$.

36. Suppose that a package of medical supplies is dropped from a helicopter straight down by parachute into a remote area. The velocity v (in feet per second) of the package t seconds after it is released is given by $v = 24.61(1 - e^{-1.3t})$.
    (a) Graph v versus t.
    (b) Show that the graph has a horizontal asymptote $v = c$.
    (c) The constant c is called the **terminal velocity**. Explain what the terminal velocity means in practical terms.
    (d) Can the package actually reach its terminal velocity? Explain.
    (e) How long does it take for the package to reach 98% of its terminal velocity?

37. A breeding group of 20 bighorn sheep is released in a protected area in Colorado. It is expected that with careful management the number of sheep, N, after t years will be given by the formula
$$N = \frac{220}{1 + 10(0.83^t)}$$
and that the sheep population will be able to maintain itself without further supervision once the population reaches a size of 80.
    (a) Graph N versus t.
    (b) How many years must the state of Colorado maintain a program to care for the sheep?
    (c) How many bighorn sheep can the environment in the protected area support? [*Hint:* Examine the graph of N versus t for large values of t.]

38. An oven is preheated and then remains at a constant temperature. A potato is placed in the oven to bake. Suppose that the temperature T (in °F) of the potato t minutes later is given by $T = 400 - 325(0.97^t)$. The potato will be considered done when its temperature is anywhere between 260°F and 280°F.
    (a) During what interval of time would the potato be considered done?
    (b) How long does it take for the difference between the potato and oven temperatures to be cut in half?

39. (a) Show that the graphs of $y = \ln x$ and $y = x^{0.2}$ intersect.
    (b) Approximate the solution(s) of the equation $\ln x = x^{0.2}$ to three decimal places.

40. (a) Show that for $x > 0$ and $k \ne 0$ the equations
$$x^k = e^x \quad \text{and} \quad \frac{\ln x}{x} = \frac{1}{k}$$
have the same solutions.

(cont.)

(b) Use the graph of $y = (\ln x)/x$ to determine the values of k for which the equation $x^k = e^x$ has two distinct positive solutions.

(c) Estimate the positive solution(s) of $x^8 = e^x$.

41. Consider $f(x) = x^2 \tan x + \ln x,\ 0 < x < \pi/2$.

(a) Explain why f is one-to-one.

(b) Use a graphing utility to generate the graph of f. Then sketch the graphs of f and f^{-1} together. What are the asymptotes for each graph?

Joe McBride/Stone/Getty Images

1

LIMITS AND CONTINUITY

Air resistance prevents the velocity of a skydiver from increasing indefinitely. The velocity approaches a limit, called the "terminal velocity."

The development of calculus in the seventeenth century by Newton and Leibniz provided scientists with their first real understanding of what is meant by an "instantaneous rate of change" such as velocity and acceleration. Once the idea was understood conceptually, efficient computational methods followed, and science took a quantum leap forward. The fundamental building block on which rates of change rest is the concept of a "limit," an idea that is so important that all other calculus concepts are now based on it.

In this chapter we will develop the concept of a limit in stages, proceeding from an informal, intuitive notion to a precise mathematical definition. We will also develop theorems and procedures for calculating limits, and we will conclude the chapter by using the limits to study "continuous" curves.

LIMITS (AN INTUITIVE APPROACH)

The concept of a "limit" is the fundamental building block on which all calculus concepts are based. In this section we will study limits informally, with the goal of developing an intuitive feel for the basic ideas. In the next three sections we will focus on computational methods and precise definitions.

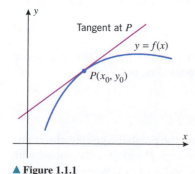

▲ **Figure 1.1.1**

Many of the ideas of calculus originated with the following two geometric problems:

THE TANGENT LINE PROBLEM   Given a function f and a point $P(x_0, y_0)$ on its graph, find an equation of the line that is tangent to the graph at P (Figure 1.1.1).

THE AREA PROBLEM   Given a function f, find the area between the graph of f and an interval $[a, b]$ on the x-axis (Figure 1.1.2).

Traditionally, that portion of calculus arising from the tangent line problem is called *differential calculus* and that arising from the area problem is called *integral calculus*. However, we will see later that the tangent line and area problems are so closely related that the distinction between differential and integral calculus is somewhat artificial.

67

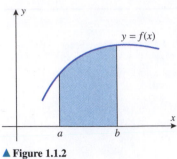

▲ **Figure 1.1.2**

■ TANGENT LINES AND LIMITS

In plane geometry, a line is called *tangent* to a circle if it meets the circle at precisely one point (Figure 1.1.3a). Although this definition is adequate for circles, it is not appropriate for more general curves. For example, in Figure 1.1.3b, the line meets the curve exactly once but is obviously not what we would regard to be a tangent line; and in Figure 1.1.3c, the line appears to be tangent to the curve, yet it intersects the curve more than once.

To obtain a definition of a tangent line that applies to curves other than circles, we must view tangent lines another way. For this purpose, suppose that we are interested in the tangent line at a point P on a curve in the xy-plane and that Q is any point that lies on the curve and is different from P. The line through P and Q is called a *secant line* for the curve at P. Intuition suggests that if we move the point Q along the curve toward P, then the secant line will rotate toward a *limiting position*. The line in this limiting position is what we will consider to be the *tangent line* at P (Figure 1.1.4a). As suggested by Figure 1.1.4b, this new concept of a tangent line coincides with the traditional concept when applied to circles.

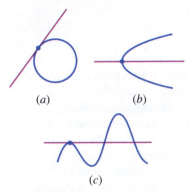

(a)   (b)

(c)

▲ **Figure 1.1.3**

▶ **Figure 1.1.4**

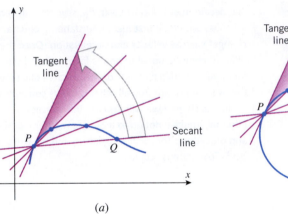

(a)   (b)

▶ **Example 1** Find an equation for the tangent line to the parabola $y = x^2$ at the point $P(1, 1)$.

Solution. If we can find the slope $m_{\tan}$ of the tangent line at P, then we can use the point P and the point-slope formula for a line (Web Appendix G) to write the equation of the tangent line as

$$y - 1 = m_{\tan}(x - 1) \tag{1}$$

To find the slope $m_{\tan}$, consider the secant line through P and a point $Q(x, x^2)$ on the parabola that is distinct from P. The slope $m_{\sec}$ of this secant line is

> Why are we requiring that P and Q be distinct?

$$m_{\sec} = \frac{x^2 - 1}{x - 1} \tag{2}$$

Figure 1.1.4a suggests that if we now let Q move along the parabola, getting closer and closer to P, then the limiting position of the secant line through P and Q will coincide with that of the tangent line at P. This in turn suggests that the value of $m_{\sec}$ will get closer and closer to the value of $m_{\tan}$ as P moves toward Q along the curve. However, to say that $Q(x, x^2)$ gets closer and closer to $P(1, 1)$ is algebraically equivalent to saying that x gets closer and closer to 1. Thus, the problem of finding $m_{\tan}$ reduces to finding the "limiting value" of $m_{\sec}$ in Formula (2) as x gets closer and closer to 1 (but with $x \neq 1$ to ensure that P and Q remain distinct).

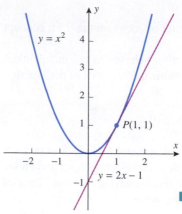

▲ **Figure 1.1.5**

We can rewrite (2) as

$$m_{\sec} = \frac{x^2 - 1}{x - 1} = \frac{(x - 1)(x + 1)}{(x - 1)} = x + 1$$

where the cancellation of the factor $(x - 1)$ is allowed because $x \neq 1$. It is now evident that $m_{\sec}$ gets closer and closer to 2 as x gets closer and closer to 1. Thus, $m_{\tan} = 2$ and (1) implies that the equation of the tangent line is

$$y - 1 = 2(x - 1) \quad \text{or equivalently} \quad y = 2x - 1$$

Figure 1.1.5 shows the graph of $y = x^2$ and this tangent line. ◀

■ AREAS AND LIMITS

Just as the general notion of a tangent line leads to the concept of *limit*, so does the general notion of area. For plane regions with straight-line boundaries, areas can often be calculated by subdividing the region into rectangles or triangles and adding the areas of the constituent parts (Figure 1.1.6). However, for regions with curved boundaries, such as that in Figure 1.1.7*a*, a more general approach is needed. One such approach is to begin by approximating the area of the region by inscribing a number of rectangles of equal width under the curve and adding the areas of these rectangles (Figure 1.1.7*b*). Intuition suggests that if we repeat that approximation process using more and more rectangles, then the rectangles will tend to fill in the gaps under the curve, and the approximations will get closer and closer to the exact area under the curve (Figure 1.1.7*c*). This suggests that we can define the area under the curve to be the limiting value of these approximations. This idea will be considered in detail later, but the point to note here is that once again the concept of a limit comes into play.

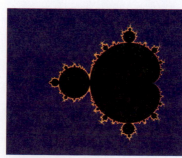

▲ **Figure 1.1.6**

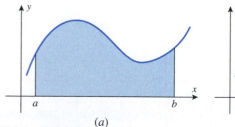

(a)

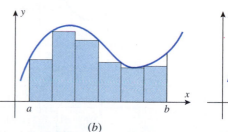

(b)

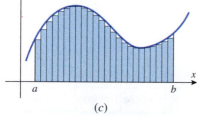

(c)

▲ **Figure 1.1.7**

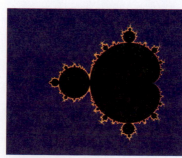

This figure shows a region called the **Mandelbrot Set**. *It illustrates how complicated a region in the plane can be and why the notion of area requires careful definition.*

■ DECIMALS AND LIMITS

Limits also arise in the familiar context of decimals. For example, the decimal expansion of the fraction $\frac{1}{3}$ is

$$\frac{1}{3} = 0.33333\ldots \tag{3}$$

in which the dots indicate that the digit 3 repeats indefinitely. Although you may not have thought about decimals in this way, we can write (3) as

$$\frac{1}{3} = 0.33333\ldots = 0.3 + 0.03 + 0.003 + 0.0003 + 0.00003 + \cdots \tag{4}$$

which is a sum with "infinitely many" terms. As we will discuss in more detail later, we interpret (4) to mean that the succession of finite sums

$$0.3, \quad 0.3 + 0.03, \quad 0.3 + 0.03 + 0.003, \quad 0.3 + 0.03 + 0.003 + 0.0003, \ldots$$

gets closer and closer to a limiting value of $\frac{1}{3}$ as more and more terms are included. Thus, limits even occur in the familiar context of decimal representations of real numbers.

■ LIMITS

Now that we have seen how limits arise in various ways, let us focus on the limit concept itself.

The most basic use of limits is to describe how a function behaves as the independent variable approaches a given value. For example, let us examine the behavior of the function

$$f(x) = x^2 - x + 1$$

for x-values closer and closer to 2. It is evident from the graph and table in Figure 1.1.8 that the values of $f(x)$ get closer and closer to 3 as values of x are selected closer and closer to 2 on either the left or the right side of 2. We describe this by saying that the "limit of $x^2 - x + 1$ is 3 as x approaches 2 from either side," and we write

$$\lim_{x \to 2} (x^2 - x + 1) = 3 \tag{5}$$

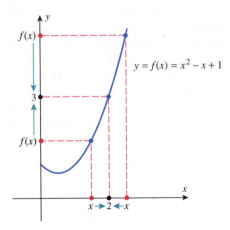

x	1.0	1.5	1.9	1.95	1.99	1.995	1.999	2	2.001	2.005	2.01	2.05	2.1	2.5	3.0
$f(x)$	1.000000	1.750000	2.710000	2.852500	2.970100	2.985025	2.997001		3.003001	3.015025	3.030100	3.152500	3.310000	4.750000	7.000000

Left side   →   ←   Right side

▲ **Figure 1.1.8**

This leads us to the following general idea.

1.1.1   LIMITS (AN INFORMAL VIEW)   If the values of $f(x)$ can be made as close as we like to L by taking values of x sufficiently close to a (but not equal to a), then we write

$$\lim_{x \to a} f(x) = L \tag{6}$$

which is read "the limit of $f(x)$ as x approaches a is L" or "$f(x)$ approaches L as x approaches a." The expression in (6) can also be written as

$$f(x) \to L \quad \text{as} \quad x \to a \tag{7}$$

Since x is required to be different from a in (6), the value of f at a, or even whether f is defined at a, has no bearing on the limit L. The limit describes the behavior of f *close to* a but not *at* a.

▶ **Example 2** Use numerical evidence to make a conjecture about the value of

$$\lim_{x \to 1} \frac{x-1}{\sqrt{x}-1} \tag{8}$$

Solution. Although the function

$$f(x) = \frac{x-1}{\sqrt{x}-1} \tag{9}$$

is undefined at $x = 1$, this has no bearing on the limit. Table 1.1.1 shows sample x-values approaching 1 from the left side and from the right side. In both cases the corresponding values of $f(x)$, calculated to six decimal places, appear to get closer and closer to 2, and hence we conjecture that

$$\lim_{x \to 1} \frac{x-1}{\sqrt{x}-1} = 2$$

This is consistent with the graph of f shown in Figure 1.1.9. In the next section we will show how to obtain this result algebraically. ◀

Table 1.1.1

x	0.99	0.999	0.9999	0.99999		1.00001	1.0001	1.001	1.01
$f(x)$	1.994987	1.999500	1.999950	1.999995		2.000005	2.000050	2.000500	2.004988

Left side ⟶ ⟵ Right side

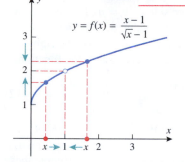

$y = f(x) = \dfrac{x-1}{\sqrt{x}-1}$

▲ **Figure 1.1.9**

▶ **Example 3** Use numerical evidence to make a conjecture about the value of

$$\lim_{x \to 0} \frac{\sin x}{x} \tag{10}$$

Solution. With the help of a calculating utility set in radian mode, we obtain Table 1.1.2. The data in the table suggest that

$$\lim_{x \to 0} \frac{\sin x}{x} = 1 \tag{11}$$

The result is consistent with the graph of $f(x) = (\sin x)/x$ shown in Figure 1.1.10. Later in this chapter we will give a geometric argument to prove that our conjecture is correct. ◀

Table 1.1.2

x (RADIANS)	$y = \dfrac{\sin x}{x}$
±1.0	0.84147
±0.9	0.87036
±0.8	0.89670
±0.7	0.92031
±0.6	0.94107
±0.5	0.95885
±0.4	0.97355
±0.3	0.98507
±0.2	0.99335
±0.1	0.99833
±0.01	0.99998

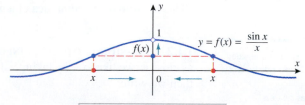

$y = f(x) = \dfrac{\sin x}{x}$

As x approaches 0 from the left or right, $f(x)$ approaches 1.

▶ **Figure 1.1.10**

■ **SAMPLING PITFALLS**

Numerical evidence can sometimes lead to incorrect conclusions about limits because of roundoff error or because the sample values chosen do not reveal the true limiting behavior. For example, one might *incorrectly* conclude from Table 1.1.3 that

$$\lim_{x \to 0} \sin\left(\frac{\pi}{x}\right) = 0$$

The fact that this is not correct is evidenced by the graph of f in Figure 1.1.11. The graph reveals that the values of f oscillate between -1 and 1 with increasing rapidity as $x \to 0$ and hence do not approach a limit. The data in the table deceived us because the x-values selected all happened to be x-intercepts for $f(x)$. This points out the need for having alternative methods for corroborating limits conjectured from numerical evidence.

Table 1.1.3

x	$\dfrac{\pi}{x}$	$f(x) = \sin\left(\dfrac{\pi}{x}\right)$
$x = \pm 1$	$\pm \pi$	$\sin(\pm \pi) = 0$
$x = \pm 0.1$	$\pm 10\pi$	$\sin(\pm 10\pi) = 0$
$x = \pm 0.01$	$\pm 100\pi$	$\sin(\pm 100\pi) = 0$
$x = \pm 0.001$	$\pm 1000\pi$	$\sin(\pm 1000\pi) = 0$
$x = \pm 0.0001$	$\pm 10,000\pi$	$\sin(\pm 10,000\pi) = 0$
$\vdots$	$\vdots$	$\vdots$

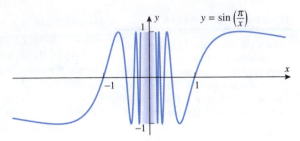

▲ Figure 1.1.11

■ ONE-SIDED LIMITS

The limit in (6) is called a **two-sided limit** because it requires the values of $f(x)$ to get closer and closer to L as values of x are taken from *either* side of $x = a$. However, some functions exhibit different behaviors on the two sides of an x-value a, in which case it is necessary to distinguish whether values of x near a are on the left side or on the right side of a for purposes of investigating limiting behavior. For example, consider the function

$$f(x) = \frac{|x|}{x} = \begin{cases} 1, & x > 0 \\ -1, & x < 0 \end{cases} \tag{12}$$

which is graphed in Figure 1.1.12. As x approaches 0 from the *right*, the values of $f(x)$ approach a limit of 1 [in fact, the values of $f(x)$ are exactly 1 for all such x], and similarly, as x approaches 0 from the *left*, the values of $f(x)$ approach a limit of -1. We denote these limits by writing

$$\lim_{x \to 0^+} \frac{|x|}{x} = 1 \quad \text{and} \quad \lim_{x \to 0^-} \frac{|x|}{x} = -1 \tag{13}$$

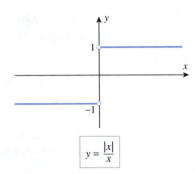

▲ Figure 1.1.12

With this notation, the superscript "$+$" indicates a limit from the right and the superscript "$-$" indicates a limit from the left.

This leads to the general idea of a **one-sided limit**.

As with two-sided limits, the one-sided limits in (14) and (15) can also be written as

$$f(x) \to L \quad \text{as} \quad x \to a^+$$

and

$$f(x) \to L \quad \text{as} \quad x \to a^-$$

respectively.

1.1.2   ONE-SIDED LIMITS (AN INFORMAL VIEW)   If the values of $f(x)$ can be made as close as we like to L by taking values of x sufficiently close to a (but greater than a), then we write

$$\lim_{x \to a^+} f(x) = L \tag{14}$$

and if the values of $f(x)$ can be made as close as we like to L by taking values of x sufficiently close to a (but less than a), then we write

$$\lim_{x \to a^-} f(x) = L \tag{15}$$

Expression (14) is read "the limit of $f(x)$ as x approaches a from the right is L" or "$f(x)$ approaches L as x approaches a from the right." Similarly, expression (15) is read "the limit of $f(x)$ as x approaches a from the left is L" or "$f(x)$ approaches L as x approaches a from the left."

■ **THE RELATIONSHIP BETWEEN ONE-SIDED LIMITS AND TWO-SIDED LIMITS**

In general, there is no guarantee that a function f will have a two-sided limit at a given point a; that is, the values of $f(x)$ may not get closer and closer to any *single* real number L as $x \to a$. In this case we say that

$$\lim_{x \to a} f(x) \quad \textit{does not exist}$$

Similarly, the values of $f(x)$ may not get closer and closer to a single real number L as $x \to a^+$ or as $x \to a^-$. In these cases we say that

$$\lim_{x \to a^+} f(x) \quad \textit{does not exist}$$

or that

$$\lim_{x \to a^-} f(x) \quad \textit{does not exist}$$

In order for the two-sided limit of a function $f(x)$ to exist at a point a, the values of $f(x)$ must approach some real number L as x approaches a, and this number must be the same regardless of whether x approaches a from the left or the right. This suggests the following result, which we state without formal proof.

1.1.3   **THE RELATIONSHIP BETWEEN ONE-SIDED AND TWO-SIDED LIMITS**   The two-sided limit of a function $f(x)$ exists at a if and only if both of the one-sided limits exist at a and have the same value; that is,

$$\lim_{x \to a} f(x) = L \quad \text{if and only if} \quad \lim_{x \to a^-} f(x) = L = \lim_{x \to a^+} f(x)$$

▶ **Example 4**   Explain why

$$\lim_{x \to 0} \frac{|x|}{x}$$

does not exist.

Solution.   As x approaches 0, the values of $f(x) = |x|/x$ approach -1 from the left and approach 1 from the right [see (13)]. Thus, the one-sided limits at 0 are not the same. ◀

▶ **Example 5**   For the functions in Figure 1.1.13, find the one-sided and two-sided limits at $x = a$ if they exist.

Solution.   The functions in all three figures have the same one-sided limits as $x \to a$, since the functions are identical, except at $x = a$. These limits are

$$\lim_{x \to a^+} f(x) = 3 \quad \text{and} \quad \lim_{x \to a^-} f(x) = 1$$

In all three cases the two-sided limit does not exist as $x \to a$ because the one-sided limits are not equal. ◀

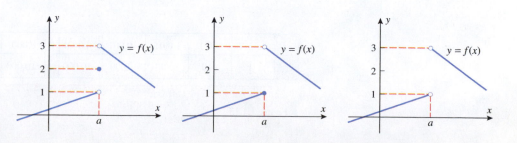

▶ **Figure 1.1.13**

▶ **Example 6** For the functions in Figure 1.1.14, find the one-sided and two-sided limits at $x = a$ if they exist.

Solution. As in the preceding example, the value of f at $x = a$ has no bearing on the limits as $x \to a$, so in all three cases we have

$$\lim_{x \to a^+} f(x) = 2 \quad \text{and} \quad \lim_{x \to a^-} f(x) = 2$$

Since the one-sided limits are equal, the two-sided limit exists and

$$\lim_{x \to a} f(x) = 2 \quad ◀$$

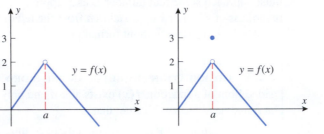

▲ **Figure 1.1.14**

■ INFINITE LIMITS

Sometimes one-sided or two-sided limits fail to exist because the values of the function increase or decrease without bound. For example, consider the behavior of $f(x) = 1/x$ for values of x near 0. It is evident from the table and graph in Figure 1.1.15 that as x-values are taken closer and closer to 0 from the right, the values of $f(x) = 1/x$ are positive and increase without bound; and as x-values are taken closer and closer to 0 from the left, the values of $f(x) = 1/x$ are negative and decrease without bound. We describe these limiting behaviors by writing

$$\lim_{x \to 0^+} \frac{1}{x} = +\infty \quad \text{and} \quad \lim_{x \to 0^-} \frac{1}{x} = -\infty$$

The symbols $+\infty$ and $-\infty$ here are *not* real numbers; they simply describe particular ways in which the limits fail to exist. Do not make the mistake of manipulating these symbols using rules of algebra. For example, it is *incorrect* to write $(+\infty) - (+\infty) = 0$.

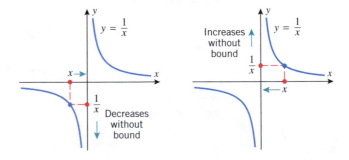

x	-1	-0.1	-0.01	-0.001	-0.0001	0	0.0001	0.001	0.01	0.1	1
$\frac{1}{x}$	-1	-10	-100	-1000	$-10,000$		10,000	1000	100	10	1

Left side   Right side

▲ **Figure 1.1.15**

1.1.4 **INFINITE LIMITS (AN INFORMAL VIEW)** The expressions

$$\lim_{x \to a^-} f(x) = +\infty \quad \text{and} \quad \lim_{x \to a^+} f(x) = +\infty$$

denote that $f(x)$ increases without bound as x approaches a from the left and from the right, respectively. If both are true, then we write

$$\lim_{x \to a} f(x) = +\infty$$

Similarly, the expressions

$$\lim_{x \to a^-} f(x) = -\infty \quad \text{and} \quad \lim_{x \to a^+} f(x) = -\infty$$

denote that $f(x)$ decreases without bound as x approaches a from the left and from the right, respectively. If both are true, then we write

$$\lim_{x \to a} f(x) = -\infty$$

▶ **Example 7** For the functions in Figure 1.1.16, describe the limits at $x = a$ in appropriate limit notation.

Solution (a). In Figure 1.1.16a, the function increases without bound as x approaches a from the right and decreases without bound as x approaches a from the left. Thus,

$$\lim_{x \to a^+} \frac{1}{x-a} = +\infty \quad \text{and} \quad \lim_{x \to a^-} \frac{1}{x-a} = -\infty$$

Solution (b). In Figure 1.1.16b, the function increases without bound as x approaches a from both the left and right. Thus,

$$\lim_{x \to a} \frac{1}{(x-a)^2} = \lim_{x \to a^+} \frac{1}{(x-a)^2} = \lim_{x \to a^-} \frac{1}{(x-a)^2} = +\infty$$

Solution (c). In Figure 1.1.16c, the function decreases without bound as x approaches a from the right and increases without bound as x approaches a from the left. Thus,

$$\lim_{x \to a^+} \frac{-1}{x-a} = -\infty \quad \text{and} \quad \lim_{x \to a^-} \frac{-1}{x-a} = +\infty$$

Solution (d). In Figure 1.1.16d, the function decreases without bound as x approaches a from both the left and right. Thus,

$$\lim_{x \to a} \frac{-1}{(x-a)^2} = \lim_{x \to a^+} \frac{-1}{(x-a)^2} = \lim_{x \to a^-} \frac{-1}{(x-a)^2} = -\infty \quad ◀$$

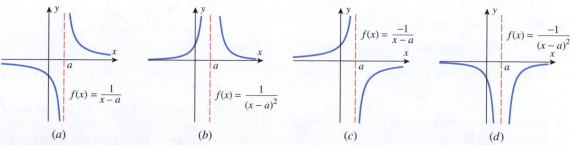

(a) (b) (c) (d)

▲ **Figure 1.1.16**

■ VERTICAL ASYMPTOTES

Figure 1.1.17 illustrates geometrically what happens when any of the following situations occur:

$$\lim_{x \to a^-} f(x) = +\infty, \quad \lim_{x \to a^+} f(x) = +\infty, \quad \lim_{x \to a^-} f(x) = -\infty, \quad \lim_{x \to a^+} f(x) = -\infty$$

In each case the graph of $y = f(x)$ either rises or falls without bound, squeezing closer and closer to the vertical line $x = a$ as x approaches a from the side indicated in the limit. The line $x = a$ is called a ***vertical asymptote*** of the curve $y = f(x)$ (from the Greek word *asymptotos*, meaning "nonintersecting").

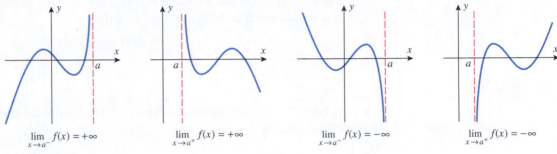

$$\lim_{x \to a^-} f(x) = +\infty \qquad \lim_{x \to a^+} f(x) = +\infty \qquad \lim_{x \to a^-} f(x) = -\infty \qquad \lim_{x \to a^+} f(x) = -\infty$$

▲ **Figure 1.1.17**

▶ **Example 8**   Referring to Figure 0.5.7 we see that the y-axis is a vertical asymptote for $y = \log_b x$ if $b > 1$ since

$$\lim_{x \to 0^+} \log_b x = -\infty$$

> For the function in (16), find expressions for the left- and right-hand limits at each asymptote.

and referring to Figure 0.3.11 we see that $x = -1$ and $x = 1$ are vertical asymptotes of the graph of

$$f(x) = \frac{x^2 + 2x}{x^2 - 1} \quad \blacktriangleleft \tag{16}$$

✔ QUICK CHECK EXERCISES 1.1   (See page 80 for answers.)

1. We write $\lim_{x \to a} f(x) = L$ provided the values of _____ can be made as close to _____ as desired, by taking values of _____ sufficiently close to _____ but not _____.

2. We write $\lim_{x \to a^-} f(x) = +\infty$ provided _____ increases without bound, as _____ approaches _____ from the left.

3. State what must be true about

$$\lim_{x \to a^-} f(x) \quad \text{and} \quad \lim_{x \to a^+} f(x)$$

in order for it to be the case that

$$\lim_{x \to a} f(x) = L$$

4. Use the accompanying graph of $y = f(x)$ $(-\infty < x < 3)$ to determine the limits.

(a) $\lim_{x \to 0} f(x) =$ _____

(b) $\lim_{x \to 2^-} f(x) =$ _____

(c) $\lim_{x \to 2^+} f(x) =$ _____

(d) $\lim_{x \to 3^-} f(x) =$ _____

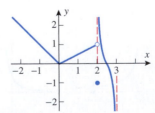

◀ **Figure Ex-4**

5. The slope of the secant line through $P(2, 4)$ and $Q(x, x^2)$ on the parabola $y = x^2$ is $m_{\text{sec}} = x + 2$. It follows that the slope of the tangent line to this parabola at the point P is _____.

EXERCISE SET 1.1   Graphing Utility   [c] CAS

1–10 In these exercises, make reasonable assumptions about the graph of the indicated function outside of the region depicted. ∎

1. For the function g graphed in the accompanying figure, find
   (a) $\lim\limits_{x \to 0^-} g(x)$ (b) $\lim\limits_{x \to 0^+} g(x)$
   (c) $\lim\limits_{x \to 0} g(x)$ (d) $g(0)$.

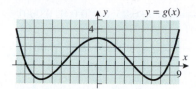

◀ **Figure Ex-1**

2. For the function G graphed in the accompanying figure, find
   (a) $\lim\limits_{x \to 0^-} G(x)$ (b) $\lim\limits_{x \to 0^+} G(x)$
   (c) $\lim\limits_{x \to 0} G(x)$ (d) $G(0)$.

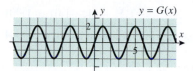

◀ **Figure Ex-2**

3. For the function f graphed in the accompanying figure, find
   (a) $\lim\limits_{x \to 3^-} f(x)$ (b) $\lim\limits_{x \to 3^+} f(x)$
   (c) $\lim\limits_{x \to 3} f(x)$ (d) $f(3)$.

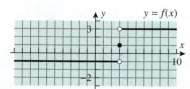

◀ **Figure Ex-3**

4. For the function f graphed in the accompanying figure, find
   (a) $\lim\limits_{x \to 2^-} f(x)$ (b) $\lim\limits_{x \to 2^+} f(x)$
   (c) $\lim\limits_{x \to 2} f(x)$ (d) $f(2)$.

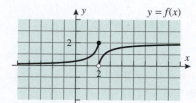

◀ **Figure Ex-4**

5. For the function F graphed in the accompanying figure, find
   (a) $\lim\limits_{x \to -2^-} F(x)$ (b) $\lim\limits_{x \to -2^+} F(x)$
   (c) $\lim\limits_{x \to -2} F(x)$ (d) $F(-2)$.

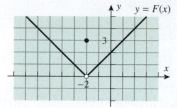

◀ **Figure Ex-5**

6. For the function G graphed in the accompanying figure, find
   (a) $\lim\limits_{x \to 0^-} G(x)$ (b) $\lim\limits_{x \to 0^+} G(x)$
   (c) $\lim\limits_{x \to 0} G(x)$ (d) $G(0)$.

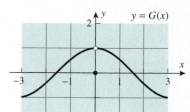

◀ **Figure Ex-6**

7. For the function f graphed in the accompanying figure, find
   (a) $\lim\limits_{x \to 3^-} f(x)$ (b) $\lim\limits_{x \to 3^+} f(x)$
   (c) $\lim\limits_{x \to 3} f(x)$ (d) $f(3)$.

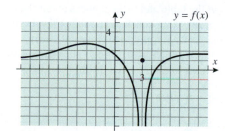

◀ **Figure Ex-7**

8. For the function ϕ graphed in the accompanying figure, find
   (a) $\lim\limits_{x \to 4^-} \phi(x)$ (b) $\lim\limits_{x \to 4^+} \phi(x)$
   (c) $\lim\limits_{x \to 4} \phi(x)$ (d) $\phi(4)$.

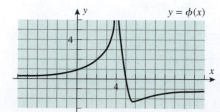

◀ **Figure Ex-8**

9. For the function f graphed in the accompanying figure on the next page, find
   (a) $\lim\limits_{x \to 0^-} f(x)$ (b) $\lim\limits_{x \to 0^+} f(x)$
   (c) $\lim\limits_{x \to 0} f(x)$ (d) $f(0)$.

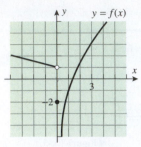

◀ **Figure Ex-9**

10. For the function g graphed in the accompanying figure, find
(a) $\lim\limits_{x \to 1^-} g(x)$
(b) $\lim\limits_{x \to 1^+} g(x)$
(c) $\lim\limits_{x \to 1} g(x)$
(d) $g(1)$.

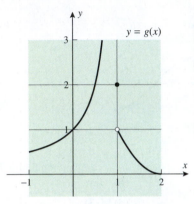

◀ **Figure Ex-10**

~ **11–12** (i) Complete the table and make a guess about the limit indicated. (ii) Confirm your conclusions about the limit by graphing a function over an appropriate interval. [*Note:* For the inverse trigonometric function, be sure to put your calculating and graphing utilities in radian mode.] ■

11. $f(x) = \dfrac{e^x - 1}{x}$; $\lim\limits_{x \to 0} f(x)$

x	-0.01	-0.001	-0.0001	0.0001	0.001	0.01
$f(x)$						

▲ **Table Ex-11**

12. $f(x) = \dfrac{\sin^{-1} 2x}{x}$; $\lim\limits_{x \to 0} f(x)$

x	-0.1	-0.01	-0.001	0.001	0.01	0.1
$f(x)$						

▲ **Table Ex-12**

C **13–16** (i) Make a guess at the limit (if it exists) by evaluating the function at the specified x-values. (ii) Confirm your conclusions about the limit by graphing the function over an appropriate interval. (iii) If you have a CAS, then use it to find the limit. [*Note:* For the trigonometric functions, be sure to put your calculating and graphing utilities in radian mode.] ■

13. (a) $\lim\limits_{x \to 1} \dfrac{x-1}{x^3-1}$; $x = 2, 1.5, 1.1, 1.01, 1.001, 0, 0.5, 0.9,$ $0.99, 0.999$

(b) $\lim\limits_{x \to 1^+} \dfrac{x+1}{x^3-1}$; $x = 2, 1.5, 1.1, 1.01, 1.001, 1.0001$

(c) $\lim\limits_{x \to 1^-} \dfrac{x+1}{x^3-1}$; $x = 0, 0.5, 0.9, 0.99, 0.999, 0.9999$

14. (a) $\lim\limits_{x \to 0} \dfrac{\sqrt{x+1}-1}{x}$; $x = \pm0.25, \pm0.1, \pm0.001,$ ±0.0001

(b) $\lim\limits_{x \to 0^+} \dfrac{\sqrt{x+1}+1}{x}$; $x = 0.25, 0.1, 0.001, 0.0001$

(c) $\lim\limits_{x \to 0^-} \dfrac{\sqrt{x+1}+1}{x}$; $x = -0.25, -0.1, -0.001,$ -0.0001

15. (a) $\lim\limits_{x \to 0} \dfrac{\sin 3x}{x}$; $x = \pm0.25, \pm0.1, \pm0.001, \pm0.0001$

(b) $\lim\limits_{x \to -1} \dfrac{\cos x}{x+1}$; $x = 0, -0.5, -0.9, -0.99, -0.999,$ $-1.5, -1.1, -1.01, -1.001$

16. (a) $\lim\limits_{x \to -1} \dfrac{\tan(x+1)}{x+1}$; $x = 0, -0.5, -0.9, -0.99, -0.999,$ $-1.5, -1.1, -1.01, -1.001$

(b) $\lim\limits_{x \to 0} \dfrac{\sin(5x)}{\sin(2x)}$; $x = \pm0.25, \pm0.1, \pm0.001, \pm0.0001$

17–20 True–False Determine whether the statement is true or false. Explain your answer. ■

17. If $f(a) = L$, then $\lim_{x \to a} f(x) = L$.

18. If $\lim_{x \to a} f(x)$ exists, then so do $\lim_{x \to a^-} f(x)$ and $\lim_{x \to a^+} f(x)$.

19. If $\lim_{x \to a^-} f(x)$ and $\lim_{x \to a^+} f(x)$ exist, then so does $\lim_{x \to a} f(x)$.

20. If $\lim_{x \to a^+} f(x) = +\infty$, then $f(a)$ is undefined.

21–26 Sketch a possible graph for a function f with the specified properties. (Many different solutions are possible.) ■

21. (i) the domain of f is $[-1, 1]$
(ii) $f(-1) = f(0) = f(1) = 0$
(iii) $\lim\limits_{x \to -1^+} f(x) = \lim\limits_{x \to 0} f(x) = \lim\limits_{x \to 1^-} f(x) = 1$

22. (i) the domain of f is $[-2, 1]$
(ii) $f(-2) = f(0) = f(1) = 0$
(iii) $\lim\limits_{x \to -2^+} f(x) = 2, \lim\limits_{x \to 0} f(x) = 0,$ and $\lim_{x \to 1^-} f(x) = 1$

23. (i) the domain of f is $(-\infty, 0]$
(ii) $f(-2) = f(0) = 1$
(iii) $\lim\limits_{x \to -2} f(x) = +\infty$

24. (i) the domain of f is $(0, +\infty)$
(ii) $f(1) = 0$
(iii) the y-axis is a vertical asymptote for the graph of f
(iv) $f(x) < 0$ if $0 < x < 1$

25. (i) $f(-3) = f(0) = f(2) = 0$

(ii) $\lim\limits_{x \to -2^-} f(x) = +\infty$ and $\lim\limits_{x \to -2^+} f(x) = -\infty$

(iii) $\lim\limits_{x \to 1} f(x) = +\infty$

26. (i) $f(-1) = 0$, $f(0) = 1$, $f(1) = 0$

(ii) $\lim\limits_{x \to -1^-} f(x) = 0$ and $\lim\limits_{x \to -1^+} f(x) = +\infty$

(iii) $\lim\limits_{x \to 1^-} f(x) = 1$ and $\lim\limits_{x \to 1^+} f(x) = +\infty$

27–30 Modify the argument of Example 1 to find the equation of the tangent line to the specified graph at the point given. ■

27. the graph of $y = x^2$ at $(-1, 1)$

28. the graph of $y = x^2$ at $(0, 0)$

29. the graph of $y = x^4$ at $(1, 1)$

30. the graph of $y = x^4$ at $(-1, 1)$

FOCUS ON CONCEPTS

31. In the special theory of relativity the length l of a narrow rod moving longitudinally is a function $l = l(v)$ of the rod's speed v. The accompanying figure, in which c denotes the speed of light, displays some of the qualitative features of this function.

(a) What is the physical interpretation of l_0?

(b) What is $\lim\limits_{v \to c^-} l(v)$? What is the physical significance of this limit?

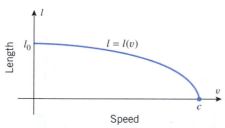

▲ Figure Ex-31

32. In the special theory of relativity the mass m of a moving object is a function $m = m(v)$ of the object's speed v. The accompanying figure, in which c denotes the speed of light, displays some of the qualitative features of this function.

(a) What is the physical interpretation of m_0?

(b) What is $\lim\limits_{v \to c^-} m(v)$? What is the physical significance of this limit?

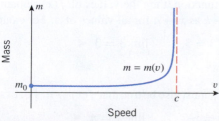

▲ Figure Ex-32

33. What do the graphs in Figure 0.5.4 imply about the value of

$$\lim_{x \to 0} \frac{e^x - 1}{x}$$

Explain your answer.

$\boxed{c}$ **34.** Let

$$f(x) = \frac{x - \sin x}{x^3}$$

(a) Make a conjecture about the limit of f as $x \to 0^+$ by completing the table.

x	0.5	0.1	0.05	0.01
$f(x)$				

(b) Make another conjecture about the limit of f as $x \to 0^+$ by evaluating $f(x)$ at $x = 0.0001, 0.00001, 0.000001,$ $0.0000001, 0.00000001, 0.000000001$.

(c) The phenomenon exhibited in part (b) is called *catastrophic subtraction*. What do you think causes catastrophic subtraction? How does it put restrictions on the use of numerical evidence to make conjectures about limits?

(d) If you have a CAS, use it to show that the exact value of the limit is $\frac{1}{6}$.

$\boxed{\sim}$ **35.** Let

$$f(x) = \left(1 + x^2\right)^{1.1/x^2}$$

(a) Graph f in the window

$$[-1, 1] \times [2.5, 3.5]$$

and use the calculator's trace feature to make a conjecture about the limit of $f(x)$ as $x \to 0$.

(b) Graph f in the window

$$[-0.001, 0.001] \times [2.5, 3.5]$$

and use the calculator's trace feature to make a conjecture about the limit of $f(x)$ as $x \to 0$.

(c) Graph f in the window

$$[-0.000001, 0.000001] \times [2.5, 3.5]$$

and use the calculator's trace feature to make a conjecture about the limit of $f(x)$ as $x \to 0$.

(d) Later we will be able to show that

$$\lim_{x \to 0} \left(1 + x^2\right)^{1.1/x^2} \approx 3.00416602$$

What flaw do your graphs reveal about using numerical evidence (as revealed by the graphs you obtained) to make conjectures about limits?

36. Writing Two students are discussing the limit of $\sqrt{x}$ as x approaches 0. One student maintains that the limit is 0, while the other claims that the limit does not exist. Write a short paragraph that discusses the pros and cons of each student's position.

37. Writing Given a function f and a real number a, explain informally why

$$\lim_{x \to 0} f(x + a) = \lim_{x \to a} f(x)$$

(Here "equality" means that either both limits exist and are equal or that both limits fail to exist.)

1. $f(x)$; L; x; a **2.** $f(x)$; x; a **3.** Both one-sided limits must exist and equal L.   **4.** (a) 0 (b) 1 (c) $+\infty$ (d) $-\infty$ **5.** 4

1.2   COMPUTING LIMITS

In this section we will discuss techniques for computing limits of many functions. We base these results on the informal development of the limit concept discussed in the preceding section. A more formal derivation of these results is possible after Section 1.4.

■ SOME BASIC LIMITS

Our strategy for finding limits algebraically has two parts:

- First we will obtain the limits of some simple functions.
- Then we will develop a repertoire of theorems that will enable us to use the limits of those simple functions as building blocks for finding limits of more complicated functions.

We start with the following basic results, which are illustrated in Figure 1.2.1.

1.2.1   THEOREM   *Let a and k be real numbers.*

(*a*) $\displaystyle\lim_{x \to a} k = k$ (*b*) $\displaystyle\lim_{x \to a} x = a$ (*c*) $\displaystyle\lim_{x \to 0^-} \frac{1}{x} = -\infty$ (*d*) $\displaystyle\lim_{x \to 0^+} \frac{1}{x} = +\infty$

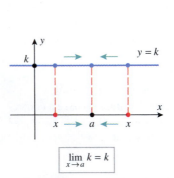
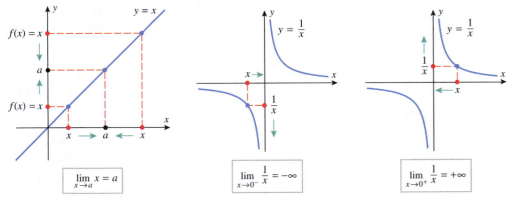

▲ **Figure 1.2.1**

The following examples explain these results further.

▶ **Example 1**   If $f(x) = k$ is a constant function, then the values of $f(x)$ remain fixed at k as x varies, which explains why $f(x) \to k$ as $x \to a$ for all values of a. For example,

$$\lim_{x \to -25} 3 = 3, \qquad \lim_{x \to 0} 3 = 3, \qquad \lim_{x \to \pi} 3 = 3 \quad ◀$$

▶ **Example 2**   If $f(x) = x$, then as $x \to a$ it must also be true that $f(x) \to a$. For example,

$$\lim_{x \to 0} x = 0, \qquad \lim_{x \to -2} x = -2, \qquad \lim_{x \to \pi} x = \pi \quad ◀$$

▶ **Example 3** You should know from your experience with fractions that for a fixed nonzero numerator, the closer the denominator is to zero, the larger the absolute value of the fraction. This fact and the data in Table 1.2.1 suggest why $1/x \to +\infty$ as $x \to 0^+$ and why $1/x \to -\infty$ as $x \to 0^-$. ◄

Table 1.2.1

	VALUES						CONCLUSION
x	-1	-0.1	-0.01	-0.001	-0.0001	$\cdots$	As $x \to 0^-$ the value of $1/x$
$1/x$	-1	-10	-100	-1000	$-10{,}000$	$\cdots$	decreases without bound.
x	1	0.1	0.01	0.001	0.0001	$\cdots$	As $x \to 0^+$ the value of $1/x$
$1/x$	1	10	100	1000	$10{,}000$	$\cdots$	increases without bound.

The following theorem, parts of which are proved in Appendix D, will be our basic tool for finding limits algebraically.

1.2.2 THEOREM *Let a be a real number, and suppose that*

$$\lim_{x \to a} f(x) = L_1 \quad and \quad \lim_{x \to a} g(x) = L_2$$

That is, the limits exist and have values L_1 and L_2, respectively. Then:

(a) $\displaystyle \lim_{x \to a} [f(x) + g(x)] = \lim_{x \to a} f(x) + \lim_{x \to a} g(x) = L_1 + L_2$

(b) $\displaystyle \lim_{x \to a} [f(x) - g(x)] = \lim_{x \to a} f(x) - \lim_{x \to a} g(x) = L_1 - L_2$

(c) $\displaystyle \lim_{x \to a} [f(x)g(x)] = \left(\lim_{x \to a} f(x) \right) \left(\lim_{x \to a} g(x) \right) = L_1 L_2$

(d) $\displaystyle \lim_{x \to a} \frac{f(x)}{g(x)} = \frac{\lim_{x \to a} f(x)}{\lim_{x \to a} g(x)} = \frac{L_1}{L_2}, \quad provided\ L_2 \neq 0$

(e) $\displaystyle \lim_{x \to a} \sqrt[n]{f(x)} = \sqrt[n]{\lim_{x \to a} f(x)} = \sqrt[n]{L_1}, \quad provided\ L_1 > 0\ if\ n\ is\ even.$

Moreover, these statements are also true for the one-sided limits as $x \to a^-$ or as $x \to a^+$.

Theorem 1.2.2(e) remains valid for n even and $L_1 = 0$, provided $f(x)$ is nonnegative for x near a with $x \neq a$.

This theorem can be stated informally as follows:

(a) The limit of a sum is the sum of the limits.

(b) The limit of a difference is the difference of the limits.

(c) The limit of a product is the product of the limits.

(d) The limit of a quotient is the quotient of the limits, provided the limit of the denominator is not zero.

(e) The limit of an nth root is the nth root of the limit.

For the special case of part (c) in which $f(x) = k$ is a constant function, we have

$$\lim_{x \to a} (kg(x)) = \lim_{x \to a} k \cdot \lim_{x \to a} g(x) = k \lim_{x \to a} g(x) \tag{1}$$

and similarly for one-sided limits. This result can be rephrased as follows:

> *A constant factor can be moved through a limit symbol.*

Although parts (a) and (c) of Theorem 1.2.2 are stated for two functions, the results hold for any finite number of functions. Moreover, the various parts of the theorem can be used in combination to reformulate expressions involving limits.

▶ **Example 4**

$$\lim_{x \to a}[f(x) - g(x) + 2h(x)] = \lim_{x \to a} f(x) - \lim_{x \to a} g(x) + 2 \lim_{x \to a} h(x)$$

$$\lim_{x \to a}[f(x)g(x)h(x)] = \left(\lim_{x \to a} f(x) \right) \left(\lim_{x \to a} g(x) \right) \left(\lim_{x \to a} h(x) \right)$$

$$\lim_{x \to a}[f(x)]^3 = \left(\lim_{x \to a} f(x) \right)^3$$ Take $g(x) = h(x) = f(x)$ in the last equation.

$$\lim_{x \to a}[f(x)]^n = \left(\lim_{x \to a} f(x) \right)^n$$ The extension of Theorem 1.2.2(c) in which there are n factors, each of which is $f(x)$

$$\lim_{x \to a} x^n = \left(\lim_{x \to a} x \right)^n = a^n$$ Apply the previous result with $f(x) = x$. ◀

■ **LIMITS OF POLYNOMIALS AND RATIONAL FUNCTIONS AS $x \to a$**

▶ **Example 5**   Find $\lim\limits_{x \to 5} (x^2 - 4x + 3)$.

Solution.

$$\lim_{x \to 5} (x^2 - 4x + 3) = \lim_{x \to 5} x^2 - \lim_{x \to 5} 4x + \lim_{x \to 5} 3$$ Theorem 1.2.2(a), (b)

$$= \lim_{x \to 5} x^2 - 4 \lim_{x \to 5} x + \lim_{x \to 5} 3$$ A constant can be moved through a limit symbol.

$$= 5^2 - 4(5) + 3$$ The last part of Example 4

$$= 8 \blacktriangleleft$$

Observe that in Example 5 the limit of the polynomial $p(x) = x^2 - 4x + 3$ as $x \to 5$ turned out to be the same as $p(5)$. This is not an accident. The next result shows that, in general, the limit of a polynomial $p(x)$ as $x \to a$ is the same as the value of the polynomial at a. Knowing this fact allows us to reduce the computation of limits of polynomials to simply evaluating the polynomial at the appropriate point.

1.2.3   **THEOREM**   *For any polynomial*

$$p(x) = c_0 + c_1 x + \cdots + c_n x^n$$

and any real number a,

$$\lim_{x \to a} p(x) = c_0 + c_1 a + \cdots + c_n a^n = p(a)$$

PROOF

$$\lim_{x \to a} p(x) = \lim_{x \to a} \left(c_0 + c_1 x + \cdots + c_n x^n\right)$$

$$= \lim_{x \to a} c_0 + \lim_{x \to a} c_1 x + \cdots + \lim_{x \to a} c_n x^n$$

$$= \lim_{x \to a} c_0 + c_1 \lim_{x \to a} x + \cdots + c_n \lim_{x \to a} x^n$$

$$= c_0 + c_1 a + \cdots + c_n a^n = p(a) \quad \blacksquare$$

▶ **Example 6**   Find $\lim_{x \to 1} (x^7 - 2x^5 + 1)^{35}$.

Solution.   The function involved is a polynomial (why?), so the limit can be obtained by evaluating this polynomial at $x = 1$. This yields

$$\lim_{x \to 1} (x^7 - 2x^5 + 1)^{35} = 0 \quad \blacktriangleleft$$

Recall that a rational function is a ratio of two polynomials. The following example illustrates how Theorems 1.2.2(*d*) and 1.2.3 can sometimes be used in combination to compute limits of rational functions.

▶ **Example 7**   Find $\lim_{x \to 2} \dfrac{5x^3 + 4}{x - 3}$.

Solution.

$$\lim_{x \to 2} \frac{5x^3 + 4}{x - 3} = \frac{\lim_{x \to 2} (5x^3 + 4)}{\lim_{x \to 2} (x - 3)} \qquad \boxed{\text{Theorem 1.2.2(}d\text{)}}$$

$$= \frac{5 \cdot 2^3 + 4}{2 - 3} = -44 \qquad \boxed{\text{Theorem 1.2.3}} \quad \blacktriangleleft$$

The method used in the last example will not work for rational functions in which the limit of the denominator is zero because Theorem 1.2.2(*d*) is not applicable. There are two cases of this type to be considered—the case where the limit of the denominator is zero and the limit of the numerator is not, and the case where the limits of the numerator and denominator are both zero. If the limit of the denominator is zero but the limit of the numerator is not, then one can prove that the limit of the rational function does not exist and that one of the following situations occurs:

- The limit may be $-\infty$ from one side and $+\infty$ from the other.
- The limit may be $+\infty$.
- The limit may be $-\infty$.

Figure 1.2.2 illustrates these three possibilities graphically for rational functions of the form $1/(x - a)$, $1/(x - a)^2$, and $-1/(x - a)^2$.

▶ **Example 8**   Find

(a) $\lim_{x \to 4^+} \dfrac{2 - x}{(x - 4)(x + 2)}$ (b) $\lim_{x \to 4^-} \dfrac{2 - x}{(x - 4)(x + 2)}$ (c) $\lim_{x \to 4} \dfrac{2 - x}{(x - 4)(x + 2)}$

Solution.   In all three parts the limit of the numerator is -2, and the limit of the denominator is 0, so the limit of the ratio does not exist. To be more specific than this, we need

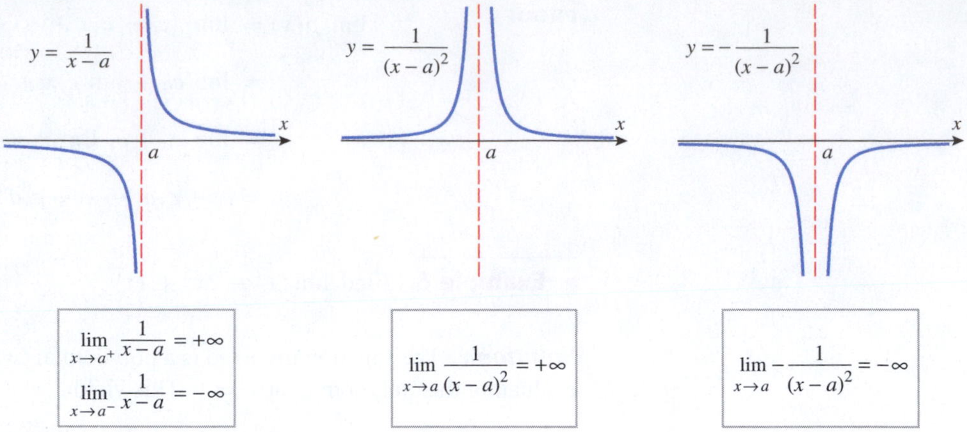

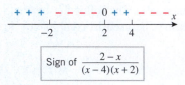

▲ Figure 1.2.2

▲ Figure 1.2.3

to analyze the sign of the ratio. The sign of the ratio, which is given in Figure 1.2.3, is determined by the signs of $2 - x$, $x - 4$, and $x + 2$. (The method of test points, discussed in Web Appendix E, provides a way of finding the sign of the ratio here.) It follows from this figure that as x approaches 4 from the right, the ratio is always negative; and as x approaches 4 from the left, the ratio is eventually positive. Thus,

$$\lim_{x \to 4^+} \frac{2 - x}{(x - 4)(x + 2)} = -\infty \quad \text{and} \quad \lim_{x \to 4^-} \frac{2 - x}{(x - 4)(x + 2)} = +\infty$$

Because the one-sided limits have opposite signs, all we can say about the two-sided limit is that it does not exist. ◄

In the case where $p(x)/q(x)$ is a rational function for which $p(a) = 0$ and $q(a) = 0$, the numerator and denominator must have one or more common factors of $x - a$. In this case the limit of $p(x)/q(x)$ as $x \to a$ can be found by canceling all common factors of $x - a$ and using one of the methods already considered to find the limit of the simplified function. Here is an example.

> In Example 9(a), the simplified function $x - 3$ is defined at $x = 3$, but the original function is not. However, this has no effect on the limit as x *approaches* 3 since the two functions are identical if $x \neq 3$ (Exercise 50).

► **Example 9**   Find

(a) $\displaystyle\lim_{x \to 3} \frac{x^2 - 6x + 9}{x - 3}$ (b) $\displaystyle\lim_{x \to -4} \frac{2x + 8}{x^2 + x - 12}$ (c) $\displaystyle\lim_{x \to 5} \frac{x^2 - 3x - 10}{x^2 - 10x + 25}$

Solution (a).   The numerator and the denominator both have a zero at $x = 3$, so there is a common factor of $x - 3$. Then

$$\lim_{x \to 3} \frac{x^2 - 6x + 9}{x - 3} = \lim_{x \to 3} \frac{(x - 3)^2}{x - 3} = \lim_{x \to 3} (x - 3) = 0$$

Solution (b).   The numerator and the denominator both have a zero at $x = -4$, so there is a common factor of $x - (-4) = x + 4$. Then

$$\lim_{x \to -4} \frac{2x + 8}{x^2 + x - 12} = \lim_{x \to -4} \frac{2(x + 4)}{(x + 4)(x - 3)} = \lim_{x \to -4} \frac{2}{x - 3} = -\frac{2}{7}$$

Solution (c).   The numerator and the denominator both have a zero at $x = 5$, so there is a common factor of $x - 5$. Then

$$\lim_{x \to 5} \frac{x^2 - 3x - 10}{x^2 - 10x + 25} = \lim_{x \to 5} \frac{(x - 5)(x + 2)}{(x - 5)(x - 5)} = \lim_{x \to 5} \frac{x + 2}{x - 5}$$

However,

$$\lim_{x \to 5}(x+2) = 7 \neq 0 \quad \text{and} \quad \lim_{x \to 5}(x-5) = 0$$

so

$$\lim_{x \to 5}\frac{x^2 - 3x - 10}{x^2 - 10x + 25} = \lim_{x \to 5}\frac{x+2}{x-5}$$

does not exist. More precisely, the sign analysis in Figure 1.2.4 implies that

$$\lim_{x \to 5^+}\frac{x^2 - 3x - 10}{x^2 - 10x + 25} = \lim_{x \to 5^+}\frac{x+2}{x-5} = +\infty$$

and

$$\lim_{x \to 5^-}\frac{x^2 - 3x - 10}{x^2 - 10x + 25} = \lim_{x \to 5^-}\frac{x+2}{x-5} = -\infty \quad \blacktriangleleft$$

+ + + 0 – – – – – – – – – – + + x

-2 5

Sign of $\dfrac{x+2}{x-5}$

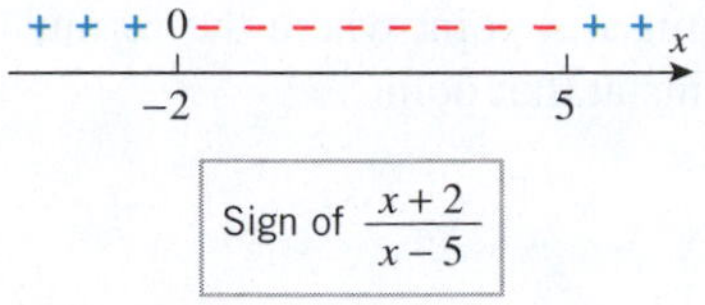

▲ **Figure 1.2.4**

Discuss the logical errors in the following statement: An indeterminate form of type $0/0$ must have a limit of zero because zero divided by anything is zero.

A quotient $f(x)/g(x)$ in which the numerator and denominator both have a limit of zero as $x \to a$ is called an ***indeterminate form of type*** $\boldsymbol{0/0}$. The problem with such limits is that it is difficult to tell by inspection whether the limit exists, and, if so, its value. Informally stated, this is because there are two conflicting influences at work. The value of $f(x)/g(x)$ would tend to zero as $f(x)$ approached zero if $g(x)$ were to remain at some fixed nonzero value, whereas the value of this ratio would tend to increase or decrease without bound as $g(x)$ approached zero if $f(x)$ were to remain at some fixed nonzero value. But with both $f(x)$ and $g(x)$ approaching zero, the behavior of the ratio depends on precisely how these conflicting tendencies offset one another for the particular f and g.

Sometimes, limits of indeterminate forms of type $0/0$ can be found by algebraic simplification, as in the last example, but frequently this will not work and other methods must be used. We will study such methods in later sections.

The following theorem summarizes our observations about limits of rational functions.

1.2.4 **THEOREM** *Let*

$$f(x) = \frac{p(x)}{q(x)}$$

be a rational function, and let a be any real number.

(a) *If $q(a) \neq 0$, then $\displaystyle\lim_{x \to a} f(x) = f(a)$.*

(b) *If $q(a) = 0$ but $p(a) \neq 0$, then $\displaystyle\lim_{x \to a} f(x)$ does not exist.*

■ **LIMITS INVOLVING RADICALS**

▶ **Example 10**   Find $\displaystyle\lim_{x \to 1}\frac{x-1}{\sqrt{x}-1}$.

Solution. In Example 2 of Section 1.1 we used numerical evidence to conjecture that this limit is 2. Here we will confirm this algebraically. Since this limit is an indeterminate form of type $0/0$, we will need to devise some strategy for making the limit (if it exists) evident. One such strategy is to rationalize the denominator of the function. This yields

$$\frac{x-1}{\sqrt{x}-1} = \frac{(x-1)(\sqrt{x}+1)}{(\sqrt{x}-1)(\sqrt{x}+1)} = \frac{(x-1)(\sqrt{x}+1)}{x-1} = \sqrt{x}+1 \quad (x \neq 1)$$

Therefore,

$$\lim_{x \to 1} \frac{x-1}{\sqrt{x}-1} = \lim_{x \to 1} (\sqrt{x}+1) = 2 \;\blacktriangleleft$$

> Confirm the limit in Example 10 by factoring the numerator.

■ LIMITS OF PIECEWISE-DEFINED FUNCTIONS

For functions that are defined piecewise, a two-sided limit at a point where the formula changes is best obtained by first finding the one-sided limits at that point.

> ▶ **Example 11**   Let

$$f(x) = \begin{cases} 1/(x+2), & x < -2 \\ x^2 - 5, & -2 < x \le 3 \\ \sqrt{x+13}, & x > 3 \end{cases}$$

Find

(a) $\displaystyle\lim_{x \to -2} f(x)$ (b) $\displaystyle\lim_{x \to 0} f(x)$ (c) $\displaystyle\lim_{x \to 3} f(x)$

Solution (a).   We will determine the stated two-sided limit by first considering the corresponding one-sided limits. For each one-sided limit, we must use that part of the formula that is applicable on the interval over which x varies. For example, as x approaches -2 from the left, the applicable part of the formula is

$$f(x) = \frac{1}{x+2}$$

and as x approaches -2 from the right, the applicable part of the formula near -2 is

$$f(x) = x^2 - 5$$

Thus,

$$\lim_{x \to -2^-} f(x) = \lim_{x \to -2^-} \frac{1}{x+2} = -\infty$$

$$\lim_{x \to -2^+} f(x) = \lim_{x \to -2^+} (x^2 - 5) = (-2)^2 - 5 = -1$$

from which it follows that $\displaystyle\lim_{x \to -2} f(x)$ does not exist.

Solution (b).   The applicable part of the formula is $f(x) = x^2 - 5$ on both sides of 0, so there is no need to consider one-sided limits here. We see directly that

$$\lim_{x \to 0} f(x) = \lim_{x \to 0} (x^2 - 5) = 0^2 - 5 = -5$$

Solution (c).   Using the applicable parts of the formula for $f(x)$, we obtain

$$\lim_{x \to 3^-} f(x) = \lim_{x \to 3^-} (x^2 - 5) = 3^2 - 5 = 4$$

$$\lim_{x \to 3^+} f(x) = \lim_{x \to 3^+} \sqrt{x+13} = \sqrt{\lim_{x \to 3^+} (x+13)} = \sqrt{3+13} = 4$$

Since the one-sided limits are equal, we have

$$\lim_{x \to 3} f(x) = 4$$

We note that the limit calculations in parts (a), (b), and (c) are consistent with the graph of f shown in Figure 1.2.5. ◀

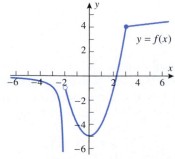

▲ **Figure 1.2.5**

✔ **QUICK CHECK EXERCISES 1.2**   *(See page 88 for answers.)*

1. In each part, find the limit by inspection.

   (a) $\lim\limits_{x \to 8} 7 =$ _____   (b) $\lim\limits_{y \to 3^+} 12y =$ _____

   (c) $\lim\limits_{x \to 0^-} \dfrac{x}{|x|} =$ _____   (d) $\lim\limits_{w \to 5} \dfrac{w}{|w|} =$ _____

   (e) $\lim\limits_{z \to 1^-} \dfrac{1}{1 - z} =$ _____

2. Given that $\lim_{x \to a} f(x) = 1$ and $\lim_{x \to a} g(x) = 2$, find the limits.

   (a) $\lim\limits_{x \to a} [3f(x) + 2g(x)] =$ _____

   (b) $\lim\limits_{x \to a} \dfrac{2f(x) + 1}{1 - f(x)g(x)} =$ _____ .

   (c) $\lim\limits_{x \to a} \dfrac{\sqrt{f(x) + 3}}{g(x)} =$ _____

3. Find the limits.

   (a) $\lim\limits_{x \to -1} (x^3 + x^2 + x)^{101} =$ _____

   (b) $\lim\limits_{x \to 2^-} \dfrac{(x - 1)(x - 2)}{x + 1} =$ _____

   (c) $\lim\limits_{x \to -1^+} \dfrac{(x - 1)(x - 2)}{x + 1} =$ _____

   (d) $\lim\limits_{x \to 4} \dfrac{x^2 - 16}{x - 4} =$ _____

4. Let
   $$f(x) = \begin{cases} x + 1, & x \le 1 \\ x - 1, & x > 1 \end{cases}$$

   Find the limits that exist.

   (a) $\lim\limits_{x \to 1^-} f(x) =$ _____

   (b) $\lim\limits_{x \to 1^+} f(x) =$ _____

   (c) $\lim\limits_{x \to 1} f(x) =$ _____

EXERCISE SET 1.2

1. Given that
   $$\lim_{x \to a} f(x) = 2, \quad \lim_{x \to a} g(x) = -4, \quad \lim_{x \to a} h(x) = 0$$
   find the limits.

   (a) $\lim\limits_{x \to a} [f(x) + 2g(x)]$

   (b) $\lim\limits_{x \to a} [h(x) - 3g(x) + 1]$

   (c) $\lim\limits_{x \to a} [f(x)g(x)]$ (d) $\lim\limits_{x \to a} [g(x)]^2$

   (e) $\lim\limits_{x \to a} \sqrt[3]{6 + f(x)}$ (f) $\lim\limits_{x \to a} \dfrac{2}{g(x)}$

2. Use the graphs of f and g in the accompanying figure to find the limits that exist. If the limit does not exist, explain why.

   (a) $\lim\limits_{x \to 2} [f(x) + g(x)]$ (b) $\lim\limits_{x \to 0} [f(x) + g(x)]$

   (c) $\lim\limits_{x \to 0^+} [f(x) + g(x)]$ (d) $\lim\limits_{x \to 0^-} [f(x) + g(x)]$

   (e) $\lim\limits_{x \to 2} \dfrac{f(x)}{1 + g(x)}$ (f) $\lim\limits_{x \to 2} \dfrac{1 + g(x)}{f(x)}$

   (g) $\lim\limits_{x \to 0^+} \sqrt{f(x)}$ (h) $\lim\limits_{x \to 0^-} \sqrt{f(x)}$

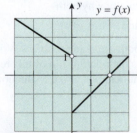
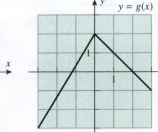

▲ **Figure Ex-2**

3–30 Find the limits. ■

3. $\lim\limits_{x \to 2} x(x - 1)(x + 1)$

4. $\lim\limits_{x \to 3} x^3 - 3x^2 + 9x$

5. $\lim\limits_{x \to 3} \dfrac{x^2 - 2x}{x + 1}$

6. $\lim\limits_{x \to 0} \dfrac{6x - 9}{x^3 - 12x + 3}$

7. $\lim\limits_{x \to 1^+} \dfrac{x^4 - 1}{x - 1}$

8. $\lim\limits_{t \to -2} \dfrac{t^3 + 8}{t + 2}$

9. $\lim\limits_{x \to -1} \dfrac{x^2 + 6x + 5}{x^2 - 3x - 4}$

10. $\lim\limits_{x \to 2} \dfrac{x^2 - 4x + 4}{x^2 + x - 6}$

11. $\lim\limits_{x \to -1} \dfrac{2x^2 + x - 1}{x + 1}$

12. $\lim\limits_{x \to 1} \dfrac{3x^2 - x - 2}{2x^2 + x - 3}$

13. $\lim\limits_{t \to 2} \dfrac{t^3 + 3t^2 - 12t + 4}{t^3 - 4t}$

14. $\lim\limits_{t \to 1} \dfrac{t^3 + t^2 - 5t + 3}{t^3 - 3t + 2}$

15. $\lim\limits_{x \to 3^+} \dfrac{x}{x - 3}$

16. $\lim\limits_{x \to 3^-} \dfrac{x}{x - 3}$

17. $\lim\limits_{x \to 3} \dfrac{x}{x - 3}$

18. $\lim\limits_{x \to 2^+} \dfrac{x}{x^2 - 4}$

19. $\lim\limits_{x \to 2^-} \dfrac{x}{x^2 - 4}$

20. $\lim\limits_{x \to 2} \dfrac{x}{x^2 - 4}$

21. $\lim\limits_{y \to 6^+} \dfrac{y + 6}{y^2 - 36}$

22. $\lim\limits_{y \to 6^-} \dfrac{y + 6}{y^2 - 36}$

23. $\lim\limits_{y \to 6} \dfrac{y + 6}{y^2 - 36}$

24. $\lim\limits_{x \to 4^+} \dfrac{3 - x}{x^2 - 2x - 8}$

25. $\lim\limits_{x \to 4^-} \dfrac{3 - x}{x^2 - 2x - 8}$

26. $\lim\limits_{x \to 4} \dfrac{3 - x}{x^2 - 2x - 8}$

27. $\lim\limits_{x \to 2^+} \dfrac{1}{|2 - x|}$

28. $\lim\limits_{x \to 3^-} \dfrac{1}{|x - 3|}$

29. $\lim\limits_{x \to 9} \dfrac{x - 9}{\sqrt{x} - 3}$

30. $\lim\limits_{y \to 4} \dfrac{4 - y}{2 - \sqrt{y}}$

31. Let
   $$f(x) = \begin{cases} x - 1, & x \le 3 \\ 3x - 7, & x > 3 \end{cases}$$

   (cont.)

Find

(a) $\lim\limits_{x \to 3^-} f(x)$ (b) $\lim\limits_{x \to 3^+} f(x)$ (c) $\lim\limits_{x \to 3} f(x)$.

32. Let

$$g(t) = \begin{cases} t - 2, & t < 0 \\ t^2, & 0 \le t \le 2 \\ 2t, & t > 2 \end{cases}$$

Find

(a) $\lim\limits_{t \to 0} g(t)$ (b) $\lim\limits_{t \to 1} g(t)$ (c) $\lim\limits_{t \to 2} g(t)$.

33–36 True–False Determine whether the statement is true or false. Explain your answer. ■

33. If $\lim\limits_{x \to a} f(x)$ and $\lim\limits_{x \to a} g(x)$ exist, then so does $\lim\limits_{x \to a}[f(x) + g(x)]$.

34. If $\lim\limits_{x \to a} g(x) = 0$ and $\lim\limits_{x \to a} f(x)$ exists, then $\lim\limits_{x \to a}[f(x)/g(x)]$ does not exist.

35. If $\lim\limits_{x \to a} f(x)$ and $\lim\limits_{x \to a} g(x)$ both exist and are equal, then $\lim\limits_{x \to a}[f(x)/g(x)] = 1$.

36. If $f(x)$ is a rational function and $x = a$ is in the domain of f, then $\lim\limits_{x \to a} f(x) = f(a)$.

37–38 First rationalize the numerator and then find the limit.

37. $\lim\limits_{x \to 0} \dfrac{\sqrt{x+4} - 2}{x}$ **38.** $\lim\limits_{x \to 0} \dfrac{\sqrt{x^2+4} - 2}{x}$

39. Let

$$f(x) = \frac{x^3 - 1}{x - 1}$$

(a) Find $\lim\limits_{x \to 1} f(x)$.

(b) Sketch the graph of $y = f(x)$.

40. Let

$$f(x) = \begin{cases} \dfrac{x^2 - 9}{x + 3}, & x \ne -3 \\ k, & x = -3 \end{cases}$$

(a) Find k so that $f(-3) = \lim\limits_{x \to -3} f(x)$.

(b) With k assigned the value $\lim\limits_{x \to -3} f(x)$, show that $f(x)$ can be expressed as a polynomial.

FOCUS ON CONCEPTS

41. (a) Explain why the following calculation is incorrect.

$$\lim_{x \to 0^+}\left(\frac{1}{x} - \frac{1}{x^2}\right) = \lim_{x \to 0^+}\frac{1}{x} - \lim_{x \to 0^+}\frac{1}{x^2}$$
$$= +\infty - (+\infty) = 0$$

(b) Show that $\lim\limits_{x \to 0^+}\left(\dfrac{1}{x} - \dfrac{1}{x^2}\right) = -\infty$.

42. (a) Explain why the following argument is incorrect.

$$\lim_{x \to 0}\left(\frac{1}{x} - \frac{2}{x^2 + 2x}\right) = \lim_{x \to 0}\frac{1}{x}\left(1 - \frac{2}{x + 2}\right)$$
$$= \infty \cdot 0 = 0$$

(b) Show that $\lim\limits_{x \to 0}\left(\dfrac{1}{x} - \dfrac{2}{x^2 + 2x}\right) = \dfrac{1}{2}$.

43. Find all values of a such that

$$\lim_{x \to 1}\left(\frac{1}{x - 1} - \frac{a}{x^2 - 1}\right)$$

exists and is finite.

44. (a) Explain informally why

$$\lim_{x \to 0^-}\left(\frac{1}{x} + \frac{1}{x^2}\right) = +\infty$$

(b) Verify the limit in part (a) algebraically.

45. Let $p(x)$ and $q(x)$ be polynomials, with $q(x_0) = 0$. Discuss the behavior of the graph of $y = p(x)/q(x)$ in the vicinity of $x = x_0$. Give examples to support your conclusions.

46. Suppose that f and g are two functions such that $\lim\limits_{x \to a} f(x)$ exists but $\lim\limits_{x \to a}[f(x) + g(x)]$ does not exist. Use Theorem 1.2.2. to prove that $\lim\limits_{x \to a} g(x)$ does not exist.

47. Suppose that f and g are two functions such that both $\lim\limits_{x \to a} f(x)$ and $\lim\limits_{x \to a}[f(x) + g(x)]$ exist. Use Theorem 1.2.2 to prove that $\lim\limits_{x \to a} g(x)$ exists.

48. Suppose that f and g are two functions such that

$$\lim_{x \to a} g(x) = 0 \quad \text{and} \quad \lim_{x \to a}\frac{f(x)}{g(x)}$$

exists. Use Theorem 1.2.2 to prove that $\lim\limits_{x \to a} f(x) = 0$.

49. Writing According to Newton's Law of Universal Gravitation, the gravitational force of attraction between two masses is inversely proportional to the square of the distance between them. What results of this section are useful in describing the gravitational force of attraction between the masses as they get closer and closer together?

50. Writing Suppose that f and g are two functions that are equal except at a finite number of points and that a denotes a real number. Explain informally why both

$$\lim_{x \to a} f(x) \quad \text{and} \quad \lim_{x \to a} g(x)$$

exist and are equal, or why both limits fail to exist. Write a short paragraph that explains the relationship of this result to the use of "algebraic simplification" in the evaluation of a limit.

✓ QUICK CHECK ANSWERS 1.2

1. (a) 7 (b) 36 (c) −1 (d) 1 (e) $+\infty$ **2.** (a) 7 (b) −3 (c) 1   **3.** (a) −1 (b) 0 (c) $+\infty$ (d) 8
4. (a) 2 (b) 0 (c) does not exist

1.3 LIMITS AT INFINITY; END BEHAVIOR OF A FUNCTION

Up to now we have been concerned with limits that describe the behavior of a function $f(x)$ as x approaches some real number a. In this section we will be concerned with the behavior of $f(x)$ as x increases or decreases without bound.

■ LIMITS AT INFINITY AND HORIZONTAL ASYMPTOTES

If the values of a variable x increase without bound, then we write $x \to +\infty$, and if the values of x decrease without bound, then we write $x \to -\infty$. The behavior of a function $f(x)$ as x increases without bound or decreases without bound is sometimes called the **end behavior** of the function. For example,

$$\lim_{x \to -\infty} \frac{1}{x} = 0 \quad \text{and} \quad \lim_{x \to +\infty} \frac{1}{x} = 0 \tag{1–2}$$

are illustrated numerically in Table 1.3.1 and geometrically in Figure 1.3.1.

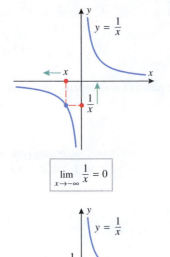

$$\lim_{x \to -\infty} \frac{1}{x} = 0$$

$$\lim_{x \to +\infty} \frac{1}{x} = 0$$

▲ **Figure 1.3.1**

Table 1.3.1

	VALUES						CONCLUSION
x	-1	-10	-100	-1000	$-10,000$	$\cdots$	As $x \to -\infty$ the value of $1/x$
$1/x$	-1	-0.1	-0.01	-0.001	-0.0001	$\cdots$	increases toward zero.
x	1	10	100	1000	$10,000$	$\cdots$	As $x \to +\infty$ the value of $1/x$
$1/x$	1	0.1	0.01	0.001	0.0001	$\cdots$	decreases toward zero.

In general, we will use the following notation.

> **1.3.1 LIMITS AT INFINITY (AN INFORMAL VIEW)**   If the values of $f(x)$ eventually get as close as we like to a number L as x increases without bound, then we write
>
> $$\lim_{x \to +\infty} f(x) = L \quad \text{or} \quad f(x) \to L \text{ as } x \to +\infty \tag{3}$$
>
> Similarly, if the values of $f(x)$ eventually get as close as we like to a number L as x decreases without bound, then we write
>
> $$\lim_{x \to -\infty} f(x) = L \quad \text{or} \quad f(x) \to L \text{ as } x \to -\infty \tag{4}$$

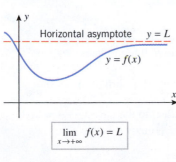

$$\lim_{x \to +\infty} f(x) = L$$

Figure 1.3.2 illustrates the end behavior of a function f when

$$\lim_{x \to +\infty} f(x) = L \quad \text{or} \quad \lim_{x \to -\infty} f(x) = L$$

In the first case the graph of f eventually comes as close as we like to the line $y = L$ as x increases without bound, and in the second case it eventually comes as close as we like to the line $y = L$ as x decreases without bound. If either limit holds, we call the line $y = L$ a **horizontal asymptote** for the graph of f.

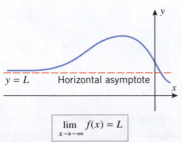

$$\lim_{x \to -\infty} f(x) = L$$

▲ **Figure 1.3.2**

▶ **Example 1**   It follows from (1) and (2) that $y = 0$ is a horizontal asymptote for the graph of $f(x) = 1/x$ in both the positive and negative directions. This is consistent with the graph of $y = 1/x$ shown in Figure 1.3.1. ◀

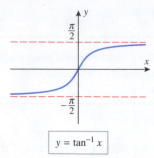

$y = \tan^{-1} x$

▲ **Figure 1.3.3**

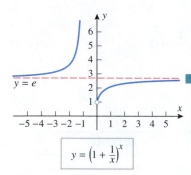

$y = \left(1 + \dfrac{1}{x}\right)^x$

▲ **Figure 1.3.4**

▶ **Example 2**   Figure 1.3.3 is the graph of $f(x) = \tan^{-1} x$. As suggested by this graph,

$$\lim_{x \to +\infty} \tan^{-1} x = \frac{\pi}{2} \quad \text{and} \quad \lim_{x \to -\infty} \tan^{-1} x = -\frac{\pi}{2} \tag{5--6}$$

so the line $y = \pi/2$ is a horizontal asymptote for f in the positive direction and the line $y = -\pi/2$ is a horizontal asymptote in the negative direction. ◀

▶ **Example 3**   Figure 1.3.4 is the graph of $f(x) = (1 + 1/x)^x$. As suggested by this graph,

$$\lim_{x \to +\infty} \left(1 + \frac{1}{x}\right)^x = e \quad \text{and} \quad \lim_{x \to -\infty} \left(1 + \frac{1}{x}\right)^x = e \tag{7--8}$$

so the line $y = e$ is a horizontal asymptote for f in both the positive and negative directions.
◀

■ **LIMIT LAWS FOR LIMITS AT INFINITY**

It can be shown that the limit laws in Theorem 1.2.2 carry over without change to limits at $+\infty$ and $-\infty$. Moreover, it follows by the same argument used in Section 1.2 that if n is a positive integer, then

$$\lim_{x \to +\infty} (f(x))^n = \left(\lim_{x \to +\infty} f(x)\right)^n \qquad \lim_{x \to -\infty} (f(x))^n = \left(\lim_{x \to -\infty} f(x)\right)^n \tag{9--10}$$

provided the indicated limit of $f(x)$ exists. It also follows that constants can be moved through the limit symbols for limits at infinity:

$$\lim_{x \to +\infty} kf(x) = k \lim_{x \to +\infty} f(x) \qquad \lim_{x \to -\infty} kf(x) = k \lim_{x \to -\infty} f(x) \tag{11--12}$$

provided the indicated limit of $f(x)$ exists.

Finally, if $f(x) = k$ is a constant function, then the values of f do not change as $x \to +\infty$ or as $x \to -\infty$, so

$$\lim_{x \to +\infty} k = k \qquad \lim_{x \to -\infty} k = k \tag{13--14}$$

▶ **Example 4**

(a) It follows from (1), (2), (9), and (10) that if n is a positive integer, then

$$\lim_{x \to +\infty} \frac{1}{x^n} = \left(\lim_{x \to +\infty} \frac{1}{x}\right)^n = 0 \quad \text{and} \quad \lim_{x \to -\infty} \frac{1}{x^n} = \left(\lim_{x \to -\infty} \frac{1}{x}\right)^n = 0$$

(b) It follows from (7) and the extension of Theorem 1.2.2(e) to the case $x \to +\infty$ that

$$\lim_{x \to +\infty} \left(1 + \frac{1}{2x}\right)^x = \lim_{x \to +\infty} \left[\left(1 + \frac{1}{2x}\right)^{2x}\right]^{1/2}$$

$$= \left[\lim_{x \to +\infty} \left(1 + \frac{1}{2x}\right)^{2x}\right]^{1/2} = e^{1/2} = \sqrt{e} \quad ◀$$

■ **INFINITE LIMITS AT INFINITY**

Limits at infinity, like limits at a real number a, can fail to exist for various reasons. One such possibility is that the values of $f(x)$ increase or decrease without bound as $x \to +\infty$ or as $x \to -\infty$. We will use the following notation to describe this situation.

1.3.2 **INFINITE LIMITS AT INFINITY (AN INFORMAL VIEW)** If the values of $f(x)$ increase without bound as $x \to +\infty$ or as $x \to -\infty$, then we write

$$\lim_{x \to +\infty} f(x) = +\infty \quad \text{or} \quad \lim_{x \to -\infty} f(x) = +\infty$$

as appropriate; and if the values of $f(x)$ decrease without bound as $x \to +\infty$ or as $x \to -\infty$, then we write

$$\lim_{x \to +\infty} f(x) = -\infty \quad \text{or} \quad \lim_{x \to -\infty} f(x) = -\infty$$

as appropriate.

◼ LIMITS OF x^n AS $x \to \pm\infty$

Figure 1.3.5 illustrates the end behavior of the polynomials x^n for $n = 1, 2, 3$, and 4. These are special cases of the following general results:

$$\lim_{x \to +\infty} x^n = +\infty, \quad n = 1, 2, 3, \dots \qquad \lim_{x \to -\infty} x^n = \begin{cases} -\infty, & n = 1, 3, 5, \dots \\ +\infty, & n = 2, 4, 6, \dots \end{cases} \qquad (15\text{--}16)$$

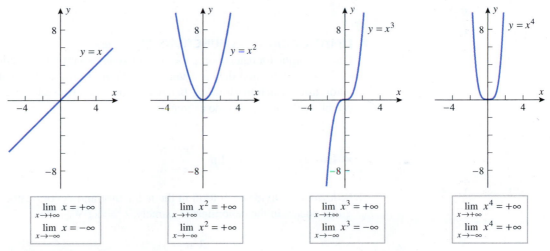

$$\lim_{x \to +\infty} x = +\infty$$
$$\lim_{x \to -\infty} x = -\infty$$

$$\lim_{x \to +\infty} x^2 = +\infty$$
$$\lim_{x \to -\infty} x^2 = +\infty$$

$$\lim_{x \to +\infty} x^3 = +\infty$$
$$\lim_{x \to -\infty} x^3 = -\infty$$

$$\lim_{x \to +\infty} x^4 = +\infty$$
$$\lim_{x \to -\infty} x^4 = +\infty$$

▲ **Figure 1.3.5**

Multiplying x^n by a positive real number does not affect limits (15) and (16), but multiplying by a negative real number reverses the sign.

▶ **Example 5**

$$\lim_{x \to +\infty} 2x^5 = +\infty, \qquad \lim_{x \to -\infty} 2x^5 = -\infty$$

$$\lim_{x \to +\infty} -7x^6 = -\infty, \qquad \lim_{x \to -\infty} -7x^6 = -\infty \quad ◀$$

◼ LIMITS OF POLYNOMIALS AS $x \to \pm\infty$

There is a useful principle about polynomials which, expressed informally, states:

The end behavior of a polynomial matches the end behavior of its highest degree term.

More precisely, if $c_n \neq 0$, then

$$\lim_{x \to -\infty} \left(c_0 + c_1 x + \cdots + c_n x^n\right) = \lim_{x \to -\infty} c_n x^n \tag{17}$$

$$\lim_{x \to +\infty} \left(c_0 + c_1 x + \cdots + c_n x^n\right) = \lim_{x \to +\infty} c_n x^n \tag{18}$$

We can motivate these results by factoring out the highest power of x from the polynomial and examining the limit of the factored expression. Thus,

$$c_0 + c_1 x + \cdots + c_n x^n = x^n \left(\frac{c_0}{x^n} + \frac{c_1}{x^{n-1}} + \cdots + c_n\right)$$

As $x \to -\infty$ or $x \to +\infty$, it follows from Example 4(a) that all of the terms with positive powers of x in the denominator approach 0, so (17) and (18) are certainly plausible.

▶ **Example 6**

$$\lim_{x \to -\infty} (7x^5 - 4x^3 + 2x - 9) = \lim_{x \to -\infty} 7x^5 = -\infty$$

$$\lim_{x \to -\infty} (-4x^8 + 17x^3 - 5x + 1) = \lim_{x \to -\infty} -4x^8 = -\infty \quad ◀$$

■ **LIMITS OF RATIONAL FUNCTIONS AS $x \to \pm\infty$**

One technique for determining the end behavior of a rational function is to divide each term in the numerator and denominator by the highest power of x that occurs in the denominator, after which the limiting behavior can be determined using results we have already established. Here are some examples.

▶ **Example 7**   Find $\displaystyle \lim_{x \to +\infty} \frac{3x + 5}{6x - 8}$.

Solution.   Divide each term in the numerator and denominator by the highest power of x that occurs in the denominator, namely, $x^1 = x$. We obtain

$$\lim_{x \to +\infty} \frac{3x + 5}{6x - 8} = \lim_{x \to +\infty} \frac{3 + \dfrac{5}{x}}{6 - \dfrac{8}{x}} \qquad \text{Divide each term by } x.$$

$$= \frac{\displaystyle\lim_{x \to +\infty} \left(3 + \frac{5}{x}\right)}{\displaystyle\lim_{x \to +\infty} \left(6 - \frac{8}{x}\right)} \qquad \begin{array}{l}\text{Limit of a quotient is the}\\\text{quotient of the limits.}\end{array}$$

$$= \frac{\displaystyle\lim_{x \to +\infty} 3 + \lim_{x \to +\infty} \frac{5}{x}}{\displaystyle\lim_{x \to +\infty} 6 - \lim_{x \to +\infty} \frac{8}{x}} \qquad \begin{array}{l}\text{Limit of a sum is the}\\\text{sum of the limits.}\end{array}$$

$$= \frac{3 + 5 \displaystyle\lim_{x \to +\infty} \frac{1}{x}}{6 - 8 \displaystyle\lim_{x \to +\infty} \frac{1}{x}} = \frac{3 + 0}{6 + 0} = \frac{1}{2} \qquad \begin{array}{l}\text{A constant can be moved through a}\\\text{limit symbol; Formulas (2) and (13).}\end{array} \quad ◀$$

▶ **Example 8** Find

$$\text{(a) } \lim_{x \to -\infty} \frac{4x^2 - x}{2x^3 - 5} \qquad \text{(b) } \lim_{x \to +\infty} \frac{5x^3 - 2x^2 + 1}{1 - 3x}$$

Solution (a).   Divide each term in the numerator and denominator by the highest power of x that occurs in the denominator, namely, x^3. We obtain

$$\lim_{x \to -\infty} \frac{4x^2 - x}{2x^3 - 5} = \lim_{x \to -\infty} \frac{\dfrac{4}{x} - \dfrac{1}{x^2}}{2 - \dfrac{5}{x^3}} \qquad \boxed{\text{Divide each term by } x^3.}$$

$$= \frac{\lim\limits_{x \to -\infty} \left(\dfrac{4}{x} - \dfrac{1}{x^2} \right)}{\lim\limits_{x \to -\infty} \left(2 - \dfrac{5}{x^3} \right)} \qquad \boxed{\begin{array}{l}\text{Limit of a quotient is the}\\ \text{quotient of the limits.}\end{array}}$$

$$= \frac{\lim\limits_{x \to -\infty} \dfrac{4}{x} - \lim\limits_{x \to -\infty} \dfrac{1}{x^2}}{\lim\limits_{x \to -\infty} 2 - \lim\limits_{x \to -\infty} \dfrac{5}{x^3}} \qquad \boxed{\begin{array}{l}\text{Limit of a difference is the}\\ \text{difference of the limits.}\end{array}}$$

$$= \frac{4 \lim\limits_{x \to -\infty} \dfrac{1}{x} - \lim\limits_{x \to -\infty} \dfrac{1}{x^2}}{2 - 5 \lim\limits_{x \to -\infty} \dfrac{1}{x^3}} = \frac{0 - 0}{2 - 0} = 0 \qquad \boxed{\begin{array}{l}\text{A constant can be moved through}\\ \text{a limit symbol; Formula (14) and}\\ \text{Example 4.}\end{array}}$$

Solution (b).   Divide each term in the numerator and denominator by the highest power of x that occurs in the denominator, namely, $x^1 = x$. We obtain

$$\lim_{x \to +\infty} \frac{5x^3 - 2x^2 + 1}{1 - 3x} = \lim_{x \to +\infty} \frac{5x^2 - 2x + \dfrac{1}{x}}{\dfrac{1}{x} - 3} \qquad (19)$$

In this case we cannot argue that the limit of the quotient is the quotient of the limits because the limit of the numerator does not exist. However, we have

$$\lim_{x \to +\infty} 5x^2 - 2x = +\infty, \qquad \lim_{x \to +\infty} \frac{1}{x} = 0, \qquad \lim_{x \to +\infty} \left(\frac{1}{x} - 3 \right) = -3$$

Thus, the numerator on the right side of (19) approaches $+\infty$ and the denominator has a finite *negative* limit. We conclude from this that the quotient approaches $-\infty$; that is,

$$\lim_{x \to +\infty} \frac{5x^3 - 2x^2 + 1}{1 - 3x} = \lim_{x \to +\infty} \frac{5x^2 - 2x + \dfrac{1}{x}}{\dfrac{1}{x} - 3} = -\infty \quad ◀$$

■ **A QUICK METHOD FOR FINDING LIMITS OF RATIONAL FUNCTIONS AS $x \to +\infty$ OR $x \to -\infty$**

Since the end behavior of a polynomial matches the end behavior of its highest degree term, one can reasonably conclude:

> *The end behavior of a rational function matches the end behavior of the quotient of the highest degree term in the numerator divided by the highest degree term in the denominator.*

▶ **Example 9** Use the preceding observation to compute the limits in Examples 7 and 8.

Solution.

$$\lim_{x \to +\infty} \frac{3x + 5}{6x - 8} = \lim_{x \to +\infty} \frac{3x}{6x} = \lim_{x \to +\infty} \frac{1}{2} = \frac{1}{2}$$

$$\lim_{x \to -\infty} \frac{4x^2 - x}{2x^3 - 5} = \lim_{x \to -\infty} \frac{4x^2}{2x^3} = \lim_{x \to -\infty} \frac{2}{x} = 0$$

$$\lim_{x \to +\infty} \frac{5x^3 - 2x^2 + 1}{1 - 3x} = \lim_{x \to +\infty} \frac{5x^3}{(-3x)} = \lim_{x \to +\infty} \left(-\frac{5}{3} x^2 \right) = -\infty \blacktriangleleft$$

■ LIMITS INVOLVING RADICALS

▶ **Example 10** Find

(a) $\displaystyle \lim_{x \to +\infty} \frac{\sqrt{x^2 + 2}}{3x - 6}$ (b) $\displaystyle \lim_{x \to -\infty} \frac{\sqrt{x^2 + 2}}{3x - 6}$

In both parts it would be helpful to manipulate the function so that the powers of x are transformed to powers of $1/x$. This can be achieved in both cases by dividing the numerator and denominator by $|x|$ and using the fact that $\sqrt{x^2} = |x|$.

Solution (a). As $x \to +\infty$, the values of x under consideration are positive, so we can replace $|x|$ by x where helpful. We obtain

$$\lim_{x \to +\infty} \frac{\sqrt{x^2 + 2}}{3x - 6} = \lim_{x \to +\infty} \frac{\dfrac{\sqrt{x^2 + 2}}{|x|}}{\dfrac{3x - 6}{|x|}} = \lim_{x \to +\infty} \frac{\dfrac{\sqrt{x^2 + 2}}{\sqrt{x^2}}}{\dfrac{3x - 6}{x}}$$

$$= \lim_{x \to +\infty} \frac{\sqrt{1 + \dfrac{2}{x^2}}}{3 - \dfrac{6}{x}} = \frac{\displaystyle \lim_{x \to +\infty} \sqrt{1 + \dfrac{2}{x^2}}}{\displaystyle \lim_{x \to +\infty} \left(3 - \dfrac{6}{x} \right)}$$

$$= \frac{\sqrt{\displaystyle \lim_{x \to +\infty} \left(1 + \dfrac{2}{x^2} \right)}}{\displaystyle \lim_{x \to +\infty} \left(3 - \dfrac{6}{x} \right)} = \frac{\sqrt{\left(\displaystyle \lim_{x \to +\infty} 1 \right) + \left(2 \displaystyle \lim_{x \to +\infty} \dfrac{1}{x^2} \right)}}{\left(\displaystyle \lim_{x \to +\infty} 3 \right) - \left(6 \displaystyle \lim_{x \to +\infty} \dfrac{1}{x} \right)}$$

$$= \frac{\sqrt{1 + (2 \cdot 0)}}{3 - (6 \cdot 0)} = \frac{1}{3}$$

TECHNOLOGY MASTERY

It follows from Example 10 that the function

$$f(x) = \frac{\sqrt{x^2 + 2}}{3x - 6}$$

has an asymptote of $y = \frac{1}{3}$ in the positive direction and an asymptote of $y = -\frac{1}{3}$ in the negative direction. Confirm this using a graphing utility.

Solution (b). As $x \to -\infty$, the values of x under consideration are negative, so we can replace $|x|$ by $-x$ where helpful. We obtain

$$\lim_{x \to -\infty} \frac{\sqrt{x^2 + 2}}{3x - 6} = \lim_{x \to -\infty} \frac{\dfrac{\sqrt{x^2 + 2}}{|x|}}{\dfrac{3x - 6}{|x|}} = \lim_{x \to -\infty} \frac{\dfrac{\sqrt{x^2 + 2}}{\sqrt{x^2}}}{\dfrac{3x - 6}{(-x)}}$$

$$= \lim_{x \to -\infty} \frac{\sqrt{1 + \dfrac{2}{x^2}}}{-3 + \dfrac{6}{x}} = -\frac{1}{3} \blacktriangleleft$$

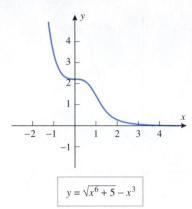

$y = \sqrt{x^6 + 5} - x^3$

(a)

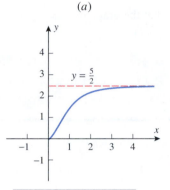

$y = \frac{5}{2}$

$y = \sqrt{x^6 + 5x^3} - x^3,\ x \geq 0$

(b)

▲ **Figure 1.3.6**

We noted in Section 1.1 that the standard rules of algebra do not apply to the symbols $+\infty$ and $-\infty$. Part (b) of Example 11 illustrates this. The terms $\sqrt{x^6 + 5x^3}$ and x^3 both approach $+\infty$ as $x \to +\infty$, but their difference does not approach 0.

▶ **Example 11** Find

$$\text{(a) } \lim_{x \to +\infty} (\sqrt{x^6 + 5} - x^3) \qquad \text{(b) } \lim_{x \to +\infty} (\sqrt{x^6 + 5x^3} - x^3)$$

Solution. Graphs of the functions $f(x) = \sqrt{x^6 + 5} - x^3$, and $g(x) = \sqrt{x^6 + 5x^3} - x^3$ for $x \geq 0$, are shown in Figure 1.3.6. From the graphs we might conjecture that the requested limits are 0 and $\frac{5}{2}$, respectively. To confirm this, we treat each function as a fraction with a denominator of 1 and rationalize the numerator.

$$\lim_{x \to +\infty} (\sqrt{x^6 + 5} - x^3) = \lim_{x \to +\infty} (\sqrt{x^6 + 5} - x^3)\left(\frac{\sqrt{x^6 + 5} + x^3}{\sqrt{x^6 + 5} + x^3}\right)$$

$$= \lim_{x \to +\infty} \frac{(x^6 + 5) - x^6}{\sqrt{x^6 + 5} + x^3} = \lim_{x \to +\infty} \frac{5}{\sqrt{x^6 + 5} + x^3}$$

$$= \lim_{x \to +\infty} \frac{\dfrac{5}{x^3}}{\sqrt{1 + \dfrac{5}{x^6}} + 1} \qquad \boxed{\sqrt{x^6} = x^3 \text{ for } x > 0}$$

$$= \frac{0}{\sqrt{1 + 0} + 1} = 0$$

$$\lim_{x \to +\infty} (\sqrt{x^6 + 5x^3} - x^3) = \lim_{x \to +\infty} (\sqrt{x^6 + 5x^3} - x^3)\left(\frac{\sqrt{x^6 + 5x^3} + x^3}{\sqrt{x^6 + 5x^3} + x^3}\right)$$

$$= \lim_{x \to +\infty} \frac{(x^6 + 5x^3) - x^6}{\sqrt{x^6 + 5x^3} + x^3} = \lim_{x \to +\infty} \frac{5x^3}{\sqrt{x^6 + 5x^3} + x^3}$$

$$= \lim_{x \to +\infty} \frac{5}{\sqrt{1 + \dfrac{5}{x^3}} + 1} \qquad \boxed{\sqrt{x^6} = x^3 \text{ for } x > 0}$$

$$= \frac{5}{\sqrt{1 + 0} + 1} = \frac{5}{2} \quad \blacktriangleleft$$

■ **END BEHAVIOR OF TRIGONOMETRIC, EXPONENTIAL, AND LOGARITHMIC FUNCTIONS**

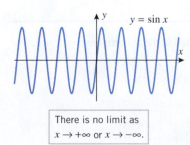

$y = \sin x$

There is no limit as $x \to +\infty$ or $x \to -\infty$.

▲ **Figure 1.3.7**

Consider the function $f(x) = \sin x$ that is graphed in Figure 1.3.7. For this function the limits as $x \to +\infty$ and as $x \to -\infty$ fail to exist not because $f(x)$ increases or decreases without bound, but rather because the values vary between -1 and 1 without approaching some specific real number. In general, the trigonometric functions fail to have limits as $x \to +\infty$ and as $x \to -\infty$ because of periodicity. There is no specific notation to denote this kind of behavior.

In Section 0.5 we showed that the functions e^x and $\ln x$ both increase without bound as $x \to +\infty$ (Figures 0.5.8 and 0.5.9). Thus, in limit notation we have

$$\lim_{x \to +\infty} \ln x = +\infty \qquad \lim_{x \to +\infty} e^x = +\infty \qquad (20\text{–}21)$$

For reference, we also list the following limits, which are consistent with the graphs in Figure 1.3.8:

$$\lim_{x \to -\infty} e^x = 0 \qquad \lim_{x \to 0^+} \ln x = -\infty \qquad (22\text{–}23)$$

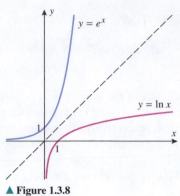

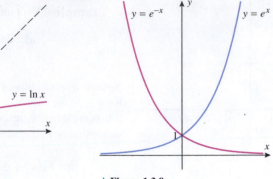

▲ Figure 1.3.8   ▲ Figure 1.3.9

Finally, the following limits can be deduced by noting that the graph of $y = e^{-x}$ is the reflection about the y-axis of the graph of $y = e^x$ (Figure 1.3.9).

$$\lim_{x \to +\infty} e^{-x} = 0 \qquad\qquad \lim_{x \to -\infty} e^{-x} = +\infty \qquad\qquad (24\text{--}25)$$

✔ QUICK CHECK EXERCISES 1.3   (See page 100 for answers.)

1. Find the limits.

(a) $\displaystyle\lim_{x \to -\infty} (3 - x) = \underline{\hspace{1cm}}$

(b) $\displaystyle\lim_{x \to +\infty} \left(5 - \frac{1}{x}\right) = \underline{\hspace{1cm}}$

(c) $\displaystyle\lim_{x \to +\infty} \ln\left(\frac{1}{x}\right) = \underline{\hspace{1cm}}$

(d) $\displaystyle\lim_{x \to +\infty} \frac{1}{e^x} = \underline{\hspace{1cm}}$

2. Find the limits that exist.

(a) $\displaystyle\lim_{x \to -\infty} \frac{2x^2 + x}{4x^2 - 3} = \underline{\hspace{1cm}}$

(b) $\displaystyle\lim_{x \to +\infty} \frac{1}{2 + \sin x} = \underline{\hspace{1cm}}$

(c) $\displaystyle\lim_{x \to +\infty} \left(1 + \frac{1}{x}\right)^x = \underline{\hspace{1cm}}$

3. Given that

$$\lim_{x \to +\infty} f(x) = 2 \quad \text{and} \quad \lim_{x \to +\infty} g(x) = -3$$

find the limits that exist.

(a) $\displaystyle\lim_{x \to +\infty} [3f(x) - g(x)] = \underline{\hspace{1cm}}$

(b) $\displaystyle\lim_{x \to +\infty} \frac{f(x)}{g(x)} = \underline{\hspace{1cm}}$

(c) $\displaystyle\lim_{x \to +\infty} \frac{2f(x) + 3g(x)}{3f(x) + 2g(x)} = \underline{\hspace{1cm}}$

(d) $\displaystyle\lim_{x \to +\infty} \sqrt{10 - f(x)g(x)} = \underline{\hspace{1cm}}$

4. Consider the graphs of $1/x$, $\sin x$, $\ln x$, e^x, and e^{-x}. Which of these graphs has a horizontal asymptote?

EXERCISE SET 1.3   Graphing Utility

1–4 In these exercises, make reasonable assumptions about the end behavior of the indicated function. ■

1. For the function g graphed in the accompanying figure, find

(a) $\displaystyle\lim_{x \to -\infty} g(x)$ (b) $\displaystyle\lim_{x \to +\infty} g(x)$.

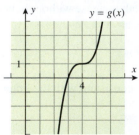

◀ Figure Ex-1

2. For the function ϕ graphed in the accompanying figure, find

(a) $\displaystyle\lim_{x \to -\infty} \phi(x)$

(b) $\displaystyle\lim_{x \to +\infty} \phi(x)$.

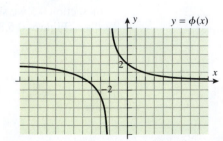

◀ Figure Ex-2

3. For the function ϕ graphed in the accompanying figure, find
   (a) $\lim\limits_{x \to -\infty} \phi(x)$ (b) $\lim\limits_{x \to +\infty} \phi(x)$.

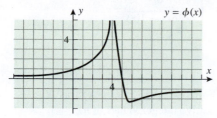

◀ **Figure Ex-3**

4. For the function G graphed in the accompanying figure, find
   (a) $\lim\limits_{x \to -\infty} G(x)$ (b) $\lim\limits_{x \to +\infty} G(x)$.

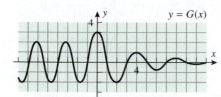

◀ **Figure Ex-4**

5. Given that
$$\lim_{x \to +\infty} f(x) = 3, \quad \lim_{x \to +\infty} g(x) = -5, \quad \lim_{x \to +\infty} h(x) = 0$$
   find the limits that exist. If the limit does not exist, explain why.
   (a) $\lim\limits_{x \to +\infty} [f(x) + 3g(x)]$
   (b) $\lim\limits_{x \to +\infty} [h(x) - 4g(x) + 1]$
   (c) $\lim\limits_{x \to +\infty} [f(x)g(x)]$ (d) $\lim\limits_{x \to +\infty} [g(x)]^2$
   (e) $\lim\limits_{x \to +\infty} \sqrt[3]{5 + f(x)}$ (f) $\lim\limits_{x \to +\infty} \dfrac{3}{g(x)}$
   (g) $\lim\limits_{x \to +\infty} \dfrac{3h(x) + 4}{x^2}$ (h) $\lim\limits_{x \to +\infty} \dfrac{6f(x)}{5f(x) + 3g(x)}$

6. Given that
$$\lim_{x \to -\infty} f(x) = 7 \quad \text{and} \quad \lim_{x \to -\infty} g(x) = -6$$
   find the limits that exist. If the limit does not exist, explain why.
   (a) $\lim\limits_{x \to -\infty} [2f(x) - g(x)]$ (b) $\lim\limits_{x \to -\infty} [6f(x) + 7g(x)]$
   (c) $\lim\limits_{x \to -\infty} [x^2 + g(x)]$ (d) $\lim\limits_{x \to -\infty} [x^2 g(x)]$
   (e) $\lim\limits_{x \to -\infty} \sqrt[3]{f(x)g(x)}$ (f) $\lim\limits_{x \to -\infty} \dfrac{g(x)}{f(x)}$
   (g) $\lim\limits_{x \to -\infty} \left[f(x) + \dfrac{g(x)}{x} \right]$ (h) $\lim\limits_{x \to -\infty} \dfrac{xf(x)}{(2x + 3)g(x)}$

7. (a) Complete the table and make a guess about the limit indicated.
$$f(x) = \tan^{-1}\left(\frac{1}{x}\right) \qquad \lim_{x \to 0^+} f(x)$$

x	0.1	0.01	0.001	0.0001	0.00001	0.000001
$f(x)$						

(b) Use Figure 1.3.3 to find the exact value of the limit in part (a).

8. Complete the table and make a guess about the limit indicated.
$$f(x) = x^{1/x} \qquad \lim_{x \to +\infty} f(x)$$

x	10	100	1000	10,000	100,000	1,000,000
$f(x)$						

9–40 Find the limits. ■

9. $\lim\limits_{x \to +\infty} (1 + 2x - 3x^5)$ 10. $\lim\limits_{x \to +\infty} (2x^3 - 100x + 5)$

11. $\lim\limits_{x \to +\infty} \sqrt{x}$ 12. $\lim\limits_{x \to -\infty} \sqrt{5 - x}$

13. $\lim\limits_{x \to +\infty} \dfrac{3x + 1}{2x - 5}$ 14. $\lim\limits_{x \to +\infty} \dfrac{5x^2 - 4x}{2x^2 + 3}$

15. $\lim\limits_{y \to -\infty} \dfrac{3}{y + 4}$ 16. $\lim\limits_{x \to +\infty} \dfrac{1}{x - 12}$

17. $\lim\limits_{x \to -\infty} \dfrac{x - 2}{x^2 + 2x + 1}$ 18. $\lim\limits_{x \to +\infty} \dfrac{5x^2 + 7}{3x^2 - x}$

19. $\lim\limits_{x \to +\infty} \dfrac{7 - 6x^5}{x + 3}$ 20. $\lim\limits_{t \to -\infty} \dfrac{5 - 2t^3}{t^2 + 1}$

21. $\lim\limits_{t \to +\infty} \dfrac{6 - t^3}{7t^3 + 3}$ 22. $\lim\limits_{x \to -\infty} \dfrac{x + 4x^3}{1 - x^2 + 7x^3}$

23. $\lim\limits_{x \to +\infty} \sqrt[3]{\dfrac{2 + 3x - 5x^2}{1 + 8x^2}}$ 24. $\lim\limits_{s \to +\infty} \sqrt[3]{\dfrac{3s^7 - 4s^5}{2s^7 + 1}}$

25. $\lim\limits_{x \to -\infty} \dfrac{\sqrt{5x^2 - 2}}{x + 3}$ 26. $\lim\limits_{x \to +\infty} \dfrac{\sqrt{5x^2 - 2}}{x + 3}$

27. $\lim\limits_{y \to -\infty} \dfrac{2 - y}{\sqrt{7 + 6y^2}}$ 28. $\lim\limits_{y \to +\infty} \dfrac{2 - y}{\sqrt{7 + 6y^2}}$

29. $\lim\limits_{x \to -\infty} \dfrac{\sqrt{3x^4 + x}}{x^2 - 8}$ 30. $\lim\limits_{x \to +\infty} \dfrac{\sqrt{3x^4 + x}}{x^2 - 8}$

31. $\lim\limits_{x \to +\infty} (\sqrt{x^2 + 3} - x)$ 32. $\lim\limits_{x \to +\infty} (\sqrt{x^2 - 3x} - x)$

33. $\lim\limits_{x \to -\infty} \dfrac{1 - e^x}{1 + e^x}$ 34. $\lim\limits_{x \to +\infty} \dfrac{1 - e^x}{1 + e^x}$

35. $\lim\limits_{x \to +\infty} \dfrac{e^x + e^{-x}}{e^x - e^{-x}}$ 36. $\lim\limits_{x \to -\infty} \dfrac{e^x + e^{-x}}{e^x - e^{-x}}$

37. $\lim\limits_{x \to +\infty} \ln\left(\dfrac{2}{x^2}\right)$ 38. $\lim\limits_{x \to 0^+} \ln\left(\dfrac{2}{x^2}\right)$

39. $\lim\limits_{x \to +\infty} \dfrac{(x + 1)^x}{x^x}$ 40. $\lim\limits_{x \to +\infty} \left(1 + \dfrac{1}{x}\right)^{-x}$

41–44 True–False Determine whether the statement is true or false. Explain your answer. ■

41. We have $\lim\limits_{x \to +\infty} \left(1 + \dfrac{1}{x}\right)^{2x} = (1 + 0)^{+\infty} = 1^{+\infty} = 1$.

42. If $y = L$ is a horizontal asymptote for the curve $y = f(x)$, then

$$\lim_{x \to -\infty} f(x) = L \quad \text{and} \quad \lim_{x \to +\infty} f(x) = L$$

43. If $y = L$ is a horizontal asymptote for the curve $y = f(x)$, then it is possible for the graph of f to intersect the line $y = L$ infinitely many times.

44. If a rational function $p(x)/q(x)$ has a horizontal asymptote, then the degree of $p(x)$ must equal the degree of $q(x)$.

FOCUS ON CONCEPTS

45. Assume that a particle is accelerated by a constant force. The two curves $v = n(t)$ and $v = e(t)$ in the accompanying figure provide velocity versus time curves for the particle as predicted by classical physics and by the special theory of relativity, respectively. The parameter c represents the speed of light. Using the language of limits, describe the differences in the long-term predictions of the two theories.

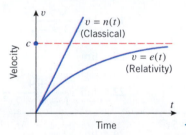

Time   ◀ **Figure Ex-45**

46. Let $T = f(t)$ denote the temperature of a baked potato t minutes after it has been removed from a hot oven. The accompanying figure shows the temperature versus time curve for the potato, where r is the temperature of the room.
   (a) What is the physical significance of $\lim_{t \to 0^+} f(t)$?
   (b) What is the physical significance of $\lim_{t \to +\infty} f(t)$?

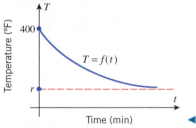

Time (min)   ◀ **Figure Ex-46**

47. Let

$$f(x) = \begin{cases} 2x^2 + 5, & x < 0 \\ \dfrac{3 - 5x^3}{1 + 4x + x^3}, & x \geq 0 \end{cases}$$

Find
   (a) $\lim_{x \to -\infty} f(x)$
   (b) $\lim_{x \to +\infty} f(x)$.

48. Let

$$g(t) = \begin{cases} \dfrac{2 + 3t}{5t^2 + 6}, & t < 1{,}000{,}000 \\ \dfrac{\sqrt{36t^2 - 100}}{5 - t}, & t > 1{,}000{,}000 \end{cases}$$

Find
   (a) $\lim_{t \to -\infty} g(t)$
   (b) $\lim_{t \to +\infty} g(t)$.

49. Discuss the limits of $p(x) = (1 - x)^n$ as $x \to +\infty$ and $x \to -\infty$ for positive integer values of n.

50. In each part, find examples of polynomials $p(x)$ and $q(x)$ that satisfy the stated condition and such that $p(x) \to +\infty$ and $q(x) \to +\infty$ as $x \to +\infty$.
   (a) $\lim_{x \to +\infty} \dfrac{p(x)}{q(x)} = 1$
   (b) $\lim_{x \to +\infty} \dfrac{p(x)}{q(x)} = 0$
   (c) $\lim_{x \to +\infty} \dfrac{p(x)}{q(x)} = +\infty$
   (d) $\lim_{x \to +\infty} [p(x) - q(x)] = 3$

51. (a) Do any of the trigonometric functions $\sin x$, $\cos x$, $\tan x$, $\cot x$, $\sec x$, and $\csc x$ have horizontal asymptotes?
   (b) Do any of the trigonometric functions have vertical asymptotes? Where?

52. Find

$$\lim_{x \to +\infty} \frac{c_0 + c_1 x + \cdots + c_n x^n}{d_0 + d_1 x + \cdots + d_m x^m}$$

where $c_n \neq 0$ and $d_m \neq 0$. [*Hint:* Your answer will depend on whether $m < n$, $m = n$, or $m > n$.]

FOCUS ON CONCEPTS

53–54 These exercises develop some versions of the *substitution principle*, a useful tool for the evaluation of limits.

53. (a) Explain why we can evaluate $\lim_{x \to +\infty} e^{x^2}$ by making the substitution $t = x^2$ and writing

$$\lim_{x \to +\infty} e^{x^2} = \lim_{t \to +\infty} e^t = +\infty$$

   (b) Suppose $g(x) \to +\infty$ as $x \to +\infty$. Given any function $f(x)$, explain why we can evaluate $\lim_{x \to +\infty} f[g(x)]$ by substituting $t = g(x)$ and writing

$$\lim_{x \to +\infty} f[g(x)] = \lim_{t \to +\infty} f(t)$$

   (Here, "equality" is interpreted to mean that either both limits exist and are equal or that both limits fail to exist.)
   (c) Why does the result in part (b) remain valid if $\lim_{x \to +\infty}$ is replaced everywhere by one of $\lim_{x \to -\infty}$, $\lim_{x \to c}$, $\lim_{x \to c^-}$, or $\lim_{x \to c^+}$?

54. (a) Explain why we can evaluate $\lim_{x \to +\infty} e^{-x^2}$ by making the substitution $t = -x^2$ and writing

$$\lim_{x \to +\infty} e^{-x^2} = \lim_{t \to -\infty} e^t = 0 \qquad \text{(cont.)}$$

(b) Suppose $g(x) \to -\infty$ as $x \to +\infty$. Given any function $f(x)$, explain why we can evaluate $\lim_{x \to +\infty} f[g(x)]$ by substituting $t = g(x)$ and writing

$$\lim_{x \to +\infty} f[g(x)] = \lim_{t \to -\infty} f(t)$$

(Here, "equality" is interpreted to mean that either both limits exist and are equal or that both limits fail to exist.)

(c) Why does the result in part (b) remain valid if $\lim_{x \to +\infty}$ is replaced everywhere by one of $\lim_{x \to -\infty}$, $\lim_{x \to c}$, $\lim_{x \to c^-}$, or $\lim_{x \to c^+}$?

55–62 Evaluate the limit using an appropriate substitution. ■

55. $\lim_{x \to 0^+} e^{1/x}$

56. $\lim_{x \to 0^-} e^{1/x}$

57. $\lim_{x \to 0^+} e^{\csc x}$

58. $\lim_{x \to 0^-} e^{\csc x}$

59. $\lim_{x \to +\infty} \dfrac{\ln 2x}{\ln 3x}$ [*Hint:* $t = \ln x$]

60. $\lim_{x \to +\infty} [\ln(x^2 - 1) - \ln(x + 1)]$ [*Hint:* $t = x - 1$]

61. $\lim_{x \to +\infty} \left(1 - \dfrac{1}{x}\right)^{-x}$ [*Hint:* $t = -x$]

62. $\lim_{x \to +\infty} \left(1 + \dfrac{2}{x}\right)^{x}$ [*Hint:* $t = x/2$]

63. Let $f(x) = b^x$, where $0 < b$. Use the substitution principle to verify the asymptotic behavior of f that is illustrated in Figure 0.5.1. [*Hint:* $f(x) = b^x = (e^{\ln b})^x = e^{(\ln b)x}$]

64. Prove that $\lim_{x \to 0}(1 + x)^{1/x} = e$ by completing parts (a) and (b).
(a) Use Equation (7) and the substitution $t = 1/x$ to prove that $\lim_{x \to 0^+}(1 + x)^{1/x} = e$.
(b) Use Equation (8) and the substitution $t = 1/x$ to prove that $\lim_{x \to 0^-}(1 + x)^{1/x} = e$.

65. Suppose that the speed v (in ft/s) of a skydiver t seconds after leaping from a plane is given by the equation $v = 190(1 - e^{-0.168t})$.
(a) Graph v versus t.
(b) By evaluating an appropriate limit, show that the graph of v versus t has a horizontal asymptote $v = c$ for an appropriate constant c.
(c) What is the physical significance of the constant c in part (b)?

66. The population p of the United States (in millions) in year t may be modeled by the function

$$p = \dfrac{50371.7}{151.3 + 181.626e^{-0.031636(t-1950)}}$$

(a) Based on this model, what was the U.S. population in 1950?
(b) Plot p versus t for the 200-year period from 1950 to 2150.

(c) By evaluating an appropriate limit, show that the graph of p versus t has a horizontal asymptote $p = c$ for an appropriate constant c.
(d) What is the significance of the constant c in part (b) for population predicted by this model?

67. (a) Compute the (approximate) values of the terms in the sequence

$$1.01^{101}, 1.001^{1001}, 1.0001^{10001}, 1.00001^{100001},$$
$$1.000001^{1000001}, 1.0000001^{10000001} \ldots$$

What number do these terms appear to be approaching?
(b) Use Equation (7) to verify your answer in part (a).
(c) Let $1 \leq a \leq 9$ denote a positive integer. What number is approached more and more closely by the terms in the following sequence?

$$1.01^{a0a}, 1.001^{a00a}, 1.0001^{a000a}, 1.00001^{a0000a},$$
$$1.000001^{a00000a}, 1.0000001^{a000000a} \ldots$$

(The powers are positive integers that begin and end with the digit a and have 0's in the remaining positions).

68. Let $f(x) = \left(1 + \dfrac{1}{x}\right)^x$.
(a) Prove the identity

$$f(-x) = \dfrac{x}{x-1} \cdot f(x-1)$$

(b) Use Equation (7) and the identity from part (a) to prove Equation (8).

69–73 The notion of an asymptote can be extended to include curves as well as lines. Specifically, we say that curves $y = f(x)$ and $y = g(x)$ are *asymptotic as $x \to +\infty$* provided

$$\lim_{x \to +\infty} [f(x) - g(x)] = 0$$

and are *asymptotic as $x \to -\infty$* provided

$$\lim_{x \to -\infty} [f(x) - g(x)] = 0$$

In these exercises, determine a simpler function $g(x)$ such that $y = f(x)$ is asymptotic to $y = g(x)$ as $x \to +\infty$ or $x \to -\infty$. Use a graphing utility to generate the graphs of $y = f(x)$ and $y = g(x)$ and identify all vertical asymptotes. ■

69. $f(x) = \dfrac{x^2 - 2}{x - 2}$ [*Hint:* Divide $x - 2$ into $x^2 - 2$.]

70. $f(x) = \dfrac{x^3 - x + 3}{x}$

71. $f(x) = \dfrac{-x^3 + 3x^2 + x - 1}{x - 3}$

72. $f(x) = \dfrac{x^5 - x^3 + 3}{x^2 - 1}$

73. $f(x) = \sin x + \dfrac{1}{x - 1}$

74. Writing In some models for learning a skill (e.g., juggling), it is assumed that the skill level for an individual increases with practice but cannot become arbitrarily high. How do concepts of this section apply to such a model?

75. Writing In some population models it is assumed that a given ecological system possesses a *carrying capacity* L. Populations greater than the carrying capacity tend to decline toward L, while populations less than the carrying capacity tend to increase toward L. Explain why these assumptions are reasonable, and discuss how the concepts of this section apply to such a model.

QUICK CHECK ANSWERS 1.3

1. (a) $+\infty$ (b) 5 (c) $-\infty$ (d) 0   **2.** (a) $\frac{1}{2}$ (b) does not exist (c) e **3.** (a) 9 (b) $-\frac{2}{3}$ (c) does not exist (d) 4
4. $1/x$, e^x, and e^{-x} each has a horizontal asymptote.

1.4   LIMITS (DISCUSSED MORE RIGOROUSLY)

In the previous sections of this chapter we focused on the discovery of values of limits, either by sampling selected x-values or by applying limit theorems that were stated without proof. Our main goal in this section is to define the notion of a limit precisely, thereby making it possible to establish limits with certainty and to prove theorems about them. This will also provide us with a deeper understanding of some of the more subtle properties of functions.

■ **MOTIVATION FOR THE DEFINITION OF A TWO-SIDED LIMIT**
The statement $\lim_{x \to a} f(x) = L$ can be interpreted informally to mean that we can make the value of $f(x)$ as close as we like to the real number L by making the value of x sufficiently close to a. It is our goal to make the informal phrases "as close as we like to L" and "sufficiently close to a" mathematically precise.

To do this, consider the function f graphed in Figure 1.4.1a for which $f(x) \to L$ as $x \to a$. For visual simplicity we have drawn the graph of f to be increasing on an open interval containing a, and we have intentionally placed a hole in the graph at $x = a$ to emphasize that f need not be defined at $x = a$ to have a limit there.

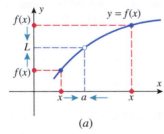
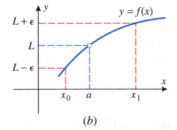
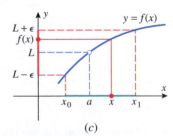

(a)   (b)   (c)

▲ **Figure 1.4.1**

Next, let us choose any positive number ϵ and ask how close x must be to a in order for the values of $f(x)$ to be within ϵ units of L. We can answer this geometrically by drawing horizontal lines from the points $L + \epsilon$ and $L - \epsilon$ on the y-axis until they meet the curve $y = f(x)$, and then drawing vertical lines from those points on the curve to the x-axis (Figure 1.4.1b). As indicated in the figure, let x_0 and x_1 be the points where those vertical lines intersect the x-axis.

Now imagine that x gets closer and closer to a (from either side). Eventually, x will lie inside the interval (x_0, x_1), which is marked in green in Figure 1.4.1c; and when this happens, the value of $f(x)$ will fall between $L - \epsilon$ and $L + \epsilon$, marked in red in the figure. Thus, we conclude:

If $f(x) \to L$ as $x \to a$, then for any positive number ϵ, we can find an open interval (x_0, x_1) on the x-axis that contains a and has the property that for each x in that interval (except possibly for $x = a$), the value of $f(x)$ is between $L - \epsilon$ and $L + \epsilon$.

What is important about this result is that it holds no matter how small we make ϵ. However, making ϵ smaller and smaller forces $f(x)$ *closer and closer* to L—which is precisely the concept we were trying to capture mathematically.

Observe that in Figure 1.4.1 the interval (x_0, x_1) extends farther on the right side of a than on the left side. However, for many purposes it is preferable to have an interval that extends the same distance on both sides of a. For this purpose, let us choose any positive number δ that is smaller than both $x_1 - a$ and $a - x_0$, and consider the interval

$$(a - \delta, a + \delta)$$

This interval extends the same distance δ on both sides of a and lies inside of the interval (x_0, x_1) (Figure 1.4.2). Moreover, the condition

$$L - \epsilon < f(x) < L + \epsilon \tag{1}$$

holds for every x in this interval (except possibly $x = a$), since this condition holds on the larger interval (x_0, x_1).

Since (1) can be expressed as

$$|f(x) - L| < \epsilon$$

and the condition that x lies in the interval $(a - \delta, a + \delta)$, but $x \neq a$, can be expressed as

$$0 < |x - a| < \delta$$

we are led to the following precise definition of a two-sided limit.

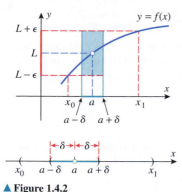

▲ **Figure 1.4.2**

The definitions of one-sided limits require minor adjustments to Definition 1.4.1. For example, for a limit from the right we need only assume that $f(x)$ is defined on an interval (a, b) extending to the right of a and that the ϵ condition is met for x in an interval $a < x < a + \delta$ extending to the right of a. A similar adjustment must be made for a limit from the left. (See Exercise 27.)

1.4.1 LIMIT DEFINITION   Let $f(x)$ be defined for all x in some open interval containing the number a, with the possible exception that $f(x)$ need not be defined at a. We will write

$$\lim_{x \to a} f(x) = L$$

if given any number $\epsilon > 0$ we can find a number $\delta > 0$ such that

$$|f(x) - L| < \epsilon \quad \text{if} \quad 0 < |x - a| < \delta$$

This definition, which is attributed to the German mathematician Karl Weierstrass and is commonly called the "epsilon-delta" definition of a two-sided limit, makes the transition from an informal concept of a limit to a precise definition. Specifically, the informal phrase "as close as we like to L" is given quantitative meaning by our ability to choose the positive number ϵ arbitrarily, and the phrase "sufficiently close to a" is quantified by the positive number δ.

In the preceding sections we illustrated various numerical and graphical methods for *guessing* at limits. Now that we have a precise definition to work with, we can actually confirm the validity of those guesses with mathematical proof. Here is a typical example of such a proof.

▶ **Example 1**   Use Definition 1.4.1 to prove that $\lim\limits_{x \to 2} (3x - 5) = 1$.

Solution.   We must show that given any positive number ϵ, we can find a positive number δ such that

$$| \underbrace{(3x - 5)}_{f(x)} - \underbrace{1}_{L} | < \epsilon \quad \text{if} \quad 0 < | x - \underbrace{2}_{a} | < \delta \tag{2}$$

There are two things to do. First, we must *discover* a value of δ for which this statement holds, and then we must *prove* that the statement holds for that δ. For the discovery part we begin by simplifying (2) and writing it as

$$|3x - 6| < \epsilon \quad \text{if} \quad 0 < |x - 2| < \delta$$

Next we will rewrite this statement in a form that will facilitate the discovery of an appropriate δ:

$$3|x - 2| < \epsilon \quad \text{if} \quad 0 < |x - 2| < \delta$$
$$|x - 2| < \epsilon/3 \quad \text{if} \quad 0 < |x - 2| < \delta \tag{3}$$

It should be self-evident that this last statement holds if $\delta = \epsilon/3$, which completes the discovery portion of our work. Now we need to prove that (2) holds for this choice of δ. However, statement (2) is equivalent to (3), and (3) holds with $\delta = \epsilon/3$, so (2) also holds with $\delta = \epsilon/3$. This proves that $\lim\limits_{x \to 2} (3x - 5) = 1$. ◀

This example illustrates the general form of a limit proof: We *assume* that we are given a positive number ϵ, and we try to *prove* that we can find a positive number δ such that

$$|f(x) - L| < \epsilon \quad \text{if} \quad 0 < |x - a| < \delta \tag{4}$$

This is done by first discovering δ, and then proving that the discovered δ works. Since the argument has to be general enough to work for all positive values of ϵ, the quantity δ has to be expressed as a function of ϵ. In Example 1 we found the function $\delta = \epsilon/3$ by some simple algebra; however, most limit proofs require a little more algebraic and logical ingenuity. Thus, if you find our ensuing discussion of "ϵ-δ" proofs challenging, do not become discouraged; the concepts and techniques are intrinsically difficult. In fact, a precise understanding of limits evaded the finest mathematical minds for more than 150 years after the basic concepts of calculus were discovered.

Karl Weierstrass (1815–1897) Weierstrass, the son of a customs officer, was born in Ostenfelde, Germany. As a youth Weierstrass showed outstanding skills in languages and mathematics. However, at the urging of his dominant father, Weierstrass entered the law and commerce program at the University of Bonn. To the chagrin of his family, the rugged and congenial young man concentrated instead on fencing and beer drinking. Four years later he returned home without a degree. In 1839 Weierstrass entered the Academy of Münster to study for a career in secondary education, and he met and studied under an excellent mathematician named Christof Gudermann. Gudermann's ideas greatly influenced the work of Weierstrass. After receiving his teaching certificate, Weierstrass spent the next 15 years in secondary education teaching German, geography, and mathematics. In addition, he taught handwriting to small children. During this period much of Weierstrass's mathematical work was ignored because he was a secondary schoolteacher and not a college professor. Then, in 1854, he published a paper of major importance that created a sensation in the mathematics world and catapulted him to international fame overnight. He was immediately given an honorary Doctorate at the University of Königsberg and began a new career in college teaching at the University of Berlin in 1856. In 1859 the strain of his mathematical research caused a temporary nervous breakdown and led to spells of dizziness that plagued him for the rest of his life. Weierstrass was a brilliant teacher and his classes overflowed with multitudes of auditors. In spite of his fame, he never lost his early beer-drinking congeniality and was always in the company of students, both ordinary and brilliant. Weierstrass was acknowledged as the leading mathematical analyst in the world. He and his students opened the door to the modern school of mathematical analysis.

▶ **Example 2**   Prove that $\lim\limits_{x \to 0^+} \sqrt{x} = 0$.

Solution.   Note that the domain of $\sqrt{x}$ is $0 \le x$, so it is valid to discuss the limit as $x \to 0^+$. We must show that given $\epsilon > 0$, there exists a $\delta > 0$ such that

$$|\sqrt{x} - 0| < \epsilon \quad \text{if} \quad 0 < x - 0 < \delta$$

or more simply,

$$\sqrt{x} < \epsilon \quad \text{if} \quad 0 < x < \delta \tag{5}$$

But, by squaring both sides of the inequality $\sqrt{x} < \epsilon$, we can rewrite (5) as

$$x < \epsilon^2 \quad \text{if} \quad 0 < x < \delta \tag{6}$$

It should be self-evident that (6) is true if $\delta = \epsilon^2$; and since (6) is a reformulation of (5), we have shown that (5) holds with $\delta = \epsilon^2$. This proves that $\lim\limits_{x \to 0^+} \sqrt{x} = 0$. ◀

> In Example 2 the limit from the left and the two-sided limit do not exist at $x = 0$ because $\sqrt{x}$ is defined only for nonnegative values of x.

▪ THE VALUE OF δ IS NOT UNIQUE

In preparation for our next example, we note that the value of δ in Definition 1.4.1 is not unique; once we have found a value of δ that fulfills the requirements of the definition, then any *smaller* positive number δ_1 will also fulfill those requirements. That is, if it is true that

$$|f(x) - L| < \epsilon \quad \text{if} \quad 0 < |x - a| < \delta$$

then it will also be true that

$$|f(x) - L| < \epsilon \quad \text{if} \quad 0 < |x - a| < \delta_1$$

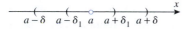

▲ **Figure 1.4.3**

This is because $\{x : 0 < |x - a| < \delta_1\}$ is a subset of $\{x : 0 < |x - a| < \delta\}$ (Figure 1.4.3), and hence if $|f(x) - L| < \epsilon$ is satisfied for all x in the larger set, then it will automatically be satisfied for all x in the subset. Thus, in Example 1, where we used $\delta = \epsilon/3$, we could have used any smaller value of δ such as $\delta = \epsilon/4$, $\delta = \epsilon/5$, or $\delta = \epsilon/6$.

▶ **Example 3**   Prove that $\lim\limits_{x \to 3} x^2 = 9$.

Solution.   We must show that given any positive number ϵ, we can find a positive number δ such that

$$|x^2 - 9| < \epsilon \quad \text{if} \quad 0 < |x - 3| < \delta \tag{7}$$

Because $|x - 3|$ occurs on the right side of this "if statement," it will be helpful to factor the left side to introduce a factor of $|x - 3|$. This yields the following alternative form of (7):

$$|x + 3||x - 3| < \epsilon \quad \text{if} \quad 0 < |x - 3| < \delta \tag{8}$$

> If you are wondering how we knew to make the restriction $\delta \le 1$, as opposed to $\delta \le 5$ or $\delta \le \frac{1}{2}$, for example, the answer is that 1 is merely a convenient choice—any restriction of the form $\delta \le c$ would work equally well.

We wish to bound the factor $|x + 3|$. If we knew, for example, that $\delta \le 1$, then we would have $-1 < x - 3 < 1$, so $5 < x + 3 < 7$, and consequently $|x + 3| < 7$. Thus, if $\delta \le 1$ and $0 < |x - 3| < \delta$, then

$$|x + 3||x - 3| < 7\delta$$

It follows that (8) will be satisfied for any positive δ such that $\delta \le 1$ and $7\delta < \epsilon$. We can achieve this by taking δ to be the minimum of the numbers 1 and $\epsilon/7$, which is sometimes written as $\delta = \min(1, \epsilon/7)$. This proves that $\lim\limits_{x \to 3} x^2 = 9$. ◀

▪ LIMITS AS $x \to \pm\infty$

In Section 1.3 we discussed the limits

$$\lim_{x \to +\infty} f(x) = L \quad \text{and} \quad \lim_{x \to -\infty} f(x) = L$$

from an intuitive point of view. The first limit can be interpreted to mean that we can make the value of $f(x)$ as close as we like to L by taking x sufficiently large, and the second can be interpreted to mean that we can make the value of $f(x)$ as close as we like to L by taking x sufficiently far to the left of 0. These ideas are captured in the following definitions and are illustrated in Figure 1.4.4.

1.4.2   DEFINITION   Let $f(x)$ be defined for all x in some infinite open interval extending in the positive x-direction. We will write

$$\lim_{x \to +\infty} f(x) = L$$

if given any number $\epsilon > 0$, there corresponds a positive number N such that

$$|f(x) - L| < \epsilon \quad \text{if} \quad x > N$$

1.4.3   DEFINITION   Let $f(x)$ be defined for all x in some infinite open interval extending in the negative x-direction. We will write

$$\lim_{x \to -\infty} f(x) = L$$

if given any number $\epsilon > 0$, there corresponds a negative number N such that

$$|f(x) - L| < \epsilon \quad \text{if} \quad x < N$$

To see how these definitions relate to our informal concepts of these limits, suppose that $f(x) \to L$ as $x \to +\infty$, and for a given ϵ let N be the positive number described in Definition 1.4.2. If x is allowed to increase indefinitely, then eventually x will lie in the interval $(N, +\infty)$, which is marked in green in Figure 1.4.4a; when this happens, the value of $f(x)$ will fall between $L - \epsilon$ and $L + \epsilon$, marked in red in the figure. Since this is true for all positive values of ϵ (no matter how small), we can force the values of $f(x)$ as close as we like to L by making N sufficiently large. This agrees with our informal concept of this limit. Similarly, Figure 1.4.4b illustrates Definition 1.4.3.

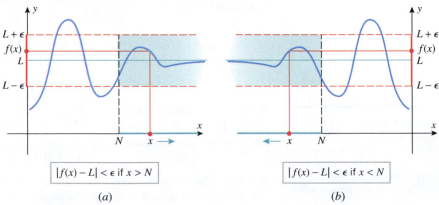

$$|f(x) - L| < \epsilon \text{ if } x > N \qquad\qquad |f(x) - L| < \epsilon \text{ if } x < N$$

$$(a) \qquad\qquad\qquad\qquad (b)$$

▲ **Figure 1.4.4**

▶ **Example 4**   Prove that $\displaystyle\lim_{x \to +\infty} \frac{1}{x} = 0$.

Solution.   Applying Definition 1.4.2 with $f(x) = 1/x$ and $L = 0$, we must show that given $\epsilon > 0$, we can find a number $N > 0$ such that

$$\left| \frac{1}{x} - 0 \right| < \epsilon \quad \text{if} \quad x > N \tag{9}$$

Because $x \to +\infty$ we can assume that $x > 0$. Thus, we can eliminate the absolute values in this statement and rewrite it as

$$\frac{1}{x} < \epsilon \quad \text{if} \quad x > N$$

or, on taking reciprocals,

$$x > \frac{1}{\epsilon} \quad \text{if} \quad x > N \tag{10}$$

It is self-evident that $N = 1/\epsilon$ satisfies this requirement, and since (10) and (9) are equivalent for $x > 0$, the proof is complete. ◀

■ INFINITE LIMITS

In Section 1.1 we discussed limits of the following type from an intuitive viewpoint:

$$\lim_{x \to a} f(x) = +\infty, \qquad \lim_{x \to a} f(x) = -\infty \tag{11}$$

$$\lim_{x \to a^+} f(x) = +\infty, \qquad \lim_{x \to a^+} f(x) = -\infty \tag{12}$$

$$\lim_{x \to a^-} f(x) = +\infty, \qquad \lim_{x \to a^-} f(x) = -\infty \tag{13}$$

Recall that each of these expressions describes a particular way in which the limit fails to exist. The $+\infty$ indicates that the limit fails to exist because $f(x)$ increases without bound, and the $-\infty$ indicates that the limit fails to exist because $f(x)$ decreases without bound. These ideas are captured more precisely in the following definitions and are illustrated in Figure 1.4.5.

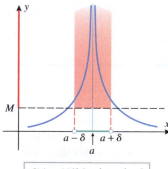

$$f(x) > M \text{ if } 0 < |x - a| < \delta$$

(a)

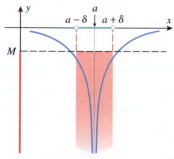

$$f(x) < M \text{ if } 0 < |x - a| < \delta$$

(b)

▲ **Figure 1.4.5**

How would you define these limits?

$$\lim_{x \to a^+} f(x) = +\infty \quad \lim_{x \to a^+} f(x) = -\infty$$
$$\lim_{x \to a^-} f(x) = +\infty \quad \lim_{x \to a^-} f(x) = -\infty$$
$$\lim_{x \to +\infty} f(x) = +\infty \quad \lim_{x \to +\infty} f(x) = -\infty$$
$$\lim_{x \to -\infty} f(x) = +\infty \quad \lim_{x \to -\infty} f(x) = -\infty$$

> **1.4.4   DEFINITION**   Let $f(x)$ be defined for all x in some open interval containing a, except that $f(x)$ need not be defined at a. We will write
>
> $$\lim_{x \to a} f(x) = +\infty$$
>
> if given any positive number M, we can find a number $\delta > 0$ such that $f(x)$ satisfies
>
> $$f(x) > M \quad \text{if} \quad 0 < |x - a| < \delta$$

> **1.4.5   DEFINITION**   Let $f(x)$ be defined for all x in some open interval containing a, except that $f(x)$ need not be defined at a. We will write
>
> $$\lim_{x \to a} f(x) = -\infty$$
>
> if given any negative number M, we can find a number $\delta > 0$ such that $f(x)$ satisfies
>
> $$f(x) < M \quad \text{if} \quad 0 < |x - a| < \delta$$

To see how these definitions relate to our informal concepts of these limits, suppose that $f(x) \to +\infty$ as $x \to a$, and for a given M let δ be the corresponding positive number described in Definition 1.4.4. Next, imagine that x gets closer and closer to a (from either side). Eventually, x will lie in the interval $(a - \delta, a + \delta)$, which is marked in green in Figure 1.4.5a; when this happens the value of $f(x)$ will be greater than M, marked in red in the figure. Since this is true for any positive value of M (no matter how large), we can force the values of $f(x)$ to be as large as we like by making x sufficiently close to a. This agrees with our informal concept of this limit. Similarly, Figure 1.4.5b illustrates Definition 1.4.5.

▶ **Example 5** Prove that $\lim\limits_{x \to 0} \dfrac{1}{x^2} = +\infty$.

Solution. Applying Definition 1.4.4 with $f(x) = 1/x^2$ and $a = 0$, we must show that given a number $M > 0$, we can find a number $\delta > 0$ such that

$$\frac{1}{x^2} > M \quad \text{if} \quad 0 < |x - 0| < \delta \tag{14}$$

or, on taking reciprocals and simplifying,

$$x^2 < \frac{1}{M} \quad \text{if} \quad 0 < |x| < \delta \tag{15}$$

But $x^2 < 1/M$ if $|x| < 1/\sqrt{M}$, so that $\delta = 1/\sqrt{M}$ satisfies (15). Since (14) is equivalent to (15), the proof is complete. ◀

✔ **QUICK CHECK EXERCISES 1.4**   *(See page 109 for answers.)*

1. The definition of a two-sided limit states: $\lim_{x \to a} f(x) = L$ if given any number _____ there is a number _____ such that $|f(x) - L| < \epsilon$ if _____.

2. Suppose that $f(x)$ is a function such that for any given $\epsilon > 0$, the condition $0 < |x - 1| < \epsilon/2$ guarantees that $|f(x) - 5| < \epsilon$. What limit results from this property?

3. Suppose that ϵ is any positive number. Find the largest value of δ such that $|5x - 10| < \epsilon$ if $0 < |x - 2| < \delta$.

4. The definition of limit at $+\infty$ states: $\lim_{x \to +\infty} f(x) = L$ if given any number _____ there is a positive number _____ such that $|f(x) - L| < \epsilon$ if _____.

5. Find the smallest positive number N such that for each $x > N$, the value of $f(x) = 1/\sqrt{x}$ is within 0.01 of 0.

EXERCISE SET 1.4   📈 Graphing Utility

1. (a) Find the largest open interval, centered at the origin on the x-axis, such that for each x in the interval the value of the function $f(x) = x + 2$ is within 0.1 unit of the number $f(0) = 2$.
   (b) Find the largest open interval, centered at $x = 3$, such that for each x in the interval the value of the function $f(x) = 4x - 5$ is within 0.01 unit of the number $f(3) = 7$.
   (c) Find the largest open interval, centered at $x = 4$, such that for each x in the interval the value of the function $f(x) = x^2$ is within 0.001 unit of the number $f(4) = 16$.

2. In each part, find the largest open interval, centered at $x = 0$, such that for each x in the interval the value of $f(x) = 2x + 3$ is within ϵ units of the number $f(0) = 3$.
   (a) $\epsilon = 0.1$ (b) $\epsilon = 0.01$
   (c) $\epsilon = 0.0012$

3. (a) Find the values of x_0 and x_1 in the accompanying figure.
   (b) Find a positive number δ such that $|\sqrt{x} - 2| < 0.05$ if $0 < |x - 4| < \delta$.

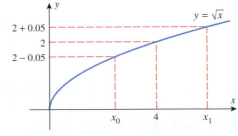

Not drawn to scale

▲ **Figure Ex-3**

4. (a) Find the values of x_0 and x_1 in the accompanying figure on the next page.
   (b) Find a positive number δ such that $|(1/x) - 1| < 0.1$ if $0 < |x - 1| < \delta$.

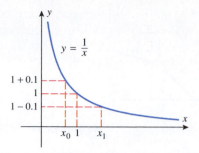

$y = \frac{1}{x}$

Not drawn to scale ◀ **Figure Ex-4**

5. Generate the graph of $f(x) = x^3 - 4x + 5$ with a graphing utility, and use the graph to find a number δ such that $|f(x) - 2| < 0.05$ if $0 < |x - 1| < \delta$. [*Hint:* Show that the inequality $|f(x) - 2| < 0.05$ can be rewritten as $1.95 < x^3 - 4x + 5 < 2.05$, and estimate the values of x for which $x^3 - 4x + 5 = 1.95$ and $x^3 - 4x + 5 = 2.05$.]

6. Use the method of Exercise 5 to find a number δ such that $|\sqrt{5x + 1} - 4| < 0.5$ if $0 < |x - 3| < \delta$.

7. Let $f(x) = x + \sqrt{x}$ with $L = \lim_{x \to 1} f(x)$ and let $\epsilon = 0.2$. Use a graphing utility and its trace feature to find a positive number δ such that $|f(x) - L| < \epsilon$ if $0 < |x - 1| < \delta$.

8. Let $f(x) = (\sin 2x)/x$ and use a graphing utility to conjecture the value of $L = \lim_{x \to 0} f(x)$. Then let $\epsilon = 0.1$ and use the graphing utility and its trace feature to find a positive number δ such that $|f(x) - L| < \epsilon$ if $0 < |x| < \delta$.

FOCUS ON CONCEPTS

9. What is wrong with the following "proof" that $\lim_{x \to 3} 2x = 6$? Suppose that $\epsilon = 1$ and $\delta = \frac{1}{2}$. Then if $|x - 3| < \frac{1}{2}$, we have

$$|2x - 6| = 2|x - 3| < 2\left(\tfrac{1}{2}\right) = 1 = \epsilon$$

Therefore, $\lim_{x \to 3} 2x = 6$.

10. What is wrong with the following "proof" that $\lim_{x \to 3} 2x = 6$? Given any $\delta > 0$, choose $\epsilon = 2\delta$. Then if $|x - 3| < \delta$, we have

$$|2x - 6| = 2|x - 3| < 2\delta = \epsilon$$

Therefore, $\lim_{x \to 3} 2x = 6$.

11. Recall from Example 1 that the creation of a limit proof involves two stages. The first is a *discovery* stage in which δ is found, and the second is the *proof* stage in which the discovered δ is shown to work. Fill in the blanks to give an explicit proof that the choice of $\delta = \epsilon/3$ in Example 1 works. Suppose that $\epsilon > 0$. Set $\delta = \epsilon/3$ and assume that $0 < |x - 2| < \delta$. Then

$$|(3x - 5) - 1| = | \underline{\hspace{2cm}} |$$
$$= 3 \cdot | \underline{\hspace{1.5cm}} | < 3 \cdot \underline{\hspace{1.5cm}} = \epsilon$$

12. Suppose that $f(x) = c$ is a constant function and that a is some fixed real number. Explain why *any* choice of $\delta > 0$ (e.g., $\delta = 1$) works to prove $\lim_{x \to a} f(x) = c$.

13–22 Use Definition 1.4.1 to prove that the limit is correct. ■

13. $\lim_{x \to 2} 3 = 3$

14. $\lim_{x \to 4} (x + 2) = 6$

15. $\lim_{x \to 5} 3x = 15$

16. $\lim_{x \to -1} (7x + 5) = -2$

17. $\lim_{x \to 0} \frac{2x^2 + x}{x} = 1$

18. $\lim_{x \to -3} \frac{x^2 - 9}{x + 3} = -6$

19. $\lim_{x \to 1} f(x) = 3$, where $f(x) = \begin{cases} x + 2, & x \neq 1 \\ 10, & x = 1 \end{cases}$

20. $\lim_{x \to 2} f(x) = 5$, where $f(x) = \begin{cases} 9 - 2x, & x \neq 2 \\ 49, & x = 2 \end{cases}$

21. $\lim_{x \to 0} |x| = 0$

22. $\lim_{x \to 2} f(x) = 5$, where $f(x) = \begin{cases} 9 - 2x, & x < 2 \\ 3x - 1, & x > 2 \end{cases}$

23–26 True–False Determine whether the statement is true or false. Explain your answer. ■

23. Suppose that $f(x) = mx + b$, $m \neq 0$. To prove that $\lim_{x \to a} f(x) = f(a)$, we can take $\delta = \epsilon/|m|$.

24. Suppose that $f(x) = mx + b$, $m \neq 0$. To prove that $\lim_{x \to a} f(x) = f(a)$, we can take $\delta = \epsilon/(2|m|)$.

25. For certain functions, the *same* δ will work for *all* $\epsilon > 0$ in a limit proof.

26. Suppose that $f(x) > 0$ for all x in the interval $(-1, 1)$. If $\lim_{x \to 0} f(x) = L$, then $L > 0$.

FOCUS ON CONCEPTS

27. Give rigorous definitions of $\lim_{x \to a^+} f(x) = L$ and $\lim_{x \to a^-} f(x) = L$.

28. Consider the statement that $\lim_{x \to a} |f(x) - L| = 0$.
    (a) Using Definition 1.4.1, write down precisely what this limit statement means.
    (b) Explain why your answer to part (a) shows that
    $$\lim_{x \to a} |f(x) - L| = 0 \quad \text{if and only if} \quad \lim_{x \to a} f(x) = L$$

29. (a) Show that
    $$|(3x^2 + 2x - 20) - 300| = |3x + 32| \cdot |x - 10|$$
    (b) Find an upper bound for $|3x + 32|$ if x satisfies $|x - 10| < 1$.
    (c) Fill in the blanks to complete a proof that
    $$\lim_{x \to 10} [3x^2 + 2x - 20] = 300$$
    Suppose that $\epsilon > 0$. Set $\delta = \min(1, \underline{\hspace{1cm}})$ and assume that $0 < |x - 10| < \delta$. Then
    $$|(3x^2 + 2x - 20) - 300| = |3x + 32| \cdot |x - 10|$$
    $$< \underline{\hspace{1cm}} \cdot |x - 10|$$
    $$< \underline{\hspace{1cm}} \cdot \underline{\hspace{1cm}}$$
    $$= \epsilon$$

30. (a) Show that

$$\left|\frac{28}{3x+1} - 4\right| = \left|\frac{12}{3x+1}\right| \cdot |x-2|$$

(b) Is $\left|12/(3x+1)\right|$ bounded if $|x-2| < 4$? If not, explain; if so, give a bound.

(c) Is $\left|12/(3x+1)\right|$ bounded if $|x-2| < 1$? If not, explain; if so, give a bound.

(d) Fill in the blanks to complete a proof that

$$\lim_{x\to 2}\left[\frac{28}{3x+1}\right] = 4$$

Suppose that $\epsilon > 0$. Set $\delta = \min(1, \underline{\hspace{1cm}})$ and assume that $0 < |x-2| < \delta$. Then

$$\left|\frac{28}{3x+1} - 4\right| = \left|\frac{12}{3x+1}\right| \cdot |x-2|$$
$$< \underline{\hspace{1cm}} \cdot |x-2|$$
$$< \underline{\hspace{1cm}} \cdot \underline{\hspace{1cm}}$$
$$= \epsilon$$

31–36 Use Definition 1.4.1 to prove that the stated limit is correct. In each case, to show that $\lim_{x\to a} f(x) = L$, factor $|f(x) - L|$ in the form

$$|f(x) - L| = |\text{"something"}| \cdot |x - a|$$

and then bound the size of $|\text{"something"}|$ by putting restrictions on the size of δ. ■

31. $\lim_{x\to 1} 2x^2 = 2$ [*Hint:* Assume $\delta \le 1$.]

32. $\lim_{x\to 3}(x^2 + x) = 12$ [*Hint:* Assume $\delta \le 1$.]

33. $\lim_{x\to -2}\dfrac{1}{x+1} = -1$ **34.** $\lim_{x\to 1/2}\dfrac{2x+3}{x} = 8$

35. $\lim_{x\to 4}\sqrt{x} = 2$ **36.** $\lim_{x\to 2} x^3 = 8$

37. Let

$$f(x) = \begin{cases} 0, & \text{if } x \text{ is rational} \\ x, & \text{if } x \text{ is irrational} \end{cases}$$

Use Definition 1.4.1 to prove that $\lim_{x\to 0} f(x) = 0$.

38. Let

$$f(x) = \begin{cases} 0, & \text{if } x \text{ is rational} \\ 1, & \text{if } x \text{ is irrational} \end{cases}$$

Use Definition 1.4.1 to prove that $\lim_{x\to 0} f(x)$ does not exist. [*Hint:* Assume $\lim_{x\to 0} f(x) = L$ and apply Definition 1.4.1 with $\epsilon = \frac{1}{2}$ to conclude that $|1 - L| < \frac{1}{2}$ and $|L| = |0 - L| < \frac{1}{2}$. Then show $1 \le |1 - L| + |L|$ and derive a contradiction.]

39. (a) Find the values of x_1 and x_2 in the accompanying figure.

(b) Find a positive number N such that

$$\left|\frac{x^2}{1+x^2} - 1\right| < \epsilon$$

for $x > N$.

(c) Find a negative number N such that

$$\left|\frac{x^2}{1+x^2} - 1\right| < \epsilon$$

for $x < N$.

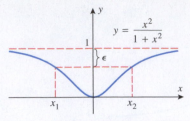

Not drawn to scale   ◀ **Figure Ex-39**

40. (a) Find the values of x_1 and x_2 in the accompanying figure.

(b) Find a positive number N such that

$$\left|\frac{1}{\sqrt[3]{x}} - 0\right| = \left|\frac{1}{\sqrt[3]{x}}\right| < \epsilon$$

for $x > N$.

(c) Find a negative number N such that

$$\left|\frac{1}{\sqrt[3]{x}} - 0\right| = \left|\frac{1}{\sqrt[3]{x}}\right| < \epsilon$$

for $x < N$.

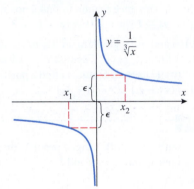

◀ **Figure Ex-40**

41–44 A positive number ϵ and the limit L of a function f at $+\infty$ are given. Find a positive number N such that $|f(x) - L| < \epsilon$ if $x > N$. ■

41. $\lim_{x\to +\infty}\dfrac{1}{x^2} = 0$; $\epsilon = 0.01$

42. $\lim_{x\to +\infty}\dfrac{1}{x+2} = 0$; $\epsilon = 0.005$

43. $\lim_{x\to +\infty}\dfrac{x}{x+1} = 1$; $\epsilon = 0.001$

44. $\lim_{x\to +\infty}\dfrac{4x-1}{2x+5} = 2$; $\epsilon = 0.1$

45–48 A positive number ϵ and the limit L of a function f at $-\infty$ are given. Find a negative number N such that $|f(x) - L| < \epsilon$ if $x < N$. ■

45. $\lim_{x\to -\infty}\dfrac{1}{x+2} = 0$; $\epsilon = 0.005$

46. $\lim_{x\to -\infty}\dfrac{1}{x^2} = 0$; $\epsilon = 0.01$

47. $\lim_{x\to -\infty}\dfrac{4x-1}{2x+5} = 2$; $\epsilon = 0.1$

48. $\lim\limits_{x \to -\infty} \dfrac{x}{x+1} = 1$; $\epsilon = 0.001$

68. $\lim\limits_{x \to 2^-} f(x) = 6$, where $f(x) = \begin{cases} x, & x > 2 \\ 3x, & x \le 2 \end{cases}$

49–54 Use Definition 1.4.2 or 1.4.3 to prove that the stated limit is correct. ■

49. $\lim\limits_{x \to +\infty} \dfrac{1}{x^2} = 0$

50. $\lim\limits_{x \to +\infty} \dfrac{1}{x+2} = 0$

51. $\lim\limits_{x \to -\infty} \dfrac{4x-1}{2x+5} = 2$

52. $\lim\limits_{x \to -\infty} \dfrac{x}{x+1} = 1$

53. $\lim\limits_{x \to +\infty} \dfrac{2\sqrt{x}}{\sqrt{x}-1} = 2$

54. $\lim\limits_{x \to -\infty} 2^x = 0$

55. (a) Find the largest open interval, centered at the origin on the x-axis, such that for each x in the interval, other than the center, the values of $f(x) = 1/x^2$ are greater than 100.

(b) Find the largest open interval, centered at $x = 1$, such that for each x in the interval, other than the center, the values of the function $f(x) = 1/|x-1|$ are greater than 1000.

(c) Find the largest open interval, centered at $x = 3$, such that for each x in the interval, other than the center, the values of the function $f(x) = -1/(x-3)^2$ are less than -1000.

(d) Find the largest open interval, centered at the origin on the x-axis, such that for each x in the interval, other than the center, the values of $f(x) = -1/x^4$ are less than $-10,000$.

56. In each part, find the largest open interval centered at $x = 1$, such that for each x in the interval, other than the center, the value of $f(x) = 1/(x-1)^2$ is greater than M.

(a) $M = 10$ (b) $M = 1000$ (c) $M = 100,000$

57–62 Use Definition 1.4.4 or 1.4.5 to prove that the stated limit is correct. ■

57. $\lim\limits_{x \to 3} \dfrac{1}{(x-3)^2} = +\infty$

58. $\lim\limits_{x \to 3} \dfrac{-1}{(x-3)^2} = -\infty$

59. $\lim\limits_{x \to 0} \dfrac{1}{|x|} = +\infty$

60. $\lim\limits_{x \to 1} \dfrac{1}{|x-1|} = +\infty$

61. $\lim\limits_{x \to 0} \left(-\dfrac{1}{x^4}\right) = -\infty$

62. $\lim\limits_{x \to 0} \dfrac{1}{x^4} = +\infty$

63–68 Use the definitions in Exercise 27 to prove that the stated one-sided limit is correct. ■

63. $\lim\limits_{x \to 2^+} (x+1) = 3$

64. $\lim\limits_{x \to 1^-} (3x+2) = 5$

65. $\lim\limits_{x \to 4^+} \sqrt{x-4} = 0$

66. $\lim\limits_{x \to 0^-} \sqrt{-x} = 0$

67. $\lim\limits_{x \to 2^+} f(x) = 2$, where $f(x) = \begin{cases} x, & x > 2 \\ 3x, & x \le 2 \end{cases}$

69–72 Write out the definition for the corresponding limit in the marginal note on page 105, and use your definition to prove that the stated limit is correct. ■

69. (a) $\lim\limits_{x \to 1^+} \dfrac{1}{1-x} = -\infty$ (b) $\lim\limits_{x \to 1^-} \dfrac{1}{1-x} = +\infty$

70. (a) $\lim\limits_{x \to 0^+} \dfrac{1}{x} = +\infty$ (b) $\lim\limits_{x \to 0^-} \dfrac{1}{x} = -\infty$

71. (a) $\lim\limits_{x \to +\infty} (x+1) = +\infty$ (b) $\lim\limits_{x \to -\infty} (x+1) = -\infty$

72. (a) $\lim\limits_{x \to +\infty} (x^2-3) = +\infty$ (b) $\lim\limits_{x \to -\infty} (x^3+5) = -\infty$

73. According to Ohm's law, when a voltage of V volts is applied across a resistor with a resistance of R ohms, a current of $I = V/R$ amperes flows through the resistor.

(a) How much current flows if a voltage of 3.0 volts is applied across a resistance of 7.5 ohms?

(b) If the resistance varies by ± 0.1 ohm, and the voltage remains constant at 3.0 volts, what is the resulting range of values for the current?

(c) If temperature variations cause the resistance to vary by $\pm \delta$ from its value of 7.5 ohms, and the voltage remains constant at 3.0 volts, what is the resulting range of values for the current?

(d) If the current is not allowed to vary by more than $\epsilon = \pm 0.001$ ampere at a voltage of 3.0 volts, what variation of $\pm \delta$ from the value of 7.5 ohms is allowable?

(e) Certain alloys become **superconductors** as their temperature approaches absolute zero ($-273°C$), meaning that their resistance approaches zero. If the voltage remains constant, what happens to the current in a superconductor as $R \to 0^+$?

74. Writing Compare informal Definition 1.1.1 with Definition 1.4.1.

(a) What portions of Definition 1.4.1 correspond to the expression "values of $f(x)$ can be made as close as we like to L" in Definition 1.1.1? Explain.

(b) What portions of Definition 1.4.1 correspond to the expression "taking values of x sufficiently close to a (but not equal to a)" in Definition 1.1.1? Explain.

75. Writing Compare informal Definition 1.3.1 with Definition 1.4.2.

(a) What portions of Definition 1.4.2 correspond to the expression "values of $f(x)$ eventually get as close as we like to a number L" in Definition 1.3.1? Explain.

(b) What portions of Definition 1.4.2 correspond to the expression "as x increases without bound" in Definition 1.3.1? Explain.

✔ **QUICK CHECK ANSWERS 1.4**

1. $\epsilon > 0$; $\delta > 0$; $0 < |x-a| < \delta$ **2.** $\lim\limits_{x \to 1} f(x) = 5$ **3.** $\delta = \epsilon/5$ **4.** $\epsilon > 0$; N; $x > N$ **5.** $N = 10,000$

1.5 CONTINUITY

A thrown baseball cannot vanish at some point and reappear someplace else to continue its motion. Thus, we perceive the path of the ball as an unbroken curve. In this section, we translate "unbroken curve" into a precise mathematical formulation called continuity, and develop some fundamental properties of continuous curves.

Joseph Helfenberger/iStockphoto

A baseball moves along a "continuous" trajectory after leaving the pitcher's hand.

■ DEFINITION OF CONTINUITY

Intuitively, the graph of a function can be described as a "continuous curve" if it has no breaks or holes. To make this idea more precise we need to understand what properties of a function can cause breaks or holes. Referring to Figure 1.5.1, we see that the graph of a function has a break or hole if any of the following conditions occur:

- The function f is undefined at c (Figure 1.5.1a).
- The limit of $f(x)$ does not exist as x approaches c (Figures 1.5.1b, 1.5.1c).
- The value of the function and the value of the limit at c are different (Figure 1.5.1d).

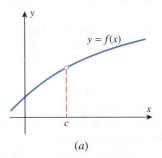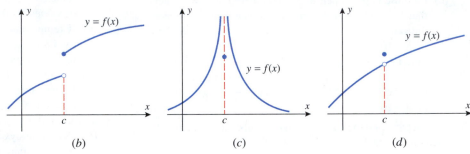

▲ **Figure 1.5.1**

This suggests the following definition.

The third condition in Definition 1.5.1 actually implies the first two, since it is tacitly understood in the statement

$$\lim_{x \to c} f(x) = f(c)$$

that the limit exists and the function is defined at c. Thus, when we want to establish continuity at c our usual procedure will be to verify the third condition only.

1.5.1 DEFINITION   A function f is said to be **_continuous at $x = c$_** provided the following conditions are satisfied:

1. $f(c)$ is defined.

2. $\lim\limits_{x \to c} f(x)$ exists.

3. $\lim\limits_{x \to c} f(x) = f(c)$.

If one or more of the conditions of this definition fails to hold, then we will say that f has a **_discontinuity at $x = c$_**. Each function drawn in Figure 1.5.1 illustrates a discontinuity at $x = c$. In Figure 1.5.1a, the function is not defined at c, violating the first condition of Definition 1.5.1. In Figure 1.5.1b, the one-sided limits of $f(x)$ as x approaches c both exist but are not equal. Thus, $\lim_{x \to c} f(x)$ does not exist, and this violates the second condition of Definition 1.5.1. We will say that a function like that in Figure 1.5.1b has a **_jump discontinuity_** at c. In Figure 1.5.1c, the one-sided limits of $f(x)$ as x approaches c are infinite. Thus, $\lim_{x \to c} f(x)$ does not exist, and this violates the second condition of Definition 1.5.1. We will say that a function like that in Figure 1.5.1c has an **_infinite discontinuity_** at c. In Figure 1.5.1d, the function is defined at c and $\lim_{x \to c} f(x)$ exists, but these two values are not equal, violating the third condition of Definition 1.5.1. We will say that a function like that in Figure 1.5.1d has a ***removable discontinuity*** at c. Exercises 33 and 34 help to explain why discontinuities of this type are given this name.

▶ **Example 1**   Determine whether the following functions are continuous at $x = 2$.

$$f(x) = \frac{x^2 - 4}{x - 2}, \qquad g(x) = \begin{cases} \dfrac{x^2 - 4}{x - 2}, & x \neq 2 \\ 3, & x = 2, \end{cases} \qquad h(x) = \begin{cases} \dfrac{x^2 - 4}{x - 2}, & x \neq 2 \\ 4, & x = 2 \end{cases}$$

Solution.   In each case we must determine whether the limit of the function as $x \to 2$ is the same as the value of the function at $x = 2$. In all three cases the functions are identical, except at $x = 2$, and hence all three have the same limit at $x = 2$, namely,

$$\lim_{x \to 2} f(x) = \lim_{x \to 2} g(x) = \lim_{x \to 2} h(x) = \lim_{x \to 2} \frac{x^2 - 4}{x - 2} = \lim_{x \to 2} (x + 2) = 4$$

The function f is undefined at $x = 2$, and hence is not continuous at $x = 2$ (Figure 1.5.2a). The function g is defined at $x = 2$, but its value there is $g(2) = 3$, which is not the same as the limit as x approaches 2; hence, g is also not continuous at $x = 2$ (Figure 1.5.2b). The value of the function h at $x = 2$ is $h(2) = 4$, which is the same as the limit as x approaches 2; hence, h is continuous at $x = 2$ (Figure 1.5.2c). (Note that the function h could have been written more simply as $h(x) = x + 2$, but we wrote it in piecewise form to emphasize its relationship to f and g.) ◀

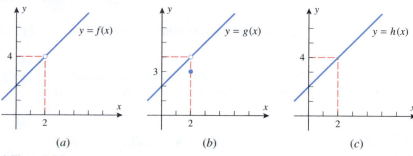

▲ **Figure 1.5.2**

Chris Hondros/Getty Images

A poor connection in a transmission cable can cause a discontinuity in the electrical signal it carries.

■ CONTINUITY IN APPLICATIONS

In applications, discontinuities often signal the occurrence of important physical events. For example, Figure 1.5.3a is a graph of voltage versus time for an underground cable that is accidentally cut by a work crew at time $t = t_0$ (the voltage drops to zero when the line is cut). Figure 1.5.3b shows the graph of inventory versus time for a company that restocks its warehouse to y_1 units when the inventory falls to y_0 units. The discontinuities occur at those times when restocking occurs.

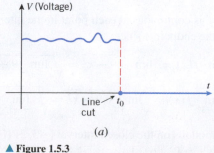

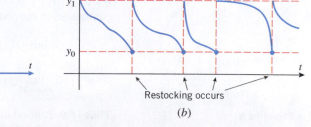

▲ **Figure 1.5.3**

■ CONTINUITY ON AN INTERVAL

If a function f is continuous at each number in an open interval (a, b), then we say that f is *continuous on* (a, b). This definition applies to infinite open intervals of the form $(a, +\infty)$, $(-\infty, b)$, and $(-\infty, +\infty)$. In the case where f is continuous on $(-\infty, +\infty)$, we will say that f is *continuous everywhere*.

Because Definition 1.5.1 involves a two-sided limit, that definition does not generally apply at the endpoints of a closed interval $[a, b]$ or at the endpoint of an interval of the form $[a, b)$, $(a, b]$, $(-\infty, b]$, or $[a, +\infty)$. To remedy this problem, we will agree that a function is continuous at an endpoint of an interval if its value at the endpoint is equal to the appropriate one-sided limit at that endpoint. For example, the function graphed in Figure 1.5.4 is continuous at the right endpoint of the interval $[a, b]$ because

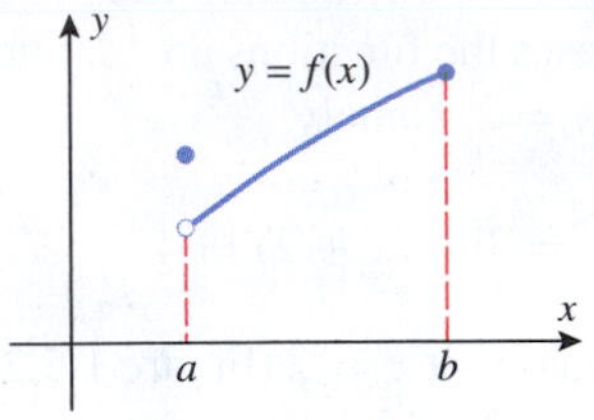

$$\lim_{x \to b^-} f(x) = f(b)$$

but it is not continuous at the left endpoint because

$$\lim_{x \to a^+} f(x) \neq f(a)$$

In general, we will say a function f is *continuous from the left* at c if

$$\lim_{x \to c^-} f(x) = f(c)$$

and is *continuous from the right* at c if

$$\lim_{x \to c^+} f(x) = f(c)$$

Using this terminology we define continuity on a closed interval as follows.

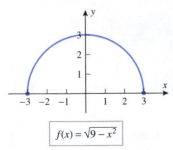

▲ **Figure 1.5.4**

1.5.2 **DEFINITION** A function f is said to be *continuous on a closed interval* $[a, b]$ if the following conditions are satisfied:

1. f is continuous on (a, b).

2. f is continuous from the right at a.

3. f is continuous from the left at b.

> Modify Definition 1.5.2 appropriately so that it applies to intervals of the form $[a, +\infty)$, $(-\infty, b]$, $(a, b]$, and $[a, b)$.

▶ **Example 2** What can you say about the continuity of the function $f(x) = \sqrt{9 - x^2}$?

Solution. Because the natural domain of this function is the closed interval $[-3, 3]$, we will need to investigate the continuity of f on the open interval $(-3, 3)$ and at the two endpoints. If c is any point in the interval $(-3, 3)$, then it follows from Theorem 1.2.2(e) that

$$\lim_{x \to c} f(x) = \lim_{x \to c} \sqrt{9 - x^2} = \sqrt{\lim_{x \to c} (9 - x^2)} = \sqrt{9 - c^2} = f(c)$$

which proves f is continuous at each point in the interval $(-3, 3)$. The function f is also continuous at the endpoints since

$$\lim_{x \to 3^-} f(x) = \lim_{x \to 3^-} \sqrt{9 - x^2} = \sqrt{\lim_{x \to 3^-} (9 - x^2)} = 0 = f(3)$$

$$\lim_{x \to -3^+} f(x) = \lim_{x \to -3^+} \sqrt{9 - x^2} = \sqrt{\lim_{x \to -3^+} (9 - x^2)} = 0 = f(-3)$$

Thus, f is continuous on the closed interval $[-3, 3]$ (Figure 1.5.5). ◀

$$f(x) = \sqrt{9 - x^2}$$

▲ **Figure 1.5.5**

■ **SOME PROPERTIES OF CONTINUOUS FUNCTIONS**

The following theorem, which is a consequence of Theorem 1.2.2, will enable us to reach conclusions about the continuity of functions that are obtained by adding, subtracting, multiplying, and dividing continuous functions.

1.5.3 **THEOREM** *If the functions f and g are continuous at c, then*

(a) *$f + g$ is continuous at c.*

(b) *$f - g$ is continuous at c.*

(c) *fg is continuous at c.*

(d) *f/g is continuous at c if $g(c) \neq 0$ and has a discontinuity at c if $g(c) = 0$.*

We will prove part (d). The remaining proofs are similar and will be left to the exercises.

PROOF First, consider the case where $g(c) = 0$. In this case $f(c)/g(c)$ is undefined, so the function f/g has a discontinuity at c.

Next, consider the case where $g(c) \neq 0$. To prove that f/g is continuous at c, we must show that

$$\lim_{x \to c} \frac{f(x)}{g(x)} = \frac{f(c)}{g(c)} \tag{1}$$

Since f and g are continuous at c,

$$\lim_{x \to c} f(x) = f(c) \quad \text{and} \quad \lim_{x \to c} g(x) = g(c)$$

Thus, by Theorem 1.2.2(d)

$$\lim_{x \to c} \frac{f(x)}{g(x)} = \frac{\lim\limits_{x \to c} f(x)}{\lim\limits_{x \to c} g(x)} = \frac{f(c)}{g(c)}$$

which proves (1). ■

■ **CONTINUITY OF POLYNOMIALS AND RATIONAL FUNCTIONS**

The general procedure for showing that a function is continuous everywhere is to show that it is continuous at an *arbitrary* point. For example, we know from Theorem 1.2.3 that if $p(x)$ is a polynomial and a is *any* real number, then

$$\lim_{x \to a} p(x) = p(a)$$

This shows that polynomials are continuous everywhere. Moreover, since rational functions are ratios of polynomials, it follows from part (d) of Theorem 1.5.3 that rational functions are continuous at points other than the zeros of the denominator, and at these zeros they have discontinuities. Thus, we have the following result.

1.5.4 **THEOREM**

(a) *A polynomial is continuous everywhere.*

(b) *A rational function is continuous at every point where the denominator is nonzero, and has discontinuities at the points where the denominator is zero.*

▶ **Example 3**   For what values of x is there a discontinuity in the graph of

$$y = \frac{x^2 - 9}{x^2 - 5x + 6}?$$

Solution.   The function being graphed is a rational function, and hence is continuous at every number where the denominator is nonzero. Solving the equation

$$x^2 - 5x + 6 = 0$$

yields discontinuities at $x = 2$ and at $x = 3$ (Figure 1.5.6). ◀

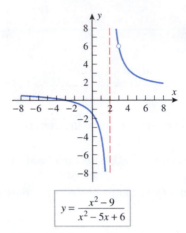

$$y = \frac{x^2 - 9}{x^2 - 5x + 6}$$

▲ **Figure 1.5.6**

▶ **Example 4**   Show that $|x|$ is continuous everywhere (Figure 0.1.9).

Solution.   We can write $|x|$ as

$$|x| = \begin{cases} x & \text{if} \quad x > 0 \\ 0 & \text{if} \quad x = 0 \\ -x & \text{if} \quad x < 0 \end{cases}$$

so $|x|$ is the same as the polynomial x on the interval $(0, +\infty)$ and is the same as the polynomial $-x$ on the interval $(-\infty, 0)$. But polynomials are continuous everywhere, so $x = 0$ is the only possible discontinuity for $|x|$. Since $|0| = 0$, to prove the continuity at $x = 0$ we must show that

$$\lim_{x \to 0} |x| = 0 \qquad (2)$$

Because the piecewise formula for $|x|$ changes at 0, it will be helpful to consider the one-sided limits at 0 rather than the two-sided limit. We obtain

$$\lim_{x \to 0^+} |x| = \lim_{x \to 0^+} x = 0 \quad \text{and} \quad \lim_{x \to 0^-} |x| = \lim_{x \to 0^-} (-x) = 0$$

Thus, (2) holds and $|x|$ is continuous at $x = 0$. ◀

■ **CONTINUITY OF COMPOSITIONS**

The following theorem, whose proof is given in Appendix J, will be useful for calculating limits of compositions of functions.

In words, Theorem 1.5.5 states that a limit symbol can be moved through a function sign provided the limit of the expression inside the function sign exists and the function is continuous at this limit.

1.5.5   THEOREM   *If* $\lim_{x \to c} g(x) = L$ *and if the function* f *is continuous at* L, *then* $\lim_{x \to c} f(g(x)) = f(L)$. *That is,*

$$\lim_{x \to c} f(g(x)) = f\left(\lim_{x \to c} g(x) \right)$$

This equality remains valid if $\lim_{x \to c}$ *is replaced everywhere by one of* $\lim_{x \to c^+}$, $\lim_{x \to c^-}$, $\lim_{x \to +\infty}$, *or* $\lim_{x \to -\infty}$.

In the special case of this theorem where $f(x) = |x|$, the fact that $|x|$ is continuous everywhere allows us to write

$$\lim_{x \to c} |g(x)| = \left| \lim_{x \to c} g(x) \right| \qquad (3)$$

provided $\lim_{x \to c} g(x)$ exists. Thus, for example,

$$\lim_{x \to 3} |5 - x^2| = \left| \lim_{x \to 3} (5 - x^2) \right| = |-4| = 4$$

The following theorem is concerned with the continuity of compositions of functions; the first part deals with continuity at a specific number and the second with continuity everywhere.

1.5.6 THEOREM

(a) *If the function g is continuous at c, and the function f is continuous at g(c), then the composition f ∘ g is continuous at c.*

(b) *If the function g is continuous everywhere and the function f is continuous everywhere, then the composition f ∘ g is continuous everywhere.*

PROOF We will prove part (*a*) only; the proof of part (*b*) can be obtained by applying part (*a*) at an arbitrary number *c*. To prove that $f \circ g$ is continuous at *c*, we must show that the value of $f \circ g$ and the value of its limit are the same at $x = c$. But this is so, since we can write

$$\lim_{x \to c} (f \circ g)(x) = \lim_{x \to c} f(g(x)) = f\left(\lim_{x \to c} g(x)\right) = f(g(c)) = (f \circ g)(c) \ \blacksquare$$

| Theorem 1.5.5 | g is continuous at c. |

Can the absolute value of a function that is not continuous everywhere be continuous everywhere? Justify your answer.

We know from Example 4 that the function $|x|$ is continuous everywhere. Thus, if $g(x)$ is continuous at *c*, then by part (*a*) of Theorem 1.5.6, the function $|g(x)|$ must also be continuous at *c*; and, more generally, if $g(x)$ is continuous everywhere, then so is $|g(x)|$. Stated informally:

The absolute value of a continuous function is continuous.

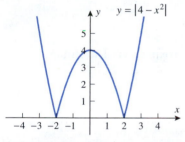

▲ **Figure 1.5.7**

For example, the polynomial $g(x) = 4 - x^2$ is continuous everywhere, so we can conclude that the function $|4 - x^2|$ is also continuous everywhere (Figure 1.5.7).

■ **THE INTERMEDIATE-VALUE THEOREM**
Figure 1.5.8 shows the graph of a function that is continuous on the closed interval $[a, b]$. The figure suggests that if we draw any horizontal line $y = k$, where *k* is between $f(a)$ and $f(b)$, then that line will cross the curve $y = f(x)$ at least once over the interval $[a, b]$. Stated in numerical terms, if *f* is continuous on $[a, b]$, then the function *f* must take on every value *k* between $f(a)$ and $f(b)$ at least once as *x* varies from *a* to *b*. For example, the polynomial $p(x) = x^5 - x + 3$ has a value of 3 at $x = 1$ and a value of 33 at $x = 2$. Thus, it follows from the continuity of *p* that the equation $x^5 - x + 3 = k$ has at least one solution in the interval $[1, 2]$ for every value of *k* between 3 and 33. This idea is stated more precisely in the following theorem.

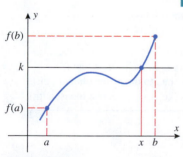

▲ **Figure 1.5.8**

1.5.7 THEOREM (*Intermediate-Value Theorem*) *If f is continuous on a closed interval $[a, b]$ and k is any number between $f(a)$ and $f(b)$, inclusive, then there is at least one number x in the interval $[a, b]$ such that $f(x) = k$.*

Although this theorem is intuitively obvious, its proof depends on a mathematically precise development of the real number system, which is beyond the scope of this text.

■ APPROXIMATING ROOTS USING THE INTERMEDIATE-VALUE THEOREM

A variety of problems can be reduced to solving an equation $f(x) = 0$ for its roots. Sometimes it is possible to solve for the roots exactly using algebra, but often this is not possible and one must settle for decimal approximations of the roots. One procedure for approximating roots is based on the following consequence of the Intermediate-Value Theorem.

1.5.8   THEOREM   *If f is continuous on $[a, b]$, and if $f(a)$ and $f(b)$ are nonzero and have opposite signs, then there is at least one solution of the equation $f(x) = 0$ in the interval (a, b).*

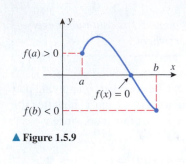

▲ **Figure 1.5.9**

This result, which is illustrated in Figure 1.5.9, can be proved as follows.

PROOF   Since $f(a)$ and $f(b)$ have opposite signs, 0 is between $f(a)$ and $f(b)$. Thus, by the Intermediate-Value Theorem there is at least one number x in the interval $[a, b]$ such that $f(x) = 0$. However, $f(a)$ and $f(b)$ are nonzero, so x must lie in the interval (a, b), which completes the proof. ■

Before we illustrate how this theorem can be used to approximate roots, it will be helpful to discuss some standard terminology for describing errors in approximations. If x is an approximation to a quantity x_0, then we call

$$\epsilon = |x - x_0|$$

the **absolute error** or (less precisely) the **error** in the approximation. The terminology in Table 1.5.1 is used to describe the size of such errors.

Table 1.5.1

ERROR	DESCRIPTION
$|x - x_0| \leq 0.1$	x approximates x_0 with an error of at most 0.1.
$|x - x_0| \leq 0.01$	x approximates x_0 with an error of at most 0.01.
$|x - x_0| \leq 0.001$	x approximates x_0 with an error of at most 0.001.
$|x - x_0| \leq 0.0001$	x approximates x_0 with an error of at most 0.0001.
$|x - x_0| \leq 0.5$	x approximates x_0 to the nearest integer.
$|x - x_0| \leq 0.05$	x approximates x_0 to 1 decimal place (i.e., to the nearest tenth).
$|x - x_0| \leq 0.005$	x approximates x_0 to 2 decimal places (i.e., to the nearest hundredth).
$|x - x_0| \leq 0.0005$	x approximates x_0 to 3 decimal places (i.e., to the nearest thousandth).

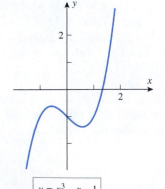

▲ **Figure 1.5.10**

▶ **Example 5**   The equation

$$x^3 - x - 1 = 0$$

cannot be solved algebraically very easily because the left side has no simple factors. However, if we graph $p(x) = x^3 - x - 1$ with a graphing utility (Figure 1.5.10), then we are led to conjecture that there is one real root and that this root lies inside the interval $[1, 2]$. The existence of a root in this interval is also confirmed by Theorem 1.5.8, since $p(1) = -1$ and $p(2) = 5$ have opposite signs. Approximate this root to two decimal-place accuracy.

Solution. Our objective is to approximate the unknown root x_0 with an error of at most 0.005. It follows that if we can find an interval of length 0.01 that contains the root, then the midpoint of that interval will approximate the root with an error of at most $\frac{1}{2}(0.01) = 0.005$, which will achieve the desired accuracy.

We know that the root x_0 lies in the interval $[1, 2]$. However, this interval has length 1, which is too large. We can pinpoint the location of the root more precisely by dividing the interval $[1, 2]$ into 10 equal parts and evaluating p at the points of subdivision using a calculating utility (Table 1.5.2). In this table $p(1.3)$ and $p(1.4)$ have opposite signs, so we know that the root lies in the interval $[1.3, 1.4]$. This interval has length 0.1, which is still too large, so we repeat the process by dividing the interval $[1.3, 1.4]$ into 10 parts and evaluating p at the points of subdivision; this yields Table 1.5.3, which tells us that the root is inside the interval $[1.32, 1.33]$ (Figure 1.5.11). Since this interval has length 0.01, its midpoint 1.325 will approximate the root with an error of at most 0.005. Thus, $x_0 \approx 1.325$ to two decimal-place accuracy. ◄

Table 1.5.2

x	1	1.1	1.2	1.3	1.4	1.5	1.6	1.7	1.8	1.9	2
$p(x)$	-1	-0.77	-0.47	-0.10	0.34	0.88	1.50	2.21	3.03	3.96	5

Table 1.5.3

x	1.3	1.31	1.32	1.33	1.34	1.35	1.36	1.37	1.38	1.39	1.4
$p(x)$	-0.103	-0.062	-0.020	0.023	0.066	0.110	0.155	0.201	0.248	0.296	0.344

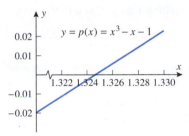

▲ **Figure 1.5.11**

REMARK To say that x approximates x_0 to n decimal places does *not* mean that the first n decimal places of x and x_0 will be the same when the numbers are rounded to n decimal places. For example, $x = 1.084$ approximates $x_0 = 1.087$ to two decimal places because $|x - x_0| = 0.003 \, (< 0.005)$. However, if we round these values to two decimal places, then we obtain $x \approx 1.08$ and $x_0 \approx 1.09$. Thus, if you approximate a number to n decimal places, then you should display that approximation to at least $n + 1$ decimal places to preserve the accuracy.

TECHNOLOGY MASTERY

Use a graphing or calculating utility to show that the root x_0 in Example 5 can be approximated as $x_0 \approx 1.3245$ to three decimal-place accuracy.

✔ **QUICK CHECK EXERCISES 1.5**   *(See page 120 for answers.)*

1. What three conditions are satisfied if f is continuous at $x = c$?

2. Suppose that f and g are continuous functions such that $f(2) = 1$ and $\lim\limits_{x \to 2} [f(x) + 4g(x)] = 13$. Find
   (a) $g(2)$
   (b) $\lim\limits_{x \to 2} g(x)$.

3. Suppose that f and g are continuous functions such that $\lim\limits_{x \to 3} g(x) = 5$ and $f(3) = -2$. Find $\lim\limits_{x \to 3} [f(x)/g(x)]$.

4. For what values of x, if any, is the function
$$f(x) = \frac{x^2 - 16}{x^2 - 5x + 4}$$
discontinuous?

5. Suppose that a function f is continuous everywhere and that $f(-2) = 3$, $f(-1) = -1$, $f(0) = -4$, $f(1) = 1$, and $f(2) = 5$. Does the Intermediate-Value Theorem guarantee that f has a root on the following intervals?
   (a) $[-2, -1]$ (b) $[-1, 0]$ (c) $[-1, 1]$ (d) $[0, 2]$

EXERCISE SET 1.5   Graphing Utility

1–4 Let f be the function whose graph is shown. On which of the following intervals, if any, is f continuous?
(a) $[1, 3]$ (b) $(1, 3)$ (c) $[1, 2]$
(d) $(1, 2)$ (e) $[2, 3]$ (f) $(2, 3)$
For each interval on which f is not continuous, indicate which conditions for the continuity of f do not hold. ■

1.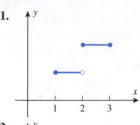

2.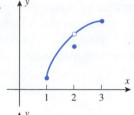

3.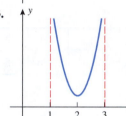

4.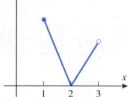

5. Consider the functions
$$f(x) = \begin{cases} 1, & x \neq 4 \\ -1, & x = 4 \end{cases} \quad \text{and} \quad g(x) = \begin{cases} 4x - 10, & x \neq 4 \\ -6, & x = 4 \end{cases}$$

In each part, is the given function continuous at $x = 4$?
(a) $f(x)$ (b) $g(x)$ (c) $-g(x)$ (d) $|f(x)|$
(e) $f(x)g(x)$ (f) $g(f(x))$ (g) $g(x) - 6f(x)$

6. Consider the functions
$$f(x) = \begin{cases} 1, & 0 \leq x \\ 0, & x < 0 \end{cases} \quad \text{and} \quad g(x) = \begin{cases} 0, & 0 \leq x \\ 1, & x < 0 \end{cases}$$

In each part, is the given function continuous at $x = 0$?
(a) $f(x)$ (b) $g(x)$ (c) $f(-x)$ (d) $|g(x)|$
(e) $f(x)g(x)$ (f) $g(f(x))$ (g) $f(x) + g(x)$

FOCUS ON CONCEPTS

7. In each part sketch the graph of a function f that satisfies the stated conditions.
(a) f is continuous everywhere except at $x = 3$, at which point it is continuous from the right.
(b) f has a two-sided limit at $x = 3$, but it is not continuous at $x = 3$.
(c) f is not continuous at $x = 3$, but if its value at $x = 3$ is changed from $f(3) = 1$ to $f(3) = 0$, it becomes continuous at $x = 3$.
(d) f is continuous on the interval $[0, 3)$ and is defined on the closed interval $[0, 3]$; but f is not continuous on the interval $[0, 3]$.

8. Assume that a function f is defined at $x = c$, and, with the aid of Definition 1.4.1, write down precisely what

condition (involving ϵ and δ) must be satisfied for f to be continuous at $x = c$. Explain why the condition $0 < |x - c| < \delta$ can be replaced by $|x - c| < \delta$.

9. A student parking lot at a university charges $2.00 for the first half hour (or any part) and $1.00 for each subsequent half hour (or any part) up to a daily maximum of $10.00.
(a) Sketch a graph of cost as a function of the time parked.
(b) Discuss the significance of the discontinuities in the graph to a student who parks there.

10. In each part determine whether the function is continuous or not, and explain your reasoning.
(a) The Earth's population as a function of time.
(b) Your exact height as a function of time.
(c) The cost of a taxi ride in your city as a function of the distance traveled.
(d) The volume of a melting ice cube as a function of time.

11–22 Find values of x, if any, at which f is not continuous. ■

11. $f(x) = 5x^4 - 3x + 7$ **12.** $f(x) = \sqrt[3]{x - 8}$

13. $f(x) = \dfrac{x + 2}{x^2 + 4}$ **14.** $f(x) = \dfrac{x + 2}{x^2 - 4}$

15. $f(x) = \dfrac{x}{2x^2 + x}$ **16.** $f(x) = \dfrac{2x + 1}{4x^2 + 4x + 5}$

17. $f(x) = \dfrac{3}{x} + \dfrac{x - 1}{x^2 - 1}$ **18.** $f(x) = \dfrac{5}{x} + \dfrac{2x}{x + 4}$

19. $f(x) = \dfrac{x^2 + 6x + 9}{|x| + 3}$ **20.** $f(x) = \left| 4 - \dfrac{8}{x^4 + x} \right|$

21. $f(x) = \begin{cases} 2x + 3, & x \leq 4 \\ 7 + \dfrac{16}{x}, & x > 4 \end{cases}$

22. $f(x) = \begin{cases} \dfrac{3}{x - 1}, & x \neq 1 \\ 3, & x = 1 \end{cases}$

23–28 True–False Determine whether the statement is true or false. Explain your answer. ■

23. If $f(x)$ is continuous at $x = c$, then so is $|f(x)|$.

24. If $|f(x)|$ is continuous at $x = c$, then so is $f(x)$.

25. If f and g are discontinuous at $x = c$, then so is $f + g$.

26. If f and g are discontinuous at $x = c$, then so is fg.

27. If $\sqrt{f(x)}$ is continuous at $x = c$, then so is $f(x)$.

28. If $f(x)$ is continuous at $x = c$, then so is $\sqrt{f(x)}$.

29–30 Find a value of the constant k, if possible, that will make the function continuous everywhere. ■

29. (a) $f(x) = \begin{cases} 7x - 2, & x \le 1 \\ kx^2, & x > 1 \end{cases}$

(b) $f(x) = \begin{cases} kx^2, & x \le 2 \\ 2x + k, & x > 2 \end{cases}$

30. (a) $f(x) = \begin{cases} 9 - x^2, & x \ge -3 \\ k/x^2, & x < -3 \end{cases}$

(b) $f(x) = \begin{cases} 9 - x^2, & x \ge 0 \\ k/x^2, & x < 0 \end{cases}$

31. Find values of the constants k and m, if possible, that will make the function f continuous everywhere.

$$f(x) = \begin{cases} x^2 + 5, & x > 2 \\ m(x+1) + k, & -1 < x \le 2 \\ 2x^3 + x + 7, & x \le -1 \end{cases}$$

32. On which of the following intervals is

$$f(x) = \frac{1}{\sqrt{x - 2}}$$

continuous?
(a) $[2, +\infty)$ (b) $(-\infty, +\infty)$ (c) $(2, +\infty)$ (d) $[1, 2)$

33–36 A function f is said to have a ***removable discontinuity*** at $x = c$ if $\lim_{x \to c} f(x)$ exists but f is not continuous at $x = c$, either because f is not defined at c or because the definition for $f(c)$ differs from the value of the limit. This terminology will be needed in these exercises. ■

33. (a) Sketch the graph of a function with a removable discontinuity at $x = c$ for which $f(c)$ is undefined.
(b) Sketch the graph of a function with a removable discontinuity at $x = c$ for which $f(c)$ is defined.

34. (a) The terminology *removable discontinuity* is appropriate because a removable discontinuity of a function f at $x = c$ can be "removed" by redefining the value of f appropriately at $x = c$. What value for $f(c)$ removes the discontinuity?
(b) Show that the following functions have removable discontinuities at $x = 1$, and sketch their graphs.

$$f(x) = \frac{x^2 - 1}{x - 1} \quad \text{and} \quad g(x) = \begin{cases} 1, & x > 1 \\ 0, & x = 1 \\ 1, & x < 1 \end{cases}$$

(c) What values should be assigned to $f(1)$ and $g(1)$ to remove the discontinuities?

35–36 Find the values of x (if any) at which f is not continuous, and determine whether each such value is a removable discontinuity. ■

35. (a) $f(x) = \frac{|x|}{x}$ (b) $f(x) = \frac{x^2 + 3x}{x + 3}$

(c) $f(x) = \frac{x - 2}{|x| - 2}$

36. (a) $f(x) = \frac{x^2 - 4}{x^3 - 8}$ (b) $f(x) = \begin{cases} 2x - 3, & x \le 2 \\ x^2, & x > 2 \end{cases}$

(c) $f(x) = \begin{cases} 3x^2 + 5, & x \ne 1 \\ 6, & x = 1 \end{cases}$

37. (a) Use a graphing utility to generate the graph of the function $f(x) = (x + 3)/(2x^2 + 5x - 3)$, and then use the graph to make a conjecture about the number and locations of all discontinuities.
(b) Check your conjecture by factoring the denominator.

38. (a) Use a graphing utility to generate the graph of the function $f(x) = x/(x^3 - x + 2)$, and then use the graph to make a conjecture about the number and locations of all discontinuities.
(b) Use the Intermediate-Value Theorem to approximate the locations of all discontinuities to two decimal places.

39. Prove that $f(x) = x^{3/5}$ is continuous everywhere, carefully justifying each step.

40. Prove that $f(x) = 1/\sqrt{x^4 + 7x^2 + 1}$ is continuous everywhere, carefully justifying each step.

41. Prove:
(a) part (*a*) of Theorem 1.5.3
(b) part (*b*) of Theorem 1.5.3
(c) part (*c*) of Theorem 1.5.3.

42. Prove part (*b*) of Theorem 1.5.4.

43. (a) Use Theorem 1.5.5 to prove that if f is continuous at $x = c$, then $\lim_{h \to 0} f(c + h) = f(c)$.
(b) Prove that if $\lim_{h \to 0} f(c + h) = f(c)$, then f is continuous at $x = c$. [*Hint:* What does this limit tell you about the continuity of $g(h) = f(c + h)$?]
(c) Conclude from parts (a) and (b) that f is continuous at $x = c$ if and only if $\lim_{h \to 0} f(c + h) = f(c)$.

44. Prove: If f and g are continuous on $[a, b]$, and $f(a) > g(a)$, $f(b) < g(b)$, then there is at least one solution of the equation $f(x) = g(x)$ in (a, b). [*Hint:* Consider $f(x) - g(x)$.]

FOCUS ON CONCEPTS

45. Give an example of a function f that is defined on a closed interval, and whose values at the endpoints have opposite signs, but for which the equation $f(x) = 0$ has no solution in the interval.

46. Let f be the function whose graph is shown in Exercise 2. For each interval, determine (i) whether the hypothesis of the Intermediate-Value Theorem is satisfied, and (ii) whether the conclusion of the Intermediate-Value Theorem is satisfied.
(a) $[1, 2]$ (b) $[2, 3]$ (c) $[1, 3]$

47. Show that the equation $x^3 + x^2 - 2x = 1$ has at least one solution in the interval $[-1, 1]$.

48. Prove: If $p(x)$ is a polynomial of odd degree, then the equation $p(x) = 0$ has at least one real solution.

49. The accompanying figure shows the graph of the equation $y = x^4 + x - 1$. Use the method of Example 5 to approximate the x-intercepts with an error of at most 0.05.

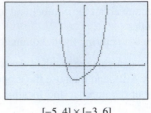

[−5, 4] × [−3, 6]
xScl = 1, yScl = 1   ◀ **Figure Ex-49**

50. The accompanying figure shows the graph of the equation $y = 5 - x - x^4$. Use the method of Example 5 to approximate the roots of the equation $5 - x - x^4 = 0$ to two decimal-place accuracy.

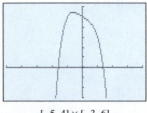

[−5, 4] × [−3, 6]
xScl = 1, yScl = 1   ◀ **Figure Ex-50**

51. Use the fact that $\sqrt{5}$ is a solution of $x^2 - 5 = 0$ to approximate $\sqrt{5}$ with an error of at most 0.005.

52. A sprinter, who is timed with a stopwatch, runs a hundred yard dash in 10 s. The stopwatch is reset to 0, and the sprinter is timed jogging back to the starting block. Show that there is at least one point on the track at which the reading on the stopwatch during the sprint is the same as the reading during the return jog. [*Hint:* Use the result in Exercise 44.]

53. Prove that there exist points on opposite sides of the equator that are at the same temperature. [*Hint:* Consider the accompanying figure, which shows a view of the equator from a point above the North Pole. Assume that the temperature $T(\theta)$ is a continuous function of the angle θ, and consider the function $f(\theta) = T(\theta + \pi) - T(\theta).$]

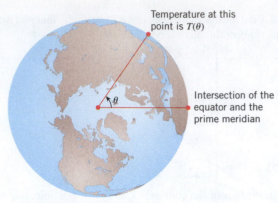

Temperature at this point is $T(\theta)$

Intersection of the equator and the prime meridian

▲ **Figure Ex-53**

54. Let R denote an elliptical region in the xy-plane, and define $f(z)$ to be the area within R that is on, or to the left of, the vertical line $x = z$. Prove that f is a continuous function of z. [*Hint:* Assume the ellipse is between the horizontal lines $y = a$ and $y = b$, $a < b$. Argue that $|f(z_1) - f(z_2)| \leq (b - a) \cdot |z_1 - z_2|.$]

55. Let R denote an elliptical region in the plane. For any line L, prove there is a line perpendicular to L that divides R in half by area. [*Hint:* Introduce coordinates so that L is the x-axis. Use the result in Exercise 54 and the Intermediate-Value Theorem.]

56. Suppose that f is continuous on the interval [0, 1] and that $0 \leq f(x) \leq 1$ for all x in this interval.
  (a) Sketch the graph of $y = x$ together with a possible graph for f over the interval [0, 1].
  (b) Use the Intermediate-Value Theorem to help prove that there is at least one number c in the interval [0, 1] such that $f(c) = c$.

57. Writing It is often assumed that changing physical quantities such as the height of a falling object or the weight of a melting snowball, are continuous functions of time. Use specific examples to discuss the merits of this assumption.

58. Writing The Intermediate-Value Theorem (Theorem 1.5.7) is an example of what is known as an "existence theorem." In your own words, describe how to recognize an existence theorem, and discuss some of the ways in which an existence theorem can be useful.

✔ **QUICK CHECK ANSWERS 1.5**

1. $f(c)$ is defined; $\lim_{x \to c} f(x)$ exists; $\lim_{x \to c} f(x) = f(c)$ **2.** (a) 3 (b) 3   **3.** −2/5   **4.** $x = 1, 4$
5. (a) yes (b) no (c) yes (d) yes

1.6

CONTINUITY OF TRIGONOMETRIC, EXPONENTIAL, AND INVERSE FUNCTIONS

In this section we will discuss the continuity properties of trigonometric functions, exponential functions, and inverses of various continuous functions. We will also discuss some important limits involving such functions.

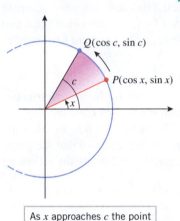

As x approaches c the point P approaches the point Q.

▲ **Figure 1.6.1**

Theorem 1.6.1 implies that the six basic trigonometric functions are continuous on their domains. In particular, $\sin x$ and $\cos x$ are continuous everywhere.

■ CONTINUITY OF TRIGONOMETRIC FUNCTIONS

Recall from trigonometry that the graphs of $\sin x$ and $\cos x$ are drawn as continuous curves. We will not formally prove that these functions are continuous, but we can motivate this fact by letting c be a fixed angle in radian measure and x a variable angle in radian measure. If, as illustrated in Figure 1.6.1, the angle x approaches the angle c, then the point $P(\cos x, \sin x)$ moves along the unit circle toward $Q(\cos c, \sin c)$, and the coordinates of P approach the corresponding coordinates of Q. This implies that

$$\lim_{x \to c} \sin x = \sin c \quad \text{and} \quad \lim_{x \to c} \cos x = \cos c \tag{1}$$

Thus, $\sin x$ and $\cos x$ are continuous at the arbitrary point c; that is, these functions are continuous everywhere.

The formulas in (1) can be used to find limits of the remaining trigonometric functions by expressing them in terms of $\sin x$ and $\cos x$; for example, if $\cos c \neq 0$, then

$$\lim_{x \to c} \tan x = \lim_{x \to c} \frac{\sin x}{\cos x} = \frac{\sin c}{\cos c} = \tan c$$

Thus, we are led to the following theorem.

1.6.1   THEOREM   *If c is any number in the natural domain of the stated trigonometric function, then*

$$\lim_{x \to c} \sin x = \sin c \qquad \lim_{x \to c} \cos x = \cos c \qquad \lim_{x \to c} \tan x = \tan c$$

$$\lim_{x \to c} \csc x = \csc c \qquad \lim_{x \to c} \sec x = \sec c \qquad \lim_{x \to c} \cot x = \cot c$$

▶ **Example 1**   Find the limit

$$\lim_{x \to 1} \cos \left(\frac{x^2 - 1}{x - 1} \right)$$

Solution.   Since the cosine function is continuous everywhere, it follows from Theorem 1.5.5 that

$$\lim_{x \to 1} \cos(g(x)) = \cos \left(\lim_{x \to 1} g(x) \right)$$

provided $\lim_{x \to 1} g(x)$ exists. Thus,

$$\lim_{x \to 1} \cos \left(\frac{x^2 - 1}{x - 1} \right) = \lim_{x \to 1} \cos(x + 1) = \cos \left(\lim_{x \to 1} (x + 1) \right) = \cos 2 \ ◀$$

■ CONTINUITY OF INVERSE FUNCTIONS

Since the graphs of a one-to-one function f and its inverse f^{-1} are reflections of one another about the line $y = x$, it is clear geometrically that if the graph of f has no breaks or holes in it, then neither does the graph of f^{-1}. This, and the fact that the range of f is the domain of f^{-1}, suggests the following result, which we state without formal proof.

To paraphrase Theorem 1.6.2, *the inverse of a continuous function is continuous.*

1.6.2   THEOREM   *If f is a one-to-one function that is continuous at each point of its domain, then f^{-1} is continuous at each point of its domain; that is, f^{-1} is continuous at each point in the range of f.*

▶ **Example 2**   Use Theorem 1.6.2 to prove that $\sin^{-1} x$ is continuous on the interval $[-1, 1]$.

Solution.   Recall that $\sin^{-1} x$ is the inverse of the restricted sine function whose domain is the interval $[-\pi/2, \pi/2]$ and whose range is the interval $[-1, 1]$ (Definition 0.4.6 and Figure 0.4.13). Since $\sin x$ is continuous on the interval $[-\pi/2, \pi/2]$, Theorem 1.6.2 implies $\sin^{-1} x$ is continuous on the interval $[-1, 1]$. ◀

Arguments similar to the solution of Example 2 show that each of the inverse trigonometric functions defined in Section 0.4 is continuous at each point of its domain.

When we introduced the exponential function $f(x) = b^x$ in Section 0.5, we assumed that its graph is a curve without breaks, gaps, or holes; that is, we assumed that the graph of $y = b^x$ is a continuous curve. This assumption and Theorem 1.6.2 imply the following theorem, which we state without formal proof.

1.6.3   THEOREM   *Let $b > 0, b \neq 1$.*

(*a*)   *The function b^x is continuous on $(-\infty, +\infty)$.*

(*b*)   *The function $\log_b x$ is continuous on $(0, +\infty)$.*

▶ **Example 3**   Where is the function $f(x) = \dfrac{\tan^{-1} x + \ln x}{x^2 - 4}$ continuous?

Solution.   The fraction will be continuous at all points where the numerator and denominator are both continuous and the denominator is nonzero. Since $\tan^{-1} x$ is continuous everywhere and $\ln x$ is continuous if $x > 0$, the numerator is continuous if $x > 0$. The denominator, being a polynomial, is continuous everywhere, so the fraction will be continuous at all points where $x > 0$ and the denominator is nonzero. Thus, f is continuous on the intervals $(0, 2)$ and $(2, +\infty)$. ◀

◼ OBTAINING LIMITS BY SQUEEZING

In Section 1.1 we used numerical evidence to conjecture that

$$\lim_{x \to 0} \frac{\sin x}{x} = 1 \tag{2}$$

However, this limit is not easy to establish with certainty. The limit is an indeterminate form of type $0/0$, and there is no simple algebraic manipulation that one can perform to obtain the limit. Later in the text we will develop general methods for finding limits of indeterminate forms, but in this particular case we can use a technique called *squeezing*.

The method of squeezing is used to prove that $f(x) \to L$ as $x \to c$ by "trapping" or "squeezing" f between two functions, g and h, whose limits as $x \to c$ are known with *certainty* to be L. As illustrated in Figure 1.6.2, this forces f to have a limit of L as well. This is the idea behind the following theorem, which we state without proof.

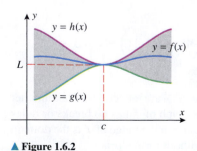

▲ **Figure 1.6.2**

1.6.4 THEOREM (*The Squeezing Theorem*) *Let f, g, and h be functions satisfying*

$$g(x) \le f(x) \le h(x)$$

for all x in some open interval containing the number c, with the possible exception that the inequalities need not hold at c. If g and h have the same limit as x approaches c, say

$$\lim_{x \to c} g(x) = \lim_{x \to c} h(x) = L$$

then f also has this limit as x approaches c, that is,

$$\lim_{x \to c} f(x) = L$$

The Squeezing Theorem also holds for one-sided limits and limits at $+\infty$ and $-\infty$. How do you think the hypotheses would change in those cases?

To illustrate how the Squeezing Theorem works, we will prove the following results, which are illustrated in Figure 1.6.3.

1.6.5 THEOREM

$$(a) \quad \lim_{x \to 0} \frac{\sin x}{x} = 1 \qquad (b) \quad \lim_{x \to 0} \frac{1 - \cos x}{x} = 0$$

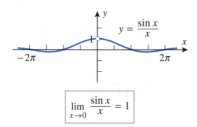

$y = \dfrac{\sin x}{x}$

$\lim_{x \to 0} \dfrac{\sin x}{x} = 1$

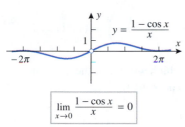

$y = \dfrac{1 - \cos x}{x}$

$\lim_{x \to 0} \dfrac{1 - \cos x}{x} = 0$

▲ **Figure 1.6.3**

PROOF (*a*) In this proof we will interpret x as an angle in radian measure, and we will assume to start that $0 < x < \pi/2$. As illustrated in Figure 1.6.4, the area of a sector with central angle x and radius 1 lies between the areas of two triangles, one with area $\frac{1}{2} \tan x$ and the other with area $\frac{1}{2} \sin x$. Since the sector has area $\frac{1}{2} x$ (see marginal note), it follows that

$$\frac{1}{2} \tan x \ge \frac{1}{2} x \ge \frac{1}{2} \sin x$$

Multiplying through by $2/(\sin x)$ and using the fact that $\sin x > 0$ for $0 < x < \pi/2$, we obtain

$$\frac{1}{\cos x} \ge \frac{x}{\sin x} \ge 1$$

Next, taking reciprocals reverses the inequalities, so we obtain

$$\cos x \le \frac{\sin x}{x} \le 1 \tag{3}$$

which squeezes the function $(\sin x)/x$ between the functions $\cos x$ and 1. Although we derived these inequalities by assuming that $0 < x < \pi/2$, they also hold for $-\pi/2 < x < 0$ [since replacing x by $-x$ and using the identities $\sin(-x) = -\sin x$, and $\cos(-x) = \cos x$

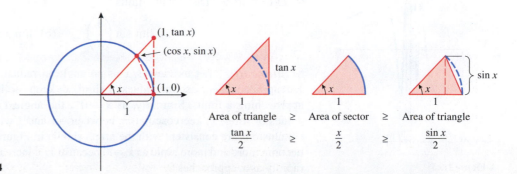

▶ **Figure 1.6.4**

Recall that the area A of a sector of radius r and central angle θ is

$$A = \frac{1}{2}r^2\theta$$

This can be derived from the relationship

$$\frac{A}{\pi r^2} = \frac{\theta}{2\pi}$$

which states that the area of the sector is to the area of the circle as the central angle of the sector is to the central angle of the circle.

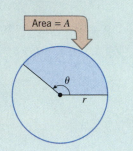

Area = A

leaves (3) unchanged]. Finally, since

$$\lim_{x\to 0}\cos x = 1 \quad\text{and}\quad \lim_{x\to 0} 1 = 1$$

the Squeezing Theorem implies that

$$\lim_{x\to 0}\frac{\sin x}{x} = 1$$

PROOF (b)   For this proof we will use the limit in part (a), the continuity of the sine function, and the trigonometric identity $\sin^2 x = 1 - \cos^2 x$. We obtain

$$\lim_{x\to 0}\frac{1 - \cos x}{x} = \lim_{x\to 0}\left[\frac{1 - \cos x}{x}\cdot\frac{1 + \cos x}{1 + \cos x}\right] = \lim_{x\to 0}\frac{\sin^2 x}{(1 + \cos x)x}$$

$$= \left(\lim_{x\to 0}\frac{\sin x}{x}\right)\left(\lim_{x\to 0}\frac{\sin x}{1 + \cos x}\right) = (1)\left(\frac{0}{1 + 1}\right) = 0 \quad\blacksquare$$

▶ **Example 4**   Find

(a) $\displaystyle\lim_{x\to 0}\frac{\tan x}{x}$ (b) $\displaystyle\lim_{\theta\to 0}\frac{\sin 2\theta}{\theta}$ (c) $\displaystyle\lim_{x\to 0}\frac{\sin 3x}{\sin 5x}$

Solution (a).

$$\lim_{x\to 0}\frac{\tan x}{x} = \lim_{x\to 0}\left(\frac{\sin x}{x}\cdot\frac{1}{\cos x}\right) = \left(\lim_{x\to 0}\frac{\sin x}{x}\right)\left(\lim_{x\to 0}\frac{1}{\cos x}\right) = (1)(1) = 1$$

Solution (b).   The trick is to multiply and divide by 2, which will make the denominator the same as the argument of the sine function [just as in Theorem 1.6.5(a)]:

$$\lim_{\theta\to 0}\frac{\sin 2\theta}{\theta} = \lim_{\theta\to 0} 2\cdot\frac{\sin 2\theta}{2\theta} = 2\lim_{\theta\to 0}\frac{\sin 2\theta}{2\theta}$$

Now make the substitution $x = 2\theta$, and use the fact that $x\to 0$ as $\theta\to 0$. This yields

$$\lim_{\theta\to 0}\frac{\sin 2\theta}{\theta} = 2\lim_{\theta\to 0}\frac{\sin 2\theta}{2\theta} = 2\lim_{x\to 0}\frac{\sin x}{x} = 2(1) = 2$$

TECHNOLOGY MASTERY

Use a graphing utility to confirm the limits in Example 4, and if you have a CAS, use it to obtain the limits.

Solution (c).

$$\lim_{x\to 0}\frac{\sin 3x}{\sin 5x} = \lim_{x\to 0}\frac{\dfrac{\sin 3x}{x}}{\dfrac{\sin 5x}{x}} = \lim_{x\to 0}\frac{3\cdot\dfrac{\sin 3x}{3x}}{5\cdot\dfrac{\sin 5x}{5x}} = \frac{3\cdot 1}{5\cdot 1} = \frac{3}{5} \quad◀$$

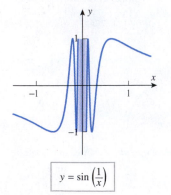

$y = \sin\left(\dfrac{1}{x}\right)$

▲ **Figure 1.6.5**

▶ **Example 5**   Discuss the limits

(a) $\displaystyle\lim_{x\to 0}\sin\left(\frac{1}{x}\right)$ (b) $\displaystyle\lim_{x\to 0} x\sin\left(\frac{1}{x}\right)$

Solution (a).   Let us view $1/x$ as an angle in radian measure. As $x\to 0^+$, the angle $1/x$ approaches $+\infty$, so the values of $\sin(1/x)$ keep oscillating between -1 and 1 without approaching a limit. Similarly, as $x\to 0^-$, the angle $1/x$ approaches $-\infty$, so again the values of $\sin(1/x)$ keep oscillating between -1 and 1 without approaching a limit. These conclusions are consistent with the graph shown in Figure 1.6.5. Note that the oscillations become more and more rapid as $x\to 0$ because $1/x$ increases (or decreases) more and more rapidly as x approaches 0.

Solution (b).   Since

$$-1 \leq \sin\left(\frac{1}{x}\right) \leq 1$$

it follows that if $x \neq 0$, then

$$-|x| \leq x \sin\left(\frac{1}{x}\right) \leq |x| \tag{4}$$

Since $|x| \to 0$ as $x \to 0$, the inequalities in (4) and the Squeezing Theorem imply that

$$\lim_{x \to 0} x \sin\left(\frac{1}{x}\right) = 0$$

This is consistent with the graph shown in Figure 1.6.6. ◄

Confirm (4) by considering the cases $x > 0$ and $x < 0$ separately.

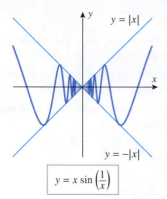

$y = x \sin\left(\frac{1}{x}\right)$

▲ Figure 1.6.6

REMARK   It follows from part (b) of this example that the function

$$f(x) = \begin{cases} x \sin(1/x), & x \neq 0 \\ 0, & x = 0 \end{cases}$$

is continuous at $x = 0$, since the value of the function and the value of the limit are the same at 0. This shows that the behavior of a function can be very complex in the vicinity of $x = c$, even though the function is continuous at c.

✔ **QUICK CHECK EXERCISES 1.6**   *(See page 128 for answers.)*

1. In each part, is the given function continuous on the interval $[0, \pi/2]$?
   (a) $\sin x$ (b) $\cos x$ (c) $\tan x$ (d) $\csc x$

2. Evaluate
   (a) $\displaystyle\lim_{x \to 0} \frac{\sin x}{x}$
   (b) $\displaystyle\lim_{x \to 0} \frac{1 - \cos x}{x}$.

3. Suppose a function f has the property that for all real numbers x
   $$3 - |x| \leq f(x) \leq 3 + |x|$$
   From this we can conclude that $f(x) \to$ _____ as $x \to$ _____.

4. In each part, give the largest interval on which the function is continuous.
   (a) e^x (b) $\ln x$ (c) $\sin^{-1} x$ (d) $\tan^{-1} x$

EXERCISE SET 1.6    Graphing Utility

1–8 Find the discontinuities, if any. ■

1. $f(x) = \sin(x^2 - 2)$

2. $f(x) = \cos\left(\dfrac{x}{x - \pi}\right)$

3. $f(x) = |\cot x|$

4. $f(x) = \sec x$

5. $f(x) = \csc x$

6. $f(x) = \dfrac{1}{1 + \sin^2 x}$

7. $f(x) = \dfrac{1}{1 - 2\sin x}$

8. $f(x) = \sqrt{2 + \tan^2 x}$

9–14 Determine where f is continuous. ■

9. $f(x) = \sin^{-1} 2x$

10. $f(x) = \cos^{-1}(\ln x)$

11. $f(x) = \dfrac{\ln(\tan^{-1} x)}{x^2 - 9}$

12. $f(x) = \exp\left(\dfrac{\sin x}{x}\right)$

13. $f(x) = \dfrac{\sin^{-1}(1/x)}{x}$

14. $f(x) = \ln|x| - 2\ln(x + 3)$

15–16 In each part, use Theorem 1.5.6(*b*) to show that the function is continuous everywhere. ■

15. (a) $\sin(x^3 + 7x + 1)$ (b) $|\sin x|$ $\cos^3(x+1)$

16. (a) $|3 + \sin 2x|$ (b) $\sin(\sin x)$
     (c) $\cos^5 x - 2\cos^3 x + 1$

17–42 Find the limits. ■

17. $\lim\limits_{x \to +\infty} \cos\left(\dfrac{1}{x}\right)$

18. $\lim\limits_{x \to +\infty} \sin\left(\dfrac{\pi x}{2 - 3x}\right)$

19. $\lim\limits_{x \to +\infty} \sin^{-1}\left(\dfrac{x}{1 - 2x}\right)$

20. $\lim\limits_{x \to +\infty} \ln\left(\dfrac{x+1}{x}\right)$

21. $\lim\limits_{x \to 0} e^{\sin x}$

22. $\lim\limits_{x \to +\infty} \cos(2\tan^{-1} x)$

23. $\lim\limits_{\theta \to 0} \dfrac{\sin 3\theta}{\theta}$

24. $\lim\limits_{h \to 0} \dfrac{\sin h}{2h}$

25. $\lim\limits_{\theta \to 0^+} \dfrac{\sin \theta}{\theta^2}$

26. $\lim\limits_{\theta \to 0} \dfrac{\sin^2 \theta}{\theta}$

27. $\lim\limits_{x \to 0} \dfrac{\tan 7x}{\sin 3x}$

28. $\lim\limits_{x \to 0} \dfrac{\sin 6x}{\sin 8x}$

29. $\lim\limits_{x \to 0^+} \dfrac{\sin x}{5\sqrt{x}}$

30. $\lim\limits_{x \to 0} \dfrac{\sin^2 x}{3x^2}$

31. $\lim\limits_{x \to 0} \dfrac{\sin x^2}{x}$

32. $\lim\limits_{h \to 0} \dfrac{\sin h}{1 - \cos h}$

33. $\lim\limits_{t \to 0} \dfrac{t^2}{1 - \cos^2 t}$

34. $\lim\limits_{x \to 0} \dfrac{x}{\cos\left(\frac{1}{2}\pi - x\right)}$

35. $\lim\limits_{\theta \to 0} \dfrac{\theta^2}{1 - \cos \theta}$

36. $\lim\limits_{h \to 0} \dfrac{1 - \cos 3h}{\cos^2 5h - 1}$

37. $\lim\limits_{x \to 0^+} \sin\left(\dfrac{1}{x}\right)$

38. $\lim\limits_{x \to 0} \dfrac{x^2 - 3\sin x}{x}$

39. $\lim\limits_{x \to 0} \dfrac{2 - \cos 3x - \cos 4x}{x}$

40. $\lim\limits_{x \to 0} \dfrac{\tan 3x^2 + \sin^2 5x}{x^2}$

41–42 (a) Complete the table and make a guess about the limit indicated. (b) Find the exact value of the limit. ■

41. $f(x) = \dfrac{\sin(x-5)}{x^2 - 25}$; $\lim\limits_{x \to 5} f(x)$

x	4	4.5	4.9	5.1	5.5	6
$f(x)$						

◀ **Table Ex-41**

42. $f(x) = \dfrac{\sin(x^2 + 3x + 2)}{x + 2}$; $\lim\limits_{x \to -2} f(x)$

x	−2.1	−2.01	−2.001	−1.999	−1.99	−1.9
$f(x)$						

▲ **Table Ex-42**

43–46 True–False Determine whether the statement is true or false. Explain your answer. ■

43. Suppose that for all real numbers x, a function f satisfies

$$|f(x) + 5| \le |x + 1|$$

Then $\lim\limits_{x \to -1} f(x) = -5$.

44. For $0 < x < \pi/2$, the graph of $y = \sin x$ lies below the graph of $y = x$ and above the graph of $y = x \cos x$.

45. If an invertible function f is continuous everywhere, then its inverse f^{-1} is also continuous everywhere.

46. Suppose that M is a positive number and that for all real numbers x, a function f satisfies

$$-M \le f(x) \le M$$

Then

$$\lim\limits_{x \to 0} xf(x) = 0 \quad \text{and} \quad \lim\limits_{x \to +\infty} \dfrac{f(x)}{x} = 0$$

FOCUS ON CONCEPTS

47. In an attempt to verify that $\lim\limits_{x \to 0} (\sin x)/x = 1$, a student constructs the accompanying table.
    (a) What mistake did the student make?
    (b) What is the exact value of the limit illustrated by this table?

x	−0.01	−0.001	0.001	0.01
$\sin x/x$	0.017453	0.017453	0.017453	0.017453

▲ **Table Ex-47**

48. Consider $\lim\limits_{x \to 0} (1 - \cos x)/x$, where x is in degrees. Why is it possible to evaluate this limit with little or no computation?

49. In the circle in the accompanying figure, a central angle of measure θ radians subtends a chord of length $c(\theta)$ and a circular arc of length $s(\theta)$. Based on your intuition, what would you conjecture is the value of $\lim\limits_{\theta \to 0^+} c(\theta)/s(\theta)$? Verify your conjecture by computing the limit.

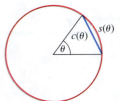

◀ **Figure Ex-49**

50. What is wrong with the following "proof" that $\lim\limits_{x \to 0}[(\sin 2x)/x] = 1$? Since

$$\lim\limits_{x \to 0} (\sin 2x - x) = \lim\limits_{x \to 0} \sin 2x - \lim\limits_{x \to 0} x = 0 - 0 = 0$$

if x is close to 0, then $\sin 2x - x \approx 0$ or, equivalently, $\sin 2x \approx x$. Dividing both sides of this approximate equality by x yields $(\sin 2x)/x \approx 1$. That is, $\lim_{x \to 0}[(\sin 2x)/x] = 1$.

51. Find a nonzero value for the constant k that makes

$$f(x) = \begin{cases} \dfrac{\tan kx}{x}, & x < 0 \\ 3x + 2k^2, & x \geq 0 \end{cases}$$

continuous at $x = 0$.

52. Is

$$f(x) = \begin{cases} \dfrac{\sin x}{|x|}, & x \neq 0 \\ 1, & x = 0 \end{cases}$$

continuous at $x = 0$? Explain.

53. In parts (a)–(c), find the limit by making the indicated substitution.

(a) $\lim\limits_{x \to +\infty} x \sin \dfrac{1}{x}$; $\quad t = \dfrac{1}{x}$

(b) $\lim\limits_{x \to -\infty} x \left(1 - \cos \dfrac{1}{x}\right)$; $\quad t = \dfrac{1}{x}$

(c) $\lim\limits_{x \to \pi} \dfrac{\pi - x}{\sin x}$; $\quad t = \pi - x$

54. Find $\lim\limits_{x \to 2} \dfrac{\cos(\pi/x)}{x - 2}$. $\quad \left[Hint: \text{ Let } t = \dfrac{\pi}{2} - \dfrac{\pi}{x}. \right]$

55. Find $\lim\limits_{x \to 1} \dfrac{\sin(\pi x)}{x - 1}$.   **56.** Find $\lim\limits_{x \to \pi/4} \dfrac{\tan x - 1}{x - \pi/4}$.

57. Find $\lim\limits_{x \to \pi/4} \dfrac{\cos x - \sin x}{x - \pi/4}$.

58. Suppose that f is an invertible function, $f(0) = 0$, f is continuous at 0, and $\lim_{x \to 0}(f(x)/x)$ exists. Given that $L = \lim_{x \to 0}(f(x)/x)$, show

$$\lim_{x \to 0} \dfrac{x}{f^{-1}(x)} = L$$

[*Hint:* Apply Theorem 1.5.5 to the composition $h \circ g$, where

$$h(x) = \begin{cases} f(x)/x, & x \neq 0 \\ L, & x = 0 \end{cases}$$

and $g(x) = f^{-1}(x)$.]

59–62 Apply the result of Exercise 58, if needed, to find the limits. ■

59. $\lim\limits_{x \to 0} \dfrac{x}{\sin^{-1} x}$ **60.** $\lim\limits_{x \to 0} \dfrac{\tan^{-1} x}{x}$

61. $\lim\limits_{x \to 0} \dfrac{\sin^{-1} 5x}{x}$ **62.** $\lim\limits_{x \to 1} \dfrac{\sin^{-1}(x - 1)}{x^2 - 1}$

FOCUS ON CONCEPTS

63. In Example 5 we used the Squeezing Theorem to prove that

$$\lim_{x \to 0} x \sin\left(\frac{1}{x}\right) = 0$$

Why couldn't we have obtained the same result by writing

$$\lim_{x \to 0} x \sin\left(\frac{1}{x}\right) = \lim_{x \to 0} x \cdot \lim_{x \to 0} \sin\left(\frac{1}{x}\right)$$
$$= 0 \cdot \lim_{x \to 0} \sin\left(\frac{1}{x}\right) = 0?$$

64. Sketch the graphs of the curves $y = 1 - x^2$, $y = \cos x$, and $y = f(x)$, where f is a function that satisfies the inequalities

$$1 - x^2 \leq f(x) \leq \cos x$$

for all x in the interval $(-\pi/2, \pi/2)$. What can you say about the limit of $f(x)$ as $x \to 0$? Explain.

65. Sketch the graphs of the curves $y = 1/x$, $y = -1/x$, and $y = f(x)$, where f is a function that satisfies the inequalities

$$-\frac{1}{x} \leq f(x) \leq \frac{1}{x}$$

for all x in the interval $[1, +\infty)$. What can you say about the limit of $f(x)$ as $x \to +\infty$? Explain your reasoning.

66. Draw pictures analogous to Figure 1.6.2 that illustrate the Squeezing Theorem for limits of the forms $\lim_{x \to +\infty} f(x)$ and $\lim_{x \to -\infty} f(x)$.

67. (a) Use the Intermediate-Value Theorem to show that the equation $x = \cos x$ has at least one solution in the interval $[0, \pi/2]$.
(b) Show graphically that there is exactly one solution in the interval.
(c) Approximate the solution to three decimal places.

68. (a) Use the Intermediate-Value Theorem to show that the equation $x + \sin x = 1$ has at least one solution in the interval $[0, \pi/6]$.
(b) Show graphically that there is exactly one solution in the interval.
(c) Approximate the solution to three decimal places.

69. In the study of falling objects near the surface of the Earth, the *acceleration g due to gravity* is commonly taken to be a constant 9.8 m/s². However, the elliptical shape of the Earth and other factors cause variations in this value that depend on latitude. The following formula, known as the World Geodetic System 1984 (WGS 84) Ellipsoidal Gravity Formula, is used to predict the value of g at a latitude of ϕ degrees (either north or south of the equator):

$$g = 9.7803253359 \dfrac{1 + 0.0019318526461 \sin^2 \phi}{\sqrt{1 - 0.0066943799901 \sin^2 \phi}} \text{ m/s}^2$$

(*cont.*)

(a) Use a graphing utility to graph the curve $y = g(\phi)$ for $0° \le \phi \le 90°$. What do the values of g at $\phi = 0°$ and at $\phi = 90°$ tell you about the WGS 84 ellipsoid model for the Earth?

(b) Show that $g = 9.8$ m/s^2 somewhere between latitudes of 38° and 39°.

70. Writing In your own words, explain the *practical value* of the Squeezing Theorem.

71. Writing A careful examination of the proof of Theorem 1.6.5 raises the issue of whether the proof might actually be a circular argument! Read the article "A Circular Argument" by Fred Richman in the March 1993 issue of *The College Mathematics Journal*, and write a short report on the author's principal points.

✔ QUICK CHECK ANSWERS 1.6

1. (a) yes (b) yes (c) yes (d) no   **2.** (a) 1 (b) 0   **3.** 3; 0   **4.** (a) $(-\infty, +\infty)$ (b) $(0, +\infty)$ (c) $[-1, 1]$ (d) $(-\infty, +\infty)$

CHAPTER 1 REVIEW EXERCISES   ∿ Graphing Utility   [C] CAS

1. For the function f graphed in the accompanying figure, find the limit if it exists.

(a) $\lim\limits_{x \to 1} f(x)$ (b) $\lim\limits_{x \to 2} f(x)$ (c) $\lim\limits_{x \to 3} f(x)$

(d) $\lim\limits_{x \to 4} f(x)$ (e) $\lim\limits_{x \to +\infty} f(x)$ (f) $\lim\limits_{x \to -\infty} f(x)$

(g) $\lim\limits_{x \to 3^+} f(x)$ (h) $\lim\limits_{x \to 3^-} f(x)$ (i) $\lim\limits_{x \to 0} f(x)$

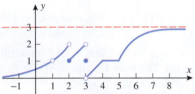

◀ **Figure Ex-1**

2. In each part, complete the table and make a conjecture about the value of the limit indicated. Confirm your conjecture by finding the limit analytically.

(a) $f(x) = \dfrac{x-2}{x^2-4}$; $\lim\limits_{x \to 2^+} f(x)$

x	2.00001	2.0001	2.001	2.01	2.1	2.5
$f(x)$						

(b) $f(x) = \dfrac{\tan 4x}{x}$; $\lim\limits_{x \to 0} f(x)$

x	−0.01	−0.001	−0.0001	0.0001	0.001	0.01
$f(x)$						

∿ **3.** (a) Approximate the value for the limit

$$\lim_{x \to 0} \frac{3^x - 2^x}{x}$$

to three decimal places by constructing an appropriate table of values.

(b) Confirm your approximation using graphical evidence.

[C] **4.** Approximate

$$\lim_{x \to 3} \frac{2^x - 8}{x - 3}$$

both by looking at a graph and by calculating values for some appropriate choices of x. Compare your answer with the value produced by a CAS.

5–10 Find the limits. ■

5. $\lim\limits_{x \to -1} \dfrac{x^3 - x^2}{x - 1}$

6. $\lim\limits_{x \to 1} \dfrac{x^3 - x^2}{x - 1}$

7. $\lim\limits_{x \to -3} \dfrac{3x + 9}{x^2 + 4x + 3}$

8. $\lim\limits_{x \to 2^-} \dfrac{x + 2}{x - 2}$

9. $\lim\limits_{x \to +\infty} \dfrac{(2x - 1)^5}{(3x^2 + 2x - 7)(x^3 - 9x)}$

10. $\lim\limits_{x \to 0} \dfrac{\sqrt{x^2 + 4} - 2}{x^2}$

11. In each part, find the horizontal asymptotes, if any.

(a) $y = \dfrac{2x - 7}{x^2 - 4x}$

(b) $y = \dfrac{x^3 - x^2 + 10}{3x^2 - 4x}$

(c) $y = \dfrac{2x^2 - 6}{x^2 + 5x}$

12. In each part, find $\lim\limits_{x \to a} f(x)$, if it exists, where a is replaced by $0, 5^+, -5^-, -5, 5, -\infty$, and $+\infty$.

(a) $f(x) = \sqrt{5 - x}$

(b) $f(x) = \begin{cases} (x - 5)/|x - 5|, & x \ne 5 \\ 0, & x = 5 \end{cases}$

13–20 Find the limits. ■

13. $\lim\limits_{x \to 0} \dfrac{\sin 3x}{\tan 3x}$

14. $\lim\limits_{x \to 0} \dfrac{x \sin x}{1 - \cos x}$

15. $\lim\limits_{x \to 0} \dfrac{3x - \sin(kx)}{x}$, $k \ne 0$

16. $\lim\limits_{\theta \to 0} \tan\left(\dfrac{1 - \cos\theta}{\theta}\right)$

17. $\lim\limits_{t \to \pi/2^+} e^{\tan t}$

18. $\lim\limits_{\theta \to 0^+} \ln(\sin 2\theta) - \ln(\tan\theta)$

19. $\displaystyle\lim_{x \to +\infty} \left(1 + \frac{3}{x}\right)^{-x}$

20. $\displaystyle\lim_{x \to +\infty} \left(1 + \frac{a}{x}\right)^{bx}$, $\quad a, b > 0$

21. If \$1000 is invested in an account that pays 7% interest compounded n times each year, then in 10 years there will be $1000(1 + 0.07/n)^{10n}$ dollars in the account. How much money will be in the account in 10 years if the interest is compounded quarterly ($n = 4$)? Monthly ($n = 12$)? Daily ($n = 365$)? Determine the amount of money that will be in the account in 10 years if the interest is compounded *continuously*, that is, as $n \to +\infty$.

22. (a) Write a paragraph or two that describes how the limit of a function can fail to exist at $x = a$, and accompany your description with some specific examples.
   (b) Write a paragraph or two that describes how the limit of a function can fail to exist as $x \to +\infty$ or $x \to -\infty$, and accompany your description with some specific examples.
   (c) Write a paragraph or two that describes how a function can fail to be continuous at $x = a$, and accompany your description with some specific examples.

23. (a) Find a formula for a rational function that has a vertical asymptote at $x = 1$ and a horizontal asymptote at $y = 2$.
   (b) Check your work by using a graphing utility to graph the function.

24. Paraphrase the ϵ-δ definition for $\lim_{x \to a} f(x) = L$ in terms of a graphing utility viewing window centered at the point (a, L).

25. Suppose that $f(x)$ is a function and that for any given $\epsilon > 0$, the condition $0 < |x - 2| < \frac{3}{4}\epsilon$ guarantees that $|f(x) - 5| < \epsilon$.
   (a) What limit is described by this statement?
   (b) Find a value of δ such that $0 < |x - 2| < \delta$ guarantees that $|8f(x) - 40| < 0.048$.

26. The limit
$$\lim_{x \to 0} \frac{\sin x}{x} = 1$$
ensures that there is a number δ such that
$$\left|\frac{\sin x}{x} - 1\right| < 0.001$$
if $0 < |x| < \delta$. Estimate the largest such δ.

27. In each part, a positive number ϵ and the limit L of a function f at a are given. Find a number δ such that $|f(x) - L| < \epsilon$ if $0 < |x - a| < \delta$.
   (a) $\lim_{x \to 2} (4x - 7) = 1$; $\epsilon = 0.01$
   (b) $\lim_{x \to 3/2} \dfrac{4x^2 - 9}{2x - 3} = 6$; $\epsilon = 0.05$
   (c) $\lim_{x \to 4} x^2 = 16$; $\epsilon = 0.001$

28. Use Definition 1.4.1 to prove the stated limits are correct.
   (a) $\lim_{x \to 2} (4x - 7) = 1$
   (b) $\lim_{x \to 3/2} \dfrac{4x^2 - 9}{2x - 3} = 6$

29. Suppose that f is continuous at x_0 and that $f(x_0) > 0$. Give either an ϵ-δ proof or a convincing verbal argument to show that there must be an open interval containing x_0 on which $f(x) > 0$.

30. (a) Let
$$f(x) = \frac{\sin x - \sin 1}{x - 1}$$
   Approximate $\lim_{x \to 1} f(x)$ by graphing f and calculating values for some appropriate choices of x.
   (b) Use the identity
$$\sin \alpha - \sin \beta = 2 \sin \frac{\alpha - \beta}{2} \cos \frac{\alpha + \beta}{2}$$
   to find the exact value of $\lim_{x \to 1} f(x)$.

31. Find values of x, if any, at which the given function is not continuous.
   (a) $f(x) = \dfrac{x}{x^2 - 1}$
   (b) $f(x) = |x^3 - 2x^2|$
   (c) $f(x) = \dfrac{x + 3}{|x^2 + 3x|}$

32. Determine where f is continuous.
   (a) $f(x) = \dfrac{x}{|x| - 3}$
   (b) $f(x) = \cos^{-1}\left(\dfrac{1}{x}\right)$
   (c) $f(x) = e^{\ln x}$

33. Suppose that
$$f(x) = \begin{cases} -x^4 + 3, & x \leq 2 \\ x^2 + 9, & x > 2 \end{cases}$$
Is f continuous everywhere? Justify your conclusion.

34. One dictionary describes a continuous function as "one whose value at each point is closely approached by its values at neighboring points."
   (a) How would you explain the meaning of the terms "neighboring points" and "closely approached" to a nonmathematician?
   (b) Write a paragraph that explains why the dictionary definition is consistent with Definition 1.5.1.

35. Show that the conclusion of the Intermediate-Value Theorem may be false if f is not continuous on the interval $[a, b]$.

36. Suppose that f is continuous on the interval $[0, 1]$, that $f(0) = 2$, and that f has no zeros in the interval. Prove that $f(x) > 0$ for all x in $[0, 1]$.

37. Show that the equation $x^4 + 5x^3 + 5x - 1 = 0$ has at least two real solutions in the interval $[-6, 2]$.

CHAPTER 1 MAKING CONNECTIONS

In Section 1.1 we developed the notion of a tangent line to a graph at a given point by considering it as a limiting position of secant lines through that point (Figure 1.1.4a). In these exercises we will develop an analogous idea in which secant lines are replaced by "secant circles" and the tangent line is replaced by a "tangent circle" (called the *osculating circle*). We begin with the graph of $y = x^2$.

1. Recall that there is a unique circle through any three noncollinear points in the plane. For any positive real number x, consider the unique "secant circle" that passes through the fixed point $O(0, 0)$ and the variable points $Q(-x, x^2)$ and $P(x, x^2)$ (see the accompanying figure). Use plane geometry to explain why the center of this circle is the intersection of the y-axis and the perpendicular bisector of segment OP.

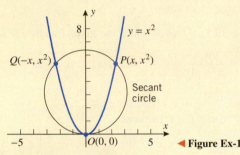

◀ **Figure Ex-1**

2. (a) Let $(0, C(x))$ denote the center of the circle in Exercise 1 and show that

   $$C(x) = \tfrac{1}{2}x^2 + \tfrac{1}{2}$$

   (b) Show that as $x \to 0^+$, the secant circles approach a limiting position given by the circle that passes through the origin and is centered at $\left(0, \tfrac{1}{2}\right)$. As shown in the accom-

panying figure, this circle is the osculating circle to the graph of $y = x^2$ at the origin.

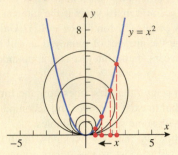

◀ **Figure Ex-2**

3. Show that if we replace the curve $y = x^2$ by the curve $y = f(x)$, where f is an even function, then the formula for $C(x)$ becomes

   $$C(x) = \frac{1}{2}\left[f(0) + f(x) + \frac{x^2}{f(x) - f(0)} \right]$$

   [Here we assume that $f(x) \neq f(0)$ for positive values of x close to 0.] If $\lim_{x \to 0^+} C(x) = L \neq f(0)$, then we define the osculating circle to the curve $y = f(x)$ at $(0, f(0))$ to be the unique circle through $(0, f(0))$ with center $(0, L)$. If $C(x)$ does not have a finite limit different from $f(0)$ as $x \to 0^+$, then we say that the curve has no osculating circle at $(0, f(0))$.

4. In each part, determine the osculating circle to the curve $y = f(x)$ at $(0, f(0))$, if it exists.
   (a) $f(x) = 4x^2$ (b) $f(x) = x^2 \cos x$
   (c) $f(x) = |x|$ (d) $f(x) = x \sin x$
   (e) $f(x) = \cos x$
   (f) $f(x) = x^2 g(x)$, where $g(x)$ is an even continuous function with $g(0) \neq 0$
   (g) $f(x) = x^4$

Photo by Kirby Lee/WireImage/Getty Images

2

THE DERIVATIVE

One of the crowning achievements of calculus is its ability to capture continuous motion mathematically, allowing that motion to be analyzed instant by instant.

Many real-world phenomena involve changing quantities—the speed of a rocket, the inflation of currency, the number of bacteria in a culture, the shock intensity of an earthquake, the voltage of an electrical signal, and so forth. In this chapter we will develop the concept of a "derivative," which is the mathematical tool for studying the rate at which one quantity changes relative to another. The study of rates of change is closely related to the geometric concept of a tangent line to a curve, so we will also be discussing the general definition of a tangent line and methods for finding its slope and equation.

2.1 TANGENT LINES AND RATES OF CHANGE

In this section we will discuss three ideas: tangent lines to curves, the velocity of an object moving along a line, and the rate at which one variable changes relative to another. Our goal is to show how these seemingly unrelated ideas are, in actuality, closely linked.

■ TANGENT LINES

In Example 1 of Section 1.1, we showed how the notion of a limit could be used to find an equation of a tangent line to a curve. At that stage in the text we did not have precise definitions of tangent lines and limits to work with, so the argument was intuitive and informal. However, now that limits have been defined precisely, we are in a position to give a mathematical definition of the tangent line to a curve $y = f(x)$ at a point $P(x_0, f(x_0))$ on the curve. As illustrated in Figure 2.1.1, consider a point $Q(x, f(x))$ on the curve that is distinct from P, and compute the slope m_{PQ} of the secant line through P and Q:

$$m_{PQ} = \frac{f(x) - f(x_0)}{x - x_0}$$

If we let x approach x_0, then the point Q will move along the curve and approach the point P. If the secant line through P and Q approaches a limiting position as $x \to x_0$, then we will regard that position to be the position of the tangent line at P. Stated another way, if the slope m_{PQ} of the secant line through P and Q approaches a limit as $x \to x_0$, then we regard that limit to be the slope $m_{\tan}$ of the tangent line at P. Thus, we make the following definition.

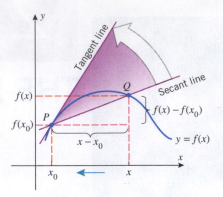

2.1.1   DEFINITION   Suppose that x_0 is in the domain of the function f. The ***tangent line*** to the curve $y = f(x)$ at the point $P(x_0, f(x_0))$ is the line with equation

$$y - f(x_0) = m_{tan}(x - x_0)$$

where

$$m_{tan} = \lim_{x \to x_0} \frac{f(x) - f(x_0)}{x - x_0} \tag{1}$$

provided the limit exists.   For simplicity, we will also call this the tangent line to $y = f(x)$ at x_0.

▶ **Example 1**   Use Definition 2.1.1 to find an equation for the tangent line to the parabola $y = x^2$ at the point $P(1, 1)$, and confirm the result agrees with that obtained in Example 1 of Section 1.1.

Solution.   Applying Formula (1) with $f(x) = x^2$ and $x_0 = 1$, we have

$$m_{tan} = \lim_{x \to 1} \frac{f(x) - f(1)}{x - 1}$$

$$= \lim_{x \to 1} \frac{x^2 - 1}{x - 1}$$

$$= \lim_{x \to 1} \frac{(x - 1)(x + 1)}{x - 1} = \lim_{x \to 1} (x + 1) = 2$$

Thus, the tangent line to $y = x^2$ at $(1, 1)$ has equation

$$y - 1 = 2(x - 1) \quad \text{or equivalently} \quad y = 2x - 1$$

which agrees with Example 1 of Section 1.1.   ◀

There is an alternative way of expressing Formula (1) that is commonly used. If we let h denote the difference

$$h = x - x_0$$

then the statement that $x \to x_0$ is equivalent to the statement $h \to 0$, so we can rewrite (1) in terms of x_0 and h as

$$m_{tan} = \lim_{h \to 0} \frac{f(x_0 + h) - f(x_0)}{h} \tag{2}$$

Figure 2.1.2 shows how Formula (2) expresses the slope of the tangent line as a limit of slopes of secant lines.

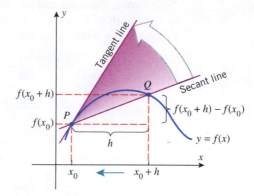

▶ **Figure 2.1.2**

▶ **Example 2** Compute the slope in Example 1 using Formula (2).

Solution. Applying Formula (2) with $f(x) = x^2$ and $x_0 = 1$, we obtain

$$m_{\tan} = \lim_{h \to 0} \frac{f(1 + h) - f(1)}{h}$$

$$= \lim_{h \to 0} \frac{(1 + h)^2 - 1^2}{h}$$

$$= \lim_{h \to 0} \frac{1 + 2h + h^2 - 1}{h} = \lim_{h \to 0} (2 + h) = 2$$

which agrees with the slope found in Example 1. ◀

> Formulas (1) and (2) for $m_{\tan}$ usually lead to indeterminate forms of type $0/0$, so you will generally need to perform algebraic simplifications or use other methods to determine limits of such indeterminate forms.

▶ **Example 3** Find an equation for the tangent line to the curve $y = 2/x$ at the point $(2, 1)$ on this curve.

Solution. First, we will find the slope of the tangent line by applying Formula (2) with $f(x) = 2/x$ and $x_0 = 2$. This yields

$$m_{\tan} = \lim_{h \to 0} \frac{f(2 + h) - f(2)}{h}$$

$$= \lim_{h \to 0} \frac{\dfrac{2}{2 + h} - 1}{h} = \lim_{h \to 0} \frac{\left(\dfrac{2 - (2 + h)}{2 + h} \right)}{h}$$

$$= \lim_{h \to 0} \frac{-h}{h(2 + h)} = - \left(\lim_{h \to 0} \frac{1}{2 + h} \right) = -\frac{1}{2}$$

Thus, an equation of the tangent line at $(2, 1)$ is

$$y - 1 = -\tfrac{1}{2}(x - 2) \quad \text{or equivalently} \quad y = -\tfrac{1}{2}x + 2$$

(see Figure 2.1.3). ◀

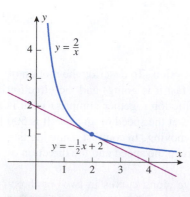

▲ **Figure 2.1.3**

▶ **Example 4** Find the slopes of the tangent lines to the curve $y = \sqrt{x}$ at $x_0 = 1$, $x_0 = 4$, and $x_0 = 9$.

Solution. We could compute each of these slopes separately, but it will be more efficient to find the slope for a general value of x_0 and then substitute the specific numerical values. Proceeding in this way we obtain

$$m_{\tan} = \lim_{h \to 0} \frac{f(x_0 + h) - f(x_0)}{h}$$

$$= \lim_{h \to 0} \frac{\sqrt{x_0 + h} - \sqrt{x_0}}{h}$$

$$= \lim_{h \to 0} \frac{\sqrt{x_0 + h} - \sqrt{x_0}}{h} \cdot \frac{\sqrt{x_0 + h} + \sqrt{x_0}}{\sqrt{x_0 + h} + \sqrt{x_0}}$$

> Rationalize the numerator to help eliminate the indeterminate form of the limit.

$$= \lim_{h \to 0} \frac{x_0 + h - x_0}{h(\sqrt{x_0 + h} + \sqrt{x_0})}$$

$$= \lim_{h \to 0} \frac{h}{h(\sqrt{x_0 + h} + \sqrt{x_0})}$$

$$= \lim_{h \to 0} \frac{1}{\sqrt{x_0 + h} + \sqrt{x_0}} = \frac{1}{2\sqrt{x_0}}$$

The slopes at $x_0 = 1, 4$, and 9 can now be obtained by substituting these values into our general formula for $m_{\tan}$. Thus,

$$\text{slope at } x_0 = 1: \frac{1}{2\sqrt{1}} = \frac{1}{2}$$

$$\text{slope at } x_0 = 4: \frac{1}{2\sqrt{4}} = \frac{1}{4}$$

$$\text{slope at } x_0 = 9: \frac{1}{2\sqrt{9}} = \frac{1}{6}$$

(see Figure 2.1.4). ◀

▶ **Figure 2.1.4**

■ VELOCITY

One of the important themes in calculus is the study of motion. To describe the motion of an object completely, one must specify its *speed* (how fast it is going) and the direction in which it is moving. The speed and the direction of motion together comprise what is called the *velocity* of the object. For example, knowing that the speed of an aircraft is 500 mi/h tells us how fast it is going, but not which way it is moving. In contrast, knowing that the velocity of the aircraft is 500 mi/h *due south* pins down the speed and the direction of motion.

Later, we will study the motion of objects that move along curves in two- or three-dimensional space, but for now we will only consider motion along a line; this is called *rectilinear motion*. Some examples are a piston moving up and down in a cylinder, a race

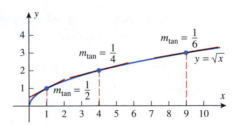

The velocity of an airplane describes its speed and direction.

car moving along a straight track, an object dropped from the top of a building and falling straight down, a ball thrown straight up and then falling down along the same line, and so forth.

For computational purposes, we will assume that a particle in rectilinear motion moves along a coordinate line, which we will call the s-axis. A graphical description of rectilinear motion along an s-axis can be obtained by making a plot of the s-coordinate of the particle versus the elapsed time t from starting time $t = 0$. This is called the *position versus time curve* for the particle. Figure 2.1.5 shows two typical position versus time curves. The first is for a car that starts at the origin and moves only in the positive direction of the s-axis. In this case s increases as t increases. The second is for a ball that is thrown straight up in the positive direction of an s-axis from some initial height s_0 and then falls straight down in the negative direction. In this case s increases as the ball moves up and decreases as it moves down.

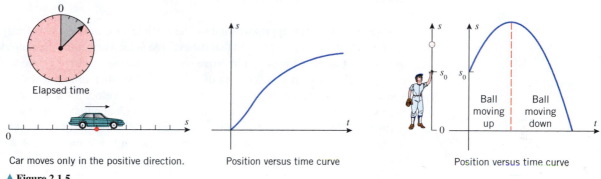

Car moves only in the positive direction.

Position versus time curve

Ball moving up | Ball moving down

Position versus time curve

▲ **Figure 2.1.5**

If a particle in rectilinear motion moves along an s-axis so that its position coordinate function of the elapsed time t is

$$s = f(t) \tag{3}$$

then f is called the *position function of the particle*; the graph of (3) is the position versus time curve. The *average velocity* of the particle over a time interval $[t_0, t_0 + h]$, $h > 0$, is defined to be

$$v_{\text{ave}} = \frac{\text{change in position}}{\text{time elapsed}} = \frac{f(t_0 + h) - f(t_0)}{h} \tag{4}$$

Show that (4) is also correct for a time interval $[t_0 + h, t_0]$, $h < 0$.

The change in position

$$f(t_0 + h) - f(t_0)$$

is also called the *displacement* of the particle over the time interval between t_0 and $t_0 + h$.

▶ **Example 5** Suppose that $s = f(t) = 1 + 5t - 2t^2$ is the position function of a particle, where s is in meters and t is in seconds. Find the average velocities of the particle over the time intervals (a) $[0, 2]$ and (b) $[2, 3]$.

Solution (a). Applying (4) with $t_0 = 0$ and $h = 2$, we see that the average velocity is

$$v_{\text{ave}} = \frac{f(t_0 + h) - f(t_0)}{h} = \frac{f(2) - f(0)}{2} = \frac{3 - 1}{2} = \frac{2}{2} = 1 \text{ m/s}$$

Solution (b). Applying (4) with $t_0 = 2$ and $h = 1$, we see that the average velocity is

$$v_{\text{ave}} = \frac{f(t_0 + h) - f(t_0)}{h} = \frac{f(3) - f(2)}{1} = \frac{-2 - 3}{1} = \frac{-5}{1} = -5 \text{ m/s} \blacktriangleleft$$

For a particle in rectilinear motion, average velocity describes its behavior over an *interval* of time. We are interested in the particle's "instantaneous velocity," which describes

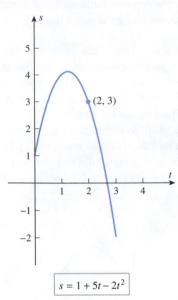

$s = 1 + 5t - 2t^2$

▲ **Figure 2.1.6**

Table 2.1.1

TIME INTERVAL	AVERAGE VELOCITY (m/s)
$2.0 \leq t \leq 3.0$	-5
$2.0 \leq t \leq 2.1$	-3.2
$2.0 \leq t \leq 2.01$	-3.02
$2.0 \leq t \leq 2.001$	-3.002
$2.0 \leq t \leq 2.0001$	-3.0002

Note the negative values for the velocities in Example 6. This is consistent with the fact that the object is moving in the negative direction along the s-axis.

its behavior at a specific *instant* in time. Formula (4) is not directly applicable for computing instantaneous velocity because the "time elapsed" at a specific instant is zero, so (4) is undefined. One way to circumvent this problem is to compute average velocities for small time intervals between $t = t_0$ and $t = t_0 + h$. These average velocities may be viewed as approximations to the "instantaneous velocity" of the particle at time t_0. If these average velocities have a limit as h approaches zero, then we can take that limit to be the *instantaneous velocity* of the particle at time t_0. Here is an example.

▶ **Example 6**   Consider the particle in Example 5, whose position function is

$$s = f(t) = 1 + 5t - 2t^2$$

The position of the particle at time $t = 2$ s is $s = 3$ m (Figure 2.1.6). Find the particle's instantaneous velocity at time $t = 2$ s.

Solution.   As a first approximation to the particle's instantaneous velocity at time $t = 2$ s, let us recall from Example 5(b) that the average velocity over the time interval from $t = 2$ to $t = 3$ is $v_{ave} = -5$ m/s. To improve on this initial approximation we will compute the average velocity over a succession of smaller and smaller time intervals. We leave it to you to verify the results in Table 2.1.1. The average velocities in this table appear to be approaching a limit of -3 m/s, providing strong evidence that the instantaneous velocity at time $t = 2$ s is -3 m/s. To confirm this analytically, we start by computing the object's average velocity over a general time interval between $t = 2$ and $t = 2 + h$ using Formula (4):

$$v_{ave} = \frac{f(2 + h) - f(2)}{h} = \frac{[1 + 5(2 + h) - 2(2 + h)^2] - 3}{h}$$

The object's instantaneous velocity at time $t = 2$ is calculated as a limit as $h \to 0$:

$$\text{instantaneous velocity} = \lim_{h \to 0} \frac{[1 + 5(2 + h) - 2(2 + h)^2] - 3}{h}$$

$$= \lim_{h \to 0} \frac{-2 + (10 + 5h) - (8 + 8h + 2h^2)}{h}$$

$$= \lim_{h \to 0} \frac{-3h - 2h^2}{h} = \lim_{h \to 0} (-3 - 2h) = -3$$

This confirms our numerical conjecture that the instantaneous velocity after 2 s is -3 m/s.   ◀

Consider a particle in rectilinear motion with position function $s = f(t)$. Motivated by Example 6, we define the instantaneous velocity v_{inst} of the particle at time t_0 to be the limit as $h \to 0$ of its average velocities v_{ave} over time intervals between $t = t_0$ and $t = t_0 + h$. Thus, from (4) we obtain

$$v_{inst} = \lim_{h \to 0} \frac{f(t_0 + h) - f(t_0)}{h} \tag{5}$$

Confirm the solution to Example 5(b) by computing the slope of an appropriate secant line.

Geometrically, the average velocity v_{ave} between $t = t_0$ and $t = t_0 + h$ is the slope of the secant line through points $P(t_0, f(t_0))$ and $Q(t_0 + h, f(t_0 + h))$ on the position versus time curve, and the instantaneous velocity v_{inst} at time t_0 is the slope of the tangent line to the position versus time curve at the point $P(t_0, f(t_0))$ (Figure 2.1.7).

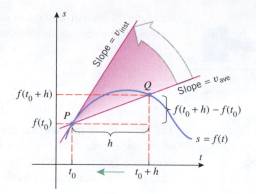

◀ **Figure 2.1.7**

SLOPES AND RATES OF CHANGE

Velocity can be viewed as *rate of change*—the rate of change of position with respect to time. Rates of change occur in other applications as well. For example:

- A microbiologist might be interested in the rate at which the number of bacteria in a colony changes with time.
- An engineer might be interested in the rate at which the length of a metal rod changes with temperature.
- An economist might be interested in the rate at which production cost changes with the quantity of a product that is manufactured.
- A medical researcher might be interested in the rate at which the radius of an artery changes with the concentration of alcohol in the bloodstream.

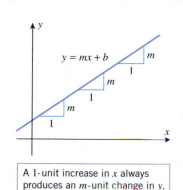

A 1-unit increase in *x* always produces an *m*-unit change in *y*.

▲ **Figure 2.1.8**

Our next objective is to define precisely what is meant by the "rate of change of y with respect to x" when y is a function of x. In the case where y is a linear function of x, say $y = mx + b$, the slope m is the natural measure of the rate of change of y with respect to x. As illustrated in Figure 2.1.8, each 1-unit increase in x anywhere along the line produces an m-unit change in y, so we see that y changes at a constant rate with respect to x along the line and that m measures this rate of change.

▶ **Example 7**   Find the rate of change of y with respect to x if

$$\text{(a)} \ \ y = 2x - 1 \qquad \text{(b)} \ \ y = -5x + 1$$

Solution.   In part (a) the rate of change of y with respect to x is $m = 2$, so each 1-unit increase in x produces a 2-unit increase in y. In part (b) the rate of change of y with respect to x is $m = -5$, so each 1-unit increase in x produces a 5-unit decrease in y. ◀

In applied problems, changing the units of measurement can change the slope of a line, so it is essential to include the units when calculating the slope and describing rates of change. The following example illustrates this.

▶ **Example 8**   Suppose that a uniform rod of length 40 cm ($= 0.4$ m) is thermally insulated around the lateral surface and that the exposed ends of the rod are held at constant temperatures of 25°C and 5°C, respectively (Figure 2.1.9*a*). It is shown in physics that under appropriate conditions the graph of the temperature T versus the distance x from the left-hand end of the rod will be a straight line. Parts (*b*) and (*c*) of Figure 2.1.9 show two

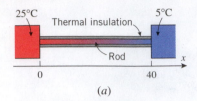

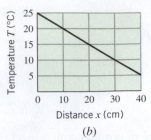

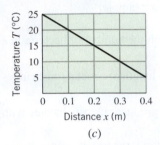

(a)

(b)

(c)

▲ **Figure 2.1.9**

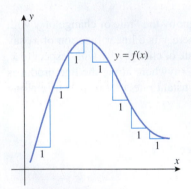

$y = f(x)$

▲ **Figure 2.1.10**

such graphs: one in which x is measured in centimeters and one in which it is measured in meters. The slopes in the two cases are

$$m = \frac{5 - 25}{40 - 0} = \frac{-20}{40} = -0.5 \tag{6}$$

$$m = \frac{5 - 25}{0.4 - 0} = \frac{-20}{0.4} = -50 \tag{7}$$

The slope in (6) implies that the temperature *decreases* at a rate of $0.5°C$ per centimeter of distance from the left end of the rod, and the slope in (7) implies that the temperature decreases at a rate of $50°C$ per meter of distance from the left end of the rod. The two statements are equivalent physically, even though the slopes differ. ◄

Although the rate of change of y with respect to x is constant along a nonvertical line $y = mx + b$, this is not true for a general curve $y = f(x)$. For example, in Figure 2.1.10 the change in y that results from a 1-unit increase in x tends to have greater magnitude in regions where the curve rises or falls rapidly than in regions where it rises or falls slowly. As with velocity, we will distinguish between the average rate of change over an interval and the instantaneous rate of change at a specific point.

If $y = f(x)$, then we define the ***average rate of change of y with respect to x over the interval*** $[x_0, x_1]$ to be

$$r_{\text{ave}} = \frac{f(x_1) - f(x_0)}{x_1 - x_0} \tag{8}$$

and we define the ***instantaneous rate of change of y with respect to x at*** x_0 to be

$$r_{\text{inst}} = \lim_{x_1 \to x_0} \frac{f(x_1) - f(x_0)}{x_1 - x_0} \tag{9}$$

Geometrically, the average rate of change of y with respect to x over the interval $[x_0, x_1]$ is the slope of the secant line through the points $P(x_0, f(x_0))$ and $Q(x_1, f(x_1))$ (Figure 2.1.11), and the instantaneous rate of change of y with respect to x at x_0 is the slope of the tangent line at the point $P(x_0, f(x_0))$ (since it is the limit of the slopes of the secant lines through P).

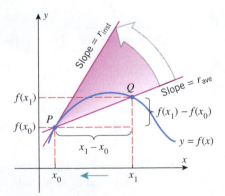

► **Figure 2.1.11**

If desired, we can let $h = x_1 - x_0$, and rewrite (8) and (9) as

$$r_{\text{ave}} = \frac{f(x_0 + h) - f(x_0)}{h} \tag{10}$$

$$r_{\text{inst}} = \lim_{h \to 0} \frac{f(x_0 + h) - f(x_0)}{h} \tag{11}$$

▶ **Example 9**   Let $y = x^2 + 1$.

(a) Find the average rate of change of y with respect to x over the interval $[3, 5]$.

(b) Find the instantaneous rate of change of y with respect to x when $x = -4$.

Solution (a).   We will apply Formula (8) with $f(x) = x^2 + 1$, $x_0 = 3$, and $x_1 = 5$. This yields

$$r_{\text{ave}} = \frac{f(x_1) - f(x_0)}{x_1 - x_0} = \frac{f(5) - f(3)}{5 - 3} = \frac{26 - 10}{2} = 8$$

Thus, y increases an average of 8 units per unit increase in x over the interval $[3, 5]$.

Solution (b).   We will apply Formula (9) with $f(x) = x^2 + 1$ and $x_0 = -4$. This yields

$$r_{\text{inst}} = \lim_{x_1 \to x_0} \frac{f(x_1) - f(x_0)}{x_1 - x_0} = \lim_{x_1 \to -4} \frac{f(x_1) - f(-4)}{x_1 - (-4)} = \lim_{x_1 \to -4} \frac{(x_1^2 + 1) - 17}{x_1 + 4}$$

$$= \lim_{x_1 \to -4} \frac{x_1^2 - 16}{x_1 + 4} = \lim_{x_1 \to -4} \frac{(x_1 + 4)(x_1 - 4)}{x_1 + 4} = \lim_{x_1 \to -4} (x_1 - 4) = -8$$

Thus, a small increase in x from $x = -4$ will produce approximately an 8-fold decrease in y. ◀

Perform the calculations in Example 9 using Formulas (10) and (11).

■ RATES OF CHANGE IN APPLICATIONS

In applied problems, average and instantaneous rates of change must be accompanied by appropriate units. In general, the units for a rate of change of y with respect to x are obtained by "dividing" the units of y by the units of x and then simplifying according to the standard rules of algebra. Here are some examples:

• If y is in degrees Fahrenheit (°F) and x is in inches (in), then a rate of change of y with respect to x has units of degrees Fahrenheit per inch (°F/in).

• If y is in feet per second (ft/s) and x is in seconds (s), then a rate of change of y with respect to x has units of feet per second per second (ft/s/s), which would usually be written as ft/s².

• If y is in newton-meters (N·m) and x is in meters (m), then a rate of change of y with respect to x has units of newtons (N), since N·m/m = N.

• If y is in foot-pounds (ft·lb) and x is in hours (h), then a rate of change of y with respect to x has units of foot-pounds per hour (ft·lb/h).

▶ **Example 10**   The limiting factor in athletic endurance is cardiac output, that is, the volume of blood that the heart can pump per unit of time during an athletic competition. Figure 2.1.12 shows a stress-test graph of cardiac output V in liters (L) of blood versus workload W in kilogram-meters (kg·m) for 1 minute of weight lifting. This graph illustrates the known medical fact that cardiac output increases with the workload, but after reaching a peak value begins to decrease.

(a) Use the secant line shown in Figure 2.1.13a to estimate the average rate of change of cardiac output with respect to workload as the workload increases from 300 to 1200 kg·m.

(b) Use the line segment shown in Figure 2.1.13b to estimate the instantaneous rate of change of cardiac output with respect to workload at the point where the workload is 300 kg·m.

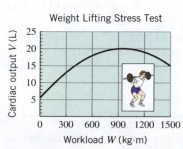

▲ **Figure 2.1.12**

Solution (a). Using the estimated points (300, 13) and (1200, 19) to find the slope of the secant line, we obtain

$$r_{\text{ave}} \approx \frac{19 - 13}{1200 - 300} \approx 0.0067 \frac{\text{L}}{\text{kg}\cdot\text{m}}$$

This means that on average a 1-unit increase in workload produced a 0.0067 L increase in cardiac output over the interval.

Solution (b). We estimate the slope of the cardiac output curve at $W = 300$ by sketching a line that appears to meet the curve at $W = 300$ with slope equal to that of the curve (Figure 2.1.13b). Estimating points (0, 7) and (900, 25) on this line, we obtain

$$r_{\text{inst}} \approx \frac{25 - 7}{900 - 0} = 0.02 \frac{\text{L}}{\text{kg}\cdot\text{m}} \quad \blacktriangleleft$$

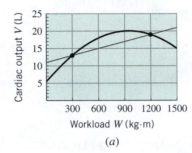

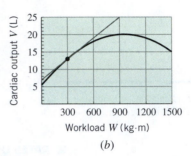

▶ **Figure 2.1.13**

(a) (b)

✔ **QUICK CHECK EXERCISES 2.1**   *(See page 143 for answers.)*

1. The slope m_{tan} of the tangent line to the curve $y = f(x)$ at the point $P(x_0, f(x_0))$ is given by

$$m_{\text{tan}} = \lim_{x \to x_0} \underline{\hspace{1cm}} = \lim_{h \to 0} \underline{\hspace{1cm}}$$

2. The tangent line to the curve $y = (x - 1)^2$ at the point $(-1, 4)$ has equation $4x + y = 0$. Thus, the value of the limit

$$\lim_{x \to -1} \frac{x^2 - 2x - 3}{x + 1}$$

is _____.

3. A particle is moving along an s-axis, where s is in feet. During the first 5 seconds of motion, the position of the particle is given by

$$s = 10 - (3 - t)^2, \quad 0 \le t \le 5$$

Use this position function to complete each part.

(a) Initially, the particle moves a distance of _____ ft in the (positive/negative) _____ direction; then it reverses direction, traveling a distance of _____ ft during the remainder of the 5-second period.

(b) The average velocity of the particle over the 5-second period is _____.

4. Let $s = f(t)$ be the equation of a position versus time curve for a particle in rectilinear motion, where s is in meters and t is in seconds. Assume that $s = -1$ when $t = 2$ and that the instantaneous velocity of the particle at this instant is 3 m/s. The equation of the tangent line to the position versus time curve at time $t = 2$ is _____.

5. Suppose that $y = x^2 + x$.
(a) The average rate of change of y with respect to x over the interval $2 \le x \le 5$ is _____.
(b) The instantaneous rate of change of y with respect to x at $x = 2$, r_{inst}, is given by the limit _____.

EXERCISE SET 2.1

1. The accompanying figure on the next page shows the position versus time curve for an elevator that moves upward a distance of 60 m and then discharges its passengers.

(a) Estimate the instantaneous velocity of the elevator at $t = 10$ s.
(b) Sketch a velocity versus time curve for the motion of the elevator for $0 \le t \le 20$.

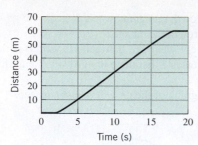

Figure Ex-1

2. The accompanying figure shows the position versus time curve for an automobile over a period of time of 10 s. Use the line segments shown in the figure to estimate the instantaneous velocity of the automobile at time $t = 4$ s and again at time $t = 8$ s.

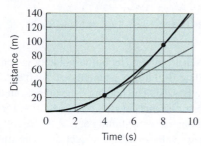

Figure Ex-2

3. The accompanying figure shows the position versus time curve for a certain particle moving along a straight line. Estimate each of the following from the graph:
   (a) the average velocity over the interval $0 \le t \le 3$
   (b) the values of t at which the instantaneous velocity is zero
   (c) the values of t at which the instantaneous velocity is either a maximum or a minimum
   (d) the instantaneous velocity when $t = 3$ s.

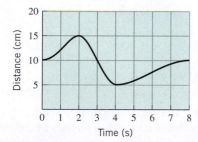

Figure Ex-3

4. The accompanying figure shows the position versus time curves of four different particles moving on a straight line. For each particle, determine whether its instantaneous velocity is increasing or decreasing with time.

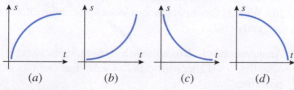

▲ **Figure Ex-4**

FOCUS ON CONCEPTS

5. If a particle moves at constant velocity, what can you say about its position versus time curve?

6. An automobile, initially at rest, begins to move along a straight track. The velocity increases steadily until suddenly the driver sees a concrete barrier in the road and applies the brakes sharply at time t_0. The car decelerates rapidly, but it is too late—the car crashes into the barrier at time t_1 and instantaneously comes to rest. Sketch a position versus time curve that might represent the motion of the car. Indicate how characteristics of your curve correspond to the events of this scenario.

7–10 For each exercise, sketch a curve and a line L satisfying the stated conditions. ■

7. L is tangent to the curve and intersects the curve in at least two points.

8. L intersects the curve in exactly one point, but L is not tangent to the curve.

9. L is tangent to the curve at two different points.

10. L is tangent to the curve at two different points and intersects the curve at a third point.

11–14 A function $y = f(x)$ and values of x_0 and x_1 are given.
   (a) Find the average rate of change of y with respect to x over the interval $[x_0, x_1]$.
   (b) Find the instantaneous rate of change of y with respect to x at the specified value of x_0.
   (c) Find the instantaneous rate of change of y with respect to x at an arbitrary value of x_0.
   (d) The average rate of change in part (a) is the slope of a certain secant line, and the instantaneous rate of change in part (b) is the slope of a certain tangent line. Sketch the graph of $y = f(x)$ together with those two lines. ■

11. $y = 2x^2$; $x_0 = 0$, $x_1 = 1$ **12.** $y = x^3$; $x_0 = 1$, $x_1 = 2$

13. $y = 1/x$; $x_0 = 2$, $x_1 = 3$ **14.** $y = 1/x^2$; $x_0 = 1$, $x_1 = 2$

15–18 A function $y = f(x)$ and an x-value x_0 are given.
   (a) Find a formula for the slope of the tangent line to the graph of f at a general point $x = x_0$.
   (b) Use the formula obtained in part (a) to find the slope of the tangent line for the given value of x_0. ■

15. $f(x) = x^2 - 1$; $x_0 = -1$

16. $f(x) = x^2 + 3x + 2$; $x_0 = 2$

17. $f(x) = x + \sqrt{x}$; $x_0 = 1$

18. $f(x) = 1/\sqrt{x}$; $x_0 = 4$

19–22 True–False Determine whether the statement is true or false. Explain your answer. ■

19. If $\lim\limits_{x \to 1} \dfrac{f(x) - f(1)}{x - 1} = 3$, then $\lim\limits_{h \to 0} \dfrac{f(1 + h) - f(1)}{h} = 3$.

20. A tangent line to a curve $y = f(x)$ is a particular kind of secant line to the curve.

21. The velocity of an object represents a change in the object's position.

22. A 50-foot horizontal metal beam is supported on either end by concrete pillars and a weight is placed on the middle of the beam. If $f(x)$ models how many inches the center of the beam sags when the weight measures x tons, then the units of the rate of change of $y = f(x)$ with respect to x are inches/ton.

23. Suppose that the outside temperature versus time curve over a 24-hour period is as shown in the accompanying figure.
   (a) Estimate the maximum temperature and the time at which it occurs.
   (b) The temperature rise is fairly linear from 8 A.M. to 2 P.M. Estimate the rate at which the temperature is increasing during this time period.
   (c) Estimate the time at which the temperature is decreasing most rapidly. Estimate the instantaneous rate of change of temperature with respect to time at this instant.

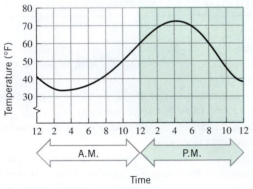

▲ **Figure Ex-23**

24. The accompanying figure shows the graph of the pressure p in atmospheres (atm) versus the volume V in liters (L) of 1 mole of an ideal gas at a constant temperature of 300 K (kelvins). Use the line segments shown in the figure to estimate the rate of change of pressure with respect to volume at the points where $V = 10$ L and $V = 25$ L.

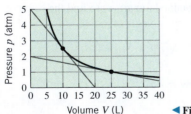

◀ **Figure Ex-24**

25. The accompanying figure shows the graph of the height h in centimeters versus the age t in years of an individual from birth to age 20.

   (a) When is the growth rate greatest?
   (b) Estimate the growth rate at age 5.
   (c) At approximately what age between 10 and 20 is the growth rate greatest? Estimate the growth rate at this age.
   (d) Draw a rough graph of the growth rate versus age.

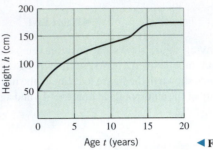

◀ **Figure Ex-25**

26. An object is released from rest (its initial velocity is zero) from the Empire State Building at a height of 1250 ft above street level (Figure Ex-26). The height of the object can be modeled by the position function $s = f(t) = 1250 - 16t^2$.
   (a) Verify that the object is still falling at $t = 5$ s.
   (b) Find the average velocity of the object over the time interval from $t = 5$ to $t = 6$ s.
   (c) Find the object's instantaneous velocity at time $t = 5$ s.

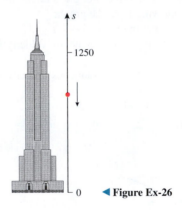

◀ **Figure Ex-26**

27. During the first 40 s of a rocket flight, the rocket is propelled straight up so that in t seconds it reaches a height of $s = 0.3t^3$ ft.
   (a) How high does the rocket travel in 40 s?
   (b) What is the average velocity of the rocket during the first 40 s?
   (c) What is the average velocity of the rocket during the first 1000 ft of its flight?
   (d) What is the instantaneous velocity of the rocket at the end of 40 s?

28. An automobile is driven down a straight highway such that after $0 \le t \le 12$ seconds it is $s = 4.5t^2$ feet from its initial position. *(cont.)*

(a) Find the average velocity of the car over the interval [0, 12].

(b) Find the instantaneous velocity of the car at $t = 6$.

29. Writing Discuss how the tangent line to the graph of a function $y = f(x)$ at a point $P(x_0, f(x_0))$ is defined in terms of secant lines to the graph through point P.

30. Writing A particle is in rectilinear motion during the time interval $0 \leq t \leq 2$. Explain the connection between the instantaneous velocity of the particle at time $t = 1$ and the average velocities of the particle during portions of the interval $0 \leq t \leq 2$.

✔ QUICK CHECK ANSWERS 2.1

1. $\dfrac{f(x) - f(x_0)}{x - x_0}$; $\dfrac{f(x_0 + h) - f(x_0)}{h}$ 2. -4 3. (a) 9; positive; 4 (b) 1 ft/s   4. $s = 3t - 7$

5. (a) 8 (b) $\lim\limits_{x \to 2} \dfrac{(x^2 + x) - 6}{x - 2}$ or $\lim\limits_{h \to 0} \dfrac{[(2 + h)^2 + (2 + h)] - 6}{h}$.

2.2 THE DERIVATIVE FUNCTION

In this section we will discuss the concept of a "derivative," which is the primary mathematical tool that is used to calculate and study rates of change.

■ **DEFINITION OF THE DERIVATIVE FUNCTION**

In the last section we showed that if the limit

$$\lim_{h \to 0} \frac{f(x_0 + h) - f(x_0)}{h}$$

exists, then it can be interpreted either as the slope of the tangent line to the curve $y = f(x)$ at $x = x_0$ or as the instantaneous rate of change of y with respect to x at $x = x_0$ [see Formulas (2) and (11) of that section]. This limit is so important that it has a special notation:

$$f'(x_0) = \lim_{h \to 0} \frac{f(x_0 + h) - f(x_0)}{h} \tag{1}$$

You can think of f' (read "f prime") as a function whose input is x_0 and whose output is the number $f'(x_0)$ that represents either the slope of the tangent line to $y = f(x)$ at $x = x_0$ or the instantaneous rate of change of y with respect to x at $x = x_0$. To emphasize this function point of view, we will replace x_0 by x in (1) and make the following definition.

The expression
$$\frac{f(x + h) - f(x)}{h}$$
that appears in (2) is commonly called the *difference quotient*.

2.2.1 DEFINITION The function f' defined by the formula

$$f'(x) = \lim_{h \to 0} \frac{f(x + h) - f(x)}{h} \tag{2}$$

is called the *derivative of f with respect to x*. The domain of f' consists of all x in the domain of f for which the limit exists.

The term "derivative" is used because the function f' is *derived* from the function f by a limiting process.

▶ **Example 1** Find the derivative with respect to x of $f(x) = x^2$, and use it to find the equation of the tangent line to $y = x^2$ at $x = 2$.

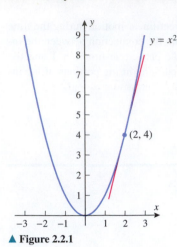

▲ Figure 2.2.1

Solution.   It follows from (2) that

$$f'(x) = \lim_{h \to 0} \frac{f(x+h) - f(x)}{h} = \lim_{h \to 0} \frac{(x+h)^2 - x^2}{h}$$

$$= \lim_{h \to 0} \frac{x^2 + 2xh + h^2 - x^2}{h} = \lim_{h \to 0} \frac{2xh + h^2}{h}$$

$$= \lim_{h \to 0} (2x + h) = 2x$$

Thus, the slope of the tangent line to $y = x^2$ at $x = 2$ is $f'(2) = 4$. Since $y = 4$ if $x = 2$, the point-slope form of the tangent line is

$$y - 4 = 4(x - 2)$$

which we can rewrite in slope-intercept form as $y = 4x - 4$ (Figure 2.2.1). ◄

You can think of f' as a "slope-producing function" in the sense that the value of $f'(x)$ at $x = x_0$ is the slope of the tangent line to the graph of f at $x = x_0$. This aspect of the derivative is illustrated in Figure 2.2.2, which shows the graphs of $f(x) = x^2$ and its derivative $f'(x) = 2x$ (obtained in Example 1). The figure illustrates that the values of $f'(x) = 2x$ at $x = -2, 0$, and 2 correspond to the slopes of the tangent lines to the graph of $f(x) = x^2$ at those values of x.

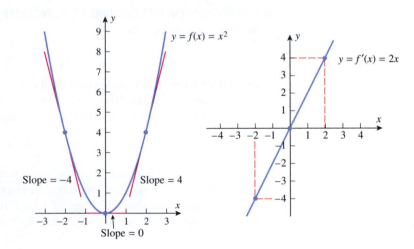

▶ Figure 2.2.2

In general, if $f'(x)$ is defined at $x = x_0$, then the point-slope form of the equation of the tangent line to the graph of $y = f(x)$ at $x = x_0$ may be found using the following steps.

Finding an Equation for the Tangent Line to $y = f(x)$ at $x = x_0$.

Step 1. Evaluate $f(x_0)$; the point of tangency is $(x_0, f(x_0))$.

Step 2. Find $f'(x)$ and evaluate $f'(x_0)$, which is the slope m of the line.

Step 3. Substitute the value of the slope m and the point $(x_0, f(x_0))$ into the point-slope form of the line

$$y - f(x_0) = f'(x_0)(x - x_0)$$

or, equivalently,

$$y = f(x_0) + f'(x_0)(x - x_0) \tag{3}$$

▶ **Example 2**

(a) Find the derivative with respect to x of $f(x) = x^3 - x$.

(b) Graph f and f' together, and discuss the relationship between the two graphs.

Solution (a).

$$
\begin{aligned}
f'(x) &= \lim_{h \to 0} \frac{f(x+h) - f(x)}{h} \\
&= \lim_{h \to 0} \frac{[(x+h)^3 - (x+h)] - [x^3 - x]}{h} \\
&= \lim_{h \to 0} \frac{[x^3 + 3x^2h + 3xh^2 + h^3 - x - h] - [x^3 - x]}{h} \\
&= \lim_{h \to 0} \frac{3x^2h + 3xh^2 + h^3 - h}{h} \\
&= \lim_{h \to 0} [3x^2 + 3xh + h^2 - 1] = 3x^2 - 1
\end{aligned}
$$

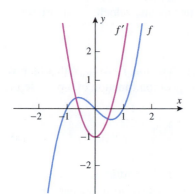

▲ Figure 2.2.3

Solution (b). Since $f'(x)$ can be interpreted as the slope of the tangent line to the graph of $y = f(x)$ at x, it follows that $f'(x)$ is positive where the tangent line has positive slope, is negative where the tangent line has negative slope, and is zero where the tangent line is horizontal. We leave it for you to verify that this is consistent with the graphs of $f(x) = x^3 - x$ and $f'(x) = 3x^2 - 1$ shown in Figure 2.2.3. ◄

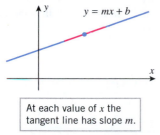

At each value of x the tangent line has slope m.

▲ Figure 2.2.4

▶ **Example 3** At each value of x, the tangent line to a line $y = mx + b$ coincides with the line itself (Figure 2.2.4), and hence all tangent lines have slope m. This suggests geometrically that if $f(x) = mx + b$, then $f'(x) = m$ for all x. This is confirmed by the following computations:

$$
\begin{aligned}
f'(x) &= \lim_{h \to 0} \frac{f(x+h) - f(x)}{h} \\
&= \lim_{h \to 0} \frac{[m(x+h) + b] - [mx + b]}{h} \\
&= \lim_{h \to 0} \frac{mh}{h} = \lim_{h \to 0} m = m \quad ◄
\end{aligned}
$$

▶ **Example 4**

(a) Find the derivative with respect to x of $f(x) = \sqrt{x}$.

(b) Find the slope of the tangent line to $y = \sqrt{x}$ at $x = 9$.

(c) Find the limits of $f'(x)$ as $x \to 0^+$ and as $x \to +\infty$, and explain what those limits say about the graph of f.

Solution (a). Recall from Example 4 of Section 2.1 that the slope of the tangent line to $y = \sqrt{x}$ at $x = x_0$ is given by $m_{\tan} = 1/(2\sqrt{x_0})$. Thus, $f'(x) = 1/(2\sqrt{x})$.

Solution (b). The slope of the tangent line at $x = 9$ is $f'(9)$. From part (a), this slope is $f'(9) = 1/(2\sqrt{9}) = \frac{1}{6}$.

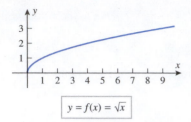

$$y = f(x) = \sqrt{x}$$

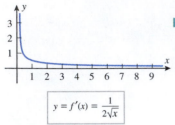

$$y = f'(x) = \frac{1}{2\sqrt{x}}$$

▲ **Figure 2.2.5**

Solution (c). The graphs of $f(x) = \sqrt{x}$ and $f'(x) = 1/(2\sqrt{x})$ are shown in Figure 2.2.5. Observe that $f'(x) > 0$ if $x > 0$, which means that all tangent lines to the graph of $y = \sqrt{x}$ have positive slope at all points in this interval. Since

$$\lim_{x \to 0^+} \frac{1}{2\sqrt{x}} = +\infty \quad \text{and} \quad \lim_{x \to +\infty} \frac{1}{2\sqrt{x}} = 0$$

the graph of f becomes more and more vertical as $x \to 0^+$ and more and more horizontal as $x \to +\infty$. ◄

■ COMPUTING INSTANTANEOUS VELOCITY

It follows from Formula (5) of Section 2.1 (with t replacing t_0) that if $s = f(t)$ is the position function of a particle in rectilinear motion, then the instantaneous velocity at an arbitrary time t is given by

$$v_{\text{inst}} = \lim_{h \to 0} \frac{f(t+h) - f(t)}{h}$$

Since the right side of this equation is the derivative of the function f (with t rather than x as the independent variable), it follows that if $f(t)$ is the position function of a particle in rectilinear motion, then the function

$$v(t) = f'(t) = \lim_{h \to 0} \frac{f(t+h) - f(t)}{h} \tag{4}$$

represents the instantaneous velocity of the particle at time t. Accordingly, we call (4) the ***instantaneous velocity function*** or, more simply, the ***velocity function*** of the particle.

▶ **Example 5** Recall the particle from Example 5 of Section 2.1 with position function $s = f(t) = 1 + 5t - 2t^2$. Here $f(t)$ is measured in meters and t is measured in seconds. Find the velocity function of the particle.

Solution. It follows from (4) that the velocity function is

$$v(t) = \lim_{h \to 0} \frac{f(t+h) - f(t)}{h} = \lim_{h \to 0} \frac{[1 + 5(t+h) - 2(t+h)^2] - [1 + 5t - 2t^2]}{h}$$

$$= \lim_{h \to 0} \frac{-2[t^2 + 2th + h^2 - t^2] + 5h}{h} = \lim_{h \to 0} \frac{-4th - 2h^2 + 5h}{h}$$

$$= \lim_{h \to 0} (-4t - 2h + 5) = 5 - 4t$$

where the units of velocity are meters per second. ◄

■ DIFFERENTIABILITY

It is possible that the limit that defines the derivative of a function f may not exist at certain points in the domain of f. At such points the derivative is undefined. To account for this possibility we make the following definition.

2.2.2 **DEFINITION** A function f is said to be ***differentiable at x_0*** if the limit

$$f'(x_0) = \lim_{h \to 0} \frac{f(x_0 + h) - f(x_0)}{h} \tag{5}$$

exists. If f is differentiable at each point of the open interval (a, b), then we say that it is ***differentiable on (a, b)***, and similarly for open intervals of the form $(a, +\infty)$, $(-\infty, b)$, and $(-\infty, +\infty)$. In the last case we say that f is ***differentiable everywhere***.

Geometrically, a function f is differentiable at x_0 if the graph of f has a tangent line at x_0. Thus, f is not differentiable at any point x_0 where the secant lines from $P(x_0, f(x_0))$ to points $Q(x, f(x))$ distinct from P do not approach a unique *nonvertical* limiting position as $x \to x_0$. Figure 2.2.6 illustrates two common ways in which a function that is continuous at x_0 can fail to be differentiable at x_0. These can be described informally as

- corner points
- points of vertical tangency

At a corner point, the slopes of the secant lines have different limits from the left and from the right, and hence the *two-sided* limit that defines the derivative does not exist (Figure 2.2.7). At a point of vertical tangency the slopes of the secant lines approach $+\infty$ or $-\infty$ from the left and from the right (Figure 2.2.8), so again the limit that defines the derivative does not exist.

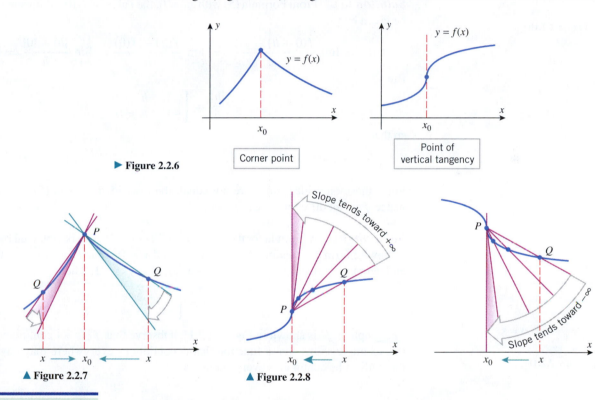

▶ **Figure 2.2.6**

Corner point

Point of vertical tangency

▲ **Figure 2.2.7**   ▲ **Figure 2.2.8**

There are other less obvious circumstances under which a function may fail to be differentiable. (See Exercise 49, for example.)

Differentiability at x_0 can also be described informally in terms of the behavior of the graph of f under increasingly stronger magnification at the point $P(x_0, f(x_0))$ (Figure 2.2.9). If f is differentiable at x_0, then under sufficiently strong magnification at P the

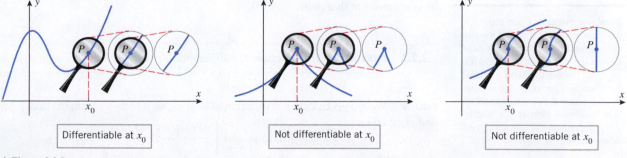

Differentiable at x_0

Not differentiable at x_0

Not differentiable at x_0

▲ **Figure 2.2.9**

graph looks like a nonvertical line (the tangent line); if a corner point occurs at x_0, then no matter how great the magnification at P the corner persists and the graph never looks like a nonvertical line; and if vertical tangency occurs at x_0, then the graph of f looks like a vertical line under sufficiently strong magnification at P.

▶ **Example 6** The graph of $y = |x|$ in Figure 2.2.10 has a corner at $x = 0$, which implies that $f(x) = |x|$ is not differentiable at $x = 0$.

(a) Prove that $f(x) = |x|$ is not differentiable at $x = 0$ by showing that the limit in Definition 2.2.2 does not exist at $x = 0$.

(b) Find a formula for $f'(x)$.

Solution (a). From Formula (5) with $x_0 = 0$, the value of $f'(0)$, if it were to exist, would be given by

$$f'(0) = \lim_{h \to 0} \frac{f(0+h) - f(0)}{h} = \lim_{h \to 0} \frac{f(h) - f(0)}{h} = \lim_{h \to 0} \frac{|h| - |0|}{h} = \lim_{h \to 0} \frac{|h|}{h} \qquad (6)$$

But

$$\frac{|h|}{h} = \begin{cases} 1, & h > 0 \\ -1, & h < 0 \end{cases}$$

so that

$$\lim_{h \to 0^-} \frac{|h|}{h} = -1 \quad \text{and} \quad \lim_{h \to 0^+} \frac{|h|}{h} = 1$$

Since these one-sided limits are not equal, the two-sided limit in (5) does not exist, and hence f is not differentiable at $x = 0$.

Solution (b). A formula for the derivative of $f(x) = |x|$ can be obtained by writing $|x|$ in piecewise form and treating the cases $x > 0$ and $x < 0$ separately. If $x > 0$, then $f(x) = x$ and $f'(x) = 1$; if $x < 0$, then $f(x) = -x$ and $f'(x) = -1$. Thus,

$$f'(x) = \begin{cases} 1, & x > 0 \\ -1, & x < 0 \end{cases}$$

The graph of f' is shown in Figure 2.2.11. Observe that f' is not continuous at $x = 0$, so this example shows that a function that is continuous everywhere may have a derivative that fails to be continuous everywhere. ◀

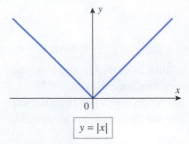

$y = |x|$

▲ **Figure 2.2.10**

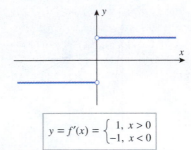

$y = f'(x) = \begin{cases} 1, & x > 0 \\ -1, & x < 0 \end{cases}$

▲ **Figure 2.2.11**

■ **THE RELATIONSHIP BETWEEN DIFFERENTIABILITY AND CONTINUITY**

We already know that functions are not differentiable at corner points and points of vertical tangency. The next theorem shows that functions are not differentiable at points of discontinuity. We will do this by proving that if f is differentiable at a point, then it must be continuous at that point.

A theorem that says "If statement A is true, then statement B is true" is equivalent to the theorem that says "If statement B is not true, then statement A is not true." The two theorems are called *contrapositive forms* of one another. Thus, Theorem 2.2.3 can be rewritten in contrapositive form as "If a function f is not continuous at x_0, then f is not differentiable at x_0."

2.2.3 THEOREM *If a function f is differentiable at x_0, then f is continuous at x_0.*

PROOF We are given that f is differentiable at x_0, so it follows from (5) that $f'(x_0)$ exists and is given by

$$f'(x_0) = \lim_{h \to 0} \left[\frac{f(x_0 + h) - f(x_0)}{h} \right] \qquad (7)$$

To show that f is continuous at x_0, we must show that $\lim_{x \to x_0} f(x) = f(x_0)$ or, equivalently,

$$\lim_{x \to x_0} [f(x) - f(x_0)] = 0$$

Expressing this in terms of the variable $h = x - x_0$, we must prove that

$$\lim_{h \to 0} [f(x_0 + h) - f(x_0)] = 0$$

However, this can be proved using (7) as follows:

$$\lim_{h \to 0} [f(x_0 + h) - f(x_0)] = \lim_{h \to 0} \left[\frac{f(x_0 + h) - f(x_0)}{h} \cdot h \right]$$

$$= \lim_{h \to 0} \left[\frac{f(x_0 + h) - f(x_0)}{h} \right] \cdot \lim_{h \to 0} h$$

$$= f'(x_0) \cdot 0 = 0 \quad \blacksquare$$

WARNING

The converse of Theorem 2.2.3 is false; that is, *a function may be continuous at a point but not differentiable at that point*. This occurs, for example, at corner points of continuous functions. For instance, $f(x) = |x|$ is continuous at $x = 0$ but not differentiable there (Example 6).

The relationship between continuity and differentiability was of great historical significance in the development of calculus. In the early nineteenth century mathematicians believed that if a continuous function had many points of nondifferentiability, these points, like the tips of a sawblade, would have to be separated from one another and joined by smooth curve segments (Figure 2.2.12). This misconception was corrected by a series of discoveries beginning in 1834. In that year a Bohemian priest, philosopher, and mathematician named Bernhard Bolzano discovered a procedure for constructing a continuous function that is not differentiable at any point. Later, in 1860, the great German mathematician Karl Weierstrass (biography on p. 102) produced the first formula for such a function. The graphs of such functions are impossible to draw; it is as if the corners are so numerous that any segment of the curve, when suitably enlarged, reveals more corners. The discovery of these functions was important in that it made mathematicians distrustful of their geometric intuition and more reliant on precise mathematical proof. Recently, such functions have started to play a fundamental role in the study of geometric objects called *fractals*. Fractals have revealed an order to natural phenomena that were previously dismissed as random and chaotic.

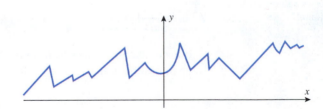

▶ **Figure 2.2.12**

Bernhard Bolzano (**1781–1848**) Bolzano, the son of an art dealer, was born in Prague, Bohemia (Czech Republic). He was educated at the University of Prague, and eventually won enough mathematical fame to be recommended for a mathematics chair there. However, Bolzano became an ordained Roman Catholic priest, and in 1805 he was appointed to a chair of Philosophy at the University of Prague. Bolzano was a man of great human compassion; he spoke out for educational reform, he voiced the right of individual conscience over government demands, and he lectured on the absurdity of war and militarism. His views so disenchanted Emperor Franz I of Austria that the emperor pressed the Archbishop of Prague to have Bolzano recant his statements. Bolzano refused and was then forced to retire in 1824 on a small pension. Bolzano's main contribution to mathematics was philosophical. His work helped convince mathematicians that sound mathematics must ultimately rest on rigorous proof rather than intuition. In addition to his work in mathematics, Bolzano investigated problems concerning space, force, and wave propagation.

■ DERIVATIVES AT THE ENDPOINTS OF AN INTERVAL

If a function f is defined on a closed interval $[a, b]$ but not outside that interval, then f' is not defined at the endpoints of the interval because derivatives are two-sided limits. To deal with this we define *left-hand derivatives* and *right-hand derivatives* by

$$f'_-(x) = \lim_{h \to 0^-} \frac{f(x+h) - f(x)}{h} \quad \text{and} \quad f'_+(x) = \lim_{h \to 0^+} \frac{f(x+h) - f(x)}{h}$$

respectively. These are called *one-sided derivatives*. Geometrically, $f'_-(x)$ is the limit of the slopes of the secant lines as x is approached from the left and $f'_+(x)$ is the limit of the slopes of the secant lines as x is approached from the right. For a closed interval $[a, b]$, we will understand the derivative at the left endpoint to be $f'_+(a)$ and at the right endpoint to be $f'_-(b)$ (Figure 2.2.13).

In general, we will say that f is *differentiable* on an interval of the form $[a, b]$, $[a, +\infty)$, $(-\infty, b]$, $[a, b)$, or $(a, b]$ if it is differentiable at all points inside the interval and the appropriate one-sided derivative exists at each included endpoint.

It can be proved that a function f is continuous from the left at those points where the left-hand derivative exists and is continuous from the right at those points where the right-hand derivative exists.

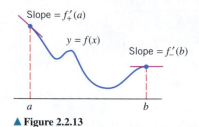

Slope $= f'_+(a)$

$y = f(x)$

Slope $= f'_-(b)$

a b

▲ Figure 2.2.13

■ OTHER DERIVATIVE NOTATIONS

The process of finding a derivative is called *differentiation*. You can think of differentiation as an *operation* on functions that associates a function f' with a function f. When the independent variable is x, the differentiation operation is also commonly denoted by

$$f'(x) = \frac{d}{dx}[f(x)] \quad \text{or} \quad f'(x) = D_x[f(x)]$$

In the case where there is a dependent variable $y = f(x)$, the derivative is also commonly denoted by

$$f'(x) = y'(x) \quad \text{or} \quad f'(x) = \frac{dy}{dx}$$

Later, the symbols dy and dx will be given specific meanings. However, for the time being do not regard dy/dx as a ratio, but rather as a single symbol denoting the derivative.

With the above notations, the value of the derivative at a point x_0 can be expressed as

$$f'(x_0) = \frac{d}{dx}[f(x)]\Big|_{x=x_0}, \quad f'(x_0) = D_x[f(x)]\big|_{x=x_0}, \quad f'(x_0) = y'(x_0), \quad f'(x_0) = \frac{dy}{dx}\Big|_{x=x_0}$$

If a variable w changes from some initial value w_0 to some final value w_1, then the final value minus the initial value is called an *increment* in w and is denoted by

$$\Delta w = w_1 - w_0 \tag{8}$$

Increments can be positive or negative, depending on whether the final value is larger or smaller than the initial value. The increment symbol in (8) should not be interpreted as a product; rather, Δw should be regarded as a single symbol representing the change in the value of w.

It is common to regard the variable h in the derivative formula

$$f'(x) = \lim_{h \to 0} \frac{f(x+h) - f(x)}{h} \tag{9}$$

as an increment Δx in x and write (9) as

$$f'(x) = \lim_{\Delta x \to 0} \frac{f(x + \Delta x) - f(x)}{\Delta x} \tag{10}$$

Moreover, if $y = f(x)$, then the numerator in (10) can be regarded as the increment

$$\Delta y = f(x + \Delta x) - f(x) \tag{11}$$

in which case

$$\frac{dy}{dx} = \lim_{\Delta x \to 0} \frac{\Delta y}{\Delta x} = \lim_{\Delta x \to 0} \frac{f(x + \Delta x) - f(x)}{\Delta x} \tag{12}$$

The geometric interpretations of Δx and Δy are shown in Figure 2.2.14.

Sometimes it is desirable to express derivatives in a form that does not use increments at all. For example, if we let $w = x + h$ in Formula (9), then $w \to x$ as $h \to 0$, so we can rewrite that formula as

$$f'(x) = \lim_{w \to x} \frac{f(w) - f(x)}{w - x} \tag{13}$$

(Compare Figures 2.2.14 and 2.2.15.)

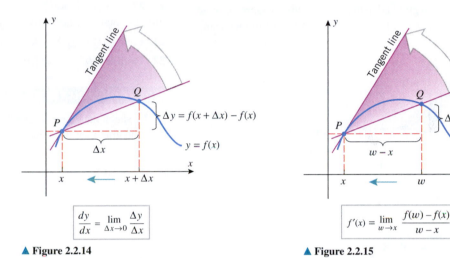

$$\frac{dy}{dx} = \lim_{\Delta x \to 0} \frac{\Delta y}{\Delta x}$$

▲ **Figure 2.2.14**

$$f'(x) = \lim_{w \to x} \frac{f(w) - f(x)}{w - x}$$

▲ **Figure 2.2.15**

When letters other than x and y are used for the independent and dependent variables, the derivative notations must be adjusted accordingly. Thus, for example, if $s = f(t)$ is the position function for a particle in rectilinear motion, then the velocity function $v(t)$ in (4) can be expressed as

$$v(t) = \frac{ds}{dt} = \lim_{\Delta t \to 0} \frac{\Delta s}{\Delta t} = \lim_{\Delta t \to 0} \frac{f(t + \Delta t) - f(t)}{\Delta t} \tag{14}$$

✔ **QUICK CHECK EXERCISES 2.2**   *(See page 155 for answers.)*

1. The function $f'(x)$ is defined by the formula

$$f'(x) = \lim_{h \to 0} \underline{\hspace{2cm}}$$

2. (a) The derivative of $f(x) = x^2$ is $f'(x) = \underline{\hspace{1.5cm}}$.
   (b) The derivative of $f(x) = \sqrt{x}$ is $f'(x) = \underline{\hspace{1.5cm}}$.

3. Suppose that the line $2x + 3y = 5$ is tangent to the graph of $y = f(x)$ at $x = 1$. The value of $f(1)$ is $\underline{\hspace{1.5cm}}$ and the value of $f'(1)$ is $\underline{\hspace{1.5cm}}$.

4. Which theorem guarantees us that if

$$\lim_{h \to 0} \frac{f(x_0 + h) - f(x_0)}{h}$$

exists, then $\lim_{x \to x_0} f(x) = f(x_0)$?

EXERCISE SET 2.2   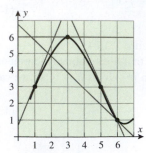 Graphing Utility

1. Use the graph of $y = f(x)$ in the accompanying figure to estimate the value of $f'(1)$, $f'(3)$, $f'(5)$, and $f'(6)$.

◀ **Figure Ex-1**

2. For the function graphed in the accompanying figure, arrange the numbers 0, $f'(-3)$, $f'(0)$, $f'(2)$, and $f'(4)$ in increasing order.

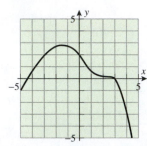

◀ **Figure Ex-2**

FOCUS ON CONCEPTS

3. (a) If you are given an equation for the tangent line at the point $(a, f(a))$ on a curve $y = f(x)$, how would you go about finding $f'(a)$?
   (b) Given that the tangent line to the graph of $y = f(x)$ at the point $(2, 5)$ has the equation $y = 3x - 1$, find $f'(2)$.
   (c) For the function $y = f(x)$ in part (b), what is the instantaneous rate of change of y with respect to x at $x = 2$?

4. Given that the tangent line to $y = f(x)$ at the point $(1, 2)$ passes through the point $(-1, -1)$, find $f'(1)$.

5. Sketch the graph of a function f for which $f(0) = -1$, $f'(0) = 0$, $f'(x) < 0$ if $x < 0$, and $f'(x) > 0$ if $x > 0$.

6. Sketch the graph of a function f for which $f(0) = 0$, $f'(0) = 0$, and $f'(x) > 0$ if $x < 0$ or $x > 0$.

7. Given that $f(3) = -1$ and $f'(3) = 5$, find an equation for the tangent line to the graph of $y = f(x)$ at $x = 3$.

8. Given that $f(-2) = 3$ and $f'(-2) = -4$, find an equation for the tangent line to the graph of $y = f(x)$ at $x = -2$.

9–14 Use Definition 2.2.1 to find $f'(x)$, and then find the tangent line to the graph of $y = f(x)$ at $x = a$. ■

9. $f(x) = 2x^2$; $a = 1$ 10. $f(x) = 1/x^2$; $a = -1$

11. $f(x) = x^3$; $a = 0$ 12. $f(x) = 2x^3 + 1$; $a = -1$

13. $f(x) = \sqrt{x + 1}$; $a = 8$ 14. $f(x) = \sqrt{2x + 1}$; $a = 4$

15–20 Use Formula (12) to find dy/dx. ■

15. $y = \dfrac{1}{x}$ 16. $y = \dfrac{1}{x + 1}$ 17. $y = x^2 - x$

18. $y = x^4$ 19. $y = \dfrac{1}{\sqrt{x}}$ 20. $y = \dfrac{1}{\sqrt{x - 1}}$

21–22 Use Definition 2.2.1 (with appropriate change in notation) to obtain the derivative requested. ■

21. Find $f'(t)$ if $f(t) = 4t^2 + t$.

22. Find dV/dr if $V = \frac{4}{3}\pi r^3$.

FOCUS ON CONCEPTS

23. Match the graphs of the functions shown in (a)–(f) with the graphs of their derivatives in (A)–(F).

(a)

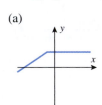

(b)

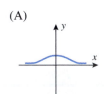

(c)

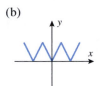

(d)

(e)

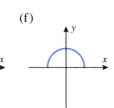

(f)

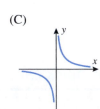

(A)

(B)

(C)

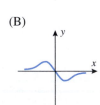

(D)

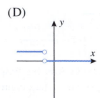

(E)

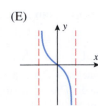

(F)

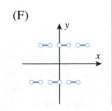

24. Let $f(x) = \sqrt{1 - x^2}$. Use a geometric argument to find $f'(\sqrt{2}/2)$.

25–26 Sketch the graph of the derivative of the function whose graph is shown. ■

25. (a)   (b)   (c)

26. (a)   (b)   (c)

27–30 True–False Determine whether the statement is true or false. Explain your answer. ■

27. If a curve $y = f(x)$ has a horizontal tangent line at $x = a$, then $f'(a)$ is not defined.

28. If the tangent line to the graph of $y = f(x)$ at $x = -2$ has negative slope, then $f'(-2) < 0$.

29. If a function f is continuous at $x = 0$, then f is differentiable at $x = 0$.

30. If a function f is differentiable at $x = 0$, then f is continuous at $x = 0$.

31–32 The given limit represents $f'(a)$ for some function f and some number a. Find $f(x)$ and a in each case. ■

31. (a) $\lim\limits_{\Delta x \to 0} \dfrac{\sqrt{1 + \Delta x} - 1}{\Delta x}$ (b) $\lim\limits_{x_1 \to 3} \dfrac{x_1^2 - 9}{x_1 - 3}$

32. (a) $\lim\limits_{h \to 0} \dfrac{\cos(\pi + h) + 1}{h}$ (b) $\lim\limits_{x \to 1} \dfrac{x^7 - 1}{x - 1}$

33. Find $dy/dx|_{x=1}$, given that $y = 1 - x^2$.

34. Find $dy/dx|_{x=-2}$, given that $y = (x + 2)/x$.

35. Find an equation for the line that is tangent to the curve $y = x^3 - 2x + 1$ at the point $(0, 1)$, and use a graphing utility to graph the curve and its tangent line on the same screen.

36. Use a graphing utility to graph the following on the same screen: the curve $y = x^2/4$, the tangent line to this curve at $x = 1$, and the secant line joining the points $(0, 0)$ and $(2, 1)$ on this curve.

37. Let $f(x) = 2^x$. Estimate $f'(1)$ by
(a) using a graphing utility to zoom in at an appropriate point until the graph looks like a straight line, and then estimating the slope
(b) using a calculating utility to estimate the limit in Formula (13) by making a table of values for a succession of values of w approaching 1.

38. Let $f(x) = \sin x$. Estimate $f'(\pi/4)$ by
(a) using a graphing utility to zoom in at an appropriate point until the graph looks like a straight line, and then estimating the slope
(b) using a calculating utility to estimate the limit in Formula (13) by making a table of values for a succession of values of w approaching $\pi/4$.

39–40 The function f whose graph is shown below has values as given in the accompanying table.

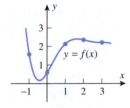

x	-1	0	1	2	3
$f(x)$	1.56	0.58	2.12	2.34	2.2

■

39. (a) Use data from the table to calculate the difference quotients
$$\frac{f(3) - f(1)}{3 - 1}, \quad \frac{f(2) - f(1)}{2 - 1}, \quad \frac{f(2) - f(0)}{2 - 0}$$
(b) Using the graph of $y = f(x)$, indicate which difference quotient in part (a) best approximates $f'(1)$ and which difference quotient gives the worst approximation to $f'(1)$.

40. Use data from the table to approximate the derivative values.
(a) $f'(0.5)$ (b) $f'(2.5)$

FOCUS ON CONCEPTS

41. Suppose that the cost of drilling x feet for an oil well is $C = f(x)$ dollars.
(a) What are the units of $f'(x)$?
(b) In practical terms, what does $f'(x)$ mean in this case?
(c) What can you say about the sign of $f'(x)$?
(d) Estimate the cost of drilling an additional foot, starting at a depth of 300 ft, given that $f'(300) = 1000$.

42. A paint manufacturing company estimates that it can sell $g = f(p)$ gallons of paint at a price of p dollars per gallon.
(a) What are the units of dg/dp?
(b) In practical terms, what does dg/dp mean in this case?
(c) What can you say about the sign of dg/dp?
(d) Given that $dg/dp|_{p=10} = -100$, what can you say about the effect of increasing the price from \$10 per gallon to \$11 per gallon?

43. It is a fact that when a flexible rope is wrapped around a rough cylinder, a small force of magnitude F_0 at one end can resist a large force of magnitude F at the other end. The size of F depends on the angle θ through which the rope is wrapped around the cylinder (see the

accompanying figure). The figure shows the graph of F (in pounds) versus θ (in radians), where F is the magnitude of the force that can be resisted by a force with magnitude $F_0 = 10$ lb for a certain rope and cylinder.
(a) Estimate the values of F and $dF/d\theta$ when the angle $\theta = 10$ radians.
(b) It can be shown that the force F satisfies the equation $dF/d\theta = \mu F$, where the constant μ is called the *coefficient of friction*. Use the results in part (a) to estimate the value of μ.

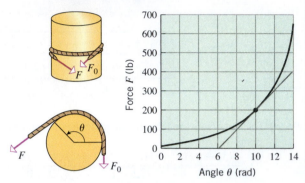

▲ **Figure Ex-43**

44. The accompanying figure shows the velocity versus time curve for a rocket in outer space where the only significant force on the rocket is from its engines. It can be shown that the mass $M(t)$ (in slugs) of the rocket at time t seconds satisfies the equation

$$M(t) = \frac{T}{dv/dt}$$

where T is the thrust (in lb) of the rocket's engines and v is the velocity (in ft/s) of the rocket. The thrust of the first stage of a *Saturn V* rocket is $T = 7,680,982$ lb. Use this value of T and the line segment in the figure to estimate the mass of the rocket at time $t = 100$.

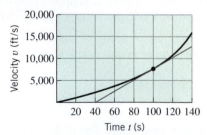

◀ **Figure Ex-44**

45. According to *Newton's Law of Cooling*, the rate of change of an object's temperature is proportional to the difference between the temperature of the object and that of the surrounding medium. The accompanying figure shows the graph of the temperature T (in degrees Fahrenheit) versus time t (in minutes) for a cup of coffee, initially with a temperature of $200°$F, that is allowed to cool in a room with a constant temperature of $75°$F.
(a) Estimate T and dT/dt when $t = 10$ min.

(b) Newton's Law of Cooling can be expressed as

$$\frac{dT}{dt} = k(T - T_0)$$

where k is the constant of proportionality and T_0 is the temperature (assumed constant) of the surrounding medium. Use the results in part (a) to estimate the value of k.

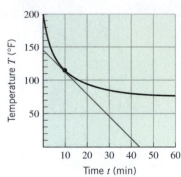

◀ **Figure Ex-45**

46. Show that $f(x)$ is continuous but not differentiable at the indicated point. Sketch the graph of f.
(a) $f(x) = \sqrt[3]{x}$, $x = 0$
(b) $f(x) = \sqrt[3]{(x-2)^2}$, $x = 2$

47. Show that

$$f(x) = \begin{cases} x^2 + 1, & x \le 1 \\ 2x, & x > 1 \end{cases}$$

is continuous and differentiable at $x = 1$. Sketch the graph of f.

48. Show that

$$f(x) = \begin{cases} x^2 + 2, & x \le 1 \\ x + 2, & x > 1 \end{cases}$$

is continuous but not differentiable at $x = 1$. Sketch the graph of f.

49. Show that

$$f(x) = \begin{cases} x \sin(1/x), & x \ne 0 \\ 0, & x = 0 \end{cases}$$

is continuous but not differentiable at $x = 0$. Sketch the graph of f near $x = 0$. (See Figure 1.6.6 and the remark following Example 5 in Section 1.6.)

50. Show that

$$f(x) = \begin{cases} x^2 \sin(1/x), & x \ne 0 \\ 0, & x = 0 \end{cases}$$

is continuous and differentiable at $x = 0$. Sketch the graph of f near $x = 0$.

FOCUS ON CONCEPTS

51. Suppose that a function f is differentiable at x_0 and that $f'(x_0) > 0$. Prove that there exists an open interval containing x_0 such that if x_1 and x_2 are any two points in this interval with $x_1 < x_0 < x_2$, then $f(x_1) < f(x_0) < f(x_2)$.

52. Suppose that a function f is differentiable at x_0 and define $g(x) = f(mx + b)$, where m and b are constants. Prove that if x_1 is a point at which $mx_1 + b = x_0$, then $g(x)$ is differentiable at x_1 and $g'(x_1) = mf'(x_0)$.

53. Suppose that a function f is differentiable at $x = 0$ with $f(0) = f'(0) = 0$, and let $y = mx$, $m \neq 0$, denote any line of nonzero slope through the origin.
  (a) Prove that there exists an open interval containing 0 such that for all nonzero x in this interval $|f(x)| < \left|\frac{1}{2}mx\right|$. [*Hint:* Let $\epsilon = \frac{1}{2}|m|$ and apply Definition 1.4.1 to (5) with $x_0 = 0$.]
  (b) Conclude from part (a) and the triangle inequality that there exists an open interval containing 0 such that $|f(x)| < |f(x) - mx|$ for all x in this interval.
  (c) Explain why the result obtained in part (b) may be interpreted to mean that the tangent line to the graph

of f at the origin is the best *linear* approximation to f at that point.

54. Suppose that f is differentiable at x_0. Modify the argument of Exercise 53 to prove that the tangent line to the graph of f at the point $P(x_0, f(x_0))$ provides the best linear approximation to f at P. [*Hint:* Suppose that $y = f(x_0) + m(x - x_0)$ is any line through $P(x_0, f(x_0))$ with slope $m \neq f'(x_0)$. Apply Definition 1.4.1 to (5) with $x = x_0 + h$ and $\epsilon = \frac{1}{2}|f'(x_0) - m|$.]

55. Writing Write a paragraph that explains what it means for a function to be differentiable. Include examples of functions that are not differentiable as well as examples of functions that are differentiable.

56. Writing Explain the relationship between continuity and differentiability.

✔ **QUICK CHECK ANSWERS 2.2**

1. $\dfrac{f(x + h) - f(x)}{h}$ **2.** (a) $2x$ (b) $\dfrac{1}{2\sqrt{x}}$ **3.** 1; $-\frac{2}{3}$

4. Theorem 2.2.3: If f is differentiable at x_0, then f is continuous at x_0.

2.3 INTRODUCTION TO TECHNIQUES OF DIFFERENTIATION

In the last section we defined the derivative of a function f as a limit, and we used that limit to calculate a few simple derivatives. In this section we will develop some important theorems that will enable us to calculate derivatives more efficiently.

■ DERIVATIVE OF A CONSTANT

The simplest kind of function is a constant function $f(x) = c$. Since the graph of f is a horizontal line of slope 0, the tangent line to the graph of f has slope 0 for every x; and hence we can see geometrically that $f'(x) = 0$ (Figure 2.3.1). We can also see this algebraically since

$$f'(x) = \lim_{h \to 0} \frac{f(x + h) - f(x)}{h} = \lim_{h \to 0} \frac{c - c}{h} = \lim_{h \to 0} 0 = 0$$

Thus, we have established the following result.

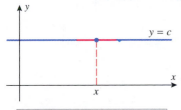

The tangent line to the graph of $f(x) = c$ has slope 0 for all x.

▲ **Figure 2.3.1**

2.3.1 THEOREM *The derivative of a constant function is 0; that is, if c is any real number, then*

$$\frac{d}{dx}[c] = 0 \tag{1}$$

▶ **Example 1**

$$\frac{d}{dx}[1] = 0, \quad \frac{d}{dx}[-3] = 0, \quad \frac{d}{dx}[\pi] = 0, \quad \frac{d}{dx}\left[-\sqrt{2}\right] = 0 \;◀$$

DERIVATIVES OF POWER FUNCTIONS

The simplest power function is $f(x) = x$. Since the graph of f is a line of slope 1, it follows from Example 3 of Section 2.2 that $f'(x) = 1$ for all x (Figure 2.3.2). In other words,

$$\frac{d}{dx}[x] = 1 \tag{2}$$

Example 1 of Section 2.2 shows that the power function $f(x) = x^2$ has derivative $f'(x) = 2x$. From Example 2 in that section one can infer that the power function $f(x) = x^3$ has derivative $f'(x) = 3x^2$. That is,

$$\frac{d}{dx}[x^2] = 2x \quad \text{and} \quad \frac{d}{dx}[x^3] = 3x^2 \tag{3–4}$$

These results are special cases of the following more general result.

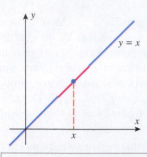

The tangent line to the graph of $f(x) = x$ has slope 1 for all x.

▲ Figure 2.3.2

2.3.2 THEOREM (*The Power Rule*) *If n is a positive integer, then*

$$\frac{d}{dx}[x^n] = nx^{n-1} \tag{5}$$

Verify that Formulas (2), (3), and (4) are the special cases of (5) in which $n = 1, 2,$ and 3.

The binomial formula can be found on the front endpaper of the text. Replacing y by h in this formula yields the identity used in the proof of Theorem 2.3.2.

PROOF Let $f(x) = x^n$. Thus, from the definition of a derivative and the binomial formula for expanding the expression $(x + h)^n$, we obtain

$$\frac{d}{dx}[x^n] = f'(x) = \lim_{h \to 0} \frac{f(x+h) - f(x)}{h} = \lim_{h \to 0} \frac{(x+h)^n - x^n}{h}$$

$$= \lim_{h \to 0} \frac{\left[x^n + nx^{n-1}h + \frac{n(n-1)}{2!}x^{n-2}h^2 + \cdots + nxh^{n-1} + h^n \right] - x^n}{h}$$

$$= \lim_{h \to 0} \frac{nx^{n-1}h + \frac{n(n-1)}{2!}x^{n-2}h^2 + \cdots + nxh^{n-1} + h^n}{h}$$

$$= \lim_{h \to 0} \left[nx^{n-1} + \frac{n(n-1)}{2!}x^{n-2}h + \cdots + nxh^{n-2} + h^{n-1} \right]$$

$$= nx^{n-1} + 0 + \cdots + 0 + 0$$

$$= nx^{n-1} \quad \blacksquare$$

► **Example 2**

$$\frac{d}{dx}[x^4] = 4x^3, \quad \frac{d}{dx}[x^5] = 5x^4, \quad \frac{d}{dt}[t^{12}] = 12t^{11} \quad ◄$$

Although our proof of the power rule in Formula (5) applies only to *positive* integer powers of x, it is not difficult to show that the same formula holds for all integer powers of x (Exercise 82). Also, we saw in Example 4 of Section 2.2 that

$$\frac{d}{dx}[\sqrt{x}] = \frac{1}{2\sqrt{x}} \tag{6}$$

which can be expressed as

$$\frac{d}{dx}[x^{1/2}] = \frac{1}{2}x^{-1/2} = \frac{1}{2}x^{(1/2)-1}$$

Thus, Formula (5) is valid for $n = \frac{1}{2}$, as well. In fact, it can be shown that this formula holds for any real exponent. We state this more general result for our use now, although we won't be prepared to prove it until Chapter 3.

2.3.3 **THEOREM** (*Extended Power Rule*) *If r is any real number, then*

$$\frac{d}{dx}[x^r] = rx^{r-1} \tag{7}$$

In words, *to differentiate a power function, decrease the constant exponent by one and multiply the resulting power function by the original exponent.*

▶ **Example 3**

$$\frac{d}{dx}[x^\pi] = \pi x^{\pi-1}$$

$$\frac{d}{dx}\left[\frac{1}{x}\right] = \frac{d}{dx}[x^{-1}] = (-1)x^{-1-1} = -x^{-2} = -\frac{1}{x^2}$$

$$\frac{d}{dw}\left[\frac{1}{w^{100}}\right] = \frac{d}{dw}[w^{-100}] = -100w^{-101} = -\frac{100}{w^{101}}$$

$$\frac{d}{dx}[x^{4/5}] = \frac{4}{5}x^{(4/5)-1} = \frac{4}{5}x^{-1/5}$$

$$\frac{d}{dx}[\sqrt[3]{x}] = \frac{d}{dx}[x^{1/3}] = \frac{1}{3}x^{-2/3} = \frac{1}{3\sqrt[3]{x^2}} \blacktriangleleft$$

■ **DERIVATIVE OF A CONSTANT TIMES A FUNCTION**

Formula (8) can also be expressed in function notation as

$$(cf)' = cf'$$

2.3.4 **THEOREM** (*Constant Multiple Rule*) *If f is differentiable at x and c is any real number, then cf is also differentiable at x and*

$$\frac{d}{dx}[cf(x)] = c\frac{d}{dx}[f(x)] \tag{8}$$

PROOF

$$\frac{d}{dx}[cf(x)] = \lim_{h \to 0} \frac{cf(x+h) - cf(x)}{h}$$

$$= \lim_{h \to 0} c\left[\frac{f(x+h) - f(x)}{h}\right]$$

$$= c \lim_{h \to 0} \frac{f(x+h) - f(x)}{h} \qquad \boxed{\text{A constant factor can be moved through a limit sign.}}$$

$$= c\frac{d}{dx}[f(x)] \quad \blacksquare$$

In words, *a constant factor can be moved through a derivative sign.*

▶ **Example 4**

$$\frac{d}{dx}[4x^8] = 4\frac{d}{dx}[x^8] = 4[8x^7] = 32x^7$$

$$\frac{d}{dx}[-x^{12}] = (-1)\frac{d}{dx}[x^{12}] = -12x^{11}$$

$$\frac{d}{dx}\left[\frac{\pi}{x}\right] = \pi\frac{d}{dx}[x^{-1}] = \pi(-x^{-2}) = -\frac{\pi}{x^2} \quad \blacktriangleleft$$

■ **DERIVATIVES OF SUMS AND DIFFERENCES**

Formulas (9) and (10) can also be expressed as

$$(f+g)' = f' + g'$$
$$(f-g)' = f' - g'$$

2.3.5   THEOREM (*Sum and Difference Rules*)   *If f and g are differentiable at x, then so are $f+g$ and $f-g$ and*

$$\frac{d}{dx}[f(x) + g(x)] = \frac{d}{dx}[f(x)] + \frac{d}{dx}[g(x)] \tag{9}$$

$$\frac{d}{dx}[f(x) - g(x)] = \frac{d}{dx}[f(x)] - \frac{d}{dx}[g(x)] \tag{10}$$

PROOF   Formula (9) can be proved as follows:

$$\frac{d}{dx}[f(x) + g(x)] = \lim_{h \to 0}\frac{[f(x+h) + g(x+h)] - [f(x) + g(x)]}{h}$$

$$= \lim_{h \to 0}\frac{[f(x+h) - f(x)] + [g(x+h) - g(x)]}{h}$$

$$= \lim_{h \to 0}\frac{f(x+h) - f(x)}{h} + \lim_{h \to 0}\frac{g(x+h) - g(x)}{h} \qquad \boxed{\text{The limit of a sum is the sum of the limits.}}$$

$$= \frac{d}{dx}[f(x)] + \frac{d}{dx}[g(x)]$$

Formula (10) can be proved in a similar manner or, alternatively, by writing $f(x) - g(x)$ as $f(x) + (-1)g(x)$ and then applying Formulas (8) and (9). ■

In words, *the derivative of a sum equals the sum of the derivatives*, and *the derivative of a difference equals the difference of the derivatives.*

▶ **Example 5**

$$\frac{d}{dx}[2x^6 + x^{-9}] = \frac{d}{dx}[2x^6] + \frac{d}{dx}[x^{-9}] = 12x^5 + (-9)x^{-10} = 12x^5 - 9x^{-10}$$

$$\frac{d}{dx}\left[\frac{\sqrt{x} - 2x}{\sqrt{x}}\right] = \frac{d}{dx}[1 - 2\sqrt{x}]$$

$$= \frac{d}{dx}[1] - \frac{d}{dx}[2\sqrt{x}] = 0 - 2\left(\frac{1}{2\sqrt{x}}\right) = -\frac{1}{\sqrt{x}} \qquad \boxed{\text{See Formula (6).}} \quad \blacktriangleleft$$

Although Formulas (9) and (10) are stated for sums and differences of two functions, they can be extended to any finite number of functions. For example, by grouping and applying Formula (9) twice we obtain

$$(f + g + h)' = [(f + g) + h]' = (f + g)' + h' = f' + g' + h'$$

As illustrated in the following example, the constant multiple rule together with the extended versions of the sum and difference rules can be used to differentiate any polynomial.

▶ **Example 6** Find dy/dx if $y = 3x^8 - 2x^5 + 6x + 1$.

Solution.

$$\frac{dy}{dx} = \frac{d}{dx}[3x^8 - 2x^5 + 6x + 1]$$

$$= \frac{d}{dx}[3x^8] - \frac{d}{dx}[2x^5] + \frac{d}{dx}[6x] + \frac{d}{dx}[1]$$

$$= 24x^7 - 10x^4 + 6 \quad ◀$$

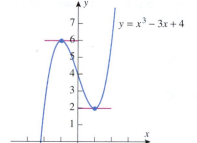

▲ Figure 2.3.3

▶ **Example 7** At what points, if any, does the graph of $y = x^3 - 3x + 4$ have a horizontal tangent line?

Solution. Horizontal tangent lines have slope zero, so we must find those values of x for which $y'(x) = 0$. Differentiating yields

$$y'(x) = \frac{d}{dx}[x^3 - 3x + 4] = 3x^2 - 3$$

Thus, horizontal tangent lines occur at those values of x for which $3x^2 - 3 = 0$, that is, if $x = -1$ or $x = 1$. The corresponding points on the curve $y = x^3 - 3x + 4$ are $(-1, 6)$ and $(1, 2)$ (see Figure 2.3.3). ◀

▶ **Example 8** Find the area of the triangle formed from the coordinate axes and the tangent line to the curve $y = 5x^{-1} - \frac{1}{5}x$ at the point $(5, 0)$.

Solution. Since the derivative of y with respect to x is

$$y'(x) = \frac{d}{dx}\left[5x^{-1} - \frac{1}{5}x\right] = \frac{d}{dx}[5x^{-1}] - \frac{d}{dx}\left[\frac{1}{5}x\right] = -5x^{-2} - \frac{1}{5}$$

the slope of the tangent line at the point $(5, 0)$ is $y'(5) = -\frac{2}{5}$. Thus, the equation of the tangent line at this point is

$$y - 0 = -\frac{2}{5}(x - 5) \quad \text{or equivalently} \quad y = -\frac{2}{5}x + 2$$

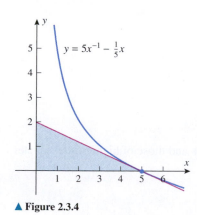

▲ Figure 2.3.4

Since the y-intercept of this line is 2, the right triangle formed from the coordinate axes and the tangent line has legs of length 5 and 2, so its area is $\frac{1}{2}(5)(2) = 5$ (Figure 2.3.4). ◀

■ **HIGHER DERIVATIVES**

The derivative f' of a function f is itself a function and hence may have a derivative of its own. If f' is differentiable, then its derivative is denoted by f'' and is called the *second derivative* of f. As long as we have differentiability, we can continue the process of differentiating to obtain third, fourth, fifth, and even higher derivatives of f. These successive derivatives are denoted by

$$f', \quad f'' = (f')', \quad f''' = (f'')', \quad f^{(4)} = (f''')', \quad f^{(5)} = (f^{(4)})', \dots$$

If $y = f(x)$, then successive derivatives can also be denoted by

$$y', \quad y'', \quad y''', \quad y^{(4)}, \quad y^{(5)}, \dots$$

Other common notations are

$$y' = \frac{dy}{dx} = \frac{d}{dx}[f(x)]$$

$$y'' = \frac{d^2 y}{dx^2} = \frac{d}{dx}\left[\frac{d}{dx}[f(x)]\right] = \frac{d^2}{dx^2}[f(x)]$$

$$y''' = \frac{d^3 y}{dx^3} = \frac{d}{dx}\left[\frac{d^2}{dx^2}[f(x)]\right] = \frac{d^3}{dx^3}[f(x)]$$

$$\vdots \qquad\qquad\qquad \vdots$$

These are called, in succession, the *first derivative*, the *second derivative*, the *third derivative*, and so forth. The number of times that f is differentiated is called the **order** of the derivative. A general nth order derivative can be denoted by

$$\frac{d^n y}{dx^n} = f^{(n)}(x) = \frac{d^n}{dx^n}[f(x)] \tag{11}$$

and the value of a general nth order derivative at a specific point $x = x_0$ can be denoted by

$$\left.\frac{d^n y}{dx^n}\right|_{x=x_0} = f^{(n)}(x_0) = \left.\frac{d^n}{dx^n}[f(x)]\right|_{x=x_0} \tag{12}$$

▶ **Example 9**   If $f(x) = 3x^4 - 2x^3 + x^2 - 4x + 2$, then

$$f'(x) = 12x^3 - 6x^2 + 2x - 4$$
$$f''(x) = 36x^2 - 12x + 2$$
$$f'''(x) = 72x - 12$$
$$f^{(4)}(x) = 72$$
$$f^{(5)}(x) = 0$$
$$\vdots$$
$$f^{(n)}(x) = 0 \quad (n \geq 5) \blacktriangleleft$$

We will discuss the significance of second derivatives and those of higher order in later sections.

✔ **QUICK CHECK EXERCISES 2.3**   *(See page 163 for answers.)*

1. In each part, determine $f'(x)$.
   (a) $f(x) = \sqrt{6}$
   (b) $f(x) = \sqrt{6x}$
   (c) $f(x) = 6\sqrt{x}$
   (d) $f(x) = \sqrt{6x}$

2. In parts (a)–(d), determine $f'(x)$.
   (a) $f(x) = x^3 + 5$
   (b) $f(x) = x^2(x^3 + 5)$
   (c) $f(x) = \dfrac{x^3 + 5}{2}$
   (d) $f(x) = \dfrac{x^3 + 5}{x^2}$

3. The slope of the tangent line to the curve $y = x^2 + 4x + 7$ at $x = 1$ is _____.

4. If $f(x) = 3x^3 - 3x^2 + x + 1$, then $f''(x) = $ _____.

EXERCISE SET 2.3  Graphing Utility

1–8 Find dy/dx. ■

1. $y = 4x^7$

2. $y = -3x^{12}$

3. $y = 3x^8 + 2x + 1$

4. $y = \frac{1}{2}(x^4 + 7)$

5. $y = \pi^3$

6. $y = \sqrt{2}x + (1/\sqrt{2})$

7. $y = -\frac{1}{3}(x^7 + 2x - 9)$

8. $y = \frac{x^2 + 1}{5}$

9–16 Find $f'(x)$. ■

9. $f(x) = x^{-3} + \dfrac{1}{x^7}$

10. $f(x) = \sqrt{x} + \dfrac{1}{x}$

11. $f(x) = -3x^{-8} + 2\sqrt{x}$

12. $f(x) = 7x^{-6} - 5\sqrt{x}$

13. $f(x) = x^e + \dfrac{1}{x^{\sqrt{10}}}$

14. $f(x) = \sqrt[3]{\dfrac{8}{x}}$

15. $f(x) = ax^3 + bx^2 + cx + d$ $(a, b, c, d$ constant$)$

16. $f(x) = \dfrac{1}{a}\left(x^2 + \dfrac{1}{b}x + c\right)$ $(a, b, c$ constant$)$

17–18 Find $y'(1)$. ■

17. $y = 5x^2 - 3x + 1$

18. $y = \dfrac{x^{3/2} + 2}{x}$

19–20 Find dx/dt. ■

19. $x = t^2 - t$

20. $x = \dfrac{t^2 + 1}{3t}$

21–24 Find $dy/dx|_{x=1}$. ■

21. $y = 1 + x + x^2 + x^3 + x^4 + x^5$

22. $y = \dfrac{1 + x + x^2 + x^3 + x^4 + x^5 + x^6}{x^3}$

23. $y = (1 - x)(1 + x)(1 + x^2)(1 + x^4)$

24. $y = x^{24} + 2x^{12} + 3x^8 + 4x^6$

25–26 Approximate $f'(1)$ by considering the difference quotient

$$\frac{f(1 + h) - f(1)}{h}$$

for values of h near 0, and then find the exact value of $f'(1)$ by differentiating. ■

25. $f(x) = x^3 - 3x + 1$

26. $f(x) = \dfrac{1}{x^2}$

27–28 Use a graphing utility to estimate the value of $f'(1)$ by zooming in on the graph of f, and then compare your estimate to the exact value obtained by differentiating. ■

27. $f(x) = \dfrac{x^2 + 1}{x}$

28. $f(x) = \dfrac{x + 2x^{3/2}}{\sqrt{x}}$

29–32 Find the indicated derivative. ■

29. $\dfrac{d}{dt}[16t^2]$

30. $\dfrac{dC}{dr}$, where $C = 2\pi r$

31. $V'(r)$, where $V = \pi r^3$

32. $\dfrac{d}{d\alpha}[2\alpha^{-1} + \alpha]$

33–36 True–False Determine whether the statement is true or false. Explain your answer. ■

33. If f and g are differentiable at $x = 2$, then

$$\frac{d}{dx}[f(x) - 8g(x)]\Big|_{x=2} = f'(2) - 8g'(2)$$

34. If $f(x)$ is a cubic polynomial, then $f'(x)$ is a quadratic polynomial.

35. If $f'(2) = 5$, then

$$\frac{d}{dx}[4f(x) + x^3]\Big|_{x=2} = \frac{d}{dx}[4f(x) + 8]\Big|_{x=2} = 4f'(2) = 20$$

36. If $f(x) = x^2(x^4 - x)$, then

$$f''(x) = \frac{d}{dx}[x^2] \cdot \frac{d}{dx}[x^4 - x] = 2x(4x^3 - 1)$$

37. A spherical balloon is being inflated.
(a) Find a general formula for the instantaneous rate of change of the volume V with respect to the radius r, given that $V = \frac{4}{3}\pi r^3$.
(b) Find the rate of change of V with respect to r at the instant when the radius is $r = 5$.

38. Find $\dfrac{d}{d\lambda}\left[\dfrac{\lambda\lambda_0 + \lambda^6}{2 - \lambda_0}\right]$ $(\lambda_0$ is constant$)$.

39. Find an equation of the tangent line to the graph of $y = f(x)$ at $x = -3$ if $f(-3) = 2$ and $f'(-3) = 5$.

40. Find an equation of the tangent line to the graph of $y = f(x)$ at $x = 2$ if $f(2) = -2$ and $f'(2) = -1$.

41–42 Find d^2y/dx^2. ■

41. (a) $y = 7x^3 - 5x^2 + x$ (b) $y = 12x^2 - 2x + 3$
(c) $y = \dfrac{x + 1}{x}$ (d) $y = (5x^2 - 3)(7x^3 + x)$

42. (a) $y = 4x^7 - 5x^3 + 2x$ (b) $y = 3x + 2$
(c) $y = \dfrac{3x - 2}{5x}$ (d) $y = (x^3 - 5)(2x + 3)$

43–44 Find y'''. ■

43. (a) $y = x^{-5} + x^5$ (b) $y = 1/x$
(c) $y = ax^3 + bx + c$ $(a, b, c$ constant$)$

44. (a) $y = 5x^2 - 4x + 7$ (b) $y = 3x^{-2} + 4x^{-1} + x$
(c) $y = ax^4 + bx^2 + c$ $(a, b, c$ constant$)$

45. Find
(a) $f'''(2)$, where $f(x) = 3x^2 - 2$
(b) $\dfrac{d^2y}{dx^2}\Big|_{x=1}$, where $y = 6x^5 - 4x^2$
(c) $\dfrac{d^4}{dx^4}[x^{-3}]\Big|_{x=1}$.

46. Find
(a) $y'''(0)$, where $y = 4x^4 + 2x^3 + 3$
(b) $\dfrac{d^4 y}{dx^4}\bigg|_{x=1}$, where $y = \dfrac{6}{x^4}$.

47. Show that $y = x^3 + 3x + 1$ satisfies $y''' + xy'' - 2y' = 0$.

48. Show that if $x \neq 0$, then $y = 1/x$ satisfies the equation $x^3 y'' + x^2 y' - xy = 0$.

~ 49–50 Use a graphing utility to make rough estimates of the locations of all horizontal tangent lines, and then find their exact locations by differentiating. ∎

49. $y = \frac{1}{3}x^3 - \frac{3}{2}x^2 + 2x$ **50.** $y = \dfrac{x^2 + 9}{x}$

FOCUS ON CONCEPTS

51. Find a function $y = ax^2 + bx + c$ whose graph has an x-intercept of 1, a y-intercept of -2, and a tangent line with a slope of -1 at the y-intercept.

52. Find k if the curve $y = x^2 + k$ is tangent to the line $y = 2x$.

53. Find the x-coordinate of the point on the graph of $y = x^2$ where the tangent line is parallel to the secant line that cuts the curve at $x = -1$ and $x = 2$.

54. Find the x-coordinate of the point on the graph of $y = \sqrt{x}$ where the tangent line is parallel to the secant line that cuts the curve at $x = 1$ and $x = 4$.

55. Find the coordinates of all points on the graph of $y = 1 - x^2$ at which the tangent line passes through the point $(2, 0)$.

56. Show that any two tangent lines to the parabola $y = ax^2$, $a \neq 0$, intersect at a point that is on the vertical line halfway between the points of tangency.

---

57. Suppose that L is the tangent line at $x = x_0$ to the graph of the cubic equation $y = ax^3 + bx$. Find the x-coordinate of the point where L intersects the graph a second time.

58. Show that the segment of the tangent line to the graph of $y = 1/x$ that is cut off by the coordinate axes is bisected by the point of tangency.

59. Show that the triangle that is formed by any tangent line to the graph of $y = 1/x$, $x > 0$, and the coordinate axes has an area of 2 square units.

60. Find conditions on a, b, c, and d so that the graph of the polynomial $f(x) = ax^3 + bx^2 + cx + d$ has
(a) exactly two horizontal tangents
(b) exactly one horizontal tangent
(c) no horizontal tangents.

61. Newton's Law of Universal Gravitation states that the magnitude F of the force exerted by a point with mass M on a point with mass m is
$$F = \frac{GmM}{r^2}$$
where G is a constant and r is the distance between the bodies. Assuming that the points are moving, find a formula for the instantaneous rate of change of F with respect to r.

62. In the temperature range between $0°C$ and $700°C$ the resistance R [in ohms (Ω)] of a certain platinum resistance thermometer is given by
$$R = 10 + 0.04124T - 1.779 \times 10^{-5}T^2$$
where T is the temperature in degrees Celsius. Where in the interval from $0°C$ to $700°C$ is the resistance of the thermometer most sensitive and least sensitive to temperature changes? [*Hint:* Consider the size of dR/dT in the interval $0 \leq T \leq 700$.]

~ 63–64 Use a graphing utility to make rough estimates of the intervals on which $f'(x) > 0$, and then find those intervals exactly by differentiating. ∎

63. $f(x) = x - \dfrac{1}{x}$ **64.** $f(x) = x^3 - 3x$

65–68 You are asked in these exercises to determine whether a piecewise-defined function f is differentiable at a value $x = x_0$, where f is defined by different formulas on different sides of x_0. You may use without proof the following result, which is a consequence of the Mean-Value Theorem (discussed in Section 4.8). ***Theorem.*** *Let f be continuous at x_0 and suppose that $\lim_{x \to x_0} f'(x)$ exists. Then f is differentiable at x_0, and $f'(x_0) = \lim_{x \to x_0} f'(x)$.* ∎

65. Show that
$$f(x) = \begin{cases} x^2 + x + 1, & x \leq 1 \\ 3x, & x > 1 \end{cases}$$
is continuous at $x = 1$. Determine whether f is differentiable at $x = 1$. If so, find the value of the derivative there. Sketch the graph of f.

66. Let
$$f(x) = \begin{cases} x^2 - 16x, & x < 9 \\ \sqrt{x}, & x \geq 9 \end{cases}$$
Is f continuous at $x = 9$? Determine whether f is differentiable at $x = 9$. If so, find the value of the derivative there.

67. Let
$$f(x) = \begin{cases} x^2, & x \leq 1 \\ \sqrt{x}, & x > 1 \end{cases}$$
Determine whether f is differentiable at $x = 1$. If so, find the value of the derivative there.

68. Let
$$f(x) = \begin{cases} x^3 + \frac{1}{16}, & x < \frac{1}{2} \\ \frac{3}{4}x^2, & x \geq \frac{1}{2} \end{cases}$$
Determine whether f is differentiable at $x = \frac{1}{2}$. If so, find the value of the derivative there.

69. Find all points where f fails to be differentiable. Justify your answer.
(a) $f(x) = |3x - 2|$ (b) $f(x) = |x^2 - 4|$

70. In each part, compute f', f'', f''', and then state the formula for $f^{(n)}$.

(a) $f(x) = 1/x$ (b) $f(x) = 1/x^2$

[*Hint:* The expression $(-1)^n$ has a value of 1 if n is even and -1 if n is odd. Use this expression in your answer.]

71. (a) Prove:

$$\frac{d^2}{dx^2}[cf(x)] = c\frac{d^2}{dx^2}[f(x)]$$

$$\frac{d^2}{dx^2}[f(x) + g(x)] = \frac{d^2}{dx^2}[f(x)] + \frac{d^2}{dx^2}[g(x)]$$

(b) Do the results in part (a) generalize to nth derivatives? Justify your answer.

72. Let $f(x) = x^8 - 2x + 3$; find

$$\lim_{w \to 2} \frac{f'(w) - f'(2)}{w - 2}$$

73. (a) Find $f^{(n)}(x)$ if $f(x) = x^n$, $n = 1, 2, 3, \ldots$.

(b) Find $f^{(n)}(x)$ if $f(x) = x^k$ and $n > k$, where k is a positive integer.

(c) Find $f^{(n)}(x)$ if

$$f(x) = a_0 + a_1 x + a_2 x^2 + \cdots + a_n x^n$$

74. (a) Prove: If $f''(x)$ exists for each x in (a, b), then both f and f' are continuous on (a, b).

(b) What can be said about the continuity of f and its derivatives if $f^{(n)}(x)$ exists for each x in (a, b)?

75. Let $f(x) = (mx + b)^n$, where m and b are constants and n is an integer. Use the result of Exercise 52 in Section 2.2 to prove that $f'(x) = nm(mx + b)^{n-1}$.

76–77 Verify the result of Exercise 75 for $f(x)$. ■

76. $f(x) = (2x + 3)^2$ **77.** $f(x) = (3x - 1)^3$

78–81 Use the result of Exercise 75 to compute the derivative of the given function $f(x)$. ■

78. $f(x) = \dfrac{1}{x - 1}$

79. $f(x) = \dfrac{3}{(2x + 1)^2}$

80. $f(x) = \dfrac{x}{x + 1}$

81. $f(x) = \dfrac{2x^2 + 4x + 3}{x^2 + 2x + 1}$

82. The purpose of this exercise is to extend the power rule (Theorem 2.3.2) to any integer exponent. Let $f(x) = x^n$, where n is any integer. If $n > 0$, then $f'(x) = nx^{n-1}$ by Theorem 2.3.2.

(a) Show that the conclusion of Theorem 2.3.2 holds in the case $n = 0$.

(b) Suppose that $n < 0$ and set $m = -n$ so that

$$f(x) = x^n = x^{-m} = \frac{1}{x^m}$$

Use Definition 2.2.1 and Theorem 2.3.2 to show that

$$\frac{d}{dx}\left[\frac{1}{x^m}\right] = -mx^{m-1} \cdot \frac{1}{x^{2m}}$$

and conclude that $f'(x) = nx^{n-1}$.

✔ **QUICK CHECK ANSWERS 2.3**

1. (a) 0 (b) $\sqrt{6}$ (c) $3/\sqrt{x}$ (d) $\sqrt{6}/(2\sqrt{x})$ **2.** (a) $3x^2$ (b) $5x^4 + 10x$ (c) $\frac{3}{2}x^2$ (d) $1 - 10x^{-3}$ **3.** 6   **4.** $18x - 6$

2.4 THE PRODUCT AND QUOTIENT RULES

In this section we will develop techniques for differentiating products and quotients of functions whose derivatives are known.

■ DERIVATIVE OF A PRODUCT

You might be tempted to conjecture that the derivative of a product of two functions is the product of their derivatives. However, a simple example will show this to be false. Consider the functions

$$f(x) = x \quad \text{and} \quad g(x) = x^2$$

The product of their derivatives is

$$f'(x)g'(x) = (1)(2x) = 2x$$

but their product is $h(x) = f(x)g(x) = x^3$, so the derivative of the product is

$$h'(x) = 3x^2$$

Thus, the derivative of the product is not equal to the product of the derivatives. The correct relationship, which is credited to Leibniz, is given by the following theorem.

2.4.1   THEOREM (*The Product Rule*)   *If f and g are differentiable at x, then so is the product $f \cdot g$, and*

Formula (1) can also be expressed as

$$(f \cdot g)' = f \cdot g' + g \cdot f'$$

$$\frac{d}{dx}[f(x)g(x)] = f(x)\frac{d}{dx}[g(x)] + g(x)\frac{d}{dx}[f(x)] \tag{1}$$

PROOF   Whereas the proofs of the derivative rules in the last section were straightforward applications of the derivative definition, a key step in this proof involves adding and subtracting the quantity $f(x + h)g(x)$ to the numerator in the derivative definition. This yields

$$\frac{d}{dx}[f(x)g(x)] = \lim_{h \to 0} \frac{f(x+h) \cdot g(x+h) - f(x) \cdot g(x)}{h}$$

$$= \lim_{h \to 0} \frac{f(x+h)g(x+h) - f(x+h)g(x) + f(x+h)g(x) - f(x)g(x)}{h}$$

$$= \lim_{h \to 0} \left[f(x+h) \cdot \frac{g(x+h) - g(x)}{h} + g(x) \cdot \frac{f(x+h) - f(x)}{h} \right]$$

$$= \lim_{h \to 0} f(x+h) \cdot \lim_{h \to 0} \frac{g(x+h) - g(x)}{h} + \lim_{h \to 0} g(x) \cdot \lim_{h \to 0} \frac{f(x+h) - f(x)}{h}$$

$$= \left[\lim_{h \to 0} f(x+h) \right] \frac{d}{dx}[g(x)] + \left[\lim_{h \to 0} g(x) \right] \frac{d}{dx}[f(x)]$$

$$= f(x)\frac{d}{dx}[g(x)] + g(x)\frac{d}{dx}[f(x)]$$

[*Note:* In the last step $f(x + h) \to f(x)$ as $h \to 0$ because f is continuous at x by Theorem 2.2.3. Also, $g(x) \to g(x)$ as $h \to 0$ because $g(x)$ does not involve h and hence is treated as constant for the limit.] ■

In words, *the derivative of a product of two functions is the first function times the derivative of the second plus the second function times the derivative of the first.*

▶ **Example 1**   Find dy/dx if $y = (4x^2 - 1)(7x^3 + x)$.

Solution.   There are two methods that can be used to find dy/dx. We can either use the product rule or we can multiply out the factors in y and then differentiate. We will give both methods.

Method 1. (*Using the Product Rule*)

$$\frac{dy}{dx} = \frac{d}{dx}[(4x^2 - 1)(7x^3 + x)]$$

$$= (4x^2 - 1)\frac{d}{dx}[7x^3 + x] + (7x^3 + x)\frac{d}{dx}[4x^2 - 1]$$

$$= (4x^2 - 1)(21x^2 + 1) + (7x^3 + x)(8x) = 140x^4 - 9x^2 - 1$$

Method 2. (*Multiplying First*)

$$y = (4x^2 - 1)(7x^3 + x) = 28x^5 - 3x^3 - x$$

Thus,

$$\frac{dy}{dx} = \frac{d}{dx}[28x^5 - 3x^3 - x] = 140x^4 - 9x^2 - 1$$

which agrees with the result obtained using the product rule. ◄

▶ **Example 2**   Find ds/dt if $s = (1 + t)\sqrt{t}$.

Solution.   Applying the product rule yields

$$\frac{ds}{dt} = \frac{d}{dt}[(1 + t)\sqrt{t}]$$

$$= (1 + t)\frac{d}{dt}[\sqrt{t}] + \sqrt{t}\frac{d}{dt}[1 + t]$$

$$= \frac{1 + t}{2\sqrt{t}} + \sqrt{t} = \frac{1 + 3t}{2\sqrt{t}} ◄$$

■ **DERIVATIVE OF A QUOTIENT**

Just as the derivative of a product is not generally the product of the derivatives, so the derivative of a quotient is not generally the quotient of the derivatives. The correct relationship is given by the following theorem.

Formula (2) can also be expressed as

$$\left(\frac{f}{g}\right)' = \frac{g \cdot f' - f \cdot g'}{g^2}$$

2.4.2   THEOREM (*The Quotient Rule*)   *If f and g are both differentiable at x and if $g(x) \neq 0$, then f/g is differentiable at x and*

$$\frac{d}{dx}\left[\frac{f(x)}{g(x)}\right] = \frac{g(x)\dfrac{d}{dx}[f(x)] - f(x)\dfrac{d}{dx}[g(x)]}{[g(x)]^2} \tag{2}$$

PROOF

$$\frac{d}{dx}\left[\frac{f(x)}{g(x)}\right] = \lim_{h \to 0} \frac{\dfrac{f(x + h)}{g(x + h)} - \dfrac{f(x)}{g(x)}}{h} = \lim_{h \to 0} \frac{f(x + h) \cdot g(x) - f(x) \cdot g(x + h)}{h \cdot g(x) \cdot g(x + h)}$$

Adding and subtracting $f(x) \cdot g(x)$ in the numerator yields

$$\frac{d}{dx}\left[\frac{f(x)}{g(x)}\right] = \lim_{h \to 0} \frac{f(x+h) \cdot g(x) - f(x) \cdot g(x) - f(x) \cdot g(x+h) + f(x) \cdot g(x)}{h \cdot g(x) \cdot g(x+h)}$$

$$= \lim_{h \to 0} \frac{\left[g(x) \cdot \dfrac{f(x+h) - f(x)}{h}\right] - \left[f(x) \cdot \dfrac{g(x+h) - g(x)}{h}\right]}{g(x) \cdot g(x+h)}$$

$$= \frac{\displaystyle\lim_{h \to 0} g(x) \cdot \lim_{h \to 0} \frac{f(x+h) - f(x)}{h} - \lim_{h \to 0} f(x) \cdot \lim_{h \to 0} \frac{g(x+h) - g(x)}{h}}{\displaystyle\lim_{h \to 0} g(x) \cdot \lim_{h \to 0} g(x+h)}$$

$$= \frac{\left[\displaystyle\lim_{h \to 0} g(x)\right] \cdot \dfrac{d}{dx}[f(x)] - \left[\displaystyle\lim_{h \to 0} f(x)\right] \cdot \dfrac{d}{dx}[g(x)]}{\displaystyle\lim_{h \to 0} g(x) \cdot \lim_{h \to 0} g(x+h)}$$

$$= \frac{g(x)\dfrac{d}{dx}[f(x)] - f(x)\dfrac{d}{dx}[g(x)]}{[g(x)]^2}$$

[See the note at the end of the proof of Theorem 2.4.1 for an explanation of the last step.] ■

In words, *the derivative of a quotient of two functions is the denominator times the derivative of the numerator minus the numerator times the derivative of the denominator, all divided by the denominator squared.*

Sometimes it is better to simplify a function first than to apply the quotient rule immediately. For example, it is easier to differentiate

$$f(x) = \frac{x^{3/2} + x}{\sqrt{x}}$$

by rewriting it as

$$f(x) = x + \sqrt{x}$$

as opposed to using the quotient rule.

▶ **Example 3**   Find $y'(x)$ for $y = \dfrac{x^3 + 2x^2 - 1}{x + 5}$.

Solution.   Applying the quotient rule yields

$$\frac{dy}{dx} = \frac{d}{dx}\left[\frac{x^3 + 2x^2 - 1}{x + 5}\right] = \frac{(x+5)\dfrac{d}{dx}[x^3 + 2x^2 - 1] - (x^3 + 2x^2 - 1)\dfrac{d}{dx}[x+5]}{(x+5)^2}$$

$$= \frac{(x+5)(3x^2 + 4x) - (x^3 + 2x^2 - 1)(1)}{(x+5)^2}$$

$$= \frac{(3x^3 + 19x^2 + 20x) - (x^3 + 2x^2 - 1)}{(x+5)^2}$$

$$= \frac{2x^3 + 17x^2 + 20x + 1}{(x+5)^2} \quad ◀$$

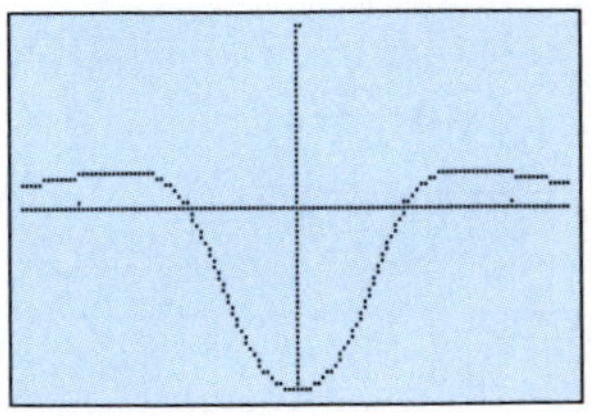

▶ **Example 4**   Let $f(x) = \dfrac{x^2 - 1}{x^4 + 1}$.

(a) Graph $y = f(x)$, and use your graph to make rough estimates of the locations of all horizontal tangent lines.

(b) By differentiating, find the exact locations of the horizontal tangent lines.

Solution (a).   In Figure 2.4.1 we have shown the graph of the equation $y = f(x)$ in the window $[-2.5, 2.5] \times [-1, 1]$. This graph suggests that horizontal tangent lines occur at $x = 0$, $x \approx 1.5$, and $x \approx -1.5$.

$[-2.5, 2.5] \times [-1, 1]$
$x\text{Scl} = 1, y\text{Scl} = 1$

$$y = \frac{x^2 - 1}{x^4 + 1}$$

▲ **Figure 2.4.1**

Solution (b). To find the exact locations of the horizontal tangent lines, we must find the points where $dy/dx = 0$. We start by finding dy/dx:

$$\frac{dy}{dx} = \frac{d}{dx}\left[\frac{x^2-1}{x^4+1}\right] = \frac{(x^4+1)\dfrac{d}{dx}[x^2-1] - (x^2-1)\dfrac{d}{dx}[x^4+1]}{(x^4+1)^2}$$

$$= \frac{(x^4+1)(2x) - (x^2-1)(4x^3)}{(x^4+1)^2}$$

> The differentiation is complete. The rest is simplification.

$$= \frac{-2x^5 + 4x^3 + 2x}{(x^4+1)^2} = -\frac{2x(x^4 - 2x^2 - 1)}{(x^4+1)^2}$$

Now we will set $dy/dx = 0$ and solve for x. We obtain

$$-\frac{2x(x^4 - 2x^2 - 1)}{(x^4+1)^2} = 0$$

The solutions of this equation are the values of x for which the numerator is 0, that is,

$$2x(x^4 - 2x^2 - 1) = 0$$

The first factor yields the solution $x = 0$. Other solutions can be found by solving the equation

$$x^4 - 2x^2 - 1 = 0$$

This can be treated as a quadratic equation in x^2 and solved by the quadratic formula. This yields

$$x^2 = \frac{2 \pm \sqrt{8}}{2} = 1 \pm \sqrt{2}$$

The minus sign yields imaginary values of x, which we ignore since they are not relevant to the problem. The plus sign yields the solutions

$$x = \pm\sqrt{1 + \sqrt{2}}$$

In summary, horizontal tangent lines occur at

$$x = 0, \quad x = \sqrt{1 + \sqrt{2}} \approx 1.55, \quad \text{and} \quad x = -\sqrt{1 + \sqrt{2}} \approx -1.55$$

which is consistent with the rough estimates that we obtained graphically in part (a). ◄

> Derive the following rule for differentiating a reciprocal:
>
> $$\left(\frac{1}{g}\right)' = -\frac{g'}{g^2}$$
>
> Use it to find the derivative of
>
> $$f(x) = \frac{1}{x^2 + 1}$$

■ SUMMARY OF DIFFERENTIATION RULES

The following table summarizes the differentiation rules that we have encountered thus far.

Table 2.4.1

RULES FOR DIFFERENTIATION

$\dfrac{d}{dx}[c] = 0$	$(f+g)' = f' + g'$	$(f \cdot g)' = f \cdot g' + g \cdot f'$	$\left(\dfrac{1}{g}\right)' = -\dfrac{g'}{g^2}$
$(cf)' = cf'$	$(f-g)' = f' - g'$	$\left(\dfrac{f}{g}\right)' = \dfrac{g \cdot f' - f \cdot g'}{g^2}$	$\dfrac{d}{dx}[x^r] = rx^{r-1}$

✔ QUICK CHECK EXERCISES 2.4 *(See page 169 for answers.)*

1. (a) $\dfrac{d}{dx}[x^2 f(x)] = $ _____ (b) $\dfrac{d}{dx}\left[\dfrac{f(x)}{x^2+1}\right] = $ _____

(c) $\dfrac{d}{dx}\left[\dfrac{x^2+1}{f(x)}\right] = $ _____

2. Find $F'(1)$ given that $f(1) = -1$, $f'(1) = 2$, $g(1) = 3$, and $g'(1) = -1$.

(a) $F(x) = 2f(x) - 3g(x)$ (b) $F(x) = [f(x)]^2$

(c) $F(x) = f(x)g(x)$ (d) $F(x) = f(x)/g(x)$

EXERCISE SET 2.4 ◰ Graphing Utility

1–4 Compute the derivative of the given function $f(x)$ by (a) multiplying and then differentiating and (b) using the product rule. Verify that (a) and (b) yield the same result. ■

1. $f(x) = (x + 1)(2x - 1)$ **2.** $f(x) = (3x^2 - 1)(x^2 + 2)$

3. $f(x) = (x^2 + 1)(x^2 - 1)$

4. $f(x) = (x + 1)(x^2 - x + 1)$

5–20 Find $f'(x)$. ■

5. $f(x) = (3x^2 + 6)\left(2x - \frac{1}{4}\right)$

6. $f(x) = (2 - x - 3x^3)(7 + x^5)$

7. $f(x) = (x^3 + 7x^2 - 8)(2x^{-3} + x^{-4})$

8. $f(x) = \left(\frac{1}{x} + \frac{1}{x^2}\right)(3x^3 + 27)$

9. $f(x) = (x - 2)(x^2 + 2x + 4)$

10. $f(x) = (x^2 + x)(x^2 - x)$

11. $f(x) = \dfrac{3x + 4}{x^2 + 1}$ **12.** $f(x) = \dfrac{x - 2}{x^4 + x + 1}$

13. $f(x) = \dfrac{x^2}{3x - 4}$ **14.** $f(x) = \dfrac{2x^2 + 5}{3x - 4}$

15. $f(x) = \dfrac{(2\sqrt{x} + 1)(x - 1)}{x + 3}$

16. $f(x) = (2\sqrt{x} + 1)\left(\dfrac{2 - x}{x^2 + 3x}\right)$

17. $f(x) = (2x + 1)\left(1 + \dfrac{1}{x}\right)(x^{-3} + 7)$

18. $f(x) = x^{-5}(x^2 + 2x)(4 - 3x)(2x^9 + 1)$

19. $f(x) = (x^7 + 2x - 3)^3$ **20.** $f(x) = (x^2 + 1)^4$

21–22 Find $dy/dx|_{x=1}$. ■

21. $y = \left(\dfrac{3x + 2}{x}\right)(x^{-5} + 1)$ **22.** $y = (2x^7 - x^2)\left(\dfrac{x - 1}{x + 1}\right)$

◰ **23–24** Use a graphing utility to estimate the value of $f'(1)$ by zooming in on the graph of f, and then compare your estimate to the exact value obtained by differentiating. ■

23. $f(x) = \dfrac{x}{x^2 + 1}$ **24.** $f(x) = \dfrac{x^2 - 1}{x^2 + 1}$

25. Find $g'(4)$ given that $f(4) = 3$ and $f'(4) = -5$.
  (a) $g(x) = \sqrt{x}\, f(x)$ (b) $g(x) = \dfrac{f(x)}{x}$

26. Find $g'(3)$ given that $f(3) = -2$ and $f'(3) = 4$.
  (a) $g(x) = 3x^2 - 5f(x)$ (b) $g(x) = \dfrac{2x + 1}{f(x)}$

27. In parts (a)–(d), $F(x)$ is expressed in terms of $f(x)$ and $g(x)$. Find $F'(2)$ given that $f(2) = -1$, $f'(2) = 4$, $g(2) = 1$, and $g'(2) = -5$.

  (a) $F(x) = 5f(x) + 2g(x)$ (b) $F(x) = f(x) - 3g(x)$
  (c) $F(x) = f(x)g(x)$ (d) $F(x) = f(x)/g(x)$

28. Find $F'(\pi)$ given that $f(\pi) = 10$, $f'(\pi) = -1$, $g(\pi) = -3$, and $g'(\pi) = 2$.

  (a) $F(x) = 6f(x) - 5g(x)$ (b) $F(x) = x(f(x) + g(x))$
  (c) $F(x) = 2f(x)g(x)$ (d) $F(x) = \dfrac{f(x)}{4 + g(x)}$

29–34 Find all values of x at which the tangent line to the given curve satisfies the stated property. ■

29. $y = \dfrac{x^2 - 1}{x + 2}$; horizontal **30.** $y = \dfrac{x^2 + 1}{x - 1}$; horizontal

31. $y = \dfrac{x^2 + 1}{x + 1}$; parallel to the line $y = x$

32. $y = \dfrac{x + 3}{x + 2}$; perpendicular to the line $y = x$

33. $y = \dfrac{1}{x + 4}$; passes through the origin

34. $y = \dfrac{2x + 5}{x + 2}$; y-intercept 2

FOCUS ON CONCEPTS

35. (a) What should it mean to say that two curves intersect at right angles?
  (b) Show that the curves $y = 1/x$ and $y = 1/(2 - x)$ intersect at right angles.

36. Find all values of a such that the curves $y = a/(x - 1)$ and $y = x^2 - 2x + 1$ intersect at right angles.

37. Find a general formula for $F''(x)$ if $F(x) = xf(x)$ and f and f' are differentiable at x.

38. Suppose that the function f is differentiable everywhere and $F(x) = xf(x)$.
  (a) Express $F'''(x)$ in terms of x and derivatives of f.
  (b) For $n \geq 2$, conjecture a formula for $F^{(n)}(x)$.

39. A manufacturer of athletic footwear finds that the sales of their ZipStride brand running shoes is a function $f(p)$ of the selling price p (in dollars) for a pair of shoes. Suppose that $f(120) = 9000$ pairs of shoes and $f'(120) = -60$ pairs of shoes per dollar. The revenue that the manufacturer will receive for selling $f(p)$ pairs of shoes at p dollars per pair is $R(p) = p \cdot f(p)$. Find $R'(120)$. What impact would a small increase in price have on the manufacturer's revenue?

40. Solve the problem in Exercise 39 under the assumption that $f(120) = 9000$ and $f'(120) = -80$.

41. Use the quotient rule (Theorem 2.4.2) to derive the formula for the derivative of $f(x) = x^{-n}$, where n is a positive integer.

1. (a) $x^2 f'(x) + 2x f(x)$ (b) $\dfrac{(x^2 + 1) f'(x) - 2x f(x)}{(x^2 + 1)^2}$ (c) $\dfrac{2x f(x) - (x^2 + 1) f'(x)}{[f(x)^2]}$ **2.** (a) 7 (b) −4 (c) 7 (d) $\frac{5}{9}$

2.5 DERIVATIVES OF TRIGONOMETRIC FUNCTIONS

The main objective of this section is to obtain formulas for the derivatives of the six basic trigonometric functions. If needed, you will find a review of trigonometric functions in Appendix B.

We will assume in this section that the variable x in the trigonometric functions $\sin x$, $\cos x$, $\tan x$, $\cot x$, $\sec x$, and $\csc x$ is measured in radians. Also, we will need the limits in Theorem 1.6.5, but restated as follows using h rather than x as the variable:

$$\lim_{h \to 0} \frac{\sin h}{h} = 1 \quad \text{and} \quad \lim_{h \to 0} \frac{1 - \cos h}{h} = 0 \qquad (1\text{–}2)$$

Let us start with the problem of differentiating $f(x) = \sin x$. Using the definition of the derivative we obtain

$$f'(x) = \lim_{h \to 0} \frac{f(x + h) - f(x)}{h}$$

$$= \lim_{h \to 0} \frac{\sin(x + h) - \sin x}{h}$$

$$= \lim_{h \to 0} \frac{\sin x \cos h + \cos x \sin h - \sin x}{h} \qquad \boxed{\text{By the addition formula for sine}}$$

$$= \lim_{h \to 0} \left[\sin x \left(\frac{\cos h - 1}{h} \right) + \cos x \left(\frac{\sin h}{h} \right) \right]$$

$$= \lim_{h \to 0} \left[\cos x \left(\frac{\sin h}{h} \right) - \sin x \left(\frac{1 - \cos h}{h} \right) \right] \qquad \boxed{\text{Algebraic reorganization}}$$

$$= \lim_{h \to 0} \cos x \cdot \lim_{h \to 0} \frac{\sin h}{h} - \lim_{h \to 0} \sin x \cdot \lim_{h \to 0} \frac{1 - \cos h}{h}$$

$$= \left(\lim_{h \to 0} \cos x \right) (1) - \left(\lim_{h \to 0} \sin x \right) (0) \qquad \boxed{\text{Formulas (1) and (2)}}$$

$$= \lim_{h \to 0} \cos x = \cos x \qquad \boxed{\begin{array}{l}\cos x \text{ does not involve the variable } h \text{ and hence}\\ \text{is treated as a constant in the limit computation.}\end{array}}$$

Thus, we have shown that

Formulas (1) and (2) and the derivation of Formulas (3) and (4) are only valid if h and x are in radians. See Exercise 49 for how Formulas (3) and (4) change when x is measured in degrees.

$$\frac{d}{dx}[\sin x] = \cos x \qquad (3)$$

In the exercises we will ask you to use the same method to derive the following formula for the derivative of $\cos x$:

$$\frac{d}{dx}[\cos x] = -\sin x \qquad (4)$$

▶ **Example 1**   Find dy/dx if $y = x \sin x$.

Solution.   Using Formula (3) and the product rule we obtain

$$\frac{dy}{dx} = \frac{d}{dx}[x \sin x]$$

$$= x \frac{d}{dx}[\sin x] + \sin x \frac{d}{dx}[x]$$

$$= x \cos x + \sin x \blacktriangleleft$$

▶ **Example 2**   Find dy/dx if $y = \dfrac{\sin x}{1 + \cos x}$.

Solution.   Using the quotient rule together with Formulas (3) and (4) we obtain

$$\frac{dy}{dx} = \frac{(1 + \cos x) \cdot \dfrac{d}{dx}[\sin x] - \sin x \cdot \dfrac{d}{dx}[1 + \cos x]}{(1 + \cos x)^2}$$

$$= \frac{(1 + \cos x)(\cos x) - (\sin x)(-\sin x)}{(1 + \cos x)^2}$$

$$= \frac{\cos x + \cos^2 x + \sin^2 x}{(1 + \cos x)^2} = \frac{\cos x + 1}{(1 + \cos x)^2} = \frac{1}{1 + \cos x} \blacktriangleleft$$

The derivatives of the remaining trigonometric functions are

$\dfrac{d}{dx}[\tan x] = \sec^2 x$	$\dfrac{d}{dx}[\sec x] = \sec x \tan x$ (5–6)
$\dfrac{d}{dx}[\cot x] = -\csc^2 x$	$\dfrac{d}{dx}[\csc x] = -\csc x \cot x$ (7–8)

> Since Formulas (3) and (4) are valid only if x is in radians, the same is true for Formulas (5)–(8).

These can all be obtained using the definition of the derivative, but it is easier to use Formulas (3) and (4) and apply the quotient rule to the relationships

$$\tan x = \frac{\sin x}{\cos x}, \quad \cot x = \frac{\cos x}{\sin x}, \quad \sec x = \frac{1}{\cos x}, \quad \csc x = \frac{1}{\sin x}$$

For example,

$$\frac{d}{dx}[\tan x] = \frac{d}{dx}\left[\frac{\sin x}{\cos x}\right] = \frac{\cos x \cdot \dfrac{d}{dx}[\sin x] - \sin x \cdot \dfrac{d}{dx}[\cos x]}{\cos^2 x}$$

$$= \frac{\cos x \cdot \cos x - \sin x \cdot (-\sin x)}{\cos^2 x} = \frac{\cos^2 x + \sin^2 x}{\cos^2 x} = \frac{1}{\cos^2 x} = \sec^2 x$$

> When finding the value of a derivative at a specific point $x = x_0$, it is important to substitute x_0 *after* the derivative is obtained. Thus, in Example 3 we made the substitution $x = \pi/4$ after f'' was calculated. What would have happened had we *incorrectly* substituted $x = \pi/4$ into $f'(x)$ before calculating f''?

▶ **Example 3**   Find $f''(\pi/4)$ if $f(x) = \sec x$.

$$f'(x) = \sec x \tan x$$

$$f''(x) = \sec x \cdot \frac{d}{dx}[\tan x] + \tan x \cdot \frac{d}{dx}[\sec x]$$

$$= \sec x \cdot \sec^2 x + \tan x \cdot \sec x \tan x$$

$$= \sec^3 x + \sec x \tan^2 x$$

Thus,

$$f''(\pi/4) = \sec^3(\pi/4) + \sec(\pi/4)\tan^2(\pi/4)$$
$$= (\sqrt{2})^3 + (\sqrt{2})(1)^2 = 3\sqrt{2} \blacktriangleleft$$

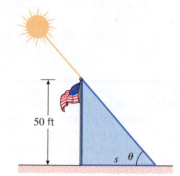

▶ **Example 4**   On a sunny day, a 50 ft flagpole casts a shadow that changes with the angle of elevation of the Sun. Let s be the length of the shadow and θ the angle of elevation of the Sun (Figure 2.5.1). Find the rate at which the length of the shadow is changing with respect to θ when $\theta = 45°$. Express your answer in units of feet/degree.

Solution.   The variables s and θ are related by $\tan\theta = 50/s$ or, equivalently,

$$s = 50\cot\theta \tag{9}$$

If θ is measured in radians, then Formula (7) is applicable, which yields

$$\frac{ds}{d\theta} = -50\csc^2\theta$$

which is the rate of change of shadow length with respect to the elevation angle θ in units of feet/radian. When $\theta = 45°$ (or equivalently $\theta = \pi/4$ radians), we obtain

$$\left.\frac{ds}{d\theta}\right|_{\theta=\pi/4} = -50\csc^2(\pi/4) = -100 \text{ feet/radian}$$

Converting radians (rad) to degrees (deg) yields

$$-100\frac{\text{ft}}{\text{rad}} \cdot \frac{\pi}{180}\frac{\text{rad}}{\text{deg}} = -\frac{5}{9}\pi\frac{\text{ft}}{\text{deg}} \approx -1.75 \text{ ft/deg}$$

Thus, when $\theta = 45°$, the shadow length is decreasing (because of the minus sign) at an approximate rate of 1.75 ft/deg increase in the angle of elevation. ◀

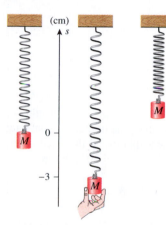

▲ **Figure 2.5.2**

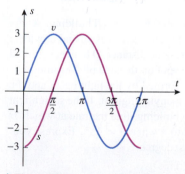

▶ **Example 5**   As illustrated in Figure 2.5.2, suppose that a spring with an attached mass is stretched 3 cm beyond its rest position and released at time $t = 0$. Assuming that the position function of the top of the attached mass is

$$s = -3\cos t \tag{10}$$

where s is in centimeters and t is in seconds, find the velocity function and discuss the motion of the attached mass.

Solution.   The velocity function is

$$v = \frac{ds}{dt} = \frac{d}{dt}[-3\cos t] = 3\sin t$$

Figure 2.5.3 shows the graphs of the position and velocity functions. The position function tells us that the top of the mass oscillates between a low point of $s = -3$ and a high point of $s = 3$ with one complete oscillation occuring every 2π seconds [the period of (10)]. The top of the mass is moving up (the positive s-direction) when v is positive, is moving down when v is negative, and is at a high or low point when $v = 0$. Thus, for example, the top of the mass moves up from time $t = 0$ to time $t = \pi$, at which time it reaches the high point $s = 3$ and then moves down until time $t = 2\pi$, at which time it reaches the low point of $s = -3$. The motion then repeats periodically. ◀

▲ **Figure 2.5.3**

In Example 5, the top of the mass has its maximum speed when it passes through its rest position. Why? What is that maximum speed?

✔ **QUICK CHECK EXERCISES 2.5** *(See page 174 for answers.)*

1. Find dy/dx.
(a) $y = \sin x$ 　　　　(b) $y = \cos x$
(c) $y = \tan x$ 　　　　(d) $y = \sec x$

2. Find $f'(x)$ and $f'(\pi/3)$ if $f(x) = \sin x \cos x$.

3. Use a derivative to evaluate each limit.
(a) $\displaystyle \lim_{h \to 0} \frac{\sin\left(\frac{\pi}{2} + h\right) - 1}{h}$ 　　(b) $\displaystyle \lim_{h \to 0} \frac{\csc(x + h) - \csc x}{h}$

EXERCISE SET 2.5 Graphing Utility

1–18 Find $f'(x)$. ■

1. $f(x) = 4\cos x + 2\sin x$ 　　**2.** $f(x) = \dfrac{5}{x^2} + \sin x$

3. $f(x) = -4x^2 \cos x$ 　　**4.** $f(x) = 2\sin^2 x$

5. $f(x) = \dfrac{5 - \cos x}{5 + \sin x}$ 　　**6.** $f(x) = \dfrac{\sin x}{x^2 + \sin x}$

7. $f(x) = \sec x - \sqrt{2}\tan x$ 　　**8.** $f(x) = (x^2 + 1)\sec x$

9. $f(x) = 4\csc x - \cot x$ 　　**10.** $f(x) = \cos x - x\csc x$

11. $f(x) = \sec x \tan x$ 　　**12.** $f(x) = \csc x \cot x$

13. $f(x) = \dfrac{\cot x}{1 + \csc x}$ 　　**14.** $f(x) = \dfrac{\sec x}{1 + \tan x}$

15. $f(x) = \sin^2 x + \cos^2 x$ 　　**16.** $f(x) = \sec^2 x - \tan^2 x$

17. $f(x) = \dfrac{\sin x \sec x}{1 + x \tan x}$ 　　**18.** $f(x) = \dfrac{(x^2 + 1)\cot x}{3 - \cos x \csc x}$

19–24 Find d^2y/dx^2. ■

19. $y = x \cos x$ 　　**20.** $y = \csc x$

21. $y = x \sin x - 3 \cos x$ 　　**22.** $y = x^2 \cos x + 4 \sin x$

23. $y = \sin x \cos x$ 　　**24.** $y = \tan x$

25. Find the equation of the line tangent to the graph of $\tan x$ at
(a) $x = 0$ 　　(b) $x = \pi/4$ 　　(c) $x = -\pi/4$.

26. Find the equation of the line tangent to the graph of $\sin x$ at
(a) $x = 0$ 　　(b) $x = \pi$ 　　(c) $x = \pi/4$.

27. (a) Show that $y = x \sin x$ is a solution to $y'' + y = 2 \cos x$.
(b) Show that $y = x \sin x$ is a solution of the equation $y^{(4)} + y'' = -2 \cos x$.

28. (a) Show that $y = \cos x$ and $y = \sin x$ are solutions of the equation $y'' + y = 0$.
(b) Show that $y = A \sin x + B \cos x$ is a solution of the equation $y'' + y = 0$ for all constants A and B.

29. Find all values in the interval $[-2\pi, 2\pi]$ at which the graph of f has a horizontal tangent line.
(a) $f(x) = \sin x$ 　　(b) $f(x) = x + \cos x$
(c) $f(x) = \tan x$ 　　(d) $f(x) = \sec x$

 30. (a) Use a graphing utility to make rough estimates of the values in the interval $[0, 2\pi]$ at which the graph of $y = \sin x \cos x$ has a horizontal tangent line.
(b) Find the exact locations of the points where the graph has a horizontal tangent line.

31. A 10 ft ladder leans against a wall at an angle θ with the horizontal, as shown in the accompanying figure. The top of the ladder is x feet above the ground. If the bottom of the ladder is pushed toward the wall, find the rate at which x changes with respect to θ when $\theta = 60°$. Express the answer in units of feet/degree.

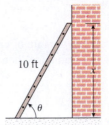

◄ **Figure Ex-31**

32. An airplane is flying on a horizontal path at a height of 3800 ft, as shown in the accompanying figure. At what rate is the distance s between the airplane and the fixed point P changing with respect to θ when $\theta = 30°$? Express the answer in units of feet/degree.

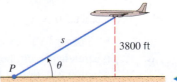

◄ **Figure Ex-32**

33. A searchlight is trained on the side of a tall building. As the light rotates, the spot it illuminates moves up and down the side of the building. That is, the distance D between ground level and the illuminated spot on the side of the building is a function of the angle θ formed by the light beam and the horizontal (see the accompanying figure). If the searchlight is located 50 m from the building, find the rate at which D is changing with respect to θ when $\theta = 45°$. Express your answer in units of meters/degree.

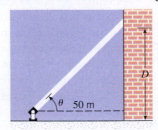

◄ **Figure Ex-33**

34. An Earth-observing satellite can see only a portion of the Earth's surface. The satellite has horizon sensors that can detect the angle θ shown in the accompanying figure. Let r be the radius of the Earth (assumed spherical) and h the distance of the satellite from the Earth's surface.

(a) Show that $h = r(\csc\theta - 1)$.

(b) Using $r = 6378$ km, find the rate at which h is changing with respect to θ when $\theta = 30°$. Express the answer in units of kilometers/degree.

Source: Adapted from *Space Mathematics*, NASA, 1985.

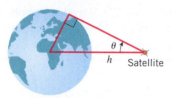

Earth   ◀ **Figure Ex-34**

35–38 True–False Determine whether the statement is true or false. Explain your answer. ■

35. If $g(x) = f(x)\sin x$, then $g'(x) = f'(x)\cos x$.

36. If $g(x) = f(x)\sin x$, then $g'(0) = f(0)$.

37. If $f(x)\cos x = \sin x$, then $f'(x) = \sec^2 x$.

38. Suppose that $g(x) = f(x)\sec x$, where $f(0) = 8$ and $f'(0) = -2$. Then

$$g'(0) = \lim_{h \to 0} \frac{f(h)\sec h - f(0)}{h} = \lim_{h \to 0} \frac{8(\sec h - 1)}{h}$$

$$= 8 \cdot \frac{d}{dx}[\sec x]\Big|_{x=0} = 8\sec 0\tan 0 = 0$$

39–40 Make a conjecture about the derivative by calculating the first few derivatives and observing the resulting pattern. ■

39. $\dfrac{d^{87}}{dx^{87}}[\sin x]$

40. $\dfrac{d^{100}}{dx^{100}}[\cos x]$

41. Let $f(x) = \cos x$. Find all positive integers n for which $f^{(n)}(x) = \sin x$.

42. Let $f(x) = \sin x$. Find all positive integers n for which $f^{(n)}(x) = \sin x$.

FOCUS ON CONCEPTS

43. In each part, determine where f is differentiable.

(a) $f(x) = \sin x$ (b) $f(x) = \cos x$

(c) $f(x) = \tan x$ (d) $f(x) = \cot x$

(e) $f(x) = \sec x$ (f) $f(x) = \csc x$

(g) $f(x) = \dfrac{1}{1 + \cos x}$ (h) $f(x) = \dfrac{1}{\sin x \cos x}$

(i) $f(x) = \dfrac{\cos x}{2 - \sin x}$

44. (a) Derive Formula (4) using the definition of a derivative.

(b) Use Formulas (3) and (4) to obtain (7).

(c) Use Formula (4) to obtain (6).

(d) Use Formula (3) to obtain (8).

45. Use Formula (1), the alternative form for the definition of derivative given in Formula (13) of Section 2.2, that is,

$$f'(x) = \lim_{w \to x} \frac{f(w) - f(x)}{w - x}$$

and the difference identity

$$\sin\alpha - \sin\beta = 2\sin\left(\frac{\alpha - \beta}{2}\right)\cos\left(\frac{\alpha + \beta}{2}\right)$$

to show that $\dfrac{d}{dx}[\sin x] = \cos x$.

46. Follow the directions of Exercise 45 using the difference identity

$$\cos\alpha - \cos\beta = -2\sin\left(\frac{\alpha - \beta}{2}\right)\sin\left(\frac{\alpha + \beta}{2}\right)$$

to show that $\dfrac{d}{dx}[\cos x] = -\sin x$.

47. (a) Show that $\lim\limits_{h \to 0} \dfrac{\tan h}{h} = 1$.

(b) Use the result in part (a) to help derive the formula for the derivative of $\tan x$ directly from the definition of a derivative.

48. Without using any trigonometric identities, find

$$\lim_{x \to 0} \frac{\tan(x + y) - \tan y}{x}$$

[*Hint:* Relate the given limit to the definition of the derivative of an appropriate function of y.]

49. The derivative formulas for $\sin x$, $\cos x$, $\tan x$, $\cot x$, $\sec x$, and $\csc x$ were obtained under the assumption that x is measured in radians. If x is measured in degrees, then

$$\lim_{x \to 0} \frac{\sin x}{x} = \frac{\pi}{180}$$

(See Exercise 49 of Section 1.6). Use this result to prove that if x is measured in degrees, then

(a) $\dfrac{d}{dx}[\sin x] = \dfrac{\pi}{180}\cos x$

(b) $\dfrac{d}{dx}[\cos x] = -\dfrac{\pi}{180}\sin x$.

50. Writing Suppose that f is a function that is differentiable everywhere. Explain the relationship, if any, between the periodicity of f and that of f'. That is, if f is periodic, must f' also be periodic? If f' is periodic, must f also be periodic?

✔ **QUICK CHECK ANSWERS 2.5**

1. (a) $\cos x$ (b) $-\sin x$ (c) $\sec^2 x$ (d) $\sec x \tan x$ **2.** $f'(x) = \cos^2 x - \sin^2 x$, $f'(\pi/3) = -\frac{1}{2}$

3. (a) $\dfrac{d}{dx}[\sin x]\bigg|_{x=\pi/2} = 0$ (b) $\dfrac{d}{dx}[\csc x] = -\csc x \cot x$

2.6 THE CHAIN RULE

In this section we will derive a formula that expresses the derivative of a composition $f \circ g$ in terms of the derivatives of f and g. This formula will enable us to differentiate complicated functions using known derivatives of simpler functions.

■ DERIVATIVES OF COMPOSITIONS

Mike Brinson/Getty Images
The cost of a car trip is a combination of fuel efficiency and the cost of gasoline.

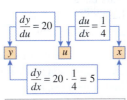

▲ **Figure 2.6.1**

Suppose you are traveling to school in your car, which gets 20 miles per gallon of gasoline. The number of miles you can travel in your car without refueling is a function of the number of gallons of gas you have in the gas tank. In symbols, if y is the number of miles you can travel and u is the number of gallons of gas you have initially, then y is a function of u, or $y = f(u)$. As you continue your travels, you note that your local service station is selling gasoline for \$4 per gallon. The number of gallons of gas you have initially is a function of the amount of money you spend for that gas. If x is the number of dollars you spend on gas, then $u = g(x)$. Now 20 miles per gallon is the rate at which your mileage changes with respect to the amount of gasoline you use, so

$$f'(u) = \frac{dy}{du} = 20 \text{ miles per gallon}$$

Similarly, since gasoline costs \$4 per gallon, each dollar you spend will give you 1/4 of a gallon of gas, and

$$g'(x) = \frac{du}{dx} = \frac{1}{4} \text{ gallons per dollar}$$

Notice that the number of miles you can travel is also a function of the number of dollars you spend on gasoline. This fact is expressible as the composition of functions

$$y = f(u) = f(g(x))$$

You might be interested in how many miles you can travel per dollar, which is dy/dx. Intuition suggests that rates of change multiply in this case (see Figure 2.6.1), so

$$\frac{dy}{dx} = \frac{dy}{du} \cdot \frac{du}{dx} = \frac{20 \text{ miles}}{1 \text{ gallon}} \cdot \frac{1 \text{ gallons}}{4 \text{ dollars}} = \frac{20 \text{ miles}}{4 \text{ dollars}} = 5 \text{ miles per dollar}$$

The following theorem, the proof of which is given in Appendix J, formalizes the preceding ideas.

The name "chain rule" is appropriate because the desired derivative is obtained by a two-link "chain" of simpler derivatives.

2.6.1 THEOREM (*The Chain Rule*) *If g is differentiable at x and f is differentiable at $g(x)$, then the composition $f \circ g$ is differentiable at x. Moreover, if*

$$y = f(g(x)) \quad \text{and} \quad u = g(x)$$

then $y = f(u)$ and

$$\frac{dy}{dx} = \frac{dy}{du} \cdot \frac{du}{dx} \tag{1}$$

> **Example 1**   Find dy/dx if $y = \cos(x^3)$.

Solution.   Let $u = x^3$ and express y as $y = \cos u$. Applying Formula (1) yields

$$\frac{dy}{dx} = \frac{dy}{du} \cdot \frac{du}{dx}$$

$$= \frac{d}{du}[\cos u] \cdot \frac{d}{dx}[x^3]$$

$$= (-\sin u) \cdot (3x^2)$$

$$= (-\sin(x^3)) \cdot (3x^2) = -3x^2 \sin(x^3) \; \blacktriangleleft$$

Formula (1) is easy to remember because the left side is exactly what results if we "cancel" the du's on the right side. This "canceling" device provides a good way of deducing the correct form of the chain rule when different variables are used. For example, if w is a function of x and x is a function of t, then the chain rule takes the form

$$\frac{dw}{dt} = \frac{dw}{dx} \cdot \frac{dx}{dt}$$

> **Example 2**   Find dw/dt if $w = \tan x$ and $x = 4t^3 + t$.

Solution.   In this case the chain rule computations take the form

$$\frac{dw}{dt} = \frac{dw}{dx} \cdot \frac{dx}{dt}$$

$$= \frac{d}{dx}[\tan x] \cdot \frac{d}{dt}[4t^3 + t]$$

$$= (\sec^2 x) \cdot (12t^2 + 1)$$

$$= [\sec^2(4t^3 + t)] \cdot (12t^2 + 1) = (12t^2 + 1)\sec^2(4t^3 + t) \; \blacktriangleleft$$

■ **AN ALTERNATIVE VERSION OF THE CHAIN RULE**

Formula (1) for the chain rule can be unwieldy in some problems because it involves so many variables. As you become more comfortable with the chain rule, you may want to dispense with writing out the dependent variables by expressing (1) in the form

Confirm that (2) is an alternative version of (1) by letting $y = f(g(x))$ and $u = g(x)$.

$$\frac{d}{dx}[f(g(x))] = (f \circ g)'(x) = f'(g(x))g'(x) \tag{2}$$

A convenient way to remember this formula is to call f the "outside function" and g the "inside function" in the composition $f(g(x))$ and then express (2) in words as:

The derivative of $f(g(x))$ is the derivative of the outside function evaluated at the inside function times the derivative of the inside function.

$$\frac{d}{dx}[f(g(x))] = \underbrace{f'(g(x))}_{} \cdot \underbrace{g'(x)}_{}$$

Derivative of the outside function evaluated at the inside function

Derivative of the inside function

▶ **Example 3** (*Example 1 revisited*)   Find $h'(x)$ if $h(x) = \cos(x^3)$.

Solution.   We can think of h as a composition $f(g(x))$ in which $g(x) = x^3$ is the inside function and $f(x) = \cos x$ is the outside function. Thus, Formula (2) yields

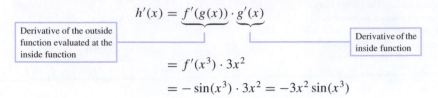

$$h'(x) = \underbrace{f'(g(x))}_{} \cdot \underbrace{g'(x)}_{}$$

Derivative of the outside function evaluated at the inside function

Derivative of the inside function

$$= f'(x^3) \cdot 3x^2$$

$$= -\sin(x^3) \cdot 3x^2 = -3x^2 \sin(x^3)$$

which agrees with the result obtained in Example 1.   ◀

▶ **Example 4**

$$\frac{d}{dx}[\tan^2 x] = \frac{d}{dx}[(\tan x)^2] = \underbrace{(2\tan x)}_{} \cdot \underbrace{(\sec^2 x)}_{} = 2\tan x \sec^2 x$$

Derivative of the outside function evaluated at the inside function

Derivative of the inside function

$$\frac{d}{dx}[\sqrt{x^2 + 1}] = \frac{1}{2\sqrt{x^2 + 1}} \cdot \underbrace{2x}_{} = \frac{x}{\sqrt{x^2 + 1}}$$

See Formula (6) of Section 2.3.   ◀

Derivative of the outside function evaluated at the inside function

Derivative of the inside function

■ GENERALIZED DERIVATIVE FORMULAS

There is a useful third variation of the chain rule that strikes a middle ground between Formulas (1) and (2). If we let $u = g(x)$ in (2), then we can rewrite that formula as

$$\frac{d}{dx}[f(u)] = f'(u)\frac{du}{dx} \tag{3}$$

This result, called the ***generalized derivative formula*** for f, provides a way of using the derivative of $f(x)$ to produce the derivative of $f(u)$, where u is a function of x. Table 2.6.1 gives some examples of this formula.

Table 2.6.1

GENERALIZED DERIVATIVE FORMULAS

$$\frac{d}{dx}[u^r] = ru^{r-1}\frac{du}{dx}$$

$$\frac{d}{dx}[\sin u] = \cos u \,\frac{du}{dx} \qquad\qquad \frac{d}{dx}[\cos u] = -\sin u \,\frac{du}{dx}$$

$$\frac{d}{dx}[\tan u] = \sec^2 u \,\frac{du}{dx} \qquad\qquad \frac{d}{dx}[\cot u] = -\csc^2 u \,\frac{du}{dx}$$

$$\frac{d}{dx}[\sec u] = \sec u \tan u \,\frac{du}{dx} \qquad\qquad \frac{d}{dx}[\csc u] = -\csc u \cot u \,\frac{du}{dx}$$

▶ **Example 5**   Find

(a) $\dfrac{d}{dx}[\sin(2x)]$ (b) $\dfrac{d}{dx}[\tan(x^2+1)]$ (c) $\dfrac{d}{dx}\left[\sqrt{x^3+\csc x}\right]$

(d) $\dfrac{d}{dx}[x^2-x+2]^{3/4}$ (e) $\dfrac{d}{dx}\left[(1+x^5\cot x)^{-8}\right]$

Solution (a).   Taking $u=2x$ in the generalized derivative formula for $\sin u$ yields

$$\frac{d}{dx}[\sin(2x)]=\frac{d}{dx}[\sin u]=\cos u\frac{du}{dx}=\cos 2x\cdot\frac{d}{dx}[2x]=\cos 2x\cdot 2=2\cos 2x$$

Solution (b).   Taking $u=x^2+1$ in the generalized derivative formula for $\tan u$ yields

$$\frac{d}{dx}[\tan(x^2+1)]=\frac{d}{dx}[\tan u]=\sec^2 u\frac{du}{dx}$$

$$=\sec^2(x^2+1)\cdot\frac{d}{dx}[x^2+1]=\sec^2(x^2+1)\cdot 2x$$

$$=2x\sec^2(x^2+1)$$

Solution (c).   Taking $u=x^3+\csc x$ in the generalized derivative formula for $\sqrt{u}$ yields

$$\frac{d}{dx}\left[\sqrt{x^3+\csc x}\right]=\frac{d}{dx}[\sqrt{u}]=\frac{1}{2\sqrt{u}}\frac{du}{dx}=\frac{1}{2\sqrt{x^3+\csc x}}\cdot\frac{d}{dx}[x^3+\csc x]$$

$$=\frac{1}{2\sqrt{x^3+\csc x}}\cdot(3x^2-\csc x\cot x)=\frac{3x^2-\csc x\cot x}{2\sqrt{x^3+\csc x}}$$

Solution (d).   Taking $u=x^2-x+2$ in the generalized derivative formula for $u^{3/4}$ yields

$$\frac{d}{dx}[x^2-x+2]^{3/4}=\frac{d}{dx}[u^{3/4}]=\frac{3}{4}u^{-1/4}\frac{du}{dx}$$

$$=\frac{3}{4}(x^2-x+2)^{-1/4}\cdot\frac{d}{dx}[x^2-x+2]$$

$$=\frac{3}{4}(x^2-x+2)^{-1/4}(2x-1)$$

Solution (e).   Taking $u=1+x^5\cot x$ in the generalized derivative formula for u^{-8} yields

$$\frac{d}{dx}\left[(1+x^5\cot x)^{-8}\right]=\frac{d}{dx}[u^{-8}]=-8u^{-9}\frac{du}{dx}$$

$$=-8(1+x^5\cot x)^{-9}\cdot\frac{d}{dx}[1+x^5\cot x]$$

$$=-8(1+x^5\cot x)^{-9}\cdot\left[x^5(-\csc^2 x)+5x^4\cot x\right]$$

$$=(8x^5\csc^2 x-40x^4\cot x)(1+x^5\cot x)^{-9}\ \blacktriangleleft$$

Sometimes you will have to make adjustments in notation or apply the chain rule more than once to calculate a derivative.

▶ **Example 6**   Find

(a) $\dfrac{d}{dx}\left[\sin(\sqrt{1+\cos x}\,)\right]$ (b) $\dfrac{d\mu}{dt}$ if $\mu=\sec\sqrt{\omega t}$ (ω constant)

Solution (a). Taking $u = \sqrt{1 + \cos x}$ in the generalized derivative formula for $\sin u$ yields

$$\frac{d}{dx}\left[\sin(\sqrt{1 + \cos x}\,)\right] = \frac{d}{dx}[\sin u] = \cos u \frac{du}{dx}$$

$$= \cos(\sqrt{1 + \cos x}) \cdot \frac{d}{dx}\left[\sqrt{1 + \cos x}\right]$$

$$= \cos(\sqrt{1 + \cos x}) \cdot \frac{-\sin x}{2\sqrt{1 + \cos x}} \quad \boxed{\text{We used the generalized derivative formula for } \sqrt{u} \text{ with } u = 1 + \cos x.}$$

$$= -\frac{\sin x \cos(\sqrt{1 + \cos x})}{2\sqrt{1 + \cos x}}$$

Solution (b).

$$\frac{d\mu}{dt} = \frac{d}{dt}[\sec\sqrt{\omega t}] = \sec\sqrt{\omega t}\,\tan\sqrt{\omega t}\,\frac{d}{dt}[\sqrt{\omega t}] \quad \boxed{\text{We used the generalized derivative formula for } \sec u \text{ with } u = \sqrt{\omega t}.}$$

$$= \sec\sqrt{\omega t}\,\tan\sqrt{\omega t}\,\frac{\omega}{2\sqrt{\omega t}} \quad \boxed{\text{We used the generalized derivative formula for } \sqrt{u} \text{ with } u = \omega t.} \quad \blacktriangleleft$$

■ DIFFERENTIATING USING COMPUTER ALGEBRA SYSTEMS

Even with the chain rule and other differentiation rules, some derivative computations can be tedious to perform. For complicated derivatives, engineers and scientists often use computer algebra systems such as *Mathematica*, *Maple*, or *Sage*. For example, although we have all the mathematical tools to compute

$$\frac{d}{dx}\left[\frac{(x^2 + 1)^{10}\sin^3(\sqrt{x}\,)}{\sqrt{1 + \csc x}}\right] \tag{4}$$

TECHNOLOGY MASTERY

If you have a CAS, use it to perform the differentiation in (4).

by hand, the computation is sufficiently involved that it may be more efficient (and less error-prone) to use a computer algebra system.

✔ **QUICK CHECK EXERCISES 2.6**   *(See page 181 for answers.)*

1. The chain rule states that the derivative of the composition of two functions is the derivative of the _____ function evaluated at the _____ function times the derivative of the _____ function.

2. If y is a differentiable function of u, and u is a differentiable function of x, then
$$\frac{dy}{dx} = \underline{\qquad} \cdot \underline{\qquad}$$

3. Find dy/dx.
   (a) $y = (x^2 + 5)^{10}$ (b) $y = \sqrt{1 + 6x}$

4. Find dy/dx.
   (a) $y = \sin(3x + 2)$ (b) $y = (x^2 \tan x)^4$

5. Suppose that $f(2) = 3$, $f'(2) = 4$, $g(3) = 6$, and $g'(3) = -5$. Evaluate
   (a) $h'(2)$, where $h(x) = g(f(x))$
   (b) $k'(3)$, where $k(x) = f\left(\frac{1}{3}g(x)\right)$.

EXERCISE SET 2.6   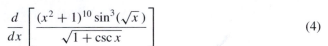 Graphing Utility   [C] CAS

---

1. Given that
$$f'(0) = 2,\ g(0) = 0 \quad \text{and} \quad g'(0) = 3$$
find $(f \circ g)'(0)$.

2. Given that
$$f'(9) = 5,\ g(2) = 9 \quad \text{and} \quad g'(2) = -3$$
find $(f \circ g)'(2)$.

3. Let $f(x) = x^5$ and $g(x) = 2x - 3$.
   (a) Find $(f \circ g)(x)$ and $(f \circ g)'(x)$.
   (b) Find $(g \circ f)(x)$ and $(g \circ f)'(x)$.

4. Let $f(x) = 5\sqrt{x}$ and $g(x) = 4 + \cos x$.
   (a) Find $(f \circ g)(x)$ and $(f \circ g)'(x)$.
   (b) Find $(g \circ f)(x)$ and $(g \circ f)'(x)$.

FOCUS ON CONCEPTS

5. Given the following table of values, find the indicated derivatives in parts (a) and (b).

x	$f(x)$	$f'(x)$	$g(x)$	$g'(x)$
3	5	−2	5	7
5	3	−1	12	4

(a) $F'(3)$, where $F(x) = f(g(x))$
(b) $G'(3)$, where $G(x) = g(f(x))$

6. Given the following table of values, find the indicated derivatives in parts (a) and (b).

x	$f(x)$	$f'(x)$	$g(x)$	$g'(x)$
−1	2	3	2	−3
2	0	4	1	−5

(a) $F'(-1)$, where $F(x) = f(g(x))$
(b) $G'(-1)$, where $G(x) = g(f(x))$

7–26 Find $f'(x)$. ■

7. $f(x) = (x^3 + 2x)^{37}$
8. $f(x) = (3x^2 + 2x - 1)^6$
9. $f(x) = \left(x^3 - \dfrac{7}{x}\right)^{-2}$
10. $f(x) = \dfrac{1}{(x^5 - x + 1)^9}$
11. $f(x) = \dfrac{4}{(3x^2 - 2x + 1)^3}$
12. $f(x) = \sqrt{x^3 - 2x + 5}$
13. $f(x) = \sqrt{4 + \sqrt{3x}}$
14. $f(x) = \sqrt[4]{x} \quad (= \sqrt{\sqrt{x}})$
15. $f(x) = \sin\left(\dfrac{1}{x^2}\right)$
16. $f(x) = \tan\sqrt{x}$
17. $f(x) = 4\cos^5 x$
18. $f(x) = 4x + 5\sin^4 x$
19. $f(x) = \cos^2(3\sqrt{x})$
20. $f(x) = \tan^4(x^3)$
21. $f(x) = 2\sec^2(x^7)$
22. $f(x) = \cos^3\left(\dfrac{x}{x+1}\right)$
23. $f(x) = \sqrt{\cos(5x)}$
24. $f(x) = \sqrt{3x - \sin^2(4x)}$
25. $f(x) = [x + \csc(x^3 + 3)]^{-3}$
26. $f(x) = [x^4 - \sec(4x^2 - 2)]^{-4}$

27–40 Find dy/dx. ■

27. $y = x^3 \sin^2(5x)$
28. $y = \sqrt{x}\tan^3(\sqrt{x})$
29. $y = x^5 \sec(1/x)$
30. $y = \dfrac{\sin x}{\sec(3x+1)}$
31. $y = \cos(\cos x)$
32. $y = \sin(\tan 3x)$
33. $y = \cos^3(\sin 2x)$
34. $y = \dfrac{1 + \csc(x^2)}{1 - \cot(x^2)}$
35. $y = (5x+8)^7 \left(1 - \sqrt{x}\right)^6$
36. $y = (x^2 + x)^5 \sin^8 x$
37. $y = \left(\dfrac{x-5}{2x+1}\right)^3$
38. $y = \left(\dfrac{1+x^2}{1-x^2}\right)^{17}$
39. $y = \dfrac{(2x+3)^3}{(4x^2 - 1)^8}$
40. $y = [1 + \sin^3(x^5)]^{12}$

C **41–42** Use a CAS to find dy/dx. ■

41. $y = [x\sin 2x + \tan^4(x^7)]^5$
42. $y = \tan^4\left(2 + \dfrac{(7-x)\sqrt{3x^2 + 5}}{x^3 + \sin x}\right)$

43–50 Find an equation for the tangent line to the graph at the specified value of x. ■

43. $y = x\cos 3x, \quad x = \pi$
44. $y = \sin(1 + x^3), \quad x = -3$
45. $y = \sec^3\left(\dfrac{\pi}{2} - x\right), \quad x = -\dfrac{\pi}{2}$
46. $y = \left(x - \dfrac{1}{x}\right)^3, \quad x = 2$
47. $y = \tan(4x^2), \quad x = \sqrt{\pi}$
48. $y = 3\cot^4 x, \quad x = \dfrac{\pi}{4}$
49. $y = x^2\sqrt{5 - x^2}, \quad x = 1$
50. $y = \dfrac{x}{\sqrt{1 - x^2}}, \quad x = 0$

51–54 Find d^2y/dx^2. ■

51. $y = x\cos(5x) - \sin^2 x$
52. $y = \sin(3x^2)$
53. $y = \dfrac{1+x}{1-x}$
54. $y = x\tan\left(\dfrac{1}{x}\right)$

55–58 Find the indicated derivative. ■

55. $y = \cot^3(\pi - \theta)$; find $\dfrac{dy}{d\theta}$.
56. $\lambda = \left(\dfrac{au+b}{cu+d}\right)^6$; find $\dfrac{d\lambda}{du}$ (a, b, c, d constants).
57. $\dfrac{d}{d\omega}[a\cos^2 \pi\omega + b\sin^2 \pi\omega]$ (a, b constants)
58. $x = \csc^2\left(\dfrac{\pi}{3} - y\right)$; find $\dfrac{dx}{dy}$.

59. (a) Use a graphing utility to obtain the graph of the function $f(x) = x\sqrt{4 - x^2}$.
    (b) Use the graph in part (a) to make a rough sketch of the graph of f'.
    (c) Find $f'(x)$, and then check your work in part (b) by using the graphing utility to obtain the graph of f'.
    (d) Find the equation of the tangent line to the graph of f at $x = 1$, and graph f and the tangent line together.

60. (a) Use a graphing utility to obtain the graph of the function $f(x) = \sin x^2 \cos x$ over the interval $[-\pi/2, \pi/2]$.
    (b) Use the graph in part (a) to make a rough sketch of the graph of f' over the interval.
    (c) Find $f'(x)$, and then check your work in part (b) by using the graphing utility to obtain the graph of f' over the interval.
    (d) Find the equation of the tangent line to the graph of f at $x = 1$, and graph f and the tangent line together over the interval.

61–64 True–False Determine whether the statement is true or false. Explain your answer. ■

61. If $y = f(x)$, then $\dfrac{d}{dx}[\sqrt{y}] = \sqrt{f'(x)}$.

62. If $y = f(u)$ and $u = g(x)$, then $dy/dx = f'(x) \cdot g'(x)$.

63. If $y = \cos[g(x)]$, then $dy/dx = -\sin[g'(x)]$.

64. If $y = \sin^3(3x^3)$, then $dy/dx = 27x^2 \sin^2(3x^3)\cos(3x^3)$.

65. If an object suspended from a spring is displaced vertically from its equilibrium position by a small amount and released, and if the air resistance and the mass of the spring are ignored, then the resulting oscillation of the object is called **simple harmonic motion**. Under appropriate conditions the displacement y from equilibrium in terms of time t is given by
$$y = A \cos \omega t$$
where A is the initial displacement at time $t = 0$, and ω is a constant that depends on the mass of the object and the stiffness of the spring (see the accompanying figure). The constant $|A|$ is called the **amplitude** of the motion and ω the **angular frequency**.
(a) Show that
$$\frac{d^2y}{dt^2} = -\omega^2 y$$
(b) The **period** T is the time required to make one complete oscillation. Show that $T = 2\pi/\omega$.
(c) The **frequency** f of the vibration is the number of oscillations per unit time. Find f in terms of the period T.
(d) Find the amplitude, period, and frequency of an object that is executing simple harmonic motion given by $y = 0.6 \cos 15t$, where t is in seconds and y is in centimeters.

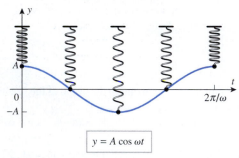

$y = A \cos \omega t$

▲ **Figure Ex-65**

66. Find the value of the constant A so that $y = A \sin 3t$ satisfies the equation
$$\frac{d^2y}{dt^2} + 2y = 4 \sin 3t$$

FOCUS ON CONCEPTS

67. Use the graph of the function f in the accompanying figure to evaluate
$$\frac{d}{dx}\left[\sqrt{x + f(x)}\right]\Bigg|_{x=-1}$$

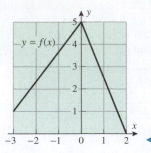

◄ **Figure Ex-67**

68. Using the function f in Exercise 67, evaluate
$$\frac{d}{dx}[f(2 \sin x)]\Bigg|_{x=\pi/6}$$

69. The accompanying figure shows the graph of atmospheric pressure p (lb/in²) versus the altitude h (mi) above sea level.
(a) From the graph and the tangent line at $h = 2$ shown on the graph, estimate the values of p and dp/dh at an altitude of 2 mi.
(b) If the altitude of a space vehicle is increasing at the rate of 0.3 mi/s at the instant when it is 2 mi above sea level, how fast is the pressure changing with time at this instant?

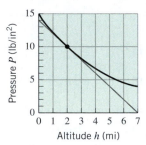

Altitude h (mi) ◄ **Figure Ex-69**

70. The force F (in pounds) acting at an angle θ with the horizontal that is needed to drag a crate weighing W pounds along a horizontal surface at a constant velocity is given by
$$F = \frac{\mu W}{\cos \theta + \mu \sin \theta}$$
where μ is a constant called the **coefficient of sliding friction** between the crate and the surface (see the accompanying figure). Suppose that the crate weighs 150 lb and that $\mu = 0.3$.
(a) Find $dF/d\theta$ when $\theta = 30°$. Express the answer in units of pounds/degree.
(b) Find dF/dt when $\theta = 30°$ if θ is decreasing at the rate of 0.5°/s at this instant.

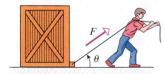

◄ **Figure Ex-70**

71. Recall that

$$\frac{d}{dx}(|x|) = \begin{cases} 1, & x > 0 \\ -1, & x < 0 \end{cases}$$

Use this result and the chain rule to find

$$\frac{d}{dx}(|\sin x|)$$

for nonzero x in the interval $(-\pi, \pi)$.

72. Use the derivative formula for $\sin x$ and the identity

$$\cos x = \sin\left(\frac{\pi}{2} - x\right)$$

to obtain the derivative formula for $\cos x$.

73. Let

$$f(x) = \begin{cases} x \sin \dfrac{1}{x}, & x \neq 0 \\ 0, & x = 0 \end{cases}$$

(a) Show that f is continuous at $x = 0$.
(b) Use Definition 2.2.1 to show that $f'(0)$ does not exist.
(c) Find $f'(x)$ for $x \neq 0$.
(d) Determine whether $\lim\limits_{x \to 0} f'(x)$ exists.

74. Let

$$f(x) = \begin{cases} x^2 \sin \dfrac{1}{x}, & x \neq 0 \\ 0, & x = 0 \end{cases}$$

(a) Show that f is continuous at $x = 0$.
(b) Use Definition 2.2.1 to find $f'(0)$.
(c) Find $f'(x)$ for $x \neq 0$.
(d) Show that f' is not continuous at $x = 0$.

75. Given the following table of values, find the indicated derivatives in parts (a) and (b).

x	$f(x)$	$f'(x)$
2	1	7
8	5	-3

(a) $g'(2)$, where $g(x) = [f(x)]^3$
(b) $h'(2)$, where $h(x) = f(x^3)$

76. Given that $f'(x) = \sqrt{3x + 4}$ and $g(x) = x^2 - 1$, find $F'(x)$ if $F(x) = f(g(x))$.

77. Given that $f'(x) = \dfrac{x}{x^2 + 1}$ and $g(x) = \sqrt{3x - 1}$, find $F'(x)$ if $F(x) = f(g(x))$.

78. Find $f'(x^2)$ if $\dfrac{d}{dx}[f(x^2)] = x^2$.

79. Find $\dfrac{d}{dx}[f(x)]$ if $\dfrac{d}{dx}[f(3x)] = 6x$.

80. Recall that a function f is **even** if $f(-x) = f(x)$ and **odd** if $f(-x) = -f(x)$, for all x in the domain of f. Assuming that f is differentiable, prove:
(a) f' is odd if f is even
(b) f' is even if f is odd

81. Draw some pictures to illustrate the results in Exercise 80, and write a paragraph that gives an informal explanation of why the results are true.

82. Let $y = f_1(u)$, $u = f_2(v)$, $v = f_3(w)$, and $w = f_4(x)$. Express dy/dx in terms of dy/du, dw/dx, du/dv, and dv/dw.

83. Find a formula for

$$\frac{d}{dx}[f(g(h(x)))]$$

84. Writing The "co" in "cosine" comes from "complementary," since the cosine of an angle is the sine of the complementary angle, and vice versa:

$$\cos x = \sin\left(\frac{\pi}{2} - x\right) \quad \text{and} \quad \sin x = \cos\left(\frac{\pi}{2} - x\right)$$

Suppose that we define a function g to be a *cofunction* of a function f if

$$g(x) = f\left(\frac{\pi}{2} - x\right) \quad \text{for all } x$$

Thus, cosine and sine are cofunctions of each other, as are cotangent and tangent, and also cosecant and secant. If g is the cofunction of f, state a formula that relates g' and the cofunction of f'. Discuss how this relationship is exhibited by the derivatives of the cosine, cotangent, and cosecant functions.

✔ **QUICK CHECK ANSWERS 2.6**

1. outside; inside; inside **2.** $\dfrac{dy}{du} \cdot \dfrac{du}{dx}$ **3.** (a) $10(x^2 + 5)^9 \cdot 2x = 20x(x^2 + 5)^9$ (b) $\dfrac{1}{2\sqrt{1 + 6x}} \cdot 6 = \dfrac{3}{\sqrt{1 + 6x}}$

4. (a) $3\cos(3x + 2)$ (b) $4(x^2 \tan x)^3(2x \tan x + x^2 \sec^2 x)$ **5.** (a) $g'(f(2))f'(2) = -20$ (b) $f'\left(\dfrac{1}{3}g(3)\right) \cdot \dfrac{1}{3}g'(3) = -\dfrac{20}{3}$

CHAPTER 2 REVIEW EXERCISES  Graphing Utility [C] CAS

---

1. Explain the difference between average and instantaneous rates of change, and discuss how they are calculated.

2. In parts (a)–(d), use the function $y = \frac{1}{2}x^2$.

(a) Find the average rate of change of y with respect to x over the interval $[3, 4]$.
(b) Find the instantaneous rate of change of y with respect to x at $x = 3$. *(cont.)*

(c) Find the instantaneous rate of change of y with respect to x at a general x-value.

(d) Sketch the graph of $y = \frac{1}{2}x^2$ together with the secant line whose slope is given by the result in part (a), and indicate graphically the slope of the tangent line that corresponds to the result in part (b).

3. Complete each part for the function $f(x) = x^2 + 1$.

(a) Find the slope of the tangent line to the graph of f at a general x-value.

(b) Find the slope of the tangent line to the graph of f at $x = 2$.

4. A car is traveling on a straight road that is 120 mi long. For the first 100 mi the car travels at an average velocity of 50 mi/h. Show that no matter how fast the car travels for the final 20 mi it cannot bring the average velocity up to 60 mi/h for the entire trip.

5. At time $t = 0$ a car moves into the passing lane to pass a slow-moving truck. The average velocity of the car from $t = 1$ to $t = 1 + h$ is

$$v_{\text{ave}} = \frac{3(h+1)^{2.5} + 580h - 3}{10h}$$

Estimate the instantaneous velocity of the car at $t = 1$, where time is in seconds and distance is in feet.

6. A skydiver jumps from an airplane. Suppose that the distance she falls during the first t seconds before her parachute opens is $s(t) = 976((0.835)^t - 1) + 176t$, where s is in feet. Graph s versus t for $0 \le t \le 20$, and use your graph to estimate the instantaneous velocity at $t = 15$.

7. A particle moves on a line away from its initial position so that after t hours it is $s = 3t^2 + t$ miles from its initial position.

(a) Find the average velocity of the particle over the interval $[1, 3]$.

(b) Find the instantaneous velocity at $t = 1$.

8. State the definition of a derivative, and give two interpretations of it.

9. Use the definition of a derivative to find dy/dx, and check your answer by calculating the derivative using appropriate derivative formulas.

(a) $y = \sqrt{9 - 4x}$ (b) $y = \dfrac{x}{x+1}$

10. Suppose that $f(x) = \begin{cases} x^2 - 1, & x \le 1 \\ k(x - 1), & x > 1. \end{cases}$

For what values of k is f

(a) continuous?   (b) differentiable?

11. The accompanying figure shows the graph of $y = f'(x)$ for an unspecified function f.

(a) For what values of x does the curve $y = f(x)$ have a horizontal tangent line?

(b) Over what intervals does the curve $y = f(x)$ have tangent lines with positive slope?

(c) Over what intervals does the curve $y = f(x)$ have tangent lines with negative slope?

(d) Given that $g(x) = f(x) \sin x$, find $g''(0)$.

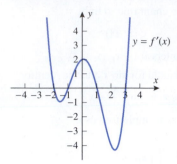

◀ **Figure Ex-11**

12. Sketch the graph of a function f for which $f(0) = 1$, $f'(0) = 0$, $f'(x) > 0$ if $x < 0$, and $f'(x) < 0$ if $x > 0$.

13. According to the U.S. Bureau of the Census, the estimated and projected midyear world population, N, in billions for the years 1950, 1975, 2000, 2025, and 2050 was 2.555, 4.088, 6.080, 7.841, and 9.104, respectively. Although the increase in population is not a continuous function of the time t, we can apply the ideas in this section if we are willing to approximate the graph of N versus t by a continuous curve, as shown in the accompanying figure.

(a) Use the tangent line at $t = 2000$ shown in the figure to approximate the value of dN/dt there. Interpret your result as a rate of change.

(b) The instantaneous **growth rate** is defined as

$$\frac{dN/dt}{N}$$

Use your answer to part (a) to approximate the instantaneous growth rate at the start of the year 2000. Express the result as a percentage and include the proper units.

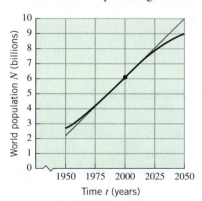

◀ **Figure Ex-13**

14. Use a graphing utility to graph the function

$$f(x) = |x^4 - x - 1| - x$$

and estimate the values of x where the derivative of this function does not exist.

c **15–18** (a) Use a CAS to find $f'(x)$ via Definition 2.2.1; (b) check the result by finding the derivative by hand; (c) use the CAS to find $f''(x)$. ■

15. $f(x) = x^2 \sin x$

16. $f(x) = \sqrt{x} + \cos^2 x$

17. $f(x) = \dfrac{2x^2 - x + 5}{3x + 2}$

18. $f(x) = \dfrac{\tan x}{1 + x^2}$

19. The amount of water in a tank t minutes after it has started to drain is given by $W = 100(t - 15)^2$ gal.
  (a) At what rate is the water running out at the end of 5 min?
  (b) What is the average rate at which the water flows out during the first 5 min?

20. Use the formula $V = l^3$ for the volume of a cube of side l to find
  (a) the average rate at which the volume of a cube changes with l as l increases from $l = 2$ to $l = 4$
  (b) the instantaneous rate at which the volume of a cube changes with l when $l = 5$.

∿ **21–22** Zoom in on the graph of f on an interval containing $x = x_0$ until the graph looks like a straight line. Estimate the slope of this line and then check your answer by finding the exact value of $f'(x_0)$. ■

21. (a) $f(x) = x^2 - 1$, $x_0 = 1.8$

  (b) $f(x) = \dfrac{x^2}{x - 2}$, $x_0 = 3.5$

22. (a) $f(x) = x^3 - x^2 + 1$, $x_0 = 2.3$

  (b) $f(x) = \dfrac{x}{x^2 + 1}$, $x_0 = -0.5$

23. Suppose that a function f is differentiable at $x = 1$ and

$$\lim_{h \to 0} \frac{f(1 + h)}{h} = 5$$

Find $f(1)$ and $f'(1)$.

24. Suppose that a function f is differentiable at $x = 2$ and

$$\lim_{x \to 2} \frac{x^3 f(x) - 24}{x - 2} = 28$$

Find $f(2)$ and $f'(2)$.

25. Find the equations of all lines through the origin that are tangent to the curve $y = x^3 - 9x^2 - 16x$.

26. Find all values of x for which the tangent line to the curve $y = 2x^3 - x^2$ is perpendicular to the line $x + 4y = 10$.

27. Let $f(x) = x^2$. Show that for any distinct values of a and b, the slope of the tangent line to $y = f(x)$ at $x = \frac{1}{2}(a + b)$ is equal to the slope of the secant line through the points (a, a^2) and (b, b^2). Draw a picture to illustrate this result.

28. In each part, evaluate the expression given that $f(1) = 1$, $g(1) = -2$, $f'(1) = 3$, and $g'(1) = -1$.
  (a) $\dfrac{d}{dx}[f(x)g(x)]\Big|_{x=1}$
  (b) $\dfrac{d}{dx}\left[\dfrac{f(x)}{g(x)}\right]\Big|_{x=1}$
  (c) $\dfrac{d}{dx}\left[\sqrt{f(x)}\right]\Big|_{x=1}$
  (d) $\dfrac{d}{dx}[f(1)g'(1)]$

29–32 Find $f'(x)$. ■

29. (a) $f(x) = x^8 - 3\sqrt{x} + 5x^{-3}$
  (b) $f(x) = (2x + 1)^{101}(5x^2 - 7)$

30. (a) $f(x) = \sin x + 2\cos^3 x$
  (b) $f(x) = (1 + \sec x)(x^2 - \tan x)$

31. (a) $f(x) = \sqrt{3x + 1}(x - 1)^2$
  (b) $f(x) = \left(\dfrac{3x + 1}{x^2}\right)^3$

32. (a) $f(x) = \cot\left(\dfrac{\csc 2x}{x^3 + 5}\right)$ (b) $f(x) = \dfrac{1}{2x + \sin^3 x}$

33–34 Find the values of x at which the curve $y = f(x)$ has a horizontal tangent line. ■

33. $f(x) = (2x + 7)^6(x - 2)^5$ **34.** $f(x) = \dfrac{(x - 3)^4}{x^2 + 2x}$

35. Find all lines that are simultaneously tangent to the graph of $y = x^2 + 1$ and to the graph of $y = -x^2 - 1$.

36. (a) Let n denote an even positive integer. Generalize the result of Exercise 35 by finding all lines that are simultaneously tangent to the graph of $y = x^n + n - 1$ and to the graph of $y = -x^n - n + 1$.
  (b) Let n denote an odd positive integer. Are there any lines that are simultaneously tangent to the graph of $y = x^n + n - 1$ and to the graph of $y = -x^n - n + 1$? Explain.

37. Find all values of x for which the line that is tangent to $y = 3x - \tan x$ is parallel to the line $y - x = 2$.

∿ **38.** Approximate the values of x at which the tangent line to the graph of $y = x^3 - \sin x$ is horizontal.

39. Suppose that $f(x) = M \sin x + N \cos x$ for some constants M and N. If $f(\pi/4) = 3$ and $f'(\pi/4) = 1$, find an equation for the tangent line to $y = f(x)$ at $x = 3\pi/4$.

40. Suppose that $f(x) = M \tan x + N \sec x$ for some constants M and N. If $f(\pi/4) = 2$ and $f'(\pi/4) = 0$, find an equation for the tangent line to $y = f(x)$ at $x = 0$.

41. Suppose that $f'(x) = 2x \cdot f(x)$ and $f(2) = 5$.
  (a) Find $g'(\pi/3)$ if $g(x) = f(\sec x)$.
  (b) Find $h'(2)$ if $h(x) = [f(x)/(x - 1)]^4$.

CHAPTER 2 MAKING CONNECTIONS

1. Suppose that f is a function with the properties (i) f is differentiable everywhere, (ii) $f(x + y) = f(x)f(y)$ for all values of x and y, (iii) $f(0) \neq 0$, and (iv) $f'(0) = 1$.
   (a) Show that $f(0) = 1$. [*Hint:* Consider $f(0 + 0)$.]
   (b) Show that $f(x) > 0$ for all values of x. [*Hint:* First show that $f(x) \neq 0$ for any x by considering $f(x - x)$.]
   (c) Use the definition of derivative (Definition 2.2.1) to show that $f'(x) = f(x)$ for all values of x.

2. Suppose that f and g are functions each of which has the properties (i)–(iv) in Exercise 1.
   (a) Show that $y = f(2x)$ satisfies the equation $y' = 2y$ in two ways: using property (ii), and by directly applying the chain rule (Theorem 2.6.1).
   (b) If k is any constant, show that $y = f(kx)$ satisfies the equation $y' = ky$.
   (c) Find a value of k such that $y = f(x)g(x)$ satisfies the equation $y' = ky$.
   (d) If $h = f/g$, find $h'(x)$. Make a conjecture about the relationship between f and g.

3. (a) Apply the product rule (Theorem 2.4.1) twice to show that if f, g, and h are differentiable functions, then $f \cdot g \cdot h$ is differentiable and
   $$(f \cdot g \cdot h)' = f' \cdot g \cdot h + f \cdot g' \cdot h + f \cdot g \cdot h'$$
   (b) Suppose that $f, g, h,$ and k are differentiable functions. Derive a formula for $(f \cdot g \cdot h \cdot k)'$.

   (c) Based on the result in part (a), make a conjecture about a formula differentiating a product of n functions. Prove your formula using induction.

4. (a) Apply the quotient rule (Theorem 2.4.2) twice to show that if f, g, and h are differentiable functions, then $(f/g)/h$ is differentiable where it is defined and
   $$[(f/g)/h]' = \frac{f' \cdot g \cdot h - f \cdot g' \cdot h - f \cdot g \cdot h'}{g^2 h^2}$$

   (b) Derive the derivative formula of part (a) by first simplifying $(f/g)/h$ and then applying the quotient and product rules.
   (c) Apply the quotient rule (Theorem 2.4.2) twice to derive a formula for $[f/(g/h)]'$.
   (d) Derive the derivative formula of part (c) by first simplifying $f/(g/h)$ and then applying the quotient and product rules.

5. Assume that $h(x) = f(x)/g(x)$ is differentiable. Derive the quotient rule formula for $h'(x)$ (Theorem 2.4.2) in two ways:
   (a) Write $h(x) = f(x) \cdot [g(x)]^{-1}$ and use the product and chain rules (Theorems 2.4.1 and 2.6.1) to differentiate h.
   (b) Write $f(x) = h(x) \cdot g(x)$ and use the product rule to derive a formula for $h'(x)$.

EXPANDING THE CALCULUS HORIZON

To learn how derivatives can be used in the field of robotics, see the module entitled **Robotics** at:

www.wiley.com/college/anton

3

TOPICS IN DIFFERENTIATION

Craig Lovell/Corbis Images

The growth and decline of animal populations and natural resources can be modeled using basic functions studied in calculus.

We begin this chapter by extending the process of differentiation to functions that are either difficult or impossible to differentiate directly. We will discuss a combination of direct and indirect methods of differentiation that will allow us to develop a number of new derivative formulas that include the derivatives of logarithmic, exponential, and inverse trigonometric functions. Later in the chapter, we will consider some applications of the derivative. These will include ways in which different rates of change can be related as well as the use of linear functions to approximate nonlinear functions. Finally, we will discuss L'Hôpital's rule, a powerful tool for evaluating limits.

3.1 IMPLICIT DIFFERENTIATION

Up to now we have been concerned with differentiating functions that are given by equations of the form $y = f(x)$. In this section we will consider methods for differentiating functions for which it is inconvenient or impossible to express them in this form.

■ FUNCTIONS DEFINED EXPLICITLY AND IMPLICITLY
An equation of the form $y = f(x)$ is said to define y *explicitly* as a function of x because the variable y appears alone on one side of the equation and does not appear at all on the other side. However, sometimes functions are defined by equations in which y is not alone on one side; for example, the equation

$$yx + y + 1 = x \qquad (1)$$

is not of the form $y = f(x)$, but it still defines y as a function of x since it can be rewritten as

$$y = \frac{x - 1}{x + 1}$$

Thus, we say that (1) defines y *implicitly* as a function of x, the function being

$$f(x) = \frac{x - 1}{x + 1}$$

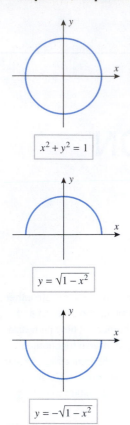

$x^2 + y^2 = 1$

$y = \sqrt{1 - x^2}$

$y = -\sqrt{1 - x^2}$

▲ **Figure 3.1.1**

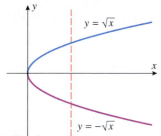

▲ **Figure 3.1.2** The graph of $x = y^2$ does not pass the vertical line test, but the graphs of $y = \sqrt{x}$ and $y = -\sqrt{x}$ do.

An equation in x and y can implicitly define more than one function of x. This can occur when the graph of the equation fails the vertical line test, so it is not the graph of a function of x. For example, if we solve the equation of the circle

$$x^2 + y^2 = 1 \tag{2}$$

for y in terms of x, we obtain $y = \pm\sqrt{1 - x^2}$, so we have found two functions that are defined implicitly by (2), namely,

$$f_1(x) = \sqrt{1 - x^2} \quad \text{and} \quad f_2(x) = -\sqrt{1 - x^2} \tag{3}$$

The graphs of these functions are the upper and lower semicircles of the circle $x^2 + y^2 = 1$ (Figure 3.1.1). This leads us to the following definition.

3.1.1 DEFINITION We will say that a given equation in x and y defines the function f *implicitly* if the graph of $y = f(x)$ coincides with a portion of the graph of the equation.

▶ **Example 1** The graph of $x = y^2$ is not the graph of a function of x, since it does not pass the vertical line test (Figure 3.1.2). However, if we solve this equation for y in terms of x, we obtain the equations $y = \sqrt{x}$ and $y = -\sqrt{x}$, whose graphs pass the vertical line test and are portions of the graph of $x = y^2$ (Figure 3.1.2). Thus, the equation $x = y^2$ implicitly defines the functions

$$f_1(x) = \sqrt{x} \quad \text{and} \quad f_2(x) = -\sqrt{x} \blacktriangleleft$$

Although it was a trivial matter in the last example to solve the equation $x = y^2$ for y in terms of x, it is difficult or impossible to do this for some equations. For example, the equation

$$x^3 + y^3 = 3xy \tag{4}$$

can be solved for y in terms of x, but the resulting formulas are too complicated to be practical. Other equations, such as $\sin(xy) = y$, cannot be solved for y by any elementary method. Thus, even though an equation may define one or more functions of x, it may not be possible or practical to find explicit formulas for those functions.

Fortunately, CAS programs, such as *Mathematica* and *Maple*, have "implicit plotting" capabilities that can graph equations such as (4). The graph of this equation, which is called the *Folium of Descartes*, is shown in Figure 3.1.3*a*. Parts (*b*) and (*c*) of the figure show the graphs (in blue) of two functions that are defined implicitly by (4).

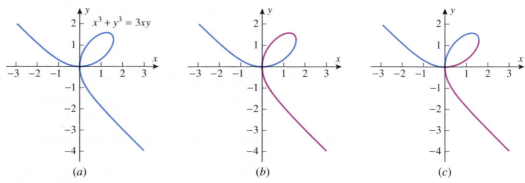

▲ **Figure 3.1.3**

■ IMPLICIT DIFFERENTIATION

In general, it is not necessary to solve an equation for y in terms of x in order to differentiate the functions defined implicitly by the equation. To illustrate this, let us consider the simple equation

$$xy = 1 \tag{5}$$

One way to find dy/dx is to rewrite this equation as

$$y = \frac{1}{x} \tag{6}$$

from which it follows that

$$\frac{dy}{dx} = -\frac{1}{x^2} \tag{7}$$

Another way to obtain this derivative is to differentiate both sides of (5) *before* solving for y in terms of x, treating y as a (temporarily unspecified) differentiable function of x. With this approach we obtain

$$\frac{d}{dx}[xy] = \frac{d}{dx}[1]$$

$$x\frac{d}{dx}[y] + y\frac{d}{dx}[x] = 0$$

$$x\frac{dy}{dx} + y = 0$$

$$\frac{dy}{dx} = -\frac{y}{x}$$

If we now substitute (6) into the last expression, we obtain

$$\frac{dy}{dx} = -\frac{1}{x^2}$$

which agrees with Equation (7). This method of obtaining derivatives is called *implicit differentiation*.

▶ **Example 2** Use implicit differentiation to find dy/dx if $5y^2 + \sin y = x^2$.

$$\frac{d}{dx}[5y^2 + \sin y] = \frac{d}{dx}[x^2]$$

$$5\frac{d}{dx}[y^2] + \frac{d}{dx}[\sin y] = 2x$$

$$5\left(2y\frac{dy}{dx}\right) + (\cos y)\frac{dy}{dx} = 2x \qquad \boxed{\text{The chain rule was used here because } y \text{ is a function of } x.}$$

$$10y\frac{dy}{dx} + (\cos y)\frac{dy}{dx} = 2x$$

René Descartes (1596–1650) Descartes, a French aristocrat, was the son of a government official. He graduated from the University of Poitiers with a law degree at age 20. After a brief probe into the pleasures of Paris he became a military engineer, first for the Dutch Prince of Nassau and then for the German Duke of Bavaria. It was during his service as a soldier that Descartes began to pursue mathematics seriously and develop his analytic geometry. After the wars, he returned to Paris where he stalked the city as an eccentric, wearing a sword in his belt and a plumed hat. He lived in leisure, seldom arose before 11 A.M., and dabbled in the study of human physiology, philosophy, glaciers, meteors, and rainbows. He eventually moved to Holland, where he published his *Discourse on the Method*, and finally to Sweden where he died while serving as tutor to Queen Christina. Descartes is regarded as a genius of the first magnitude. In addition to major contributions in mathematics and philosophy he is considered, along with William Harvey, to be a founder of modern physiology.

Solving for dy/dx we obtain

$$\frac{dy}{dx} = \frac{2x}{10y + \cos y} \qquad (8)$$

Note that this formula involves both x and y. In order to obtain a formula for dy/dx that involves x alone, we would have to solve the original equation for y in terms of x and then substitute in (8). However, it is impossible to do this, so we are forced to leave the formula for dy/dx in terms of x and y. ◄

▶ **Example 3**   Use implicit differentiation to find d^2y/dx^2 if $4x^2 - 2y^2 = 9$.

Solution.   Differentiating both sides of $4x^2 - 2y^2 = 9$ with respect to x yields

$$8x - 4y\frac{dy}{dx} = 0$$

from which we obtain

$$\frac{dy}{dx} = \frac{2x}{y} \qquad (9)$$

Differentiating both sides of (9) yields

$$\frac{d^2y}{dx^2} = \frac{(y)(2) - (2x)(dy/dx)}{y^2} \qquad (10)$$

Substituting (9) into (10) and simplifying using the original equation, we obtain

$$\frac{d^2y}{dx^2} = \frac{2y - 2x(2x/y)}{y^2} = \frac{2y^2 - 4x^2}{y^3} = -\frac{9}{y^3} \blacktriangleleft$$

In Examples 2 and 3, the resulting formulas for dy/dx involved both x and y. Although it is usually more desirable to have the formula for dy/dx expressed in terms of x alone, having the formula in terms of x and y is not an impediment to finding slopes and equations of tangent lines provided the x- and y-coordinates of the point of tangency are known. This is illustrated in the following example.

▶ **Example 4**   Find the slopes of the tangent lines to the curve $y^2 - x + 1 = 0$ at the points $(2, -1)$ and $(2, 1)$.

Solution.   We could proceed by solving the equation for y in terms of x, and then evaluating the derivative of $y = \sqrt{x-1}$ at $(2, 1)$ and the derivative of $y = -\sqrt{x-1}$ at $(2, -1)$ (Figure 3.1.4). However, implicit differentiation is more efficient since it can be used for the slopes of *both* tangent lines. Differentiating implicitly yields

$$\frac{d}{dx}[y^2 - x + 1] = \frac{d}{dx}[0]$$

$$\frac{d}{dx}[y^2] - \frac{d}{dx}[x] + \frac{d}{dx}[1] = \frac{d}{dx}[0]$$

$$2y\frac{dy}{dx} - 1 = 0$$

$$\frac{dy}{dx} = \frac{1}{2y}$$

At $(2, -1)$ we have $y = -1$, and at $(2, 1)$ we have $y = 1$, so the slopes of the tangent lines to the curve at those points are

$$\frac{dy}{dx}\bigg|_{\substack{x=2 \\ y=-1}} = -\frac{1}{2} \quad \text{and} \quad \frac{dy}{dx}\bigg|_{\substack{x=2 \\ y=1}} = \frac{1}{2} \blacktriangleleft$$

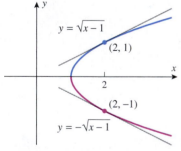

$y = \sqrt{x-1}$

$(2, 1)$

$(2, -1)$

$y = -\sqrt{x-1}$

▲ Figure 3.1.4

▶ **Example 5**

(a) Use implicit differentiation to find dy/dx for the Folium of Descartes $x^3 + y^3 = 3xy$.

(b) Find an equation for the tangent line to the Folium of Descartes at the point $\left(\frac{3}{2}, \frac{3}{2}\right)$.

(c) At what point(s) in the first quadrant is the tangent line to the Folium of Descartes horizontal?

Solution (a).   Differentiating implicitly yields

$$\frac{d}{dx}[x^3 + y^3] = \frac{d}{dx}[3xy]$$

$$3x^2 + 3y^2\frac{dy}{dx} = 3x\frac{dy}{dx} + 3y$$

$$x^2 + y^2\frac{dy}{dx} = x\frac{dy}{dx} + y$$

$$(y^2 - x)\frac{dy}{dx} = y - x^2$$

$$\frac{dy}{dx} = \frac{y - x^2}{y^2 - x} \tag{11}$$

Formula (11) cannot be evaluated at $(0, 0)$ and hence provides no information about the nature of the Folium of Descartes at the origin. Based on the graphs in Figure 3.1.3, what can you say about the differentiability of the implicitly defined functions graphed in blue in parts (b) and (c) of the figure?

Solution (b).   At the point $\left(\frac{3}{2}, \frac{3}{2}\right)$, we have $x = \frac{3}{2}$ and $y = \frac{3}{2}$, so from (11) the slope m_{tan} of the tangent line at this point is

$$m_{\text{tan}} = \frac{dy}{dx}\bigg|_{\substack{x=3/2 \\ y=3/2}} = \frac{(3/2) - (3/2)^2}{(3/2)^2 - (3/2)} = -1$$

Thus, the equation of the tangent line at the point $\left(\frac{3}{2}, \frac{3}{2}\right)$ is

$$y - \tfrac{3}{2} = -1\left(x - \tfrac{3}{2}\right) \quad \text{or} \quad x + y = 3$$

which is consistent with Figure 3.1.5.

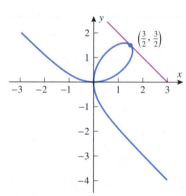

▲ **Figure 3.1.5**

Solution (c).   The tangent line is horizontal at the points where $dy/dx = 0$, and from (11) this occurs only where $y - x^2 = 0$ or

$$y = x^2 \tag{12}$$

Substituting this expression for y in the equation $x^3 + y^3 = 3xy$ for the curve yields

$$x^3 + (x^2)^3 = 3x^3$$
$$x^6 - 2x^3 = 0$$
$$x^3(x^3 - 2) = 0$$

whose solutions are $x = 0$ and $x = 2^{1/3}$. From (12), the solutions $x = 0$ and $x = 2^{1/3}$ yield the points $(0, 0)$ and $(2^{1/3}, 2^{2/3})$, respectively. Of these two, only $(2^{1/3}, 2^{2/3})$ is in the first quadrant. Substituting $x = 2^{1/3}$, $y = 2^{2/3}$ into (11) yields

$$\frac{dy}{dx}\bigg|_{\substack{x=2^{1/3} \\ y=2^{2/3}}} = \frac{0}{2^{4/3} - 2^{2/3}} = 0$$

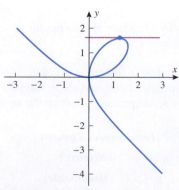

▲ **Figure 3.1.6**

We conclude that $(2^{1/3}, 2^{2/3}) \approx (1.26, 1.59)$ is the only point on the Folium of Descartes in the first quadrant at which the tangent line is horizontal (Figure 3.1.6). ◀

■ DIFFERENTIABILITY OF FUNCTIONS DEFINED IMPLICITLY

When differentiating implicitly, it is assumed that y represents a differentiable function of x. If this is not so, then the resulting calculations may be nonsense. For example, if we differentiate the equation

$$x^2 + y^2 + 1 = 0 \tag{13}$$

we obtain

$$2x + 2y\frac{dy}{dx} = 0 \quad \text{or} \quad \frac{dy}{dx} = -\frac{x}{y}$$

However, this derivative is meaningless because there are no real values of x and y that satisfy (13) (why?); and hence (13) does not define any real functions implicitly.

The nonsensical conclusion of these computations conveys the importance of knowing whether an equation in x and y that is to be differentiated implicitly actually defines some differentiable function of x implicitly. Unfortunately, this can be a difficult problem, so we will leave the discussion of such matters for more advanced courses in analysis.

✔ **QUICK CHECK EXERCISES 3.1**   *(See page 192 for answers.)*

1. The equation $xy + 2y = 1$ defines implicitly the function $y = \underline{\hspace{1cm}}$.

2. Use implicit differentiation to find dy/dx for $x^2 - y^3 = xy$.

3. The slope of the tangent line to the graph of $x + y + xy = 3$ at $(1, 1)$ is $\underline{\hspace{1cm}}$.

4. Use implicit differentiation to find d^2y/dx^2 for $\sin y = x$.

EXERCISE SET 3.1   ☐c CAS

1–2
(a) Find dy/dx by differentiating implicitly.
(b) Solve the equation for y as a function of x, and find dy/dx from that equation.
(c) Confirm that the two results are consistent by expressing the derivative in part (a) as a function of x alone. ■

1. $x + xy - 2x^3 = 2$ 2. $\sqrt{y} - \sin x = 2$

3–12 Find dy/dx by implicit differentiation. ■

3. $x^2 + y^2 = 100$ 4. $x^3 + y^3 = 3xy^2$

5. $x^2 y + 3xy^3 - x = 3$ 6. $x^3 y^2 - 5x^2 y + x = 1$

7. $\dfrac{1}{\sqrt{x}} + \dfrac{1}{\sqrt{y}} = 1$ 8. $x^2 = \dfrac{x + y}{x - y}$

9. $\sin(x^2 y^2) = x$ 10. $\cos(xy^2) = y$

11. $\tan^3(xy^2 + y) = x$ 12. $\dfrac{xy^3}{1 + \sec y} = 1 + y^4$

13–18 Find d^2y/dx^2 by implicit differentiation. ■

13. $2x^2 - 3y^2 = 4$ 14. $x^3 + y^3 = 1$

15. $x^3 y^3 - 4 = 0$ 16. $xy + y^2 = 2$

17. $y + \sin y = x$ 18. $x \cos y = y$

19–20 Find the slope of the tangent line to the curve at the given points in two ways: first by solving for y in terms of x and differentiating and then by implicit differentiation. ■

19. $x^2 + y^2 = 1$; $(1/2, \sqrt{3}/2), (1/2, -\sqrt{3}/2)$

20. $y^2 - x + 1 = 0$; $(10, 3), (10, -3)$

21–24 True–False Determine whether the statement is true or false. Explain your answer. ■

21. If an equation in x and y defines a function $y = f(x)$ implicitly, then the graph of the equation and the graph of f are identical.

22. The function

$$f(x) = \begin{cases} \sqrt{1 - x^2}, & 0 < x \le 1 \\ -\sqrt{1 - x^2}, & -1 \le x \le 0 \end{cases}$$

is defined implicitly by the equation $x^2 + y^2 = 1$.

23. The function $|x|$ is not defined implicitly by the equation $(x + y)(x - y) = 0$.

24. If y is defined implicitly as a function of x by the equation $x^2 + y^2 = 1$, then $dy/dx = -x/y$.

25–28 Use implicit differentiation to find the slope of the tangent line to the curve at the specified point, and check that your answer is consistent with the accompanying graph on the next page. ■

25. $x^4 + y^4 = 16$; $(1, \sqrt[4]{15})$ [*Lamé's special quartic*]

26. $y^3 + yx^2 + x^2 - 3y^2 = 0$; $(0, 3)$ [*trisectrix*]

27. $2(x^2 + y^2)^2 = 25(x^2 - y^2)$; $(3, 1)$ [*lemniscate*]

28. $x^{2/3} + y^{2/3} = 4$; $(-1, 3\sqrt{3})$ [*four-cusped hypocycloid*]

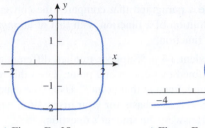

▲ Figure Ex-25

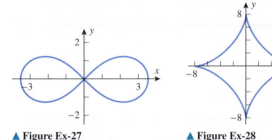

▲ Figure Ex-26

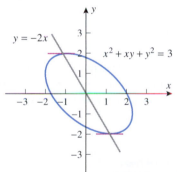

▲ Figure Ex-27   ▲ Figure Ex-28

33–34 These exercises deal with the rotated ellipse C whose equation is $x^2 - xy + y^2 = 4$. ∎

33. Show that the line $y = x$ intersects C at two points P and Q and that the tangent lines to C at P and Q are parallel.

34. Prove that if $P(a, b)$ is a point on C, then so is $Q(-a, -b)$ and that the tangent lines to C through P and through Q are parallel.

35. Find the values of a and b for the curve $x^2 y + ay^2 = b$ if the point $(1, 1)$ is on its graph and the tangent line at $(1, 1)$ has the equation $4x + 3y = 7$.

36. At what point(s) is the tangent line to the curve $y^3 = 2x^2$ perpendicular to the line $x + 2y - 2 = 0$?

37–38 Two curves are said to be **orthogonal** if their tangent lines are perpendicular at each point of intersection, and two families of curves are said to be **orthogonal trajectories** of one another if each member of one family is orthogonal to each member of the other family. This terminology is used in these exercises. ∎

37. The accompanying figure shows some typical members of the families of circles $x^2 + (y - c)^2 = c^2$ (black curves) and $(x - k)^2 + y^2 = k^2$ (gray curves). Show that these families are orthogonal trajectories of one another. [*Hint:* For the tangent lines to be perpendicular at a point of intersection, the slopes of those tangent lines must be negative reciprocals of one another.]

38. The accompanying figure shows some typical members of the families of hyperbolas $xy = c$ (black curves) and $x^2 - y^2 = k$ (gray curves), where $c \neq 0$ and $k \neq 0$. Use the hint in Exercise 37 to show that these families are orthogonal trajectories of one another.

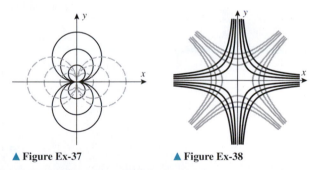

▲ Figure Ex-37   ▲ Figure Ex-38

FOCUS ON CONCEPTS

29. In the accompanying figure, it appears that the ellipse $x^2 + xy + y^2 = 3$ has horizontal tangent lines at the points of intersection of the ellipse and the line $y = -2x$. Use implicit differentiation to explain why this is the case.

◀ Figure Ex-29

30. (a) A student claims that the ellipse $x^2 - xy + y^2 = 1$ has a horizontal tangent line at the point $(1, 1)$. Without doing any computations, explain why the student's claim must be incorrect.
(b) Find all points on the ellipse $x^2 - xy + y^2 = 1$ at which the tangent line is horizontal.

C 31. (a) Use the implicit plotting capability of a CAS to graph the equation $y^4 + y^2 = x(x - 1)$.
(b) Use implicit differentiation to help explain why the graph in part (a) has no horizontal tangent lines.
(c) Solve the equation $y^4 + y^2 = x(x - 1)$ for x in terms of y and explain why the graph in part (a) consists of two parabolas.

32. Use implicit differentiation to find all points on the graph of $y^4 + y^2 = x(x - 1)$ at which the tangent line is vertical.

C 39. (a) Use the implicit plotting capability of a CAS to graph the curve C whose equation is $x^3 - 2xy + y^3 = 0$.
(b) Use the graph in part (a) to estimate the x-coordinates of a point in the first quadrant that is on C and at which the tangent line to C is parallel to the x-axis.
(c) Find the exact value of the x-coordinate in part (b).

C 40. (a) Use the implicit plotting capability of a CAS to graph the curve C whose equation is $x^3 - 2xy + y^3 = 0$.
(b) Use the graph to guess the coordinates of a point in the first quadrant that is on C and at which the tangent line to C is parallel to the line $y = -x$.   *(cont.)*

(c) Use implicit differentiation to verify your conjecture in part (b).

41. Prove that for every nonzero rational number r, the tangent line to the graph of $x^r + y^r = 2$ at the point $(1, 1)$ has slope -1.

42. Find equations for two lines through the origin that are tangent to the ellipse $2x^2 - 4x + y^2 + 1 = 0$.

43. **Writing** Write a paragraph that compares the concept of an *explicit* definition of a function with that of an *implicit* definition of a function.

44. **Writing** A student asks: "Suppose implicit differentiation yields an undefined expression at a point. Does this mean that dy/dx is undefined at that point?" Using the equation $x^2 - 2xy + y^2 = 0$ as a basis for your discussion, write a paragraph that answers the student's question.

✔ **QUICK CHECK ANSWERS 3.1**

1. $\dfrac{1}{x+2}$ 2. $\dfrac{dy}{dx} = \dfrac{2x - y}{x + 3y^2}$ 3. -1 4. $\dfrac{d^2y}{dx^2} = \sec^2 y \tan y$

3.2 DERIVATIVES OF LOGARITHMIC FUNCTIONS

In this section we will obtain derivative formulas for logarithmic functions, and we will explain why the natural logarithm function is preferred over logarithms with other bases in calculus.

■ DERIVATIVES OF LOGARITHMIC FUNCTIONS

We will establish that $f(x) = \ln x$ is differentiable for $x > 0$ by applying the derivative definition to $f(x)$. To evaluate the resulting limit, we will need the fact that $\ln x$ is continuous for $x > 0$ (Theorem 1.6.3), and we will need the limit

$$\lim_{v \to 0} (1 + v)^{1/v} = e \tag{1}$$

This limit can be obtained from limits (7) and (8) of Section 1.3 by making the substitution $v = 1/x$ and using the fact that $v \to 0^+$ as $x \to +\infty$ and $v \to 0^-$ as $x \to -\infty$. This produces two equal one-sided limits that together imply (1) (see Exercise 64 of Section 1.3).

$$\frac{d}{dx}[\ln x] = \lim_{h \to 0} \frac{\ln(x + h) - \ln x}{h}$$

$$= \lim_{h \to 0} \frac{1}{h} \ln\left(\frac{x + h}{x}\right) \qquad \boxed{\text{The quotient property of logarithms in Theorem 0.5.2}}$$

$$= \lim_{h \to 0} \frac{1}{h} \ln\left(1 + \frac{h}{x}\right)$$

$$= \lim_{v \to 0} \frac{1}{vx} \ln(1 + v) \qquad \boxed{\text{Let } v = h/x \text{ and note that } v \to 0 \text{ if and only if } h \to 0.}$$

$$= \frac{1}{x} \lim_{v \to 0} \frac{1}{v} \ln(1 + v) \qquad \boxed{x \text{ is fixed in this limit computation, so } 1/x \text{ can be moved through the limit sign.}}$$

$$= \frac{1}{x} \lim_{v \to 0} \ln(1 + v)^{1/v} \qquad \boxed{\text{The power property of logarithms in Theorem 0.5.2}}$$

$$= \frac{1}{x} \ln\left[\lim_{v \to 0} (1 + v)^{1/v}\right] \qquad \boxed{\ln x \text{ is continuous on } (0, +\infty) \text{ so we can move the limit through the function symbol.}}$$

$$= \frac{1}{x} \ln e$$

$$= \frac{1}{x} \qquad \boxed{\text{Since } \ln e = 1}$$

Thus,

$$\frac{d}{dx}[\ln x] = \frac{1}{x}, \quad x > 0 \tag{2}$$

A derivative formula for the general logarithmic function $\log_b x$ can be obtained from (2) by using Formula (6) of Section 0.5 to write

$$\frac{d}{dx}[\log_b x] = \frac{d}{dx}\left[\frac{\ln x}{\ln b}\right] = \frac{1}{\ln b}\frac{d}{dx}[\ln x]$$

It follows from this that

$$\frac{d}{dx}[\log_b x] = \frac{1}{x \ln b}, \quad x > 0 \tag{3}$$

> Note that, among all possible bases, the base $b = e$ produces the simplest formula for the derivative of $\log_b x$. This is one of the reasons why the natural logarithm function is preferred over other logarithms in calculus.

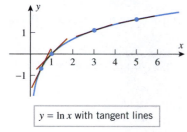

$y = \ln x$ with tangent lines

▲ **Figure 3.2.1**

▶ **Example 1**

(a) Figure 3.2.1 shows the graph of $y = \ln x$ and its tangent lines at the points $x = \frac{1}{2}, 1, 3$, and 5. Find the slopes of those tangent lines.

(b) Does the graph of $y = \ln x$ have any horizontal tangent lines? Use the derivative of $\ln x$ to justify your answer.

Solution (a). From (2), the slopes of the tangent lines at the points $x = \frac{1}{2}, 1, 3$, and 5 are $1/x = 2, 1, \frac{1}{3}$, and $\frac{1}{5}$, respectively, which is consistent with Figure 3.2.1.

Solution (b). It does not appear from the graph of $y = \ln x$ that there are any horizontal tangent lines. This is confirmed by the fact that $dy/dx = 1/x$ is not equal to zero for any real value of x. ◀

If u is a differentiable function of x, and if $u(x) > 0$, then applying the chain rule to (2) and (3) produces the following generalized derivative formulas:

$$\frac{d}{dx}[\ln u] = \frac{1}{u} \cdot \frac{du}{dx} \quad \text{and} \quad \frac{d}{dx}[\log_b u] = \frac{1}{u \ln b} \cdot \frac{du}{dx} \tag{4–5}$$

▶ **Example 2** Find $\dfrac{d}{dx}[\ln(x^2 + 1)]$.

Solution. Using (4) with $u = x^2 + 1$ we obtain

$$\frac{d}{dx}[\ln(x^2 + 1)] = \frac{1}{x^2 + 1} \cdot \frac{d}{dx}[x^2 + 1] = \frac{1}{x^2 + 1} \cdot 2x = \frac{2x}{x^2 + 1} \quad ◀$$

When possible, the properties of logarithms in Theorem 0.5.2 should be used to convert products, quotients, and exponents into sums, differences, and constant multiples *before* differentiating a function involving logarithms.

▶ **Example 3**

$$\frac{d}{dx}\left[\ln\left(\frac{x^2 \sin x}{\sqrt{1 + x}}\right)\right] = \frac{d}{dx}\left[2 \ln x + \ln(\sin x) - \frac{1}{2}\ln(1 + x)\right]$$

$$= \frac{2}{x} + \frac{\cos x}{\sin x} - \frac{1}{2(1 + x)}$$

$$= \frac{2}{x} + \cot x - \frac{1}{2 + 2x} \quad ◀$$

Figure 3.2.2 shows the graph of $f(x) = \ln|x|$. This function is important because it "extends" the domain of the natural logarithm function in the sense that the values of $\ln|x|$ and $\ln x$ are the same for $x > 0$, but $\ln|x|$ is defined for all nonzero values of x, and $\ln x$ is only defined for positive values of x.

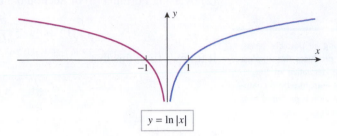

▶ **Figure 3.2.2**

The derivative of $\ln|x|$ for $x \neq 0$ can be obtained by considering the cases $x > 0$ and $x < 0$ separately:

Case $x > 0$. In this case $|x| = x$, so

$$\frac{d}{dx}[\ln|x|] = \frac{d}{dx}[\ln x] = \frac{1}{x}$$

Case $x < 0$. In this case $|x| = -x$, so it follows from (4) that

$$\frac{d}{dx}[\ln|x|] = \frac{d}{dx}[\ln(-x)] = \frac{1}{(-x)} \cdot \frac{d}{dx}[-x] = \frac{1}{x}$$

Since the same formula results in both cases, we have shown that

$$\frac{d}{dx}[\ln|x|] = \frac{1}{x} \quad \text{if } x \neq 0 \tag{6}$$

▶ **Example 4** From (6) and the chain rule,

$$\frac{d}{dx}[\ln|\sin x|] = \frac{1}{\sin x} \cdot \frac{d}{dx}[\sin x] = \frac{\cos x}{\sin x} = \cot x \;\blacktriangleleft$$

■ LOGARITHMIC DIFFERENTIATION

We now consider a technique called ***logarithmic differentiation*** that is useful for differentiating functions that are composed of products, quotients, and powers.

▶ **Example 5** The derivative of

$$y = \frac{x^2 \sqrt[3]{7x - 14}}{(1 + x^2)^4} \tag{7}$$

is messy to calculate directly. However, if we first take the natural logarithm of both sides and then use its properties, we can write

$$\ln y = 2\ln x + \tfrac{1}{3}\ln(7x - 14) - 4\ln(1 + x^2)$$

Differentiating both sides with respect to x yields

$$\frac{1}{y}\frac{dy}{dx} = \frac{2}{x} + \frac{7/3}{7x - 14} - \frac{8x}{1 + x^2}$$

Thus, on solving for dy/dx and using (7) we obtain

$$\frac{dy}{dx} = \frac{x^2 \sqrt[3]{7x-14}}{(1+x^2)^4}\left[\frac{2}{x} + \frac{1}{3x-6} - \frac{8x}{1+x^2}\right] \blacktriangleleft$$

REMARK | Since $\ln y$ is only defined for $y > 0$, the computations in Example 5 are only valid for $x > 2$ (verify). However, because the derivative of $\ln y$ is the same as the derivative of $\ln|y|$, and because $\ln|y|$ is defined for $y < 0$ as well as $y > 0$, it follows that the formula obtained for dy/dx is valid for $x < 2$ as well as $x > 2$. In general, whenever a derivative dy/dx is obtained by logarithmic differentiation, the resulting derivative formula will be valid for all values of x for which $y \neq 0$. It may be valid at those points as well, but it is not guaranteed.

■ DERIVATIVES OF REAL POWERS OF x

We know from Theorem 2.3.2 and Exercise 82 in Section 2.3, that the differentiation formula

$$\frac{d}{dx}[x^r] = rx^{r-1} \tag{8}$$

holds for constant integer values of r. We will now use logarithmic differentiation to show that this formula holds if r is *any* real number (rational or irrational). In our computations we will assume that x^r is a differentiable function and that the familiar laws of exponents hold for real exponents.

> In the next section we will discuss differentiating functions that have exponents which are not constant.

Let $y = x^r$, where r is a real number. The derivative dy/dx can be obtained by logarithmic differentiation as follows:

$$\ln|y| = \ln|x^r| = r\ln|x|$$

$$\frac{d}{dx}[\ln|y|] = \frac{d}{dx}[r\ln|x|]$$

$$\frac{1}{y}\frac{dy}{dx} = \frac{r}{x}$$

$$\frac{dy}{dx} = \frac{r}{x}y = \frac{r}{x}x^r = rx^{r-1}$$

✔ QUICK CHECK EXERCISES 3.2   *(See page 196 for answers.)*

1. The equation of the tangent line to the graph of $y = \ln x$ at $x = e^2$ is _____.

2. Find dy/dx.
   (a) $y = \ln 3x$ (b) $y = \ln\sqrt{x}$
   (c) $y = \log(1/|x|)$

3. Use logarithmic differentiation to find the derivative of

$$f(x) = \frac{\sqrt{x+1}}{\sqrt[3]{x-1}}$$

4. $\displaystyle\lim_{h\to 0} \frac{\ln(1+h)}{h} = $ _____

EXERCISE SET 3.2

1–26 Find dy/dx. ■

1. $y = \ln 5x$
2. $y = \ln\dfrac{x}{3}$
3. $y = \ln|1+x|$
4. $y = \ln(2+\sqrt{x})$
5. $y = \ln|x^2 - 1|$
6. $y = \ln|x^3 - 7x^2 - 3|$
7. $y = \ln\left(\dfrac{x}{1+x^2}\right)$
8. $y = \ln\left|\dfrac{1+x}{1-x}\right|$
9. $y = \ln x^2$
10. $y = (\ln x)^3$
11. $y = \sqrt{\ln x}$
12. $y = \ln\sqrt{x}$
13. $y = x\ln x$
14. $y = x^3\ln x$
15. $y = x^2\log_2(3-2x)$
16. $y = x[\log_2(x^2 - 2x)]^3$
17. $y = \dfrac{x^2}{1+\log x}$
18. $y = \dfrac{\log x}{1+\log x}$

19. $y = \ln(\ln x)$ **20.** $y = \ln(\ln(\ln x))$

21. $y = \ln(\tan x)$ **22.** $y = \ln(\cos x)$

23. $y = \cos(\ln x)$ **24.** $y = \sin^2(\ln x)$

25. $y = \log(\sin^2 x)$ **26.** $y = \log(1 - \sin^2 x)$

27–30 Use the method of Example 3 to help perform the indicated differentiation. ■

27. $\dfrac{d}{dx}[\ln((x-1)^3(x^2+1)^4)]$

28. $\dfrac{d}{dx}[\ln((\cos^2 x)\sqrt{1+x^4})]$

29. $\dfrac{d}{dx}\left[\ln\dfrac{\cos x}{\sqrt{4-3x^2}}\right]$ **30.** $\dfrac{d}{dx}\left[\ln\sqrt{\dfrac{x-1}{x+1}}\right]$

31–34 True–False Determine whether the statement is true or false. Explain your answer. ■

31. The slope of the tangent line to the graph of $y = \ln x$ at $x = a$ approaches infinity as $a \to 0^+$.

32. If $\lim_{x \to +\infty} f'(x) = 0$, then the graph of $y = f(x)$ has a horizontal asymptote.

33. The derivative of $\ln|x|$ is an odd function.

34. We have

$$\frac{d}{dx}((\ln x)^2) = \frac{d}{dx}(2(\ln x)) = \frac{2}{x}$$

35–38 Find dy/dx using logarithmic differentiation. ■

35. $y = x\sqrt[3]{1+x^2}$ **36.** $y = \sqrt[5]{\dfrac{x-1}{x+1}}$

37. $y = \dfrac{(x^2-8)^{1/3}\sqrt{x^3+1}}{x^6-7x+5}$ **38.** $y = \dfrac{\sin x \cos x \tan^3 x}{\sqrt{x}}$

39. Find

  (a) $\dfrac{d}{dx}[\log_x e]$ (b) $\dfrac{d}{dx}[\log_x 2]$.

40. Find

  (a) $\dfrac{d}{dx}[\log_{(1/x)} e]$ (b) $\dfrac{d}{dx}[\log_{(\ln x)} e]$.

41–44 Find the equation of the tangent line to the graph of $y = f(x)$ at $x = x_0$. ■

41. $f(x) = \ln x$; $x_0 = e^{-1}$ **42.** $f(x) = \log x$; $x_0 = 10$

43. $f(x) = \ln(-x)$; $x_0 = -e$ **44.** $f(x) = \ln|x|$; $x_0 = -2$

FOCUS ON CONCEPTS

45. (a) Find the equation of a line through the origin that is tangent to the graph of $y = \ln x$.
    (b) Explain why the y-intercept of a tangent line to the curve $y = \ln x$ must be 1 unit less than the y-coordinate of the point of tangency.

46. Use logarithmic differentiation to verify the product and quotient rules. Explain what properties of $\ln x$ are important for this verification.

47. Find a formula for the area $A(w)$ of the triangle bounded by the tangent line to the graph of $y = \ln x$ at $P(w, \ln w)$, the horizontal line through P, and the y-axis.

48. Find a formula for the area $A(w)$ of the triangle bounded by the tangent line to the graph of $y = \ln x^2$ at $P(w, \ln w^2)$, the horizontal line through P, and the y-axis.

49. Verify that $y = \ln(x + e)$ satisfies $dy/dx = e^{-y}$, with $y = 1$ when $x = 0$.

50. Verify that $y = -\ln(e^2 - x)$ satisfies $dy/dx = e^y$, with $y = -2$ when $x = 0$.

51. Find a function f such that $y = f(x)$ satisfies $dy/dx = e^{-y}$, with $y = 0$ when $x = 0$.

52. Find a function f such that $y = f(x)$ satisfies $dy/dx = e^y$, with $y = -\ln 2$ when $x = 0$.

53–55 Find the limit by interpreting the expression as an appropriate derivative. ■

53. (a) $\lim_{x \to 0} \dfrac{\ln(1+3x)}{x}$ (b) $\lim_{x \to 0} \dfrac{\ln(1-5x)}{x}$

54. (a) $\lim_{\Delta x \to 0} \dfrac{\ln(e^2 + \Delta x) - 2}{\Delta x}$ (b) $\lim_{w \to 1} \dfrac{\ln w}{w - 1}$

55. (a) $\lim_{x \to 0} \dfrac{\ln(\cos x)}{x}$ (b) $\lim_{h \to 0} \dfrac{(1+h)^{\sqrt{2}} - 1}{h}$

56. Modify the derivation of Equation (2) to give another proof of Equation (3).

57. Writing Review the derivation of the formula

$$\frac{d}{dx}[\ln x] = \frac{1}{x}$$

and then write a paragraph that discusses all the ingredients (theorems, limit properties, etc.) that are needed for this derivation.

58. Writing Write a paragraph that explains how logarithmic differentiation can replace a difficult differentiation computation with a simpler computation.

✓ **QUICK CHECK ANSWERS 3.2**

1. $y = \dfrac{x}{e^2} + 1$ **2.** (a) $\dfrac{dy}{dx} = \dfrac{1}{x}$ (b) $\dfrac{dy}{dx} = \dfrac{1}{2x}$ (c) $\dfrac{dy}{dx} = -\dfrac{1}{x \ln 10}$ **3.** $\dfrac{\sqrt{x+1}}{\sqrt[3]{x-1}}\left[\dfrac{1}{2(x+1)} - \dfrac{1}{3(x-1)}\right]$ **4.** 1

3.3 DERIVATIVES OF EXPONENTIAL AND INVERSE TRIGONOMETRIC FUNCTIONS

In this section we will show how the derivative of a one-to-one function can be used to obtain the derivative of its inverse function. This will provide the tools we need to obtain derivative formulas for exponential functions from the derivative formulas for logarithmic functions and to obtain derivative formulas for inverse trigonometric functions from the derivative formulas for trigonometric functions.

See Section 0.4 for a review of one-to-one functions and inverse functions.

Our first goal in this section is to obtain a formula relating the derivative of the inverse function f^{-1} to the derivative of the function f.

▶ **Example 1** Suppose that f is a one-to-one differentiable function such that $f(2) = 1$ and $f'(2) = \frac{3}{4}$. Then the tangent line to $y = f(x)$ at the point $(2, 1)$ has equation

$$y - 1 = \tfrac{3}{4}(x - 2)$$

The tangent line to $y = f^{-1}(x)$ at the point $(1, 2)$ is the reflection about the line $y = x$ of the tangent line to $y = f(x)$ at the point $(2, 1)$ (Figure 3.3.1), and its equation can be obtained by interchanging x and y:

$$x - 1 = \tfrac{3}{4}(y - 2) \quad \text{or} \quad y - 2 = \tfrac{4}{3}(x - 1)$$

Notice that the slope of the tangent line to $y = f^{-1}(x)$ at $x = 1$ is the reciprocal of the slope of the tangent line to $y = f(x)$ at $x = 2$. That is,

$$(f^{-1})'(1) = \frac{1}{f'(2)} = \frac{4}{3} \quad ◀ \tag{1}$$

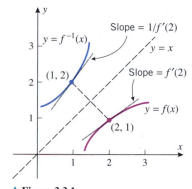

▲ **Figure 3.3.1**

Since $2 = f^{-1}(1)$ for the function f in Example 1, it follows that $f'(2) = f'(f^{-1}(1))$. Thus, Formula (1) can also be expressed as

$$(f^{-1})'(1) = \frac{1}{f'(f^{-1}(1))}$$

In general, if f is a differentiable and one-to-one function, then

$$(f^{-1})'(x) = \frac{1}{f'(f^{-1}(x))} \tag{2}$$

provided $f'(f^{-1}(x)) \neq 0$.

Formula (2) can be confirmed using implicit differentiation. The equation $y = f^{-1}(x)$ is equivalent to $x = f(y)$. Differentiating with respect to x we obtain

$$1 = \frac{d}{dx}[x] = \frac{d}{dx}[f(y)] = f'(y) \cdot \frac{dy}{dx}$$

so that

$$\frac{dy}{dx} = \frac{1}{f'(y)} = \frac{1}{f'(f^{-1}(x))}$$

Also from $x = f(y)$ we have $dx/dy = f'(y)$, which gives the following alternative version of Formula (2):

$$\frac{dy}{dx} = \frac{1}{dx/dy} \tag{3}$$

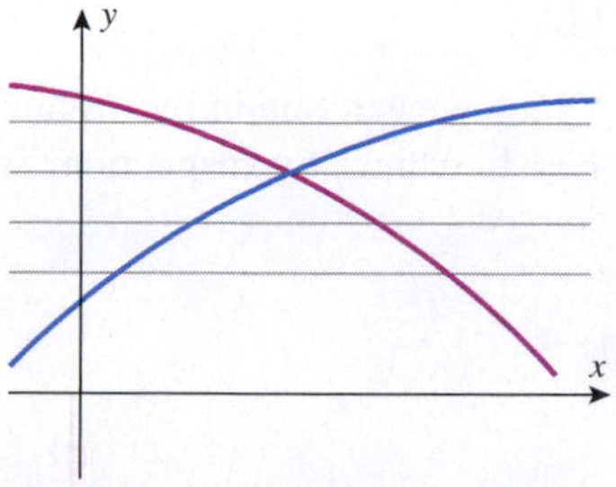

▲ **Figure 3.3.2** The graph of an increasing function (blue) or a decreasing function (purple) is cut at most once by any horizontal line.

■ INCREASING OR DECREASING FUNCTIONS ARE ONE-TO-ONE

If the graph of a function f is always increasing or always decreasing over the domain of f, then a horizontal line will cut the graph of f in at most one point (Figure 3.3.2), so f

must have an inverse function (see Section 0.4). We will prove in the next chapter that f is increasing on any interval on which $f'(x) > 0$ (since the graph has positive slope) and that f is decreasing on any interval on which $f'(x) < 0$ (since the graph has negative slope). These intuitive observations, together with Formula (2), suggest the following theorem, which we state without formal proof.

3.3.1   THEOREM   *Suppose that the domain of a function f is an open interval on which $f'(x) > 0$ or on which $f'(x) < 0$. Then f is one-to-one, $f^{-1}(x)$ is differentiable at all values of x in the range of f, and the derivative of $f^{-1}(x)$ is given by Formula (2).*

► **Example 2**   Consider the function $f(x) = x^5 + x + 1$.

(a) Show that f is one-to-one on the interval $(-\infty, +\infty)$.

(b) Find a formula for the derivative of f^{-1}.

(c) Compute $(f^{-1})'(1)$.

> In general, once it is established that f^{-1} is differentiable, one has the option of calculating the derivative of f^{-1} using Formula (2) or (3), or by differentiating implicitly, as in Example 2.

Solution (a).   Since

$$f'(x) = 5x^4 + 1 > 0$$

for all real values of x, it follows from Theorem 3.3.1 that f is one-to-one on the interval $(-\infty, +\infty)$.

Solution (b).   Let $y = f^{-1}(x)$. Differentiating $x = f(y) = y^5 + y + 1$ implicitly with respect to x yields

$$\frac{d}{dx}[x] = \frac{d}{dx}[y^5 + y + 1]$$

$$1 = (5y^4 + 1)\frac{dy}{dx}$$

$$\frac{dy}{dx} = \frac{1}{5y^4 + 1} \tag{4}$$

We cannot solve $x = y^5 + y + 1$ for y in terms of x, so we leave the expression for dy/dx in Equation (4) in terms of y.

Solution (c).   From Equation (4),

$$(f^{-1})'(1) = \left.\frac{dy}{dx}\right|_{x=1} = \left.\frac{1}{5y^4 + 1}\right|_{x=1}$$

Thus, we need to know the value of $y = f^{-1}(x)$ at $x = 1$, which we can obtain by solving the equation $f(y) = 1$ for y. This equation is $y^5 + y + 1 = 1$, which, by inspection, is satisfied by $y = 0$. Thus,

$$(f^{-1})'(1) = \left.\frac{1}{5y^4 + 1}\right|_{y=0} = 1 \quad ◄$$

■ **DERIVATIVES OF EXPONENTIAL FUNCTIONS**

Our next objective is to show that the general exponential function b^x ($b > 0, b \neq 1$) is differentiable everywhere and to find its derivative. To do this, we will use the fact that

b^x is the inverse of the function $f(x) = \log_b x$. We will assume that $b > 1$. With this assumption we have $\ln b > 0$, so

$$f'(x) = \frac{d}{dx}[\log_b x] = \frac{1}{x \ln b} > 0 \quad \text{for all } x \text{ in the interval } (0, +\infty)$$

It now follows from Theorem 3.3.1 that $f^{-1}(x) = b^x$ is differentiable for all x in the range of $f(x) = \log_b x$. But we know from Table 0.5.3 that the range of $\log_b x$ is $(-\infty, +\infty)$, so we have established that b^x is differentiable everywhere.

To obtain a derivative formula for b^x we rewrite $y = b^x$ as

$$x = \log_b y$$

and differentiate implicitly using Formula (5) of Section 3.2 to obtain

$$1 = \frac{1}{y \ln b} \cdot \frac{dy}{dx}$$

Solving for dy/dx and replacing y by b^x we have

$$\frac{dy}{dx} = y \ln b = b^x \ln b$$

Thus, we have shown that

> How does the derivation of Formula (5) change if $0 < b < 1$?

$$\frac{d}{dx}[b^x] = b^x \ln b \tag{5}$$

In the special case where $b = e$ we have $\ln e = 1$, so that (5) becomes

$$\frac{d}{dx}[e^x] = e^x \tag{6}$$

> In Section 0.5 we stated that $b = e$ is the only base for which the slope of the tangent line to the curve $y = b^x$ at any point P on the curve is the y-coordinate at P (see page 54). Verify this statement.

Moreover, if u is a differentiable function of x, then it follows from (5) and (6) that

$$\frac{d}{dx}[b^u] = b^u \ln b \cdot \frac{du}{dx} \quad \text{and} \quad \frac{d}{dx}[e^u] = e^u \cdot \frac{du}{dx} \tag{7–8}$$

> It is important to distinguish between differentiating an exponential function b^x (variable exponent and constant base) and a power function x^b (variable base and constant exponent). For example, compare the derivative
>
> $$\frac{d}{dx}[x^2] = 2x$$
>
> to the derivative of 2^x in Example 3.

▶ **Example 3** The following computations use Formulas (7) and (8).

$$\frac{d}{dx}[2^x] = 2^x \ln 2$$

$$\frac{d}{dx}[e^{-2x}] = e^{-2x} \cdot \frac{d}{dx}[-2x] = -2e^{-2x}$$

$$\frac{d}{dx}[e^{x^3}] = e^{x^3} \cdot \frac{d}{dx}[x^3] = 3x^2 e^{x^3}$$

$$\frac{d}{dx}[e^{\cos x}] = e^{\cos x} \cdot \frac{d}{dx}[\cos x] = -(\sin x)e^{\cos x} \quad ◀$$

Functions of the form $f(x) = u^v$ in which u and v are *nonconstant* functions of x are neither exponential functions nor power functions. Functions of this form can be differentiated using logarithmic differentiation.

▶ **Example 4** Use logarithmic differentiation to find $\dfrac{d}{dx}[(x^2 + 1)^{\sin x}]$.

Solution. Setting $y = (x^2 + 1)^{\sin x}$ we have

$$\ln y = \ln[(x^2 + 1)^{\sin x}] = (\sin x) \ln(x^2 + 1)$$

Differentiating both sides with respect to x yields

$$\frac{1}{y}\frac{dy}{dx} = \frac{d}{dx}[(\sin x)\ln(x^2 + 1)]$$

$$= (\sin x)\frac{1}{x^2 + 1}(2x) + (\cos x)\ln(x^2 + 1)$$

Thus,

$$\frac{dy}{dx} = y\left[\frac{2x\sin x}{x^2 + 1} + (\cos x)\ln(x^2 + 1)\right]$$

$$= (x^2 + 1)^{\sin x}\left[\frac{2x\sin x}{x^2 + 1} + (\cos x)\ln(x^2 + 1)\right] \quad \blacktriangleleft$$

■ DERIVATIVES OF THE INVERSE TRIGONOMETRIC FUNCTIONS

To obtain formulas for the derivatives of the inverse trigonometric functions, we will need to use some of the identities given in Formulas (11) to (17) of Section 0.4. Rather than memorize those identities, we recommend that you review the "triangle technique" that we used to obtain them.

To begin, consider the function $\sin^{-1} x$. If we let $f(x) = \sin x\ (-\pi/2 \leq x \leq \pi/2)$, then it follows from Formula (2) that $f^{-1}(x) = \sin^{-1} x$ will be differentiable at any point x where $\cos(\sin^{-1} x) \neq 0$. This is equivalent to the condition

$$\sin^{-1} x \neq -\frac{\pi}{2} \quad \text{and} \quad \sin^{-1} x \neq \frac{\pi}{2}$$

so it follows that $\sin^{-1} x$ is differentiable on the interval $(-1, 1)$.

A derivative formula for $\sin^{-1} x$ on $(-1, 1)$ can be obtained by using Formula (2) or (3) or by differentiating implicitly. We will use the latter method. Rewriting the equation $y = \sin^{-1} x$ as $x = \sin y$ and differentiating implicitly with respect to x, we obtain

$$\frac{d}{dx}[x] = \frac{d}{dx}[\sin y]$$

$$1 = \cos y \cdot \frac{dy}{dx}$$

$$\frac{dy}{dx} = \frac{1}{\cos y} = \frac{1}{\cos(\sin^{-1} x)}$$

> Observe that $\sin^{-1} x$ is only differentiable on the interval $(-1, 1)$, even though its domain is $[-1, 1]$. This is because the graph of $y = \sin x$ has horizontal tangent lines at the points $(\pi/2, 1)$ and $(-\pi/2, -1)$, so the graph of $y = \sin^{-1} x$ has vertical tangent lines at $x = \pm 1$.

At this point we have succeeded in obtaining the derivative; however, this derivative formula can be simplified using the identity indicated in Figure 3.3.3. This yields

$$\frac{dy}{dx} = \frac{1}{\sqrt{1 - x^2}}$$

Thus, we have shown that

$$\frac{d}{dx}[\sin^{-1} x] = \frac{1}{\sqrt{1 - x^2}} \qquad (-1 < x < 1)$$

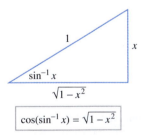

$$\cos(\sin^{-1} x) = \sqrt{1 - x^2}$$

▲ Figure 3.3.3

More generally, if u is a differentiable function of x, then the chain rule produces the following generalized version of this formula:

$$\frac{d}{dx}[\sin^{-1} u] = \frac{1}{\sqrt{1 - u^2}}\frac{du}{dx} \qquad (-1 < u < 1)$$

The method used to derive this formula can be used to obtain generalized derivative formulas for the remaining inverse trigonometric functions. The following is a complete list of these formulas, each of which is valid on the natural domain of the function that multiplies du/dx.

$$\frac{d}{dx}[\sin^{-1}u] = \frac{1}{\sqrt{1-u^2}}\frac{du}{dx} \qquad \frac{d}{dx}[\cos^{-1}u] = -\frac{1}{\sqrt{1-u^2}}\frac{du}{dx} \qquad (9\text{–}10)$$

$$\frac{d}{dx}[\tan^{-1}u] = \frac{1}{1+u^2}\frac{du}{dx} \qquad \frac{d}{dx}[\cot^{-1}u] = -\frac{1}{1+u^2}\frac{du}{dx} \qquad (11\text{–}12)$$

$$\frac{d}{dx}[\sec^{-1}u] = \frac{1}{|u|\sqrt{u^2-1}}\frac{du}{dx} \qquad \frac{d}{dx}[\csc^{-1}u] = -\frac{1}{|u|\sqrt{u^2-1}}\frac{du}{dx} \qquad (13\text{–}14)$$

The appearance of $|u|$ in (13) and (14) will be explained in Exercise 58.

▶ **Example 5** Find dy/dx if

$$\text{(a) } y = \sin^{-1}(x^3) \qquad \text{(b) } y = \sec^{-1}(e^x)$$

Solution (a). From (9)

$$\frac{dy}{dx} = \frac{1}{\sqrt{1-(x^3)^2}}(3x^2) = \frac{3x^2}{\sqrt{1-x^6}}$$

Solution (b). From (13)

$$\frac{dy}{dx} = \frac{1}{e^x\sqrt{(e^x)^2-1}}(e^x) = \frac{1}{\sqrt{e^{2x}-1}} \quad ◀$$

✔ QUICK CHECK EXERCISES 3.3   *(See page 203 for answers.)*

1. Suppose that a one-to-one function f has tangent line $y = 5x + 3$ at the point $(1, 8)$. Evaluate $(f^{-1})'(8)$.

2. In each case, from the given derivative, determine whether the function f is invertible.
   (a) $f'(x) = x^2 + 1$ (b) $f'(x) = x^2 - 1$
   (c) $f'(x) = \sin x$ (d) $f'(x) = \dfrac{\pi}{2} + \tan^{-1}x$

3. Evaluate the derivative.
   (a) $\dfrac{d}{dx}[e^x]$ (b) $\dfrac{d}{dx}[7^x]$
   (c) $\dfrac{d}{dx}[\cos(e^x + 1)]$ (d) $\dfrac{d}{dx}[e^{3x-2}]$

4. Let $f(x) = e^{x^3+x}$. Use $f'(x)$ to verify that f is one-to-one.

EXERCISE SET 3.3   Graphing Utility

FOCUS ON CONCEPTS

1. Let $f(x) = x^5 + x^3 + x$.
   (a) Show that f is one-to-one and confirm that $f(1) = 3$.
   (b) Find $(f^{-1})'(3)$.

2. Let $f(x) = x^3 + 2e^x$.
   (a) Show that f is one-to-one and confirm that $f(0) = 2$.
   (b) Find $(f^{-1})'(2)$.

3–4 Find $(f^{-1})'(x)$ using Formula (2), and check your answer by differentiating f^{-1} directly. ■

3. $f(x) = 2/(x+3)$ 4. $f(x) = \ln(2x+1)$

5–6 Determine whether the function f is one-to-one by examining the sign of $f'(x)$. ■

5. (a) $f(x) = x^2 + 8x + 1$
   (b) $f(x) = 2x^5 + x^3 + 3x + 2$
   (c) $f(x) = 2x + \sin x$
   (d) $f(x) = \left(\frac{1}{2}\right)^x$

6. (a) $f(x) = x^3 + 3x^2 - 8$
   (b) $f(x) = x^5 + 8x^3 + 2x - 1$
   (c) $f(x) = \dfrac{x}{x+1}$
   (d) $f(x) = \log_b x, \quad 0 < b < 1$

7–10 Find the derivative of f^{-1} by using Formula (3), and check your result by differentiating implicitly. ■

7. $f(x) = 5x^3 + x - 7$

8. $f(x) = 1/x^2, \quad x > 0$

9. $f(x) = 2x^5 + x^3 + 1$

10. $f(x) = 5x - \sin 2x, \quad -\dfrac{\pi}{4} < x < \dfrac{\pi}{4}$

FOCUS ON CONCEPTS

11. Figure 0.4.8 is a "proof by picture" that the reflection of a point $P(a, b)$ about the line $y = x$ is the point $Q(b, a)$. Establish this result rigorously by completing each part.
   (a) Prove that if P is not on the line $y = x$, then P and Q are distinct, and the line $\overleftrightarrow{PQ}$ is perpendicular to the line $y = x$.
   (b) Prove that if P is not on the line $y = x$, the midpoint of segment PQ is on the line $y = x$.
   (c) Carefully explain what it means geometrically to reflect P about the line $y = x$.
   (d) Use the results of parts (a)–(c) to prove that Q is the reflection of P about the line $y = x$.

12. Prove that the reflection about the line $y = x$ of a line with slope m, $m \neq 0$, is a line with slope $1/m$. [*Hint:* Apply the result of the previous exercise to a pair of points on the line of slope m and to a corresponding pair of points on the reflection of this line about the line $y = x$.]

13. Suppose that f and g are increasing functions. Determine which of the functions $f(x) + g(x)$, $f(x)g(x)$, and $f(g(x))$ must also be increasing.

14. Suppose that f and g are one-to-one functions. Determine which of the functions $f(x) + g(x)$, $f(x)g(x)$, and $f(g(x))$ must also be one-to-one.

15–26 Find dy/dx. ■

15. $y = e^{7x}$

16. $y = e^{-5x^2}$

17. $y = x^3 e^x$

18. $y = e^{1/x}$

19. $y = \dfrac{e^x - e^{-x}}{e^x + e^{-x}}$

20. $y = \sin(e^x)$

21. $y = e^x \tan x$

22. $y = \dfrac{e^x}{\ln x}$

23. $y = e^{(x - e^{3x})}$

24. $y = \exp(\sqrt{1 + 5x^3})$

25. $y = \ln(1 - xe^{-x})$

26. $y = \ln(\cos e^x)$

27–30 Find $f'(x)$ by Formula (7) and then by logarithmic differentiation. ■

27. $f(x) = 2^x$

28. $f(x) = 3^{-x}$

29. $f(x) = \pi^{\sin x}$

30. $f(x) = \pi^{x \tan x}$

31–35 Find dy/dx using the method of logarithmic differentiation. ■

31. $y = (x^3 - 2x)^{\ln x}$

32. $y = x^{\sin x}$

33. $y = (\ln x)^{\tan x}$

34. $y = (x^2 + 3)^{\ln x}$

35. $y = (\ln x)^{\ln x}$

36. (a) Explain why Formula (5) cannot be used to find $(d/dx)[x^x]$.
   (b) Find this derivative by logarithmic differentiation.

37–52 Find dy/dx. ■

37. $y = \sin^{-1}(3x)$

38. $y = \cos^{-1}\left(\dfrac{x + 1}{2}\right)$

39. $y = \sin^{-1}(1/x)$

40. $y = \cos^{-1}(\cos x)$

41. $y = \tan^{-1}(x^3)$

42. $y = \sec^{-1}(x^5)$

43. $y = (\tan x)^{-1}$

44. $y = \dfrac{1}{\tan^{-1} x}$

45. $y = e^x \sec^{-1} x$

46. $y = \ln(\cos^{-1} x)$

47. $y = \sin^{-1} x + \cos^{-1} x$

48. $y = x^2 (\sin^{-1} x)^3$

49. $y = \sec^{-1} x + \csc^{-1} x$

50. $y = \csc^{-1}(e^x)$

51. $y = \cot^{-1}(\sqrt{x})$

52. $y = \sqrt{\cot^{-1} x}$

53–56 True–False Determine whether the statement is true or false. Explain your answer. ■

53. If a function $y = f(x)$ satisfies $dy/dx = y$, then $y = e^x$.

54. If $y = f(x)$ is a function such that dy/dx is a rational function, then $f(x)$ is also a rational function.

55. $\dfrac{d}{dx}(\log_b |x|) = \dfrac{1}{x \ln b}$

56. We can conclude from the derivatives of $\sin^{-1} x$ and $\cos^{-1} x$ that $\sin^{-1} x + \cos^{-1} x$ is constant.

57. (a) Use Formula (2) to prove that
$$\frac{d}{dx}[\cot^{-1} x]\Big|_{x=0} = -1$$
   (b) Use part (a) above, part (a) of Exercise 48 in Section 0.4, and the chain rule to show that
$$\frac{d}{dx}[\cot^{-1} x] = -\frac{1}{1 + x^2}$$
   for $-\infty < x < +\infty$.
   (c) Conclude from part (b) that
$$\frac{d}{dx}[\cot^{-1} u] = -\frac{1}{1 + u^2}\frac{du}{dx}$$
   for $-\infty < u < +\infty$.

58. (a) Use part (c) of Exercise 48 in Section 0.4 and the chain rule to show that
$$\frac{d}{dx}[\csc^{-1} x] = -\frac{1}{|x|\sqrt{x^2 - 1}}$$
   for $1 < |x|$.
   (b) Conclude from part (a) that
$$\frac{d}{dx}[\csc^{-1} u] = -\frac{1}{|u|\sqrt{u^2 - 1}}\frac{du}{dx}$$
   for $1 < |u|$.

(cont.)

(c) Use Equation (11) in Section 0.4 and parts (b) and (c) of Exercise 48 in that section to show that if $|x| \geq 1$ then, $\sec^{-1} x + \csc^{-1} x = \pi/2$. Conclude from part (a) that

$$\frac{d}{dx}[\sec^{-1} x] = \frac{1}{|x|\sqrt{x^2 - 1}}$$

(d) Conclude from part (c) that

$$\frac{d}{dx}[\sec^{-1} u] = \frac{1}{|u|\sqrt{x^2 - 1}}\frac{du}{dx}$$

59–60 Find dy/dx by implicit differentiation. ■

59. $x^3 + x\tan^{-1} y = e^y$ **60.** $\sin^{-1}(xy) = \cos^{-1}(x - y)$

61. (a) Show that $f(x) = x^3 - 3x^2 + 2x$ is not one-to-one on $(-\infty, +\infty)$.
(b) Find the largest value of k such that f is one-to-one on the interval $(-k, k)$.

62. (a) Show that the function $f(x) = x^4 - 2x^3$ is not one-to-one on $(-\infty, +\infty)$.
(b) Find the smallest value of k such that f is one-to-one on the interval $[k, +\infty)$.

63. Let $f(x) = x^4 + x^3 + 1, 0 \leq x \leq 2$.
(a) Show that f is one-to-one.
(b) Let $g(x) = f^{-1}(x)$ and define $F(x) = f(2g(x))$. Find an equation for the tangent line to $y = F(x)$ at $x = 3$.

64. Let $f(x) = \dfrac{\exp(4 - x^2)}{x}, x > 0$.
(a) Show that f is one-to-one.
(b) Let $g(x) = f^{-1}(x)$ and define $F(x) = f([g(x)]^2)$. Find $F'\left(\frac{1}{2}\right)$.

65. Show that for any constants A and k, the function $y = Ae^{kt}$ satisfies the equation $dy/dt = ky$.

66. Show that for any constants A and B, the function

$$y = Ae^{2x} + Be^{-4x}$$

satisfies the equation

$$y'' + 2y' - 8y = 0$$

67. Show that
(a) $y = xe^{-x}$ satisfies the equation $xy' = (1 - x)y$
(b) $y = xe^{-x^2/2}$ satisfies the equation $xy' = (1 - x^2)y$.

68. Show that the rate of change of $y = 100e^{-0.2x}$ with respect to x is proportional to y.

69. Show that

$$y = \frac{60}{5 + 7e^{-t}} \quad \text{satisfies} \quad \frac{dy}{dt} = r\left(1 - \frac{y}{K}\right)y$$

for some constants r and K, and determine the values of these constants.

70. Suppose that the population of oxygen-dependent bacteria in a pond is modeled by the equation

$$P(t) = \frac{60}{5 + 7e^{-t}}$$

where $P(t)$ is the population (in billions) t days after an initial observation at time $t = 0$.
(a) Use a graphing utility to graph the function $P(t)$.
(b) In words, explain what happens to the population over time. Check your conclusion by finding $\lim_{t \to +\infty} P(t)$.
(c) In words, what happens to the *rate* of population growth over time? Check your conclusion by graphing $P'(t)$.

71–76 Find the limit by interpreting the expression as an appropriate derivative. ■

71. $\lim\limits_{x \to 0} \dfrac{e^{3x} - 1}{x}$ **72.** $\lim\limits_{x \to 0} \dfrac{\exp(x^2) - 1}{x}$

73. $\lim\limits_{h \to 0} \dfrac{10^h - 1}{h}$ **74.** $\lim\limits_{h \to 0} \dfrac{\tan^{-1}(1 + h) - \pi/4}{h}$

75. $\lim\limits_{\Delta x \to 0} \dfrac{9\left[\sin^{-1}\left(\frac{\sqrt{3}}{2} + \Delta x\right)\right]^2 - \pi^2}{\Delta x}$

76. $\lim\limits_{w \to 2} \dfrac{3\sec^{-1} w - \pi}{w - 2}$

77. Writing Let G denote the graph of an invertible function f and consider G as a fixed set of points in the plane. Suppose we relabel the coordinate axes so that the x-axis becomes the y-axis and vice versa. Carefully explain why now the same set of points G becomes the graph of f^{-1} (with the coordinate axes in a nonstandard position). Use this result to explain Formula (2).

78. Writing Suppose that f has an inverse function. Carefully explain the connection between Formula (2) and implicit differentiation of the equation $x = f(y)$.

✔ QUICK CHECK ANSWERS 3.3

1. $\frac{1}{5}$ **2.** (a) yes (b) no (c) no (d) yes   **3.** (a) e^x (b) $7^x \ln 7$ (c) $-e^x \sin(e^x + 1)$ (d) $3e^{3x-2}$
4. $f'(x) = e^{x^3+x} \cdot (3x^2 + 1) > 0$ for all x

3.4 RELATED RATES

In this section we will study related rates problems. In such problems one tries to find the rate at which some quantity is changing by relating the quantity to other quantities whose rates of change are known.

■ **DIFFERENTIATING EQUATIONS TO RELATE RATES**

Figure 3.4.1 shows a liquid draining through a conical filter. As the liquid drains, its volume V, height h, and radius r are functions of the elapsed time t, and at each instant these variables are related by the equation

$$V = \frac{\pi}{3}r^2 h$$

If we were interested in finding the rate of change of the volume V with respect to the time t, we could begin by differentiating both sides of this equation with respect to t to obtain

$$\frac{dV}{dt} = \frac{\pi}{3}\left[r^2\frac{dh}{dt} + h\left(2r\frac{dr}{dt}\right)\right] = \frac{\pi}{3}\left(r^2\frac{dh}{dt} + 2rh\frac{dr}{dt}\right)$$

Thus, to find dV/dt at a specific time t from this equation we would need to have values for r, h, dh/dt, and dr/dt at that time. This is called a ***related rates problem*** because the goal is to find an unknown rate of change by *relating* it to other variables whose values and whose rates of change at time t are known or can be found in some way. Let us begin with a simple example.

▶ **Figure 3.4.1**

▶ **Example 1** Suppose that x and y are differentiable functions of t and are related by the equation $y = x^3$. Find dy/dt at time $t = 1$ if $x = 2$ and $dx/dt = 4$ at time $t = 1$.

Solution. Using the chain rule to differentiate both sides of the equation $y = x^3$ with respect to t yields

$$\frac{dy}{dt} = \frac{d}{dt}[x^3] = 3x^2\frac{dx}{dt}$$

Thus, the value of dy/dt at time $t = 1$ is

$$\left.\frac{dy}{dt}\right|_{t=1} = 3(2)^2\left.\frac{dx}{dt}\right|_{t=1} = 12 \cdot 4 = 48 \quad ◀$$

Arni Katz/Phototake

Oil spill from a ruptured tanker.

▶ **Example 2**   Assume that oil spilled from a ruptured tanker spreads in a circular pattern whose radius increases at a constant rate of 2 ft/s. How fast is the area of the spill increasing when the radius of the spill is 60 ft?

Solution.   Let

$$t = \text{number of seconds elapsed from the time of the spill}$$
$$r = \text{radius of the spill in feet after } t \text{ seconds}$$
$$A = \text{area of the spill in square feet after } t \text{ seconds}$$

(Figure 3.4.2). We know the rate at which the radius is increasing, and we want to find the rate at which the area is increasing at the instant when $r = 60$; that is, we want to find

$$\left.\frac{dA}{dt}\right|_{r=60} \qquad \text{given that} \qquad \frac{dr}{dt} = 2 \text{ ft/s}$$

This suggests that we look for an equation relating A and r that we can differentiate with respect to t to produce a relationship between dA/dt and dr/dt. But A is the area of a circle of radius r, so

$$A = \pi r^2 \tag{1}$$

Differentiating both sides of (1) with respect to t yields

$$\frac{dA}{dt} = 2\pi r \frac{dr}{dt} \tag{2}$$

Thus, when $r = 60$ the area of the spill is increasing at the rate of

$$\left.\frac{dA}{dt}\right|_{r=60} = 2\pi(60)(2) = 240\pi \text{ ft}^2/\text{s} \approx 754 \text{ ft}^2/\text{s} \blacktriangleleft$$

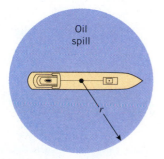

▲ **Figure 3.4.2**

With some minor variations, the method used in Example 2 can be used to solve a variety of related rates problems. We can break the method down into five steps.

WARNING

We have italicized the word "After" in Step 5 because it is a common error to substitute numerical values before performing the differentiation. For instance, in Example 2 had we substituted the known value of $r = 60$ in (1) before differentiating, we would have obtained $dA/dt = 0$, which is obviously incorrect.

A Strategy for Solving Related Rates Problems

Step 1. Assign letters to all quantities that vary with time and any others that seem relevant to the problem. Give a definition for each letter.

Step 2. Identify the rates of change that are known and the rate of change that is to be found. Interpret each rate as a derivative.

Step 3. Find an equation that relates the variables whose rates of change were identified in Step 2. To do this, it will often be helpful to draw an appropriately labeled figure that illustrates the relationship.

Step 4. Differentiate both sides of the equation obtained in Step 3 with respect to time to produce a relationship between the known rates of change and the unknown rate of change.

Step 5. *After* completing Step 4, substitute all known values for the rates of change and the variables, and then solve for the unknown rate of change.

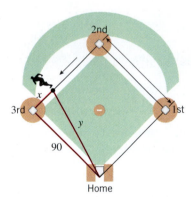

▲ **Figure 3.4.3**

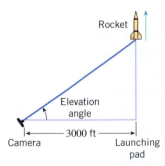

▲ **Figure 3.4.4**

▶ **Example 3**   A baseball diamond is a square whose sides are 90 ft long (Figure 3.4.3). Suppose that a player running from second base to third base has a speed of 30 ft/s at the instant when he is 20 ft from third base. At what rate is the player's distance from home plate changing at that instant?

Solution.   We are given a constant speed with which the player is approaching third base, and we want to find the rate of change of the distance between the player and home plate at a particular instant. Thus, let

$$t = \text{number of seconds since the player left second base}$$
$$x = \text{distance in feet from the player to third base}$$
$$y = \text{distance in feet from the player to home plate}$$

(Figure 3.4.4). Thus, we want to find

$$\left.\frac{dy}{dt}\right|_{x=20} \qquad \text{given that} \qquad \left.\frac{dx}{dt}\right|_{x=20} = -30 \text{ ft/s}$$

As suggested by Figure 3.4.4, an equation relating the variables x and y can be obtained using the Theorem of Pythagoras:

$$x^2 + 90^2 = y^2 \tag{3}$$

Differentiating both sides of this equation with respect to t yields

$$2x\frac{dx}{dt} = 2y\frac{dy}{dt}$$

from which we obtain

$$\frac{dy}{dt} = \frac{x}{y}\frac{dx}{dt} \tag{4}$$

When $x = 20$, it follows from (3) that

$$y = \sqrt{20^2 + 90^2} = \sqrt{8500} = 10\sqrt{85}$$

so that (4) yields

$$\left.\frac{dy}{dt}\right|_{x=20} = \frac{20}{10\sqrt{85}}(-30) = -\frac{60}{\sqrt{85}} \approx -6.51 \text{ ft/s}$$

The negative sign in the answer tells us that y is decreasing, which makes sense physically from Figure 3.4.4. ◀

> The quantity
> $$\left.\frac{dx}{dt}\right|_{x=20}$$
> is negative because x is decreasing with respect to t.

▶ **Example 4**   In Figure 3.4.5 we have shown a camera mounted at a point 3000 ft from the base of a rocket launching pad. If the rocket is rising vertically at 880 ft/s when it is 4000 ft above the launching pad, how fast must the camera elevation angle change at that instant to keep the camera aimed at the rocket?

Solution.   Let

$$t = \text{number of seconds elapsed from the time of launch}$$
$$\phi = \text{camera elevation angle in radians after } t \text{ seconds}$$
$$h = \text{height of the rocket in feet after } t \text{ seconds}$$

(Figure 3.4.6). At each instant the rate at which the camera elevation angle must change

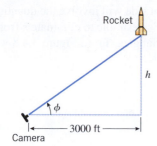

▲ Figure 3.4.6

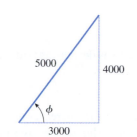

▲ Figure 3.4.7

is $d\phi/dt$, and the rate at which the rocket is rising is dh/dt. We want to find

$$\left.\frac{d\phi}{dt}\right|_{h=4000} \qquad \text{given that} \qquad \left.\frac{dh}{dt}\right|_{h=4000} = 880 \text{ ft/s}$$

From Figure 3.4.6 we see that

$$\tan\phi = \frac{h}{3000} \tag{5}$$

Differentiating both sides of (5) with respect to t yields

$$(\sec^2\phi)\frac{d\phi}{dt} = \frac{1}{3000}\frac{dh}{dt} \tag{6}$$

When $h = 4000$, it follows that

$$(\sec\phi)\big|_{h=4000} = \frac{5000}{3000} = \frac{5}{3}$$

(see Figure 3.4.7), so that from (6)

$$\left(\frac{5}{3}\right)^2 \left.\frac{d\phi}{dt}\right|_{h=4000} = \frac{1}{3000}\cdot 880 = \frac{22}{75}$$

$$\left.\frac{d\phi}{dt}\right|_{h=4000} = \frac{22}{75}\cdot\frac{9}{25} = \frac{66}{625} \approx 0.11 \text{ rad/s} \approx 6.05 \text{ deg/s} \blacktriangleleft$$

▶ **Example 5** Suppose that liquid is to be cleared of sediment by allowing it to drain through a conical filter that is 16 cm high and has a radius of 4 cm at the top (Figure 3.4.8). Suppose also that the liquid is forced out of the cone at a constant rate of 2 cm³/min.

(a) Do you think that the depth of the liquid will decrease at a constant rate? Give a verbal argument that justifies your conclusion.

(b) Find a formula that expresses the rate at which the depth of the liquid is changing in terms of the depth, and use that formula to determine whether your conclusion in part (a) is correct.

(c) At what rate is the depth of the liquid changing at the instant when the liquid in the cone is 8 cm deep?

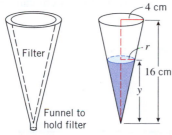

▲ Figure 3.4.8

Solution (a). For the volume of liquid to decrease by a *fixed amount*, it requires a greater decrease in depth when the cone is close to empty than when it is almost full (Figure 3.4.9). This suggests that for the volume to decrease at a constant rate, the depth must decrease at an increasing rate.

Solution (b). Let

$$t = \text{time elapsed from the initial observation (min)}$$
$$V = \text{volume of liquid in the cone at time } t \text{ (cm}^3)$$
$$y = \text{depth of the liquid in the cone at time } t \text{ (cm)}$$
$$r = \text{radius of the liquid surface at time } t \text{ (cm)}$$

(Figure 3.4.8). At each instant the rate at which the volume of liquid is changing is dV/dt, and the rate at which the depth is changing is dy/dt. We want to express dy/dt in terms of y given that dV/dt has a constant value of $dV/dt = -2$. (We must use a minus sign here because V *decreases* as t increases.)

From the formula for the volume of a cone, the volume V, the radius r, and the depth y are related by

$$V = \tfrac{1}{3}\pi r^2 y \tag{7}$$

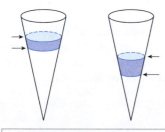

The same volume has drained, but the change in height is greater near the bottom than near the top.

▲ Figure 3.4.9

If we differentiate both sides of (7) with respect to t, the right side will involve the quantity dr/dt. Since we have no direct information about dr/dt, it is desirable to eliminate r from (7) before differentiating. This can be done using similar triangles. From Figure 3.4.8 we see that

$$\frac{r}{y} = \frac{4}{16} \quad \text{or} \quad r = \frac{1}{4}y$$

Substituting this expression in (7) gives

$$V = \frac{\pi}{48}y^3 \tag{8}$$

Differentiating both sides of (8) with respect to t we obtain

$$\frac{dV}{dt} = \frac{\pi}{48}\left(3y^2\frac{dy}{dt}\right)$$

or

$$\frac{dy}{dt} = \frac{16}{\pi y^2}\frac{dV}{dt} = \frac{16}{\pi y^2}(-2) = -\frac{32}{\pi y^2} \tag{9}$$

which expresses dy/dt in terms of y. The minus sign tells us that y is decreasing with time, and

$$\left|\frac{dy}{dt}\right| = \frac{32}{\pi y^2}$$

tells us how fast y is decreasing. From this formula we see that $|dy/dt|$ increases as y decreases, which confirms our conjecture in part (a) that the depth of the liquid decreases more quickly as the liquid drains through the filter.

Solution (c). The rate at which the depth is changing when the depth is 8 cm can be obtained from (9) with $y = 8$:

$$\frac{dy}{dt}\bigg|_{y=8} = -\frac{32}{\pi(8^2)} = -\frac{1}{2\pi} \approx -0.16 \text{ cm/min} \blacktriangleleft$$

✔ **QUICK CHECK EXERCISES 3.4**   *(See page 211 for answers.)*

1. If $A = x^2$ and $\dfrac{dx}{dt} = 3$, find $\dfrac{dA}{dt}\bigg|_{x=10}$.

2. If $A = x^2$ and $\dfrac{dA}{dt} = 3$, find $\dfrac{dx}{dt}\bigg|_{x=10}$.

3. A 10-foot ladder stands on a horizontal floor and leans against a vertical wall. Use x to denote the distance along the floor from the wall to the foot of the ladder, and use y to denote the distance along the wall from the floor to the top of the ladder. If the foot of the ladder is dragged away from the wall, find an equation that relates rates of change of x and y with respect to time.

4. Suppose that a block of ice in the shape of a right circular cylinder melts so that it retains its cylindrical shape. Find an equation that relates the rates of change of the volume (V), height (h), and radius (r) of the block of ice.

EXERCISE SET 3.4

1–4 Both x and y denote functions of t that are related by the given equation. Use this equation and the given derivative information to find the specified derivative. ■

1. Equation: $y = 3x + 5$.
   (a) Given that $dx/dt = 2$, find dy/dt when $x = 1$.
   (b) Given that $dy/dt = -1$, find dx/dt when $x = 0$.

2. Equation: $x + 4y = 3$.
   (a) Given that $dx/dt = 1$, find dy/dt when $x = 2$.
   (b) Given that $dy/dt = 4$, find dx/dt when $x = 3$.

3. Equation: $4x^2 + 9y^2 = 1$.
   (a) Given that $dx/dt = 3$, find dy/dt when
   $$(x, y) = \left(\frac{1}{2\sqrt{2}}, \frac{1}{3\sqrt{2}}\right).$$ *(cont.)*

(b) Given that $dy/dt = 8$, find dx/dt when
$(x, y) = \left(\frac{1}{3}, -\frac{\sqrt{5}}{9}\right)$.

4. Equation: $x^2 + y^2 = 2x + 4y$.
   (a) Given that $dx/dt = -5$, find dy/dt when
       $(x, y) = (3, 1)$.
   (b) Given that $dy/dt = 6$, find dx/dt when
       $(x, y) = (1 + \sqrt{2}, 2 + \sqrt{3})$.

FOCUS ON CONCEPTS

5. Let A be the area of a square whose sides have length
   x, and assume that x varies with the time t.
   (a) Draw a picture of the square with the labels A and
       x placed appropriately.
   (b) Write an equation that relates A and x.
   (c) Use the equation in part (b) to find an equation that
       relates dA/dt and dx/dt.
   (d) At a certain instant the sides are 3 ft long and in-
       creasing at a rate of 2 ft/min. How fast is the area
       increasing at that instant?

6. In parts (a)–(d), let A be the area of a circle of radius r,
   and assume that r increases with the time t.
   (a) Draw a picture of the circle with the labels A and r
       placed appropriately.
   (b) Write an equation that relates A and r.
   (c) Use the equation in part (b) to find an equation that
       relates dA/dt and dr/dt.
   (d) At a certain instant the radius is 5 cm and increasing
       at the rate of 2 cm/s. How fast is the area increasing
       at that instant?

7. Let V be the volume of a cylinder having height h and
   radius r, and assume that h and r vary with time.
   (a) How are dV/dt, dh/dt, and dr/dt related?
   (b) At a certain instant, the height is 6 in and increasing
       at 1 in/s, while the radius is 10 in and decreasing
       at 1 in/s. How fast is the volume changing at that
       instant? Is the volume increasing or decreasing at
       that instant?

8. Let l be the length of a diagonal of a rectangle whose
   sides have lengths x and y, and assume that x and y vary
   with time.
   (a) How are dl/dt, dx/dt, and dy/dt related?
   (b) If x increases at a constant rate of $\frac{1}{2}$ ft/s and y de-
       creases at a constant rate of $\frac{1}{4}$ ft/s, how fast is the
       size of the diagonal changing when $x = 3$ ft and
       $y = 4$ ft? Is the diagonal increasing or decreasing
       at that instant?

9. Let θ (in radians) be an acute angle in a right triangle,
   and let x and y, respectively, be the lengths of the sides
   adjacent to and opposite θ. Suppose also that x and y
   vary with time.
   (a) How are $d\theta/dt$, dx/dt, and dy/dt related?
   (b) At a certain instant, $x = 2$ units and is increasing at

1 unit/s, while $y = 2$ units and is decreasing at $\frac{1}{4}$
unit/s. How fast is θ changing at that instant? Is θ
increasing or decreasing at that instant?

10. Suppose that $z = x^3 y^2$, where both x and y are changing
    with time. At a certain instant when $x = 1$ and $y = 2$, x is
    decreasing at the rate of 2 units/s, and y is increasing at the
    rate of 3 units/s. How fast is z changing at this instant? Is
    z increasing or decreasing?

11. The minute hand of a certain clock is 4 in long. Starting
    from the moment when the hand is pointing straight up,
    how fast is the area of the sector that is swept out by the
    hand increasing at any instant during the next revolution of
    the hand?

12. A stone dropped into a still pond sends out a circular ripple
    whose radius increases at a constant rate of 3 ft/s. How
    rapidly is the area enclosed by the ripple increasing at the
    end of 10 s?

13. Oil spilled from a ruptured tanker spreads in a circle whose
    area increases at a constant rate of 6 mi^2/h. How fast is the
    radius of the spill increasing when the area is 9 mi^2?

14. A spherical balloon is inflated so that its volume is increas-
    ing at the rate of 3 ft^3/min. How fast is the diameter of the
    balloon increasing when the radius is 1 ft?

15. A spherical balloon is to be deflated so that its radius
    decreases at a constant rate of 15 cm/min. At what rate
    must air be removed when the radius is 9 cm?

16. A 17 ft ladder is leaning against a wall. If the bottom of the
    ladder is pulled along the ground away from the wall at a
    constant rate of 5 ft/s, how fast will the top of the ladder be
    moving down the wall when it is 8 ft above the ground?

17. A 13 ft ladder is leaning against a wall. If the top of the
    ladder slips down the wall at a rate of 2 ft/s, how fast will
    the foot be moving away from the wall when the top is 5 ft
    above the ground?

18. A 10 ft plank is leaning against a wall. If at a certain instant
    the bottom of the plank is 2 ft from the wall and is being
    pushed toward the wall at the rate of 6 in/s, how fast is the
    acute angle that the plank makes with the ground increasing?

19. A softball diamond is a square whose sides are 60 ft long.
    Suppose that a player running from first to second base has a
    speed of 25 ft/s at the instant when she is 10 ft from second
    base. At what rate is the player's distance from home plate
    changing at that instant?

20. A rocket, rising vertically, is tracked by a radar station that
    is on the ground 5 mi from the launchpad. How fast is the
    rocket rising when it is 4 mi high and its distance from the
    radar station is increasing at a rate of 2000 mi/h?

21. For the camera and rocket shown in Figure 3.4.5, at what rate
    is the camera-to-rocket distance changing when the rocket
    is 4000 ft up and rising vertically at 880 ft/s?

22. For the camera and rocket shown in Figure 3.4.5, at what rate is the rocket rising when the elevation angle is $\pi/4$ radians and increasing at a rate of 0.2 rad/s?

23. A satellite is in an elliptical orbit around the Earth. Its distance r (in miles) from the center of the Earth is given by

$$r = \frac{4995}{1 + 0.12\cos\theta}$$

where θ is the angle measured from the point on the orbit nearest the Earth's surface (see the accompanying figure).
   (a) Find the altitude of the satellite at *perigee* (the point nearest the surface of the Earth) and at *apogee* (the point farthest from the surface of the Earth). Use 3960 mi as the radius of the Earth.
   (b) At the instant when θ is 120°, the angle θ is increasing at the rate of 2.7°/min. Find the altitude of the satellite and the rate at which the altitude is changing at this instant. Express the rate in units of mi/min.

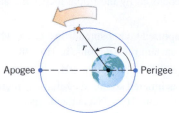

◀ **Figure Ex-23**

24. An aircraft is flying horizontally at a constant height of 4000 ft above a fixed observation point (see the accompanying figure). At a certain instant the angle of elevation θ is 30° and decreasing, and the speed of the aircraft is 300 mi/h.
   (a) How fast is θ decreasing at this instant? Express the result in units of deg/s.
   (b) How fast is the distance between the aircraft and the observation point changing at this instant? Express the result in units of ft/s. Use 1 mi = 5280 ft.

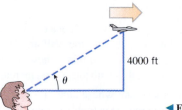

◀ **Figure Ex-24**

25. A conical water tank with vertex down has a radius of 10 ft at the top and is 24 ft high. If water flows into the tank at a rate of 20 ft³/min, how fast is the depth of the water increasing when the water is 16 ft deep?

26. Grain pouring from a chute at the rate of 8 ft³/min forms a conical pile whose height is always twice its radius. How fast is the height of the pile increasing at the instant when the pile is 6 ft high?

27. Sand pouring from a chute forms a conical pile whose height is always equal to the diameter. If the height increases at a constant rate of 5 ft/min, at what rate is sand pouring from the chute when the pile is 10 ft high?

28. Wheat is poured through a chute at the rate of 10 ft³/min and falls in a conical pile whose bottom radius is always half the altitude. How fast will the circumference of the base be increasing when the pile is 8 ft high?

29. An aircraft is climbing at a 30° angle to the horizontal. How fast is the aircraft gaining altitude if its speed is 500 mi/h?

30. A boat is pulled into a dock by means of a rope attached to a pulley on the dock (see the accompanying figure). The rope is attached to the bow of the boat at a point 10 ft below the pulley. If the rope is pulled through the pulley at a rate of 20 ft/min, at what rate will the boat be approaching the dock when 125 ft of rope is out?

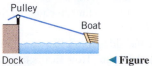

◀ **Figure Ex-30**

31. For the boat in Exercise 30, how fast must the rope be pulled if we want the boat to approach the dock at a rate of 12 ft/min at the instant when 125 ft of rope is out?

32. A man 6 ft tall is walking at the rate of 3 ft/s toward a streetlight 18 ft high (see the accompanying figure).
   (a) At what rate is his shadow length changing?
   (b) How fast is the tip of his shadow moving?

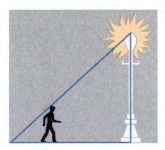

◀ **Figure Ex-32**

33. A beacon that makes one revolution every 10 s is located on a ship anchored 4 kilometers from a straight shoreline. How fast is the beam moving along the shoreline when it makes an angle of 45° with the shore?

34. An aircraft is flying at a constant altitude with a constant speed of 600 mi/h. An antiaircraft missile is fired on a straight line perpendicular to the flight path of the aircraft so that it will hit the aircraft at a point P (see the accompanying figure). At the instant the aircraft is 2 mi from the impact point P the missile is 4 mi from P and flying at 1200 mi/h. At that instant, how rapidly is the distance between missile and aircraft decreasing?

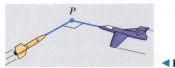

◀ **Figure Ex-34**

35. Solve Exercise 34 under the assumption that the angle between the flight paths is 120° instead of the assumption that the paths are perpendicular. [*Hint:* Use the law of cosines.]

36. A police helicopter is flying due north at 100 mi/h and at a constant altitude of $\frac{1}{2}$ mi. Below, a car is traveling west on a highway at 75 mi/h. At the moment the helicopter crosses over the highway the car is 2 mi east of the helicopter.
  (a) How fast is the distance between the car and helicopter changing at the moment the helicopter crosses the highway?
  (b) Is the distance between the car and helicopter increasing or decreasing at that moment?

37. A particle is moving along the curve whose equation is
$$\frac{xy^3}{1+y^2} = \frac{8}{5}$$
Assume that the x-coordinate is increasing at the rate of 6 units/s when the particle is at the point $(1, 2)$.
  (a) At what rate is the y-coordinate of the point changing at that instant?
  (b) Is the particle rising or falling at that instant?

38. A point P is moving along the curve whose equation is $y = \sqrt{x^3 + 17}$. When P is at $(2, 5)$, y is increasing at the rate of 2 units/s. How fast is x changing?

39. A point P is moving along the line whose equation is $y = 2x$. How fast is the distance between P and the point $(3, 0)$ changing at the instant when P is at $(3, 6)$ if x is decreasing at the rate of 2 units/s at that instant?

40. A point P is moving along the curve whose equation is $y = \sqrt{x}$. Suppose that x is increasing at the rate of 4 units/s when $x = 3$.
  (a) How fast is the distance between P and the point $(2, 0)$ changing at this instant?
  (b) How fast is the angle of inclination of the line segment from P to $(2, 0)$ changing at this instant?

41. A particle is moving along the curve $y = x/(x^2 + 1)$. Find all values of x at which the rate of change of x with respect to time is three times that of y. [Assume that dx/dt is never zero.]

42. A particle is moving along the curve $16x^2 + 9y^2 = 144$. Find all points (x, y) at which the rates of change of x and y with respect to time are equal. [Assume that dx/dt and dy/dt are never both zero at the same point.]

43. The *thin lens equation* in physics is
$$\frac{1}{s} + \frac{1}{S} = \frac{1}{f}$$
where s is the object distance from the lens, S is the image distance from the lens, and f is the focal length of the lens. Suppose that a certain lens has a focal length of 6 cm and that an object is moving toward the lens at the rate of 2 cm/s. How fast is the image distance changing at the instant when the object is 10 cm from the lens? Is the image moving away from the lens or toward the lens?

44. Water is stored in a cone-shaped reservoir (vertex down). Assuming the water evaporates at a rate proportional to the surface area exposed to the air, show that the depth of the water will decrease at a constant rate that does not depend on the dimensions of the reservoir.

45. A meteor enters the Earth's atmosphere and burns up at a rate that, at each instant, is proportional to its surface area. Assuming that the meteor is always spherical, show that the radius decreases at a constant rate.

46. On a certain clock the minute hand is 4 in long and the hour hand is 3 in long. How fast is the distance between the tips of the hands changing at 9 o'clock?

47. Coffee is poured at a uniform rate of 20 cm³/s into a cup whose inside is shaped like a truncated cone (see the accompanying figure). If the upper and lower radii of the cup are 4 cm and 2 cm and the height of the cup is 6 cm, how fast will the coffee level be rising when the coffee is halfway up? [*Hint:* Extend the cup downward to form a cone.]

◄ **Figure Ex-47**

✔ **QUICK CHECK ANSWERS 3.4**

1. 60 **2.** $\dfrac{3}{20}$ **3.** $x\dfrac{dx}{dt} + y\dfrac{dy}{dt} = 0$ **4.** $\dfrac{dV}{dt} = 2\pi r h\dfrac{dr}{dt} + \pi r^2\dfrac{dh}{dt}$

3.5   LOCAL LINEAR APPROXIMATION; DIFFERENTIALS

In this section we will show how derivatives can be used to approximate nonlinear functions by linear functions. Also, up to now we have been interpreting dy/dx as a single entity representing the derivative. In this section we will define the quantities dx and dy themselves, thereby allowing us to interpret dy/dx as an actual ratio.

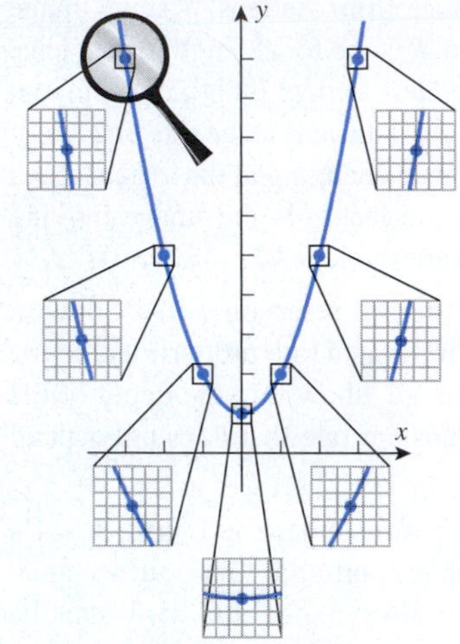

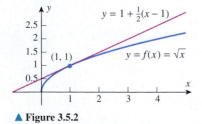

Magnifying portions of the graph of $y = x^2 + 1$

▲ **Figure 3.5.1**

Recall from Section 2.2 that if a function f is differentiable at x_0, then a sufficiently magnified portion of the graph of f centered at the point $P(x_0, f(x_0))$ takes on the appearance of a straight line segment. Figure 3.5.1 illustrates this at several points on the graph of $y = x^2 + 1$. For this reason, a function that is differentiable at x_0 is sometimes said to be *locally linear* at x_0.

The line that best approximates the graph of f in the vicinity of $P(x_0, f(x_0))$ is the tangent line to the graph of f at x_0, given by the equation

$$y = f(x_0) + f'(x_0)(x - x_0)$$

[see Formula (3) of Section 2.2]. Thus, for values of x near x_0 we can approximate values of $f(x)$ by

$$f(x) \approx f(x_0) + f'(x_0)(x - x_0) \tag{1}$$

This is called the *local linear approximation* of f at x_0. This formula can also be expressed in terms of the increment $\Delta x = x - x_0$ as

$$f(x_0 + \Delta x) \approx f(x_0) + f'(x_0)\Delta x \tag{2}$$

▶ **Example 1**

(a) Find the local linear approximation of $f(x) = \sqrt{x}$ at $x_0 = 1$.

(b) Use the local linear approximation obtained in part (a) to approximate $\sqrt{1.1}$, and compare your approximation to the result produced directly by a calculating utility.

Solution (a).   Since $f'(x) = 1/(2\sqrt{x})$, it follows from (1) that the local linear approximation of $\sqrt{x}$ at a point x_0 is

$$\sqrt{x} \approx \sqrt{x_0} + \frac{1}{2\sqrt{x_0}}(x - x_0)$$

Thus, the local linear approximation at $x_0 = 1$ is

$$\sqrt{x} \approx 1 + \tfrac{1}{2}(x - 1) \tag{3}$$

The graphs of $y = \sqrt{x}$ and the local linear approximation $y = 1 + \tfrac{1}{2}(x - 1)$ are shown in Figure 3.5.2.

Solution (b).   Applying (3) with $x = 1.1$ yields

$$\sqrt{1.1} \approx 1 + \tfrac{1}{2}(1.1 - 1) = 1.05$$

Since the tangent line $y = 1 + \tfrac{1}{2}(x - 1)$ in Figure 3.5.2 lies above the graph of $f(x) = \sqrt{x}$, we would expect this approximation to be slightly too large. This expectation is confirmed by the calculator approximation $\sqrt{1.1} \approx 1.04881$. ◀

$y = 1 + \tfrac{1}{2}(x - 1)$

$(1, 1)$

$y = f(x) = \sqrt{x}$

▲ **Figure 3.5.2**

Examples 1 and 2 illustrate important ideas and are not meant to suggest that you should use local linear approximations for computations that your calculating utility can perform. The main application of local linear approximation is in modeling problems where it is useful to replace complicated functions by simpler ones.

▶ **Example 2**

(a) Find the local linear approximation of $f(x) = \sin x$ at $x_0 = 0$.

(b) Use the local linear approximation obtained in part (a) to approximate $\sin 2°$, and compare your approximation to the result produced directly by your calculating device.

Solution (a). Since $f'(x) = \cos x$, it follows from (1) that the local linear approximation of $\sin x$ at a point x_0 is

$$\sin x \approx \sin x_0 + (\cos x_0)(x - x_0)$$

Thus, the local linear approximation at $x_0 = 0$ is

$$\sin x \approx \sin 0 + (\cos 0)(x - 0)$$

which simplifies to
$$\sin x \approx x \tag{4}$$

Solution (b). The variable x in (4) is in radian measure, so we must first convert $2°$ to radians before we can apply this approximation. Since

$$2° = 2\left(\frac{\pi}{180}\right) = \frac{\pi}{90} \approx 0.0349066 \text{ radian}$$

it follows from (4) that $\sin 2° \approx 0.0349066$. Comparing the two graphs in Figure 3.5.3, we would expect this approximation to be slightly larger than the exact value. The calculator approximation $\sin 2° \approx 0.0348995$ shows that this is indeed the case. ◀

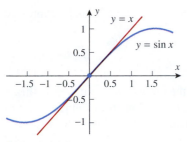

▲ **Figure 3.5.3**

■ **ERROR IN LOCAL LINEAR APPROXIMATIONS**

As a general rule, the accuracy of the local linear approximation to $f(x)$ at x_0 will deteriorate as x gets progressively farther from x_0. To illustrate this for the approximation $\sin x \approx x$ in Example 2, let us graph the function

$$E(x) = |\sin x - x|$$

which is the absolute value of the error in the approximation (Figure 3.5.4).

In Figure 3.5.4, the graph shows how the absolute error in the local linear approximation of $\sin x$ increases as x moves progressively farther from 0 in either the positive or negative direction. The graph also tells us that for values of x between the two vertical lines, the absolute error does not exceed 0.01. Thus, for example, we could use the local linear approximation $\sin x \approx x$ for all values of x in the interval $-0.35 < x < 0.35$ (radians) with confidence that the approximation is within ± 0.01 of the exact value.

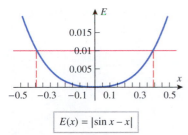

$$E(x) = |\sin x - x|$$

▲ **Figure 3.5.4**

■ **DIFFERENTIALS**

Newton and Leibniz each used a different notation when they published their discoveries of calculus, thereby creating a notational divide between Britain and the European continent that lasted for more than 50 years. The ***Leibniz notation*** dy/dx eventually prevailed because it suggests correct formulas in a natural way, the chain rule

$$\frac{dy}{dx} = \frac{dy}{du} \cdot \frac{du}{dx}$$

being a good example.

Up to now we have interpreted dy/dx as a single entity representing the derivative of y with respect to x; the symbols "dy" and "dx," which are called ***differentials***, have had no meanings attached to them. Our next goal is to define these symbols in such a way that dy/dx can be treated as an actual ratio. To do this, assume that f is differentiable at a point x, *define dx* to be an independent variable that can have any real value, and *define dy* by the formula

$$dy = f'(x)\,dx \tag{5}$$

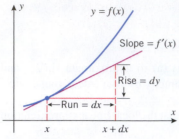

▲ **Figure 3.5.5**

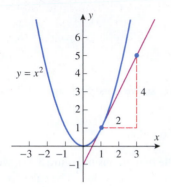

▲ **Figure 3.5.6**

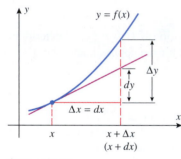

▲ **Figure 3.5.7**

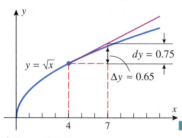

▲ **Figure 3.5.8**

If $dx \neq 0$, then we can divide both sides of (5) by dx to obtain

$$\frac{dy}{dx} = f'(x) \tag{6}$$

Thus, we have achieved our goal of defining dy and dx so their ratio is $f'(x)$. Formula (5) is said to express (6) in **differential form**.

To interpret (5) geometrically, note that $f'(x)$ is the slope of the tangent line to the graph of f at x. The differentials dy and dx can be viewed as a corresponding rise and run of this tangent line (Figure 3.5.5).

▶ **Example 3** Express the derivative with respect to x of $y = x^2$ in differential form, and discuss the relationship between dy and dx at $x = 1$.

Solution. The derivative of y with respect to x is $dy/dx = 2x$, which can be expressed in differential form as
$$dy = 2x \, dx$$

When $x = 1$ this becomes
$$dy = 2 \, dx$$

This tells us that if we travel along the tangent line to the curve $y = x^2$ at $x = 1$, then a change of dx units in x produces a change of $2 \, dx$ units in y. Thus, for example, a run of $dx = 2$ units produces a rise of $dy = 4$ units along the tangent line (Figure 3.5.6). ◀

It is important to understand the distinction between the increment Δy and the differential dy. To see the difference, let us assign the independent variables dx and Δx the same value, so $dx = \Delta x$. Then Δy represents the change in y that occurs when we start at x and travel *along the curve* $y = f(x)$ until we have moved $\Delta x \, (= dx)$ units in the x-direction, while dy represents the change in y that occurs if we start at x and travel *along the tangent line* until we have moved $dx \, (= \Delta x)$ units in the x-direction (Figure 3.5.7).

▶ **Example 4** Let $y = \sqrt{x}$. Find dy and Δy at $x = 4$ with $dx = \Delta x = 3$. Then make a sketch of $y = \sqrt{x}$, showing dy and Δy in the picture.

Solution. With $f(x) = \sqrt{x}$ we obtain
$$\Delta y = f(x + \Delta x) - f(x) = \sqrt{x + \Delta x} - \sqrt{x} = \sqrt{7} - \sqrt{4} \approx 0.65$$

If $y = \sqrt{x}$, then
$$\frac{dy}{dx} = \frac{1}{2\sqrt{x}}, \quad \text{so} \quad dy = \frac{1}{2\sqrt{x}} \, dx = \frac{1}{2\sqrt{4}}(3) = \frac{3}{4} = 0.75$$

Figure 3.5.8 shows the curve $y = \sqrt{x}$ together with dy and Δy. ◀

■ **LOCAL LINEAR APPROXIMATION FROM THE DIFFERENTIAL POINT OF VIEW**

Although Δy and dy are generally different, the differential dy will nonetheless be a good approximation of Δy provided $dx = \Delta x$ is close to 0. To see this, recall from Section 2.2 that
$$f'(x) = \lim_{\Delta x \to 0} \frac{\Delta y}{\Delta x}$$

It follows that if Δx is close to 0, then we will have $f'(x) \approx \Delta y / \Delta x$ or, equivalently,
$$\Delta y \approx f'(x) \Delta x$$

If we agree to let $dx = \Delta x$, then we can rewrite this as
$$\Delta y \approx f'(x) \, dx = dy \tag{7}$$

In words, this states that for values of dx near zero the differential dy closely approximates the increment Δy (Figure 3.5.7). But this is to be expected since the graph of the tangent line at x is the local linear approximation of the graph of f.

■ ERROR PROPAGATION

© Michael Newman/PhotoEdit
Real-world measurements inevitably have small errors.

In real-world applications, small errors in measured quantities will invariably occur. These measurement errors are of importance in scientific research—all scientific measurements come with measurement errors included. For example, your height might be measured as 170 ± 0.5 cm, meaning that your exact height lies somewhere between 169.5 and 170.5 cm. Researchers often must use these inexactly measured quantities to compute other quantities, thereby *propagating* the errors from the measured quantitites to the computed quantities. This phenomenon is called ***error propagation***. Researchers must be able to estimate errors in the computed quantities. Our goal is to show how to estimate these errors using local linear approximation and differentials. For this purpose, suppose

x_0 is the exact value of the quantity being measured
$y_0 = f(x_0)$ is the exact value of the quantity being computed
x is the measured value of x_0
$y = f(x)$ is the computed value of y

We define $dx\ (= \Delta x) = x - x_0$ to be the ***measurement error*** of x
$\Delta y = f(x) - f(x_0)$ to be the ***propagated error*** of y

It follows from (7) with x_0 replacing x that the propagated error Δy can be approximated by

$$\Delta y \approx dy = f'(x_0)\, dx \tag{8}$$

Unfortunately, there is a practical difficulty in applying this formula since the value of x_0 is unknown. (Keep in mind that only the measured value x is known to the researcher.) This being the case, it is standard practice in research to use the measured value x in place of x_0 in (8) and use the approximation

$$\Delta y \approx dy = f'(x)\, dx \tag{9}$$

for the propagated error.

> Note that measurement error is positive if the measured value is greater than the exact value and is negative if it is less than the exact value. The sign of the propagated error conveys similar information.

> Explain why an error estimate of at most $\pm\frac{1}{32}$ inch is reasonable for a ruler that is calibrated in sixteenths of an inch.

▶ **Example 5**   Suppose that the side of a square is measured with a ruler to be 10 inches with a measurement error of at most $\pm\frac{1}{32}$ in. Estimate the error in the computed area of the square.

Solution.   Let x denote the exact length of a side and y the exact area so that $y = x^2$. It follows from (9) with $f(x) = x^2$ that if dx is the measurement error, then the propagated error Δy can be approximated as

$$\Delta y \approx dy = 2x\, dx$$

Substituting the measured value $x = 10$ into this equation yields

$$dy = 20\, dx \tag{10}$$

But to say that the measurement error is at most $\pm\frac{1}{32}$ means that

$$-\frac{1}{32} \le dx \le \frac{1}{32}$$

Multiplying these inequalities through by 20 and applying (10) yields

$$20\left(-\tfrac{1}{32}\right) \le dy \le 20\left(\tfrac{1}{32}\right) \quad \text{or equivalently} \quad -\tfrac{5}{8} \le dy \le \tfrac{5}{8}$$

Thus, the propagated error in the area is estimated to be within $\pm\frac{5}{8}$ in^2. ◀

If the true value of a quantity is q and a measurement or calculation produces an error Δq, then $\Delta q/q$ is called the **relative error** in the measurement or calculation; when expressed as a percentage, $\Delta q/q$ is called the **percentage error**. As a practical matter, the true value q is usually unknown, so that the measured or calculated value of q is used instead; and the relative error is approximated by dq/q.

▶ **Example 6** The radius of a sphere is measured with a percentage error within $\pm 0.04\%$. Estimate the percentage error in the calculated volume of the sphere.

Solution. The volume V of a sphere is $V = \frac{4}{3}\pi r^3$, so

$$\frac{dV}{dr} = 4\pi r^2$$

from which it follows that $dV = 4\pi r^2 \, dr$. Thus, the relative error in V is approximately

$$\frac{dV}{V} = \frac{4\pi r^2 \, dr}{\frac{4}{3}\pi r^3} = 3\frac{dr}{r} \tag{11}$$

We are given that the relative error in the measured value of r is $\pm 0.04\%$, which means that

$$-0.0004 \leq \frac{dr}{r} \leq 0.0004$$

Multiplying these inequalities through by 3 and applying (11) yields

$$3(-0.0004) \leq \frac{dV}{V} \leq 3(0.0004) \quad \text{or equivalently} \quad -0.0012 \leq \frac{dV}{V} \leq 0.0012$$

Thus, we estimate the percentage error in the calculated value of V to be within $\pm 0.12\%$. ◀

> Formula (11) tells us that, as a rule of thumb, the percentage error in the computed volume of a sphere is approximately 3 times the percentage error in the measured value of its radius. As a rule of thumb, how is the percentage error in the computed area of a square related to the percentage error in the measured value of a side?

■ MORE NOTATION; DIFFERENTIAL FORMULAS

The symbol df is another common notation for the differential of a function $y = f(x)$. For example, if $f(x) = \sin x$, then we can write $df = \cos x \, dx$. We can also view the symbol "d" as an *operator* that acts on a function to produce the corresponding differential. For example, $d[x^2] = 2x \, dx$, $d[\sin x] = \cos x \, dx$, and so on. All of the general rules of differentiation then have corresponding differential versions:

DERIVATIVE FORMULA	DIFFERENTIAL FORMULA
$\dfrac{d}{dx}[c] = 0$	$d[c] = 0$
$\dfrac{d}{dx}[cf] = c\dfrac{df}{dx}$	$d[cf] = c\,df$
$\dfrac{d}{dx}[f + g] = \dfrac{df}{dx} + \dfrac{dg}{dx}$	$d[f + g] = df + dg$
$\dfrac{d}{dx}[fg] = f\dfrac{dg}{dx} + g\dfrac{df}{dx}$	$d[fg] = f\,dg + g\,df$
$\dfrac{d}{dx}\left[\dfrac{f}{g}\right] = \dfrac{g\dfrac{df}{dx} - f\dfrac{dg}{dx}}{g^2}$	$d\left[\dfrac{f}{g}\right] = \dfrac{g\,df - f\,dg}{g^2}$

For example,

$$\begin{aligned} d[x^2 \sin x] &= (x^2 \cos x + 2x \sin x)\,dx \\ &= x^2(\cos x \, dx) + (2x \, dx)\sin x \\ &= x^2 d[\sin x] + (\sin x)\,d[x^2] \end{aligned}$$

illustrates the differential version of the product rule.

✔ **QUICK CHECK EXERCISES 3.5**   *(See page 219 for answers.)*

1. The local linear approximation of f at x_0 uses the _____ line to the graph of $y = f(x)$ at $x = x_0$ to approximate values of _____ for values of x near _____.

2. Find an equation for the local linear approximation to $y = 5 - x^2$ at $x_0 = 2$.

3. Let $y = 5 - x^2$. Find dy and Δy at $x = 2$ with $dx = \Delta x = 0.1$.

4. The intensity of light from a light source is a function $I = f(x)$ of the distance x from the light source. Suppose that a small gemstone is measured to be 10 m from a light source, $f(10) = 0.2 \text{ W/m}^2$, and $f'(10) = -0.04 \text{ W/m}^3$. If the distance $x = 10$ m was obtained with a measurement error within ± 0.05 m, estimate the percentage error in the calculated intensity of the light on the gemstone.

EXERCISE SET 3.5    Graphing Utility

---

1. (a) Use Formula (1) to obtain the local linear approximation of x^3 at $x_0 = 1$.
   (b) Use Formula (2) to rewrite the approximation obtained in part (a) in terms of Δx.
   (c) Use the result obtained in part (a) to approximate $(1.02)^3$, and confirm that the formula obtained in part (b) produces the same result.

2. (a) Use Formula (1) to obtain the local linear approximation of $1/x$ at $x_0 = 2$.
   (b) Use Formula (2) to rewrite the approximation obtained in part (a) in terms of Δx.
   (c) Use the result obtained in part (a) to approximate $1/2.05$, and confirm that the formula obtained in part (b) produces the same result.

FOCUS ON CONCEPTS

3. (a) Find the local linear approximation of the function $f(x) = \sqrt{1+x}$ at $x_0 = 0$, and use it to approximate $\sqrt{0.9}$ and $\sqrt{1.1}$.
   (b) Graph f and its tangent line at x_0 together, and use the graphs to illustrate the relationship between the exact values and the approximations of $\sqrt{0.9}$ and $\sqrt{1.1}$.

4. A student claims that whenever a local linear approximation is used to approximate the square root of a number, the approximation is too large.
   (a) Write a few sentences that make the student's claim precise, and justify this claim geometrically.
   (b) Verify the student's claim algebraically using approximation (1).

5–10 Confirm that the stated formula is the local linear approximation at $x_0 = 0$. ■

5. $(1+x)^{15} \approx 1 + 15x$

6. $\dfrac{1}{\sqrt{1-x}} \approx 1 + \dfrac{1}{2}x$

7. $\tan x \approx x$

8. $\dfrac{1}{1+x} \approx 1 - x$

9. $e^x \approx 1 + x$

10. $\ln(1+x) \approx x$

11–16 Confirm that the stated formula is the local linear approximation of f at $x_0 = 1$, where $\Delta x = x - 1$. ■

11. $f(x) = x^4;\ (1 + \Delta x)^4 \approx 1 + 4\Delta x$

12. $f(x) = \sqrt{x};\ \sqrt{1 + \Delta x} \approx 1 + \tfrac{1}{2}\Delta x$

13. $f(x) = \dfrac{1}{2+x};\ \dfrac{1}{3 + \Delta x} \approx \dfrac{1}{3} - \dfrac{1}{9}\Delta x$

14. $f(x) = (4+x)^3;\ (5 + \Delta x)^3 \approx 125 + 75\Delta x$

15. $\tan^{-1} x;\ \tan^{-1}(1 + \Delta x) \approx \dfrac{\pi}{4} + \dfrac{1}{2}\Delta x$

16. $\sin^{-1}\left(\dfrac{x}{2}\right);\ \sin^{-1}\left(\dfrac{1}{2} + \dfrac{1}{2}\Delta x\right) \approx \dfrac{\pi}{6} + \dfrac{1}{\sqrt{3}}\Delta x$

~ **17–20** Confirm that the formula is the local linear approximation at $x_0 = 0$, and use a graphing utility to estimate an interval of x-values on which the error is at most ± 0.1. ■

17. $\sqrt{x+3} \approx \sqrt{3} + \dfrac{1}{2\sqrt{3}}x$

18. $\dfrac{1}{\sqrt{9-x}} \approx \dfrac{1}{3} + \dfrac{1}{54}x$

19. $\tan 2x \approx 2x$

20. $\dfrac{1}{(1+2x)^5} \approx 1 - 10x$

21. (a) Use the local linear approximation of $\sin x$ at $x_0 = 0$ obtained in Example 2 to approximate $\sin 1°$, and compare the approximation to the result produced directly by your calculating device.
   (b) How would you choose x_0 to approximate $\sin 44°$?
   (c) Approximate $\sin 44°$; compare the approximation to the result produced directly by your calculating device.

22. (a) Use the local linear approximation of $\tan x$ at $x_0 = 0$ to approximate $\tan 2°$, and compare the approximation to the result produced directly by your calculating device.
   (b) How would you choose x_0 to approximate $\tan 61°$?
   (c) Approximate $\tan 61°$; compare the approximation to the result produced directly by your calculating device.

23–31 Use an appropriate local linear approximation to estimate the value of the given quantity. ■

23. $(3.02)^4$ 24. $(1.97)^3$ 25. $\sqrt{65}$

26. $\sqrt{24}$ **27.** $\sqrt{80.9}$ **28.** $\sqrt{36.03}$

29. $\sin 0.1$ **30.** $\tan 0.2$ **31.** $\cos 31°$

32. $\ln(1.01)$ **33.** $\tan^{-1}(0.99)$

FOCUS ON CONCEPTS

34. The approximation $(1+x)^k \approx 1 + kx$ is commonly used by engineers for quick calculations.
  (a) Derive this result, and use it to make a rough estimate of $(1.001)^{37}$.
  (b) Compare your estimate to that produced directly by your calculating device.
  (c) If k is a positive integer, how is the approximation $(1+x)^k \approx 1+kx$ related to the expansion of $(1+x)^k$ using the binomial theorem?

35. Use the approximation $(1+x)^k \approx 1+kx$, along with some mental arithmetic to show that $\sqrt[3]{8.24} \approx 2.02$ and $4.08^{3/2} \approx 8.24$.

36. Referring to the accompanying figure, suppose that the angle of elevation of the top of the building, as measured from a point 500 ft from its base, is found to be $\theta = 6°$. Use an appropriate local linear approximation, along with some mental arithmetic to show that the building is about 52 ft high.

|←——— 500 ft ———→| ◄ **Figure Ex-36**

37. (a) Let $y = x^2$. Find dy and Δy at $x = 2$ with $dx = \Delta x = 1$.
  (b) Sketch the graph of $y = x^2$, showing dy and Δy in the picture.

38. (a) Let $y = x^3$. Find dy and Δy at $x = 1$ with $dx = \Delta x = 1$.
  (b) Sketch the graph of $y = x^3$, showing dy and Δy in the picture.

39–42 Find formulas for dy and Δy. ■

39. $y = x^3$ **40.** $y = 8x - 4$

41. $y = x^2 - 2x + 1$ **42.** $y = \sin x$

43–46 Find the differential dy. ■

43. (a) $y = 4x^3 - 7x^2$ (b) $y = x \cos x$

44. (a) $y = 1/x$ (b) $y = 5 \tan x$

45. (a) $y = x\sqrt{1-x}$ (b) $y = (1+x)^{-17}$

46. (a) $y = \dfrac{1}{x^3 - 1}$ (b) $y = \dfrac{1-x^3}{2-x}$

47–50 True–False Determine whether the statement is true or false. Explain your answer. ■

47. A differential dy is defined to be a very small change in y.

48. The error in approximation (2) is the same as the error in approximation (7).

49. A local linear approximation to a function can never be identically equal to the function.

50. A local linear approximation to a nonconstant function can never be constant.

51–54 Use the differential dy to approximate Δy when x changes as indicated. ■

51. $y = \sqrt{3x-2}$; from $x = 2$ to $x = 2.03$

52. $y = \sqrt{x^2+8}$; from $x = 1$ to $x = 0.97$

53. $y = \dfrac{x}{x^2+1}$; from $x = 2$ to $x = 1.96$

54. $y = x\sqrt{8x+1}$; from $x = 3$ to $x = 3.05$

55. The side of a square is measured to be 10 ft, with a possible error of ± 0.1 ft.
  (a) Use differentials to estimate the error in the calculated area.
  (b) Estimate the percentage errors in the side and the area.

56. The side of a cube is measured to be 25 cm, with a possible error of ± 1 cm.
  (a) Use differentials to estimate the error in the calculated volume.
  (b) Estimate the percentage errors in the side and volume.

57. The hypotenuse of a right triangle is known to be 10 in exactly, and one of the acute angles is measured to be 30°, with a possible error of $\pm 1°$.
  (a) Use differentials to estimate the errors in the sides opposite and adjacent to the measured angle.
  (b) Estimate the percentage errors in the sides.

58. One side of a right triangle is known to be 25 cm exactly. The angle opposite to this side is measured to be 60°, with a possible error of $\pm 0.5°$.
  (a) Use differentials to estimate the errors in the adjacent side and the hypotenuse.
  (b) Estimate the percentage errors in the adjacent side and hypotenuse.

59. The electrical resistance R of a certain wire is given by $R = k/r^2$, where k is a constant and r is the radius of the wire. Assuming that the radius r has a possible error of $\pm 5\%$, use differentials to estimate the percentage error in R. (Assume k is exact.)

60. A 12-foot ladder leaning against a wall makes an angle θ with the floor. If the top of the ladder is h feet up the wall, express h in terms of θ and then use dh to estimate the change in h if θ changes from 60° to 59°.

61. The area of a right triangle with a hypotenuse of H is calculated using the formula $A = \frac{1}{4}H^2 \sin 2\theta$, where θ is one of the acute angles. Use differentials to approximate the error in calculating A if $H = 4$ cm (exactly) and θ is measured to be 30°, with a possible error of $\pm 15'$.

62. The side of a square is measured with a possible percentage error of ±1%. Use differentials to estimate the percentage error in the area.

63. The side of a cube is measured with a possible percentage error of ±2%. Use differentials to estimate the percentage error in the volume.

64. The volume of a sphere is to be computed from a measured value of its radius. Estimate the maximum permissible percentage error in the measurement if the percentage error in the volume must be kept within ±3%. ($V = \frac{4}{3}\pi r^3$ is the volume of a sphere of radius r.)

65. The area of a circle is to be computed from a measured value of its diameter. Estimate the maximum permissible percentage error in the measurement if the percentage error in the area must be kept within ±1%.

66. A steel cube with 1-inch sides is coated with 0.01 inch of copper. Use differentials to estimate the volume of copper in the coating. [*Hint:* Let ΔV be the change in the volume of the cube.]

67. A metal rod 15 cm long and 5 cm in diameter is to be covered (except for the ends) with insulation that is 0.1 cm thick. Use differentials to estimate the volume of insulation. [*Hint:* Let ΔV be the change in volume of the rod.]

68. The time required for one complete oscillation of a pendulum is called its *period*. If L is the length of the pendulum and the oscillation is small, then the period is given by $P = 2\pi\sqrt{L/g}$, where g is the constant acceleration due to gravity. Use differentials to show that the percentage error in P is approximately half the percentage error in L.

69. If the temperature T of a metal rod of length L is changed by an amount ΔT, then the length will change by the amount $\Delta L = \alpha L \Delta T$, where α is called the **coefficient of linear expansion**. For moderate changes in temperature α is taken as constant.
 (a) Suppose that a rod 40 cm long at 20°C is found to be 40.006 cm long when the temperature is raised to 30°C. Find α.
 (b) If an aluminum pole is 180 cm long at 15°C, how long is the pole if the temperature is raised to 40°C? [Take $\alpha = 2.3 \times 10^{-5}/°C$.]

70. If the temperature T of a solid or liquid of volume V is changed by an amount ΔT, then the volume will change by the amount $\Delta V = \beta V \Delta T$, where β is called the **coefficient of volume expansion**. For moderate changes in temperature β is taken as constant. Suppose that a tank truck loads 4000 gallons of ethyl alcohol at a temperature of 35°C and delivers its load sometime later at a temperature of 15°C. Using $\beta = 7.5 \times 10^{-4}/°C$ for ethyl alcohol, find the number of gallons delivered.

71. Writing Explain why the local linear approximation of a function value is equivalent to the use of a differential to approximate a change in the function.

72. Writing The local linear approximation
$$\sin x \approx x$$
is known as the *small angle approximation* and has both practical and theoretical applications. Do some research on some of these applications, and write a short report on the results of your investigations.

✔ **QUICK CHECK ANSWERS 3.5**

1. tangent; $f(x)$; x_0 **2.** $y = 1 + (-4)(x - 2)$ or $y = -4x + 9$ **3.** $dy = -0.4, \Delta y = -0.41$ **4.** within ±1%

3.6 L'HÔPITAL'S RULE; INDETERMINATE FORMS

In this section we will discuss a general method for using derivatives to find limits. This method will enable us to establish limits with certainty that earlier in the text we were only able to conjecture using numerical or graphical evidence. The method that we will discuss in this section is an extremely powerful tool that is used internally by many computer programs to calculate limits of various types.

■ **INDETERMINATE FORMS OF TYPE 0/0**
Recall that a limit of the form
$$\lim_{x \to a} \frac{f(x)}{g(x)} \tag{1}$$
in which $f(x) \to 0$ and $g(x) \to 0$ as $x \to a$ is called an **indeterminate form of type 0/0**. Some examples encountered earlier in the text are
$$\lim_{x \to 1} \frac{x^2 - 1}{x - 1} = 2, \quad \lim_{x \to 0} \frac{\sin x}{x} = 1, \quad \lim_{x \to 0} \frac{1 - \cos x}{x} = 0$$

The first limit was obtained algebraically by factoring the numerator and canceling the common factor of $x - 1$, and the second two limits were obtained using geometric methods. However, there are many indeterminate forms for which neither algebraic nor geometric methods will produce the limit, so we need to develop a more general method.

To motivate such a method, suppose that (1) is an indeterminate form of type $0/0$ in which f' and g' are continuous at $x = a$ and $g'(a) \neq 0$. Since f and g can be closely approximated by their local linear approximations near a, it is reasonable to expect that

$$\lim_{x \to a} \frac{f(x)}{g(x)} = \lim_{x \to a} \frac{f(a) + f'(a)(x - a)}{g(a) + g'(a)(x - a)} \tag{2}$$

Since we are assuming that f' and g' are continuous at $x = a$, we have

$$\lim_{x \to a} f'(x) = f'(a) \quad \text{and} \quad \lim_{x \to a} g'(x) = g'(a)$$

and since the differentiability of f and g at $x = a$ implies the continuity of f and g at $x = a$, we have

$$f(a) = \lim_{x \to a} f(x) = 0 \quad \text{and} \quad g(a) = \lim_{x \to a} g(x) = 0$$

Thus, we can rewrite (2) as

$$\lim_{x \to a} \frac{f(x)}{g(x)} = \lim_{x \to a} \frac{f'(a)(x - a)}{g'(a)(x - a)} = \lim_{x \to a} \frac{f'(a)}{g'(a)} = \lim_{x \to a} \frac{f'(x)}{g'(x)} \tag{3}$$

This result, called **L'Hôpital's rule**, converts the given indeterminate form into a limit involving derivatives that is often easier to evaluate.

Although we motivated (3) by assuming that f and g have continuous derivatives at $x = a$ and that $g'(a) \neq 0$, the result is true under less stringent conditions and is also valid for one-sided limits and limits at $+\infty$ and $-\infty$. The proof of the following precise statement of L'Hôpital's rule is omitted.

3.6.1 **THEOREM** (*L'Hôpital's Rule for Form* $0/0$) *Suppose that f and g are differentiable functions on an open interval containing $x = a$, except possibly at $x = a$, and that*

$$\lim_{x \to a} f(x) = 0 \quad \text{and} \quad \lim_{x \to a} g(x) = 0$$

If $\lim_{x \to a} [f'(x)/g'(x)]$ *exists, or if this limit is* $+\infty$ *or* $-\infty$, *then*

$$\lim_{x \to a} \frac{f(x)}{g(x)} = \lim_{x \to a} \frac{f'(x)}{g'(x)}$$

Moreover, this statement is also true in the case of a limit as $x \to a^-$, $x \to a^+$, $x \to -\infty$, *or as* $x \to +\infty$.

WARNING

Note that in L'Hôpital's rule the numerator and denominator are differentiated individually. This is *not* the same as differentiating $f(x)/g(x)$.

In the examples that follow we will apply L'Hôpital's rule using the following three-step process:

Applying L'Hôpital's Rule

Step 1. Check that the limit of $f(x)/g(x)$ is an indeterminate form of type $0/0$.

Step 2. Differentiate f and g separately.

Step 3. Find the limit of $f'(x)/g'(x)$. If this limit is finite, $+\infty$, or $-\infty$, then it is equal to the limit of $f(x)/g(x)$.

▶ **Example 1**   Find the limit

$$\lim_{x \to 2} \frac{x^2 - 4}{x - 2}$$

using L'Hôpital's rule, and check the result by factoring.

Solution.   The numerator and denominator have a limit of 0, so the limit is an indeterminate form of type 0/0. Applying L'Hôpital's rule yields

$$\lim_{x \to 2} \frac{x^2 - 4}{x - 2} = \lim_{x \to 2} \frac{\dfrac{d}{dx}[x^2 - 4]}{\dfrac{d}{dx}[x - 2]} = \lim_{x \to 2} \frac{2x}{1} = 4$$

This agrees with the computation

> The limit in Example 1 can be interpreted as the limit form of a certain derivative. Use that derivative to evaluate the limit.

$$\lim_{x \to 2} \frac{x^2 - 4}{x - 2} = \lim_{x \to 2} \frac{(x - 2)(x + 2)}{x - 2} = \lim_{x \to 2} (x + 2) = 4 \ ◀$$

▶ **Example 2**   In each part confirm that the limit is an indeterminate form of type 0/0, and evaluate it using L'Hôpital's rule.

(a) $\displaystyle\lim_{x \to 0} \frac{\sin 2x}{x}$ (b) $\displaystyle\lim_{x \to \pi/2} \frac{1 - \sin x}{\cos x}$ (c) $\displaystyle\lim_{x \to 0} \frac{e^x - 1}{x^3}$

(d) $\displaystyle\lim_{x \to 0^-} \frac{\tan x}{x^2}$ (e) $\displaystyle\lim_{x \to 0} \frac{1 - \cos x}{x^2}$ (f) $\displaystyle\lim_{x \to +\infty} \frac{x^{-4/3}}{\sin(1/x)}$

Solution (a).   The numerator and denominator have a limit of 0, so the limit is an indeterminate form of type 0/0. Applying L'Hôpital's rule yields

Applying L'Hôpital's rule to limits that are not indeterminate forms can produce incorrect results. For example, the computation

$$\lim_{x \to 0} \frac{x + 6}{x + 2} = \lim_{x \to 0} \frac{\dfrac{d}{dx}[x + 6]}{\dfrac{d}{dx}[x + 2]}$$

$$= \lim_{x \to 0} \frac{1}{1} = 1$$

is *not valid*, since the limit is not an indeterminate form. The correct result is

$$\lim_{x \to 0} \frac{x + 6}{x + 2} = \frac{0 + 6}{0 + 2} = 3$$

$$\lim_{x \to 0} \frac{\sin 2x}{x} = \lim_{x \to 0} \frac{\dfrac{d}{dx}[\sin 2x]}{\dfrac{d}{dx}[x]} = \lim_{x \to 0} \frac{2\cos 2x}{1} = 2$$

Observe that this result agrees with that obtained by substitution in Example 4(b) of Section 1.6.

Solution (b).   The numerator and denominator have a limit of 0, so the limit is an indeterminate form of type 0/0. Applying L'Hôpital's rule yields

$$\lim_{x \to \pi/2} \frac{1 - \sin x}{\cos x} = \lim_{x \to \pi/2} \frac{\dfrac{d}{dx}[1 - \sin x]}{\dfrac{d}{dx}[\cos x]} = \lim_{x \to \pi/2} \frac{-\cos x}{-\sin x} = \frac{0}{-1} = 0$$

Guillaume François Antoine de L'Hôpital (1661–1704) French mathematician. L'Hôpital, born to parents of the French high nobility, held the title of Marquis de Sainte-Mesme Comte d'Autrement. He showed mathematical talent quite early and at age 15 solved a difficult problem about cycloids posed by Pascal. As a young man he served briefly as a cavalry officer, but resigned because of near-sightedness. In his own time he gained fame as the author of the first textbook ever published on differential calculus, *L'Analyse des* *Infiniment Petits pour l'Intelligence des Lignes Courbes* (1696). L'Hôpital's rule appeared for the first time in that book. Actually, L'Hôpital's rule and most of the material in the calculus text were due to John Bernoulli, who was L'Hôpital's teacher. L'Hôpital dropped his plans for a book on integral calculus when Leibniz informed him that he intended to write such a text. L'Hôpital was apparently generous and personable, and his many contacts with major mathematicians provided the vehicle for disseminating major discoveries in calculus throughout Europe.

Solution (c). The numerator and denominator have a limit of 0, so the limit is an indeterminate form of type $0/0$. Applying L'Hôpital's rule yields

$$\lim_{x \to 0} \frac{e^x - 1}{x^3} = \lim_{x \to 0} \frac{\dfrac{d}{dx}[e^x - 1]}{\dfrac{d}{dx}[x^3]} = \lim_{x \to 0} \frac{e^x}{3x^2} = +\infty$$

Solution (d). The numerator and denominator have a limit of 0, so the limit is an indeterminate form of type $0/0$. Applying L'Hôpital's rule yields

$$\lim_{x \to 0^-} \frac{\tan x}{x^2} = \lim_{x \to 0^-} \frac{\sec^2 x}{2x} = -\infty$$

Solution (e). The numerator and denominator have a limit of 0, so the limit is an indeterminate form of type $0/0$. Applying L'Hôpital's rule yields

$$\lim_{x \to 0} \frac{1 - \cos x}{x^2} = \lim_{x \to 0} \frac{\sin x}{2x}$$

Since the new limit is another indeterminate form of type $0/0$, we apply L'Hôpital's rule again:

$$\lim_{x \to 0} \frac{1 - \cos x}{x^2} = \lim_{x \to 0} \frac{\sin x}{2x} = \lim_{x \to 0} \frac{\cos x}{2} = \frac{1}{2}$$

Solution (f). The numerator and denominator have a limit of 0, so the limit is an indeterminate form of type $0/0$. Applying L'Hôpital's rule yields

$$\lim_{x \to +\infty} \frac{x^{-4/3}}{\sin(1/x)} = \lim_{x \to +\infty} \frac{-\frac{4}{3}x^{-7/3}}{(-1/x^2)\cos(1/x)} = \lim_{x \to +\infty} \frac{\frac{4}{3}x^{-1/3}}{\cos(1/x)} = \frac{0}{1} = 0 \blacktriangleleft$$

■ INDETERMINATE FORMS OF TYPE ∞/∞

When we want to indicate that the limit (or a one-sided limit) of a function is $+\infty$ or $-\infty$ without being specific about the sign, we will say that the limit is ∞. For example,

$$\lim_{x \to a^+} f(x) = \infty \quad \text{means} \quad \lim_{x \to a^+} f(x) = +\infty \quad \text{or} \quad \lim_{x \to a^+} f(x) = -\infty$$

$$\lim_{x \to +\infty} f(x) = \infty \quad \text{means} \quad \lim_{x \to +\infty} f(x) = +\infty \quad \text{or} \quad \lim_{x \to +\infty} f(x) = -\infty$$

$$\lim_{x \to a} f(x) = \infty \quad \text{means} \quad \lim_{x \to a^+} f(x) = \pm\infty \quad \text{and} \quad \lim_{x \to a^-} f(x) = \pm\infty$$

The limit of a ratio, $f(x)/g(x)$, in which the numerator has limit ∞ and the denominator has limit ∞ is called an *indeterminate form of type ∞/∞*. The following version of L'Hôpital's rule, which we state without proof, can often be used to evaluate limits of this type.

3.6.2 **THEOREM** (*L'Hôpital's Rule for Form ∞/∞*) *Suppose that f and g are differentiable functions on an open interval containing $x = a$, except possibly at $x = a$, and that*

$$\lim_{x \to a} f(x) = \infty \quad \text{and} \quad \lim_{x \to a} g(x) = \infty$$

If $\lim_{x \to a} [f'(x)/g'(x)]$ exists, or if this limit is $+\infty$ or $-\infty$, then

$$\lim_{x \to a} \frac{f(x)}{g(x)} = \lim_{x \to a} \frac{f'(x)}{g'(x)}$$

Moreover, this statement is also true in the case of a limit as $x \to a^-$, $x \to a^+$, $x \to -\infty$, or as $x \to +\infty$.

▶ **Example 3** In each part confirm that the limit is an indeterminate form of type ∞/∞ and apply L'Hôpital's rule.

$$\text{(a)} \lim_{x \to +\infty} \frac{x}{e^x} \qquad \text{(b)} \lim_{x \to 0^+} \frac{\ln x}{\csc x}$$

Solution (a). The numerator and denominator both have a limit of $+\infty$, so we have an indeterminate form of type ∞/∞. Applying L'Hôpital's rule yields

$$\lim_{x \to +\infty} \frac{x}{e^x} = \lim_{x \to +\infty} \frac{1}{e^x} = 0$$

Solution (b). The numerator has a limit of $-\infty$ and the denominator has a limit of $+\infty$, so we have an indeterminate form of type ∞/∞. Applying L'Hôpital's rule yields

$$\lim_{x \to 0^+} \frac{\ln x}{\csc x} = \lim_{x \to 0^+} \frac{1/x}{-\csc x \cot x} \tag{4}$$

This last limit is again an indeterminate form of type ∞/∞. Moreover, any additional applications of L'Hôpital's rule will yield powers of $1/x$ in the numerator and expressions involving $\csc x$ and $\cot x$ in the denominator; thus, repeated application of L'Hôpital's rule simply produces new indeterminate forms. We must try something else. The last limit in (4) can be rewritten as

$$\lim_{x \to 0^+} \left(-\frac{\sin x}{x} \tan x \right) = -\lim_{x \to 0^+} \frac{\sin x}{x} \cdot \lim_{x \to 0^+} \tan x = -(1)(0) = 0$$

Thus,

$$\lim_{x \to 0^+} \frac{\ln x}{\csc x} = 0 \ \blacktriangleleft$$

■ **ANALYZING THE GROWTH OF EXPONENTIAL FUNCTIONS USING L'HÔPITAL'S RULE**
If n is any positive integer, then $x^n \to +\infty$ as $x \to +\infty$. Such integer powers of x are sometimes used as "measuring sticks" to describe how rapidly other functions grow. For example, we know that $e^x \to +\infty$ as $x \to +\infty$ and that the growth of e^x is very rapid (Table 0.5.5); however, the growth of x^n is also rapid when n is a high power, so it is reasonable to ask whether high powers of x grow more or less rapidly than e^x. One way to investigate this is to examine the behavior of the ratio x^n/e^x as $x \to +\infty$. For example, Figure 3.6.1a shows the graph of $y = x^5/e^x$. This graph suggests that $x^5/e^x \to 0$ as $x \to +\infty$, and this implies that the growth of the function e^x is sufficiently rapid that its values eventually overtake those of x^5 and force the ratio toward zero. Stated informally, "e^x eventually grows more rapidly than x^5." The same conclusion could have been reached by putting e^x on top and examining the behavior of e^x/x^5 as $x \to +\infty$ (Figure 3.6.1b). In this case the values of e^x eventually overtake those of x^5 and force the ratio toward $+\infty$. More generally, we can use L'Hôpital's rule to show that e^x *eventually grows more rapidly than any positive integer power of x*, that is,

$$\lim_{x \to +\infty} \frac{x^n}{e^x} = 0 \quad \text{and} \quad \lim_{x \to +\infty} \frac{e^x}{x^n} = +\infty \tag{5–6}$$

Both limits are indeterminate forms of type ∞/∞ that can be evaluated using L'Hôpital's rule. For example, to establish (5), we will need to apply L'Hôpital's rule n times. For this purpose, observe that successive differentiations of x^n reduce the exponent by 1 each time, thus producing a constant for the nth derivative. For example, the successive derivatives

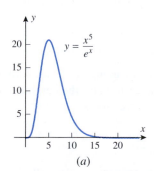

(a)

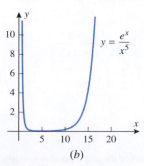

(b)

▲ **Figure 3.6.1**

of x^3 are $3x^2$, $6x$, and 6. In general, the nth derivative of x^n is $n(n-1)(n-2)\cdots 1 = n!$ (verify).[*] Thus, applying L'Hôpital's rule n times to (5) yields

$$\lim_{x \to +\infty} \frac{x^n}{e^x} = \lim_{x \to +\infty} \frac{n!}{e^x} = 0$$

Limit (6) can be established similarly.

■ **INDETERMINATE FORMS OF TYPE $0 \cdot \infty$**

Thus far we have discussed indeterminate forms of type $0/0$ and ∞/∞. However, these are not the only possibilities; in general, the limit of an expression that has one of the forms

$$\frac{f(x)}{g(x)}, \quad f(x) \cdot g(x), \quad f(x)^{g(x)}, \quad f(x) - g(x), \quad f(x) + g(x)$$

is called an *indeterminate form* if the limits of $f(x)$ and $g(x)$ individually exert conflicting influences on the limit of the entire expression. For example, the limit

$$\lim_{x \to 0^+} x \ln x$$

is an *indeterminate form of type $0 \cdot \infty$* because the limit of the first factor is 0, the limit of the second factor is $-\infty$, and these two limits exert conflicting influences on the product. On the other hand, the limit

$$\lim_{x \to +\infty} [\sqrt{x}(1 - x^2)]$$

is not an indeterminate form because the first factor has a limit of $+\infty$, the second factor has a limit of $-\infty$, and these influences work together to produce a limit of $-\infty$ for the product.

Indeterminate forms of type $0 \cdot \infty$ can sometimes be evaluated by rewriting the product as a ratio, and then applying L'Hôpital's rule for indeterminate forms of type $0/0$ or ∞/∞.

WARNING

It is tempting to argue that an indeterminate form of type $0 \cdot \infty$ has value 0 since "zero times anything is zero." However, this is fallacious since $0 \cdot \infty$ is not a product of numbers, but rather a statement about limits. For example, here are two indeterminate forms of type $0 \cdot \infty$ whose limits are *not* zero:

$$\lim_{x \to 0} \left(x \cdot \frac{1}{x} \right) = \lim_{x \to 0} 1 = 1$$

$$\lim_{x \to 0^+} \left(\sqrt{x} \cdot \frac{1}{x} \right) = \lim_{x \to 0^+} \left(\frac{1}{\sqrt{x}} \right)$$

$$= +\infty$$

▶ **Example 4**   Evaluate

(a) $\displaystyle\lim_{x \to 0^+} x \ln x$ (b) $\displaystyle\lim_{x \to \pi/4} (1 - \tan x) \sec 2x$

Solution (a).   The factor x has a limit of 0 and the factor $\ln x$ has a limit of $-\infty$, so the stated problem is an indeterminate form of type $0 \cdot \infty$. There are two possible approaches: we can rewrite the limit as

$$\lim_{x \to 0^+} \frac{\ln x}{1/x} \quad \text{or} \quad \lim_{x \to 0^+} \frac{x}{1/\ln x}$$

the first being an indeterminate form of type ∞/∞ and the second an indeterminate form of type $0/0$. However, the first form is the preferred initial choice because the derivative of $1/x$ is less complicated than the derivative of $1/\ln x$. That choice yields

$$\lim_{x \to 0^+} x \ln x = \lim_{x \to 0^+} \frac{\ln x}{1/x} = \lim_{x \to 0^+} \frac{1/x}{-1/x^2} = \lim_{x \to 0^+} (-x) = 0$$

Solution (b).   The stated problem is an indeterminate form of type $0 \cdot \infty$. We will convert it to an indeterminate form of type $0/0$:

$$\lim_{x \to \pi/4} (1 - \tan x) \sec 2x = \lim_{x \to \pi/4} \frac{1 - \tan x}{1/\sec 2x} = \lim_{x \to \pi/4} \frac{1 - \tan x}{\cos 2x}$$

$$= \lim_{x \to \pi/4} \frac{-\sec^2 x}{-2 \sin 2x} = \frac{-2}{-2} = 1 \blacktriangleleft$$

---

[*]Recall that for $n \geq 1$ the expression $n!$, read ***n-factorial***, denotes the product of the first n positive integers.

■ **INDETERMINATE FORMS OF TYPE** $\infty - \infty$

A limit problem that leads to one of the expressions

$$(+\infty) - (+\infty), \quad (-\infty) - (-\infty),$$
$$(+\infty) + (-\infty), \quad (-\infty) + (+\infty)$$

is called an *indeterminate form of type* $\infty - \infty$. Such limits are indeterminate because the two terms exert conflicting influences on the expression: one pushes it in the positive direction and the other pushes it in the negative direction. However, limit problems that lead to one of the expressions

$$(+\infty) + (+\infty), \quad (+\infty) - (-\infty),$$
$$(-\infty) + (-\infty), \quad (-\infty) - (+\infty)$$

are not indeterminate, since the two terms work together (those on the top produce a limit of $+\infty$ and those on the bottom produce a limit of $-\infty$).

Indeterminate forms of type $\infty - \infty$ can sometimes be evaluated by combining the terms and manipulating the result to produce an indeterminate form of type $0/0$ or ∞/∞.

▶ **Example 5** Evaluate $\lim\limits_{x \to 0^+} \left(\dfrac{1}{x} - \dfrac{1}{\sin x} \right)$.

Solution. Both terms have a limit of $+\infty$, so the stated problem is an indeterminate form of type $\infty - \infty$. Combining the two terms yields

$$\lim_{x \to 0^+} \left(\frac{1}{x} - \frac{1}{\sin x} \right) = \lim_{x \to 0^+} \frac{\sin x - x}{x \sin x}$$

which is an indeterminate form of type $0/0$. Applying L'Hôpital's rule twice yields

$$\lim_{x \to 0^+} \frac{\sin x - x}{x \sin x} = \lim_{x \to 0^+} \frac{\cos x - 1}{\sin x + x \cos x}$$
$$= \lim_{x \to 0^+} \frac{-\sin x}{\cos x + \cos x - x \sin x} = \frac{0}{2} = 0 \quad ◀$$

■ **INDETERMINATE FORMS OF TYPE** $0^0, \infty^0, 1^\infty$

Limits of the form

$$\lim f(x)^{g(x)}$$

can give rise to *indeterminate forms of the types* $0^0, \infty^0,$ *and* 1^∞. (The interpretations of these symbols should be clear.) For example, the limit

$$\lim_{x \to 0^+} (1 + x)^{1/x}$$

whose value we know to be e [see Formula (1) of Section 3.2] is an indeterminate form of type 1^∞. It is indeterminate because the expressions $1 + x$ and $1/x$ exert two conflicting influences: the first approaches 1, which drives the expression toward 1, and the second approaches $+\infty$, which drives the expression toward $+\infty$.

Indeterminate forms of types $0^0, \infty^0,$ and 1^∞ can sometimes be evaluated by first introducing a dependent variable

$$y = f(x)^{g(x)}$$

and then computing the limit of $\ln y$. Since

$$\ln y = \ln[f(x)^{g(x)}] = g(x) \cdot \ln[f(x)]$$

the limit of $\ln y$ will be an indeterminate form of type $0 \cdot \infty$ (verify), which can be evaluated by methods we have already studied. Once the limit of $\ln y$ is known, it is a straightforward matter to determine the limit of $y = f(x)^{g(x)}$, as we will illustrate in the next example.

▶ **Example 6** Find $\lim\limits_{x \to 0} (1 + \sin x)^{1/x}$.

Solution. As discussed above, we begin by introducing a dependent variable

$$y = (1 + \sin x)^{1/x}$$

and taking the natural logarithm of both sides:

$$\ln y = \ln(1 + \sin x)^{1/x} = \frac{1}{x} \ln(1 + \sin x) = \frac{\ln(1 + \sin x)}{x}$$

Thus,

$$\lim_{x \to 0} \ln y = \lim_{x \to 0} \frac{\ln(1 + \sin x)}{x}$$

which is an indeterminate form of type $0/0$, so by L'Hôpital's rule

$$\lim_{x \to 0} \ln y = \lim_{x \to 0} \frac{\ln(1 + \sin x)}{x} = \lim_{x \to 0} \frac{(\cos x)/(1 + \sin x)}{1} = 1$$

Since we have shown that $\ln y \to 1$ as $x \to 0$, the continuity of the exponential function implies that $e^{\ln y} \to e^1$ as $x \to 0$, and this implies that $y \to e$ as $x \to 0$. Thus,

$$\lim_{x \to 0} (1 + \sin x)^{1/x} = e \quad ◀$$

✔**QUICK CHECK EXERCISES 3.6**   *(See page 228 for answers.)*

1. In each part, does L'Hôpital's rule apply to the given limit?

   (a) $\lim\limits_{x \to 1} \dfrac{2x - 2}{x^3 + x - 2}$ (b) $\lim\limits_{x \to 0} \dfrac{\cos x}{x}$

   (c) $\lim\limits_{x \to 0} \dfrac{e^{2x} - 1}{\tan x}$

2. Evaluate each of the limits in Quick Check Exercise 1.

3. Using L'Hôpital's rule, $\lim\limits_{x \to +\infty} \dfrac{e^x}{500x^2} = \underline{\qquad}$.

EXERCISE SET 3.6   Graphing Utility   CAS

1–2 Evaluate the given limit without using L'Hôpital's rule, and then check that your answer is correct using L'Hôpital's rule. ■

1. (a) $\lim\limits_{x \to 2} \dfrac{x^2 - 4}{x^2 + 2x - 8}$ (b) $\lim\limits_{x \to +\infty} \dfrac{2x - 5}{3x + 7}$

2. (a) $\lim\limits_{x \to 0} \dfrac{\sin x}{\tan x}$ (b) $\lim\limits_{x \to 1} \dfrac{x^2 - 1}{x^3 - 1}$

3–6 True–False Determine whether the statement is true or false. Explain your answer. ■

3. L'Hôpital's rule does not apply to $\lim\limits_{x \to -\infty} \dfrac{\ln x}{x}$.

4. For any polynomial $p(x)$, $\lim\limits_{x \to +\infty} \dfrac{p(x)}{e^x} = 0$.

5. If n is chosen sufficiently large, then $\lim\limits_{x \to +\infty} \dfrac{(\ln x)^n}{x} = +\infty$.

6. $\lim\limits_{x \to 0^+} (\sin x)^{1/x} = 0$

7–45 Find the limits. ■

7. $\lim\limits_{x \to 0} \dfrac{e^x - 1}{\sin x}$

8. $\lim\limits_{x \to 0} \dfrac{\sin 2x}{\sin 5x}$

9. $\lim\limits_{\theta \to 0} \dfrac{\tan \theta}{\theta}$

10. $\lim\limits_{t \to 0} \dfrac{te^t}{1 - e^t}$

11. $\lim\limits_{x \to \pi^+} \dfrac{\sin x}{x - \pi}$

12. $\lim\limits_{x \to 0^+} \dfrac{\sin x}{x^2}$

13. $\lim\limits_{x \to +\infty} \dfrac{\ln x}{x}$

14. $\lim\limits_{x \to +\infty} \dfrac{e^{3x}}{x^2}$

15. $\lim\limits_{x \to 0^+} \dfrac{\cot x}{\ln x}$

16. $\lim\limits_{x \to 0^+} \dfrac{1 - \ln x}{e^{1/x}}$

17. $\lim\limits_{x \to +\infty} \dfrac{x^{100}}{e^x}$

18. $\lim\limits_{x \to 0^+} \dfrac{\ln(\sin x)}{\ln(\tan x)}$

19. $\lim\limits_{x \to 0} \dfrac{\sin^{-1} 2x}{x}$

20. $\lim\limits_{x \to 0} \dfrac{x - \tan^{-1} x}{x^3}$

21. $\lim\limits_{x \to +\infty} xe^{-x}$

22. $\lim\limits_{x \to \pi^-} (x - \pi) \tan \tfrac{1}{2}x$

23. $\lim\limits_{x \to +\infty} x \sin \dfrac{\pi}{x}$

24. $\lim\limits_{x \to 0^+} \tan x \ln x$

25. $\lim\limits_{x \to \pi/2^-} \sec 3x \cos 5x$

26. $\lim\limits_{x \to \pi} (x - \pi) \cot x$

27. $\lim\limits_{x \to +\infty} (1 - 3/x)^x$

28. $\lim\limits_{x \to 0} (1 + 2x)^{-3/x}$

29. $\lim\limits_{x \to 0} (e^x + x)^{1/x}$

30. $\lim\limits_{x \to +\infty} (1 + a/x)^{bx}$

31. $\lim\limits_{x \to 1} (2 - x)^{\tan[(\pi/2)x]}$

32. $\lim\limits_{x \to +\infty} [\cos(2/x)]^{x^2}$

33. $\lim\limits_{x \to 0} (\csc x - 1/x)$

34. $\lim\limits_{x \to 0} \left(\dfrac{1}{x^2} - \dfrac{\cos 3x}{x^2} \right)$

35. $\lim\limits_{x \to +\infty} (\sqrt{x^2 + x} - x)$

36. $\lim\limits_{x \to 0} \left(\dfrac{1}{x} - \dfrac{1}{e^x - 1} \right)$

37. $\lim\limits_{x \to +\infty} [x - \ln(x^2 + 1)]$

38. $\lim\limits_{x \to +\infty} [\ln x - \ln(1 + x)]$

39. $\lim\limits_{x \to 0^+} x^{\sin x}$

40. $\lim\limits_{x \to 0^+} (e^{2x} - 1)^x$

41. $\lim\limits_{x \to 0^+} \left[-\dfrac{1}{\ln x} \right]^x$

42. $\lim\limits_{x \to +\infty} x^{1/x}$

43. $\lim\limits_{x \to +\infty} (\ln x)^{1/x}$

44. $\lim\limits_{x \to 0^+} (-\ln x)^x$

45. $\lim\limits_{x \to \pi/2^-} (\tan x)^{(\pi/2) - x}$

46. Show that for any positive integer n

(a) $\lim\limits_{x \to +\infty} \dfrac{\ln x}{x^n} = 0$ (b) $\lim\limits_{x \to +\infty} \dfrac{x^n}{\ln x} = +\infty$.

FOCUS ON CONCEPTS

47. (a) Find the error in the following calculation:

$$\lim_{x \to 1} \frac{x^3 - x^2 + x - 1}{x^3 - x^2} = \lim_{x \to 1} \frac{3x^2 - 2x + 1}{3x^2 - 2x}$$
$$= \lim_{x \to 1} \frac{6x - 2}{6x - 2} = 1$$

(b) Find the correct limit.

48. (a) Find the error in the following calculation:

$$\lim_{x \to 2} \frac{e^{3x^2 - 12x + 12}}{x^4 - 16} = \lim_{x \to 2} \frac{(6x - 12)e^{3x^2 - 12x + 12}}{4x^3} = 0$$

(b) Find the correct limit.

49–52 Make a conjecture about the limit by graphing the function involved with a graphing utility; then check your conjecture using L'Hôpital's rule. ■

49. $\lim\limits_{x \to +\infty} \dfrac{\ln(\ln x)}{\sqrt{x}}$

50. $\lim\limits_{x \to 0^+} x^x$

51. $\lim\limits_{x \to 0^+} (\sin x)^{3/\ln x}$

52. $\lim\limits_{x \to (\pi/2)^-} \dfrac{4 \tan x}{1 + \sec x}$

53–56 Make a conjecture about the equations of horizontal asymptotes, if any, by graphing the equation with a graphing utility; then check your answer using L'Hôpital's rule. ■

53. $y = \ln x - e^x$

54. $y = x - \ln(1 + 2e^x)$

55. $y = (\ln x)^{1/x}$

56. $y = \left(\dfrac{x + 1}{x + 2} \right)^x$

57. Limits of the type

$$0/\infty, \quad \infty/0, \quad 0^\infty, \quad \infty \cdot \infty, \quad +\infty + (+\infty),$$
$$+\infty - (-\infty), \quad -\infty + (-\infty), \quad -\infty - (+\infty)$$

are *not* indeterminate forms. Find the following limits by inspection.

(a) $\lim\limits_{x \to 0^+} \dfrac{x}{\ln x}$

(b) $\lim\limits_{x \to +\infty} \dfrac{x^3}{e^{-x}}$

(c) $\lim\limits_{x \to (\pi/2)^-} (\cos x)^{\tan x}$

(d) $\lim\limits_{x \to 0^+} (\ln x) \cot x$

(e) $\lim\limits_{x \to 0^+} \left(\dfrac{1}{x} - \ln x \right)$

(f) $\lim\limits_{x \to -\infty} (x + x^3)$

58. There is a myth that circulates among beginning calculus students which states that all indeterminate forms of types 0^0, ∞^0, and 1^∞ have value 1 because "anything to the zero power is 1" and "1 to any power is 1." The fallacy is that 0^0, ∞^0, and 1^∞ are not powers of numbers, but rather descriptions of limits. The following examples, which were suggested by Prof. Jack Staib of Drexel University, show that such indeterminate forms can have any positive real value:

(a) $\lim\limits_{x \to 0^+} [x^{(\ln a)/(1 + \ln x)}] = a$ (form 0^0)

(b) $\lim\limits_{x \to +\infty} [x^{(\ln a)/(1 + \ln x)}] = a$ (form ∞^0)

(c) $\lim\limits_{x \to 0} [(x + 1)^{(\ln a)/x}] = a$ (form 1^∞).

Verify these results.

59–62 Verify that L'Hôpital's rule is of no help in finding the limit; then find the limit, if it exists, by some other method. ■

59. $\lim\limits_{x \to +\infty} \dfrac{x + \sin 2x}{x}$

60. $\lim\limits_{x \to +\infty} \dfrac{2x - \sin x}{3x + \sin x}$

61. $\lim\limits_{x \to +\infty} \dfrac{x(2 + \sin 2x)}{x + 1}$

62. $\lim\limits_{x \to +\infty} \dfrac{x(2 + \sin x)}{x^2 + 1}$

63. The accompanying schematic diagram represents an electrical circuit consisting of an electromotive force that produces a voltage V, a resistor with resistance R, and an inductor with inductance L. It is shown in electrical circuit theory that if the voltage is first applied at time $t = 0$, then the current I flowing through the circuit at time t is given by

$$I = \frac{V}{R}(1 - e^{-Rt/L})$$

What is the effect on the current at a fixed time t if the resistance approaches 0 (i.e., $R \to 0^+$)?

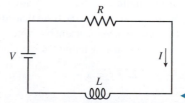

◀ **Figure Ex-63**

64. (a) Show that $\lim\limits_{x \to \pi/2} (\pi/2 - x) \tan x = 1$.

(b) Show that

$$\lim_{x \to \pi/2} \left(\frac{1}{\pi/2 - x} - \tan x \right) = 0$$

(c) It follows from part (b) that the approximation

$$\tan x \approx \frac{1}{\pi/2 - x}$$

should be good for values of x near $\pi/2$. Use a calculator to find $\tan x$ and $1/(\pi/2 - x)$ for $x = 1.57$; compare the results.

C **65.** (a) Use a CAS to show that if k is a positive constant, then

$$\lim_{x \to +\infty} x(k^{1/x} - 1) = \ln k$$

(b) Confirm this result using L'Hôpital's rule. [*Hint:* Express the limit in terms of $t = 1/x$.]

(c) If n is a positive integer, then it follows from part (a) with $x = n$ that the approximation

$$n(\sqrt[n]{k} - 1) \approx \ln k$$

should be good when n is large. Use this result and the square root key on a calculator to approximate the values of $\ln 0.3$ and $\ln 2$ with $n = 1024$, then compare the values obtained with values of the logarithms generated directly from the calculator. [*Hint:* The nth roots for which n is a power of 2 can be obtained as successive square roots.]

66. Find all values of k and l such that

$$\lim_{x \to 0} \frac{k + \cos lx}{x^2} = -4$$

FOCUS ON CONCEPTS

 67. Let $f(x) = x^2 \sin(1/x)$.

(a) Are the limits $\lim_{x \to 0^+} f(x)$ and $\lim_{x \to 0^-} f(x)$ indeterminate forms?

(b) Use a graphing utility to generate the graph of f, and use the graph to make conjectures about the limits in part (a).

(c) Use the Squeezing Theorem (1.6.4) to confirm that your conjectures in part (b) are correct.

68. (a) Explain why L'Hôpital's rule does not apply to the problem

$$\lim_{x \to 0} \frac{x^2 \sin(1/x)}{\sin x}$$

(b) Find the limit.

69. Find $\displaystyle\lim_{x \to 0^+} \frac{x \sin(1/x)}{\sin x}$ if it exists.

70. Suppose that functions f and g are differentiable at $x = a$ and that $f(a) = g(a) = 0$. If $g'(a) \neq 0$, show that

$$\lim_{x \to a} \frac{f(x)}{g(x)} = \frac{f'(a)}{g'(a)}$$

without using L'Hôpital's rule. [*Hint:* Divide the numerator and denominator of $f(x)/g(x)$ by $x - a$ and use the definitions for $f'(a)$ and $g'(a)$.]

71. Writing Were we to use L'Hôpital's rule to evaluate either

$$\lim_{x \to 0} \frac{\sin x}{x} \quad \text{or} \quad \lim_{x \to +\infty} \left(1 + \frac{1}{x}\right)^x$$

we could be accused of circular reasoning. Explain why.

72. Writing Exercise 58 shows that the indeterminate forms 0^0 and ∞^0 can assume any positive real value. However, it is often the case that these indeterminate forms have value 1. Read the article "Indeterminate Forms of Exponential Type" by John Baxley and Elmer Hayashi in the June–July 1978 issue of *The American Mathematical Monthly*, and write a short report on why this is the case.

✔ QUICK CHECK ANSWERS 3.6

1. (a) yes (b) no (c) yes   **2.** (a) $\frac{1}{2}$ (b) does not exist (c) 2   **3.** $+\infty$

CHAPTER 3 REVIEW EXERCISES   Graphing Utility

1–2 (a) Find dy/dx by differentiating implicitly. (b) Solve the equation for y as a function of x, and find dy/dx from that equation. (c) Confirm that the two results are consistent by expressing the derivative in part (a) as a function of x alone. ■

1. $x^3 + xy - 2x = 1$ **2.** $xy = x - y$

3–6 Find dy/dx by implicit differentiation. ■

3. $\dfrac{1}{y} + \dfrac{1}{x} = 1$ **4.** $x^3 - y^3 = 6xy$

5. $\sec(xy) = y$ **6.** $x^2 = \dfrac{\cot y}{1 + \csc y}$

7–8 Find d^2y/dx^2 by implicit differentiation. ■

7. $3x^2 - 4y^2 = 7$ **8.** $2xy - y^2 = 3$

9. Use implicit differentiation to find the slope of the tangent line to the curve $y = x \tan(\pi y/2)$, $x > 0$, $y > 0$ (*the quadratrix of Hippias*) at the point $\left(\frac{1}{2}, \frac{1}{2}\right)$.

10. At what point(s) is the tangent line to the curve $y^2 = 2x^3$ perpendicular to the line $4x - 3y + 1 = 0$?

11. Prove that if P and Q are two distinct points on the rotated ellipse $x^2 + xy + y^2 = 4$ such that P, Q, and the origin are collinear, then the tangent lines to the ellipse at P and Q are parallel.

12. Find the coordinates of the point in the first quadrant at which the tangent line to the curve $x^3 - xy + y^3 = 0$ is parallel to the x-axis.

13. Find the coordinates of the point in the first quadrant at which the tangent line to the curve $x^3 - xy + y^3 = 0$ is parallel to the y-axis.

14. Use implicit differentiation to show that the equation of the tangent line to the curve $y^2 = kx$ at (x_0, y_0) is

$$y_0 y = \tfrac{1}{2} k(x + x_0)$$

15–16 Find dy/dx by first using algebraic properties of the natural logarithm function. ■

15. $y = \ln \left(\dfrac{(x + 1)(x + 2)^2}{(x + 3)^3 (x + 4)^4} \right)$ **16.** $y = \ln \left(\dfrac{\sqrt{x}\,\sqrt[3]{x + 1}}{\sin x \sec x} \right)$

17–34 Find dy/dx. ■

17. $y = \ln 2x$ **18.** $y = (\ln x)^2$

19. $y = \sqrt[3]{\ln x + 1}$ **20.** $y = \ln(\sqrt[3]{x + 1})$

21. $y = \log(\ln x)$ **22.** $y = \dfrac{1 + \log x}{1 - \log x}$

23. $y = \ln(x^{3/2}\sqrt{1 + x^4})$ **24.** $y = \ln \left(\dfrac{\sqrt{x} \cos x}{1 + x^2} \right)$

25. $y = e^{\ln(x^2 + 1)}$ **26.** $y = \ln \left(\dfrac{1 + e^x + e^{2x}}{1 - e^{3x}} \right)$

27. $y = 2xe^{\sqrt{x}}$ **28.** $y = \dfrac{a}{1 + be^{-x}}$

29. $y = \dfrac{1}{\pi} \tan^{-1} 2x$ **30.** $y = 2^{\sin^{-1} x}$

31. $y = x^{(e^x)}$ **32.** $y = (1 + x)^{1/x}$

33. $y = \sec^{-1}(2x + 1)$ **34.** $y = \sqrt{\cos^{-1} x^2}$

35–36 Find dy/dx using logarithmic differentiation. ■

35. $y = \dfrac{x^3}{\sqrt{x^2 + 1}}$ **36.** $y = \sqrt[3]{\dfrac{x^2 - 1}{x^2 + 1}}$

37. (a) Make a conjecture about the shape of the graph of $y = \tfrac{1}{2} x - \ln x$, and draw a rough sketch.
     (b) Check your conjecture by graphing the equation over the interval $0 < x < 5$ with a graphing utility.
     (c) Show that the slopes of the tangent lines to the curve at $x = 1$ and $x = e$ have opposite signs.
     (d) What does part (c) imply about the existence of a horizontal tangent line to the curve? Explain.
     (e) Find the exact x-coordinates of all horizontal tangent lines to the curve.

38. Recall from Section 0.5 that the loudness β of a sound in decibels (dB) is given by $\beta = 10 \log(I/I_0)$, where I is the intensity of the sound in watts per square meter (W/m^2) and I_0 is a constant that is approximately the intensity of a sound at the threshold of human hearing. Find the rate of change of β with respect to I at the point where
     (a) $I/I_0 = 10$ (b) $I/I_0 = 100$ (c) $I/I_0 = 1000$.

39. A particle is moving along the curve $y = x \ln x$. Find all values of x at which the rate of change of y with respect to time is three times that of x. [Assume that dx/dt is never zero.]

40. Find the equation of the tangent line to the graph of $y = \ln(5 - x^2)$ at $x = 2$.

41. Find the value of b so that the line $y = x$ is tangent to the graph of $y = \log_b x$. Confirm your result by graphing both $y = x$ and $y = \log_b x$ in the same coordinate system.

42. In each part, find the value of k for which the graphs of $y = f(x)$ and $y = \ln x$ share a common tangent line at their point of intersection. Confirm your result by graphing $y = f(x)$ and $y = \ln x$ in the same coordinate system.
     (a) $f(x) = \sqrt{x} + k$ (b) $f(x) = k\sqrt{x}$

43. If f and g are inverse functions and f is differentiable on its domain, must g be differentiable on its domain? Give a reasonable informal argument to support your answer.

44. In each part, find $(f^{-1})'(x)$ using Formula (2) of Section 3.3, and check your answer by differentiating f^{-1} directly.
     (a) $f(x) = 3/(x + 1)$ (b) $f(x) = \sqrt{e^x}$

45. Find a point on the graph of $y = e^{3x}$ at which the tangent line passes through the origin.

46. Show that the rate of change of $y = 5000e^{1.07x}$ is proportional to y.

47. Show that the rate of change of $y = 3^{2x} 5^{7x}$ is proportional to y.

48. The equilibrium constant k of a balanced chemical reaction changes with the absolute temperature T according to the law

$$k = k_0 \exp \left(-\dfrac{q(T - T_0)}{2 T_0 T} \right)$$

where k_0, q, and T_0 are constants. Find the rate of change of k with respect to T.

49. Show that the function $y = e^{ax} \sin bx$ satisfies

$$y'' - 2ay' + (a^2 + b^2)y = 0$$

for any real constants a and b.

50. Show that the function $y = \tan^{-1} x$ satisfies

$$y'' = -2 \sin y \cos^3 y$$

51. Suppose that the population of deer on an island is modeled by the equation

$$P(t) = \dfrac{95}{5 - 4e^{-t/4}}$$

where $P(t)$ is the number of deer t weeks after an initial observation at time $t = 0$.
     (a) Use a graphing utility to graph the function $P(t)$.
     (b) In words, explain what happens to the population over time. Check your conclusion by finding $\lim_{t \to +\infty} P(t)$.
     (c) In words, what happens to the *rate* of population growth over time? Check your conclusion by graphing $P'(t)$.

52. In each part, find each limit by interpreting the expression as an appropriate derivative.
     (a) $\lim\limits_{h \to 0} \dfrac{(1 + h)^{\pi} - 1}{h}$ (b) $\lim\limits_{x \to e} \dfrac{1 - \ln x}{(x - e) \ln x}$

53. Suppose that $\lim f(x) = \pm\infty$ and $\lim g(x) = \pm\infty$. In each of the four possible cases, state whether $\lim[f(x) - g(x)]$ is an indeterminate form, and give a reasonable informal argument to support your answer.

54. (a) Under what conditions will a limit of the form
$$\lim_{x \to a}[f(x)/g(x)]$$
be an indeterminate form?

(b) If $\lim_{x \to a} g(x) = 0$, must $\lim_{x \to a}[f(x)/g(x)]$ be an indeterminate form? Give some examples to support your answer.

55–58 Evaluate the given limit. ■

55. $\lim\limits_{x \to +\infty} (e^x - x^2)$

56. $\lim\limits_{x \to 1} \sqrt{\dfrac{\ln x}{x^4 - 1}}$

57. $\lim\limits_{x \to 0} \dfrac{x^2 e^x}{\sin^2 3x}$

58. $\lim\limits_{x \to 0} \dfrac{a^x - 1}{x}, \quad a > 0$

59. An oil slick on a lake is surrounded by a floating circular containment boom. As the boom is pulled in, the circular containment area shrinks. If the boom is pulled in at the rate of 5 m/min, at what rate is the containment area shrinking when the containment area has a diameter of 100 m?

60. The hypotenuse of a right triangle is growing at a constant rate of a centimeters per second and one leg is decreasing t at a constant rate of b centimeters per second. How fast is the acute angle between the hypotenuse and the other leg changing at the instant when both legs are 1 cm?

61. In each part, use the given information to find Δx, Δy, and dy.
(a) $y = 1/(x - 1)$; x decreases from 2 to 1.5.
(b) $y = \tan x$; x increases from $-\pi/4$ to 0.
(c) $y = \sqrt{25 - x^2}$; x increases from 0 to 3.

62. Use an appropriate local linear approximation to estimate the value of $\cot 46°$, and compare your answer to the value obtained with a calculating device.

63. The base of the Great Pyramid at Giza is a square that is 230 m on each side.
(a) As illustrated in the accompanying figure, suppose that an archaeologist standing at the center of a side measures the angle of elevation of the apex to be $\phi = 51°$ with an error of $\pm 0.5°$. What can the archaeologist reasonably say about the height of the pyramid?
(b) Use differentials to estimate the allowable error in the elevation angle that will ensure that the error in calculating the height is at most ± 5 m.

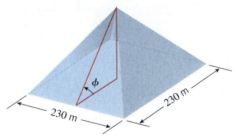

▲ **Figure Ex-63**

CHAPTER 3 MAKING CONNECTIONS

In these exercises we explore an application of exponential functions to radioactive decay, and we consider another approach to computing the derivative of the natural exponential function.

1. Consider a simple model of radioactive decay. We assume that given any quantity of a radioactive element, the fraction of the quantity that decays over a period of time will be a constant that depends on only the particular element and the length of the time period. We choose a time parameter $-\infty < t < +\infty$ and let $A = A(t)$ denote the amount of the element remaining at time t. We also choose units of measure such that the initial amount of the element is $A(0) = 1$, and we let $b = A(1)$ denote the amount at time $t = 1$. Prove that the function $A(t)$ has the following properties.

(a) $A(-t) = \dfrac{1}{A(t)}$ [*Hint:* For $t > 0$, you can interpret $A(t)$ as the fraction of any given amount that remains after a time period of length t.]

(b) $A(s + t) = A(s) \cdot A(t)$ [*Hint:* First consider positive s and t. For the other cases use the property in part (a).]

(c) If n is any nonzero integer, then
$$A\left(\frac{1}{n}\right) = (A(1))^{1/n} = b^{1/n}$$

(d) If m and n are integers with $n \neq 0$, then
$$A\left(\frac{m}{n}\right) = (A(1))^{m/n} = b^{m/n}$$

(e) Assuming that $A(t)$ is a continuous function of t, then $A(t) = b^t$. [*Hint:* Prove that if two continuous functions agree on the set of rational numbers, then they are equal.]

(f) If we replace the assumption that $A(0) = 1$ by the condition $A(0) = A_0$, prove that $A = A_0 b^t$.

2. Refer to Figure 1.3.4.

  (a) Make the substitution $h = 1/x$ and conclude that

$$(1 + h)^{1/h} < e < (1 - h)^{-1/h} \quad \text{for } h > 0$$

and

$$(1 - h)^{-1/h} < e < (1 + h)^{1/h} \quad \text{for } h < 0$$

  (b) Use the inequalities in part (a) and the Squeezing Theorem to prove that

$$\lim_{h \to 0} \frac{e^h - 1}{h} = 1$$

  (c) Explain why the limit in part (b) confirms Figure 0.5.4.

  (d) Use the limit in part (b) to prove that

$$\frac{d}{dx}(e^x) = e^x$$

Stone/Getty Images

4

THE DERIVATIVE IN GRAPHING AND APPLICATIONS

Derivatives can help to find the most cost-effective location for an offshore oil-drilling rig.

In this chapter we will study various applications of the derivative. For example, we will use methods of calculus to analyze functions and their graphs. In the process, we will show how calculus and graphing utilities, working together, can provide most of the important information about the behavior of functions. Another important application of the derivative will be in the solution of optimization problems. For example, if time is the main consideration in a problem, we might be interested in finding the quickest way to perform a task, and if cost is the main consideration, we might be interested in finding the least expensive way to perform a task. Mathematically, optimization problems can be reduced to finding the largest or smallest value of a function on some interval, and determining where the largest or smallest value occurs. Using the derivative, we will develop the mathematical tools necessary for solving such problems. We will also use the derivative to study the motion of a particle moving along a line, and we will show how the derivative can help us to approximate solutions of equations.

4.1 ANALYSIS OF FUNCTIONS I: INCREASE, DECREASE, AND CONCAVITY

Although graphing utilities are useful for determining the general shape of a graph, many problems require more precision than graphing utilities are capable of producing. The purpose of this section is to develop mathematical tools that can be used to determine the exact shape of a graph and the precise locations of its key features.

■ INCREASING AND DECREASING FUNCTIONS
The terms *increasing*, *decreasing*, and *constant* are used to describe the behavior of a function as we travel left to right along its graph. For example, the function graphed in Figure 4.1.1 can be described as increasing to the left of $x = 0$, decreasing from $x = 0$ to $x = 2$, increasing from $x = 2$ to $x = 4$, and constant to the right of $x = 4$.

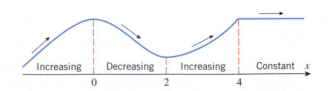

▶ **Figure 4.1.1**

The following definition, which is illustrated in Figure 4.1.2, expresses these intuitive ideas precisely.

The definitions of "increasing," "decreasing," and "constant" describe the behavior of a function on an *interval* and not at a point. In particular, it is not inconsistent to say that the function in Figure 4.1.1 is decreasing on the interval [0, 2] and increasing on the interval [2, 4].

4.1.1   DEFINITION   Let f be defined on an interval, and let x_1 and x_2 denote points in that interval.

(a) f is **increasing** on the interval if $f(x_1) < f(x_2)$ whenever $x_1 < x_2$.

(b) f is **decreasing** on the interval if $f(x_1) > f(x_2)$ whenever $x_1 < x_2$.

(c) f is **constant** on the interval if $f(x_1) = f(x_2)$ for all points x_1 and x_2.

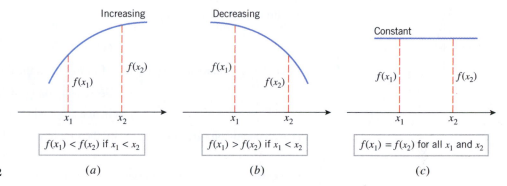

▶ **Figure 4.1.2**

(*a*)   (*b*)   (*c*)

Figure 4.1.3 suggests that a differentiable function f is increasing on any interval where each tangent line to its graph has positive slope, is decreasing on any interval where each tangent line to its graph has negative slope, and is constant on any interval where each tangent line to its graph has zero slope. This intuitive observation suggests the following important theorem that will be proved in Section 4.8.

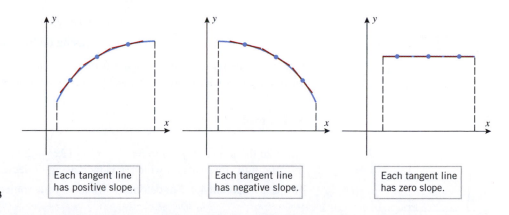

▶ **Figure 4.1.3**

Observe that the derivative conditions in Theorem 4.1.2 are only required to hold *inside* the interval [a, b], even though the conclusions apply to the entire interval.

4.1.2   THEOREM   *Let f be a function that is continuous on a closed interval $[a, b]$ and differentiable on the open interval (a, b).*

(a)   *If $f'(x) > 0$ for every value of x in (a, b), then f is increasing on $[a, b]$.*

(b)   *If $f'(x) < 0$ for every value of x in (a, b), then f is decreasing on $[a, b]$.*

(c)   *If $f'(x) = 0$ for every value of x in (a, b), then f is constant on $[a, b]$.*

Although stated for closed intervals, Theorem 4.1.2 is applicable on any interval on which f is continuous. For example, if f is continuous on $[a, +\infty)$ and $f'(x) > 0$ on $(a, +\infty)$, then f is increasing on $[a, +\infty)$; and if f is continuous on $(-\infty, +\infty)$ and $f'(x) < 0$ on $(-\infty, +\infty)$, then f is decreasing on $(-\infty, +\infty)$.

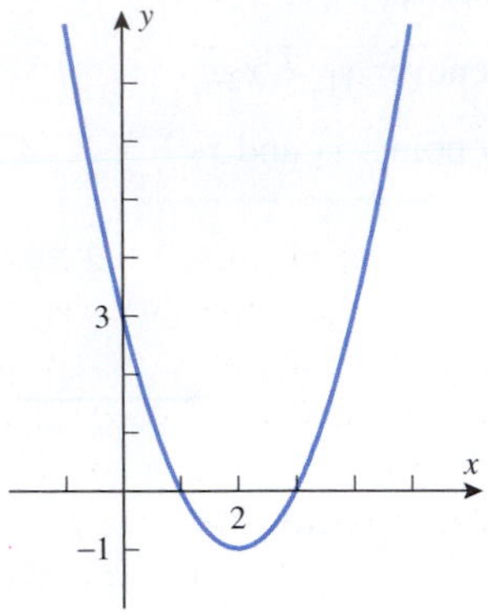

▶ **Example 1** Find the intervals on which $f(x) = x^2 - 4x + 3$ is increasing and the intervals on which it is decreasing.

Solution. The graph of f in Figure 4.1.4 suggests that f is decreasing for $x \leq 2$ and increasing for $x \geq 2$. To confirm this, we analyze the sign of f'. The derivative of f is

$$f'(x) = 2x - 4 = 2(x - 2)$$

It follows that

$$f'(x) < 0 \quad \text{if} \quad x < 2$$
$$f'(x) > 0 \quad \text{if} \quad 2 < x$$

Since f is continuous everywhere, it follows from the comment after Theorem 4.1.2 that

$$f \text{ is decreasing on } (-\infty, 2]$$
$$f \text{ is increasing on } [2, +\infty)$$

These conclusions are consistent with the graph of f in Figure 4.1.4. ◀

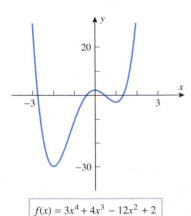

$$f(x) = x^2 - 4x + 3$$

▶ **Figure 4.1.4**

▶ **Example 2** Find the intervals on which $f(x) = x^3$ is increasing and the intervals on which it is decreasing.

Solution. The graph of f in Figure 4.1.5 suggests that f is increasing over the entire x-axis. To confirm this, we differentiate f to obtain $f'(x) = 3x^2$. Thus,

$$f'(x) > 0 \quad \text{if} \quad x < 0$$
$$f'(x) > 0 \quad \text{if} \quad 0 < x$$

Since f is continuous everywhere,

$$f \text{ is increasing on } (-\infty, 0]$$
$$f \text{ is increasing on } [0, +\infty)$$

Since f is increasing on the adjacent intervals $(-\infty, 0]$ and $[0, +\infty)$, it follows that f is increasing on their union $(-\infty, +\infty)$ (see Exercise 59). ◀

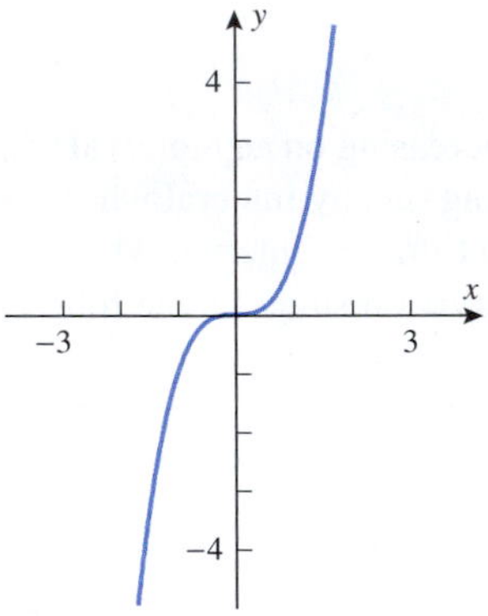

$$f(x) = x^3$$

▶ **Figure 4.1.5**

▶ **Example 3**

(a) Use the graph of $f(x) = 3x^4 + 4x^3 - 12x^2 + 2$ in Figure 4.1.6 to make a conjecture about the intervals on which f is increasing or decreasing.

(b) Use Theorem 4.1.2 to determine whether your conjecture is correct.

Solution (a). The graph suggests that the function f is decreasing if $x \leq -2$, increasing if $-2 \leq x \leq 0$, decreasing if $0 \leq x \leq 1$, and increasing if $x \geq 1$.

Solution (b). Differentiating f we obtain

$$f'(x) = 12x^3 + 12x^2 - 24x = 12x(x^2 + x - 2) = 12x(x + 2)(x - 1)$$

The sign analysis of f' in Table 4.1.1 can be obtained using the method of test points discussed in Web Appendix E. The conclusions in Table 4.1.1 confirm the conjecture in part (a). ◀

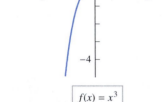

$$f(x) = 3x^4 + 4x^3 - 12x^2 + 2$$

▶ **Figure 4.1.6**

Table 4.1.1

INTERVAL	$(12x)(x+2)(x-1)$	$f'(x)$	CONCLUSION
$x < -2$	$(-)(-)(-)$	$-$	f is decreasing on $(-\infty, -2]$
$-2 < x < 0$	$(-)(+)(-)$	$+$	f is increasing on $[-2, 0]$
$0 < x < 1$	$(+)(+)(-)$	$-$	f is decreasing on $[0, 1]$
$1 < x$	$(+)(+)(+)$	$+$	f is increasing on $[1, +\infty)$

■ CONCAVITY

Although the sign of the derivative of f reveals where the graph of f is increasing or decreasing, it does not reveal the direction of *curvature*. For example, the graph is increasing on both sides of the point in Figure 4.1.7, but on the left side it has an upward curvature ("holds water") and on the right side it has a downward curvature ("spills water"). On intervals where the graph of f has upward curvature we say that f is *concave up*, and on intervals where the graph has downward curvature we say that f is *concave down*.

Figure 4.1.8 suggests two ways to characterize the concavity of a differentiable function f on an open interval:

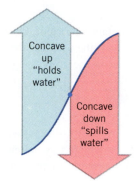

• f is concave up on an open interval if its tangent lines have increasing slopes on that interval and is concave down if they have decreasing slopes.

• f is concave up on an open interval if its graph lies above its tangent lines on that interval and is concave down if it lies below its tangent lines.

Our formal definition for "concave up" and "concave down" corresponds to the first of these characterizations.

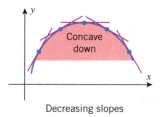

▶ **Figure 4.1.7**

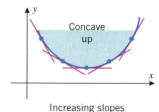

Increasing slopes

4.1.3   DEFINITION   If f is differentiable on an open interval, then f is said to be *concave up* on the open interval if f' is increasing on that interval, and f is said to be *concave down* on the open interval if f' is decreasing on that interval.

Since the slopes of the tangent lines to the graph of a differentiable function f are the values of its derivative f', it follows from Theorem 4.1.2 (applied to f' rather than f) that f' will be increasing on intervals where f'' is positive and that f' will be decreasing on intervals where f'' is negative. Thus, we have the following theorem.

Decreasing slopes

▶ **Figure 4.1.8**

4.1.4   THEOREM   *Let f be twice differentiable on an open interval.*

(a)   *If $f''(x) > 0$ for every value of x in the open interval, then f is concave up on that interval.*

(b)   *If $f''(x) < 0$ for every value of x in the open interval, then f is concave down on that interval.*

▶ **Example 4**   Figure 4.1.4 suggests that the function $f(x) = x^2 - 4x + 3$ is concave up on the interval $(-\infty, +\infty)$. This is consistent with Theorem 4.1.4, since $f'(x) = 2x - 4$ and $f''(x) = 2$, so

$$f''(x) > 0 \quad \text{on the interval } (-\infty, +\infty)$$

Also, Figure 4.1.5 suggests that $f(x) = x^3$ is concave down on the interval $(-\infty, 0)$ and concave up on the interval $(0, +\infty)$. This agrees with Theorem 4.1.4, since $f'(x) = 3x^2$ and $f''(x) = 6x$, so

$$f''(x) < 0 \quad \text{if } x < 0 \quad \text{and} \quad f''(x) > 0 \quad \text{if } x > 0 \blacktriangleleft$$

■ INFLECTION POINTS

We see from Example 4 and Figure 4.1.5 that the graph of $f(x) = x^3$ changes from concave down to concave up at $x = 0$. Points where a curve changes from concave up to concave down or vice versa are of special interest, so there is some terminology associated with them.

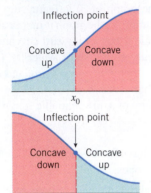

> **4.1.5 DEFINITION** If f is continuous on an open interval containing a value x_0, and if f changes the direction of its concavity at the point $(x_0, f(x_0))$, then we say that f has an **inflection point at x_0**, and we call the point $(x_0, f(x_0))$ on the graph of f an **inflection point** of f (Figure 4.1.9).

► **Figure 4.1.9**

▶ **Example 5** Figure 4.1.10 shows the graph of the function $f(x) = x^3 - 3x^2 + 1$. Use the first and second derivatives of f to determine the intervals on which f is increasing, decreasing, concave up, and concave down. Locate all inflection points and confirm that your conclusions are consistent with the graph.

Solution. Calculating the first two derivatives of f we obtain

$$f'(x) = 3x^2 - 6x = 3x(x - 2)$$
$$f''(x) = 6x - 6 = 6(x - 1)$$

The sign analysis of these derivatives is shown in the following tables:

INTERVAL	$(3x)(x-2)$	$f'(x)$	CONCLUSION
$x < 0$	$(-)(-)$	$+$	f is increasing on $(-\infty, 0]$
$0 < x < 2$	$(+)(-)$	$-$	f is decreasing on $[0, 2]$
$x > 2$	$(+)(+)$	$+$	f is increasing on $[2, +\infty)$

INTERVAL	$6(x-1)$	$f''(x)$	CONCLUSION
$x < 1$	$(-)$	$-$	f is concave down on $(-\infty, 1)$
$x > 1$	$(+)$	$+$	f is concave up on $(1, +\infty)$

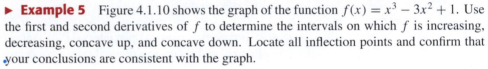

$$f(x) = x^3 - 3x^2 + 1$$

► **Figure 4.1.10**

The second table shows that there is an inflection point at $x = 1$, since f changes from concave down to concave up at that point. The inflection point is $(1, f(1)) = (1, -1)$. All of these conclusions are consistent with the graph of f. ◀

One can correctly guess from Figure 4.1.10 that the function $f(x) = x^3 - 3x^2 + 1$ has an inflection point at $x = 1$ without actually computing derivatives. However, sometimes changes in concavity are so subtle that calculus is essential to confirm their existence and identify their location. Here is an example.

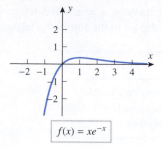

$f(x) = xe^{-x}$

▶ **Figure 4.1.11**

▶ **Example 6** Figure 4.1.11 suggests that the function $f(x) = xe^{-x}$ has an inflection point but its exact location is not evident from the graph in this figure. Use the first and second derivatives of f to determine the intervals on which f is increasing, decreasing, concave up, and concave down. Locate all inflection points.

Solution. Calculating the first two derivatives of f we obtain (verify)

$$f'(x) = (1 - x)e^{-x}$$
$$f''(x) = (x - 2)e^{-x}$$

Keeping in mind that e^{-x} is positive for all x, the sign analysis of these derivatives is easily determined:

INTERVAL	$(1-x)(e^{-x})$	$f'(x)$	CONCLUSION
$x < 1$	$(+)(+)$	$+$	f is increasing on $(-\infty, 1]$
$x > 1$	$(-)(+)$	$-$	f is decreasing on $[1, +\infty)$

INTERVAL	$(x-2)(e^{-x})$	$f''(x)$	CONCLUSION
$x < 2$	$(-)(+)$	$-$	f is concave down on $(-\infty, 2)$
$x > 2$	$(+)(+)$	$+$	f is concave up on $(2, +\infty)$

The second table shows that there is an inflection point at $x = 2$, since f changes from concave down to concave up at that point. All of these conclusions are consistent with the graph of f. ◄

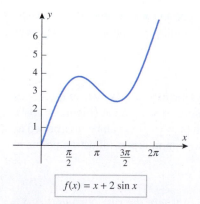

$f(x) = x + 2 \sin x$

▶ **Figure 4.1.12**

▶ **Example 7** Figure 4.1.12 shows the graph of the function $f(x) = x + 2 \sin x$ over the interval $[0, 2\pi]$. Use the first and second derivatives of f to determine where f is increasing, decreasing, concave up, and concave down. Locate all inflection points and confirm that your conclusions are consistent with the graph.

Solution. Calculating the first two derivatives of f we obtain

$$f'(x) = 1 + 2 \cos x$$
$$f''(x) = -2 \sin x$$

Since f' is a continuous function, it changes sign on the interval $(0, 2\pi)$ only at points where $f'(x) = 0$ (why?). These values are solutions of the equation

$$1 + 2 \cos x = 0 \quad \text{or equivalently} \quad \cos x = -\tfrac{1}{2}$$

There are two solutions of this equation in the interval $(0, 2\pi)$, namely, $x = 2\pi/3$ and $x = 4\pi/3$ (verify). Similarly, f'' is a continuous function, so its sign changes in the interval $(0, 2\pi)$ will occur only at values of x for which $f''(x) = 0$. These values are solutions of the equation

$$-2 \sin x = 0$$

There is one solution of this equation in the interval $(0, 2\pi)$, namely, $x = \pi$. With the help of these "sign transition points" we obtain the sign analysis shown in the following tables:

INTERVAL	$f'(x) = 1 + 2\cos x$	CONCLUSION
$0 < x < 2\pi/3$	$+$	f is increasing on $[0, 2\pi/3]$
$2\pi/3 < x < 4\pi/3$	$-$	f is decreasing on $[2\pi/3, 4\pi/3]$
$4\pi/3 < x < 2\pi$	$+$	f is increasing on $[4\pi/3, 2\pi]$

INTERVAL	$f''(x) = -2\sin x$	CONCLUSION
$0 < x < \pi$	$-$	f is concave down on $(0, \pi)$
$\pi < x < 2\pi$	$+$	f is concave up on $(\pi, 2\pi)$

> The signs in the two tables of Example 7 can be obtained either using the method of test points or using the unit circle definition of the sine and cosine functions.

The second table shows that there is an inflection point at $x = \pi$, since f changes from concave down to concave up at that point. All of these conclusions are consistent with the graph of f. ◄

In the preceding examples the inflection points of f occurred wherever $f''(x) = 0$. However, this is not always the case. Here is a specific example.

▶ **Example 8**   Find the inflection points, if any, of $f(x) = x^4$.

Solution.   Calculating the first two derivatives of f we obtain

$$f'(x) = 4x^3$$

$$f''(x) = 12x^2$$

Since $f''(x)$ is positive for $x < 0$ and for $x > 0$, the function f is concave up on the interval $(-\infty, 0)$ and on the interval $(0, +\infty)$. Thus, there is no change in concavity and hence no inflection point at $x = 0$, even though $f''(0) = 0$ (Figure 4.1.13). ◄

We will see later that if a function f has an inflection point at $x = x_0$ and $f''(x_0)$ exists, then $f''(x_0) = 0$. Also, we will see in Section 4.3 that an inflection point may also occur where $f''(x)$ is not defined.

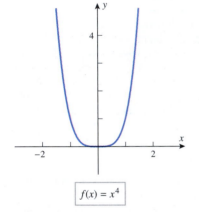

$f(x) = x^4$

▶ **Figure 4.1.13**

> Give an argument to show that the function $f(x) = x^4$ graphed in Figure 4.1.13 is concave up on the interval $(-\infty, +\infty)$.

■ **INFLECTION POINTS IN APPLICATIONS**

Inflection points of a function f are those points on the graph of $y = f(x)$ where the slopes of the tangent lines change from increasing to decreasing or vice versa (Figure 4.1.14). Since the slope of the tangent line at a point on the graph of $y = f(x)$ can be interpreted as the rate of change of y with respect to x at that point, we can interpret inflection points in the following way:

> *Inflection points mark the places on the curve $y = f(x)$ where the rate of change of y with respect to x changes from increasing to decreasing, or vice versa.*

This is a subtle idea, since we are dealing with a change in a rate of change. It can help with your understanding of this idea to realize that inflection points may have interpretations in more familiar contexts. For example, consider the statement "Oil prices rose sharply during the first half of the year but have since begun to level off." If the price of oil is plotted as a function of time of year, this statement suggests the existence of an inflection point

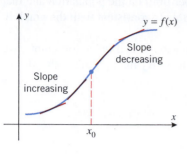

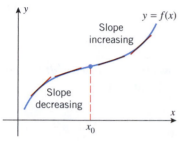

on the graph near the end of June. (Why?) To give a more visual example, consider the flask shown in Figure 4.1.15. Suppose that water is added to the flask so that the volume increases at a constant rate with respect to the time t, and let us examine the rate at which the water level y rises with respect to t. Initially, the level y will rise at a slow rate because of the wide base. However, as the diameter of the flask narrows, the rate at which the level y rises will increase until the level is at the narrow point in the neck. From that point on the rate at which the level rises will decrease as the diameter gets wider and wider. Thus, the narrow point in the neck is the point at which the rate of change of y with respect to t changes from increasing to decreasing.

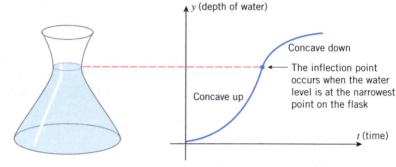

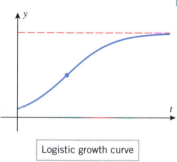

▶ **Figure 4.1.14**

▶ **Figure 4.1.15**

■ LOGISTIC CURVES

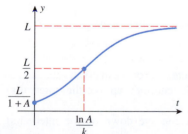

Logistic growth curve

▲ **Figure 4.1.16**

When a population grows in an environment in which space or food is limited, the graph of population versus time is typically an S-shaped curve of the form shown in Figure 4.1.16. The scenario described by this curve is a population that grows slowly at first and then more and more rapidly as the number of individuals producing offspring increases. However, at a certain point in time (where the inflection point occurs) the environmental factors begin to show their effect, and the growth rate begins a steady decline. Over an extended period of time the population approaches a limiting value that represents the upper limit on the number of individuals that the available space or food can sustain. Population growth curves of this type are called *logistic growth curves*.

▶ **Example 9**   We will see in a later chapter that logistic growth curves arise from equations of the form

$$y = \frac{L}{1 + Ae^{-kt}} \tag{1}$$

where y is the population at time t ($t \geq 0$) and A, k, and L are positive constants. Show that Figure 4.1.17 correctly describes the graph of this equation when $A > 1$.

Solution.   It follows from (1) that at time $t = 0$ the value of y is

$$y = \frac{L}{1 + A}$$

and it follows from (1) and the fact that $0 < e^{-kt} \leq 1$ for $t \geq 0$ that

$$\frac{L}{1 + A} \leq y < L \tag{2}$$

(verify). This is consistent with the graph in Figure 4.1.17. The horizontal asymptote at $y = L$ is confirmed by the limit

$$\lim_{t \to +\infty} y = \lim_{t \to +\infty} \frac{L}{1 + Ae^{-kt}} = \frac{L}{1 + 0} = L \tag{3}$$

▲ **Figure 4.1.17**

Physically, Formulas (2) and (3) tell us that L is an upper limit on the population and that the population approaches this limit over time. Again, this is consistent with the graph in Figure 4.1.17.

To investigate intervals of increase and decrease, concavity, and inflection points, we need the first and second derivatives of y with respect to t. By multiplying both sides of Equation (1) by $e^{kt}(1 + Ae^{-kt})$, we can rewrite (1) as

$$ye^{kt} + Ay = Le^{kt}$$

Using implicit differentiation, we can derive that

$$\frac{dy}{dt} = \frac{k}{L}y(L - y) \tag{4}$$

$$\frac{d^2y}{dt^2} = \frac{k^2}{L^2}y(L - y)(L - 2y) \tag{5}$$

(Exercise 70). Since $k > 0$, $y > 0$, and $L - y > 0$, it follows from (4) that $dy/dt > 0$ for all t. Thus, y is always increasing, which is consistent with Figure 4.1.17.

Since $y > 0$ and $L - y > 0$, it follows from (5) that

$$\frac{d^2y}{dt^2} > 0 \quad \text{if} \quad L - 2y > 0$$

$$\frac{d^2y}{dt^2} < 0 \quad \text{if} \quad L - 2y < 0$$

Thus, the graph of y versus t is concave up if $y < L/2$, concave down if $y > L/2$, and has an inflection point where $y = L/2$, all of which is consistent with Figure 4.1.17.

Finally, we leave it for you to solve the equation

$$\frac{L}{2} = \frac{L}{1 + Ae^{-kt}}$$

for t to show that the inflection point occurs at

$$t = \frac{1}{k}\ln A = \frac{\ln A}{k} \quad \blacktriangleleft \tag{6}$$

✓ **QUICK CHECK EXERCISES 4.1**   *(See page 244 for answers.)*

1. (a) A function f is increasing on (a, b) if _____ whenever $a < x_1 < x_2 < b$.
   (b) A function f is decreasing on (a, b) if _____ whenever $a < x_1 < x_2 < b$.
   (c) A function f is concave up on (a, b) if f' is _____ on (a, b).
   (d) If $f''(a)$ exists and f has an inflection point at $x = a$, then $f''(a)$ _____.

2. Let $f(x) = 0.1(x^3 - 3x^2 - 9x)$. Then

   $$f'(x) = 0.1(3x^2 - 6x - 9) = 0.3(x + 1)(x - 3)$$

   $$f''(x) = 0.6(x - 1)$$

   (a) Solutions to $f'(x) = 0$ are $x =$ _____.
   (b) The function f is increasing on the interval(s) _____.

(c) The function f is concave down on the interval(s) _____.
(d) _____ is an inflection point on the graph of f.

3. Suppose that $f(x)$ has derivative $f'(x) = (x - 4)^2e^{-x/2}$. Then $f''(x) = -\frac{1}{2}(x - 4)(x - 8)e^{-x/2}$.
   (a) The function f is increasing on the interval(s) _____.
   (b) The function f is concave up on the interval(s) _____.
   (c) The function f is concave down on the interval(s) _____.

4. Consider the statement "The rise in the cost of living slowed during the first half of the year." If we graph the cost of living versus time for the first half of the year, how does the graph reflect this statement?

EXERCISE SET 4.1   ∿ Graphing Utility   [C] CAS

FOCUS ON CONCEPTS

1. In each part, sketch the graph of a function f with the stated properties, and discuss the signs of f' and f''.
   (a) The function f is concave up and increasing on the interval $(-\infty, +\infty)$.
   (b) The function f is concave down and increasing on the interval $(-\infty, +\infty)$.
   (c) The function f is concave up and decreasing on the interval $(-\infty, +\infty)$.
   (d) The function f is concave down and decreasing on the interval $(-\infty, +\infty)$.

2. In each part, sketch the graph of a function f with the stated properties.
   (a) f is increasing on $(-\infty, +\infty)$, has an inflection point at the origin, and is concave up on $(0, +\infty)$.
   (b) f is increasing on $(-\infty, +\infty)$, has an inflection point at the origin, and is concave down on $(0, +\infty)$.
   (c) f is decreasing on $(-\infty, +\infty)$, has an inflection point at the origin, and is concave up on $(0, +\infty)$.
   (d) f is decreasing on $(-\infty, +\infty)$, has an inflection point at the origin, and is concave down on $(0, +\infty)$.

3. Use the graph of the equation $y = f(x)$ in the accompanying figure to find the signs of dy/dx and d^2y/dx^2 at the points A, B, and C.

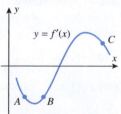

◀ Figure Ex-3

4. Use the graph of the equation $y = f'(x)$ in the accompanying figure to find the signs of dy/dx and d^2y/dx^2 at the points A, B, and C.

5. Use the graph of $y = f''(x)$ in the accompanying figure to determine the x-coordinates of all inflection points of f. Explain your reasoning.

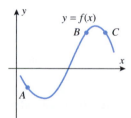

▲ Figure Ex-4   ▲ Figure Ex-5

6. Use the graph of $y = f'(x)$ in the accompanying figure to replace the question mark with $<$, $=$, or $>$, as appropriate. Explain your reasoning.
   (a) $f(0)$? $f(1)$ (b) $f(1)$? $f(2)$ (c) $f'(0)$? 0
   (d) $f'(1)$? 0 (e) $f''(0)$? 0 (f) $f''(2)$? 0

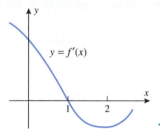

◀ Figure Ex-6

7. In each part, use the graph of $y = f(x)$ in the accompanying figure to find the requested information.
   (a) Find the intervals on which f is increasing.
   (b) Find the intervals on which f is decreasing.
   (c) Find the open intervals on which f is concave up.
   (d) Find the open intervals on which f is concave down.
   (e) Find all values of x at which f has an inflection point.

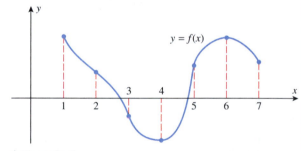

▲ Figure Ex-7

8. Use the graph in Exercise 7 to make a table that shows the signs of f' and f'' over the intervals $(1, 2)$, $(2, 3)$, $(3, 4)$, $(4, 5)$, $(5, 6)$, and $(6, 7)$.

9–10 A sign chart is presented for the first and second derivatives of a function f. Assuming that f is continuous everywhere, find: (a) the intervals on which f is increasing, (b) the intervals on which f is decreasing, (c) the open intervals on which f is concave up, (d) the open intervals on which f is concave down, and (e) the x-coordinates of all inflection points. ∎

9.

INTERVAL	SIGN OF $f'(x)$	SIGN OF $f''(x)$
$x < 1$	$-$	$+$
$1 < x < 2$	$+$	$+$
$2 < x < 3$	$+$	$-$
$3 < x < 4$	$-$	$-$
$4 < x$	$-$	$+$

10.

INTERVAL	SIGN OF $f'(x)$	SIGN OF $f''(x)$
$x < 1$	$+$	$+$
$1 < x < 3$	$+$	$-$
$3 < x$	$+$	$+$

11–14 True–False Assume that f is differentiable everywhere. Determine whether the statement is true or false. Explain your answer. ■

11. If f is decreasing on $[0, 2]$, then $f(0) > f(1) > f(2)$.

12. If $f'(1) > 0$, then f is increasing on $[0, 2]$.

13. If f is increasing on $[0, 2]$, then $f'(1) > 0$.

14. If f' is increasing on $[0, 1]$ and f' is decreasing on $[1, 2]$, then f has an inflection point at $x = 1$.

15–32 Find: (a) the intervals on which f is increasing, (b) the intervals on which f is decreasing, (c) the open intervals on which f is concave up, (d) the open intervals on which f is concave down, and (e) the x-coordinates of all inflection points. ■

15. $f(x) = x^2 - 3x + 8$
16. $f(x) = 5 - 4x - x^2$
17. $f(x) = (2x + 1)^3$
18. $f(x) = 5 + 12x - x^3$
19. $f(x) = 3x^4 - 4x^3$
20. $f(x) = x^4 - 5x^3 + 9x^2$
21. $f(x) = \dfrac{x - 2}{(x^2 - x + 1)^2}$
22. $f(x) = \dfrac{x}{x^2 + 2}$
23. $f(x) = \sqrt[3]{x^2 + x + 1}$
24. $f(x) = x^{4/3} - x^{1/3}$
25. $f(x) = (x^{2/3} - 1)^2$
26. $f(x) = x^{2/3} - x$
27. $f(x) = e^{-x^2/2}$
28. $f(x) = xe^{x^2}$
29. $f(x) = \ln \sqrt{x^2 + 4}$
30. $f(x) = x^3 \ln x$
31. $f(x) = \tan^{-1}(x^2 - 1)$
32. $f(x) = \sin^{-1} x^{2/3}$

33–38 Analyze the trigonometric function f over the specified interval, stating where f is increasing, decreasing, concave up, and concave down, and stating the x-coordinates of all inflection points. Confirm that your results are consistent with the graph of f generated with a graphing utility. ■

33. $f(x) = \sin x - \cos x$; $[-\pi, \pi]$
34. $f(x) = \sec x \tan x$; $(-\pi/2, \pi/2)$
35. $f(x) = 1 - \tan(x/2)$; $(-\pi, \pi)$
36. $f(x) = 2x + \cot x$; $(0, \pi)$
37. $f(x) = (\sin x + \cos x)^2$; $[-\pi, \pi]$
38. $f(x) = \sin^2 2x$; $[0, \pi]$

FOCUS ON CONCEPTS

39. In parts (a)–(c), sketch a continuous curve $y = f(x)$ with the stated properties.
(a) $f(2) = 4$, $f'(2) = 0$, $f''(x) > 0$ for all x
(b) $f(2) = 4$, $f'(2) = 0$, $f''(x) < 0$ for $x < 2$, $f''(x) > 0$ for $x > 2$

(c) $f(2) = 4$, $f''(x) < 0$ for $x \neq 2$ and $\lim_{x \to 2^+} f'(x) = +\infty$, $\lim_{x \to 2^-} f'(x) = -\infty$

40. In each part sketch a continuous curve $y = f(x)$ with the stated properties.
(a) $f(2) = 4$, $f'(2) = 0$, $f''(x) < 0$ for all x
(b) $f(2) = 4$, $f'(2) = 0$, $f''(x) < 0$ for $x < 2$, $f''(x) > 0$ for $x > 2$
(c) $f(2) = 4$, $f''(x) > 0$ for $x \neq 2$ and $\lim_{x \to 2^+} f'(x) = -\infty$, $\lim_{x \to 2^-} f'(x) = +\infty$

41–46 If f is increasing on an interval $[0, b)$, then it follows from Definition 4.1.1 that $f(0) < f(x)$ for each x in the interval $(0, b)$. Use this result in these exercises. ■

41. Show that $\sqrt[3]{1 + x} < 1 + \frac{1}{3}x$ if $x > 0$, and confirm the inequality with a graphing utility. [*Hint:* Show that the function $f(x) = 1 + \frac{1}{3}x - \sqrt[3]{1 + x}$ is increasing on $[0, +\infty)$.]

42. Show that $x < \tan x$ if $0 < x < \pi/2$, and confirm the inequality with a graphing utility. [*Hint:* Show that the function $f(x) = \tan x - x$ is increasing on $[0, \pi/2)$.]

43. Use a graphing utility to make a conjecture about the relative sizes of x and $\sin x$ for $x \geq 0$, and prove your conjecture.

44. Use a graphing utility to make a conjecture about the relative sizes of $1 - x^2/2$ and $\cos x$ for $x \geq 0$, and prove your conjecture. [*Hint:* Use the result of Exercise 43.]

45. (a) Show that $\ln(x + 1) \leq x$ if $x \geq 0$.
(b) Show that $\ln(x + 1) \geq x - \frac{1}{2}x^2$ if $x \geq 0$.
(c) Confirm the inequalities in parts (a) and (b) with a graphing utility.

46. (a) Show that $e^x \geq 1 + x$ if $x \geq 0$.
(b) Show that $e^x \geq 1 + x + \frac{1}{2}x^2$ if $x \geq 0$.
(c) Confirm the inequalities in parts (a) and (b) with a graphing utility.

47–48 Use a graphing utility to generate the graphs of f' and f'' over the stated interval; then use those graphs to estimate the x-coordinates of the inflection points of f, the intervals on which f is concave up or down, and the intervals on which f is increasing or decreasing. Check your estimates by graphing f. ■

47. $f(x) = x^4 - 24x^2 + 12x$, $-5 \leq x \leq 5$
48. $f(x) = \dfrac{1}{1 + x^2}$, $-5 \leq x \leq 5$

49–50 Use a CAS to find f'' and to approximate the x-coordinates of the inflection points to six decimal places. Confirm that your answer is consistent with the graph of f. ■

49. $f(x) = \dfrac{10x - 3}{3x^2 - 5x + 8}$
50. $f(x) = \dfrac{x^3 - 8x + 7}{\sqrt{x^2 + 1}}$

51. Use Definition 4.1.1 to prove that $f(x) = x^2$ is increasing on $[0, +\infty)$.

52. Use Definition 4.1.1 to prove that $f(x) = 1/x$ is decreasing on $(0, +\infty)$.

53–54 Determine whether the statements are true or false. If a statement is false, find functions for which the statement fails to hold. ■

53. (a) If f and g are increasing on an interval, then so is $f + g$.
(b) If f and g are increasing on an interval, then so is $f \cdot g$.

54. (a) If f and g are concave up on an interval, then so is $f + g$.
(b) If f and g are concave up on an interval, then so is $f \cdot g$.

55. In each part, find functions f and g that are increasing on $(-\infty, +\infty)$ and for which $f - g$ has the stated property.
(a) $f - g$ is decreasing on $(-\infty, +\infty)$.
(b) $f - g$ is constant on $(-\infty, +\infty)$.
(c) $f - g$ is increasing on $(-\infty, +\infty)$.

56. In each part, find functions f and g that are positive and increasing on $(-\infty, +\infty)$ and for which f/g has the stated property.
(a) f/g is decreasing on $(-\infty, +\infty)$.
(b) f/g is constant on $(-\infty, +\infty)$.
(c) f/g is increasing on $(-\infty, +\infty)$.

57. (a) Prove that a general cubic polynomial

$$f(x) = ax^3 + bx^2 + cx + d \quad (a \neq 0)$$

has exactly one inflection point.
(b) Prove that if a cubic polynomial has three x-intercepts, then the inflection point occurs at the average value of the intercepts.
(c) Use the result in part (b) to find the inflection point of the cubic polynomial $f(x) = x^3 - 3x^2 + 2x$, and check your result by using f'' to determine where f is concave up and concave down.

58. From Exercise 57, the polynomial $f(x) = x^3 + bx^2 + 1$ has one inflection point. Use a graphing utility to reach a conclusion about the effect of the constant b on the location of the inflection point. Use f'' to explain what you have observed graphically.

59. Use Definition 4.1.1 to prove:
(a) If f is increasing on the intervals $(a, c]$ and $[c, b)$, then f is increasing on (a, b).
(b) If f is decreasing on the intervals $(a, c]$ and $[c, b)$, then f is decreasing on (a, b).

60. Use part (a) of Exercise 59 to show that $f(x) = x + \sin x$ is increasing on the interval $(-\infty, +\infty)$.

61. Use part (b) of Exercise 59 to show that $f(x) = \cos x - x$ is decreasing on the interval $(-\infty, +\infty)$.

62. Let $y = 1/(1 + x^2)$. Find the values of x for which y is increasing most rapidly or decreasing most rapidly.

63–66 Suppose that water is flowing at a constant rate into the container shown. Make a rough sketch of the graph of the water level y versus the time t. Make sure that your sketch conveys where the graph is concave up and concave down, and label the y-coordinates of the inflection points. ■

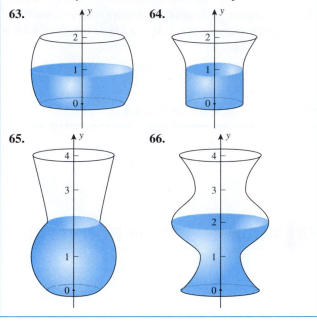

63.

64.

65.

66.

67. Suppose that a population y grows according to the logistic model given by Formula (1).
(a) At what rate is y increasing at time $t = 0$?
(b) In words, describe how the rate of growth of y varies with time.
(c) At what time is the population growing most rapidly?

68. Suppose that the number of individuals at time t in a certain wildlife population is given by

$$N(t) = \frac{340}{1 + 9(0.77)^t}, \quad t \geq 0$$

where t is in years. Use a graphing utility to estimate the time at which the size of the population is increasing most rapidly.

69. Suppose that the spread of a flu virus on a college campus is modeled by the function

$$y(t) = \frac{1000}{1 + 999e^{-0.9t}}$$

where $y(t)$ is the number of infected students at time t (in days, starting with $t = 0$). Use a graphing utility to estimate the day on which the virus is spreading most rapidly.

70. The logistic growth model given in Formula (1) is equivalent to

$$ye^{kt} + Ay = Le^{kt}$$

where y is the population at time t ($t \geq 0$) and A, k, and L

are positive constants. Use implicit differentiation to verify that

$$\frac{dy}{dt} = \frac{k}{L} y(L - y)$$

$$\frac{d^2y}{dt^2} = \frac{k^2}{L^2} y(L - y)(L - 2y)$$

71. Assuming that A, k, and L are positive constants, verify that the graph of $y = L/(1 + Ae^{-kt})$ has an inflection point at $\left(\frac{1}{k} \ln A, \frac{1}{2}L\right)$.

72. Writing An approaching storm causes the air temperature to fall. Make a statement that indicates there is an inflection point in the graph of temperature versus time. Explain how the existence of an inflection point follows from your statement.

73. Writing Explain what the sign analyses of $f'(x)$ and $f''(x)$ tell us about the graph of $y = f(x)$.

✔ **QUICK CHECK ANSWERS 4.1**

1. (a) $f(x_1) < f(x_2)$ (b) $f(x_1) > f(x_2)$ (c) increasing (d) $= 0$ **2.** (a) $-1, 3$ (b) $(-\infty, -1]$ and $[3, +\infty)$ (c) $(-\infty, 1)$ (d) $(1, -1.1)$ **3.** (a) $(-\infty, +\infty)$ (b) $(4, 8)$ (c) $(-\infty, 4), (8, +\infty)$ **4.** The graph is increasing and concave down.

4.2 ANALYSIS OF FUNCTIONS II: RELATIVE EXTREMA; GRAPHING POLYNOMIALS

In this section we will develop methods for finding the high and low points on the graph of a function and we will discuss procedures for analyzing the graphs of polynomials.

■ RELATIVE MAXIMA AND MINIMA

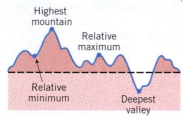

▲ **Figure 4.2.1**

If we imagine the graph of a function f to be a two-dimensional mountain range with hills and valleys, then the tops of the hills are called "relative maxima," and the bottoms of the valleys are called "relative minima" (Figure 4.2.1). The relative maxima are the high points in their *immediate vicinity*, and the relative minima are the low points. A relative maximum need not be the highest point in the entire mountain range, and a relative minimum need not be the lowest point—they are just high and low points *relative* to the nearby terrain. These ideas are captured in the following definition.

4.2.1   DEFINITION   A function f is said to have a *relative maximum* at x_0 if there is an open interval containing x_0 on which $f(x_0)$ is the largest value, that is, $f(x_0) \geq f(x)$ for all x in the interval. Similarly, f is said to have a *relative minimum* at x_0 if there is an open interval containing x_0 on which $f(x_0)$ is the smallest value, that is, $f(x_0) \leq f(x)$ for all x in the interval. If f has either a relative maximum or a relative minimum at x_0, then f is said to have a *relative extremum* at x_0.

▶ **Example 1**   We can see from Figure 4.2.2 that:

- $f(x) = x^2$ has a relative minimum at $x = 0$ but no relative maxima.
- $f(x) = x^3$ has no relative extrema.
- $f(x) = x^3 - 3x + 3$ has a relative maximum at $x = -1$ and a relative minimum at $x = 1$.
- $f(x) = \frac{1}{2}x^4 - \frac{4}{3}x^3 - x^2 + 4x + 1$ has relative minima at $x = -1$ and $x = 2$ and a relative maximum at $x = 1$.

- $f(x) = \cos x$ has relative maxima at all even multiples of π and relative minima at all odd multiples of π. ◄

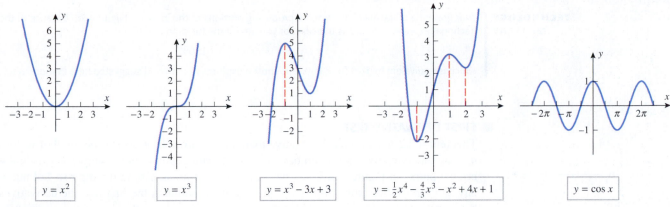

$\boxed{y = x^2}$ $\boxed{y = x^3}$ $\boxed{y = x^3 - 3x + 3}$ $\boxed{y = \frac{1}{2}x^4 - \frac{4}{3}x^3 - x^2 + 4x + 1}$ $\boxed{y = \cos x}$

▲ **Figure 4.2.2**

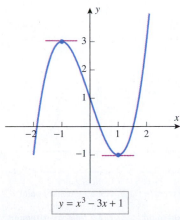

▲ **Figure 4.2.3** The points x_1, x_2, x_3, x_4, and x_5 are critical points. Of these, x_1, x_2, and x_5 are stationary points.

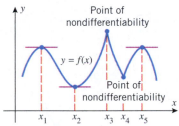

$\boxed{y = x^3 - 3x + 1}$

▲ **Figure 4.2.4**

What is the maximum number of critical points that a polynomial of degree n can have? Why?

The relative extrema for the five functions in Example 1 occur at points where the graphs of the functions have horizontal tangent lines. Figure 4.2.3 illustrates that a relative extremum can also occur at a point where a function is not differentiable. In general, we define a ***critical point*** for a function f to be a point in the domain of f at which either the graph of f has a horizontal tangent line or f is not differentiable. To distinguish between the two types of critical points we call x a ***stationary point*** of f if $f'(x) = 0$. The following theorem, which is proved in Appendix J, states that the critical points for a function form a complete set of candidates for relative extrema on the interior of the domain of the function.

4.2.2 THEOREM *Suppose that f is a function defined on an open interval containing the point x_0. If f has a relative extremum at $x = x_0$, then $x = x_0$ is a critical point of f; that is, either $f'(x_0) = 0$ or f is not differentiable at x_0.*

▶ **Example 2** Find all critical points of $f(x) = x^3 - 3x + 1$.

Solution. The function f, being a polynomial, is differentiable everywhere, so its critical points are all stationary points. To find these points we must solve the equation $f'(x) = 0$. Since

$$f'(x) = 3x^2 - 3 = 3(x + 1)(x - 1)$$

we conclude that the critical points occur at $x = -1$ and $x = 1$. This is consistent with the graph of f in Figure 4.2.4. ◄

▶ **Example 3** Find all critical points of $f(x) = 3x^{5/3} - 15x^{2/3}$.

Solution. The function f is continuous everywhere and its derivative is

$$f'(x) = 5x^{2/3} - 10x^{-1/3} = 5x^{-1/3}(x - 2) = \frac{5(x - 2)}{x^{1/3}}$$

We see from this that $f'(x) = 0$ if $x = 2$ and $f'(x)$ is undefined if $x = 0$. Thus $x = 0$ and $x = 2$ are critical points and $x = 2$ is a stationary point. This is consistent with the graph of f shown in Figure 4.2.5. ◄

TECHNOLOGY MASTERY

Your graphing utility may have trouble producing portions of the graph in Figure 4.2.5 because of the fractional exponents. If this is the case for you, graph the function

$$y = 3(|x|/x)|x|^{5/3} - 15|x|^{2/3}$$

which is equivalent to $f(x)$ for $x \neq 0$. Appendix A explores the method suggested here in more detail.

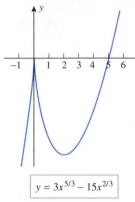

$$y = 3x^{5/3} - 15x^{2/3}$$

▲ Figure 4.2.5

■ FIRST DERIVATIVE TEST

Theorem 4.2.2 asserts that the relative extrema must occur at critical points, but it does *not* say that a relative extremum occurs at *every* critical point. For example, for the eight critical points in Figure 4.2.6, relative extrema occur at each x_0 in the top row but not at any x_0 in the bottom row. Moreover, at the critical points in the first row the derivatives have opposite signs on the two sides of x_0, whereas at the critical points in the second row the signs of the derivatives are the same on both sides. This suggests:

> *A function f has a relative extremum at those critical points where f' changes sign.*

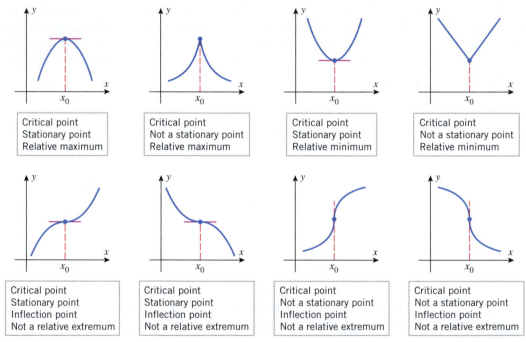

▲ Figure 4.2.6

We can actually take this a step further. At the two relative maxima in Figure 4.2.6 the derivative is positive on the left side and negative on the right side, and at the two relative minima the derivative is negative on the left side and positive on the right side. All of this is summarized more precisely in the following theorem.

4.2.3   THEOREM (*First Derivative Test*)   *Suppose that f is continuous at a critical point x_0.*

(a)   *If $f'(x) > 0$ on an open interval extending left from x_0 and $f'(x) < 0$ on an open interval extending right from x_0, then f has a relative maximum at x_0.*

(b)   *If $f'(x) < 0$ on an open interval extending left from x_0 and $f'(x) > 0$ on an open interval extending right from x_0, then f has a relative minimum at x_0.*

(c)   *If $f'(x)$ has the same sign on an open interval extending left from x_0 as it does on an open interval extending right from x_0, then f does not have a relative extremum at x_0.*

Informally stated, parts (*a*) and (*b*) of Theorem 4.2.3 tell us that for a continuous function, relative maxima occur at critical points where the derivative changes from $+$ to $-$ and relative minima where it changes from $-$ to $+$.

Use the first derivative test to confirm the behavior at x_0 of each graph in Figure 4.2.6.

PROOF   We will prove part (*a*) and leave parts (*b*) and (*c*) as exercises. We are assuming that $f'(x) > 0$ on the interval (a, x_0) and that $f'(x) < 0$ on the interval (x_0, b), and we want to show that

$$f(x_0) \geq f(x)$$

for all x in the interval (a, b). However, the two hypotheses, together with Theorem 4.1.2 and its associated marginal note imply that f is increasing on the interval $(a, x_0]$ and decreasing on the interval $[x_0, b)$. Thus, $f(x_0) \geq f(x)$ for all x in (a, b) with equality only at x_0. ∎

▶ **Example 4**   We showed in Example 3 that the function $f(x) = 3x^{5/3} - 15x^{2/3}$ has critical points at $x = 0$ and $x = 2$. Figure 4.2.5 suggests that f has a relative maximum at $x = 0$ and a relative minimum at $x = 2$. Confirm this using the first derivative test.

Solution.   We showed in Example 3 that

$$f'(x) = \frac{5(x-2)}{x^{1/3}}$$

A sign analysis of this derivative is shown in Table 4.2.1. The sign of f' changes from $+$ to $-$ at $x = 0$, so there is a relative maximum at that point. The sign changes from $-$ to $+$ at $x = 2$, so there is a relative minimum at that point. ◀

Table 4.2.1

INTERVAL	$5(x-2)/x^{1/3}$	$f'(x)$
$x < 0$	$(-)/(-)$	$+$
$0 < x < 2$	$(-)/(+)$	$-$
$x > 2$	$(+)/(+)$	$+$

SECOND DERIVATIVE TEST

There is another test for relative extrema that is based on the following geometric observation: A function f has a relative maximum at a stationary point if the graph of f is concave down on an open interval containing that point, and it has a relative minimum if it is concave up (Figure 4.2.7).

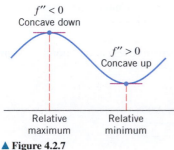

$f'' < 0$
Concave down

$f'' > 0$
Concave up

Relative maximum   Relative minimum

▲ **Figure 4.2.7**

4.2.4   THEOREM (*Second Derivative Test*)   *Suppose that f is twice differentiable at the point x_0.*

(a)   *If $f'(x_0) = 0$ and $f''(x_0) > 0$, then f has a relative minimum at x_0.*

(b)   *If $f'(x_0) = 0$ and $f''(x_0) < 0$, then f has a relative maximum at x_0.*

(c)   *If $f'(x_0) = 0$ and $f''(x_0) = 0$, then the test is inconclusive; that is, f may have a relative maximum, a relative minimum, or neither at x_0.*

We will prove parts (a) and (c) and leave part (b) as an exercise.

PROOF (a)   We are given that $f'(x_0) = 0$ and $f''(x_0) > 0$, and we want to show that f has a relative minimum at x_0. Expressing $f''(x_0)$ as a limit and using the two given conditions we obtain

$$f''(x_0) = \lim_{x \to x_0} \frac{f'(x) - f'(x_0)}{x - x_0} = \lim_{x \to x_0} \frac{f'(x)}{x - x_0} > 0$$

This implies that for x sufficiently close to but different from x_0 we have

$$\frac{f'(x)}{x - x_0} > 0 \tag{1}$$

Thus, there is an open interval extending left from x_0 and an open interval extending right from x_0 on which (1) holds. On the open interval extending left the denominator in (1) is negative, so $f'(x) < 0$, and on the open interval extending right the denominator is positive, so $f'(x) > 0$. It now follows from part (b) of the first derivative test (Theorem 4.2.3) that f has a relative minimum at x_0.

PROOF (c)   To prove this part of the theorem we need only provide functions for which $f'(x_0) = 0$ and $f''(x_0) = 0$ at some point x_0, but with one having a relative minimum at x_0, one having a relative maximum at x_0, and one having neither at x_0. We leave it as an exercise for you to show that three such functions are $f(x) = x^4$ (relative minimum at $x = 0$), $f(x) = -x^4$ (relative maximum at $x = 0$), and $f(x) = x^3$ (neither a relative maximum nor a relative minimum at x_0). ∎

▶ **Example 5**   Find the relative extrema of $f(x) = 3x^5 - 5x^3$.

Solution.   We have

$$f'(x) = 15x^4 - 15x^2 = 15x^2(x^2 - 1) = 15x^2(x + 1)(x - 1)$$
$$f''(x) = 60x^3 - 30x = 30x(2x^2 - 1)$$

Solving $f'(x) = 0$ yields the stationary points $x = 0$, $x = -1$, and $x = 1$. As shown in the following table, we can conclude from the second derivative test that f has a relative maximum at $x = -1$ and a relative minimum at $x = 1$.

STATIONARY POINT	$30x(2x^2 - 1)$	$f''(x)$	SECOND DERIVATIVE TEST
$x = -1$	-30	$-$	f has a relative maximum
$x = 0$	0	0	Inconclusive
$x = 1$	30	$+$	f has a relative minimum

The test is inconclusive at $x = 0$, so we will try the first derivative test at that point. A sign analysis of f' is given in the following table:

INTERVAL	$15x^2(x + 1)(x - 1)$	$f'(x)$
$-1 < x < 0$	$(+)(+)(-)$	$-$
$0 < x < 1$	$(+)(+)(-)$	$-$

Since there is no sign change in f' at $x = 0$, there is neither a relative maximum nor a relative minimum at that point. All of this is consistent with the graph of f shown in Figure 4.2.8. ◀

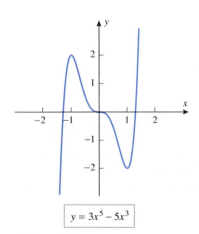

$y = 3x^5 - 5x^3$

▲ **Figure 4.2.8**

■ GEOMETRIC IMPLICATIONS OF MULTIPLICITY

Our final goal in this section is to outline a general procedure that can be used to analyze and graph polynomials. To do so, it will be helpful to understand how the graph of a polynomial behaves in the vicinity of its roots. For example, it would be nice to know what property of the polynomial in Example 5 produced the inflection point and horizontal tangent at the root $x = 0$.

Recall that a root $x = r$ of a polynomial $p(x)$ has ***multiplicity m*** if $(x - r)^m$ divides $p(x)$ but $(x - r)^{m+1}$ does not. A root of multiplicity 1 is called a ***simple root***. Figure 4.2.9 and the following theorem show that the behavior of a polynomial in the vicinity of a real root is determined by the multiplicity of that root (we omit the proof).

4.2.5   THE GEOMETRIC IMPLICATIONS OF MULTIPLICITY   *Suppose that $p(x)$ is a polynomial with a root of multiplicity m at $x = r$.*

(a)   *If m is even, then the graph of $y = p(x)$ is tangent to the x-axis at $x = r$, does not cross the x-axis there, and does not have an inflection point there.*

(b)   *If m is odd and greater than 1, then the graph is tangent to the x-axis at $x = r$, crosses the x-axis there, and also has an inflection point there.*

(c)   *If $m = 1$ (so that the root is simple), then the graph is not tangent to the x-axis at $x = r$, crosses the x-axis there, and may or may not have an inflection point there.*

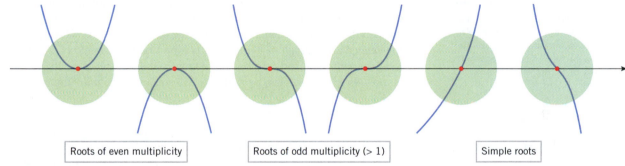

| Roots of even multiplicity | Roots of odd multiplicity (> 1) | Simple roots |

▲ **Figure 4.2.9**

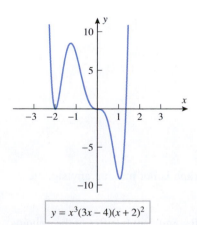

$y = x^3(3x - 4)(x + 2)^2$

▲ **Figure 4.2.10**

▶ **Example 6**   Make a conjecture about the behavior of the graph of

$$y = x^3(3x - 4)(x + 2)^2$$

in the vicinity of its x-intercepts, and test your conjecture by generating the graph.

Solution.   The x-intercepts occur at $x = 0$, $x = \frac{4}{3}$, and $x = -2$. The root $x = 0$ has multiplicity 3, which is odd, so at that point the graph should be tangent to the x-axis, cross the x-axis, and have an inflection point there. The root $x = -2$ has multiplicity 2, which is even, so the graph should be tangent to but not cross the x-axis there. The root $x = \frac{4}{3}$ is simple, so at that point the curve should cross the x-axis without being tangent to it. All of this is consistent with the graph in Figure 4.2.10. ◀

■ ANALYSIS OF POLYNOMIALS

Historically, the term "curve sketching" meant using calculus to help draw the graph of a function by hand—the graph was the goal. Since graphs can now be produced with great precision using calculators and computers, the purpose of curve sketching has changed. Today, we typically start with a graph produced by a calculator or computer, then use curve sketching to identify important features of the graph that the calculator or computer might have missed. Thus, the goal of curve sketching is no longer the graph itself, but rather the information it reveals about the function.

Polynomials are among the simplest functions to graph and analyze. Their significant features are symmetry, intercepts, relative extrema, inflection points, and the behavior as $x \to +\infty$ and as $x \to -\infty$. Figure 4.2.11 shows the graphs of four polynomials in x. The graphs in Figure 4.2.11 have properties that are common to all polynomials:

- The natural domain of a polynomial is $(-\infty, +\infty)$.

- Polynomials are continuous everywhere.

- Polynomials are differentiable everywhere, so their graphs have no corners or vertical tangent lines.

- The graph of a nonconstant polynomial eventually increases or decreases without bound as $x \to +\infty$ and as $x \to -\infty$. This is because the limit of a nonconstant polynomial as $x \to +\infty$ or as $x \to -\infty$ is $\pm\infty$, depending on the sign of the term of highest degree and whether the polynomial has even or odd degree [see Formulas (17) and (18) of Section 1.3 and the related discussion].

- The graph of a polynomial of degree $n \, (> 2)$ has at most n x-intercepts, at most $n - 1$ relative extrema, and at most $n - 2$ inflection points. This is because the x-intercepts, relative extrema, and inflection points of a polynomial $p(x)$ are among the real solutions of the equations $p(x) = 0$, $p'(x) = 0$, and $p''(x) = 0$, and the polynomials in these equations have degree n, $n - 1$, and $n - 2$, respectively. Thus, for example, the graph of a quadratic polynomial has at most two x-intercepts, one relative extremum, and no inflection points; and the graph of a cubic polynomial has at most three x-intercepts, two relative extrema, and one inflection point.

> For each of the graphs in Figure 4.2.11, count the number of x-intercepts, relative extrema, and inflection points, and confirm that your count is consistent with the degree of the polynomial.

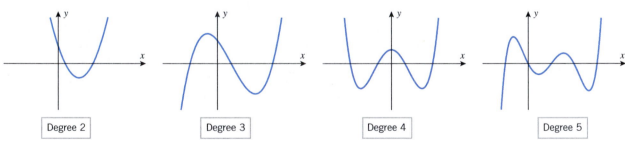

| Degree 2 | Degree 3 | Degree 4 | Degree 5 |

▲ **Figure 4.2.11**

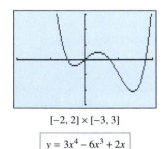

$[-2, 2] \times [-3, 3]$

$y = 3x^4 - 6x^3 + 2x$

▲ **Figure 4.2.12**

▶ **Example 7**   Figure 4.2.12 shows the graph of

$$y = 3x^4 - 6x^3 + 2x$$

produced on a graphing calculator. Confirm that the graph is not missing any significant features.

Solution.   We can be confident that the graph shows all significant features of the polynomial because the polynomial has degree 4 and we can account for four roots, three relative extrema, and two inflection points. Moreover, the graph suggests the correct behavior as

$x \to +\infty$ and as $x \to -\infty$, since

$$\lim_{x \to +\infty} (3x^4 - 6x^3 + 2x) = \lim_{x \to +\infty} 3x^4 = +\infty$$

$$\lim_{x \to -\infty} (3x^4 - 6x^3 + 2x) = \lim_{x \to -\infty} 3x^4 = +\infty \blacktriangleleft$$

▶ **Example 8** Sketch the graph of the equation

$$y = x^3 - 3x + 2$$

and identify the locations of the intercepts, relative extrema, and inflection points.

Solution. The following analysis will produce the information needed to sketch the graph:

> A review of polynomial factoring is given in Appendix C.

- *x-intercepts:* Factoring the polynomial yields

$$x^3 - 3x + 2 = (x + 2)(x - 1)^2$$

which tells us that the *x*-intercepts are $x = -2$ and $x = 1$.

- *y-intercept:* Setting $x = 0$ yields $y = 2$.

- *End behavior:* We have

$$\lim_{x \to +\infty} (x^3 - 3x + 2) = \lim_{x \to +\infty} x^3 = +\infty$$

$$\lim_{x \to -\infty} (x^3 - 3x + 2) = \lim_{x \to -\infty} x^3 = -\infty$$

so the graph increases without bound as $x \to +\infty$ and decreases without bound as $x \to -\infty$.

- *Derivatives:*

$$\frac{dy}{dx} = 3x^2 - 3 = 3(x - 1)(x + 1)$$

$$\frac{d^2y}{dx^2} = 6x$$

- *Increase, decrease, relative extrema, inflection points:* Figure 4.2.13 gives a sign analysis of the first and second derivatives and indicates its geometric significance. There are stationary points at $x = -1$ and $x = 1$. Since the sign of dy/dx changes from $+$ to $-$ at $x = -1$, there is a relative maximum there, and since it changes from $-$ to $+$ at $x = 1$, there is a relative minimum there. The sign of d^2y/dx^2 changes from $-$ to $+$ at $x = 0$, so there is an inflection point there.

- *Final sketch:* Figure 4.2.14 shows the final sketch with the coordinates of the intercepts, relative extrema, and inflection point labeled. ◀

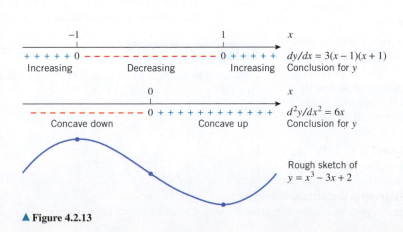

▲ **Figure 4.2.13**

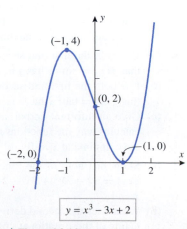

▲ **Figure 4.2.14**

✔ **QUICK CHECK EXERCISES 4.2**   *(See page 254 for answers.)*

1. A function f has a relative maximum at x_0 if there is an open interval containing x_0 on which $f(x)$ is _____ $f(x_0)$ for every x in the interval.

2. Suppose that f is defined everywhere and $x = 2, 3, 5, 7$ are critical points for f. If $f'(x)$ is positive on the intervals $(-\infty, 2)$ and $(5, 7)$, and if $f'(x)$ is negative on the intervals $(2, 3)$, $(3, 5)$, and $(7, +\infty)$, then f has relative maxima at $x =$ _____ and f has relative minima at $x =$ _____.

3. Suppose that f is defined everywhere and $x = -2$ and $x = 1$ are critical points for f. If $f''(x) = 2x + 1$, then f has a relative _____ at $x = -2$ and f has a relative _____ at $x = 1$.

4. Let $f(x) = (x^2 - 4)^2$. Then $f'(x) = 4x(x^2 - 4)$ and $f''(x) = 4(3x^2 - 4)$. Identify the locations of the (a) relative maxima, (b) relative minima, and (c) inflection points on the graph of f.

EXERCISE SET 4.2   Graphing Utility   [C] CAS

1. In each part, sketch the graph of a continuous function f with the stated properties.
   (a) f is concave up on the interval $(-\infty, +\infty)$ and has exactly one relative extremum.
   (b) f is concave up on the interval $(-\infty, +\infty)$ and has no relative extrema.
   (c) The function f has exactly two relative extrema on the interval $(-\infty, +\infty)$, and $f(x) \to +\infty$ as $x \to +\infty$.
   (d) The function f has exactly two relative extrema on the interval $(-\infty, +\infty)$, and $f(x) \to -\infty$ as $x \to +\infty$.

2. In each part, sketch the graph of a continuous function f with the stated properties.
   (a) f has exactly one relative extremum on $(-\infty, +\infty)$, and $f(x) \to 0$ as $x \to +\infty$ and as $x \to -\infty$.
   (b) f has exactly two relative extrema on $(-\infty, +\infty)$, and $f(x) \to 0$ as $x \to +\infty$ and as $x \to -\infty$.
   (c) f has exactly one inflection point and one relative extremum on $(-\infty, +\infty)$.
   (d) f has infinitely many relative extrema, and $f(x) \to 0$ as $x \to +\infty$ and as $x \to -\infty$.

3. (a) Use both the first and second derivative tests to show that $f(x) = 3x^2 - 6x + 1$ has a relative minimum at $x = 1$.
   (b) Use both the first and second derivative tests to show that $f(x) = x^3 - 3x + 3$ has a relative minimum at $x = 1$ and a relative maximum at $x = -1$.

4. (a) Use both the first and second derivative tests to show that $f(x) = \sin^2 x$ has a relative minimum at $x = 0$.
   (b) Use both the first and second derivative tests to show that $g(x) = \tan^2 x$ has a relative minimum at $x = 0$.
   (c) Give an informal verbal argument to explain without calculus why the functions in parts (a) and (b) have relative minima at $x = 0$.

5. (a) Show that both of the functions $f(x) = (x - 1)^4$ and $g(x) = x^3 - 3x^2 + 3x - 2$ have stationary points at $x = 1$.
   (b) What does the second derivative test tell you about the nature of these stationary points?

   (c) What does the first derivative test tell you about the nature of these stationary points?

6. (a) Show that $f(x) = 1 - x^5$ and $g(x) = 3x^4 - 8x^3$ both have stationary points at $x = 0$.
   (b) What does the second derivative test tell you about the nature of these stationary points?
   (c) What does the first derivative test tell you about the nature of these stationary points?

7–14 Locate the critical points and identify which critical points are stationary points. ■

7. $f(x) = 4x^4 - 16x^2 + 17$
8. $f(x) = 3x^4 + 12x$
9. $f(x) = \dfrac{x + 1}{x^2 + 3}$
10. $f(x) = \dfrac{x^2}{x^3 + 8}$
11. $f(x) = \sqrt[3]{x^2 - 25}$
12. $f(x) = x^2(x - 1)^{2/3}$
13. $f(x) = |\sin x|$
14. $f(x) = \sin|x|$

15–18 True–False Assume that f is continuous everywhere. Determine whether the statement is true or false. Explain your answer. ■

15. If f has a relative maximum at $x = 1$, then $f(1) \geq f(2)$.
16. If f has a relative maximum at $x = 1$, then $x = 1$ is a critical point for f.
17. If $f''(x) > 0$, then f has a relative minimum at $x = 1$.
18. If $p(x)$ is a polynomial such that $p'(x)$ has a simple root at $x = 1$, then p has a relative extremum at $x = 1$.

19–20 The graph of a function $f(x)$ is given. Sketch graphs of $y = f'(x)$ and $y = f''(x)$. ■

19.

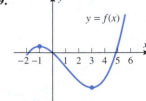

20.

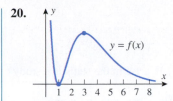

21–24 Use the graph of f' shown in the figure to estimate all values of x at which f has (a) relative minima, (b) relative maxima, and (c) inflection points. (d) Draw a rough sketch of the graph of a function f with the given derivative. ■

21.   **22.**

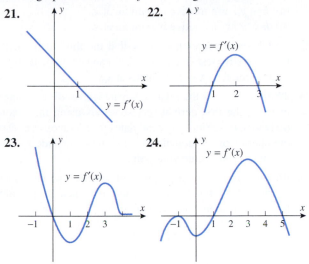

23.   **24.**

25–32 Use the given derivative to find all critical points of f, and at each critical point determine whether a relative maximum, relative minimum, or neither occurs. Assume in each case that f is continuous everywhere. ■

25. $f'(x) = x^2(x^3 - 5)$ **26.** $f'(x) = 4x^3 - 9x$

27. $f'(x) = \dfrac{2 - 3x}{\sqrt[3]{x + 2}}$ **28.** $f'(x) = \dfrac{x^2 - 7}{\sqrt[3]{x^2 + 4}}$

29. $f'(x) = xe^{1-x^2}$ **30.** $f'(x) = x^4(e^x - 3)$

31. $f'(x) = \ln\left(\dfrac{2}{1 + x^2}\right)$ **32.** $f'(x) = e^{2x} - 5e^x + 6$

33–36 Find the relative extrema using both first and second derivative tests. ■

33. $f(x) = 1 + 8x - 3x^2$ **34.** $f(x) = x^4 - 12x^3$

35. $f(x) = \sin 2x, \quad 0 < x < \pi$ **36.** $f(x) = (x - 3)e^x$

37–50 Use any method to find the relative extrema of the function f. ■

37. $f(x) = x^4 - 4x^3 + 4x^2$ **38.** $f(x) = x(x - 4)^3$

39. $f(x) = x^3(x + 1)^2$ **40.** $f(x) = x^2(x + 1)^3$

41. $f(x) = 2x + 3x^{2/3}$ **42.** $f(x) = 2x + 3x^{1/3}$

43. $f(x) = \dfrac{x + 3}{x - 2}$ **44.** $f(x) = \dfrac{x^2}{x^4 + 16}$

45. $f(x) = \ln(2 + x^2)$ **46.** $f(x) = \ln|2 + x^3|$

47. $f(x) = e^{2x} - e^x$ **48.** $f(x) = (xe^x)^2$

49. $f(x) = |3x - x^2|$ **50.** $f(x) = |1 + \sqrt[3]{x}|$

51–60 Give a graph of the polynomial and label the coordinates of the intercepts, stationary points, and inflection points. Check your work with a graphing utility. ■

51. $p(x) = x^2 - 3x - 4$ **52.** $p(x) = 1 + 8x - x^2$

53. $p(x) = 2x^3 - 3x^2 - 36x + 5$

54. $p(x) = 2 - x + 2x^2 - x^3$

55. $p(x) = (x + 1)^2(2x - x^2)$

56. $p(x) = x^4 - 6x^2 + 5$

57. $p(x) = x^4 - 2x^3 + 2x - 1$ **58.** $p(x) = 4x^3 - 9x^4$

59. $p(x) = x(x^2 - 1)^2$ **60.** $p(x) = x(x^2 - 1)^3$

61. In each part: (i) Make a conjecture about the behavior of the graph in the vicinity of its x-intercepts. (ii) Make a rough sketch of the graph based on your conjecture and the limits of the polynomial as $x \to +\infty$ and as $x \to -\infty$. (iii) Compare your sketch to the graph generated with a graphing utility.
(a) $y = x(x - 1)(x + 1)$ (b) $y = x^2(x - 1)^2(x + 1)^2$
(c) $y = x^2(x - 1)^2(x + 1)^3$ (d) $y = x(x - 1)^5(x + 1)^4$

62. Sketch the graph of $y = (x - a)^m(x - b)^n$ for the stated values of m and n, assuming that $a < b$ (six graphs in total).
(a) $m = 1, \ n = 1, 2, 3$ (b) $m = 2, \ n = 2, 3$
(c) $m = 3, \ n = 3$

63–66 Find the relative extrema in the interval $0 < x < 2\pi$, and confirm that your results are consistent with the graph of f generated with a graphing utility. ■

63. $f(x) = |\sin 2x|$ **64.** $f(x) = \sqrt{3}x + 2\sin x$

65. $f(x) = \cos^2 x$ **66.** $f(x) = \dfrac{\sin x}{2 - \cos x}$

67–70 Use a graphing utility to make a conjecture about the relative extrema of f, and then check your conjecture using either the first or second derivative test. ■

67. $f(x) = x \ln x$ **68.** $f(x) = \dfrac{2}{e^x + e^{-x}}$

69. $f(x) = x^2 e^{-2x}$ **70.** $f(x) = 10 \ln x - x$

71–72 Use a graphing utility to generate the graphs of f' and f'' over the stated interval, and then use those graphs to estimate the x-coordinates of the relative extrema of f. Check that your estimates are consistent with the graph of f. ■

71. $f(x) = x^4 - 24x^2 + 12x, \quad -5 \le x \le 5$

72. $f(x) = \sin \tfrac{1}{2}x \cos x, \quad -\pi/2 \le x \le \pi/2$

[c] **73–76** Use a CAS to graph f' and f'', and then use those graphs to estimate the x-coordinates of the relative extrema of f. Check that your estimates are consistent with the graph of f. ■

73. $f(x) = \dfrac{10x^3 - 3}{3x^2 - 5x + 8}$ **74.** $f(x) = \dfrac{\tan^{-1}(x^2 - x)}{x^2 + 4}$

75. $f(x) = \sqrt{x^4 + \cos^2 x}$ **76.** $f(x) = x^2(e^{2x} - e^x)$

77. In each part, find k so that f has a relative extremum at the point where $x = 3$.

  (a) $f(x) = x^2 + \dfrac{k}{x}$ (b) $f(x) = \dfrac{x}{x^2 + k}$

c 78. (a) Use a CAS to graph the function

$$f(x) = \frac{x^4 + 1}{x^2 + 1}$$

  and use the graph to estimate the x-coordinates of the relative extrema.

  (b) Find the exact x-coordinates by using the CAS to solve the equation $f'(x) = 0$.

79. Functions similar to

$$f(x) = \frac{1}{\sqrt{2\pi}} e^{-x^2/2}$$

arise in a wide variety of statistical problems.

  (a) Use the first derivative test to show that f has a relative maximum at $x = 0$, and confirm this by using a graphing utility to graph f.

  (b) Sketch the graph of

$$f(x) = \frac{1}{\sqrt{2\pi}} e^{-(x-\mu)^2/2}$$

  where μ is a constant, and label the coordinates of the relative extrema.

80. Functions of the form

$$f(x) = \frac{x^n e^{-x}}{n!}, \quad x > 0$$

where n is a positive integer, arise in the statistical study of traffic flow.

  (a) Use a graphing utility to generate the graph of f for $n = 2, 3, 4,$ and 5, and make a conjecture about the number and locations of the relative extrema of f.

  (b) Confirm your conjecture using the first derivative test.

81. Let h and g have relative maxima at x_0. Prove or disprove:

  (a) $h + g$ has a relative maximum at x_0.

  (b) $h - g$ has a relative maximum at x_0.

82. Sketch some curves that show that the three parts of the first derivative test (Theorem 4.2.3) can be false without the assumption that f is continuous at x_0.

83. Writing Discuss the relative advantages or disadvantages of using the first derivative test versus using the second derivative test to classify candidates for relative extrema on the interior of the domain of a function. Include specific examples to illustrate your points.

84. Writing If $p(x)$ is a polynomial, discuss the usefulness of knowing zeros for p, p', and p'' when determining information about the graph of p.

✔ **QUICK CHECK ANSWERS 4.2**

1. less than or equal to   **2.** 2, 7; 5   **3.** maximum; minimum   **4.** (a) $(0, 16)$ (b) $(-2, 0)$ and $(2, 0)$
(c) $(-2/\sqrt{3}, 64/9)$ and $(2/\sqrt{3}, 64/9)$

4.3 ANALYSIS OF FUNCTIONS III: RATIONAL FUNCTIONS, CUSPS, AND VERTICAL TANGENTS

In this section we will discuss procedures for graphing rational functions and other kinds of curves. We will also discuss the interplay between calculus and technology in curve sketching.

■ **PROPERTIES OF GRAPHS**

In many problems, the properties of interest in the graph of a function are:

- symmetries
- x-intercepts
- relative extrema
- intervals of increase and decrease
- asymptotes

- periodicity
- y-intercepts
- concavity
- inflection points
- behavior as $x \to +\infty$ or as $x \to -\infty$

Some of these properties may not be relevant in certain cases; for example, asymptotes are characteristic of rational functions but not of polynomials, and periodicity is characteristic of trigonometric functions but not of polynomial or rational functions. Thus, when analyzing the graph of a function f, it helps to know something about the general properties of the family to which it belongs.

In a given problem you will usually have a definite objective for your analysis of a graph. For example, you may be interested in showing all of the important characteristics of the function, you may only be interested in the behavior of the graph as $x \to +\infty$ or as $x \to -\infty$, or you may be interested in some specific feature such as a particular inflection point. Thus, your objectives in the problem will dictate those characteristics on which you want to focus.

■ **GRAPHING RATIONAL FUNCTIONS**

Recall that a rational function is a function of the form $f(x) = P(x)/Q(x)$ in which $P(x)$ and $Q(x)$ are polynomials. Graphs of rational functions are more complicated than those of polynomials because of the possibility of asymptotes and discontinuities (see Figure 0.3.11, for example). If $P(x)$ and $Q(x)$ have no common factors, then the information obtained in the following steps will usually be sufficient to obtain an accurate sketch of the graph of a rational function.

Graphing a Rational Function $f(x) = P(x)/Q(x)$ if $P(x)$ and $Q(x)$ have no Common Factors

Step 1. (symmetries). Determine whether there is symmetry about the y-axis or the origin.

Step 2. (x- and y-intercepts). Find the x- and y-intercepts.

Step 3. (vertical asymptotes). Find the values of x for which $Q(x) = 0$. The graph has a vertical asymptote at each such value.

Step 4. (sign of $f(x)$). The only places where $f(x)$ can change sign are at the x-intercepts or vertical asymptotes. Mark the points on the x-axis at which these occur and calculate a sample value of $f(x)$ in each of the open intervals determined by these points. This will tell you whether $f(x)$ is positive or negative over that interval.

Step 5. (end behavior). Determine the end behavior of the graph by computing the limits of $f(x)$ as $x \to +\infty$ and as $x \to -\infty$. If either limit has a finite value L, then the line $y = L$ is a horizontal asymptote.

Step 6. (derivatives). Find $f'(x)$ and $f''(x)$.

Step 7. (conclusions and graph). Analyze the sign changes of $f'(x)$ and $f''(x)$ to determine the intervals where $f(x)$ is increasing, decreasing, concave up, and concave down. Determine the locations of all stationary points, relative extrema, and inflection points. Use the sign analysis of $f(x)$ to determine the behavior of the graph in the vicinity of the vertical asymptotes. Sketch a graph of f that exhibits these conclusions.

▶ **Example 1** Sketch a graph of the equation

$$y = \frac{2x^2 - 8}{x^2 - 16}$$

and identify the locations of the intercepts, relative extrema, inflection points, and asymptotes.

Solution. The numerator and denominator have no common factors, so we will use the procedure just outlined.

- *Symmetries:* Replacing x by $-x$ does not change the equation, so the graph is symmetric about the y-axis.

- *x- and y-intercepts:* Setting $y = 0$ yields the x-intercepts $x = -2$ and $x = 2$. Setting $x = 0$ yields the y-intercept $y = \frac{1}{2}$.

- *Vertical asymptotes:* We observed above that the numerator and denominator of y have no common factors, so the graph has vertical asymptotes at the points where the denominator of y is zero, namely, at $x = -4$ and $x = 4$.

- *Sign of y:* The set of points where x-intercepts or vertical asymptotes occur is $\{-4, -2, 2, 4\}$. These points divide the x-axis into the open intervals

$$(-\infty, -4), \quad (-4, -2), \quad (-2, 2), \quad (2, 4), \quad (4, +\infty)$$

We can find the sign of y on each interval by choosing an arbitrary test point in the interval and evaluating $y = f(x)$ at the test point (Table 4.3.1). This analysis is summarized on the first line of Figure 4.3.1a.

- *End behavior:* The limits

$$\lim_{x \to +\infty} \frac{2x^2 - 8}{x^2 - 16} = \lim_{x \to +\infty} \frac{2 - (8/x^2)}{1 - (16/x^2)} = 2$$

$$\lim_{x \to -\infty} \frac{2x^2 - 8}{x^2 - 16} = \lim_{x \to -\infty} \frac{2 - (8/x^2)}{1 - (16/x^2)} = 2$$

yield the horizontal asymptote $y = 2$.

- *Derivatives:*

$$\frac{dy}{dx} = \frac{(x^2 - 16)(4x) - (2x^2 - 8)(2x)}{(x^2 - 16)^2} = -\frac{48x}{(x^2 - 16)^2}$$

$$\frac{d^2y}{dx^2} = \frac{48(16 + 3x^2)}{(x^2 - 16)^3} \quad \text{(verify)}$$

Table 4.3.1

SIGN ANALYSIS OF $y = \dfrac{2x^2 - 8}{x^2 - 16}$

INTERVAL	TEST POINT	VALUE OF y	SIGN OF y
$(-\infty, -4)$	-5	14/3	$+$
$(-4, -2)$	-3	$-10/7$	$-$
$(-2, 2)$	0	1/2	$+$
$(2, 4)$	3	$-10/7$	$-$
$(4, +\infty)$	5	14/3	$+$

Conclusions and graph:

- The sign analysis of y in Figure 4.3.1a reveals the behavior of the graph in the vicinity of the vertical asymptotes: The graph increases without bound as $x \to -4^-$ and decreases without bound as $x \to -4^+$; and the graph decreases without bound as $x \to 4^-$ and increases without bound as $x \to 4^+$ (Figure 4.3.1b).

- The sign analysis of dy/dx in Figure 4.3.1a shows that the graph is increasing to the left of $x = 0$ and is decreasing to the right of $x = 0$. Thus, there is a relative maximum at the stationary point $x = 0$. There are no relative minima.

- The sign analysis of d^2y/dx^2 in Figure 4.3.1a shows that the graph is concave up to the left of $x = -4$, is concave down between $x = -4$ and $x = 4$, and is concave up to the right of $x = 4$. There are no inflection points.

The procedure we stated for graphing a rational function $P(x)/Q(x)$ applies only if the polynomials $P(x)$ and $Q(x)$ have no common factors. How would you find the graph if those polynomials have common factors?

The graph is shown in Figure 4.3.1c. ◄

▶ **Example 2** Sketch a graph of

$$y = \frac{x^2 - 1}{x^3}$$

and identify the locations of all asymptotes, intercepts, relative extrema, and inflection points.

Solution. The numerator and denominator have no common factors, so we will use the procedure outlined previously.

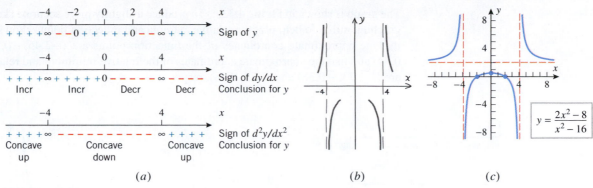

▲ **Figure 4.3.1**

Table 4.3.2

SIGN ANALYSIS OF $y = \dfrac{x^2 - 1}{x^3}$

INTERVAL	TEST POINT	VALUE OF y	SIGN OF y
$(-\infty, -1)$	-2	$-\frac{3}{8}$	$-$
$(-1, 0)$	$-\frac{1}{2}$	6	$+$
$(0, 1)$	$\frac{1}{2}$	-6	$-$
$(1, +\infty)$	2	$\frac{3}{8}$	$+$

- *Symmetries:* Replacing x by $-x$ and y by $-y$ yields an equation that simplifies to the original equation, so the graph is symmetric about the origin.

- *x- and y-intercepts:* Setting $y = 0$ yields the x-intercepts $x = -1$ and $x = 1$. Setting $x = 0$ leads to a division by zero, so there is no y-intercept.

- *Vertical asymptotes:* Setting $x^3 = 0$ yields the solution $x = 0$. This is not a root of $x^2 - 1$, so $x = 0$ is a vertical asymptote.

- *Sign of y:* The set of points where x-intercepts or vertical asymptotes occur is $\{-1, 0, 1\}$. These points divide the x-axis into the open intervals

$$(-\infty, -1), \quad (-1, 0), \quad (0, 1), \quad (1, +\infty)$$

Table 4.3.2 uses the method of test points to produce the sign of y on each of these intervals.

- *End behavior:* The limits

$$\lim_{x \to +\infty} \frac{x^2 - 1}{x^3} = \lim_{x \to +\infty} \left(\frac{1}{x} - \frac{1}{x^3} \right) = 0$$

$$\lim_{x \to -\infty} \frac{x^2 - 1}{x^3} = \lim_{x \to -\infty} \left(\frac{1}{x} - \frac{1}{x^3} \right) = 0$$

yield the horizontal asymptote $y = 0$.

- *Derivatives:*

$$\frac{dy}{dx} = \frac{x^3(2x) - (x^2 - 1)(3x^2)}{(x^3)^2} = \frac{3 - x^2}{x^4} = \frac{(\sqrt{3} + x)(\sqrt{3} - x)}{x^4}$$

$$\frac{d^2y}{dx^2} = \frac{x^4(-2x) - (3 - x^2)(4x^3)}{(x^4)^2} = \frac{2(x^2 - 6)}{x^5} = \frac{2(x - \sqrt{6})(x + \sqrt{6})}{x^5}$$

Conclusions and graph:

- The sign analysis of y in Figure 4.3.2a reveals the behavior of the graph in the vicinity of the vertical asymptote $x = 0$: The graph increases without bound as $x \to 0^-$ and decreases without bound as $x \to 0^+$ (Figure 4.3.2b).

- The sign analysis of dy/dx in Figure 4.3.2a shows that there is a relative minimum at $x = -\sqrt{3}$ and a relative maximum at $x = \sqrt{3}$.

- The sign analysis of d^2y/dx^2 in Figure 4.3.2a shows that the graph changes concavity at the vertical asymptote $x = 0$ and that there are inflection points at $x = -\sqrt{6}$ and $x = \sqrt{6}$.

The graph is shown in Figure 4.3.2c. To produce a slightly more accurate sketch, we used a graphing utility to help plot the relative extrema and inflection points. You should confirm that the approximate coordinates of the inflection points are $(-2.45, -0.34)$ and $(2.45, 0.34)$ and that the approximate coordinates of the relative minimum and relative maximum are $(-1.73, -0.38)$ and $(1.73, 0.38)$, respectively. ◄

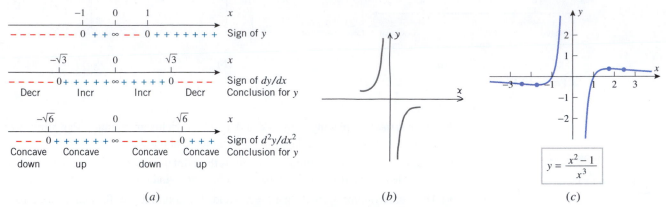

(a)   (b)   (c)

▲ **Figure 4.3.2**

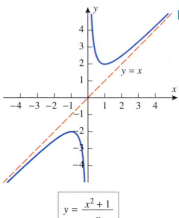

▲ **Figure 4.3.3**

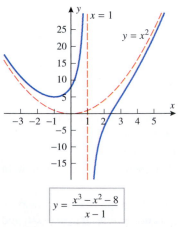

▲ **Figure 4.3.4**

RATIONAL FUNCTIONS WITH OBLIQUE OR CURVILINEAR ASYMPTOTES

In the rational functions of Examples 1 and 2, the degree of the numerator did not exceed the degree of the denominator, and the asymptotes were either vertical or horizontal. If the numerator of a rational function has greater degree than the denominator, then other kinds of "asymptotes" are possible. For example, consider the rational functions

$$f(x) = \frac{x^2 + 1}{x} \quad \text{and} \quad g(x) = \frac{x^3 - x^2 - 8}{x - 1} \tag{1}$$

By division we can rewrite these as

$$f(x) = x + \frac{1}{x} \quad \text{and} \quad g(x) = x^2 - \frac{8}{x - 1}$$

Since the second terms both approach 0 as $x \to +\infty$ or as $x \to -\infty$, it follows that

$$(f(x) - x) \to 0 \quad \text{as } x \to +\infty \text{ or as } x \to -\infty$$
$$(g(x) - x^2) \to 0 \quad \text{as } x \to +\infty \text{ or as } x \to -\infty$$

Geometrically, this means that the graph of $y = f(x)$ eventually gets closer and closer to the line $y = x$ as $x \to +\infty$ or as $x \to -\infty$. The line $y = x$ is called an **oblique** or **slant asymptote** of f. Similarly, the graph of $y = g(x)$ eventually gets closer and closer to the parabola $y = x^2$ as $x \to +\infty$ or as $x \to -\infty$. The parabola is called a **curvilinear asymptote** of g. The graphs of the functions in (1) are shown in Figures 4.3.3 and 4.3.4.

In general, if $f(x) = P(x)/Q(x)$ is a rational function, then we can find quotient and remainder polynomials $q(x)$ and $r(x)$ such that

$$f(x) = q(x) + \frac{r(x)}{Q(x)}$$

and the degree of $r(x)$ is less than the degree of $Q(x)$. Then $r(x)/Q(x) \to 0$ as $x \to +\infty$ and as $x \to -\infty$, so $y = q(x)$ is an asymptote of f. This asymptote will be an oblique line if the degree of $P(x)$ is one greater than the degree of $Q(x)$, and it will be curvilinear if the degree of $P(x)$ exceeds that of $Q(x)$ by two or more. Problems involving these kinds of asymptotes are given in the exercises (Exercises 17 and 18).

■ GRAPHS WITH VERTICAL TANGENTS AND CUSPS

Figure 4.3.5 shows four curve elements that are commonly found in graphs of functions that involve radicals or fractional exponents. In all four cases, the function is not differentiable at x_0 because the secant line through $(x_0, f(x_0))$ and $(x, f(x))$ approaches a vertical position as x approaches x_0 from either side. Thus, in each case, the curve has a vertical tangent line at $(x_0, f(x_0))$. In parts (a) and (b) of the figure, there is an inflection point at x_0 because there is a change in concavity at that point. In parts (c) and (d), where $f'(x)$ approaches $+\infty$ from one side of x_0 and $-\infty$ from the other side, we say that the graph has a ***cusp*** at x_0.

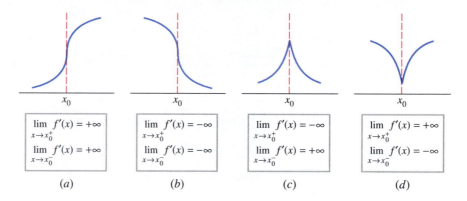

$$\lim_{x \to x_0^+} f'(x) = +\infty$$
$$\lim_{x \to x_0^-} f'(x) = +\infty$$

$$\lim_{x \to x_0^+} f'(x) = -\infty$$
$$\lim_{x \to x_0^-} f'(x) = -\infty$$

$$\lim_{x \to x_0^+} f'(x) = -\infty$$
$$\lim_{x \to x_0^-} f'(x) = +\infty$$

$$\lim_{x \to x_0^+} f'(x) = +\infty$$
$$\lim_{x \to x_0^-} f'(x) = -\infty$$

▶ **Figure 4.3.5** (a) (b) (c) (d)

> The steps that are used to sketch the graph of a rational function can serve as guidelines for sketching graphs of other types of functions. This is illustrated in Examples 3, 4, and 5.

▶ **Example 3**   Sketch the graph of $y = (x - 4)^{2/3}$.

- *Symmetries:* There are no symmetries about the coordinate axes or the origin (verify). However, the graph of $y = (x - 4)^{2/3}$ is symmetric about the line $x = 4$ since it is a translation (4 units to the right) of the graph of $y = x^{2/3}$, which is symmetric about the y-axis.

- *x- and y-intercepts:* Setting $y = 0$ yields the x-intercept $x = 4$. Setting $x = 0$ yields the y-intercept $y = \sqrt[3]{16} \approx 2.5$.

- *Vertical asymptotes:* None, since $f(x) = (x - 4)^{2/3}$ is continuous everywhere.

- *End behavior:* The graph has no horizontal asymptotes since

$$\lim_{x \to +\infty} (x - 4)^{2/3} = +\infty \quad \text{and} \quad \lim_{x \to -\infty} (x - 4)^{2/3} = +\infty$$

- *Derivatives:*

$$\frac{dy}{dx} = f'(x) = \frac{2}{3}(x - 4)^{-1/3} = \frac{2}{3(x - 4)^{1/3}}$$

$$\frac{d^2 y}{dx^2} = f''(x) = -\frac{2}{9}(x - 4)^{-4/3} = -\frac{2}{9(x - 4)^{4/3}}$$

- *Vertical tangent lines:* There is a vertical tangent line and cusp at $x = 4$ of the type in Figure 4.3.5d since $f(x) = (x - 4)^{2/3}$ is continuous at $x = 4$ and

$$\lim_{x \to 4^+} f'(x) = \lim_{x \to 4^+} \frac{2}{3(x - 4)^{1/3}} = +\infty$$

$$\lim_{x \to 4^-} f'(x) = \lim_{x \to 4^-} \frac{2}{3(x - 4)^{1/3}} = -\infty$$

Conclusions and graph:

- The function $f(x) = (x - 4)^{2/3} = ((x - 4)^{1/3})^2$ is nonnegative for all x. There is a zero for f at $x = 4$.

- There is a critical point at $x = 4$ since f is not differentiable there. We saw above that a cusp occurs at this point. The sign analysis of dy/dx in Figure 4.3.6a and the first derivative test show that there is a relative minimum at this cusp since $f'(x) < 0$ if $x < 4$ and $f'(x) > 0$ if $x > 4$.

- The sign analysis of d^2y/dx^2 in Figure 4.3.6a shows that the graph is concave down on both sides of the cusp.

The graph is shown in Figure 4.3.6b. ◄

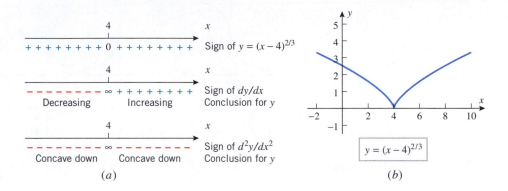

▶ **Figure 4.3.6**   (a)   (b)

▶ **Example 4**   Sketch the graph of $y = 6x^{1/3} + 3x^{4/3}$.

Solution.   It will help in our analysis to write

$$f(x) = 6x^{1/3} + 3x^{4/3} = 3x^{1/3}(2 + x)$$

- *Symmetries:* There are no symmetries about the coordinate axes or the origin (verify).

- *x- and y-intercepts:* Setting $y = 3x^{1/3}(2 + x) = 0$ yields the x-intercepts $x = 0$ and $x = -2$. Setting $x = 0$ yields the y-intercept $y = 0$.

- *Vertical asymptotes:* None, since $f(x) = 6x^{1/3} + 3x^{4/3}$ is continuous everywhere.

- *End behavior:* The graph has no horizontal asymptotes since

$$\lim_{x \to +\infty} (6x^{1/3} + 3x^{4/3}) = \lim_{x \to +\infty} 3x^{1/3}(2 + x) = +\infty$$

$$\lim_{x \to -\infty} (6x^{1/3} + 3x^{4/3}) = \lim_{x \to -\infty} 3x^{1/3}(2 + x) = +\infty$$

- *Derivatives:*

$$\frac{dy}{dx} = f'(x) = 2x^{-2/3} + 4x^{1/3} = 2x^{-2/3}(1 + 2x) = \frac{2(2x + 1)}{x^{2/3}}$$

$$\frac{d^2y}{dx^2} = f''(x) = -\frac{4}{3}x^{-5/3} + \frac{4}{3}x^{-2/3} = \frac{4}{3}x^{-5/3}(-1 + x) = \frac{4(x - 1)}{3x^{5/3}}$$

- *Vertical tangent lines:* There is a vertical tangent line at $x = 0$ since f is continuous there and

$$\lim_{x \to 0^+} f'(x) = \lim_{x \to 0^+} \frac{2(2x + 1)}{x^{2/3}} = +\infty$$

$$\lim_{x \to 0^-} f'(x) = \lim_{x \to 0^-} \frac{2(2x + 1)}{x^{2/3}} = +\infty$$

This and the change in concavity at $x = 0$ mean that $(0, 0)$ is an inflection point of the type in Figure 4.3.5a.

Conclusions and graph:

- From the sign analysis of y in Figure 4.3.7a, the graph is below the x-axis between the x-intercepts $x = -2$ and $x = 0$ and is above the x-axis if $x < -2$ or $x > 0$.

- From the formula for dy/dx we see that there is a stationary point at $x = -\frac{1}{2}$ and a critical point at $x = 0$ at which f is not differentiable. We saw above that a vertical tangent line and inflection point are at that critical point.

- The sign analysis of dy/dx in Figure 4.3.7a and the first derivative test show that there is a relative minimum at the stationary point at $x = -\frac{1}{2}$ (verify).

- The sign analysis of d^2y/dx^2 in Figure 4.3.7a shows that in addition to the inflection point at the vertical tangent there is an inflection point at $x = 1$ at which the graph changes from concave down to concave up.

The graph is shown in Figure 4.3.7b. ◄

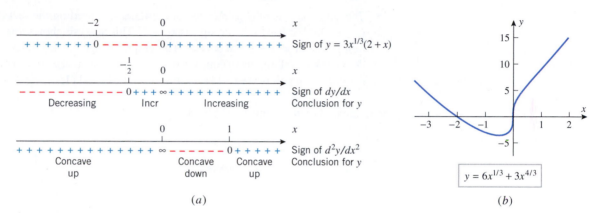

▲ **Figure 4.3.7**

■ GRAPHING OTHER KINDS OF FUNCTIONS

We have discussed methods for graphing polynomials, rational functions, and functions with cusps and vertical tangent lines. The same calculus tools that we used to analyze these functions can also be used to analyze and graph trigonometric functions, logarithmic and exponential functions, and an endless variety of other kinds of functions.

▶ **Example 5** Sketch the graph of $y = e^{-x^2/2}$ and identify the locations of all relative extrema and inflection points.

Solution.

- *Symmetries:* Replacing x by $-x$ does not change the equation, so the graph is symmetric about the y-axis.

- *x- and y-intercepts:* Setting $y = 0$ leads to the equation $e^{-x^2/2} = 0$, which has no solutions since all powers of e have positive values. Thus, there are no x-intercepts. Setting $x = 0$ yields the y-intercept $y = 1$.

- *Vertical asymptotes:* There are no vertical asymptotes since $e^{-x^2/2}$ is continuous on $(-\infty, +\infty)$.

- *End behavior:* The x-axis ($y = 0$) is a horizontal asymptote since

$$\lim_{x \to -\infty} e^{-x^2/2} = \lim_{x \to +\infty} e^{-x^2/2} = 0$$

- *Derivatives:*

$$\frac{dy}{dx} = e^{-x^2/2}\frac{d}{dx}\left[-\frac{x^2}{2}\right] = -xe^{-x^2/2}$$

$$\frac{d^2y}{dx^2} = -x\frac{d}{dx}[e^{-x^2/2}] + e^{-x^2/2}\frac{d}{dx}[-x]$$

$$= x^2 e^{-x^2/2} - e^{-x^2/2} = (x^2 - 1)e^{-x^2/2}$$

Conclusions and graph:

- The sign analysis of y in Figure 4.3.8a is based on the fact that $e^{-x^2/2} > 0$ for all x. This shows that the graph is always above the x-axis.
- The sign analysis of dy/dx in Figure 4.3.8a is based on the fact that $dy/dx = -xe^{-x^2/2}$ has the same sign as $-x$. This analysis and the first derivative test show that there is a stationary point at $x = 0$ at which there is a relative maximum. The value of y at the relative maximum is $y = e^0 = 1$.
- The sign analysis of d^2y/dx^2 in Figure 4.3.8a is based on the fact that $d^2y/dx^2 = (x^2 - 1)e^{-x^2/2}$ has the same sign as $x^2 - 1$. This analysis shows that there are inflection points at $x = -1$ and $x = 1$. The graph changes from concave up to concave down at $x = -1$ and from concave down to concave up at $x = 1$. The coordinates of the inflection points are $(-1, e^{-1/2}) \approx (-1, 0.61)$ and $(1, e^{-1/2}) \approx (1, 0.61)$.

The graph is shown in Figure 4.3.8b. ◄

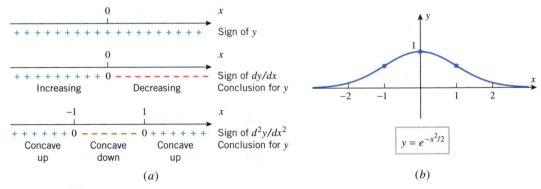

▲ **Figure 4.3.8**

■ GRAPHING USING CALCULUS AND TECHNOLOGY TOGETHER

Thus far in this chapter we have used calculus to produce graphs of functions; the graph was the end result. Now we will work in the reverse direction by *starting* with a graph produced by a graphing utility. Our goal will be to use the tools of calculus to determine the exact locations of relative extrema, inflection points, and other features suggested by that graph and to determine whether the graph may be missing some important features that we would like to see.

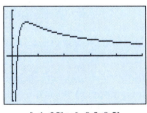

$[-1, 25] \times [-0.5, 0.5]$
xScl = 5, yScl = 0.2

$$y = \frac{\ln x}{x}$$

▲ **Figure 4.3.9**

▶ **Example 6**   Use a graphing utility to generate the graph of $f(x) = (\ln x)/x$, and discuss what it tells you about relative extrema, inflection points, asymptotes, and end behavior. Use calculus to find the locations of all key features of the graph.

Solution.   Figure 4.3.9 shows a graph of f produced by a graphing utility. The graph suggests that there is an x-intercept near $x = 1$, a relative maximum somewhere between

$x = 0$ and $x = 5$, an inflection point near $x = 5$, a vertical asymptote at $x = 0$, and possibly a horizontal asymptote $y = 0$. For a more precise analysis of this information we need to consider the derivatives

$$f'(x) = \frac{x\left(\dfrac{1}{x}\right) - (\ln x)(1)}{x^2} = \frac{1 - \ln x}{x^2}$$

$$f''(x) = \frac{x^2\left(-\dfrac{1}{x}\right) - (1 - \ln x)(2x)}{x^4} = \frac{2x \ln x - 3x}{x^4} = \frac{2 \ln x - 3}{x^3}$$

- *Relative extrema:* Solving $f'(x) = 0$ yields the stationary point $x = e$ (verify). Since

$$f''(e) = \frac{2 - 3}{e^3} = -\frac{1}{e^3} < 0$$

there is a relative maximum at $x = e \approx 2.7$ by the second derivative test.

- *Inflection points:* Since $f(x) = (\ln x)/x$ is only defined for positive values of x, the second derivative $f''(x)$ has the same sign as $2 \ln x - 3$. We leave it for you to use the inequalities $(2 \ln x - 3) < 0$ and $(2 \ln x - 3) > 0$ to show that $f''(x) < 0$ if $x < e^{3/2}$ and $f''(x) > 0$ if $x > e^{3/2}$. Thus, there is an inflection point at $x = e^{3/2} \approx 4.5$.

- *Asymptotes:* Applying L'Hôpital's rule we have

$$\lim_{x \to +\infty} \frac{\ln x}{x} = \lim_{x \to +\infty} \frac{(1/x)}{1} = \lim_{x \to +\infty} \frac{1}{x} = 0$$

so that $y = 0$ is a horizontal asymptote. Also, there is a vertical asymptote at $x = 0$ since

$$\lim_{x \to 0^+} \frac{\ln x}{x} = -\infty$$

(why?).

- *Intercepts:* Setting $f(x) = 0$ yields $(\ln x)/x = 0$. The only real solution of this equation is $x = 1$, so there is an x-intercept at this point. ◄

✓ **QUICK CHECK EXERCISES 4.3**   *(See page 266 for answers.)*

1. Let $f(x) = \dfrac{3(x + 1)(x - 3)}{(x + 2)(x - 4)}$. Given that

$$f'(x) = \frac{-30(x - 1)}{(x + 2)^2(x - 4)^2}, \qquad f''(x) = \frac{90(x^2 - 2x + 4)}{(x + 2)^3(x - 4)^3}$$

determine the following properties of the graph of f.
   (a) The x- and y-intercepts are _____.
   (b) The vertical asymptotes are _____.
   (c) The horizontal asymptote is _____.
   (d) The graph is above the x-axis on the intervals _____.
   (e) The graph is increasing on the intervals _____.
   (f) The graph is concave up on the intervals _____.
   (g) The relative maximum point on the graph is _____.

2. Let $f(x) = \dfrac{x^2 - 4}{x^{8/3}}$. Given that

$$f'(x) = \frac{-2(x^2 - 16)}{3x^{11/3}}, \qquad f''(x) = \frac{2(5x^2 - 176)}{9x^{14/3}}$$

determine the following properties of the graph of f.

   (a) The x-intercepts are _____.
   (b) The vertical asymptote is _____.
   (c) The horizontal asymptote is _____.
   (d) The graph is above the x-axis on the intervals _____.
   (e) The graph is increasing on the intervals _____.
   (f) The graph is concave up on the intervals _____.
   (g) Inflection points occur at $x =$ _____.

3. Let $f(x) = (x - 2)^2 e^{x/2}$. Given that

$$f'(x) = \tfrac{1}{2}(x^2 - 4)e^{x/2}, \qquad f''(x) = \tfrac{1}{4}(x^2 + 4x - 4)e^{x/2}$$

determine the following properties of the graph of f.
   (a) The horizontal asymptote is _____.
   (b) The graph is above the x-axis on the intervals _____.
   (c) The graph is increasing on the intervals _____.
   (d) The graph is concave up on the intervals _____.
   (e) The relative minimum point on the graph is _____.
   (f) The relative maximum point on the graph is _____.
   (g) Inflection points occur at $x =$ _____.

EXERCISE SET 4.3 Graphing Utility

1–14 Give a graph of the rational function and label the co-ordinates of the stationary points and inflection points. Show the horizontal and vertical asymptotes and label them with their equations. Label point(s), if any, where the graph crosses a horizontal asymptote. Check your work with a graphing utility.

1. $\dfrac{2x-6}{4-x}$

2. $\dfrac{8}{x^2-4}$

3. $\dfrac{x}{x^2-4}$

4. $\dfrac{x^2}{x^2-4}$

5. $\dfrac{x^2}{x^2+4}$

6. $\dfrac{(x^2-1)^2}{x^4+1}$

7. $\dfrac{x^3+1}{x^3-1}$

8. $2-\dfrac{1}{3x^2+x^3}$

9. $\dfrac{4}{x^2}-\dfrac{2}{x}+3$

10. $\dfrac{3(x+1)^2}{(x-1)^2}$

11. $\dfrac{(3x+1)^2}{(x-1)^2}$

12. $3+\dfrac{x+1}{(x-1)^4}$

13. $\dfrac{x^2+x}{1-x^2}$

14. $\dfrac{x^2}{1-x^3}$

15–16 In each part, make a rough sketch of the graph using asymptotes and appropriate limits but no derivatives. Compare your graph to that generated with a graphing utility.

15. (a) $y=\dfrac{3x^2-8}{x^2-4}$ (b) $y=\dfrac{x^2+2x}{x^2-1}$

16. (a) $y=\dfrac{2x-x^2}{x^2+x-2}$ (b) $y=\dfrac{x^2}{x^2-x-2}$

17. Show that $y=x+3$ is an oblique asymptote of the graph of $f(x)=x^2/(x-3)$. Sketch the graph of $y=f(x)$ showing this asymptotic behavior.

18. Show that $y=3-x^2$ is a curvilinear asymptote of the graph of $f(x)=(2+3x-x^3)/x$. Sketch the graph of $y=f(x)$ showing this asymptotic behavior.

19–24 Sketch a graph of the rational function and label the co-ordinates of the stationary points and inflection points. Show the horizontal, vertical, oblique, and curvilinear asymptotes and label them with their equations. Label point(s), if any, where the graph crosses an asymptote. Check your work with a graphing utility.

19. $x^2-\dfrac{1}{x}$

20. $\dfrac{x^2-2}{x}$

21. $\dfrac{(x-2)^3}{x^2}$

22. $x-\dfrac{1}{x}-\dfrac{1}{x^2}$

23. $\dfrac{x^3-4x-8}{x+2}$

24. $\dfrac{x^5}{x^2+1}$

FOCUS ON CONCEPTS

25. In each part, match the function with graphs I–VI.
    (a) $x^{1/3}$ (b) $x^{1/4}$ (c) $x^{1/5}$
    (d) $x^{2/5}$ (e) $x^{4/3}$ (f) $x^{-1/3}$

I 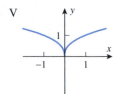 II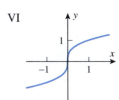

III   IV

V   VI

▲ **Figure Ex-25**

26. Sketch the general shape of the graph of $y=x^{1/n}$, and then explain in words what happens to the shape of the graph as n increases if
    (a) n is a positive even integer
    (b) n is a positive odd integer.

27–30 True–False Determine whether the statement is true or false. Explain your answer.

27. Suppose that $f(x)=P(x)/Q(x)$, where P and Q are polynomials with no common factors. If $y=5$ is a horizontal asymptote for the graph of f, then P and Q have the same degree.

28. If the graph of f has a vertical asymptote at $x=1$, then f cannot be continuous at $x=1$.

29. If the graph of f' has a vertical asymptote at $x=1$, then f cannot be continuous at $x=1$.

30. If the graph of f has a cusp at $x=1$, then f cannot have an inflection point at $x=1$.

31–38 Give a graph of the function and identify the locations of all critical points and inflection points. Check your work with a graphing utility.

31. $\sqrt{4x^2-1}$

32. $\sqrt[3]{x^2-4}$

33. $2x+3x^{2/3}$

34. $2x^2-3x^{4/3}$

35. $4x^{1/3}-x^{4/3}$

36. $5x^{2/3}+x^{5/3}$

37. $\dfrac{8+x}{2+\sqrt[3]{x}}$

38. $\dfrac{8(\sqrt{x}-1)}{x}$

39–44 Give a graph of the function and identify the locations of all relative extrema and inflection points. Check your work with a graphing utility. ■

39. $x + \sin x$ **40.** $x - \tan x$

41. $\sqrt{3}\cos x + \sin x$ **42.** $\sin x + \cos x$

43. $\sin^2 x - \cos x, \quad -\pi \le x \le 3\pi$

44. $\sqrt{\tan x}, \quad 0 \le x < \pi/2$

 45–54 Using L'Hôpital's rule (Section 3.6) one can verify that

$$\lim_{x \to +\infty} \frac{e^x}{x} = +\infty, \quad \lim_{x \to +\infty} \frac{x}{e^x} = 0, \quad \lim_{x \to -\infty} xe^x = 0$$

In these exercises: (a) Use these results, as necessary, to find the limits of $f(x)$ as $x \to +\infty$ and as $x \to -\infty$. (b) Sketch a graph of $f(x)$ and identify all relative extrema, inflection points, and asymptotes (as appropriate). Check your work with a graphing utility. ■

45. $f(x) = xe^x$ **46.** $f(x) = xe^{-x}$

47. $f(x) = x^2 e^{-2x}$ **48.** $f(x) = x^2 e^{2x}$

49. $f(x) = x^2 e^{-x^2}$ **50.** $f(x) = e^{-1/x^2}$

51. $f(x) = \dfrac{e^x}{1-x}$ **52.** $f(x) = x^{2/3} e^x$

53. $f(x) = x^2 e^{1-x}$ **54.** $f(x) = x^3 e^{x-1}$

 55–60 Using L'Hôpital's rule (Section 3.6) one can verify that

$$\lim_{x \to +\infty} \frac{\ln x}{x^r} = 0, \quad \lim_{x \to +\infty} \frac{x^r}{\ln x} = +\infty, \quad \lim_{x \to 0^+} x^r \ln x = 0$$

for any positive real number r. In these exercises: (a) Use these results, as necessary, to find the limits of $f(x)$ as $x \to +\infty$ and as $x \to 0^+$. (b) Sketch a graph of $f(x)$ and identify all relative extrema, inflection points, and asymptotes (as appropriate). Check your work with a graphing utility. ■

55. $f(x) = x \ln x$ **56.** $f(x) = x^2 \ln x$

57. $f(x) = x^2 \ln(2x)$ **58.** $f(x) = \ln(x^2 + 1)$

59. $f(x) = x^{2/3} \ln x$ **60.** $f(x) = x^{-1/3} \ln x$

FOCUS ON CONCEPTS

61. Consider the family of curves $y = xe^{-bx}$ $(b > 0)$.
  (a) Use a graphing utility to generate some members of this family.
  (b) Discuss the effect of varying b on the shape of the graph, and discuss the locations of the relative extrema and inflection points.

62. Consider the family of curves $y = e^{-bx^2}$ $(b > 0)$.
  (a) Use a graphing utility to generate some members of this family.
  (b) Discuss the effect of varying b on the shape of the graph, and discuss the locations of the relative extrema and inflection points.

63. (a) Determine whether the following limits exist, and if so, find them:
$$\lim_{x \to +\infty} e^x \cos x, \quad \lim_{x \to -\infty} e^x \cos x$$

  (b) Sketch the graphs of the equations $y = e^x$, $y = -e^x$, and $y = e^x \cos x$ in the same coordinate system, and label any points of intersection.
  (c) Use a graphing utility to generate some members of the family $y = e^{ax} \cos bx$ $(a > 0$ and $b > 0)$, and discuss the effect of varying a and b on the shape of the curve.

64. Consider the family of curves $y = x^n e^{-x^2/n}$, where n is a positive integer.
  (a) Use a graphing utility to generate some members of this family.
  (b) Discuss the effect of varying n on the shape of the graph, and discuss the locations of the relative extrema and inflection points.

65. The accompanying figure shows the graph of the *derivative* of a function h that is defined and continuous on the interval $(-\infty, +\infty)$. Assume that the graph of h' has a vertical asymptote at $x = 3$ and that
$$h'(x) \to 0^+ \text{ as } x \to -\infty$$
$$h'(x) \to -\infty \text{ as } x \to +\infty$$

  (a) What are the critical points for $h(x)$?
  (b) Identify the intervals on which $h(x)$ is increasing.
  (c) Identify the x-coordinates of relative extrema for $h(x)$ and classify each as a relative maximum or relative minimum.
  (d) Estimate the x-coordinates of inflection points for $h(x)$.

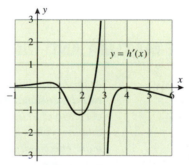

$y = h'(x)$

◀ **Figure Ex-65**

66. Let $f(x) = (1 - 2x)h(x)$, where $h(x)$ is as given in Exercise 65. Suppose that $x = 5$ is a critical point for $f(x)$.
  (a) Estimate $h(5)$.
  (b) Use the second derivative test to determine whether $f(x)$ has a relative maximum or a relative minimum at $x = 5$.

67. A rectangular plot of land is to be fenced off so that the area enclosed will be 400 ft². Let L be the length of fencing needed and x the length of one side of the rectangle. Show that $L = 2x + 800/x$ for $x > 0$, and sketch the graph of L versus x for $x > 0$.

68. A box with a square base and open top is to be made from sheet metal so that its volume is 500 in³. Let S be the area of the surface of the box and x the length of a side of the square base. Show that $S = x^2 + 2000/x$ for $x > 0$, and sketch the graph of S versus x for $x > 0$.

69. The accompanying figure shows a computer-generated graph of the polynomial $y = 0.1x^5(x - 1)$ using a viewing window of $[-2, 2.5] \times [-1, 5]$. Show that the choice of the vertical scale caused the computer to miss important features of the graph. Find the features that were missed and make your own sketch of the graph that shows the missing features.

70. The accompanying figure shows a computer-generated graph of the polynomial $y = 0.1x^5(x + 1)^2$ using a viewing window of $[-2, 1.5] \times [-0.2, 0.2]$. Show that the choice of the vertical scale caused the computer to miss important features of the graph. Find the features that were missed and make your own sketch of the graph that shows the missing features.

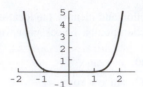

Generated by Mathematica

▲ **Figure Ex-69**

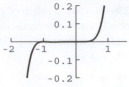

Generated by Mathematica

▲ **Figure Ex-70**

71. Writing Suppose that $x = x_0$ is a point at which a function f is continuous but not differentiable and that $f'(x)$ approaches different finite limits as x approaches x_0 from either side. Invent your own term to describe the graph of f at such a point and discuss the appropriateness of your term.

72. Writing Suppose that the graph of a function f is obtained using a graphing utility. Discuss the information that calculus techniques can provide about f to add to what can already be inferred about f from the graph as shown on your utility's display.

✔ **QUICK CHECK ANSWERS 4.3**

1. (a) $(-1, 0), (3, 0), (0, \frac{9}{8})$ (b) $x = -2$ and $x = 4$ (c) $y = 3$ (d) $(-\infty, -2), (-1, 3),$ and $(4, +\infty)$ (e) $(-\infty, -2)$ and $(-2, 1]$
(f) $(-\infty, -2)$ and $(4, +\infty)$ (g) $\left(1, \frac{4}{3}\right)$ **2.** (a) $(-2, 0), (2, 0)$ (b) $x = 0$ (c) $y = 0$ (d) $(-\infty, -2)$ and $(2, +\infty)$
(e) $(-\infty, -4]$ and $(0, 4]$ (f) $(-\infty, -4\sqrt{11/5})$ and $(4\sqrt{11/5}, +\infty)$ (g) $\pm 4\sqrt{11/5} \approx \pm 5.93$ **3.** (a) $y = 0$ (as $x \to -\infty$)
(b) $(-\infty, 2)$ and $(2, +\infty)$ (c) $(-\infty, -2]$ and $[2, +\infty)$ (d) $(-\infty, -2 - 2\sqrt{2})$ and $(-2 + 2\sqrt{2}, +\infty)$ (e) $(2, 0)$
(f) $(-2, 16e^{-1}) \approx (-2, 5.89)$ (g) $-2 \pm 2\sqrt{2}$

4.4 ABSOLUTE MAXIMA AND MINIMA

At the beginning of Section 4.2 we observed that if the graph of a function f is viewed as a two-dimensional mountain range (Figure 4.2.1), then the relative maxima and minima correspond to the tops of the hills and the bottoms of the valleys; that is, they are the high and low points in their immediate vicinity. In this section we will be concerned with the more encompassing problem of finding the highest and lowest points over the entire mountain range, that is, we will be looking for the top of the highest hill and the bottom of the deepest valley. In mathematical terms, we will be looking for the largest and smallest values of a function over an interval.

■ **ABSOLUTE EXTREMA**

We will begin with some terminology for describing the largest and smallest values of a function on an interval.

4.4.1 DEFINITION Consider an interval in the domain of a function f and a point x_0 in that interval. We say that f has an **absolute maximum** at x_0 if $f(x) \le f(x_0)$ for all x in the interval, and we say that f has an **absolute minimum** at x_0 if $f(x_0) \le f(x)$ for all x in the interval. We say that f has an **absolute extremum** at x_0 if it has either an absolute maximum or an absolute minimum at that point.

If f has an absolute maximum at the point x_0 on an interval, then $f(x_0)$ is the largest value of f on the interval, and if f has an absolute minimum at x_0, then $f(x_0)$ is the smallest value of f on the interval. In general, there is no guarantee that a function will actually have an absolute maximum or minimum on a given interval (Figure 4.4.1).

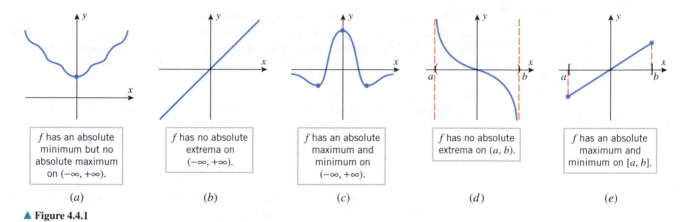

f has an absolute minimum but no absolute maximum on $(-\infty, +\infty)$.

(a)

f has no absolute extrema on $(-\infty, +\infty)$.

(b)

f has an absolute maximum and minimum on $(-\infty, +\infty)$.

(c)

f has no absolute extrema on (a, b).

(d)

f has an absolute maximum and minimum on $[a, b]$.

(e)

▲ **Figure 4.4.1**

■ THE EXTREME VALUE THEOREM

Parts (a)–(d) of Figure 4.4.1 show that a continuous function may or may not have absolute maxima or minima on an infinite interval or on a finite open interval. However, the following theorem shows that a continuous function must have both an absolute maximum and an absolute minimum on every *finite closed* interval [see part (e) of Figure 4.4.1].

> The hypotheses in the Extreme-Value Theorem are essential. That is, if either the interval is not closed or f is not continuous on the interval, then f need not have absolute extrema on the interval (Exercises 4–6).

4.4.2 **THEOREM** (*Extreme-Value Theorem*) *If a function f is continuous on a finite closed interval $[a, b]$, then f has both an absolute maximum and an absolute minimum on $[a, b]$.*

REMARK Although the proof of this theorem is too difficult to include here, you should be able to convince yourself of its validity with a little experimentation—try graphing various continuous functions over the interval $[0, 1]$, and convince yourself that there is no way to avoid having a highest and lowest point on a graph. As a physical analogy, if you imagine the graph to be a roller-coaster track starting at $x = 0$ and ending at $x = 1$, the roller coaster will have to pass through a highest point and a lowest point during the trip.

The Extreme-Value Theorem is an example of what mathematicians call an ***existence theorem***. Such theorems state conditions under which certain objects exist, in this case absolute extrema. However, knowing that an object exists and finding it are two separate things. We will now address methods for determining the locations of absolute extrema under the conditions of the Extreme-Value Theorem.

If f is continuous on the finite closed interval $[a, b]$, then the absolute extrema of f occur either at the endpoints of the interval or inside on the open interval (a, b). If the absolute extrema happen to fall inside, then the following theorem tells us that they must occur at critical points of f.

> Theorem 4.4.3 is also valid on infinite open intervals, that is, intervals of the form $(-\infty, +\infty)$, $(a, +\infty)$, and $(-\infty, b)$.

4.4.3 **THEOREM** *If f has an absolute extremum on an open interval (a, b), then it must occur at a critical point of f.*

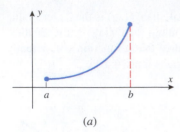

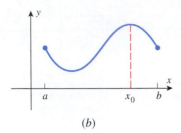

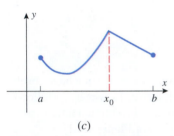

▲ Figure 4.4.2   In part (*a*) the absolute maximum occurs at an endpoint of [*a*, *b*], in part (*b*) it occurs at a stationary point in (*a*, *b*), and in part (*c*) it occurs at a critical point in (*a*, *b*) where *f* is not differentiable.

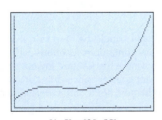

[1, 5] × [20, 55]
*x*Scl = 1, *y*Scl = 10

$$y = 2x^3 - 15x^2 + 36x$$

▲ Figure 4.4.3

Table 4.4.1

x	-1	0	$\frac{1}{8}$	1
$f(x)$	9	0	$-\frac{9}{8}$	3

PROOF   If f has an absolute maximum on (a, b) at x_0, then $f(x_0)$ is also a relative maximum for f; for if $f(x_0)$ is the largest value of f on all (a, b), then $f(x_0)$ is certainly the largest value for f in the immediate vicinity of x_0. Thus, x_0 is a critical point of f by Theorem 4.2.2. The proof for absolute minima is similar. ■

It follows from this theorem that if f is continuous on the finite closed interval $[a, b]$, then the absolute extrema occur either at the endpoints of the interval or at critical points inside the interval (Figure 4.4.2). Thus, we can use the following procedure to find the absolute extrema of a continuous function on a finite closed interval $[a, b]$.

> *A Procedure for Finding the Absolute Extrema of a Continuous Function f on a Finite Closed Interval* [*a*, *b*]
>
> **Step 1.**   Find the critical points of f in (a, b).
>
> **Step 2.**   Evaluate f at all the critical points and at the endpoints a and b.
>
> **Step 3.**   The largest of the values in Step 2 is the absolute maximum value of f on $[a, b]$ and the smallest value is the absolute minimum.

▶ **Example 1**   Find the absolute maximum and minimum values of the function $f(x) = 2x^3 - 15x^2 + 36x$ on the interval $[1, 5]$, and determine where these values occur.

Solution.   Since f is continuous and differentiable everywhere, the absolute extrema must occur either at endpoints of the interval or at solutions to the equation $f'(x) = 0$ in the open interval $(1, 5)$. The equation $f'(x) = 0$ can be written as

$$6x^2 - 30x + 36 = 6(x^2 - 5x + 6) = 6(x - 2)(x - 3) = 0$$

Thus, there are stationary points at $x = 2$ and at $x = 3$. Evaluating f at the endpoints, at $x = 2$, and at $x = 3$ yields

$$f(1) = 2(1)^3 - 15(1)^2 + 36(1) = 23$$
$$f(2) = 2(2)^3 - 15(2)^2 + 36(2) = 28$$
$$f(3) = 2(3)^3 - 15(3)^2 + 36(3) = 27$$
$$f(5) = 2(5)^3 - 15(5)^2 + 36(5) = 55$$

from which we conclude that the absolute minimum of f on $[1, 5]$ is 23, occurring at $x = 1$, and the absolute maximum of f on $[1, 5]$ is 55, occurring at $x = 5$. This is consistent with the graph of f in Figure 4.4.3. ◀

▶ **Example 2**   Find the absolute extrema of $f(x) = 6x^{4/3} - 3x^{1/3}$ on the interval $[-1, 1]$, and determine where these values occur.

Solution.   Note that f is continuous everywhere and therefore the Extreme-Value Theorem guarantees that f has a maximum and a minimum value in the interval $[-1, 1]$. Differentiating, we obtain

$$f'(x) = 8x^{1/3} - x^{-2/3} = x^{-2/3}(8x - 1) = \frac{8x - 1}{x^{2/3}}$$

Thus, $f'(x) = 0$ at $x = \frac{1}{8}$, and $f'(x)$ is undefined at $x = 0$. Evaluating f at these critical points and endpoints yields Table 4.4.1, from which we conclude that an absolute minimum value of $-\frac{9}{8}$ occurs at $x = \frac{1}{8}$, and an absolute maximum value of 9 occurs at $x = -1$. ◀

■ ABSOLUTE EXTREMA ON INFINITE INTERVALS

We observed earlier that a continuous function may or may not have absolute extrema on an infinite interval (see Figure 4.4.1). However, certain conclusions about the existence of absolute extrema of a continuous function f on $(-\infty, +\infty)$ can be drawn from the behavior of $f(x)$ as $x \to -\infty$ and as $x \to +\infty$ (Table 4.4.2).

Table 4.4.2
ABSOLUTE EXTREMA ON INFINITE INTERVALS

LIMITS	$\lim\limits_{x \to -\infty} f(x) = +\infty$ $\lim\limits_{x \to +\infty} f(x) = +\infty$	$\lim\limits_{x \to -\infty} f(x) = -\infty$ $\lim\limits_{x \to +\infty} f(x) = -\infty$	$\lim\limits_{x \to -\infty} f(x) = -\infty$ $\lim\limits_{x \to +\infty} f(x) = +\infty$	$\lim\limits_{x \to -\infty} f(x) = +\infty$ $\lim\limits_{x \to +\infty} f(x) = -\infty$
CONCLUSION IF f **IS CONTINUOUS EVERYWHERE**	f has an absolute minimum but no absolute maximum on $(-\infty, +\infty)$.	f has an absolute maximum but no absolute minimum on $(-\infty, +\infty)$.	f has neither an absolute maximum nor an absolute minimum on $(-\infty, +\infty)$.	f has neither an absolute maximum nor an absolute minimum on $(-\infty, +\infty)$.
GRAPH				

▶ **Example 3**   What can you say about the existence of absolute extrema on $(-\infty, +\infty)$ for polynomials?

Solution.   If $p(x)$ is a polynomial of odd degree, then

$$\lim_{x \to +\infty} p(x) \quad \text{and} \quad \lim_{x \to -\infty} p(x) \tag{1}$$

have opposite signs (one is $+\infty$ and the other is $-\infty$), so there are no absolute extrema. On the other hand, if $p(x)$ has even degree, then the limits in (1) have the same sign (both $+\infty$ or both $-\infty$). If the leading coefficient is positive, then both limits are $+\infty$, and there is an absolute minimum but no absolute maximum; if the leading coefficient is negative, then both limits are $-\infty$, and there is an absolute maximum but no absolute minimum. ◄

▶ **Example 4**   Determine by inspection whether $p(x) = 3x^4 + 4x^3$ has any absolute extrema. If so, find them and state where they occur.

Solution.   Since $p(x)$ has even degree and the leading coefficient is positive, $p(x) \to +\infty$ as $x \to \pm\infty$. Thus, there is an absolute minimum but no absolute maximum. From Theorem 4.4.3 [applied to the interval $(-\infty, +\infty)$], the absolute minimum must occur at a critical point of p. Since p is differentiable everywhere, we can find all critical points by solving the equation $p'(x) = 0$. This equation is

$$12x^3 + 12x^2 = 12x^2(x + 1) = 0$$

from which we conclude that the critical points are $x = 0$ and $x = -1$. Evaluating p at these critical points yields

$$p(0) = 0 \quad \text{and} \quad p(-1) = -1$$

Therefore, p has an absolute minimum of -1 at $x = -1$ (Figure 4.4.4). ◄

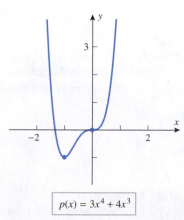

$p(x) = 3x^4 + 4x^3$

▲ **Figure 4.4.4**

■ ABSOLUTE EXTREMA ON OPEN INTERVALS

We know that a continuous function may or may not have absolute extrema on an open interval. However, certain conclusions about the existence of absolute extrema of a continuous function f on a finite open interval (a, b) can be drawn from the behavior of $f(x)$ as $x \to a^+$ and as $x \to b^-$ (Table 4.4.3). Similar conclusions can be drawn for intervals of the form $(-\infty, b)$ or $(a, +\infty)$.

Table 4.4.3
ABSOLUTE EXTREMA ON OPEN INTERVALS

LIMITS	$\lim\limits_{x \to a^+} f(x) = +\infty$ $\lim\limits_{x \to b^-} f(x) = +\infty$	$\lim\limits_{x \to a^+} f(x) = -\infty$ $\lim\limits_{x \to b^-} f(x) = -\infty$	$\lim\limits_{x \to a^+} f(x) = -\infty$ $\lim\limits_{x \to b^-} f(x) = +\infty$	$\lim\limits_{x \to a^+} f(x) = +\infty$ $\lim\limits_{x \to b^-} f(x) = -\infty$
CONCLUSION IF ***f* IS CONTINUOUS** **ON (a, b)**	f has an absolute minimum but no absolute maximum on (a, b).	f has an absolute maximum but no absolute minimum on (a, b).	f has neither an absolute maximum nor an absolute minimum on (a, b).	f has neither an absolute maximum nor an absolute minimum on (a, b).
GRAPH				

▶ **Example 5** Determine whether the function

$$f(x) = \frac{1}{x^2 - x}$$

has any absolute extrema on the interval $(0, 1)$. If so, find them and state where they occur.

Solution. Since f is continuous on the interval $(0, 1)$ and

$$\lim_{x \to 0^+} f(x) = \lim_{x \to 0^+} \frac{1}{x^2 - x} = \lim_{x \to 0^+} \frac{1}{x(x - 1)} = -\infty$$

$$\lim_{x \to 1^-} f(x) = \lim_{x \to 1^-} \frac{1}{x^2 - x} = \lim_{x \to 1^-} \frac{1}{x(x - 1)} = -\infty$$

the function f has an absolute maximum but no absolute minimum on the interval $(0, 1)$. By Theorem 4.4.3 the absolute maximum must occur at a critical point of f in the interval $(0, 1)$. We have

$$f'(x) = -\frac{2x - 1}{\left(x^2 - x\right)^2}$$

so the only solution of the equation $f'(x) = 0$ is $x = \frac{1}{2}$. Although f is not differentiable at $x = 0$ or at $x = 1$, these values are doubly disqualified since they are neither in the domain of f nor in the interval $(0, 1)$. Thus, the absolute maximum occurs at $x = \frac{1}{2}$, and this absolute maximum is

$$f\left(\tfrac{1}{2}\right) = \frac{1}{\left(\tfrac{1}{2}\right)^2 - \tfrac{1}{2}} = -4$$

(Figure 4.4.5). ◀

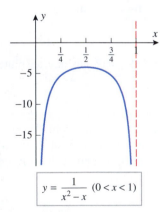

$$y = \frac{1}{x^2 - x} \quad (0 < x < 1)$$

▲ **Figure 4.4.5**

■ ABSOLUTE EXTREMA OF FUNCTIONS WITH ONE RELATIVE EXTREMUM

If a continuous function has only one relative extremum on a finite or infinite interval, then that relative extremum must of necessity also be an absolute extremum. To understand why

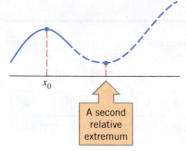

▲ Figure 4.4.6

this is so, suppose that f has a relative maximum at x_0 in an interval, and there are no other relative extrema of f on the interval. If $f(x_0)$ is *not* the absolute maximum of f on the interval, then the graph of f has to make an upward turn somewhere on the interval to rise above $f(x_0)$. However, this cannot happen because in the process of making an upward turn it would produce a second relative extremum (Figure 4.4.6). Thus, $f(x_0)$ must be the absolute maximum as well as a relative maximum. This idea is captured in the following theorem, which we state without proof.

4.4.4 THEOREM *Suppose that f is continuous and has exactly one relative extremum on an interval, say at x_0.*

(a) *If f has a relative minimum at x_0, then $f(x_0)$ is the absolute minimum of f on the interval.*

(b) *If f has a relative maximum at x_0, then $f(x_0)$ is the absolute maximum of f the interval.*

This theorem is often helpful in situations where other methods are difficult or tedious to apply.

▶ **Example 6** Find the absolute extrema, if any, of the function $f(x) = e^{(x^3-3x^2)}$ on the interval $(0, +\infty)$.

Solution. We have

$$\lim_{x \to +\infty} f(x) = +\infty$$

(verify), so f does not have an absolute maximum on the interval $(0, +\infty)$. However, the continuity of f together with the fact that

$$\lim_{x \to 0^+} f(x) = e^0 = 1$$

is finite allow for the possibility that f has an absolute minimum on $(0, +\infty)$. If so, it would have to occur at a critical point of f, so we consider

$$f'(x) = e^{(x^3-3x^2)}(3x^2 - 6x) = 3x(x - 2)e^{(x^3-3x^2)}$$

Since $e^{(x^3-3x^2)} > 0$ for all values of x, we see that $x = 0$ and $x = 2$ are the only critical points of f. Of these, only $x = 2$ is in the interval $(0, +\infty)$, so this is the point at which an absolute minimum could occur. To see whether an absolute minimum actually does occur at this point, we can apply part (a) of Theorem 4.4.4. Since

$$f''(x) = e^{(x^3-3x^2)}(3x^2 - 6x)^2 + e^{(x^3-3x^2)}(6x - 6)$$
$$= [(3x^2 - 6x)^2 + (6x - 6)]e^{(x^3-3x^2)}$$

we have

$$f''(2) = (0 + 6)e^{-4} = 6e^{-4} > 0$$

so a relative minimum occurs at $x = 2$ by the second derivative test. Thus, $f(x)$ has an absolute minimum at $x = 2$, and this absolute minimum is $f(2) = e^{-4} \approx 0.0183$ (Figure 4.4.7). ◀

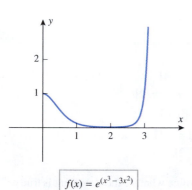

$f(x) = e^{(x^3-3x^2)}$

▲ Figure 4.4.7

Does the function in Example 6 have an absolute minimum on the interval $(-\infty, +\infty)$?

✔ **QUICK CHECK EXERCISES 4.4** *(See page 274 for answers.)*

1. Use the accompanying graph to find the *x*-coordinates of the relative extrema and absolute extrema of *f* on [0, 6].

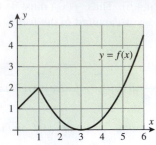

◄ **Figure Ex-1**

2. Suppose that a function *f* is continuous on [−4, 4] and has critical points at *x* = −3, 0, 2. Use the accompanying table to determine the absolute maximum and absolute minimum values, if any, for *f* on the indicated intervals.
(a) [1, 4] (b) [−2, 2] (c) [−4, 4] (d) (−4, 4)

x	−4	−3	−2	−1	0	1	2	3	4
f(*x*)	2224	−1333	0	1603	2096	2293	2400	2717	6064

3. Let $f(x) = x^3 - 3x^2 - 9x + 25$. Use the derivative $f'(x) = 3(x + 1)(x - 3)$ to determine the absolute maximum and absolute minimum values, if any, for *f* on each of the given intervals.
(a) [0, 4] (b) [−2, 4] (c) [−4, 2]
(d) [−5, 10] (e) (−5, 4)

EXERCISE SET 4.4 ⌁ Graphing Utility 🄲 CAS

FOCUS ON CONCEPTS

1–2 Use the graph to find *x*-coordinates of the relative extrema and absolute extrema of *f* on [0, 7]. ◾

1. 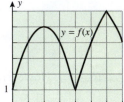 **2.**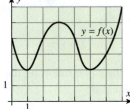

3. In each part, sketch the graph of a continuous function *f* with the stated properties on the interval [0, 10].
(a) *f* has an absolute minimum at *x* = 0 and an absolute maximum at *x* = 10.
(b) *f* has an absolute minimum at *x* = 2 and an absolute maximum at *x* = 7.
(c) *f* has relative minima at *x* = 1 and *x* = 8, has relative maxima at *x* = 3 and *x* = 7, has an absolute minimum at *x* = 5, and has an absolute maximum at *x* = 10.

4. In each part, sketch the graph of a continuous function *f* with the stated properties on the interval (−∞, +∞).
(a) *f* has no relative extrema or absolute extrema.
(b) *f* has an absolute minimum at *x* = 0 but no absolute maximum.
(c) *f* has an absolute maximum at *x* = −5 and an absolute minimum at *x* = 5.

5. Let
$$f(x) = \begin{cases} \dfrac{1}{1-x}, & 0 \le x < 1 \\ 0, & x = 1 \end{cases}$$
Explain why *f* has a minimum value but no maximum value on the closed interval [0, 1].

6. Let
$$f(x) = \begin{cases} x, & 0 < x < 1 \\ \frac{1}{2}, & x = 0, 1 \end{cases}$$
Explain why *f* has neither a minimum value nor a maximum value on the closed interval [0, 1].

7–16 Find the absolute maximum and minimum values of *f* on the given closed interval, and state where those values occur. ◾

7. $f(x) = 4x^2 - 12x + 10$; [1, 2]

8. $f(x) = 8x - x^2$; [0, 6]

9. $f(x) = (x - 2)^3$; [1, 4]

10. $f(x) = 2x^3 + 3x^2 - 12x$; [−3, 2]

11. $f(x) = \dfrac{3x}{\sqrt{4x^2 + 1}}$; [−1, 1]

12. $f(x) = (x^2 + x)^{2/3}$; [−2, 3]

13. $f(x) = x - 2\sin x$; [−π/4, π/2]

14. $f(x) = \sin x - \cos x$; [0, π]

15. $f(x) = 1 + |9 - x^2|$; [−5, 1]

16. $f(x) = |6 - 4x|$; [−3, 3]

17–20 True–False Determine whether the statement is true or false. Explain your answer. ◾

17. If a function *f* is continuous on [*a*, *b*], then *f* has an absolute maximum on [*a*, *b*].

18. If a function *f* is continuous on (*a*, *b*), then *f* has an absolute minimum on (*a*, *b*).

19. If a function *f* has an absolute minimum on (*a*, *b*), then there is a critical point of *f* in (*a*, *b*).

20. If a function *f* is continuous on [*a*, *b*] and *f* has no relative extreme values in (*a*, *b*), then the absolute maximum value of *f* exists and occurs either at *x* = *a* or at *x* = *b*.

21–28 Find the absolute maximum and minimum values of f, if any, on the given interval, and state where those values occur. ■

21. $f(x) = x^2 - x - 2$; $(-\infty, +\infty)$

22. $f(x) = 3 - 4x - 2x^2$; $(-\infty, +\infty)$

23. $f(x) = 4x^3 - 3x^4$; $(-\infty, +\infty)$

24. $f(x) = x^4 + 4x$; $(-\infty, +\infty)$

25. $f(x) = 2x^3 - 6x + 2$; $(-\infty, +\infty)$

26. $f(x) = x^3 - 9x + 1$; $(-\infty, +\infty)$

27. $f(x) = \dfrac{x^2 + 1}{x + 1}$; $(-5, -1)$

28. $f(x) = \dfrac{x - 2}{x + 1}$; $(-1, 5]$

29–42 Use a graphing utility to estimate the absolute maximum and minimum values of f, if any, on the stated interval, and then use calculus methods to find the exact values. ■

29. $f(x) = (x^2 - 2x)^2$; $(-\infty, +\infty)$

30. $f(x) = (x - 1)^2(x + 2)^2$; $(-\infty, +\infty)$

31. $f(x) = x^{2/3}(20 - x)$; $[-1, 20]$

32. $f(x) = \dfrac{x}{x^2 + 2}$; $[-1, 4]$

33. $f(x) = 1 + \dfrac{1}{x}$; $(0, +\infty)$

34. $f(x) = \dfrac{2x^2 - 3x + 3}{x^2 - 2x + 2}$; $[1, +\infty)$

35. $f(x) = \dfrac{2 - \cos x}{\sin x}$; $[\pi/4, 3\pi/4]$

36. $f(x) = \sin^2 x + \cos x$; $[-\pi, \pi]$

37. $f(x) = x^3 e^{-2x}$; $[1, 4]$

38. $f(x) = \dfrac{\ln(2x)}{x}$; $[1, e]$

39. $f(x) = 5\ln(x^2 + 1) - 3x$; $[0, 4]$

40. $f(x) = (x^2 - 1)e^x$; $[-2, 2]$

41. $f(x) = \sin(\cos x)$; $[0, 2\pi]$

42. $f(x) = \cos(\sin x)$; $[0, \pi]$

43. Find the absolute maximum and minimum values of
$$f(x) = \begin{cases} 4x - 2, & x < 1 \\ (x - 2)(x - 3), & x \ge 1 \end{cases}$$
on $\left[\frac{1}{2}, \frac{7}{2}\right]$.

44. Let $f(x) = x^2 + px + q$. Find the values of p and q such that $f(1) = 3$ is an extreme value of f on $[0, 2]$. Is this value a maximum or minimum?

45–46 If f is a periodic function, then the locations of all absolute extrema on the interval $(-\infty, +\infty)$ can be obtained by finding the locations of the absolute extrema for one period and using the periodicity to locate the rest. Use this idea in these exercises to find the absolute maximum and minimum values of the function, and state the x-values at which they occur. ■

45. $f(x) = 2\cos x + \cos 2x$ **46.** $f(x) = 3\cos\dfrac{x}{3} + 2\cos\dfrac{x}{2}$

47–48 One way of proving that $f(x) \le g(x)$ for all x in a given interval is to show that $0 \le g(x) - f(x)$ for all x in the interval; and one way of proving the latter inequality is to show that the absolute minimum value of $g(x) - f(x)$ on the interval is nonnegative. Use this idea to prove the inequalities in these exercises. ■

47. Prove that $\sin x \le x$ for all x in the interval $[0, 2\pi]$.

48. Prove that $\cos x \ge 1 - (x^2/2)$ for all x in the interval $[0, 2\pi]$.

49. What is the smallest possible slope for a tangent to the graph of the equation $y = x^3 - 3x^2 + 5x$?

50. (a) Show that $f(x) = \sec x + \csc x$ has a minimum value but no maximum value on the interval $(0, \pi/2)$.
(b) Find the minimum value in part (a).

C **51.** Show that the absolute minimum value of
$$f(x) = x^2 + \dfrac{x^2}{(8 - x)^2}, \quad x > 8$$
occurs at $x = 10$ by using a CAS to find $f'(x)$ and to solve the equation $f'(x) = 0$.

52. The concentration $C(t)$ of a drug in the bloodstream t hours after it has been injected is commonly modeled by an equation of the form
$$C(t) = \dfrac{K(e^{-bt} - e^{-at})}{a - b}$$
where $K > 0$ and $a > b > 0$.
(a) At what time does the maximum concentration occur?
(b) Let $K = 1$ for simplicity, and use a graphing utility to check your result in part (a) by graphing $C(t)$ for various values of a and b.

53. Suppose that the equations of motion of a paper airplane during the first 12 seconds of flight are
$$x = t - 2\sin t, \quad y = 2 - 2\cos t \quad (0 \le t \le 12)$$
What are the highest and lowest points in the trajectory, and when is the airplane at those points?

54. The accompanying figure shows the path of a fly whose equations of motion are
$$x = \dfrac{\cos t}{2 + \sin t}, \quad y = 3 + \sin(2t) - 2\sin^2 t \quad (0 \le t \le 2\pi)$$
(a) How high and low does it fly?
(b) How far left and right of the origin does it fly?

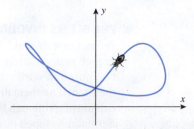

◀ **Figure Ex-54**

55. Let $f(x) = ax^2 + bx + c$, where $a > 0$. Prove that $f(x) \geq 0$ for all x if and only if $b^2 - 4ac \leq 0$. [*Hint:* Find the minimum of $f(x)$.]

56. Prove Theorem 4.4.3 in the case where the extreme value is a minimum.

57. Writing Suppose that f is continuous and positive-valued everywhere and that the x-axis is an asymptote for the graph of f, both as $x \to -\infty$ and as $x \to +\infty$. Explain why f cannot have an absolute minimum but may have a relative minimum.

58. Writing Explain the difference between a relative maximum and an absolute maximum. Sketch a graph that illustrates a function with a relative maximum that is not an absolute maximum, and sketch another graph illustrating an absolute maximum that is not a relative maximum. Explain how these graphs satisfy the given conditions.

✔ **QUICK CHECK ANSWERS 4.4**

1. There is a relative minimum at $x = 3$, a relative maximum at $x = 1$, an absolute minimum at $x = 3$, and an absolute maximum at $x = 6$.   **2.** (a) max, 6064; min, 2293 (b) max, 2400; min, 0 (c) max, 6064; min, -1333 (d) no max; min, -1333
3. (a) max, $f(0) = 25$; min, $f(3) = -2$ (b) max, $f(-1) = 30$; min, $f(3) = -2$ (c) max, $f(-1) = 30$; min, $f(-4) = -51$
(d) max, $f(10) = 635$; min, $f(-5) = -130$ (e) max, $f(-1) = 30$; no min

4.5 APPLIED MAXIMUM AND MIMIMUM PROBLEMS

In this section we will show how the methods discussed in the last section can be used to solve various applied optimization problems.

CLASSIFICATION OF OPTIMIZATION PROBLEMS

The applied optimization problems that we will consider in this section fall into the following two categories:

- Problems that reduce to maximizing or minimizing a continuous function over a finite closed interval.
- Problems that reduce to maximizing or minimizing a continuous function over an infinite interval or a finite interval that is not closed.

For problems of the first type the Extreme-Value Theorem (4.4.2) guarantees that the problem has a solution, and we know that the solution can be obtained by examining the values of the function at the critical points and at the endpoints. However, for problems of the second type there may or may not be a solution. If the function is continuous and has exactly one relative extremum of the appropriate type on the interval, then Theorem 4.4.4 guarantees the existence of a solution and provides a method for finding it. In cases where this theorem is not applicable some ingenuity may be required to solve the problem.

PROBLEMS INVOLVING FINITE CLOSED INTERVALS

In his *On a Method for the Evaluation of Maxima and Minima*, the seventeenth century French mathematician Pierre de Fermat solved an optimization problem very similar to the one posed in our first example. Fermat's work on such optimization problems prompted the French mathematician Laplace to proclaim Fermat the "true inventor of the differential calculus." Although this honor must still reside with Newton and Leibniz, it is the case that Fermat developed procedures that anticipated parts of differential calculus.

▶ **Example 1**   A garden is to be laid out in a rectangular area and protected by a chicken wire fence. What is the largest possible area of the garden if only 100 running feet of chicken wire is available for the fence?

Solution.   Let

$$x = \text{length of the rectangle (ft)}$$
$$y = \text{width of the rectangle (ft)}$$
$$A = \text{area of the rectangle (ft}^2)$$

Then

$$A = xy \tag{1}$$

Since the perimeter of the rectangle is 100 ft, the variables x and y are related by the equation

$$2x + 2y = 100 \quad \text{or} \quad y = 50 - x \tag{2}$$

(See Figure 4.5.1.) Substituting (2) in (1) yields

$$A = x(50 - x) = 50x - x^2 \tag{3}$$

Because x represents a length, it cannot be negative, and because the two sides of length x cannot have a combined length exceeding the total perimeter of 100 ft, the variable x must satisfy

$$0 \leq x \leq 50 \tag{4}$$

Thus, we have reduced the problem to that of finding the value (or values) of x in $[0, 50]$, for which A is maximum. Since A is a polynomial in x, it is continuous on $[0, 50]$, and so the maximum must occur at an endpoint of this interval or at a critical point.

From (3) we obtain

$$\frac{dA}{dx} = 50 - 2x$$

Setting $dA/dx = 0$ we obtain

$$50 - 2x = 0$$

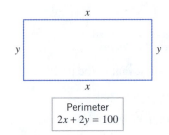

Perimeter
$2x + 2y = 100$

▲ **Figure 4.5.1**

Pierre de Fermat (1601–1665) Fermat, the son of a successful French leather merchant, was a lawyer who practiced mathematics as a hobby. He received a Bachelor of Civil Laws degree from the University of Orleans in 1631 and subsequently held various government positions, including a post as councillor to the Toulouse parliament. Although he was apparently financially successful, confidential documents of that time suggest that his performance in office and as a lawyer was poor, perhaps because he devoted so much time to mathematics. Throughout his life, Fermat fought all efforts to have his mathematical results published. He had the unfortunate habit of scribbling his work in the margins of books and often sent his results to friends without keeping copies for himself. As a result, he never received credit for many major achievements until his name was raised from obscurity in the mid-nineteenth century. It is now known that Fermat, simultaneously and independently of Descartes, developed analytic geometry. Unfortunately, Descartes and Fermat argued bitterly over various problems so that there was never any real cooperation between these two great geniuses.

Fermat solved many fundamental calculus problems. He obtained the first procedure for differentiating polynomials, and solved many important maximization, minimization, area, and tangent problems. His work served to inspire Isaac Newton. Fermat is best known for his work in number theory, the study of properties of and relationships between whole numbers. He was the first mathematician to make substantial contributions to this field after the ancient Greek mathematician Diophantus. Unfortunately, none of Fermat's contemporaries appreciated his work in this area, a fact that eventually pushed Fermat into isolation and obscurity in later life. In addition to his work in calculus and number theory, Fermat was one of the founders of probability theory and made major contributions to the theory of optics. Outside mathematics, Fermat was a classical scholar of some note, was fluent in French, Italian, Spanish, Latin, and Greek, and he composed a considerable amount of Latin poetry.

One of the great mysteries of mathematics is shrouded in Fermat's work in number theory. In the margin of a book by Diophantus, Fermat scribbled that for integer values of n greater than 2, the equation $x^n + y^n = z^n$ has no nonzero integer solutions for x, y, and z. He stated, "I have discovered a truly marvelous proof of this, which however the margin is not large enough to contain." This result, which became known as "Fermat's last theorem," appeared to be true, but its proof evaded the greatest mathematical geniuses for 300 years until Professor Andrew Wiles of Princeton University presented a proof in June 1993 in a dramatic series of three lectures that drew international media attention (see *New York Times*, June 27, 1993). As it turned out, that proof had a serious gap that Wiles and Richard Taylor fixed and published in 1995. A prize of 100,000 German marks was offered in 1908 for the solution, but it is worthless today because of inflation.

Table 4.5.1

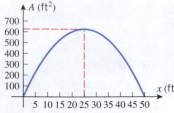

x	0	25	50
A	0	625	0

A (ft²) graph

▲ **Figure 4.5.2**

In Example 1 we included $x = 0$ and $x = 50$ as possible values of x, even though these correspond to rectangles with two sides of length zero. If we view this as a purely mathematical problem, then there is nothing wrong with this. However, if we view this as an applied problem in which the rectangle will be formed from physical material, then it would make sense to exclude these values.

or $x = 25$. Thus, the maximum occurs at one of the values

$$x = 0, \quad x = 25, \quad x = 50$$

Substituting these values in (3) yields Table 4.5.1, which tells us that the maximum area of 625 ft² occurs at $x = 25$, which is consistent with the graph of (3) in Figure 4.5.2. From (2) the corresponding value of y is 25, so the rectangle of perimeter 100 ft with greatest area is a square with sides of length 25 ft. ◄

Example 1 illustrates the following five-step procedure that can be used for solving many applied maximum and minimum problems.

A Procedure for Solving Applied Maximum and Minimum Problems

Step 1. Draw an appropriate figure and label the quantities relevant to the problem.

Step 2. Find a formula for the quantity to be maximized or minimized.

Step 3. Using the conditions stated in the problem to eliminate variables, express the quantity to be maximized or minimized as a function of one variable.

Step 4. Find the interval of possible values for this variable from the physical restrictions in the problem.

Step 5. If applicable, use the techniques of the preceding section to obtain the maximum or minimum.

▶ **Example 2** An open box is to be made from a 16-inch by 30-inch piece of cardboard by cutting out squares of equal size from the four corners and bending up the sides (Figure 4.5.3). What size should the squares be to obtain a box with the largest volume?

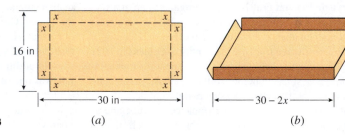

▶ **Figure 4.5.3**   (a)   (b)

Solution. For emphasis, we explicitly list the steps of the five-step problem-solving procedure given above as an outline for the solution of this problem. (In later examples we will follow these guidelines without listing the steps.)

- *Step 1:* Figure 4.5.3a illustrates the cardboard piece with squares removed from its corners. Let

   x = length (in inches) of the sides of the squares to be cut out
   V = volume (in cubic inches) of the resulting box

- *Step 2:* Because we are removing a square of side x from each corner, the resulting box will have dimensions $16 - 2x$ by $30 - 2x$ by x (Figure 4.5.3b). Since the volume of a box is the product of its dimensions, we have

$$V = (16 - 2x)(30 - 2x)x = 480x - 92x^2 + 4x^3 \tag{5}$$

- *Step 3:* Note that our volume expression is already in terms of the single variable x.
- *Step 4:* The variable x in (5) is subject to certain restrictions. Because x represents a length, it cannot be negative, and because the width of the cardboard is 16 inches, we cannot cut out squares whose sides are more than 8 inches long. Thus, the variable x in (5) must satisfy

$$0 \leq x \leq 8$$

and hence we have reduced our problem to finding the value (or values) of x in the interval [0, 8] for which (5) is a maximum.

- *Step 5:* From (5) we obtain

$$\frac{dV}{dx} = 480 - 184x + 12x^2 = 4(120 - 46x + 3x^2)$$
$$= 4(x - 12)(3x - 10)$$

Setting $dV/dx = 0$ yields

$$x = \tfrac{10}{3} \quad \text{and} \quad x = 12$$

Since $x = 12$ falls outside the interval [0, 8], the maximum value of V occurs either at the critical point $x = \tfrac{10}{3}$ or at the endpoints $x = 0$, $x = 8$. Substituting these values into (5) yields Table 4.5.2, which tells us that the greatest possible volume $V = \tfrac{19{,}600}{27}$ in$^3 \approx 726$ in^3 occurs when we cut out squares whose sides have length $\tfrac{10}{3}$ inches. This is consistent with the graph of (5) shown in Figure 4.5.4. ◄

Table 4.5.2

x	0	$\frac{10}{3}$	8
V	0	$\frac{19{,}600}{27} \approx 726$	0

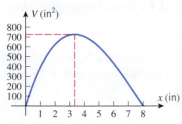

▲ **Figure 4.5.4**

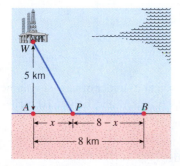

▲ **Figure 4.5.5**

► **Example 3** Figure 4.5.5 shows an offshore oil well located at a point W that is 5 km from the closest point A on a straight shoreline. Oil is to be piped from W to a shore point B that is 8 km from A by piping it on a straight line under water from W to some shore point P between A and B and then on to B via pipe along the shoreline. If the cost of laying pipe is $1,000,000/km under water and $500,000/km over land, where should the point P be located to minimize the cost of laying the pipe?

Solution. Let

$$x = \text{distance (in kilometers) between } A \text{ and } P$$
$$c = \text{cost (in millions of dollars) for the entire pipeline}$$

From Figure 4.5.5 the length of pipe under water is the distance between W and P. By the Theorem of Pythagoras that length is

$$\sqrt{x^2 + 25} \qquad (6)$$

Also from Figure 4.5.5, the length of pipe over land is the distance between P and B, which is

$$8 - x \qquad (7)$$

From (6) and (7) it follows that the total cost c (in millions of dollars) for the pipeline is

$$c = 1(\sqrt{x^2 + 25}) + \tfrac{1}{2}(8 - x) = \sqrt{x^2 + 25} + \tfrac{1}{2}(8 - x) \qquad (8)$$

Because the distance between A and B is 8 km, the distance x between A and P must satisfy

$$0 \leq x \leq 8$$

We have thus reduced our problem to finding the value (or values) of x in the interval [0, 8] for which c is a minimum. Since c is a continuous function of x on the closed interval [0, 8], we can use the methods developed in the preceding section to find the minimum.

From (8) we obtain

$$\frac{dc}{dx} = \frac{x}{\sqrt{x^2 + 25}} - \frac{1}{2}$$

Setting $dc/dx = 0$ and solving for x yields

$$\frac{x}{\sqrt{x^2 + 25}} = \frac{1}{2} \tag{9}$$

$$x^2 = \frac{1}{4}(x^2 + 25)$$

$$x = \pm\frac{5}{\sqrt{3}}$$

The number $-5/\sqrt{3}$ is not a solution of (9) and must be discarded, leaving $x = 5/\sqrt{3}$ as the only critical point. Since this point lies in the interval $[0, 8]$, the minimum must occur at one of the values

$$x = 0, \quad x = 5/\sqrt{3}, \quad x = 8$$

Substituting these values into (8) yields Table 4.5.3, which tells us that the least possible cost of the pipeline (to the nearest dollar) is $c = \$8,330,127$, and this occurs when the point P is located at a distance of $5/\sqrt{3} \approx 2.89$ km from A. ◄

Table 4.5.3

x	0	$\frac{5}{\sqrt{3}}$	8
c	9	$\frac{10}{\sqrt{3}} + \left(4 - \frac{5}{2\sqrt{3}}\right) \approx 8.330127$	$\sqrt{89} \approx 9.433981$

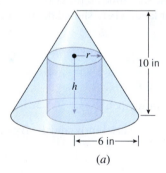

(a)

(b)

▲ **Figure 4.5.6**

► **Example 4** Find the radius and height of the right circular cylinder of largest volume that can be inscribed in a right circular cone with radius 6 inches and height 10 inches (Figure 4.5.6a).

Solution. Let

$$r = \text{radius (in inches) of the cylinder}$$
$$h = \text{height (in inches) of the cylinder}$$
$$V = \text{volume (in cubic inches) of the cylinder}$$

The formula for the volume of the inscribed cylinder is

$$V = \pi r^2 h \tag{10}$$

To eliminate one of the variables in (10) we need a relationship between r and h. Using similar triangles (Figure 4.5.6b) we obtain

$$\frac{10 - h}{r} = \frac{10}{6} \quad \text{or} \quad h = 10 - \frac{5}{3}r \tag{11}$$

Substituting (11) into (10) we obtain

$$V = \pi r^2 \left(10 - \frac{5}{3}r\right) = 10\pi r^2 - \frac{5}{3}\pi r^3 \tag{12}$$

which expresses V in terms of r alone. Because r represents a radius, it cannot be negative, and because the radius of the inscribed cylinder cannot exceed the radius of the cone, the variable r must satisfy

$$0 \leq r \leq 6$$

Thus, we have reduced the problem to that of finding the value (or values) of r in $[0, 6]$ for which (12) is a maximum. Since V is a continuous function of r on $[0, 6]$, the methods developed in the preceding section apply.

From (12) we obtain

$$\frac{dV}{dr} = 20\pi r - 5\pi r^2 = 5\pi r(4 - r)$$

Setting $dV/dr = 0$ gives

$$5\pi r(4 - r) = 0$$

so $r = 0$ and $r = 4$ are critical points. Since these lie in the interval $[0, 6]$, the maximum must occur at one of the values

$$r = 0, \quad r = 4, \quad r = 6$$

Table 4.5.4

r	0	4	6
V	0	$\frac{160}{3}\pi$	0

Substituting these values into (12) yields Table 4.5.4, which tells us that the maximum volume $V = \frac{160}{3}\pi \approx 168$ in^3 occurs when the inscribed cylinder has radius 4 in. When $r = 4$ it follows from (11) that $h = \frac{10}{3}$. Thus, the inscribed cylinder of largest volume has radius $r = 4$ in and height $h = \frac{10}{3}$ in. ◄

■ PROBLEMS INVOLVING INTERVALS THAT ARE NOT BOTH FINITE AND CLOSED

► **Example 5** A closed cylindrical can is to hold 1 liter (1000 cm^3) of liquid. How should we choose the height and radius to minimize the amount of material needed to manufacture the can?

Solution.   Let

$$h = \text{height (in cm) of the can}$$
$$r = \text{radius (in cm) of the can}$$
$$S = \text{surface area (in cm}^2\text{) of the can}$$

Assuming there is no waste or overlap, the amount of material needed for manufacture will be the same as the surface area of the can. Since the can consists of two circular disks of radius r and a rectangular sheet with dimensions h by $2\pi r$ (Figure 4.5.7), the surface area will be

$$S = 2\pi r^2 + 2\pi r h \tag{13}$$

Since S depends on two variables, r and h, we will look for some condition in the problem that will allow us to express one of these variables in terms of the other. For this purpose,

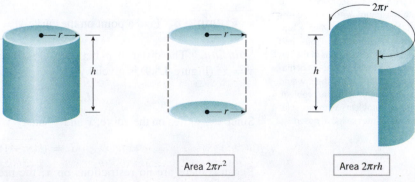

Area $2\pi r^2$ Area $2\pi r h$

▲ **Figure 4.5.7**

observe that the volume of the can is 1000 cm³, so it follows from the formula $V = \pi r^2 h$ for the volume of a cylinder that

$$1000 = \pi r^2 h \quad \text{or} \quad h = \frac{1000}{\pi r^2} \tag{14–15}$$

Substituting (15) in (13) yields

$$S = 2\pi r^2 + \frac{2000}{r} \tag{16}$$

Thus, we have reduced the problem to finding a value of r in the interval $(0, +\infty)$ for which S is minimum. Since S is a continuous function of r on the interval $(0, +\infty)$ and

$$\lim_{r \to 0^+} \left(2\pi r^2 + \frac{2000}{r} \right) = +\infty \quad \text{and} \quad \lim_{r \to +\infty} \left(2\pi r^2 + \frac{2000}{r} \right) = +\infty$$

the analysis in Table 4.4.3 implies that S does have a minimum on the interval $(0, +\infty)$. Since this minimum must occur at a critical point, we calculate

$$\frac{dS}{dr} = 4\pi r - \frac{2000}{r^2} \tag{17}$$

Setting $dS/dr = 0$ gives

$$r = \frac{10}{\sqrt[3]{2\pi}} \approx 5.4 \tag{18}$$

Since (18) is the only critical point in the interval $(0, +\infty)$, this value of r yields the minimum value of S. From (15) the value of h corresponding to this r is

$$h = \frac{1000}{\pi (10/\sqrt[3]{2\pi})^2} = \frac{20}{\sqrt[3]{2\pi}} = 2r$$

It is not an accident here that the minimum occurs when the height of the can is equal to the diameter of its base (Exercise 29).

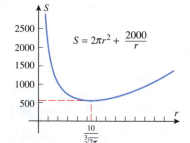

$$S = 2\pi r^2 + \frac{2000}{r}$$

▲ **Figure 4.5.8**

Second Solution.   The conclusion that a minimum occurs at the value of r in (18) can be deduced from Theorem 4.4.4 and the second derivative test by noting that

$$\frac{d^2 S}{dr^2} = 4\pi + \frac{4000}{r^3}$$

is positive if $r > 0$ and hence is positive if $r = 10/\sqrt[3]{2\pi}$. This implies that a relative minimum, and therefore a minimum, occurs at the critical point $r = 10/\sqrt[3]{2\pi}$.

Third Solution.   An alternative justification that the critical point $r = 10/\sqrt[3]{2\pi}$ corresponds to a minimum for S is to view the graph of S versus r (Figure 4.5.8). ◄

In Example 5, the surface area S has no absolute maximum, since S increases without bound as the radius r approaches 0 (Figure 4.5.8). Thus, had we asked for the dimensions of the can requiring the *maximum* amount of material for its manufacture, there would have been no solution to the problem. Optimization problems with no solution are sometimes called *ill posed*.

▶ **Example 6**   Find a point on the curve $y = x^2$ that is closest to the point $(18, 0)$.

Solution.   The distance L between $(18, 0)$ and an arbitrary point (x, y) on the curve $y = x^2$ (Figure 4.5.9) is given by

$$L = \sqrt{(x - 18)^2 + (y - 0)^2}$$

Since (x, y) lies on the curve, x and y satisfy $y = x^2$; thus,

$$L = \sqrt{(x - 18)^2 + x^4} \tag{19}$$

Because there are no restrictions on x, the problem reduces to finding a value of x in $(-\infty, +\infty)$ for which (19) is a minimum. The distance L and the square of the distance L^2

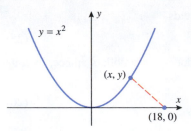

▲ **Figure 4.5.9**

are minimized at the same value (see Exercise 66). Thus, the minimum value of L in (19) and the minimum value of

$$S = L^2 = (x - 18)^2 + x^4 \tag{20}$$

occur at the same x-value.

From (20),

$$\frac{dS}{dx} = 2(x - 18) + 4x^3 = 4x^3 + 2x - 36 \tag{21}$$

so the critical points satisfy $4x^3 + 2x - 36 = 0$ or, equivalently,

$$2x^3 + x - 18 = 0 \tag{22}$$

To solve for x we will begin by checking the divisors of -18 to see whether the polynomial on the left side has any integer roots (see Appendix C). These divisors are $\pm 1, \pm 2, \pm 3, \pm 6,$ $\pm 9,$ and ± 18. A check of these values shows that $x = 2$ is a root, so $x - 2$ is a factor of the polynomial. After dividing the polynomial by this factor we can rewrite (22) as

$$(x - 2)(2x^2 + 4x + 9) = 0$$

Thus, the remaining solutions of (22) satisfy the quadratic equation

$$2x^2 + 4x + 9 = 0$$

But this equation has no real solutions (using the quadratic formula), so $x = 2$ is the only critical point of S. To determine the nature of this critical point we will use the second derivative test. From (21),

$$\frac{d^2 S}{dx^2} = 12x^2 + 2, \quad \text{so} \quad \left. \frac{d^2 S}{dx^2} \right|_{x=2} = 50 > 0$$

which shows that a relative minimum occurs at $x = 2$. Since $x = 2$ yields the only relative extremum for L, it follows from Theorem 4.4.4 that an absolute minimum value of L also occurs at $x = 2$. Thus, the point on the curve $y = x^2$ closest to $(18, 0)$ is

$$(x, y) = (x, x^2) = (2, 4) \blacktriangleleft$$

◼ AN APPLICATION TO ECONOMICS

Three functions of importance to an economist or a manufacturer are

$C(x) = $ total cost of producing x units of a product during some time period

$R(x) = $ total revenue from selling x units of the product during the time period

$P(x) = $ total profit obtained by selling x units of the product during the time period

These are called, respectively, the ***cost function***, ***revenue function***, and ***profit function***. If all units produced are sold, then these are related by

$$P(x) = R(x) - C(x) \tag{23}$$

[profit] = [revenue] − [cost]

The total cost $C(x)$ of producing x units can be expressed as a sum

$$C(x) = a + M(x) \tag{24}$$

where a is a constant, called ***overhead***, and $M(x)$ is a function representing ***manufacturing cost***. The overhead, which includes such fixed costs as rent and insurance, does not depend on x; it must be paid even if nothing is produced. On the other hand, the manufacturing cost $M(x)$, which includes such items as cost of materials and labor, depends on the number of items manufactured. It is shown in economics that with suitable simplifying assumptions, $M(x)$ can be expressed in the form

$$M(x) = bx + cx^2$$

where b and c are constants. Substituting this in (24) yields

$$C(x) = a + bx + cx^2 \qquad (25)$$

If a manufacturing firm can sell all the items it produces for p dollars apiece, then its total revenue $R(x)$ (in dollars) will be

$$R(x) = px \qquad (26)$$

and its total profit $P(x)$ (in dollars) will be

$$P(x) = [\text{total revenue}] - [\text{total cost}] = R(x) - C(x) = px - C(x)$$

Thus, if the cost function is given by (25),

$$P(x) = px - (a + bx + cx^2) \qquad (27)$$

Depending on such factors as number of employees, amount of machinery available, economic conditions, and competition, there will be some upper limit l on the number of items a manufacturer is capable of producing and selling. Thus, during a fixed time period the variable x in (27) will satisfy

$$0 \leq x \leq l$$

By determining the value or values of x in $[0, l]$ that maximize (27), the firm can determine how many units of its product must be manufactured and sold to yield the greatest profit. This is illustrated in the following numerical example.

▶ **Example 7**   A liquid form of antibiotic manufactured by a pharmaceutical firm is sold in bulk at a price of $200 per unit. If the total production cost (in dollars) for x units is

$$C(x) = 500{,}000 + 80x + 0.003x^2$$

and if the production capacity of the firm is at most 30,000 units in a specified time, how many units of antibiotic must be manufactured and sold in that time to maximize the profit?

Jim Karageorge/Getty Images

A pharmaceutical firm's profit is a function of the number of units produced.

Solution.   Since the total revenue for selling x units is $R(x) = 200x$, the profit $P(x)$ on x units will be

$$P(x) = R(x) - C(x) = 200x - (500{,}000 + 80x + 0.003x^2) \qquad (28)$$

Since the production capacity is at most 30,000 units, x must lie in the interval $[0, 30{,}000]$. From (28)

$$\frac{dP}{dx} = 200 - (80 + 0.006x) = 120 - 0.006x$$

Setting $dP/dx = 0$ gives

$$120 - 0.006x = 0 \quad \text{or} \quad x = 20{,}000$$

Since this critical point lies in the interval $[0, 30{,}000]$, the maximum profit must occur at one of the values

$$x = 0, \quad x = 20{,}000, \quad \text{or} \quad x = 30{,}000$$

Substituting these values in (28) yields Table 4.5.5, which tells us that the maximum profit $P = \$700{,}000$ occurs when $x = 20{,}000$ units are manufactured and sold in the specified time. ◀

Table 4.5.5

x	0	20,000	30,000
$P(x)$	−500,000	700,000	400,000

■ MARGINAL ANALYSIS

Economists call $P'(x)$, $R'(x)$, and $C'(x)$ the **marginal profit**, **marginal revenue**, and **marginal cost**, respectively; and they interpret these quantities as the *additional* profit, revenue, and cost that result from producing and selling one additional unit of the product when the production and sales levels are at x units. These interpretations follow from the local linear approximations of the profit, revenue, and cost functions. For example, it follows from Formula (2) of Section 3.5 that when the production and sales levels are at x units the local linear approximation of the profit function is

$$P(x + \Delta x) \approx P(x) + P'(x)\Delta x$$

Thus, if $\Delta x = 1$ (one additional unit produced and sold), this formula implies

$$P(x + 1) \approx P(x) + P'(x)$$

and hence the *additional* profit that results from producing and selling one additional unit can be approximated as

$$P(x + 1) - P(x) \approx P'(x)$$

Similarly, $R(x + 1) - R(x) \approx R'(x)$ and $C(x + 1) - C(x) \approx C'(x)$.

■ A BASIC PRINCIPLE OF ECONOMICS

It follows from (23) that $P'(x) = 0$ has the same solution as $C'(x) = R'(x)$, and this implies that the maximum profit must occur at a point where the marginal revenue is equal to the marginal cost; that is:

> *If profit is maximum, then the cost of manufacturing and selling an additional unit of a product is approximately equal to the revenue generated by the additional unit.*

In Example 7, the maximum profit occurs when $x = 20{,}000$ units. Note that

$$C(20{,}001) - C(20{,}000) = \$200.003 \quad \text{and} \quad R(20{,}001) - R(20{,}000) = \$200$$

which is consistent with this basic economic principle.

✔ QUICK CHECK EXERCISES 4.5   (See page 288 for answers.)

1. A positive number x and its reciprocal are added together. The smallest possible value of this sum is obtained by minimizing $f(x) =$ _____ for x in the interval _____.

2. Two nonnegative numbers, x and y, have a sum equal to 10. The largest possible product of the two numbers is obtained by maximizing $f(x) =$ _____ for x in the interval _____.

3. A rectangle in the xy-plane has one corner at the origin, an adjacent corner at the point $(x, 0)$, and a third corner at a point on the line segment from $(0, 4)$ to $(3, 0)$. The largest possible area of the rectangle is obtained by maximizing $A(x) =$ _____ for x in the interval _____.

4. An open box is to be made from a 20-inch by 32-inch piece of cardboard by cutting out x-inch by x-inch squares from the four corners and bending up the sides. The largest possible volume of the box is obtained by maximizing $V(x) =$ _____ for x in the interval _____.

EXERCISE SET 4.5

1. Find a number in the closed interval $\left[\frac{1}{2}, \frac{3}{2}\right]$ such that the sum of the number and its reciprocal is
   (a) as small as possible
   (b) as large as possible.

2. How should two nonnegative numbers be chosen so that their sum is 1 and the sum of their squares is
   (a) as large as possible
   (b) as small as possible?

3. A rectangular field is to be bounded by a fence on three sides and by a straight stream on the fourth side. Find the dimensions of the field with maximum area that can be enclosed using 1000 ft of fence.

4. The boundary of a field is a right triangle with a straight stream along its hypotenuse and with fences along its other two sides. Find the dimensions of the field with maximum area that can be enclosed using 1000 ft of fence.

5. A rectangular plot of land is to be fenced in using two kinds of fencing. Two opposite sides will use heavy-duty fencing selling for $3 a foot, while the remaining two sides will use standard fencing selling for $2 a foot. What are the dimensions of the rectangular plot of greatest area that can be fenced in at a cost of $6000?

6. A rectangle is to be inscribed in a right triangle having sides of length 6 in, 8 in, and 10 in. Find the dimensions of the rectangle with greatest area assuming the rectangle is positioned as in Figure Ex-6.

7. Solve the problem in Exercise 6 assuming the rectangle is positioned as in Figure Ex-7.

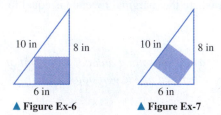

| 10 in | 8 in | 10 in | 8 in |

▲ **Figure Ex-6**   ▲ **Figure Ex-7**

8. A rectangle has its two lower corners on the x-axis and its two upper corners on the curve $y = 16 - x^2$. For all such rectangles, what are the dimensions of the one with largest area?

9. Find the dimensions of the rectangle with maximum area that can be inscribed in a circle of radius 10.

10. Find the point P in the first quadrant on the curve $y = x^{-2}$ such that a rectangle with sides on the coordinate axes and a vertex at P has the smallest possible perimeter.

11. A rectangular area of 3200 ft² is to be fenced off. Two opposite sides will use fencing costing $1 per foot and the remaining sides will use fencing costing $2 per foot. Find the dimensions of the rectangle of least cost.

12. Show that among all rectangles with perimeter p, the square has the maximum area.

13. Show that among all rectangles with area A, the square has the minimum perimeter.

14. A wire of length 12 in can be bent into a circle, bent into a square, or cut into two pieces to make both a circle and a square. How much wire should be used for the circle if the total area enclosed by the figure(s) is to be
 (a) a maximum   (b) a minimum?

15. A rectangle R in the plane has corners at $(\pm 8, \pm 12)$, and a 100 by 100 square S is positioned in the plane so that its

sides are parallel to the coordinate axes and the lower left corner of S is on the line $y = -3x$. What is the largest possible area of a region in the plane that is contained in both R and S?

16. Solve the problem in Exercise 15 if S is a 16 by 16 square.

17. Solve the problem in Exercise 15 if S is positioned with its lower left corner on the line $y = -6x$.

18. A rectangular page is to contain 42 square inches of printable area. The margins at the top and bottom of the page are each 1 inch, one side margin is 1 inch, and the other side margin is 2 inches. What should the dimensions of the page be so that the least amount of paper is used?

19. A box with a square base is taller than it is wide. In order to send the box through the U.S. mail, the height of the box and the perimeter of the base can sum to no more than 108 in. What is the maximum volume for such a box?

20. A box with a square base is wider than it is tall. In order to send the box through the U.S. mail, the width of the box and the perimeter of one of the (nonsquare) sides of the box can sum to no more than 108 in. What is the maximum volume for such a box?

21. An open box is to be made from a 3 ft by 8 ft rectangular piece of sheet metal by cutting out squares of equal size from the four corners and bending up the sides. Find the maximum volume that the box can have.

22. A closed rectangular container with a square base is to have a volume of 2250 in³. The material for the top and bottom of the container will cost $2 per in², and the material for the sides will cost $3 per in². Find the dimensions of the container of least cost.

23. A closed rectangular container with a square base is to have a volume of 2000 cm³. It costs twice as much per square centimeter for the top and bottom as it does for the sides. Find the dimensions of the container of least cost.

24. A container with square base, vertical sides, and open top is to be made from 1000 ft² of material. Find the dimensions of the container with greatest volume.

25. A rectangular container with two square sides and an open top is to have a volume of V cubic units. Find the dimensions of the container with minimum surface area.

26. A church window consisting of a rectangle topped by a semicircle is to have a perimeter p. Find the radius of the semicircle if the area of the window is to be maximum.

27. Find the dimensions of the right circular cylinder of largest volume that can be inscribed in a sphere of radius R.

28. Find the dimensions of the right circular cylinder of greatest surface area that can be inscribed in a sphere of radius R.

29. A closed, cylindrical can is to have a volume of V cubic units. Show that the can of minimum surface area is achieved when the height is equal to the diameter of the base.

30. A closed cylindrical can is to have a surface area of S square units. Show that the can of maximum volume is achieved when the height is equal to the diameter of the base.

31. A cylindrical can, open at the top, is to hold 500 cm^3 of liquid. Find the height and radius that minimize the amount of material needed to manufacture the can.

32. A soup can in the shape of a right circular cylinder of radius r and height h is to have a prescribed volume V. The top and bottom are cut from squares as shown in Figure Ex-32. If the shaded corners are wasted, but there is no other waste, find the ratio r/h for the can requiring the least material (including waste).

33. A box-shaped wire frame consists of two identical wire squares whose vertices are connected by four straight wires of equal length (Figure Ex-33). If the frame is to be made from a wire of length L, what should the dimensions be to obtain a box of greatest volume?

▲ **Figure Ex-32**   ▲ **Figure Ex-33**

34. Suppose that the sum of the surface areas of a sphere and a cube is a constant.
 (a) Show that the sum of their volumes is smallest when the diameter of the sphere is equal to the length of an edge of the cube.
 (b) When will the sum of their volumes be greatest?

35. Find the height and radius of the cone of slant height L whose volume is as large as possible.

36. A cone is made from a circular sheet of radius R by cutting out a sector and gluing the cut edges of the remaining piece together (Figure Ex-36). What is the maximum volume attainable for the cone?

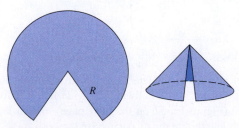

▲ **Figure Ex-36**

37. A cone-shaped paper drinking cup is to hold 100 cm^3 of water. Find the height and radius of the cup that will require the least amount of paper.

38. Find the dimensions of the isosceles triangle of least area that can be circumscribed about a circle of radius R.

39. Find the height and radius of the right circular cone with least volume that can be circumscribed about a sphere of radius R.

40. A commercial cattle ranch currently allows 20 steers per acre of grazing land; on the average its steers weigh 2000 lb at market. Estimates by the Agriculture Department indicate that the average market weight per steer will be reduced by 50 lb for each additional steer added per acre of grazing land. How many steers per acre should be allowed in order for the ranch to get the largest possible total market weight for its cattle?

41. A company mines low-grade nickel ore. If the company mines x tons of ore, it can sell the ore for $p = 225 - 0.25x$ dollars per ton. Find the revenue and marginal revenue functions. At what level of production would the company obtain the maximum revenue?

42. A fertilizer producer finds that it can sell its product at a price of $p = 300 - 0.1x$ dollars per unit when it produces x units of fertilizer. The total production cost (in dollars) for x units is

$$C(x) = 15{,}000 + 125x + 0.025x^2$$

If the production capacity of the firm is at most 1000 units of fertilizer in a specified time, how many units must be manufactured and sold in that time to maximize the profit?

43. (a) A chemical manufacturer sells sulfuric acid in bulk at a price of \$100 per unit. If the daily total production cost in dollars for x units is

$$C(x) = 100{,}000 + 50x + 0.0025x^2$$

and if the daily production capacity is at most 7000 units, how many units of sulfuric acid must be manufactured and sold daily to maximize the profit?
 (b) Would it benefit the manufacturer to expand the daily production capacity?
 (c) Use marginal analysis to approximate the effect on profit if daily production could be increased from 7000 to 7001 units.

44. A firm determines that x units of its product can be sold daily at p dollars per unit, where

$$x = 1000 - p$$

The cost of producing x units per day is

$$C(x) = 3000 + 20x$$

 (a) Find the revenue function $R(x)$.
 (b) Find the profit function $P(x)$.
 (c) Assuming that the production capacity is at most 500 units per day, determine how many units the company must produce and sell each day to maximize the profit.
 (d) Find the maximum profit.
 (e) What price per unit must be charged to obtain the maximum profit?

45. In a certain chemical manufacturing process, the daily weight y of defective chemical output depends on the total weight x of all output according to the empirical formula

$$y = 0.01x + 0.00003x^2$$

where x and y are in pounds. If the profit is \$100 per pound of nondefective chemical produced and the loss is \$20 per pound of defective chemical produced, how many pounds of chemical should be produced daily to maximize the total daily profit?

46. An independent truck driver charges a client \$15 for each hour of driving, plus the cost of fuel. At highway speeds of v miles per hour, the trucker's rig gets $10 - 0.07v$ miles per gallon of diesel fuel. If diesel fuel costs \$2.50 per gallon, what speed v will minimize the cost to the client?

47. A trapezoid is inscribed in a semicircle of radius 2 so that one side is along the diameter (Figure Ex-47). Find the maximum possible area for the trapezoid. [*Hint:* Express the area of the trapezoid in terms of θ.]

48. A drainage channel is to be made so that its cross section is a trapezoid with equally sloping sides (Figure Ex-48). If the sides and bottom all have a length of 5 ft, how should the angle θ ($0 \le \theta \le \pi/2$) be chosen to yield the greatest cross-sectional area of the channel?

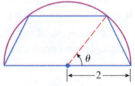

▲ **Figure Ex-47**   ▲ **Figure Ex-48**

49. A lamp is suspended above the center of a round table of radius r. How high above the table should the lamp be placed to achieve maximum illumination at the edge of the table? [Assume that the illumination I is directly proportional to the cosine of the angle of incidence ϕ of the light rays and inversely proportional to the square of the distance l from the light source (Figure Ex-49).]

50. A plank is used to reach over a fence 8 ft high to support a wall that is 1 ft behind the fence (Figure Ex-50). What is the length of the shortest plank that can be used? [*Hint:* Express the length of the plank in terms of the angle θ shown in the figure.]

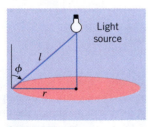

▲ **Figure Ex-49**   ▲ **Figure Ex-50**

51. Find the coordinates of the point P on the curve

$$y = \frac{1}{x^2} \quad (x > 0)$$

where the segment of the tangent line at P that is cut off by the coordinate axes has its shortest length.

52. Find the x-coordinate of the point P on the parabola

$$y = 1 - x^2 \quad (0 < x \le 1)$$

where the triangle that is enclosed by the tangent line at P and the coordinate axes has the smallest area.

53. Where on the curve $y = (1 + x^2)^{-1}$ does the tangent line have the greatest slope?

54. Suppose that the number of bacteria in a culture at time t is given by $N = 5000(25 + te^{-t/20})$.
   (a) Find the largest and smallest number of bacteria in the culture during the time interval $0 \le t \le 100$.
   (b) At what time during the time interval in part (a) is the number of bacteria decreasing most rapidly?

55. The shoreline of Circle Lake is a circle with diameter 2 mi. Nancy's training routine begins at point E on the eastern shore of the lake. She jogs along the north shore to a point P and then swims the straight line distance, if any, from P to point W diametrically opposite E (Figure Ex-55). Nancy swims at a rate of 2 mi/h and jogs at 8 mi/h. How far should Nancy jog in order to complete her training routine in
   (a) the least amount of time
   (b) the greatest amount of time?

56. A man is floating in a rowboat 1 mile from the (straight) shoreline of a large lake. A town is located on the shoreline 1 mile from the point on the shoreline closest to the man. As suggested in Figure Ex-56, he intends to row in a straight line to some point P on the shoreline and then walk the remaining distance to the town. To what point should he row in order to reach his destination in the least time if
   (a) he can walk 5 mi/h and row 3 mi/h
   (b) he can walk 5 mi/h and row 4 mi/h?

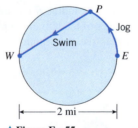
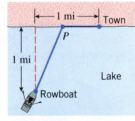

▲ **Figure Ex-55**   ▲ **Figure Ex-56**

57. A pipe of negligible diameter is to be carried horizontally around a corner from a hallway 8 ft wide into a hallway 4 ft wide (Figure Ex-57 on the next page). What is the maximum length that the pipe can have?

Source: An interesting discussion of this problem in the case where the diameter of the pipe is not neglected is given by Norman Miller in the *American Mathematical Monthly*, Vol. 56, 1949, pp. 177–179.

58. A concrete barrier whose cross section is an isosceles triangle runs parallel to a wall. The height of the barrier is 3 ft, the width of the base of a cross section is 8 ft, and the barrier is positioned on level ground with its base 1 ft from the wall. A straight, stiff metal rod of negligible diameter has one end on the ground, the other end against the wall, and touches the top of the barrier (Figure Ex-58). What is the minimum length the rod can have?

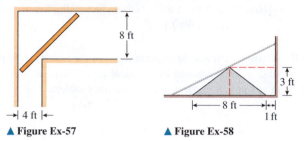

▲ **Figure Ex-57**   ▲ **Figure Ex-58**

59. Suppose that the intensity of a point light source is directly proportional to the strength of the source and inversely proportional to the square of the distance from the source. Two point light sources with strengths of S and $8S$ are separated by a distance of 90 cm. Where on the line segment between the two sources is the total intensity a minimum?

60. Given points $A(2, 1)$ and $B(5, 4)$, find the point P in the interval $[2, 5]$ on the x-axis that maximizes angle APB.

61. The lower edge of a painting, 10 ft in height, is 2 ft above an observer's eye level. Assuming that the best view is obtained when the angle subtended at the observer's eye by the painting is maximum, how far from the wall should the observer stand?

FOCUS ON CONCEPTS

62. *Fermat's principle* (biography on p. 275) in optics states that light traveling from one point to another follows that path for which the total travel time is minimum. In a uniform medium, the paths of "minimum time" and "shortest distance" turn out to be the same, so that light, if unobstructed, travels along a straight line. Assume that we have a light source, a flat mirror, and an observer in a uniform medium. If a light ray leaves the source, bounces off the mirror, and travels on to the observer, then its path will consist of two line segments, as shown in Figure Ex-62. According to Fermat's principle, the path will be such that the total travel time t is minimum or, since the medium is uniform, the path will be such that the total distance traveled from A to P to B is as small as possible. Assuming the minimum occurs when $dt/dx = 0$, show that the light ray will strike the mirror at the point P where the "angle of incidence" θ_1 equals the "angle of reflection" θ_2.

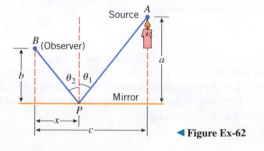

◀ **Figure Ex-62**

63. Fermat's principle (Exercise 62) also explains why light rays traveling between air and water undergo bending (refraction). Imagine that we have two uniform media (such as air and water) and a light ray traveling from a source A in one medium to an observer B in the other medium (Figure Ex-63). It is known that light travels at a constant speed in a uniform medium, but more slowly in a dense medium (such as water) than in a thin medium (such as air). Consequently, the path of shortest time from A to B is not necessarily a straight line, but rather some broken line path A to P to B allowing the light to take greatest advantage of its higher speed through the thin medium. *Snell's law of refraction* (biography on p. 288) states that the path of the light ray will be such that

$$\frac{\sin \theta_1}{v_1} = \frac{\sin \theta_2}{v_2}$$

where v_1 is the speed of light in the first medium, v_2 is the speed of light in the second medium, and θ_1 and θ_2 are the angles shown in Figure Ex-63. Show that this follows from the assumption that the path of minimum time occurs when $dt/dx = 0$.

64. A farmer wants to walk at a constant rate from her barn to a straight river, fill her pail, and carry it to her house in the least time.

(a) Explain how this problem relates to Fermat's principle and the light-reflection problem in Exercise 62.

(b) Use the result of Exercise 62 to describe geometrically the best path for the farmer to take.

(c) Use part (b) to determine where the farmer should fill her pail if her house and barn are located as in Figure Ex-64.

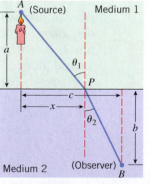 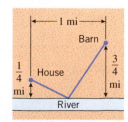

▲ **Figure Ex-63**   ▲ **Figure Ex-64**

65. If an unknown physical quantity x is measured n times, the measurements $x_1, x_2, \ldots, x_n$ often vary because of uncontrollable factors such as temperature, atmospheric pressure, and so forth. Thus, a scientist is often faced with the problem of using n different observed measurements to obtain an estimate $\bar{x}$ of an unknown quantity x. One method for making such an estimate is based on the *least squares principle*, which states that the estimate $\bar{x}$

should be chosen to minimize

$$s = (x_1 - \bar{x})^2 + (x_2 - \bar{x})^2 + \cdots + (x_n - \bar{x})^2$$

which is the sum of the squares of the deviations between the estimate $\bar{x}$ and the measured values. Show that the estimate resulting from the least squares principle is

$$\bar{x} = \frac{1}{n}(x_1 + x_2 + \cdots + x_n)$$

that is, $\bar{x}$ is the arithmetic average of the observed values.

66. Prove: If $f(x) \geq 0$ on an interval and if $f(x)$ has a maximum value on that interval at x_0, then $\sqrt{f(x)}$ also has a maximum value at x_0. Similarly for minimum values. [*Hint:* Use the fact that $\sqrt{x}$ is an increasing function on the interval $[0, +\infty)$.]

67. Writing Discuss the importance of finding intervals of possible values imposed by physical restrictions on variables in an applied maximum or minimum problem.

✔ **QUICK CHECK ANSWERS 4.5**

1. $x + \dfrac{1}{x}$; $(0, +\infty)$ **2.** $x(10 - x)$; $[0, 10]$ **3.** $x\left(-\frac{4}{3}x + 4\right) = -\frac{4}{3}x^2 + 4x$; $[0, 3]$
4. $x(20 - 2x)(32 - 2x) = 4x^3 - 104x^2 + 640x$; $[0, 10]$

4.6   RECTILINEAR MOTION

In this section we will continue the study of rectilinear motion that we began in Section 2.1. We will define the notion of "acceleration" mathematically, and we will show how the tools of calculus developed earlier in this chapter can be used to analyze rectilinear motion in more depth.

■ REVIEW OF TERMINOLOGY

Recall from Section 2.1 that a particle that can move in either direction along a coordinate line is said to be in ***rectilinear motion***. The line might be an x-axis, a y-axis, or a coordinate line inclined at some angle. In general discussions we will designate the coordinate line as the s-axis. We will assume that units are chosen for measuring distance and time and that we begin observing the motion of the particle at time $t = 0$. As the particle moves along the s-axis, its coordinate s will be some function of time, say $s = s(t)$. We call $s(t)$ the ***position function*** of the particle,[*] and we call the graph of s versus t the ***position versus time curve***. If the coordinate of a particle at time t_1 is $s(t_1)$ and the coordinate at a later time t_2 is $s(t_2)$, then $s(t_2) - s(t_1)$ is called the ***displacement*** of the particle over the time interval $[t_1, t_2]$. The displacement describes the change in position of the particle.

Figure 4.6.1 shows a typical position versus time curve for a particle in rectilinear motion. We can tell from that graph that the coordinate of the particle at time $t = 0$ is s_0, and we can tell from the sign of s when the particle is on the negative or the positive side of the origin as it moves along the coordinate line.

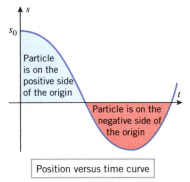

Particle is on the positive side of the origin

Particle is on the negative side of the origin

Position versus time curve

▲ **Figure 4.6.1**

---
[*] In writing $s = s(t)$, rather than the more familiar $s = f(t)$, we are using the letter s both as the dependent variable and the name of the function. This is common practice in engineering and physics.

Willebrord van Roijen Snell (1591–1626) Dutch mathematician. Snell, who succeeded his father to the post of Professor of Mathematics at the University of Leiden in 1613, is most famous for the result of light refraction that bears his name. Although this phenomenon was studied as far back as the ancient Greek astronomer Ptolemy, until Snell's work the relationship was incorrectly thought to be $\theta_1/v_1 = \theta_2/v_2$. Snell's law was published by Descartes in 1638 without giving proper credit to Snell. Snell also discovered a method for determining distances by triangulation that founded the modern technique of mapmaking.

▶ **Example 1** Figure 4.6.2*a* shows the position versus time curve for a particle moving along an *s*-axis. In words, describe how the position of the particle changes with time.

Solution. The particle is at $s = -3$ at time $t = 0$. It moves in the positive direction until time $t = 4$, since s is increasing. At time $t = 4$ the particle is at position $s = 3$. At that time it turns around and travels in the negative direction until time $t = 7$, since s is decreasing. At time $t = 7$ the particle is at position $s = -1$, and it remains stationary thereafter, since s is constant for $t > 7$. This is illustrated schematically in Figure 4.6.2*b*. ◀

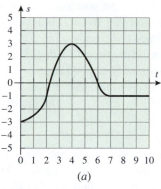

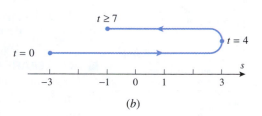

(a)   (b)

▲ **Figure 4.6.2**

■ **VELOCITY AND SPEED**

We should more properly call $v(t)$ the ***instantaneous velocity function*** to distinguish instantaneous velocity from average velocity. However, we will follow the standard practice of referring to it as the "velocity function," leaving it understood that it describes instantaneous velocity.

Recall from Formula (5) of Section 2.1 and Formula (4) of Section 2.2 that the instantaneous velocity of a particle in rectilinear motion is the derivative of the position function. Thus, if a particle in rectilinear motion has position function $s(t)$, then we define its ***velocity function*** $v(t)$ to be

$$v(t) = s'(t) = \frac{ds}{dt} \tag{1}$$

The sign of the velocity tells which way the particle is moving—a positive value for $v(t)$ means that s is increasing with time, so the particle is moving in the positive direction, and a negative value for $v(t)$ means that s is decreasing with time, so the particle is moving in the negative direction. If $v(t) = 0$, then the particle has momentarily stopped.

For a particle in rectilinear motion it is important to distinguish between its ***velocity***, which describes how fast and in what direction the particle is moving, and its ***speed***, which describes only how fast the particle is moving. We make this distinction by defining speed to be the absolute value of velocity. Thus a particle with a velocity of 2 m/s has a speed of 2 m/s and is moving in the positive direction, while a particle with a velocity of -2 m/s also has a speed of 2 m/s but is moving in the negative direction.

Since the instantaneous speed of a particle is the absolute value of its instantaneous velocity, we define its ***speed function*** to be

$$|v(t)| = |s'(t)| = \left| \frac{ds}{dt} \right| \tag{2}$$

The speed function, which is always nonnegative, tells us how fast the particle is moving but not its direction of motion.

▶ **Example 2** Let $s(t) = t^3 - 6t^2$ be the position function of a particle moving along an *s*-axis, where *s* is in meters and *t* is in seconds. Find the velocity and speed functions, and show the graphs of position, velocity, and speed versus time.

Solution. From (1) and (2), the velocity and speed functions are given by

$$v(t) = \frac{ds}{dt} = 3t^2 - 12t \quad \text{and} \quad |v(t)| = |3t^2 - 12t|$$

The graphs of position, velocity, and speed versus time are shown in Figure 4.6.3. Observe that velocity and speed both have units of meters per second (m/s), since s is in meters (m) and time is in seconds (s). ◄

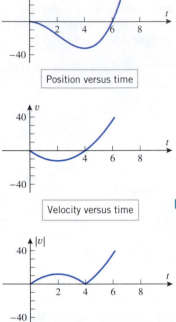

Position versus time

Velocity versus time

Speed versus time

▲ **Figure 4.6.3**

The graphs in Figure 4.6.3 provide a wealth of visual information about the motion of the particle. For example, the position versus time curve tells us that the particle is on the negative side of the origin for $0 < t < 6$, is on the positive side of the origin for $t > 6$, and is at the origin at times $t = 0$ and $t = 6$. The velocity versus time curve tells us that the particle is moving in the negative direction if $0 < t < 4$, is moving in the positive direction if $t > 4$, and is momentarily stopped at times $t = 0$ and $t = 4$ (the velocity is zero at those times). The speed versus time curve tells us that the speed of the particle is increasing for $0 < t < 2$, decreasing for $2 < t < 4$, and increasing again for $t > 4$.

■ **ACCELERATION**

In rectilinear motion, the rate at which the instantaneous velocity of a particle changes with time is called its ***instantaneous acceleration***. Thus, if a particle in rectilinear motion has velocity function $v(t)$, then we define its ***acceleration function*** to be

$$a(t) = v'(t) = \frac{dv}{dt} \tag{3}$$

Alternatively, we can use the fact that $v(t) = s'(t)$ to express the acceleration function in terms of the position function as

$$a(t) = s''(t) = \frac{d^2s}{dt^2} \tag{4}$$

▶ **Example 3** Let $s(t) = t^3 - 6t^2$ be the position function of a particle moving along an s-axis, where s is in meters and t is in seconds. Find the acceleration function $a(t)$, and show the graph of acceleration versus time.

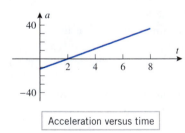

Acceleration versus time

▲ **Figure 4.6.4**

Solution. From Example 2, the velocity function of the particle is $v(t) = 3t^2 - 12t$, so the acceleration function is

$$a(t) = \frac{dv}{dt} = 6t - 12$$

and the acceleration versus time curve is the line shown in Figure 4.6.4. Note that in this example the acceleration has units of m/s^2, since v is in meters per second (m/s) and time is in seconds (s). ◄

■ **SPEEDING UP AND SLOWING DOWN**

We will say that a particle in rectilinear motion is ***speeding up*** when its speed is increasing and is ***slowing down*** when its speed is decreasing. In everyday language an object that is speeding up is said to be "accelerating" and an object that is slowing down is said to be "decelerating"; thus, one might expect that a particle in rectilinear motion will be speeding up when its acceleration is positive and slowing down when it is negative. Although this is true for a particle moving in the positive direction, it is *not* true for a particle moving in the negative direction—a particle with negative velocity is speeding up when its acceleration is negative and slowing down when its acceleration is positive. This is because a positive acceleration implies an increasing velocity, and increasing a negative velocity decreases its absolute value; similarly, a negative acceleration implies a decreasing velocity, and decreasing a negative velocity increases its absolute value.

The preceding informal discussion can be summarized as follows (Exercise 41):

> If $a(t) = 0$ over a certain time interval, what does this tell you about the motion of the particle during that time?

INTERPRETING THE SIGN OF ACCELERATION *A particle in rectilinear motion is speeding up when its velocity and acceleration have the same sign and slowing down when they have opposite signs.*

▶ **Example 4** In Examples 2 and 3 we found the velocity versus time curve and the acceleration versus time curve for a particle with position function $s(t) = t^3 - 6t^2$. Use those curves to determine when the particle is speeding up and slowing down, and confirm that your results are consistent with the speed versus time curve obtained in Example 2.

Solution. Over the time interval $0 < t < 2$ the velocity and acceleration are negative, so the particle is speeding up. This is consistent with the speed versus time curve, since the speed is increasing over this time interval. Over the time interval $2 < t < 4$ the velocity is negative and the acceleration is positive, so the particle is slowing down. This is also consistent with the speed versus time curve, since the speed is decreasing over this time interval. Finally, on the time interval $t > 4$ the velocity and acceleration are positive, so the particle is speeding up, which again is consistent with the speed versus time curve. ◀

■ ANALYZING THE POSITION VERSUS TIME CURVE

The position versus time curve contains all of the significant information about the position and velocity of a particle in rectilinear motion:

- If $s(t) > 0$, the particle is on the positive side of the s-axis.
- If $s(t) < 0$, the particle is on the negative side of the s-axis.
- The slope of the curve at any time is equal to the instantaneous velocity at that time.
- Where the curve has positive slope, the velocity is positive and the particle is moving in the positive direction.
- Where the curve has negative slope, the velocity is negative and the particle is moving in the negative direction.
- Where the slope of the curve is zero, the velocity is zero, and the particle is momentarily stopped.

Information about the acceleration of a particle in rectilinear motion can also be deduced from the position versus time curve by examining its concavity. For example, we know that the position versus time curve will be concave up on intervals where $s''(t) > 0$ and will be concave down on intervals where $s''(t) < 0$. But we know from (4) that $s''(t)$ is the acceleration, so that on intervals where the position versus time curve is concave up the particle has a positive acceleration, and on intervals where it is concave down the particle has a negative acceleration.

Table 4.6.1 summarizes our observations about the position versus time curve.

Table 4.6.1

ANALYSIS OF PARTICLE MOTION

POSITION VERSUS TIME CURVE	CHARACTERISTICS OF THE CURVE AT $t = t_0$	BEHAVIOR OF THE PARTICLE AT TIME $t = t_0$
	• $s(t_0) > 0$ • Curve has positive slope. • Curve is concave down.	• Particle is on the positive side of the origin. • Particle is moving in the positive direction. • Velocity is decreasing. • Particle is slowing down.
	• $s(t_0) > 0$ • Curve has negative slope. • Curve is concave down.	• Particle is on the positive side of the origin. • Particle is moving in the negative direction. • Velocity is decreasing. • Particle is speeding up.
	• $s(t_0) < 0$ • Curve has negative slope. • Curve is concave up.	• Particle is on the negative side of the origin. • Particle is moving in the negative direction. • Velocity is increasing. • Particle is slowing down.
	• $s(t_0) > 0$ • Curve has zero slope. • Curve is concave down.	• Particle is on the positive side of the origin. • Particle is momentarily stopped. • Velocity is decreasing.

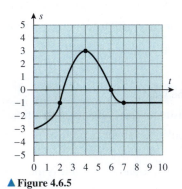

▲ **Figure 4.6.5**

▶ **Example 5** Use the position versus time curve in Figure 4.6.5 to determine when the particle in Example 1 is speeding up and slowing down.

Solution. From $t = 0$ to $t = 2$, the acceleration and velocity are positive, so the particle is speeding up. From $t = 2$ to $t = 4$, the acceleration is negative and the velocity is positive, so the particle is slowing down. At $t = 4$, the velocity is zero, so the particle has momentarily stopped. From $t = 4$ to $t = 6$, the acceleration is negative and the velocity is negative, so the particle is speeding up. From $t = 6$ to $t = 7$, the acceleration is positive and the velocity is negative, so the particle is slowing down. Thereafter, the velocity is zero, so the particle has stopped. ◀

▶ **Example 6** Suppose that the position function of a particle moving on a coordinate line is given by $s(t) = 2t^3 - 21t^2 + 60t + 3$. Analyze the motion of the particle for $t \geq 0$.

Solution. The velocity and acceleration functions are

$$v(t) = s'(t) = 6t^2 - 42t + 60 = 6(t - 2)(t - 5)$$
$$a(t) = v'(t) = 12t - 42 = 12 \left(t - \tfrac{7}{2} \right)$$

• *Direction of motion:* The sign analysis of the velocity function in Figure 4.6.6 shows that the particle is moving in the positive direction over the time interval $0 \leq t < 2$, stops momentarily at time $t = 2$, moves in the negative direction over the time interval $2 < t < 5$, stops momentarily at time $t = 5$, and then moves in the positive direction thereafter.

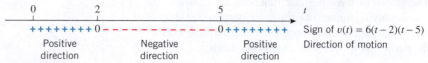

▲ **Figure 4.6.6**

- *Change in speed:* A comparison of the signs of the velocity and acceleration functions is shown in Figure 4.6.7. Since the particle is speeding up when the signs are the same and is slowing down when they are opposite, we see that the particle is slowing down over the time interval $0 \leq t < 2$ and stops momentarily at time $t = 2$. It is then speeding up over the time interval $2 < t < \frac{7}{2}$. At time $t = \frac{7}{2}$ the instantaneous acceleration is zero, so the particle is neither speeding up nor slowing down. It is then slowing down over the time interval $\frac{7}{2} < t < 5$ and stops momentarily at time $t = 5$. Thereafter, it is speeding up.

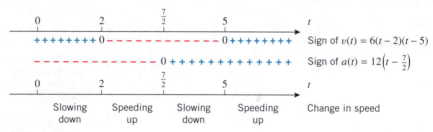

▲ **Figure 4.6.7**

Conclusions: The diagram in Figure 4.6.8 summarizes the above information schematically. The curved line is descriptive only; the actual path is back and forth on the coordinate line. The coordinates of the particle at times $t = 0$, $t = 2$, $t = \frac{7}{2}$, and $t = 5$ were computed from $s(t)$. Segments in red indicate that the particle is speeding up and segments in blue indicate that it is slowing down. ◄

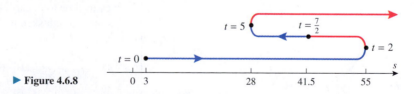

▶ **Figure 4.6.8**

✔ **QUICK CHECK EXERCISES 4.6**   *(See page 296 for answers.)*

---

1. For a particle in rectilinear motion, the velocity and position functions $v(t)$ and $s(t)$ are related by the equation _____, and the acceleration and velocity functions $a(t)$ and $v(t)$ are related by the equation _____.

2. Suppose that a particle moving along the s-axis has position function $s(t) = 7t - 2t^2$. At time $t = 3$, the particle's position is _____, its velocity is _____, its speed is _____, and its acceleration is _____.

3. A particle in rectilinear motion is speeding up if the signs of its velocity and acceleration are _____, and it is slowing down if these signs are _____.

4. Suppose that a particle moving along the s-axis has position function $s(t) = t^4 - 24t^2$ over the time interval $t \geq 0$. The particle slows down over the time interval(s) _____.

EXERCISE SET 4.6   Graphing Utility

FOCUS ON CONCEPTS

1. The graphs of three position functions are shown in the accompanying figure. In each case determine the signs of the velocity and acceleration, and then determine whether the particle is speeding up or slowing down.

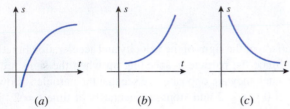

(a)   (b)   (c)

▲ **Figure Ex-1**

2. The graphs of three velocity functions are shown in the accompanying figure. In each case determine the sign of the acceleration, and then determine whether the particle is speeding up or slowing down.

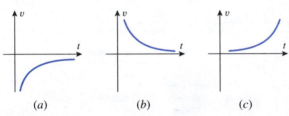

(a)   (b)   (c)

▲ **Figure Ex-2**

3. The graph of the position function of a particle moving on a horizontal line is shown in the accompanying figure.
   (a) Is the particle moving left or right at time t_0?
   (b) Is the acceleration positive or negative at time t_0?
   (c) Is the particle speeding up or slowing down at time t_0?
   (d) Is the particle speeding up or slowing down at time t_1?

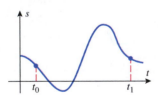

◄ **Figure Ex-3**

4. For the graphs in the accompanying figure, match the position functions (a)–(c) with their corresponding velocity functions (I)–(III).

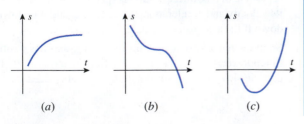

(a)   (b)   (c)

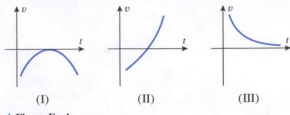

(I)   (II)   (III)

▲ **Figure Ex-4**

5. Sketch a reasonable graph of s versus t for a mouse that is trapped in a narrow corridor (an s-axis with the positive direction to the right) and scurries back and forth as follows. It runs right with a constant speed of 1.2 m/s for a while, then gradually slows down to 0.6 m/s, then quickly speeds up to 2.0 m/s, then gradually slows to a stop but immediately reverses direction and quickly speeds up to 1.2 m/s.

6. The accompanying figure shows the position versus time curve for an ant that moves along a narrow vertical pipe, where t is measured in seconds and the s-axis is along the pipe with the positive direction up.
   (a) When, if ever, is the ant above the origin?
   (b) When, if ever, does the ant have velocity zero?
   (c) When, if ever, is the ant moving down the pipe?

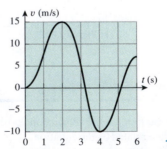

◄ **Figure Ex-6**

7. The accompanying figure shows the graph of velocity versus time for a particle moving along a coordinate line. Make a rough sketch of the graphs of speed versus time and acceleration versus time.

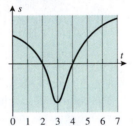

◄ **Figure Ex-7**

8. The accompanying figure (on the next page) shows the position versus time graph for an elevator that ascends 40 m from one stop to the next.
   (a) Estimate the velocity when the elevator is halfway up to the top.   *(cont.)*

(b) Sketch rough graphs of the velocity versus time curve and the acceleration versus time curve.

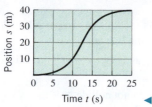

◀ **Figure Ex-8**

9–12 True–False Determine whether the statement is true or false. Explain your answer. ■

9. A particle is speeding up when its position versus time graph is increasing.

10. Velocity is the derivative of position with respect to time.

11. Acceleration is the absolute value of velocity.

12. If the position versus time curve is increasing and concave down, then the particle is slowing down.

13. The accompanying figure shows the velocity versus time graph for a test run on a Pontiac Grand Prix GTP. Using this graph, estimate
(a) the acceleration at 60 mi/h (in ft/s²)
(b) the time at which the maximum acceleration occurs.
Source: Data from *Car and Driver Magazine*, July 2003.

14. The accompanying figure shows the velocity versus time graph for a test run on a Chevrolet Malibu. Using this graph, estimate
(a) the acceleration at 60 mi/h (in ft/s²)
(b) the time at which the maximum acceleration occurs.
Source: Data from *Car and Driver Magazine*, November 2003.

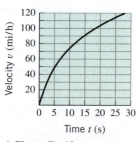

▲ **Figure Ex-13**

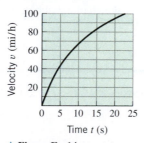

▲ **Figure Ex-14**

15–16 The function $s(t)$ describes the position of a particle moving along a coordinate line, where s is in meters and t is in seconds.
(a) Make a table showing the position, velocity, and acceleration to two decimal places at times $t = 1, 2, 3, 4, 5$.
(b) At each of the times in part (a), determine whether the particle is stopped; if it is not, state its direction of motion.
(c) At each of the times in part (a), determine whether the particle is speeding up, slowing down, or neither. ■

15. $s(t) = \sin \dfrac{\pi t}{4}$ 16. $s(t) = t^4 e^{-t}$, $t \geq 0$

17–22 The function $s(t)$ describes the position of a particle moving along a coordinate line, where s is in feet and t is in seconds.
(a) Find the velocity and acceleration functions.
(b) Find the position, velocity, speed, and acceleration at time $t = 1$.
(c) At what times is the particle stopped?
(d) When is the particle speeding up? Slowing down?
(e) Find the total distance traveled by the particle from time $t = 0$ to time $t = 5$. ■

17. $s(t) = t^3 - 3t^2$, $t \geq 0$

18. $s(t) = t^4 - 4t^2 + 4$, $t \geq 0$

19. $s(t) = 9 - 9\cos(\pi t/3)$, $0 \leq t \leq 5$

20. $s(t) = \dfrac{t}{t^2 + 4}$, $t \geq 0$

21. $s(t) = (t^2 + 8)e^{-t/3}$, $t \geq 0$

22. $s(t) = \frac{1}{4}t^2 - \ln(t + 1)$, $t \geq 0$

23. Let $s(t) = t/(t^2 + 5)$ be the position function of a particle moving along a coordinate line, where s is in meters and t is in seconds. Use a graphing utility to generate the graphs of $s(t)$, $v(t)$, and $a(t)$ for $t \geq 0$, and use those graphs where needed.
(a) Use the appropriate graph to make a rough estimate of the time at which the particle first reverses the direction of its motion; and then find the time exactly.
(b) Find the exact position of the particle when it first reverses the direction of its motion.
(c) Use the appropriate graphs to make a rough estimate of the time intervals on which the particle is speeding up and on which it is slowing down; and then find those time intervals exactly.

24. Let $s(t) = t/e^t$ be the position function of a particle moving along a coordinate line, where s is in meters and t is in seconds. Use a graphing utility to generate the graphs of $s(t)$, $v(t)$, and $a(t)$ for $t \geq 0$, and use those graphs where needed.
(a) Use the appropriate graph to make a rough estimate of the time at which the particle first reverses the direction of its motion; and then find the time exactly.
(b) Find the exact position of the particle when it first reverses the direction of its motion.
(c) Use the appropriate graphs to make a rough estimate of the time intervals on which the particle is speeding up and on which it is slowing down; and then find those time intervals exactly.

25–32 A position function of a particle moving along a coordinate line is given. Use the method of Example 6 to analyze the motion of the particle for $t \geq 0$, and give a schematic picture of the motion (as in Figure 4.6.8). ■

25. $s = -4t + 3$ 26. $s = 5t^2 - 20t$

27. $s = t^3 - 9t^2 + 24t$ 28. $s = t^3 - 6t^2 + 9t + 1$

29. $s = 16te^{-(t^2/8)}$

30. $s = t + \dfrac{25}{t+2}$

31. $s = \begin{cases} \cos t, & 0 \le t < 2\pi \\ 1, & t \ge 2\pi \end{cases}$

32. $s = \begin{cases} 2t(t-2)^2, & 0 \le t < 3 \\ 13 - 7(t-4)^2, & t \ge 3 \end{cases}$

33. Let $s(t) = 5t^2 - 22t$ be the position function of a particle moving along a coordinate line, where s is in feet and t is in seconds.
 (a) Find the maximum speed of the particle during the time interval $1 \le t \le 3$.
 (b) When, during the time interval $1 \le t \le 3$, is the particle farthest from the origin? What is its position at that instant?

34. Let $s = 100/(t^2 + 12)$ be the position function of a particle moving along a coordinate line, where s is in feet and t is in seconds. Find the maximum speed of the particle for $t \ge 0$, and find the direction of motion of the particle when it has its maximum speed.

35–36 A position function of a particle moving along a coordinate line is provided. (a) Evaluate s and v when $a = 0$. (b) Evaluate s and a when $v = 0$. ■

35. $s = \ln(3t^2 - 12t + 13)$ **36.** $s = t^3 - 6t^2 + 1$

37. Let $s = \sqrt{2t^2 + 1}$ be the position function of a particle moving along a coordinate line.
 (a) Use a graphing utility to generate the graph of v versus t, and make a conjecture about the velocity of the particle as $t \to +\infty$.
 (b) Check your conjecture by finding $\lim\limits_{t \to +\infty} v$.

38. (a) Use the chain rule to show that for a particle in rectilinear motion $a = v(dv/ds)$.

(b) Let $s = \sqrt{3t + 7}, t \ge 0$. Find a formula for v in terms of s and use the equation in part (a) to find the acceleration when $s = 5$.

39. Suppose that the position functions of two particles, P_1 and P_2, in motion along the same line are

$$s_1 = \tfrac{1}{2}t^2 - t + 3 \quad \text{and} \quad s_2 = -\tfrac{1}{4}t^2 + t + 1$$

respectively, for $t \ge 0$.
 (a) Prove that P_1 and P_2 do not collide.
 (b) How close do P_1 and P_2 get to each other?
 (c) During what intervals of time are they moving in opposite directions?

40. Let $s_A = 15t^2 + 10t + 20$ and $s_B = 5t^2 + 40t, t \ge 0$, be the position functions of cars A and B that are moving along parallel straight lanes of a highway.
 (a) How far is car A ahead of car B when $t = 0$?
 (b) At what instants of time are the cars next to each other?
 (c) At what instant of time do they have the same velocity? Which car is ahead at this instant?

41. Prove that a particle is speeding up if the velocity and acceleration have the same sign, and slowing down if they have opposite signs. [*Hint:* Let $r(t) = |v(t)|$ and find $r'(t)$ using the chain rule.]

42. **Writing** A speedometer on a bicycle calculates the bicycle's speed by measuring the time per rotation for one of the bicycle's wheels. Explain how this measurement can be used to calculate an average velocity for the bicycle, and discuss how well it approximates the instantaneous velocity for the bicycle.

43. **Writing** A toy rocket is launched into the air and falls to the ground after its fuel runs out. Describe the rocket's acceleration and when the rocket is speeding up or slowing down during its flight. Accompany your description with a sketch of a graph of the rocket's acceleration versus time.

✔ **QUICK CHECK ANSWERS 4.6**

1. $v(t) = s'(t); \ a(t) = v'(t)$ **2.** 3; -5; 5; -4 **3.** the same; opposite   **4.** $2 < t < 2\sqrt{3}$

4.7 NEWTON'S METHOD

In Section 1.5 we showed how to approximate the roots of an equation $f(x) = 0$ using the Intermediate-Value Theorem. In this section we will study a technique, called "Newton's Method," that is usually more efficient than that method. Newton's Method is the technique used by many commercial and scientific computer programs for finding roots.

■ **NEWTON'S METHOD**

In beginning algebra one learns that the solution of a first-degree equation $ax + b = 0$ is given by the formula $x = -b/a$, and the solutions of a second-degree equation

$$ax^2 + bx + c = 0$$

are given by the quadratic formula. Formulas also exist for the solutions of all third- and fourth-degree equations, although they are too complicated to be of practical use. In 1826 it was shown by the Norwegian mathematician Niels Henrik Abel that it is impossible to construct a similar formula for the solutions of a *general* fifth-degree equation or higher. Thus, for a *specific* fifth-degree polynomial equation such as

$$x^5 - 9x^4 + 2x^3 - 5x^2 + 17x - 8 = 0$$

it may be difficult or impossible to find exact values for all of the solutions. Similar difficulties occur for nonpolynomial equations such as

$$x - \cos x = 0$$

For such equations the solutions are generally approximated in some way, often by the method we will now discuss.

Suppose that we are trying to find a root r of the equation $f(x) = 0$, and suppose that by some method we are able to obtain an initial rough estimate, x_1, of r, say by generating the graph of $y = f(x)$ with a graphing utility and examining the x-intercept. If $f(x_1) = 0$, then $r = x_1$. If $f(x_1) \neq 0$, then we consider an easier problem, that of finding a root to a linear equation. The best linear approximation to $y = f(x)$ near $x = x_1$ is given by the tangent line to the graph of f at x_1, so it might be reasonable to expect that the x-intercept to this tangent line provides an improved approximation to r. Call this intercept x_2 (Figure 4.7.1). We can now treat x_2 in the same way we did x_1. If $f(x_2) = 0$, then $r = x_2$. If $f(x_2) \neq 0$, then construct the tangent line to the graph of f at x_2, and take x_3 to be the x-intercept of this tangent line. Continuing in this way we can generate a succession of values $x_1, x_2, x_3, x_4, \dots$ that will usually approach r. This procedure for approximating r is called **Newton's Method**.

To implement Newton's Method analytically, we must derive a formula that will tell us how to calculate each improved approximation from the preceding approximation. For this purpose, we note that the point-slope form of the tangent line to $y = f(x)$ at the initial

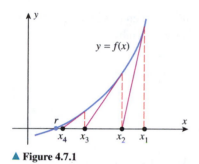

▲ **Figure 4.7.1**

Niels Henrik Abel (1802–1829) Norwegian mathematician. Abel was the son of a poor Lutheran minister and a remarkably beautiful mother from whom he inherited strikingly good looks. In his brief life of 26 years Abel lived in virtual poverty and suffered a succession of adversities, yet he managed to prove major results that altered the mathematical landscape forever. At the age of thirteen he was sent away from home to a school whose better days had long passed. By a stroke of luck the school had just hired a teacher named Bernt Michael Holmboe, who quickly discovered that Abel had extraordinary mathematical ability. Together, they studied the calculus texts of Euler and works of Newton and the later French mathematicians. By the time he graduated, Abel was familar with most of the great mathematical literature. In 1820 his father died, leaving the family in dire financial straits. Abel was able to enter the University of Christiania in Oslo only because he was granted a free room and several professors supported him directly from their salaries. The University had no advanced courses in mathematics, so Abel took a preliminary degree in 1822 and then continued to study mathematics on his own. In 1824 he published at his own expense the proof that it is impossible to solve the general fifth-degree polynomial equation algebraically. With the hope that this landmark paper would lead to his recognition and acceptance by the European mathematical community, Abel sent the paper to the great German mathematician Gauss, who casually declared it to be a "monstrosity" and tossed it aside. However, in 1826 Abel's paper on the fifth-degree equation and other work was published in the first issue of a new journal, founded by his friend, Leopold Crelle. In the summer of 1826 he completed a landmark work on transcendental functions, which he submitted to the French Academy of Sciences. He hoped to establish himself as a major mathematician, for many young mathematicians had gained quick distinction by having their work accepted by the Academy. However, Abel waited in vain because the paper was either ignored or misplaced by one of the referees, and it did not surface again until two years after his death. That paper was later described by one major mathematician as "…the most important mathematical discovery that has been made in our century…." After submitting his paper, Abel returned to Norway, ill with tuberculosis and in heavy debt. While eking out a meager living as a tutor, he continued to produce great work and his fame spread. Soon great efforts were being made to secure a suitable mathematical position for him. Fearing that his great work had been lost by the Academy, he mailed a proof of the main results to Crelle in January of 1829. In April he suffered a violent hemorrhage and died. Two days later Crelle wrote to inform him that an appointment had been secured for him in Berlin and his days of poverty were over! Abel's great paper was finally published by the Academy twelve years after his death.

approximation x_1 is

$$y - f(x_1) = f'(x_1)(x - x_1) \tag{1}$$

If $f'(x_1) \neq 0$, then this line is not parallel to the x-axis and consequently it crosses the x-axis at some point $(x_2, 0)$. Substituting the coordinates of this point in (1) yields

$$-f(x_1) = f'(x_1)(x_2 - x_1)$$

Solving for x_2 we obtain

$$x_2 = x_1 - \frac{f(x_1)}{f'(x_1)} \tag{2}$$

The next approximation can be obtained more easily. If we view x_2 as the starting approximation and x_3 the new approximation, we can simply apply (2) with x_2 in place of x_1 and x_3 in place of x_2. This yields

$$x_3 = x_2 - \frac{f(x_2)}{f'(x_2)} \tag{3}$$

provided $f'(x_2) \neq 0$. In general, if x_n is the nth approximation, then it is evident from the pattern in (2) and (3) that the improved approximation x_{n+1} is given by

> **Newton's Method**
>
> $$x_{n+1} = x_n - \frac{f(x_n)}{f'(x_n)}, \quad n = 1, 2, 3, \ldots$$
> $$\tag{4}$$

▶ **Example 1** Use Newton's Method to approximate the real solutions of

$$x^3 - x - 1 = 0$$

Solution. Let $f(x) = x^3 - x - 1$, so $f'(x) = 3x^2 - 1$ and (4) becomes

$$x_{n+1} = x_n - \frac{x_n^3 - x_n - 1}{3x_n^2 - 1} \tag{5}$$

From the graph of f in Figure 4.7.2, we see that the given equation has only one real solution. This solution lies between 1 and 2 because $f(1) = -1 < 0$ and $f(2) = 5 > 0$. We will use $x_1 = 1.5$ as our first approximation ($x_1 = 1$ or $x_1 = 2$ would also be reasonable choices).

Letting $n = 1$ in (5) and substituting $x_1 = 1.5$ yields

$$x_2 = 1.5 - \frac{(1.5)^3 - 1.5 - 1}{3(1.5)^2 - 1} \approx 1.34782609 \tag{6}$$

(We used a calculator that displays nine digits.) Next, we let $n = 2$ in (5) and substitute x_2 to obtain

$$x_3 = x_2 - \frac{x_2^3 - x_2 - 1}{3x_2^2 - 1} \approx 1.32520040 \tag{7}$$

If we continue this process until two identical approximations are generated in succession, we obtain

$$x_1 = 1.5$$
$$x_2 \approx 1.34782609$$
$$x_3 \approx 1.32520040$$
$$x_4 \approx 1.32471817$$
$$x_5 \approx 1.32471796$$
$$x_6 \approx 1.32471796$$

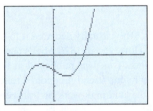

$[-2, 4] \times [-3, 3]$
$x\text{Scl} = 1, \ y\text{Scl} = 1$

$$\boxed{y = x^3 - x - 1}$$

▲ **Figure 4.7.2**

TECHNOLOGY MASTERY

Many calculators and computer programs calculate internally with more digits than they display. Where possible, you should use stored calculated values rather than values displayed from earlier calculations. Thus, in Example 1 the value of x_2 used in (7) should be the stored value, not the value in (6).

At this stage there is no need to continue further because we have reached the display accuracy limit of our calculator, and all subsequent approximations that the calculator generates will likely be the same. Thus, the solution is approximately $x \approx 1.32471796$. ◄

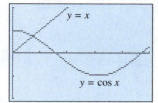

$[0, 5] \times [-2, 2]$
xScl $= 1$, yScl $= 1$

▲ **Figure 4.7.3**

▶ **Example 2**   It is evident from Figure 4.7.3 that if x is in radians, then the equation

$$\cos x = x$$

has a solution between 0 and 1. Use Newton's Method to approximate it.

Solution.   Rewrite the equation as

$$x - \cos x = 0$$

and apply (4) with $f(x) = x - \cos x$. Since $f'(x) = 1 + \sin x$, (4) becomes

$$x_{n+1} = x_n - \frac{x_n - \cos x_n}{1 + \sin x_n} \tag{8}$$

From Figure 4.7.3, the solution seems closer to $x = 1$ than $x = 0$, so we will use $x_1 = 1$ (radian) as our initial approximation. Letting $n = 1$ in (8) and substituting $x_1 = 1$ yields

$$x_2 = 1 - \frac{1 - \cos 1}{1 + \sin 1} \approx 0.750363868$$

Next, letting $n = 2$ in (8) and substituting this value of x_2 yields

$$x_3 = x_2 - \frac{x_2 - \cos x_2}{1 + \sin x_2} \approx 0.739112891$$

If we continue this process until two identical approximations are generated in succession, we obtain

$$x_1 = 1$$
$$x_2 \approx 0.750363868$$
$$x_3 \approx 0.739112891$$
$$x_4 \approx 0.739085133$$
$$x_5 \approx 0.739085133$$

Thus, to the accuracy limit of our calculator, the solution of the equation $\cos x = x$ is $x \approx 0.739085133$. ◄

SOME DIFFICULTIES WITH NEWTON'S METHOD

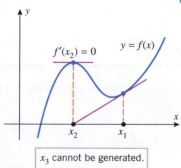

x_3 cannot be generated.

▲ **Figure 4.7.4**

When Newton's Method works, the approximations usually converge toward the solution with dramatic speed. However, there are situations in which the method fails. For example, if $f'(x_n) = 0$ for some n, then (4) involves a division by zero, making it impossible to generate x_{n+1}. However, this is to be expected because the tangent line to $y = f(x)$ is parallel to the x-axis where $f'(x_n) = 0$, and hence this tangent line does not cross the x-axis to generate the next approximation (Figure 4.7.4).

Newton's Method can fail for other reasons as well; sometimes it may overlook the root you are trying to find and converge to a different root, and sometimes it may fail to converge altogether. For example, consider the equation

$$x^{1/3} = 0$$

which has $x = 0$ as its only solution, and try to approximate this solution by Newton's Method with a starting value of $x_0 = 1$. Letting $f(x) = x^{1/3}$, Formula (4) becomes

$$x_{n+1} = x_n - \frac{(x_n)^{1/3}}{\frac{1}{3}(x_n)^{-2/3}} = x_n - 3x_n = -2x_n$$

Beginning with $x_1 = 1$, the successive values generated by this formula are

$$x_1 = 1, \quad x_2 = -2, \quad x_3 = 4, \quad x_4 = -8, \ldots$$

which obviously do not converge to $x = 0$. Figure 4.7.5 illustrates what is happening geometrically in this situation.

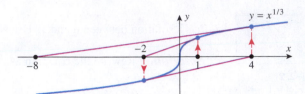

▶ **Figure 4.7.5**

To learn more about the conditions under which Newton's Method converges and for a discussion of error questions, you should consult a book on numerical analysis. For a more in-depth discussion of Newton's Method and its relationship to contemporary studies of chaos and fractals, you may want to read the article, "Newton's Method and Fractal Patterns," by Philip Straffin, which appears in *Applications of Calculus*, MAA Notes, Vol. 3, No. 29, 1993, published by the Mathematical Association of America.

✔ **QUICK CHECK EXERCISES 4.7**   *(See page 302 for answers.)*

1. Use the accompanying graph to estimate x_2 and x_3 if Newton's Method is applied to the equation $y = f(x)$ with $x_1 = 8$.

2. Suppose that $f(1) = 2$ and $f'(1) = 4$. If Newton's Method is applied to $y = f(x)$ with $x_1 = 1$, then $x_2 = $ _____.

3. Suppose we are given that $f(0) = 3$ and that $x_2 = 3$ when Newton's Method is applied to $y = f(x)$ with $x_1 = 0$. Then $f'(0) = $ _____.

4. If Newton's Method is applied to $y = e^x - 1$ with $x_1 = \ln 2$, then $x_2 = $ _____.

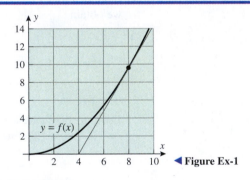

◀ **Figure Ex-1**

EXERCISE SET 4.7   ⌣ Graphing Utility

In this exercise set, express your answers with as many decimal digits as your calculating utility can display, but use the procedure in the Technology Mastery on p. 298. ■

1. Approximate $\sqrt{2}$ by applying Newton's Method to the equation $x^2 - 2 = 0$.

2. Approximate $\sqrt{5}$ by applying Newton's Method to the equation $x^2 - 5 = 0$.

3. Approximate $\sqrt[3]{6}$ by applying Newton's Method to the equation $x^3 - 6 = 0$.

4. To what equation would you apply Newton's Method to approximate the nth root of a?

5–8 The given equation has one real solution. Approximate it by Newton's Method. ■

5. $x^3 - 2x - 2 = 0$ 6. $x^3 + x - 1 = 0$

7. $x^5 + x^4 - 5 = 0$ 8. $x^5 - 3x + 3 = 0$

⌣ **9–14** Use a graphing utility to determine how many solutions the equation has, and then use Newton's Method to approximate the solution that satisfies the stated condition. ■

9. $x^4 + x^2 - 4 = 0; \quad x < 0$

10. $x^5 - 5x^3 - 2 = 0; \quad x > 0$

11. $2\cos x = x; \quad x > 0$ 12. $\sin x = x^2; \quad x > 0$

13. $x - \tan x = 0$; $\pi/2 < x < 3\pi/2$

14. $1 + e^x \sin x = 0$; $\pi/2 < x < 3\pi/2$

15–20 Use a graphing utility to determine the number of times the curves intersect; and then apply Newton's Method, where needed, to approximate the x-coordinates of all intersections. ■

15. $y = x^3$ and $y = 1 - x$

16. $y = \sin x$ and $y = x^3 - 2x^2 + 1$

17. $y = x^2$ and $y = \sqrt{2x + 1}$

18. $y = \frac{1}{8}x^3 - 1$ and $y = \cos x - 2$

19. $y = 1$ and $y = e^x \sin x$; $0 < x < \pi$

20. $y = e^{-x}$ and $y = \ln x$

21–24 True–False Determine whether the statement is true or false. Explain your answer. ■

21. Newton's Method uses the tangent line to $y = f(x)$ at $x = x_n$ to compute x_{n+1}.

22. Newton's Method is a process to find exact solutions to $f(x) = 0$.

23. If $f(x) = 0$ has a root, then Newton's Method starting at $x = x_1$ will approximate the root nearest x_1.

24. Newton's Method can be used to appoximate a point of intersection of two curves.

25. The *mechanic's rule* for approximating square roots states that $\sqrt{a} \approx x_{n+1}$, where

$$x_{n+1} = \frac{1}{2}\left(x_n + \frac{a}{x_n}\right), \quad n = 1, 2, 3, \ldots$$

and x_1 is any positive approximation to $\sqrt{a}$.
(a) Apply Newton's Method to

$$f(x) = x^2 - a$$

to derive the mechanic's rule.
(b) Use the mechanic's rule to approximate $\sqrt{10}$.

26. Many calculators compute reciprocals using the approximation $1/a \approx x_{n+1}$, where

$$x_{n+1} = x_n(2 - ax_n), \quad n = 1, 2, 3, \ldots$$

and x_1 is an initial approximation to $1/a$. This formula makes it possible to perform divisions using multiplications and subtractions, which is a faster procedure than dividing directly.
(a) Apply Newton's Method to

$$f(x) = \frac{1}{x} - a$$

to derive this approximation.
(b) Use the formula to approximate $\frac{1}{17}$.

27. Use Newton's Method to approximate the absolute minimum of $f(x) = \frac{1}{4}x^4 + x^2 - 5x$.

28. Use Newton's Method to approximate the absolute maximum of $f(x) = x \sin x$ on the interval $[0, \pi]$.

29. For the function

$$f(x) = \frac{e^{-x}}{1 + x^2}$$

use Newton's Method to approximate the x-coordinates of the inflection points to two decimal places.

30. Use Newton's Method to approximate the absolute maximum of $f(x) = (1 - 2x)\tan^{-1} x$.

31. Use Newton's Method to approximate the coordinates of the point on the parabola $y = x^2$ that is closest to the point $(1, 0)$.

32. Use Newton's Method to approximate the dimensions of the rectangle of largest area that can be inscribed under the curve $y = \cos x$ for $0 \le x \le \pi/2$ (Figure Ex-32).

33. (a) Show that on a circle of radius r, the central angle θ that subtends an arc whose length is 1.5 times the length L of its chord satisfies the equation $\theta = 3\sin(\theta/2)$ (Figure Ex-33).
(b) Use Newton's Method to approximate θ.

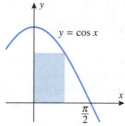

▲ **Figure Ex-32**   ▲ **Figure Ex-33**

34. A *segment* of a circle is the region enclosed by an arc and its chord (Figure Ex-34). If r is the radius of the circle and θ the angle subtended at the center of the circle, then it can be shown that the area A of the segment is $A = \frac{1}{2}r^2(\theta - \sin \theta)$, where θ is in radians. Find the value of θ for which the area of the segment is one-fourth the area of the circle. Give θ to the nearest degree.

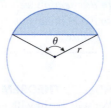

◀ **Figure Ex-34**

35–36 Use Newton's Method to approximate all real values of y satisfying the given equation for the indicated value of x. ■

35. $xy^4 + x^3y = 1$; $x = 1$ **36.** $xy - \cos\left(\frac{1}{2}xy\right) = 0$; $x = 2$

37. An *annuity* is a sequence of equal payments that are paid or received at regular time intervals. For example, you may want to deposit equal amounts at the end of each year into an interest-bearing account for the purpose of accumulating a lump sum at some future time. If, at the end of each year, interest of $i \times 100\%$ on the account balance for that year is added to the account, then the account is said to pay $i \times 100\%$ interest, *compounded annually*. It can be shown that if payments of Q dollars are deposited at the end of each year into an account that pays $i \times 100\%$ compounded annually, then at the time when the nth payment and the accrued interest for the past year are deposited, the amount $S(n)$ in the account is given by the formula

$$S(n) = \frac{Q}{i}[(1+i)^n - 1]$$

Suppose that you can invest \$5000 in an interest-bearing account at the end of each year, and your objective is to have \$250,000 on the 25th payment. Approximately what annual compound interest rate must the account pay for you to achieve your goal? [*Hint:* Show that the interest rate i satisfies the equation $50i = (1+i)^{25} - 1$, and solve it using Newton's Method.]

FOCUS ON CONCEPTS

 38. (a) Use a graphing utility to generate the graph of

$$f(x) = \frac{x}{x^2 + 1}$$

and use it to explain what happens if you apply Newton's Method with a starting value of $x_1 = 2$. Check your conclusion by computing x_2, x_3, x_4, and x_5.

(b) Use the graph generated in part (a) to explain what happens if you apply Newton's Method with a start-

ing value of $x_1 = 0.5$. Check your conclusion by computing x_2, x_3, x_4, and x_5.

39. (a) Apply Newton's Method to $f(x) = x^2 + 1$ with a starting value of $x_1 = 0.5$, and determine if the values of $x_2, \ldots, x_{10}$ appear to converge.

(b) Explain what is happening.

40. In each part, explain what happens if you apply Newton's Method to a function f when the given condition is satisfied for some value of n.

(a) $f(x_n) = 0$ (b) $x_{n+1} = x_n$

(c) $x_{n+2} = x_n \neq x_{n+1}$

41. Writing Compare Newton's Method and the Intermediate-Value Theorem (1.5.7; see Example 5 in Section 1.5) as methods to locate solutions to $f(x) = 0$.

42. Writing Newton's Method uses a local linear approximation to $y = f(x)$ at $x = x_n$ to find an "improved" approximation x_{n+1} to a zero of f. Your friend proposes a process that uses a local quadratic approximation to $y = f(x)$ at $x = x_n$ (that is, matching values for the function and its first two derivatives) to obtain x_{n+1}. Discuss the pros and cons of this proposal. Support your statements with some examples.

✔ **QUICK CHECK ANSWERS 4.7**

1. $x_2 \approx 4, x_3 \approx 2$ **2.** $\frac{1}{2}$ **3.** -1 **4.** $\ln 2 - \frac{1}{2} \approx 0.193147$

4.8   ROLLE'S THEOREM; MEAN-VALUE THEOREM

In this section we will discuss a result called the Mean-Value Theorem. This theorem has so many important consequences that it is regarded as one of the major principles in calculus.

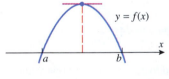

▲ **Figure 4.8.1**

■ **ROLLE'S THEOREM**

We will begin with a special case of the Mean-Value Theorem, called Rolle's Theorem, in honor of the mathematician Michel Rolle. This theorem states the geometrically obvious fact that if the graph of a differentiable function intersects the x-axis at two places, a and b, then somewhere between a and b there must be at least one place where the tangent line is horizontal (Figure 4.8.1). The precise statement of the theorem is as follows.

4.8.1   THEOREM (*Rolle's Theorem*)   *Let f be continuous on the closed interval $[a, b]$ and differentiable on the open interval (a, b). If*

$$f(a) = 0 \quad and \quad f(b) = 0$$

then there is at least one point c in the interval (a, b) such that $f'(c) = 0$.

PROOF We will divide the proof into three cases: the case where $f(x) = 0$ for all x in (a, b), the case where $f(x) > 0$ at some point in (a, b), and the case where $f(x) < 0$ at some point in (a, b).

CASE 1 If $f(x) = 0$ for all x in (a, b), then $f'(c) = 0$ at every point c in (a, b) because f is a constant function on that interval.

CASE 2 Assume that $f(x) > 0$ at some point in (a, b). Since f is continuous on $[a, b]$, it follows from the Extreme-Value Theorem (4.4.2) that f has an absolute maximum on $[a, b]$. The absolute maximum value cannot occur at an endpoint of $[a, b]$ because we have assumed that $f(a) = f(b) = 0$, and that $f(x) > 0$ at some point in (a, b). Thus, the absolute maximum must occur at some point c in (a, b). It follows from Theorem 4.4.3 that c is a critical point of f, and since f is differentiable on (a, b), this critical point must be a stationary point; that is, $f'(c) = 0$.

CASE 3 Assume that $f(x) < 0$ at some point in (a, b). The proof of this case is similar to Case 2 and will be omitted. ∎

▶ **Example 1** Find the two x-intercepts of the function $f(x) = x^2 - 5x + 4$ and confirm that $f'(c) = 0$ at some point c between those intercepts.

Solution. The function f can be factored as

$$x^2 - 5x + 4 = (x - 1)(x - 4)$$

so the x-intercepts are $x = 1$ and $x = 4$. Since the polynomial f is continuous and differentiable everywhere, the hypotheses of Rolle's Theorem are satisfied on the interval $[1, 4]$. Thus, we are guaranteed the existence of at least one point c in the interval $(1, 4)$ such that $f'(c) = 0$. Differentiating f yields

$$f'(x) = 2x - 5$$

Solving the equation $f'(x) = 0$ yields $x = \frac{5}{2}$, so $c = \frac{5}{2}$ is a point in the interval $(1, 4)$ at which $f'(c) = 0$ (Figure 4.8.2). ◀

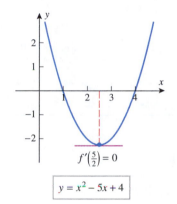

$$f'\left(\tfrac{5}{2}\right) = 0$$

$$y = x^2 - 5x + 4$$

▲ **Figure 4.8.2**

▶ **Example 2** The differentiability requirement in Rolle's Theorem is critical. If f fails to be differentiable at even one place in the interval (a, b), then the conclusion of the

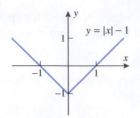

▲ **Figure 4.8.3**

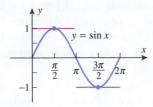

▲ **Figure 4.8.4**

In Examples 1 and 3 we were able to find exact values of c because the equation $f'(x) = 0$ was easy to solve. However, in the applications of Rolle's Theorem it is usually the *existence of* c that is important and not its actual value.

theorem may not hold. For example, the function $f(x) = |x| - 1$ graphed in Figure 4.8.3 has roots at $x = -1$ and $x = 1$, yet there is no horizontal tangent to the graph of f over the interval $(-1, 1)$. ◄

▶ **Example 3**   If f satisfies the conditions of Rolle's Theorem on $[a, b]$, then the theorem guarantees the existence of *at least* one point c in (a, b) at which $f'(c) = 0$. There may, however, be more than one such c. For example, the function $f(x) = \sin x$ is continuous and differentiable everywhere, so the hypotheses of Rolle's Theorem are satisfied on the interval $[0, 2\pi]$ whose endpoints are roots of f. As indicated in Figure 4.8.4, there are two points in the interval $[0, 2\pi]$ at which the graph of f has a horizontal tangent, $c_1 = \pi/2$ and $c_2 = 3\pi/2$. ◄

■ **THE MEAN-VALUE THEOREM**

Rolle's Theorem is a special case of a more general result, called the *Mean-Value Theorem*. Geometrically, this theorem states that between any two points $A(a, f(a))$ and $B(b, f(b))$ on the graph of a differentiable function f, there is at least one place where the tangent line to the graph is parallel to the secant line joining A and B (Figure 4.8.5).

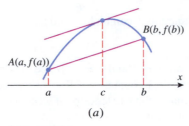

(a) (b)

▲ **Figure 4.8.5**

Note that the slope of the secant line joining $A(a, f(a))$ and $B(b, f(b))$ is

$$\frac{f(b) - f(a)}{b - a}$$

and that the slope of the tangent line at c in Figure 4.8.5a is $f'(c)$. Similarly, in Figure 4.8.5b the slopes of the tangent lines at c_1 and c_2 are $f'(c_1)$ and $f'(c_2)$, respectively. Since nonvertical parallel lines have the same slope, the Mean-Value Theorem can be stated precisely as follows.

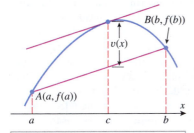

The tangent line is parallel to the secant line where the vertical distance $v(x)$ between the secant line and the graph of f is maximum.

▲ **Figure 4.8.6**

4.8.2   THEOREM (*Mean-Value Theorem*)   *Let f be continuous on the closed interval $[a, b]$ and differentiable on the open interval (a, b). Then there is at least one point c in (a, b) such that*

$$f'(c) = \frac{f(b) - f(a)}{b - a} \tag{1}$$

MOTIVATION FOR THE PROOF OF THEOREM 4.8.2   Figure 4.8.6 suggests that (1) will hold (i.e., the tangent line will be parallel to the secant line) at a point c where the vertical distance between the curve and the secant line is maximum. Thus, to prove the Mean-Value Theorem it is natural to begin by looking for a formula for the vertical distance $v(x)$ between the curve $y = f(x)$ and the secant line joining $(a, f(a))$ and $(b, f(b))$.

PROOF OF THEOREM 4.8.2 Since the two-point form of the equation of the secant line joining $(a, f(a))$ and $(b, f(b))$ is

$$y - f(a) = \frac{f(b) - f(a)}{b - a}(x - a)$$

or, equivalently,

$$y = \frac{f(b) - f(a)}{b - a}(x - a) + f(a)$$

the difference $v(x)$ between the height of the graph of f and the height of the secant line is

$$v(x) = f(x) - \left[\frac{f(b) - f(a)}{b - a}(x - a) + f(a)\right] \tag{2}$$

Since $f(x)$ is continuous on $[a, b]$ and differentiable on (a, b), so is $v(x)$. Moreover,

$$v(a) = 0 \quad \text{and} \quad v(b) = 0$$

so that $v(x)$ satisfies the hypotheses of Rolle's Theorem on the interval $[a, b]$. Thus, there is a point c in (a, b) such that $v'(c) = 0$. But from Equation (2)

$$v'(x) = f'(x) - \frac{f(b) - f(a)}{b - a}$$

so

$$v'(c) = f'(c) - \frac{f(b) - f(a)}{b - a}$$

Since $v'(c) = 0$, we have

$$f'(c) = \frac{f(b) - f(a)}{b - a} \quad \blacksquare$$

▶ **Example 4** Show that the function $f(x) = \frac{1}{4}x^3 + 1$ satisfies the hypotheses of the Mean-Value Theorem over the interval $[0, 2]$, and find all values of c in the interval $(0, 2)$ at which the tangent line to the graph of f is parallel to the secant line joining the points $(0, f(0))$ and $(2, f(2))$.

Solution. The function f is continuous and differentiable everywhere because it is a polynomial. In particular, f is continuous on $[0, 2]$ and differentiable on $(0, 2)$, so the hypotheses of the Mean-Value Theorem are satisfied with $a = 0$ and $b = 2$. But

$$f(a) = f(0) = 1, \quad f(b) = f(2) = 3$$

$$f'(x) = \frac{3x^2}{4}, \quad f'(c) = \frac{3c^2}{4}$$

so in this case Equation (1) becomes

$$\frac{3c^2}{4} = \frac{3 - 1}{2 - 0} \quad \text{or} \quad 3c^2 = 4$$

which has the two solutions $c = \pm 2/\sqrt{3} \approx \pm 1.15$. However, only the positive solution lies in the interval $(0, 2)$; this value of c is consistent with Figure 4.8.7. ◀

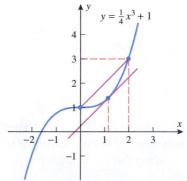

$y = \frac{1}{4}x^3 + 1$

▲ **Figure 4.8.7**

■ **VELOCITY INTERPRETATION OF THE MEAN-VALUE THEOREM**

There is a nice interpretation of the Mean-Value Theorem in the situation where $x = f(t)$ is the position versus time curve for a car moving along a straight road. In this case, the right side of (1) is the average velocity of the car over the time interval from $a \leq t \leq b$, and the left side is the instantaneous velocity at time $t = c$. Thus, the Mean-Value Theorem implies that at least once during the time interval the instantaneous velocity must equal the average velocity. This agrees with our real-world experience—if the average velocity for a trip is 40 mi/h, then sometime during the trip the speedometer has to read 40 mi/h.

▶ **Example 5** You are driving on a straight highway on which the speed limit is 55 mi/h. At 8:05 A.M. a police car clocks your velocity at 50 mi/h and at 8:10 A.M. a second police car posted 5 mi down the road clocks your velocity at 55 mi/h. Explain why the police have a right to charge you with a speeding violation.

Solution. You traveled 5 mi in 5 min $\left(= \frac{1}{12} \text{ h}\right)$, so your average velocity was 60 mi/h. Therefore, the Mean-Value Theorem guarantees the police that your instantaneous velocity was 60 mi/h at least once over the 5 mi section of highway. ◀

■ CONSEQUENCES OF THE MEAN-VALUE THEOREM

We stated at the beginning of this section that the Mean-Value Theorem is the starting point for many important results in calculus. As an example of this, we will use it to prove Theorem 4.1.2, which was one of our fundamental tools for analyzing graphs of functions.

4.1.2 **THEOREM** (*Revisited*) *Let f be a function that is continuous on a closed interval $[a, b]$ and differentiable on the open interval (a, b).*

(a) *If $f'(x) > 0$ for every value of x in (a, b), then f is increasing on $[a, b]$.*

(b) *If $f'(x) < 0$ for every value of x in (a, b), then f is decreasing on $[a, b]$.*

(c) *If $f'(x) = 0$ for every value of x in (a, b), then f is constant on $[a, b]$.*

PROOF (*a*) Suppose that x_1 and x_2 are points in $[a, b]$ such that $x_1 < x_2$. We must show that $f(x_1) < f(x_2)$. Because the hypotheses of the Mean-Value Theorem are satisfied on the entire interval $[a, b]$, they are satisfied on the subinterval $[x_1, x_2]$. Thus, there is some point c in the open interval (x_1, x_2) such that

$$f'(c) = \frac{f(x_2) - f(x_1)}{x_2 - x_1}$$

or, equivalently,

$$f(x_2) - f(x_1) = f'(c)(x_2 - x_1) \tag{3}$$

Since c is in the open interval (x_1, x_2), it follows that $a < c < b$; thus, $f'(c) > 0$. However, $x_2 - x_1 > 0$ since we assumed that $x_1 < x_2$. It follows from (3) that $f(x_2) - f(x_1) > 0$ or, equivalently, $f(x_1) < f(x_2)$, which is what we were to prove. The proofs of parts (*b*) and (*c*) are similar and are left as exercises. ■

■ THE CONSTANT DIFFERENCE THEOREM

We know from our earliest study of derivatives that the derivative of a constant is zero. Part (*c*) of Theorem 4.1.2 is the converse of that result; that is, a function whose derivative is zero on an interval must be constant on that interval. If we apply this to the difference of two functions, we obtain the following useful theorem.

4.8.3 **THEOREM** (*Constant Difference Theorem*) *If f and g are differentiable on an interval, and if $f'(x) = g'(x)$ for all x in that interval, then $f - g$ is constant on the interval; that is, there is a constant k such that $f(x) - g(x) = k$ or, equivalently,*

$$f(x) = g(x) + k$$

for all x in the interval.

PROOF Let x_1 and x_2 be any points in the interval such that $x_1 < x_2$. Since the functions f and g are differentiable on the interval, they are continuous on the interval. Since $[x_1, x_2]$ is a subinterval, it follows that f and g are continuous on $[x_1, x_2]$ and differentiable on (x_1, x_2). Moreover, it follows from the basic properties of derivatives and continuity that the same is true of the function

$$F(x) = f(x) - g(x)$$

Since

$$F'(x) = f'(x) - g'(x) = 0$$

it follows from part (c) of Theorem 4.1.2 that $F(x) = f(x) - g(x)$ is constant on the interval $[x_1, x_2]$. This means that $f(x) - g(x)$ has the same value at any two points x_1 and x_2 in the interval, and this implies that $f - g$ is constant on the interval. ∎

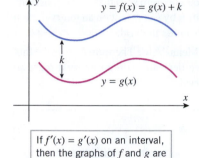

y = f(x) = g(x) + k

y = g(x)

If $f'(x) = g'(x)$ on an interval, then the graphs of f and g are vertical translations of each other.

▲ **Figure 4.8.8**

Geometrically, the Constant Difference Theorem tells us that if f and g have the same derivative on an interval, then the graphs of f and g are vertical translations of each other over that interval (Figure 4.8.8).

▶ **Example 6** Part (c) of Theorem 4.1.2 is sometimes useful for establishing identities. For example, although we do not need calculus to prove the identity

$$\sin^{-1} x + \cos^{-1} x = \frac{\pi}{2} \qquad (-1 \le x \le 1) \tag{4}$$

it can be done by letting $f(x) = \sin^{-1} x + \cos^{-1} x$. It follows from Formulas (9) and (10) of Section 3.3 that

$$f'(x) = \frac{d}{dx}[\sin^{-1} x] + \frac{d}{dx}[\cos^{-1} x] = \frac{1}{\sqrt{1 - x^2}} - \frac{1}{\sqrt{1 - x^2}} = 0$$

so $f(x) = \sin^{-1} x + \cos^{-1} x$ is constant on the interval $[-1, 1]$. We can find this constant by evaluating f at any convenient point in this interval. For example, using $x = 0$ we obtain

$$f(0) = \sin^{-1} 0 + \cos^{-1} 0 = 0 + \frac{\pi}{2} = \frac{\pi}{2}$$

which proves (4). ◀

✔ QUICK CHECK EXERCISES 4.8   (See page 310 for answers.)

1. Let $f(x) = x^2 - x$.
   (a) An interval on which f satisfies the hypotheses of Rolle's Theorem is _____.
   (b) Find all values of c that satisfy the conclusion of Rolle's Theorem for the function f on the interval in part (a).

2. Use the accompanying graph of f to find an interval $[a, b]$ on which Rolle's Theorem applies, and find all values of c in that interval that satisfy the conclusion of the theorem.

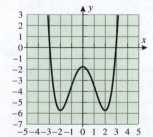

◀ **Figure Ex-2**

3. Let $f(x) = x^2 - x$.
   (a) Find a point b such that the slope of the secant line through $(0, 0)$ and $(b, f(b))$ is 1.
   (b) Find all values of c that satisfy the conclusion of the Mean-Value Theorem for the function f on the interval $[0, b]$, where b is the point found in part (a).

4. Use the graph of f in the accompanying figure to estimate all values of c that satisfy the conclusion of the Mean-Value Theorem on the interval
   (a) $[0, 8]$ (b) $[0, 4]$.

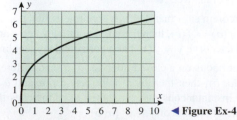

◀ **Figure Ex-4**

5. Find a function f such that the graph of f contains the point $(1, 5)$ and such that for every value of x_0 the tangent line to the graph of f at x_0 is parallel to the tangent line to the graph of $y = x^2$ at x_0.

EXERCISE SET 4.8   Graphing Utility

1–4 Verify that the hypotheses of Rolle's Theorem are satisfied on the given interval, and find all values of c in that interval that satisfy the conclusion of the theorem. ■

1. $f(x) = x^2 - 8x + 15$; $[3, 5]$

2. $f(x) = x^3 - 3x^2 + 2x$; $[0, 2]$

3. $f(x) = \cos x$; $[\pi/2, 3\pi/2]$

4. $f(x) = \ln(4 + 2x - x^2)$; $[-1, 3]$

5–8 Verify that the hypotheses of the Mean-Value Theorem are satisfied on the given interval, and find all values of c in that interval that satisfy the conclusion of the theorem. ■

5. $f(x) = x^2 - x$; $[-3, 5]$

6. $f(x) = x^3 + x - 4$; $[-1, 2]$

7. $f(x) = \sqrt{x + 1}$; $[0, 3]$

8. $f(x) = x - \dfrac{1}{x}$; $[3, 4]$

9. (a) Find an interval $[a, b]$ on which

$$f(x) = x^4 + x^3 - x^2 + x - 2$$

satisfies the hypotheses of Rolle's Theorem.

(b) Generate the graph of $f'(x)$, and use it to make rough estimates of all values of c in the interval obtained in part (a) that satisfy the conclusion of Rolle's Theorem.

(c) Use Newton's Method to improve on the rough estimates obtained in part (b).

10. Let $f(x) = x^3 - 4x$.

(a) Find the equation of the secant line through the points $(-2, f(-2))$ and $(1, f(1))$.

(b) Show that there is only one point c in the interval $(-2, 1)$ that satisfies the conclusion of the Mean-Value Theorem for the secant line in part (a).

(c) Find the equation of the tangent line to the graph of f at the point $(c, f(c))$.

(d) Use a graphing utility to generate the secant line in part (a) and the tangent line in part (c) in the same coordinate system, and confirm visually that the two lines seem parallel.

11–14 True–False Determine whether the statement is true or false. Explain your answer. ■

11. Rolle's Theorem says that if f is a continuous function on $[a, b]$ and $f(a) = f(b)$, then there is a point between a and b at which the curve $y = f(x)$ has a horizontal tangent line.

12. If f is continuous on a closed interval $[a, b]$ and differentiable on (a, b), then there is a point between a and b at which the instantaneous rate of change of f matches the average rate of change of f over $[a, b]$.

13. The Constant Difference Theorem says that if two functions have derivatives that differ by a constant on an interval, then the functions are equal on the interval.

14. One application of the Mean-Value Theorem is to prove that a function with positive derivative on an interval must be increasing on that interval.

FOCUS ON CONCEPTS

15. Let $f(x) = \tan x$.
(a) Show that there is no point c in the interval $(0, \pi)$ such that $f'(c) = 0$, even though $f(0) = f(\pi) = 0$.
(b) Explain why the result in part (a) does not contradict Rolle's Theorem.

16. Let $f(x) = x^{2/3}$, $a = -1$, and $b = 8$.
(a) Show that there is no point c in (a, b) such that

$$f'(c) = \frac{f(b) - f(a)}{b - a}$$

(b) Explain why the result in part (a) does not contradict the Mean-Value Theorem.

17. (a) Show that if f is differentiable on $(-\infty, +\infty)$, and if $y = f(x)$ and $y = f'(x)$ are graphed in the same coordinate system, then between any two x-intercepts of f there is at least one x-intercept of f'.
(b) Give some examples that illustrate this.

18. Review Formulas (8) and (9) in Section 2.1 and use the Mean-Value Theorem to show that if f is differentiable on $(-\infty, +\infty)$, then for any interval $[x_0, x_1]$ there is at least one point in (x_0, x_1) where the instantaneous rate of change of y with respect to x is equal to the average rate of change over the interval.

19–21 Use the result of Exercise 18 in these exercises. ■

19. An automobile travels 4 mi along a straight road in 5 min. Show that the speedometer reads exactly 48 mi/h at least once during the trip.

20. At 11 A.M. on a certain morning the outside temperature was $76°$F. At 11 P.M. that evening it had dropped to $52°$F.
(a) Show that at some instant during this period the temperature was decreasing at the rate of $2°$F/h.
(b) Suppose that you know the temperature reached a high of $88°$F sometime between 11 A.M. and 11 P.M. Show that at some instant during this period the temperature was decreasing at a rate greater than $3°$F/h.

21. Suppose that two runners in a 100 m dash finish in a tie. Show that they had the same velocity at least once during the race.

22. Use the fact that

$$\frac{d}{dx}[x\ln(2-x)] = \ln(2-x) - \frac{x}{2-x}$$

to show that the equation $x = (2-x)\ln(2-x)$ has at least one solution in the interval $(0, 1)$.

23. (a) Use the Constant Difference Theorem (4.8.3) to show that if $f'(x) = g'(x)$ for all x in the interval $(-\infty, +\infty)$, and if f and g have the same value at some point x_0, then $f(x) = g(x)$ for all x in $(-\infty, +\infty)$.

(b) Use the result in part (a) to confirm the trigonometric identity $\sin^2 x + \cos^2 x = 1$.

24. (a) Use the Constant Difference Theorem (4.8.3) to show that if $f'(x) = g'(x)$ for all x in $(-\infty, +\infty)$, and if $f(x_0) - g(x_0) = c$ at some point x_0, then

$$f(x) - g(x) = c$$

for all x in $(-\infty, +\infty)$.

(b) Use the result in part (a) to show that the function

$$h(x) = (x-1)^3 - (x^2+3)(x-3)$$

is constant for all x in $(-\infty, +\infty)$, and find the constant.

(c) Check the result in part (b) by multiplying out and simplifying the formula for $h(x)$.

25. Let $g(x) = xe^x - e^x$. Find $f(x)$ so that $f'(x) = g'(x)$ and $f(1) = 2$.

26. Let $g(x) = \tan^{-1} x$. Find $f(x)$ so that $f'(x) = g'(x)$ and $f(1) = 2$.

27. (a) Use the Mean-Value Theorem to show that if f is differentiable on an interval, and if $|f'(x)| \le M$ for all values of x in the interval, then

$$|f(x) - f(y)| \le M|x - y|$$

for all values of x and y in the interval.

(b) Use the result in part (a) to show that

$$|\sin x - \sin y| \le |x - y|$$

for all real values of x and y.

28. (a) Use the Mean-Value Theorem to show that if f is differentiable on an open interval, and if $|f'(x)| \ge M$ for all values of x in the interval, then

$$|f(x) - f(y)| \ge M|x - y|$$

for all values of x and y in the interval.

(b) Use the result in part (a) to show that

$$|\tan x - \tan y| \ge |x - y|$$

for all values of x and y in the interval $(-\pi/2, \pi/2)$.

(c) Use the result in part (b) to show that

$$|\tan x + \tan y| \ge |x + y|$$

for all values of x and y in the interval $(-\pi/2, \pi/2)$.

29. (a) Use the Mean-Value Theorem to show that

$$\sqrt{y} - \sqrt{x} < \frac{y-x}{2\sqrt{x}}$$

if $0 < x < y$.

(b) Use the result in part (a) to show that if $0 < x < y$, then $\sqrt{xy} < \frac{1}{2}(x+y)$.

30. Show that if f is differentiable on an open interval and $f'(x) \ne 0$ on the interval, the equation $f(x) = 0$ can have at most one real root in the interval.

31. Use the result in Exercise 30 to show the following:

(a) The equation $x^3 + 4x - 1 = 0$ has exactly one real root.

(b) If $b^2 - 3ac < 0$ and if $a \ne 0$, then the equation

$$ax^3 + bx^2 + cx + d = 0$$

has exactly one real root.

32. Use the inequality $\sqrt{3} < 1.8$ to prove that

$$1.7 < \sqrt{3} < 1.75$$

[*Hint:* Let $f(x) = \sqrt{x}, a = 3$, and $b = 4$ in the Mean-Value Theorem.]

33. Use the Mean-Value Theorem to prove that

$$\frac{x}{1+x^2} < \tan^{-1} x < x \quad (x > 0)$$

34. (a) Show that if f and g are functions for which

$$f'(x) = g(x) \quad \text{and} \quad g'(x) = f(x)$$

for all x, then $f^2(x) - g^2(x)$ is a constant.

(b) Show that the function $f(x) = \frac{1}{2}(e^x + e^{-x})$ and the function $g(x) = \frac{1}{2}(e^x - e^{-x})$ have this property.

35. (a) Show that if f and g are functions for which

$$f'(x) = g(x) \quad \text{and} \quad g'(x) = -f(x)$$

for all x, then $f^2(x) + g^2(x)$ is a constant.

(b) Give an example of functions f and g with this property.

36. Let f and g be continuous on $[a, b]$ and differentiable on (a, b). Prove: If $f(a) = g(a)$ and $f(b) = g(b)$, then there is a point c in (a, b) such that $f'(c) = g'(c)$.

37. Illustrate the result in Exercise 36 by drawing an appropriate picture.

38. (a) Prove that if $f''(x) > 0$ for all x in (a, b), then $f'(x) = 0$ at most once in (a, b).

(b) Give a geometric interpretation of the result in (a).

39. (a) Prove part (b) of Theorem 4.1.2.

(b) Prove part (c) of Theorem 4.1.2.

40. Use the Mean-Value Theorem to prove the following result: Let f be continuous at x_0 and suppose that $\lim_{x \to x_0} f'(x)$ exists. Then f is differentiable at x_0, and

$$f'(x_0) = \lim_{x \to x_0} f'(x)$$

[*Hint:* The derivative $f'(x_0)$ is given by

$$f'(x_0) = \lim_{x \to x_0} \frac{f(x) - f(x_0)}{x - x_0}$$

provided this limit exists.]

41. Let
$$f(x) = \begin{cases} 3x^2, & x \le 1 \\ ax + b, & x > 1 \end{cases}$$
Find the values of a and b so that f will be differentiable at $x = 1$.

42. (a) Let
$$f(x) = \begin{cases} x^2, & x \le 0 \\ x^2 + 1, & x > 0 \end{cases}$$
Show that
$$\lim_{x \to 0^-} f'(x) = \lim_{x \to 0^+} f'(x)$$
but that $f'(0)$ does not exist.

(b) Let
$$f(x) = \begin{cases} x^2, & x \le 0 \\ x^3, & x > 0 \end{cases}$$
Show that $f'(0)$ exists but $f''(0)$ does not.

43. Use the Mean-Value Theorem to prove the following result: The graph of a function f has a point of vertical tangency at $(x_0, f(x_0))$ if f is continuous at x_0 and $f'(x)$ approaches either $+\infty$ or $-\infty$ as $x \to x_0^+$ and as $x \to x_0^-$.

44. Writing Suppose that $p(x)$ is a nonconstant polynomial with zeros at $x = a$ and $x = b$. Explain how both the Extreme-Value Theorem (4.4.2) and Rolle's Theorem can be used to show that p has a critical point between a and b.

45. Writing Find and describe a physical situation that illustrates the Mean-Value Theorem.

✔ **QUICK CHECK ANSWERS 4.8**

1. (a) $[0, 1]$ (b) $c = \frac{1}{2}$ **2.** $[-3, 3]$; $c = -2, 0, 2$ **3.** (a) $b = 2$ (b) $c = 1$ **4.** (a) 1.5 (b) 0.8 **5.** $f(x) = x^2 + 4$

CHAPTER 4 REVIEW EXERCISES ⌇ Graphing Utility ⊂ CAS

1. (a) If $x_1 < x_2$, what relationship must hold between $f(x_1)$ and $f(x_2)$ if f is increasing on an interval containing x_1 and x_2? Decreasing? Constant?

(b) What condition on f' ensures that f is increasing on an interval $[a, b]$? Decreasing? Constant?

2. (a) What condition on f' ensures that f is concave up on an open interval? Concave down?

(b) What condition on f'' ensures that f is concave up on an open interval? Concave down?

(c) In words, what is an inflection point of f?

3–10 Find: (a) the intervals on which f is increasing, (b) the intervals on which f is decreasing, (c) the open intervals on which f is concave up, (d) the open intervals on which f is concave down, and (e) the x-coordinates of all inflection points. ■

3. $f(x) = x^2 - 5x + 6$ **4.** $f(x) = x^4 - 8x^2 + 16$

5. $f(x) = \dfrac{x^2}{x^2 + 2}$ **6.** $f(x) = \sqrt[3]{x + 2}$

7. $f(x) = x^{1/3}(x + 4)$ **8.** $f(x) = x^{4/3} - x^{1/3}$

9. $f(x) = 1/e^{x^2}$ **10.** $f(x) = \tan^{-1} x^2$

⌇ **11–14** Analyze the trigonometric function f over the specified interval, stating where f is increasing, decreasing, concave up, and concave down, and stating the x-coordinates of all inflection points. Confirm that your results are consistent with the graph of f generated with a graphing utility. ■

11. $f(x) = \cos x$; $[0, 2\pi]$

12. $f(x) = \tan x$; $(-\pi/2, \pi/2)$

13. $f(x) = \sin x \cos x$; $[0, \pi]$

14. $f(x) = \cos^2 x - 2 \sin x$; $[0, 2\pi]$

15. In each part, sketch a continuous curve $y = f(x)$ with the stated properties.

(a) $f(2) = 4$, $f'(2) = 1$, $f''(x) < 0$ for $x < 2$, $f''(x) > 0$ for $x > 2$

(b) $f(2) = 4$, $f''(x) > 0$ for $x < 2$, $f''(x) < 0$ for $x > 2$, $\lim_{x \to 2^-} f'(x) = +\infty$, $\lim_{x \to 2^+} f'(x) = +\infty$

(c) $f(2) = 4$, $f''(x) < 0$ for $x \ne 2$, $\lim_{x \to 2^-} f'(x) = 1$, $\lim_{x \to 2^+} f'(x) = -1$

⌇ **16.** In parts (a)–(d), the graph of a polynomial with degree at most 6 is given. Find equations for polynomials that produce graphs with these shapes, and check your answers with a graphing utility.

(a)
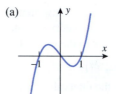

(b)
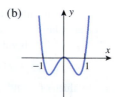

(c)
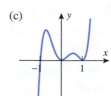

(d)
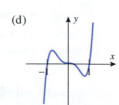

17. For a general quadratic polynomial

$$f(x) = ax^2 + bx + c \quad (a \neq 0)$$

find conditions on a, b, and c to ensure that f is always increasing or always decreasing on $[0, +\infty)$.

18. For the general cubic polynomial

$$f(x) = ax^3 + bx^2 + cx + d \quad (a \neq 0)$$

find conditions on a, b, c, and d to ensure that f is always increasing or always decreasing on $(-\infty, +\infty)$.

~ 19. Use a graphing utility to estimate the value of x at which

$$f(x) = \frac{2^x}{1 + 2^{x+1}}$$

is increasing most rapidly.

20. Prove that for any positive constants a and k, the graph of

$$y = \frac{a^x}{1 + a^{x+k}}$$

has an inflection point at $x = -k$.

21. (a) Where on the graph of $y = f(x)$ would you expect y to be increasing or decreasing most rapidly with respect to x?

(b) In words, what is a relative extremum?

(c) State a procedure for determining where the relative extrema of f occur.

22. Determine whether the statement is true or false. If it is false, give an example for which the statement fails.

(a) If f has a relative maximum at x_0, then $f(x_0)$ is the largest value that $f(x)$ can have.

(b) If the largest value for f on the interval (a, b) is at x_0, then f has a relative maximum at x_0.

(c) A function f has a relative extremum at each of its critical points.

23. (a) According to the first derivative test, what conditions ensure that f has a relative maximum at x_0? A relative minimum?

(b) According to the second derivative test, what conditions ensure that f has a relative maximum at x_0? A relative minimum?

24–26 Locate the critical points and identify which critical points correspond to stationary points. ■

24. (a) $f(x) = x^3 + 3x^2 - 9x + 1$

(b) $f(x) = x^4 - 6x^2 - 3$

25. (a) $f(x) = \dfrac{x}{x^2 + 2}$ (b) $f(x) = \dfrac{x^2 - 3}{x^2 + 1}$

26. (a) $f(x) = x^{1/3}(x - 4)$ (b) $f(x) = x^{4/3} - 6x^{1/3}$

27. In each part, find all critical points, and use the first derivative test to classify them as relative maxima, relative minima, or neither.

(a) $f(x) = x^{1/3}(x - 7)^2$

(b) $f(x) = 2\sin x - \cos 2x, \quad 0 \leq x \leq 2\pi$

(c) $f(x) = 3x - (x - 1)^{3/2}$

28. In each part, find all critical points, and use the second derivative test (where possible) to classify them as relative maxima, relative minima, or neither.

(a) $f(x) = x^{-1/2} + \frac{1}{9}x^{1/2}$

(b) $f(x) = x^2 + 8/x$

(c) $f(x) = \sin^2 x - \cos x, \quad 0 \leq x \leq 2\pi$

29–36 Give a graph of the function f, and identify the limits as $x \to \pm\infty$, as well as locations of all relative extrema, inflection points, and asymptotes (as appropriate). ■

29. $f(x) = x^4 - 3x^3 + 3x^2 + 1$

30. $f(x) = x^5 - 4x^4 + 4x^3$

31. $f(x) = \tan(x^2 + 1)$ **32.** $f(x) = x - \cos x$

33. $f(x) = \dfrac{x^2}{x^2 + 2x + 5}$ **34.** $f(x) = \dfrac{25 - 9x^2}{x^3}$

35. $f(x) = \begin{cases} \frac{1}{2}x^2, & x \leq 0 \\ -x^2, & x > 0 \end{cases}$

36. $f(x) = (1 + x)^{2/3}(3 - x)^{1/3}$

37–44 Use any method to find the relative extrema of the function f. ■

37. $f(x) = x^3 + 5x - 2$ **38.** $f(x) = x^4 - 2x^2 + 7$

39. $f(x) = x^{4/5}$ **40.** $f(x) = 2x + x^{2/3}$

41. $f(x) = \dfrac{x^2}{x^2 + 1}$ **42.** $f(x) = \dfrac{x}{x + 2}$

43. $f(x) = \ln(1 + x^2)$ **44.** $f(x) = x^2 e^x$

~ 45–46 When using a graphing utility, important features of a graph may be missed if the viewing window is not chosen appropriately. This is illustrated in Exercises 45 and 46. ■

45. (a) Generate the graph of $f(x) = \frac{1}{3}x^3 - \frac{1}{400}x$ over the interval $[-5, 5]$, and make a conjecture about the locations and nature of all critical points.

(b) Find the exact locations of all the critical points, and classify them as relative maxima, relative minima, or neither.

(c) Confirm the results in part (b) by graphing f over an appropriate interval.

46. (a) Generate the graph of

$$f(x) = \frac{1}{5}x^5 - \frac{7}{8}x^4 + \frac{1}{3}x^3 + \frac{7}{2}x^2 - 6x$$

over the interval $[-5, 5]$, and make a conjecture about the locations and nature of all critical points.

(b) Find the exact locations of all the critical points, and classify them as relative maxima, relative minima, or neither.

(c) Confirm the results in part (b) by graphing portions of f over appropriate intervals. [*Note:* It will not be possible to find a single window in which all of the critical points are discernible.]

47. (a) Use a graphing utility to generate the graphs of $y = x$ and $y = (x^3 - 8)/(x^2 + 1)$ together over the interval $[-5, 5]$, and make a conjecture about the relationship between the two graphs.

(b) Confirm your conjecture in part (a).

48. Use implicit differentiation to show that a function defined implicitly by $\sin x + \cos y = 2y$ has a critical point whenever $\cos x = 0$. Then use either the first or second derivative test to classify these critical points as relative maxima or minima.

49. Let

$$f(x) = \frac{2x^3 + x^2 - 15x + 7}{(2x - 1)(3x^2 + x - 1)}$$

Graph $y = f(x)$, and find the equations of all horizontal and vertical asymptotes. Explain why there is no vertical asymptote at $x = \frac{1}{2}$, even though the denominator of f is zero at that point.

c 50. Let

$$f(x) = \frac{x^5 - x^4 - 3x^3 + 2x + 4}{x^7 - 2x^6 - 3x^5 + 6x^4 + 4x - 8}$$

(a) Use a CAS to factor the numerator and denominator of f, and use the results to determine the locations of all vertical asymptotes.

(b) Confirm that your answer is consistent with the graph of f.

51. (a) What inequality must $f(x)$ satisfy for the function f to have an absolute maximum on an interval I at x_0?

(b) What inequality must $f(x)$ satisfy for f to have an absolute minimum on an interval I at x_0?

(c) What is the difference between an absolute extremum and a relative extremum?

52. According to the Extreme-Value Theorem, what conditions on a function f and a given interval guarantee that f will have both an absolute maximum and an absolute minimum on the interval?

53. In each part, determine whether the statement is true or false, and justify your answer.

(a) If f is differentiable on the open interval (a, b), and if f has an absolute extremum on that interval, then it must occur at a stationary point of f.

(b) If f is continuous on the open interval (a, b), and if f has an absolute extremum on that interval, then it must occur at a stationary point of f.

54–56 In each part, find the absolute minimum m and the absolute maximum M of f on the given interval (if they exist), and state where the absolute extrema occur. ■

54. (a) $f(x) = 1/x$; $[-2, -1]$
(b) $f(x) = x^3 - x^4$; $[-1, \frac{3}{2}]$
(c) $f(x) = x - \tan x$; $[-\pi/4, \pi/4]$
(d) $f(x) = -|x^2 - 2x|$; $[1, 3]$

55. (a) $f(x) = x^2 - 3x - 1$; $(-\infty, +\infty)$
(b) $f(x) = x^3 - 3x - 2$; $(-\infty, +\infty)$

(c) $f(x) = e^x/x^2$; $(0, +\infty)$
(d) $f(x) = x^x$; $(0, +\infty)$

56. (a) $f(x) = 2x^5 - 5x^4 + 7$; $(-1, 3)$
(b) $f(x) = (3 - x)/(2 - x)$; $(0, 2)$
(c) $f(x) = 2x/(x^2 + 3)$; $(0, 2]$
(d) $f(x) = x^2(x - 2)^{1/3}$; $(0, 3]$

57. In each part, use a graphing utility to estimate the absolute maximum and minimum values of f, if any, on the stated interval, and then use calculus methods to find the exact values.

(a) $f(x) = (x^2 - 1)^2$; $(-\infty, +\infty)$
(b) $f(x) = x/(x^2 + 1)$; $[0, +\infty)$
(c) $f(x) = 2 \sec x - \tan x$; $[0, \pi/4]$
(d) $f(x) = x/2 + \ln(x^2 + 1)$; $[-4, 0]$

58. Prove that $x \le \sin^{-1} x$ for all x in $[0, 1]$.

c 59. Let

$$f(x) = \frac{x^3 + 2}{x^4 + 1}$$

(a) Generate the graph of $y = f(x)$, and use the graph to make rough estimates of the coordinates of the absolute extrema.

(b) Use a CAS to solve the equation $f'(x) = 0$ and then use it to make more accurate approximations of the coordinates in part (a).

60. A church window consists of a blue semicircular section surmounting a clear rectangular section as shown in the accompanying figure. The blue glass lets through half as much light per unit area as the clear glass. Find the radius r of the window that admits the most light if the perimeter of the entire window is to be P feet.

61. Find the dimensions of the rectangle of maximum area that can be inscribed inside the ellipse $(x/4)^2 + (y/3)^2 = 1$ (see the accompanying figure).

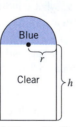

▲ **Figure Ex-60**

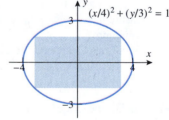

▲ **Figure Ex-61**

c 62. As shown in the accompanying figure on the next page, suppose that a boat enters the river at the point $(1, 0)$ and maintains a heading toward the origin. As a result of the strong current, the boat follows the path

$$y = \frac{x^{10/3} - 1}{2x^{2/3}}$$

where x and y are in miles.

(a) Graph the path taken by the boat.

(b) Can the boat reach the origin? If not, discuss its fate and find how close it comes to the origin.

◀ **Figure Ex-62**

63. A sheet of cardboard 12 in square is used to make an open box by cutting squares of equal size from the four corners and folding up the sides. What size squares should be cut to obtain a box with largest possible volume?

64. Is it true or false that a particle in rectilinear motion is speeding up when its velocity is increasing and slowing down when its velocity is decreasing? Justify your answer.

65. (a) Can an object in rectilinear motion reverse direction if its acceleration is constant? Justify your answer using a velocity versus time curve.
(b) Can an object in rectilinear motion have increasing speed and decreasing acceleration? Justify your answer using a velocity versus time curve.

66. Suppose that the position function of a particle in rectilinear motion is given by the formula $s(t) = t/(2t^2 + 8)$ for $t \geq 0$.
(a) Use a graphing utility to generate the position, velocity, and acceleration versus time curves.
(b) Use the appropriate graph to make a rough estimate of the time when the particle reverses direction, and then find that time exactly.
(c) Find the position, velocity, and acceleration at the instant when the particle reverses direction.
(d) Use the appropriate graphs to make rough estimates of the time intervals on which the particle is speeding up and the time intervals on which it is slowing down, and then find those time intervals exactly.
(e) When does the particle have its maximum and minimum velocities?

67. For parts (a)–(f), suppose that the position function of a particle in rectilinear motion is given by the formula

$$s(t) = \frac{t^2 + 1}{t^4 + 1}, \quad t \geq 0$$

(a) Use a CAS to find simplified formulas for the velocity function $v(t)$ and the acceleration function $a(t)$.
(b) Graph the position, velocity, and acceleration versus time curves.
(c) Use the appropriate graph to make a rough estimate of the time at which the particle is farthest from the origin and its distance from the origin at that time.
(d) Use the appropriate graph to make a rough estimate of the time interval during which the particle is moving in the positive direction.

(e) Use the appropriate graphs to make rough estimates of the time intervals during which the particle is speeding up and the time intervals during which it is slowing down.
(f) Use the appropriate graph to make a rough estimate of the maximum speed of the particle and the time at which the maximum speed occurs.

68. Draw an appropriate picture, and describe the basic idea of Newton's Method without using any formulas.

69. Use Newton's Method to approximate all three solutions of $x^3 - 4x + 1 = 0$.

70. Use Newton's Method to approximate the smallest positive solution of $\sin x + \cos x = 0$.

71. Use a graphing utility to determine the number of times the curve $y = x^3$ intersects the curve $y = (x/2) - 1$. Then apply Newton's Method to approximate the x-coordinates of all intersections.

72. According to **Kepler's law**, the planets in our solar system move in elliptical orbits around the Sun. If a planet's closest approach to the Sun occurs at time $t = 0$, then the distance r from the center of the planet to the center of the Sun at some later time t can be determined from the equation

$$r = a(1 - e\cos\phi)$$

where a is the average distance between centers, e is a positive constant that measures the "flatness" of the elliptical orbit, and ϕ is the solution of **Kepler's equation**

$$\frac{2\pi t}{T} = \phi - e\sin\phi$$

in which T is the time it takes for one complete orbit of the planet. Estimate the distance from the Earth to the Sun when $t = 90$ days. [First find ϕ from Kepler's equation, and then use this value of ϕ to find the distance. Use $a = 150 \times 10^6$ km, $e = 0.0167$, and $T = 365$ days.]

73. Using the formulas in Exercise 72, find the distance from the planet Mars to the Sun when $t = 1$ year. For Mars use $a = 228 \times 10^6$ km, $e = 0.0934$, and $T = 1.88$ years.

74. Suppose that f is continuous on the closed interval $[a, b]$ and differentiable on the open interval (a, b), and suppose that $f(a) = f(b)$. Is it true or false that f must have at least one stationary point in (a, b)? Justify your answer.

75. In each part, determine whether all of the hypotheses of Rolle's Theorem are satisfied on the stated interval. If not, state which hypotheses fail; if so, find all values of c guaranteed in the conclusion of the theorem.
(a) $f(x) = \sqrt{4 - x^2}$ on $[-2, 2]$
(b) $f(x) = x^{2/3} - 1$ on $[-1, 1]$
(c) $f(x) = \sin(x^2)$ on $[0, \sqrt{\pi}]$

76. In each part, determine whether all of the hypotheses of the Mean-Value Theorem are satisfied on the stated interval. If not, state which hypotheses fail; if so, find all values of c guaranteed in the conclusion of the theorem.
(a) $f(x) = |x - 1|$ on $[-2, 2]$ *(cont.)*

(b) $f(x) = \dfrac{x+1}{x-1}$ on $[2, 3]$

(c) $f(x) = \begin{cases} 3 - x^2 & \text{if } x \le 1 \\ 2/x & \text{if } x > 1 \end{cases}$ on $[0, 2]$

77. Use the fact that

$$\frac{d}{dx}(x^6 - 2x^2 + x) = 6x^5 - 4x + 1$$

to show that the equation $6x^5 - 4x + 1 = 0$ has at least one solution in the interval $(0, 1)$.

78. Let $g(x) = x^3 - 4x + 6$. Find $f(x)$ so that $f'(x) = g'(x)$ and $f(1) = 2$.

CHAPTER 4 MAKING CONNECTIONS

1. Suppose that $g(x)$ is a function that is defined and differentiable for all real numbers x and that $g(x)$ has the following properties:

(i) $g(0) = 2$ and $g'(0) = -\frac{2}{3}$.

(ii) $g(4) = 3$ and $g'(4) = 3$.

(iii) $g(x)$ is concave up for $x < 4$ and concave down for $x > 4$.

(iv) $g(x) \ge -10$ for all x.

Use these properties to answer the following questions.

(a) How many zeros does g have?

(b) How many zeros does g' have?

(c) Exactly one of the following limits is possible:

$$\lim_{x \to +\infty} g'(x) = -5, \quad \lim_{x \to +\infty} g'(x) = 0, \quad \lim_{x \to +\infty} g'(x) = 5$$

Identify which of these results is possible and draw a rough sketch of the graph of such a function $g(x)$. Explain why the other two results are impossible.

2. The two graphs in the accompanying figure depict a function $r(x)$ and its derivative $r'(x)$.

(a) Approximate the coordinates of each inflection point on the graph of $y = r(x)$.

(b) Suppose that $f(x)$ is a function that is continuous everywhere and whose *derivative* satisfies

$$f'(x) = (x^2 - 4) \cdot r(x)$$

What are the critical points for $f(x)$? At each critical point, identify whether $f(x)$ has a (relative) maximum, minimum, or neither a maximum or minimum. Approximate $f''(1)$.

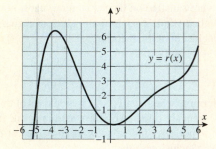

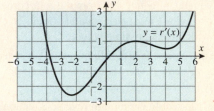

◀ **Figure Ex-2**

3. With the function $r(x)$ as provided in Exercise 2, let $g(x)$ be a function that is continuous everywhere such that $g'(x) = x - r(x)$. For which values of x does $g(x)$ have an inflection point?

4. Suppose that f is a function whose derivative is continuous everywhere. Assume that there exists a real number c such that when Newton's Method is applied to f, the inequality

$$|x_n - c| < \frac{1}{n}$$

is satisfied for all values of $n = 1, 2, 3, \dots$.

(a) Explain why

$$|x_{n+1} - x_n| < \frac{2}{n}$$

for all values of $n = 1, 2, 3, \dots$.

(b) Show that there exists a positive constant M such that

$$|f(x_n)| \le M|x_{n+1} - x_n| < \frac{2M}{n}$$

for all values of $n = 1, 2, 3, \dots$.

(c) Prove that if $f(c) \ne 0$, then there exists a positive integer N such that

$$\frac{|f(c)|}{2} < |f(x_n)|$$

if $n > N$. [*Hint:* Argue that $f(x) \to f(c)$ as $x \to c$ and then apply Definition 1.4.1 with $\epsilon = \frac{1}{2}|f(c)|$.]

(d) What can you conclude from parts (b) and (c)?

5. What are the important elements in the argument suggested by Exercise 4? Can you extend this argument to a wider collection of functions?

6. A bug crawling on a linoleum floor along the edge of a plush carpet encounters an irregularity in the form of a 2 in by 3 in rectangular section of carpet that juts out into the linoleum as illustrated in Figure Ex-6a on the next page.

The bug crawls at 0.7 in/s on the linoleum, but only at 0.3 in/s through the carpet, and its goal is to travel from point A to point B. Four possible routes from A to B are as follows: (i) crawl on linoleum along the edge of the carpet; (ii) crawl through the carpet to a point on the wider side of the rectangle, and finish the journey on linoleum along the edge of the carpet; (iii) crawl through the carpet to a point on the shorter side of the rectangle, and finish the journey on linoleum along the edge of the carpet; or (iv) crawl through the carpet directly to point B. (See Figure Ex-6b.)

(a) Calculate the times it would take the bug to crawl from A to B via routes (i) and (iv).

(b) Suppose the bug follows route (ii) and use x to represent the total distance the bug crawls on linoleum. Identify the appropriate interval for x in this case, and determine the shortest time for the bug to complete the journey using route (ii).

(c) Suppose the bug follows route (iii) and again use x to represent the total distance the bug crawls on linoleum. Identify the appropriate interval for x in this case, and determine the shortest time for the bug to complete the journey using route (iii).

(d) Which of routes (i), (ii), (iii), or (iv) is quickest? What is the shortest time for the bug to complete the journey?

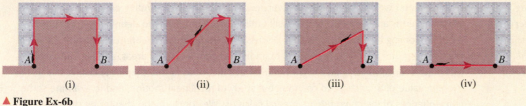

(i)   (ii)   (iii)   (iv)

▲ Figure Ex-6b

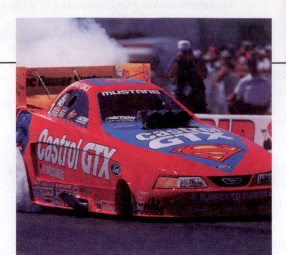

Jon Ferrey/Allsport/Getty Images

5

INTEGRATION

If a dragster moves with varying velocity over a certain time interval, it is possible to find the distance it travels during that time interval using techniques of calculus.

In this chapter we will begin with an overview of the problem of finding areas—we will discuss what the term "area" means, and we will outline two approaches to defining and calculating areas. Following this overview, we will discuss the Fundamental Theorem of Calculus, which is the theorem that relates the problems of finding tangent lines and areas, and we will discuss techniques for calculating areas. We will then use the ideas in this chapter to define the average value of a function, to continue our study of rectilinear motion, and to examine some consequences of the chain rule in integral calculus. We conclude the chapter by studying functions defined by integrals, with a focus on the natural logarithm function.

5.1 AN OVERVIEW OF THE AREA PROBLEM

In this introductory section we will consider the problem of calculating areas of plane regions with curvilinear boundaries. All of the results in this section will be reexamined in more detail later in this chapter. Our purpose here is simply to introduce and motivate the fundamental concepts.

■ THE AREA PROBLEM

Formulas for the areas of polygons, such as squares, rectangles, triangles, and trapezoids, were well known in many early civilizations. However, the problem of finding formulas for regions with curved boundaries (a circle being the simplest example) caused difficulties for early mathematicians.

The first real progress in dealing with the general area problem was made by the Greek mathematician Archimedes, who obtained areas of regions bounded by circular arcs, parabolas, spirals, and various other curves using an ingenious procedure that was later called the *method of exhaustion*. The method, when applied to a circle, consists of inscribing a succession of regular polygons in the circle and allowing the number of sides to increase indefinitely (Figure 5.1.1). As the number of sides increases, the polygons tend to "exhaust" the region inside the circle, and the areas of the polygons become better and better approximations of the exact area of the circle.

To see how this works numerically, let $A(n)$ denote the area of a regular n-sided polygon inscribed in a circle of radius 1. Table 5.1.1 shows the values of $A(n)$ for various choices of n. Note that for large values of n the area $A(n)$ appears to be close to π (square units),

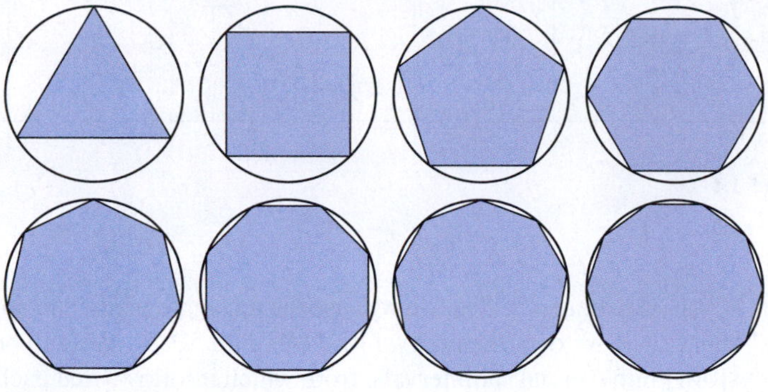

Table 5.1.1

n	$A(n)$
100	3.13952597647
200	3.14107590781
300	3.14136298250
400	3.14146346236
500	3.14150997084
1000	3.14157198278
2000	3.14158748588
3000	3.14159035683
4000	3.14159136166
5000	3.14159182676
10,000	3.14159244688

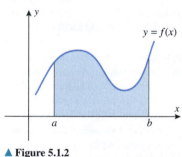

▲ **Figure 5.1.1**

as one would expect. This suggests that for a circle of radius 1, the method of exhaustion is equivalent to an equation of the form

$$\lim_{n \to \infty} A(n) = \pi$$

Since Greek mathematicians were suspicious of the concept of "infinity," they avoided its use in mathematical arguments. As a result, computation of area using the method of exhaustion was a very cumbersome procedure. It remained for Newton and Leibniz to obtain a general method for finding areas that explicitly used the notion of a limit. We will discuss their method in the context of the following problem.

▲ **Figure 5.1.2**

5.1.1 THE AREA PROBLEM Given a function f that is continuous and nonnegative on an interval $[a, b]$, find the area between the graph of f and the interval $[a, b]$ on the x-axis (Figure 5.1.2).

■ **THE RECTANGLE METHOD FOR FINDING AREAS**

One approach to the area problem is to use Archimedes' method of exhaustion in the following way:

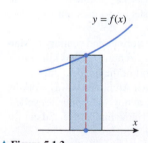

▲ **Figure 5.1.3**

- Divide the interval $[a, b]$ into n equal subintervals, and over each subinterval construct a rectangle that extends from the x-axis to any point on the curve $y = f(x)$ that is above the subinterval; the particular point does not matter—it can be above the center, above an endpoint, or above any other point in the subinterval. In Figure 5.1.3 it is above the center.

- For each n, the total area of the rectangles can be viewed as an *approximation* to the exact area under the curve over the interval $[a, b]$. Moreover, it is evident intuitively that as n increases these approximations will get better and better and will approach the exact area as a limit (Figure 5.1.4). That is, if A denotes the exact area under the curve and A_n denotes the approximation to A using n rectangles, then

$$A = \lim_{n \to +\infty} A_n$$

We will call this the *rectangle method* for computing A.

Logically speaking, we cannot really talk about computing areas without a precise mathematical definition of the term "area." Later in this chapter we will give such a definition, but for now we will treat the concept intuitively.

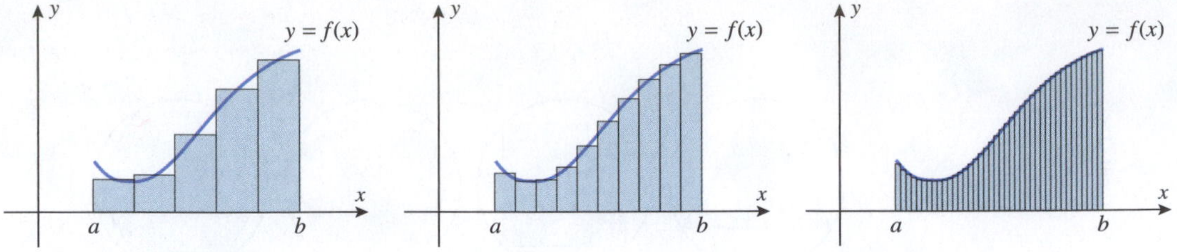

▲ Figure 5.1.4

To illustrate this idea, we will use the rectangle method to approximate the area under the curve $y = x^2$ over the interval $[0, 1]$ (Figure 5.1.5). We will begin by dividing the interval $[0, 1]$ into n equal subintervals, from which it follows that each subinterval has length $1/n$; the endpoints of the subintervals occur at

$$0, \; \frac{1}{n}, \; \frac{2}{n}, \; \frac{3}{n}, \ldots, \; \frac{n-1}{n}, \; 1$$

Archimedes (**287** B.C.–**212** B.C.) Greek mathematician and scientist. Born in Syracuse, Sicily, Archimedes was the son of the astronomer Pheidias and possibly related to Heiron II, king of Syracuse. Most of the facts about his life come from the Roman biographer, Plutarch, who inserted a few tantalizing pages about him in the massive biography of the Roman soldier, Marcellus. In the words of one writer, "the account of Archimedes is slipped like a tissue-thin shaving of ham in a bull-choking sandwich."

Archimedes ranks with Newton and Gauss as one of the three greatest mathematicians who ever lived, and he is certainly the greatest mathematician of antiquity. His mathematical work is so modern in spirit and technique that it is barely distinguishable from that of a seventeenth-century mathematician, yet it was all done without benefit of algebra or a convenient number system. Among his mathematical achievements, Archimedes developed a general method (exhaustion) for finding areas and volumes, and he used the method to find areas bounded by parabolas and spirals and to find volumes of cylinders, paraboloids, and segments of spheres. He gave a procedure for approximating π and bounded its value between $3\frac{10}{71}$ and $3\frac{1}{7}$. In spite of the limitations of the Greek numbering system, he devised methods for finding square roots and invented a method based on the Greek myriad (10,000) for representing numbers as large as 1 followed by 80 million billion zeros.

Of all his mathematical work, Archimedes was most proud of his discovery of a method for finding the volume of a sphere—he showed that the volume of a sphere is two-thirds the volume of the smallest cylinder that can contain it. At his request, the figure of a sphere and cylinder was engraved on his tombstone.

In addition to mathematics, Archimedes worked extensively in mechanics and hydrostatics. Nearly every schoolchild knows Archimedes as the absent-minded scientist who, on realizing that a floating object displaces its weight of liquid, leaped from his bath and ran naked through the streets of Syracuse shouting, "Eureka, Eureka!"—(meaning, "I have found it!"). Archimedes actually created the discipline of hydrostatics and used it to find equilibrium positions for various floating bodies. He laid down the fundamental postulates of mechanics, discovered the laws of levers, and calculated centers of gravity for various flat surfaces and solids. In the excitement of discovering the mathematical laws of the lever, he is said to have declared, "Give me a place to stand and I will move the earth."

Although Archimedes was apparently more interested in pure mathematics than its applications, he was an engineering genius. During the second Punic war, when Syracuse was attacked by the Roman fleet under the command of Marcellus, it was reported by Plutarch that Archimedes' military inventions held the fleet at bay for three years. He invented super catapults that showered the Romans with rocks weighing a quarter ton or more, and fearsome mechanical devices with iron "beaks and claws" that reached over the city walls, grasped the ships, and spun them against the rocks. After the first repulse, Marcellus called Archimedes a "geometrical Briareus (a hundred-armed mythological monster) who uses our ships like cups to ladle water from the sea."

Eventually the Roman army was victorious and contrary to Marcellus' specific orders the 75-year-old Archimedes was killed by a Roman soldier. According to one report of the incident, the soldier cast a shadow across the sand in which Archimedes was working on a mathematical problem. When the annoyed Archimedes yelled, "Don't disturb my circles," the soldier flew into a rage and cut the old man down.

Although there is no known likeness or statue of this great man, nine works of Archimedes have survived to the present day. Especially important is his treatise, *The Method of Mechanical Theorems*, which was part of a palimpsest found in Constantinople in 1906. In this treatise Archimedes explains how he made some of his discoveries, using reasoning that anticipated ideas of the integral calculus. Thought to be lost, the Archimedes palimpsest later resurfaced in 1998, when it was purchased by an anonymous private collector for two million dollars.

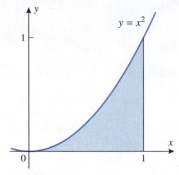

▲ **Figure 5.1.5**

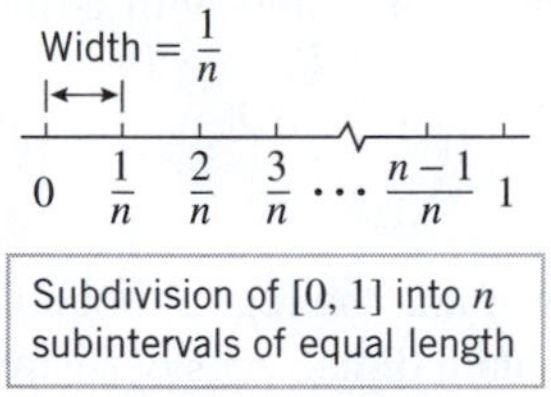

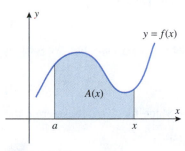

▲ **Figure 5.1.6**

(Figure 5.1.6). We want to construct a rectangle over each of these subintervals whose height is the value of the function $f(x) = x^2$ at some point in the subinterval. To be specific, let us use the right endpoints, in which case the heights of our rectangles will be

$$\left(\frac{1}{n}\right)^2, \left(\frac{2}{n}\right)^2, \left(\frac{3}{n}\right)^2, \ldots, 1^2$$

and since each rectangle has a base of width $1/n$, the total area A_n of the n rectangles will be

$$A_n = \left[\left(\frac{1}{n}\right)^2 + \left(\frac{2}{n}\right)^2 + \left(\frac{3}{n}\right)^2 + \cdots + 1^2\right]\left(\frac{1}{n}\right) \tag{1}$$

For example, if $n = 4$, then the total area of the four approximating rectangles would be

$$A_4 = \left[\left(\tfrac{1}{4}\right)^2 + \left(\tfrac{2}{4}\right)^2 + \left(\tfrac{3}{4}\right)^2 + 1^2\right]\left(\tfrac{1}{4}\right) = \tfrac{15}{32} = 0.46875$$

Table 5.1.2 shows the result of evaluating (1) on a computer for some increasingly large values of n. These computations suggest that the exact area is close to $\frac{1}{3}$. Later in this chapter we will prove that this area is exactly $\frac{1}{3}$ by showing that

$$\lim_{n\to\infty} A_n = \tfrac{1}{3}$$

Table 5.1.2

n	4	10	100	1000	10,000	100,000
A_n	0.468750	0.385000	0.338350	0.333834	0.333383	0.333338

TECHNOLOGY MASTERY

Use a calculating utility to compute the value of A_{10} in Table 5.1.2. Some calculating utilities have special commands for computing sums such as that in (1) for any specified value of n. If your utility has this feature, use it to compute A_{100} as well.

THE ANTIDERIVATIVE METHOD FOR FINDING AREAS

Although the rectangle method is appealing intuitively, the limits that result can only be evaluated in certain cases. For this reason, progress on the area problem remained at a rudimentary level until the latter part of the seventeenth century when Isaac Newton and Gottfried Leibniz independently discovered a fundamental relationship between areas and derivatives. Briefly stated, they showed that if f is a nonnegative continuous function on the interval $[a, b]$, and if $A(x)$ denotes the area under the graph of f over the interval $[a, x]$, where x is any point in the interval $[a, b]$ (Figure 5.1.7), then

$$A'(x) = f(x) \tag{2}$$

▲ **Figure 5.1.7**

The following example confirms Formula (2) in some cases where a formula for $A(x)$ can be found using elementary geometry.

► **Example 1** For each of the functions f, find the area $A(x)$ between the graph of f and the interval $[a, x] = [-1, x]$, and find the derivative $A'(x)$ of this area function.

(a) $f(x) = 2$ (b) $f(x) = x + 1$ (c) $f(x) = 2x + 3$

Solution (a). From Figure 5.1.8*a* we see that

$$A(x) = 2(x - (-1)) = 2(x + 1) = 2x + 2$$

is the area of a rectangle of height 2 and base $x + 1$. For this area function,

$$A'(x) = 2 = f(x)$$

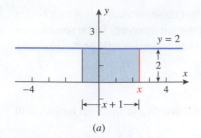

(a)

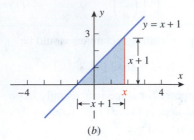

(b)

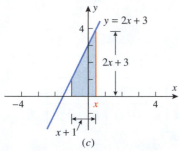

(c)

▲ **Figure 5.1.8**

How does the solution to Example 2 change if the interval $[0, 1]$ is replaced by the interval $[-1, 1]$?

Solution (b).   From Figure 5.1.8b we see that

$$A(x) = \frac{1}{2}(x + 1)(x + 1) = \frac{x^2}{2} + x + \frac{1}{2}$$

is the area of an isosceles right triangle with base and height equal to $x + 1$. For this area function,

$$A'(x) = x + 1 = f(x)$$

Solution (c).   Recall that the formula for the area of a trapezoid is $A = \frac{1}{2}(b + b')h$, where b and b' denote the lengths of the parallel sides of the trapezoid, and the altitude h denotes the distance between the parallel sides. From Figure 5.1.8c we see that

$$A(x) = \frac{1}{2}((2x + 3) + 1)(x - (-1)) = x^2 + 3x + 2$$

is the area of a trapezoid with parallel sides of lengths 1 and $2x + 3$ and with altitude $x - (-1) = x + 1$. For this area function,

$$A'(x) = 2x + 3 = f(x) \quad \blacktriangleleft$$

Formula (2) is important because it relates the area function A and the region-bounding function f. Although a formula for $A(x)$ may be difficult to obtain directly, its derivative, $f(x)$, is given. If a formula for $A(x)$ can be recovered from the given formula for $A'(x)$, then the area under the graph of f over the interval $[a, b]$ can be obtained by computing $A(b)$.

The process of finding a function from its derivative is called **antidifferentiation**, and a procedure for finding areas via antidifferentiation is called the **antiderivative method**. To illustrate this method, let us revisit the problem of finding the area in Figure 5.1.5.

▶ **Example 2**   Use the antiderivative method to find the area under the graph of $y = x^2$ over the interval $[0, 1]$.

Solution.   Let x be any point in the interval $[0, 1]$, and let $A(x)$ denote the area under the graph of $f(x) = x^2$ over the interval $[0, x]$. It follows from (2) that

$$A'(x) = x^2 \tag{3}$$

To find $A(x)$ we must look for a function whose derivative is x^2. By guessing, we see that one such function is $\frac{1}{3}x^3$, so by Theorem 4.8.3

$$A(x) = \frac{1}{3}x^3 + C \tag{4}$$

for some real constant C. We can determine the specific value for C by considering the case where $x = 0$. In this case (4) implies that

$$A(0) = C \tag{5}$$

But if $x = 0$, then the interval $[0, x]$ reduces to a single point. If we agree that the area above a single point should be taken as zero, then $A(0) = 0$ and (5) implies that $C = 0$. Thus, it follows from (4) that

$$A(x) = \frac{1}{3}x^3$$

is the area function we are seeking. This implies that the area under the graph of $y = x^2$ over the interval $[0, 1]$ is

$$A(1) = \frac{1}{3}(1^3) = \frac{1}{3}$$

This is consistent with the result that we previously obtained numerically. ◀

As Example 2 illustrates, antidifferentiation is a process in which one tries to "undo" a differentiation. One of the objectives in this chapter is to develop efficient antidifferentiation procedures.

■ **THE RECTANGLE METHOD AND THE ANTIDERIVATIVE METHOD COMPARED**

The rectangle method and the antiderivative method provide two very different approaches to the area problem, each of which is important. The antiderivative method is usually the more efficient way to *compute* areas, but it is the rectangle method that is used to formally *define* the notion of area, thereby allowing us to prove mathematical results about areas. The underlying idea of the rectangle approach is also important because it can be adapted readily to such diverse problems as finding the volume of a solid, the length of a curve, the mass of an object, and the work done in pumping water out of a tank, to name a few.

✔ **QUICK CHECK EXERCISES 5.1**   *(See page 322 for answers.)*

1. Let R denote the region below the graph of $f(x) = \sqrt{1 - x^2}$ and above the interval $[-1, 1]$.
   (a) Use a geometric argument to find the area of R.
   (b) What estimate results if the area of R is approximated by the total area within the rectangles of the accompanying figure?

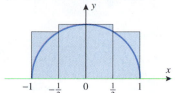

◀ **Figure Ex-1**

2. Suppose that when the area A between the graph of a function $y = f(x)$ and an interval $[a, b]$ is approximated by the areas of n rectangles, the total area of the rectangles is $A_n = 2 + (2/n)$, $n = 1, 2, \ldots$. Then, $A = $ _____.

3. The area under the graph of $y = x^2$ over the interval $[0, 3]$ is _____.

4. Find a formula for the area $A(x)$ between the graph of the function $f(x) = x$ and the interval $[0, x]$, and verify that $A'(x) = f(x)$.

5. The area under the graph of $y = f(x)$ over the interval $[0, x]$ is $A(x) = x + e^x - 1$. It follows that $f(x) = $ _____.

EXERCISE SET 5.1

1–12 Estimate the area between the graph of the function f and the interval $[a, b]$. Use an approximation scheme with n rectangles similar to our treatment of $f(x) = x^2$ in this section. If your calculating utility will perform automatic summations, estimate the specified area using $n = 10, 50,$ and 100 rectangles. Otherwise, estimate this area using $n = 2, 5,$ and 10 rectangles. ■

1. $f(x) = \sqrt{x}$; $[a, b] = [0, 1]$

2. $f(x) = \dfrac{1}{x + 1}$; $[a, b] = [0, 1]$

3. $f(x) = \sin x$; $[a, b] = [0, \pi]$

4. $f(x) = \cos x$; $[a, b] = [0, \pi/2]$

5. $f(x) = \dfrac{1}{x}$; $[a, b] = [1, 2]$

6. $f(x) = \cos x$; $[a, b] = [-\pi/2, \pi/2]$

7. $f(x) = \sqrt{1 - x^2}$; $[a, b] = [0, 1]$

8. $f(x) = \sqrt{1 - x^2}$; $[a, b] = [-1, 1]$

9. $f(x) = e^x$; $[a, b] = [-1, 1]$

10. $f(x) = \ln x$; $[a, b] = [1, 2]$

11. $f(x) = \sin^{-1} x$; $[a, b] = [0, 1]$

12. $f(x) = \tan^{-1} x$; $[a, b] = [0, 1]$

13–18 Graph each function over the specified interval. Then use simple area formulas from geometry to find the area function $A(x)$ that gives the area between the graph of the specified function f and the interval $[a, x]$. Confirm that $A'(x) = f(x)$ in every case. ■

13. $f(x) = 3$; $[a, x] = [1, x]$

14. $f(x) = 5$; $[a, x] = [2, x]$

15. $f(x) = 2x + 2$; $[a, x] = [0, x]$

16. $f(x) = 3x - 3$; $[a, x] = [1, x]$

17. $f(x) = 2x + 2$; $[a, x] = [1, x]$

18. $f(x) = 3x - 3$; $[a, x] = [2, x]$

19–22 True–False Determine whether the statement is true or false. Explain your answer. ■

19. If $A(n)$ denotes the area of a regular n-sided polygon inscribed in a circle of radius 2, then $\lim_{n \to +\infty} A(n) = 2\pi$.

20. If the area under the curve $y = x^2$ over an interval is approximated by the total area of a collection of rectangles, the approximation will be too large.

21. If $A(x)$ is the area under the graph of a nonnegative continuous function f over an interval $[a, x]$, then $A'(x) = f(x)$.

22. If $A(x)$ is the area under the graph of a nonnegative continuous function f over an interval $[a, x]$, then $A(x)$ will be a continuous function.

23. Explain how to use the formula for $A(x)$ found in the solution to Example 2 to determine the area between the graph of $y = x^2$ and the interval $[3, 6]$.

24. Repeat Exercise 23 for the interval $[-3, 9]$.

25. Let A denote the area between the graph of $f(x) = \sqrt{x}$ and the interval $[0, 1]$, and let B denote the area between the graph of $f(x) = x^2$ and the interval $[0, 1]$. Explain geometrically why $A + B = 1$.

26. Let A denote the area between the graph of $f(x) = 1/x$ and the interval $[1, 2]$, and let B denote the area between the graph of f and the interval $\left[\frac{1}{2}, 1\right]$. Explain geometrically why $A = B$.

27–28 The area $A(x)$ under the graph of f and over the interval $[a, x]$ is given. Find the function f and the value of a. ■

27. $A(x) = x^2 - 4$ **28.** $A(x) = x^2 - x$

29. Writing Compare and contrast the rectangle method and the antiderivative method.

30. Writing Suppose that f is a nonnegative continuous function on an interval $[a, b]$ and that $g(x) = f(x) + C$, where C is a positive constant. What will be the area of the region *between* the graphs of f and g?

✔ **QUICK CHECK ANSWERS 5.1**

1. (a) $\dfrac{\pi}{2}$ (b) $1 + \dfrac{\sqrt{3}}{2}$ **2.** 2   **3.** 9   **4.** $A(x) = \dfrac{x^2}{2}$; $A'(x) = \dfrac{2x}{2} = x = f(x)$ **5.** $e^x + 1$

5.2 THE INDEFINITE INTEGRAL

In the last section we saw how antidifferentiation could be used to find exact areas. In this section we will develop some fundamental results about antidifferentiation.

■ ANTIDERIVATIVES

5.2.1   DEFINITION   A function F is called an *antiderivative* of a function f on a given open interval if $F'(x) = f(x)$ for all x in the interval.

For example, the function $F(x) = \frac{1}{3}x^3$ is an antiderivative of $f(x) = x^2$ on the interval $(-\infty, +\infty)$ because for each x in this interval

$$F'(x) = \frac{d}{dx}\left[\tfrac{1}{3}x^3\right] = x^2 = f(x)$$

However, $F(x) = \frac{1}{3}x^3$ is not the only antiderivative of f on this interval. If we add any constant C to $\frac{1}{3}x^3$, then the function $G(x) = \frac{1}{3}x^3 + C$ is also an antiderivative of f on $(-\infty, +\infty)$, since

$$G'(x) = \frac{d}{dx}\left[\tfrac{1}{3}x^3 + C\right] = x^2 + 0 = f(x)$$

In general, once any single antiderivative is known, other antiderivatives can be obtained by adding constants to the known antiderivative. Thus,

$$\tfrac{1}{3}x^3, \quad \tfrac{1}{3}x^3 + 2, \quad \tfrac{1}{3}x^3 - 5, \quad \tfrac{1}{3}x^3 + \sqrt{2}$$

are all antiderivatives of $f(x) = x^2$.

It is reasonable to ask if there are antiderivatives of a function f that cannot be obtained by adding some constant to a known antiderivative F. The answer is *no*—once a single antiderivative of f on an open interval is known, all other antiderivatives on that interval are obtainable by adding constants to the known antiderivative. This is so because Theorem 4.8.3 tells us that if two functions have the same derivative on an open interval, then the functions differ by a constant on the interval. The following theorem summarizes these observations.

> **5.2.2 THEOREM** *If $F(x)$ is any antiderivative of $f(x)$ on an open interval, then for any constant C the function $F(x) + C$ is also an antiderivative on that interval. Moreover, each antiderivative of $f(x)$ on the interval can be expressed in the form $F(x) + C$ by choosing the constant C appropriately.*

■ THE INDEFINITE INTEGRAL

The process of finding antiderivatives is called ***antidifferentiation*** or ***integration***. Thus, if

$$\frac{d}{dx}[F(x)] = f(x) \tag{1}$$

then ***integrating*** (or ***antidifferentiating***) the function $f(x)$ produces an antiderivative of the form $F(x) + C$. To emphasize this process, Equation (1) is recast using ***integral notation***,

$$\int f(x)\,dx = F(x) + C \tag{2}$$

where C is understood to represent an arbitrary constant. It is important to note that (1) and (2) are just different notations to express the same fact. For example,

$$\int x^2\,dx = \tfrac{1}{3}x^3 + C \quad \text{is equivalent to} \quad \frac{d}{dx}\left[\tfrac{1}{3}x^3\right] = x^2$$

Note that if we differentiate an antiderivative of $f(x)$, we obtain $f(x)$ back again. Thus,

$$\frac{d}{dx}\left[\int f(x)\,dx\right] = f(x) \tag{3}$$

The expression $\int f(x)\,dx$ is called an ***indefinite integral***. The adjective "indefinite" emphasizes that the result of antidifferentiation is a "generic" function, described only up to a constant term. The "elongated s" that appears on the left side of (2) is called an ***integral sign***,[*] the function $f(x)$ is called the ***integrand***, and the constant C is called the ***constant of integration***. Equation (2) should be read as:

The integral of $f(x)$ with respect to x is equal to $F(x)$ plus a constant.

The differential symbol, dx, in the differentiation and antidifferentiation operations

$$\frac{d}{dx}[\ \ ] \quad \text{and} \quad \int [\ \ ]\,dx$$

Reproduced from C. I. Gerhardt's "Briefwechsel von G. W. Leibniz mit Mathematikern (1899)."

Extract from the manuscript of Leibniz dated October 29, 1675 in which the integral sign first appeared (see yellow highlight).

---

[*]This notation was devised by Leibniz. In his early papers Leibniz used the notation "omn." (an abbreviation for the Latin word "omnes") to denote integration. Then on October 29, 1675 he wrote, "It will be useful to write $\int$ for omn., thus $\int l$ for omn. $l\ldots$." Two or three weeks later he refined the notation further and wrote $\int [\ \ ]\,dx$ rather than $\int$ alone. This notation is so useful and so powerful that its development by Leibniz must be regarded as a major milestone in the history of mathematics and science.

serves to identify the independent variable. If an independent variable other than x is used, say t, then the notation must be adjusted appropriately. Thus,

$$\frac{d}{dt}[F(t)] = f(t) \quad \text{and} \quad \int f(t)\, dt = F(t) + C$$

are equivalent statements. Here are some examples of derivative formulas and their equivalent integration formulas:

DERIVATIVE FORMULA	EQUIVALENT INTEGRATION FORMULA
$\dfrac{d}{dx}[x^3] = 3x^2$	$\displaystyle\int 3x^2\, dx = x^3 + C$
$\dfrac{d}{dx}[\sqrt{x}] = \dfrac{1}{2\sqrt{x}}$	$\displaystyle\int \dfrac{1}{2\sqrt{x}}\, dx = \sqrt{x} + C$
$\dfrac{d}{dt}[\tan t] = \sec^2 t$	$\displaystyle\int \sec^2 t\, dt = \tan t + C$
$\dfrac{d}{du}[u^{3/2}] = \dfrac{3}{2}u^{1/2}$	$\displaystyle\int \dfrac{3}{2}u^{1/2}\, du = u^{3/2} + C$

For simplicity, the dx is sometimes absorbed into the integrand. For example,

$$\int 1\, dx \quad \text{can be written as} \quad \int dx$$

$$\int \frac{1}{x^2}\, dx \quad \text{can be written as} \quad \int \frac{dx}{x^2}$$

■ **INTEGRATION FORMULAS**

Integration is essentially educated guesswork—given the derivative f of a function F, one tries to guess what the function F is. However, many basic integration formulas can be obtained directly from their companion differentiation formulas. Some of the most important are given in Table 5.2.1.

Table 5.2.1
INTEGRATION FORMULAS

DIFFERENTIATION FORMULA	INTEGRATION FORMULA	DIFFERENTIATION FORMULA	INTEGRATION FORMULA
1. $\dfrac{d}{dx}[x] = 1$	$\displaystyle\int dx = x + C$	8. $\dfrac{d}{dx}[-\csc x] = \csc x \cot x$	$\displaystyle\int \csc x \cot x\, dx = -\csc x + C$
2. $\dfrac{d}{dx}\left[\dfrac{x^{r+1}}{r+1}\right] = x^r \ (r \neq -1)$	$\displaystyle\int x^r\, dx = \dfrac{x^{r+1}}{r+1} + C \ (r \neq -1)$	9. $\dfrac{d}{dx}[e^x] = e^x$	$\displaystyle\int e^x\, dx = e^x + C$
3. $\dfrac{d}{dx}[\sin x] = \cos x$	$\displaystyle\int \cos x\, dx = \sin x + C$	10. $\dfrac{d}{dx}\left[\dfrac{b^x}{\ln b}\right] = b^x \ (0 < b, b \neq 1)$	$\displaystyle\int b^x\, dx = \dfrac{b^x}{\ln b} + C \ (0 < b, b \neq 1)$
4. $\dfrac{d}{dx}[-\cos x] = \sin x$	$\displaystyle\int \sin x\, dx = -\cos x + C$	11. $\dfrac{d}{dx}[\ln |x|] = \dfrac{1}{x}$	$\displaystyle\int \dfrac{1}{x}\, dx = \ln |x| + C$
5. $\dfrac{d}{dx}[\tan x] = \sec^2 x$	$\displaystyle\int \sec^2 x\, dx = \tan x + C$	12. $\dfrac{d}{dx}[\tan^{-1} x] = \dfrac{1}{1+x^2}$	$\displaystyle\int \dfrac{1}{1+x^2}\, dx = \tan^{-1} x + C$
6. $\dfrac{d}{dx}[-\cot x] = \csc^2 x$	$\displaystyle\int \csc^2 x\, dx = -\cot x + C$	13. $\dfrac{d}{dx}[\sin^{-1} x] = \dfrac{1}{\sqrt{1-x^2}}$	$\displaystyle\int \dfrac{1}{\sqrt{1-x^2}}\, dx = \sin^{-1} x + C$
7. $\dfrac{d}{dx}[\sec x] = \sec x \tan x$	$\displaystyle\int \sec x \tan x\, dx = \sec x + C$	14. $\dfrac{d}{dx}[\sec^{-1} |x|] = \dfrac{1}{x\sqrt{x^2-1}}$	$\displaystyle\int \dfrac{1}{x\sqrt{x^2-1}}\, dx = \sec^{-1} |x| + C$

See Exercise 72 for a justification of Formula 14 in Table 5.2.1.

▶ **Example 1** The second integration formula in Table 5.2.1 will be easier to remember if you express it in words:

To integrate a power of x (other than −1), add 1 to the exponent and divide by the new exponent.

Although Formula 2 in Table 5.2.1 is not applicable to integrating x^{-1}, this function can be integrated by rewriting the integral in Formula 11 as

$$\int \frac{1}{x}\,dx = \int x^{-1}\,dx = \ln |x| + C$$

Here are some examples:

$$\int x^2\,dx = \frac{x^3}{3} + C \qquad \boxed{r=2}$$

$$\int x^3\,dx = \frac{x^4}{4} + C \qquad \boxed{r=3}$$

$$\int \frac{1}{x^5}\,dx = \int x^{-5}\,dx = \frac{x^{-5+1}}{-5+1} + C = -\frac{1}{4x^4} + C \qquad \boxed{r=-5}$$

$$\int \sqrt{x}\,dx = \int x^{\frac{1}{2}}\,dx = \frac{x^{\frac{1}{2}+1}}{\frac{1}{2}+1} + C = \frac{2}{3}x^{\frac{3}{2}} + C = \frac{2}{3}(\sqrt{x})^3 + C \qquad \boxed{r=\frac{1}{2}} \quad ◀$$

■ **PROPERTIES OF THE INDEFINITE INTEGRAL**

Our first properties of antiderivatives follow directly from the simple constant factor, sum, and difference rules for derivatives.

5.2.3 THEOREM *Suppose that $F(x)$ and $G(x)$ are antiderivatives of $f(x)$ and $g(x)$, respectively, and that c is a constant. Then:*

(a) A constant factor can be moved through an integral sign; that is,

$$\int cf(x)\,dx = cF(x) + C$$

(b) An antiderivative of a sum is the sum of the antiderivatives; that is,

$$\int [f(x) + g(x)]\,dx = F(x) + G(x) + C$$

(c) An antiderivative of a difference is the difference of the antiderivatives; that is,

$$\int [f(x) - g(x)]\,dx = F(x) - G(x) + C$$

PROOF In general, to establish the validity of an equation of the form

$$\int h(x)\,dx = H(x) + C$$

one must show that

$$\frac{d}{dx}[H(x)] = h(x)$$

We are given that $F(x)$ and $G(x)$ are antiderivatives of $f(x)$ and $g(x)$, respectively, so we know that

$$\frac{d}{dx}[F(x)] = f(x) \quad \text{and} \quad \frac{d}{dx}[G(x)] = g(x)$$

Thus,

$$\frac{d}{dx}[cF(x)] = c\frac{d}{dx}[F(x)] = cf(x)$$

$$\frac{d}{dx}[F(x) + G(x)] = \frac{d}{dx}[F(x)] + \frac{d}{dx}[G(x)] = f(x) + g(x)$$

$$\frac{d}{dx}[F(x) - G(x)] = \frac{d}{dx}[F(x)] - \frac{d}{dx}[G(x)] = f(x) - g(x)$$

which proves the three statements of the theorem. ∎

The statements in Theorem 5.2.3 can be summarized by the following formulas:

$$\int cf(x)\,dx = c\int f(x)\,dx \tag{4}$$

$$\int [f(x) + g(x)]\,dx = \int f(x)\,dx + \int g(x)\,dx \tag{5}$$

$$\int [f(x) - g(x)]\,dx = \int f(x)\,dx - \int g(x)\,dx \tag{6}$$

However, these equations must be applied carefully to avoid errors and unnecessary complexities arising from the constants of integration. For example, if you use (4) to integrate $2x$ by writing

$$\int 2x\,dx = 2\int x\,dx = 2\left(\frac{x^2}{2} + C\right) = x^2 + 2C$$

then you will have an unnecessarily complicated form of the arbitrary constant. This kind of problem can be avoided by inserting the constant of integration in the final result rather than in intermediate calculations. Exercises 65 and 66 explore how careless application of these formulas can lead to errors.

▶ **Example 2**   Evaluate

$$\text{(a)}\ \int 4\cos x\,dx \qquad \text{(b)}\ \int (x + x^2)\,dx$$

Solution (a).   Since $F(x) = \sin x$ is an antiderivative for $f(x) = \cos x$ (Table 5.2.1), we obtain

$$\int 4\cos x\,dx = 4\underset{(4)}{\int} \cos x\,dx = 4\sin x + C$$

Solution (b).   From Table 5.2.1 we obtain

$$\int (x + x^2)\,dx = \underset{(5)}{\int} x\,dx + \int x^2\,dx = \frac{x^2}{2} + \frac{x^3}{3} + C \ \blacktriangleleft$$

Parts (*b*) and (*c*) of Theorem 5.2.3 can be extended to more than two functions, which in combination with part (*a*) results in the following general formula:

$$\int [c_1 f_1(x) + c_2 f_2(x) + \cdots + c_n f_n(x)]\,dx$$

$$= c_1\int f_1(x)\,dx + c_2\int f_2(x)\,dx + \cdots + c_n\int f_n(x)\,dx \tag{7}$$

► **Example 3**

$$\int (3x^6 - 2x^2 + 7x + 1)\,dx = 3\int x^6\,dx - 2\int x^2\,dx + 7\int x\,dx + \int 1\,dx$$

$$= \frac{3x^7}{7} - \frac{2x^3}{3} + \frac{7x^2}{2} + x + C \quad \blacktriangleleft$$

Sometimes it is useful to rewrite an integrand in a different form before performing the integration. This is illustrated in the following example.

► **Example 4** Evaluate

$$\text{(a)} \int \frac{\cos x}{\sin^2 x}\,dx \qquad \text{(b)} \int \frac{t^2 - 2t^4}{t^4}\,dt \qquad \text{(c)} \int \frac{x^2}{x^2 + 1}\,dx$$

Solution (a).

$$\int \frac{\cos x}{\sin^2 x}\,dx = \int \frac{1}{\sin x}\frac{\cos x}{\sin x}\,dx = \int \csc x \cot x\,dx = -\csc x + C$$

> Formula 8 in Table 5.2.1

Solution (b).

$$\int \frac{t^2 - 2t^4}{t^4}\,dt = \int \left(\frac{1}{t^2} - 2\right)dt = \int (t^{-2} - 2)\,dt$$

$$= \frac{t^{-1}}{-1} - 2t + C = -\frac{1}{t} - 2t + C$$

Solution (c). By adding and subtracting 1 from the numerator of the integrand, we can rewrite the integral in a form in which Formulas 1 and 12 of Table 5.2.1 can be applied:

$$\int \frac{x^2}{x^2 + 1}\,dx = \int \left(\frac{x^2 + 1}{x^2 + 1} - \frac{1}{x^2 + 1}\right)dx$$

$$= \int \left(1 - \frac{1}{x^2 + 1}\right)dx = x - \tan^{-1}x + C \quad \blacktriangleleft$$

Perform the integration in part (c) by first performing a long division on the integrand.

■ **INTEGRAL CURVES**

Graphs of antiderivatives of a function f are called **integral curves** of f. We know from Theorem 5.2.2 that if $y = F(x)$ is any integral curve of $f(x)$, then all other integral curves are vertical translations of this curve, since they have equations of the form $y = F(x) + C$. For example, $y = \frac{1}{3}x^3$ is one integral curve for $f(x) = x^2$, so all the other integral curves have equations of the form $y = \frac{1}{3}x^3 + C$; conversely, the graph of any equation of this form is an integral curve (Figure 5.2.1).

In many problems one is interested in finding a function whose derivative satisfies specified conditions. The following example illustrates a geometric problem of this type.

► **Example 5** Suppose that a curve $y = f(x)$ in the xy-plane has the property that at each point (x, y) on the curve, the tangent line has slope x^2. Find an equation for the curve given that it passes through the point $(2, 1)$.

Solution. Since the slope of the line tangent to $y = f(x)$ is dy/dx, we have $dy/dx = x^2$, and

$$y = \int x^2\,dx = \frac{1}{3}x^3 + C$$

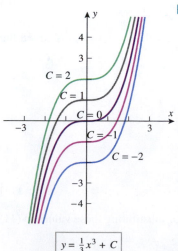

$y = \frac{1}{3}x^3 + C$

▲ **Figure 5.2.1**

In Example 5, the requirement that the graph of f pass through the point $(2, 1)$ selects the single integral curve $y = \frac{1}{3}x^3 - \frac{5}{3}$ from the family of curves $y = \frac{1}{3}x^3 + C$ (Figure 5.2.2).

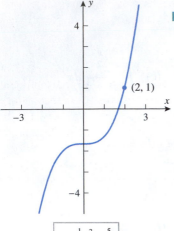

$y = \frac{1}{3}x^3 - \frac{5}{3}$

▲ **Figure 5.2.2**

Since the curve passes through $(2, 1)$, a specific value for C can be found by using the fact that $y = 1$ if $x = 2$. Substituting these values in the above equation yields

$$1 = \tfrac{1}{3}(2^3) + C \quad \text{or} \quad C = -\tfrac{5}{3}$$

so an equation of the curve is

$$y = \tfrac{1}{3}x^3 - \tfrac{5}{3}$$

(Figure 5.2.2). ◄

■ INTEGRATION FROM THE VIEWPOINT OF DIFFERENTIAL EQUATIONS

We will now consider another way of looking at integration that will be useful in our later work. Suppose that $f(x)$ is a known function and we are interested in finding a function $F(x)$ such that $y = F(x)$ satisfies the equation

$$\frac{dy}{dx} = f(x) \tag{8}$$

The solutions of this equation are the antiderivatives of $f(x)$, and we know that these can be obtained by integrating $f(x)$. For example, the solutions of the equation

$$\frac{dy}{dx} = x^2 \tag{9}$$

are

$$y = \int x^2 \, dx = \frac{x^3}{3} + C$$

Equation (8) is called a **differential equation** because it involves a derivative of an unknown function. Differential equations are different from the kinds of equations we have encountered so far in that the unknown is a *function* and not a *number* as in an equation such as $x^2 + 5x - 6 = 0$.

Sometimes we will not be interested in finding all of the solutions of (8), but rather we will want only the solution whose graph passes through a specified point (x_0, y_0). For example, in Example 5 we solved (9) for the integral curve that passed through the point $(2, 1)$.

For simplicity, it is common in the study of differential equations to denote a solution of $dy/dx = f(x)$ as $y(x)$ rather than $F(x)$, as earlier. With this notation, the problem of finding a function $y(x)$ whose derivative is $f(x)$ and whose graph passes through the point (x_0, y_0) is expressed as

$$\frac{dy}{dx} = f(x), \quad y(x_0) = y_0 \tag{10}$$

This is called an **initial-value problem**, and the requirement that $y(x_0) = y_0$ is called the **initial condition** for the problem.

▶ **Example 6** Solve the initial-value problem

$$\frac{dy}{dx} = \cos x, \quad y(0) = 1$$

Solution. The solution of the differential equation is

$$y = \int \cos x \, dx = \sin x + C \tag{11}$$

The initial condition $y(0) = 1$ implies that $y = 1$ if $x = 0$; substituting these values in (11) yields

$$1 = \sin(0) + C \quad \text{or} \quad C = 1$$

Thus, the solution of the initial-value problem is $y = \sin x + 1$. ◄

■ **SLOPE FIELDS**

If we interpret dy/dx as the slope of a tangent line, then at a point (x, y) on an integral curve of the equation $dy/dx = f(x)$, the slope of the tangent line is $f(x)$. What is interesting about this is that the slopes of the tangent lines to the integral curves can be obtained without actually solving the differential equation. For example, if

$$\frac{dy}{dx} = \sqrt{x^2 + 1}$$

then we know without solving the equation that at the point where $x = 1$ the tangent line to an integral curve has slope $\sqrt{1^2 + 1} = \sqrt{2}$; and more generally, at a point where $x = a$, the tangent line to an integral curve has slope $\sqrt{a^2 + 1}$.

A geometric description of the integral curves of a differential equation $dy/dx = f(x)$ can be obtained by choosing a rectangular grid of points in the xy-plane, calculating the slopes of the tangent lines to the integral curves at the gridpoints, and drawing small portions of the tangent lines through those points. The resulting picture, which is called a *slope field* or *direction field* for the equation, shows the "direction" of the integral curves at the gridpoints. With sufficiently many gridpoints it is often possible to visualize the integral curves themselves; for example, Figure 5.2.3*a* shows a slope field for the differential equation $dy/dx = x^2$, and Figure 5.2.3*b* shows that same field with the integral curves imposed on it—the more gridpoints that are used, the more completely the slope field reveals the shape of the integral curves. However, the amount of computation can be considerable, so computers are usually used when slope fields with many gridpoints are needed.

Slope fields will be studied in more detail later in the text.

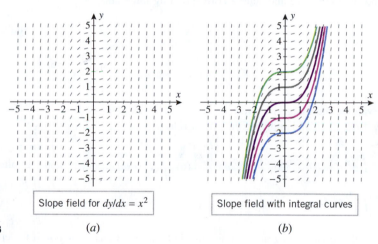

Slope field for $dy/dx = x^2$ Slope field with integral curves

▶ **Figure 5.2.3**   (*a*)   (*b*)

✔ **QUICK CHECK EXERCISES 5.2**   *(See page 332 for answers.)*

1. A function F is an antiderivative of a function f on an interval if _____ for all x in the interval.

2. Write an equivalent integration formula for each given derivative formula.

   (a) $\dfrac{d}{dx}[\sqrt{x}] = \dfrac{1}{2\sqrt{x}}$ (b) $\dfrac{d}{dx}[e^{4x}] = 4e^{4x}$

3. Evaluate the integrals.

   (a) $\displaystyle\int [x^3 + x + 5]\, dx$ (b) $\displaystyle\int [\sec^2 x - \csc x \cot x]\, dx$

4. The graph of $y = x^2 + x$ is an integral curve for the func-

tion $f(x) = $ _____. If G is a function whose graph is also an integral curve for f, and if $G(1) = 5$, then $G(x) = $ _____.

5. A slope field for the differential equation

$$\frac{dy}{dx} = \frac{2x}{x^2 - 4}$$

has a line segment with slope _____ through the point $(0, 5)$ and has a line segment with slope _____ through the point $(-4, 1)$.

EXERCISE SET 5.2   Graphing Utility   CAS

1. In each part, confirm that the formula is correct, and state a corresponding integration formula.

   (a) $\dfrac{d}{dx}[\sqrt{1+x^2}] = \dfrac{x}{\sqrt{1+x^2}}$

   (b) $\dfrac{d}{dx}[xe^x] = (x+1)e^x$

2. In each part, confirm that the stated formula is correct by differentiating.

   (a) $\displaystyle\int x \sin x \, dx = \sin x - x \cos x + C$

   (b) $\displaystyle\int \dfrac{dx}{(1-x^2)^{3/2}} = \dfrac{x}{\sqrt{1-x^2}} + C$

FOCUS ON CONCEPTS

3. What is a *constant of integration*? Why does an answer to an integration problem involve a constant of integration?

4. What is an *integral curve* of a function f? How are two integral curves of a function f related?

5–8 Find the derivative and state a corresponding integration formula. ▨

5. $\dfrac{d}{dx}[\sqrt{x^3+5}]$

6. $\dfrac{d}{dx}\left[\dfrac{x}{x^2+3}\right]$

7. $\dfrac{d}{dx}[\sin(2\sqrt{x})]$

8. $\dfrac{d}{dx}[\sin x - x \cos x]$

9–10 Evaluate the integral by rewriting the integrand appropriately, if required, and applying the power rule (Formula 2 in Table 5.2.1). ▨

9. (a) $\displaystyle\int x^8 \, dx$ (b) $\displaystyle\int x^{5/7} \, dx$ (c) $\displaystyle\int x^3\sqrt{x}\, dx$

10. (a) $\displaystyle\int \sqrt[3]{x^2}\, dx$ (b) $\displaystyle\int \dfrac{1}{x^6}\, dx$ (c) $\displaystyle\int x^{-7/8}\, dx$

11–14 Evaluate each integral by applying Theorem 5.2.3 and Formula 2 in Table 5.2.1 appropriately. ▨

11. $\displaystyle\int \left[5x + \dfrac{2}{3x^5}\right] dx$

12. $\displaystyle\int \left[x^{-1/2} - 3x^{7/5} + \tfrac{1}{9}\right] dx$

13. $\displaystyle\int [x^{-3} - 3x^{1/4} + 8x^2]\, dx$

14. $\displaystyle\int \left[\dfrac{10}{y^{3/4}} - \sqrt[3]{y} + \dfrac{4}{\sqrt{y}}\right] dy$

15–34 Evaluate the integral and check your answer by differentiating. ▨

15. $\displaystyle\int x(1+x^3)\, dx$

16. $\displaystyle\int (2+y^2)^2\, dy$

17. $\displaystyle\int x^{1/3}(2-x)^2\, dx$

18. $\displaystyle\int (1+x^2)(2-x)\, dx$

19. $\displaystyle\int \dfrac{x^5+2x^2-1}{x^4}\, dx$

20. $\displaystyle\int \dfrac{1-2t^3}{t^3}\, dt$

21. $\displaystyle\int \left[\dfrac{2}{x} + 3e^x\right] dx$

22. $\displaystyle\int \left[\dfrac{1}{2t} - \sqrt{2}e^t\right] dt$

23. $\displaystyle\int [3\sin x - 2\sec^2 x]\, dx$

24. $\displaystyle\int [\csc^2 t - \sec t \tan t]\, dt$

25. $\displaystyle\int \sec x (\sec x + \tan x)\, dx$

26. $\displaystyle\int \csc x (\sin x + \cot x)\, dx$

27. $\displaystyle\int \dfrac{\sec \theta}{\cos \theta}\, d\theta$

28. $\displaystyle\int \dfrac{dy}{\csc y}$

29. $\displaystyle\int \dfrac{\sin x}{\cos^2 x}\, dx$

30. $\displaystyle\int \left[\phi + \dfrac{2}{\sin^2 \phi}\right] d\phi$

31. $\displaystyle\int [1 + \sin^2 \theta \csc \theta]\, d\theta$

32. $\displaystyle\int \dfrac{\sec x + \cos x}{2 \cos x}\, dx$

33. $\displaystyle\int \left[\dfrac{1}{2\sqrt{1-x^2}} - \dfrac{3}{1+x^2}\right] dx$

34. $\displaystyle\int \left[\dfrac{4}{x\sqrt{x^2-1}} + \dfrac{1+x+x^3}{1+x^2}\right] dx$

35. Evaluate the integral

$$\int \dfrac{1}{1+\sin x}\, dx$$

by multiplying the numerator and denominator by an appropriate expression.

36. Use the double-angle formula $\cos 2x = 2\cos^2 x - 1$ to evaluate the integral

$$\int \dfrac{1}{1+\cos 2x}\, dx$$

37–40 True–False Determine whether the statement is true or false. Explain your answer. ▨

37. If $F(x)$ is an antiderivative of $f(x)$, then

$$\int f(x)\, dx = F(x) + C$$

38. If C denotes a constant of integration, the two formulas

$$\int \cos x \, dx = \sin x + C$$

$$\int \cos x \, dx = (\sin x + \pi) + C$$

are both correct equations.

39. The function $f(x) = e^{-x} + 1$ is a solution to the initial-value problem

$$\dfrac{dy}{dx} = -\dfrac{1}{e^x}, \quad y(0) = 1$$

40. Every integral curve of the slope field

$$\dfrac{dy}{dx} = \dfrac{1}{\sqrt{x^2+1}}$$

is the graph of an increasing function of x.

41. Use a graphing utility to generate some representative integral curves of the function $f(x) = 5x^4 - \sec^2 x$ over the interval $(-\pi/2, \pi/2)$.

42. Use a graphing utility to generate some representative integral curves of the function $f(x) = (x - 1)/x$ over the interval $(0, 5)$.

43–46 Solve the initial-value problems. ■

43. (a) $\dfrac{dy}{dx} = \sqrt[3]{x},\ y(1) = 2$

(b) $\dfrac{dy}{dt} = \sin t + 1,\ y\left(\dfrac{\pi}{3}\right) = \dfrac{1}{2}$

(c) $\dfrac{dy}{dx} = \dfrac{x + 1}{\sqrt{x}},\ y(1) = 0$

44. (a) $\dfrac{dy}{dx} = \dfrac{1}{(2x)^3},\ y(1) = 0$

(b) $\dfrac{dy}{dt} = \sec^2 t - \sin t,\ y\left(\dfrac{\pi}{4}\right) = 1$

(c) $\dfrac{dy}{dx} = x^2\sqrt{x^3},\ y(0) = 0$

45. (a) $\dfrac{dy}{dx} = 4e^x,\ y(0) = 1$ (b) $\dfrac{dy}{dt} = \dfrac{1}{t},\ y(-1) = 5$

46. (a) $\dfrac{dy}{dt} = \dfrac{3}{\sqrt{1 - t^2}},\ y\left(\dfrac{\sqrt{3}}{2}\right) = 0$

(b) $\dfrac{dy}{dx} = \dfrac{x^2 - 1}{x^2 + 1},\ y(1) = \dfrac{\pi}{2}$

47–50 A particle moves along an s-axis with position function $s = s(t)$ and velocity function $v(t) = s'(t)$. Use the given information to find $s(t)$. ■

47. $v(t) = 32t;\ s(0) = 20$ **48.** $v(t) = \cos t;\ s(0) = 2$

49. $v(t) = 3\sqrt{t};\ s(4) = 1$ **50.** $v(t) = 3e^t;\ s(1) = 0$

51. Find the general form of a function whose second derivative is $\sqrt{x}$. [*Hint:* Solve the equation $f''(x) = \sqrt{x}$ for $f(x)$ by integrating both sides twice.]

52. Find a function f such that $f''(x) = x + \cos x$ and such that $f(0) = 1$ and $f'(0) = 2$. [*Hint:* Integrate both sides of the equation twice.]

53–57 Find an equation of the curve that satisfies the given conditions. ■

53. At each point (x, y) on the curve the slope is $2x + 1$; the curve passes through the point $(-3, 0)$.

54. At each point (x, y) on the curve the slope is $(x + 1)^2$; the curve passes through the point $(-2, 8)$.

55. At each point (x, y) on the curve the slope is $-\sin x$; the curve passes through the point $(0, 2)$.

56. At each point (x, y) on the curve the slope equals the square of the distance between the point and the y-axis; the point $(-1, 2)$ is on the curve.

57. At each point (x, y) on the curve, y satisfies the condition $d^2y/dx^2 = 6x$; the line $y = 5 - 3x$ is tangent to the curve at the point where $x = 1$.

C 58. In each part, use a CAS to solve the initial-value problem.

(a) $\dfrac{dy}{dx} = x^2 \cos 3x,\ y(\pi/2) = -1$

(b) $\dfrac{dy}{dx} = \dfrac{x^3}{(4 + x^2)^{3/2}},\ y(0) = -2$

59. (a) Use a graphing utility to generate a slope field for the differential equation $dy/dx = x$ in the region $-5 \le x \le 5$ and $-5 \le y \le 5$.

(b) Graph some representative integral curves of the function $f(x) = x$.

(c) Find an equation for the integral curve that passes through the point $(2, 1)$.

60. (a) Use a graphing utility to generate a slope field for the differential equation $dy/dx = e^x/2$ in the region $-1 \le x \le 4$ and $-1 \le y \le 4$.

(b) Graph some representative integral curves of the function $f(x) = e^x/2$.

(c) Find an equation for the integral curve that passes through the point $(0, 1)$.

61–64 The given slope field figure corresponds to one of the differential equations below. Identify the differential equation that matches the figure, and sketch solution curves through the highlighted points.

(a) $\dfrac{dy}{dx} = 2$ (b) $\dfrac{dy}{dx} = -x$

(c) $\dfrac{dy}{dx} = x^2 - 4$ (d) $\dfrac{dy}{dx} = e^{x/3}$ ■

61.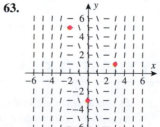

62.

63.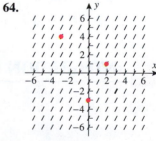

64.

65. Critique the following "proof" that an arbitrary constant must be zero:

$$C = \int 0\, dx = \int 0 \cdot 0\, dx = 0 \int 0\, dx = 0$$

66. Critique the following "proof" that an arbitrary constant must be zero:

$$0 = \left(\int x \, dx \right) - \left(\int x \, dx \right)$$

$$= \int (x - x) \, dx = \int 0 \, dx = C$$

67. (a) Show that

$$F(x) = \tan^{-1} x \quad \text{and} \quad G(x) = -\tan^{-1}(1/x)$$

differ by a constant on the interval $(0, +\infty)$ by showing that they are antiderivatives of the same function.

(b) Find the constant C such that $F(x) - G(x) = C$ by evaluating the functions $F(x)$ and $G(x)$ at a particular value of x.

(c) Check your answer to part (b) by using trigonometric identities.

68. Let F and G be the functions defined by

$$F(x) = \frac{x^2 + 3x}{x} \quad \text{and} \quad G(x) = \begin{cases} x + 3, & x > 0 \\ x, & x < 0 \end{cases}$$

(a) Show that F and G have the same derivative.

(b) Show that $G(x) \neq F(x) + C$ for any constant C.

(c) Do parts (a) and (b) contradict Theorem 5.2.2? Explain.

69–70 Use a trigonometric identity to evaluate the integral. ■

69. $\displaystyle\int \tan^2 x \, dx$ **70.** $\displaystyle\int \cot^2 x \, dx$

71. Use the identities $\cos 2\theta = 1 - 2\sin^2 \theta = 2\cos^2 \theta - 1$ to help evaluate the integrals

(a) $\displaystyle\int \sin^2(x/2) \, dx$ (b) $\displaystyle\int \cos^2(x/2) \, dx$

72. Recall that

$$\frac{d}{dx}[\sec^{-1} x] = \frac{1}{|x|\sqrt{x^2 - 1}}$$

Use this to verify Formula 14 in Table 5.2.1.

73. The speed of sound in air at $0°C$ (or 273 K on the Kelvin scale) is 1087 ft/s, but the speed v increases as the temperature T rises. Experimentation has shown that the rate of change of v with respect to T is

$$\frac{dv}{dT} = \frac{1087}{2\sqrt{273}} T^{-1/2}$$

where v is in feet per second and T is in kelvins (K). Find a formula that expresses v as a function of T.

74. Suppose that a uniform metal rod 50 cm long is insulated laterally, and the temperatures at the exposed ends are maintained at $25°C$ and $85°C$, respectively. Assume that an x-axis is chosen as in the accompanying figure and that the temperature $T(x)$ satisfies the equation

$$\frac{d^2 T}{dx^2} = 0$$

Find $T(x)$ for $0 \leq x \leq 50$.

◀ **Figure Ex-74**

75. **Writing**   What is an *initial-value problem*? Describe the sequence of steps for solving an initial-value problem.

76. **Writing**   What is a *slope field*? How are slope fields and integral curves related?

✔ **QUICK CHECK ANSWERS 5.2**

1. $F'(x) = f(x)$ **2.** (a) $\displaystyle\int \frac{1}{2\sqrt{x}} \, dx = \sqrt{x} + C$ (b) $\displaystyle\int 4e^{4x} \, dx = e^{4x} + C$

3. (a) $\frac{1}{4}x^4 + \frac{1}{2}x^2 + 5x + C$ (b) $\tan x + \csc x + C$ **4.** $2x + 1$; $x^2 + x + 3$ **5.** 0; $-\frac{2}{3}$

5.3   INTEGRATION BY SUBSTITUTION

*In this section we will study a technique, called **substitution**, that can often be used to transform complicated integration problems into simpler ones.*

■ *u*-SUBSTITUTION

The method of substitution can be motivated by examining the chain rule from the viewpoint of antidifferentiation. For this purpose, suppose that F is an antiderivative of f and that g is a differentiable function. The chain rule implies that the derivative of $F(g(x))$ can be expressed as

$$\frac{d}{dx}[F(g(x))] = F'(g(x))g'(x)$$

which we can write in integral form as

$$\int F'(g(x))g'(x)\,dx = F(g(x)) + C \tag{1}$$

or since F is an antiderivative of f,

$$\int f(g(x))g'(x)\,dx = F(g(x)) + C \tag{2}$$

For our purposes it will be useful to let $u = g(x)$ and to write $du/dx = g'(x)$ in the differential form $du = g'(x)\,dx$. With this notation (2) can be expressed as

$$\int f(u)\,du = F(u) + C \tag{3}$$

The process of evaluating an integral of form (2) by converting it into form (3) with the substitution

$$u = g(x) \quad \text{and} \quad du = g'(x)\,dx$$

is called the **method of u-substitution**. Here our emphasis is *not* on the interpretation of the expression $du = g'(x)\,dx$. Rather, the differential notation serves primarily as a useful "bookkeeping" device for the method of u-substitution. The following example illustrates how the method works.

▶ **Example 1** Evaluate $\int (x^2 + 1)^{50} \cdot 2x\,dx$.

Solution. If we let $u = x^2 + 1$, then $du/dx = 2x$, which implies that $du = 2x\,dx$. Thus, the given integral can be written as

$$\int (x^2 + 1)^{50} \cdot 2x\,dx = \int u^{50}\,du = \frac{u^{51}}{51} + C = \frac{(x^2 + 1)^{51}}{51} + C \blacktriangleleft$$

It is important to realize that in the method of u-substitution you have control over the choice of u, but once you make that choice you have no control over the resulting expression for du. Thus, in the last example we *chose* $u = x^2 + 1$ but $du = 2x\,dx$ was *computed*. Fortunately, our choice of u, combined with the computed du, worked out perfectly to produce an integral involving u that was easy to evaluate. However, in general, the method of u-substitution will fail if the chosen u and the computed du cannot be used to produce an integrand in which no expressions involving x remain, or if you cannot evaluate the resulting integral. Thus, for example, the substitution $u = x^2$, $du = 2x\,dx$ will not work for the integral

$$\int 2x \sin x^4\,dx$$

because this substitution results in the integral

$$\int \sin u^2\,du$$

which still cannot be evaluated in terms of familiar functions.

In general, there are no hard and fast rules for choosing u, and in some problems no choice of u will work. In such cases other methods need to be used, some of which will be discussed later. Making appropriate choices for u will come with experience, but you may find the following guidelines, combined with a mastery of the basic integrals in Table 5.2.1, helpful.

Guidelines for u-Substitution

Step 1. Look for some composition $f(g(x))$ within the integrand for which the substitution

$$u = g(x), \quad du = g'(x)\, dx$$

produces an integral that is expressed entirely in terms of u and its differential du. This may or may not be possible.

Step 2. If you are successful in Step 1, then try to evaluate the resulting integral in terms of u. Again, this may or may not be possible.

Step 3. If you are successful in Step 2, then replace u by $g(x)$ to express your final answer in terms of x.

■ EASY TO RECOGNIZE SUBSTITUTIONS

The easiest substitutions occur when the integrand is the derivative of a known function, except for a constant added to or subtracted from the independent variable.

▶ **Example 2**

$$\int \sin(x+9)\, dx = \int \sin u\, du = -\cos u + C = -\cos(x+9) + C$$

$$\boxed{\begin{array}{l} u = x + 9 \\ du = 1 \cdot dx = dx \end{array}}$$

$$\int (x-8)^{23}\, dx = \int u^{23}\, du = \frac{u^{24}}{24} + C = \frac{(x-8)^{24}}{24} + C \;\blacktriangleleft$$

$$\boxed{\begin{array}{l} u = x - 8 \\ du = 1 \cdot dx = dx \end{array}}$$

Another easy u-substitution occurs when the integrand is the derivative of a known function, except for a constant that multiplies or divides the independent variable. The following example illustrates two ways to evaluate such integrals.

▶ **Example 3**   Evaluate $\displaystyle\int \cos 5x\, dx$.

Solution.

$$\int \cos 5x\, dx = \int (\cos u) \cdot \frac{1}{5}\, du = \frac{1}{5}\int \cos u\, du = \frac{1}{5}\sin u + C = \frac{1}{5}\sin 5x + C$$

$$\boxed{\begin{array}{l} u = 5x \\ du = 5\, dx \text{ or } dx = \tfrac{1}{5}\, du \end{array}}$$

Alternative Solution.   There is a variation of the preceding method that some people prefer. The substitution $u = 5x$ requires $du = 5\, dx$. If there were a factor of 5 in the integrand, then we could group the 5 and dx together to form the du required by the substitution. Since there is no factor of 5, we will insert one and compensate by putting a factor of $\frac{1}{5}$ in front of the integral. The computations are as follows:

$$\int \cos 5x\, dx = \frac{1}{5}\int \cos 5x \cdot 5\, dx = \frac{1}{5}\int \cos u\, du = \frac{1}{5}\sin u + C = \frac{1}{5}\sin 5x + C \;\blacktriangleleft$$

$$\boxed{\begin{array}{l} u = 5x \\ du = 5\, dx \end{array}}$$

More generally, if the integrand is a composition of the form $f(ax + b)$, where $f(x)$ is an easy to integrate function, then the substitution $u = ax + b$, $du = a\,dx$ will work.

▶ **Example 4**

$$\int \frac{dx}{\left(\frac{1}{3}x - 8\right)^5} = \int \frac{3\,du}{u^5} = 3\int u^{-5}\,du = -\frac{3}{4}u^{-4} + C = -\frac{3}{4}\left(\frac{1}{3}x - 8\right)^{-4} + C \quad ◀$$

$$\boxed{\begin{array}{l} u = \frac{1}{3}x - 8 \\ du = \frac{1}{3}\,dx \text{ or } dx = 3\,du \end{array}}$$

▶ **Example 5** Evaluate $\displaystyle\int \frac{dx}{1 + 3x^2}$.

Solution. Substituting

$$u = \sqrt{3}x, \quad du = \sqrt{3}\,dx$$

yields

$$\int \frac{dx}{1 + 3x^2} = \frac{1}{\sqrt{3}}\int \frac{du}{1 + u^2} = \frac{1}{\sqrt{3}}\tan^{-1} u + C = \frac{1}{\sqrt{3}}\tan^{-1}(\sqrt{3}x) + C \quad ◀$$

With the help of Theorem 5.2.3, a complicated integral can sometimes be computed by expressing it as a sum of simpler integrals.

▶ **Example 6**

$$\int \left(\frac{1}{x} + \sec^2 \pi x\right) dx = \int \frac{dx}{x} + \int \sec^2 \pi x\,dx$$

$$= \ln|x| + \int \sec^2 \pi x\,dx$$

$$= \ln|x| + \frac{1}{\pi}\int \sec^2 u\,du$$

$$\boxed{\begin{array}{l} u = \pi x \\ du = \pi\,dx \text{ or } dx = \frac{1}{\pi}\,du \end{array}}$$

$$= \ln|x| + \frac{1}{\pi}\tan u + C = \ln|x| + \frac{1}{\pi}\tan \pi x + C \quad ◀$$

The next four examples illustrate a substitution $u = g(x)$ where $g(x)$ is a nonlinear function.

▶ **Example 7** Evaluate $\displaystyle\int \sin^2 x \cos x\,dx$.

Solution. If we let $u = \sin x$, then

$$\frac{du}{dx} = \cos x, \quad \text{so} \quad du = \cos x\,dx$$

Thus,

$$\int \sin^2 x \cos x\,dx = \int u^2\,du = \frac{u^3}{3} + C = \frac{\sin^3 x}{3} + C \quad ◀$$

▶ **Example 8**   Evaluate $\displaystyle\int \frac{e^{\sqrt{x}}}{\sqrt{x}}\,dx$.

Solution.   If we let $u = \sqrt{x}$, then

$$\frac{du}{dx} = \frac{1}{2\sqrt{x}}, \quad \text{so} \quad du = \frac{1}{2\sqrt{x}}\,dx \quad \text{or} \quad 2\,du = \frac{1}{\sqrt{x}}\,dx$$

Thus,

$$\int \frac{e^{\sqrt{x}}}{\sqrt{x}}\,dx = \int 2e^u\,du = 2\int e^u\,du = 2e^u + C = 2e^{\sqrt{x}} + C \blacktriangleleft$$

▶ **Example 9**   Evaluate $\displaystyle\int t^4 \sqrt[3]{3 - 5t^5}\,dt$.

Solution.

$$\int t^4 \sqrt[3]{3 - 5t^5}\,dt = -\frac{1}{25}\int \sqrt[3]{u}\,du = -\frac{1}{25}\int u^{1/3}\,du$$

$$\boxed{\begin{array}{l} u = 3 - 5t^5 \\ du = -25t^4\,dt \text{ or } -\frac{1}{25}\,du = t^4\,dt \end{array}}$$

$$= -\frac{1}{25}\frac{u^{4/3}}{4/3} + C = -\frac{3}{100}\left(3 - 5t^5\right)^{4/3} + C \blacktriangleleft$$

▶ **Example 10**   Evaluate $\displaystyle\int \frac{e^x}{\sqrt{1 - e^{2x}}}\,dx$.

Solution.   Substituting

$$u = e^x, \quad du = e^x\,dx$$

yields

$$\int \frac{e^x}{\sqrt{1 - e^{2x}}}\,dx = \int \frac{du}{\sqrt{1 - u^2}} = \sin^{-1}u + C = \sin^{-1}(e^x) + C \blacktriangleleft$$

■ **LESS APPARENT SUBSTITUTIONS**

The method of substitution is relatively straightforward, provided the integrand contains an easily recognized composition $f(g(x))$ and the remainder of the integrand is a constant multiple of $g'(x)$. If this is not the case, the method may still apply but may require more computation.

▶ **Example 11**   Evaluate $\displaystyle\int x^2\sqrt{x - 1}\,dx$.

Solution.   The composition $\sqrt{x - 1}$ suggests the substitution

$$u = x - 1 \quad \text{so that} \quad du = dx \tag{4}$$

From the first equality in (4)

$$x^2 = (u + 1)^2 = u^2 + 2u + 1$$

so that

$$\int x^2\sqrt{x - 1}\,dx = \int (u^2 + 2u + 1)\sqrt{u}\,du = \int (u^{5/2} + 2u^{3/2} + u^{1/2})\,du$$

$$= \tfrac{2}{7}u^{7/2} + \tfrac{4}{5}u^{5/2} + \tfrac{2}{3}u^{3/2} + C$$

$$= \tfrac{2}{7}(x - 1)^{7/2} + \tfrac{4}{5}(x - 1)^{5/2} + \tfrac{2}{3}(x - 1)^{3/2} + C \blacktriangleleft$$

► **Example 12** Evaluate $\int \cos^3 x \, dx$.

Solution. The only compositions in the integrand that suggest themselves are

$$\cos^3 x = (\cos x)^3 \quad \text{and} \quad \cos^2 x = (\cos x)^2$$

However, neither the substitution $u = \cos x$ nor the substitution $u = \cos^2 x$ work (verify). In this case, an appropriate substitution is not suggested by the composition contained in the integrand. On the other hand, note from Equation (2) that the derivative $g'(x)$ appears as a factor in the integrand. This suggests that we write

$$\int \cos^3 x \, dx = \int \cos^2 x \cos x \, dx$$

and solve the equation $du = \cos x \, dx$ for $u = \sin x$. Since $\sin^2 x + \cos^2 x = 1$, we then have

$$\int \cos^3 x \, dx = \int \cos^2 x \cos x \, dx = \int (1 - \sin^2 x) \cos x \, dx = \int (1 - u^2) \, du$$

$$= u - \frac{u^3}{3} + C = \sin x - \frac{1}{3} \sin^3 x + C \quad ◄$$

► **Example 13** Evaluate $\int \dfrac{dx}{a^2 + x^2} \, dx$, where $a \neq 0$ is a constant.

Solution. Some simple algebra and an appropriate u-substitution will allow us to use Formula 12 in Table 5.2.1.

$$\int \frac{dx}{a^2 + x^2} = \int \frac{a(dx/a)}{a^2(1 + (x/a)^2)} = \frac{1}{a} \int \frac{dx/a}{1 + (x/a)^2} \qquad \boxed{\begin{array}{l} u = x/a \\ du = dx/a \end{array}}$$

$$= \frac{1}{a} \int \frac{du}{1 + u^2} = \frac{1}{a} \tan^{-1} u + C = \frac{1}{a} \tan^{-1} \frac{x}{a} + C \quad ◄$$

The method of Example 13 leads to the following generalizations of Formulas 12, 13, and 14 in Table 5.2.1 for $a > 0$:

$$\int \frac{du}{a^2 + u^2} = \frac{1}{a} \tan^{-1} \frac{u}{a} + C \tag{5}$$

$$\int \frac{du}{\sqrt{a^2 - u^2}} = \sin^{-1} \frac{u}{a} + C \tag{6}$$

$$\int \frac{du}{u\sqrt{u^2 - a^2}} = \frac{1}{a} \sec^{-1} \left| \frac{u}{a} \right| + C \tag{7}$$

► **Example 14** Evaluate $\int \dfrac{dx}{\sqrt{2 - x^2}}$.

Solution. Applying (6) with $u = x$ and $a = \sqrt{2}$ yields

$$\int \frac{dx}{\sqrt{2 - x^2}} = \sin^{-1} \frac{x}{\sqrt{2}} + C \quad ◄$$

■ INTEGRATION USING COMPUTER ALGEBRA SYSTEMS

The advent of computer algebra systems has made it possible to evaluate many kinds of integrals that would be laborious to evaluate by hand. For example, a handheld calculator evaluated the integral

$$\int \frac{5x^2}{(1+x)^{1/3}}\,dx = \frac{3(x+1)^{2/3}(5x^2 - 6x + 9)}{8} + C$$

in about a second. The computer algebra system *Mathematica*, running on a personal computer, required even less time to evaluate this same integral. However, just as one would not want to rely on a calculator to compute $2 + 2$, so one would not want to use a CAS to integrate a simple function such as $f(x) = x^2$. Thus, even if you have a CAS, you will want to develop a reasonable level of competence in evaluating basic integrals. Moreover, the mathematical techniques that we will introduce for evaluating basic integrals are precisely the techniques that computer algebra systems use to evaluate more complicated integrals.

✔ QUICK CHECK EXERCISES 5.3   *(See page 340 for answers.)*

1. Indicate the u-substitution.

   (a) $\displaystyle\int 3x^2(1+x^3)^{25}\,dx = \int u^{25}\,du$ if $u = $ _____ and $du = $ _____.

   (b) $\displaystyle\int 2x \sin x^2\,dx = \int \sin u\,du$ if $u = $ _____ and $du = $ _____.

   (c) $\displaystyle\int \frac{18x}{1+9x^2}\,dx = \int \frac{1}{u}\,du$ if $u = $ _____ and $du = $ _____.

   (d) $\displaystyle\int \frac{3}{1+9x^2}\,dx = \int \frac{1}{1+u^2}\,du$ if $u = $ _____ and $du = $ _____.

2. Supply the missing integrand corresponding to the indicated u-substitution.

   (a) $\displaystyle\int 5(5x - 3)^{-1/3}\,dx = \int $ _____ du; $u = 5x - 3$

   (b) $\displaystyle\int (3 - \tan x) \sec^2 x\,dx = \int $ _____ du; $u = 3 - \tan x$

   (c) $\displaystyle\int \frac{\sqrt[3]{8+\sqrt{x}}}{\sqrt{x}}\,dx = \int $ _____ du; $u = 8 + \sqrt{x}$

   (d) $\displaystyle\int e^{3x}\,dx = \int $ _____ du; $u = 3x$

EXERCISE SET 5.3   ~ Graphing Utility   [c] CAS

1–12 Evaluate the integrals using the indicated substitutions.

1. (a) $\displaystyle\int 2x(x^2 + 1)^{23}\,dx$; $u = x^2 + 1$

   (b) $\displaystyle\int \cos^3 x \sin x\,dx$; $u = \cos x$

2. (a) $\displaystyle\int \frac{1}{\sqrt{x}} \sin \sqrt{x}\,dx$; $u = \sqrt{x}$

   (b) $\displaystyle\int \frac{3x\,dx}{\sqrt{4x^2 + 5}}$; $u = 4x^2 + 5$

3. (a) $\displaystyle\int \sec^2(4x + 1)\,dx$; $u = 4x + 1$

   (b) $\displaystyle\int y\sqrt{1 + 2y^2}\,dy$; $u = 1 + 2y^2$

4. (a) $\displaystyle\int \sqrt{\sin \pi\theta} \cos \pi\theta\,d\theta$; $u = \sin \pi\theta$

   (b) $\displaystyle\int (2x + 7)(x^2 + 7x + 3)^{4/5}\,dx$; $u = x^2 + 7x + 3$

5. (a) $\displaystyle\int \cot x \csc^2 x\,dx$; $u = \cot x$

   (b) $\displaystyle\int (1 + \sin t)^9 \cos t\,dt$; $u = 1 + \sin t$

6. (a) $\displaystyle\int \cos 2x\,dx$; $u = 2x$ (b) $\displaystyle\int x \sec^2 x^2\,dx$; $u = x^2$

7. (a) $\displaystyle\int x^2\sqrt{1 + x}\,dx$; $u = 1 + x$

   (b) $\displaystyle\int [\csc(\sin x)]^2 \cos x\,dx$; $u = \sin x$

8. (a) $\displaystyle\int \sin(x - \pi)\,dx$; $u = x - \pi$

   (b) $\displaystyle\int \frac{5x^4}{(x^5 + 1)^2}\,dx$; $u = x^5 + 1$

9. (a) $\displaystyle\int \frac{dx}{x \ln x}$; $u = \ln x$

   (b) $\displaystyle\int e^{-5x}\,dx$; $u = -5x$

10. (a) $\displaystyle\int \frac{\sin 3\theta}{1 + \cos 3\theta}\, d\theta;\ u = 1 + \cos 3\theta$

(b) $\displaystyle\int \frac{e^x}{1 + e^x}\, dx;\ u = 1 + e^x$

11. (a) $\displaystyle\int \frac{x^2\, dx}{1 + x^6};\ u = x^3$

(b) $\displaystyle\int \frac{dx}{x\sqrt{1 - (\ln x)^2}};\ u = \ln x$

12. (a) $\displaystyle\int \frac{dx}{x\sqrt{9x^2 - 1}};\ u = 3x$

(b) $\displaystyle\int \frac{dx}{\sqrt{x}(1 + x)};\ u = \sqrt{x}$

FOCUS ON CONCEPTS

13. Explain the connection between the chain rule for differentiation and the method of u-substitution for integration.

14. Explain how the substitution $u = ax + b$ helps to perform an integration in which the integrand is $f(ax + b)$, where $f(x)$ is an easy to integrate function.

15–56 Evaluate the integrals using appropriate substitutions. ■

15. $\displaystyle\int (4x - 3)^9\, dx$

16. $\displaystyle\int x^3\sqrt{5 + x^4}\, dx$

17. $\displaystyle\int \sin 7x\, dx$

18. $\displaystyle\int \cos \frac{x}{3}\, dx$

19. $\displaystyle\int \sec 4x \tan 4x\, dx$

20. $\displaystyle\int \sec^2 5x\, dx$

21. $\displaystyle\int e^{2x}\, dx$

22. $\displaystyle\int \frac{dx}{2x}$

23. $\displaystyle\int \frac{dx}{\sqrt{1 - 4x^2}}$

24. $\displaystyle\int \frac{dx}{1 + 16x^2}$

25. $\displaystyle\int t\sqrt{7t^2 + 12}\, dt$

26. $\displaystyle\int \frac{x}{\sqrt{4 - 5x^2}}\, dx$

27. $\displaystyle\int \frac{6}{(1 - 2x)^3}\, dx$

28. $\displaystyle\int \frac{x^2 + 1}{\sqrt{x^3 + 3x}}\, dx$

29. $\displaystyle\int \frac{x^3}{(5x^4 + 2)^3}\, dx$

30. $\displaystyle\int \frac{\sin(1/x)}{3x^2}\, dx$

31. $\displaystyle\int e^{\sin x} \cos x\, dx$

32. $\displaystyle\int x^3 e^{x^4}\, dx$

33. $\displaystyle\int x^2 e^{-2x^3}\, dx$

34. $\displaystyle\int \frac{e^x + e^{-x}}{e^x - e^{-x}}\, dx$

35. $\displaystyle\int \frac{e^x}{1 + e^{2x}}\, dx$

36. $\displaystyle\int \frac{t}{t^4 + 1}\, dt$

37. $\displaystyle\int \frac{\sin(5/x)}{x^2}\, dx$

38. $\displaystyle\int \frac{\sec^2(\sqrt{x})}{\sqrt{x}}\, dx$

39. $\displaystyle\int \cos^4 3t \sin 3t\, dt$

40. $\displaystyle\int \cos 2t \sin^5 2t\, dt$

41. $\displaystyle\int x \sec^2(x^2)\, dx$

42. $\displaystyle\int \frac{\cos 4\theta}{(1 + 2\sin 4\theta)^4}\, d\theta$

43. $\displaystyle\int \cos 4\theta\sqrt{2 - \sin 4\theta}\, d\theta$

44. $\displaystyle\int \tan^3 5x \sec^2 5x\, dx$

45. $\displaystyle\int \frac{\sec^2 x\, dx}{\sqrt{1 - \tan^2 x}}$

46. $\displaystyle\int \frac{\sin\theta}{\cos^2\theta + 1}\, d\theta$

47. $\displaystyle\int \sec^3 2x \tan 2x\, dx$

48. $\displaystyle\int [\sin(\sin\theta)] \cos\theta\, d\theta$

49. $\displaystyle\int \frac{dx}{e^x}$

50. $\displaystyle\int \sqrt{e^x}\, dx$

51. $\displaystyle\int \frac{dx}{\sqrt{x}\, e^{(2\sqrt{x})}}$

52. $\displaystyle\int \frac{e^{\sqrt{2y+1}}}{\sqrt{2y + 1}}\, dy$

53. $\displaystyle\int \frac{y}{\sqrt{2y + 1}}\, dy$

54. $\displaystyle\int x\sqrt{4 - x}\, dx$

55. $\displaystyle\int \sin^3 2\theta\, d\theta$

56. $\displaystyle\int \sec^4 3\theta\, d\theta$ [*Hint:* Apply a trigonometric identity.]

57–60 Evaluate each integral by first modifying the form of the integrand and then making an appropriate substitution, if needed. ■

57. $\displaystyle\int \frac{t + 1}{t}\, dt$

58. $\displaystyle\int e^{2\ln x}\, dx$

59. $\displaystyle\int [\ln(e^x) + \ln(e^{-x})]\, dx$

60. $\displaystyle\int \cot x\, dx$

61–62 Evaluate the integrals with the aid of Formulas (5), (6), and (7). ■

61. (a) $\displaystyle\int \frac{dx}{\sqrt{9 - x^2}}$ (b) $\displaystyle\int \frac{dx}{5 + x^2}$ (c) $\displaystyle\int \frac{dx}{x\sqrt{x^2 - \pi}}$

62. (a) $\displaystyle\int \frac{e^x}{4 + e^{2x}}\, dx$ (b) $\displaystyle\int \frac{dx}{\sqrt{9 - 4x^2}}$ (c) $\displaystyle\int \frac{dy}{y\sqrt{5y^2 - 3}}$

63–65 Evaluate the integrals assuming that n is a positive integer and $b \neq 0$. ■

63. $\displaystyle\int (a + bx)^n\, dx$

64. $\displaystyle\int \sqrt[n]{a + bx}\, dx$

65. $\displaystyle\int \sin^n(a + bx)\cos(a + bx)\, dx$

C **66.** Use a CAS to check the answers you obtained in Exercises 63–65. If the answer produced by the CAS does not match yours, show that the two answers are equivalent. [*Suggestion: Mathematica* users may find it helpful to apply the Simplify command to the answer.]

FOCUS ON CONCEPTS

67. (a) Evaluate the integral $\int \sin x \cos x\, dx$ by two methods: first by letting $u = \sin x$, and then by letting $u = \cos x$.

(b) Explain why the two apparently different answers obtained in part (a) are really equivalent.

68. (a) Evaluate the integral $\int (5x-1)^2\, dx$ by two methods: first square and integrate, then let $u = 5x - 1$.
(b) Explain why the two apparently different answers obtained in part (a) are really equivalent.

69–72 Solve the initial-value problems. ■

69. $\dfrac{dy}{dx} = \sqrt{5x+1}, \quad y(3) = -2$

70. $\dfrac{dy}{dx} = 2 + \sin 3x, \quad y(\pi/3) = 0$

71. $\dfrac{dy}{dt} = -e^{2t}, \quad y(0) = 6$

72. $\dfrac{dy}{dt} = \dfrac{1}{25 + 9t^2}, \quad y\left(-\dfrac{5}{3}\right) = \dfrac{\pi}{30}$

73. (a) Evaluate $\int [x/\sqrt{x^2+1}]\, dx$.
(b) Use a graphing utility to generate some typical integral curves of $f(x) = x/\sqrt{x^2+1}$ over the interval $(-5, 5)$.

74. (a) Evaluate $\int [x/(x^2+1)]\, dx$.
(b) Use a graphing utility to generate some typical integral curves of $f(x) = x/(x^2+1)$ over the interval $(-5, 5)$.

75. Find a function f such that the slope of the tangent line at a point (x, y) on the curve $y = f(x)$ is $\sqrt{3x+1}$ and the curve passes through the point $(0, 1)$.

76. A population of minnows in a lake is estimated to be 100,000 at the beginning of the year 2005. Suppose that t years after the beginning of 2005 the rate of growth of the population $p(t)$ (in thousands) is given by $p'(t) = (3 + 0.12t)^{3/2}$. Estimate the projected population at the beginning of the year 2010.

77. Derive integration Formula (6).

78. Derive integration Formula (7).

79. Writing If you want to evaluate an integral by u-substitution, how do you decide what part of the integrand to choose for u?

80. Writing The evaluation of an integral can sometimes result in apparently different answers (Exercises 67 and 68). Explain why this occurs and give an example. How might you show that two apparently different answers are actually equivalent?

✔**QUICK CHECK ANSWERS 5.3**

1. (a) $1 + x^3$; $3x^2\, dx$ (b) x^2; $2x\, dx$ (c) $1 + 9x^2$; $18x\, dx$ (d) $3x$; $3\, dx$ **2.** (a) $u^{-1/3}$ (b) $-u$ (c) $2\sqrt[3]{u}$ (d) $\frac{1}{3}e^u$

5.4 THE DEFINITION OF AREA AS A LIMIT; SIGMA NOTATION

Our main goal in this section is to use the rectangle method to give a precise mathematical definition of the "area under a curve."

■ SIGMA NOTATION

To simplify our computations, we will begin by discussing a useful notation for expressing lengthy sums in a compact form. This notation is called *sigma notation* or *summation notation* because it uses the uppercase Greek letter Σ (sigma) to denote various kinds of sums. To illustrate how this notation works, consider the sum

$$1^2 + 2^2 + 3^2 + 4^2 + 5^2$$

in which each term is of the form k^2, where k is one of the integers from 1 to 5. In sigma notation this sum can be written as

$$\sum_{k=1}^{5} k^2$$

which is read "the summation of k^2, where k runs from 1 to 5." The notation tells us to form the sum of the terms that result when we substitute successive integers for k in the expression k^2, starting with $k = 1$ and ending with $k = 5$.

More generally, if $f(k)$ is a function of k, and if m and n are integers such that $m \leq n$, then

$$\sum_{k=m}^{n} f(k) \tag{1}$$

denotes the sum of the terms that result when we substitute successive integers for k, starting with $k = m$ and ending with $k = n$ (Figure 5.4.1).

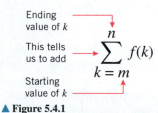

Ending value of k

This tells us to add $\quad \displaystyle\sum_{k=m}^{n} f(k)$

Starting value of k

▲ **Figure 5.4.1**

▶ **Example 1**

$$\sum_{k=4}^{8} k^3 = 4^3 + 5^3 + 6^3 + 7^3 + 8^3$$

$$\sum_{k=1}^{5} 2k = 2 \cdot 1 + 2 \cdot 2 + 2 \cdot 3 + 2 \cdot 4 + 2 \cdot 5 = 2 + 4 + 6 + 8 + 10$$

$$\sum_{k=0}^{5} (2k + 1) = 1 + 3 + 5 + 7 + 9 + 11$$

$$\sum_{k=0}^{5} (-1)^k (2k + 1) = 1 - 3 + 5 - 7 + 9 - 11$$

$$\sum_{k=-3}^{1} k^3 = (-3)^3 + (-2)^3 + (-1)^3 + 0^3 + 1^3 = -27 - 8 - 1 + 0 + 1$$

$$\sum_{k=1}^{3} k \sin\left(\frac{k\pi}{5}\right) = \sin\frac{\pi}{5} + 2\sin\frac{2\pi}{5} + 3\sin\frac{3\pi}{5} \quad \blacktriangleleft$$

The numbers m and n in (1) are called, respectively, the **lower** and **upper limits of summation**; and the letter k is called the **index of summation**. It is not essential to use k as the index of summation; any letter not reserved for another purpose will do. For example,

$$\sum_{i=1}^{6} \frac{1}{i}, \quad \sum_{j=1}^{6} \frac{1}{j}, \quad \text{and} \quad \sum_{n=1}^{6} \frac{1}{n}$$

all denote the sum

$$1 + \frac{1}{2} + \frac{1}{3} + \frac{1}{4} + \frac{1}{5} + \frac{1}{6}$$

If the upper and lower limits of summation are the same, then the "sum" in (1) reduces to a single term. For example,

$$\sum_{k=2}^{2} k^3 = 2^3 \quad \text{and} \quad \sum_{i=1}^{1} \frac{1}{i+2} = \frac{1}{1+2} = \frac{1}{3}$$

In the sums

$$\sum_{i=1}^{5} 2 \quad \text{and} \quad \sum_{j=0}^{2} x^3$$

the expression to the right of the Σ sign does not involve the index of summation. In such cases, we take all the terms in the sum to be the same, with one term for each allowable value of the summation index. Thus,

$$\sum_{i=1}^{5} 2 = 2 + 2 + 2 + 2 + 2 \quad \text{and} \quad \sum_{j=0}^{2} x^3 = x^3 + x^3 + x^3$$

■ **CHANGING THE LIMITS OF SUMMATION**

A sum can be written in more than one way using sigma notation with different limits of summation and correspondingly different summands. For example,

$$\sum_{i=1}^{5} 2i = 2 + 4 + 6 + 8 + 10 = \sum_{j=0}^{4} (2j + 2) = \sum_{k=3}^{7} (2k - 4)$$

On occasion we will want to change the sigma notation for a given sum to a sigma notation with different limits of summation.

■ PROPERTIES OF SUMS

When stating general properties of sums it is often convenient to use a subscripted letter such as a_k in place of the function notation $f(k)$. For example,

$$\sum_{k=1}^{5} a_k = a_1 + a_2 + a_3 + a_4 + a_5 = \sum_{j=1}^{5} a_j = \sum_{k=-1}^{3} a_{k+2}$$

$$\sum_{k=1}^{n} a_k = a_1 + a_2 + \cdots + a_n = \sum_{j=1}^{n} a_j = \sum_{k=-1}^{n-2} a_{k+2}$$

Our first properties provide some basic rules for manipulating sums.

5.4.1 THEOREM

(a) $\displaystyle\sum_{k=1}^{n} ca_k = c \sum_{k=1}^{n} a_k$ (*if c does not depend on k*)

(b) $\displaystyle\sum_{k=1}^{n} (a_k + b_k) = \sum_{k=1}^{n} a_k + \sum_{k=1}^{n} b_k$

(c) $\displaystyle\sum_{k=1}^{n} (a_k - b_k) = \sum_{k=1}^{n} a_k - \sum_{k=1}^{n} b_k$

We will prove parts (a) and (b) and leave part (c) as an exercise.

PROOF (a)

$$\sum_{k=1}^{n} ca_k = ca_1 + ca_2 + \cdots + ca_n = c(a_1 + a_2 + \cdots + a_n) = c \sum_{k=1}^{n} a_k$$

PROOF (b)

$$\sum_{k=1}^{n} (a_k + b_k) = (a_1 + b_1) + (a_2 + b_2) + \cdots + (a_n + b_n)$$

$$= (a_1 + a_2 + \cdots + a_n) + (b_1 + b_2 + \cdots + b_n) = \sum_{k=1}^{n} a_k + \sum_{k=1}^{n} b_k \quad ■$$

Restating Theorem 5.4.1 in words:

(a) *A constant factor can be moved through a sigma sign.*

(b) *Sigma distributes across sums.*

(c) *Sigma distributes across differences.*

■ SUMMATION FORMULAS

The following theorem lists some useful formulas for sums of powers of integers. The derivations of these formulas are given in Appendix D.

5.4.2   THEOREM

(a) $\displaystyle\sum_{k=1}^{n} k = 1 + 2 + \cdots + n = \dfrac{n(n+1)}{2}$

(b) $\displaystyle\sum_{k=1}^{n} k^2 = 1^2 + 2^2 + \cdots + n^2 = \dfrac{n(n+1)(2n+1)}{6}$

(c) $\displaystyle\sum_{k=1}^{n} k^3 = 1^3 + 2^3 + \cdots + n^3 = \left[\dfrac{n(n+1)}{2}\right]^2$

► **Example 2**   Evaluate $\displaystyle\sum_{k=1}^{30} k(k+1)$.

Solution.

$$\sum_{k=1}^{30} k(k+1) = \sum_{k=1}^{30} (k^2 + k) = \sum_{k=1}^{30} k^2 + \sum_{k=1}^{30} k$$

$$= \frac{30(31)(61)}{6} + \frac{30(31)}{2} = 9920 \qquad \boxed{\text{Theorem 5.4.2}(a),\,(b)} \quad \blacktriangleleft$$

In formulas such as

$$\sum_{k=1}^{n} k = \frac{n(n+1)}{2} \quad \text{or} \quad 1 + 2 + \cdots + n = \frac{n(n+1)}{2}$$

the left side of the equality is said to express the sum in **open form** and the right side is said to express it in **closed form**. The open form indicates the summands and the closed form is an explicit formula for the sum.

► **Example 3**   Express $\displaystyle\sum_{k=1}^{n} (3+k)^2$ in closed form.

Solution.

$$\sum_{k=1}^{n} (3+k)^2 = 4^2 + 5^2 + \cdots + (3+n)^2$$

$$= [1^2 + 2^2 + 3^3 + 4^2 + 5^2 + \cdots + (3+n)^2] - [1^2 + 2^2 + 3^2]$$

$$= \left(\sum_{k=1}^{3+n} k^2\right) - 14$$

$$= \frac{(3+n)(4+n)(7+2n)}{6} - 14 = \frac{1}{6}(73n + 21n^2 + 2n^3) \quad \blacktriangleleft$$

■ A DEFINITION OF AREA

We now turn to the problem of giving a precise definition of what is meant by the "area under a curve." Specifically, suppose that the function f is continuous and nonnegative on the interval $[a, b]$, and let R denote the region bounded below by the x-axis, bounded on the sides by the vertical lines $x = a$ and $x = b$, and bounded above by the curve $y = f(x)$ (Figure 5.4.2). Using the rectangle method of Section 5.1, we can motivate a definition for the area of R as follows:

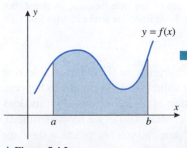

▲ **Figure 5.4.2**

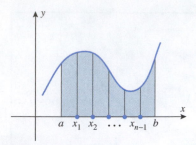

▲ **Figure 5.4.3**

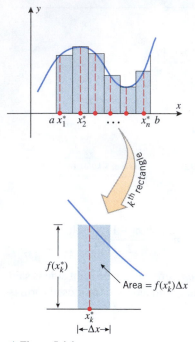

▲ **Figure 5.4.4**

The limit in (2) is interpreted to mean that given any number $\epsilon > 0$ the inequality

$$\left| A - \sum_{k=1}^{n} f(x_k^*)\Delta x \right| < \epsilon$$

holds when n is sufficiently large, no matter how the points x_k^* are selected.

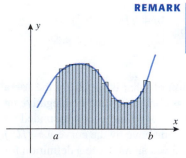

▲ **Figure 5.4.5** $\text{area}(R_n) \approx \text{area}(R)$

- Divide the interval $[a, b]$ into n equal subintervals by inserting $n - 1$ equally spaced points between a and b, and denote those points by

$$x_1, x_2, \ldots, x_{n-1}$$

(Figure 5.4.3). Each of these subintervals has width $(b - a)/n$, which is customarily denoted by

$$\Delta x = \frac{b - a}{n}$$

- Over each subinterval construct a rectangle whose height is the value of f at an arbitrarily selected point in the subinterval. Thus, if

$$x_1^*, x_2^*, \ldots, x_n^*$$

denote the points selected in the subintervals, then the rectangles will have heights $f(x_1^*), f(x_2^*), \ldots, f(x_n^*)$ and areas

$$f(x_1^*)\Delta x, \quad f(x_2^*)\Delta x, \ldots, \quad f(x_n^*)\Delta x$$

(Figure 5.4.4).

- The union of the rectangles forms a region R_n whose area can be regarded as an approximation to the area A of the region R; that is,

$$A = \text{area}(R) \approx \text{area}(R_n) = f(x_1^*)\Delta x + f(x_2^*)\Delta x + \cdots + f(x_n^*)\Delta x$$

(Figure 5.4.5). This can be expressed more compactly in sigma notation as

$$A \approx \sum_{k=1}^{n} f(x_k^*)\Delta x$$

- Repeat the process using more and more subdivisions, and define the area of R to be the "limit" of the areas of the approximating regions R_n as n increases without bound. That is, we define the area A as

$$A = \lim_{n \to +\infty} \sum_{k=1}^{n} f(x_k^*)\Delta x$$

In summary, we make the following definition.

5.4.3   **DEFINITION** (*Area Under a Curve*)   If the function f is continuous on $[a, b]$ and if $f(x) \geq 0$ for all x in $[a, b]$, then the **area** A under the curve $y = f(x)$ over the interval $[a, b]$ is defined by

$$A = \lim_{n \to +\infty} \sum_{k=1}^{n} f(x_k^*)\Delta x \qquad (2)$$

REMARK | There is a difference in interpretation between $\lim_{n \to +\infty}$ and $\lim_{x \to +\infty}$, where n represents a positive integer and x represents a real number. Later we will study limits of the type $\lim_{n \to +\infty}$ in detail, but for now suffice it to say that the computational techniques we have used for limits of type $\lim_{x \to +\infty}$ will also work for $\lim_{n \to +\infty}$.

The values of $x_1^*, x_2^*, \ldots, x_n^*$ in (2) can be chosen arbitrarily, so it is conceivable that different choices of these values might produce different values of A. Were this to happen, then Definition 5.4.3 would not be an acceptable definition of area. Fortunately, this does not happen; it is proved in advanced courses that if f is continuous (as we have assumed), then the same value of A results no matter how the x_k^* are chosen. In practice they are chosen in some systematic fashion, some common choices being

- the left endpoint of each subinterval
- the right endpoint of each subinterval
- the midpoint of each subinterval

To be more specific, suppose that the interval $[a, b]$ is divided into n equal parts of length $\Delta x = (b - a)/n$ by the points $x_1, x_2, \ldots, x_{n-1}$, and let $x_0 = a$ and $x_n = b$ (Figure 5.4.6). Then,

$$x_k = a + k\Delta x \quad \text{for } k = 0, 1, 2, \ldots, n$$

Thus, the left endpoint, right endpoint, and midpoint choices for $x_1^*, x_2^*, \ldots, x_n^*$ are given by

$$x_k^* = x_{k-1} = a + (k - 1)\Delta x \qquad \boxed{\text{Left endpoint}} \tag{3}$$

$$x_k^* = x_k = a + k\Delta x \qquad \boxed{\text{Right endpoint}} \tag{4}$$

$$x_k^* = \tfrac{1}{2}(x_{k-1} + x_k) = a + \left(k - \tfrac{1}{2}\right)\Delta x \qquad \boxed{\text{Midpoint}} \tag{5}$$

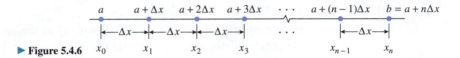

▶ **Figure 5.4.6**

When applicable, the antiderivative method will be the method of choice for finding exact areas. However, the following examples will help to reinforce the ideas that we have just discussed.

▶ **Example 4**   Use Definition 5.4.3 with x_k^* as the right endpoint of each subinterval to find the area between the graph of $f(x) = x^2$ and the interval $[0, 1]$.

Solution.   The length of each subinterval is

$$\Delta x = \frac{b - a}{n} = \frac{1 - 0}{n} = \frac{1}{n}$$

so it follows from (4) that

$$x_k^* = a + k\Delta x = \frac{k}{n}$$

Thus,

$$\sum_{k=1}^{n} f(x_k^*)\Delta x = \sum_{k=1}^{n}(x_k^*)^2\Delta x = \sum_{k=1}^{n}\left(\frac{k}{n}\right)^2 \frac{1}{n} = \frac{1}{n^3}\sum_{k=1}^{n}k^2$$

$$= \frac{1}{n^3}\left[\frac{n(n+1)(2n+1)}{6}\right] \qquad \boxed{\text{Part } (b) \text{ of Theorem 5.4.2}}$$

$$= \frac{1}{6}\left(\frac{n}{n} \cdot \frac{n+1}{n} \cdot \frac{2n+1}{n}\right) = \frac{1}{6}\left(1 + \frac{1}{n}\right)\left(2 + \frac{1}{n}\right)$$

from which it follows that

$$A = \lim_{n \to +\infty}\sum_{k=1}^{n} f(x_k^*)\Delta x = \lim_{n \to +\infty}\left[\frac{1}{6}\left(1 + \frac{1}{n}\right)\left(2 + \frac{1}{n}\right)\right] = \frac{1}{3}$$

Observe that this is consistent with the results in Table 5.1.2 and the related discussion in Section 5.1. ◀

In the solution to Example 4 we made use of one of the "closed form" summation formulas from Theorem 5.4.2. The next result collects some consequences of Theorem 5.4.2 that can facilitate computations of area using Definition 5.4.3.

What pattern is revealed by parts (b)–(d) of Theorem 5.4.4? Does part (a) also fit this pattern? What would you conjecture to be the value of

$$\lim_{n \to +\infty} \frac{1}{n^m} \sum_{k=1}^{n} k^{m-1}$$

5.4.4   THEOREM

(a) $\displaystyle \lim_{n \to +\infty} \frac{1}{n} \sum_{k=1}^{n} 1 = 1$ (b) $\displaystyle \lim_{n \to +\infty} \frac{1}{n^2} \sum_{k=1}^{n} k = \frac{1}{2}$

(c) $\displaystyle \lim_{n \to +\infty} \frac{1}{n^3} \sum_{k=1}^{n} k^2 = \frac{1}{3}$ (d) $\displaystyle \lim_{n \to +\infty} \frac{1}{n^4} \sum_{k=1}^{n} k^3 = \frac{1}{4}$

The proof of Theorem 5.4.4 is left as an exercise for the reader.

▶ **Example 5**   Use Definition 5.4.3 with x_k^* as the midpoint of each subinterval to find the area under the parabola $y = f(x) = 9 - x^2$ and over the interval $[0, 3]$.

Solution.   Each subinterval has length

$$\Delta x = \frac{b - a}{n} = \frac{3 - 0}{n} = \frac{3}{n}$$

so it follows from (5) that

$$x_k^* = a + \left(k - \frac{1}{2}\right)\Delta x = \left(k - \frac{1}{2}\right)\left(\frac{3}{n}\right)$$

Thus,

$$f(x_k^*)\Delta x = [9 - (x_k^*)^2]\Delta x = \left[9 - \left(k - \frac{1}{2}\right)^2 \left(\frac{3}{n}\right)^2\right]\left(\frac{3}{n}\right)$$

$$= \left[9 - \left(k^2 - k + \frac{1}{4}\right)\left(\frac{9}{n^2}\right)\right]\left(\frac{3}{n}\right)$$

$$= \frac{27}{n} - \frac{27}{n^3}k^2 + \frac{27}{n^3}k - \frac{27}{4n^3}$$

from which it follows that

$$A = \lim_{n \to +\infty} \sum_{k=1}^{n} f(x_k^*)\Delta x$$

$$= \lim_{n \to +\infty} \sum_{k=1}^{n} \left(\frac{27}{n} - \frac{27}{n^3}k^2 + \frac{27}{n^3}k - \frac{27}{4n^3}\right)$$

$$= \lim_{n \to +\infty} 27 \left[\frac{1}{n}\sum_{k=1}^{n} 1 - \frac{1}{n^3}\sum_{k=1}^{n} k^2 + \frac{1}{n}\left(\frac{1}{n^2}\sum_{k=1}^{n} k\right) - \frac{1}{4n^2}\left(\frac{1}{n}\sum_{k=1}^{n} 1\right)\right]$$

$$= 27\left[1 - \frac{1}{3} + 0 \cdot \frac{1}{2} - 0 \cdot 1\right] = 18 \quad \boxed{\text{Theorem 5.4.4}} \quad ◀$$

■ NUMERICAL APPROXIMATIONS OF AREA

The antiderivative method discussed in Section 5.1 (and to be studied in more detail later) is an appropriate tool for finding the exact area under a curve when an antiderivative of the integrand can be found. However, if an antiderivative cannot be found, then we must resort to *approximating* the area. Definition 5.4.3 provides a way of doing this. It follows from this definition that if n is large, then

$$\sum_{k=1}^{n} f(x_k^*)\Delta x = \Delta x \sum_{k=1}^{n} f(x_k^*) = \Delta x[f(x_1^*) + f(x_2^*) + \cdots + f(x_n^*)] \tag{6}$$

will be a good approximation to the area A. If one of Formulas (3), (4), or (5) is used to choose the x_k^* in (6), then the result is called the *left endpoint approximation*, the *right endpoint approximation*, or the *midpoint approximation*, respectively (Figure 5.4.7).

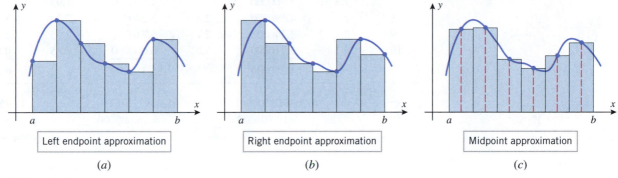

Left endpoint approximation

(a)

Right endpoint approximation

(b)

Midpoint approximation

(c)

▲ **Figure 5.4.7**

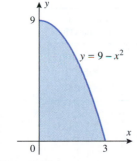

▲ **Figure 5.4.8**

▶ **Example 6** Find the left endpoint, right endpoint, and midpoint approximations of the area under the curve $y = 9 - x^2$ over the interval $[0, 3]$ with $n = 10$, $n = 20$, and $n = 50$ (Figure 5.4.8). Compare the accuracies of these three methods.

Solution. Details of the computations for the case $n = 10$ are shown to six decimal places in Table 5.4.1 and the results of all the computations are given in Table 5.4.2. We showed in Example 5 that the exact area is 18 (i.e., 18 square units), so in this case the midpoint approximation is more accurate than the endpoint approximations. This is also evident geometrically from Figure 5.4.9. You can also see from the figure that in this case the left endpoint approximation overestimates the area and the right endpoint approximation underestimates it. Later in the text we will investigate the error that results when an area is approximated by the midpoint rule. ◀

■ NET SIGNED AREA

In Definition 5.4.3 we assumed that f is continuous and nonnegative on the interval $[a, b]$. If f is continuous and attains both positive and negative values on $[a, b]$, then the limit

$$\lim_{n \to +\infty} \sum_{k=1}^{n} f(x_k^*)\Delta x \tag{7}$$

no longer represents the area between the curve $y = f(x)$ and the interval $[a, b]$ on the x-axis; rather, it represents a difference of areas—the area of the region that is above the interval $[a, b]$ and below the curve $y = f(x)$ minus the area of the region that is below the interval $[a, b]$ and above the curve $y = f(x)$. We call this the *net signed area*

Table 5.4.1

$n = 10, \Delta x = (b - a)/n = (3 - 0)/10 = 0.3$

	LEFT ENDPOINT APPROXIMATION		RIGHT ENDPOINT APPROXIMATION		MIDPOINT APPROXIMATION	
k	x_k^*	$9 - (x_k^*)^2$	x_k^*	$9 - (x_k^*)^2$	x_k^*	$9 - (x_k^*)^2$
1	0.0	9.000000	0.3	8.910000	0.15	8.977500
2	0.3	8.910000	0.6	8.640000	0.45	8.797500
3	0.6	8.640000	0.9	8.190000	0.75	8.437500
4	0.9	8.190000	1.2	7.560000	1.05	7.897500
5	1.2	7.560000	1.5	6.750000	1.35	7.177500
6	1.5	6.750000	1.8	5.760000	1.65	6.277500
7	1.8	5.760000	2.1	4.590000	1.95	5.197500
8	2.1	4.590000	2.4	3.240000	2.25	3.937500
9	2.4	3.240000	2.7	1.710000	2.55	2.497500
10	2.7	1.710000	3.0	0.000000	2.85	0.877500
		64.350000		55.350000		60.075000

$$\Delta x \sum_{k=1}^{n} f(x_k^*) \qquad \begin{matrix}(0.3)(64.350000)\\= 19.305000\end{matrix} \qquad \begin{matrix}(0.3)(55.350000)\\= 16.605000\end{matrix} \qquad \begin{matrix}(0.3)(60.075000)\\= 18.022500\end{matrix}$$

Table 5.4.2

n	LEFT ENDPOINT APPROXIMATION	RIGHT ENDPOINT APPROXIMATION	MIDPOINT APPROXIMATION
10	19.305000	16.605000	18.022500
20	18.663750	17.313750	18.005625
50	18.268200	17.728200	18.000900

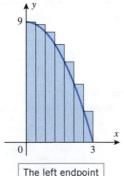

The left endpoint approximation overestimates the area.

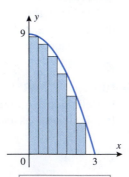

The right endpoint approximation underestimates the area.

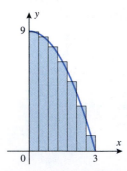

The midpoint approximation is better than the endpoint approximations.

▲ **Figure 5.4.9**

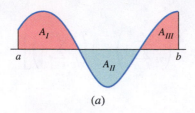

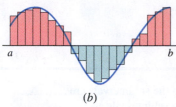

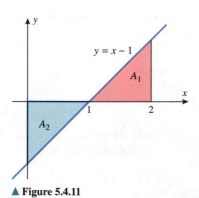

▲ Figure 5.4.10

between the graph of $y = f(x)$ and the interval $[a, b]$. For example, in Figure 5.4.10a, the net signed area between the curve $y = f(x)$ and the interval $[a, b]$ is

$$(A_I + A_{III}) - A_{II} = [\text{area above } [a, b]] - [\text{area below } [a, b]]$$

To explain why the limit in (7) represents this net signed area, let us subdivide the interval $[a, b]$ in Figure 5.4.10a into n equal subintervals and examine the terms in the sum

$$\sum_{k=1}^{n} f(x_k^*)\Delta x \tag{8}$$

If $f(x_k^*)$ is positive, then the product $f(x_k^*)\Delta x$ represents the area of the rectangle with height $f(x_k^*)$ and base Δx (the pink rectangles in Figure 5.4.10b). However, if $f(x_k^*)$ is negative, then the product $f(x_k^*)\Delta x$ is the *negative* of the area of the rectangle with height $|f(x_k^*)|$ and base Δx (the green rectangles in Figure 5.4.10b). Thus, (8) represents the total area of the pink rectangles minus the total area of the green rectangles. As n increases, the pink rectangles fill out the regions with areas A_I and A_{III} and the green rectangles fill out the region with area A_{II}, which explains why the limit in (7) represents the signed area between $y = f(x)$ and the interval $[a, b]$. We formalize this in the following definition.

As with Definition 5.4.3, it can be proved that the limit in (9) always exists and that the same value of A results no matter how the points in the subintervals are chosen.

5.4.5 **DEFINITION** (*Net Signed Area*)   If the function f is continuous on $[a, b]$, then the **net signed area** A between $y = f(x)$ and the interval $[a, b]$ is defined by

$$A = \lim_{n \to +\infty} \sum_{k=1}^{n} f(x_k^*)\Delta x \tag{9}$$

Figure 5.4.11 shows the graph of $f(x) = x - 1$ over the interval $[0, 2]$. It is geometrically evident that the areas A_1 and A_2 in that figure are equal, so we expect the net signed area between the graph of f and the interval $[0, 2]$ to be zero.

▶ **Example 7**   Confirm that the net signed area between the graph of $f(x) = x - 1$ and the interval $[0, 2]$ is zero by using Definition 5.4.5 with x_k^* chosen to be the left endpoint of each subinterval.

Solution.   Each subinterval has length

$$\Delta x = \frac{b - a}{n} = \frac{2 - 0}{n} = \frac{2}{n}$$

so it follows from (3) that

$$x_k^* = a + (k - 1)\Delta x = (k - 1)\left(\frac{2}{n}\right)$$

Thus,

$$\sum_{k=1}^{n} f(x_k^*)\Delta x = \sum_{k=1}^{n} (x_k^* - 1)\Delta x = \sum_{k=1}^{n} \left[(k - 1)\left(\frac{2}{n}\right) - 1\right]\left(\frac{2}{n}\right)$$

$$= \sum_{k=1}^{n} \left[\left(\frac{4}{n^2}\right)k - \frac{4}{n^2} - \frac{2}{n}\right]$$

▲ Figure 5.4.11

(The graph shows $y = x - 1$ with region A_1 above the x-axis and A_2 below, with markings at 1 and 2.)

from which it follows that

$$A = \lim_{n \to +\infty} \sum_{k=1}^{n} f(x_k^*) \Delta x = \lim_{n \to +\infty} \left[4 \left(\frac{1}{n^2} \sum_{k=1}^{n} k \right) - \frac{4}{n} \left(\frac{1}{n} \sum_{k=1}^{n} 1 \right) - 2 \left(\frac{1}{n} \sum_{k=1}^{n} 1 \right) \right]$$

$$= 4 \left(\frac{1}{2} \right) - 0 \cdot 1 - 2 \cdot 1 = 0 \quad \boxed{\text{Theorem 5.4.4}}$$

This confirms that the net signed area is zero. ◄

✔ **QUICK CHECK EXERCISES 5.4** (*See page 352 for answers.*)

1. (a) Write the sum in two ways:

$$\frac{1}{2} + \frac{1}{4} + \frac{1}{6} + \frac{1}{8} = \sum_{k=1}^{4} \underline{\qquad} = \sum_{j=0}^{3} \underline{\qquad}$$

   (b) Express the sum $10 + 10^2 + 10^3 + 10^4 + 10^5$ using sigma notation.

2. Express the sums in closed form.

   (a) $\sum_{k=1}^{n} k$ (b) $\sum_{k=1}^{n} (6k+1)$ (c) $\sum_{k=1}^{n} k^2$

3. Divide the interval $[1, 3]$ into $n = 4$ subintervals of equal length.

   (a) Each subinterval has width _____.

   (b) The left endpoints of the subintervals are _____.
   (c) The midpoints of the subintervals are _____.
   (d) The right endpoints of the subintervals are _____.

4. Find the left endpoint approximation for the area between the curve $y = x^2$ and the interval $[1, 3]$ using $n = 4$ equal subdivisions of the interval.

5. The right endpoint approximation for the net signed area between $y = f(x)$ and an interval $[a, b]$ is given by

$$\sum_{k=1}^{n} \frac{6k+1}{n^2}$$

Find the exact value of this net signed area.

EXERCISE SET 5.4 c CAS

1. Evaluate.

   (a) $\sum_{k=1}^{3} k^3$ (b) $\sum_{j=2}^{6} (3j-1)$ (c) $\sum_{i=-4}^{1} (i^2 - i)$

   (d) $\sum_{n=0}^{5} 1$ (e) $\sum_{k=0}^{4} (-2)^k$ (f) $\sum_{n=1}^{6} \sin n\pi$

2. Evaluate.

   (a) $\sum_{k=1}^{4} k \sin \frac{k\pi}{2}$ (b) $\sum_{j=0}^{5} (-1)^j$ (c) $\sum_{i=7}^{20} \pi^2$

   (d) $\sum_{m=3}^{5} 2^{m+1}$ (e) $\sum_{n=1}^{6} \sqrt{n}$ (f) $\sum_{k=0}^{10} \cos k\pi$

3–8 Write each expression in sigma notation but do not evaluate. ■

3. $1 + 2 + 3 + \cdots + 10$

4. $3 \cdot 1 + 3 \cdot 2 + 3 \cdot 3 + \cdots + 3 \cdot 20$

5. $2 + 4 + 6 + 8 + \cdots + 20$ 6. $1 + 3 + 5 + 7 + \cdots + 15$

7. $1 - 3 + 5 - 7 + 9 - 11$ 8. $1 - \frac{1}{2} + \frac{1}{3} - \frac{1}{4} + \frac{1}{5}$

9. (a) Express the sum of the even integers from 2 to 100 in sigma notation.

   (b) Express the sum of the odd integers from 1 to 99 in sigma notation.

10. Express in sigma notation.

    (a) $a_1 - a_2 + a_3 - a_4 + a_5$
    (b) $-b_0 + b_1 - b_2 + b_3 - b_4 + b_5$
    (c) $a_0 + a_1 x + a_2 x^2 + \cdots + a_n x^n$
    (d) $a^5 + a^4 b + a^3 b^2 + a^2 b^3 + ab^4 + b^5$

11–16 Use Theorem 5.4.2 to evaluate the sums. Check your answers using the summation feature of a calculating utility. ■

11. $\sum_{k=1}^{100} k$ 12. $\sum_{k=1}^{100} (7k+1)$ 13. $\sum_{k=1}^{20} k^2$

14. $\sum_{k=4}^{20} k^2$ 15. $\sum_{k=1}^{30} k(k-2)(k+2)$

16. $\sum_{k=1}^{6} (k - k^3)$

17–20 Express the sums in closed form. ■

17. $\sum_{k=1}^{n} \frac{3k}{n}$ 18. $\sum_{k=1}^{n-1} \frac{k^2}{n}$ 19. $\sum_{k=1}^{n-1} \frac{k^3}{n^2}$

20. $\sum_{k=1}^{n} \left(\frac{5}{n} - \frac{2k}{n} \right)$

21–24 True–False Determine whether the statement is true or false. Explain your answer. ■

21. For all positive integers n
$$1^3 + 2^3 + \cdots + n^3 = (1 + 2 + \cdots + n)^2$$

22. The midpoint approximation is the average of the left endpoint approximation and the right endpoint approximation.

23. Every right endpoint approximation for the area under the graph of $y = x^2$ over an interval $[a, b]$ will be an overestimate.

24. For any continuous function f, the area between the graph of f and an interval $[a, b]$ (on which f is defined) is equal to the absolute value of the net signed area between the graph of f and the interval $[a, b]$.

FOCUS ON CONCEPTS

25. (a) Write the first three and final two summands in the sum
$$\sum_{k=1}^{n} \left(2 + k \cdot \frac{3}{n}\right)^4 \frac{3}{n}$$
Explain why this sum gives the right endpoint approximation for the area under the curve $y = x^4$ over the interval $[2, 5]$.
(b) Show that a change in the index range of the sum in part (a) can produce the left endpoint approximation for the area under the curve $y = x^4$ over the interval $[2, 5]$.

26. For a function f that is continuous on $[a, b]$, Definition 5.4.5 says that the net signed area A between $y = f(x)$ and the interval $[a, b]$ is
$$A = \lim_{n \to +\infty} \sum_{k=1}^{n} f(x_k^*) \Delta x$$
Give geometric interpretations for the symbols n, x_k^*, and Δx. Explain how to interpret the limit in this definition.

27–30 Divide the specified interval into $n = 4$ subintervals of equal length and then compute
$$\sum_{k=1}^{4} f(x_k^*) \Delta x$$
with x_k^* as (a) the left endpoint of each subinterval, (b) the midpoint of each subinterval, and (c) the right endpoint of each subinterval. Illustrate each part with a graph of f that includes the rectangles whose areas are represented in the sum. ■

27. $f(x) = 3x + 1$; $[2, 6]$ **28.** $f(x) = 1/x$; $[1, 9]$

29. $f(x) = \cos x$; $[0, \pi]$ **30.** $f(x) = 2x - x^2$; $[-1, 3]$

C **31–34** Use a calculating utility with summation capabilities or a CAS to obtain an approximate value for the area between the curve $y = f(x)$ and the specified interval with $n = 10, 20$, and 50 subintervals using the (a) left endpoint, (b) midpoint, and (c) right endpoint approximations. ■

31. $f(x) = 1/x$; $[1, 2]$ **32.** $f(x) = 1/x^2$; $[1, 3]$

33. $f(x) = \sqrt{x}$; $[0, 4]$ **34.** $f(x) = \sin x$; $[0, \pi/2]$

35–40 Use Definition 5.4.3 with x_k^* as the *right* endpoint of each subinterval to find the area under the curve $y = f(x)$ over the specified interval. ■

35. $f(x) = x/2$; $[1, 4]$ **36.** $f(x) = 5 - x$; $[0, 5]$

37. $f(x) = 9 - x^2$; $[0, 3]$ **38.** $f(x) = 4 - \frac{1}{4}x^2$; $[0, 3]$

39. $f(x) = x^3$; $[2, 6]$ **40.** $f(x) = 1 - x^3$; $[-3, -1]$

41–44 Use Definition 5.4.3 with x_k^* as the *left* endpoint of each subinterval to find the area under the curve $y = f(x)$ over the specified interval. ■

41. $f(x) = x/2$; $[1, 4]$ **42.** $f(x) = 5 - x$; $[0, 5]$

43. $f(x) = 9 - x^2$; $[0, 3]$ **44.** $f(x) = 4 - \frac{1}{4}x^2$; $[0, 3]$

45–48 Use Definition 5.4.3 with x_k^* as the *midpoint* of each subinterval to find the area under the curve $y = f(x)$ over the specified interval. ■

45. $f(x) = 2x$; $[0, 4]$ **46.** $f(x) = 6 - x$; $[1, 5]$

47. $f(x) = x^2$; $[0, 1]$ **48.** $f(x) = x^2$; $[-1, 1]$

49–52 Use Definition 5.4.5 with x_k^* as the *right* endpoint of each subinterval to find the net signed area between the curve $y = f(x)$ and the specified interval. ■

49. $f(x) = x$; $[-1, 1]$. Verify your answer with a simple geometric argument.

50. $f(x) = x$; $[-1, 2]$. Verify your answer with a simple geometric argument.

51. $f(x) = x^2 - 1$; $[0, 2]$ **52.** $f(x) = x^3$; $[-1, 1]$

53. (a) Show that the area under the graph of $y = x^3$ and over the interval $[0, b]$ is $b^4/4$.
(b) Find a formula for the area under $y = x^3$ over the interval $[a, b]$, where $a \geq 0$.

54. Find the area between the graph of $y = \sqrt{x}$ and the interval $[0, 1]$. [*Hint:* Use the result of Exercise 25 of Section 5.1.]

55. An artist wants to create a rough triangular design using uniform square tiles glued edge to edge. She places n tiles in a row to form the base of the triangle and then makes each successive row two tiles shorter than the preceding row. Find a formula for the number of tiles used in the design. [*Hint:* Your answer will depend on whether n is even or odd.]

56. An artist wants to create a sculpture by gluing together uniform spheres. She creates a rough rectangular base that has 50 spheres along one edge and 30 spheres along the other. She then creates successive layers by gluing spheres in the grooves of the preceding layer. How many spheres will there be in the sculpture?

57–60 Consider the sum

$$\sum_{k=1}^{4}[(k+1)^3 - k^3] = [5^3 - 4^3] + [4^3 - 3^3]$$
$$+ [3^3 - 2^3] + [2^3 - 1^3]$$
$$= 5^3 - 1^3 = 124$$

For convenience, the terms are listed in reverse order. Note how cancellation allows the entire sum to collapse like a telescope. A sum is said to *telescope* when part of each term cancels part of an adjacent term, leaving only portions of the first and last terms uncanceled. Evaluate the telescoping sums in these exercises. ∎

57. $\displaystyle\sum_{k=5}^{17}(3^k - 3^{k-1})$

58. $\displaystyle\sum_{k=1}^{50}\left(\frac{1}{k} - \frac{1}{k+1}\right)$

59. $\displaystyle\sum_{k=2}^{20}\left(\frac{1}{k^2} - \frac{1}{(k-1)^2}\right)$

60. $\displaystyle\sum_{k=1}^{100}(2^{k+1} - 2^k)$

61. (a) Show that

$$\frac{1}{1\cdot 3} + \frac{1}{3\cdot 5} + \cdots + \frac{1}{(2n-1)(2n+1)} = \frac{n}{2n+1}$$

$$\left[Hint: \frac{1}{(2n-1)(2n+1)} = \frac{1}{2}\left(\frac{1}{2n-1} - \frac{1}{2n+1}\right)\right]$$

(b) Use the result in part (a) to find

$$\lim_{n\to +\infty}\sum_{k=1}^{n}\frac{1}{(2k-1)(2k+1)}$$

62. (a) Show that

$$\frac{1}{1\cdot 2} + \frac{1}{2\cdot 3} + \frac{1}{3\cdot 4} + \cdots + \frac{1}{n(n+1)} = \frac{n}{n+1}$$

$$\left[Hint: \frac{1}{n(n+1)} = \frac{1}{n} - \frac{1}{n+1}\right]$$

(b) Use the result in part (a) to find

$$\lim_{n\to +\infty}\sum_{k=1}^{n}\frac{1}{k(k+1)}$$

63. Let $\bar{x}$ denote the arithmetic average of the n numbers $x_1, x_2, \ldots, x_n$. Use Theorem 5.4.1 to prove that

$$\sum_{i=1}^{n}(x_i - \bar{x}) = 0$$

64. Let

$$S = \sum_{k=0}^{n}ar^k$$

Show that $S - rS = a - ar^{n+1}$ and hence that

$$\sum_{k=0}^{n}ar^k = \frac{a - ar^{n+1}}{1-r} \quad (r \neq 1)$$

(A sum of this form is called a *geometric sum*.)

65. By writing out the sums, determine whether the following are valid identities.

(a) $\displaystyle\int\left[\sum_{i=1}^{n}f_i(x)\right]dx = \sum_{i=1}^{n}\left[\int f_i(x)\,dx\right]$

(b) $\displaystyle\frac{d}{dx}\left[\sum_{i=1}^{n}f_i(x)\right] = \sum_{i=1}^{n}\left[\frac{d}{dx}[f_i(x)]\right]$

66. Which of the following are valid identities?

(a) $\displaystyle\sum_{i=1}^{n}a_ib_i = \sum_{i=1}^{n}a_i\sum_{i=1}^{n}b_i$ (b) $\displaystyle\sum_{i=1}^{n}a_i^2 = \left(\sum_{i=1}^{n}a_i\right)^2$

(c) $\displaystyle\sum_{i=1}^{n}\frac{a_i}{b_i} = \frac{\displaystyle\sum_{i=1}^{n}a_i}{\displaystyle\sum_{i=1}^{n}b_i}$ (d) $\displaystyle\sum_{i=1}^{n}a_i = \sum_{j=0}^{n-1}a_{j+1}$

67. Prove part (c) of Theorem 5.4.1.

68. Prove Theorem 5.4.4.

69. **Writing** What is *net signed area*? How does this concept expand our application of the rectangle method?

70. **Writing** Based on Example 6, one might conjecture that the midpoint approximation always provides a better approximation than either endpoint approximation. Discuss the merits of this conjecture.

✔ **QUICK CHECK ANSWERS 5.4**

1. (a) $\dfrac{1}{2k};\ \dfrac{1}{2(j+1)}$ (b) $\displaystyle\sum_{k=1}^{5}10^k$ **2.** (a) $\dfrac{n(n+1)}{2}$ (b) $3n(n+1)+n$ (c) $\dfrac{n(n+1)(2n+1)}{6}$ **3.** (a) 0.5 (b) 1, 1.5, 2, 2.5

(c) 1.25, 1.75, 2.25, 2.75 (d) 1.5, 2, 2.5, 3 **4.** 6.75 **5.** $\displaystyle\lim_{n\to +\infty}\dfrac{3n^2 + 4n}{n^2} = 3$

5.5   THE DEFINITE INTEGRAL

In this section we will introduce the concept of a "definite integral," which will link the concept of area to other important concepts such as length, volume, density, probability, and work.

■ RIEMANN SUMS AND THE DEFINITE INTEGRAL

In our definition of net signed area (Definition 5.4.5), we assumed that for each positive number n, the interval $[a, b]$ was subdivided into n subintervals of equal length to create bases for the approximating rectangles. For some functions it may be more convenient to use rectangles with different widths (see Making Connections Exercises 2 and 3); however, if we are to "exhaust" an area with rectangles of different widths, then it is important that successive subdivisions be constructed in such a way that the widths of all the rectangles approach zero as n increases (Figure 5.5.1). Thus, we must preclude the kind of situation that occurs in Figure 5.5.2 in which the right half of the interval is never subdivided. If this kind of subdivision were allowed, the error in the approximation would not approach zero as n increased.

A **partition** of the interval $[a, b]$ is a collection of points

$$a = x_0 < x_1 < x_2 < \cdots < x_{n-1} < x_n = b$$

that divides $[a, b]$ into n subintervals of lengths

$$\Delta x_1 = x_1 - x_0, \quad \Delta x_2 = x_2 - x_1, \quad \Delta x_3 = x_3 - x_2, \ldots, \quad \Delta x_n = x_n - x_{n-1}$$

The partition is said to be **regular** provided the subintervals all have the same length

$$\Delta x_k = \Delta x = \frac{b - a}{n}$$

For a regular partition, the widths of the approximating rectangles approach zero as n is made large. Since this need not be the case for a general partition, we need some way to measure the "size" of these widths. One approach is to let $\max \Delta x_k$ denote the largest of the subinterval widths. The magnitude $\max \Delta x_k$ is called the **mesh size** of the partition. For example, Figure 5.5.3 shows a partition of the interval $[0, 6]$ into four subintervals with a mesh size of 2.

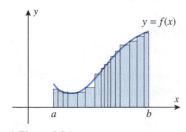

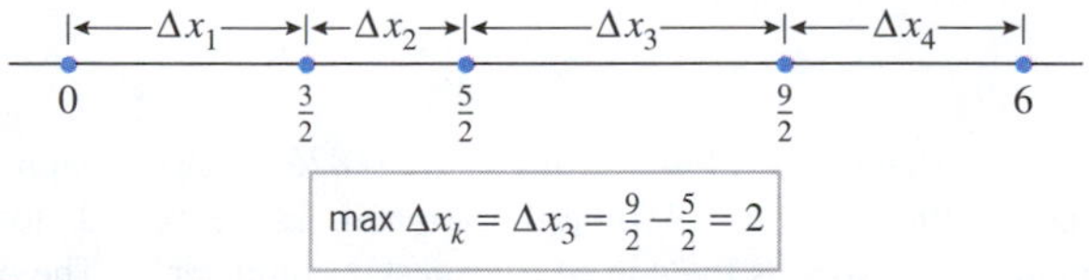

▲ Figure 5.5.1

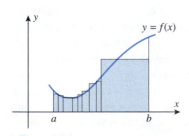

▲ Figure 5.5.2

$$|\!\leftarrow\!\Delta x_1\!\rightarrow\!|\!\leftarrow\!\Delta x_2\!\rightarrow\!|\!\leftarrow\!\text{———}\Delta x_3\text{———}\!\rightarrow\!|\!\leftarrow\!\Delta x_4\!\rightarrow\!|$$

$$0 \qquad \frac{3}{2} \qquad \frac{5}{2} \qquad \frac{9}{2} \qquad 6$$

$$\boxed{\max \Delta x_k = \Delta x_3 = \tfrac{9}{2} - \tfrac{5}{2} = 2}$$

▶ Figure 5.5.3

If we are to generalize Definition 5.4.5 so that it allows for unequal subinterval widths, we must replace the constant length Δx by the variable length Δx_k. When this is done the sum

$$\sum_{k=1}^{n} f(x_k^*)\Delta x \quad \text{is replaced by} \quad \sum_{k=1}^{n} f(x_k^*)\Delta x_k$$

We also need to replace the expression $n \to +\infty$ by an expression that guarantees us that the lengths of all subintervals approach zero. We will use the expression $\max \Delta x_k \to 0$ for this purpose. Based on our intuitive concept of area, we would then expect the net signed area A between the graph of f and the interval $[a, b]$ to satisfy the equation

Some writers use the symbol $\|\Delta\|$ rather than $\max \Delta x_k$ for the mesh size of the partition, in which case $\max \Delta x_k \to 0$ would be replaced by $\|\Delta\| \to 0$.

$$A = \lim_{\max \Delta x_k \to 0} \sum_{k=1}^{n} f(x_k^*)\Delta x_k$$

(We will see in a moment that this is the case.) The limit that appears in this expression is one of the fundamental concepts of integral calculus and forms the basis for the following definition.

5.5.1 DEFINITION A function f is said to be *integrable* on a finite closed interval $[a, b]$ if the limit

$$\lim_{\max \Delta x_k \to 0} \sum_{k=1}^{n} f(x_k^*) \Delta x_k$$

exists and does not depend on the choice of partitions or on the choice of the points x_k^* in the subintervals. When this is the case we denote the limit by the symbol

$$\int_a^b f(x) \, dx = \lim_{\max \Delta x_k \to 0} \sum_{k=1}^{n} f(x_k^*) \Delta x_k$$

which is called the *definite integral* of f from a to b. The numbers a and b are called the *lower limit of integration* and the *upper limit of integration*, respectively, and $f(x)$ is called the *integrand*.

The notation used for the definite integral deserves some comment. Historically, the expression "$f(x) \, dx$" was interpreted to be the "infinitesimal area" of a rectangle with height $f(x)$ and "infinitesimal" width dx. By "summing" these infinitesimal areas, the entire area under the curve was obtained. The integral symbol "$\int$" is an "elongated s" that was used to indicate this summation. For us, the integral symbol "$\int$" and the symbol "dx" can serve as reminders that the definite integral is actually a limit of a *summation* as $\Delta x_k \to 0$. The sum that appears in Definition 5.5.1 is called a ***Riemann sum***, and the definite integral is sometimes called the ***Riemann integral*** in honor of the German mathematician Bernhard Riemann who formulated many of the basic concepts of integral calculus. (The reason for the similarity in notation between the definite integral and the indefinite integral will become clear in the next section, where we will establish a link between the two types of "integration.")

Georg Friedrich Bernhard Riemann (1826–1866) German mathematician. Bernhard Riemann, as he is commonly known, was the son of a Protestant minister. He received his elementary education from his father and showed brilliance in arithmetic at an early age. In 1846 he enrolled at Göttingen University to study theology and philology, but he soon transferred to mathematics. He studied physics under W. E. Weber and mathematics under Carl Friedrich Gauss, whom some people consider to be the greatest mathematician who ever lived. In 1851 Riemann received his Ph.D. under Gauss, after which he remained at Göttingen to teach. In 1862, one month after his marriage, Riemann suffered an attack of pleuritis, and for the remainder of his life was an extremely sick man. He finally succumbed to tuberculosis in 1866 at age 39.

An interesting story surrounds Riemann's work in geometry. For his introductory lecture prior to becoming an associate professor, Riemann submitted three possible topics to Gauss. Gauss surprised Riemann by choosing the topic Riemann liked the least, the foundations of geometry. The lecture was like a scene from a movie. The old and failing Gauss, a giant in his day, watching intently as his brilliant and youthful protégé skillfully pieced together portions of the old man's own work into a complete and beautiful system. Gauss is said to have gasped with delight as the lecture neared its end, and on the way home he marveled at his student's brilliance. Gauss died shortly thereafter. The results presented by Riemann that day eventually evolved into a fundamental tool that Einstein used some 50 years later to develop relativity theory.

In addition to his work in geometry, Riemann made major contributions to the theory of complex functions and mathematical physics. The notion of the definite integral, as it is presented in most basic calculus courses, is due to him. Riemann's early death was a great loss to mathematics, for his mathematical work was brilliant and of fundamental importance.

The limit that appears in Definition 5.5.1 is somewhat different from the kinds of limits discussed in Chapter 1. Loosely phrased, the expression

$$\lim_{\max \Delta x_k \to 0} \sum_{k=1}^{n} f(x_k^*) \Delta x_k = L$$

is intended to convey the idea that we can force the Riemann sums to be as close as we please to L, regardless of how the values of x_k^* are chosen, by making the mesh size of the partition sufficiently small. While it is possible to give a more formal definition of this limit, we will simply rely on intuitive arguments when applying Definition 5.5.1.

Although a function need not be continuous on an interval to be integrable on that interval (Exercise 42), we will be interested primarily in definite integrals of continuous functions. The following theorem, which we will state without proof, says that if a function is continuous on a finite closed interval, then it is integrable on that interval, and its definite integral is the net signed area between the graph of the function and the interval.

5.5.2 THEOREM *If a function f is continuous on an interval $[a, b]$, then f is integrable on $[a, b]$, and the net signed area A between the graph of f and the interval $[a, b]$ is*

$$A = \int_a^b f(x)\,dx \tag{1}$$

REMARK Formula (1) follows from the integrability of f, since the integrability allows us to use any partitions to evaluate the integral. In particular, if we use *regular* partitions of $[a, b]$, then

$$\Delta x_k = \Delta x = \frac{b - a}{n}$$

for all values of k. This implies that $\max \Delta x_k = (b - a)/n$, from which it follows that $\max \Delta x_k \to 0$ if and only if $n \to +\infty$. Thus,

$$\int_a^b f(x)\,dx = \lim_{\max \Delta x_k \to 0} \sum_{k=1}^{n} f(x_k^*) \Delta x_k = \lim_{n \to +\infty} \sum_{k=1}^{n} f(x_k^*) \Delta x = A$$

In the simplest cases, definite integrals of continuous functions can be calculated using formulas from plane geometry to compute signed areas.

▶ **Example 1** Sketch the region whose area is represented by the definite integral, and evaluate the integral using an appropriate formula from geometry.

> In Example 1, it is understood that the units of area are the squared units of length, even though we have not stated the units of length explicitly.

(a) $\displaystyle\int_1^4 2\,dx$ (b) $\displaystyle\int_{-1}^2 (x + 2)\,dx$ (c) $\displaystyle\int_0^1 \sqrt{1 - x^2}\,dx$

Solution (a). The graph of the integrand is the horizontal line $y = 2$, so the region is a rectangle of height 2 extending over the interval from 1 to 4 (Figure 5.5.4a). Thus,

$$\int_1^4 2\,dx = (\text{area of rectangle}) = 2(3) = 6$$

Solution (b). The graph of the integrand is the line $y = x + 2$, so the region is a trapezoid whose base extends from $x = -1$ to $x = 2$ (Figure 5.5.4b). Thus,

$$\int_{-1}^2 (x + 2)\,dx = (\text{area of trapezoid}) = \frac{1}{2}(1 + 4)(3) = \frac{15}{2}$$

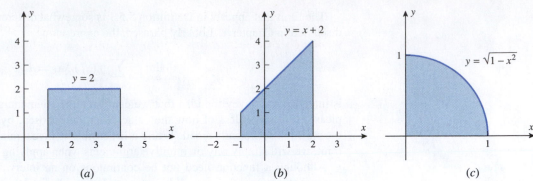

▲ **Figure 5.5.4**

Solution (c).   The graph of $y = \sqrt{1 - x^2}$ is the upper semicircle of radius 1, centered at the origin, so the region is the right quarter-circle extending from $x = 0$ to $x = 1$ (Figure 5.5.4c). Thus,

$$\int_0^1 \sqrt{1 - x^2}\, dx = \text{(area of quarter-circle)} = \frac{1}{4}\pi(1^2) = \frac{\pi}{4} \blacktriangleleft$$

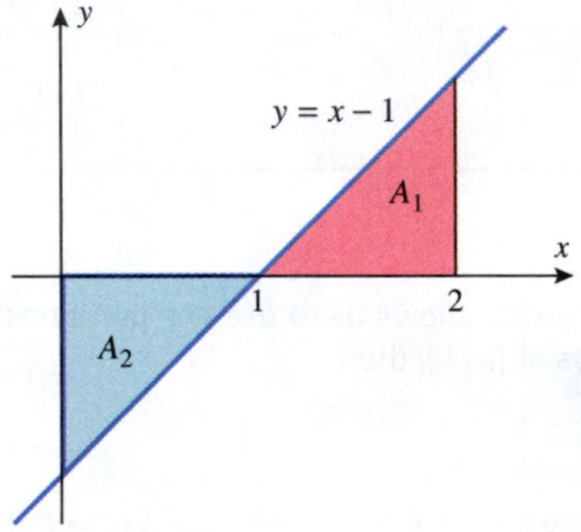

▶ **Example 2**   Evaluate

$$\text{(a)} \quad \int_0^2 (x - 1)\, dx \qquad \text{(b)} \quad \int_0^1 (x - 1)\, dx$$

Solution.   The graph of $y = x - 1$ is shown in Figure 5.5.5, and we leave it for you to verify that the shaded triangular regions both have area $\frac{1}{2}$. Over the interval $[0, 2]$ the net signed area is $A_1 - A_2 = \frac{1}{2} - \frac{1}{2} = 0$, and over the interval $[0, 1]$ the net signed area is $-A_2 = -\frac{1}{2}$. Thus,

$$\int_0^2 (x - 1)\, dx = 0 \quad \text{and} \quad \int_0^1 (x - 1)\, dx = -\frac{1}{2}$$

(Recall that in Example 7 of Section 5.4, we used Definition 5.4.5 to show that the net signed area between the graph of $y = x - 1$ and the interval $[0, 2]$ is zero.) ◀

▲ **Figure 5.5.5**

PROPERTIES OF THE DEFINITE INTEGRAL

It is assumed in Definition 5.5.1 that $[a, b]$ is a finite closed interval with $a < b$, and hence the upper limit of integration in the definite integral is greater than the lower limit of integration. However, it will be convenient to extend this definition to allow for cases in which the upper and lower limits of integration are equal or the lower limit of integration is greater than the upper limit of integration. For this purpose we make the following special definitions.

5.5.3   DEFINITION

(a)   If a is in the domain of f, we define

$$\int_a^a f(x)\, dx = 0$$

(b)   If f is integrable on $[a, b]$, then we define

$$\int_b^a f(x)\, dx = -\int_a^b f(x)\, dx$$

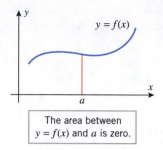

The area between
$y = f(x)$ and a is zero.

▲ **Figure 5.5.6**

Part (a) of this definition is consistent with the intuitive idea that the area between a point on the x-axis and a curve $y = f(x)$ should be zero (Figure 5.5.6). Part (b) of the definition is simply a useful convention; it states that interchanging the limits of integration reverses the sign of the integral.

▶ **Example 3**

(a) $\displaystyle\int_1^1 x^2\,dx = 0$

(b) $\displaystyle\int_1^0 \sqrt{1 - x^2}\,dx = -\int_0^1 \sqrt{1 - x^2}\,dx = -\frac{\pi}{4}$ ◀

Example 1(c)

Because definite integrals are defined as limits, they inherit many of the properties of limits. For example, we know that constants can be moved through limit signs and that the limit of a sum or difference is the sum or difference of the limits. Thus, you should not be surprised by the following theorem, which we state without formal proof.

5.5.4 THEOREM *If f and g are integrable on $[a, b]$ and if c is a constant, then cf, $f + g$, and $f - g$ are integrable on $[a, b]$ and*

(a) $\displaystyle\int_a^b cf(x)\,dx = c\int_a^b f(x)\,dx$

(b) $\displaystyle\int_a^b [f(x) + g(x)]\,dx = \int_a^b f(x)\,dx + \int_a^b g(x)\,dx$

(c) $\displaystyle\int_a^b [f(x) - g(x)]\,dx = \int_a^b f(x)\,dx - \int_a^b g(x)\,dx$

Part (b) of this theorem can be extended to more than two functions. More precisely,

$$\int_a^b [f_1(x) + f_2(x) + \cdots + f_n(x)]\,dx$$
$$= \int_a^b f_1(x)\,dx + \int_a^b f_2(x)\,dx + \cdots + \int_a^b f_n(x)\,dx$$

Some properties of definite integrals can be motivated by interpreting the integral as an area. For example, if f is continuous and nonnegative on the interval $[a, b]$, and if c is a point between a and b, then the area under $y = f(x)$ over the interval $[a, b]$ can be split into two parts and expressed as the area under the graph from a to c plus the area under the graph from c to b (Figure 5.5.7), that is,

$$\int_a^b f(x)\,dx = \int_a^c f(x)\,dx + \int_c^b f(x)\,dx$$

This is a special case of the following theorem about definite integrals, which we state without proof.

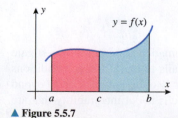

▲ **Figure 5.5.7**

> **5.5.5**   **THEOREM**   *If f is integrable on a closed interval containing the three points a, b, and c, then*
>
> $$\int_a^b f(x)\,dx = \int_a^c f(x)\,dx + \int_c^b f(x)\,dx$$
>
> *no matter how the points are ordered.*

The following theorem, which we state without formal proof, can also be motivated by interpreting definite integrals as areas.

> **5.5.6**   **THEOREM**
>
> (a)   *If f is integrable on [a, b] and f(x) ≥ 0 for all x in [a, b], then*
>
> $$\int_a^b f(x)\,dx \geq 0$$
>
> (b)   *If f and g are integrable on [a, b] and f(x) ≥ g(x) for all x in [a, b], then*
>
> $$\int_a^b f(x)\,dx \geq \int_a^b g(x)\,dx$$

Part (*b*) of Theorem 5.5.6 states that the direction (sometimes called the *sense*) of the inequality $f(x) \geq g(x)$ is unchanged if one integrates both sides. Moreover, if $b > a$, then both parts of the theorem remain true if ≥ is replaced by ≤, >, or < throughout.

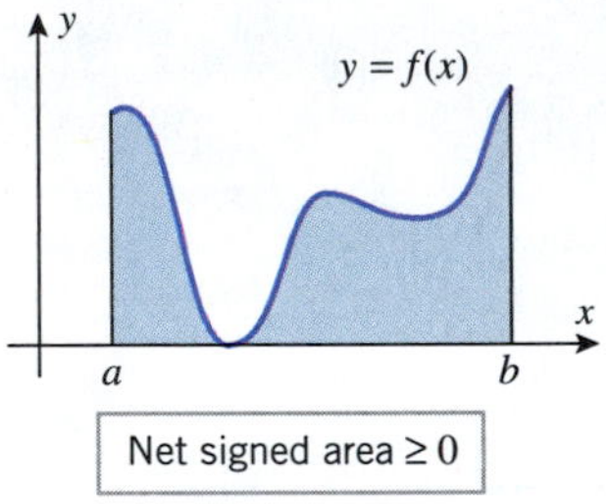

Geometrically, part (*a*) of this theorem states the obvious fact that if f is nonnegative on $[a, b]$, then the net signed area between the graph of f and the interval $[a, b]$ is also nonnegative (Figure 5.5.8). Part (*b*) has its simplest interpretation when f and g are nonnegative on $[a, b]$, in which case the theorem states that if the graph of f does not go below the graph of g, then the area under the graph of f is at least as large as the area under the graph of g (Figure 5.5.9).

Net signed area ≥ 0

▲ **Figure 5.5.8**

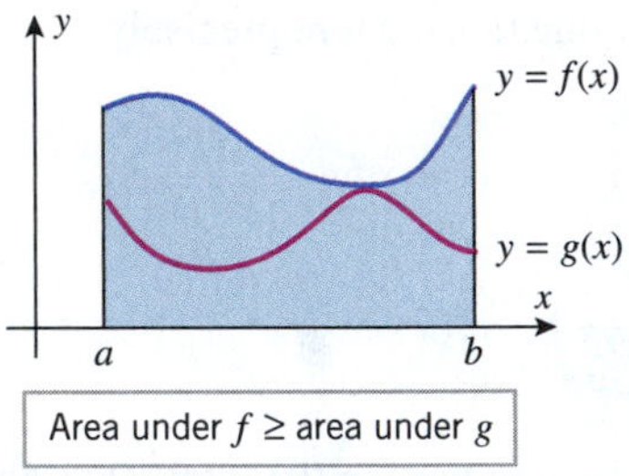

Area under f ≥ area under g

▲ **Figure 5.5.9**

▶ **Example 4**   Evaluate

$$\int_0^1 (5 - 3\sqrt{1 - x^2})\,dx$$

Solution.   From parts (*a*) and (*c*) of Theorem 5.5.4 we can write

$$\int_0^1 (5 - 3\sqrt{1 - x^2})\,dx = \int_0^1 5\,dx - \int_0^1 3\sqrt{1 - x^2}\,dx = \int_0^1 5\,dx - 3\int_0^1 \sqrt{1 - x^2}\,dx$$

The first integral in this difference can be interpreted as the area of a rectangle of height 5 and base 1, so its value is 5, and from Example 1 the value of the second integral is $\pi/4$. Thus,

$$\int_0^1 (5 - 3\sqrt{1 - x^2})\,dx = 5 - 3\left(\frac{\pi}{4}\right) = 5 - \frac{3\pi}{4} \blacktriangleleft$$

■ DISCONTINUITIES AND INTEGRABILITY

In the late nineteenth and early twentieth centuries, mathematicians began to investigate conditions under which the limit that defines an integral fails to exist, that is, conditions under which a function fails to be integrable. The matter is quite complex and beyond the scope of this text. However, there are a few basic results about integrability that are important to know; we begin with a definition.

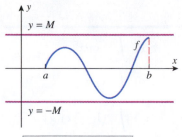

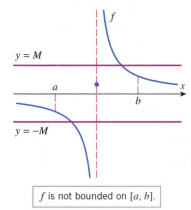

y = M

f

y = M

a

b

x

y = -M

f is not bounded on [a, b].

▲ Figure 5.5.11

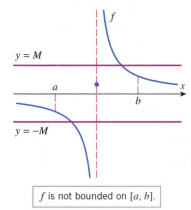
y = M

f is bounded on $[a, b]$.

▲ Figure 5.5.10

5.5.7 **DEFINITION** A function f that is defined on an interval is said to be **bounded** on the interval if there is a positive number M such that

$$-M \leq f(x) \leq M$$

for all x in the interval. Geometrically, this means that the graph of f over the interval lies between the lines $y = -M$ and $y = M$.

For example, a continuous function f is bounded on *every* finite closed interval because the Extreme-Value Theorem (4.4.2) implies that f has an absolute maximum and an absolute minimum on the interval; hence, its graph will lie between the lines $y = -M$ and $y = M$, provided we make M large enough (Figure 5.5.10). In contrast, a function that has a vertical asymptote inside of an interval is not bounded on that interval because its graph over the interval cannot be made to lie between the lines $y = -M$ and $y = M$, no matter how large we make the value of M (Figure 5.5.11).

The following theorem, which we state without proof, provides some facts about integrability for functions with discontinuities. In the exercises we have included some problems that are concerned with this theorem (Exercises 42, 43, and 44).

5.5.8 **THEOREM** *Let f be a function that is defined on the finite closed interval $[a, b]$.*

(a) *If f has finitely many discontinuities in $[a, b]$ but is bounded on $[a, b]$, then f is integrable on $[a, b]$.*

(b) *If f is not bounded on $[a, b]$, then f is not integrable on $[a, b]$.*

✔ **QUICK CHECK EXERCISES 5.5**   *(See page 362 for answers.)*

---

1. In each part, use the partition of $[2, 7]$ in the accompanying figure.

2   3   4.5   6.5   7

▲ Figure Ex-1

(a) What is n, the number of subintervals in this partition?

(b) $x_0 = $ _____; $x_1 = $ _____; $x_2 = $ _____;
$x_3 = $ _____; $x_4 = $ _____

(c) $\Delta x_1 = $ _____; $\Delta x_2 = $ _____; $\Delta x_3 = $ _____;
$\Delta x_4 = $ _____

(d) The mesh of this partition is _____.

2. Let $f(x) = 2x - 8$. Use the partition of $[2, 7]$ in Quick Check Exercise 1 and the choices $x_1^* = 2$, $x_2^* = 4$, $x_3^* = 5$, and $x_4^* = 7$ to evaluate the Riemann sum

$$\sum_{k=1}^{4} f(x_k^*)\Delta x_k$$

3. Use the accompanying figure to evaluate

$$\int_{2}^{7} (2x - 8)\, dx$$

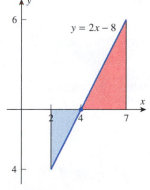

y = 2x - 8

◄ Figure Ex-3

4. Suppose that $g(x)$ is a function for which

$$\int_{-2}^{1} g(x)\, dx = 5 \quad \text{and} \quad \int_{1}^{2} g(x)\, dx = -2$$

Use this information to evaluate the definite integrals.

(a) $\displaystyle\int_{1}^{2} 5g(x)\, dx$

(b) $\displaystyle\int_{-2}^{2} g(x)\, dx$

(c) $\displaystyle\int_{1}^{1} [g(x)]^2\, dx$

(d) $\displaystyle\int_{2}^{-2} 4g(x)\, dx$

EXERCISE SET 5.5

1–4 Find the value of

(a) $\displaystyle\sum_{k=1}^{n} f(x_k^*)\,\Delta x_k$ (b) max Δx_k. ■

1. $f(x) = x + 1$; $a = 0, b = 4$; $n = 3$;
   $\Delta x_1 = 1, \Delta x_2 = 1, \Delta x_3 = 2$;
   $x_1^* = \frac{1}{3}, x_2^* = \frac{3}{2}, x_3^* = 3$

2. $f(x) = \cos x$; $a = 0, b = 2\pi$; $n = 4$;
   $\Delta x_1 = \pi/2, \Delta x_2 = 3\pi/4, \Delta x_3 = \pi/2, \Delta x_4 = \pi/4$;
   $x_1^* = \pi/4, x_2^* = \pi, x_3^* = 3\pi/2, x_4^* = 7\pi/4$

3. $f(x) = 4 - x^2$; $a = -3, b = 4$; $n = 4$;
   $\Delta x_1 = 1, \Delta x_2 = 2, \Delta x_3 = 1, \Delta x_4 = 3$;
   $x_1^* = -\frac{5}{2}, x_2^* = -1, x_3^* = \frac{1}{4}, x_4^* = 3$

4. $f(x) = x^3$; $a = -3, b = 3$; $n = 4$;
   $\Delta x_1 = 2, \Delta x_2 = 1, \Delta x_3 = 1, \Delta x_4 = 2$;
   $x_1^* = -2, x_2^* = 0, x_3^* = 0, x_4^* = 2$

5–8 Use the given values of a and b to express the following limits as integrals. (Do not evaluate the integrals.) ■

5. $\displaystyle\lim_{\max \Delta x_k \to 0}\sum_{k=1}^{n}(x_k^*)^2 \Delta x_k$; $a = -1, b = 2$

6. $\displaystyle\lim_{\max \Delta x_k \to 0}\sum_{k=1}^{n}(x_k^*)^3 \Delta x_k$; $a = 1, b = 2$

7. $\displaystyle\lim_{\max \Delta x_k \to 0}\sum_{k=1}^{n}4x_k^*(1 - 3x_k^*)\Delta x_k$; $a = -3, b = 3$

8. $\displaystyle\lim_{\max \Delta x_k \to 0}\sum_{k=1}^{n}(\sin^2 x_k^*)\Delta x_k$; $a = 0, b = \pi/2$

9–10 Use Definition 5.5.1 to express the integrals as limits of Riemann sums. (Do not evaluate the integrals.) ■

9. (a) $\displaystyle\int_1^2 2x\,dx$ (b) $\displaystyle\int_0^1 \frac{x}{x+1}\,dx$

10. (a) $\displaystyle\int_1^2 \sqrt{x}\,dx$ (b) $\displaystyle\int_{-\pi/2}^{\pi/2} (1 + \cos x)\,dx$

FOCUS ON CONCEPTS

11. Explain informally why Theorem 5.5.4(a) follows from Definition 5.5.1.

12. Explain informally why Theorem 5.5.6(a) follows from Definition 5.5.1.

13–16 Sketch the region whose signed area is represented by the definite integral, and evaluate the integral using an appropriate formula from geometry, where needed. ■

13. (a) $\displaystyle\int_0^3 x\,dx$ (b) $\displaystyle\int_{-2}^{-1} x\,dx$
    (c) $\displaystyle\int_{-1}^4 x\,dx$ (d) $\displaystyle\int_{-5}^5 x\,dx$

14. (a) $\displaystyle\int_0^2 \left(1 - \tfrac{1}{2}x\right) dx$ (b) $\displaystyle\int_{-1}^1 \left(1 - \tfrac{1}{2}x\right) dx$
    (c) $\displaystyle\int_2^3 \left(1 - \tfrac{1}{2}x\right) dx$ (d) $\displaystyle\int_0^3 \left(1 - \tfrac{1}{2}x\right) dx$

15. (a) $\displaystyle\int_0^5 2\,dx$ (b) $\displaystyle\int_0^\pi \cos x\,dx$
    (c) $\displaystyle\int_{-1}^2 |2x - 3|\,dx$ (d) $\displaystyle\int_{-1}^1 \sqrt{1 - x^2}\,dx$

16. (a) $\displaystyle\int_{-10}^{-5} 6\,dx$ (b) $\displaystyle\int_{-\pi/3}^{\pi/3} \sin x\,dx$
    (c) $\displaystyle\int_0^3 |x - 2|\,dx$ (d) $\displaystyle\int_0^2 \sqrt{4 - x^2}\,dx$

17. In each part, evaluate the integral, given that

$$f(x) = \begin{cases} |x - 2|, & x \ge 0 \\ x + 2, & x < 0 \end{cases}$$

(a) $\displaystyle\int_{-2}^0 f(x)\,dx$ (b) $\displaystyle\int_{-2}^2 f(x)\,dx$
(c) $\displaystyle\int_0^6 f(x)\,dx$ (d) $\displaystyle\int_{-4}^6 f(x)\,dx$

18. In each part, evaluate the integral, given that

$$f(x) = \begin{cases} 2x, & x \le 1 \\ 2, & x > 1 \end{cases}$$

(a) $\displaystyle\int_0^1 f(x)\,dx$ (b) $\displaystyle\int_{-1}^1 f(x)\,dx$
(c) $\displaystyle\int_1^{10} f(x)\,dx$ (d) $\displaystyle\int_{1/2}^5 f(x)\,dx$

FOCUS ON CONCEPTS

19–20 Use the areas shown in the figure to find

(a) $\displaystyle\int_a^b f(x)\,dx$ (b) $\displaystyle\int_b^c f(x)\,dx$
(c) $\displaystyle\int_a^c f(x)\,dx$ (d) $\displaystyle\int_a^d f(x)\,dx$. ■

19.

20.

21. Find $\displaystyle\int_{-1}^{2} [f(x) + 2g(x)] \, dx$ if

$$\int_{-1}^{2} f(x) \, dx = 5 \quad \text{and} \quad \int_{-1}^{2} g(x) \, dx = -3$$

22. Find $\displaystyle\int_{1}^{4} [3f(x) - g(x)] \, dx$ if

$$\int_{1}^{4} f(x) \, dx = 2 \quad \text{and} \quad \int_{1}^{4} g(x) \, dx = 10$$

23. Find $\displaystyle\int_{1}^{5} f(x) \, dx$ if

$$\int_{0}^{1} f(x) \, dx = -2 \quad \text{and} \quad \int_{0}^{5} f(x) \, dx = 1$$

24. Find $\displaystyle\int_{3}^{-2} f(x) \, dx$ if

$$\int_{-2}^{1} f(x) \, dx = 2 \quad \text{and} \quad \int_{1}^{3} f(x) \, dx = -6$$

25–28 Use Theorem 5.5.4 and appropriate formulas from geometry to evaluate the integrals. ■

25. $\displaystyle\int_{-1}^{3} (4 - 5x) \, dx$ **26.** $\displaystyle\int_{-2}^{2} (1 - 3|x|) \, dx$

27. $\displaystyle\int_{0}^{1} (x + 2\sqrt{1 - x^2}) \, dx$ **28.** $\displaystyle\int_{-3}^{0} (2 + \sqrt{9 - x^2}) \, dx$

29–32 True–False Determine whether the statement is true or false. Explain your answer. ■

29. If $f(x)$ is integrable on $[a, b]$, then $f(x)$ is continuous on $[a, b]$.

30. It is the case that

$$0 < \int_{-1}^{1} \frac{\cos x}{\sqrt{1 + x^2}} \, dx$$

31. If the integral of $f(x)$ over the interval $[a, b]$ is negative, then $f(x) \le 0$ for $a \le x \le b$.

32. The function

$$f(x) = \begin{cases} 0, & x \le 0 \\ x^2, & x > 0 \end{cases}$$

is integrable over every closed interval $[a, b]$.

33–34 Use Theorem 5.5.6 to determine whether the value of the integral is positive or negative. ■

33. (a) $\displaystyle\int_{2}^{3} \frac{\sqrt{x}}{1 - x} \, dx$ (b) $\displaystyle\int_{0}^{4} \frac{x^2}{3 - \cos x} \, dx$

34. (a) $\displaystyle\int_{-3}^{-1} \frac{x^4}{\sqrt{3 - x}} \, dx$ (b) $\displaystyle\int_{-2}^{2} \frac{x^3 - 9}{|x| + 1} \, dx$

35. Prove that if f is continuous and if $m \le f(x) \le M$ on $[a, b]$, then

$$m(b - a) \le \int_{a}^{b} f(x) \, dx \le M(b - a)$$

36. Find the maximum and minimum values of $\sqrt{x^3 + 2}$ for $0 \le x \le 3$. Use these values, and the inequalities in Exercise 35, to find bounds on the value of the integral

$$\int_{0}^{3} \sqrt{x^3 + 2} \, dx$$

37–38 Evaluate the integrals by completing the square and applying appropriate formulas from geometry. ■

37. $\displaystyle\int_{0}^{10} \sqrt{10x - x^2} \, dx$ **38.** $\displaystyle\int_{0}^{3} \sqrt{6x - x^2} \, dx$

39–40 Evaluate the limit by expressing it as a definite integral over the interval $[a, b]$ and applying appropriate formulas from geometry. ■

39. $\displaystyle\lim_{\max \Delta x_k \to 0} \sum_{k=1}^{n} (3x_k^* + 1) \Delta x_k$; $a = 0, b = 1$

40. $\displaystyle\lim_{\max \Delta x_k \to 0} \sum_{k=1}^{n} \sqrt{4 - (x_k^*)^2} \, \Delta x_k$; $a = -2, b = 2$

FOCUS ON CONCEPTS

41. Let $f(x) = C$ be a constant function.
(a) Use a formula from geometry to show that

$$\int_{a}^{b} f(x) \, dx = C(b - a)$$

(b) Show that any Riemann sum for $f(x)$ over $[a, b]$ evaluates to $C(b - a)$. Use Definition 5.5.1 to show that

$$\int_{a}^{b} f(x) \, dx = C(b - a)$$

42. Define a function f on $[0, 1]$ by

$$f(x) = \begin{cases} 1, & 0 < x \le 1 \\ 0, & x = 0 \end{cases}$$

Use Definition 5.5.1 to show that

$$\int_{0}^{1} f(x) \, dx = 1$$

43. It can be shown that every interval contains both rational and irrational numbers. Accepting this to be so, do you believe that the function

$$f(x) = \begin{cases} 1 & \text{if} \quad x \text{ is rational} \\ 0 & \text{if} \quad x \text{ is irrational} \end{cases}$$

is integrable on a closed interval $[a, b]$? Explain your reasoning.

44. Define the function f by

$$f(x) = \begin{cases} \dfrac{1}{x}, & x \neq 0 \\ 0, & x = 0 \end{cases}$$

It follows from Theorem 5.5.8(b) that f is not integrable on the interval $[0, 1]$. Prove this to be the case by applying Definition 5.5.1. [*Hint:* Argue that no matter how small the mesh size is for a partition of $[0, 1]$, there will always be a choice of x_1^* that will make the Riemann sum in Definition 5.5.1 as large as we like.]

45. In each part, use Theorems 5.5.2 and 5.5.8 to determine whether the function f is integrable on the interval $[-1, 1]$.

(a) $f(x) = \cos x$

(b) $f(x) = \begin{cases} x/|x|, & x \neq 0 \\ 0, & x = 0 \end{cases}$

(c) $f(x) = \begin{cases} 1/x^2, & x \neq 0 \\ 0, & x = 0 \end{cases}$

(d) $f(x) = \begin{cases} \sin 1/x, & x \neq 0 \\ 0, & x = 0 \end{cases}$

46. Writing Write a short paragraph that discusses the similarities and differences between indefinite integrals and definite integrals.

47. Writing Write a paragraph that explains informally what it means for a function to be "integrable."

✔ **QUICK CHECK ANSWERS 5.5**

1. (a) $n = 4$ (b) $2, 3, 4.5, 6.5, 7$ (c) $1, 1.5, 2, 0.5$ (d) 2 **2.** 3 **3.** 5 **4.** (a) -10 (b) 3 (c) 0 (d) -12

5.6 THE FUNDAMENTAL THEOREM OF CALCULUS

In this section we will establish two basic relationships between definite and indefinite integrals that together constitute a result called the "Fundamental Theorem of Calculus." One part of this theorem will relate the rectangle and antiderivative methods for calculating areas, and the second part will provide a powerful method for evaluating definite integrals using antiderivatives.

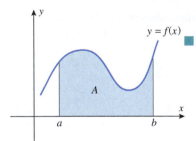

▲ **Figure 5.6.1**

■ **THE FUNDAMENTAL THEOREM OF CALCULUS**

As in earlier sections, let us begin by assuming that f is nonnegative and continuous on an interval $[a, b]$, in which case the area A under the graph of f over the interval $[a, b]$ is represented by the definite integral

$$A = \int_a^b f(x)\,dx \tag{1}$$

(Figure 5.6.1).

Recall that our discussion of the antiderivative method in Section 5.1 suggested that if $A(x)$ is the area under the graph of f from a to x (Figure 5.6.2), then

- $A'(x) = f(x)$

- $A(a) = 0$ The area under the curve from a to a is the area above the single point a, and hence is zero.

- $A(b) = A$ The area under the curve from a to b is A.

The formula $A'(x) = f(x)$ states that $A(x)$ is an antiderivative of $f(x)$, which implies that every other antiderivative of $f(x)$ on $[a, b]$ can be obtained by adding a constant to $A(x)$. Accordingly, let

$$F(x) = A(x) + C$$

be any antiderivative of $f(x)$, and consider what happens when we subtract $F(a)$ from $F(b)$:

$$F(b) - F(a) = [A(b) + C] - [A(a) + C] = A(b) - A(a) = A - 0 = A$$

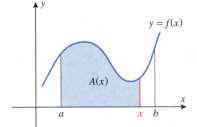

▲ **Figure 5.6.2**

Hence (1) can be expressed as

$$\int_a^b f(x)\,dx = F(b) - F(a)$$

In words, this equation states:

The definite integral can be evaluated by finding any antiderivative of the integrand and then subtracting the value of this antiderivative at the lower limit of integration from its value at the upper limit of integration.

Although our evidence for this result assumed that f is nonnegative on $[a, b]$, this assumption is not essential.

5.6.1   THEOREM (*The Fundamental Theorem of Calculus, Part 1*)   *If f is continuous on $[a, b]$ and F is any antiderivative of f on $[a, b]$, then*

$$\int_a^b f(x)\,dx = F(b) - F(a) \tag{2}$$

PROOF   Let $x_1, x_2, \ldots, x_{n-1}$ be any points in $[a, b]$ such that

$$a < x_1 < x_2 < \cdots < x_{n-1} < b$$

These values divide $[a, b]$ into n subintervals

$$[a, x_1], [x_1, x_2], \ldots, [x_{n-1}, b] \tag{3}$$

whose lengths, as usual, we denote by

$$\Delta x_1, \Delta x_2, \ldots, \Delta x_n$$

(see Figure 5.6.3). By hypothesis, $F'(x) = f(x)$ for all x in $[a, b]$, so F satisfies the hypotheses of the Mean-Value Theorem (4.8.2) on each subinterval in (3). Hence, we can find points $x_1^*, x_2^*, \ldots, x_n^*$ in the respective subintervals in (3) such that

$$F(x_1) - F(a) = F'(x_1^*)(x_1 - a) = f(x_1^*)\Delta x_1$$
$$F(x_2) - F(x_1) = F'(x_2^*)(x_2 - x_1) = f(x_2^*)\Delta x_2$$
$$F(x_3) - F(x_2) = F'(x_3^*)(x_3 - x_2) = f(x_3^*)\Delta x_3$$
$$\vdots$$
$$F(b) - F(x_{n-1}) = F'(x_n^*)(b - x_{n-1}) = f(x_n^*)\Delta x_n$$

Adding the preceding equations yields

$$F(b) - F(a) = \sum_{k=1}^{n} f(x_k^*)\Delta x_k \tag{4}$$

Let us now increase n in such a way that max $\Delta x_k \to 0$. Since f is assumed to be continuous, the right side of (4) approaches $\int_a^b f(x)\,dx$ by Theorem 5.5.2 and Definition 5.5.1. However,

▶ **Figure 5.6.3**

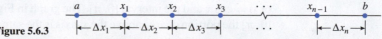

the left side of (4) is independent of n; that is, the left side of (4) remains constant as n increases. Thus,

$$F(b) - F(a) = \lim_{\max \Delta x_k \to 0} \sum_{k=1}^{n} f(x_k^*) \Delta x_k = \int_a^b f(x)\, dx \quad \blacksquare$$

It is standard to denote the difference $F(b) - F(a)$ as

$$F(x)\Big]_a^b = F(b) - F(a) \quad \text{or} \quad \Big[F(x)\Big]_a^b = F(b) - F(a)$$

For example, using the first of these notations we can express (2) as

$$\int_a^b f(x)\, dx = F(x)\Big]_a^b \tag{5}$$

We will sometimes write

$$F(x)\Big]_{x=a}^b = F(b) - F(a)$$

when it is important to emphasize that a and b are values for the variable x.

▶ **Example 1**   Evaluate $\displaystyle\int_1^2 x\, dx$.

The integral in Example 1 represents the area of a certain trapezoid. Sketch the trapezoid, and find its area using geometry.

Solution.   The function $F(x) = \frac{1}{2}x^2$ is an antiderivative of $f(x) = x$; thus, from (2)

$$\int_1^2 x\, dx = \frac{1}{2}x^2\Big]_1^2 = \frac{1}{2}(2)^2 - \frac{1}{2}(1)^2 = 2 - \frac{1}{2} = \frac{3}{2} \quad \blacktriangleleft$$

▶ **Example 2**   In Example 5 of Section 5.4 we used the definition of area to show that the area under the graph of $y = 9 - x^2$ over the interval $[0, 3]$ is 18 (square units). We can now solve that problem much more easily using the Fundamental Theorem of Calculus:

$$A = \int_0^3 (9 - x^2)\, dx = \left[9x - \frac{x^3}{3}\right]_0^3 = \left(27 - \frac{27}{3}\right) - 0 = 18 \quad \blacktriangleleft$$

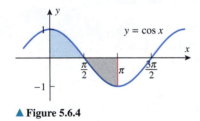

▲ **Figure 5.6.4**

▶ **Example 3**

(a) Find the area under the curve $y = \cos x$ over the interval $[0, \pi/2]$ (Figure 5.6.4).

(b) Make a conjecture about the value of the integral

$$\int_0^\pi \cos x\, dx$$

and confirm your conjecture using the Fundamental Theorem of Calculus.

Solution (a).   Since $\cos x \geq 0$ over the interval $[0, \pi/2]$, the area A under the curve is

$$A = \int_0^{\pi/2} \cos x\, dx = \sin x\Big]_0^{\pi/2} = \sin\frac{\pi}{2} - \sin 0 = 1$$

Solution (b).   The given integral can be interpreted as the signed area between the graph of $y = \cos x$ and the interval $[0, \pi]$. The graph in Figure 5.6.4 suggests that over the interval $[0, \pi]$ the portion of area above the x-axis is the same as the portion of area below the x-axis, so we conjecture that the signed area is zero; this implies that the value of the integral is zero. This is confirmed by the computations

$$\int_0^\pi \cos x \, dx = \sin x \Big]_0^\pi = \sin \pi - \sin 0 = 0 \quad \blacktriangleleft$$

■ THE RELATIONSHIP BETWEEN DEFINITE AND INDEFINITE INTEGRALS

Observe that in the preceding examples we did not include a constant of integration in the antiderivatives. In general, when applying the Fundamental Theorem of Calculus there is no need to include a constant of integration because it will drop out anyway. To see that this is so, let F be any antiderivative of the integrand on $[a, b]$, and let C be any constant; then

$$\int_a^b f(x) \, dx = \big[F(x) + C\big]_a^b = [F(b) + C] - [F(a) + C] = F(b) - F(a)$$

Thus, for purposes of evaluating a definite integral we can omit the constant of integration in

$$\int_a^b f(x) \, dx = \big[F(x) + C\big]_a^b$$

and express (5) as

$$\int_a^b f(x) \, dx = \int f(x) \, dx \bigg]_a^b \tag{6}$$

which relates the definite and indefinite integrals.

► **Example 4**

$$\int_1^9 \sqrt{x} \, dx = \int x^{1/2} \, dx \bigg]_1^9 = \frac{2}{3} x^{3/2} \bigg]_1^9 = \frac{2}{3}(27 - 1) = \frac{52}{3} \quad \blacktriangleleft$$

► **Example 5** Table 5.2.1 will be helpful for the following computations.

$$\int_4^9 x^2 \sqrt{x} \, dx = \int_4^9 x^{5/2} \, dx = \frac{2}{7} x^{7/2} \bigg]_4^9 = \frac{2}{7}(2187 - 128) = \frac{4118}{7} = 588\frac{2}{7}$$

$$\int_0^{\pi/2} \frac{\sin x}{5} \, dx = -\frac{1}{5} \cos x \bigg]_0^{\pi/2} = -\frac{1}{5}\left[\cos\left(\frac{\pi}{2}\right) - \cos 0\right] = -\frac{1}{5}[0 - 1] = \frac{1}{5}$$

$$\int_0^{\pi/3} \sec^2 x \, dx = \tan x \bigg]_0^{\pi/3} = \tan\left(\frac{\pi}{3}\right) - \tan 0 = \sqrt{3} - 0 = \sqrt{3}$$

$$\int_0^{\ln 3} 5e^x \, dx = 5e^x \bigg]_0^{\ln 3} = 5[e^{\ln 3} - e^0] = 5[3 - 1] = 10$$

$$\int_{-e}^{-1} \frac{1}{x} \, dx = \ln |x| \bigg]_{-e}^{-1} = \ln |-1| - \ln |-e| = 0 - 1 = -1$$

$$\int_{-1/2}^{1/2} \frac{1}{\sqrt{1 - x^2}} \, dx = \sin^{-1} x \bigg]_{-1/2}^{1/2} = \sin^{-1}\left(\frac{1}{2}\right) - \sin^{-1}\left(-\frac{1}{2}\right) = \frac{\pi}{6} - \left(-\frac{\pi}{6}\right) = \frac{\pi}{3} \quad \blacktriangleleft$$

TECHNOLOGY MASTERY

If you have a CAS, read the documentation on evaluating definite integrals and then check the results in Example 5.

WARNING The requirements in the Fundamental Theorem of Calculus that f be continuous on $[a, b]$ and that F be an antiderivative for f over the entire interval $[a, b]$ are important to keep in mind. Disregarding these assumptions will likely lead to incorrect results. For example, the function $f(x) = 1/x^2$ fails on two counts to be continuous at $x = 0$: $f(x)$ is not defined at $x = 0$ and $\lim_{x \to 0} f(x)$ does not exist. Thus, the Fundamental Theorem of Calculus should not be used to integrate f on any interval that contains $x = 0$. However, if we ignore this and mistakenly apply Formula (2) over the interval $[-1, 1]$, we might *incorrectly* compute $\int_{-1}^{1} (1/x^2) \, dx$ by evaluating an antiderivative, $-1/x$, at the endpoints, arriving at the answer

$$-\frac{1}{x}\Bigg]_{-1}^{1} = -[1 - (-1)] = -2$$

But $f(x) = 1/x^2$ is a nonnegative function, so clearly a negative value for the definite integral is impossible.

The Fundamental Theorem of Calculus can be applied without modification to definite integrals in which the lower limit of integration is greater than or equal to the upper limit of integration.

▶ **Example 6**

$$\int_{1}^{1} x^2 \, dx = \frac{x^3}{3}\Bigg]_{1}^{1} = \frac{1}{3} - \frac{1}{3} = 0$$

$$\int_{4}^{0} x \, dx = \frac{x^2}{2}\Bigg]_{4}^{0} = \frac{0}{2} - \frac{16}{2} = -8$$

The latter result is consistent with the result that would be obtained by first reversing the limits of integration in accordance with Definition 5.5.3(b):

$$\int_{4}^{0} x \, dx = -\int_{0}^{4} x \, dx = -\frac{x^2}{2}\Bigg]_{0}^{4} = -\left[\frac{16}{2} - \frac{0}{2}\right] = -8 \blacktriangleleft$$

To integrate a continuous function that is defined piecewise on an interval $[a, b]$, split this interval into subintervals at the breakpoints of the function, and integrate separately over each subinterval in accordance with Theorem 5.5.5.

▶ **Example 7** Evaluate $\int_{0}^{3} f(x) \, dx$ if

$$f(x) = \begin{cases} x^2, & x < 2 \\ 3x - 2, & x \geq 2 \end{cases}$$

Solution. See Figure 5.6.5. From Theorem 5.5.5 we can integrate from 0 to 2 and from 2 to 3 separately and add the results. This yields

$$\int_{0}^{3} f(x) \, dx = \int_{0}^{2} f(x) \, dx + \int_{2}^{3} f(x) \, dx = \int_{0}^{2} x^2 \, dx + \int_{2}^{3} (3x - 2) \, dx$$

$$= \frac{x^3}{3}\Bigg]_{0}^{2} + \left[\frac{3x^2}{2} - 2x\right]_{2}^{3} = \left(\frac{8}{3} - 0\right) + \left(\frac{15}{2} - 2\right) = \frac{49}{6} \blacktriangleleft$$

If f is a continuous function on the interval $[a, b]$, then we define the **total area** between the curve $y = f(x)$ and the interval $[a, b]$ to be

$$\text{total area} = \int_{a}^{b} |f(x)| \, dx \tag{7}$$

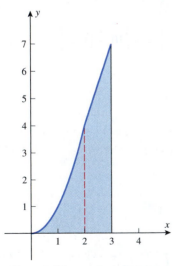

▲ **Figure 5.6.5**

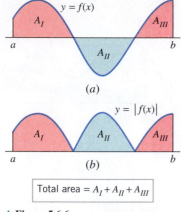

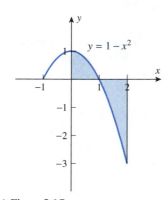

Total area = $A_I + A_{II} + A_{III}$

▲ **Figure 5.6.6**

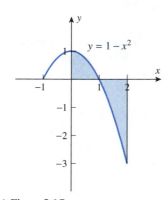

▲ **Figure 5.6.7**

(Figure 5.6.6). To compute total area using Formula (7), begin by dividing the interval of integration into subintervals on which $f(x)$ does not change sign. On the subintervals for which $0 \leq f(x)$ replace $|f(x)|$ by $f(x)$, and on the subintervals for which $f(x) \leq 0$ replace $|f(x)|$ by $-f(x)$. Adding the resulting integrals then yields the total area.

▶ **Example 8** Find the total area between the curve $y = 1 - x^2$ and the x-axis over the interval $[0, 2]$ (Figure 5.6.7).

Solution. The area A is given by

$$A = \int_0^2 |1 - x^2|\, dx = \int_0^1 (1 - x^2)\, dx + \int_1^2 -(1 - x^2)\, dx$$

$$= \left[x - \frac{x^3}{3} \right]_0^1 - \left[x - \frac{x^3}{3} \right]_1^2$$

$$= \frac{2}{3} - \left(-\frac{4}{3} \right) = 2 \; \blacktriangleleft$$

■ **DUMMY VARIABLES**

To evaluate a definite integral using the Fundamental Theorem of Calculus, one needs to be able to find an antiderivative of the integrand; thus, it is important to know what kinds of functions have antiderivatives. It is our next objective to show that all continuous functions have antiderivatives, but to do this we will need some preliminary results.

Formula (6) shows that there is a close relationship between the integrals

$$\int_a^b f(x)\, dx \quad \text{and} \quad \int f(x)\, dx$$

However, the definite and indefinite integrals differ in some important ways. For one thing, the two integrals are different kinds of objects—the definite integral is a *number* (the net signed area between the graph of $y = f(x)$ and the interval $[a, b]$), whereas the indefinite integral is a *function*, or more accurately a family of functions [the antiderivatives of $f(x)$]. However, the two types of integrals also differ in the role played by the variable of integration. In an indefinite integral, the variable of integration is "passed through" to the antiderivative in the sense that integrating a function of x produces a function of x, integrating a function of t produces a function of t, and so forth. For example,

$$\int x^2\, dx = \frac{x^3}{3} + C \quad \text{and} \quad \int t^2\, dt = \frac{t^3}{3} + C$$

In contrast, the variable of integration in a definite integral is not passed through to the end result, since the end result is a number. Thus, integrating a function of x over an interval and integrating the same function of t over the same interval of integration produce the same value for the integral. For example,

$$\int_1^3 x^2\, dx = \frac{x^3}{3} \bigg]_{x=1}^3 = \frac{27}{3} - \frac{1}{3} = \frac{26}{3} \quad \text{and} \quad \int_1^3 t^2\, dt = \frac{t^3}{3} \bigg]_{t=1}^3 = \frac{27}{3} - \frac{1}{3} = \frac{26}{3}$$

However, this latter result should not be surprising, since the area under the graph of the curve $y = f(x)$ over an interval $[a, b]$ on the x-axis is the same as the area under the graph of the curve $y = f(t)$ over the interval $[a, b]$ on the t-axis (Figure 5.6.8).

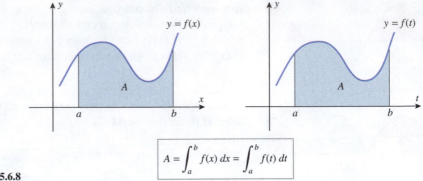

$$A = \int_a^b f(x)\,dx = \int_a^b f(t)\,dt$$

▶ **Figure 5.6.8**

Because the variable of integration in a definite integral plays no role in the end result, it is often referred to as a ***dummy variable***. In summary:

> *Whenever you find it convenient to change the letter used for the variable of integration in a definite integral, you can do so without changing the value of the integral.*

■ THE MEAN-VALUE THEOREM FOR INTEGRALS

To reach our goal of showing that continuous functions have antiderivatives, we will need to develop a basic property of definite integrals, known as the *Mean-Value Theorem for Integrals*. In Section 5.8 we will interpret this theorem using the concept of the "average value" of a continuous function over an interval. Here we will need it as a tool for developing other results.

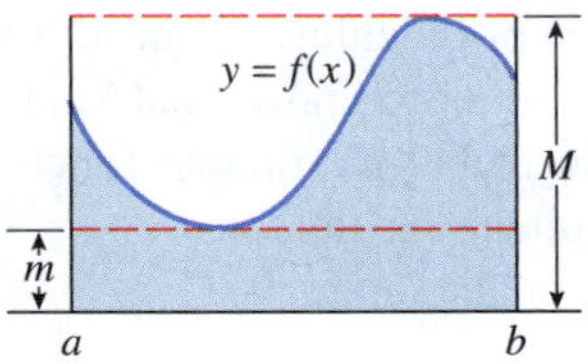

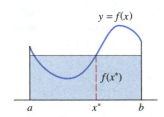

▲ **Figure 5.6.9**

Let f be a continuous nonnegative function on $[a, b]$, and let m and M be the minimum and maximum values of $f(x)$ on this interval. Consider the rectangles of heights m and M over the interval $[a, b]$ (Figure 5.6.9). It is clear geometrically from this figure that the area

$$A = \int_a^b f(x)\,dx$$

under $y = f(x)$ is at least as large as the area of the rectangle of height m and no larger than the area of the rectangle of height M. It seems reasonable, therefore, that there is a rectangle over the interval $[a, b]$ of some appropriate height $f(x^*)$ between m and M whose area is precisely A; that is,

$$\int_a^b f(x)\,dx = f(x^*)(b - a)$$

(Figure 5.6.10). This is a special case of the following result.

The area of the shaded rectangle is equal to the area of the shaded region in Figure 5.6.9.

▲ **Figure 5.6.10**

5.6.2 **THEOREM** (*The Mean-Value Theorem for Integrals*) *If f is continuous on a closed interval $[a, b]$, then there is at least one point x^* in $[a, b]$ such that*

$$\int_a^b f(x)\,dx = f(x^*)(b - a) \tag{8}$$

PROOF By the Extreme-Value Theorem (4.4.2), f assumes a maximum value M and a minimum value m on $[a, b]$. Thus, for all x in $[a, b]$,

$$m \le f(x) \le M$$

and from Theorem 5.5.6(b)

$$\int_a^b m \, dx \leq \int_a^b f(x) \, dx \leq \int_a^b M \, dx$$

or

$$m(b - a) \leq \int_a^b f(x) \, dx \leq M(b - a) \tag{9}$$

or

$$m \leq \frac{1}{b - a} \int_a^b f(x) \, dx \leq M$$

This implies that

$$\frac{1}{b - a} \int_a^b f(x) \, dx \tag{10}$$

is a number between m and M, and since $f(x)$ assumes the values m and M on $[a, b]$, it follows from the Intermediate-Value Theorem (1.5.7) that $f(x)$ must assume the value (10) at some x^* in $[a, b]$; that is,

$$\frac{1}{b - a} \int_a^b f(x) \, dx = f(x^*) \quad \text{or} \quad \int_a^b f(x) \, dx = f(x^*)(b - a) \quad \blacksquare$$

▶ **Example 9** Since $f(x) = x^2$ is continuous on the interval $[1, 4]$, the Mean-Value Theorem for Integrals guarantees that there is a point x^* in $[1, 4]$ such that

$$\int_1^4 x^2 \, dx = f(x^*)(4 - 1) = (x^*)^2(4 - 1) = 3(x^*)^2$$

But

$$\int_1^4 x^2 \, dx = \frac{x^3}{3} \Big]_1^4 = 21$$

so that

$$3(x^*)^2 = 21 \quad \text{or} \quad (x^*)^2 = 7 \quad \text{or} \quad x^* = \pm\sqrt{7}$$

Thus, $x^* = \sqrt{7} \approx 2.65$ is the point in the interval $[1, 4]$ whose existence is guaranteed by the Mean-Value Theorem for Integrals. ◀

■ **PART 2 OF THE FUNDAMENTAL THEOREM OF CALCULUS**

In Section 5.1 we suggested that if f is continuous and nonnegative on $[a, b]$, and if $A(x)$ is the area under the graph of $y = f(x)$ over the interval $[a, x]$ (Figure 5.6.2), then $A'(x) = f(x)$. But $A(x)$ can be expressed as the definite integral

$$A(x) = \int_a^x f(t) \, dt$$

(where we have used t rather than x as the variable of integration to avoid confusion with the x that appears as the upper limit of integration). Thus, the relationship $A'(x) = f(x)$ can be expressed as

$$\frac{d}{dx} \left[\int_a^x f(t) \, dt \right] = f(x)$$

This is a special case of the following more general result, which applies even if f has negative values.

> **5.6.3   THEOREM** (*The Fundamental Theorem of Calculus, Part 2*)   *If f is continuous on an interval, then f has an antiderivative on that interval. In particular, if a is any point in the interval, then the function F defined by*
>
> $$F(x) = \int_a^x f(t)\,dt$$
>
> *is an antiderivative of f; that is, $F'(x) = f(x)$ for each x in the interval, or in an alternative notation*
>
> $$\frac{d}{dx}\left[\int_a^x f(t)\,dt\right] = f(x) \tag{11}$$

PROOF   We will show first that $F(x)$ is defined at each x in the interval. If $x > a$ and x is in the interval, then Theorem 5.5.2 applied to the interval $[a, x]$ and the continuity of f ensure that $F(x)$ is defined; and if x is in the interval and $x \le a$, then Definition 5.5.3 combined with Theorem 5.5.2 ensures that $F(x)$ is defined. Thus, $F(x)$ is defined for all x in the interval.

Next we will show that $F'(x) = f(x)$ for each x in the interval. If x is not an endpoint, then it follows from the definition of a derivative that

$$F'(x) = \lim_{h\to 0} \frac{F(x+h) - F(x)}{h}$$

$$= \lim_{h\to 0} \frac{1}{h}\left[\int_a^{x+h} f(t)\,dt - \int_a^x f(t)\,dt\right]$$

$$= \lim_{h\to 0} \frac{1}{h}\left[\int_a^{x+h} f(t)\,dt + \int_x^a f(t)\,dt\right]$$

$$= \lim_{h\to 0} \frac{1}{h}\int_x^{x+h} f(t)\,dt \qquad \boxed{\text{Theorem 5.5.5}} \tag{12}$$

Applying the Mean-Value Theorem for Integrals (5.6.2) to the integral in (12) we obtain

$$\frac{1}{h}\int_x^{x+h} f(t)\,dt = \frac{1}{h}[f(t^*)\cdot h] = f(t^*) \tag{13}$$

where t^* is some number between x and $x + h$. Because t^* is trapped between x and $x + h$, it follows that $t^* \to x$ as $h \to 0$. Thus, the continuity of f at x implies that $f(t^*) \to f(x)$ as $h \to 0$. Therefore, it follows from (12) and (13) that

$$F'(x) = \lim_{h\to 0}\left(\frac{1}{h}\int_x^{x+h} f(t)\,dt\right) = \lim_{h\to 0} f(t^*) = f(x)$$

If x is an endpoint of the interval, then the two-sided limits in the proof must be replaced by the appropriate one-sided limits, but otherwise the arguments are identical. ■

In words, Formula (11) states:

> *If a definite integral has a variable upper limit of integration, a constant lower limit of integration, and a continuous integrand, then the derivative of the integral with respect to its upper limit is equal to the integrand evaluated at the upper limit.*

▶ **Example 10**   Find

$$\frac{d}{dx}\left[\int_1^x t^3\,dt\right]$$

by applying Part 2 of the Fundamental Theorem of Calculus, and then confirm the result by performing the integration and then differentiating.

Solution. The integrand is a continuous function, so from (11)

$$\frac{d}{dx}\left[\int_1^x t^3\,dt\right] = x^3$$

Alternatively, evaluating the integral and then differentiating yields

$$\int_1^x t^3\,dt = \frac{t^4}{4}\Bigg]_{t=1}^x = \frac{x^4}{4} - \frac{1}{4}, \quad \frac{d}{dx}\left[\frac{x^4}{4} - \frac{1}{4}\right] = x^3$$

so the two methods for differentiating the integral agree. ◀

▶ **Example 11** Since

$$f(x) = \frac{\sin x}{x}$$

is continuous on any interval that does not contain the origin, it follows from (11) that on the interval $(0, +\infty)$ we have

$$\frac{d}{dx}\left[\int_1^x \frac{\sin t}{t}\,dt\right] = \frac{\sin x}{x}$$

Unlike the preceding example, there is no way to evaluate the integral in terms of familiar functions, so Formula (11) provides the only simple method for finding the derivative. ◀

■ DIFFERENTIATION AND INTEGRATION ARE INVERSE PROCESSES

The two parts of the Fundamental Theorem of Calculus, when taken together, tell us that differentiation and integration are inverse processes in the sense that each undoes the effect of the other. To see why this is so, note that Part 1 of the Fundamental Theorem of Calculus (5.6.1) implies that

$$\int_a^x f'(t)\,dt = f(x) - f(a)$$

which tells us that if the value of $f(a)$ is known, then the function f can be recovered from its derivative f' by integrating. Conversely, Part 2 of the Fundamental Theorem of Calculus (5.6.3) states that

$$\frac{d}{dx}\left[\int_a^x f(t)\,dt\right] = f(x)$$

which tells us that the function f can be recovered from its integral by differentiating. Thus, differentiation and integration can be viewed as inverse processes.

It is common to treat parts 1 and 2 of the Fundamental Theorem of Calculus as a single theorem and refer to it simply as the *Fundamental Theorem of Calculus*. This theorem ranks as one of the greatest discoveries in the history of science, and its formulation by Newton and Leibniz is generally regarded to be the "discovery of calculus."

■ INTEGRATING RATES OF CHANGE

The Fundamental Theorem of Calculus

$$\int_a^b f(x)\,dx = F(b) - F(a) \tag{14}$$

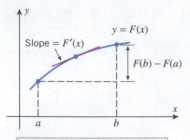

Integrating the slope of $y = F(x)$ over the interval $[a, b]$ produces the change $F(b) - F(a)$ in the value of $F(x)$.

▲ **Figure 5.6.11**

has a useful interpretation that can be seen by rewriting it in a slightly different form. Since F is an antiderivative of f on the interval $[a, b]$, we can use the relationship $F'(x) = f(x)$ to rewrite (14) as

$$\int_a^b F'(x)\, dx = F(b) - F(a) \tag{15}$$

In this formula we can view $F'(x)$ as the rate of change of $F(x)$ with respect to x, and we can view $F(b) - F(a)$ as the *change* in the value of $F(x)$ as x increases from a to b (Figure 5.6.11). Thus, we have the following useful principle.

5.6.4 **INTEGRATING A RATE OF CHANGE** Integrating the rate of change of $F(x)$ with respect to x over an interval $[a, b]$ produces the change in the value of $F(x)$ that occurs as x increases from a to b.

Mitchell Funk/Getty Images

Mathematical analysis plays an important role in understanding human population growth.

Here are some examples of this idea:

- If $s(t)$ is the position of a particle in rectilinear motion, then $s'(t)$ is the instantaneous velocity of the particle at time t, and

$$\int_{t_1}^{t_2} s'(t)\, dt = s(t_2) - s(t_1)$$

is the displacement (or the change in the position) of the particle between the times t_1 and t_2.

- If $P(t)$ is a population (e.g., plants, animals, or people) at time t, then $P'(t)$ is the rate at which the population is changing at time t, and

$$\int_{t_1}^{t_2} P'(t)\, dt = P(t_2) - P(t_1)$$

is the change in the population between times t_1 and t_2.

- If $A(t)$ is the area of an oil spill at time t, then $A'(t)$ is the rate at which the area of the spill is changing at time t, and

$$\int_{t_1}^{t_2} A'(t)\, dt = A(t_2) - A(t_1)$$

is the change in the area of the spill between times t_1 and t_2.

- If $P'(x)$ is the marginal profit that results from producing and selling x units of a product (see Section 4.5), then

$$\int_{x_1}^{x_2} P'(x)\, dx = P(x_2) - P(x_1)$$

is the change in the profit that results when the production level increases from x_1 units to x_2 units.

✔ **QUICK CHECK EXERCISES 5.6** *(See page 376 for answers.)*

1. (a) If $F(x)$ is an antiderivative for $f(x)$, then

$$\int_a^b f(x)\, dx = \underline{\qquad}$$

(b) $\displaystyle\int_a^b F'(x)\, dx = \underline{\qquad}$

(c) $\displaystyle\frac{d}{dx}\left[\int_a^x f(t)\, dt\right] = \underline{\qquad}$

2. (a) $\displaystyle\int_0^2 (3x^2 - 2x)\,dx = $ _____

(b) $\displaystyle\int_{-\pi}^{\pi} \cos x\,dx = $ _____

(c) $\displaystyle\int_0^{\frac{1}{2}\ln 5} e^x\,dx = $ _____

(d) $\displaystyle\int_{-1/2}^{1/2} \frac{1}{\sqrt{1-x^2}}\,dx = $ _____

3. For the function $f(x) = 3x^2 - 2x$ and an interval $[a, b]$, the point x^* guaranteed by the Mean-Value Theorem for Integrals is $x^* = \frac{2}{3}$. It follows that

$$\int_a^b (3x^2 - 2x)\,dx = \text{_____}$$

4. The area of an oil spill is increasing at a rate of $25t$ ft^2/s t seconds after the spill. Between times $t = 2$ and $t = 4$ the area of the spill increases by _____ ft^2.

EXERCISE SET 5.6    Graphing Utility   CAS

---

1. In each part, use a definite integral to find the area of the region, and check your answer using an appropriate formula from geometry.

(a)   (b)   (c)

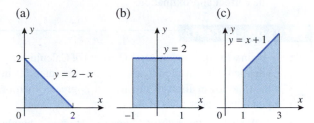

2. In each part, use a definite integral to find the area under the curve $y = f(x)$ over the stated interval, and check your answer using an appropriate formula from geometry.
(a) $f(x) = x$; $[0, 5]$
(b) $f(x) = 5$; $[3, 9]$
(c) $f(x) = x + 3$; $[-1, 2]$

3. In each part, sketch the analogue of Figure 5.6.10 for the specified region. [Let $y = f(x)$ denote the upper boundary of the region. If x^* is unique, label both it and $f(x^*)$ on your sketch. Otherwise, label $f(x^*)$ on your sketch, and determine all values of x^* that satisfy Equation (8).]
(a) The region in part (a) of Exercise 1.
(b) The region in part (b) of Exercise 1.
(c) The region in part (c) of Exercise 1.

4. In each part, sketch the analogue of Figure 5.6.10 for the function and interval specified. [If x^* is unique, label both it and $f(x^*)$ on your sketch. Otherwise, label $f(x^*)$ on your sketch, and determine all values of x^* that satisfy Equation (8).]
(a) The function and interval in part (a) of Exercise 2.
(b) The function and interval in part (b) of Exercise 2.
(c) The function and interval in part (c) of Exercise 2.

5–10 Find the area under the curve $y = f(x)$ over the stated interval. ■

5. $f(x) = x^3$; $[2, 3]$ **6.** $f(x) = x^4$; $[-1, 1]$

7. $f(x) = 3\sqrt{x}$; $[1, 4]$ **8.** $f(x) = x^{-2/3}$; $[1, 27]$

9. $f(x) = e^{2x}$; $[0, \ln 2]$ **10.** $f(x) = \dfrac{1}{x}$; $[1, 5]$

11–12 Find all values of x^* in the stated interval that satisfy Equation (8) in the Mean-Value Theorem for Integrals (5.6.2), and explain what these numbers represent. ■

11. (a) $f(x) = \sqrt{x}$; $[0, 3]$
(b) $f(x) = x^2 + x$; $[-12, 0]$

12. (a) $f(x) = \sin x$; $[-\pi, \pi]$ (b) $f(x) = 1/x^2$; $[1, 3]$

13–30 Evaluate the integrals using Part 1 of the Fundamental Theorem of Calculus. ■

13. $\displaystyle\int_{-2}^1 (x^2 - 6x + 12)\,dx$ **14.** $\displaystyle\int_{-1}^2 4x(1 - x^2)\,dx$

15. $\displaystyle\int_1^4 \frac{4}{x^2}\,dx$ **16.** $\displaystyle\int_1^2 \frac{1}{x^6}\,dx$

17. $\displaystyle\int_4^9 2x\sqrt{x}\,dx$ **18.** $\displaystyle\int_1^4 \frac{1}{x\sqrt{x}}\,dx$

19. $\displaystyle\int_{-\pi/2}^{\pi/2} \sin\theta\,d\theta$ **20.** $\displaystyle\int_0^{\pi/4} \sec^2\theta\,d\theta$

21. $\displaystyle\int_{-\pi/4}^{\pi/4} \cos x\,dx$ **22.** $\displaystyle\int_0^{\pi/3} (2x - \sec x \tan x)\,dx$

23. $\displaystyle\int_{\ln 2}^3 5e^x\,dx$ **24.** $\displaystyle\int_{1/2}^1 \frac{1}{2x}\,dx$

25. $\displaystyle\int_0^{1/\sqrt{2}} \frac{dx}{\sqrt{1-x^2}}$ **26.** $\displaystyle\int_{-1}^1 \frac{dx}{1+x^2}$

27. $\displaystyle\int_{\sqrt{2}}^2 \frac{dx}{x\sqrt{x^2-1}}$ **28.** $\displaystyle\int_{-\sqrt{2}}^{-2/\sqrt{3}} \frac{dx}{x\sqrt{x^2-1}}$

29. $\displaystyle\int_1^4 \left(\frac{1}{\sqrt{t}} - 3\sqrt{t}\right)dt$ **30.** $\displaystyle\int_{\pi/6}^{\pi/2} \left(x + \frac{2}{\sin^2 x}\right)dx$

31–34 Use Theorem 5.5.5 to evaluate the given integrals. ■

31. (a) $\displaystyle\int_{-1}^1 |2x - 1|\,dx$ (b) $\displaystyle\int_0^{3\pi/4} |\cos x|\,dx$

32. (a) $\displaystyle\int_{-1}^2 \sqrt{2 + |x|}\,dx$ (b) $\displaystyle\int_0^{\pi/2} \left|\tfrac{1}{2} - \cos x\right|\,dx$

33. (a) $\int_{-1}^{1} |e^x - 1| \, dx$ (b) $\int_{1}^{4} \frac{|2 - x|}{x} \, dx$

34. (a) $\int_{-3}^{3} \left| x^2 - 1 - \frac{15}{x^2 + 1} \right| \, dx$

(b) $\int_{0}^{\sqrt{3}/2} \left| \frac{1}{\sqrt{1 - x^2}} - \sqrt{2} \right| \, dx$

35–36 A function $f(x)$ is defined piecewise on an interval. In these exercises: (a) Use Theorem 5.5.5 to find the integral of $f(x)$ over the interval. (b) Find an antiderivative of $f(x)$ on the interval. (c) Use parts (a) and (b) to verify Part 1 of the Fundamental Theorem of Calculus. ∎

35. $f(x) = \begin{cases} x, & 0 \le x \le 1 \\ x^2, & 1 < x \le 2 \end{cases}$

36. $f(x) = \begin{cases} \sqrt{x}, & 0 \le x < 1 \\ 1/x^2, & 1 \le x \le 4 \end{cases}$

37–40 True–False Determine whether the statement is true or false. Explain your answer. ∎

37. There does not exist a differentiable function $F(x)$ such that $F'(x) = |x|$.

38. If $f(x)$ is continuous on the interval $[a, b]$, and if the definite integral of $f(x)$ over this interval has value 0, then the equation $f(x) = 0$ has at least one solution in the interval $[a, b]$.

39. If $F(x)$ is an antiderivative of $f(x)$ and $G(x)$ is an antiderivative of $g(x)$, then

$$\int_{a}^{b} f(x) \, dx = \int_{a}^{b} g(x) \, dx$$

if and only if

$$G(a) + F(b) = F(a) + G(b)$$

40. If $f(x)$ is continuous everywhere and

$$F(x) = \int_{0}^{x} f(t) \, dt$$

then the equation $F(x) = 0$ has at least one solution.

41–44 Use a calculating utility to find the midpoint approximation of the integral using $n = 20$ subintervals, and then find the exact value of the integral using Part 1 of the Fundamental Theorem of Calculus. ∎

41. $\int_{1}^{3} \frac{1}{x^2} \, dx$ **42.** $\int_{0}^{\pi/2} \sin x \, dx$

43. $\int_{-1}^{1} \sec^2 x \, dx$ **44.** $\int_{1}^{3} \frac{1}{x} \, dx$

45–48 Sketch the region described and find its area. ∎

45. The region under the curve $y = x^2 + 1$ and over the interval $[0, 3]$.

46. The region below the curve $y = x - x^2$ and above the x-axis.

47. The region under the curve $y = 3 \sin x$ and over the interval $[0, 2\pi/3]$.

48. The region below the interval $[-2, -1]$ and above the curve $y = x^3$.

49–52 Sketch the curve and find the total area between the curve and the given interval on the x-axis. ∎

49. $y = x^2 - x$; $[0, 2]$ **50.** $y = \sin x$; $[0, 3\pi/2]$

51. $y = e^x - 1$; $[-1, 1]$ **52.** $y = \frac{x^2 - 1}{x^2}$; $[\frac{1}{2}, 2]$

53. A student wants to find the area enclosed by the graphs of $y = 1/\sqrt{1 - x^2}$, $y = 0$, $x = 0$, and $x = 0.8$.
(a) Show that the exact area is $\sin^{-1} 0.8$.
(b) The student uses a calculator to approximate the result in part (a) to two decimal places and obtains an incorrect answer of 53.13. What was the student's error? Find the correct approximation.

FOCUS ON CONCEPTS

54. Explain why the Fundamental Theorem of Calculus may be applied without modification to definite integrals in which the lower limit of integration is greater than or equal to the upper limit of integration.

55. (a) If $h'(t)$ is the rate of change of a child's height measured in inches per year, what does the integral $\int_{0}^{10} h'(t) \, dt$ represent, and what are its units?
(b) If $r'(t)$ is the rate of change of the radius of a spherical balloon measured in centimeters per second, what does the integral $\int_{1}^{2} r'(t) \, dt$ represent, and what are its units?
(c) If $H(t)$ is the rate of change of the speed of sound with respect to temperature measured in ft/s per °F, what does the integral $\int_{32}^{100} H(t) \, dt$ represent, and what are its units?
(d) If $v(t)$ is the velocity of a particle in rectilinear motion, measured in cm/h, what does the integral $\int_{t_1}^{t_2} v(t) \, dt$ represent, and what are its units?

56. (a) Use a graphing utility to generate the graph of
$$f(x) = \frac{1}{100}(x + 2)(x + 1)(x - 3)(x - 5)$$
and use the graph to make a conjecture about the sign of the integral
$$\int_{-2}^{5} f(x) \, dx$$
(b) Check your conjecture by evaluating the integral.

57. Define $F(x)$ by
$$F(x) = \int_{1}^{x} (3t^2 - 3) \, dt$$
(a) Use Part 2 of the Fundamental Theorem of Calculus to find $F'(x)$.
(b) Check the result in part (a) by first integrating and then differentiating.

58. Define $F(x)$ by

$$F(x) = \int_{\pi/4}^{x} \cos 2t \, dt$$

(a) Use Part 2 of the Fundamental Theorem of Calculus to find $F'(x)$.

(b) Check the result in part (a) by first integrating and then differentiating.

59–62 Use Part 2 of the Fundamental Theorem of Calculus to find the derivatives. ■

59. (a) $\dfrac{d}{dx} \displaystyle\int_{1}^{x} \sin(t^2) \, dt$ (b) $\dfrac{d}{dx} \displaystyle\int_{0}^{x} e^{\sqrt{t}} \, dt$

60. (a) $\dfrac{d}{dx} \displaystyle\int_{0}^{x} \dfrac{dt}{1 + \sqrt{t}}$ (b) $\dfrac{d}{dx} \displaystyle\int_{1}^{x} \ln t \, dt$

61. $\dfrac{d}{dx} \displaystyle\int_{x}^{0} t \sec t \, dt$ [*Hint*: Use Definition 5.5.3(b).]

62. $\dfrac{d}{du} \displaystyle\int_{0}^{u} |x| \, dx$

63. Let $F(x) = \displaystyle\int_{4}^{x} \sqrt{t^2 + 9} \, dt$. Find

(a) $F(4)$ (b) $F'(4)$ (c) $F''(4)$.

64. Let $F(x) = \displaystyle\int_{\sqrt{3}}^{x} \tan^{-1} t \, dt$. Find

(a) $F(\sqrt{3})$ (b) $F'(\sqrt{3})$ (c) $F''(\sqrt{3})$.

65. Let $F(x) = \displaystyle\int_{0}^{x} \dfrac{t - 3}{t^2 + 7} \, dt$ for $-\infty < x < +\infty$.

(a) Find the value of x where F attains its minimum value.

(b) Find intervals over which F is only increasing or only decreasing.

(c) Find open intervals over which F is only concave up or only concave down.

C 66. Use the plotting and numerical integration commands of a CAS to generate the graph of the function F in Exercise 65 over the interval $-20 \leq x \leq 20$, and confirm that the graph is consistent with the results obtained in that exercise.

67. (a) Over what open interval does the formula

$$F(x) = \int_{1}^{x} \dfrac{dt}{t}$$

represent an antiderivative of $f(x) = 1/x$?

(b) Find a point where the graph of F crosses the x-axis.

68. (a) Over what open interval does the formula

$$F(x) = \int_{1}^{x} \dfrac{1}{t^2 - 9} \, dt$$

represent an antiderivative of

$$f(x) = \dfrac{1}{x^2 - 9}?$$

(b) Find a point where the graph of F crosses the x-axis.

69. (a) Suppose that a reservoir supplies water to an industrial park at a constant rate of $r = 4$ gallons per minute (gal/min) between 8:30 A.M. and 9:00 A.M. How much water does the reservoir supply during that time period?

(b) Suppose that one of the industrial plants increases its water consumption between 9:00 A.M. and 10:00 A.M. and that the rate at which the reservoir supplies water increases linearly, as shown in the accompanying figure. How much water does the reservoir supply during that 1-hour time period?

(c) Suppose that from 10:00 A.M. to 12 noon the rate at which the reservoir supplies water is given by the formula $r(t) = 10 + \sqrt{t}$ gal/min, where t is the time (in minutes) since 10:00 A.M. How much water does the reservoir supply during that 2-hour time period?

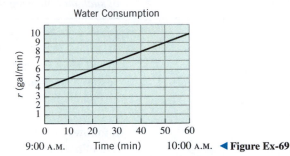

Water Consumption

◀ Figure Ex-69

70. A traffic engineer monitors the rate at which cars enter the main highway during the afternoon rush hour. From her data she estimates that between 4:30 P.M. and 5:30 P.M. the rate $R(t)$ at which cars enter the highway is given by the formula $R(t) = 100(1 - 0.0001t^2)$ cars per minute, where t is the time (in minutes) since 4:30 P.M.

(a) When does the peak traffic flow into the highway occur?

(b) Estimate the number of cars that enter the highway during the rush hour.

71–72 Evaluate each limit by interpreting it as a Riemann sum in which the given interval is divided into n subintervals of equal width. ■

71. $\displaystyle\lim_{n \to +\infty} \sum_{k=1}^{n} \dfrac{\pi}{4n} \sec^2\left(\dfrac{\pi k}{4n}\right); \quad \left[0, \dfrac{\pi}{4}\right]$

72. $\displaystyle\lim_{n \to +\infty} \sum_{k=1}^{n} \dfrac{n}{n^2 + k^2}; \quad [0, 1]$

73. Prove the Mean-Value Theorem for Integrals (Theorem 5.6.2) by applying the Mean-Value Theorem (4.8.2) to an antiderivative F for f.

74. Writing Write a short paragraph that describes the various ways in which integration and differentiation may be viewed as inverse processes. (Be sure to discuss both definite and indefinite integrals.)

75. Writing Let f denote a function that is continuous on an interval $[a, b]$, and let x^* denote the point guaranteed by the Mean-Value Theorem for Integrals. Explain geometrically why $f(x^*)$ may be interpreted as a "mean" or average value of $f(x)$ over $[a, b]$. (In Section 5.8 we will discuss the concept of "average value" in more detail.)

5.7 RECTILINEAR MOTION REVISITED USING INTEGRATION

In Section 4.6 we used the derivative to define the notions of instantaneous velocity and acceleration for a particle in rectilinear motion. In this section we will resume the study of such motion using the tools of integration.

■ FINDING POSITION AND VELOCITY BY INTEGRATION

Recall from Formulas (1) and (3) of Section 4.6 that if a particle in rectilinear motion has position function $s(t)$, then its instantaneous velocity and acceleration are given by the formulas

$$v(t) = s'(t) \quad \text{and} \quad a(t) = v'(t)$$

It follows from these formulas that $s(t)$ is an antiderivative of $v(t)$ and $v(t)$ is an antiderivative of $a(t)$; that is,

$$s(t) = \int v(t)\, dt \qquad \text{and} \qquad v(t) = \int a(t)\, dt \qquad (1\text{–}2)$$

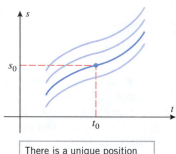

There is a unique position function such that $s(t_0) = s_0$.

▲ **Figure 5.7.1**

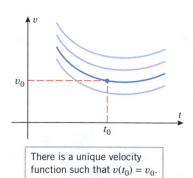

There is a unique velocity function such that $v(t_0) = v_0$.

▲ **Figure 5.7.2**

By Formula (1), if we know the velocity function $v(t)$ of a particle in rectilinear motion, then by integrating $v(t)$ we can produce a family of position functions with that velocity function. If, in addition, we know the position s_0 of the particle at any time t_0, then we have sufficient information to find the constant of integration and determine a unique position function (Figure 5.7.1). Similarly, if we know the acceleration function $a(t)$ of the particle, then by integrating $a(t)$ we can produce a family of velocity functions with that acceleration function. If, in addition, we know the velocity v_0 of the particle at any time t_0, then we have sufficient information to find the constant of integration and determine a unique velocity function (Figure 5.7.2).

▶ **Example 1**   Suppose that a particle moves with velocity $v(t) = \cos \pi t$ along a coordinate line. Assuming that the particle has coordinate $s = 4$ at time $t = 0$, find its position function.

Solution.   The position function is

$$s(t) = \int v(t)\, dt = \int \cos \pi t\, dt = \frac{1}{\pi} \sin \pi t + C$$

Since $s = 4$ when $t = 0$, it follows that

$$4 = s(0) = \frac{1}{\pi} \sin 0 + C = C$$

Thus,

$$s(t) = \frac{1}{\pi} \sin \pi t + 4 \quad ◀$$

■ COMPUTING DISPLACEMENT AND DISTANCE TRAVELED BY INTEGRATION

Recall that the displacement over a time interval of a particle in rectilinear motion is its final coordinate minus its initial coordinate. Thus, if the position function of the particle is $s(t)$, then its displacement (or change in position) over the time interval $[t_0, t_1]$ is $s(t_1) - s(t_0)$. This can be written in integral form as

Recall that Formula (3) is a special case of the formula

$$\int_a^b F'(x)\,dx = F(b) - F(a)$$

for integrating a rate of change.

$$\begin{bmatrix} \text{displacement} \\ \text{over the time} \\ \text{interval } [t_0, t_1] \end{bmatrix} = \int_{t_0}^{t_1} v(t)\,dt = \int_{t_0}^{t_1} s'(t)\,dt = s(t_1) - s(t_0) \tag{3}$$

In contrast, to find the distance traveled by the particle over the time interval $[t_0, t_1]$ (distance traveled in the positive direction plus the distance traveled in the negative direction), we must integrate the absolute value of the velocity function; that is,

$$\begin{bmatrix} \text{distance traveled} \\ \text{during time} \\ \text{interval } [t_0, t_1] \end{bmatrix} = \int_{t_0}^{t_1} |v(t)|\,dt \tag{4}$$

Since the absolute value of velocity is speed, Formulas (3) and (4) can be summarized informally as follows:

> *Integrating velocity over a time interval produces displacement, and integrating speed over a time interval produces distance traveled.*

▶ **Example 2** Suppose that a particle moves on a coordinate line so that its velocity at time t is $v(t) = t^2 - 2t$ m/s (Figure 5.7.3).

(a) Find the displacement of the particle during the time interval $0 \le t \le 3$.

(b) Find the distance traveled by the particle during the time interval $0 \le t \le 3$.

Solution (a). From (3) the displacement is

$$\int_0^3 v(t)\,dt = \int_0^3 (t^2 - 2t)\,dt = \left[\frac{t^3}{3} - t^2\right]_0^3 = 0$$

Thus, the particle is at the same position at time $t = 3$ as at $t = 0$.

Solution (b). The velocity can be written as $v(t) = t^2 - 2t = t(t - 2)$, from which we see that $v(t) \le 0$ for $0 \le t \le 2$ and $v(t) \ge 0$ for $2 \le t \le 3$. Thus, it follows from (4) that the distance traveled is

$$\int_0^3 |v(t)|\,dt = \int_0^2 -v(t)\,dt + \int_2^3 v(t)\,dt$$

$$= \int_0^2 -(t^2 - 2t)\,dt + \int_2^3 (t^2 - 2t)\,dt$$

$$= -\left[\frac{t^3}{3} - t^2\right]_0^2 + \left[\frac{t^3}{3} - t^2\right]_2^3 = \frac{4}{3} + \frac{4}{3} = \frac{8}{3} \text{ m} \quad ◀$$

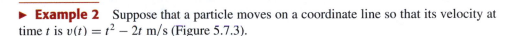

Figure 5.7.3

The figure shows a graph with the curve $v = t^2 - 2t$.

In physical problems it is important to associate correct units with definite integrals. In general, the units for

$$\int_a^b f(x)\,dx$$

are units of $f(x)$ times units of x, since the integral is the limit of Riemann sums, each of whose terms has these units. For example, if $v(t)$ is in meters per second (m/s) and t is in seconds (s), then

$$\int_a^b v(t)\,dt$$

is in meters since

$$\text{(m/s)} \times \text{s} = \text{m}$$

■ ANALYZING THE VELOCITY VERSUS TIME CURVE

In Section 4.6 we showed how to use the position versus time curve to obtain information about the behavior of a particle in rectilinear motion (Table 4.6.1). Similarly, there is valuable information that can be obtained from the velocity versus time curve. For example, the integral in (3) can be interpreted geometrically as the *net signed area* between the graph of $v(t)$ and the interval $[t_0, t_1]$, and the integral in (4) can be interpreted as the *total area* between the graph of $v(t)$ and the interval $[t_0, t_1]$. Thus we have the following result.

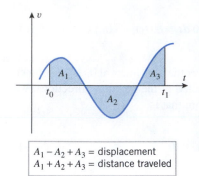

$A_1 - A_2 + A_3$ = displacement
$A_1 + A_2 + A_3$ = distance traveled

▲ **Figure 5.7.4**

5.7.1 FINDING DISPLACEMENT AND DISTANCE TRAVELED FROM THE VELOCITY VERSUS TIME CURVE   For a particle in rectilinear motion, the net signed area between the velocity versus time curve and the interval $[t_0, t_1]$ on the t-axis represents the displacement of the particle over that time interval, and the total area between the velocity versus time curve and the interval $[t_0, t_1]$ on the t-axis represents the distance traveled by the particle over that time interval (Figure 5.7.4).

▶ **Example 3**   Figure 5.7.5 shows three velocity versus time curves for a particle in rectilinear motion along a horizontal line with the positive direction to the right. In each case find the displacement and the distance traveled over the time interval $0 \le t \le 4$, and explain what that information tells you about the motion of the particle.

Solution (a).   In part (*a*) of the figure the area and the net signed area over the interval are both 2. Thus, at the end of the time period the particle is 2 units to the right of its starting point and has traveled a distance of 2 units.

Solution (b).   In part (*b*) of the figure the net signed area is -2, and the total area is 2. Thus, at the end of the time period the particle is 2 units to the left of its starting point and has traveled a distance of 2 units.

Solution (c).   In part (*c*) of the figure the net signed area is 0, and the total area is 2. Thus, at the end of the time period the particle is back at its starting point and has traveled a distance of 2 units. More specifically, it traveled 1 unit to the right over the time interval $0 \le t \le 1$ and then 1 unit to the left over the time interval $1 \le t \le 2$ (why?). ◀

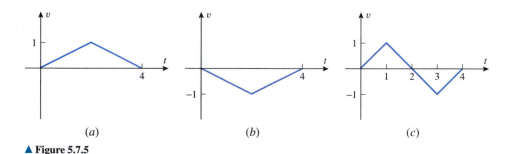

(*a*)   (*b*)   (*c*)

▲ **Figure 5.7.5**

■ **CONSTANT ACCELERATION**

One of the most important cases of rectilinear motion occurs when a particle has *constant acceleration*. We will show that if a particle moves with constant acceleration along an s-axis, and if the position and velocity of the particle are known at some point in time, say when $t = 0$, then it is possible to derive formulas for the position $s(t)$ and the velocity $v(t)$ at any time t. To see how this can be done, suppose that the particle has constant acceleration

$$a(t) = a \tag{5}$$

and

$$s = s_0 \quad \text{when} \quad t = 0 \tag{6}$$
$$v = v_0 \quad \text{when} \quad t = 0 \tag{7}$$

where s_0 and v_0 are known. We call (6) and (7) the *initial conditions*.

With (5) as a starting point, we can integrate $a(t)$ to obtain $v(t)$, and we can integrate $v(t)$ to obtain $s(t)$, using an initial condition in each case to determine the constant of integration. The computations are as follows:

$$v(t) = \int a(t)\, dt = \int a\, dt = at + C_1 \tag{8}$$

To determine the constant of integration C_1 we apply initial condition (7) to this equation to obtain

$$v_0 = v(0) = a \cdot 0 + C_1 = C_1$$

Substituting this in (8) and putting the constant term first yields

$$v(t) = v_0 + at$$

Since v_0 is constant, it follows that

$$s(t) = \int v(t)\, dt = \int (v_0 + at)\, dt = v_0 t + \tfrac{1}{2}at^2 + C_2 \tag{9}$$

To determine the constant C_2 we apply initial condition (6) to this equation to obtain

$$s_0 = s(0) = v_0 \cdot 0 + \tfrac{1}{2}a \cdot 0 + C_2 = C_2$$

Substituting this in (9) and putting the constant term first yields

$$s(t) = s_0 + v_0 t + \tfrac{1}{2}at^2$$

In summary, we have the following result.

5.7.2 CONSTANT ACCELERATION If a particle moves with constant acceleration a along an s-axis, and if the position and velocity at time $t = 0$ are s_0 and v_0, respectively, then the position and velocity functions of the particle are

$$s(t) = s_0 + v_0 t + \tfrac{1}{2}at^2 \tag{10}$$

$$v(t) = v_0 + at \tag{11}$$

How can you tell from the graph of the velocity versus time curve whether a particle moving along a line has constant acceleration?

▶ **Example 4** Suppose that an intergalactic spacecraft uses a sail and the "solar wind" to produce a constant acceleration of 0.032 m/s^2. Assuming that the spacecraft has a velocity of 10,000 m/s when the sail is first raised, how far will the spacecraft travel in 1 hour, and what will its velocity be at the end of this hour?

Solution. In this problem the choice of a coordinate axis is at our discretion, so we will choose it to make the computations as simple as possible. Accordingly, let us introduce an s-axis whose positive direction is in the direction of motion, and let us take the origin to coincide with the position of the spacecraft at the time $t = 0$ when the sail is raised. Thus, Formulas (10) and (11) apply with

$$s_0 = s(0) = 0, \quad v_0 = v(0) = 10{,}000, \quad \text{and} \quad a = 0.032$$

Since 1 hour corresponds to $t = 3600$ s, it follows from (10) that in 1 hour the spacecraft travels a distance of

$$s(3600) = 10{,}000(3600) + \tfrac{1}{2}(0.032)(3600)^2 \approx 36{,}200{,}000 \text{ m}$$

and it follows from (11) that after 1 hour its velocity is

$$v(3600) = 10{,}000 + (0.032)(3600) \approx 10{,}100 \text{ m/s} \blacktriangleleft$$

▶ **Example 5** A bus has stopped to pick up riders, and a woman is running at a constant velocity of 5 m/s to catch it. When she is 11 m behind the front door the bus pulls away with a constant acceleration of 1 m/s². From that point in time, how long will it take for the woman to reach the front door of the bus if she keeps running with a velocity of 5 m/s?

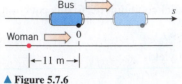

▲ **Figure 5.7.6**

Solution. As shown in Figure 5.7.6, choose the s-axis so that the bus and the woman are moving in the positive direction, and the front door of the bus is at the origin at the time $t = 0$ when the bus begins to pull away. To catch the bus at some later time t, the woman will have to cover a distance $s_w(t)$ that is equal to 11 m plus the distance $s_b(t)$ traveled by the bus; that is, the woman will catch the bus when

$$s_w(t) = s_b(t) + 11 \tag{12}$$

Since the woman has a constant velocity of 5 m/s, the distance she travels in t seconds is $s_w(t) = 5t$. Thus, (12) can be written as

$$s_b(t) = 5t - 11 \tag{13}$$

Since the bus has a constant acceleration of $a = 1$ m/s², and since $s_0 = v_0 = 0$ at time $t = 0$ (why?), it follows from (10) that

$$s_b(t) = \tfrac{1}{2}t^2$$

Substituting this equation into (13) and reorganizing the terms yields the quadratic equation

$$\tfrac{1}{2}t^2 - 5t + 11 = 0 \quad \text{or} \quad t^2 - 10t + 22 = 0$$

Solving this equation for t using the quadratic formula yields two solutions:

$$t = 5 - \sqrt{3} \approx 3.3 \quad \text{and} \quad t = 5 + \sqrt{3} \approx 6.7$$

(verify). Thus, the woman can reach the door at two different times, $t = 3.3$ s and $t = 6.7$ s. The reason that there are two solutions can be explained as follows: When the woman first reaches the door, she is running faster than the bus and can run past it if the driver does not see her. However, as the bus speeds up, it eventually catches up to her, and she has another chance to flag it down. ◀

■ FREE-FALL MODEL

Motion that occurs when an object near the Earth is imparted some initial velocity (up or down) and thereafter moves along a vertical line is called *free-fall motion*. In modeling free-fall motion we assume that the only force acting on the object is the Earth's gravity and that the object stays sufficiently close to the Earth that the gravitational force is constant. In particular, air resistance and the gravitational pull of other celestial bodies are neglected.

In our model we will ignore the physical size of the object by treating it as a particle, and we will assume that it moves along an s-axis whose origin is at the surface of the Earth and whose positive direction is up. With this convention, the s-coordinate of the particle is the height of the particle above the surface of the Earth (Figure 5.7.7).

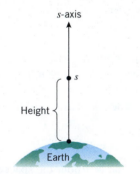

▲ **Figure 5.7.7**

It is a fact of physics that a particle with free-fall motion has constant acceleration. The magnitude of this constant, denoted by the letter g, is called the *acceleration due to gravity* and is approximately 9.8 m/s² or 32 ft/s², depending on whether distance is measured in meters or feet.[*]

Recall that a particle is speeding up when its velocity and acceleration have the same sign and is slowing down when they have opposite signs. Thus, because we have chosen

---

[*]Strictly speaking, the constant g varies with the latitude and the distance from the Earth's center. However, for motion at a fixed latitude and near the surface of the Earth, the assumption of a constant g is satisfactory for many applications.

the positive direction to be up, it follows that the acceleration $a(t)$ of a particle in free fall is negative for all values of t. To see that this is so, observe that an upward-moving particle (positive velocity) is slowing down, so its acceleration must be negative; and a downward-moving particle (negative velocity) is speeding up, so its acceleration must also be negative. Thus, we conclude that

$$a(t) = -g \tag{14}$$

It now follows from this and Formulas (10) and (11) that the position and velocity functions for a particle in free-fall motion are

$$s(t) = s_0 + v_0 t - \tfrac{1}{2}gt^2 \tag{15}$$

$$v(t) = v_0 - gt \tag{16}$$

How would Formulas (14), (15), and (16) change if we choose the direction of the positive s-axis to be down?

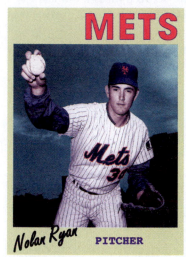

Corbis.Bettmann
Nolan Ryan's rookie baseball card.

In Example 6 the ball is moving up when the velocity is positive and is moving down when the velocity is negative, so it makes sense physically that the velocity is zero when the ball reaches its peak.

▶ **Example 6** Nolan Ryan, a member of the Baseball Hall of Fame and one of the fastest baseball pitchers of all time, was able to throw a baseball 150 ft/s (over 102 mi/h). During his career, he had the opportunity to pitch in the Houston Astrodome, home to the Houston Astros Baseball Team from 1965 to 1999. The Astrodome was an indoor stadium with a ceiling 208 ft high. Could Nolan Ryan have hit the ceiling of the Astrodome if he were capable of giving a baseball an upward velocity of 100 ft/s from a height of 7 ft?

Solution. Since distance is in feet, we take $g = 32$ ft/s^2. Initially, we have $s_0 = 7$ ft and $v_0 = 100$ ft/s, so from (15) and (16) we have

$$s(t) = 7 + 100t - 16t^2$$
$$v(t) = 100 - 32t$$

The ball will rise until $v(t) = 0$, that is, until $100 - 32t = 0$. Solving this equation we see that the ball is at its maximum height at time $t = \frac{25}{8}$. To find the height of the ball at this instant we substitute this value of t into the position function to obtain

$$s\left(\tfrac{25}{8}\right) = 7 + 100\left(\tfrac{25}{8}\right) - 16\left(\tfrac{25}{8}\right)^2 = 163.25 \text{ ft}$$

which is roughly 45 ft short of hitting the ceiling. ◀

▶ **Example 7** A penny is released from rest near the top of the Empire State Building at a point that is 1250 ft above the ground (Figure 5.7.8). Assuming that the free-fall model applies, how long does it take for the penny to hit the ground, and what is its speed at the time of impact?

Solution. Since distance is in feet, we take $g = 32$ ft/s^2. Initially, we have $s_0 = 1250$ and $v_0 = 0$, so from (15)

$$s(t) = 1250 - 16t^2 \tag{17}$$

Impact occurs when $s(t) = 0$. Solving this equation for t, we obtain

$$1250 - 16t^2 = 0$$

$$t^2 = \frac{1250}{16} = \frac{625}{8}$$

$$t = \pm\frac{25}{\sqrt{8}} \approx \pm 8.8 \text{ s}$$

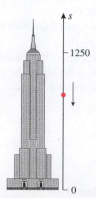

▲ **Figure 5.7.8**

Since $t \geq 0$, we can discard the negative solution and conclude that it takes $25/\sqrt{8} \approx 8.8$ s for the penny to hit the ground. To obtain the velocity at the time of impact, we substitute $t = 25/\sqrt{8}$, $v_0 = 0$, and $g = 32$ in (16) to obtain

$$v\left(\frac{25}{\sqrt{8}}\right) = 0 - 32\left(\frac{25}{\sqrt{8}}\right) = -200\sqrt{2} \approx -282.8 \text{ ft/s}$$

Thus, the speed at the time of impact is

$$\left|v\left(\frac{25}{\sqrt{8}}\right)\right| = 200\sqrt{2} \approx 282.8 \text{ ft/s}$$

which is more than 192 mi/h. ◄

✔ QUICK CHECK EXERCISES 5.7   *(See page 385 for answers.)*

1. Suppose that a particle is moving along an s-axis with velocity $v(t) = 2t + 1$. If at time $t = 0$ the particle is at position $s = 2$, the position function of the particle is $s(t) = $ _____.

2. Let $v(t)$ denote the velocity function of a particle that is moving along an s-axis with constant acceleration $a = -2$. If $v(1) = 4$, then $v(t) = $ _____.

3. Let $v(t)$ denote the velocity function of a particle in rectilinear motion. Suppose that $v(0) = -1$, $v(3) = 2$, and the

velocity versus time curve is a straight line. The displacement of the particle between times $t = 0$ and $t = 3$ is _____, and the distance traveled by the particle over this period of time is _____.

4. Based on the free-fall model, from what height must a coin be dropped so that it strikes the ground with speed 48 ft/s?

EXERCISE SET 5.7   Graphing Utility   ⒸCAS

FOCUS ON CONCEPTS

1. In each part, the velocity versus time curve is given for a particle moving along a line. Use the curve to find the displacement and the distance traveled by the particle over the time interval $0 \le t \le 3$.

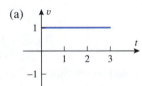

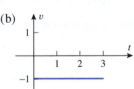

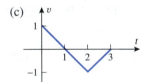

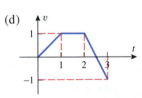

2. Sketch a velocity versus time curve for a particle that travels a distance of 5 units along a coordinate line during the time interval $0 \le t \le 10$ and has a displacement of 0 units.

3. The accompanying figure shows the acceleration versus time curve for a particle moving along a coordinate line. If the initial velocity of the particle is 20 m/s, estimate
(a) the velocity at time $t = 4$ s
(b) the velocity at time $t = 6$ s.

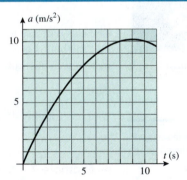

◄ **Figure Ex-3**

4. The accompanying figure shows the velocity versus time curve over the time interval $1 \le t \le 5$ for a particle moving along a horizontal coordinate line.
(a) What can you say about the sign of the acceleration over the time interval?
(b) When is the particle speeding up? Slowing down?
(c) What can you say about the location of the particle at time $t = 5$ relative to its location at time $t = 1$? Explain your reasoning.

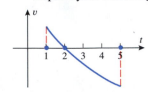

◄ **Figure Ex-4**

5–8 A particle moves along an s-axis. Use the given information to find the position function of the particle. ■

5. (a) $v(t) = 3t^2 - 2t$; $s(0) = 1$
   (b) $a(t) = 3 \sin 3t$; $v(0) = 3$; $s(0) = 3$

6. (a) $v(t) = 1 + \sin t$; $s(0) = -3$
   (b) $a(t) = t^2 - 3t + 1$; $v(0) = 0$; $s(0) = 0$

7. (a) $v(t) = 3t + 1$; $s(2) = 4$
   (b) $a(t) = t^{-2}$; $v(1) = 0$; $s(1) = 2$

8. (a) $v(t) = t^{2/3}$; $s(8) = 0$
   (b) $a(t) = \sqrt{t}$; $v(4) = 1$; $s(4) = -5$

9–12 A particle moves with a velocity of $v(t)$ m/s along an s-axis. Find the displacement and the distance traveled by the particle during the given time interval. ■

9. (a) $v(t) = \sin t$; $0 \le t \le \pi/2$
   (b) $v(t) = \cos t$; $\pi/2 \le t \le 2\pi$

10. (a) $v(t) = 3t - 2$; $0 \le t \le 2$
    (b) $v(t) = |1 - 2t|$; $0 \le t \le 2$

11. (a) $v(t) = t^3 - 3t^2 + 2t$; $0 \le t \le 3$
    (b) $v(t) = \sqrt{t} - 2$; $0 \le t \le 3$

12. (a) $v(t) = t - \sqrt{t}$; $0 \le t \le 4$
    (b) $v(t) = \dfrac{1}{\sqrt{t+1}}$; $0 < t < 3$

13–16 A particle moves with acceleration $a(t)$ m/s^2 along an s-axis and has velocity v_0 m/s at time $t = 0$. Find the displacement and the distance traveled by the particle during the given time interval. ■

13. $a(t) = 3$; $v_0 = -1$; $0 \le t \le 2$

14. $a(t) = t - 2$; $v_0 = 0$; $1 \le t \le 5$

15. $a(t) = 1/\sqrt{3t + 1}$; $v_0 = \frac{4}{3}$; $1 \le t \le 5$

16. $a(t) = \sin t$; $v_0 = 1$; $\pi/4 \le t \le \pi/2$

17. In each part, use the given information to find the position, velocity, speed, and acceleration at time $t = 1$.
    (a) $v = \sin \frac{1}{2}\pi t$; $s = 0$ when $t = 0$
    (b) $a = -3t$; $s = 1$ and $v = 0$ when $t = 0$

18. In each part, use the given information to find the position, velocity, speed, and acceleration at time $t = 1$.
    (a) $v = \cos \frac{1}{3}\pi t$; $s = 0$ when $t = \frac{3}{2}$
    (b) $a = 4e^{2t-2}$; $s = 1/e^2$ and $v = (2/e^2) - 3$ when $t = 0$

19. Suppose that a particle moves along a line so that its velocity v at time t is given by

$$v(t) = \begin{cases} 5t, & 0 \le t < 1 \\ 6\sqrt{t} - \dfrac{1}{t}, & 1 \le t \end{cases}$$

where t is in seconds and v is in centimeters per second (cm/s). Estimate the time(s) at which the particle is 4 cm from its starting position.

20. Suppose that a particle moves along a line so that its velocity v at time t is given by

$$v(t) = \frac{3}{t^2 + 1} - 0.5t, \quad t \ge 0$$

where t is in seconds and v is in centimeters per second (cm/s). Estimate the time(s) at which the particle is 2 cm from its starting position.

21. Suppose that the velocity function of a particle moving along an s-axis is $v(t) = 20t^2 - 110t + 120$ ft/s and that the particle is at the origin at time $t = 0$. Use a graphing utility to generate the graphs of $s(t)$, $v(t)$, and $a(t)$ for the first 6 s of motion.

22. Suppose that the acceleration function of a particle moving along an s-axis is $a(t) = 4t - 30$ m/s^2 and that the position and velocity at time $t = 0$ are $s_0 = -5$ m and $v_0 = 3$ m/s. Use a graphing utility to generate the graphs of $s(t)$, $v(t)$, and $a(t)$ for the first 25 s of motion.

23–26 True–False Determine whether the statement is true or false. Explain your answer. Each question refers to a particle in rectilinear motion. ■

23. If the particle has constant acceleration, the velocity versus time graph will be a straight line.

24. If the particle has constant nonzero acceleration, its position versus time curve will be a parabola.

25. If the total area between the velocity versus time curve and a time interval $[a, b]$ is positive, then the displacement of the particle over this time interval will be nonzero.

26. If $D(t)$ denotes the distance traveled by the particle over the time interval $[0, t]$, then $D(t)$ is an antiderivative for the speed of the particle.

[C] **27–30** For the given velocity function $v(t)$:
(a) Generate the velocity versus time curve, and use it to make a conjecture about the sign of the displacement over the given time interval.
(b) Use a CAS to find the displacement. ■

27. $v(t) = 0.5 - t \sin t$; $0 \le t \le 5$

28. $v(t) = 0.5 - t \cos \pi t$; $0 \le t \le 1$

29. $v(t) = 0.5 - te^{-t}$; $0 \le t \le 5$

30. $v(t) = t \ln(t + 0.1)$; $0 \le t \le 1$

31. Suppose that at time $t = 0$ a particle is at the origin of an x-axis and has a velocity of $v_0 = 25$ cm/s. For the first 4 s thereafter it has no acceleration, and then it is acted on by a retarding force that produces a constant negative acceleration of $a = -10$ cm/s^2.
    (a) Sketch the acceleration versus time curve over the interval $0 \le t \le 12$.
    (b) Sketch the velocity versus time curve over the time interval $0 \le t \le 12$.
    (c) Find the x-coordinate of the particle at times $t = 8$ s and $t = 12$ s. *(cont.)*

(d) What is the maximum x-coordinate of the particle over the time interval $0 \le t \le 12$?

32–36 In these exercises assume that the object is moving with constant acceleration in the positive direction of a coordinate line, and apply Formulas (10) and (11) as appropriate. In some of these problems you will need the fact that 88 ft/s = 60 mi/h.

32. A car traveling 60 mi/h along a straight road decelerates at a constant rate of 11 ft/s^2.
(a) How long will it take until the speed is 45 mi/h?
(b) How far will the car travel before coming to a stop?

33. Spotting a police car, you hit the brakes on your new Porsche to reduce your speed from 90 mi/h to 60 mi/h at a constant rate over a distance of 200 ft.
(a) Find the acceleration in ft/s^2.
(b) How long does it take for you to reduce your speed to 55 mi/h?
(c) At the acceleration obtained in part (a), how long would it take for you to bring your Porsche to a complete stop from 90 mi/h?

34. A particle moving along a straight line is accelerating at a constant rate of 5 m/s^2. Find the initial velocity if the particle moves 60 m in the first 4 s.

35. A car that has stopped at a toll booth leaves the booth with a constant acceleration of 4 ft/s^2. At the time the car leaves the booth it is 2500 ft behind a truck traveling with a constant velocity of 50 ft/s. How long will it take for the car to catch the truck, and how far will the car be from the toll booth at that time?

36. In the final sprint of a rowing race the challenger is rowing at a constant speed of 12 m/s. At the point where the leader is 100 m from the finish line and the challenger is 15 m behind, the leader is rowing at 8 m/s but starts accelerating at a constant 0.5 m/s^2. Who wins?

37–46 Assume that a free-fall model applies. Solve these exercises by applying Formulas (15) and (16). In these exercises take $g = 32$ ft/s^2 or $g = 9.8$ m/s^2, depending on the units. ■

37. A projectile is launched vertically upward from ground level with an initial velocity of 112 ft/s.
(a) Find the velocity at $t = 3$ s and $t = 5$ s.
(b) How high will the projectile rise?
(c) Find the speed of the projectile when it hits the ground.

38. A projectile fired downward from a height of 112 ft reaches the ground in 2 s. What is its initial velocity?

39. A projectile is fired vertically upward from ground level with an initial velocity of 16 ft/s.
(a) How long will it take for the projectile to hit the ground?
(b) How long will the projectile be moving upward?

40. In 1939, Joe Sprinz of the San Francisco Seals Baseball Club attempted to catch a ball dropped from a blimp at a height of 800 ft (for the purpose of breaking the record for catching a

ball dropped from the greatest height set the preceding year by members of the Cleveland Indians).
(a) How long does it take for a ball to drop 800 ft?
(b) What is the velocity of a ball in miles per hour after an 800 ft drop (88 ft/s = 60 mi/h)?
[*Note:* As a practical matter, it is unrealistic to ignore wind resistance in this problem; however, even with the slowing effect of wind resistance, the impact of the ball slammed Sprinz's glove hand into his face, fractured his upper jaw in 12 places, broke five teeth, and knocked him unconscious. He dropped the ball!]

41. A projectile is launched upward from ground level with an initial speed of 60 m/s.
(a) How long does it take for the projectile to reach its highest point?
(b) How high does the projectile go?
(c) How long does it take for the projectile to drop back to the ground from its highest point?
(d) What is the speed of the projectile when it hits the ground?

42. (a) Use the results in Exercise 41 to make a conjecture about the relationship between the initial and final speeds of a projectile that is launched upward from ground level and returns to ground level.
(b) Prove your conjecture.

43. A projectile is fired vertically upward with an initial velocity of 49 m/s from a tower 150 m high.
(a) How long will it take for the projectile to reach its maximum height?
(b) What is the maximum height?
(c) How long will it take for the projectile to pass its starting point on the way down?
(d) What is the velocity when it passes the starting point on the way down?
(e) How long will it take for the projectile to hit the ground?
(f) What will be its speed at impact?

44. A man drops a stone from a bridge. What is the height of the bridge if
(a) the stone hits the water 4 s later
(b) the sound of the splash reaches the man 4 s later? [Take 1080 ft/s as the speed of sound.]

45. In Example 6, how fast would Nolan Ryan have to throw a ball upward from a height of 7 ft in order to hit the ceiling of the Astrodome?

46. A rock thrown downward with an unknown initial velocity from a height of 1000 ft reaches the ground in 5 s. Find the velocity of the rock when it hits the ground.

47. Writing Make a list of important features of a velocity versus time curve, and interpret each feature in terms of the motion.

48. Writing Use Riemann sums to argue informally that integrating speed over a time interval produces the distance traveled.

5.8 AVERAGE VALUE OF A FUNCTION AND ITS APPLICATIONS

In this section we will define the notion of the "average value" of a function, and we will give various applications of this idea.

■ AVERAGE VELOCITY REVISITED

Let $s = s(t)$ denote the position function of a particle in rectilinear motion. In Section 2.1 we defined the average velocity v_{ave} of the particle over the time interval $[t_0, t_1]$ to be

$$v_{ave} = \frac{s(t_1) - s(t_0)}{t_1 - t_0}$$

Let $v(t) = s'(t)$ denote the velocity function of the particle. We saw in Section 5.7 that integrating $s'(t)$ over a time interval gives the displacement of the particle over that interval. Thus,

$$\int_{t_0}^{t_1} v(t)\, dt = \int_{t_0}^{t_1} s'(t)\, dt = s(t_1) - s(t_0)$$

It follows that

$$v_{ave} = \frac{s(t_1) - s(t_0)}{t_1 - t_0} = \frac{1}{t_1 - t_0} \int_{t_0}^{t_1} v(t)\, dt \tag{1}$$

▶ **Example 1**   Suppose that a particle moves along a coordinate line so that its velocity at time t is $v(t) = 2 + \cos t$. Find the average velocity of the particle during the time interval $0 \le t \le \pi$.

Solution.   From (1) the average velocity is

$$\frac{1}{\pi - 0} \int_0^{\pi} (2 + \cos t)\, dt = \frac{1}{\pi} \left[2t + \sin t \right]_0^{\pi} = \frac{1}{\pi}(2\pi) = 2 \blacktriangleleft$$

We will see that Formula (1) is a special case of a formula for what we will call the *average value* of a continuous function over a given interval.

■ AVERAGE VALUE OF A CONTINUOUS FUNCTION

In scientific work, numerical information is often summarized by an *average value* or *mean value* of the observed data. There are various kinds of averages, but the most common is the **arithmetic mean** or **arithmetic average**, which is formed by adding the data and dividing by the number of data points. Thus, the arithmetic average $\bar{a}$ of n numbers $a_1, a_2, \ldots, a_n$ is

$$\bar{a} = \frac{1}{n}(a_1 + a_2 + \cdots + a_n) = \frac{1}{n} \sum_{k=1}^{n} a_k$$

In the case where the a_k's are values of a function f, say,

$$a_1 = f(x_1), a_2 = f(x_2), \ldots, a_n = f(x_n)$$

then the arithmetic average $\bar{a}$ of these function values is

$$\bar{a} = \frac{1}{n} \sum_{k=1}^{n} f(x_k)$$

We will now show how to extend this concept so that we can compute not only the arithmetic average of finitely many function values but an average of *all* values of $f(x)$ as x varies over a closed interval $[a, b]$. For this purpose recall the Mean-Value Theorem for Integrals (5.6.2), which states that if f is continuous on the interval $[a, b]$, then there is at least one point x^* in this interval such that

$$\int_a^b f(x)\, dx = f(x^*)(b - a)$$

The quantity

$$f(x^*) = \frac{1}{b - a} \int_a^b f(x)\, dx$$

will be our candidate for the average value of f over the interval $[a, b]$. To explain what motivates this, divide the interval $[a, b]$ into n subintervals of equal length

$$\Delta x = \frac{b - a}{n} \tag{2}$$

and choose arbitrary points $x_1^*, x_2^*, \ldots, x_n^*$ in successive subintervals. Then the arithmetic average of the values $f(x_1^*), f(x_2^*), \ldots, f(x_n^*)$ is

$$\text{ave} = \frac{1}{n}[f(x_1^*) + f(x_2^*) + \cdots + f(x_n^*)]$$

or from (2)

$$\text{ave} = \frac{1}{b - a}[f(x_1^*)\Delta x + f(x_2^*)\Delta x + \cdots + f(x_n^*)\Delta x] = \frac{1}{b - a} \sum_{k=1}^{n} f(x_k^*)\Delta x$$

Taking the limit as $n \to +\infty$ yields

$$\lim_{n \to +\infty} \frac{1}{b - a} \sum_{k=1}^{n} f(x_k^*)\Delta x = \frac{1}{b - a} \int_a^b f(x)\, dx$$

Since this equation describes what happens when we compute the average of "more and more" values of $f(x)$, we are led to the following definition.

Note that the Mean-Value Theorem for Integrals, when expressed in form (3), ensures that there is always at least one point x^* in $[a, b]$ at which the value of f is equal to the average value of f over the interval.

5.8.1 **DEFINITION** If f is continuous on $[a, b]$, then the *average value* (or *mean value*) of f on $[a, b]$ is defined to be

$$f_{\text{ave}} = \frac{1}{b - a} \int_a^b f(x)\, dx \tag{3}$$

REMARK When f is nonnegative on $[a, b]$, the quantity f_{ave} has a simple geometric interpretation, which can be seen by writing (3) as

$$f_{\text{ave}} \cdot (b - a) = \int_a^b f(x)\, dx$$

The left side of this equation is the area of a rectangle with a height of f_{ave} and base of length $b - a$, and the right side is the area under $y = f(x)$ over $[a, b]$. Thus, f_{ave} is the height of a rectangle constructed over the interval $[a, b]$, whose area is the same as the area under the graph of f over that interval (Figure 5.8.1).

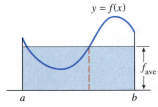

▲ **Figure 5.8.1**

▶ **Example 2** Find the average value of the function $f(x) = \sqrt{x}$ over the interval $[1, 4]$, and find all points in the interval at which the value of f is the same as the average.

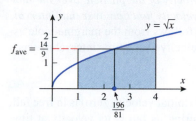

▲ Figure 5.8.2

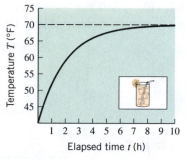

▲ Figure 5.8.3

In Example 3, the temperature T of the lemonade rises from an initial temperature of $40°$F toward the room temperature of $70°$F. Explain why the formula

$$T = 70 - 30e^{-0.5t}$$

is a good model for this situation.

Solution.

$$f_{\text{ave}} = \frac{1}{b-a} \int_a^b f(x)\,dx = \frac{1}{4-1}\int_1^4 \sqrt{x}\,dx = \frac{1}{3}\left[\frac{2x^{3/2}}{3}\right]_1^4$$

$$= \frac{1}{3}\left[\frac{16}{3} - \frac{2}{3}\right] = \frac{14}{9} \approx 1.6$$

The x-values at which $f(x) = \sqrt{x}$ is the same as this average satisfy $\sqrt{x} = 14/9$, from which we obtain $x = 196/81 \approx 2.4$ (Figure 5.8.2). ◄

▶ **Example 3** A glass of lemonade with a temperature of $40°$F is left to sit in a room whose temperature is a constant $70°$F. Using a principle of physics called *Newton's Law of Cooling*, one can show that if the temperature of the lemonade reaches $52°$F in 1 hour, then the temperature T of the lemonade as a function of the elapsed time t is modeled by the equation

$$T = 70 - 30e^{-0.5t}$$

where T is in degrees Fahrenheit and t is in hours. The graph of this equation, shown in Figure 5.8.3, conforms to our everyday experience that the temperature of the lemonade gradually approaches the temperature of the room. Find the average temperature T_{ave} of the lemonade over the first 5 hours.

Solution. From Definition 5.8.1 the average value of T over the time interval $[0, 5]$ is

$$T_{\text{ave}} = \frac{1}{5}\int_0^5 (70 - 30e^{-0.5t})\,dt \tag{4}$$

To evaluate the definite integral, we first find the indefinite integral

$$\int (70 - 30e^{-0.5t})\,dt$$

by making the substitution

$$u = -0.5t \quad \text{so that} \quad du = -0.5\,dt \quad (\text{or } dt = -2\,du)$$

Thus,

$$\int (70 - 30e^{-0.5t})\,dt = \int (70 - 30e^{u})(-2)\,du = -2(70u - 30e^{u}) + C$$

$$= -2[70(-0.5t) - 30e^{-0.5t}] + C = 70t + 60e^{-0.5t} + C$$

and (4) can be expressed as

$$T_{\text{ave}} = \frac{1}{5}\left[70t + 60e^{-0.5t}\right]_0^5 = \frac{1}{5}\left[(350 + 60e^{-2.5}) - 60\right]$$

$$= 58 + 12e^{-2.5} \approx 59°\text{F} \quad ◄$$

■ **AVERAGE VALUE AND AVERAGE VELOCITY**

We now have two ways to calculate the average velocity of a particle in rectilinear motion, since

$$\frac{s(t_1) - s(t_0)}{t_1 - t_0} = \frac{1}{t_1 - t_0}\int_{t_0}^{t_1} v(t)\,dt \tag{5}$$

and both of these expressions are equal to the average velocity. The left side of (5) gives the average rate of change of s over $[t_0, t_1]$, while the right side gives the average value of

$v = s'$ over the interval $[t_0, t_1]$. That is, *the average velocity of the particle over the time interval $[t_0, t_1]$ is the same as the average value of the velocity function over that interval.*

Since velocity functions are generally continuous, it follows from the marginal note associated with Definition 5.8.1 that a particle's average velocity over a time interval matches the particle's velocity at some time in the interval.

▶ **Example 4** Show that if a body released from rest (initial velocity zero) is in free fall, then its average velocity over a time interval $[0, T]$ during its fall is its velocity at time $t = T/2$.

Solution. It follows from Formula (16) of Section 5.7 with $v_0 = 0$ that the velocity function of the body is $v(t) = -gt$. Thus, its average velocity over a time interval $[0, T]$ is

$$v_{ave} = \frac{1}{T - 0} \int_0^T v(t)\, dt$$

$$= \frac{1}{T} \int_0^T -gt\, dt$$

$$= -\frac{g}{T} \left[\frac{1}{2} t^2 \right]_0^T = -g \cdot \frac{T}{2} = v\left(\frac{T}{2} \right) \quad ◀$$

The result of Example 4 can be generalized to show that the average velocity of a particle with constant acceleration during a time interval $[a, b]$ is the velocity at time $t = (a + b)/2$. (See Exercise 18.)

✔ **QUICK CHECK EXERCISES 5.8** *(See page 390 for answers.)*

1. The arithmetic average of n numbers, $a_1, a_2, \ldots, a_n$ is _____.

2. If f is continuous on $[a, b]$, then the average value of f on $[a, b]$ is _____.

3. If f is continuous on $[a, b]$, then the Mean-Value Theorem for Integrals guarantees that for at least one point x^* in $[a, b]$ _____ equals the average value of f on $[a, b]$.

4. The average value of $f(x) = 4x^3$ on $[1, 3]$ is _____.

EXERCISE SET 5.8 [C] CAS

1. (a) Find f_{ave} of $f(x) = 2x$ over $[0, 4]$.
   (b) Find a point x^* in $[0, 4]$ such that $f(x^*) = f_{ave}$.
   (c) Sketch a graph of $f(x) = 2x$ over $[0, 4]$, and construct a rectangle over the interval whose area is the same as the area under the graph of f over the interval.

2. (a) Find f_{ave} of $f(x) = x^2$ over $[0, 2]$.
   (b) Find a point x^* in $[0, 2]$ such that $f(x^*) = f_{ave}$.
   (c) Sketch a graph of $f(x) = x^2$ over $[0, 2]$, and construct a rectangle over the interval whose area is the same as the area under the graph of f over the interval.

3–12 Find the average value of the function over the given interval. ■

3. $f(x) = 3x$; $[1, 3]$

4. $f(x) = \sqrt[3]{x}$; $[-1, 8]$

5. $f(x) = \sin x$; $[0, \pi]$

6. $f(x) = \sec x \tan x$; $[0, \pi/3]$

7. $f(x) = 1/x$; $[1, e]$

8. $f(x) = e^x$; $[-1, \ln 5]$

9. $f(x) = \dfrac{1}{1 + x^2}$; $[1, \sqrt{3}]$

10. $f(x) = \dfrac{1}{\sqrt{1 - x^2}}$; $\left[-\frac{1}{2}, 0 \right]$

11. $f(x) = e^{-2x}$; $[0, 4]$

12. $f(x) = \sec^2 x$; $[-\pi/4, \pi/4]$

FOCUS ON CONCEPTS

13. Let $f(x) = 3x^2$.
    (a) Find the arithmetic average of the values $f(0.4)$, $f(0.8)$, $f(1.2)$, $f(1.6)$, and $f(2.0)$.
    (b) Find the arithmetic average of the values $f(0.1)$, $f(0.2)$, $f(0.3)$, \ldots, $f(2.0)$.
    (c) Find the average value of f on $[0, 2]$.
    (d) Explain why the answer to part (c) is less than the answers to parts (a) and (b).

14. In parts (a)–(d), let $f(x) = 1 + (1/x)$.
    (a) Find the arithmetic average of the values $f\left(\frac{6}{5} \right)$, $f\left(\frac{7}{5} \right)$, $f\left(\frac{8}{5} \right)$, $f\left(\frac{9}{5} \right)$, and $f(2)$.
    (b) Find the arithmetic average of the values $f(1.1)$, $f(1.2)$, $f(1.3)$, \ldots, $f(2)$.
    (cont.)

(c) Find the average value of f on $[1, 2]$.

(d) Explain why the answer to part (c) is greater than the answers to parts (a) and (b).

15. In each part, the velocity versus time curve is given for a particle moving along a line. Use the curve to find the average velocity of the particle over the time interval $0 \le t \le 3$.

(a)　　　　　　　　(b)

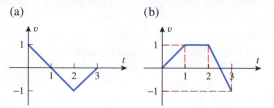

16. Suppose that a particle moving along a line starts from rest and has an average velocity of 2 ft/s over the time interval $0 \le t \le 5$. Sketch a velocity versus time curve for the particle assuming that the particle is also at rest at time $t = 5$. Explain how your curve satisfies the required properties.

17. Suppose that f is a linear function. Using the graph of f, explain why the average value of f on $[a, b]$ is

$$f\left(\frac{a+b}{2}\right)$$

18. Suppose that a particle moves along a coordinate line with constant acceleration. Show that the average velocity of the particle during a time interval $[a, b]$ matches the velocity of the particle at the midpoint of the interval.

19–22 True–False Determine whether the statement is true or false. Explain your answer. (Assume that f and g denote continuous functions on an interval $[a, b]$ and that f_{ave} and g_{ave} denote the respective average values of f and g on $[a, b]$.) ■

19. If $g_{ave} < f_{ave}$, then $g(x) \le f(x)$ on $[a, b]$.

20. The average value of a constant multiple of f is the same multiple of f_{ave}; that is, if c is any constant,

$$(c \cdot f)_{ave} = c \cdot f_{ave}$$

21. The average of the sum of two functions on an interval is the sum of the average values of the two functions on the interval; that is,

$$(f + g)_{ave} = f_{ave} + g_{ave}$$

22. The average of the product of two functions on an interval is the product of the average values of the two functions on the interval; that is

$$(f \cdot g)_{ave} = f_{ave} \cdot g_{ave}$$

23. (a) Suppose that the velocity function of a particle moving along a coordinate line is $v(t) = 3t^3 + 2$. Find the average velocity of the particle over the time interval $1 \le t \le 4$ by integrating.

(b) Suppose that the position function of a particle moving along a coordinate line is $s(t) = 6t^2 + t$. Find the average velocity of the particle over the time interval $1 \le t \le 4$ algebraically.

24. (a) Suppose that the acceleration function of a particle moving along a coordinate line is $a(t) = t + 1$. Find the average acceleration of the particle over the time interval $0 \le t \le 5$ by integrating.

(b) Suppose that the velocity function of a particle moving along a coordinate line is $v(t) = \cos t$. Find the average acceleration of the particle over the time interval $0 \le t \le \pi/4$ algebraically.

25. Water is run at a constant rate of 1 ft^3/min to fill a cylindrical tank of radius 3 ft and height 5 ft. Assuming that the tank is initially empty, make a conjecture about the average weight of the water in the tank over the time period required to fill it, and then check your conjecture by integrating. [Take the weight density of water to be 62.4 lb/ft^3.]

26. (a) The temperature of a 10 m long metal bar is 15°C at one end and 30°C at the other end. Assuming that the temperature increases linearly from the cooler end to the hotter end, what is the average temperature of the bar?

(b) Explain why there must be a point on the bar where the temperature is the same as the average, and find it.

27. A traffic engineer monitors the rate at which cars enter the main highway during the afternoon rush hour. From her data she estimates that between 4:30 P.M. and 5:30 P.M. the rate $R(t)$ at which cars enter the highway is given by the formula $R(t) = 100(1 - 0.0001t^2)$ cars per minute, where t is the time (in minutes) since 4:30 P.M. Find the average rate, in cars per minute, at which cars enter the highway during the first half-hour of rush hour.

28. Suppose that the value of a yacht in dollars after t years of use is $V(t) = 275{,}000e^{-0.17t}$. What is the average value of the yacht over its first 10 years of use?

29. A large juice glass containing 60 ml of orange juice is replenished by a server. The accompanying figure shows the rate at which orange juice is poured into the glass in milliliters per second (ml/s). Show that the average rate of change of the volume of juice in the glass during these 5 s is equal to the average value of the rate of flow of juice into the glass.

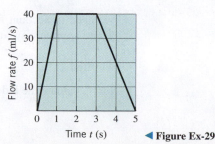

Time t (s)　　◀ **Figure Ex-29**

C **30.** The function J_0 defined by

$$J_0(x) = \frac{1}{\pi} \int_0^\pi \cos(x \sin t)\, dt$$

is called the **Bessel function of order zero**.
(a) Find a function f and an interval $[a, b]$ for which $J_0(1)$ is the average value of f over $[a, b]$.
(b) Estimate $J_0(1)$.
(c) Use a CAS to graph the equation $y = J_0(x)$ over the interval $0 \le x \le 8$.
(d) Estimate the smallest positive zero of J_0.

31. Find a positive value of k such that the average value of $f(x) = \sqrt{3x}$ over the interval $[0, k]$ is 6.

32. Suppose that a tumor grows at the rate of $r(t) = kt$ grams per week for some positive constant k, where t is the num-

ber of weeks since the tumor appeared. When, during the second 26 weeks of growth, is the mass of the tumor the same as its average mass during that period?

33. **Writing** Consider the following statement: *The average value of the rate of change of a function over an interval is equal to the average rate of change of the function over that interval.* Write a short paragraph that explains why this statement may be interpreted as a rewording of Part 1 of the Fundamental Theorem of Calculus.

34. **Writing** If an automobile gets an average of 25 miles per gallon of gasoline, then it is also the case that on average the automobile expends $1/25$ gallon of gasoline per mile. Interpret this statement using the concept of the average value of a function over an interval.

✔ **QUICK CHECK ANSWERS 5.8**

1. $\dfrac{1}{n} \displaystyle\sum_{k=1}^{n} a_k$ **2.** $\dfrac{1}{b-a} \displaystyle\int_a^b f(x)\, dx$ **3.** $f(x^*)$ **4.** 40

5.9 EVALUATING DEFINITE INTEGRALS BY SUBSTITUTION

In this section we will discuss two methods for evaluating definite integrals in which a substitution is required.

■ **TWO METHODS FOR MAKING SUBSTITUTIONS IN DEFINITE INTEGRALS**

Recall from Section 5.3 that indefinite integrals of the form

$$\int f(g(x)) g'(x)\, dx$$

can sometimes be evaluated by making the u-substitution

$$u = g(x), \quad du = g'(x)\, dx \tag{1}$$

which converts the integral to the form

$$\int f(u)\, du$$

To apply this method to a definite integral of the form

$$\int_a^b f(g(x)) g'(x)\, dx$$

we need to account for the effect that the substitution has on the x-limits of integration. There are two ways of doing this.

Method 1.

First evaluate the indefinite integral

$$\int f(g(x)) g'(x)\, dx$$

by substitution, and then use the relationship

$$\int_a^b f(g(x))g'(x)\,dx = \left[\int f(g(x))g'(x)\,dx\right]_a^b$$

to evaluate the definite integral. This procedure does not require any modification of the x-limits of integration.

Method 2.

Make the substitution (1) directly in the definite integral, and then use the relationship $u = g(x)$ to replace the x-limits, $x = a$ and $x = b$, by corresponding u-limits, $u = g(a)$ and $u = g(b)$. This produces a new definite integral

$$\int_{g(a)}^{g(b)} f(u)\,du$$

that is expressed entirely in terms of u.

▶ **Example 1**   Use the two methods above to evaluate $\displaystyle\int_0^2 x(x^2+1)^3\,dx$.

Solution by Method 1.   If we let

$$u = x^2 + 1 \quad\text{so that}\quad du = 2x\,dx \tag{2}$$

then we obtain

$$\int x(x^2+1)^3\,dx = \frac{1}{2}\int u^3\,du = \frac{u^4}{8} + C = \frac{(x^2+1)^4}{8} + C$$

Thus,

$$\int_0^2 x(x^2+1)^3\,dx = \left[\int x(x^2+1)^3\,dx\right]_{x=0}^2$$

$$= \frac{(x^2+1)^4}{8}\Bigg]_{x=0}^2 = \frac{625}{8} - \frac{1}{8} = 78$$

Solution by Method 2.   If we make the substitution $u = x^2 + 1$ in (2), then

$$\text{if}\quad x = 0,\quad u = 1$$
$$\text{if}\quad x = 2,\quad u = 5$$

Thus,

$$\int_0^2 x(x^2+1)^3\,dx = \frac{1}{2}\int_1^5 u^3\,du$$

$$= \frac{u^4}{8}\Bigg]_{u=1}^5 = \frac{625}{8} - \frac{1}{8} = 78$$

which agrees with the result obtained by Method 1. ◀

The following theorem states precise conditions under which Method 2 can be used.

5.9.1   THEOREM   *If g' is continuous on $[a, b]$ and f is continuous on an interval containing the values of $g(x)$ for $a \le x \le b$, then*

$$\int_a^b f(g(x))g'(x)\,dx = \int_{g(a)}^{g(b)} f(u)\,du$$

PROOF   Since f is continuous on an interval containing the values of $g(x)$ for $a \le x \le b$, it follows that f has an antiderivative F on that interval. If we let $u = g(x)$, then the chain rule implies that

$$\frac{d}{dx}F(g(x)) = \frac{d}{dx}F(u) = \frac{dF}{du}\frac{du}{dx} = f(u)\frac{du}{dx} = f(g(x))g'(x)$$

for each x in $[a, b]$. Thus, $F(g(x))$ is an antiderivative of $f(g(x))g'(x)$ on $[a, b]$. Therefore, by Part 1 of the Fundamental Theorem of Calculus (5.6.1)

$$\int_a^b f(g(x))g'(x)\,dx = F(g(x))\Big]_a^b = F(g(b)) - F(g(a)) = \int_{g(a)}^{g(b)} f(u)\,du \quad \blacksquare$$

The choice of methods for evaluating definite integrals by substitution is generally a matter of taste, but in the following examples we will use the second method, since the idea is new.

▶ **Example 2**   Evaluate

$$\text{(a)} \int_0^{\pi/8} \sin^5 2x \cos 2x\,dx \qquad \text{(b)} \int_2^5 (2x - 5)(x - 3)^9\,dx$$

Solution (a).   Let

$$u = \sin 2x \quad \text{so that} \quad du = 2\cos 2x\,dx \quad \left(\text{or } \tfrac{1}{2}\,du = \cos 2x\,dx\right)$$

With this substitution,

$$\text{if} \quad x = 0, \quad u = \sin(0) = 0$$
$$\text{if} \quad x = \pi/8, \quad u = \sin(\pi/4) = 1/\sqrt{2}$$

so

$$\int_0^{\pi/8} \sin^5 2x \cos 2x\,dx = \frac{1}{2}\int_0^{1/\sqrt{2}} u^5\,du$$

$$= \frac{1}{2}\cdot\frac{u^6}{6}\Big]_{u=0}^{1/\sqrt{2}} = \frac{1}{2}\left[\frac{1}{6(\sqrt{2})^6} - 0\right] = \frac{1}{96}$$

Solution (b).   Let

$$u = x - 3 \quad \text{so that} \quad du = dx$$

This leaves a factor of $2x - 5$ unresolved in the integrand. However,

$$x = u + 3, \quad \text{so} \quad 2x - 5 = 2(u + 3) - 5 = 2u + 1$$

With this substitution,

$$\text{if} \quad x = 2, \quad u = 2 - 3 = -1$$
$$\text{if} \quad x = 5, \quad u = 5 - 3 = 2$$

so

$$\int_2^5 (2x - 5)(x - 3)^9\,dx = \int_{-1}^2 (2u + 1)u^9\,du = \int_{-1}^2 (2u^{10} + u^9)\,du$$

$$= \left[\frac{2u^{11}}{11} + \frac{u^{10}}{10}\right]_{u=-1}^2 = \left(\frac{2^{12}}{11} + \frac{2^{10}}{10}\right) - \left(-\frac{2}{11} + \frac{1}{10}\right)$$

$$= \frac{52{,}233}{110} \approx 474.8 \quad \blacktriangleleft$$

▶ **Example 3** Evaluate

$$\text{(a)} \int_0^{3/4} \frac{dx}{1-x} \qquad \text{(b)} \int_0^{\ln 3} e^x (1+e^x)^{1/2} \, dx$$

Solution (a). Let

$$u = 1 - x \quad \text{so that} \quad du = -dx$$

With this substitution,

$$\text{if} \quad x = 0, \quad u = 1$$
$$\text{if} \quad x = \tfrac{3}{4}, \quad u = \tfrac{1}{4}$$

Thus,

$$\int_0^{3/4} \frac{dx}{1-x} = -\int_1^{1/4} \frac{du}{u}$$

$$= -\ln|u|\Big]_{u=1}^{1/4} = -\left[\ln\left(\frac{1}{4}\right) - \ln(1)\right] = \ln 4$$

Solution (b). Make the u-substitution

$$u = 1 + e^x, \quad du = e^x \, dx$$

and change the x-limits of integration ($x = 0$, $x = \ln 3$) to the u-limits

$$u = 1 + e^0 = 2, \quad u = 1 + e^{\ln 3} = 1 + 3 = 4$$

This yields

$$\int_0^{\ln 3} e^x (1+e^x)^{1/2} \, dx = \int_2^4 u^{1/2} \, du$$

$$= \frac{2}{3} u^{3/2}\Big]_{u=2}^4 = \frac{2}{3}[4^{3/2} - 2^{3/2}] = \frac{16 - 4\sqrt{2}}{3} \blacktriangleleft$$

> The u-substitution in Example 3(a) produces an integral in which the upper u-limit is smaller than the lower u-limit. Use Definition 5.5.3(b) to convert this integral to one whose lower limit is smaller than the upper limit and verify that it produces an integral with the same value as that in the example.

✔ **QUICK CHECK EXERCISES 5.9** *(See page 396 for answers.)*

1. Assume that g' is continuous on $[a, b]$ and that f is continuous on an interval containing the values of $g(x)$ for $a \le x \le b$. If F is an antiderivative for f, then

$$\int_a^b f(g(x))g'(x) \, dx = \underline{\hspace{2cm}}$$

2. In each part, use the substitution to replace the given integral with an integral involving the variable u. (Do not evaluate the integral.)

   (a) $\displaystyle\int_0^2 3x^2(1+x^3)^3 \, dx; \quad u = 1 + x^3$

   (b) $\displaystyle\int_0^2 \frac{x}{\sqrt{5-x^2}} \, dx; \quad u = 5 - x^2$

   (c) $\displaystyle\int_0^1 \frac{e^{\sqrt{x}}}{\sqrt{x}} \, dx; \quad u = \sqrt{x}$

3. Evaluate the integral by making an appropriate substitution.

   (a) $\displaystyle\int_{-\pi}^0 \sin(3x - \pi) \, dx = \underline{\hspace{2cm}}$

   (b) $\displaystyle\int_2^3 \frac{x}{x^2 - 2} \, dx = \underline{\hspace{2cm}}$

   (c) $\displaystyle\int_0^{\pi/2} \sqrt[3]{\sin x} \cos x \, dx = \underline{\hspace{2cm}}$

EXERCISE SET 5.9   ~ Graphing Utility   C CAS

1–4 Express the integral in terms of the variable u, but do not evaluate it. ■

1. (a) $\displaystyle\int_1^3 (2x-1)^3\,dx; \quad u=2x-1$

   (b) $\displaystyle\int_0^4 3x\sqrt{25-x^2}\,dx; \quad u=25-x^2$

   (c) $\displaystyle\int_{-1/2}^{1/2} \cos(\pi\theta)\,d\theta; \quad u=\pi\theta$

   (d) $\displaystyle\int_0^1 (x+2)(x+1)^5\,dx; \quad u=x+1$

2. (a) $\displaystyle\int_{-1}^4 (5-2x)^8\,dx; \quad u=5-2x$

   (b) $\displaystyle\int_{-\pi/3}^{2\pi/3} \frac{\sin x}{\sqrt{2+\cos x}}\,dx; \quad u=2+\cos x$

   (c) $\displaystyle\int_0^{\pi/4} \tan^2 x \sec^2 x\,dx; \quad u=\tan x$

   (d) $\displaystyle\int_0^1 x^3\sqrt{x^2+3}\,dx; \quad u=x^2+3$

3. (a) $\displaystyle\int_0^1 e^{2x-1}\,dx; \quad u=2x-1$

   (b) $\displaystyle\int_e^{e^2} \frac{\ln x}{x}\,dx; \quad u=\ln x$

4. (a) $\displaystyle\int_1^{\sqrt{3}} \frac{\sqrt{\tan^{-1}x}}{1+x^2}\,dx; \quad u=\tan^{-1}x$

   (b) $\displaystyle\int_1^{\sqrt{e}} \frac{dx}{x\sqrt{1-(\ln x)^2}}; \quad u=\ln x$

5–18 Evaluate the definite integral two ways: first by a u-substitution in the definite integral and then by a u-substitution in the corresponding indefinite integral. ■

5. $\displaystyle\int_0^1 (2x+1)^3\,dx$

6. $\displaystyle\int_1^2 (4x-2)^3\,dx$

7. $\displaystyle\int_0^1 (2x-1)^3\,dx$

8. $\displaystyle\int_1^2 (4-3x)^8\,dx$

9. $\displaystyle\int_0^8 x\sqrt{1+x}\,dx$

10. $\displaystyle\int_{-3}^0 x\sqrt{1-x}\,dx$

11. $\displaystyle\int_0^{\pi/2} 4\sin(x/2)\,dx$

12. $\displaystyle\int_0^{\pi/6} 2\cos 3x\,dx$

13. $\displaystyle\int_{-2}^{-1} \frac{x}{(x^2+2)^3}\,dx$

14. $\displaystyle\int_{1-\pi}^{1+\pi} \sec^2\left(\tfrac{1}{4}x-\tfrac{1}{4}\right)dx$

15. $\displaystyle\int_{-\ln 3}^{\ln 3} \frac{e^x}{e^x+4}\,dx$

16. $\displaystyle\int_0^{\ln 5} e^x(3-4e^x)\,dx$

17. $\displaystyle\int_1^3 \frac{dx}{\sqrt{x}\,(x+1)}$

18. $\displaystyle\int_{\ln 2}^{\ln(2/\sqrt{3})} \frac{e^{-x}\,dx}{\sqrt{1-e^{-2x}}}$

19–22 Evaluate the definite integral by expressing it in terms of u and evaluating the resulting integral using a formula from geometry. ■

19. $\displaystyle\int_{-5/3}^{5/3} \sqrt{25-9x^2}\,dx; \quad u=3x$

20. $\displaystyle\int_0^2 x\sqrt{16-x^4}\,dx; \quad u=x^2$

21. $\displaystyle\int_{\pi/3}^{\pi/2} \sin\theta\sqrt{1-4\cos^2\theta}\,d\theta; \quad u=2\cos\theta$

22. $\displaystyle\int_{e^{-3}}^{e^3} \frac{\sqrt{9-(\ln x)^2}}{x}\,dx; \quad u=\ln x$

23. A particle moves with a velocity of $v(t)=\sin\pi t$ m/s along an s-axis. Find the distance traveled by the particle over the time interval $0\le t\le 1$.

24. A particle moves with a velocity of $v(t)=3\cos 2t$ m/s along an s-axis. Find the distance traveled by the particle over the time interval $0\le t\le\pi/8$.

25. Find the area under the curve $y=9/(x+2)^2$ over the interval $[-1,1]$.

26. Find the area under the curve $y=1/(3x+1)^2$ over the interval $[0,1]$.

27. Find the area of the region enclosed by the graphs of $y=1/\sqrt{1-9x^2}$, $y=0$, $x=0$, and $x=\tfrac{1}{6}$.

28. Find the area of the region enclosed by the graphs of $y=\sin^{-1}x$, $x=0$, and $y=\pi/2$.

29–48 Evaluate the integrals by any method. ■

29. $\displaystyle\int_1^5 \frac{dx}{\sqrt{2x-1}}$

30. $\displaystyle\int_1^2 \sqrt{5x-1}\,dx$

31. $\displaystyle\int_{-1}^1 \frac{x^2\,dx}{\sqrt{x^3+9}}$

32. $\displaystyle\int_{\pi/2}^{\pi} 6\sin x(\cos x+1)^5\,dx$

33. $\displaystyle\int_1^3 \frac{x+2}{\sqrt{x^2+4x+7}}\,dx$

34. $\displaystyle\int_1^2 \frac{dx}{x^2-6x+9}$

35. $\displaystyle\int_0^{\pi/4} 4\sin x\cos x\,dx$

36. $\displaystyle\int_0^{\pi/4} \sqrt{\tan x}\,\sec^2 x\,dx$

37. $\displaystyle\int_0^{\sqrt{\pi}} 5x\cos(x^2)\,dx$

38. $\displaystyle\int_{\pi^2}^{4\pi^2} \frac{1}{\sqrt{x}}\sin\sqrt{x}\,dx$

39. $\displaystyle\int_{\pi/12}^{\pi/9} \sec^2 3\theta\,d\theta$

40. $\displaystyle\int_0^{\pi/6} \tan 2\theta\,d\theta$

41. $\displaystyle\int_0^1 \frac{y^2\,dy}{\sqrt{4-3y}}$

42. $\displaystyle\int_{-1}^4 \frac{x\,dx}{\sqrt{5+x}}$

43. $\displaystyle\int_0^e \frac{dx}{2x+e}$

44. $\displaystyle\int_1^{\sqrt{2}} xe^{-x^2}\,dx$

45. $\displaystyle\int_0^1 \frac{x}{\sqrt{4-3x^4}}\,dx$

46. $\displaystyle\int_1^2 \frac{1}{\sqrt{x}\sqrt{4-x}}\,dx$

47. $\displaystyle\int_{0}^{1/\sqrt{3}} \frac{1}{1+9x^2}\,dx$ **48.** $\displaystyle\int_{1}^{\sqrt{2}} \frac{x}{3+x^4}\,dx$

c **49.** (a) Use a CAS to find the exact value of the integral

$$\int_{0}^{\pi/6} \sin^4 x \cos^3 x\,dx$$

(b) Confirm the exact value by hand calculation.
[*Hint:* Use the identity $\cos^2 x = 1 - \sin^2 x$.]

c **50.** (a) Use a CAS to find the exact value of the integral

$$\int_{-\pi/4}^{\pi/4} \tan^4 x\,dx$$

(b) Confirm the exact value by hand calculation.
[*Hint:* Use the identity $1 + \tan^2 x = \sec^2 x$.]

51. (a) Find $\displaystyle\int_{0}^{1} f(3x + 1)\,dx$ if $\displaystyle\int_{1}^{4} f(x)\,dx = 5$.

(b) Find $\displaystyle\int_{0}^{3} f(3x)\,dx$ if $\displaystyle\int_{0}^{9} f(x)\,dx = 5$.

(c) Find $\displaystyle\int_{-2}^{0} x f(x^2)\,dx$ if $\displaystyle\int_{0}^{4} f(x)\,dx = 1$.

52. Given that m and n are positive integers, show that

$$\int_{0}^{1} x^m(1-x)^n\,dx = \int_{0}^{1} x^n(1-x)^m\,dx$$

by making a substitution. Do not attempt to evaluate the integrals.

53. Given that n is a positive integer, show that

$$\int_{0}^{\pi/2} \sin^n x\,dx = \int_{0}^{\pi/2} \cos^n x\,dx$$

by using a trigonometric identity and making a substitution. Do not attempt to evaluate the integrals.

54. Given that n is a positive integer, evaluate the integral

$$\int_{0}^{1} x(1-x)^n\,dx$$

55. Suppose that at time $t = 0$ there are 750 bacteria in a growth medium and the bacteria population $y(t)$ grows at the rate $y'(t) = 802.137e^{1.528t}$ bacteria per hour. How many bacteria will there be in 12 hours?

56. Suppose that a particle moving along a coordinate line has velocity $v(t) = 25 + 10e^{-0.05t}$ ft/s.
(a) What is the distance traveled by the particle from time $t = 0$ to time $t = 10$?
(b) Does the term $10e^{-0.05t}$ have much effect on the distance traveled by the particle over that time interval? Explain your reasoning.

57. (a) The accompanying table shows the fraction of the Moon that is illuminated (as seen from Earth) at midnight (Eastern Standard Time) for the first week of 2005. Find the average fraction of the Moon illuminated during the first week of 2005.

Source: Data from the U.S Naval Observatory Astronomical Applications Department.

(b) The function $f(x) = 0.5 + 0.5\sin(0.213x + 2.481)$ models data for illumination of the Moon for the first 60 days of 2005. Find the average value of this illumination function over the interval $[0, 7]$.

DAY	1	2	3	4	5	6	7
ILLUMINATION	0.74	0.65	0.56	0.45	0.35	0.25	0.16

▲ **Table Ex-57**

58. Electricity is supplied to homes in the form of *alternating current*, which means that the voltage has a sinusoidal waveform described by an equation of the form

$$V = V_p \sin(2\pi f t)$$

(see the accompanying figure). In this equation, V_p is called the **peak voltage** or **amplitude** of the current, f is called its **frequency**, and $1/f$ is called its **period**. The voltages V and V_p are measured in volts (V), the time t is measured in seconds (s), and the frequency is measured in hertz (Hz). (1 Hz = 1 cycle per second; a *cycle* is the electrical term for one period of the waveform.) Most alternating-current voltmeters read what is called the **rms** or **root-mean-square** value of V. By definition, this is the square root of the average value of V^2 over one period.
(a) Show that

$$V_{\text{rms}} = \frac{V_p}{\sqrt{2}}$$

[*Hint:* Compute the average over the cycle from $t = 0$ to $t = 1/f$, and use the identity $\sin^2\theta = \frac{1}{2}(1 - \cos 2\theta)$ to help evaluate the integral.]
(b) In the United States, electrical outlets supply alternating current with an rms voltage of 120 V at a frequency of 60 Hz. What is the peak voltage at such an outlet?

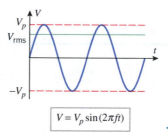

$\boxed{V = V_p \sin(2\pi f t)}$

◀ **Figure Ex-58**

59. Find a positive value of k such that the area under the graph of $y = e^{2x}$ over the interval $[0, k]$ is 3 square units.

60. Use a graphing utility to estimate the value of k ($k > 0$) so that the region enclosed by $y = 1/(1 + kx^2)$, $y = 0$, $x = 0$, and $x = 2$ has an area of 0.6 square unit.

c **61.** (a) Find the limit

$$\lim_{n \to +\infty} \sum_{k=1}^{n} \frac{\sin(k\pi/n)}{n}$$

by evaluating an appropriate definite integral over the interval $[0, 1]$.
(b) Check your answer to part (a) by evaluating the limit directly with a CAS.

62. Let

$$I = \int_{-1}^{1} \frac{1}{1+x^2}\,dx$$

(a) Explain why $I > 0$.

(b) Show that the substitution $x = 1/u$ results in

$$I = -\int_{-1}^{1} \frac{1}{1+x^2}\,dx = -I$$

Thus, $2I = 0$, which implies that $I = 0$. But this contradicts part (a). What is the error?

63. (a) Prove that if f is an odd function, then

$$\int_{-a}^{a} f(x)\,dx = 0$$

and give a geometric explanation of this result.
[*Hint:* One way to prove that a quantity q is zero is to show that $q = -q$.]

(b) Prove that if f is an even function, then

$$\int_{-a}^{a} f(x)\,dx = 2\int_{0}^{a} f(x)\,dx$$

and give a geometric explanation of this result.
[*Hint:* Split the interval of integration from $-a$ to a into two parts at 0.]

64. Show that if f and g are continuous functions, then

$$\int_{0}^{t} f(t-x)g(x)\,dx = \int_{0}^{t} f(x)g(t-x)\,dx$$

65. (a) Let

$$I = \int_{0}^{a} \frac{f(x)}{f(x)+f(a-x)}\,dx$$

Show that $I = a/2$.
[*Hint:* Let $u = a - x$, and then note the difference between the resulting integrand and 1.]

(b) Use the result of part (a) to find

$$\int_{0}^{3} \frac{\sqrt{x}}{\sqrt{x}+\sqrt{3-x}}\,dx$$

(c) Use the result of part (a) to find

$$\int_{0}^{\pi/2} \frac{\sin x}{\sin x + \cos x}\,dx$$

66. Evaluate

(a) $\displaystyle\int_{-1}^{1} x\sqrt{\cos(x^2)}\,dx$

(b) $\displaystyle\int_{0}^{\pi} \sin^8 x \cos^5 x\,dx$.

[*Hint:* Use the substitution $u = x - (\pi/2)$.]

67. Writing The two substitution methods discussed in this section yield the same result when used to evaluate a definite integral. Write a short paragraph that carefully explains why this is the case.

68. Writing In some cases, the second method for the evaluation of definite integrals has distinct advantages over the first. Provide some illustrations, and write a short paragraph that discusses the advantages of the second method in each case. [*Hint:* To get started, consider the results in Exercises 52–54, 63, and 65.]

✔ **QUICK CHECK ANSWERS 5.9**

1. $F(g(b)) - F(g(a))$ **2.** (a) $\displaystyle\int_{1}^{9} u^3\,du$ (b) $\displaystyle\int_{1}^{5} \frac{1}{2\sqrt{u}}\,du$ (c) $\displaystyle\int_{0}^{1} 2e^u\,du$ **3.** (a) $\dfrac{2}{3}$ (b) $\dfrac{1}{2}\ln\left(\dfrac{7}{2}\right)$ (c) $\dfrac{3}{4}$

5.10   LOGARITHMIC AND OTHER FUNCTIONS DEFINED BY INTEGRALS

In Section 0.5 we defined the natural logarithm function $\ln x$ to be the inverse of e^x. Although this was convenient and enabled us to deduce many properties of $\ln x$, the mathematical foundation was shaky in that we accepted the continuity of e^x and of all exponential functions without proof. In this section we will show that $\ln x$ can be defined as a certain integral, and we will use this new definition to prove that exponential functions are continuous. This integral definition is also important in applications because it provides a way of recognizing when integrals that appear in solutions of problems can be expressed as natural logarithms.

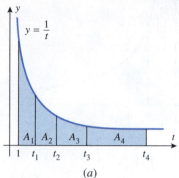

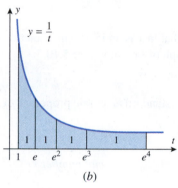

Not drawn to scale

▲ **Figure 5.10.1**

Review Theorem 5.5.8 and then explain why x is required to be positive in Definition 5.10.1.

None of the properties of $\ln x$ obtained in this section should be new, but now, for the first time, we give them a sound mathematical footing.

■ THE CONNECTION BETWEEN NATURAL LOGARITHMS AND INTEGRALS

The connection between natural logarithms and integrals was made in the middle of the seventeenth century in the course of investigating areas under the curve $y = 1/t$. The problem being considered was to find values of $t_1, t_2, t_3, \ldots, t_n, \ldots$ for which the areas $A_1, A_2, A_3, \ldots, A_n, \ldots$ in Figure 5.10.1a would be equal. Through the combined work of Isaac Newton, the Belgian Jesuit priest Gregory of St. Vincent (1584–1667), and Gregory's student Alfons A. de Sarasa (1618–1667), it was shown that by taking the points to be

$$t_1 = e, \quad t_2 = e^2, \quad t_3 = e^3, \ldots, \quad t_n = e^n, \ldots$$

each of the areas would be 1 (Figure 5.10.1b). Thus, in modern integral notation

$$\int_1^{e^n} \frac{1}{t}\, dt = n$$

which can be expressed as

$$\int_1^{e^n} \frac{1}{t}\, dt = \ln(e^n)$$

By comparing the upper limit of the integral and the expression inside the logarithm, it is a natural leap to the more general result

$$\int_1^x \frac{1}{t}\, dt = \ln x$$

which today we take as the formal definition of the natural logarithm.

5.10.1   **DEFINITION**   The ***natural logarithm*** of x is denoted by $\ln x$ and is defined by the integral

$$\ln x = \int_1^x \frac{1}{t}\, dt, \quad x > 0 \tag{1}$$

Our strategy for putting the study of logarithmic and exponential functions on a sound mathematical footing is to use (1) as a starting point and then define e^x as the inverse of $\ln x$. This is the exact opposite of our previous approach in which we defined $\ln x$ to be the inverse of e^x. However, whereas previously we had to *assume* that e^x is continuous, the continuity of e^x will now follow from our definitions as a *theorem*. Our first challenge is to demonstrate that the properties of $\ln x$ resulting from Definition 5.10.1 are consistent with those obtained earlier. To start, observe that Part 2 of the Fundamental Theorem of Calculus (5.6.3) implies that $\ln x$ is differentiable and

$$\frac{d}{dx}[\ln x] = \frac{d}{dx}\left[\int_1^x \frac{1}{t}\, dt\right] = \frac{1}{x} \quad (x > 0) \tag{2}$$

This is consistent with the derivative formula for $\ln x$ that we obtained previously. Moreover, because differentiability implies continuity, it follows that $\ln x$ is a continuous function on the interval $(0, +\infty)$.

Other properties of $\ln x$ can be obtained by interpreting the integral in (1) geometrically: In the case where $x > 1$, this integral represents the area under the curve $y = 1/t$ from $t = 1$ to $t = x$ (Figure 5.10.2a); in the case where $0 < x < 1$, the integral represents the negative of the area under the curve $y = 1/t$ from $t = x$ to $t = 1$ (Figure 5.10.2b); and in the case where $x = 1$, the integral has value 0 because its upper and lower limits of integration are the same. These geometric observations imply that

$$\ln x > 0 \quad \text{if} \quad x > 1$$
$$\ln x < 0 \quad \text{if} \quad 0 < x < 1$$
$$\ln x = 0 \quad \text{if} \quad x = 1$$

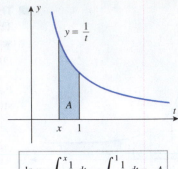

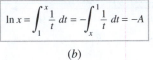

▶ **Figure 5.10.2**

(a)   (b)

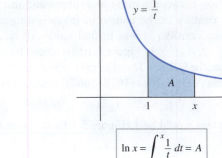

▲ **Figure 5.10.3**

Also, since $1/x$ is positive for $x > 0$, it follows from (2) that $\ln x$ is an increasing function on the interval $(0, +\infty)$. This is all consistent with the graph of $\ln x$ in Figure 5.10.3.

■ **ALGEBRAIC PROPERTIES OF ln x**

We can use (1) to show that Definition 5.10.1 produces the standard algebraic properties of logarithms.

5.10.2 **THEOREM** *For any positive numbers a and c and any rational number r:*

(a) $\ln ac = \ln a + \ln c$ (b) $\ln \dfrac{1}{c} = -\ln c$

(c) $\ln \dfrac{a}{c} = \ln a - \ln c$ (d) $\ln a^r = r \ln a$

PROOF (a) Treating a as a constant, consider the function $f(x) = \ln(ax)$. Then

$$f'(x) = \frac{1}{ax} \cdot \frac{d}{dx}(ax) = \frac{1}{ax} \cdot a = \frac{1}{x}$$

Thus, $\ln ax$ and $\ln x$ have the same derivative on $(0, +\infty)$, so these functions must differ by a constant on this interval. That is, there is a constant k such that

$$\ln ax - \ln x = k \tag{3}$$

on $(0, +\infty)$. Substituting $x = 1$ into this equation we conclude that $\ln a = k$ (verify). Thus, (3) can be written as

$$\ln ax - \ln x = \ln a$$

Setting $x = c$ establishes that

$$\ln ac - \ln c = \ln a \quad \text{or} \quad \ln ac = \ln a + \ln c$$

PROOFS (b) AND (c) Part (b) follows immediately from part (a) by substituting $1/c$ for a (verify). Then

$$\ln \frac{a}{c} = \ln\left(a \cdot \frac{1}{c}\right) = \ln a + \ln \frac{1}{c} = \ln a - \ln c$$

PROOF (d) First, we will argue that part (d) is satisfied if r is any nonnegative integer. If $r = 1$, then (d) is clearly satisfied; if $r = 0$, then (d) follows from the fact that $\ln 1 = 0$. Suppose that we know (d) is satisfied for r equal to some integer n. It then follows from part (a) that

$$\ln a^{n+1} = \ln[a \cdot a^n] = \ln a + \ln a^n = \ln a + n \ln a = (n + 1) \ln a$$

How is the proof of Theorem 5.10.2(*d*) for the case where *r* is a nonnegative integer analogous to a row of falling dominos? (This "domino" argument uses an informal version of a property of the integers known as the *principle of mathematical induction*.)

That is, if (*d*) is valid for *r* equal to some integer *n*, then it is also valid for $r = n + 1$. However, since we know (*d*) is satisfied if $r = 1$, it follows that (*d*) is valid for $r = 2$. But this implies that (*d*) is satisfied for $r = 3$, which in turn implies that (*d*) is valid for $r = 4$, and so forth. We conclude that (*d*) is satisfied if *r* is any nonnegative integer.

Next, suppose that $r = -m$ is a negative integer. Then

$$\ln a^r = \ln a^{-m} = \ln \frac{1}{a^m} = -\ln a^m$$ By part (*b*)

$$= -m \ln a$$ Part (*d*) is valid for positive powers.

$$= r \ln a$$

which shows that (*d*) is valid for any negative integer *r*. Combining this result with our previous conclusion that (*d*) is satisfied for a nonnegative integer *r* shows that (*d*) is valid if *r* is *any* integer.

Finally, suppose that $r = m/n$ is any rational number, where $m \neq 0$ and $n \neq 0$ are integers. Then

$$\ln a^r = \frac{n \ln a^r}{n} = \frac{\ln[(a^r)^n]}{n}$$ Part (*d*) is valid for integer powers.

$$= \frac{\ln a^{rn}}{n}$$ Property of exponents

$$= \frac{\ln a^m}{n}$$ Definition of *r*

$$= \frac{m \ln a}{n}$$ Part (*d*) is valid for integer powers.

$$= \frac{m}{n} \ln a = r \ln a$$

which shows that (*d*) is valid for any rational number *r*. ∎

APPROXIMATING ln *x* NUMERICALLY

For specific values of *x*, the value of ln *x* can be approximated numerically by approximating the definite integral in (1), say by using the midpoint approximation that was discussed in Section 5.4.

▶ **Example 1** Approximate ln 2 using the midpoint approximation with $n = 10$.

Solution. From (1), the exact value of ln 2 is represented by the integral

$$\ln 2 = \int_1^2 \frac{1}{t}\, dt$$

The midpoint rule is given in Formulas (5) and (6) of Section 5.4. Expressed in terms of *t*, the latter formula is

$$\int_a^b f(t)\, dt \approx \Delta t \sum_{k=1}^n f(t_k^*)$$

where Δt is the common width of the subintervals and $t_1^*, t_2^*, \ldots, t_n^*$ are the midpoints. In this case we have 10 subintervals, so $\Delta t = (2-1)/10 = 0.1$. The computations to six decimal places are shown in Table 5.10.1. By comparison, a calculator set to display six decimal places gives $\ln 2 \approx 0.693147$, so the magnitude of the error in the midpoint approximation is about 0.000311. Greater accuracy in the midpoint approximation can be obtained by increasing *n*. For example, the midpoint approximation with $n = 100$ yields $\ln 2 \approx 0.693144$, which is correct to five decimal places. ◀

Table 5.10.1

$n = 10$
$\Delta t = (b-a)/n = (2-1)/10 = 0.1$

k	t_k^*	$1/t_k^*$
1	1.05	0.952381
2	1.15	0.869565
3	1.25	0.800000
4	1.35	0.740741
5	1.45	0.689655
6	1.55	0.645161
7	1.65	0.606061
8	1.75	0.571429
9	1.85	0.540541
10	1.95	0.512821
		6.928355

$$\Delta t \sum_{k=1}^n f(t_k^*) \approx (0.1)(6.928355)$$
$$\approx 0.692836$$

■ **DOMAIN, RANGE, AND END BEHAVIOR OF ln x**

5.10.3   THEOREM

(a) *The domain of* ln x *is* $(0, +\infty)$.

(b) $\displaystyle\lim_{x \to 0^+} \ln x = -\infty$ *and* $\displaystyle\lim_{x \to +\infty} \ln x = +\infty$

(c) *The range of* ln x *is* $(-\infty, +\infty)$.

PROOFS (a) AND (b)   We have already shown that ln x is defined and increasing on the interval $(0, +\infty)$. To prove that ln $x \to +\infty$ as $x \to +\infty$, we must show that given any number $M > 0$, the value of ln x exceeds M for sufficiently large values of x. To do this, let N be any integer. If $x > 2^N$, then

$$\ln x > \ln 2^N = N \ln 2 \tag{4}$$

by Theorem 5.10.2(*d*). Since

$$\ln 2 = \int_1^2 \frac{1}{t}\, dt > 0$$

it follows that $N \ln 2$ can be made arbitrarily large by choosing N sufficiently large. In particular, we can choose N so that $N \ln 2 > M$. It now follows from (4) that if $x > 2^N$, then ln $x > M$, and this proves that

$$\lim_{x \to +\infty} \ln x = +\infty$$

Furthermore, by observing that $v = 1/x \to +\infty$ as $x \to 0^+$, we can use the preceding limit and Theorem 5.10.2(*b*) to conclude that

$$\lim_{x \to 0^+} \ln x = \lim_{v \to +\infty} \ln \frac{1}{v} = \lim_{v \to +\infty} (-\ln v) = -\infty$$

PROOF (c)   It follows from part (*a*), the continuity of ln x, and the Intermediate-Value Theorem (1.5.7) that ln x assumes every real value as x varies over the interval $(0, +\infty)$ (why?). ■

■ **DEFINITION OF e^x**

In Chapter 0 we defined ln x to be the inverse of the natural exponential function e^x. Now that we have a formal definition of ln x in terms of an integral, we will define the natural exponential function to be the inverse of ln x.

Since ln x is increasing and continuous on $(0, +\infty)$ with range $(-\infty, +\infty)$, there is exactly one (positive) solution to the equation ln $x = 1$. We *define* e to be the unique solution to ln $x = 1$, so

$$\ln e = 1 \tag{5}$$

Furthermore, if x is any real number, there is a unique positive solution y to ln $y = x$, so for irrational values of x we *define* e^x to be this solution. That is, when x is irrational, e^x is defined by

$$\ln e^x = x \tag{6}$$

Note that for rational values of x, we also have ln $e^x = x \ln e = x$ from Theorem 5.10.2(*d*). Moreover, it follows immediately that $e^{\ln x} = x$ for any $x > 0$. Thus, (6) defines the exponential function for all real values of x as the inverse of the natural logarithm function.

5.10.4   DEFINITION   The inverse of the natural logarithm function ln x is denoted by e^x and is called the ***natural exponential function***.

We can now establish the differentiability of e^x and confirm that

$$\frac{d}{dx}[e^x] = e^x$$

5.10.5 THEOREM *The natural exponential function e^x is differentiable, and hence continuous, on $(-\infty, +\infty)$, and its derivative is*

$$\frac{d}{dx}[e^x] = e^x$$

PROOF Because $\ln x$ is differentiable and

$$\frac{d}{dx}[\ln x] = \frac{1}{x} > 0$$

for all x in $(0, +\infty)$, it follows from Theorem 3.3.1, with $f(x) = \ln x$ and $f^{-1}(x) = e^x$, that e^x is differentiable on $(-\infty, +\infty)$ and its derivative is

$$\frac{d}{dx} \underbrace{[e^x]}_{f^{-1}(x)} = \underbrace{\frac{1}{1/e^x}}_{f'(f^{-1}(x))} = e^x \quad \blacksquare$$

■ IRRATIONAL EXPONENTS

Recall from Theorem 5.10.2(d) that if $a > 0$ and r is a rational number, then $\ln a^r = r \ln a$. Then $a^r = e^{\ln a^r} = e^{r \ln a}$ for any positive value of a and any rational number r. But the expression $e^{r \ln a}$ makes sense for *any* real number r, whether rational or irrational, so it is a good candidate to give meaning to a^r for any real number r.

Use Definition 5.10.6 to prove that if $a > 0$ and r is a real number, then $\ln a^r = r \ln a$.

5.10.6 DEFINITION If $a > 0$ and r is a real number, a^r is defined by

$$a^r = e^{r \ln a} \tag{7}$$

With this definition it can be shown that the standard algebraic properties of exponents, such as

$$a^p a^q = a^{p+q}, \quad \frac{a^p}{a^q} = a^{p-q}, \quad (a^p)^q = a^{pq}, \quad (a^p)(b^p) = (ab)^p$$

hold for any real values of a, b, p, and q, where a and b are positive. In addition, using (7) for a real exponent r, we can define the power function x^r whose domain consists of all positive real numbers, and for a positive base b we can define the ***base b exponential function b^x*** whose domain consists of all real numbers.

5.10.7 THEOREM

(a) *For any real number r, the power function x^r is differentiable on $(0, +\infty)$ and its derivative is*

$$\frac{d}{dx}[x^r] = rx^{r-1}$$

(b) *For $b > 0$ and $b \neq 1$, the base b exponential function b^x is differentiable on $(-\infty, +\infty)$ and its derivative is*

$$\frac{d}{dx}[b^x] = b^x \ln b$$

PROOF The differentiability of $x^r = e^{r \ln x}$ and $b^x = e^{x \ln b}$ on their domains follows from the differentiability of $\ln x$ on $(0, +\infty)$ and of e^x on $(-\infty, +\infty)$:

$$\frac{d}{dx}[x^r] = \frac{d}{dx}[e^{r \ln x}] = e^{r \ln x} \cdot \frac{d}{dx}[r \ln x] = x^r \cdot \frac{r}{x} = rx^{r-1}$$

$$\frac{d}{dx}[b^x] = \frac{d}{dx}[e^{x \ln b}] = e^{x \ln b} \cdot \frac{d}{dx}[x \ln b] = b^x \ln b \quad \blacksquare$$

We expressed e as the value of a limit in Formulas (7) and (8) of Section 1.3 and in Formula (1) of Section 3.2. We now have the mathematical tools necessary to prove the existence of these limits.

5.10.8 **THEOREM**

(a) $\displaystyle\lim_{x \to 0}(1 + x)^{1/x} = e$ *(b)* $\displaystyle\lim_{x \to +\infty}\left(1 + \frac{1}{x}\right)^x = e$ *(c)* $\displaystyle\lim_{x \to -\infty}\left(1 + \frac{1}{x}\right)^x = e$

PROOF We will prove part *(a)*; the proofs of parts *(b)* and *(c)* follow from this limit and are left as exercises. We first observe that

$$\frac{d}{dx}[\ln(x+1)]\bigg|_{x=0} = \frac{1}{x+1} \cdot 1\bigg|_{x=0} = 1$$

However, using the definition of the derivative, we obtain

$$1 = \frac{d}{dx}[\ln(x+1)]\bigg|_{x=0} = \lim_{h \to 0}\frac{\ln(0+h+1) - \ln(0+1)}{h}$$

$$= \lim_{h \to 0}\left[\frac{1}{h} \cdot \ln(1+h)\right]$$

or, equivalently,

$$\lim_{x \to 0}\frac{1}{x} \cdot \ln(1+x) = 1 \tag{8}$$

Now

$$\lim_{x \to 0}(1+x)^{1/x} = \lim_{x \to 0}e^{(\ln(1+x))/x} \qquad \boxed{\text{Definition 5.10.6}}$$

$$= e^{\lim_{x \to 0}[(\ln(1+x))/x]} \qquad \boxed{\text{Theorem 1.5.5}}$$

$$= e^1 \qquad \boxed{\text{Equation (8)}}$$

$$= e \quad \blacksquare$$

■ **GENERAL LOGARITHMS**

We note that for $b > 0$ and $b \neq 1$, the function b^x is one-to-one and so has an inverse function. Using the definition of b^x, we can solve $y = b^x$ for x as a function of y:

$$y = b^x = e^{x \ln b}$$

$$\ln y = \ln(e^{x \ln b}) = x \ln b$$

$$\frac{\ln y}{\ln b} = x$$

Thus, the inverse function for b^x is $(\ln x)/(\ln b)$.

5.10.9   DEFINITION   For $b > 0$ and $b \neq 1$, the ***base b logarithm*** function, denoted $\log_b x$, is defined by

$$\log_b x = \frac{\ln x}{\ln b} \qquad (9)$$

It follows immediately from this definition that $\log_b x$ is the inverse function for b^x and satisfies the properties in Table 0.5.3. Furthermore, $\log_b x$ is differentiable, and hence continuous, on $(0, +\infty)$, and its derivative is

$$\frac{d}{dx}[\log_b x] = \frac{1}{x \ln b}$$

As a final note of consistency, we observe that $\log_e x = \ln x$.

■ FUNCTIONS DEFINED BY INTEGRALS

The functions we have dealt with thus far in this text are called ***elementary functions***; they include polynomial, rational, power, exponential, logarithmic, trigonometric, and inverse trigonometric functions, and all other functions that can be obtained from these by addition, subtraction, multiplication, division, root extraction, and composition.

However, there are many important functions that do not fall into this category. Such functions occur in many ways, but they commonly arise in the course of solving initial-value problems of the form

$$\frac{dy}{dx} = f(x), \quad y(x_0) = y_0 \qquad (10)$$

Recall from Example 6 of Section 5.2 and the discussion preceding it that the basic method for solving (10) is to integrate $f(x)$, and then use the initial condition to determine the constant of integration. It can be proved that if f is continuous, then (10) has a unique solution and that this procedure produces it. However, there is another approach: Instead of solving each initial-value problem individually, we can find a general formula for the solution of (10), and then apply that formula to solve specific problems. We will now show that

$$y(x) = y_0 + \int_{x_0}^{x} f(t)\, dt \qquad (11)$$

is a formula for the solution of (10). To confirm this we must show that $dy/dx = f(x)$ and that $y(x_0) = y_0$. The computations are as follows:

$$\frac{dy}{dx} = \frac{d}{dx}\left[y_0 + \int_{x_0}^{x} f(t)\, dt \right] = 0 + f(x) = f(x)$$

$$y(x_0) = y_0 + \int_{x_0}^{x_0} f(t)\, dt = y_0 + 0 = y_0$$

▶ **Example 2**   In Example 6 of Section 5.2 we showed that the solution of the initial-value problem

$$\frac{dy}{dx} = \cos x, \quad y(0) = 1$$

is $y(x) = 1 + \sin x$. This initial-value problem can also be solved by applying Formula (11) with $f(x) = \cos x$, $x_0 = 0$, and $y_0 = 1$. This yields

$$y(x) = 1 + \int_{0}^{x} \cos t\, dt = 1 + \left[\sin t \right]_{t=0}^{x} = 1 + \sin x \blacktriangleleft$$

In the last example we were able to perform the integration in Formula (11) and express the solution of the initial-value problem as an elementary function. However, sometimes this will not be possible, in which case the solution of the initial-value problem must be left in terms of an "unevaluated" integral. For example, from (11), the solution of the initial-value problem

$$\frac{dy}{dx} = e^{-x^2}, \quad y(0) = 1$$

is

$$y(x) = 1 + \int_0^x e^{-t^2}\, dt$$

However, it can be shown that there is no way to express the integral in this solution as an elementary function. Thus, we have encountered a *new* function, which we regard to be *defined* by the integral. A close relative of this function, known as the **error function**, plays an important role in probability and statistics; it is denoted by erf(x) and is defined as

$$\text{erf}(x) = \frac{2}{\sqrt{\pi}} \int_0^x e^{-t^2}\, dt \tag{12}$$

Indeed, many of the most important functions in science and engineering are defined as integrals that have special names and notations associated with them. For example, the functions defined by

$$S(x) = \int_0^x \sin\left(\frac{\pi t^2}{2}\right) dt \quad \text{and} \quad C(x) = \int_0^x \cos\left(\frac{\pi t^2}{2}\right) dt \tag{13–14}$$

are called the **Fresnel sine and cosine functions**, respectively, in honor of the French physicist Augustin Fresnel (1788–1827), who first encountered them in his study of diffraction of light waves.

■ EVALUATING AND GRAPHING FUNCTIONS DEFINED BY INTEGRALS

The following values of $S(1)$ and $C(1)$ were produced by a CAS that has a built-in algorithm for approximating definite integrals:

$$S(1) = \int_0^1 \sin\left(\frac{\pi t^2}{2}\right) dt \approx 0.438259, \qquad C(1) = \int_0^1 \cos\left(\frac{\pi t^2}{2}\right) dt \approx 0.779893$$

To generate graphs of functions defined by integrals, computer programs choose a set of x-values in the domain, approximate the integral for each of those values, and then plot the resulting points. Thus, there is a lot of computation involved in generating such graphs, since each plotted point requires the approximation of an integral. The graphs of the Fresnel functions in Figure 5.10.4 were generated in this way using a CAS.

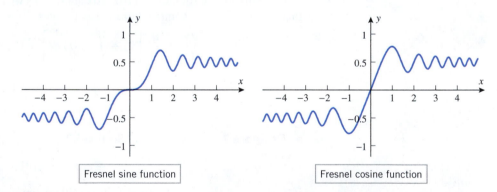

Fresnel sine function Fresnel cosine function

▶ **Figure 5.10.4**

REMARK Although it required a considerable amount of computation to generate the graphs of the Fresnel functions, the derivatives of $S(x)$ and $C(x)$ are easy to obtain using Part 2 of the Fundamental Theorem of Calculus (5.6.3); they are

$$S'(x) = \sin\left(\frac{\pi x^2}{2}\right) \quad \text{and} \quad C'(x) = \cos\left(\frac{\pi x^2}{2}\right) \tag{15–16}$$

These derivatives can be used to determine the locations of the relative extrema and inflection points and to investigate other properties of $S(x)$ and $C(x)$.

■ INTEGRALS WITH FUNCTIONS AS LIMITS OF INTEGRATION

Various applications can lead to integrals in which at least one of the limits of integration is a function of x. Some examples are

$$\int_x^1 \sqrt{\sin t}\, dt, \quad \int_{x^2}^{\sin x} \sqrt{t^3 + 1}\, dt, \quad \int_{\ln x}^{\pi} \frac{dt}{t^7 - 8}$$

We will complete this section by showing how to differentiate integrals of the form

$$\int_a^{g(x)} f(t)\, dt \tag{17}$$

where a is constant. Derivatives of other kinds of integrals with functions as limits of integration will be discussed in the exercises.

To differentiate (17) we can view the integral as a composition $F(g(x))$, where

$$F(x) = \int_a^x f(t)\, dt$$

If we now apply the chain rule, we obtain

$$\frac{d}{dx}\left[\int_a^{g(x)} f(t)\, dt \right] = \frac{d}{dx}[F(g(x))] = F'(g(x))g'(x) = \underbrace{f(g(x))g'(x)}_{\text{Theorem 5.6.3}}$$

Thus,

$$\frac{d}{dx}\left[\int_a^{g(x)} f(t)\, dt \right] = f(g(x))g'(x) \tag{18}$$

In words:

> *To differentiate an integral with a constant lower limit and a function as the upper limit, substitute the upper limit into the integrand, and multiply by the derivative of the upper limit.*

▶ **Example 3**

$$\frac{d}{dx}\left[\int_1^{\sin x} (1 - t^2)\, dt \right] = (1 - \sin^2 x) \cos x = \cos^3 x \quad ◀$$

✔ **QUICK CHECK EXERCISES 5.10**   *(See page 408 for answers.)*

1. $\displaystyle\int_1^{1/e} \frac{1}{t}\, dt =$ _____

2. Estimate $\ln 2$ using Definition 5.10.1 and
   (a) a left endpoint approximation with $n = 2$
   (b) a right endpoint approximation with $n = 2$.

3. $\pi^{1/(\ln \pi)} =$ _____

4. A solution to the initial-value problem

$$\frac{dy}{dx} = \cos x^3, \quad y(0) = 2$$

   that is defined by an integral is $y =$ _____.

5. $\displaystyle\frac{d}{dx}\left[\int_0^{e^{-x}} \frac{1}{1 + t^4}\, dt \right] =$ _____

EXERCISE SET 5.10   Graphing Utility   c CAS

1. Sketch the curve $y = 1/t$, and shade a region under the curve whose area is
   (a) $\ln 2$ (b) $-\ln 0.5$ (c) 2.

2. Sketch the curve $y = 1/t$, and shade two different regions under the curve whose areas are $\ln 1.5$.

3. Given that $\ln a = 2$ and $\ln c = 5$, find
   (a) $\displaystyle\int_1^{ac} \frac{1}{t}\, dt$ (b) $\displaystyle\int_1^{1/c} \frac{1}{t}\, dt$
   (c) $\displaystyle\int_1^{a/c} \frac{1}{t}\, dt$ (d) $\displaystyle\int_1^{a^3} \frac{1}{t}\, dt$.

4. Given that $\ln a = 9$, find
   (a) $\displaystyle\int_1^{\sqrt{a}} \frac{1}{t}\, dt$ (b) $\displaystyle\int_1^{2a} \frac{1}{t}\, dt$
   (c) $\displaystyle\int_1^{2/a} \frac{1}{t}\, dt$ (d) $\displaystyle\int_2^{a} \frac{1}{t}\, dt$.

5. Approximate $\ln 5$ using the midpoint rule with $n = 10$, and estimate the magnitude of the error by comparing your answer to that produced directly by a calculating utility.

6. Approximate $\ln 3$ using the midpoint rule with $n = 20$, and estimate the magnitude of the error by comparing your answer to that produced directly by a calculating utility.

7. Simplify the expression and state the values of x for which your simplification is valid.
   (a) $e^{-\ln x}$ (b) $e^{\ln x^2}$
   (c) $\ln\left(e^{-x^2}\right)$ (d) $\ln(1/e^x)$
   (e) $\exp(3\ln x)$ (f) $\ln(xe^x)$
   (g) $\ln\left(e^{x-\sqrt[3]{x}}\right)$ (h) $e^{x-\ln x}$

8. (a) Let $f(x) = e^{-2x}$. Find the simplest exact value of the function $f(\ln 3)$.
   (b) Let $f(x) = e^x + 3e^{-x}$. Find the simplest exact value of the function $f(\ln 2)$.

9–10 Express the given quantity as a power of e. ■

9. (a) 3^{π} (b) $2^{\sqrt{2}}$

10. (a) π^{-x} (b) $x^{2x}, \quad x > 0$

11–12 Find the limits by making appropriate substitutions in the limits given in Theorem 5.10.8. ■

11. (a) $\displaystyle\lim_{x \to +\infty} \left(1 + \frac{1}{2x}\right)^x$ (b) $\displaystyle\lim_{x \to 0} (1 + 2x)^{1/x}$

12. (a) $\displaystyle\lim_{x \to +\infty} \left(1 + \frac{3}{x}\right)^x$ (b) $\displaystyle\lim_{x \to 0} (1 + x)^{1/(3x)}$

13–14 Find $g'(x)$ using Part 2 of the Fundamental Theorem of Calculus, and check your answer by evaluating the integral and then differentiating. ■

13. $g(x) = \displaystyle\int_1^x (t^2 - t)\, dt$ 14. $g(x) = \displaystyle\int_{\pi}^x (1 - \cos t)\, dt$

15–16 Find the derivative using Formula (18), and check your answer by evaluating the integral and then differentiating the result. ■

15. (a) $\displaystyle\frac{d}{dx} \int_1^{x^3} \frac{1}{t}\, dt$ (b) $\displaystyle\frac{d}{dx} \int_1^{\ln x} e^t\, dt$

16. (a) $\displaystyle\frac{d}{dx} \int_{-1}^{x^2} \sqrt{t + 1}\, dt$ (b) $\displaystyle\frac{d}{dx} \int_{\pi}^{1/x} \sin t\, dt$

17. Let $F(x) = \displaystyle\int_0^x \frac{\sin t}{t^2 + 1}\, dt$. Find
    (a) $F(0)$ (b) $F'(0)$ (c) $F''(0)$.

18. Let $F(x) = \displaystyle\int_2^x \sqrt{3t^2 + 1}\, dt$. Find
    (a) $F(2)$ (b) $F'(2)$ (c) $F''(2)$.

19–22 True–False Determine whether the equation is true or false. Explain your answer. ■

19. $\displaystyle\int_1^{1/a} \frac{1}{t}\, dt = -\int_1^a \frac{1}{t}\, dt, \quad$ for $0 < a$

20. $\displaystyle\int_1^{\sqrt{a}} \frac{1}{t}\, dt = \frac{1}{2}\int_1^a \frac{1}{t}\, dt, \quad$ for $0 < a$

21. $\displaystyle\int_{-1}^e \frac{1}{t}\, dt = 1$

22. $\displaystyle\int \frac{2x}{1 + x^2}\, dx = \int_1^{1+x^2} \frac{1}{t}\, dt + C$

c 23. (a) Use Formula (18) to find
   $$\frac{d}{dx} \int_1^{x^2} t\sqrt{1 + t}\, dt$$
   (b) Use a CAS to evaluate the integral and differentiate the resulting function.
   (c) Use the simplification command of the CAS, if necessary, to confirm that the answers in parts (a) and (b) are the same.

24. Show that
   (a) $\displaystyle\frac{d}{dx}\left[\int_x^a f(t)\, dt\right] = -f(x)$
   (b) $\displaystyle\frac{d}{dx}\left[\int_{g(x)}^a f(t)\, dt\right] = -f(g(x))g'(x)$.

25–26 Use the results in Exercise 24 to find the derivative. ■

25. (a) $\displaystyle\frac{d}{dx} \int_x^{\pi} \cos(t^3)\, dt$ (b) $\displaystyle\frac{d}{dx} \int_{\tan x}^3 \frac{t^2}{1 + t^2}\, dt$

26. (a) $\displaystyle\frac{d}{dx} \int_x^0 \frac{1}{(t^2 + 1)^2}\, dt$ (b) $\displaystyle\frac{d}{dx} \int_{1/x}^{\pi} \cos^3 t\, dt$

27. Find
   $$\frac{d}{dx}\left[\int_{3x}^{x^2} \frac{t - 1}{t^2 + 1}\, dt\right]$$
   by writing
   $$\int_{3x}^{x^2} \frac{t - 1}{t^2 + 1}\, dt = \int_{3x}^0 \frac{t - 1}{t^2 + 1}\, dt + \int_0^{x^2} \frac{t - 1}{t^2 + 1}\, dt$$

28. Use Exercise 24(b) and the idea in Exercise 27 to show that

$$\frac{d}{dx} \int_{h(x)}^{g(x)} f(t)\, dt = f(g(x))g'(x) - f(h(x))h'(x)$$

29. Use the result obtained in Exercise 28 to perform the following differentiations:

(a) $\dfrac{d}{dx} \displaystyle\int_{x^2}^{x^3} \sin^2 t\, dt$ (b) $\dfrac{d}{dx} \displaystyle\int_{-x}^{x} \dfrac{1}{1+t}\, dt.$

30. Prove that the function

$$F(x) = \int_{x}^{5x} \frac{1}{t}\, dt$$

is constant on the interval $(0, +\infty)$ by using Exercise 28 to find $F'(x)$. What is that constant?

FOCUS ON CONCEPTS

31. Let $F(x) = \int_0^x f(t)\, dt$, where f is the function whose graph is shown in the accompanying figure.
   (a) Find $F(0)$, $F(3)$, $F(5)$, $F(7)$, and $F(10)$.
   (b) On what subintervals of the interval $[0, 10]$ is F increasing? Decreasing?
   (c) Where does F have its maximum value? Its minimum value?
   (d) Sketch the graph of F.

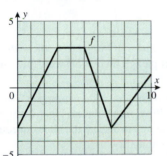

◀ **Figure Ex-31**

32. Determine the inflection point(s) for the graph of F in Exercise 31.

33–34 Express $F(x)$ in a piecewise form that does not involve an integral. ■

33. $F(x) = \displaystyle\int_{-1}^{x} |t|\, dt$

34. $F(x) = \displaystyle\int_{0}^{x} f(t)\, dt$, where $f(x) = \begin{cases} x, & 0 \le x \le 2 \\ 2, & x > 2 \end{cases}$

35–38 Use Formula (11) to solve the initial-value problem. ■

35. $\dfrac{dy}{dx} = \dfrac{2x^2 + 1}{x}$, $y(1) = 2$ **36.** $\dfrac{dy}{dx} = \dfrac{x+1}{\sqrt{x}}$, $y(1) = 0$

37. $\dfrac{dy}{dx} = \sec^2 x - \sin x$, $y(\pi/4) = 1$

38. $\dfrac{dy}{dx} = \dfrac{1}{x \ln x}$, $y(e) = 1$

39. Suppose that at time $t = 0$ there are P_0 individuals who have disease X, and suppose that a certain model for the spread of the disease predicts that the disease will spread at the rate of $r(t)$ individuals per day. Write a formula for the number of individuals who will have disease X after x days.

40. Suppose that $v(t)$ is the velocity function of a particle moving along an s-axis. Write a formula for the coordinate of the particle at time T if the particle is at s_1 at time $t = 1$.

FOCUS ON CONCEPTS

41. The accompanying figure shows the graphs of $y = f(x)$ and $y = \int_0^x f(t)\, dt$. Determine which graph is which, and explain your reasoning.

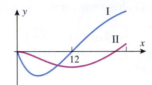

◀ **Figure Ex-41**

42. (a) Make a conjecture about the value of the limit

$$\lim_{k \to 0} \int_{1}^{b} t^{k-1}\, dt \quad (b > 0)$$

   (b) Check your conjecture by evaluating the integral and finding the limit. [*Hint:* Interpret the limit as the definition of the derivative of an exponential function.]

43. Let $F(x) = \int_0^x f(t)\, dt$, where f is the function graphed in the accompanying figure.
   (a) Where do the relative minima of F occur?
   (b) Where do the relative maxima of F occur?
   (c) Where does the absolute maximum of F on the interval $[0, 5]$ occur?
   (d) Where does the absolute minimum of F on the interval $[0, 5]$ occur?
   (e) Where is F concave up? Concave down?
   (f) Sketch the graph of F.

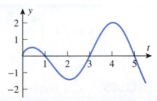

◀ **Figure Ex-43**

44. CAS programs have commands for working with most of the important nonelementary functions. Check your CAS documentation for information about the error function erf(x) [see Formula (12)], and then complete the following.
   (a) Generate the graph of erf(x).
   (b) Use the graph to make a conjecture about the existence and location of any relative maxima and minima of erf(x).
   (c) Check your conjecture in part (b) using the derivative of erf(x).

(cont.)

(d) Use the graph to make a conjecture about the existence and location of any inflection points of erf(x).

(e) Check your conjecture in part (d) using the second derivative of erf(x).

(f) Use the graph to make a conjecture about the existence of horizontal asymptotes of erf(x).

(g) Check your conjecture in part (f) by using the CAS to find the limits of erf(x) as $x \to \pm\infty$.

45. The Fresnel sine and cosine functions $S(x)$ and $C(x)$ were defined in Formulas (13) and (14) and graphed in Figure 5.10.4. Their derivatives were given in Formulas (15) and (16).

(a) At what points does $C(x)$ have relative minima? Relative maxima?

(b) Where do the inflection points of $C(x)$ occur?

(c) Confirm that your answers in parts (a) and (b) are consistent with the graph of $C(x)$.

46. Find the limit

$$\lim_{h \to 0} \frac{1}{h} \int_x^{x+h} \ln t \, dt$$

47. Find a function f and a number a such that

$$4 + \int_a^x f(t) \, dt = e^{2x}$$

48. (a) Give a geometric argument to show that

$$\frac{1}{x+1} < \int_x^{x+1} \frac{1}{t} \, dt < \frac{1}{x}, \quad x > 0$$

(b) Use the result in part (a) to prove that

$$\frac{1}{x+1} < \ln\left(1 + \frac{1}{x}\right) < \frac{1}{x}, \quad x > 0$$

(c) Use the result in part (b) to prove that

$$e^{x/(x+1)} < \left(1 + \frac{1}{x}\right)^x < e, \quad x > 0$$

and hence that

$$\lim_{x \to +\infty} \left(1 + \frac{1}{x}\right)^x = e$$

(d) Use the result in part (b) to prove that

$$\left(1 + \frac{1}{x}\right)^x < e < \left(1 + \frac{1}{x}\right)^{x+1}, \quad x > 0$$

49. Use a graphing utility to generate the graph of

$$y = \left(1 + \frac{1}{x}\right)^{x+1} - \left(1 + \frac{1}{x}\right)^x$$

in the window $[0, 100] \times [0, 0.2]$, and use that graph and part (d) of Exercise 48 to make a rough estimate of the error in the approximation

$$e \approx \left(1 + \frac{1}{50}\right)^{50}$$

50. Prove: If f is continuous on an open interval and a is any point in that interval, then

$$F(x) = \int_a^x f(t) \, dt$$

is continuous on the interval.

51. Writing A student objects that it is circular reasoning to make the definition

$$\ln x = \int_1^x \frac{1}{t} \, dt$$

since to evaluate the integral we need to know the value of $\ln x$. Write a short paragraph that answers this student's objection.

52. Writing Write a short paragraph that compares Definition 5.10.1 with the definition of the natural logarithm function given in Chapter 0. Be sure to discuss the issues surrounding continuity and differentiability.

✔**QUICK CHECK ANSWERS 5.10**

1. -1 **2.** (a) $\frac{5}{6}$ (b) $\frac{7}{12}$ **3.** e **4.** $y = 2 + \int_0^x \cos t^3 \, dt$ **5.** $-\dfrac{e^{-x}}{1 + e^{-4x}}$

CHAPTER 5 REVIEW EXERCISES    Graphing Utility   CAS

1–8 Evaluate the integrals. ■

1. $\displaystyle\int \left[\frac{1}{2x^3} + 4\sqrt{x}\right] dx$

2. $\displaystyle\int [u^3 - 2u + 7] \, du$

3. $\displaystyle\int [4\sin x + 2\cos x] \, dx$

4. $\displaystyle\int \sec x (\tan x + \cos x) \, dx$

5. $\displaystyle\int [x^{-2/3} - 5e^x] \, dx$

6. $\displaystyle\int \left[\frac{3}{4x} - \sec^2 x\right] dx$

7. $\displaystyle\int \left[\frac{1}{1+x^2} + \frac{2}{\sqrt{1-x^2}}\right] dx$

8. $\displaystyle\int \left[\frac{12}{x\sqrt{x^2-1}} + \frac{1-x^4}{1+x^2}\right] dx$

9. Solve the initial-value problems.

(a) $\dfrac{dy}{dx} = \dfrac{1-x}{\sqrt{x}}$, $y(1) = 0$

(b) $\dfrac{dy}{dx} = \cos x - 5e^x$, $y(0) = 0$

(c) $\dfrac{dy}{dx} = \sqrt[3]{x}$, $y(1) = 2$

(d) $\dfrac{dy}{dx} = xe^{x^2}$, $y(0) = 0$

10. The accompanying figure shows the slope field for a differential equation $dy/dx = f(x)$. Which of the following functions is most likely to be $f(x)$?

$$\sqrt{x}, \quad \sin x, \quad x^4, \quad x$$

Explain your reasoning.

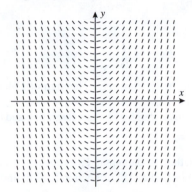

◀ **Figure Ex-10**

11. (a) Show that the substitutions $u = \sec x$ and $u = \tan x$ produce different values for the integral

$$\int \sec^2 x \tan x \, dx$$

(b) Explain why both are correct.

12. Use the two substitutions in Exercise 11 to evaluate the definite integral

$$\int_0^{\pi/4} \sec^2 x \tan x \, dx$$

and confirm that they produce the same result.

13. Evaluate the integral

$$\int \frac{x}{(x^2 - 1)\sqrt{x^4 - 2x^2}} \, dx$$

by making the substitution $u = x^2 - 1$.

14. Evaluate the integral

$$\int \sqrt{1 + x^{-2/3}} \, dx$$

by making the substitution $u = 1 + x^{2/3}$.

c **15–18** Evaluate the integrals by hand, and check your answers with a CAS if you have one. ■

15. $\displaystyle\int \frac{\cos 3x}{\sqrt{5 + 2\sin 3x}} \, dx$ **16.** $\displaystyle\int \frac{\sqrt{3 + \sqrt{x}}}{\sqrt{x}} \, dx$

17. $\displaystyle\int \frac{x^2}{(ax^3 + b)^2} \, dx$ **18.** $\displaystyle\int x \sec^2(ax^2) \, dx$

19. Express

$$\sum_{k=4}^{18} k(k - 3)$$

in sigma notation with
(a) $k = 0$ as the lower limit of summation
(b) $k = 5$ as the lower limit of summation.

20. (a) Fill in the blank:

$$1 + 3 + 5 + \cdots + (2n - 1) = \sum_{k=1}^{n} \underline{\quad\quad}$$

(b) Use part (a) to prove that the sum of the first n consecutive odd integers is a perfect square.

21. Find the area under the graph of $f(x) = 4x - x^2$ over the interval $[0, 4]$ using Definition 5.4.3 with x_k^* as the *right* endpoint of each subinterval.

22. Find the area under the graph of $f(x) = 5x - x^2$ over the interval $[0, 5]$ using Definition 5.4.3 with x_k^* as the *left* endpoint of each subinterval.

23–24 Use a calculating utility to find the left endpoint, right endpoint, and midpoint approximations to the area under the curve $y = f(x)$ over the stated interval using $n = 10$ subintervals. ■

23. $y = \ln x; \quad [1, 2]$ **24.** $y = e^x; \quad [0, 1]$

25. The *definite integral* of f over the interval $[a, b]$ is defined as the limit

$$\int_a^b f(x) \, dx = \lim_{\max \Delta x_k \to 0} \sum_{k=1}^{n} f(x_k^*) \Delta x_k$$

Explain what the various symbols on the right side of this equation mean.

26. Use a geometric argument to evaluate

$$\int_0^1 |2x - 1| \, dx$$

27. Suppose that

$$\int_0^1 f(x) \, dx = \tfrac{1}{2}, \quad \int_1^2 f(x) \, dx = \tfrac{1}{4},$$

$$\int_0^3 f(x) \, dx = -1, \quad \int_0^1 g(x) \, dx = 2$$

In each part, use this information to evaluate the given integral, if possible. If there is not enough information to evaluate the integral, then say so.

(a) $\displaystyle\int_0^2 f(x) \, dx$ (b) $\displaystyle\int_1^3 f(x) \, dx$ (c) $\displaystyle\int_2^3 5f(x) \, dx$

(d) $\displaystyle\int_1^0 g(x) \, dx$ (e) $\displaystyle\int_0^1 g(2x) \, dx$ (f) $\displaystyle\int_0^1 [g(x)]^2 \, dx$

28. In parts (a)–(d), use the information in Exercise 27 to evaluate the given integral. If there is not enough information to evaluate the integral, then say so. *(cont.)*

(a) $\displaystyle\int_0^1 [f(x) + g(x)]\,dx$ (b) $\displaystyle\int_0^1 f(x)g(x)\,dx$

(c) $\displaystyle\int_0^1 \frac{f(x)}{g(x)}\,dx$ (d) $\displaystyle\int_0^1 [4g(x) - 3f(x)]\,dx$

29. In each part, evaluate the integral. Where appropriate, you may use a geometric formula.

(a) $\displaystyle\int_{-1}^1 (1 + \sqrt{1 - x^2})\,dx$

(b) $\displaystyle\int_0^3 (x\sqrt{x^2 + 1} - \sqrt{9 - x^2})\,dx$

(c) $\displaystyle\int_0^1 x\sqrt{1 - x^4}\,dx$

30. In each part, find the limit by interpreting it as a limit of Riemann sums in which the interval $[0, 1]$ is divided into n subintervals of equal length.

(a) $\displaystyle\lim_{n \to +\infty} \frac{\sqrt{1} + \sqrt{2} + \sqrt{3} + \cdots + \sqrt{n}}{n^{3/2}}$

(b) $\displaystyle\lim_{n \to +\infty} \frac{1^4 + 2^4 + 3^4 + \cdots + n^4}{n^5}$

(c) $\displaystyle\lim_{n \to +\infty} \frac{e^{1/n} + e^{2/n} + e^{3/n} + \cdots + e^{n/n}}{n}$

31–38 Evaluate the integrals using the Fundamental Theorem of Calculus and (if necessary) properties of the definite integral. ■

31. $\displaystyle\int_{-3}^0 (x^2 - 4x + 7)\,dx$ **32.** $\displaystyle\int_{-1}^2 x(1 + x^3)\,dx$

33. $\displaystyle\int_1^3 \frac{1}{x^2}\,dx$ **34.** $\displaystyle\int_1^8 (5x^{2/3} - 4x^{-2})\,dx$

35. $\displaystyle\int_0^1 (x - \sec x \tan x)\,dx$

36. $\displaystyle\int_1^4 \left(\frac{3}{\sqrt{t}} - 5\sqrt{t} - t^{-3/2}\right)\,dt$

37. $\displaystyle\int_0^2 |2x - 3|\,dx$ **38.** $\displaystyle\int_0^{\pi/2} \left|\tfrac{1}{2} - \sin x\right|\,dx$

39–42 Find the area under the curve $y = f(x)$ over the stated interval. ■

39. $f(x) = \sqrt{x};\ [1, 9]$ **40.** $f(x) = x^{-3/5};\ [1, 4]$

41. $f(x) = e^x;\ [1, 3]$ **42.** $f(x) = \dfrac{1}{x};\ [1, e^3]$

43. Find the area that is above the x-axis but below the curve $y = (1 - x)(x - 2)$. Make a sketch of the region.

c **44.** Use a CAS to find the area of the region in the first quadrant that lies below the curve $y = x + x^2 - x^3$ and above the x-axis.

45–46 Sketch the curve and find the total area between the curve and the given interval on the x-axis. ■

45. $y = x^2 - 1;\ [0, 3]$ **46.** $y = \sqrt{x + 1} - 1;\ [-1, 1]$

47. Define $F(x)$ by
$$F(x) = \int_1^x (t^3 + 1)\,dt$$

(a) Use Part 2 of the Fundamental Theorem of Calculus to find $F'(x)$.

(b) Check the result in part (a) by first integrating and then differentiating.

48. Define $F(x)$ by
$$F(x) = \int_4^x \frac{1}{\sqrt{t}}\,dt$$

(a) Use Part 2 of the Fundamental Theorem of Calculus to find $F'(x)$.

(b) Check the result in part (a) by first integrating and then differentiating.

49–54 Use Part 2 of the Fundamental Theorem of Calculus and (where necessary) Formula (18) of Section 5.10 to find the derivatives. ■

49. $\displaystyle\frac{d}{dx}\left[\int_0^x e^{t^2}\,dt\right]$ **50.** $\displaystyle\frac{d}{dx}\left[\int_0^x \frac{t}{\cos t^2}\,dt\right]$

51. $\displaystyle\frac{d}{dx}\left[\int_0^x |t - 1|\,dt\right]$ **52.** $\displaystyle\frac{d}{dx}\left[\int_\pi^x \cos\sqrt{t}\,dt\right]$

53. $\displaystyle\frac{d}{dx}\left[\int_2^{\sin x} \frac{1}{1 + t^3}\,dt\right]$ **54.** $\displaystyle\frac{d}{dx}\left[\int_e^{\sqrt{x}} (\ln t)^2\,dt\right]$

55. State the two parts of the Fundamental Theorem of Calculus, and explain what is meant by the statement "Differentiation and integration are inverse processes."

c **56.** Let $F(x) = \displaystyle\int_0^x \frac{t^2 - 3}{t^4 + 7}\,dt$.

(a) Find the intervals on which F is increasing and those on which F is decreasing.

(b) Find the open intervals on which F is concave up and those on which F is concave down.

(c) Find the x-values, if any, at which the function F has absolute extrema.

(d) Use a CAS to graph F, and confirm that the results in parts (a), (b), and (c) are consistent with the graph.

57. (a) Use differentiation to prove that the function
$$F(x) = \int_0^x \frac{1}{1 + t^2}\,dt + \int_0^{1/x} \frac{1}{1 + t^2}\,dt$$
is constant on the interval $(0, +\infty)$.

(b) Determine the constant value of the function in part (a) and then interpret (a) as an identity involving the inverse tangent function.

58. What is the natural domain of the function
$$F(x) = \int_1^x \frac{1}{t^2 - 9}\,dt?$$
Explain your reasoning.

59. In each part, determine the values of x for which $F(x)$ is positive, negative, or zero without performing the integration; explain your reasoning.

(a) $F(x) = \int_{1}^{x} \dfrac{t^4}{t^2 + 3}\, dt$ (b) $F(x) = \int_{-1}^{x} \sqrt{4 - t^2}\, dt$

c 60. Use a CAS to approximate the largest and smallest values of the integral

$$\int_{-1}^{x} \dfrac{t}{\sqrt{2 + t^3}}\, dt$$

for $1 \le x \le 3$.

61. Find all values of x^* in the stated interval that are guaranteed to exist by the Mean-Value Theorem for Integrals, and explain what these numbers represent.

(a) $f(x) = \sqrt{x}$; $[0, 3]$ (b) $f(x) = 1/x$; $[1, e]$

62. A 10-gram tumor is discovered in a laboratory rat on March 1. The tumor is growing at a rate of $r(t) = t/7$ grams per week, where t denotes the number of weeks since March 1. What will be the mass of the tumor on June 7?

63. Use the graph of f shown in the accompanying figure to find the average value of f on the interval $[0, 10]$.

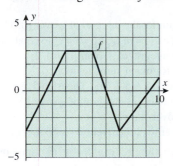

◀ Figure Ex-63

64. Find the average value of $f(x) = e^x + e^{-x}$ over the interval $\left[\ln \tfrac{1}{2}, \ln 2\right]$.

65. Derive the formulas for the position and velocity functions of a particle that moves with constant acceleration along a coordinate line.

66. The velocity of a particle moving along an s-axis is measured at 5 s intervals for 40 s, and the velocity function is modeled by a smooth curve. (The curve and the data points are shown in the accompanying figure.) Use this model in each part.

(a) Does the particle have constant acceleration? Explain your reasoning.

(b) Is there any 15 s time interval during which the acceleration is constant? Explain your reasoning.

(c) Estimate the distance traveled by the particle from time $t = 0$ to time $t = 40$.

(d) Estimate the average velocity of the particle over the 40 s time period.

(e) Is the particle ever slowing down during the 40 s time period? Explain your reasoning.

(f) Is there sufficient information for you to determine the s-coordinate of the particle at time $t = 10$? If so,

find it. If not, explain what additional information you need.

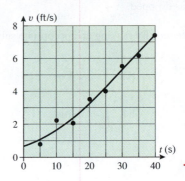

◀ Figure Ex-66

67–70 A particle moves along an s-axis. Use the given information to find the position function of the particle. ■

67. $v(t) = t^3 - 2t^2 + 1$; $s(0) = 1$

68. $a(t) = 4\cos 2t$; $v(0) = -1, s(0) = -3$

69. $v(t) = 2t - 3$; $s(1) = 5$

70. $a(t) = \cos t - 2t$; $v(0) = 0, s(0) = 0$

71–74 A particle moves with a velocity of $v(t)$ m/s along an s-axis. Find the displacement and the distance traveled by the particle during the given time interval. ■

71. $v(t) = 2t - 4$; $0 \le t \le 6$

72. $v(t) = |t - 3|$; $0 \le t \le 5$

73. $v(t) = \dfrac{1}{2} - \dfrac{1}{t^2}$; $1 \le t \le 3$

74. $v(t) = \dfrac{3}{\sqrt{t}}$; $4 \le t \le 9$

75–76 A particle moves with acceleration $a(t)$ m/s² along an s-axis and has velocity v_0 m/s at time $t = 0$. Find the displacement and the distance traveled by the particle during the given time interval. ■

75. $a(t) = -2$; $v_0 = 3$; $1 \le t \le 4$

76. $a(t) = \dfrac{1}{\sqrt{5t + 1}}$; $v_0 = 2$; $0 \le t \le 3$

77. A car traveling 60 mi/h ($= 88$ ft/s) along a straight road decelerates at a constant rate of 10 ft/s².

(a) How long will it take until the speed is 45 mi/h?

(b) How far will the car travel before coming to a stop?

78. Suppose that the velocity function of a particle moving along an s-axis is $v(t) = 20t^2 - 100t + 50$ ft/s and that the particle is at the origin at time $t = 0$. Use a graphing utility to generate the graphs of $s(t)$, $v(t)$, and $a(t)$ for the first 6 s of motion.

79. A ball is thrown vertically upward from a height of s_0 ft with an initial velocity of v_0 ft/s. If the ball is caught at height s_0, determine its average speed through the air using the free-fall model.

80. A rock, dropped from an unknown height, strikes the ground with a speed of 24 m/s. Find the height from which the rock was dropped.

81–88 Evaluate the integrals by making an appropriate substitution. ■

81. $\int_0^1 (2x+1)^4 \, dx$

82. $\int_{-5}^0 x\sqrt{4-x} \, dx$

83. $\int_0^1 \dfrac{dx}{\sqrt{3x+1}}$

84. $\int_0^{\sqrt{\pi}} x \sin x^2 \, dx$

85. $\int_0^1 \sin^2(\pi x) \cos(\pi x) \, dx$

86. $\int_e^{e^2} \dfrac{dx}{x \ln x}$

87. $\int_0^1 \dfrac{dx}{\sqrt{e^x}}$

88. $\int_0^{2/\sqrt{3}} \dfrac{1}{4+9x^2} \, dx$

89. Evaluate the limits.

(a) $\displaystyle\lim_{x \to +\infty} \left(1 + \dfrac{1}{x}\right)^{2x}$

(b) $\displaystyle\lim_{x \to +\infty} \left(1 + \dfrac{1}{3x}\right)^{x}$

90. Find a function f and a number a such that

$$2 + \int_a^x f(t) \, dt = e^{3x}$$

CHAPTER 5 MAKING CONNECTIONS

1. Consider a Riemann sum

$$\sum_{k=1}^n 2x_k^* \Delta x_k$$

for the integral of $f(x) = 2x$ over an interval $[a, b]$.

(a) Show that if x_k^* is the midpoint of the kth subinterval, the Riemann sum is a telescoping sum. (See Exercises 57–60 of Section 5.4 for other examples of telescoping sums.)

(b) Use part (a), Definition 5.5.1, and Theorem 5.5.2 to evaluate the definite integral of $f(x) = 2x$ over $[a, b]$.

2. The function $f(x) = \sqrt{x}$ is continuous on $[0, 4]$ and therefore integrable on this interval. Evaluate

$$\int_0^4 \sqrt{x} \, dx$$

by using Definition 5.5.1. Use subintervals of unequal length given by the partition

$$0 < 4(1)^2/n^2 < 4(2)^2/n^2 < \cdots < 4(n-1)^2/n^2 < 4$$

and let x_k^* be the right endpoint of the kth subinterval.

3. Make appropriate modifications and repeat Exercise 2 for

$$\int_0^8 \sqrt[3]{x} \, dx$$

4. Given a continuous function f and a positive real number m, let g denote the function defined by the composition $g(x) = f(mx)$.

(a) Suppose that

$$\sum_{k=1}^n g(x_k^*) \Delta x_k$$

is any Riemann sum for the integral of g over $[0, 1]$. Use the correspondence $u_k = mx_k$, $u_k^* = mx_k^*$ to create a Riemann sum for the integral of f over $[0, m]$. How are the values of the two Riemann sums related?

(b) Use part (a), Definition 5.5.1, and Theorem 5.5.2 to find an equation that relates the integral of g over $[0, 1]$ with the integral of f over $[0, m]$.

(c) How is your answer to part (b) related to Theorem 5.9.1?

5. Given a continuous function f, let g denote the function defined by $g(x) = 2xf(x^2)$.

(a) Suppose that

$$\sum_{k=1}^n g(x_k^*) \Delta x_k$$

is any Riemann sum for the integral of g over $[2, 3]$, with $x_k^* = (x_k + x_{k-1})/2$ the midpoint of the kth subinterval. Use the correspondence $u_k = x_k^2$, $u_k^* = (x_k^*)^2$ to create a Riemann sum for the integral of f over $[4, 9]$. How are the values of the two Riemann sums related?

(b) Use part (a), Definition 5.5.1, and Theorem 5.5.2 to find an equation that relates the integral of g over $[2, 3]$ with the integral of f over $[4, 9]$.

(c) How is your answer to part (b) related to Theorem 5.9.1?

Courtesy NASA

6

APPLICATIONS OF THE DEFINITE INTEGRAL IN GEOMETRY, SCIENCE, AND ENGINEERING

Calculus is essential for the computations required to land an astronaut on the moon.

In the last chapter we introduced the definite integral as the limit of Riemann sums in the context of finding areas. However, Riemann sums and definite integrals have applications that extend far beyond the area problem. In this chapter we will show how Riemann sums and definite integrals arise in such problems as finding the volume and surface area of a solid, finding the length of a plane curve, calculating the work done by a force, finding the center of gravity of a planar region, finding the pressure and force exerted by a fluid on a submerged object, and finding properties of suspended cables.

Although these problems are diverse, the required calculations can all be approached by the same procedure that we used to find areas—breaking the required calculation into "small parts," making an approximation for each part, adding the approximations from the parts to produce a Riemann sum that approximates the entire quantity to be calculated, and then taking the limit of the Riemann sums to produce an exact result.

6.1 AREA BETWEEN TWO CURVES

In the last chapter we showed how to find the area between a curve $y = f(x)$ and an interval on the x-axis. Here we will show how to find the area between two curves.

■ A REVIEW OF RIEMANN SUMS

Before we consider the problem of finding the area between two curves it will be helpful to review the basic principle that underlies the calculation of area as a definite integral. Recall that if f is continuous and nonnegative on $[a, b]$, then the definite integral for the area A under $y = f(x)$ over the interval $[a, b]$ is obtained in four steps (Figure 6.1.1):

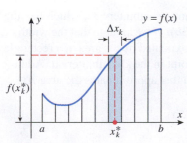

▲ Figure 6.1.1

- Divide the interval $[a, b]$ into n subintervals, and use those subintervals to divide the region under the curve $y = f(x)$ into n strips.

- Assuming that the width of the kth strip is Δx_k, approximate the area of that strip by the area $f(x_k^*)\Delta x_k$ of a rectangle of width Δx_k and height $f(x_k^*)$, where x_k^* is a point in the kth subinterval.

- Add the approximate areas of the strips to approximate the entire area A by the Riemann sum:

$$A \approx \sum_{k=1}^{n} f(x_k^*)\Delta x_k$$

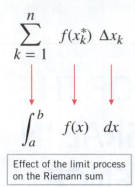

$$\sum_{k=1}^{n} f(x_k^*)\, \Delta x_k$$

$$\int_a^b \quad f(x) \quad dx$$

Effect of the limit process
on the Riemann sum

▲ **Figure 6.1.2**

• Take the limit of the Riemann sums as the number of subintervals increases and all their widths approach zero. This causes the error in the approximations to approach zero and produces the following definite integral for the exact area A:

$$A = \lim_{\max \Delta x_k \to 0} \sum_{k=1}^{n} f(x_k^*)\Delta x_k = \int_a^b f(x)\, dx$$

Figure 6.1.2 illustrates the effect that the limit process has on the various parts of the Riemann sum:

• The quantity x_k^* in the Riemann sum becomes the variable x in the definite integral.
• The interval width Δx_k in the Riemann sum becomes the dx in the definite integral.
• The interval $[a, b]$, which is the union of the subintervals with widths $\Delta x_1, \Delta x_2, \ldots, \Delta x_n$, does not appear explicitly in the Riemann sum but is represented by the upper and lower limits of integration in the definite integral.

■ AREA BETWEEN $y = f(x)$ AND $y = g(x)$

We will now consider the following extension of the area problem.

6.1.1 FIRST AREA PROBLEM   Suppose that f and g are continuous functions on an interval $[a, b]$ and

$$f(x) \geq g(x) \quad \text{for} \quad a \leq x \leq b$$

[This means that the curve $y = f(x)$ lies above the curve $y = g(x)$ and that the two can touch but not cross.] Find the area A of the region bounded above by $y = f(x)$, below by $y = g(x)$, and on the sides by the lines $x = a$ and $x = b$ (Figure 6.1.3a).

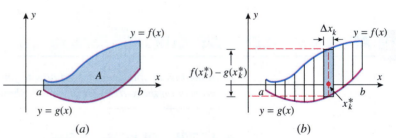

▶ **Figure 6.1.3** (a) (b)

To solve this problem we divide the interval $[a, b]$ into n subintervals, which has the effect of subdividing the region into n strips (Figure 6.1.3b). If we assume that the width of the kth strip is Δx_k, then the area of the strip can be approximated by the area of a rectangle of width Δx_k and height $f(x_k^*) - g(x_k^*)$, where x_k^* is a point in the kth subinterval. Adding these approximations yields the following Riemann sum that approximates the area A:

$$A \approx \sum_{k=1}^{n} [f(x_k^*) - g(x_k^*)]\Delta x_k$$

Taking the limit as n increases and the widths of all the subintervals approach zero yields the following definite integral for the area A between the curves:

$$A = \lim_{\max \Delta x_k \to 0} \sum_{k=1}^{n} [f(x_k^*) - g(x_k^*)]\Delta x_k = \int_a^b [f(x) - g(x)]\, dx$$

In summary, we have the following result.

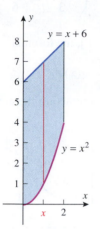

▲ Figure 6.1.4

> **6.1.2** **AREA FORMULA** If f and g are continuous functions on the interval $[a, b]$, and if $f(x) \geq g(x)$ for all x in $[a, b]$, then the area of the region bounded above by $y = f(x)$, below by $y = g(x)$, on the left by the line $x = a$, and on the right by the line $x = b$ is
>
> $$A = \int_a^b [f(x) - g(x)]\,dx \qquad (1)$$

▶ **Example 1** Find the area of the region bounded above by $y = x + 6$, bounded below by $y = x^2$, and bounded on the sides by the lines $x = 0$ and $x = 2$.

Solution. The region and a cross section are shown in Figure 6.1.4. The cross section extends from $g(x) = x^2$ on the bottom to $f(x) = x + 6$ on the top. If the cross section is moved through the region, then its leftmost position will be $x = 0$ and its rightmost position will be $x = 2$. Thus, from (1)

$$A = \int_0^2 [(x + 6) - x^2]\,dx = \left[\frac{x^2}{2} + 6x - \frac{x^3}{3} \right]_0^2 = \frac{34}{3} - 0 = \frac{34}{3} \;\blacktriangleleft$$

What does the integral in (1) represent if the graphs of f and g cross each other over the interval $[a, b]$? How would you find the area between the curves in this case?

It is possible that the upper and lower boundaries of a region may intersect at one or both endpoints, in which case the sides of the region will be points, rather than vertical line segments (Figure 6.1.5). When that occurs you will have to determine the points of intersection to obtain the limits of integration.

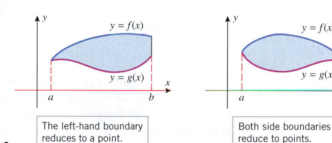

| The left-hand boundary reduces to a point. | Both side boundaries reduce to points. |

▶ Figure 6.1.5

▶ **Example 2** Find the area of the region that is enclosed between the curves $y = x^2$ and $y = x + 6$.

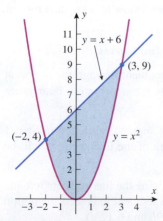

▲ Figure 6.1.6

Solution. A sketch of the region (Figure 6.1.6) shows that the lower boundary is $y = x^2$ and the upper boundary is $y = x + 6$. At the endpoints of the region, the upper and lower boundaries have the same y-coordinates; thus, to find the endpoints we equate

$$y = x^2 \quad \text{and} \quad y = x + 6 \qquad (2)$$

This yields

$$x^2 = x + 6 \quad \text{or} \quad x^2 - x - 6 = 0 \quad \text{or} \quad (x + 2)(x - 3) = 0$$

from which we obtain

$$x = -2 \quad \text{and} \quad x = 3$$

Although the y-coordinates of the endpoints are not essential to our solution, they may be obtained from (2) by substituting $x = -2$ and $x = 3$ in either equation. This yields $y = 4$ and $y = 9$, so the upper and lower boundaries intersect at $(-2, 4)$ and $(3, 9)$.

From (1) with $f(x) = x + 6$, $g(x) = x^2$, $a = -2$, and $b = 3$, we obtain the area

$$A = \int_{-2}^{3} [(x+6) - x^2]\, dx = \left[\frac{x^2}{2} + 6x - \frac{x^3}{3} \right]_{-2}^{3} = \frac{27}{2} - \left(-\frac{22}{3} \right) = \frac{125}{6} \quad \blacktriangleleft$$

In the case where f and g are *nonnegative* on the interval $[a, b]$, the formula

$$A = \int_{a}^{b} [f(x) - g(x)]\, dx = \int_{a}^{b} f(x)\, dx - \int_{a}^{b} g(x)\, dx$$

states that the area A between the curves can be obtained by subtracting the area under $y = g(x)$ from the area under $y = f(x)$ (Figure 6.1.7).

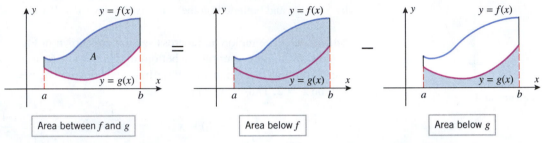

Area between f and g Area below f Area below g

▲ **Figure 6.1.7**

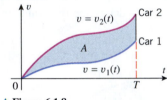

▲ **Figure 6.1.8**

▶ **Example 3**   Figure 6.1.8 shows velocity versus time curves for two race cars that move along a straight track, starting from rest at the same time. Give a physical interpretation of the area A between the curves over the interval $0 \le t \le T$.

Solution.   From (1)

$$A = \int_{0}^{T} [v_2(t) - v_1(t)]\, dt = \int_{0}^{T} v_2(t)\, dt - \int_{0}^{T} v_1(t)\, dt$$

Since v_1 and v_2 are nonnegative functions on $[0, T]$, it follows from Formula (4) of Section 5.7 that the integral of v_1 over $[0, T]$ is the distance traveled by car 1 during the time interval $0 \le t \le T$, and the integral of v_2 over $[0, T]$ is the distance traveled by car 2 during the same time interval. Since $v_1(t) \le v_2(t)$ on $[0, T]$, car 2 travels farther than car 1 does over the time interval $0 \le t \le T$, and the area A represents the distance by which car 2 is ahead of car 1 at time T. ◀

Some regions may require careful thought to determine the integrand and limits of integration in (1). Here is a systematic procedure that you can follow to set up this formula.

It is not necessary to make an extremely accurate sketch in Step 1; the only purpose of the sketch is to determine which curve is the upper boundary and which is the lower boundary.

Finding the Limits of Integration for the Area Between Two Curves

Step 1.   Sketch the region and then draw a vertical line segment through the region at an arbitrary point x on the x-axis, connecting the top and bottom boundaries (Figure 6.1.9a).

Step 2.   The y-coordinate of the top endpoint of the line segment sketched in Step 1 will be $f(x)$, the bottom one $g(x)$, and the length of the line segment will be $f(x) - g(x)$. This is the integrand in (1).

Step 3.   To determine the limits of integration, imagine moving the line segment left and then right. The leftmost position at which the line segment intersects the region is $x = a$ and the rightmost is $x = b$ (Figures 6.1.9b and 6.1.9c).

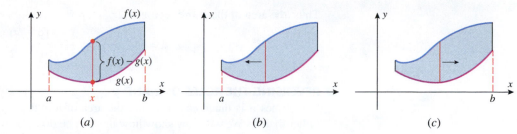

▲ **Figure 6.1.9**

There is a useful way of thinking about this procedure:

If you view the vertical line segment as the "cross section" of the region at the point x, then Formula (1) states that the area between the curves is obtained by integrating the length of the cross section over the interval $[a, b]$.

It is possible for the upper or lower boundary of a region to consist of two or more different curves, in which case it will be convenient to subdivide the region into smaller pieces in order to apply Formula (1). This is illustrated in the next example.

▶ **Example 4**   Find the area of the region enclosed by $x = y^2$ and $y = x - 2$.

Solution.   To determine the appropriate boundaries of the region, we need to know where the curves $x = y^2$ and $y = x - 2$ intersect. In Example 2 we found intersections by equating the expressions for y. Here it is easier to rewrite the latter equation as $x = y + 2$ and equate the expressions for x, namely,

$$x = y^2 \quad \text{and} \quad x = y + 2 \tag{3}$$

This yields

$$y^2 = y + 2 \quad \text{or} \quad y^2 - y - 2 = 0 \quad \text{or} \quad (y + 1)(y - 2) = 0$$

from which we obtain $y = -1$, $y = 2$. Substituting these values in either equation in (3) we see that the corresponding x-values are $x = 1$ and $x = 4$, respectively, so the points of intersection are $(1, -1)$ and $(4, 2)$ (Figure 6.1.10a).

To apply Formula (1), the equations of the boundaries must be written so that y is expressed explicitly as a function of x. The upper boundary can be written as $y = \sqrt{x}$ (rewrite $x = y^2$ as $y = \pm\sqrt{x}$ and choose the $+$ for the upper portion of the curve). The lower boundary consists of two parts:

$$y = -\sqrt{x} \quad \text{for} \quad 0 \le x \le 1 \qquad \text{and} \qquad y = x - 2 \quad \text{for} \quad 1 \le x \le 4$$

(Figure 6.1.10b). Because of this change in the formula for the lower boundary, it is necessary to divide the region into two parts and find the area of each part separately.

From (1) with $f(x) = \sqrt{x}$, $g(x) = -\sqrt{x}$, $a = 0$, and $b = 1$, we obtain

$$A_1 = \int_0^1 [\sqrt{x} - (-\sqrt{x})]\,dx = 2\int_0^1 \sqrt{x}\,dx = 2\left[\frac{2}{3}x^{3/2}\right]_0^1 = \frac{4}{3} - 0 = \frac{4}{3}$$

From (1) with $f(x) = \sqrt{x}$, $g(x) = x - 2$, $a = 1$, and $b = 4$, we obtain

$$A_2 = \int_1^4 [\sqrt{x} - (x - 2)]\,dx = \int_1^4 (\sqrt{x} - x + 2)\,dx$$

$$= \left[\frac{2}{3}x^{3/2} - \frac{1}{2}x^2 + 2x\right]_1^4 = \left(\frac{16}{3} - 8 + 8\right) - \left(\frac{2}{3} - \frac{1}{2} + 2\right) = \frac{19}{6}$$

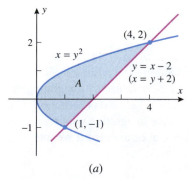

(a)

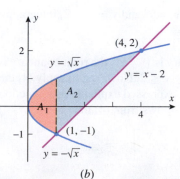

(b)

▲ **Figure 6.1.10**

Thus, the area of the entire region is

$$A = A_1 + A_2 = \frac{4}{3} + \frac{19}{6} = \frac{9}{2} \blacktriangleleft$$

■ REVERSING THE ROLES OF x AND y

Sometimes it is much easier to find the area of a region by integrating with respect to y rather than x. We will now show how this can be done.

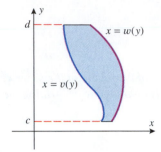

▲ **Figure 6.1.11**

6.1.3   **SECOND AREA PROBLEM**   Suppose that w and v are continuous functions of y on an interval $[c, d]$ and that

$$w(y) \geq v(y) \quad \text{for} \quad c \leq y \leq d$$

[This means that the curve $x = w(y)$ lies to the right of the curve $x = v(y)$ and that the two can touch but not cross.] Find the area A of the region bounded on the left by $x = v(y)$, on the right by $x = w(y)$, and above and below by the lines $y = d$ and $y = c$ (Figure 6.1.11).

Proceeding as in the derivation of (1), but with the roles of x and y reversed, leads to the following analog of 6.1.2.

6.1.4   **AREA FORMULA**   If w and v are continuous functions and if $w(y) \geq v(y)$ for all y in $[c, d]$, then the area of the region bounded on the left by $x = v(y)$, on the right by $x = w(y)$, below by $y = c$, and above by $y = d$ is

$$A = \int_c^d [w(y) - v(y)]\, dy \qquad (4)$$

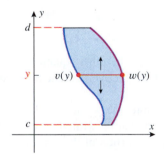

▲ **Figure 6.1.12**

The guiding principle in applying this formula is the same as with (1): The integrand in (4) can be viewed as the length of the horizontal cross section at an arbitrary point y on the y-axis, in which case Formula (4) states that the area can be obtained by integrating the length of the horizontal cross section over the interval $[c, d]$ on the y-axis (Figure 6.1.12).

In Example 4, we split the region into two parts to facilitate integrating with respect to x. In the next example we will see that splitting this region can be avoided if we integrate with respect to y.

▶ **Example 5**   Find the area of the region enclosed by $x = y^2$ and $y = x - 2$, integrating with respect to y.

Solution.   As indicated in Figure 6.1.10 the left boundary is $x = y^2$, the right boundary is $y = x - 2$, and the region extends over the interval $-1 \leq y \leq 2$. However, to apply (4) the equations for the boundaries must be written so that x is expressed explicitly as a function of y. Thus, we rewrite $y = x - 2$ as $x = y + 2$. It now follows from (4) that

$$A = \int_{-1}^2 [(y + 2) - y^2]\, dy = \left[\frac{y^2}{2} + 2y - \frac{y^3}{3} \right]_{-1}^2 = \frac{9}{2}$$

which agrees with the result obtained in Example 4. ◀

The choice between Formulas (1) and (4) is usually dictated by the shape of the region and which formula requires the least amount of splitting. However, sometimes one might choose the formula that requires more splitting because it is easier to evaluate the resulting integrals.

✔ **QUICK CHECK EXERCISES 6.1**   *(See page 421 for answers.)*

1. An integral expression for the area of the region between the curves $y = 20 - 3x^2$ and $y = e^x$ and bounded on the sides by $x = 0$ and $x = 2$ is _____.

2. An integral expression for the area of the parallelogram bounded by $y = 2x + 8$, $y = 2x - 3$, $x = -1$, and $x = 5$ is _____. The value of this integral is _____.

3. (a) The points of intersection for the circle $x^2 + y^2 = 4$ and the line $y = x + 2$ are _____ and _____.

(b) Expressed as a definite integral with respect to x, _____ gives the area of the region inside the circle $x^2 + y^2 = 4$ and above the line $y = x + 2$.

(c) Expressed as a definite integral with respect to y, _____ gives the area of the region described in part (b).

4. The area of the region enclosed by the curves $y = x^2$ and $y = \sqrt[3]{x}$ is _____.

EXERCISE SET 6.1   ⌁ Graphing Utility   C CAS

1–4 Find the area of the shaded region. ■

1.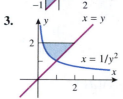
   $y = x^2 + 1$
   $y = x$

2.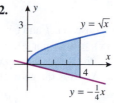
   $y = \sqrt{x}$
   $y = -\frac{1}{4}x$

3.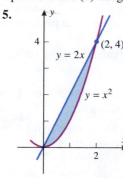
   $x = y$
   $x = 1/y^2$

4.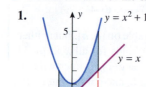
   $x = -y$
   $x = 2 - y^2$

5–6 Find the area of the shaded region by (a) integrating with respect to x and (b) integrating with respect to y. ■

5.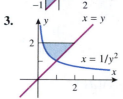
   $y = 2x$ $(2, 4)$
   $y = x^2$

6.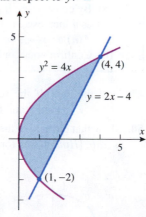
   $y^2 = 4x$ $(4, 4)$
   $y = 2x - 4$
   $(1, -2)$

7–18 Sketch the region enclosed by the curves and find its area. ■

7. $y = x^2$, $y = \sqrt{x}$, $x = \frac{1}{4}$, $x = 1$

8. $y = x^3 - 4x$, $y = 0$, $x = 0$, $x = 2$

9. $y = \cos 2x$, $y = 0$, $x = \pi/4$, $x = \pi/2$

10. $y = \sec^2 x$, $y = 2$, $x = -\pi/4$, $x = \pi/4$

11. $x = \sin y$, $x = 0$, $y = \pi/4$, $y = 3\pi/4$

12. $x^2 = y$, $x = y - 2$

13. $y = e^x$, $y = e^{2x}$, $x = 0$, $x = \ln 2$

14. $x = 1/y$, $x = 0$, $y = 1$, $y = e$

15. $y = \dfrac{2}{1 + x^2}$, $y = |x|$ 16. $y = \dfrac{1}{\sqrt{1 - x^2}}$, $y = 2$

17. $y = 2 + |x - 1|$, $y = -\frac{1}{5}x + 7$

18. $y = x$, $y = 4x$, $y = -x + 2$

⌁ **19–26** Use a graphing utility, where helpful, to find the area of the region enclosed by the curves. ■

19. $y = x^3 - 4x^2 + 3x$, $y = 0$

20. $y = x^3 - 2x^2$, $y = 2x^2 - 3x$

21. $y = \sin x$, $y = \cos x$, $x = 0$, $x = 2\pi$

22. $y = x^3 - 4x$, $y = 0$ 23. $x = y^3 - y$, $x = 0$

24. $x = y^3 - 4y^2 + 3y$, $x = y^2 - y$

25. $y = xe^{x^2}$, $y = 2|x|$

26. $y = \dfrac{1}{x\sqrt{1 - (\ln x)^2}}$, $y = \dfrac{3}{x}$

27–30 True–False Determine whether the statement is true or false. Explain your answer. [In each exercise, assume that f and g are distinct continuous functions on $[a, b]$ and that A denotes the area of the region bounded by the graphs of $y = f(x)$, $y = g(x)$, $x = a$, and $x = b$.] ■

27. If f and g differ by a positive constant c, then $A = c(b - a)$.

28. If
$$\int_a^b [f(x) - g(x)]\,dx = -3$$
then $A = 3$.

29. If
$$\int_a^b [f(x) - g(x)]\,dx = 0$$
then the graphs of $y = f(x)$ and $y = g(x)$ cross at least once on $[a, b]$.

30. If
$$A = \left| \int_a^b [f(x) - g(x)]\,dx \right|$$

then the graphs of $y = f(x)$ and $y = g(x)$ don't cross on $[a, b]$.

31. Estimate the value of k ($0 < k < 1$) so that the region enclosed by $y = 1/\sqrt{1 - x^2}$, $y = x$, $x = 0$, and $x = k$ has an area of 1 square unit.

32. Estimate the area of the region in the first quadrant enclosed by $y = \sin 2x$ and $y = \sin^{-1} x$.

C **33.** Use a CAS to find the area enclosed by $y = 3 - 2x$ and $y = x^6 + 2x^5 - 3x^4 + x^2$.

C **34.** Use a CAS to find the exact area enclosed by the curves $y = x^5 - 2x^3 - 3x$ and $y = x^3$.

35. Find a horizontal line $y = k$ that divides the area between $y = x^2$ and $y = 9$ into two equal parts.

36. Find a vertical line $x = k$ that divides the area enclosed by $x = \sqrt{y}$, $x = 2$, and $y = 0$ into two equal parts.

37. (a) Find the area of the region enclosed by the parabola $y = 2x - x^2$ and the x-axis.
(b) Find the value of m so that the line $y = mx$ divides the region in part (a) into two regions of equal area.

38. Find the area between the curve $y = \sin x$ and the line segment joining the points $(0, 0)$ and $(5\pi/6, 1/2)$ on the curve.

39–42 Use Newton's Method (Section 4.7), where needed, to approximate the x-coordinates of the intersections of the curves to at least four decimal places, and then use those approximations to approximate the area of the region. ■

39. The region that lies below the curve $y = \sin x$ and above the line $y = 0.2x$, where $x \geq 0$.

40. The region enclosed by the graphs of $y = x^2$ and $y = \cos x$.

41. The region enclosed by the graphs of $y = (\ln x)/x$ and $y = x - 2$.

42. The region enclosed by the graphs of $y = 3 - 2\cos x$ and $y = 2/(1 + x^2)$.

43. Find the area of the region that is enclosed by the curves $y = x^2 - 1$ and $y = 2\sin x$.

C **44.** Referring to the accompanying figure, use a CAS to estimate the value of k so that the areas of the shaded regions are equal.

Source: This exercise is based on Problem A1 that was posed in the Fifty-Fourth Annual William Lowell Putnam Mathematical Competition.

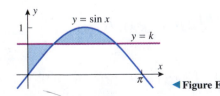

◄ **Figure Ex-44**

FOCUS ON CONCEPTS

45. Two racers in adjacent lanes move with velocity functions $v_1(t)$ m/s and $v_2(t)$ m/s, respectively. Suppose that the racers are even at time $t = 60$ s. Interpret the

value of the integral

$$\int_0^{60} [v_2(t) - v_1(t)]\, dt$$

in this context.

46. The accompanying figure shows acceleration versus time curves for two cars that move along a straight track, accelerating from rest at the starting line. What does the area A between the curves over the interval $0 \leq t \leq T$ represent? Justify your answer.

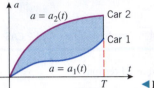

◄ **Figure Ex-46**

47. Suppose that f and g are integrable on $[a, b]$, but neither $f(x) \geq g(x)$ nor $g(x) \geq f(x)$ holds for all x in $[a, b]$ [i.e., the curves $y = f(x)$ and $y = g(x)$ are intertwined].
(a) What is the geometric significance of the integral

$$\int_a^b [f(x) - g(x)]\, dx?$$

(b) What is the geometric significance of the integral

$$\int_a^b |f(x) - g(x)|\, dx?$$

48. Let $A(n)$ be the area in the first quadrant enclosed by the curves $y = \sqrt[n]{x}$ and $y = x$.
(a) By considering how the graph of $y = \sqrt[n]{x}$ changes as n increases, make a conjecture about the limit of $A(n)$ as $n \to +\infty$.
(b) Confirm your conjecture by calculating the limit.

49. Find the area of the region enclosed between the curve $x^{1/2} + y^{1/2} = a^{1/2}$ and the coordinate axes.

50. Show that the area of the ellipse in the accompanying figure is πab. [*Hint:* Use a formula from geometry.]

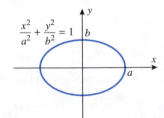

◄ **Figure Ex-50**

51. **Writing** Suppose that f and g are continuous on $[a, b]$ but that the graphs of $y = f(x)$ and $y = g(x)$ cross several times. Describe a step-by-step procedure for determining the area bounded by the graphs of $y = f(x)$, $y = g(x)$, $x = a$, and $x = b$.

52. Writing Suppose that R and S are two regions in the xy-plane that lie between a pair of lines L_1 and L_2 that are parallel to the y-axis. Assume that each line between L_1 and L_2 that is parallel to the y-axis intersects R and S in line segments of equal length. Give an informal argument that the area of R is equal to the area of S. (Make reasonable assumptions about the boundaries of R and S.)

✔ **QUICK CHECK ANSWERS 6.1**

1. $\displaystyle\int_0^2 [(20 - 3x^2) - e^x]\,dx$ **2.** $\displaystyle\int_{-1}^5 [(2x+8) - (2x-3)]\,dx;\ 66$ **3.** (a) $(-2, 0);\ (0, 2)$ (b) $\displaystyle\int_{-2}^0 [\sqrt{4 - x^2} - (x + 2)]\,dx$

(c) $\displaystyle\int_0^2 [(y - 2) + \sqrt{4 - y^2}]\,dy$ **4.** $\dfrac{5}{12}$

6.2 VOLUMES BY SLICING; DISKS AND WASHERS

In the last section we showed that the area of a plane region bounded by two curves can be obtained by integrating the length of a general cross section over an appropriate interval. In this section we will see that the same basic principle can be used to find volumes of certain three-dimensional solids.

■ VOLUMES BY SLICING

Recall that the underlying principle for finding the area of a plane region is to divide the region into thin strips, approximate the area of each strip by the area of a rectangle, add the approximations to form a Riemann sum, and take the limit of the Riemann sums to produce an integral for the area. Under appropriate conditions, the same strategy can be used to find the volume of a solid. The idea is to divide the solid into thin slabs, approximate the volume of each slab, add the approximations to form a Riemann sum, and take the limit of the Riemann sums to produce an integral for the volume (Figure 6.2.1).

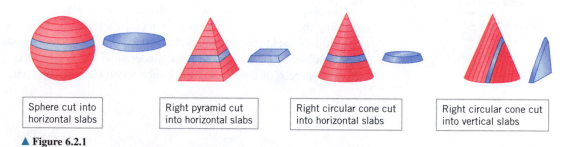

| Sphere cut into horizontal slabs | Right pyramid cut into horizontal slabs | Right circular cone cut into horizontal slabs | Right circular cone cut into vertical slabs |

▲ **Figure 6.2.1**

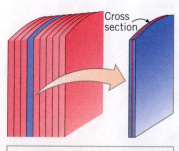

In a thin slab, the cross sections do not vary much in size and shape.

▲ **Figure 6.2.2**

What makes this method work is the fact that a *thin* slab has a cross section that does not vary much in size or shape, which, as we will see, makes its volume easy to approximate (Figure 6.2.2). Moreover, the thinner the slab, the less variation in its cross sections and the better the approximation. Thus, once we approximate the volumes of the slabs, we can set up a Riemann sum whose limit is the volume of the entire solid. We will give the details shortly, but first we need to discuss how to find the volume of a solid whose cross sections do not vary in size and shape (i.e., are congruent).

One of the simplest examples of a solid with congruent cross sections is a right circular cylinder of radius r, since all cross sections taken perpendicular to the central axis are circular regions of radius r. The volume V of a right circular cylinder of radius r and height h can be expressed in terms of the height and the area of a cross section as

$$V = \pi r^2 h = [\text{area of a cross section}] \times [\text{height}] \tag{1}$$

This is a special case of a more general volume formula that applies to solids called right cylinders. A *right cylinder* is a solid that is generated when a plane region is translated along a line or *axis* that is perpendicular to the region (Figure 6.2.3).

Some Right Cylinders

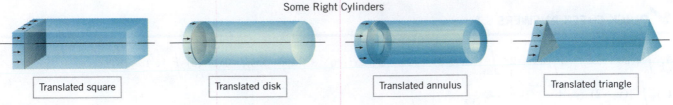

Translated square   Translated disk   Translated annulus   Translated triangle

▲ **Figure 6.2.3**

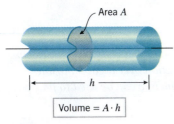

Area A

Volume = $A \cdot h$

▲ **Figure 6.2.4**

If a right cylinder is generated by translating a region of area A through a distance h, then h is called the **height** (or sometimes the **width**) of the cylinder, and the volume V of the cylinder is defined to be

$$V = A \cdot h = [\text{area of a cross section}] \times [\text{height}] \qquad (2)$$

(Figure 6.2.4). Note that this is consistent with Formula (1) for the volume of a right *circular* cylinder.

We now have all of the tools required to solve the following problem.

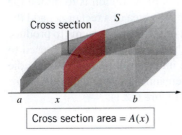

Cross section   S

Cross section area = $A(x)$

▲ **Figure 6.2.5**

6.2.1 PROBLEM   Let S be a solid that extends along the x-axis and is bounded on the left and right, respectively, by the planes that are perpendicular to the x-axis at $x = a$ and $x = b$ (Figure 6.2.5). Find the volume V of the solid, assuming that its cross-sectional area $A(x)$ is known at each x in the interval $[a, b]$.

To solve this problem we begin by dividing the interval $[a, b]$ into n subintervals, thereby dividing the solid into n slabs as shown in the left part of Figure 6.2.6. If we assume that the width of the kth subinterval is Δx_k, then the volume of the kth slab can be approximated by the volume $A(x_k^*)\Delta x_k$ of a right cylinder of width (height) Δx_k and cross-sectional area $A(x_k^*)$, where x_k^* is a point in the kth subinterval (see the right part of Figure 6.2.6).

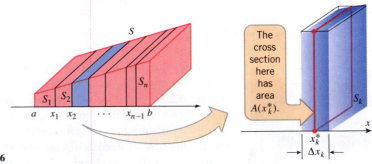

The cross section here has area $A(x_k^*)$.

▶ **Figure 6.2.6**

Adding these approximations yields the following Riemann sum that approximates the volume V:

$$V \approx \sum_{k=1}^{n} A(x_k^*)\Delta x_k$$

Taking the limit as n increases and the widths of all the subintervals approach zero yields the definite integral

$$V = \lim_{\max \Delta x_k \to 0} \sum_{k=1}^{n} A(x_k^*) \Delta x_k = \int_a^b A(x)\, dx$$

In summary, we have the following result.

It is understood in our calculations of volume that the units of volume are the cubed units of length [e.g., cubic inches (in^3) or cubic meters (m^3)].

6.2.2   VOLUME FORMULA   Let S be a solid bounded by two parallel planes perpendicular to the x-axis at $x = a$ and $x = b$. If, for each x in $[a, b]$, the cross-sectional area of S perpendicular to the x-axis is $A(x)$, then the volume of the solid is

$$V = \int_a^b A(x)\, dx \tag{3}$$

provided $A(x)$ is integrable.

There is a similar result for cross sections perpendicular to the y-axis.

6.2.3   VOLUME FORMULA   Let S be a solid bounded by two parallel planes perpendicular to the y-axis at $y = c$ and $y = d$. If, for each y in $[c, d]$, the cross-sectional area of S perpendicular to the y-axis is $A(y)$, then the volume of the solid is

$$V = \int_c^d A(y)\, dy \tag{4}$$

provided $A(y)$ is integrable.

In words, these formulas state:

The volume of a solid can be obtained by integrating the cross-sectional area from one end of the solid to the other.

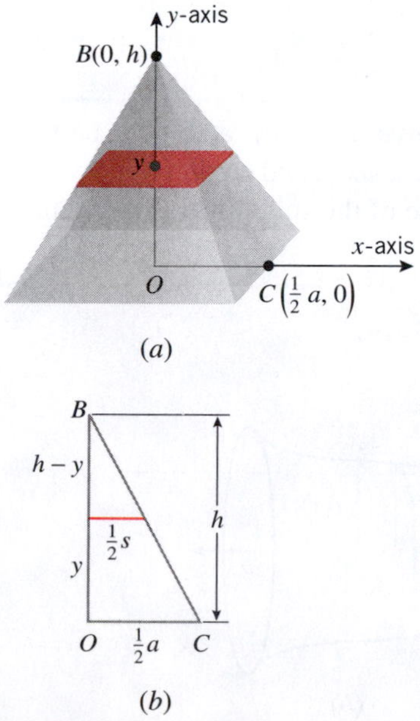

(a)

(b)

▲ **Figure 6.2.7**

▶ **Example 1**   Derive the formula for the volume of a right pyramid whose altitude is h and whose base is a square with sides of length a.

Solution.   As illustrated in Figure 6.2.7a, we introduce a rectangular coordinate system in which the y-axis passes through the apex and is perpendicular to the base, and the x-axis passes through the base and is parallel to a side of the base.

At any y in the interval $[0, h]$ on the y-axis, the cross section perpendicular to the y-axis is a square. If s denotes the length of a side of this square, then by similar triangles (Figure 6.2.7b)

$$\frac{\frac{1}{2}s}{\frac{1}{2}a} = \frac{h - y}{h} \quad \text{or} \quad s = \frac{a}{h}(h - y)$$

Thus, the area $A(y)$ of the cross section at y is

$$A(y) = s^2 = \frac{a^2}{h^2}(h - y)^2$$

and by (4) the volume is

$$V = \int_0^h A(y)\, dy = \int_0^h \frac{a^2}{h^2}(h-y)^2\, dy = \frac{a^2}{h^2}\int_0^h (h-y)^2\, dy$$

$$= \frac{a^2}{h^2}\left[-\frac{1}{3}(h-y)^3\right]_{y=0}^h = \frac{a^2}{h^2}\left[0 + \frac{1}{3}h^3\right] = \frac{1}{3}a^2 h$$

That is, the volume is $\frac{1}{3}$ of the area of the base times the altitude. ◄

■ SOLIDS OF REVOLUTION

A *solid of revolution* is a solid that is generated by revolving a plane region about a line that lies in the same plane as the region; the line is called the *axis of revolution*. Many familiar solids are of this type (Figure 6.2.8).

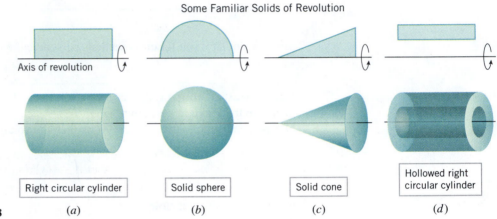

Some Familiar Solids of Revolution

Axis of revolution

| Right circular cylinder | Solid sphere | Solid cone | Hollowed right circular cylinder |

▶ **Figure 6.2.8**   (*a*)   (*b*)   (*c*)   (*d*)

■ VOLUMES BY DISKS PERPENDICULAR TO THE *x*-AXIS

We will be interested in the following general problem.

6.2.4   PROBLEM   Let f be continuous and nonnegative on $[a, b]$, and let R be the region that is bounded above by $y = f(x)$, below by the x-axis, and on the sides by the lines $x = a$ and $x = b$ (Figure 6.2.9*a*). Find the volume of the solid of revolution that is generated by revolving the region R about the x-axis.

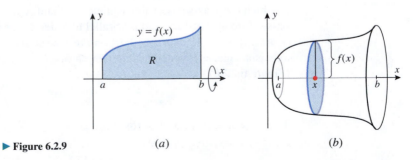

▶ **Figure 6.2.9**   (*a*)   (*b*)

We can solve this problem by slicing. For this purpose, observe that the cross section of the solid taken perpendicular to the x-axis at the point x is a circular disk of radius $f(x)$ (Figure 6.2.9b). The area of this region is

$$A(x) = \pi[f(x)]^2$$

Thus, from (3) the volume of the solid is

$$V = \int_a^b \pi[f(x)]^2\, dx \qquad (5)$$

Because the cross sections are disk shaped, the application of this formula is called the *method of disks*.

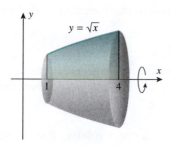

▲ Figure 6.2.10

▶ **Example 2**   Find the volume of the solid that is obtained when the region under the curve $y = \sqrt{x}$ over the interval $[1, 4]$ is revolved about the x-axis (Figure 6.2.10).

Solution.   From (5), the volume is

$$V = \int_a^b \pi[f(x)]^2\, dx = \int_1^4 \pi x\, dx = \frac{\pi x^2}{2}\Bigg]_1^4 = 8\pi - \frac{\pi}{2} = \frac{15\pi}{2} \quad \blacktriangleleft$$

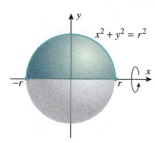

▲ Figure 6.2.11

▶ **Example 3**   Derive the formula for the volume of a sphere of radius r.

Solution.   As indicated in Figure 6.2.11, a sphere of radius r can be generated by revolving the upper semicircular disk enclosed between the x-axis and

$$x^2 + y^2 = r^2$$

about the x-axis. Since the upper half of this circle is the graph of $y = f(x) = \sqrt{r^2 - x^2}$, it follows from (5) that the volume of the sphere is

$$V = \int_a^b \pi[f(x)]^2\, dx = \int_{-r}^r \pi(r^2 - x^2)\, dx = \pi\left[r^2 x - \frac{x^3}{3}\right]_{-r}^r = \frac{4}{3}\pi r^3 \quad \blacktriangleleft$$

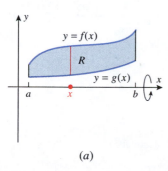

■ **VOLUMES BY WASHERS PERPENDICULAR TO THE x-AXIS**

Not all solids of revolution have solid interiors; some have holes or channels that create interior surfaces, as in Figure 6.2.8d. So we will also be interested in problems of the following type.

(a)

> **6.2.5   PROBLEM**   Let f and g be continuous and nonnegative on $[a, b]$, and suppose that $f(x) \geq g(x)$ for all x in the interval $[a, b]$. Let R be the region that is bounded above by $y = f(x)$, below by $y = g(x)$, and on the sides by the lines $x = a$ and $x = b$ (Figure 6.2.12a). Find the volume of the solid of revolution that is generated by revolving the region R about the x-axis (Figure 6.2.12b).

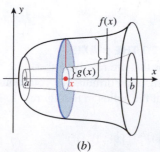

(b)

▲ Figure 6.2.12

We can solve this problem by slicing. For this purpose, observe that the cross section of the solid taken perpendicular to the x-axis at the point x is the annular or "washer-shaped" region with inner radius $g(x)$ and outer radius $f(x)$ (Figure 6.2.12b); its area is

$$A(x) = \pi[f(x)]^2 - \pi[g(x)]^2 = \pi([f(x)]^2 - [g(x)]^2)$$

Thus, from (3) the volume of the solid is

$$V = \int_a^b \pi([f(x)]^2 - [g(x)]^2) \, dx \tag{6}$$

Because the cross sections are washer shaped, the application of this formula is called the *method of washers*.

▶ **Example 4**   Find the volume of the solid generated when the region between the graphs of the equations $f(x) = \frac{1}{2} + x^2$ and $g(x) = x$ over the interval $[0, 2]$ is revolved about the x-axis.

Solution.   First sketch the region (Figure 6.2.13a); then imagine revolving it about the x-axis (Figure 6.2.13b). From (6) the volume is

$$V = \int_a^b \pi([f(x)]^2 - [g(x)]^2) \, dx = \int_0^2 \pi \left(\left[\tfrac{1}{2} + x^2 \right]^2 - x^2 \right) \, dx$$

$$= \int_0^2 \pi \left(\frac{1}{4} + x^4 \right) \, dx = \pi \left[\frac{x}{4} + \frac{x^5}{5} \right]_0^2 = \frac{69\pi}{10} \quad \blacktriangleleft$$

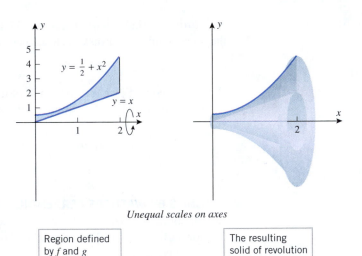

Unequal scales on axes

| Region defined by f and g | The resulting solid of revolution |

▶ **Figure 6.2.13**   (*a*)   (*b*)

■ **VOLUMES BY DISKS AND WASHERS PERPENDICULAR TO THE *y*-AXIS**

The methods of disks and washers have analogs for regions that are revolved about the y-axis (Figures 6.2.14 and 6.2.15). Using the method of slicing and Formula (4), you should be able to deduce the following formulas for the volumes of the solids in the figures.

$$V = \int_c^d \pi[u(y)]^2 \, dy \qquad V = \int_c^d \pi([w(y)]^2 - [v(y)]^2) \, dy \tag{7–8}$$

Disks   Washers

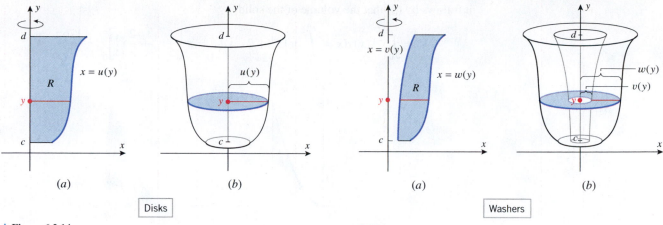

Disks

▲ **Figure 6.2.14**

Washers

▲ **Figure 6.2.15**

▶ **Example 5** Find the volume of the solid generated when the region enclosed by $y = \sqrt{x}$, $y = 2$, and $x = 0$ is revolved about the y-axis.

Solution. First sketch the region and the solid (Figure 6.2.16). The cross sections taken perpendicular to the y-axis are disks, so we will apply (7). But first we must rewrite $y = \sqrt{x}$ as $x = y^2$. Thus, from (7) with $u(y) = y^2$, the volume is

$$V = \int_c^d \pi[u(y)]^2 \, dy = \int_0^2 \pi y^4 \, dy = \frac{\pi y^5}{5}\Big]_0^2 = \frac{32\pi}{5} \quad \blacktriangleleft$$

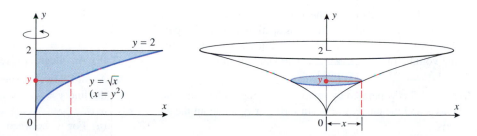

▶ **Figure 6.2.16**

■ OTHER AXES OF REVOLUTION

It is possible to use the method of disks and the method of washers to find the volume of a solid of revolution whose axis of revolution is a line other than one of the coordinate axes. Instead of developing a new formula for each situation, we will appeal to Formulas (3) and (4) and integrate an appropriate cross-sectional area to find the volume.

▶ **Example 6** Find the volume of the solid generated when the region under the curve $y = x^2$ over the interval [0, 2] is rotated about the line $y = -1$.

Solution. First sketch the region and the axis of revolution; then imagine revolving the region about the axis (Figure 6.2.17). At each x in the interval $0 \leq x \leq 2$, the cross section of the solid perpendicular to the axis $y = -1$ is a washer with outer radius $x^2 + 1$ and inner radius 1. Since the area of this washer is

$$A(x) = \pi([x^2 + 1]^2 - 1^2) = \pi(x^4 + 2x^2)$$

it follows by (3) that the volume of the solid is

$$V = \int_0^2 A(x)\,dx = \int_0^2 \pi\left(x^4 + 2x^2\right) dx = \pi\left[\frac{1}{5}x^5 + \frac{2}{3}x^3\right]_0^2 = \frac{176\pi}{15} \blacktriangleleft$$

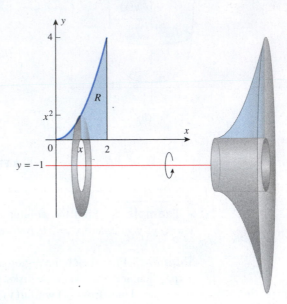

▶ **Figure 6.2.17**

✔ **QUICK CHECK EXERCISES 6.2**   *(See page 431 for answers.)*

1. A solid S extends along the x-axis from $x = 1$ to $x = 3$. For x between 1 and 3, the cross-sectional area of S perpendicular to the x-axis is $3x^2$. An integral expression for the volume of S is _____. The value of this integral is _____.

2. A solid S is generated by revolving the region between the x-axis and the curve $y = \sqrt{\sin x}$ $(0 \le x \le \pi)$ about the x-axis.
   (a) For x between 0 and π, the cross-sectional area of S perpendicular to the x-axis at x is $A(x) =$ _____.
   (b) An integral expression for the volume of S is _____.
   (c) The value of the integral in part (b) is _____.

3. A solid S is generated by revolving the region enclosed by the line $y = 2x + 1$ and the curve $y = x^2 + 1$ about the x-axis.

(a) For x between _____ and _____, the cross-sectional area of S perpendicular to the x-axis at x is $A(x) =$ _____.
   (b) An integral expression for the volume of S is _____.

4. A solid S is generated by revolving the region enclosed by the line $y = x + 1$ and the curve $y = x^2 + 1$ about the y-axis.
   (a) For y between _____ and _____, the cross-sectional area of S perpendicular to the y-axis at y is $A(y) =$ _____.
   (b) An integral expression for the volume of S is _____.

EXERCISE SET 6.2   ⒸCAS

1–8 Find the volume of the solid that results when the shaded region is revolved about the indicated axis. ■

1.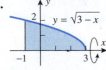
$y = \sqrt{3 - x}$

2.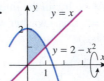
$y = x$
$y = 2 - x^2$

3.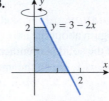
$y = 3 - 2x$

4.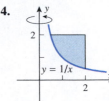
$y = 1/x$

5.

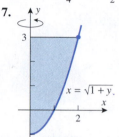

$y = \sqrt{\cos x}$

$\dfrac{\pi}{4}$ $\dfrac{\pi}{2}$

6.

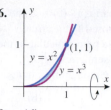

$(1, 1)$

$y = x^2$

$y = x^3$

1

7.

3

$x = \sqrt{1 + y}$

2

8.

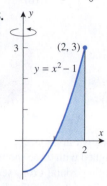

3 $(2, 3)$

$y = x^2 - 1$

2

9. Find the volume of the solid whose base is the region bounded between the curve $y = x^2$ and the x-axis from $x = 0$ to $x = 2$ and whose cross sections taken perpendicular to the x-axis are squares.

10. Find the volume of the solid whose base is the region bounded between the curve $y = \sec x$ and the x-axis from $x = \pi/4$ to $x = \pi/3$ and whose cross sections taken perpendicular to the x-axis are squares.

11–18 Find the volume of the solid that results when the region enclosed by the given curves is revolved about the x-axis. ■

11. $y = \sqrt{25 - x^2}$, $y = 3$

12. $y = 9 - x^2$, $y = 0$ **13.** $x = \sqrt{y}$, $x = y/4$

14. $y = \sin x$, $y = \cos x$, $x = 0$, $x = \pi/4$
[*Hint:* Use the identity $\cos 2x = \cos^2 x - \sin^2 x$.]

15. $y = e^x$, $y = 0$, $x = 0$, $x = \ln 3$

16. $y = e^{-2x}$, $y = 0$, $x = 0$, $x = 1$

17. $y = \dfrac{1}{\sqrt{4 + x^2}}$, $x = -2$, $x = 2$, $y = 0$

18. $y = \dfrac{e^{3x}}{\sqrt{1 + e^{6x}}}$, $x = 0$, $x = 1$, $y = 0$

19. Find the volume of the solid whose base is the region bounded between the curve $y = x^3$ and the y-axis from $y = 0$ to $y = 1$ and whose cross sections taken perpendicular to the y-axis are squares.

20. Find the volume of the solid whose base is the region enclosed between the curve $x = 1 - y^2$ and the y-axis and whose cross sections taken perpendicular to the y-axis are squares.

21–26 Find the volume of the solid that results when the region enclosed by the given curves is revolved about the y-axis. ■

21. $x = \csc y$, $y = \pi/4$, $y = 3\pi/4$, $x = 0$

22. $y = x^2$, $x = y^2$

23. $x = y^2$, $x = y + 2$

24. $x = 1 - y^2$, $x = 2 + y^2$, $y = -1$, $y = 1$

25. $y = \ln x$, $x = 0$, $y = 0$, $y = 1$

26. $y = \sqrt{\dfrac{1 - x^2}{x^2}}$ $(x > 0)$, $x = 0$, $y = 0$, $y = 2$

27–30 True–False Determine whether the statement is true or false. Explain your answer. [In these exercises, assume that a solid S of volume V is bounded by two parallel planes perpendicular to the x-axis at $x = a$ and $x = b$ and that for each x in $[a, b]$, $A(x)$ denotes the cross-sectional area of S perpendicular to the x-axis.] ■

27. If each cross section of S perpendicular to the x-axis is a square, then S is a rectangular parallelepiped (i.e., is box shaped).

28. If each cross section of S is a disk or a washer, then S is a solid of revolution.

29. If x is in centimeters (cm), then $A(x)$ must be a quadratic function of x, since units of $A(x)$ will be square centimeters (cm^2).

30. The average value of $A(x)$ on the interval $[a, b]$ is given by $V/(b - a)$.

31. Find the volume of the solid that results when the region above the x-axis and below the ellipse

$$\frac{x^2}{a^2} + \frac{y^2}{b^2} = 1 \quad (a > 0, b > 0)$$

is revolved about the x-axis.

32. Let V be the volume of the solid that results when the region enclosed by $y = 1/x$, $y = 0$, $x = 2$, and $x = b$ $(0 < b < 2)$ is revolved about the x-axis. Find the value of b for which $V = 3$.

33. Find the volume of the solid generated when the region enclosed by $y = \sqrt{x + 1}$, $y = \sqrt{2x}$, and $y = 0$ is revolved about the x-axis. [*Hint:* Split the solid into two parts.]

34. Find the volume of the solid generated when the region enclosed by $y = \sqrt{x}$, $y = 6 - x$, and $y = 0$ is revolved about the x-axis. [*Hint:* Split the solid into two parts.]

FOCUS ON CONCEPTS

35. Suppose that f is a continuous function on $[a, b]$, and let R be the region between the curve $y = f(x)$ and the line $y = k$ from $x = a$ to $x = b$. Using the method of disks, derive with explanation a formula for the volume of a solid generated by revolving R about the line $y = k$. State and explain additional assumptions, if any, that you need about f for your formula.

36. Suppose that v and w are continuous functions on $[c, d]$, and let R be the region between the curves $x = v(y)$ and $x = w(y)$ from $y = c$ to $y = d$. Using the method of washers, derive with explanation a formula for the volume of a solid generated by revolving R about the line

$x = k$. State and explain additional assumptions, if any, that you need about v and w for your formula.

37. Consider the solid generated by revolving the shaded region in Exercise 1 about the line $y = 2$.
   (a) Make a conjecture as to which is larger: the volume of this solid or the volume of the solid in Exercise 1. Explain the basis of your conjecture.
   (b) Check your conjecture by calculating this volume and comparing it to the volume obtained in Exercise 1.

38. Consider the solid generated by revolving the shaded region in Exercise 4 about the line $x = 2.5$.
   (a) Make a conjecture as to which is larger: the volume of this solid or the volume of the solid in Exercise 4. Explain the basis of your conjecture.
   (b) Check your conjecture by calculating this volume and comparing it to the volume obtained in Exercise 4.

39. Find the volume of the solid that results when the region enclosed by $y = \sqrt{x}$, $y = 0$, and $x = 9$ is revolved about the line $x = 9$.

40. Find the volume of the solid that results when the region in Exercise 39 is revolved about the line $y = 3$.

41. Find the volume of the solid that results when the region enclosed by $x = y^2$ and $x = y$ is revolved about the line $y = -1$.

42. Find the volume of the solid that results when the region in Exercise 41 is revolved about the line $x = -1$.

43. Find the volume of the solid that results when the region enclosed by $y = x^2$ and $y = x^3$ is revolved about the line $x = 1$.

44. Find the volume of the solid that results when the region in Exercise 43 is revolved about the line $y = -1$.

45. A nose cone for a space reentry vehicle is designed so that a cross section, taken x ft from the tip and perpendicular to the axis of symmetry, is a circle of radius $\frac{1}{4}x^2$ ft. Find the volume of the nose cone given that its length is 20 ft.

46. A certain solid is 1 ft high, and a horizontal cross section taken x ft above the bottom of the solid is an annulus of inner radius x^2 ft and outer radius $\sqrt{x}$ ft. Find the volume of the solid.

47. Find the volume of the solid whose base is the region bounded between the curves $y = x$ and $y = x^2$, and whose cross sections perpendicular to the x-axis are squares.

48. The base of a certain solid is the region enclosed by $y = \sqrt{x}$, $y = 0$, and $x = 4$. Every cross section perpendicular to the x-axis is a semicircle with its diameter across the base. Find the volume of the solid.

49. In parts (a)–(c) find the volume of the solid whose base is enclosed by the circle $x^2 + y^2 = 1$ and whose cross sections taken perpendicular to the x-axis are
   (a) semicircles   (b) squares
   (c) equilateral triangles.

(a)   (b)   (c)

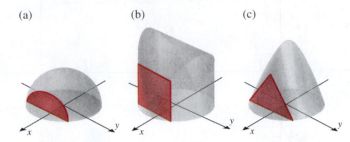

50. As shown in the accompanying figure, a cathedral dome is designed with three semicircular supports of radius r so that each horizontal cross section is a regular hexagon. Show that the volume of the dome is $r^3\sqrt{3}$.

◀ **Figure Ex-50**

C **51–54** Use a CAS to estimate the volume of the solid that results when the region enclosed by the curves is revolved about the stated axis. ■

51. $y = \sin^8 x$, $y = 2x/\pi$, $x = 0$, $x = \pi/2$; x-axis

52. $y = \pi^2 \sin x \cos^3 x$, $y = 4x^2$, $x = 0$, $x = \pi/4$; x-axis

53. $y = e^x$, $x = 1$, $y = 1$; y-axis

54. $y = x\sqrt{\tan^{-1} x}$, $y = x$; x-axis

55. The accompanying figure shows a ***spherical cap*** of radius ρ and height h cut from a sphere of radius r. Show that the volume V of the spherical cap can be expressed as
   (a) $V = \frac{1}{3}\pi h^2(3r - h)$ (b) $V = \frac{1}{6}\pi h(3\rho^2 + h^2)$.

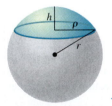

◀ **Figure Ex-55**

56. If fluid enters a hemispherical bowl with a radius of 10 ft at a rate of $\frac{1}{2}$ ft^3/min, how fast will the fluid be rising when the depth is 5 ft? [*Hint:* See Exercise 55.]

57. The accompanying figure (on the next page) shows the dimensions of a small lightbulb at 10 equally spaced points.
   (a) Use formulas from geometry to make a rough estimate of the volume enclosed by the glass portion of the bulb.

(cont.)

(b) Use the average of left and right endpoint approximations to approximate the volume.

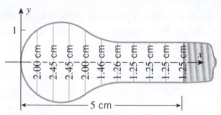

▲ **Figure Ex-57**

58. Use the result in Exercise 55 to find the volume of the solid
that remains when a hole of radius $r/2$ is drilled through the
center of a sphere of radius r, and then check your answer
by integrating.

59. As shown in the accompanying figure, a cocktail glass with
a bowl shaped like a hemisphere of diameter 8 cm contains
a cherry with a diameter of 2 cm. If the glass is filled to
a depth of h cm, what is the volume of liquid it contains?
[*Hint:* First consider the case where the cherry is partially
submerged, then the case where it is totally submerged.]

◀ **Figure Ex-59**

60. Find the volume of the torus that results when the region enclosed by the circle of radius r with center at $(h, 0)$, $h > r$,
is revolved about the y-axis. [*Hint:* Use an appropriate
formula from plane geometry to help evaluate the definite
integral.]

61. A wedge is cut from a right circular cylinder of radius r by
two planes, one perpendicular to the axis of the cylinder and
the other making an angle θ with the first. Find the volume
of the wedge by slicing perpendicular to the y-axis as shown
in the accompanying figure.

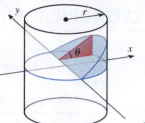

◀ **Figure Ex-61**

62. Find the volume of the wedge described in Exercise 61 by
slicing perpendicular to the x-axis.

63. Two right circular cylinders of radius r have axes that intersect at right angles. Find the volume of the solid common to
the two cylinders. [*Hint:* One-eighth of the solid is sketched
in the accompanying figure.]

64. In 1635 Bonaventura Cavalieri, a student of Galileo, stated
the following result, called **Cavalieri's principle**: *If two
solids have the same height, and if the areas of their cross
sections taken parallel to and at equal distances from their
bases are always equal, then the solids have the same volume.* Use this result to find the volume of the oblique cylinder in the accompanying figure. (See Exercise 52 of Section
6.1 for a planar version of Cavalieri's principle.)

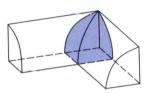

▲ **Figure Ex-63**   ▲ **Figure Ex-64**

65. Writing Use the results of this section to derive Cavalieri's
principle (Exercise 64).

66. Writing Write a short paragraph that explains how Formulas (4)–(8) may all be viewed as consequences of Formula (3).

✔**QUICK CHECK ANSWERS 6.2**

1. $\displaystyle\int_1^3 3x^2\,dx$; 26   **2.** (a) $\pi\sin x$ (b) $\displaystyle\int_0^\pi \pi\sin x\,dx$ (c) 2π **3.** (a) 0; 2; $\pi[(2x+1)^2 - (x^2+1)^2] = \pi[-x^4 + 2x^2 + 4x]$

(b) $\displaystyle\int_0^2 \pi[-x^4 + 2x^2 + 4x]\,dx$ **4.** (a) 1; 2; $\pi[(y-1) - (y-1)^2] = \pi[-y^2 + 3y - 2]$ (b) $\displaystyle\int_1^2 \pi[-y^2 + 3y - 2]\,dy$

6.3 VOLUMES BY CYLINDRICAL SHELLS

The methods for computing volumes that have been discussed so far depend on our ability to compute the cross-sectional area of the solid and to integrate that area across the solid. In this section we will develop another method for finding volumes that may be applicable when the cross-sectional area cannot be found or the integration is too difficult.

■ **CYLINDRICAL SHELLS**
In this section we will be interested in the following problem.

> **6.3.1 PROBLEM** Let f be continuous and nonnegative on $[a, b]$ $(0 \le a < b)$, and let R be the region that is bounded above by $y = f(x)$, below by the x-axis, and on the sides by the lines $x = a$ and $x = b$. Find the volume V of the solid of revolution S that is generated by revolving the region R about the y-axis (Figure 6.3.1).

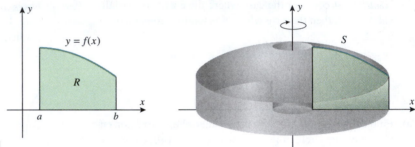

▶ **Figure 6.3.1**

Sometimes problems of the above type can be solved by the method of disks or washers perpendicular to the y-axis, but when that method is not applicable or the resulting integral is difficult, the *method of cylindrical shells*, which we will discuss here, will often work.

A *cylindrical shell* is a solid enclosed by two concentric right circular cylinders (Figure 6.3.2). The volume V of a cylindrical shell with inner radius r_1, outer radius r_2, and height h can be written as

$$V = [\text{area of cross section}] \cdot [\text{height}]$$
$$= (\pi r_2^2 - \pi r_1^2)h$$
$$= \pi(r_2 + r_1)(r_2 - r_1)h$$
$$= 2\pi \cdot \left[\tfrac{1}{2}(r_1 + r_2)\right] \cdot h \cdot (r_2 - r_1)$$

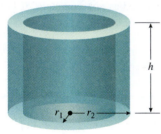

▲ **Figure 6.3.2**

But $\frac{1}{2}(r_1 + r_2)$ is the average radius of the shell and $r_2 - r_1$ is its thickness, so

$$V = 2\pi \cdot [\text{average radius}] \cdot [\text{height}] \cdot [\text{thickness}] \tag{1}$$

We will now show how this formula can be used to solve Problem 6.3.1. The underlying idea is to divide the interval $[a, b]$ into n subintervals, thereby subdividing the region R into n strips, $R_1, R_2, \ldots, R_n$ (Figure 6.3.3a). When the region R is revolved about the y-axis, these strips generate "tube-like" solids $S_1, S_2, \ldots, S_n$ that are nested one inside the other and together comprise the entire solid S (Figure 6.3.3b). Thus, the volume V of the solid can be obtained by adding together the volumes of the tubes; that is,

$$V = V(S_1) + V(S_2) + \cdots + V(S_n)$$

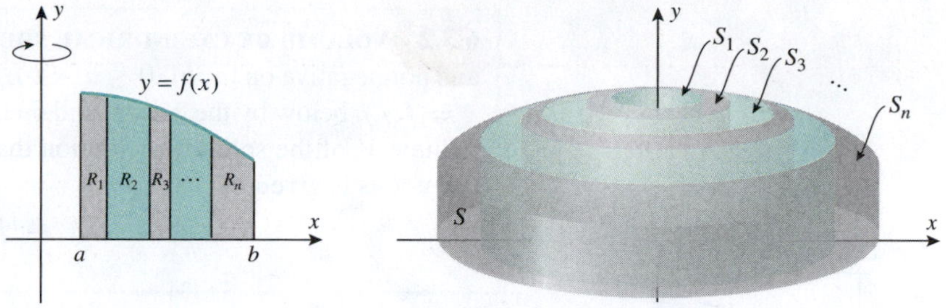

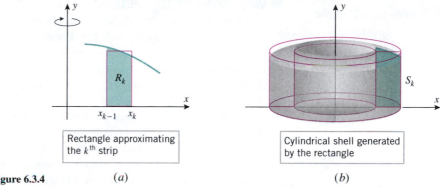

▶ Figure 6.3.3   (a)   (b)

As a rule, the tubes will have curved upper surfaces, so there will be no simple formulas for their volumes. However, if the strips are thin, then we can approximate each strip by a rectangle (Figure 6.3.4a). These rectangles, when revolved about the y-axis, will produce cylindrical shells whose volumes closely approximate the volumes of the tubes generated by the original strips (Figure 6.3.4b). We will show that by adding the volumes of the cylindrical shells we can obtain a Riemann sum that approximates the volume V, and by taking the limit of the Riemann sums we can obtain an integral for the exact volume V.

| Rectangle approximating the kth strip | Cylindrical shell generated by the rectangle |

▶ Figure 6.3.4   (a)   (b)

To implement this idea, suppose that the kth strip extends from x_{k-1} to x_k and that the width of this strip is

$$\Delta x_k = x_k - x_{k-1}$$

If we let x_k^* be the *midpoint* of the interval $[x_{k-1}, x_k]$, and if we construct a rectangle of height $f(x_k^*)$ over the interval, then revolving this rectangle about the y-axis produces a cylindrical shell of average radius x_k^*, height $f(x_k^*)$, and thickness Δx_k (Figure 6.3.5). From (1), the volume V_k of this cylindrical shell is

$$V_k = 2\pi x_k^* f(x_k^*) \Delta x_k$$

Adding the volumes of the n cylindrical shells yields the following Riemann sum that approximates the volume V:

$$V \approx \sum_{k=1}^{n} 2\pi x_k^* f(x_k^*) \Delta x_k$$

Taking the limit as n increases and the widths of all the subintervals approach zero yields the definite integral

$$V = \lim_{\max \Delta x_k \to 0} \sum_{k=1}^{n} 2\pi x_k^* f(x_k^*) \Delta x_k = \int_a^b 2\pi x f(x)\, dx$$

In summary, we have the following result.

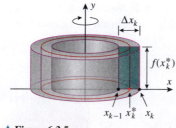

▲ Figure 6.3.5

6.3.2   VOLUME BY CYLINDRICAL SHELLS ABOUT THE y-AXIS   Let f be continuous and nonnegative on $[a, b]$ $(0 \leq a < b)$, and let R be the region that is bounded above by $y = f(x)$, below by the x-axis, and on the sides by the lines $x = a$ and $x = b$. Then the volume V of the solid of revolution that is generated by revolving the region R about the y-axis is given by

$$V = \int_a^b 2\pi x f(x)\, dx \qquad (2)$$

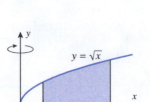

$y = \sqrt{x}$

(a)

Cutaway view of the solid

(b)

▲ **Figure 6.3.6**

▶ **Example 1**   Use cylindrical shells to find the volume of the solid generated when the region enclosed between $y = \sqrt{x}$, $x = 1$, $x = 4$, and the x-axis is revolved about the y-axis.

Solution.   First sketch the region (Figure 6.3.6a); then imagine revolving it about the y-axis (Figure 6.3.6b). Since $f(x) = \sqrt{x}$, $a = 1$, and $b = 4$, Formula (2) yields

$$V = \int_1^4 2\pi x \sqrt{x}\, dx = 2\pi \int_1^4 x^{3/2}\, dx = \left[2\pi \cdot \frac{2}{5} x^{5/2}\right]_1^4 = \frac{4\pi}{5}[32 - 1] = \frac{124\pi}{5} \quad ◀$$

■ **VARIATIONS OF THE METHOD OF CYLINDRICAL SHELLS**

The method of cylindrical shells is applicable in a variety of situations that do not fit the conditions required by Formula (2). For example, the region may be enclosed between two curves, or the axis of revolution may be some line other than the y-axis. However, rather than develop a separate formula for every possible situation, we will give a general way of thinking about the method of cylindrical shells that can be adapted to each new situation as it arises.

For this purpose, we will need to reexamine the integrand in Formula (2): At each x in the interval $[a, b]$, the vertical line segment from the x-axis to the curve $y = f(x)$ can be viewed as the cross section of the region R at x (Figure 6.3.7a). When the region R is revolved about the y-axis, the cross section at x sweeps out the *surface* of a right circular cylinder of height $f(x)$ and radius x (Figure 6.3.7b). The area of this surface is

$$2\pi x f(x)$$

(Figure 6.3.7c), which is the integrand in (2). Thus, Formula (2) can be viewed informally in the following way.

6.3.3   AN INFORMAL VIEWPOINT ABOUT CYLINDRICAL SHELLS   The volume V of a solid of revolution that is generated by revolving a region R about an axis can be obtained by integrating the area of the surface generated by an arbitrary cross section of R taken parallel to the axis of revolution.

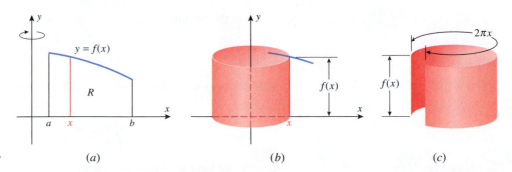

▶ **Figure 6.3.7** (a) (b) (c)

The following examples illustrate how to apply this result in situations where Formula (2) is not applicable.

▶ **Example 2**   Use cylindrical shells to find the volume of the solid generated when the region R in the first quadrant enclosed between $y = x$ and $y = x^2$ is revolved about the y-axis (Figure 6.3.8a).

Solution.   As illustrated in part (b) of Figure 6.3.8, at each x in $[0, 1]$ the cross section of R parallel to the y-axis generates a cylindrical surface of height $x - x^2$ and radius x. Since the area of this surface is

$$2\pi x (x - x^2)$$

the volume of the solid is

$$V = \int_0^1 2\pi x (x - x^2)\, dx = 2\pi \int_0^1 (x^2 - x^3)\, dx$$

$$= 2\pi \left[\frac{x^3}{3} - \frac{x^4}{4} \right]_0^1 = 2\pi \left[\frac{1}{3} - \frac{1}{4} \right] = \frac{\pi}{6} \quad \blacktriangleleft$$

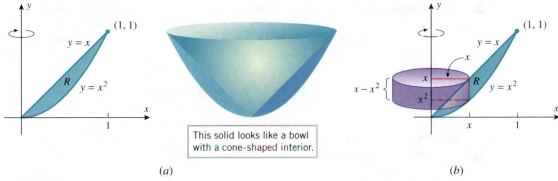

This solid looks like a bowl with a cone-shaped interior.

(a)   (b)

▲ **Figure 6.3.8**

▶ **Example 3**   Use cylindrical shells to find the volume of the solid generated when the region R under $y = x^2$ over the interval $[0, 2]$ is revolved about the line $y = -1$.

Solution.   First draw the axis of revolution; then imagine revolving the region about the axis (Figure 6.3.9a). As illustrated in Figure 6.3.9b, at each y in the interval $0 \le y \le 4$, the cross section of R parallel to the x-axis generates a cylindrical surface of height $2 - \sqrt{y}$ and radius $y + 1$. Since the area of this surface is

$$2\pi (y + 1)(2 - \sqrt{y})$$

it follows that the volume of the solid is

$$\int_0^4 2\pi (y + 1)(2 - \sqrt{y})\, dy = 2\pi \int_0^4 (2y - y^{3/2} + 2 - y^{1/2})\, dy$$

$$= 2\pi \left[y^2 - \frac{2}{5} y^{5/2} + 2y - \frac{2}{3} y^{3/2} \right]_0^4 = \frac{176\pi}{15} \quad \blacktriangleleft$$

Note that the volume found in Example 3 agrees with the volume of the same solid found by the method of washers in Example 6 of Section 6.2. Confirm that the volume in Example 2 found by the method of cylindrical shells can also be obtained by the method of washers.

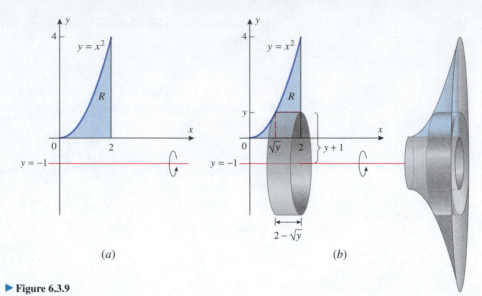

▶ **Figure 6.3.9**

---

1. Let R be the region between the x-axis and the curve $y = 1 + \sqrt{x}$ for $1 \leq x \leq 4$.
(a) For x between 1 and 4, the area of the cylindrical surface generated by revolving the vertical cross section of R at x about the y-axis is _____.
(b) Using cylindrical shells, an integral expression for the volume of the solid generated by revolving R about the y-axis is _____.

2. Let R be the region described in Quick Check Exercise 1.
(a) For x between 1 and 4, the area of the cylindrical surface generated by revolving the vertical cross section of R at x about the line $x = 5$ is _____.
(b) Using cylindrical shells, an integral expression for the volume of the solid generated by revolving R about the line $x = 5$ is _____.

3. A solid S is generated by revolving the region enclosed by the curves $x = (y - 2)^2$ and $x = 4$ about the x-axis. Using cylindrical shells, an integral expression for the volume of S is _____.

EXERCISE SET 6.3   ⊂ CAS

1–4 Use cylindrical shells to find the volume of the solid generated when the shaded region is revolved about the indicated axis. ∎

1.

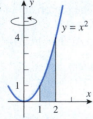

2.

3.
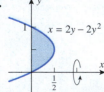

4.
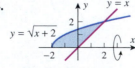

5–12 Use cylindrical shells to find the volume of the solid generated when the region enclosed by the given curves is revolved about the y-axis. ∎

5. $y = x^3$, $x = 1$, $y = 0$

6. $y = \sqrt{x}$, $x = 4$, $x = 9$, $y = 0$

7. $y = 1/x$, $y = 0$, $x = 1$, $x = 3$

8. $y = \cos(x^2)$, $x = 0$, $x = \frac{1}{2}\sqrt{\pi}$, $y = 0$

9. $y = 2x - 1$, $y = -2x + 3$, $x = 2$

10. $y = 2x - x^2$, $y = 0$

11. $y = \dfrac{1}{x^2 + 1}$, $x = 0$, $x = 1$, $y = 0$

12. $y = e^{x^2}$, $x = 1$, $x = \sqrt{3}$, $y = 0$

13–16 Use cylindrical shells to find the volume of the solid generated when the region enclosed by the given curves is revolved about the x-axis. ∎

13. $y^2 = x$, $y = 1$, $x = 0$

14. $x = 2y$, $y = 2$, $y = 3$, $x = 0$

15. $y = x^2$, $x = 1$, $y = 0$ 16. $xy = 4$, $x + y = 5$

17–20 True–False Determine whether the statement is true or false. Explain your answer. ■

17. The volume of a cylindrical shell is equal to the product of the thickness of the shell with the surface area of a cylinder whose height is that of the shell and whose radius is equal to the average of the inner and outer radii of the shell.

18. The method of cylindrical shells is a special case of the method of integration of cross-sectional area that was discussed in Section 6.2.

19. In the method of cylindrical shells, integration is over an interval on a coordinate axis that is *perpendicular* to the axis of revolution of the solid.

20. The Riemann sum approximation

$$V \approx \sum_{k=1}^{n} 2\pi x_k^* f(x_k^*) \Delta x_k \quad \left(\text{where } x_k^* = \frac{x_k + x_{k-1}}{2} \right)$$

for the volume of a solid of revolution is exact when f is a constant function.

C 21. Use a CAS to find the volume of the solid generated when the region enclosed by $y = e^x$ and $y = 0$ for $1 \le x \le 2$ is revolved about the y-axis.

C 22. Use a CAS to find the volume of the solid generated when the region enclosed by $y = \cos x$, $y = 0$, and $x = 0$ for $0 \le x \le \pi/2$ is revolved about the y-axis.

C 23. Consider the region to the right of the y-axis, to the left of the vertical line $x = k$ $(0 < k < \pi)$, and between the curve $y = \sin x$ and the x-axis. Use a CAS to estimate the value of k so that the solid generated by revolving the region about the y-axis has a volume of 8 cubic units.

FOCUS ON CONCEPTS

24. Let R_1 and R_2 be regions of the form shown in the accompanying figure. Use cylindrical shells to find a formula for the volume of the solid that results when
    (a) region R_1 is revolved about the y-axis
    (b) region R_2 is revolved about the x-axis.

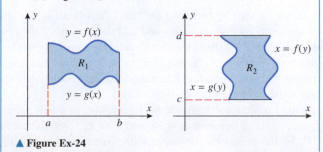

▲ **Figure Ex-24**

25. (a) Use cylindrical shells to find the volume of the solid that is generated when the region under the curve

    $$y = x^3 - 3x^2 + 2x$$

    over $[0, 1]$ is revolved about the y-axis.
    (b) For this problem, is the method of cylindrical shells easier or harder than the method of slicing discussed in the last section? Explain.

26. Let f be continuous and nonnegative on $[a, b]$, and let R be the region that is enclosed by $y = f(x)$ and $y = 0$ for $a \le x \le b$. Using the method of cylindrical shells, derive with explanation a formula for the volume of the solid generated by revolving R about the line $x = k$, where $k \le a$.

27–28 Using the method of cylindrical shells, set up but do not evaluate an integral for the volume of the solid generated when the region R is revolved about (a) the line $x = 1$ and (b) the line $y = -1$. ■

27. R is the region bounded by the graphs of $y = x$, $y = 0$, and $x = 1$.

28. R is the region in the first quadrant bounded by the graphs of $y = \sqrt{1 - x^2}$, $y = 0$, and $x = 0$.

29. Use cylindrical shells to find the volume of the solid that is generated when the region that is enclosed by $y = 1/x^3$, $x = 1$, $x = 2$, $y = 0$ is revolved about the line $x = -1$.

30. Use cylindrical shells to find the volume of the solid that is generated when the region that is enclosed by $y = x^3$, $y = 1$, $x = 0$ is revolved about the line $y = 1$.

31. Use cylindrical shells to find the volume of the cone generated when the triangle with vertices $(0, 0)$, $(0, r)$, $(h, 0)$, where $r > 0$ and $h > 0$, is revolved about the x-axis.

32. The region enclosed between the curve $y^2 = kx$ and the line $x = \frac{1}{4}k$ is revolved about the line $x = \frac{1}{2}k$. Use cylindrical shells to find the volume of the resulting solid. (Assume $k > 0$.)

33. As shown in the accompanying figure, a cylindrical hole is drilled all the way through the center of a sphere. Show that the volume of the remaining solid depends only on the length L of the hole, not on the size of the sphere.

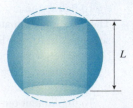

◀ **Figure Ex-33**

34. Use cylindrical shells to find the volume of the torus obtained by revolving the circle $x^2 + y^2 = a^2$ about the line

$x = b$, where $b > a > 0$. [*Hint:* It may help in the integration to think of an integral as an area.]

35. Let V_x and V_y be the volumes of the solids that result when the region enclosed by $y = 1/x$, $y = 0$, $x = \frac{1}{2}$, and $x = b$ $(b > \frac{1}{2})$ is revolved about the x-axis and y-axis, respectively. Is there a value of b for which $V_x = V_y$?

36. (a) Find the volume V of the solid generated when the region bounded by $y = 1/(1 + x^4)$, $y = 0$, $x = 1$, and $x = b$ $(b > 1)$ is revolved about the y-axis.
(b) Find $\lim\limits_{b \to +\infty} V$.

37. Writing Faced with the problem of computing the volume of a solid of revolution, how would you go about deciding whether to use the method of disks/washers or the method of cylindrical shells?

38. Writing With both the method of disks/washers and with the method of cylindrical shells, we integrate an "area" to get the volume of a solid of revolution. However, these two approaches differ in very significant ways. Write a brief paragraph that discusses these differences.

✔ **QUICK CHECK ANSWERS 6.3**

1. (a) $2\pi x(1 + \sqrt{x})$ (b) $\int_1^4 2\pi x(1 + \sqrt{x})\,dx$ **2.** (a) $2\pi(5 - x)(1 + \sqrt{x})$ (b) $\int_1^4 2\pi(5 - x)(1 + \sqrt{x})\,dx$

3. $\int_0^4 2\pi y[4 - (y - 2)^2]\,dy$

6.4 LENGTH OF A PLANE CURVE

In this section we will use the tools of calculus to study the problem of finding the length of a plane curve.

■ ARC LENGTH

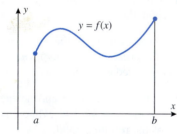

▲ Figure 6.4.1

Our first objective is to define what we mean by the **length** (also called the **arc length**) of a plane curve $y = f(x)$ over an interval $[a, b]$ (Figure 6.4.1). Once that is done we will be able to focus on the problem of computing arc lengths. To avoid some complications that would otherwise occur, we will impose the requirement that f' be continuous on $[a, b]$, in which case we will say that $y = f(x)$ is a **smooth curve** on $[a, b]$ or that f is a **smooth function** on $[a, b]$. Thus, we will be concerned with the following problem.

6.4.1 ARC LENGTH PROBLEM Suppose that $y = f(x)$ is a smooth curve on the interval $[a, b]$. Define and find a formula for the arc length L of the curve $y = f(x)$ over the interval $[a, b]$.

> Intuitively, you might think of the arc length of a curve as the number obtained by aligning a piece of string with the curve and then measuring the length of the string after it is straightened out.

To define the arc length of a curve we start by breaking the curve into small segments. Then we approximate the curve segments by line segments and add the lengths of the line segments to form a Riemann sum. Figure 6.4.2 illustrates how such line segments tend to become better and better approximations to a curve as the number of segments increases. As the number of segments increases, the corresponding Riemann sums approach a definite integral whose value we will take to be the arc length L of the curve.

To implement our idea for solving Problem 6.4.1, divide the interval $[a, b]$ into n subintervals by inserting points $x_1, x_2, \ldots, x_{n-1}$ between $a = x_0$ and $b = x_n$. As shown in Figure 6.4.3a, let $P_0, P_1, \ldots, P_n$ be the points on the curve with x-coordinates $a = x_0$,

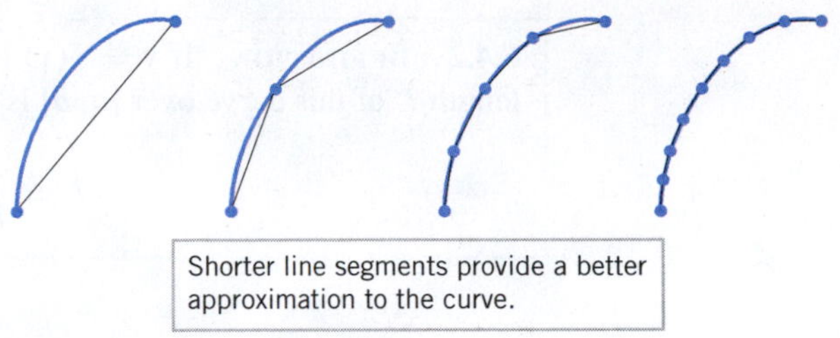

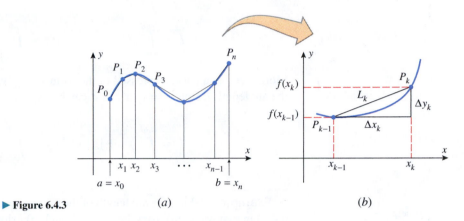

> Shorter line segments provide a better approximation to the curve.

▶ **Figure 6.4.2**

▶ **Figure 6.4.3** (*a*) (*b*)

$x_1, x_2, \ldots, x_{n-1}, b = x_n$ and join these points with straight line segments. These line segments form a ***polygonal path*** that we can regard as an approximation to the curve $y = f(x)$. As indicated in Figure 6.4.3b, the length L_k of the kth line segment in the polygonal path is

$$L_k = \sqrt{(\Delta x_k)^2 + (\Delta y_k)^2} = \sqrt{(\Delta x_k)^2 + [f(x_k) - f(x_{k-1})]^2} \tag{1}$$

If we now add the lengths of these line segments, we obtain the following approximation to the length L of the curve

$$L \approx \sum_{k=1}^{n} L_k = \sum_{k=1}^{n} \sqrt{(\Delta x_k)^2 + [f(x_k) - f(x_{k-1})]^2} \tag{2}$$

To put this in the form of a Riemann sum we will apply the Mean-Value Theorem (4.8.2). This theorem implies that there is a point x_k^* between x_{k-1} and x_k such that

$$\frac{f(x_k) - f(x_{k-1})}{x_k - x_{k-1}} = f'(x_k^*) \quad \text{or} \quad f(x_k) - f(x_{k-1}) = f'(x_k^*)\Delta x_k$$

and hence we can rewrite (2) as

$$L \approx \sum_{k=1}^{n} \sqrt{(\Delta x_k)^2 + [f'(x_k^*)]^2 (\Delta x_k)^2} = \sum_{k=1}^{n} \sqrt{1 + [f'(x_k^*)]^2} \, \Delta x_k$$

Thus, taking the limit as n increases and the widths of all the subintervals approach zero yields the following integral that defines the arc length L:

> Explain why the approximation in (2) cannot be greater than L.

$$L = \lim_{\max \Delta x_k \to 0} \sum_{k=1}^{n} \sqrt{1 + [f'(x_k^*)]^2} \, \Delta x_k = \int_a^b \sqrt{1 + [f'(x)]^2} \, dx$$

In summary, we have the following definition.

6.4.2 **DEFINITION** If $y = f(x)$ is a smooth curve on the interval $[a, b]$, then the arc length L of this curve over $[a, b]$ is defined as

$$L = \int_a^b \sqrt{1 + [f'(x)]^2}\, dx \tag{3}$$

This result provides both a definition and a formula for computing arc lengths. Where convenient, (3) can also be expressed as

$$L = \int_a^b \sqrt{1 + [f'(x)]^2}\, dx = \int_a^b \sqrt{1 + \left(\frac{dy}{dx}\right)^2}\, dx \tag{4}$$

Moreover, for a curve expressed in the form $x = g(y)$, where g' is continuous on $[c, d]$, the arc length L from $y = c$ to $y = d$ can be expressed as

$$L = \int_c^d \sqrt{1 + [g'(y)]^2}\, dy = \int_c^d \sqrt{1 + \left(\frac{dx}{dy}\right)^2}\, dy \tag{5}$$

▶ **Example 1** Find the arc length of the curve $y = x^{3/2}$ from $(1, 1)$ to $(2, 2\sqrt{2})$ (Figure 6.4.4) in two ways: (a) using Formula (4) and (b) using Formula (5).

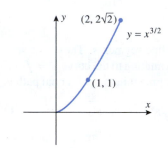

▲ **Figure 6.4.4**

Solution (a).

$$\frac{dy}{dx} = \frac{3}{2}x^{1/2}$$

and since the curve extends from $x = 1$ to $x = 2$, it follows from (4) that

$$L = \int_1^2 \sqrt{1 + \left(\tfrac{3}{2}x^{1/2}\right)^2}\, dx = \int_1^2 \sqrt{1 + \tfrac{9}{4}x}\, dx$$

To evaluate this integral we make the u-substitution

$$u = 1 + \tfrac{9}{4}x, \quad du = \tfrac{9}{4}\, dx$$

and then change the x-limits of integration $(x = 1, x = 2)$ to the corresponding u-limits $\left(u = \tfrac{13}{4}, u = \tfrac{22}{4}\right)$:

$$L = \frac{4}{9}\int_{13/4}^{22/4} u^{1/2}\, du = \frac{8}{27}u^{3/2}\Big]_{13/4}^{22/4} = \frac{8}{27}\left[\left(\frac{22}{4}\right)^{3/2} - \left(\frac{13}{4}\right)^{3/2}\right]$$

$$= \frac{22\sqrt{22} - 13\sqrt{13}}{27} \approx 2.09$$

Solution (b). To apply Formula (5) we must first rewrite the equation $y = x^{3/2}$ so that x is expressed as a function of y. This yields $x = y^{2/3}$ and

$$\frac{dx}{dy} = \frac{2}{3}y^{-1/3}$$

Since the curve extends from $y = 1$ to $y = 2\sqrt{2}$, it follows from (5) that

$$L = \int_1^{2\sqrt{2}} \sqrt{1 + \tfrac{4}{9}y^{-2/3}}\, dy = \frac{1}{3}\int_1^{2\sqrt{2}} y^{-1/3}\sqrt{9y^{2/3} + 4}\, dy$$

To evaluate this integral we make the u-substitution

$$u = 9y^{2/3} + 4, \quad du = 6y^{-1/3}\, dy$$

and change the y-limits of integration ($y = 1$, $y = 2\sqrt{2}$) to the corresponding u-limits ($u = 13$, $u = 22$). This gives

$$L = \frac{1}{18}\int_{13}^{22} u^{1/2}\, du = \frac{1}{27}u^{3/2}\Big]_{13}^{22} = \frac{1}{27}[(22)^{3/2} - (13)^{3/2}] = \frac{22\sqrt{22} - 13\sqrt{13}}{27}$$

The answer in part (b) agrees with that in part (a); however, the integration in part (b) is more tedious. In problems where there is a choice between using (4) or (5), it is often the case that one of the formulas leads to a simpler integral than the other. ◄

> The arc from the point $(1, 1)$ to the point $(2, 2\sqrt{2})$ in Figure 6.4.4 is nearly a straight line, so the arc length should be only slightly larger than the straight-line distance between these points. Show that this is so.

■ FINDING ARC LENGTH BY NUMERICAL METHODS

In the next chapter we will develop some techniques of integration that will enable us to find exact values of more integrals encountered in arc length calculations; however, generally speaking, most such integrals are impossible to evaluate in terms of elementary functions. In these cases one usually approximates the integral using a numerical method such as the midpoint rule discussed in Section 5.4.

TECHNOLOGY MASTERY

> If your calculating utility has a numerical integration capability, use it to confirm that the arc length L in Example 2 is approximately $L \approx 3.8202$.

► **Example 2** From (4), the arc length of $y = \sin x$ from $x = 0$ to $x = \pi$ is given by the integral

$$L = \int_0^\pi \sqrt{1 + (\cos x)^2}\, dx$$

This integral cannot be evaluated in terms of elementary functions; however, using a calculating utility with a numerical integration capability yields the approximation $L \approx 3.8202$. ◄

✔ QUICK CHECK EXERCISES 6.4   (See page 443 for answers.)

1. A function f is smooth on $[a, b]$ if f' is _____ on $[a, b]$.

2. If a function f is smooth on $[a, b]$, then the length of the curve $y = f(x)$ over $[a, b]$ is _____.

3. The distance between points $(1, 0)$ and $(e, 1)$ is _____.

4. Let L be the length of the curve $y = \ln x$ from $(1, 0)$ to $(e, 1)$.
   (a) Integrating with respect to x, an integral expression for L is _____.
   (b) Integrating with respect to y, an integral expression for L is _____.

EXERCISE SET 6.4   [C] CAS

1. Use the Theorem of Pythagoras to find the length of the line segment $y = 2x$ from $(1, 2)$ to $(2, 4)$, and confirm that the value is consistent with the length computed using
   (a) Formula (4)   (b) Formula (5).

2. Use the Theorem of Pythagoras to find the length of the line segment $y = 5x$ from $(0, 0)$ and $(1, 5)$, and confirm that the value is consistent with the length computed using
   (a) Formula (4)   (b) Formula (5).

3–8 Find the exact arc length of the curve over the interval. ■

3. $y = 3x^{3/2} - 1$ from $x = 0$ to $x = 1$

4. $x = \frac{1}{3}(y^2 + 2)^{3/2}$ from $y = 0$ to $y = 1$

5. $y = x^{2/3}$ from $x = 1$ to $x = 8$

6. $y = (x^6 + 8)/(16x^2)$ from $x = 2$ to $x = 3$

7. $24xy = y^4 + 48$ from $y = 2$ to $y = 4$

8. $x = \frac{1}{8}y^4 + \frac{1}{4}y^{-2}$ from $y = 1$ to $y = 4$

9–12 True–False Determine whether the statement is true or false. Explain your answer. ■

9. The graph of $y = \sqrt{1 - x^2}$ is a smooth curve on $[-1, 1]$.

10. The approximation

$$L \approx \sum_{k=1}^{n} \sqrt{(\Delta x_k)^2 + [f(x_k) - f(x_{k-1})]^2}$$

for arc length is not expressed in the form of a Riemann sum.

11. The approximation

$$L \approx \sum_{k=1}^{n} \sqrt{1 + [f'(x_k^*)]^2} \, \Delta x_k$$

for arc length is exact when f is a linear function of x.

12. In our definition of the arc length for the graph of $y = f(x)$, we need $f'(x)$ to be a continuous function in order for f to satisfy the hypotheses of the Mean-Value Theorem (4.8.2).

C **13–14** Express the exact arc length of the curve over the given interval as an integral that has been simplified to eliminate the radical, and then evaluate the integral using a CAS. ■

13. $y = \ln(\sec x)$ from $x = 0$ to $x = \pi/4$

14. $y = \ln(\sin x)$ from $x = \pi/4$ to $x = \pi/2$

FOCUS ON CONCEPTS

15. Consider the curve $y = x^{2/3}$.
    (a) Sketch the portion of the curve between $x = -1$ and $x = 8$.
    (b) Explain why Formula (4) cannot be used to find the arc length of the curve sketched in part (a).
    (c) Find the arc length of the curve sketched in part (a).

16. The curve segment $y = x^2$ from $x = 1$ to $x = 2$ may also be expressed as the graph of $x = \sqrt{y}$ from $y = 1$ to $y = 4$. Set up two integrals that give the arc length of this curve segment, one by integrating with respect to x, and the other by integrating with respect to y. Demonstrate a substitution that verifies that these two integrals are equal.

17. Consider the curve segments $y = x^2$ from $x = \frac{1}{2}$ to $x = 2$ and $y = \sqrt{x}$ from $x = \frac{1}{4}$ to $x = 4$.
    (a) Graph the two curve segments and use your graphs to explain why the lengths of these two curve segments should be equal.
    (b) Set up integrals that give the arc lengths of the curve segments by integrating with respect to x. Demonstrate a substitution that verifies that these two integrals are equal.
    (c) Set up integrals that give the arc lengths of the curve segments by integrating with respect to y.
    (d) Approximate the arc length of each curve segment using Formula (2) with $n = 10$ equal subintervals.
    (e) Which of the two approximations in part (d) is more accurate? Explain.
    (f) Use the midpoint approximation with $n = 10$ subintervals to approximate each arc length integral in part (b).

(g) Use a calculating utility with numerical integration capabilities to approximate the arc length integrals in part (b) to four decimal places.

18. Follow the directions of Exercise 17 for the curve segments $y = x^{8/3}$ from $x = 10^{-3}$ to $x = 1$ and $y = x^{3/8}$ from $x = 10^{-8}$ to $x = 1$.

19. Follow the directions of Exercise 17 for the curve segment $y = \tan x$ from $x = 0$ to $x = \pi/3$ and for the curve segment $y = \tan^{-1} x$ from $x = 0$ to $x = \sqrt{3}$.

20. Let $y = f(x)$ be a smooth curve on the closed interval $[a, b]$. Prove that if m and M are nonnegative numbers such that $m \le |f'(x)| \le M$ for all x in $[a, b]$, then the arc length L of $y = f(x)$ over the interval $[a, b]$ satisfies the inequalities

$$(b-a)\sqrt{1+m^2} \le L \le (b-a)\sqrt{1+M^2}$$

21. Use the result of Exercise 20 to show that the arc length L of $y = \sec x$ over the interval $0 \le x \le \pi/3$ satisfies

$$\frac{\pi}{3} \le L \le \frac{\pi}{3}\sqrt{13}$$

C 22. A basketball player makes a successful shot from the free throw line. Suppose that the path of the ball from the moment of release to the moment it enters the hoop is described by

$$y = 2.15 + 2.09x - 0.41x^2, \quad 0 \le x \le 4.6$$

where x is the horizontal distance (in meters) from the point of release, and y is the vertical distance (in meters) above the floor. Use a CAS or a scientific calculator with a numerical integration capability to approximate the distance the ball travels from the moment it is released to the moment it enters the hoop. Round your answer to two decimal places.

C 23. Find a positive value of k (to two decimal places) such that the curve $y = k \sin x$ has an arc length of $L = 5$ units over the interval from $x = 0$ to $x = \pi$. [*Hint:* Find an integral for the arc length L in terms of k, and then use a CAS or a scientific calculator with a numerical integration capability to find integer values of k at which the values of $L - 5$ have opposite signs. Complete the solution by using the Intermediate-Value Theorem (1.5.7) to approximate the value of k to two decimal places.]

C 24. As shown in the accompanying figure on the next page, a horizontal beam with dimensions 2 in $\times$ 6 in $\times$ 16 ft is fixed at both ends and is subjected to a uniformly distributed load of 120 lb/ft. As a result of the load, the centerline of the beam undergoes a deflection that is described by

$$y = -1.67 \times 10^{-8}(x^4 - 2Lx^3 + L^2x^2)$$

$(0 \le x \le 192)$, where $L = 192$ in is the length of the unloaded beam, x is the horizontal distance along the beam measured in inches from the left end, and y is the deflection of the centerline in inches.
    (a) Graph y versus x for $0 \le x \le 192$.
    (b) Find the maximum deflection of the centerline. *(cont.)*

(c) Use a CAS or a calculator with a numerical integration capability to find the length of the centerline of the loaded beam. Round your answer to two decimal places.

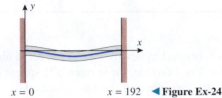

$x = 0$ $x = 192$ ◀ **Figure Ex-24**

C **25.** A golfer makes a successful chip shot to the green. Suppose that the path of the ball from the moment it is struck to the moment it hits the green is described by

$$y = 12.54x - 0.41x^2$$

where x is the horizontal distance (in yards) from the point where the ball is struck, and y is the vertical distance (in yards) above the fairway. Use a CAS or a calculating utility with a numerical integration capability to find the distance the ball travels from the moment it is struck to the moment it hits the green. Assume that the fairway and green are at the same level and round your answer to two decimal places.

26–34 These exercises assume familiarity with the basic concepts of parametric curves. If needed, an introduction to this material is provided in Web Appendix I. ■

C **26.** Assume that no segment of the curve

$$x = x(t), \quad y = y(t), \quad (a \le t \le b)$$

is traced more than once as t increases from a to b. Divide the interval $[a, b]$ into n subintervals by inserting points $t_1, t_2, \ldots, t_{n-1}$ between $a = t_0$ and $b = t_n$. Let L denote the arc length of the curve. Give an informal argument for the approximation

$$L \approx \sum_{k=1}^{n} \sqrt{[x(t_k) - x(t_{k-1})]^2 + [y(t_k) - y(t_{k-1})]^2}$$

If dx/dt and dy/dt are continuous functions for $a \le t \le b$, then it can be shown that as max $\Delta t_k \to 0$, this sum converges to

$$L = \int_a^b \sqrt{\left(\frac{dx}{dt}\right)^2 + \left(\frac{dy}{dt}\right)^2}\, dt$$

27–32 Use the arc length formula from Exercise 26 to find the arc length of the curve. ■

27. $x = \frac{1}{3}t^3, \quad y = \frac{1}{2}t^2 \quad (0 \le t \le 1)$

28. $x = (1 + t)^2, \quad y = (1 + t)^3 \quad (0 \le t \le 1)$

29. $x = \cos 2t, \quad y = \sin 2t \quad (0 \le t \le \pi/2)$

30. $x = \cos t + t \sin t, \quad y = \sin t - t \cos t \quad (0 \le t \le \pi)$

31. $x = e^t \cos t, \quad y = e^t \sin t \quad (0 \le t \le \pi/2)$

32. $x = e^t(\sin t + \cos t), \quad y = e^t(\cos t - \sin t) \quad (1 \le t \le 4)$

C **33.** (a) Show that the total arc length of the ellipse

$$x = 2\cos t, \quad y = \sin t \quad (0 \le t \le 2\pi)$$

is given by

$$4 \int_0^{\pi/2} \sqrt{1 + 3\sin^2 t}\, dt$$

(b) Use a CAS or a scientific calculator with a numerical integration capability to approximate the arc length in part (a). Round your answer to two decimal places.

(c) Suppose that the parametric equations in part (a) describe the path of a particle moving in the xy-plane, where t is time in seconds and x and y are in centimeters. Use a CAS or a scientific calculator with a numerical integration capability to approximate the distance traveled by the particle from $t = 1.5$ s to $t = 4.8$ s. Round your answer to two decimal places.

34. Show that the total arc length of the ellipse $x = a\cos t$, $y = b\sin t, 0 \le t \le 2\pi$ for $a > b > 0$ is given by

$$4a \int_0^{\pi/2} \sqrt{1 - k^2 \cos^2 t}\, dt$$

where $k = \sqrt{a^2 - b^2}/a$.

35. Writing In our discussion of Arc Length Problem 6.4.1, we derived the approximation

$$L \approx \sum_{k=1}^{n} \sqrt{1 + [f'(x_k^*)]^2}\, \Delta x_k$$

Discuss the geometric meaning of this approximation. (Be sure to address the appearance of the derivative f'.)

36. Writing Give examples in which Formula (4) for arc length cannot be applied directly, and describe how you would go about finding the arc length of the curve in each case. (Discuss both the use of alternative formulas and the use of numerical methods.)

✔ **QUICK CHECK ANSWERS 6.4**

1. continuous   **2.** $\int_a^b \sqrt{1 + [f'(x)]^2}\, dx$ **3.** $\sqrt{(e - 1)^2 + 1}$ **4.** (a) $\int_1^e \sqrt{1 + (1/x)^2}\, dx$ (b) $\int_0^1 \sqrt{1 + e^{2y}}\, dy$

6.5 AREA OF A SURFACE OF REVOLUTION

In this section we will consider the problem of finding the area of a surface that is generated by revolving a plane curve about a line.

■ SURFACE AREA

A *surface of revolution* is a surface that is generated by revolving a plane curve about an axis that lies in the same plane as the curve. For example, the surface of a sphere can be generated by revolving a semicircle about its diameter, and the lateral surface of a right circular cylinder can be generated by revolving a line segment about an axis that is parallel to it (Figure 6.5.1).

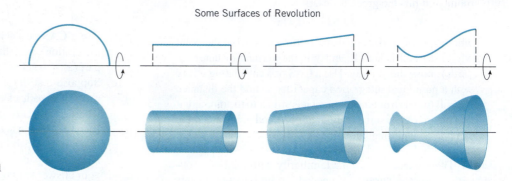

Some Surfaces of Revolution

▶ **Figure 6.5.1**

In this section we will be concerned with the following problem.

> **6.5.1 SURFACE AREA PROBLEM** Suppose that f is a smooth, nonnegative function on $[a, b]$ and that a surface of revolution is generated by revolving the portion of the curve $y = f(x)$ between $x = a$ and $x = b$ about the x-axis (Figure 6.5.2). Define what is meant by the *area S* of the surface, and find a formula for computing it.

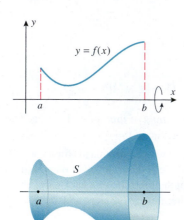

▲ **Figure 6.5.2**

To motivate an appropriate definition for the area S of a surface of revolution, we will decompose the surface into small sections whose areas can be approximated by elementary formulas, add the approximations of the areas of the sections to form a Riemann sum that approximates S, and then take the limit of the Riemann sums to obtain an integral for the exact value of S.

To implement this idea, divide the interval $[a, b]$ into n subintervals by inserting points x_1, $x_2, \ldots, x_{n-1}$ between $a = x_0$ and $b = x_n$. As illustrated in Figure 6.5.3a, the corresponding points on the graph of f define a polygonal path that approximates the curve $y = f(x)$ over the interval $[a, b]$. As illustrated in Figure 6.5.3b, when this polygonal path is revolved about the x-axis, it generates a surface consisting of n parts, each of which is a portion of a right circular cone called a *frustum* (from the Latin meaning "bit" or "piece"). Thus, the area of each part of the approximating surface can be obtained from the formula

$$S = \pi(r_1 + r_2)l \tag{1}$$

for the lateral area S of a frustum of slant height l and base radii r_1 and r_2 (Figure 6.5.4). As suggested by Figure 6.5.5, the kth frustum has radii $f(x_{k-1})$ and $f(x_k)$ and height Δx_k. Its slant height is the length L_k of the kth line segment in the polygonal path, which from Formula (1) of Section 6.4 is

$$L_k = \sqrt{(\Delta x_k)^2 + [f(x_k) - f(x_{k-1})]^2}$$

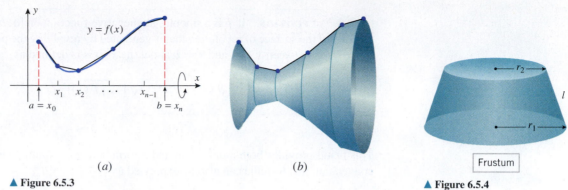

(a)

▲ **Figure 6.5.3**

(b)

Frustum

▲ **Figure 6.5.4**

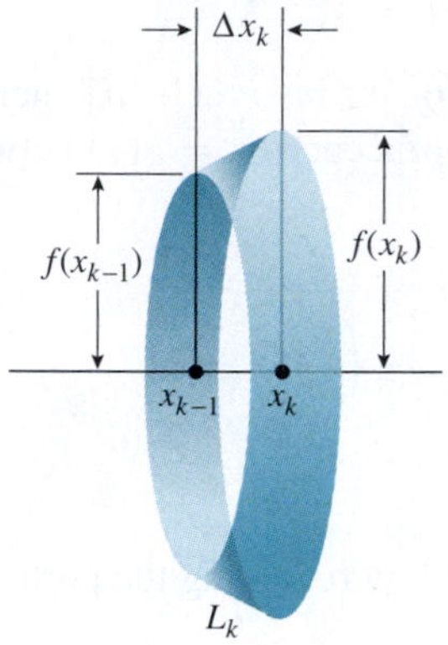

▲ **Figure 6.5.5**

This makes the lateral area S_k of the kth frustum

$$S_k = \pi[f(x_{k-1}) + f(x_k)]\sqrt{(\Delta x_k)^2 + [f(x_k) - f(x_{k-1})]^2}$$

If we add these areas, we obtain the following approximation to the area S of the entire surface:

$$S \approx \sum_{k=1}^{n} \pi[f(x_{k-1}) + f(x_k)]\sqrt{(\Delta x_k)^2 + [f(x_k) - f(x_{k-1})]^2} \qquad (2)$$

To put this in the form of a Riemann sum we will apply the Mean-Value Theorem (4.8.2). This theorem implies that there is a point x_k^* between x_{k-1} and x_k such that

$$\frac{f(x_k) - f(x_{k-1})}{x_k - x_{k-1}} = f'(x_k^*) \quad \text{or} \quad f(x_k) - f(x_{k-1}) = f'(x_k^*)\Delta x_k$$

and hence we can rewrite (2) as

$$S \approx \sum_{k=1}^{n} \pi[f(x_{k-1}) + f(x_k)]\sqrt{(\Delta x_k)^2 + [f'(x_k^*)]^2(\Delta x_k)^2}$$

$$= \sum_{k=1}^{n} \pi[f(x_{k-1}) + f(x_k)]\sqrt{1 + [f'(x_k^*)]^2}\,\Delta x_k \qquad (3)$$

However, this is not yet a Riemann sum because it involves the variables x_{k-1} and x_k. To eliminate these variables from the expression, observe that the average value of the numbers $f(x_{k-1})$ and $f(x_k)$ lies between these numbers, so the continuity of f and the Intermediate-Value Theorem (1.5.7) imply that there is a point x_k^{**} between x_{k-1} and x_k such that

$$\tfrac{1}{2}[f(x_{k-1}) + f(x_k)] = f(x_k^{**})$$

Thus, (2) can be expressed as

$$S \approx \sum_{k=1}^{n} 2\pi f(x_k^{**})\sqrt{1 + [f'(x_k^*)]^2}\,\Delta x_k$$

Although this expression is close to a Riemann sum in form, it is not a true Riemann sum because it involves two variables x_k^* and x_k^{**}, rather than x_k^* alone. However, it is proved in advanced calculus courses that this has no effect on the limit because of the continuity of f. Thus, we can assume that $x_k^{**} = x_k^*$ when taking the limit, and this suggests that S can be defined as

$$S = \lim_{\max \Delta x_k \to 0} \sum_{k=1}^{n} 2\pi f(x_k^{**})\sqrt{1 + [f'(x_k^*)]^2}\,\Delta x_k = \int_a^b 2\pi f(x)\sqrt{1 + [f'(x)]^2}\,dx$$

In summary, we have the following definition.

6.5.2   **DEFINITION**   If f is a smooth, nonnegative function on $[a, b]$, then the surface area S of the surface of revolution that is generated by revolving the portion of the curve $y = f(x)$ between $x = a$ and $x = b$ about the x-axis is defined as

$$S = \int_a^b 2\pi f(x)\sqrt{1 + [f'(x)]^2}\, dx$$

This result provides both a definition and a formula for computing surface areas. Where convenient, this formula can also be expressed as

$$S = \int_a^b 2\pi f(x)\sqrt{1 + [f'(x)]^2}\, dx = \int_a^b 2\pi y\sqrt{1 + \left(\frac{dy}{dx}\right)^2}\, dx \qquad (4)$$

Moreover, if g is nonnegative and $x = g(y)$ is a smooth curve on the interval $[c, d]$, then the area of the surface that is generated by revolving the portion of a curve $x = g(y)$ between $y = c$ and $y = d$ about the y-axis can be expressed as

$$S = \int_c^d 2\pi g(y)\sqrt{1 + [g'(y)]^2}\, dy = \int_c^d 2\pi x\sqrt{1 + \left(\frac{dx}{dy}\right)^2}\, dy \qquad (5)$$

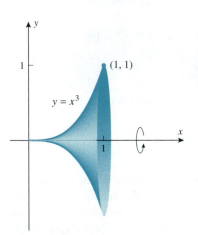

▲ **Figure 6.5.6**

▶ **Example 1**   Find the area of the surface that is generated by revolving the portion of the curve $y = x^3$ between $x = 0$ and $x = 1$ about the x-axis.

Solution.   First sketch the curve; then imagine revolving it about the x-axis (Figure 6.5.6). Since $y = x^3$, we have $dy/dx = 3x^2$, and hence from (4) the surface area S is

$$S = \int_0^1 2\pi y\sqrt{1 + \left(\frac{dy}{dx}\right)^2}\, dx$$

$$= \int_0^1 2\pi x^3\sqrt{1 + (3x^2)^2}\, dx$$

$$= 2\pi \int_0^1 x^3(1 + 9x^4)^{1/2}\, dx$$

$$= \frac{2\pi}{36} \int_1^{10} u^{1/2}\, du \qquad \boxed{\begin{array}{l} u = 1 + 9x^4 \\ du = 36x^3\, dx \end{array}}$$

$$= \frac{2\pi}{36} \cdot \frac{2}{3} u^{3/2}\Big]_{u=1}^{10} = \frac{\pi}{27}(10^{3/2} - 1) \approx 3.56 \blacktriangleleft$$

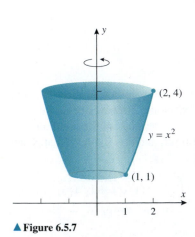

▲ **Figure 6.5.7**

▶ **Example 2**   Find the area of the surface that is generated by revolving the portion of the curve $y = x^2$ between $x = 1$ and $x = 2$ about the y-axis.

Solution.   First sketch the curve; then imagine revolving it about the y-axis (Figure 6.5.7). Because the curve is revolved about the y-axis we will apply Formula (5). Toward this end, we rewrite $y = x^2$ as $x = \sqrt{y}$ and observe that the y-values corresponding to $x = 1$ and

$x = 2$ are $y = 1$ and $y = 4$. Since $x = \sqrt{y}$, we have $dx/dy = 1/(2\sqrt{y})$, and hence from (5) the surface area S is

$$S = \int_1^4 2\pi x \sqrt{1 + \left(\frac{dx}{dy}\right)^2}\, dy$$

$$= \int_1^4 2\pi \sqrt{y} \sqrt{1 + \left(\frac{1}{2\sqrt{y}}\right)^2}\, dy$$

$$= \pi \int_1^4 \sqrt{4y + 1}\, dy$$

$$= \frac{\pi}{4} \int_5^{17} u^{1/2}\, du \qquad \boxed{\begin{array}{l} u = 4y + 1 \\ du = 4\, dy \end{array}}$$

$$= \frac{\pi}{4} \cdot \frac{2}{3} u^{3/2} \Big]_{u=5}^{17} = \frac{\pi}{6}(17^{3/2} - 5^{3/2}) \approx 30.85 \blacktriangleleft$$

✔ **QUICK CHECK EXERCISES 6.5**   (*See page 449 for answers.*)

1. If f is a smooth, nonnegative function on $[a, b]$, then the surface area S of the surface of revolution generated by revolving the portion of the curve $y = f(x)$ between $x = a$ and $x = b$ about the x-axis is _____.

2. The lateral area of the frustum with slant height $\sqrt{10}$ and base radii $r_1 = 1$ and $r_2 = 2$ is _____.

3. An integral expression for the area of the surface generated by rotating the line segment joining $(3, 1)$ and $(6, 2)$ about the x-axis is _____.

4. An integral expression for the area of the surface generated by rotating the line segment joining $(3, 1)$ and $(6, 2)$ about the y-axis is _____.

EXERCISE SET 6.5   C CAS

1–4 Find the area of the surface generated by revolving the given curve about the x-axis. ■

1. $y = 7x$, $0 \le x \le 1$

2. $y = \sqrt{x}$, $1 \le x \le 4$

3. $y = \sqrt{4 - x^2}$, $-1 \le x \le 1$

4. $x = \sqrt[3]{y}$, $1 \le y \le 8$

5–8 Find the area of the surface generated by revolving the given curve about the y-axis. ■

5. $x = 9y + 1$, $0 \le y \le 2$

6. $x = y^3$, $0 \le y \le 1$

7. $x = \sqrt{9 - y^2}$, $-2 \le y \le 2$

8. $x = 2\sqrt{1 - y}$, $-1 \le y \le 0$

C **9–12** Use a CAS to find the exact area of the surface generated by revolving the curve about the stated axis. ■

9. $y = \sqrt{x} - \frac{1}{3}x^{3/2}$, $1 \le x \le 3$; x-axis

10. $y = \frac{1}{3}x^3 + \frac{1}{4}x^{-1}$, $1 \le x \le 2$; x-axis

11. $8xy^2 = 2y^6 + 1$, $1 \le y \le 2$; y-axis

12. $x = \sqrt{16 - y}$, $0 \le y \le 15$; y-axis

C **13–16** Use a CAS or a calculating utility with a numerical integration capability to approximate the area of the surface generated by revolving the curve about the stated axis. Round your answer to two decimal places. ■

13. $y = \sin x$, $0 \le x \le \pi$; x-axis

14. $x = \tan y$, $0 \le y \le \pi/4$; y-axis

15. $y = e^x$, $0 \le x \le 1$; x-axis

16. $y = e^x$, $1 \le y \le e$; y-axis

17–20 True–False Determine whether the statement is true or false. Explain your answer. ■

17. The lateral surface area S of a right circular cone with height h and base radius r is $S = \pi r \sqrt{r^2 + h^2}$.

18. The lateral surface area of a frustum of slant height l and base radii r_1 and r_2 is equal to the lateral surface area of a right circular cylinder of height l and radius equal to the average of r_1 and r_2.

19. The approximation

$$S \approx \sum_{k=1}^{n} 2\pi f(x_k^{**})\sqrt{1+[f'(x_k^*)]^2}\,\Delta x_k$$

for surface area is exact if f is a positive-valued constant function.

20. The expression

$$\sum_{k=1}^{n} 2\pi f(x_k^{**})\sqrt{1+[f'(x_k^*)]^2}\,\Delta x_k$$

is not a true Riemann sum for

$$\int_a^b 2\pi f(x)\sqrt{1+[f'(x)]^2}\,dx$$

21–22 Approximate the area of the surface using Formula (2) with $n = 20$ subintervals of equal width. Round your answer to two decimal places. ■

21. The surface of Exercise 13.

22. The surface of Exercise 16.

FOCUS ON CONCEPTS

23. Assume that $y = f(x)$ is a smooth curve on the interval $[a, b]$ and assume that $f(x) \geq 0$ for $a \leq x \leq b$. Derive a formula for the surface area generated when the curve $y = f(x)$, $a \leq x \leq b$, is revolved about the line $y = -k$ $(k > 0)$.

24. Would it be circular reasoning to use Definition 6.5.2 to find the surface area of a frustum of a right circular cone? Explain your answer.

---

25. Show that the area of the surface of a sphere of radius r is $4\pi r^2$. [*Hint:* Revolve the semicircle $y = \sqrt{r^2 - x^2}$ about the x-axis.]

26. The accompanying figure shows a spherical cap of height h cut from a sphere of radius r. Show that the surface area S of the cap is $S = 2\pi rh$. [*Hint:* Revolve an appropriate portion of the circle $x^2 + y^2 = r^2$ about the y-axis.]

◀ **Figure Ex-26**

27. The portion of a sphere that is cut by two parallel planes is called a *zone*. Use the result of Exercise 26 to show that the surface area of a zone depends on the radius of the sphere and the distance between the planes, but not on the location of the zone.

28. Let $y = f(x)$ be a smooth curve on the interval $[a, b]$ and assume that $f(x) \geq 0$ for $a \leq x \leq b$. By the Extreme-Value

Theorem (4.4.2), the function f has a maximum value K and a minimum value k on $[a, b]$. Prove: If L is the arc length of the curve $y = f(x)$ between $x = a$ and $x = b$, and if S is the area of the surface that is generated by revolving this curve about the x-axis, then

$$2\pi kL \leq S \leq 2\pi KL$$

29. Use the results of Exercise 28 above and Exercise 21 in Section 6.4 to show that the area S of the surface generated by revolving the curve $y = \sec x$, $0 \leq x \leq \pi/3$, about the x-axis satisfies

$$\frac{2\pi^2}{3} \leq S \leq \frac{4\pi^2}{3}\sqrt{13}$$

30. Let $y = f(x)$ be a smooth curve on $[a, b]$ and assume that $f(x) \geq 0$ for $a \leq x \leq b$. Let A be the area under the curve $y = f(x)$ between $x = a$ and $x = b$, and let S be the area of the surface obtained when this section of curve is revolved about the x-axis.
(a) Prove that $2\pi A \leq S$.
(b) For what functions f is $2\pi A = S$?

31–37 These exercises assume familiarity with the basic concepts of parametric curves. If needed, an introduction to this material is provided in Web Appendix I. ■

31–32 For these exercises, divide the interval $[a, b]$ into n subintervals by inserting points $t_1, t_2, \ldots, t_{n-1}$ between $a = t_0$ and $b = t_n$, and assume that $x'(t)$ and $y'(t)$ are continuous functions and that no segment of the curve

$$x = x(t), \quad y = y(t) \quad (a \leq t \leq b)$$

is traced more than once. ■

31. Let S be the area of the surface generated by revolving the curve $x = x(t)$, $y = y(t)$ $(a \leq t \leq b)$ about the x-axis. Explain how S can be approximated by

$$S \approx \sum_{k=1}^{n} (\pi[y(t_{k-1}) + y(t_k)]$$
$$\times \sqrt{[x(t_k) - x(t_{k-1})]^2 + [y(t_k) - y(t_{k-1})]^2}\,)$$

Using results from advanced calculus, it can be shown that as max $\Delta t_k \to 0$, this sum converges to

$$S = \int_a^b 2\pi y(t)\sqrt{[x'(t)]^2 + [y'(t)]^2}\,dt \qquad \text{(A)}$$

32. Let S be the area of the surface generated by revolving the curve $x = x(t)$, $y = y(t)$ $(a \leq t \leq b)$ about the y-axis. Explain how S can be approximated by

$$S \approx \sum_{k=1}^{n} (\pi[x(t_{k-1}) + x(t_k)]$$
$$\times \sqrt{[x(t_k) - x(t_{k-1})]^2 + [y(t_k) - y(t_{k-1})]^2}\,)$$

Using results from advanced calculus, it can be shown that as max $\Delta t_k \to 0$, this sum converges to

$$S = \int_a^b 2\pi x(t)\sqrt{[x'(t)]^2 + [y'(t)]^2}\,dt \qquad \text{(B)}$$

33–37 Use Formulas (A) and (B) from Exercises 31 and 32. ∎

33. Find the area of the surface generated by revolving the parametric curve $x = t^2$, $y = 2t$ ($0 \leq t \leq 4$) about the x-axis.

C **34.** Use a CAS to find the area of the surface generated by revolving the parametric curve
$$x = \cos^2 t, \quad y = 5 \sin t \quad (0 \leq t \leq \pi/2)$$
about the x-axis.

35. Find the area of the surface generated by revolving the parametric curve $x = t$, $y = 2t^2$ ($0 \leq t \leq 1$) about the y-axis.

36. Find the area of the surface generated by revolving the parametric curve $x = \cos^2 t$, $y = \sin^2 t$ ($0 \leq t \leq \pi/2$) about the y-axis.

37. By revolving the semicircle
$$x = r \cos t, \quad y = r \sin t \quad (0 \leq t \leq \pi)$$
about the x-axis, show that the surface area of a sphere of radius r is $4\pi r^2$.

38. Writing Compare the derivation of Definition 6.5.2 with that of Definition 6.4.2. Discuss the geometric features that result in similarities in the two definitions.

39. Writing Discuss what goes wrong if we replace the frustums of right circular cones by right circular cylinders in the derivation of Definition 6.5.2.

✔ **QUICK CHECK ANSWERS 6.5**

1. $\displaystyle\int_a^b 2\pi f(x)\sqrt{1 + [f'(x)]^2}\, dx$ **2.** $3\sqrt{10}\,\pi$ **3.** $\displaystyle\int_3^6 (2\pi)\left(\frac{x}{3}\right)\sqrt{\frac{10}{9}}\, dx = \int_3^6 \frac{2\sqrt{10}\,\pi}{9} x\, dx$ **4.** $\displaystyle\int_1^2 (2\pi)(3y)\sqrt{10}\, dy$

6.6 WORK

In this section we will use the integration tools developed in the preceding chapter to study some of the basic principles of "work," which is one of the fundamental concepts in physics and engineering.

■ **THE ROLE OF WORK IN PHYSICS AND ENGINEERING**

In this section we will be concerned with two related concepts, *work* and *energy*. To put these ideas in a familiar setting, when you push a stalled car for a certain distance you are performing work, and the effect of your work is to make the car move. The energy of motion caused by the work is called the *kinetic energy* of the car. The exact connection between work and kinetic energy is governed by a principle of physics called the *work–energy relationship*. Although we will touch on this idea in this section, a detailed study of the relationship between work and energy will be left for courses in physics and engineering. Our primary goal here will be to explain the role of integration in the study of work.

■ **WORK DONE BY A CONSTANT FORCE APPLIED IN THE DIRECTION OF MOTION**

When a stalled car is pushed, the speed that the car attains depends on the force F with which it is pushed and the distance d over which that force is applied (Figure 6.6.1). Force and distance appear in the following definition of work.

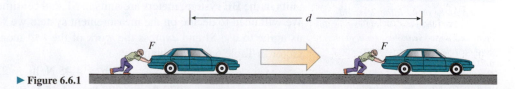

▶ **Figure 6.6.1**

6.6.1   DEFINITION   If a constant force of magnitude F is applied in the direction of motion of an object, and if that object moves a distance d, then we define the **work** W performed by the force on the object to be

$$W = F \cdot d \qquad (1)$$

> If you push against an immovable object, such as a brick wall, you may tire yourself out, but you will not perform any work. Why?

Common units for measuring force are newtons (N) in the International System of Units (SI), dynes (dyn) in the centimeter-gram-second (CGS) system, and pounds (lb) in the British Engineering (BE) system. One newton is the force required to give a mass of 1 kg an acceleration of 1 m/s^2, one dyne is the force required to give a mass of 1 g an acceleration of 1 cm/s^2, and one pound of force is the force required to give a mass of 1 slug an acceleration of 1 ft/s^2.

It follows from Definition 6.6.1 that work has units of force times distance. The most common units of work are newton-meters (N·m), dyne-centimeters (dyn·cm), and foot-pounds (ft·lb). As indicated in Table 6.6.1, one newton-meter is also called a **joule** (J), and one dyne-centimeter is also called an **erg**. One foot-pound is approximately 1.36 J.

Table 6.6.1

SYSTEM	FORCE	×	DISTANCE	=	WORK
SI	newton (N)		meter (m)		joule (J)
CGS	dyne (dyn)		centimeter (cm)		erg
BE	pound (lb)		foot (ft)		foot-pound (ft·lb)

CONVERSION FACTORS:

$1 \text{ N} = 10^5 \text{ dyn} \approx 0.225 \text{ lb}$ $\qquad$ $1 \text{ lb} \approx 4.45 \text{ N}$

$1 \text{ J} = 10^7 \text{ erg} \approx 0.738 \text{ ft·lb}$ $\qquad$ $1 \text{ ft·lb} \approx 1.36 \text{ J} = 1.36 \times 10^7 \text{ erg}$

▶ **Example 1**   An object moves 5 ft along a line while subjected to a constant force of 100 lb in its direction of motion. The work done is

$$W = F \cdot d = 100 \cdot 5 = 500 \text{ ft·lb}$$

An object moves 25 m along a line while subjected to a constant force of 4 N in its direction of motion. The work done is

$$W = F \cdot d = 4 \cdot 25 = 100 \text{ N·m} = 100 \text{ J} \blacktriangleleft$$

▶ **Example 2**   In the 1976 Olympics, Vasili Alexeev astounded the world by lifting a record-breaking 562 lb from the floor to above his head (about 2 m). Equally astounding was the feat of strongman Paul Anderson, who in 1957 braced himself on the floor and used his back to lift 6270 lb of lead and automobile parts a distance of 1 cm. Who did more work?

Solution.   To lift an object one must apply sufficient force to overcome the gravitational force that the Earth exerts on that object. The force that the Earth exerts on an object is that object's weight; thus, in performing their feats, Alexeev applied a force of 562 lb over a distance of 2 m and Anderson applied a force of 6270 lb over a distance of 1 cm. Pounds are units in the BE system, meters are units in SI, and centimeters are units in the CGS system. We will need to decide on the measurement system we want to use and be consistent. Let us agree to use SI and express the work of the two men in joules. Using the conversion factor in Table 6.6.1 we obtain

$$562 \text{ lb} \approx 562 \text{ lb} \times 4.45 \text{ N/lb} \approx 2500 \text{ N}$$

$$6270 \text{ lb} \approx 6270 \text{ lb} \times 4.45 \text{ N/lb} \approx 27{,}900 \text{ N}$$

Vasili Alexeev shown lifting a record-breaking 562 lb in the 1976 Olympics. In eight successive years he won olympic gold medals, captured six world championships, and broke 80 world records. In 1999 he was honored in Greece as the best sportsman of the 20th Century.

Using these values and the fact that 1 cm = 0.01 m we obtain

$$\text{Alexeev's work} = (2500 \text{ N}) \times (2 \text{ m}) = 5000 \text{ J}$$

$$\text{Anderson's work} = (27{,}900 \text{ N}) \times (0.01 \text{ m}) = 279 \text{ J}$$

Therefore, even though Anderson's lift required a tremendous upward force, it was applied over such a short distance that Alexeev did more work. ◄

■ WORK DONE BY A VARIABLE FORCE APPLIED IN THE DIRECTION OF MOTION

Many important problems are concerned with finding the work done by a *variable* force that is applied in the direction of motion. For example, Figure 6.6.2a shows a spring in its natural state (neither compressed nor stretched). If we want to pull the block horizontally (Figure 6.6.2b), then we would have to apply more and more force to the block to overcome the increasing force of the stretching spring. Thus, our next objective is to define what is meant by the work performed by a variable force and to find a formula for computing it. This will require calculus.

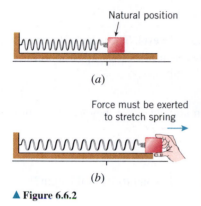

Natural position

(a)

Force must be exerted
to stretch spring

(b)

▲ Figure 6.6.2

6.6.2   PROBLEM   Suppose that an object moves in the positive direction along a coordinate line while subjected to a variable force $F(x)$ that is applied in the direction of motion. Define what is meant by the *work* W performed by the force on the object as the object moves from $x = a$ to $x = b$, and find a formula for computing the work.

The basic idea for solving this problem is to break up the interval $[a, b]$ into subintervals that are sufficiently small that the force does not vary much on each subinterval. This will allow us to treat the force as constant on each subinterval and to approximate the work on each subinterval using Formula (1). By adding the approximations to the work on the subintervals, we will obtain a Riemann sum that approximates the work W over the entire interval, and by taking the limit of the Riemann sums we will obtain an integral for W.

To implement this idea, divide the interval $[a, b]$ into n subintervals by inserting points $x_1, x_2, \ldots, x_{n-1}$ between $a = x_0$ and $b = x_n$. We can use Formula (1) to approximate the work W_k done in the kth subinterval by choosing any point x_k^* in this interval and regarding the force to have a constant value $F(x_k^*)$ throughout the interval. Since the width of the kth subinterval is $x_k - x_{k-1} = \Delta x_k$, this yields the approximation

$$W_k \approx F(x_k^*)\Delta x_k$$

Adding these approximations yields the following Riemann sum that approximates the work W done over the entire interval:

$$W \approx \sum_{k=1}^{n} F(x_k^*)\Delta x_k$$

Taking the limit as n increases and the widths of all the subintervals approach zero yields the definite integral

$$W = \lim_{\max \Delta x_k \to 0} \sum_{k=1}^{n} F(x_k^*)\Delta x_k = \int_{a}^{b} F(x)\,dx$$

In summary, we have the following result.

6.6.3 DEFINITION Suppose that an object moves in the positive direction along a coordinate line over the interval $[a, b]$ while subjected to a variable force $F(x)$ that is applied in the direction of motion. Then we define the *work* W performed by the force on the object to be

$$W = \int_a^b F(x)\,dx \qquad (2)$$

Hooke's law [Robert Hooke (1635–1703), English physicist] states that under appropriate conditions a spring that is stretched x units beyond its natural length pulls back with a force

$$F(x) = kx$$

where k is a constant (called the *spring constant* or *spring stiffness*). The value of k depends on such factors as the thickness of the spring and the material used in its composition. Since $k = F(x)/x$, the constant k has units of force per unit length.

▶ **Example 3** A spring exerts a force of 5 N when stretched 1 m beyond its natural length.

(a) Find the spring constant k.

(b) How much work is required to stretch the spring 1.8 m beyond its natural length?

Solution (a). From Hooke's law,

$$F(x) = kx$$

From the data, $F(x) = 5$ N when $x = 1$ m, so $5 = k \cdot 1$. Thus, the spring constant is $k = 5$ newtons per meter (N/m). This means that the force $F(x)$ required to stretch the spring x meters is

$$F(x) = 5x \qquad (3)$$

Solution (b). Place the spring along a coordinate line as shown in Figure 6.6.3. We want to find the work W required to stretch the spring over the interval from $x = 0$ to $x = 1.8$. From (2) and (3) the work W required is

$$W = \int_a^b F(x)\,dx = \int_0^{1.8} 5x\,dx = \left.\frac{5x^2}{2}\right]_0^{1.8} = 8.1\ \text{J} \blacktriangleleft$$

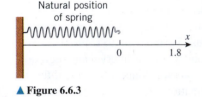

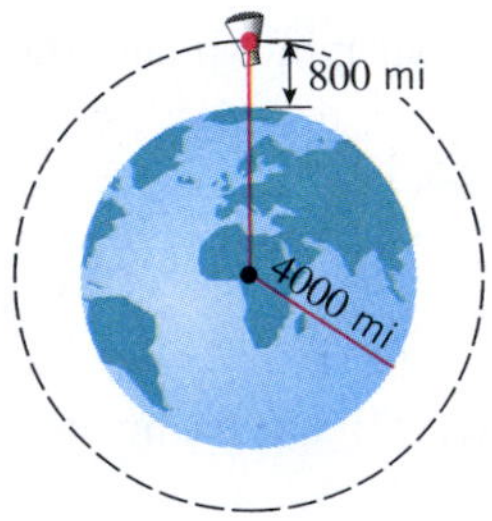

Natural position
of spring

▲ **Figure 6.6.3**

▶ **Example 4** An astronaut's *weight* (or more precisely, *Earth weight*) is the force exerted on the astronaut by the Earth's gravity. As the astronaut moves upward into space, the gravitational pull of the Earth decreases, and hence so does his or her weight. If the Earth is assumed to be a sphere of radius 4000 mi, then it can be shown using physics that an astronaut who weighs 150 lb on Earth will have a weight of

$$w(x) = \frac{2{,}400{,}000{,}000}{x^2}\ \text{lb}, \quad x \geq 4000$$

at a distance of x mi from the Earth's center (Exercise 25). Use this formula to determine the work in foot-pounds required to lift the astronaut to a point that is 800 mi above the surface of the Earth (Figure 6.6.4).

▲ **Figure 6.6.4**

Solution. Since the Earth has a radius of 4000 mi, the astronaut is lifted from a point that is 4000 mi from the Earth's center to a point that is 4800 mi from the Earth's center. Thus, from (2), the work W required to lift the astronaut is

$$W = \int_{4000}^{4800} \frac{2,400,000,000}{x^2} \, dx$$

$$= -\frac{2,400,000,000}{x} \Big]_{4000}^{4800}$$

$$= -500,000 + 600,000$$

$$= 100,000 \text{ mile-pounds}$$

$$= (100,000 \text{ mi·lb}) \times (5280 \text{ ft/mi})$$

$$= 5.28 \times 10^8 \text{ ft·lb} \blacktriangleleft$$

■ CALCULATING WORK FROM BASIC PRINCIPLES

Some problems cannot be solved by mechanically substituting into formulas, and one must return to basic principles to obtain solutions. This is illustrated in the next example.

▶ **Example 5**   Figure 6.6.5*a* shows a conical container of radius 10 ft and height 30 ft. Suppose that this container is filled with water to a depth of 15 ft. How much work is required to pump all of the water out through a hole in the top of the container?

Solution.   Our strategy will be to divide the water into thin layers, approximate the work required to move each layer to the top of the container, add the approximations for the layers to obtain a Riemann sum that approximates the total work, and then take the limit of the Riemann sums to produce an integral for the total work.

To implement this idea, introduce an x-axis as shown in Figure 6.6.5*a*, and divide the water into n layers with Δx_k denoting the thickness of the kth layer. This division induces a partition of the interval [15, 30] into n subintervals. Although the upper and lower surfaces of the kth layer are at different distances from the top, the difference will be small if the layer is thin, and we can reasonably assume that the entire layer is concentrated at a single point x_k^* (Figure 6.6.5*a*). Thus, the work W_k required to move the kth layer to the top of the container is approximately

$$W_k \approx F_k x_k^* \tag{4}$$

where F_k is the force required to lift the kth layer. But the force required to lift the kth layer is the force needed to overcome gravity, and this is the same as the weight of the layer. If the layer is very thin, we can approximate the volume of the kth layer with the volume of a cylinder of height Δx_k and radius r_k, where (by similar triangles)

$$\frac{r_k}{x_k^*} = \frac{10}{30} = \frac{1}{3}$$

or, equivalently, $r_k = x_k^*/3$ (Figure 6.6.5*b*). Therefore, the volume of the kth layer of water is approximately

$$\pi r_k^2 \Delta x_k = \pi (x_k^*/3)^2 \Delta x_k = \frac{\pi}{9}(x_k^*)^2 \Delta x_k$$

Since the weight density of water is 62.4 lb/ft^3, it follows that

$$F_k \approx \frac{62.4\pi}{9}(x_k^*)^2 \Delta x_k$$

Thus, from (4)

$$W_k \approx \left(\frac{62.4\pi}{9}(x_k^*)^2 \Delta x_k \right) x_k^* = \frac{62.4\pi}{9}(x_k^*)^3 \Delta x_k$$

and hence the work W required to move all n layers has the approximation

$$W = \sum_{k=1}^{n} W_k \approx \sum_{k=1}^{n} \frac{62.4\pi}{9}(x_k^*)^3 \Delta x_k$$

To find the *exact* value of the work we take the limit as max $\Delta x_k \to 0$. This yields

$$W = \lim_{\max \Delta x_k \to 0} \sum_{k=1}^{n} \frac{62.4\pi}{9}(x_k^*)^3 \Delta x_k = \int_{15}^{30} \frac{62.4\pi}{9} x^3 \, dx$$

$$= \frac{62.4\pi}{9} \left(\frac{x^4}{4} \right) \Big]_{15}^{30} = 1,316,250\pi \approx 4,135,000 \text{ ft·lb} \quad \blacktriangleleft$$

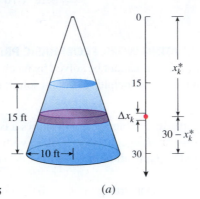
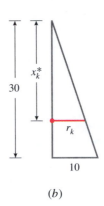

▶ **Figure 6.6.5**   (*a*)   (*b*)

■ THE WORK–ENERGY RELATIONSHIP

When you see an object in motion, you can be certain that somehow work has been expended to create that motion. For example, when you drop a stone from a building, the stone gathers speed because the force of the Earth's gravity is performing work on it, and when a hockey player strikes a puck with a hockey stick, the work performed on the puck during the brief period of contact with the stick creates the enormous speed of the puck across the ice. However, experience shows that the speed obtained by an object depends not only on the amount of work done, but also on the mass of the object. For example, the work required to throw a 5 oz baseball 50 mi/h would accelerate a 10 lb bowling ball to less than 9 mi/h.

Using the method of substitution for definite integrals, we will derive a simple equation that relates the work done on an object to the object's mass and velocity. Furthermore, this equation will allow us to motivate an appropriate definition for the "energy of motion" of an object. As in Definition 6.6.3, we will assume that an object moves in the positive direction along a coordinate line over the interval $[a, b]$ while subjected to a force $F(x)$ that is applied in the direction of motion. We let m denote the mass of the object, and we let $x = x(t)$, $v = v(t) = x'(t)$, and $a = a(t) = v'(t)$ denote the respective position, velocity, and acceleration of the object at time t. We will need the following important result from physics that relates the force acting on an object with the mass and acceleration of the object.

Mike Brinson/Getty Images
The work performed by the skater's stick in a brief interval of time produces the blinding speed of the hockey puck.

6.6.4   NEWTON'S SECOND LAW OF MOTION   If an object with mass m is subjected to a force F, then the object undergoes an acceleration a that satisfies the equation

$$F = ma \tag{5}$$

It follows from Newton's Second Law of Motion that

$$F(x(t)) = ma(t) = mv'(t)$$

Assume that

$$x(t_0) = a \quad \text{and} \quad x(t_1) = b$$

with

$$v(t_0) = v_i \quad \text{and} \quad v(t_1) = v_f$$

the initial and final velocities of the object, respectively. Then

$$W = \int_a^b F(x)\,dx = \int_{x(t_0)}^{x(t_1)} F(x)\,dx$$

$$= \int_{t_0}^{t_1} F(x(t))x'(t)\,dt \qquad \boxed{\text{By Theorem 5.9.1 with } x = x(t),\, dx = x'(t)\,dt}$$

$$= \int_{t_0}^{t_1} mv'(t)v(t)\,dt = \int_{t_0}^{t_1} mv(t)v'(t)\,dt$$

$$= \int_{v(t_0)}^{v(t_1)} mv\,dv \qquad \boxed{\text{By Theorem 5.9.1 with } v = v(t),\, dv = v'(t)\,dt}$$

$$= \int_{v_i}^{v_f} mv\,dv = \tfrac{1}{2}mv^2\Big|_{v_i}^{v_f} = \tfrac{1}{2}mv_f^2 - \tfrac{1}{2}mv_i^2$$

We see from the equation

$$W = \tfrac{1}{2}mv_f^2 - \tfrac{1}{2}mv_i^2 \tag{6}$$

that the work done on the object is equal to the change in the quantity $\tfrac{1}{2}mv^2$ from its initial value to its final value. We will refer to Equation (6) as the **work–energy relationship**. If we define the "energy of motion" or **kinetic energy** of our object to be given by

$$K = \tfrac{1}{2}mv^2 \tag{7}$$

then Equation (6) tells us that the work done on an object is equal to the *change* in the object's kinetic energy. Loosely speaking, we may think of work done on an object as being "transformed" into kinetic energy of the object. The units of kinetic energy are the same as the units of work. For example, in SI kinetic energy is measured in joules (J).

▶ **Example 6** A space probe of mass $m = 5.00 \times 10^4$ kg travels in deep space subjected only to the force of its own engine. Starting at a time when the speed of the probe is $v = 1.10 \times 10^4$ m/s, the engine is fired continuously over a distance of 2.50×10^6 m with a constant force of 4.00×10^5 N in the direction of motion. What is the final speed of the probe?

Solution.   Since the force applied by the engine is constant and in the direction of motion, the work W expended by the engine on the probe is

$$W = \text{force} \times \text{distance} = (4.00 \times 10^5 \text{ N}) \times (2.50 \times 10^6 \text{ m}) = 1.00 \times 10^{12} \text{ J}$$

From (6), the final kinetic energy $K_f = \tfrac{1}{2}mv_f^2$ of the probe can be expressed in terms of the work W and the initial kinetic energy $K_i = \tfrac{1}{2}mv_i^2$ as

$$K_f = W + K_i$$

Thus, from the known mass and initial speed we have

$$K_f = (1.00 \times 10^{12} \text{ J}) + \tfrac{1}{2}(5.00 \times 10^4 \text{ kg})(1.10 \times 10^4 \text{ m/s})^2 = 4.025 \times 10^{12} \text{ J}$$

The final kinetic energy is $K_f = \tfrac{1}{2}mv_f^2$, so the final speed of the probe is

$$v_f = \sqrt{\frac{2K_f}{m}} = \sqrt{\frac{2(4.025 \times 10^{12})}{5.00 \times 10^4}} \approx 1.27 \times 10^4 \text{ m/s} \quad \blacktriangleleft$$

✔ **QUICK CHECK EXERCISES 6.6**   *(See page 458 for answers.)*

1. If a constant force of 5 lb moves an object 10 ft, then the work done by the force on the object is _____.

2. A newton-meter is also called a _____. A dyne-centimeter is also called an _____.

3. Suppose that an object moves in the positive direction along a coordinate line over the interval $[a, b]$. The work per-

formed on the object by a variable force $F(x)$ applied in the direction of motion is $W = $ _____.

4. A force $F(x) = 10 - 2x$ N applied in the positive x-direction moves an object 3 m from $x = 2$ to $x = 5$. The work done by the force on the object is _____.

EXERCISE SET 6.6

FOCUS ON CONCEPTS

1. A variable force $F(x)$ in the positive x-direction is graphed in the accompanying figure. Find the work done by the force on a particle that moves from $x = 0$ to $x = 3$.

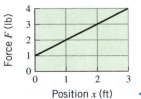
Position x (ft)   ◀ **Figure Ex-1**

2. A variable force $F(x)$ in the positive x-direction is graphed in the accompanying figure. Find the work done by the force on a particle that moves from $x = 0$ to $x = 5$.

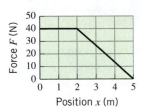
Position x (m)   ◀ **Figure Ex-2**

3. For the variable force $F(x)$ in Exercise 2, consider the distance d for which the work done by the force on the particle when the particle moves from $x = 0$ to $x = d$ is half of the work done when the particle moves from $x = 0$ to $x = 5$. By inspecting the graph of F, is d more or less than 2.5? Explain, and then find the exact value of d.

4. Suppose that a variable force $F(x)$ is applied in the positive x-direction so that an object moves from $x = a$ to $x = b$. Relate the work done by the force on the object and the average value of F over $[a, b]$, and illustrate this relationship graphically.

5. A constant force of 10 lb in the positive x-direction is applied to a particle whose velocity versus time curve is shown in the accompanying figure. Find the work done by the force on the particle from time $t = 0$ to $t = 5$.

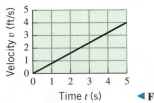
Time t (s)   ◀ **Figure Ex-5**

6. A spring exerts a force of 6 N when it is stretched from its natural length of 4 m to a length of $4\frac{1}{2}$ m. Find the work required to stretch the spring from its natural length to a length of 6 m.

7. A spring exerts a force of 100 N when it is stretched 0.2 m beyond its natural length. How much work is required to stretch the spring 0.8 m beyond its natural length?

8. A spring whose natural length is 15 cm exerts a force of 45 N when stretched to a length of 20 cm.
   (a) Find the spring constant (in newtons/meter).
   (b) Find the work that is done in stretching the spring 3 cm beyond its natural length.
   (c) Find the work done in stretching the spring from a length of 20 cm to a length of 25 cm.

9. Assume that 10 ft·lb of work is required to stretch a spring 1 ft beyond its natural length. What is the spring constant?

10–13 True–False Determine whether the statement is true or false. Explain your answer. ■

10. In order to support the weight of a parked automobile, the surface of a driveway must do work against the force of gravity on the vehicle.

11. A force of 10 lb in the direction of motion of an object that moves 5 ft in 2 s does six times the work of a force of 10 lb in the direction of motion of an object that moves 5 ft in 12 s.

12. It follows from Hooke's law that in order to double the distance a spring is stretched beyond its natural length, four times as much work is required.

13. In the International System of Units, work and kinetic energy have the same units.

14. A cylindrical tank of radius 5 ft and height 9 ft is two-thirds filled with water. Find the work required to pump all the water over the upper rim.

15. Solve Exercise 14 assuming that the tank is half-filled with water.

16. A cone-shaped water reservoir is 20 ft in diameter across the top and 15 ft deep. If the reservoir is filled to a depth of 10 ft, how much work is required to pump all the water to the top of the reservoir?

17. The vat shown in the accompanying figure contains water to a depth of 2 m. Find the work required to pump all the water to the top of the vat. [Use 9810 N/m^3 as the weight density of water.]

18. The cylindrical tank shown in the accompanying figure is filled with a liquid weighing 50 lb/ft^3. Find the work required to pump all the liquid to a level 1 ft above the top of the tank.

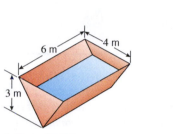

▲ Figure Ex-17

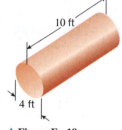

▲ Figure Ex-18

19. A swimming pool is built in the shape of a rectangular parallelepiped 10 ft deep, 15 ft wide, and 20 ft long.
   (a) If the pool is filled to 1 ft below the top, how much work is required to pump all the water into a drain at the top edge of the pool?
   (b) A one-horsepower motor can do 550 ft·lb of work per second. What size motor is required to empty the pool in 1 hour?

20. How much work is required to fill the swimming pool in Exercise 19 to 1 ft below the top if the water is pumped in through an opening located at the bottom of the pool?

21. A 100 ft length of steel chain weighing 15 lb/ft is dangling from a pulley. How much work is required to wind the chain onto the pulley?

22. A 3 lb bucket containing 20 lb of water is hanging at the end of a 20 ft rope that weighs 4 oz/ft. The other end of the rope is attached to a pulley. How much work is required to wind the length of rope onto the pulley, assuming that the rope is wound onto the pulley at a rate of 2 ft/s and that as the bucket is being lifted, water leaks from the bucket at a rate of 0.5 lb/s?

23. A rocket weighing 3 tons is filled with 40 tons of liquid fuel. In the initial part of the flight, fuel is burned off at a constant rate of 2 tons per 1000 ft of vertical height. How much work in foot-tons (ft·ton) is done lifting the rocket 3000 ft?

24. It follows from Coulomb's law in physics that two like electrostatic charges repel each other with a force inversely proportional to the square of the distance between them. Suppose that two charges A and B repel with a force of k newtons when they are positioned at points $A(-a, 0)$ and $B(a, 0)$, where a is measured in meters. Find the work W required to move charge A along the x-axis to the origin if charge B remains stationary.

25. It is a law of physics that the gravitational force exerted by the Earth on an object above the Earth's surface varies inversely as the square of its distance from the Earth's center. Thus, an object's weight $w(x)$ is related to its distance x from the Earth's center by a formula of the form

$$w(x) = \frac{k}{x^2}$$

where k is a constant of proportionality that depends on the mass of the object.
   (a) Use this fact and the assumption that the Earth is a sphere of radius 4000 mi to obtain the formula for $w(x)$ in Example 4.
   (b) Find a formula for the weight $w(x)$ of a satellite that is x mi from the Earth's surface if its weight on Earth is 6000 lb.
   (c) How much work is required to lift the satellite from the surface of the Earth to an orbital position that is 1000 mi high?

26. (a) The formula $w(x) = k/x^2$ in Exercise 25 is applicable to all celestial bodies. Assuming that the Moon is a sphere of radius 1080 mi, find the force that the Moon exerts on an astronaut who is x mi from the surface of the Moon if her weight on the Moon's surface is 20 lb.
   (b) How much work is required to lift the astronaut to a point that is 10.8 mi above the Moon's surface?

27. The world's first commercial high-speed magnetic levitation (MAGLEV) train, a 30 km double-track project connecting Shanghai, China, to Pudong International Airport, began full revenue service in 2003. Suppose that a MAGLEV train has a mass $m = 4.00 \times 10^5$ kg and that starting at a time when the train has a speed of 20 m/s the engine applies a force of 6.40×10^5 N in the direction of motion over a distance of 3.00×10^3 m. Use the work–energy relationship (6) to find the final speed of the train.

28. Assume that a Mars probe of mass $m = 2.00 \times 10^3$ kg is subjected only to the force of its own engine. Starting at a time when the speed of the probe is $v = 1.00 \times 10^4$ m/s, the engine is fired continuously over a distance of 1.50×10^5 m with a constant force of 2.00×10^5 N in the direction of motion. Use the work–energy relationship (6) to find the final speed of the probe.

29. On August 10, 1972 a meteorite with an estimated mass of 4×10^6 kg and an estimated speed of 15 km/s skipped across the atmosphere above the western United States and Canada but fortunately did not hit the Earth. *(cont.)*

(a) Assuming that the meteorite had hit the Earth with a speed of 15 km/s, what would have been its change in kinetic energy in joules (J)?

(b) Express the energy as a multiple of the explosive energy of 1 megaton of TNT, which is 4.2×10^{15} J.

(c) The energy associated with the Hiroshima atomic bomb was 13 kilotons of TNT. To how many such bombs would the meteorite impact have been equivalent?

30. Writing After reading Examples 3–5, a student classifies work problems as either "pushing/pulling" or "pumping."

Describe these categories in your own words and discuss the methods used to solve each type. Give examples to illustrate that these categories are not mutually exclusive.

31. Writing How might you recognize that a problem can be solved by means of the work–energy relationship? That is, what sort of "givens" and "unknowns" would suggest such a solution? Discuss two or three examples.

✔ **QUICK CHECK ANSWERS 6.6**

1. 50 ft·lb   **2.** joule; erg   **3.** $\displaystyle\int_a^b F(x)\,dx$ **4.** 9 J

6.7   MOMENTS, CENTERS OF GRAVITY, AND CENTROIDS

*Suppose that a rigid physical body is acted on by a constant gravitational field. Because the body is composed of many particles, each of which is affected by gravity, the action of the gravitational field on the body consists of a large number of forces distributed over the entire body. However, it is a fact of physics that these individual forces can be replaced by a single force acting at a point called the **center of gravity** of the body. In this section we will show how integrals can be used to locate centers of gravity.*

■ DENSITY AND MASS OF A LAMINA

Let us consider an idealized flat object that is thin enough to be viewed as a two-dimensional plane region (Figure 6.7.1). Such an object is called a **lamina**. A lamina is called **homogeneous** if its composition is uniform throughout and **inhomogeneous** otherwise. We will consider homogeneous laminas in this section. Inhomogeneous laminas will be discussed in Chapter 14. The **density** of a *homogeneous* lamina is defined to be its mass per unit area. Thus, the density δ of a homogeneous lamina of mass M and area A is given by $\delta = M/A$. Notice that the mass M of a homogeneous lamina can be expressed as

$$M = \delta A \tag{1}$$

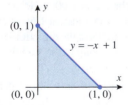

The thickness of a lamina is negligible.

▲ **Figure 6.7.1**

The units in Equation (1) are consistent since mass = (mass/area) × area.

▶ **Example 1**   A triangular lamina with vertices (0, 0), (0, 1), and (1, 0) has density $\delta = 3$. Find its total mass.

Solution.   Referring to (1) and Figure 6.7.2, the mass M of the lamina is

$$M = \delta A = 3 \cdot \frac{1}{2} = \frac{3}{2} \text{ (unit of mass)} \blacktriangleleft$$

■ CENTER OF GRAVITY OF A LAMINA

Assume that the acceleration due to the force of gravity is constant and acts downward, and suppose that a lamina occupies a region R in a horizontal xy-plane. It can be shown that there exists a unique point $(\bar{x}, \bar{y})$ (which may or may not belong to R) such that the effect

▲ **Figure 6.7.2**

(0, 1)

$y = -x + 1$

(0, 0)   (1, 0)

of gravity on the lamina is "equivalent" to that of a single force acting at the point $(\bar{x}, \bar{y})$. This point is called the *center of gravity* of the lamina, and if it is in R, then the lamina will balance horizontally on the point of a support placed at $(\bar{x}, \bar{y})$. For example, the center of gravity of a homogeneous disk is at the center of the disk, and the center of gravity of a homogeneous rectangular region is at the center of the rectangle. For an irregularly shaped homogeneous lamina, locating the center of gravity requires calculus.

6.7.1   PROBLEM   Let f be a positive continuous function on the interval $[a, b]$. Suppose that a homogeneous lamina with constant density δ occupies a region R in a horizontal xy-plane bounded by the graphs of $y = f(x)$, $y = 0$, $x = a$, and $x = b$. Find the coordinates $(\bar{x}, \bar{y})$ of the center of gravity of the lamina.

To motivate the solution, consider what happens if we try to balance the lamina on a knife-edge parallel to the x-axis. Suppose the lamina in Figure 6.7.3 is placed on a knife-edge along a line $y = c$ that does not pass through the center of gravity. Because the lamina behaves as if its entire mass is concentrated at the center of gravity $(\bar{x}, \bar{y})$, the lamina will be rotationally unstable and the force of gravity will cause a rotation about $y = c$. Similarly, the lamina will undergo a rotation if placed on a knife-edge along $y = d$. However, if the knife-edge runs along the line $y = \bar{y}$ through the center of gravity, the lamina will be in perfect balance. Similarly, the lamina will be in perfect balance on a knife-edge along the line $x = \bar{x}$ through the center of gravity. This suggests that the center of gravity of a lamina can be determined as the intersection of two lines of balance, one parallel to the x-axis and the other parallel to the y-axis. In order to find these lines of balance, we will need some preliminary results about rotations.

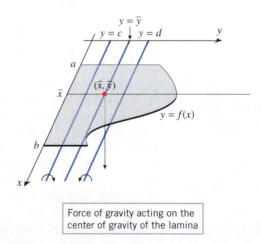

▶ **Figure 6.7.3**

Force of gravity acting on the center of gravity of the lamina

Children on a seesaw learn by experience that a lighter child can balance a heavier one by sitting farther from the fulcrum or pivot point. This is because the tendency for an object to produce rotation is proportional not only to its mass but also to the distance between the object and the fulcrum. To make this more precise, consider an x-axis, which we view as a weightless beam. If a mass m is located on the axis at x, then the tendency for that mass to produce a rotation of the beam about a point a on the axis is measured by the following quantity, called the *moment of m about x = a*:

$$\begin{bmatrix} \text{moment of } m \\ \text{about } a \end{bmatrix} = m(x - a)$$

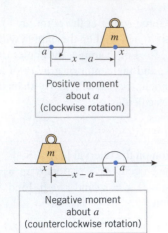

Positive moment
about a
(clockwise rotation)

Negative moment
about a
(counterclockwise rotation)

▲ **Figure 6.7.4**

The number $x - a$ is called the **lever arm**. Depending on whether the mass is to the right or left of a, the lever arm is either the distance between x and a or the negative of this distance (Figure 6.7.4). Positive lever arms result in positive moments and clockwise rotations, and negative lever arms result in negative moments and counterclockwise rotations.

Suppose that masses $m_1, m_2, \ldots, m_n$ are located at $x_1, x_2, \ldots, x_n$ on a coordinate axis and a fulcrum is positioned at the point a (Figure 6.7.5). Depending on whether the sum of the moments about a,

$$\sum_{k=1}^{n} m_k(x_k - a) = m_1(x_1 - a) + m_2(x_2 - a) + \cdots + m_n(x_n - a)$$

is positive, negative, or zero, a weightless beam along the axis will rotate clockwise about a, rotate counterclockwise about a, or balance perfectly. In the last case, the system of masses is said to be in **equilibrium**.

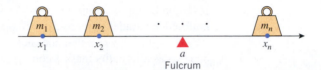

▶ **Figure 6.7.5**

The preceding ideas can be extended to masses distributed in two-dimensional space. If we imagine the xy-plane to be a weightless sheet supporting a mass m located at a point (x, y), then the tendency for the mass to produce a rotation of the sheet about the line $x = a$ is $m(x - a)$, called the **moment of m about x = a**, and the tendency for the mass to produce a rotation about the line $y = c$ is $m(y - c)$, called the **moment of m about y = c** (Figure 6.7.6). In summary,

$$\begin{bmatrix} \text{moment of } m \\ \text{about the} \\ \text{line } x = a \end{bmatrix} = m(x - a) \quad \text{and} \quad \begin{bmatrix} \text{moment of } m \\ \text{about the} \\ \text{line } y = c \end{bmatrix} = m(y - c) \quad (2\text{-}3)$$

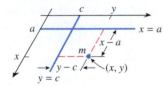

▲ **Figure 6.7.6**

If a number of masses are distributed throughout the xy-plane, then the plane (viewed as a weightless sheet) will balance on a knife-edge along the line $x = a$ if the sum of the moments about the line is zero. Similarly, the plane will balance on a knife-edge along the line $y = c$ if the sum of the moments about that line are zero.

We are now ready to solve Problem 6.7.1. The basic idea for solving this problem is to divide the lamina into strips whose areas may be approximated by the areas of rectangles. These area approximations, along with Formulas (2) and (3), will allow us to create a Riemann sum that approximates the moment of the lamina about a horizontal or vertical line. By taking the limit of Riemann sums we will then obtain an integral for the moment of a lamina about a horizontal or vertical line. We observe that since the lamina balances on the lines $x = \bar{x}$ and $y = \bar{y}$, the moment of the lamina about those lines should be zero. This observation will enable us to calculate $\bar{x}$ and $\bar{y}$.

To implement this idea, we divide the interval $[a, b]$ into n subintervals by inserting the points $x_1, x_2, \ldots, x_{n-1}$ between $a = x_0$ and $b = x_n$. This has the effect of dividing the lamina R into n strips $R_1, R_2, \ldots, R_n$ (Figure 6.7.7a). Suppose that the kth strip extends from x_{k-1} to x_k and that the width of this strip is

$$\Delta x_k = x_k - x_{k-1}$$

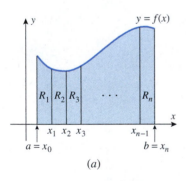

(a)

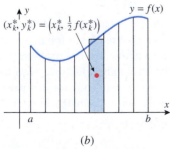

(b)

▲ **Figure 6.7.7**

We will let x_k^* be the midpoint of the kth subinterval and we will approximate R_k by a rectangle of width Δx_k and height $f(x_k^*)$. From (1), the mass ΔM_k of this rectangle is $\Delta M_k = \delta f(x_k^*)\Delta x_k$, and we will assume that the rectangle behaves as if its entire mass is concentrated at its center $(x_k^*, y_k^*) = (x_k^*, \frac{1}{2} f(x_k^*))$ (Figure 6.7.7b). It then follows from (2) and (3) that the moments of R_k about the lines $x = \bar{x}$ and $y = \bar{y}$ may be approximated by $(x_k^* - \bar{x})\Delta M_k$ and $(y_k^* - \bar{y})\Delta M_k$, respectively. Adding these approximations yields the following Riemann sums that approximate the moment of the entire lamina about the lines $x = \bar{x}$ and $y = \bar{y}$:

$$\sum_{k=1}^{n}(x_k^* - \bar{x})\Delta M_k = \sum_{k=1}^{n}(x_k^* - \bar{x})\delta f(x_k^*)\Delta x_k$$

$$\sum_{k=1}^{n}(y_k^* - \bar{y})\Delta M_k = \sum_{k=1}^{n}\left(\frac{f(x_k^*)}{2} - \bar{y}\right)\delta f(x_k^*)\Delta x_k$$

Taking the limits as n increases and the widths of all the rectangles approach zero yields the definite integrals

$$\int_a^b (x - \bar{x})\delta f(x)\,dx \quad \text{and} \quad \int_a^b \left(\frac{f(x)}{2} - \bar{y}\right)\delta f(x)\,dx$$

that represent the moments of the lamina about the lines $x = \bar{x}$ and $y = \bar{y}$. Since the lamina balances on those lines, the moments of the lamina about those lines should be zero:

$$\int_a^b (x - \bar{x})\delta f(x)\,dx = \int_a^b \left(\frac{f(x)}{2} - \bar{y}\right)\delta f(x)\,dx = 0$$

Since $\bar{x}$ and $\bar{y}$ are constant, these equations can be rewritten as

$$\int_a^b \delta x f(x)\,dx = \bar{x}\int_a^b \delta f(x)\,dx$$

$$\int_a^b \frac{1}{2}\delta(f(x))^2\,dx = \bar{y}\int_a^b \delta f(x)\,dx$$

from which we obtain the following formulas for the center of gravity of the lamina:

> **Center of Gravity $(\bar{x}, \bar{y})$ of a Lamina**
>
> $$\bar{x} = \frac{\displaystyle\int_a^b \delta x f(x)\,dx}{\displaystyle\int_a^b \delta f(x)\,dx}, \qquad \bar{y} = \frac{\displaystyle\int_a^b \frac{1}{2}\delta\,(f(x))^2\,dx}{\displaystyle\int_a^b \delta f(x)\,dx} \qquad (4\text{--}5)$$

Observe that in both formulas the denominator is the mass M of the lamina. The numerator in the formula for $\bar{x}$ is denoted by M_y and is called the *first moment of the lamina about the y-axis*; the numerator of the formula for $\bar{y}$ is denoted by M_x and is called the *first moment of the lamina about the x-axis*. Thus, we can write (4) and (5) as

> **Alternative Formulas for Center of Gravity $(\bar{x}, \bar{y})$ of a Lamina**
>
> $$\bar{x} = \frac{M_y}{M} = \frac{1}{\text{mass of } R}\int_a^b \delta x f(x)\,dx \qquad (6)$$
>
> $$\bar{y} = \frac{M_x}{M} = \frac{1}{\text{mass of } R}\int_a^b \frac{1}{2}\delta\,(f(x))^2\,dx \qquad (7)$$

▶ **Example 2** Find the center of gravity of the triangular lamina with vertices $(0, 0)$, $(0, 1)$, and $(1, 0)$ and density $\delta = 3$.

Solution. The lamina is shown in Figure 6.7.2. In Example 1 we found the mass of the lamina to be

$$M = \frac{3}{2}$$

The moment of the lamina about the y-axis is

$$M_y = \int_0^1 \delta x f(x)\, dx = \int_0^1 3x(-x+1)\, dx$$

$$= \int_0^1 (-3x^2 + 3x)\, dx = \left(-x^3 + \frac{3}{2}x^2\right)\Bigg]_0^1 = -1 + \frac{3}{2} = \frac{1}{2}$$

and the moment about the x-axis is

$$M_x = \int_0^1 \frac{1}{2}\delta(f(x))^2\, dx = \int_0^1 \frac{3}{2}(-x+1)^2\, dx$$

$$= \int_0^1 \frac{3}{2}(x^2 - 2x + 1)\, dx = \frac{3}{2}\left(\frac{1}{3}x^3 - x^2 + x\right)\Bigg]_0^1 = \frac{3}{2}\left(\frac{1}{3}\right) = \frac{1}{2}$$

From (6) and (7),

$$\bar{x} = \frac{M_y}{M} = \frac{1/2}{3/2} = \frac{1}{3}, \qquad \bar{y} = \frac{M_x}{M} = \frac{1/2}{3/2} = \frac{1}{3}$$

so the center of gravity is $(\frac{1}{3}, \frac{1}{3})$. ◀

In the case of a *homogeneous* lamina, the center of gravity of a lamina occupying the region R is called the ***centroid of the region R***. Since the lamina is homogeneous, δ is constant. The factor δ in (4) and (5) may thus be moved through the integral signs and canceled, and (4) and (5) can be expressed as

Since the density factor has canceled, we may interpret the centroid as a *geometric property* of the region, and distinguish it from the center of gravity, which is a *physical property* of an idealized object that occupies the region.

> ### Centroid of a Region R
>
> $$\bar{x} = \frac{\displaystyle\int_a^b x f(x)\, dx}{\displaystyle\int_a^b f(x)\, dx} = \frac{1}{\text{area of } R}\int_a^b x f(x)\, dx \tag{8}$$
>
> $$\bar{y} = \frac{\displaystyle\int_a^b \frac{1}{2}(f(x))^2\, dx}{\displaystyle\int_a^b f(x)\, dx} = \frac{1}{\text{area of } R}\int_a^b \frac{1}{2}(f(x))^2\, dx \tag{9}$$

▲ **Figure 6.7.8**

▶ **Example 3** Find the centroid of the semicircular region in Figure 6.7.8.

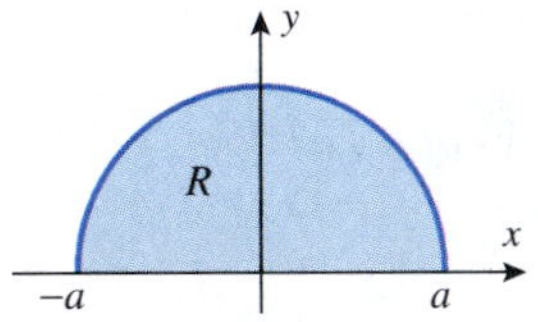

Solution. By symmetry, $\bar{x} = 0$ since the y-axis is obviously a line of balance. To find $\bar{y}$, first note that the equation of the semicircle is $y = f(x) = \sqrt{a^2 - x^2}$. From (9),

$$\bar{y} = \frac{1}{\text{area of } R}\int_{-a}^a \frac{1}{2}(f(x))^2\, dx = \frac{1}{\frac{1}{2}\pi a^2}\int_{-a}^a \frac{1}{2}(a^2 - x^2)\, dx$$

$$= \frac{1}{\pi a^2}\left(a^2 x - \frac{1}{3}x^3\right)\Bigg]_{-a}^a$$

$$= \frac{1}{\pi a^2}\left[\left(a^3 - \frac{1}{3}a^3\right) - \left(-a^3 + \frac{1}{3}a^3\right)\right]$$

$$= \frac{1}{\pi a^2}\left(\frac{4a^3}{3}\right) = \frac{4a}{3\pi}$$

so the centroid is $(0, 4a/3\pi)$. ◀

■ OTHER TYPES OF REGIONS

The strategy used to find the center of gravity of the region in Problem 6.7.1 can be used to find the center of gravity of regions that are not of that form.

Consider a homogeneous lamina that occupies the region R between two continuous functions $f(x)$ and $g(x)$ over the interval $[a, b]$, where $f(x) \geq g(x)$ for $a \leq x \leq b$. To find the center of gravity of this lamina we can subdivide it into n strips using lines parallel to the x-axis. If x_k^* is the midpoint of the kth strip, the strip can be approximated by a rectangle of width Δx_k and height $f(x_k^*) - g(x_k^*)$. We assume that the entire mass of the kth rectangle is concentrated at its center $(x_k^*, y_k^*) = (x_k^*, \frac{1}{2}(f(x_k^*) + g(x_k^*)))$ (Figure 6.7.9). Continuing the argument as in the solution of Problem 6.7.1, we find that the center of gravity of the lamina is

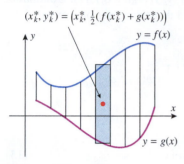

$(x_k^*, y_k^*) = \left(x_k^*, \frac{1}{2}(f(x_k^*) + g(x_k^*))\right)$

$y = f(x)$

$y = g(x)$

▲ **Figure 6.7.9**

$$\bar{x} = \frac{\displaystyle\int_a^b x(f(x) - g(x))\, dx}{\displaystyle\int_a^b (f(x) - g(x))\, dx} = \frac{1}{\text{area of } R} \int_a^b x(f(x) - g(x))\, dx \tag{10}$$

$$\bar{y} = \frac{\displaystyle\int_a^b \frac{1}{2}\left([f(x)]^2 - [g(x)]^2\right) dx}{\displaystyle\int_a^b (f(x) - g(x))\, dx} = \frac{1}{\text{area of } R} \int_a^b \frac{1}{2}\left([f(x)]^2 - [g(x)]^2\right) dx \tag{11}$$

Note that the density of the lamina does not appear in Equations (10) and (11). This reflects the fact that the centroid is a geometric property of R.

▶ **Example 4**   Find the centroid of the region R enclosed between the curves $y = x^2$ and $y = x + 6$.

Solution.   To begin, we note that the two curves intersect when $x = -2$ and $x = 3$ and that $x + 6 \geq x^2$ over that interval (Figure 6.7.10). The area of R is

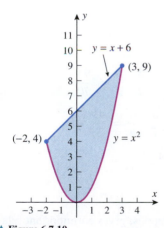

$y = x + 6$

$(3, 9)$

$(-2, 4)$ $y = x^2$

▲ **Figure 6.7.10**

$$\int_{-2}^{3} [(x + 6) - x^2]\, dx = \frac{125}{6}$$

From (10) and (11),

$$\bar{x} = \frac{1}{\text{area of } R} \int_{-2}^{3} x[(x + 6) - x^2]\, dx$$

$$= \frac{6}{125} \left(\frac{1}{3}x^3 + 3x^2 - \frac{1}{4}x^4\right)\Bigg]_{-2}^{3}$$

$$= \frac{6}{125} \cdot \frac{125}{12} = \frac{1}{2}$$

and

$$\bar{y} = \frac{1}{\text{area of } R} \int_{-2}^{3} \frac{1}{2}((x + 6)^2 - (x^2)^2)\, dx$$

$$= \frac{6}{125} \int_{-2}^{3} \frac{1}{2}(x^2 + 12x + 36 - x^4)\, dx$$

$$= \frac{6}{125} \cdot \frac{1}{2} \left(\frac{1}{3}x^3 + 6x^2 + 36x - \frac{1}{5}x^5\right)\Bigg]_{-2}^{3}$$

$$= \frac{6}{125} \cdot \frac{250}{3} = 4$$

so the centroid of R is $(\frac{1}{2}, 4)$. ◀

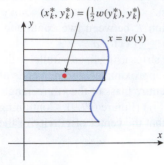

$(x_k^*, y_k^*) = \left(\tfrac{1}{2}w(y_k^*), y_k^*\right)$

$x = w(y)$

▲ **Figure 6.7.11**

Suppose that w is a continuous function of y on an interval $[c, d]$ with $w(y) \geq 0$ for $c \leq y \leq d$. Consider a lamina that occupies a region R bounded above by $y = d$, below by $y = c$, on the left by the y-axis, and on the right by $x = w(y)$ (Figure 6.7.11). To find the center of gravity of this lamina, we note that the roles of x and y in Problem 6.7.1 have been reversed. We now imagine the lamina to be subdivided into n strips using lines parallel to the x-axis. We let y_k^* be the midpoint of the kth subinterval and approximate the strip by a rectangle of width Δy_k and height $w(y_k^*)$. We assume that the entire mass of the kth rectangle is concentrated at its center $(x_k^*, y_k^*) = (\tfrac{1}{2}w(y_k^*), y_k^*)$ (Figure 6.7.11). Continuing the argument as in the solution of Problem 6.7.1, we find that the center of gravity of the lamina is

$$\bar{x} = \frac{\displaystyle\int_c^d \frac{1}{2}(w(y))^2 \, dy}{\displaystyle\int_c^d w(y) \, dy} = \frac{1}{\text{area of } R} \int_c^d \frac{1}{2}(w(y))^2 \, dy \tag{12}$$

$$\bar{y} = \frac{\displaystyle\int_c^d yw(y) \, dy}{\displaystyle\int_c^d w(y) \, dy} = \frac{1}{\text{area of } R} \int_c^d yw(y) \, dy \tag{13}$$

Once again, the absence of the density in Equations (12) and (13) reflects the geometric nature of the centroid.

▶ **Example 5**   Find the centroid of the region R enclosed between the curves $y = \sqrt{x}$, $y = 1$, $y = 2$, and the y-axis (Figure 6.7.12).

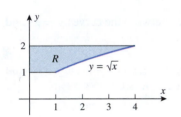

R

$y = \sqrt{x}$

▲ **Figure 6.7.12**

Solution.   Note that $x = w(y) = y^2$ and that the area of R is

$$\int_1^2 y^2 \, dy = \frac{7}{3}$$

From (12) and (13),

$$\bar{x} = \frac{1}{\text{area of } R} \int_1^2 \frac{1}{2}(y^2)^2 \, dy = \frac{3}{7} \cdot \frac{1}{10}y^5 \bigg]_1^2 = \frac{3}{7} \cdot \frac{31}{10} = \frac{93}{70}$$

$$\bar{y} = \frac{1}{\text{area of } R} \int_1^2 y(y^2) \, dy = \frac{3}{7} \cdot \frac{1}{4}y^4 \bigg]_1^2 = \frac{3}{7} \cdot \frac{15}{4} = \frac{45}{28}$$

so the centroid of R is $(93/70, 45/28) \approx (1.329, 1.607)$. ◀

THEOREM OF PAPPUS

The following theorem, due to the Greek mathematician Pappus, gives an important relationship between the centroid of a plane region R and the volume of the solid generated when the region is revolved about a line.

6.7.2   THEOREM (*Theorem of Pappus*)   *If R is a bounded plane region and L is a line that lies in the plane of R such that R is entirely on one side of L, then the volume of the solid formed by revolving R about L is given by*

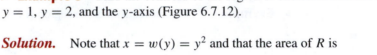

$$volume = (area\ of\ R) \cdot \binom{distance\ traveled}{by\ the\ centroid}$$

PROOF We prove this theorem in the special case where L is the y-axis, the region R is in the first quadrant, and the region R is of the form given in Problem 6.7.1. (A more general proof will be outlined in the Exercises of Section 14.8.) In this case, the volume V of the solid formed by revolving R about L can be found by the method of cylindrical shells (Section 6.3) to be

$$V = 2\pi \int_a^b x f(x)\, dx$$

Thus, it follows from (8) that

$$V = 2\pi \bar{x}[\text{area of } R]$$

This completes the proof since $2\pi \bar{x}$ is the distance traveled by the centroid when R is revolved about the y-axis. ∎

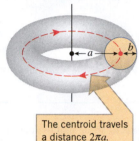

The centroid travels a distance $2\pi a$.

▲ **Figure 6.7.13**

▶ **Example 6** Use Pappus' Theorem to find the volume V of the torus generated by revolving a circular region of radius b about a line at a distance a (greater than b) from the center of the circle (Figure 6.7.13).

Solution. By symmetry, the centroid of a circular region is its center. Thus, the distance traveled by the centroid is $2\pi a$. Since the area of a circle of radius b is πb^2, it follows from Pappus' Theorem that the volume of the torus is

$$V = (2\pi a)(\pi b^2) = 2\pi^2 a b^2 \quad ◀$$

✔**QUICK CHECK EXERCISES 6.7**   *(See page 467 for answers.)*

1. The total mass of a homogeneous lamina of area A and density δ is _____.

2. A homogeneous lamina of mass M and density δ occupies a region in the xy-plane bounded by the graphs of $y = f(x)$, $y = 0, x = a$, and $x = b$, where f is a nonnegative continuous function defined on an interval $[a, b]$. The x-coordinate of the center of gravity of the lamina is M_y/M, where M_y is called the _____ and is given by the integral _____.

3. Let R be the region between the graphs of $y = x^2$ and $y = 2 - x$ for $0 \le x \le 1$. The area of R is $\frac{7}{6}$ and the centroid of R is _____.

4. If the region R in Quick Check Exercise 3 is used to generate a solid G by rotating R about a horizontal line 6 units above its centroid, then the volume of G is _____.

EXERCISE SET 6.7   boxed[C] CAS

FOCUS ON CONCEPTS

1. Masses $m_1 = 5$, $m_2 = 10$, and $m_3 = 20$ are positioned on a weightless beam as shown in the accompanying figure.

   (a) Suppose that the fulcrum is positioned at $x = 5$. Without computing the sum of moments about 5, determine whether the sum is positive, zero, or negative. Explain.

   (b) Where should the fulcrum be placed so that the beam is in equilibrium?

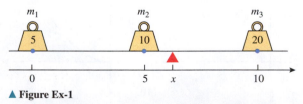

▲ **Figure Ex-1**

Pappus of Alexandria (**4th century A.D.**) Greek mathematician. Pappus lived during the early Christian era when mathematical activity was in a period of decline. His main contributions to mathematics appeared in a series of eight books called *The Collection* (written about 340 A.D.). This work, which survives only partially, contained some original results but was devoted mostly to statements, refinements, and proofs of results by earlier mathematicians. Pappus' Theorem, stated without proof in Book VII of *The Collection*, was probably known and proved in earlier times. This result is sometimes called Guldin's Theorem in recognition of the Swiss mathematician, Paul Guldin (1577–1643), who rediscovered it independently.

2. Masses $m_1 = 10$, $m_2 = 3$, $m_3 = 4$, and m are positioned on a weightless beam, with the fulcrum positioned at point 4, as shown in the accompanying figure.
   (a) Suppose that $m = 14$. Without computing the sum of the moments about 4, determine whether the sum is positive, zero, or negative. Explain.
   (b) For what value of m is the beam in equilibrium?

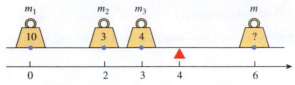

▲ **Figure Ex-2**

3–6 Find the centroid of the region by inspection and confirm your answer by integrating. ■

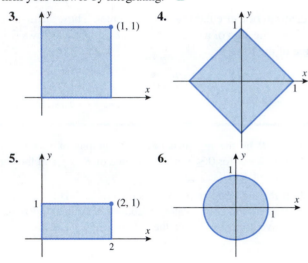

3.

4.

5.

6.

7–20 Find the centroid of the region. ■

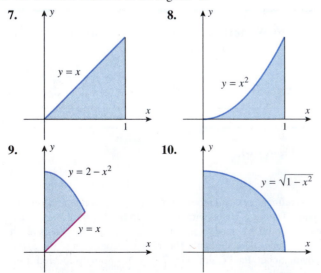

7.

8.

$y = x$

$y = x^2$

9.

10.

$y = 2 - x^2$

$y = \sqrt{1 - x^2}$

$y = x$

11. The triangle with vertices $(0, 0)$, $(2, 0)$, and $(0, 1)$.

12. The triangle with vertices $(0, 0)$, $(1, 1)$, and $(2, 0)$.

13. The region bounded by the graphs of $y = x^2$ and $x + y = 6$.

14. The region bounded on the left by the y-axis, on the right by the line $x = 2$, below by the parabola $y = x^2$, and above by the line $y = x + 6$.

15. The region bounded by the graphs of $y = x^2$ and $y = x + 2$.

16. The region bounded by the graphs of $y = x^2$ and $y = 1$.

17. The region bounded by the graphs of $y = \sqrt{x}$ and $y = x^2$.

18. The region bounded by the graphs of $x = 1/y$, $x = 0$, $y = 1$, and $y = 2$.

19. The region bounded by the graphs of $y = x$, $x = 1/y^2$, and $y = 2$.

20. The region bounded by the graphs of $xy = 4$ and $x + y = 5$.

FOCUS ON CONCEPTS

21. Use symmetry considerations to argue that the centroid of an isosceles triangle lies on the median to the base of the triangle.

22. Use symmetry considerations to argue that the centroid of an ellipse lies at the intersection of the major and minor axes of the ellipse.

23–26 Find the mass and center of gravity of the lamina with density δ. ■

23. A lamina bounded by the x-axis, the line $x = 1$, and the curve $y = \sqrt{x}$; $\delta = 2$.

24. A lamina bounded by the graph of $x = y^4$ and the line $x = 1$; $\delta = 15$.

25. A lamina bounded by the graph of $y = |x|$ and the line $y = 1$; $\delta = 3$.

26. A lamina bounded by the x-axis and the graph of the equation $y = 1 - x^2$; $\delta = 3$.

c **27–30** Use a CAS to find the mass and center of gravity of the lamina with density δ. ■

27. A lamina bounded by $y = \sin x$, $y = 0$, $x = 0$, and $x = \pi$; $\delta = 4$.

28. A lamina bounded by $y = e^x$, $y = 0$, $x = 0$, and $x = 1$; $\delta = 1/(e - 1)$.

29. A lamina bounded by the graph of $y = \ln x$, the x-axis, and the line $x = 2$; $\delta = 1$.

30. A lamina bounded by the graphs of $y = \cos x$, $y = \sin x$, $x = 0$, and $x = \pi/4$; $\delta = 1 + \sqrt{2}$.

31–34 True–False Determine whether the statement is true or false. Explain your answer. [In Exercise 34, assume that the (rotated) square lies in the xy-plane to the right of the y-axis.]

■

31. The centroid of a rectangle is the intersection of the diagonals of the rectangle.

32. The centroid of a rhombus is the intersection of the diagonals of the rhombus.

33. The centroid of an equilateral triangle is the intersection of the medians of the triangle.

34. By rotating a square about its center, it is possible to change the volume of the solid of revolution generated by revolving the square about the y-axis.

35. Find the centroid of the triangle with vertices $(0, 0)$, (a, b), and $(a, -b)$.

36. Prove that the centroid of a triangle is the point of intersection of the three medians of the triangle. [*Hint:* Choose coordinates so that the vertices of the triangle are located at $(0, -a)$, $(0, a)$, and (b, c).]

37. Find the centroid of the isosceles trapezoid with vertices $(-a, 0)$, $(a, 0)$, $(-b, c)$, and (b, c).

38. Prove that the centroid of a parallelogram is the point of intersection of the diagonals of the parallelogram. [*Hint:* Choose coordinates so that the vertices of the parallelogram are located at $(0, 0)$, $(0, a)$, (b, c), and $(b, a + c)$.]

39. Use the Theorem of Pappus and the fact that the volume of a sphere of radius a is $V = \frac{4}{3}\pi a^3$ to show that the centroid of the lamina that is bounded by the x-axis and the semicircle $y = \sqrt{a^2 - x^2}$ is $(0, 4a/(3\pi))$. (This problem was solved directly in Example 3.)

40. Use the Theorem of Pappus and the result of Exercise 39 to find the volume of the solid generated when the region

bounded by the x-axis and the semicircle $y = \sqrt{a^2 - x^2}$ is revolved about

  (a) the line $y = -a$ (b) the line $y = x - a$.

41. Use the Theorem of Pappus and the fact that the area of an ellipse with semiaxes a and b is πab to find the volume of the elliptical torus generated by revolving the ellipse

$$\frac{(x - k)^2}{a^2} + \frac{y^2}{b^2} = 1$$

about the y-axis. Assume that $k > a$.

42. Use the Theorem of Pappus to find the volume of the solid that is generated when the region enclosed by $y = x^2$ and $y = 8 - x^2$ is revolved about the x-axis.

43. Use the Theorem of Pappus to find the centroid of the triangular region with vertices $(0, 0)$, $(a, 0)$, and $(0, b)$, where $a > 0$ and $b > 0$. [*Hint:* Revolve the region about the x-axis to obtain $\bar{y}$ and about the y-axis to obtain $\bar{x}$.]

44. **Writing** Suppose that a region R in the plane is decomposed into two regions R_1 and R_2 whose areas are A_1 and A_2, respectively, and whose centroids are $(\bar{x}_1, \bar{y}_1)$ and $(\bar{x}_2, \bar{y}_2)$, respectively. Investigate the problem of expressing the centroid of R in terms of A_1, A_2, $(\bar{x}_1, \bar{y}_1)$, and $(\bar{x}_2, \bar{y}_2)$. Write a short report on your investigations, supporting your reasoning with plausible arguments. Can you extend your results to decompositions of R into more than two regions?

45. **Writing** How might you recognize that a problem can be solved by means of the Theorem of Pappus? That is, what sort of "givens" and "unknowns" would suggest such a solution? Discuss two or three examples.

✔ **QUICK CHECK ANSWERS 6.7**

1. δA **2.** first moment about the y-axis; $\displaystyle\int_a^b \delta x f(x)\, dx$ **3.** $\left(\dfrac{5}{14}, \dfrac{32}{35}\right)$ **4.** 14π

6.8 FLUID PRESSURE AND FORCE

In this section we will use the integration tools developed in the preceding chapter to study the pressures and forces exerted by fluids on submerged objects.

WHAT IS A FLUID?

A *fluid* is a substance that flows to conform to the boundaries of any container in which it is placed. Fluids include *liquids*, such as water, oil, and mercury, as well as *gases*, such as helium, oxygen, and air. The study of fluids falls into two categories: *fluid statics* (the study of fluids at rest) and *fluid dynamics* (the study of fluids in motion). In this section we will be concerned only with fluid statics; toward the end of this text we will investigate problems in fluid dynamics.

■ THE CONCEPT OF PRESSURE

Jupiter Images Corp.

Snowshoes prevent the woman from sinking by spreading her weight over a large area to reduce her pressure on the snow.

The effect that a force has on an object depends on how that force is spread over the surface of the object. For example, when you walk on soft snow with boots, the weight of your body crushes the snow and you sink into it. However, if you put on a pair of snowshoes to spread the weight of your body over a greater surface area, then the weight of your body has less of a crushing effect on the snow. The concept that accounts for both the magnitude of a force and the area over which it is applied is called *pressure*.

> **6.8.1 DEFINITION** If a force of magnitude F is applied to a surface of area A, then we define the ***pressure*** P exerted by the force on the surface to be
>
> $$P = \frac{F}{A} \tag{1}$$

It follows from this definition that pressure has units of force per unit area. The most common units of pressure are newtons per square meter (N/m^2) in SI and pounds per square inch (lb/in^2) or pounds per square foot (lb/ft^2) in the BE system. As indicated in Table 6.8.1, one newton per square meter is called a *pascal* (Pa). A pressure of 1 Pa is quite small ($1\ Pa = 1.45 \times 10^{-4}\ lb/in^2$), so in countries using SI, tire pressure gauges are usually calibrated in kilopascals (kPa), which is 1000 pascals.

Table 6.8.1

UNITS OF FORCE AND PRESSURE

SYSTEM	FORCE	÷	AREA	=	PRESSURE
SI	newton (N)		square meter (m^2)		pascal (Pa)
BE	pound (lb)		square foot (ft^2)		lb/ft^2
BE	pound (lb)		square inch (in^2)		lb/in^2 (psi)

CONVERSION FACTORS:

$1\ Pa \approx 1.45 \times 10^{-4}\ lb/in^2 \approx 2.09 \times 10^{-2}\ lb/ft^2$

$1\ lb/in^2 \approx 6.89 \times 10^3\ Pa \qquad 1\ lb/ft^2 \approx 47.9\ Pa$

Blaise Pascal (1623–1662) French mathematician and scientist. Pascal's mother died when he was three years old and his father, a highly educated magistrate, personally provided the boy's early education. Although Pascal showed an inclination for science and mathematics, his father refused to tutor him in those subjects until he mastered Latin and Greek. Pascal's sister and primary biographer claimed that he independently discovered the first thirty-two propositions of Euclid without ever reading a book on geometry. (However, it is generally agreed that the story is apocryphal.) Nevertheless, the precocious Pascal published a highly respected essay on conic sections by the time he was sixteen years old. Descartes, who read the essay, thought it so brilliant that he could not believe that it was written by such a young man. By age 18 his health began to fail and until his death he was in frequent pain. However, his creativity was unimpaired.

Pascal's contributions to physics include the discovery that air pressure decreases with altitude and the principle of fluid pressure that bears his name. However, the originality of his work is questioned by some historians. Pascal made major contributions to a branch of mathematics called "projective geometry," and he helped to develop probability theory through a series of letters with Fermat.

In 1646, Pascal's health problems resulted in a deep emotional crisis that led him to become increasingly concerned with religious matters. Although born a Catholic, he converted to a religious doctrine called Jansenism and spent most of his final years writing on religion and philosophy.

In this section we will be interested in pressures and forces on objects submerged in fluids. Pressures themselves have no directional characteristics, but the forces that they create always act perpendicular to the face of the submerged object. Thus, in Figure 6.8.1 the water pressure creates horizontal forces on the sides of the tank, vertical forces on the bottom of the tank, and forces that vary in direction, so as to be perpendicular to the different parts of the swimmer's body.

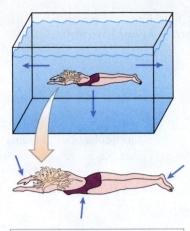

Fluid forces always act perpendicular to the surface of a submerged object.

▲ **Figure 6.8.1**

▶ **Example 1**   Referring to Figure 6.8.1, suppose that the back of the swimmer's hand has a surface area of 8.4×10^{-3} m^2 and that the pressure acting on it is 5.1×10^4 Pa (a realistic value near the bottom of a deep diving pool). Find the force that acts on the swimmer's hand.

Solution.   From (1), the force F is

$$F = PA = (5.1 \times 10^4 \text{ N/m}^2)(8.4 \times 10^{-3} \text{ m}^2) \approx 4.3 \times 10^2 \text{ N}$$

This is quite a large force (nearly 100 lb in the BE system). ◀

■ **FLUID DENSITY**

Scuba divers know that the pressure and forces on their bodies increase with the depth they dive. This is caused by the weight of the water and air above—the deeper the diver goes, the greater the weight above and so the greater the pressure and force exerted on the diver.

To calculate pressures and forces on submerged objects, we need to know something about the characteristics of the fluids in which they are submerged. For simplicity, we will assume that the fluids under consideration are *homogeneous*, by which we mean that any two samples of the fluid with the same volume have the same mass. It follows from this assumption that the mass per unit volume is a constant δ that depends on the physical characteristics of the fluid but not on the size or location of the sample; we call

Table 6.8.2

WEIGHT DENSITIES

SI	N/m^3
Machine oil	4708
Gasoline	6602
Fresh water	9810
Seawater	10,045
Mercury	133,416

BE SYSTEM	lb/ft^3
Machine oil	30.0
Gasoline	42.0
Fresh water	62.4
Seawater	64.0
Mercury	849.0

All densities are affected by variations in temperature and pressure. Weight densities are also affected by variations in g.

$$\delta = \frac{m}{V} \tag{2}$$

the *mass density* of the fluid. Sometimes it is more convenient to work with weight per unit volume than with mass per unit volume. Thus, we define the *weight density* ρ of a fluid to be

$$\rho = \frac{w}{V} \tag{3}$$

where w is the weight of a fluid sample of volume V. Thus, if the weight density of a fluid is known, then the weight w of a fluid sample of volume V can be computed from the formula $w = \rho V$. Table 6.8.2 shows some typical weight densities.

■ **FLUID PRESSURE**

To calculate fluid pressures and forces we will need to make use of an experimental observation. Suppose that a flat surface of area A is submerged in a homogeneous fluid of weight density ρ such that the entire surface lies between depths h_1 and h_2, where $h_1 \leq h_2$ (Figure 6.8.2). Experiments show that on both sides of the surface, the fluid exerts a force that is perpendicular to the surface and whose magnitude F satisfies the inequalities

$$\rho h_1 A \leq F \leq \rho h_2 A \tag{4}$$

Thus, it follows from (1) that the pressure $P = F/A$ on a given side of the surface satisfies the inequalities

$$\rho h_1 \leq P \leq \rho h_2 \tag{5}$$

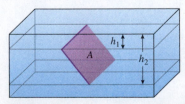

▲ **Figure 6.8.2**

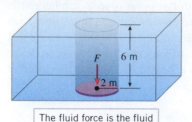

The fluid force is the fluid pressure times the area.

▲ **Figure 6.8.3**

Note that it is now a straightforward matter to calculate fluid force and pressure on a flat surface that is submerged *horizontally* at depth h, for then $h = h_1 = h_2$ and inequalities (4) and (5) become the *equalities*

$$F = \rho h A \qquad (6)$$

and

$$P = \rho h \qquad (7)$$

▶ **Example 2**   Find the fluid pressure and force on the top of a flat circular plate of radius 2 m that is submerged horizontally in water at a depth of 6 m (Figure 6.8.3).

Solution.   Since the weight density of water is $\rho = 9810 \text{ N/m}^3$, it follows from (7) that the fluid pressure is

$$P = \rho h = (9810)(6) = 58{,}860 \text{ Pa}$$

and it follows from (6) that the fluid force is

$$F = \rho h A = \rho h (\pi r^2) = (9810)(6)(4\pi) = 235{,}440\pi \approx 739{,}700 \text{ N} \blacktriangleleft$$

■ **FLUID FORCE ON A VERTICAL SURFACE**

It was easy to calculate the fluid force on the horizontal plate in Example 2 because each point on the plate was at the same depth. The problem of finding the fluid force on a vertical surface is more complicated because the depth, and hence the pressure, is not constant over the surface. To find the fluid force on a vertical surface we will need calculus.

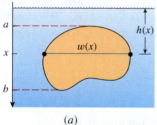

(a)

6.8.2   PROBLEM   Suppose that a flat surface is immersed vertically in a fluid of weight density ρ and that the submerged portion of the surface extends from $x = a$ to $x = b$ along an x-axis whose positive direction is down (Figure 6.8.4*a*). For $a \leq x \leq b$, suppose that $w(x)$ is the width of the surface and that $h(x)$ is the depth of the point x. Define what is meant by the *fluid force F* on the surface, and find a formula for computing it.

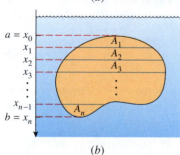

(b)

The basic idea for solving this problem is to divide the surface into horizontal strips whose areas may be approximated by areas of rectangles. These area approximations, along with inequalities (4), will allow us to create a Riemann sum that approximates the total force on the surface. By taking a limit of Riemann sums we will then obtain an integral for F.

To implement this idea, we divide the interval $[a, b]$ into n subintervals by inserting the points $x_1, x_2, \ldots, x_{n-1}$ between $a = x_0$ and $b = x_n$. This has the effect of dividing the surface into n strips of area A_k, $k = 1, 2, \ldots, n$ (Figure 6.8.4*b*). It follows from (4) that the force F_k on the kth strip satisfies the inequalities

$$\rho h(x_{k-1})A_k \leq F_k \leq \rho h(x_k)A_k$$

or, equivalently,

$$h(x_{k-1}) \leq \frac{F_k}{\rho A_k} \leq h(x_k)$$

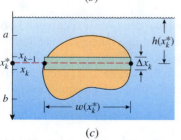

(c)

▲ **Figure 6.8.4**

Since the depth function $h(x)$ increases linearly, there must exist a point x_k^* between x_{k-1} and x_k such that

$$h(x_k^*) = \frac{F_k}{\rho A_k}$$

or, equivalently,

$$F_k = \rho h(x_k^*)A_k$$

We now approximate the area A_k of the kth strip of the surface by the area of a rectangle of width $w(x_k^*)$ and height $\Delta x_k = x_k - x_{k-1}$ (Figure 6.8.4c). It follows that F_k may be approximated as

$$F_k = \rho h(x_k^*)A_k \approx \rho h(x_k^*) \cdot \underbrace{w(x_k^*)\Delta x_k}_{\text{Area of rectangle}}$$

Adding these approximations yields the following Riemann sum that approximates the total force F on the surface:

$$F = \sum_{k=1}^{n} F_k \approx \sum_{k=1}^{n} \rho h(x_k^*)w(x_k^*)\Delta x_k$$

Taking the limit as n increases and the widths of all the subintervals approach zero yields the definite integral

$$F = \lim_{\max \Delta x_k \to 0} \sum_{k=1}^{n} \rho h(x_k^*)w(x_k^*)\Delta x_k = \int_a^b \rho h(x)w(x)\, dx$$

In summary, we have the following result.

6.8.3 DEFINITION Suppose that a flat surface is immersed vertically in a fluid of weight density ρ and that the submerged portion of the surface extends from $x = a$ to $x = b$ along an x-axis whose positive direction is down (Figure 6.8.4a). For $a \leq x \leq b$, suppose that $w(x)$ is the width of the surface and that $h(x)$ is the depth of the point x. Then we define the ***fluid force*** F on the surface to be

$$F = \int_a^b \rho h(x)w(x)\, dx \tag{8}$$

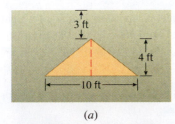

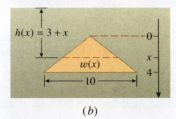

▲ **Figure 6.8.5**

▶ **Example 3** The face of a dam is a vertical rectangle of height 100 ft and width 200 ft (Figure 6.8.5a). Find the total fluid force exerted on the face when the water surface is level with the top of the dam.

Solution. Introduce an x-axis with its origin at the water surface as shown in Figure 6.8.5b. At a point x on this axis, the width of the dam in feet is $w(x) = 200$ and the depth in feet is $h(x) = x$. Thus, from (8) with $\rho = 62.4$ lb/ft^3 (the weight density of water) we obtain as the total force on the face

$$F = \int_0^{100} (62.4)(x)(200)\, dx = 12{,}480 \int_0^{100} x\, dx$$

$$= 12{,}480 \left. \frac{x^2}{2} \right]_0^{100} = 62{,}400{,}000 \text{ lb} \blacktriangleleft$$

▶ **Example 4** A plate in the form of an isosceles triangle with base 10 ft and altitude 4 ft is submerged vertically in machine oil as shown in Figure 6.8.6a. Find the fluid force F against the plate surface if the oil has weight density $\rho = 30$ lb/ft^3.

Solution. Introduce an x-axis as shown in Figure 6.8.6b. By similar triangles, the width of the plate, in feet, at a depth of $h(x) = (3 + x)$ ft satisfies

$$\frac{w(x)}{10} = \frac{x}{4}, \quad \text{so} \quad w(x) = \frac{5}{2}x$$

▲ **Figure 6.8.6**

Thus, it follows from (8) that the force on the plate is

$$F = \int_a^b \rho h(x) w(x)\, dx = \int_0^4 (30)(3+x)\left(\frac{5}{2}x\right) dx$$

$$= 75 \int_0^4 (3x + x^2)\, dx = 75\left[\frac{3x^2}{2} + \frac{x^3}{3}\right]_0^4 = 3400\ \text{lb} \blacktriangleleft$$

✔ QUICK CHECK EXERCISES 6.8   *(See page 473 for answers.)*

1. The pressure unit equivalent to a newton per square meter (N/m²) is called a _____. The pressure unit psi stands for _____.

2. Given that the weight density of water is 9810 N/m³, the fluid pressure on a rectangular 2 m × 3 m flat plate submerged horizontally in water at a depth of 10 m is _____. The fluid force on the plate is _____.

3. Suppose that a flat surface is immersed vertically in a fluid of weight density ρ and that the submerged portion of the surface extends from $x = a$ to $x = b$ along an x-axis whose positive direction is down. If, for $a \le x \le b$, the surface has width $w(x)$ and depth $h(x)$, then the fluid force on the surface is $F =$ _____.

4. A rectangular plate 2 m wide and 3 m high is submerged vertically in water so that the top of the plate is 5 m below the water surface. An integral expression for the force of the water on the plate surface is $F =$ _____.

EXERCISE SET 6.8

In this exercise set, refer to Table 6.8.2 for weight densities of fluids, where needed. ■

1. A flat rectangular plate is submerged horizontally in water.
   (a) Find the force (in lb) and the pressure (in lb/ft²) on the top surface of the plate if its area is 100 ft² and the surface is at a depth of 5 ft.
   (b) Find the force (in N) and the pressure (in Pa) on the top surface of the plate if its area is 25 m² and the surface is at a depth of 10 m.

2. (a) Find the force (in N) on the deck of a sunken ship if its area is 160 m² and the pressure acting on it is 6.0×10^5 Pa.
   (b) Find the force (in lb) on a diver's face mask if its area is 60 in² and the pressure acting on it is 100 lb/in².

3–8 The flat surfaces shown are submerged vertically in water. Find the fluid force against each surface. ■

3.
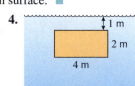
2 ft
4 ft

4.
1 m
2 m
4 m

5.
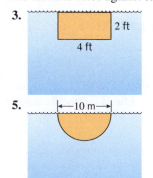
|←10 m→|

6.
|← 4 ft →|
4 ft   4 ft

7.
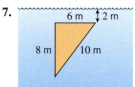
6 m   2 m
8 m   10 m

8.
4 ft   4 ft
8 ft
16 ft

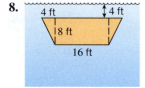

9. Suppose that a flat surface is immersed vertically in a fluid of weight density ρ. If ρ is doubled, is the force on the plate also doubled? Explain your reasoning.

10. An oil tank is shaped like a right circular cylinder of diameter 4 ft. Find the total fluid force against one end when the axis is horizontal and the tank is half filled with oil of weight density 50 lb/ft³.

11. A square plate of side a feet is dipped in a liquid of weight density ρ lb/ft³. Find the fluid force on the plate if a vertex is at the surface and a diagonal is perpendicular to the surface.

12–15 True–False Determine whether the statement is true or false. Explain your answer. ■

12. In the International System of Units, pressure and force have the same units.

13. In a cylindrical water tank (with vertical axis), the fluid force on the base of the tank is equal to the weight of water in the tank.

14. In a rectangular water tank, the fluid force on any side of the tank must be less than the fluid force on the base of the tank.

15. In any water tank with a flat base, no matter what the shape of the tank, the fluid force on the base is at most equal to the weight of water in the tank.

16–19 Formula (8) gives the fluid force on a flat surface immersed vertically in a fluid. More generally, if a flat surface is immersed so that it makes an angle of $0 \le \theta < \pi/2$ with the vertical, then the fluid force on the surface is given by

$$F = \int_a^b \rho h(x) w(x) \sec \theta \, dx$$

Use this formula in these exercises. ■

16. Derive the formula given above for the fluid force on a flat surface immersed at an angle in a fluid.

17. The accompanying figure shows a rectangular swimming pool whose bottom is an inclined plane. Find the fluid force on the bottom when the pool is filled to the top.

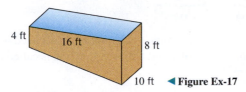

4 ft   16 ft   8 ft   10 ft   ◄ **Figure Ex-17**

18. By how many feet should the water in the pool of Exercise 17 be lowered in order for the force on the bottom to be reduced by a factor of $\frac{1}{2}$?

19. The accompanying figure shows a dam whose face is an inclined rectangle. Find the fluid force on the face when the water is level with the top of this dam.

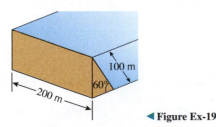

100 m   60°   200 m   ◄ **Figure Ex-19**

20. An observation window on a submarine is a square with 2 ft sides. Using ρ_0 for the weight density of seawater, find the fluid force on the window when the submarine has descended so that the window is vertical and its top is at a depth of h feet.

21. (a) Show: If the submarine in Exercise 20 descends vertically at a constant rate, then the fluid force on the window increases at a constant rate.
   (b) At what rate is the force on the window increasing if the submarine is descending vertically at 20 ft/min?

22. (a) Let $D = D_a$ denote a disk of radius a submerged in a fluid of weight density ρ such that the center of D is h units below the surface of the fluid. For each value of r in the interval $(0, a]$, let D_r denote the disk of radius r that is concentric with D. Select a side of the disk D and define $P(r)$ to be the fluid pressure on the chosen side of D_r. Use (5) to prove that

$$\lim_{r \to 0^+} P(r) = \rho h$$

   (b) Explain why the result in part (a) may be interpreted to mean that *fluid pressure at a given depth is the same in all directions*. (This statement is one version of a result known as *Pascal's Principle*.)

23. **Writing** Suppose that we model the Earth's atmosphere as a "fluid." Atmospheric pressure at sea level is $P = 14.7$ lb/in^2 and the weight density of air at sea level is about $\rho = 4.66 \times 10^{-5}$ lb/in^3. With these numbers, what would Formula (7) yield as the height of the atmosphere above the Earth? Do you think this answer is reasonable? If not, explain how we might modify our assumptions to yield a more plausible answer.

24. **Writing** Suppose that the weight density ρ of a fluid is a function $\rho = \rho(x)$ of the depth x within the fluid. How do you think that Formula (7) for fluid pressure will need to be modified? Support your answer with plausible arguments.

✔ **QUICK CHECK ANSWERS 6.8**

1. pascal; pounds per square inch   **2.** 98,100 Pa; 588,600 N   **3.** $\int_a^b \rho h(x) w(x) \, dx$ **4.** $\int_0^3 9810 \, [(5 + x)2] \, dx$

6.9 **HYPERBOLIC FUNCTIONS AND HANGING CABLES**

In this section we will study certain combinations of e^x and e^{-x}, called "hyperbolic functions." These functions, which arise in various engineering applications, have many properties in common with the trigonometric functions. This similarity is somewhat surprising, since there is little on the surface to suggest that there should be any relationship between exponential and trigonometric functions. This is because the relationship occurs within the context of complex numbers, a topic which we will leave for more advanced courses.

■ DEFINITIONS OF HYPERBOLIC FUNCTIONS

To introduce the hyperbolic functions, observe from Exercise 61 in Section 0.2 that the function e^x can be expressed in the following way as the sum of an even function and an odd function:

$$e^x = \underbrace{\frac{e^x + e^{-x}}{2}}_{\text{Even}} + \underbrace{\frac{e^x - e^{-x}}{2}}_{\text{Odd}}$$

These functions are sufficiently important that there are names and notation associated with them: the odd function is called the *hyperbolic sine* of x and the even function is called the *hyperbolic cosine* of x. They are denoted by

$$\sinh x = \frac{e^x - e^{-x}}{2} \quad \text{and} \quad \cosh x = \frac{e^x + e^{-x}}{2}$$

where sinh is pronounced "cinch" and cosh rhymes with "gosh." From these two building blocks we can create four more functions to produce the following set of six *hyperbolic functions*.

6.9.1 **DEFINITION**

Hyperbolic sine	$\sinh x = \dfrac{e^x - e^{-x}}{2}$
Hyperbolic cosine	$\cosh x = \dfrac{e^x + e^{-x}}{2}$
Hyperbolic tangent	$\tanh x = \dfrac{\sinh x}{\cosh x} = \dfrac{e^x - e^{-x}}{e^x + e^{-x}}$
Hyperbolic cotangent	$\coth x = \dfrac{\cosh x}{\sinh x} = \dfrac{e^x + e^{-x}}{e^x - e^{-x}}$
Hyperbolic secant	$\operatorname{sech} x = \dfrac{1}{\cosh x} = \dfrac{2}{e^x + e^{-x}}$
Hyperbolic cosecant	$\operatorname{csch} x = \dfrac{1}{\sinh x} = \dfrac{2}{e^x - e^{-x}}$

The terms "tanh," "sech," and "csch" are pronounced "tanch," "seech," and "coseech," respectively.

TECHNOLOGY MASTERY

Computer algebra systems have built-in capabilities for evaluating hyperbolic functions directly, but some calculators do not. However, if you need to evaluate a hyperbolic function on a calculator, you can do so by expressing it in terms of exponential functions, as in Example 1.

▶ **Example 1**

$$\sinh 0 = \frac{e^0 - e^{-0}}{2} = \frac{1 - 1}{2} = 0$$

$$\cosh 0 = \frac{e^0 + e^{-0}}{2} = \frac{1 + 1}{2} = 1$$

$$\sinh 2 = \frac{e^2 - e^{-2}}{2} \approx 3.6269 \quad ◀$$

■ GRAPHS OF THE HYPERBOLIC FUNCTIONS

The graphs of the hyperbolic functions, which are shown in Figure 6.9.1, can be generated with a graphing utility, but it is worthwhile to observe that the general shape of the graph of $y = \cosh x$ can be obtained by sketching the graphs of $y = \frac{1}{2}e^x$ and $y = \frac{1}{2}e^{-x}$ separately and adding the corresponding y-coordinates [see part (a) of the figure]. Similarly, the general shape of the graph of $y = \sinh x$ can be obtained by sketching the graphs of $y = \frac{1}{2}e^x$ and $y = -\frac{1}{2}e^{-x}$ separately and adding corresponding y-coordinates [see part (b) of the figure].

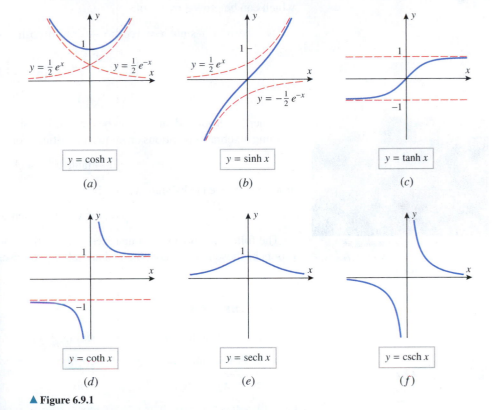

▲ Figure 6.9.1

Glen Allison/Stone/Getty Images

The design of the Gateway Arch near St. Louis is based on an inverted hyperbolic cosine curve (Exercise 73).

Observe that $\sinh x$ has a domain of $(-\infty, +\infty)$ and a range of $(-\infty, +\infty)$, whereas $\cosh x$ has a domain of $(-\infty, +\infty)$ and a range of $[1, +\infty)$. Observe also that $y = \frac{1}{2}e^x$ and $y = \frac{1}{2}e^{-x}$ are *curvilinear asymptotes* for $y = \cosh x$ in the sense that the graph of $y = \cosh x$ gets closer and closer to the graph of $y = \frac{1}{2}e^x$ as $x \to +\infty$ and gets closer and closer to the graph of $y = \frac{1}{2}e^{-x}$ as $x \to -\infty$. (See Section 4.3.) Similarly, $y = \frac{1}{2}e^x$ is a curvilinear asymptote for $y = \sinh x$ as $x \to +\infty$ and $y = -\frac{1}{2}e^{-x}$ is a curvilinear asymptote as $x \to -\infty$. Other properties of the hyperbolic functions are explored in the exercises.

■ HANGING CABLES AND OTHER APPLICATIONS

Hyperbolic functions arise in vibratory motions inside elastic solids and more generally in many problems where mechanical energy is gradually absorbed by a surrounding medium. They also occur when a homogeneous, flexible cable is suspended between two points, as with a telephone line hanging between two poles. Such a cable forms a curve, called a *catenary* (from the Latin *catena*, meaning "chain"). If, as in Figure 6.9.2, a coordinate system is introduced so that the low point of the cable lies on the y-axis, then it can be shown using principles of physics that the cable has an equation of the form

$$y = a \cosh\left(\frac{x}{a}\right) + c$$

▲ **Figure 6.9.2**

where the parameters a and c are determined by the distance between the poles and the composition of the cable.

■ HYPERBOLIC IDENTITIES

The hyperbolic functions satisfy various identities that are similar to identities for trigonometric functions. The most fundamental of these is

$$\cosh^2 x - \sinh^2 x = 1 \qquad (1)$$

which can be proved by writing

$$\cosh^2 x - \sinh^2 x = (\cosh x + \sinh x)(\cosh x - \sinh x)$$

$$= \left(\frac{e^x + e^{-x}}{2} + \frac{e^x - e^{-x}}{2} \right) \left(\frac{e^x + e^{-x}}{2} - \frac{e^x - e^{-x}}{2} \right)$$

$$= e^x \cdot e^{-x} = 1$$

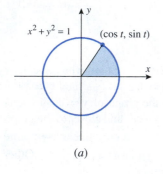

Larry Auippy/Mira.com/Digital Railroad, Inc.

A flexible cable suspended between two poles forms a catenary.

Other hyperbolic identities can be derived in a similar manner or, alternatively, by performing algebraic operations on known identities. For example, if we divide (1) by $\cosh^2 x$, we obtain

$$1 - \tanh^2 x = \operatorname{sech}^2 x$$

and if we divide (1) by $\sinh^2 x$, we obtain

$$\coth^2 x - 1 = \operatorname{csch}^2 x$$

The following theorem summarizes some of the more useful hyperbolic identities. The proofs of those not already obtained are left as exercises.

6.9.2   THEOREM

$\cosh x + \sinh x = e^x$	$\sinh(x + y) = \sinh x \cosh y + \cosh x \sinh y$
$\cosh x - \sinh x = e^{-x}$	$\cosh(x + y) = \cosh x \cosh y + \sinh x \sinh y$
$\cosh^2 x - \sinh^2 x = 1$	$\sinh(x - y) = \sinh x \cosh y - \cosh x \sinh y$
$1 - \tanh^2 x = \operatorname{sech}^2 x$	$\cosh(x - y) = \cosh x \cosh y - \sinh x \sinh y$
$\coth^2 x - 1 = \operatorname{csch}^2 x$	$\sinh 2x = 2 \sinh x \cosh x$
$\cosh(-x) = \cosh x$	$\cosh 2x = \cosh^2 x + \sinh^2 x$
$\sinh(-x) = -\sinh x$	$\cosh 2x = 2 \sinh^2 x + 1 = 2 \cosh^2 x - 1$

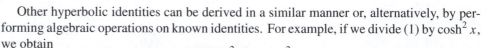

(a)

■ WHY THEY ARE CALLED HYPERBOLIC FUNCTIONS

Recall that the parametric equations

$$x = \cos t, \quad y = \sin t \qquad (0 \leq t \leq 2\pi)$$

represent the unit circle $x^2 + y^2 = 1$ (Figure 6.9.3a), as may be seen by writing

$$x^2 + y^2 = \cos^2 t + \sin^2 t = 1$$

If $0 \leq t \leq 2\pi$, then the parameter t can be interpreted as the angle in radians from the positive x-axis to the point $(\cos t, \sin t)$ or, alternatively, as twice the shaded area of the sector in Figure 6.9.3a (verify). Analogously, the parametric equations

$$x = \cosh t, \quad y = \sinh t \qquad (-\infty < t < +\infty)$$

(b)

▲ **Figure 6.9.3**

represent a portion of the curve $x^2 - y^2 = 1$, as may be seen by writing

$$x^2 - y^2 = \cosh^2 t - \sinh^2 t = 1$$

and observing that $x = \cosh t > 0$. This curve, which is shown in Figure 6.9.3b, is the right half of a larger curve called the ***unit hyperbola***; this is the reason why the functions in this section are called *hyperbolic* functions. It can be shown that if $t \geq 0$, then the parameter t can be interpreted as twice the shaded area in Figure 6.9.3b. (We omit the details.)

■ **DERIVATIVE AND INTEGRAL FORMULAS**

Derivative formulas for $\sinh x$ and $\cosh x$ can be obtained by expressing these functions in terms of e^x and e^{-x}:

$$\frac{d}{dx}[\sinh x] = \frac{d}{dx}\left[\frac{e^x - e^{-x}}{2}\right] = \frac{e^x + e^{-x}}{2} = \cosh x$$

$$\frac{d}{dx}[\cosh x] = \frac{d}{dx}\left[\frac{e^x + e^{-x}}{2}\right] = \frac{e^x - e^{-x}}{2} = \sinh x$$

Derivatives of the remaining hyperbolic functions can be obtained by expressing them in terms of sinh and cosh and applying appropriate identities. For example,

$$\frac{d}{dx}[\tanh x] = \frac{d}{dx}\left[\frac{\sinh x}{\cosh x}\right] = \frac{\cosh x \dfrac{d}{dx}[\sinh x] - \sinh x \dfrac{d}{dx}[\cosh x]}{\cosh^2 x}$$

$$= \frac{\cosh^2 x - \sinh^2 x}{\cosh^2 x} = \frac{1}{\cosh^2 x} = \text{sech}^2 x$$

The following theorem provides a complete list of the generalized derivative formulas and corresponding integration formulas for the hyperbolic functions.

6.9.3 THEOREM

$$\frac{d}{dx}[\sinh u] = \cosh u \frac{du}{dx} \qquad \int \cosh u \, du = \sinh u + C$$

$$\frac{d}{dx}[\cosh u] = \sinh u \frac{du}{dx} \qquad \int \sinh u \, du = \cosh u + C$$

$$\frac{d}{dx}[\tanh u] = \text{sech}^2 u \frac{du}{dx} \qquad \int \text{sech}^2 u \, du = \tanh u + C$$

$$\frac{d}{dx}[\coth u] = -\text{csch}^2 u \frac{du}{dx} \qquad \int \text{csch}^2 u \, du = -\coth u + C$$

$$\frac{d}{dx}[\text{sech } u] = -\text{sech } u \tanh u \frac{du}{dx} \qquad \int \text{sech } u \tanh u \, du = -\text{sech } u + C$$

$$\frac{d}{dx}[\text{csch } u] = -\text{csch } u \coth u \frac{du}{dx} \qquad \int \text{csch } u \coth u \, du = -\text{csch } u + C$$

▶ **Example 2**

$$\frac{d}{dx}[\cosh(x^3)] = \sinh(x^3) \cdot \frac{d}{dx}[x^3] = 3x^2 \sinh(x^3)$$

$$\frac{d}{dx}[\ln(\tanh x)] = \frac{1}{\tanh x} \cdot \frac{d}{dx}[\tanh x] = \frac{\text{sech}^2 x}{\tanh x} \qquad ◀$$

▶ **Example 3**

$$\int \sinh^5 x \cosh x \, dx = \tfrac{1}{6} \sinh^6 x + C \qquad \boxed{\begin{array}{l} u = \sinh x \\ du = \cosh x \, dx \end{array}}$$

$$\int \tanh x \, dx = \int \frac{\sinh x}{\cosh x} \, dx$$

$$= \ln |\cosh x| + C \qquad \boxed{\begin{array}{l} u = \cosh x \\ du = \sinh x \, dx \end{array}}$$

$$= \ln(\cosh x) + C$$

We were justified in dropping the absolute value signs since $\cosh x > 0$ for all x. ◀

▶ **Example 4** A 100 ft wire is attached at its ends to the tops of two 50 ft poles that are positioned 90 ft apart. How high above the ground is the middle of the wire?

Solution. From above, the wire forms a catenary curve with equation

$$y = a \cosh\left(\frac{x}{a}\right) + c$$

where the origin is on the ground midway between the poles. Using Formula (4) of Section 6.4 for the length of the catenary, we have

$$100 = \int_{-45}^{45} \sqrt{1 + \left(\frac{dy}{dx}\right)^2} \, dx$$

$$= 2 \int_{0}^{45} \sqrt{1 + \left(\frac{dy}{dx}\right)^2} \, dx \qquad \boxed{\begin{array}{l} \text{By symmetry} \\ \text{about the } y\text{-axis} \end{array}}$$

$$= 2 \int_{0}^{45} \sqrt{1 + \sinh^2\left(\frac{x}{a}\right)} \, dx$$

$$= 2 \int_{0}^{45} \cosh\left(\frac{x}{a}\right) \, dx \qquad \boxed{\begin{array}{l} \text{By (1) and the fact} \\ \text{that } \cosh x > 0 \end{array}}$$

$$= 2a \sinh\left(\frac{x}{a}\right) \Bigg]_{0}^{45} = 2a \sinh\left(\frac{45}{a}\right)$$

Using a calculating utility's numeric solver to solve

$$100 = 2a \sinh\left(\frac{45}{a}\right)$$

for a gives $a \approx 56.01$. Then

$$50 = y(45) = 56.01 \cosh\left(\frac{45}{56.01}\right) + c \approx 75.08 + c$$

so $c \approx -25.08$. Thus, the middle of the wire is $y(0) \approx 56.01 - 25.08 = 30.93$ ft above the ground (Figure 6.9.4). ◀

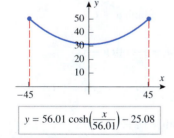

$$y = 56.01 \cosh\left(\frac{x}{56.01}\right) - 25.08$$

▲ **Figure 6.9.4**

INVERSES OF HYPERBOLIC FUNCTIONS

Referring to Figure 6.9.1, it is evident that the graphs of $\sinh x$, $\tanh x$, $\coth x$, and $\operatorname{csch} x$ pass the horizontal line test, but the graphs of $\cosh x$ and $\operatorname{sech} x$ do not. In the latter case, restricting x to be nonnegative makes the functions invertible (Figure 6.9.5). The graphs of the six inverse hyperbolic functions in Figure 6.9.6 were obtained by reflecting the graphs of the hyperbolic functions (with the appropriate restrictions) about the line $y = x$.

Table 6.9.1 summarizes the basic properties of the inverse hyperbolic functions. You should confirm that the domains and ranges listed in this table agree with the graphs in Figure 6.9.6.

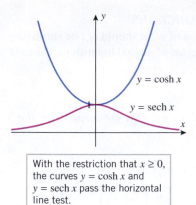

With the restriction that $x \geq 0$, the curves $y = \cosh x$ and $y = \operatorname{sech} x$ pass the horizontal line test.

▲ **Figure 6.9.5**

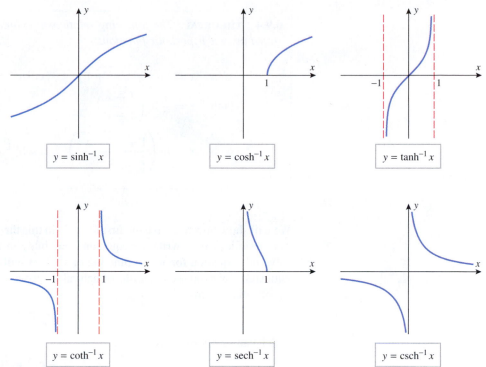

▶ **Figure 6.9.6**

Table 6.9.1
PROPERTIES OF INVERSE HYPERBOLIC FUNCTIONS

FUNCTION	DOMAIN	RANGE	BASIC RELATIONSHIPS
$\sinh^{-1} x$	$(-\infty, +\infty)$	$(-\infty, +\infty)$	$\sinh^{-1}(\sinh x) = x$ if $-\infty < x < +\infty$ $\sinh(\sinh^{-1} x) = x$ if $-\infty < x < +\infty$
$\cosh^{-1} x$	$[1, +\infty)$	$[0, +\infty)$	$\cosh^{-1}(\cosh x) = x$ if $x \geq 0$ $\cosh(\cosh^{-1} x) = x$ if $x \geq 1$
$\tanh^{-1} x$	$(-1, 1)$	$(-\infty, +\infty)$	$\tanh^{-1}(\tanh x) = x$ if $-\infty < x < +\infty$ $\tanh(\tanh^{-1} x) = x$ if $-1 < x < 1$
$\coth^{-1} x$	$(-\infty, -1) \cup (1, +\infty)$	$(-\infty, 0) \cup (0, +\infty)$	$\coth^{-1}(\coth x) = x$ if $x < 0$ or $x > 0$ $\coth(\coth^{-1} x) = x$ if $x < -1$ or $x > 1$
$\operatorname{sech}^{-1} x$	$(0, 1]$	$[0, +\infty)$	$\operatorname{sech}^{-1}(\operatorname{sech} x) = x$ if $x \geq 0$ $\operatorname{sech}(\operatorname{sech}^{-1} x) = x$ if $0 < x \leq 1$
$\operatorname{csch}^{-1} x$	$(-\infty, 0) \cup (0, +\infty)$	$(-\infty, 0) \cup (0, +\infty)$	$\operatorname{csch}^{-1}(\operatorname{csch} x) = x$ if $x < 0$ or $x > 0$ $\operatorname{csch}(\operatorname{csch}^{-1} x) = x$ if $x < 0$ or $x > 0$

■ **LOGARITHMIC FORMS OF INVERSE HYPERBOLIC FUNCTIONS**

Because the hyperbolic functions are expressible in terms of e^x, it should not be surprising that the inverse hyperbolic functions are expressible in terms of natural logarithms; the next theorem shows that this is so.

> **6.9.4** **THEOREM** *The following relationships hold for all x in the domains of the stated inverse hyperbolic functions:*
>
> $$\sinh^{-1} x = \ln(x + \sqrt{x^2 + 1}) \qquad \cosh^{-1} x = \ln(x + \sqrt{x^2 - 1})$$
>
> $$\tanh^{-1} x = \frac{1}{2} \ln\left(\frac{1+x}{1-x}\right) \qquad \coth^{-1} x = \frac{1}{2} \ln\left(\frac{x+1}{x-1}\right)$$
>
> $$\operatorname{sech}^{-1} x = \ln\left(\frac{1+\sqrt{1-x^2}}{x}\right) \qquad \operatorname{csch}^{-1} x = \ln\left(\frac{1}{x} + \frac{\sqrt{1+x^2}}{|x|}\right)$$

We will show how to derive the first formula in this theorem and leave the rest as exercises. The basic idea is to write the equation $x = \sinh y$ in terms of exponential functions and solve this equation for y as a function of x. This will produce the equation $y = \sinh^{-1} x$ with $\sinh^{-1} x$ expressed in terms of natural logarithms. Expressing $x = \sinh y$ in terms of exponentials yields

$$x = \sinh y = \frac{e^y - e^{-y}}{2}$$

which can be rewritten as

$$e^y - 2x - e^{-y} = 0$$

Multiplying this equation through by e^y we obtain

$$e^{2y} - 2xe^y - 1 = 0$$

and applying the quadratic formula yields

$$e^y = \frac{2x \pm \sqrt{4x^2 + 4}}{2} = x \pm \sqrt{x^2 + 1}$$

Since $e^y > 0$, the solution involving the minus sign is extraneous and must be discarded. Thus,

$$e^y = x + \sqrt{x^2 + 1}$$

Taking natural logarithms yields

$$y = \ln(x + \sqrt{x^2 + 1}) \quad \text{or} \quad \sinh^{-1} x = \ln(x + \sqrt{x^2 + 1})$$

▶ **Example 5**

$$\sinh^{-1} 1 = \ln(1 + \sqrt{1^2 + 1}) = \ln(1 + \sqrt{2}) \approx 0.8814$$

$$\tanh^{-1}\left(\frac{1}{2}\right) = \frac{1}{2} \ln\left(\frac{1 + \frac{1}{2}}{1 - \frac{1}{2}}\right) = \frac{1}{2} \ln 3 \approx 0.5493 \quad ◀$$

■ DERIVATIVES AND INTEGRALS INVOLVING INVERSE HYPERBOLIC FUNCTIONS

Formulas for the derivatives of the inverse hyperbolic functions can be obtained from Theorem 6.9.4. For example,

$$\frac{d}{dx}[\sinh^{-1} x] = \frac{d}{dx}[\ln(x + \sqrt{x^2 + 1})] = \frac{1}{x + \sqrt{x^2 + 1}}\left(1 + \frac{x}{\sqrt{x^2 + 1}}\right)$$

$$= \frac{\sqrt{x^2 + 1} + x}{(x + \sqrt{x^2 + 1})(\sqrt{x^2 + 1})} = \frac{1}{\sqrt{x^2 + 1}}$$

> Show that the derivative of the function $\sinh^{-1} x$ can also be obtained by letting $y = \sinh^{-1} x$ and then differentiating $x = \sinh y$ implicitly.

This computation leads to two integral formulas, a formula that involves $\sinh^{-1} x$ and an equivalent formula that involves logarithms:

$$\int \frac{dx}{\sqrt{x^2 + 1}} = \sinh^{-1} x + C = \ln(x + \sqrt{x^2 + 1}) + C$$

The following two theorems list the generalized derivative formulas and corresponding integration formulas for the inverse hyperbolic functions. Some of the proofs appear as exercises.

6.9.5   THEOREM

$$\frac{d}{dx}(\sinh^{-1} u) = \frac{1}{\sqrt{1 + u^2}}\frac{du}{dx} \qquad\qquad \frac{d}{dx}(\coth^{-1} u) = \frac{1}{1 - u^2}\frac{du}{dx}, \quad |u| > 1$$

$$\frac{d}{dx}(\cosh^{-1} u) = \frac{1}{\sqrt{u^2 - 1}}\frac{du}{dx}, \quad u > 1 \qquad\qquad \frac{d}{dx}(\operatorname{sech}^{-1} u) = -\frac{1}{u\sqrt{1 - u^2}}\frac{du}{dx}, \quad 0 < u < 1$$

$$\frac{d}{dx}(\tanh^{-1} u) = \frac{1}{1 - u^2}\frac{du}{dx}, \quad |u| < 1 \qquad\qquad \frac{d}{dx}(\operatorname{csch}^{-1} u) = -\frac{1}{|u|\sqrt{1 + u^2}}\frac{du}{dx}, \quad u \neq 0$$

6.9.6   THEOREM   *If $a > 0$, then*

$$\int \frac{du}{\sqrt{a^2 + u^2}} = \sinh^{-1}\left(\frac{u}{a}\right) + C \ \ or \ \ \ln(u + \sqrt{u^2 + a^2}) + C$$

$$\int \frac{du}{\sqrt{u^2 - a^2}} = \cosh^{-1}\left(\frac{u}{a}\right) + C \ \ or \ \ \ln(u + \sqrt{u^2 - a^2}) + C, \quad u > a$$

$$\int \frac{du}{a^2 - u^2} = \begin{cases} \dfrac{1}{a}\tanh^{-1}\left(\dfrac{u}{a}\right) + C, \ |u| < a \\[2mm] \dfrac{1}{a}\coth^{-1}\left(\dfrac{u}{a}\right) + C, \ |u| > a \end{cases} \ \ or \ \ \frac{1}{2a}\ln\left|\frac{a + u}{a - u}\right| + C, \quad |u| \neq a$$

$$\int \frac{du}{u\sqrt{a^2 - u^2}} = -\frac{1}{a}\operatorname{sech}^{-1}\left|\frac{u}{a}\right| + C \ \ or \ \ -\frac{1}{a}\ln\left(\frac{a + \sqrt{a^2 - u^2}}{|u|}\right) + C, \quad 0 < |u| < a$$

$$\int \frac{du}{u\sqrt{a^2 + u^2}} = -\frac{1}{a}\operatorname{csch}^{-1}\left|\frac{u}{a}\right| + C \ \ or \ \ -\frac{1}{a}\ln\left(\frac{a + \sqrt{a^2 + u^2}}{|u|}\right) + C, \quad u \neq 0$$

▶ **Example 6** Evaluate $\displaystyle\int \frac{dx}{\sqrt{4x^2 - 9}}, x > \frac{3}{2}$.

Solution. Let $u = 2x$. Thus, $du = 2\,dx$ and

$$\int \frac{dx}{\sqrt{4x^2 - 9}} = \frac{1}{2}\int \frac{2\,dx}{\sqrt{4x^2 - 9}} = \frac{1}{2}\int \frac{du}{\sqrt{u^2 - 3^2}}$$

$$= \frac{1}{2}\cosh^{-1}\left(\frac{u}{3}\right) + C = \frac{1}{2}\cosh^{-1}\left(\frac{2x}{3}\right) + C$$

Alternatively, we can use the logarithmic equivalent of $\cosh^{-1}(2x/3)$,

$$\cosh^{-1}\left(\frac{2x}{3}\right) = \ln(2x + \sqrt{4x^2 - 9}) - \ln 3$$

(verify), and express the answer as

$$\int \frac{dx}{\sqrt{4x^2 - 9}} = \frac{1}{2}\ln(2x + \sqrt{4x^2 - 9}) + C \quad ◀$$

✔ **QUICK CHECK EXERCISES 6.9**   (See page 485 for answers.)

1. $\cosh x =$ _____ $\sinh x =$ _____
$\tanh x =$ _____

2. Complete the table.

	$\cosh x$	$\sinh x$	$\tanh x$	$\coth x$	$\operatorname{sech} x$	$\operatorname{csch} x$
DOMAIN						
RANGE						

3. The parametric equations

$$x = \cosh t, \quad y = \sinh t \quad (-\infty < t < +\infty)$$

represent the right half of the curve called a _____. Eliminating the parameter, the equation of this curve is _____.

4. $\dfrac{d}{dx}[\cosh x] =$ _____ $\dfrac{d}{dx}[\sinh x] =$ _____
$\dfrac{d}{dx}[\tanh x] =$ _____

5. $\displaystyle\int \cosh x\,dx =$ _____ $\displaystyle\int \sinh x\,dx =$ _____
$\displaystyle\int \tanh x\,dx =$ _____

6. $\dfrac{d}{dx}[\cosh^{-1} x] =$ _____ $\dfrac{d}{dx}[\sinh^{-1} x] =$ _____
$\dfrac{d}{dx}[\tanh^{-1} x] =$ _____

EXERCISE SET 6.9   Graphing Utility

1–2 Approximate the expression to four decimal places. ■

1. (a) $\sinh 3$ (b) $\cosh(-2)$ (c) $\tanh(\ln 4)$
(d) $\sinh^{-1}(-2)$ (e) $\cosh^{-1} 3$ (f) $\tanh^{-1}\frac{3}{4}$

2. (a) $\operatorname{csch}(-1)$ (b) $\operatorname{sech}(\ln 2)$ (c) $\coth 1$
(d) $\operatorname{sech}^{-1}\frac{1}{2}$ (e) $\coth^{-1} 3$ (f) $\operatorname{csch}^{-1}(-\sqrt{3})$

3. Find the exact numerical value of each expression.
(a) $\sinh(\ln 3)$ (b) $\cosh(-\ln 2)$
(c) $\tanh(2\ln 5)$ (d) $\sinh(-3\ln 2)$

4. In each part, rewrite the expression as a ratio of polynomials.
(a) $\cosh(\ln x)$ (b) $\sinh(\ln x)$
(c) $\tanh(2\ln x)$ (d) $\cosh(-\ln x)$

5. In each part, a value for one of the hyperbolic functions is given at an unspecified positive number x_0. Use appropriate identities to find the exact values of the remaining five hyperbolic functions at x_0.
(a) $\sinh x_0 = 2$ (b) $\cosh x_0 = \frac{5}{4}$ (c) $\tanh x_0 = \frac{4}{5}$

6. Obtain the derivative formulas for $\operatorname{csch} x$, $\operatorname{sech} x$, and $\coth x$ from the derivative formulas for $\sinh x$, $\cosh x$, and $\tanh x$.

7. Find the derivatives of $\cosh^{-1} x$ and $\tanh^{-1} x$ by differentiating the formulas in Theorem 6.9.4.

8. Find the derivatives of $\sinh^{-1} x$, $\cosh^{-1} x$, and $\tanh^{-1} x$ by differentiating the equations $x = \sinh y$, $x = \cosh y$, and $x = \tanh y$ implicitly.

9–28 Find dy/dx. ■

9. $y = \sinh(4x - 8)$ **10.** $y = \cosh(x^4)$

11. $y = \coth(\ln x)$

12. $y = \ln(\tanh 2x)$

13. $y = \operatorname{csch}(1/x)$

14. $y = \operatorname{sech}(e^{2x})$

15. $y = \sqrt{4x + \cosh^2(5x)}$

16. $y = \sinh^3(2x)$

17. $y = x^3 \tanh^2(\sqrt{x})$

18. $y = \sinh(\cos 3x)$

19. $y = \sinh^{-1}\left(\frac{1}{3}x\right)$

20. $y = \sinh^{-1}(1/x)$

21. $y = \ln(\cosh^{-1} x)$

22. $y = \cosh^{-1}(\sinh^{-1} x)$

23. $y = \dfrac{1}{\tanh^{-1} x}$

24. $y = (\coth^{-1} x)^2$

25. $y = \cosh^{-1}(\cosh x)$

26. $y = \sinh^{-1}(\tanh x)$

27. $y = e^x \operatorname{sech}^{-1}\sqrt{x}$

28. $y = (1 + x \operatorname{csch}^{-1} x)^{10}$

29–44 Evaluate the integrals. ■

29. $\displaystyle\int \sinh^6 x \cosh x \, dx$

30. $\displaystyle\int \cosh(2x - 3) \, dx$

31. $\displaystyle\int \sqrt{\tanh x} \, \operatorname{sech}^2 x \, dx$

32. $\displaystyle\int \operatorname{csch}^2(3x) \, dx$

33. $\displaystyle\int \tanh x \, dx$

34. $\displaystyle\int \coth^2 x \, \operatorname{csch}^2 x \, dx$

35. $\displaystyle\int_{\ln 2}^{\ln 3} \tanh x \, \operatorname{sech}^3 x \, dx$

36. $\displaystyle\int_0^{\ln 3} \frac{e^x - e^{-x}}{e^x + e^{-x}} \, dx$

37. $\displaystyle\int \frac{dx}{\sqrt{1 + 9x^2}}$

38. $\displaystyle\int \frac{dx}{\sqrt{x^2 - 2}} \quad (x > \sqrt{2})$

39. $\displaystyle\int \frac{dx}{\sqrt{1 - e^{2x}}} \quad (x < 0)$

40. $\displaystyle\int \frac{\sin\theta \, d\theta}{\sqrt{1 + \cos^2\theta}}$

41. $\displaystyle\int \frac{dx}{x\sqrt{1 + 4x^2}}$

42. $\displaystyle\int \frac{dx}{\sqrt{9x^2 - 25}} \quad (x > 5/3)$

43. $\displaystyle\int_0^{1/2} \frac{dx}{1 - x^2}$

44. $\displaystyle\int_0^{\sqrt{3}} \frac{dt}{\sqrt{t^2 + 1}}$

45–48 True–False Determine whether the statement is true or false. Explain your answer. ■

45. The equation $\cosh x = \sinh x$ has no solutions.

46. Exactly two of the hyperbolic functions are bounded.

47. There is exactly one hyperbolic function $f(x)$ such that for all real numbers a, the equation $f(x) = a$ has a unique solution x.

48. The identities in Theorem 6.9.2 may be obtained from the corresponding trigonometric identities by replacing each trigonometric function with its hyperbolic analogue.

49. Find the area enclosed by $y = \sinh 2x$, $y = 0$, and $x = \ln 3$.

50. Find the volume of the solid that is generated when the region enclosed by $y = \operatorname{sech} x$, $y = 0$, $x = 0$, and $x = \ln 2$ is revolved about the x-axis.

51. Find the volume of the solid that is generated when the region enclosed by $y = \cosh 2x$, $y = \sinh 2x$, $x = 0$, and $x = 5$ is revolved about the x-axis.

52. Approximate the positive value of the constant a such that the area enclosed by $y = \cosh ax$, $y = 0$, $x = 0$, and $x = 1$

is 2 square units. Express your answer to at least five decimal places.

53. Find the arc length of the catenary $y = \cosh x$ between $x = 0$ and $x = \ln 2$.

54. Find the arc length of the catenary $y = a \cosh(x/a)$ between $x = 0$ and $x = x_1 \ (x_1 > 0)$.

55. In parts (a)–(f) find the limits, and confirm that they are consistent with the graphs in Figures 6.9.1 and 6.9.6.

(a) $\displaystyle\lim_{x \to +\infty} \sinh x$

(b) $\displaystyle\lim_{x \to -\infty} \sinh x$

(c) $\displaystyle\lim_{x \to +\infty} \tanh x$

(d) $\displaystyle\lim_{x \to -\infty} \tanh x$

(e) $\displaystyle\lim_{x \to +\infty} \sinh^{-1} x$

(f) $\displaystyle\lim_{x \to 1^-} \tanh^{-1} x$

FOCUS ON CONCEPTS

56. Explain how to obtain the asymptotes for $y = \tanh x$ from the curvilinear asymptotes for $y = \cosh x$ and $y = \sinh x$.

57. Prove that $\sinh x$ is an odd function of x and that $\cosh x$ is an even function of x, and check that this is consistent with the graphs in Figure 6.9.1.

58–59 Prove the identities. ■

58. (a) $\cosh x + \sinh x = e^x$

(b) $\cosh x - \sinh x = e^{-x}$

(c) $\sinh(x + y) = \sinh x \cosh y + \cosh x \sinh y$

(d) $\sinh 2x = 2 \sinh x \cosh x$

(e) $\cosh(x + y) = \cosh x \cosh y + \sinh x \sinh y$

(f) $\cosh 2x = \cosh^2 x + \sinh^2 x$

(g) $\cosh 2x = 2 \sinh^2 x + 1$

(h) $\cosh 2x = 2 \cosh^2 x - 1$

59. (a) $1 - \tanh^2 x = \operatorname{sech}^2 x$

(b) $\tanh(x + y) = \dfrac{\tanh x + \tanh y}{1 + \tanh x \tanh y}$

(c) $\tanh 2x = \dfrac{2 \tanh x}{1 + \tanh^2 x}$

60. Prove:

(a) $\cosh^{-1} x = \ln(x + \sqrt{x^2 - 1}), \quad x \geq 1$

(b) $\tanh^{-1} x = \dfrac{1}{2} \ln\left(\dfrac{1 + x}{1 - x}\right), \quad -1 < x < 1$.

61. Use Exercise 60 to obtain the derivative formulas for $\cosh^{-1} x$ and $\tanh^{-1} x$.

62. Prove:

$\operatorname{sech}^{-1} x = \cosh^{-1}(1/x), \quad 0 < x \leq 1$

$\coth^{-1} x = \tanh^{-1}(1/x), \quad |x| > 1$

$\operatorname{csch}^{-1} x = \sinh^{-1}(1/x), \quad x \neq 0$

63. Use Exercise 62 to express the integral

$$\int \frac{du}{1 - u^2}$$

entirely in terms of $\tanh^{-1}$.

64. Show that

(a) $\dfrac{d}{dx}[\text{sech}^{-1}|x|] = -\dfrac{1}{x\sqrt{1-x^2}}$

(b) $\dfrac{d}{dx}[\text{csch}^{-1}|x|] = -\dfrac{1}{x\sqrt{1+x^2}}$.

65. In each part, find the limit.

(a) $\displaystyle\lim_{x\to+\infty}(\cosh^{-1}x - \ln x)$ (b) $\displaystyle\lim_{x\to+\infty}\dfrac{\cosh x}{e^x}$

66. Use the first and second derivatives to show that the graph of $y = \tanh^{-1}x$ is always increasing and has an inflection point at the origin.

67. The integration formulas for $1/\sqrt{u^2 - a^2}$ in Theorem 6.9.6 are valid for $u > a$. Show that the following formula is valid for $u < -a$:

$$\int \dfrac{du}{\sqrt{u^2-a^2}} = -\cosh^{-1}\left(-\dfrac{u}{a}\right) + C \quad\text{or}\quad \ln\left|u + \sqrt{u^2 - a^2}\right| + C$$

68. Show that $(\sinh x + \cosh x)^n = \sinh nx + \cosh nx$.

69. Show that

$$\int_{-a}^{a} e^{tx}\,dx = \dfrac{2\sinh at}{t}$$

70. A cable is suspended between two poles as shown in Figure 6.9.2. Assume that the equation of the curve formed by the cable is $y = a\cosh(x/a)$, where a is a positive constant. Suppose that the x-coordinates of the points of support are $x = -b$ and $x = b$, where $b > 0$.

(a) Show that the length L of the cable is given by

$$L = 2a\sinh\dfrac{b}{a}$$

(b) Show that the sag S (the vertical distance between the highest and lowest points on the cable) is given by

$$S = a\cosh\dfrac{b}{a} - a$$

71–72 These exercises refer to the hanging cable described in Exercise 70. ■

71. Assuming that the poles are 400 ft apart and the sag in the cable is 30 ft, approximate the length of the cable by approximating a. Express your final answer to the nearest tenth of a foot. [*Hint:* First let $u = 200/a$.]

72. Assuming that the cable is 120 ft long and the poles are 100 ft apart, approximate the sag in the cable by approximating a. Express your final answer to the nearest tenth of a foot. [*Hint:* First let $u = 50/a$.]

73. The design of the Gateway Arch in St. Louis, Missouri, by architect Eero Saarinen was implemented using equations provided by Dr. Hannskarl Badel. The equation used for the centerline of the arch was

$$y = 693.8597 - 68.7672\cosh(0.0100333x)\text{ ft}$$

for x between -299.2239 and 299.2239.

(a) Use a graphing utility to graph the centerline of the arch.

(b) Find the length of the centerline to four decimal places.

(c) For what values of x is the height of the arch 100 ft? Round your answers to four decimal places.

(d) Approximate, to the nearest degree, the acute angle that the tangent line to the centerline makes with the ground at the ends of the arch.

74. Suppose that a hollow tube rotates with a constant angular velocity of ω rad/s about a horizontal axis at one end of the tube, as shown in the accompanying figure. Assume that an object is free to slide without friction in the tube while the tube is rotating. Let r be the distance from the object to the pivot point at time $t \geq 0$, and assume that the object is at rest and $r = 0$ when $t = 0$. It can be shown that if the tube is horizontal at time $t = 0$ and rotating as shown in the figure, then

$$r = \dfrac{g}{2\omega^2}[\sinh(\omega t) - \sin(\omega t)]$$

during the period that the object is in the tube. Assume that t is in seconds and r is in meters, and use $g = 9.8$ m/s^2 and $\omega = 2$ rad/s.

(a) Graph r versus t for $0 \leq t \leq 1$.

(b) Assuming that the tube has a length of 1 m, approximately how long does it take for the object to reach the end of the tube?

(c) Use the result of part (b) to approximate dr/dt at the instant that the object reaches the end of the tube.

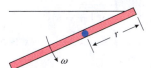

◀ **Figure Ex-74**

75. The accompanying figure (on the next page) shows a person pulling a boat by holding a rope of length a attached to the bow and walking along the edge of a dock. If we assume that the rope is always tangent to the curve traced by the bow of the boat, then this curve, which is called a *tractrix*, has the property that the segment of the tangent line between the curve and the y-axis has a constant length a. It can be proved that the equation of this tractrix is

$$y = a\,\text{sech}^{-1}\dfrac{x}{a} - \sqrt{a^2 - x^2}$$

(a) Show that to move the bow of the boat to a point (x, y), the person must walk a distance

$$D = a\,\text{sech}^{-1}\dfrac{x}{a}$$

from the origin.

(b) If the rope has a length of 15 m, how far must the person walk from the origin to bring the boat 10 m from the dock? Round your answer to two decimal places.

(c) Find the distance traveled by the bow along the tractrix as it moves from its initial position to the point where it is 5 m from the dock.

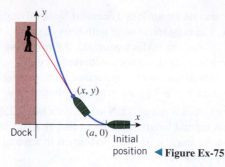

Dock | $(a, 0)$ Initial position ◀ **Figure Ex-75**

76. Writing Suppose that, by analogy with the trigonometric functions, we *define* $\cosh t$ and $\sinh t$ geometrically using Figure 6.9.3*b*:

"For any real number t, define $x = \cosh t$ and $y = \sinh t$ to be the unique values of x and y such that

(i) $P(x, y)$ is on the right branch of the unit hyperbola $x^2 - y^2 = 1$;

(ii) t and y have the same sign (or are both 0);

(iii) the area of the region bounded by the x-axis, the right branch of the unit hyperbola, and the segment from the origin to P is $|t|/2$."

Discuss what properties would first need to be verified in order for this to be a legitimate definition.

77. Writing Investigate what properties of $\cosh t$ and $\sinh t$ can be proved directly from the geometric definition in Exercise 76. Write a short description of the results of your investigation.

✔ QUICK CHECK ANSWERS 6.9

1. $\dfrac{e^x + e^{-x}}{2}; \dfrac{e^x - e^{-x}}{2}; \dfrac{e^x - e^{-x}}{e^x + e^{-x}}$

2.

	$\cosh x$	$\sinh x$	$\tanh x$	$\coth x$	$\operatorname{sech} x$	$\operatorname{csch} x$
DOMAIN	$(-\infty, +\infty)$	$(-\infty, +\infty)$	$(-\infty, +\infty)$	$(-\infty, 0) \cup (0, +\infty)$	$(-\infty, +\infty)$	$(-\infty, 0) \cup (0, +\infty)$
RANGE	$[1, +\infty)$	$(-\infty, +\infty)$	$(-1, 1)$	$(-\infty, -1) \cup (1, +\infty)$	$(0, 1]$	$(-\infty, 0) \cup (0, +\infty)$

3. unit hyperbola; $x^2 - y^2 = 1$ 4. $\sinh x$; $\cosh x$; $\operatorname{sech}^2 x$ 5. $\sinh x + C$; $\cosh x + C$; $\ln(\cosh x) + C$

6. $\dfrac{1}{\sqrt{x^2 - 1}}; \dfrac{1}{\sqrt{1 + x^2}}; \dfrac{1}{1 - x^2}$

CHAPTER 6 REVIEW EXERCISES

1. Describe the method of slicing for finding volumes, and use that method to derive an integral formula for finding volumes by the method of disks.

2. State an integral formula for finding a volume by the method of cylindrical shells, and use Riemann sums to derive the formula.

3. State an integral formula for finding the arc length of a smooth curve $y = f(x)$ over an interval $[a, b]$, and use Riemann sums to derive the formula.

4. State an integral formula for the work W done by a variable force $F(x)$ applied in the direction of motion to an object moving from $x = a$ to $x = b$, and use Riemann sums to derive the formula.

5. State an integral formula for the fluid force F exerted on a vertical flat surface immersed in a fluid of weight density ρ, and use Riemann sums to derive the formula.

6. Let R be the region in the first quadrant enclosed by $y = x^2$, $y = 2 + x$, and $x = 0$. In each part, set up, but *do not eval-*

uate, an integral or a sum of integrals that will solve the problem.

(a) Find the area of R by integrating with respect to x.

(b) Find the area of R by integrating with respect to y.

(c) Find the volume of the solid generated by revolving R about the x-axis by integrating with respect to x.

(d) Find the volume of the solid generated by revolving R about the x-axis by integrating with respect to y.

(e) Find the volume of the solid generated by revolving R about the y-axis by integrating with respect to x.

(f) Find the volume of the solid generated by revolving R about the y-axis by integrating with respect to y.

(g) Find the volume of the solid generated by revolving R about the line $y = -3$ by integrating with respect to x.

(h) Find the volume of the solid generated by revolving R about the line $x = 5$ by integrating with respect to x.

7. (a) Set up a sum of definite integrals that represents the total shaded area between the curves $y = f(x)$ and $y = g(x)$ in the accompanying figure on the next page. *(cont.)*

(b) Find the total area enclosed between $y = x^3$ and $y = x$ over the interval $[-1, 2]$.

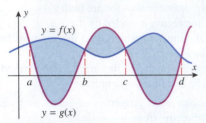

◀ **Figure Ex-7**

8. The accompanying figure shows velocity versus time curves for two cars that move along a straight track, accelerating from rest at a common starting line.
   (a) How far apart are the cars after 60 seconds?
   (b) How far apart are the cars after T seconds, where $0 \le T \le 60$?

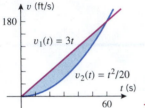

◀ **Figure Ex-8**

9. Let R be the region enclosed by the curves $y = x^2 + 4$, $y = x^3$, and the y-axis. Find and evaluate a definite integral that represents the volume of the solid generated by revolving R about the x-axis.

10. A football has the shape of the solid generated by revolving the region bounded between the x-axis and the parabola $y = 4R(x^2 - \frac{1}{4}L^2)/L^2$ about the x-axis. Find its volume.

11. Find the volume of the solid whose base is the region bounded between the curves $y = \sqrt{x}$ and $y = 1/\sqrt{x}$ for $1 \le x \le 4$ and whose cross sections perpendicular to the x-axis are squares.

12. Consider the region enclosed by $y = \sin^{-1} x$, $y = 0$, and $x = 1$. Set up, but *do not evaluate*, an integral that represents the volume of the solid generated by revolving the region about the x-axis using
   (a) disks   **(b)** cylindrical shells.

13. Find the arc length in the second quadrant of the curve $x^{2/3} + y^{2/3} = 4$ from $x = -8$ to $x = -1$.

14. Let C be the curve $y = e^x$ between $x = 0$ and $x = \ln 10$. In each part, set up, but *do not evaluate*, an integral that solves the problem.
   (a) Find the arc length of C by integrating with respect to x.
   (b) Find the arc length of C by integrating with respect to y.

15. Find the area of the surface generated by revolving the curve $y = \sqrt{25 - x}$, $9 \le x \le 16$, about the x-axis.

16. Let C be the curve $27x - y^3 = 0$ between $y = 0$ and $y = 2$. In each part, set up, but *do not evaluate*, an integral or a sum of integrals that solves the problem.

(a) Find the area of the surface generated by revolving C about the x-axis by integrating with respect to x.
   (b) Find the area of the surface generated by revolving C about the y-axis by integrating with respect to y.
   (c) Find the area of the surface generated by revolving C about the line $y = -2$ by integrating with respect to y.

17. (a) A spring exerts a force of 0.5 N when stretched 0.25 m beyond its natural length. Assuming that Hooke's law applies, how much work was performed in stretching the spring to this length?
   (b) How far beyond its natural length can the spring be stretched with 25 J of work?

18. A boat is anchored so that the anchor is 150 ft below the surface of the water. In the water, the anchor weighs 2000 lb and the chain weighs 30 lb/ft. How much work is required to raise the anchor to the surface?

19–20 Find the centroid of the region. ■

19. The region bounded by $y^2 = 4x$ and $y^2 = 8(x - 2)$.

20. The upper half of the ellipse $(x/a)^2 + (y/b)^2 = 1$.

21. In each part, set up, but *do not evaluate*, an integral that solves the problem.
   (a) Find the fluid force exerted on a side of a box that has a 3 m square base and is filled to a depth of 1 m with a liquid of weight density ρ N/m^3.
   (b) Find the fluid force exerted by a liquid of weight density ρ lb/ft^3 on a face of the vertical plate shown in part (a) of the accompanying figure.
   (c) Find the fluid force exerted on the parabolic dam in part (b) of the accompanying figure by water that extends to the top of the dam.

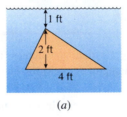
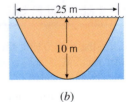

   (a) (b)

▲ **Figure Ex-21**

22. Show that for any constant a, the function $y = \sinh(ax)$ satisfies the equation $y'' = a^2 y$.

23. In each part, prove the identity.
   (a) $\cosh 3x = 4 \cosh^3 x - 3 \cosh x$
   (b) $\cosh \frac{1}{2}x = \sqrt{\frac{1}{2}(\cosh x + 1)}$
   (c) $\sinh \frac{1}{2}x = \pm\sqrt{\frac{1}{2}(\cosh x - 1)}$

CHAPTER 6 MAKING CONNECTIONS

1. Suppose that f is a nonnegative function defined on $[0, 1]$ such that the area between the graph of f and the interval $[0, 1]$ is A_1 and such that the area of the region R between the graph of $g(x) = f(x^2)$ and the interval $[0, 1]$ is A_2. In each part, express your answer in terms of A_1 and A_2.
   (a) What is the volume of the solid of revolution generated by revolving R about the y-axis?
   (b) Find a value of a such that if the xy-plane were horizontal, the region R would balance on the line $x = a$.

2. A water tank has the shape of a conical frustum with radius of the base 5 ft, radius of the top 10 ft and (vertical) height 15 ft. Suppose the tank is filled with water and consider the problem of finding the work required to pump all the water out through a hole in the top of the tank.
   (a) Solve this problem using the method of Example 5 in Section 6.6.
   (b) Solve this problem using Definition 6.6.3. [*Hint:* Think of the base as the head of a piston that expands to a water-tight fit against the sides of the tank as the piston is pushed upward. What important result about water pressure do you need to use?]

3. A disk of radius a is an inhomogeneous lamina whose density is a function $f(r)$ of the distance r to the center of the lamina.

Modify the argument used to derive the method of cylindrical shells to find a formula for the mass of the lamina.

4. Compare Formula (10) in Section 6.7 with Formula (8) in Section 6.8. Then give a plausible argument that the force on a flat surface immersed vertically in a fluid of constant weight density is equal to the product of the area of the surface and the pressure at the centroid of the surface. Conclude that the force on the surface is the same as if the surface were immersed horizontally at the depth of the centroid.

5. *Archimedes' Principle* states that a solid immersed in a fluid experiences a buoyant force equal to the weight of the fluid displaced by the solid.
   (a) Use the results of Section 6.8 to verify Archimedes' Principle in the case of (i) a box-shaped solid with a pair of faces parallel to the surface of the fluid, (ii) a solid cylinder with vertical axis, and (iii) a cylindrical shell with vertical axis.
   (b) Give a plausible argument for Archimedes' Principle in the case of a solid of revolution immersed in fluid such that the axis of revolution of the solid is vertical. [*Hint:* Approximate the solid by a union of cylindrical shells and use the result from part (a).]

© AP/Wide World Photos

7

PRINCIPLES OF INTEGRAL EVALUATION

The floating roof on the Stade de France sports complex is an ellipse. Finding the arc length of an ellipse involves numerical integration techniques introduced in this chapter.

In earlier chapters we obtained many basic integration formulas as an immediate consequence of the corresponding differentiation formulas. For example, knowing that the derivative of sin x is cos x enabled us to deduce that the integral of cos x is sin x. Subsequently, we expanded our integration repertoire by introducing the method of u-substitution. That method enabled us to integrate many functions by transforming the integrand of an unfamiliar integral into a familiar form. However, u-substitution alone is not adequate to handle the wide variety of integrals that arise in applications, so additional integration techniques are still needed. In this chapter we will discuss some of those techniques, and we will provide a more systematic procedure for attacking unfamiliar integrals. We will talk more about numerical approximations of definite integrals, and we will explore the idea of integrating over infinite intervals.

7.1 AN OVERVIEW OF INTEGRATION METHODS

In this section we will give a brief overview of methods for evaluating integrals, and we will review the integration formulas that were discussed in earlier sections.

■ METHODS FOR APPROACHING INTEGRATION PROBLEMS
There are three basic approaches for evaluating unfamiliar integrals:

- **Technology**—CAS programs such as *Mathematica*, *Maple*, and the open source program *Sage* are capable of evaluating extremely complicated integrals, and such programs are increasingly available for both computers and handheld calculators.

- **Tables**—Prior to the development of CAS programs, scientists relied heavily on tables to evaluate difficult integrals arising in applications. Such tables were compiled over many years, incorporating the skills and experience of many people. One such table appears in the endpapers of this text, but more comprehensive tables appear in various reference books such as the *CRC Standard Mathematical Tables and Formulae*, CRC Press, Inc., 2002.

- **Transformation Methods**—Transformation methods are methods for converting unfamiliar integrals into familiar integrals. These include u-substitution, algebraic manipulation of the integrand, and other methods that we will discuss in this chapter.

None of the three methods is perfect; for example, CAS programs often encounter integrals that they cannot evaluate and they sometimes produce answers that are unnecessarily complicated, tables are not exhaustive and may not include a particular integral of interest, and transformation methods rely on human ingenuity that may prove to be inadequate in difficult problems.

In this chapter we will focus on transformation methods and tables, so it will *not be necessary* to have a CAS such as *Mathematica*, *Maple*, or *Sage*. However, if you have a CAS, then you can use it to confirm the results in the examples, and there are exercises that are designed to be solved with a CAS. If you have a CAS, keep in mind that many of the algorithms that it uses are based on the methods we will discuss here, so an understanding of these methods will help you to use your technology in a more informed way.

■ A REVIEW OF FAMILIAR INTEGRATION FORMULAS

The following is a list of basic integrals that we have encountered thus far:

CONSTANTS, POWERS, EXPONENTIALS

1. $\displaystyle\int du = u + C$

2. $\displaystyle\int a\,du = a\int du = au + C$

3. $\displaystyle\int u^r\,du = \frac{u^{r+1}}{r+1} + C,\ r \neq -1$

4. $\displaystyle\int \frac{du}{u} = \ln|u| + C$

5. $\displaystyle\int e^u\,du = e^u + C$

6. $\displaystyle\int b^u\,du = \frac{b^u}{\ln b} + C,\ b > 0, b \neq 1$

TRIGONOMETRIC FUNCTIONS

7. $\displaystyle\int \sin u\,du = -\cos u + C$

8. $\displaystyle\int \cos u\,du = \sin u + C$

9. $\displaystyle\int \sec^2 u\,du = \tan u + C$

10. $\displaystyle\int \csc^2 u\,du = -\cot u + C$

11. $\displaystyle\int \sec u\tan u\,du = \sec u + C$

12. $\displaystyle\int \csc u\cot u\,du = -\csc u + C$

13. $\displaystyle\int \tan u\,du = -\ln|\cos u| + C$

14. $\displaystyle\int \cot u\,du = \ln|\sin u| + C$

HYPERBOLIC FUNCTIONS

15. $\displaystyle\int \sinh u\,du = \cosh u + C$

16. $\displaystyle\int \cosh u\,du = \sinh u + C$

17. $\displaystyle\int \operatorname{sech}^2 u\,du = \tanh u + C$

18. $\displaystyle\int \operatorname{csch}^2 u\,du = -\coth u + C$

19. $\displaystyle\int \operatorname{sech} u\tanh u\,du = -\operatorname{sech} u + C$

20. $\displaystyle\int \operatorname{csch} u\coth u\,du = -\operatorname{csch} u + C$

ALGEBRAIC FUNCTIONS ($a > 0$)

21. $\displaystyle\int \frac{du}{\sqrt{a^2 - u^2}} = \sin^{-1}\frac{u}{a} + C \qquad (|u| < a)$

22. $\displaystyle\int \frac{du}{a^2 + u^2} = \frac{1}{a}\tan^{-1}\frac{u}{a} + C$

23. $\displaystyle\int \frac{du}{u\sqrt{u^2 - a^2}} = \frac{1}{a}\sec^{-1}\left|\frac{u}{a}\right| + C \qquad (0 < a < |u|)$

24. $\int \dfrac{du}{\sqrt{a^2+u^2}} = \ln(u+\sqrt{u^2+a^2}) + C$

25. $\int \dfrac{du}{\sqrt{u^2-a^2}} = \ln\left|u+\sqrt{u^2-a^2}\right| + C \qquad (0 < a < |u|)$

26. $\int \dfrac{du}{a^2-u^2} = \dfrac{1}{2a}\ln\left|\dfrac{a+u}{a-u}\right| + C$

27. $\int \dfrac{du}{u\sqrt{a^2-u^2}} = -\dfrac{1}{a}\ln\left|\dfrac{a+\sqrt{a^2-u^2}}{u}\right| + C \qquad (0 < |u| < a)$

28. $\int \dfrac{du}{u\sqrt{a^2+u^2}} = -\dfrac{1}{a}\ln\left|\dfrac{a+\sqrt{a^2+u^2}}{u}\right| + C$

REMARK | Formula 25 is a generalization of a result in Theorem 6.9.6. Readers who did not cover Section 6.9 can ignore Formulas 24–28 for now, since we will develop other methods for obtaining them in this chapter.

✔ QUICK CHECK EXERCISES 7.1 (See page 491 for answers.)

1. Use algebraic manipulation and (if necessary) u-substitution to integrate the function.

(a) $\displaystyle\int \frac{x+1}{x}\,dx = \underline{\hspace{1cm}}$

(b) $\displaystyle\int \frac{x+2}{x+1}\,dx = \underline{\hspace{1cm}}$

(c) $\displaystyle\int \frac{2x+1}{x^2+1}\,dx = \underline{\hspace{1cm}}$

(d) $\displaystyle\int xe^{3\ln x}\,dx = \underline{\hspace{1cm}}$

2. Use trigonometric identities and (if necessary) u-substitution to integrate the function.

(a) $\displaystyle\int \frac{1}{\csc x}\,dx = \underline{\hspace{1cm}}$

(b) $\displaystyle\int \frac{1}{\cos^2 x}\,dx = \underline{\hspace{1cm}}$

(c) $\displaystyle\int (\cot^2 x + 1)\,dx = \underline{\hspace{1cm}}$

(d) $\displaystyle\int \frac{1}{\sec x + \tan x}\,dx = \underline{\hspace{1cm}}$

3. Integrate the function.

(a) $\displaystyle\int \sqrt{x-1}\,dx = \underline{\hspace{1cm}}$

(b) $\displaystyle\int e^{2x+1}\,dx = \underline{\hspace{1cm}}$

(c) $\displaystyle\int (\sin^3 x \cos x + \sin x \cos^3 x)\,dx = \underline{\hspace{1cm}}$

(d) $\displaystyle\int \frac{1}{(e^x + e^{-x})^2}\,dx = \underline{\hspace{1cm}}$

EXERCISE SET 7.1

1–30 Evaluate the integrals by making appropriate u-substitutions and applying the formulas reviewed in this section. ■

1. $\displaystyle\int (4-2x)^3\,dx$

2. $\displaystyle\int 3\sqrt{4+2x}\,dx$

3. $\displaystyle\int x\sec^2(x^2)\,dx$

4. $\displaystyle\int 4x\tan(x^2)\,dx$

5. $\displaystyle\int \frac{\sin 3x}{2+\cos 3x}\,dx$

6. $\displaystyle\int \frac{1}{9+4x^2}\,dx$

7. $\displaystyle\int e^x \sinh(e^x)\,dx$

8. $\displaystyle\int \frac{\sec(\ln x)\tan(\ln x)}{x}\,dx$

9. $\displaystyle\int e^{\tan x}\sec^2 x\,dx$

10. $\displaystyle\int \frac{x}{\sqrt{1-x^4}}\,dx$

11. $\displaystyle\int \cos^5 5x \sin 5x\,dx$

12. $\displaystyle\int \frac{\cos x}{\sin x\sqrt{\sin^2 x + 1}}\,dx$

13. $\displaystyle\int \frac{e^x}{\sqrt{4+e^{2x}}}\,dx$

14. $\displaystyle\int \frac{e^{\tan^{-1} x}}{1+x^2}\,dx$

15. $\displaystyle\int \frac{e^{\sqrt{x-1}}}{\sqrt{x-1}}\,dx$

16. $\displaystyle\int (x+1)\cot(x^2+2x)\,dx$

17. $\displaystyle\int \frac{\cosh\sqrt{x}}{\sqrt{x}}\,dx$

18. $\displaystyle\int \frac{dx}{x(\ln x)^2}$

19. $\displaystyle\int \frac{dx}{\sqrt{x}\,3^{\sqrt{x}}}$

20. $\displaystyle\int \sec(\sin\theta)\tan(\sin\theta)\cos\theta\,d\theta$

21. $\displaystyle\int \frac{\operatorname{csch}^2(2/x)}{x^2}\,dx$ **22.** $\displaystyle\int \frac{dx}{\sqrt{x^2-4}}$

23. $\displaystyle\int \frac{e^{-x}}{4-e^{-2x}}\,dx$ **24.** $\displaystyle\int \frac{\cos(\ln x)}{x}\,dx$

25. $\displaystyle\int \frac{e^x}{\sqrt{1-e^{2x}}}\,dx$ **26.** $\displaystyle\int \frac{\sinh(x^{-1/2})}{x^{3/2}}\,dx$

27. $\displaystyle\int \frac{x}{\csc(x^2)}\,dx$ **28.** $\displaystyle\int \frac{e^x}{\sqrt{4-e^{2x}}}\,dx$

29. $\displaystyle\int x4^{-x^2}\,dx$ **30.** $\displaystyle\int 2^{\pi x}\,dx$

FOCUS ON CONCEPTS

31. (a) Evaluate the integral $\int \sin x \cos x\,dx$ using the substitution $u = \sin x$.
   (b) Evaluate the integral $\int \sin x \cos x\,dx$ using the identity $\sin 2x = 2\sin x \cos x$.
   (c) Explain why your answers to parts (a) and (b) are consistent.

32. (a) Derive the identity

$$\frac{\operatorname{sech}^2 x}{1 + \tanh^2 x} = \operatorname{sech} 2x$$

   (b) Use the result in part (a) to evaluate $\int \operatorname{sech} x\,dx$.
   (c) Derive the identity

$$\operatorname{sech} x = \frac{2e^x}{e^{2x}+1}$$

   (d) Use the result in part (c) to evaluate $\int \operatorname{sech} x\,dx$.
   (e) Explain why your answers to parts (b) and (d) are consistent.

33. (a) Derive the identity

$$\frac{\sec^2 x}{\tan x} = \frac{1}{\sin x \cos x}$$

   (b) Use the identity $\sin 2x = 2\sin x \cos x$ along with the result in part (a) to evaluate $\int \csc x\,dx$.
   (c) Use the identity $\cos x = \sin[(\pi/2) - x]$ along with your answer to part (a) to evaluate $\int \sec x\,dx$.

✔ **QUICK CHECK ANSWERS 7.1**

1. (a) $x + \ln|x| + C$ (b) $x + \ln|x+1| + C$ (c) $\ln(x^2+1) + \tan^{-1}x + C$ (d) $\dfrac{x^5}{5} + C$ **2.** (a) $-\cos x + C$ (b) $\tan x + C$

(c) $-\cot x + C$ (d) $\ln(1 + \sin x) + C$ **3.** (a) $\frac{2}{3}(x-1)^{3/2} + C$ (b) $\frac{1}{2}e^{2x+1} + C$ (c) $\frac{1}{2}\sin^2 x + C$ (d) $\frac{1}{4}\tanh x + C$

7.2 INTEGRATION BY PARTS

In this section we will discuss an integration technique that is essentially an antiderivative formulation of the formula for differentiating a product of two functions.

■ THE PRODUCT RULE AND INTEGRATION BY PARTS

Our primary goal in this section is to develop a general method for attacking integrals of the form

$$\int f(x)g(x)\,dx$$

As a first step, let $G(x)$ be *any* antiderivative of $g(x)$. In this case $G'(x) = g(x)$, so the product rule for differentiating $f(x)G(x)$ can be expressed as

$$\frac{d}{dx}[f(x)G(x)] = f(x)G'(x) + f'(x)G(x) = f(x)g(x) + f'(x)G(x) \tag{1}$$

This implies that $f(x)G(x)$ is an antiderivative of the function on the right side of (1), so we can express (1) in integral form as

$$\int [f(x)g(x) + f'(x)G(x)]\,dx = f(x)G(x)$$

or, equivalently, as

$$\int f(x)g(x)\,dx = f(x)G(x) - \int f'(x)G(x)\,dx \tag{2}$$

This formula allows us to integrate $f(x)g(x)$ by integrating $f'(x)G(x)$ instead, and in many cases the net effect is to replace a difficult integration with an easier one. The application of this formula is called *integration by parts*.

In practice, we usually rewrite (2) by letting

$$u = f(x), \quad du = f'(x)\,dx$$
$$v = G(x), \quad dv = G'(x)\,dx = g(x)\,dx$$

This yields the following alternative form for (2):

$$\int u\,dv = uv - \int v\,du \tag{3}$$

Note that in Example 1 we omitted the constant of integration in calculating v from dv. Had we included a constant of integration, it would have eventually dropped out. This is always the case in integration by parts [Exercise 68(b)], so it is common to omit the constant at this stage of the computation. However, there are certain cases in which making a clever choice of a constant of integration to include with v can simplify the computation of $\int v\,du$ (Exercises 69–71).

▶ **Example 1** Use integration by parts to evaluate $\int x \cos x\,dx$.

Solution. We will apply Formula (3). The first step is to make a choice for u and dv to put the given integral in the form $\int u\,dv$. We will let

$$u = x \quad \text{and} \quad dv = \cos x\,dx$$

(Other possibilities will be considered later.) The second step is to compute du from u and v from dv. This yields

$$du = dx \quad \text{and} \quad v = \int dv = \int \cos x\,dx = \sin x$$

The third step is to apply Formula (3). This yields

$$\int \underbrace{x}_{u}\ \underbrace{\cos x\,dx}_{dv} = \underbrace{x}_{u}\ \underbrace{\sin x}_{v} - \int \underbrace{\sin x}_{v}\ \underbrace{dx}_{du}$$
$$= x \sin x - (-\cos x) + C = x \sin x + \cos x + C \blacktriangleleft$$

■ GUIDELINES FOR INTEGRATION BY PARTS

The main goal in integration by parts is to choose u and dv to obtain a new integral that is easier to evaluate than the original. In general, there are no hard and fast rules for doing this; it is mainly a matter of experience that comes from lots of practice. A strategy that often works is to choose u and dv so that u becomes "simpler" when differentiated, while leaving a dv that can be readily integrated to obtain v. Thus, for the integral $\int x \cos x\,dx$ in Example 1, both goals were achieved by letting $u = x$ and $dv = \cos x\,dx$. In contrast, $u = \cos x$ would not have been a good first choice in that example, since $du/dx = -\sin x$ is no simpler than u. Indeed, had we chosen

$$u = \cos x \qquad dv = x\,dx$$
$$du = -\sin x\,dx \qquad v = \int x\,dx = \frac{x^2}{2}$$

then we would have obtained

$$\int x \cos x\,dx = \frac{x^2}{2}\cos x - \int \frac{x^2}{2}(-\sin x)\,dx = \frac{x^2}{2}\cos x + \frac{1}{2}\int x^2 \sin x\,dx$$

For this choice of u and dv, the new integral is actually more complicated than the original.

The LIATE method is discussed in the article "A Technique for Integration by Parts," *American Mathematical Monthly*, Vol. 90, 1983, pp. 210–211, by Herbert Kasube.

There is another useful strategy for choosing u and dv that can be applied when the integrand is a product of two functions from *different* categories in the list

$$\underline{\text{L}}\text{ogarithmic, } \underline{\text{I}}\text{nverse trigonometric, } \underline{\text{A}}\text{lgebraic, } \underline{\text{T}}\text{rigonometric, } \underline{\text{E}}\text{xponential}$$

In this case you will often be successful if you take u to be the function whose category occurs earlier in the list and take dv to be the rest of the integrand. The acronym LIATE will help you to remember the order. The method does not work all the time, but it works often enough to be useful.

Note, for example, that the integrand in Example 1 consists of the product of the *algebraic* function x and the *trigonometric* function $\cos x$. Thus, the LIATE method suggests that we should let $u = x$ and $dv = \cos x\, dx$, which proved to be a successful choice.

▶ **Example 2**   Evaluate $\displaystyle\int xe^x\, dx$.

Solution.   In this case the integrand is the product of the algebraic function x with the exponential function e^x. According to LIATE we should let

$$u = x \quad \text{and} \quad dv = e^x\, dx$$

so that

$$du = dx \quad \text{and} \quad v = \int e^x\, dx = e^x$$

Thus, from (3)

$$\int xe^x\, dx = \int u\, dv = uv - \int v\, du = xe^x - \int e^x\, dx = xe^x - e^x + C \; \blacktriangleleft$$

▶ **Example 3**   Evaluate $\displaystyle\int \ln x\, dx$.

Solution.   One choice is to let $u = 1$ and $dv = \ln x\, dx$. But with this choice finding v is equivalent to evaluating $\int \ln x\, dx$ and we have gained nothing. Therefore, the only reasonable choice is to let

$$u = \ln x \qquad dv = dx$$
$$du = \frac{1}{x}\, dx \qquad v = \int dx = x$$

With this choice it follows from (3) that

$$\int \ln x\, dx = \int u\, dv = uv - \int v\, du = x \ln x - \int dx = x \ln x - x + C \; \blacktriangleleft$$

■ **REPEATED INTEGRATION BY PARTS**

It is sometimes necessary to use integration by parts more than once in the same problem.

▶ **Example 4**   Evaluate $\displaystyle\int x^2 e^{-x}\, dx$.

Solution.   Let

$$u = x^2, \quad dv = e^{-x}\, dx, \quad du = 2x\, dx, \quad v = \int e^{-x}\, dx = -e^{-x}$$

so that from (3)

$$\int x^2 e^{-x}\, dx = \int u\, dv = uv - \int v\, du$$

$$= x^2(-e^{-x}) - \int -e^{-x}(2x)\, dx$$

$$= -x^2 e^{-x} + 2\int xe^{-x}\, dx \tag{4}$$

The last integral is similar to the original except that we have replaced x^2 by x. Another integration by parts applied to $\int xe^{-x}\, dx$ will complete the problem. We let

$$u = x, \quad dv = e^{-x}\, dx, \quad du = dx, \quad v = \int e^{-x}\, dx = -e^{-x}$$

so that

$$\int xe^{-x}\, dx = x(-e^{-x}) - \int -e^{-x}\, dx = -xe^{-x} + \int e^{-x}\, dx = -xe^{-x} - e^{-x} + C$$

Finally, substituting this into the last line of (4) yields

$$\int x^2 e^{-x}\, dx = -x^2 e^{-x} + 2\int xe^{-x}\, dx = -x^2 e^{-x} + 2(-xe^{-x} - e^{-x}) + C$$

$$= -(x^2 + 2x + 2)e^{-x} + C \blacktriangleleft$$

The LIATE method suggests that integrals of the form

$$\int e^{ax} \sin bx\, dx \quad \text{and} \quad \int e^{ax} \cos bx\, dx$$

can be evaluated by letting $u = \sin bx$ or $u = \cos bx$ and $dv = e^{ax}\, dx$. However, this will require a technique that deserves special attention.

▶ **Example 5**   Evaluate $\int e^x \cos x\, dx$.

Solution.   Let

$$u = \cos x, \quad dv = e^x\, dx, \quad du = -\sin x\, dx, \quad v = \int e^x\, dx = e^x$$

Thus,

$$\int e^x \cos x\, dx = \int u\, dv = uv - \int v\, du = e^x \cos x + \int e^x \sin x\, dx \tag{5}$$

Since the integral $\int e^x \sin x\, dx$ is similar in form to the original integral $\int e^x \cos x\, dx$, it seems that nothing has been accomplished. However, let us integrate this new integral by parts. We let

$$u = \sin x, \quad dv = e^x\, dx, \quad du = \cos x\, dx, \quad v = \int e^x\, dx = e^x$$

Thus,

$$\int e^x \sin x\, dx = \int u\, dv = uv - \int v\, du = e^x \sin x - \int e^x \cos x\, dx$$

Together with Equation (5) this yields

$$\int e^x \cos x\, dx = e^x \cos x + e^x \sin x - \int e^x \cos x\, dx \tag{6}$$

which is an equation we can solve for the unknown integral. We obtain

$$2 \int e^x \cos x \, dx = e^x \cos x + e^x \sin x$$

and hence

$$\int e^x \cos x \, dx = \tfrac{1}{2} e^x \cos x + \tfrac{1}{2} e^x \sin x + C \blacktriangleleft$$

■ A TABULAR METHOD FOR REPEATED INTEGRATION BY PARTS

Integrals of the form

$$\int p(x) f(x) \, dx$$

More information on tabular integration by parts can be found in the articles "Tabular Integration by Parts," *College Mathematics Journal*, Vol. 21, 1990, pp. 307–311, by David Horowitz and "More on Tabular Integration by Parts," *College Mathematics Journal*, Vol. 22, 1991, pp. 407–410, by Leonard Gillman.

where $p(x)$ is a polynomial, can sometimes be evaluated using repeated integration by parts in which u is taken to be $p(x)$ or one of its derivatives at each stage. Since du is computed by differentiating u, the repeated differentiation of $p(x)$ will eventually produce 0, at which point you may be left with a simplified integration problem. A convenient method for organizing the computations into two columns is called *tabular integration by parts*.

> **Tabular Integration by Parts**
>
> **Step 1.** Differentiate $p(x)$ repeatedly until you obtain 0, and list the results in the first column.
>
> **Step 2.** Integrate $f(x)$ repeatedly and list the results in the second column.
>
> **Step 3.** Draw an arrow from each entry in the first column to the entry that is one row down in the second column.
>
> **Step 4.** Label the arrows with alternating $+$ and $-$ signs, starting with a $+$.
>
> **Step 5.** For each arrow, form the product of the expressions at its tip and tail and then multiply that product by $+1$ or -1 in accordance with the sign on the arrow. Add the results to obtain the value of the integral.

This process is illustrated in Figure 7.2.1 for the integral $\int (x^2 - x) \cos x \, dx$.

REPEATED DIFFERENTIATION		REPEATED INTEGRATION
$x^2 - x$	$+$	$\cos x$
$2x - 1$	$-$	$\sin x$
2	$+$	$-\cos x$
0		$-\sin x$

$$\int (x^2 - x) \cos x \, dx = (x^2 - x) \sin x + (2x - 1) \cos x - 2 \sin x + C$$
$$= (x^2 - x - 2) \sin x + (2x - 1) \cos x + C$$

▶ **Figure 7.2.1**

▶ **Example 6** In Example 11 of Section 5.3 we evaluated $\int x^2 \sqrt{x - 1} \, dx$ using u-substitution. Evaluate this integral using tabular integration by parts.

Solution.

REPEATED DIFFERENTIATION		REPEATED INTEGRATION
x^2	$+$	$(x-1)^{1/2}$
$2x$	$-$	$\frac{2}{3}(x-1)^{3/2}$
2	$+$	$\frac{4}{15}(x-1)^{5/2}$
0		$\frac{8}{105}(x-1)^{7/2}$

> The result obtained in Example 6 looks quite different from that obtained in Example 11 of Section 5.3. Show that the two answers are equivalent.

Thus, it follows that

$$\int x^2\sqrt{x-1}\,dx = \tfrac{2}{3}x^2(x-1)^{3/2} - \tfrac{8}{15}x(x-1)^{5/2} + \tfrac{16}{105}(x-1)^{7/2} + C \blacktriangleleft$$

■ INTEGRATION BY PARTS FOR DEFINITE INTEGRALS

For definite integrals the formula corresponding to (3) is

$$\int_a^b u\,dv = uv\Big]_a^b - \int_a^b v\,du \tag{7}$$

REMARK It is important to keep in mind that the variables u and v in this formula are functions of x and that the limits of integration in (7) are limits on the variable x. Sometimes it is helpful to emphasize this by writing (7) as

$$\int_{x=a}^b u\,dv = uv\Big]_{x=a}^b - \int_{x=a}^b v\,du \tag{8}$$

The next example illustrates how integration by parts can be used to integrate the inverse trigonometric functions.

▶ **Example 7** Evaluate $\displaystyle\int_0^1 \tan^{-1} x\,dx$.

Solution. Let

$$u = \tan^{-1} x, \quad dv = dx, \quad du = \frac{1}{1+x^2}\,dx, \quad v = x$$

Thus,

$$\int_0^1 \tan^{-1} x\,dx = \int_0^1 u\,dv = uv\Big]_0^1 - \int_0^1 v\,du \qquad \boxed{\text{The limits of integration refer to } x; \text{ that is, } x=0 \text{ and } x=1.}$$

$$= x\tan^{-1} x\Big]_0^1 - \int_0^1 \frac{x}{1+x^2}\,dx$$

But

$$\int_0^1 \frac{x}{1+x^2}\,dx = \frac{1}{2}\int_0^1 \frac{2x}{1+x^2}\,dx = \frac{1}{2}\ln(1+x^2)\Big]_0^1 = \frac{1}{2}\ln 2$$

so

$$\int_0^1 \tan^{-1} x\,dx = x\tan^{-1} x\Big]_0^1 - \frac{1}{2}\ln 2 = \left(\frac{\pi}{4} - 0\right) - \frac{1}{2}\ln 2 = \frac{\pi}{4} - \ln\sqrt{2} \blacktriangleleft$$

■ REDUCTION FORMULAS

Integration by parts can be used to derive *reduction formulas* for integrals. These are formulas that express an integral involving a power of a function in terms of an integral that involves a *lower* power of that function. For example, if n is a positive integer and $n \geq 2$, then integration by parts can be used to obtain the reduction formulas

$$\int \sin^n x \, dx = -\frac{1}{n} \sin^{n-1} x \cos x + \frac{n-1}{n} \int \sin^{n-2} x \, dx \tag{9}$$

$$\int \cos^n x \, dx = \frac{1}{n} \cos^{n-1} x \sin x + \frac{n-1}{n} \int \cos^{n-2} x \, dx \tag{10}$$

To illustrate how such formulas can be obtained, let us derive (10). We begin by writing $\cos^n x$ as $\cos^{n-1} x \cdot \cos x$ and letting

$$u = \cos^{n-1} x \qquad\qquad dv = \cos x \, dx$$
$$du = (n-1) \cos^{n-2} x(-\sin x) \, dx \qquad v = \sin x$$
$$= -(n-1) \cos^{n-2} x \sin x \, dx$$

so that

$$\int \cos^n x \, dx = \int \cos^{n-1} x \cos x \, dx = \int u \, dv = uv - \int v \, du$$

$$= \cos^{n-1} x \sin x + (n-1) \int \sin^2 x \cos^{n-2} x \, dx$$

$$= \cos^{n-1} x \sin x + (n-1) \int (1 - \cos^2 x) \cos^{n-2} x \, dx$$

$$= \cos^{n-1} x \sin x + (n-1) \int \cos^{n-2} x \, dx - (n-1) \int \cos^n x \, dx$$

Moving the last term on the right to the left side yields

$$n \int \cos^n x \, dx = \cos^{n-1} x \sin x + (n-1) \int \cos^{n-2} x \, dx$$

from which (10) follows. The derivation of reduction formula (9) is similar (Exercise 63).

Reduction formulas (9) and (10) reduce the exponent of sine (or cosine) by 2. Thus, if the formulas are applied repeatedly, the exponent can eventually be reduced to 0 if n is even or 1 if n is odd, at which point the integration can be completed. We will discuss this method in more detail in the next section, but for now, here is an example that illustrates how reduction formulas work.

▶ **Example 8** Evaluate $\int \cos^4 x \, dx$.

Solution. From (10) with $n = 4$

$$\int \cos^4 x \, dx = \tfrac{1}{4} \cos^3 x \sin x + \tfrac{3}{4} \int \cos^2 x \, dx \qquad \boxed{\text{Now apply (10) with } n = 2.}$$

$$= \tfrac{1}{4} \cos^3 x \sin x + \tfrac{3}{4} \left(\tfrac{1}{2} \cos x \sin x + \tfrac{1}{2} \int dx \right)$$

$$= \tfrac{1}{4} \cos^3 x \sin x + \tfrac{3}{8} \cos x \sin x + \tfrac{3}{8} x + C \quad ◀$$

✔ **QUICK CHECK EXERCISES 7.2** (See page 500 for answers.)

1. (a) If $G'(x) = g(x)$, then
$$\int f(x)g(x)\,dx = f(x)G(x) - \underline{\hspace{1cm}}$$

   (b) If $u = f(x)$ and $v = G(x)$, then the formula in part (a) can be written in the form $\int u\,dv = \underline{\hspace{1cm}}$.

2. Find an appropriate choice of u and dv for integration by parts of each integral. Do not evaluate the integral.

   (a) $\int x \ln x\,dx$; $u = \underline{\hspace{1cm}}$, $dv = \underline{\hspace{1cm}}$

   (b) $\int (x-2)\sin x\,dx$; $u = \underline{\hspace{1cm}}$, $dv = \underline{\hspace{1cm}}$

   (c) $\int \sin^{-1} x\,dx$; $u = \underline{\hspace{1cm}}$, $dv = \underline{\hspace{1cm}}$

   (d) $\int \dfrac{x}{\sqrt{x-1}}\,dx$; $u = \underline{\hspace{1cm}}$, $dv = \underline{\hspace{1cm}}$

3. Use integration by parts to evaluate the integral.

   (a) $\int x e^{2x}\,dx$ (b) $\int \ln(x-1)\,dx$

   (c) $\int_0^{\pi/6} x \sin 3x\,dx$

4. Use a reduction formula to evaluate $\int \sin^3 x\,dx$.

EXERCISE SET 7.2

1–38 Evaluate the integral. ■

1. $\int x e^{-2x}\,dx$ 2. $\int x e^{3x}\,dx$

3. $\int x^2 e^x\,dx$ 4. $\int x^2 e^{-2x}\,dx$

5. $\int x \sin 3x\,dx$ 6. $\int x \cos 2x\,dx$

7. $\int x^2 \cos x\,dx$ 8. $\int x^2 \sin x\,dx$

9. $\int x \ln x\,dx$ 10. $\int \sqrt{x} \ln x\,dx$

11. $\int (\ln x)^2\,dx$ 12. $\int \dfrac{\ln x}{\sqrt{x}}\,dx$

13. $\int \ln(3x-2)\,dx$ 14. $\int \ln(x^2+4)\,dx$

15. $\int \sin^{-1} x\,dx$ 16. $\int \cos^{-1}(2x)\,dx$

17. $\int \tan^{-1}(3x)\,dx$ 18. $\int x \tan^{-1} x\,dx$

19. $\int e^x \sin x\,dx$ 20. $\int e^{3x} \cos 2x\,dx$

21. $\int \sin(\ln x)\,dx$ 22. $\int \cos(\ln x)\,dx$

23. $\int x \sec^2 x\,dx$ 24. $\int x \tan^2 x\,dx$

25. $\int x^3 e^{x^2}\,dx$ 26. $\int \dfrac{x e^x}{(x+1)^2}\,dx$

27. $\int_0^2 x e^{2x}\,dx$ 28. $\int_0^1 x e^{-5x}\,dx$

29. $\int_1^e x^2 \ln x\,dx$ 30. $\int_{\sqrt{e}}^e \dfrac{\ln x}{x^2}\,dx$

31. $\int_{-1}^1 \ln(x+2)\,dx$ 32. $\int_0^{\sqrt{3}/2} \sin^{-1} x\,dx$

33. $\int_2^4 \sec^{-1}\sqrt{\theta}\,d\theta$ 34. $\int_1^2 x \sec^{-1} x\,dx$

35. $\int_0^\pi x \sin 2x\,dx$ 36. $\int_0^\pi (x + x\cos x)\,dx$

37. $\int_1^3 \sqrt{x} \tan^{-1}\sqrt{x}\,dx$ 38. $\int_0^2 \ln(x^2+1)\,dx$

39–42 True–False Determine whether the statement is true or false. Explain your answer. ■

39. The main goal in integration by parts is to choose u and dv to obtain a new integral that is easier to evaluate than the original.

40. Applying the LIATE strategy to evaluate $\int x^3 \ln x\,dx$, we should choose $u = x^3$ and $dv = \ln x\,dx$.

41. To evaluate $\int \ln e^x\,dx$ using integration by parts, choose $dv = e^x\,dx$.

42. Tabular integration by parts is useful for integrals of the form $\int p(x)f(x)\,dx$, where $p(x)$ is a polynomial and $f(x)$ can be repeatedly integrated.

43–44 Evaluate the integral by making a u-substitution and then integrating by parts. ■

43. $\int e^{\sqrt{x}}\,dx$ 44. $\int \cos\sqrt{x}\,dx$

45. Prove that tabular integration by parts gives the correct answer for
$$\int p(x)f(x)\,dx$$
where $p(x)$ is any *quadratic polynomial* and $f(x)$ is any function that can be repeatedly integrated.

46. The computations of any integral evaluated by repeated integration by parts can be organized using tabular integration by parts. Use this organization to evaluate $\int e^x \cos x\,dx$ in

two ways: first by repeated differentiation of $\cos x$ (compare Example 5), and then by repeated differentiation of e^x.

47–52 Evaluate the integral using tabular integration by parts. ∎

47. $\displaystyle \int (3x^2 - x + 2)e^{-x}\, dx$ **48.** $\displaystyle \int (x^2 + x + 1) \sin x\, dx$

49. $\displaystyle \int 4x^4 \sin 2x\, dx$ **50.** $\displaystyle \int x^3 \sqrt{2x + 1}\, dx$

51. $\displaystyle \int e^{ax} \sin bx\, dx$ **52.** $\displaystyle \int e^{-3\theta} \sin 5\theta\, d\theta$

53. Consider the integral $\int \sin x \cos x\, dx$.
   (a) Evaluate the integral two ways: first using integration by parts, and then using the substitution $u = \sin x$.
   (b) Show that the results of part (a) are equivalent.
   (c) Which of the two methods do you prefer? Discuss the reasons for your preference.

54. Evaluate the integral
$$\int_0^1 \frac{x^3}{\sqrt{x^2 + 1}}\, dx$$
using
   (a) integration by parts
   (b) the substitution $u = \sqrt{x^2 + 1}$.

55. (a) Find the area of the region enclosed by $y = \ln x$, the line $x = e$, and the x-axis.
   (b) Find the volume of the solid generated when the region in part (a) is revolved about the x-axis.

56. Find the area of the region between $y = x \sin x$ and $y = x$ for $0 \le x \le \pi/2$.

57. Find the volume of the solid generated when the region between $y = \sin x$ and $y = 0$ for $0 \le x \le \pi$ is revolved about the y-axis.

58. Find the volume of the solid generated when the region enclosed between $y = \cos x$ and $y = 0$ for $0 \le x \le \pi/2$ is revolved about the y-axis.

59. A particle moving along the x-axis has velocity function $v(t) = t^3 \sin t$. How far does the particle travel from time $t = 0$ to $t = \pi$?

60. The study of sawtooth waves in electrical engineering leads to integrals of the form
$$\int_{-\pi/\omega}^{\pi/\omega} t \sin(k\omega t)\, dt$$
where k is an integer and ω is a nonzero constant. Evaluate the integral.

61. Use reduction formula (9) to evaluate
   (a) $\displaystyle \int \sin^4 x\, dx$ (b) $\displaystyle \int_0^{\pi/2} \sin^5 x\, dx$.

62. Use reduction formula (10) to evaluate
   (a) $\displaystyle \int \cos^5 x\, dx$ (b) $\displaystyle \int_0^{\pi/2} \cos^6 x\, dx$.

63. Derive reduction formula (9).

64. In each part, use integration by parts or other methods to derive the reduction formula.
   (a) $\displaystyle \int \sec^n x\, dx = \frac{\sec^{n-2} x \tan x}{n - 1} + \frac{n - 2}{n - 1} \int \sec^{n-2} x\, dx$
   (b) $\displaystyle \int \tan^n x\, dx = \frac{\tan^{n-1} x}{n - 1} - \int \tan^{n-2} x\, dx$
   (c) $\displaystyle \int x^n e^x\, dx = x^n e^x - n \int x^{n-1} e^x\, dx$

65–66 Use the reduction formulas in Exercise 64 to evaluate the integrals. ∎

65. (a) $\displaystyle \int \tan^4 x\, dx$ (b) $\displaystyle \int \sec^4 x\, dx$ (c) $\displaystyle \int x^3 e^x\, dx$

66. (a) $\displaystyle \int x^2 e^{3x}\, dx$ (b) $\displaystyle \int_0^1 x e^{-\sqrt{x}}\, dx$
   [*Hint:* First make a substitution.]

67. Let f be a function whose second derivative is continuous on $[-1, 1]$. Show that
$$\int_{-1}^1 x f''(x)\, dx = f'(1) + f'(-1) - f(1) + f(-1)$$

FOCUS ON CONCEPTS

68. (a) In the integral $\int x \cos x\, dx$, let
$$u = x, \quad dv = \cos x\, dx,$$
$$du = dx, \quad v = \sin x + C_1$$
Show that the constant C_1 cancels out, thus giving the same solution obtained by omitting C_1.
   (b) Show that in general
$$uv - \int v\, du = u(v + C_1) - \int (v + C_1)\, du$$
thereby justifying the omission of the constant of integration when calculating v in integration by parts.

69. Evaluate $\int \ln(x + 1)\, dx$ using integration by parts. Simplify the computation of $\int v\, du$ by introducing a constant of integration $C_1 = 1$ when going from dv to v.

70. Evaluate $\int \ln(3x - 2)\, dx$ using integration by parts. Simplify the computation of $\int v\, du$ by introducing a constant of integration $C_1 = -\frac{2}{3}$ when going from dv to v. Compare your solution with your answer to Exercise 13.

71. Evaluate $\int x \tan^{-1} x\, dx$ using integration by parts. Simplify the computation of $\int v\, du$ by introducing a constant of integration $C_1 = \frac{1}{2}$ when going from dv to v.

72. What equation results if integration by parts is applied to the integral
$$\int \frac{1}{x \ln x}\, dx$$
with the choices
$$u = \frac{1}{\ln x} \quad \text{and} \quad dv = \frac{1}{x}\, dx?$$
In what sense is this equation true? In what sense is it false?

73. Writing Explain how the product rule for derivatives and the technique of integration by parts are related.

74. Writing For what sort of problems are the integration techniques of substitution and integration by parts "competing"

techniques? Describe situations, with examples, where each of these techniques would be preferred over the other.

✔ **QUICK CHECK ANSWERS 7.2**

1. (a) $\int f'(x)G(x)\,dx$ (b) $uv - \int v\,du$ **2.** (a) $\ln x$; $x\,dx$ (b) $x-2$; $\sin x\,dx$ (c) $\sin^{-1}x$; dx (d) x; $\dfrac{1}{\sqrt{x-1}}\,dx$

3. (a) $\left(\dfrac{x}{2} - \dfrac{1}{4}\right)e^{2x} + C$ (b) $(x-1)\ln(x-1) - x + C$ (c) $\frac{1}{9}$ **4.** $-\frac{1}{3}\sin^2 x \cos x - \frac{2}{3}\cos x + C$

7.3 INTEGRATING TRIGONOMETRIC FUNCTIONS

In the last section we derived reduction formulas for integrating positive integer powers of sine, cosine, tangent, and secant. In this section we will show how to work with those reduction formulas, and we will discuss methods for integrating other kinds of integrals that involve trigonometric functions.

■ INTEGRATING POWERS OF SINE AND COSINE

We begin by recalling two reduction formulas from the preceding section.

$$\int \sin^n x\,dx = -\frac{1}{n}\sin^{n-1}x\cos x + \frac{n-1}{n}\int \sin^{n-2}x\,dx \tag{1}$$

$$\int \cos^n x\,dx = \frac{1}{n}\cos^{n-1}x\sin x + \frac{n-1}{n}\int \cos^{n-2}x\,dx \tag{2}$$

In the case where $n = 2$, these formulas yield

$$\int \sin^2 x\,dx = -\tfrac{1}{2}\sin x \cos x + \tfrac{1}{2}\int dx = \tfrac{1}{2}x - \tfrac{1}{2}\sin x \cos x + C \tag{3}$$

$$\int \cos^2 x\,dx = \tfrac{1}{2}\cos x \sin x + \tfrac{1}{2}\int dx = \tfrac{1}{2}x + \tfrac{1}{2}\sin x \cos x + C \tag{4}$$

Alternative forms of these integration formulas can be derived from the trigonometric identities

$$\sin^2 x = \tfrac{1}{2}(1 - \cos 2x) \quad \text{and} \quad \cos^2 x = \tfrac{1}{2}(1 + \cos 2x) \tag{5–6}$$

which follow from the double-angle formulas

$$\cos 2x = 1 - 2\sin^2 x \quad \text{and} \quad \cos 2x = 2\cos^2 x - 1$$

These identities yield

$$\int \sin^2 x\,dx = \tfrac{1}{2}\int (1 - \cos 2x)\,dx = \tfrac{1}{2}x - \tfrac{1}{4}\sin 2x + C \tag{7}$$

$$\int \cos^2 x\,dx = \tfrac{1}{2}\int (1 + \cos 2x)\,dx = \tfrac{1}{2}x + \tfrac{1}{4}\sin 2x + C \tag{8}$$

Observe that the antiderivatives in Formulas (3) and (4) involve both sines and cosines, whereas those in (7) and (8) involve sines alone. However, the apparent discrepancy is easy to resolve by using the identity

$$\sin 2x = 2 \sin x \cos x$$

to rewrite (7) and (8) in forms (3) and (4), or conversely.

In the case where $n = 3$, the reduction formulas for integrating $\sin^3 x$ and $\cos^3 x$ yield

$$\int \sin^3 x \, dx = -\tfrac{1}{3} \sin^2 x \cos x + \tfrac{2}{3} \int \sin x \, dx = -\tfrac{1}{3} \sin^2 x \cos x - \tfrac{2}{3} \cos x + C \quad (9)$$

$$\int \cos^3 x \, dx = \tfrac{1}{3} \cos^2 x \sin x + \tfrac{2}{3} \int \cos x \, dx = \tfrac{1}{3} \cos^2 x \sin x + \tfrac{2}{3} \sin x + C \quad (10)$$

If desired, Formula (9) can be expressed in terms of cosines alone by using the identity $\sin^2 x = 1 - \cos^2 x$, and Formula (10) can be expressed in terms of sines alone by using the identity $\cos^2 x = 1 - \sin^2 x$. We leave it for you to do this and confirm that

$$\int \sin^3 x \, dx = \tfrac{1}{3} \cos^3 x - \cos x + C \quad (11)$$

$$\int \cos^3 x \, dx = \sin x - \tfrac{1}{3} \sin^3 x + C \quad (12)$$

We leave it as an exercise to obtain the following formulas by first applying the reduction formulas, and then using appropriate trigonometric identities.

$$\int \sin^4 x \, dx = \tfrac{3}{8}x - \tfrac{1}{4} \sin 2x + \tfrac{1}{32} \sin 4x + C \quad (13)$$

$$\int \cos^4 x \, dx = \tfrac{3}{8}x + \tfrac{1}{4} \sin 2x + \tfrac{1}{32} \sin 4x + C \quad (14)$$

▶ **Example 1**   Find the volume V of the solid that is obtained when the region under the curve $y = \sin^2 x$ over the interval $[0, \pi]$ is revolved about the x-axis (Figure 7.3.1).

Solution.   Using the method of disks, Formula (5) of Section 6.2, and Formula (13) above yields

$$V = \int_0^\pi \pi \sin^4 x \, dx = \pi \left[\tfrac{3}{8}x - \tfrac{1}{4} \sin 2x + \tfrac{1}{32} \sin 4x \right]_0^\pi = \tfrac{3}{8} \pi^2 \quad ◄$$

▲ **Figure 7.3.1**

■ **INTEGRATING PRODUCTS OF SINES AND COSINES**

If m and n are positive integers, then the integral

$$\int \sin^m x \cos^n x \, dx$$

can be evaluated by one of the three procedures stated in Table 7.3.1, depending on whether m and n are odd or even.

▶ **Example 2**   Evaluate

(a) $\displaystyle\int \sin^4 x \cos^5 x \, dx$ (b) $\displaystyle\int \sin^4 x \cos^4 x \, dx$

Table 7.3.1

INTEGRATING PRODUCTS OF SINES AND COSINES

$\int \sin^m x \cos^n x \, dx$	PROCEDURE	RELEVANT IDENTITIES
n odd	• Split off a factor of $\cos x$. • Apply the relevant identity. • Make the substitution $u = \sin x$.	$\cos^2 x = 1 - \sin^2 x$
m odd	• Split off a factor of $\sin x$. • Apply the relevant identity. • Make the substitution $u = \cos x$.	$\sin^2 x = 1 - \cos^2 x$
$\begin{cases} m \text{ even} \\ n \text{ even} \end{cases}$	• Use the relevant identities to reduce the powers on $\sin x$ and $\cos x$.	$\begin{cases} \sin^2 x = \frac{1}{2}(1 - \cos 2x) \\ \cos^2 x = \frac{1}{2}(1 + \cos 2x) \end{cases}$

Solution (a).   Since $n = 5$ is odd, we will follow the first procedure in Table 7.3.1:

$$\int \sin^4 x \cos^5 x \, dx = \int \sin^4 x \cos^4 x \cos x \, dx$$

$$= \int \sin^4 x (1 - \sin^2 x)^2 \cos x \, dx$$

$$= \int u^4 (1 - u^2)^2 \, du$$

$$= \int (u^4 - 2u^6 + u^8) \, du$$

$$= \tfrac{1}{5} u^5 - \tfrac{2}{7} u^7 + \tfrac{1}{9} u^9 + C$$

$$= \tfrac{1}{5} \sin^5 x - \tfrac{2}{7} \sin^7 x + \tfrac{1}{9} \sin^9 x + C$$

Solution (b).   Since $m = n = 4$, both exponents are even, so we will follow the third procedure in Table 7.3.1:

$$\int \sin^4 x \cos^4 x \, dx = \int (\sin^2 x)^2 (\cos^2 x)^2 \, dx$$

$$= \int \left(\tfrac{1}{2}[1 - \cos 2x] \right)^2 \left(\tfrac{1}{2}[1 + \cos 2x] \right)^2 \, dx$$

$$= \tfrac{1}{16} \int (1 - \cos^2 2x)^2 \, dx$$

$$= \tfrac{1}{16} \int \sin^4 2x \, dx \qquad \boxed{\text{Note that this can be obtained more directly from the original integral using the identity } \sin x \cos x = \tfrac{1}{2} \sin 2x.}$$

$$= \tfrac{1}{32} \int \sin^4 u \, du \qquad \boxed{\begin{array}{l} u = 2x \\ du = 2\,dx \text{ or } dx = \tfrac{1}{2}\,du \end{array}}$$

$$= \tfrac{1}{32} \left(\tfrac{3}{8} u - \tfrac{1}{4} \sin 2u + \tfrac{1}{32} \sin 4u \right) + C \qquad \boxed{\text{Formula (13)}}$$

$$= \tfrac{3}{128} x - \tfrac{1}{128} \sin 4x + \tfrac{1}{1024} \sin 8x + C \quad \blacktriangleleft$$

Integrals of the form

$$\int \sin mx \cos nx\, dx, \qquad \int \sin mx \sin nx\, dx, \qquad \int \cos mx \cos nx\, dx \qquad (15)$$

can be found by using the trigonometric identities

$$\sin \alpha \cos \beta = \tfrac{1}{2}[\sin(\alpha - \beta) + \sin(\alpha + \beta)] \qquad (16)$$

$$\sin \alpha \sin \beta = \tfrac{1}{2}[\cos(\alpha - \beta) - \cos(\alpha + \beta)] \qquad (17)$$

$$\cos \alpha \cos \beta = \tfrac{1}{2}[\cos(\alpha - \beta) + \cos(\alpha + \beta)] \qquad (18)$$

to express the integrand as a sum or difference of sines and cosines.

▶ **Example 3** Evaluate $\int \sin 7x \cos 3x\, dx$.

Solution. Using (16) yields

$$\int \sin 7x \cos 3x\, dx = \tfrac{1}{2}\int (\sin 4x + \sin 10x)\, dx = -\tfrac{1}{8}\cos 4x - \tfrac{1}{20}\cos 10x + C \quad \blacktriangleleft$$

■ INTEGRATING POWERS OF TANGENT AND SECANT

The procedures for integrating powers of tangent and secant closely parallel those for sine and cosine. The idea is to use the following reduction formulas (which were derived in Exercise 64 of Section 7.2) to reduce the exponent in the integrand until the resulting integral can be evaluated:

$$\int \tan^n x\, dx = \frac{\tan^{n-1} x}{n-1} - \int \tan^{n-2} x\, dx \qquad (19)$$

$$\int \sec^n x\, dx = \frac{\sec^{n-2} x \tan x}{n-1} + \frac{n-2}{n-1}\int \sec^{n-2} x\, dx \qquad (20)$$

In the case where n is odd, the exponent can be reduced to 1, leaving us with the problem of integrating $\tan x$ or $\sec x$. These integrals are given by

$$\int \tan x\, dx = \ln |\sec x| + C \qquad (21)$$

$$\int \sec x\, dx = \ln |\sec x + \tan x| + C \qquad (22)$$

Formula (21) can be obtained by writing

$$\int \tan x\, dx = \int \frac{\sin x}{\cos x}\, dx$$

$$= -\ln |\cos x| + C \qquad \boxed{\begin{array}{l} u = \cos x \\ du = -\sin x\, dx \end{array}}$$

$$= \ln |\sec x| + C \qquad \boxed{\ln |\cos x| = -\ln \frac{1}{|\cos x|}}$$

To obtain Formula (22) we write

$$\int \sec x\, dx = \int \sec x \left(\frac{\sec x + \tan x}{\sec x + \tan x} \right) dx = \int \frac{\sec^2 x + \sec x \tan x}{\sec x + \tan x}\, dx$$

$$= \ln |\sec x + \tan x| + C \qquad \boxed{\begin{array}{l} u = \sec x + \tan x \\ du = (\sec^2 x + \sec x \tan x)\, dx \end{array}}$$

The following basic integrals occur frequently and are worth noting:

$$\int \tan^2 x \, dx = \tan x - x + C \tag{23}$$

$$\int \sec^2 x \, dx = \tan x + C \tag{24}$$

Formula (24) is already known to us, since the derivative of $\tan x$ is $\sec^2 x$. Formula (23) can be obtained by applying reduction formula (19) with $n = 2$ (verify) or, alternatively, by using the identity

$$1 + \tan^2 x = \sec^2 x$$

to write

$$\int \tan^2 x \, dx = \int (\sec^2 x - 1) \, dx = \tan x - x + C$$

The formulas

$$\int \tan^3 x \, dx = \tfrac{1}{2} \tan^2 x - \ln |\sec x| + C \tag{25}$$

$$\int \sec^3 x \, dx = \tfrac{1}{2} \sec x \tan x + \tfrac{1}{2} \ln |\sec x + \tan x| + C \tag{26}$$

can be deduced from (21), (22), and reduction formulas (19) and (20) as follows:

$$\int \tan^3 x \, dx = \tfrac{1}{2} \tan^2 x - \int \tan x \, dx = \tfrac{1}{2} \tan^2 x - \ln |\sec x| + C$$

$$\int \sec^3 x \, dx = \tfrac{1}{2} \sec x \tan x + \tfrac{1}{2} \int \sec x \, dx = \tfrac{1}{2} \sec x \tan x + \tfrac{1}{2} \ln |\sec x + \tan x| + C$$

■ INTEGRATING PRODUCTS OF TANGENTS AND SECANTS

If m and n are positive integers, then the integral

$$\int \tan^m x \, \sec^n x \, dx$$

can be evaluated by one of the three procedures stated in Table 7.3.2, depending on whether m and n are odd or even.

Table 7.3.2

INTEGRATING PRODUCTS OF TANGENTS AND SECANTS

$\int \tan^m x \, \sec^n x \, dx$	PROCEDURE	RELEVANT IDENTITIES
n even	• Split off a factor of $\sec^2 x$. • Apply the relevant identity. • Make the substitution $u = \tan x$.	$\sec^2 x = \tan^2 x + 1$
m odd	• Split off a factor of $\sec x \tan x$. • Apply the relevant identity. • Make the substitution $u = \sec x$.	$\tan^2 x = \sec^2 x - 1$
$\begin{cases} m \text{ even} \\ n \text{ odd} \end{cases}$	• Use the relevant identities to reduce the integrand to powers of $\sec x$ alone. • Then use the reduction formula for powers of $\sec x$.	$\tan^2 x = \sec^2 x - 1$

▶ **Example 4** Evaluate

$$\text{(a) } \int \tan^2 x \sec^4 x \, dx \qquad \text{(b) } \int \tan^3 x \sec^3 x \, dx \qquad \text{(c) } \int \tan^2 x \sec x \, dx$$

Solution (a). Since $n = 4$ is even, we will follow the first procedure in Table 7.3.2:

$$\int \tan^2 x \sec^4 x \, dx = \int \tan^2 x \sec^2 x \sec^2 x \, dx$$

$$= \int \tan^2 x (\tan^2 x + 1) \sec^2 x \, dx$$

$$= \int u^2 (u^2 + 1) \, du$$

$$= \tfrac{1}{5} u^5 + \tfrac{1}{3} u^3 + C = \tfrac{1}{5} \tan^5 x + \tfrac{1}{3} \tan^3 x + C$$

Solution (b). Since $m = 3$ is odd, we will follow the second procedure in Table 7.3.2:

$$\int \tan^3 x \sec^3 x \, dx = \int \tan^2 x \sec^2 x (\sec x \tan x) \, dx$$

$$= \int (\sec^2 x - 1) \sec^2 x (\sec x \tan x) \, dx$$

$$= \int (u^2 - 1) u^2 \, du$$

$$= \tfrac{1}{5} u^5 - \tfrac{1}{3} u^3 + C = \tfrac{1}{5} \sec^5 x - \tfrac{1}{3} \sec^3 x + C$$

Solution (c). Since $m = 2$ is even and $n = 1$ is odd, we will follow the third procedure in Table 7.3.2:

$$\int \tan^2 x \sec x \, dx = \int (\sec^2 x - 1) \sec x \, dx$$

$$= \int \sec^3 x \, dx - \int \sec x \, dx \qquad \boxed{\text{See (26) and (22).}}$$

$$= \tfrac{1}{2} \sec x \tan x + \tfrac{1}{2} \ln |\sec x + \tan x| - \ln |\sec x + \tan x| + C$$

$$= \tfrac{1}{2} \sec x \tan x - \tfrac{1}{2} \ln |\sec x + \tan x| + C \ \blacktriangleleft$$

■ **AN ALTERNATIVE METHOD FOR INTEGRATING POWERS OF SINE, COSINE, TANGENT, AND SECANT**

The methods in Tables 7.3.1 and 7.3.2 can sometimes be applied if $m = 0$ or $n = 0$ to integrate positive integer powers of sine, cosine, tangent, and secant without reduction formulas. For example, instead of using the reduction formula to integrate $\sin^3 x$, we can apply the second procedure in Table 7.3.1:

$$\int \sin^3 x \, dx = \int (\sin^2 x) \sin x \, dx$$

$$= \int (1 - \cos^2 x) \sin x \, dx \qquad \boxed{\begin{array}{l} u = \cos x \\ du = -\sin x \, dx \end{array}}$$

$$= -\int (1 - u^2) \, du$$

$$= \tfrac{1}{3} u^3 - u + C = \tfrac{1}{3} \cos^3 x - \cos x + C$$

which agrees with (11).

With the aid of the identity

$$1 + \cot^2 x = \csc^2 x$$

the techniques in Table 7.3.2 can be adapted to evaluate integrals of the form

$$\int \cot^m x \csc^n x \, dx$$

It is also possible to derive reduction formulas for powers of cot and csc that are analogous to Formulas (19) and (20).

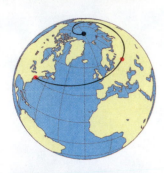

▲ Figure 7.3.2 A flight path with constant compass heading from New York City to Moscow follows a spiral toward the North Pole but is a straight line segment on a Mercator projection.

■ MERCATOR'S MAP OF THE WORLD

The integral of sec x plays an important role in the design of navigational maps for charting nautical and aeronautical courses. Sailors and pilots usually chart their courses along paths with constant compass headings; for example, the course might be 30° northeast or 135° southeast. Except for courses that are parallel to the equator or run due north or south, a course with constant compass heading spirals around the Earth toward one of the poles (as in the top part of Figure 7.3.2). In 1569 the Flemish mathematician and geographer Gerhard Kramer (1512–1594) (better known by the Latin name Mercator) devised a world map, called the *Mercator projection*, in which spirals of constant compass headings appear as straight lines. This was extremely important because it enabled sailors to determine compass headings between two points by connecting them with a straight line on a map (as in the bottom part of Figure 7.3.2).

If the Earth is assumed to be a sphere of radius 4000 mi, then the lines of latitude at 1° increments are equally spaced about 70 mi apart (why?). However, in the Mercator projection, the lines of latitude become wider apart toward the poles, so that two widely spaced latitude lines near the poles may be actually the same distance apart on the Earth as two closely spaced latitude lines near the equator. It can be proved that on a Mercator map in which the equatorial line has length L, the vertical distance D_β on the map between the equator (latitude 0°) and the line of latitude $\beta°$ is

$$D_\beta = \frac{L}{2\pi} \int_0^{\beta\pi/180} \sec x \, dx \qquad (27)$$

✔ QUICK CHECK EXERCISES 7.3   (See page 508 for answers.)

1. Complete each trigonometric identity with an expression involving cos $2x$.
   (a) $\sin^2 x =$ _____   (b) $\cos^2 x =$ _____
   (c) $\cos^2 x - \sin^2 x =$ _____

2. Evaluate the integral.
   (a) $\displaystyle\int \sec^2 x \, dx =$ _____
   (b) $\displaystyle\int \tan^2 x \, dx =$ _____
   (c) $\displaystyle\int \sec x \, dx =$ _____
   (d) $\displaystyle\int \tan x \, dx =$ _____

3. Use the indicated substitution to rewrite the integral in terms of u. Do not evaluate the integral.
   (a) $\displaystyle\int \sin^2 x \cos x \, dx$; $u = \sin x$
   (b) $\displaystyle\int \sin^3 x \cos^2 x \, dx$; $u = \cos x$
   (c) $\displaystyle\int \tan^3 x \sec^2 x \, dx$; $u = \tan x$
   (d) $\displaystyle\int \tan^3 x \sec x \, dx$; $u = \sec x$

EXERCISE SET 7.3

1–52 Evaluate the integral. ■

1. $\displaystyle\int \cos^3 x \sin x \, dx$

2. $\displaystyle\int \sin^5 3x \cos 3x \, dx$

3. $\displaystyle\int \sin^2 5\theta \, d\theta$

4. $\displaystyle\int \cos^2 3x \, dx$

5. $\displaystyle\int \sin^3 a\theta \, d\theta$

6. $\displaystyle\int \cos^3 at \, dt$

7. $\displaystyle\int \sin ax \cos ax \, dx$

8. $\displaystyle\int \sin^3 x \cos^3 x \, dx$

9. $\displaystyle\int \sin^2 t \cos^3 t \, dt$

10. $\displaystyle\int \sin^3 x \cos^2 x \, dx$

11. $\displaystyle\int \sin^2 x \cos^2 x \, dx$

12. $\displaystyle\int \sin^2 x \cos^4 x \, dx$

13. $\displaystyle\int \sin 2x \cos 3x \, dx$

14. $\displaystyle\int \sin 3\theta \cos 2\theta \, d\theta$

15. $\displaystyle\int \sin x \cos(x/2) \, dx$

16. $\displaystyle\int \cos^{1/3} x \sin x \, dx$

17. $\displaystyle\int_0^{\pi/2} \cos^3 x \, dx$

18. $\displaystyle\int_0^{\pi/2} \sin^2 \frac{x}{2} \cos^2 \frac{x}{2} \, dx$

19. $\displaystyle\int_0^{\pi/3} \sin^4 3x \cos^3 3x \, dx$

20. $\displaystyle\int_{-\pi}^{\pi} \cos^2 5\theta \, d\theta$

21. $\displaystyle\int_0^{\pi/6} \sin 4x \cos 2x \, dx$

22. $\displaystyle\int_0^{2\pi} \sin^2 kx \, dx$

23. $\displaystyle\int \sec^2(2x - 1) \, dx$

24. $\displaystyle\int \tan 5x \, dx$

25. $\displaystyle\int e^{-x} \tan(e^{-x}) \, dx$

26. $\displaystyle\int \cot 3x \, dx$

27. $\displaystyle\int \sec 4x \, dx$

28. $\displaystyle\int \frac{\sec(\sqrt{x})}{\sqrt{x}} \, dx$

29. $\displaystyle\int \tan^2 x \sec^2 x \, dx$

30. $\displaystyle\int \tan^5 x \sec^4 x \, dx$

31. $\displaystyle\int \tan 4x \sec^4 4x \, dx$

32. $\displaystyle\int \tan^4 \theta \sec^4 \theta \, d\theta$

33. $\displaystyle\int \sec^5 x \tan^3 x \, dx$

34. $\displaystyle\int \tan^5 \theta \sec \theta \, d\theta$

35. $\displaystyle\int \tan^4 x \sec x \, dx$

36. $\displaystyle\int \tan^2 x \sec^3 x \, dx$

37. $\displaystyle\int \tan t \sec^3 t \, dt$

38. $\displaystyle\int \tan x \sec^5 x \, dx$

39. $\displaystyle\int \sec^4 x \, dx$

40. $\displaystyle\int \sec^5 x \, dx$

41. $\displaystyle\int \tan^3 4x \, dx$

42. $\displaystyle\int \tan^4 x \, dx$

43. $\displaystyle\int \sqrt{\tan x} \sec^4 x \, dx$

44. $\displaystyle\int \tan x \sec^{3/2} x \, dx$

45. $\displaystyle\int_0^{\pi/8} \tan^2 2x \, dx$

46. $\displaystyle\int_0^{\pi/6} \sec^3 2\theta \tan 2\theta \, d\theta$

47. $\displaystyle\int_0^{\pi/2} \tan^5 \frac{x}{2} \, dx$

48. $\displaystyle\int_0^{1/4} \sec \pi x \tan \pi x \, dx$

49. $\displaystyle\int \cot^3 x \csc^3 x \, dx$

50. $\displaystyle\int \cot^2 3t \sec 3t \, dt$

51. $\displaystyle\int \cot^3 x \, dx$

52. $\displaystyle\int \csc^4 x \, dx$

53–56 True–False Determine whether the statement is true or false. Explain your answer. ∎

53. To evaluate $\int \sin^5 x \cos^8 x \, dx$, use the trigonometric identity $\sin^2 x = 1 - \cos^2 x$ and the substitution $u = \cos x$.

54. To evaluate $\int \sin^8 x \cos^5 x \, dx$, use the trigonometric identity $\sin^2 x = 1 - \cos^2 x$ and the substitution $u = \cos x$.

55. The trigonometric identity
$$\sin \alpha \cos \beta = \tfrac{1}{2}[\sin(\alpha - \beta) + \sin(\alpha + \beta)]$$
is often useful for evaluating integrals of the form $\int \sin^m x \cos^n x \, dx$.

56. The integral $\int \tan^4 x \sec^5 x \, dx$ is equivalent to one whose integrand is a polynomial in $\sec x$.

57. Let m, n be distinct nonnegative integers. Use Formulas (16)–(18) to prove:

(a) $\displaystyle\int_0^{2\pi} \sin mx \cos nx \, dx = 0$

(b) $\displaystyle\int_0^{2\pi} \cos mx \cos nx \, dx = 0$

(c) $\displaystyle\int_0^{2\pi} \sin mx \sin nx \, dx = 0.$

58. Evaluate the integrals in Exercise 57 when m and n denote the *same* nonnegative integer.

59. Find the arc length of the curve $y = \ln(\cos x)$ over the interval $[0, \pi/4]$.

60. Find the volume of the solid generated when the region enclosed by $y = \tan x$, $y = 1$, and $x = 0$ is revolved about the x-axis.

61. Find the volume of the solid that results when the region enclosed by $y = \cos x$, $y = \sin x$, $x = 0$, and $x = \pi/4$ is revolved about the x-axis.

62. The region bounded below by the x-axis and above by the portion of $y = \sin x$ from $x = 0$ to $x = \pi$ is revolved about the x-axis. Find the volume of the resulting solid.

63. Use Formula (27) to show that if the length of the equatorial line on a Mercator projection is L, then the vertical distance D between the latitude lines at $\alpha°$ and $\beta°$ on the same side of the equator (where $\alpha < \beta$) is
$$D = \frac{L}{2\pi} \ln \left| \frac{\sec \beta° + \tan \beta°}{\sec \alpha° + \tan \alpha°} \right|$$

64. Suppose that the equator has a length of 100 cm on a Mercator projection. In each part, use the result in Exercise 63 to answer the question.
(a) What is the vertical distance on the map between the equator and the line at 25° north latitude?
(b) What is the vertical distance on the map between New Orleans, Louisiana, at 30° north latitude and Winnipeg, Canada, at 50° north latitude?

FOCUS ON CONCEPTS

65. (a) Show that
$$\int \csc x \, dx = -\ln|\csc x + \cot x| + C$$
(b) Show that the result in part (a) can also be written as
$$\int \csc x \, dx = \ln|\csc x - \cot x| + C$$
and
$$\int \csc x \, dx = \ln \left| \tan \tfrac{1}{2}x \right| + C$$

66. Rewrite $\sin x + \cos x$ in the form

$$A \sin(x + \phi)$$

and use your result together with Exercise 65 to evaluate

$$\int \frac{dx}{\sin x + \cos x}$$

67. Use the method of Exercise 66 to evaluate

$$\int \frac{dx}{a \sin x + b \cos x} \quad (a, b \text{ not both zero})$$

68. (a) Use Formula (9) in Section 7.2 to show that

$$\int_0^{\pi/2} \sin^n x \, dx = \frac{n-1}{n} \int_0^{\pi/2} \sin^{n-2} x \, dx \quad (n \ge 2)$$

(b) Use this result to derive the **Wallis sine formulas**:

$$\int_0^{\pi/2} \sin^n x \, dx = \frac{\pi}{2} \cdot \frac{1 \cdot 3 \cdot 5 \cdots (n-1)}{2 \cdot 4 \cdot 6 \cdots n} \quad \binom{n \text{ even}}{\text{and } \ge 2}$$

$$\int_0^{\pi/2} \sin^n x \, dx = \frac{2 \cdot 4 \cdot 6 \cdots (n-1)}{3 \cdot 5 \cdot 7 \cdots n} \quad \binom{n \text{ odd}}{\text{and } \ge 3}$$

69. Use the Wallis formulas in Exercise 68 to evaluate

(a) $\displaystyle\int_0^{\pi/2} \sin^3 x \, dx$ **(b)** $\displaystyle\int_0^{\pi/2} \sin^4 x \, dx$

(c) $\displaystyle\int_0^{\pi/2} \sin^5 x \, dx$ **(d)** $\displaystyle\int_0^{\pi/2} \sin^6 x \, dx.$

70. Use Formula (10) in Section 7.2 and the method of Exercise 68 to derive the **Wallis cosine formulas**:

$$\int_0^{\pi/2} \cos^n x \, dx = \frac{\pi}{2} \cdot \frac{1 \cdot 3 \cdot 5 \cdots (n-1)}{2 \cdot 4 \cdot 6 \cdots n} \quad \binom{n \text{ even}}{\text{and } \ge 2}$$

$$\int_0^{\pi/2} \cos^n x \, dx = \frac{2 \cdot 4 \cdot 6 \cdots (n-1)}{3 \cdot 5 \cdot 7 \cdots n} \quad \binom{n \text{ odd}}{\text{and } \ge 3}$$

71. Writing Describe the various approaches for evaluating integrals of the form

$$\int \sin^m x \cos^n x \, dx$$

Into what cases do these types of integrals fall? What procedures and identities are used in each case?

72. Writing Describe the various approaches for evaluating integrals of the form

$$\int \tan^m x \sec^n x \, dx$$

Into what cases do these types of integrals fall? What procedures and identities are used in each case?

✔**QUICK CHECK ANSWERS 7.3**

1. (a) $\dfrac{1 - \cos 2x}{2}$ **(b)** $\dfrac{1 + \cos 2x}{2}$ **(c)** $\cos 2x$ **2. (a)** $\tan x + C$ **(b)** $\tan x - x + C$ **(c)** $\ln|\sec x + \tan x| + C$ **(d)** $\ln|\sec x| + C$

3. (a) $\displaystyle\int u^2 \, du$ **(b)** $\displaystyle\int (u^2 - 1)u^2 \, du$ **(c)** $\displaystyle\int u^3 \, du$ **(d)** $\displaystyle\int (u^2 - 1) \, du$

7.4 TRIGONOMETRIC SUBSTITUTIONS

In this section we will discuss a method for evaluating integrals containing radicals by making substitutions involving trigonometric functions. We will also show how integrals containing quadratic polynomials can sometimes be evaluated by completing the square.

■ THE METHOD OF TRIGONOMETRIC SUBSTITUTION

To start, we will be concerned with integrals that contain expressions of the form

$$\sqrt{a^2 - x^2}, \quad \sqrt{x^2 + a^2}, \quad \sqrt{x^2 - a^2}$$

in which a is a positive constant. The basic idea for evaluating such integrals is to make a substitution for x that will eliminate the radical. For example, to eliminate the radical in the expression $\sqrt{a^2 - x^2}$, we can make the substitution

$$x = a \sin\theta, \quad -\pi/2 \le \theta \le \pi/2 \tag{1}$$

which yields

$$\sqrt{a^2 - x^2} = \sqrt{a^2 - a^2 \sin^2\theta} = \sqrt{a^2(1 - \sin^2\theta)}$$

$$= a\sqrt{\cos^2\theta} = a|\cos\theta| = a\cos\theta \quad \boxed{\cos\theta \ge 0 \text{ since } -\pi/2 \le \theta \le \pi/2}$$

The restriction on θ in (1) serves two purposes—it enables us to replace $|\cos\theta|$ by $\cos\theta$ to simplify the calculations, and it also ensures that the substitutions can be rewritten as $\theta = \sin^{-1}(x/a)$, if needed.

▶ **Example 1** Evaluate $\displaystyle\int \frac{dx}{x^2\sqrt{4 - x^2}}$.

Solution. To eliminate the radical we make the substitution

$$x = 2\sin\theta, \quad dx = 2\cos\theta\, d\theta$$

This yields

$$\int \frac{dx}{x^2\sqrt{4 - x^2}} = \int \frac{2\cos\theta\, d\theta}{(2\sin\theta)^2\sqrt{4 - 4\sin^2\theta}}$$

$$= \int \frac{2\cos\theta\, d\theta}{(2\sin\theta)^2(2\cos\theta)} = \frac{1}{4}\int \frac{d\theta}{\sin^2\theta}$$

$$= \frac{1}{4}\int \csc^2\theta\, d\theta = -\frac{1}{4}\cot\theta + C \qquad (2)$$

At this point we have completed the integration; however, because the original integral was expressed in terms of x, it is desirable to express $\cot\theta$ in terms of x as well. This can be done using trigonometric identities, but the expression can also be obtained by writing the substitution $x = 2\sin\theta$ as $\sin\theta = x/2$ and representing it geometrically as in Figure 7.4.1. From that figure we obtain

$$\cot\theta = \frac{\sqrt{4 - x^2}}{x}$$

Substituting this in (2) yields

$$\int \frac{dx}{x^2\sqrt{4 - x^2}} = -\frac{1}{4}\frac{\sqrt{4 - x^2}}{x} + C \quad \blacktriangleleft$$

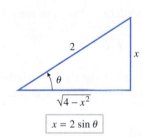

$x = 2\sin\theta$

▲ **Figure 7.4.1**

▶ **Example 2** Evaluate $\displaystyle\int_1^{\sqrt{2}} \frac{dx}{x^2\sqrt{4 - x^2}}$.

Solution. There are two possible approaches: we can make the substitution in the indefinite integral (as in Example 1) and then evaluate the definite integral using the x-limits of integration, or we can make the substitution in the definite integral and convert the x-limits to the corresponding θ-limits.

Method 1.

Using the result from Example 1 with the x-limits of integration yields

$$\int_1^{\sqrt{2}} \frac{dx}{x^2\sqrt{4 - x^2}} = -\frac{1}{4}\left[\frac{\sqrt{4 - x^2}}{x}\right]_1^{\sqrt{2}} = -\frac{1}{4}\left[1 - \sqrt{3}\right] = \frac{\sqrt{3} - 1}{4}$$

Method 2.

The substitution $x = 2\sin\theta$ can be expressed as $x/2 = \sin\theta$ or $\theta = \sin^{-1}(x/2)$, so the θ-limits that correspond to $x = 1$ and $x = \sqrt{2}$ are

$$x = 1: \quad \theta = \sin^{-1}(1/2) = \pi/6$$

$$x = \sqrt{2}: \quad \theta = \sin^{-1}(\sqrt{2}/2) = \pi/4$$

Thus, from (2) in Example 1 we obtain

$$\int_1^{\sqrt{2}} \frac{dx}{x^2\sqrt{4-x^2}} = \frac{1}{4}\int_{\pi/6}^{\pi/4} \csc^2\theta\, d\theta \qquad \boxed{\text{Convert } x\text{-limits to } \theta\text{-limits.}}$$

$$= -\frac{1}{4}\big[\cot\theta\big]_{\pi/6}^{\pi/4} = -\frac{1}{4}\Big[1 - \sqrt{3}\Big] = \frac{\sqrt{3}-1}{4} \quad \blacktriangleleft$$

▶ **Example 3**   Find the area of the ellipse

$$\frac{x^2}{a^2} + \frac{y^2}{b^2} = 1$$

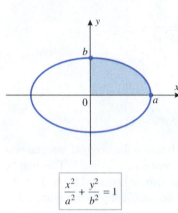

$$\frac{x^2}{a^2} + \frac{y^2}{b^2} = 1$$

▲ **Figure 7.4.2**

Solution.   Because the ellipse is symmetric about both axes, its area A is four times the area in the first quadrant (Figure 7.4.2). If we solve the equation of the ellipse for y in terms of x, we obtain

$$y = \pm\frac{b}{a}\sqrt{a^2 - x^2}$$

where the positive square root gives the equation of the upper half. Thus, the area A is given by

$$A = 4\int_0^a \frac{b}{a}\sqrt{a^2 - x^2}\, dx = \frac{4b}{a}\int_0^a \sqrt{a^2 - x^2}\, dx$$

To evaluate this integral, we will make the substitution $x = a\sin\theta$ (so $dx = a\cos\theta\, d\theta$) and convert the x-limits of integration to θ-limits. Since the substitution can be expressed as $\theta = \sin^{-1}(x/a)$, the θ-limits of integration are

$$x = 0: \quad \theta = \sin^{-1}(0) = 0$$

$$x = a: \quad \theta = \sin^{-1}(1) = \pi/2$$

Thus, we obtain

$$A = \frac{4b}{a}\int_0^a \sqrt{a^2 - x^2}\, dx = \frac{4b}{a}\int_0^{\pi/2}\sqrt{a^2 - a^2\sin^2\theta}\cdot a\cos\theta\, d\theta$$

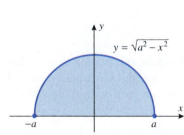

$$y = \sqrt{a^2 - x^2}$$

▲ **Figure 7.4.3**

$$= \frac{4b}{a}\int_0^{\pi/2} a\cos\theta \cdot a\cos\theta\, d\theta$$

$$= 4ab\int_0^{\pi/2}\cos^2\theta\, d\theta = 4ab\int_0^{\pi/2}\frac{1}{2}(1 + \cos 2\theta)\, d\theta$$

$$= 2ab\Big[\theta + \frac{1}{2}\sin 2\theta\Big]_0^{\pi/2} = 2ab\Big[\frac{\pi}{2} - 0\Big] = \pi ab \quad \blacktriangleleft$$

REMARK | In the special case where $a = b$, the ellipse becomes a circle of radius a, and the area formula becomes $A = \pi a^2$, as expected. It is worth noting that

$$\int_{-a}^a \sqrt{a^2 - x^2}\, dx = \tfrac{1}{2}\pi a^2 \tag{3}$$

since this integral represents the area of the upper semicircle (Figure 7.4.3).

TECHNOLOGY MASTERY

If you have a calculating utility with a numerical integration capability, use it and Formula (3) to approximate π to three decimal places.

Thus far, we have focused on using the substitution $x = a\sin\theta$ to evaluate integrals involving radicals of the form $\sqrt{a^2 - x^2}$. Table 7.4.1 summarizes this method and describes some other substitutions of this type.

Table 7.4.1
TRIGONOMETRIC SUBSTITUTIONS

EXPRESSION IN THE INTEGRAND	SUBSTITUTION	RESTRICTION ON θ	SIMPLIFICATION
$\sqrt{a^2 - x^2}$	$x = a\sin\theta$	$-\pi/2 \le \theta \le \pi/2$	$a^2 - x^2 = a^2 - a^2\sin^2\theta = a^2\cos^2\theta$
$\sqrt{a^2 + x^2}$	$x = a\tan\theta$	$-\pi/2 < \theta < \pi/2$	$a^2 + x^2 = a^2 + a^2\tan^2\theta = a^2\sec^2\theta$
$\sqrt{x^2 - a^2}$	$x = a\sec\theta$	$\begin{cases} 0 \le \theta < \pi/2 & \text{(if } x \ge a) \\ \pi/2 < \theta \le \pi & \text{(if } x \le -a) \end{cases}$	$x^2 - a^2 = a^2\sec^2\theta - a^2 = a^2\tan^2\theta$

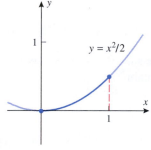

▲ **Figure 7.4.4**

▶ **Example 4** Find the arc length of the curve $y = x^2/2$ from $x = 0$ to $x = 1$ (Figure 7.4.4).

Solution. From Formula (4) of Section 6.4 the arc length L of the curve is

$$L = \int_0^1 \sqrt{1 + \left(\frac{dy}{dx}\right)^2}\, dx = \int_0^1 \sqrt{1 + x^2}\, dx$$

The integrand involves a radical of the form $\sqrt{a^2 + x^2}$ with $a = 1$, so from Table 7.4.1 we make the substitution

$$x = \tan\theta, \quad -\pi/2 < \theta < \pi/2$$

$$\frac{dx}{d\theta} = \sec^2\theta \quad \text{or} \quad dx = \sec^2\theta\, d\theta$$

Since this substitution can be expressed as $\theta = \tan^{-1} x$, the θ-limits of integration that correspond to the x-limits, $x = 0$ and $x = 1$, are

$$x = 0: \quad \theta = \tan^{-1} 0 = 0$$
$$x = 1: \quad \theta = \tan^{-1} 1 = \pi/4$$

Thus,

$$L = \int_0^1 \sqrt{1 + x^2}\, dx = \int_0^{\pi/4} \sqrt{1 + \tan^2\theta}\, \sec^2\theta\, d\theta$$

$$= \int_0^{\pi/4} \sqrt{\sec^2\theta}\, \sec^2\theta\, d\theta \qquad \boxed{1 + \tan^2\theta = \sec^2\theta}$$

$$= \int_0^{\pi/4} |\sec\theta|\sec^2\theta\, d\theta$$

$$= \int_0^{\pi/4} \sec^3\theta\, d\theta \qquad \boxed{\sec\theta > 0 \text{ since } -\pi/2 < \theta < \pi/2}$$

$$= \left[\tfrac{1}{2}\sec\theta\tan\theta + \tfrac{1}{2}\ln|\sec\theta + \tan\theta|\right]_0^{\pi/4} \qquad \boxed{\text{Formula (26) of Section 7.3}}$$

$$= \tfrac{1}{2}\left[\sqrt{2} + \ln(\sqrt{2} + 1)\right] \approx 1.148 \quad \blacktriangleleft$$

▶ **Example 5** Evaluate $\displaystyle\int \frac{\sqrt{x^2 - 25}}{x}\, dx$, assuming that $x \ge 5$.

Solution. The integrand involves a radical of the form $\sqrt{x^2 - a^2}$ with $a = 5$, so from Table 7.4.1 we make the substitution

$$x = 5 \sec \theta, \quad 0 \le \theta < \pi/2$$

$$\frac{dx}{d\theta} = 5 \sec \theta \tan \theta \quad \text{or} \quad dx = 5 \sec \theta \tan \theta \, d\theta$$

Thus,

$$\int \frac{\sqrt{x^2 - 25}}{x} \, dx = \int \frac{\sqrt{25 \sec^2 \theta - 25}}{5 \sec \theta} (5 \sec \theta \tan \theta) \, d\theta$$

$$= \int \frac{5|\tan \theta|}{5 \sec \theta} (5 \sec \theta \tan \theta) \, d\theta$$

$$= 5 \int \tan^2 \theta \, d\theta \qquad \boxed{\tan \theta \ge 0 \text{ since } 0 \le \theta < \pi/2}$$

$$= 5 \int (\sec^2 \theta - 1) \, d\theta = 5 \tan \theta - 5\theta + C$$

To express the solution in terms of x, we will represent the substitution $x = 5 \sec \theta$ geometrically by the triangle in Figure 7.4.5, from which we obtain

$$\tan \theta = \frac{\sqrt{x^2 - 25}}{5}$$

From this and the fact that the substitution can be expressed as $\theta = \sec^{-1}(x/5)$, we obtain

$$\int \frac{\sqrt{x^2 - 25}}{x} \, dx = \sqrt{x^2 - 25} - 5 \sec^{-1}\left(\frac{x}{5}\right) + C \blacktriangleleft$$

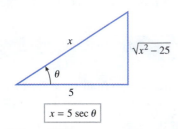

$x = 5 \sec \theta$

▲ **Figure 7.4.5**

■ **INTEGRALS INVOLVING** $ax^2 + bx + c$

Integrals that involve a quadratic expression $ax^2 + bx + c$, where $a \ne 0$ and $b \ne 0$, can often be evaluated by first completing the square, then making an appropriate substitution. The following example illustrates this idea.

▶ **Example 6** Evaluate $\displaystyle\int \frac{x}{x^2 - 4x + 8} \, dx$.

Solution. Completing the square yields

$$x^2 - 4x + 8 = (x^2 - 4x + 4) + 8 - 4 = (x - 2)^2 + 4$$

Thus, the substitution $\qquad u = x - 2, \quad du = dx$

yields

$$\int \frac{x}{x^2 - 4x + 8} \, dx = \int \frac{x}{(x - 2)^2 + 4} \, dx = \int \frac{u + 2}{u^2 + 4} \, du$$

$$= \int \frac{u}{u^2 + 4} \, du + 2 \int \frac{du}{u^2 + 4}$$

$$= \frac{1}{2} \int \frac{2u}{u^2 + 4} \, du + 2 \int \frac{du}{u^2 + 4}$$

$$= \frac{1}{2} \ln(u^2 + 4) + 2 \left(\frac{1}{2}\right) \tan^{-1} \frac{u}{2} + C$$

$$= \frac{1}{2} \ln[(x - 2)^2 + 4] + \tan^{-1}\left(\frac{x - 2}{2}\right) + C \blacktriangleleft$$

✔ **QUICK CHECK EXERCISES 7.4**   *(See page 514 for answers.)*

1. For each expression, give a trigonometric substitution that will eliminate the radical.
   (a) $\sqrt{a^2 - x^2}$ _____   (b) $\sqrt{a^2 + x^2}$ _____
   (c) $\sqrt{x^2 - a^2}$ _____

2. If $x = 2\sec\theta$ and $0 < \theta < \pi/2$, then
   (a) $\sin\theta =$ _____   (b) $\cos\theta =$ _____
   (c) $\tan\theta =$ _____.

3. In each part, state the trigonometric substitution that you would try first to evaluate the integral. Do not evaluate the integral.
   (a) $\int \sqrt{9 + x^2}\, dx$ _____

   (b) $\int \sqrt{9 - x^2}\, dx$ _____

   (c) $\int \sqrt{1 - 9x^2}\, dx$ _____

   (d) $\int \sqrt{x^2 - 9}\, dx$ _____

   (e) $\int \sqrt{9 + 3x^2}\, dx$ _____

   (f) $\int \sqrt{1 + (9x)^2}\, dx$ _____

4. In each part, determine the substitution u.
   (a) $\int \dfrac{1}{x^2 - 2x + 10}\, dx = \int \dfrac{1}{u^2 + 3^2}\, du$;

   $u =$ _____

   (b) $\int \sqrt{x^2 - 6x + 8}\, dx = \int \sqrt{u^2 - 1}\, du$;

   $u =$ _____

   (c) $\int \sqrt{12 - 4x - x^2}\, dx = \int \sqrt{4^2 - u^2}\, du$;

   $u =$ _____

EXERCISE SET 7.4   **C** CAS

1–26 Evaluate the integral.

1. $\displaystyle\int \sqrt{4 - x^2}\, dx$

2. $\displaystyle\int \sqrt{1 - 4x^2}\, dx$

3. $\displaystyle\int \frac{x^2}{\sqrt{16 - x^2}}\, dx$

4. $\displaystyle\int \frac{dx}{x^2\sqrt{9 - x^2}}$

5. $\displaystyle\int \frac{dx}{(4 + x^2)^2}$

6. $\displaystyle\int \frac{x^2}{\sqrt{5 + x^2}}\, dx$

7. $\displaystyle\int \frac{\sqrt{x^2 - 9}}{x}\, dx$

8. $\displaystyle\int \frac{dx}{x^2\sqrt{x^2 - 16}}$

9. $\displaystyle\int \frac{3x^3}{\sqrt{1 - x^2}}\, dx$

10. $\displaystyle\int x^3\sqrt{5 - x^2}\, dx$

11. $\displaystyle\int \frac{dx}{x^2\sqrt{9x^2 - 4}}$

12. $\displaystyle\int \frac{\sqrt{1 + t^2}}{t}\, dt$

13. $\displaystyle\int \frac{dx}{(1 - x^2)^{3/2}}$

14. $\displaystyle\int \frac{dx}{x^2\sqrt{x^2 + 25}}$

15. $\displaystyle\int \frac{dx}{\sqrt{x^2 - 9}}$

16. $\displaystyle\int \frac{dx}{1 + 2x^2 + x^4}$

17. $\displaystyle\int \frac{dx}{(4x^2 - 9)^{3/2}}$

18. $\displaystyle\int \frac{3x^3}{\sqrt{x^2 - 25}}\, dx$

19. $\displaystyle\int e^x\sqrt{1 - e^{2x}}\, dx$

20. $\displaystyle\int \frac{\cos\theta}{\sqrt{2 - \sin^2\theta}}\, d\theta$

21. $\displaystyle\int_0^1 5x^3\sqrt{1 - x^2}\, dx$

22. $\displaystyle\int_0^{1/2} \frac{dx}{(1 - x^2)^2}$

23. $\displaystyle\int_{\sqrt{2}}^2 \frac{dx}{x^2\sqrt{x^2 - 1}}$

24. $\displaystyle\int_{\sqrt{2}}^2 \frac{\sqrt{2x^2 - 4}}{x}\, dx$

25. $\displaystyle\int_1^3 \frac{dx}{x^4\sqrt{x^2 + 3}}$

26. $\displaystyle\int_0^3 \frac{x^3}{(3 + x^2)^{5/2}}\, dx$

27–30 True–False Determine whether the statement is true or false. Explain your answer.

27. An integrand involving a radical of the form $\sqrt{a^2 - x^2}$ suggests the substitution $x = a\sin\theta$.

28. The trigonometric substitution $x = a\sin\theta$ is made with the restriction $0 \le \theta \le \pi$.

29. An integrand involving a radical of the form $\sqrt{x^2 - a^2}$ suggests the substitution $x = a\cos\theta$.

30. The area enclosed by the ellipse $x^2 + 4y^2 = 1$ is $\pi/2$.

FOCUS ON CONCEPTS

31. The integral
   $$\int \frac{x}{x^2 + 4}\, dx$$
   can be evaluated either by a trigonometric substitution or by the substitution $u = x^2 + 4$. Do it both ways and show that the results are equivalent.

32. The integral
   $$\int \frac{x^2}{x^2 + 4}\, dx$$
   can be evaluated either by a trigonometric substitution or by algebraically rewriting the numerator of the integrand as $(x^2 + 4) - 4$. Do it both ways and show that the results are equivalent.

33. Find the arc length of the curve $y = \ln x$ from $x = 1$ to $x = 2$.

34. Find the arc length of the curve $y = x^2$ from $x = 0$ to $x = 1$.

35. Find the area of the surface generated when the curve in Exercise 34 is revolved about the x-axis.

36. Find the volume of the solid generated when the region enclosed by $x = y(1 - y^2)^{1/4}$, $y = 0$, $y = 1$, and $x = 0$ is revolved about the y-axis.

37–48 Evaluate the integral. ■

37. $\displaystyle \int \frac{dx}{x^2 - 4x + 5}$

38. $\displaystyle \int \frac{dx}{\sqrt{2x - x^2}}$

39. $\displaystyle \int \frac{dx}{\sqrt{3 + 2x - x^2}}$

40. $\displaystyle \int \frac{dx}{16x^2 + 16x + 5}$

41. $\displaystyle \int \frac{dx}{\sqrt{x^2 - 6x + 10}}$

42. $\displaystyle \int \frac{x}{x^2 + 2x + 2}\, dx$

43. $\displaystyle \int \sqrt{3 - 2x - x^2}\, dx$

44. $\displaystyle \int \frac{e^x}{\sqrt{1 + e^x + e^{2x}}}\, dx$

45. $\displaystyle \int \frac{dx}{2x^2 + 4x + 7}$

46. $\displaystyle \int \frac{2x + 3}{4x^2 + 4x + 5}\, dx$

47. $\displaystyle \int_1^2 \frac{dx}{\sqrt{4x - x^2}}$

48. $\displaystyle \int_0^4 \sqrt{x(4 - x)}\, dx$

C **49–50** There is a good chance that your CAS will not be able to evaluate these integrals as stated. If this is so, make a substitution that converts the integral into one that your CAS can evaluate. ■

49. $\displaystyle \int \cos x \sin x \sqrt{1 - \sin^4 x}\, dx$

50. $\displaystyle \int (x \cos x + \sin x)\sqrt{1 + x^2 \sin^2 x}\, dx$

51. (a) Use the **hyperbolic substitution** $x = 3 \sinh u$, the identity $\cosh^2 u - \sinh^2 u = 1$, and Theorem 6.9.4 to evaluate

$$\int \frac{dx}{\sqrt{x^2 + 9}}$$

(b) Evaluate the integral in part (a) using a trigonometric substitution and show that the result agrees with that obtained in part (a).

52. Use the hyperbolic substitution $x = \cosh u$, the identity $\sinh^2 u = \frac{1}{2}(\cosh 2u - 1)$, and the results referenced in Exercise 51 to evaluate

$$\int \sqrt{x^2 - 1}\, dx, \quad x \geq 1$$

53. **Writing** The trigonometric substitution $x = a \sin\theta$, $-\pi/2 \leq \theta \leq \pi/2$, is suggested for an integral whose integrand involves $\sqrt{a^2 - x^2}$. Discuss the implications of restricting θ to $\pi/2 \leq \theta \leq 3\pi/2$, and explain why the restriction $-\pi/2 \leq \theta \leq \pi/2$ should be preferred.

54. **Writing** The trigonometric substitution $x = a \cos\theta$ could also be used for an integral whose integrand involves $\sqrt{a^2 - x^2}$. Determine an appropriate restriction for θ with the substitution $x = a \cos\theta$, and discuss how to apply this substitution in appropriate integrals. Illustrate your discussion by evaluating the integral in Example 1 using a substitution of this type.

✔ **QUICK CHECK ANSWERS 7.4**

1. (a) $x = a \sin\theta$ (b) $x = a \tan\theta$ (c) $x = a \sec\theta$ **2.** (a) $\dfrac{\sqrt{x^2 - 4}}{x}$ (b) $\dfrac{2}{x}$ (c) $\dfrac{\sqrt{x^2 - 4}}{2}$ **3.** (a) $x = 3 \tan\theta$ (b) $x = 3 \sin\theta$
(c) $x = \frac{1}{3}\sin\theta$ (d) $x = 3 \sec\theta$ (e) $x = \sqrt{3} \tan\theta$ (f) $x = \frac{1}{9}\tan\theta$ **4.** (a) $x - 1$ (b) $x - 3$ (c) $x + 2$

7.5 **INTEGRATING RATIONAL FUNCTIONS BY PARTIAL FRACTIONS**

Recall that a rational function is a ratio of two polynomials. In this section we will give a general method for integrating rational functions that is based on the idea of decomposing a rational function into a sum of simple rational functions that can be integrated by the methods studied in earlier sections.

■ **PARTIAL FRACTIONS**

In algebra, one learns to combine two or more fractions into a single fraction by finding a common denominator. For example,

$$\frac{2}{x - 4} + \frac{3}{x + 1} = \frac{2(x + 1) + 3(x - 4)}{(x - 4)(x + 1)} = \frac{5x - 10}{x^2 - 3x - 4} \tag{1}$$

However, for purposes of integration, the left side of (1) is preferable to the right side since each of the terms is easy to integrate:

$$\int \frac{5x - 10}{x^2 - 3x - 4}\,dx = \int \frac{2}{x - 4}\,dx + \int \frac{3}{x + 1}\,dx = 2\ln|x - 4| + 3\ln|x + 1| + C$$

Thus, it is desirable to have some method that will enable us to obtain the left side of (1), starting with the right side. To illustrate how this can be done, we begin by noting that on the left side the numerators are constants and the denominators are the factors of the denominator on the right side. Thus, to find the left side of (1), starting from the right side, we could factor the denominator of the right side and look for constants A and B such that

$$\frac{5x - 10}{(x - 4)(x + 1)} = \frac{A}{x - 4} + \frac{B}{x + 1} \tag{2}$$

One way to find the constants A and B is to multiply (2) through by $(x - 4)(x + 1)$ to clear fractions. This yields

$$5x - 10 = A(x + 1) + B(x - 4) \tag{3}$$

This relationship holds for all x, so it holds in particular if $x = 4$ or $x = -1$. Substituting $x = 4$ in (3) makes the second term on the right drop out and yields the equation $10 = 5A$ or $A = 2$; and substituting $x = -1$ in (3) makes the first term on the right drop out and yields the equation $-15 = -5B$ or $B = 3$. Substituting these values in (2) we obtain

$$\frac{5x - 10}{(x - 4)(x + 1)} = \frac{2}{x - 4} + \frac{3}{x + 1} \tag{4}$$

which agrees with (1).

A second method for finding the constants A and B is to multiply out the right side of (3) and collect like powers of x to obtain

$$5x - 10 = (A + B)x + (A - 4B)$$

Since the polynomials on the two sides are identical, their corresponding coefficients must be the same. Equating the corresponding coefficients on the two sides yields the following system of equations in the unknowns A and B:

$$\begin{aligned} A + B &= 5 \\ A - 4B &= -10 \end{aligned}$$

Solving this system yields $A = 2$ and $B = 3$ as before (verify).

The terms on the right side of (4) are called **partial fractions** of the expression on the left side because they each constitute *part* of that expression. To find those partial fractions we first had to make a guess about their form, and then we had to find the unknown constants. Our next objective is to extend this idea to general rational functions. For this purpose, suppose that $P(x)/Q(x)$ is a **proper rational function**, by which we mean that the degree of the numerator is less than the degree of the denominator. There is a theorem in advanced algebra which states that every proper rational function can be expressed as a sum

$$\frac{P(x)}{Q(x)} = F_1(x) + F_2(x) + \cdots + F_n(x)$$

where $F_1(x), F_2(x), \ldots, F_n(x)$ are rational functions of the form

$$\frac{A}{(ax + b)^k} \quad \text{or} \quad \frac{Ax + B}{(ax^2 + bx + c)^k}$$

in which the denominators are factors of $Q(x)$. The sum is called the **partial fraction decomposition** of $P(x)/Q(x)$, and the terms are called **partial fractions**. As in our opening example, there are two parts to finding a partial fraction decomposition: determining the exact form of the decomposition and finding the unknown constants.

■ FINDING THE FORM OF A PARTIAL FRACTION DECOMPOSITION

The first step in finding the form of the partial fraction decomposition of a proper rational function $P(x)/Q(x)$ is to factor $Q(x)$ completely into linear and irreducible quadratic factors, and then collect all repeated factors so that $Q(x)$ is expressed as a product of *distinct* factors of the form

$$(ax + b)^m \quad \text{and} \quad (ax^2 + bx + c)^m$$

From these factors we can determine the form of the partial fraction decomposition using two rules that we will now discuss.

■ LINEAR FACTORS

If all of the factors of $Q(x)$ are linear, then the partial fraction decomposition of $P(x)/Q(x)$ can be determined by using the following rule:

LINEAR FACTOR RULE   For each factor of the form $(ax + b)^m$, the partial fraction decomposition contains the following sum of m partial fractions:

$$\frac{A_1}{ax + b} + \frac{A_2}{(ax + b)^2} + \cdots + \frac{A_m}{(ax + b)^m}$$

where $A_1, A_2, \ldots, A_m$ are constants to be determined. In the case where $m = 1$, only the first term in the sum appears.

▶ **Example 1**   Evaluate $\displaystyle\int \frac{dx}{x^2 + x - 2}$.

Solution.   The integrand is a proper rational function that can be written as

$$\frac{1}{x^2 + x - 2} = \frac{1}{(x - 1)(x + 2)}$$

The factors $x - 1$ and $x + 2$ are both linear and appear to the first power, so each contributes one term to the partial fraction decomposition by the linear factor rule. Thus, the decomposition has the form

$$\frac{1}{(x - 1)(x + 2)} = \frac{A}{x - 1} + \frac{B}{x + 2} \tag{5}$$

where A and B are constants to be determined. Multiplying this expression through by $(x - 1)(x + 2)$ yields

$$1 = A(x + 2) + B(x - 1) \tag{6}$$

As discussed earlier, there are two methods for finding A and B: we can substitute values of x that are chosen to make terms on the right drop out, or we can multiply out on the right and equate corresponding coefficients on the two sides to obtain a system of equations that can be solved for A and B. We will use the first approach.

Setting $x = 1$ makes the second term in (6) drop out and yields $1 = 3A$ or $A = \frac{1}{3}$; and setting $x = -2$ makes the first term in (6) drop out and yields $1 = -3B$ or $B = -\frac{1}{3}$. Substituting these values in (5) yields the partial fraction decomposition

$$\frac{1}{(x - 1)(x + 2)} = \frac{\frac{1}{3}}{x - 1} + \frac{-\frac{1}{3}}{x + 2}$$

The integration can now be completed as follows:

$$\int \frac{dx}{(x-1)(x+2)} = \frac{1}{3}\int \frac{dx}{x-1} - \frac{1}{3}\int \frac{dx}{x+2}$$

$$= \frac{1}{3}\ln|x-1| - \frac{1}{3}\ln|x+2| + C = \frac{1}{3}\ln\left|\frac{x-1}{x+2}\right| + C \blacktriangleleft$$

If the factors of $Q(x)$ are linear and none are repeated, as in the last example, then the recommended method for finding the constants in the partial fraction decomposition is to substitute appropriate values of x to make terms drop out. However, if some of the linear factors are repeated, then it will not be possible to find all of the constants in this way. In this case the recommended procedure is to find as many constants as possible by substitution and then find the rest by equating coefficients. This is illustrated in the next example.

▶ **Example 2**   Evaluate $\displaystyle\int \frac{2x+4}{x^3-2x^2}\,dx$.

Solution.   The integrand can be rewritten as

$$\frac{2x+4}{x^3-2x^2} = \frac{2x+4}{x^2(x-2)}$$

Although x^2 is a quadratic factor, it is *not* irreducible since $x^2 = xx$. Thus, by the linear factor rule, x^2 introduces two terms (since $m = 2$) of the form

$$\frac{A}{x} + \frac{B}{x^2}$$

and the factor $x - 2$ introduces one term (since $m = 1$) of the form

$$\frac{C}{x-2}$$

so the partial fraction decomposition is

$$\frac{2x+4}{x^2(x-2)} = \frac{A}{x} + \frac{B}{x^2} + \frac{C}{x-2} \tag{7}$$

Multiplying by $x^2(x-2)$ yields

$$2x+4 = Ax(x-2) + B(x-2) + Cx^2 \tag{8}$$

which, after multiplying out and collecting like powers of x, becomes

$$2x+4 = (A+C)x^2 + (-2A+B)x - 2B \tag{9}$$

Setting $x = 0$ in (8) makes the first and third terms drop out and yields $B = -2$, and setting $x = 2$ in (8) makes the first and second terms drop out and yields $C = 2$ (verify). However, there is no substitution in (8) that produces A directly, so we look to Equation (9) to find this value. This can be done by equating the coefficients of x^2 on the two sides to obtain

$$A + C = 0 \quad \text{or} \quad A = -C = -2$$

Substituting the values $A = -2$, $B = -2$, and $C = 2$ in (7) yields the partial fraction decomposition

$$\frac{2x+4}{x^2(x-2)} = \frac{-2}{x} + \frac{-2}{x^2} + \frac{2}{x-2}$$

Thus,

$$\int \frac{2x+4}{x^2(x-2)}\,dx = -2\int \frac{dx}{x} - 2\int \frac{dx}{x^2} + 2\int \frac{dx}{x-2}$$

$$= -2\ln|x| + \frac{2}{x} + 2\ln|x-2| + C = 2\ln\left|\frac{x-2}{x}\right| + \frac{2}{x} + C \blacktriangleleft$$

■ **QUADRATIC FACTORS**

If some of the factors of $Q(x)$ are irreducible quadratics, then the contribution of those factors to the partial fraction decomposition of $P(x)/Q(x)$ can be determined from the following rule:

QUADRATIC FACTOR RULE   For each factor of the form $(ax^2 + bx + c)^m$, the partial fraction decomposition contains the following sum of m partial fractions:

$$\frac{A_1 x + B_1}{ax^2 + bx + c} + \frac{A_2 x + B_2}{(ax^2 + bx + c)^2} + \cdots + \frac{A_m x + B_m}{(ax^2 + bx + c)^m}$$

where $A_1, A_2, \ldots, A_m, B_1, B_2, \ldots, B_m$ are constants to be determined. In the case where $m = 1$, only the first term in the sum appears.

▶ **Example 3**   Evaluate $\displaystyle\int \frac{x^2 + x - 2}{3x^3 - x^2 + 3x - 1}\,dx$.

Solution.   The denominator in the integrand can be factored by grouping:

$$3x^3 - x^2 + 3x - 1 = x^2(3x - 1) + (3x - 1) = (3x - 1)(x^2 + 1)$$

By the linear factor rule, the factor $3x - 1$ introduces one term, namely,

$$\frac{A}{3x - 1}$$

and by the quadratic factor rule, the factor $x^2 + 1$ introduces one term, namely,

$$\frac{Bx + C}{x^2 + 1}$$

Thus, the partial fraction decomposition is

$$\frac{x^2 + x - 2}{(3x - 1)(x^2 + 1)} = \frac{A}{3x - 1} + \frac{Bx + C}{x^2 + 1} \tag{10}$$

Multiplying by $(3x - 1)(x^2 + 1)$ yields

$$x^2 + x - 2 = A(x^2 + 1) + (Bx + C)(3x - 1) \tag{11}$$

We could find A by substituting $x = \frac{1}{3}$ to make the last term drop out, and then find the rest of the constants by equating corresponding coefficients. However, in this case it is just as easy to find *all* of the constants by equating coefficients and solving the resulting system. For this purpose we multiply out the right side of (11) and collect like terms:

$$x^2 + x - 2 = (A + 3B)x^2 + (-B + 3C)x + (A - C)$$

Equating corresponding coefficients gives

$$
\begin{aligned}
A + 3B & & = 1 \\
- B + 3C & = 1 \\
A \quad - C & = -2
\end{aligned}
$$

To solve this system, subtract the third equation from the first to eliminate A. Then use the resulting equation together with the second equation to solve for B and C. Finally, determine A from the first or third equation. This yields (verify)

$$A = -\frac{7}{5}, \quad B = \frac{4}{5}, \quad C = \frac{3}{5}$$

Thus, (10) becomes

$$\frac{x^2 + x - 2}{(3x - 1)(x^2 + 1)} = \frac{-\frac{7}{5}}{3x - 1} + \frac{\frac{4}{5}x + \frac{3}{5}}{x^2 + 1}$$

and

$$\int \frac{x^2 + x - 2}{(3x - 1)(x^2 + 1)}\, dx = -\frac{7}{5} \int \frac{dx}{3x - 1} + \frac{4}{5} \int \frac{x}{x^2 + 1}\, dx + \frac{3}{5} \int \frac{dx}{x^2 + 1}$$

$$= -\frac{7}{15} \ln |3x - 1| + \frac{2}{5} \ln(x^2 + 1) + \frac{3}{5} \tan^{-1} x + C \blacktriangleleft$$

TECHNOLOGY MASTERY

Computer algebra systems have built-in capabilities for finding partial fraction decompositions. If you have a CAS, use it to find the decompositions in Examples 1, 2, and 3.

▶ **Example 4** Evaluate $\displaystyle\int \frac{3x^4 + 4x^3 + 16x^2 + 20x + 9}{(x + 2)(x^2 + 3)^2}\, dx.$

Solution. Observe that the integrand is a proper rational function since the numerator has degree 4 and the denominator has degree 5. Thus, the method of partial fractions is applicable. By the linear factor rule, the factor $x + 2$ introduces the single term

$$\frac{A}{x + 2}$$

and by the quadratic factor rule, the factor $(x^2 + 3)^2$ introduces two terms (since $m = 2$):

$$\frac{Bx + C}{x^2 + 3} + \frac{Dx + E}{(x^2 + 3)^2}$$

Thus, the partial fraction decomposition of the integrand is

$$\frac{3x^4 + 4x^3 + 16x^2 + 20x + 9}{(x + 2)(x^2 + 3)^2} = \frac{A}{x + 2} + \frac{Bx + C}{x^2 + 3} + \frac{Dx + E}{(x^2 + 3)^2} \tag{12}$$

Multiplying by $(x + 2)(x^2 + 3)^2$ yields

$$3x^4 + 4x^3 + 16x^2 + 20x + 9$$
$$= A(x^2 + 3)^2 + (Bx + C)(x^2 + 3)(x + 2) + (Dx + E)(x + 2) \tag{13}$$

which, after multiplying out and collecting like powers of x, becomes

$$3x^4 + 4x^3 + 16x^2 + 20x + 9$$
$$= (A + B)x^4 + (2B + C)x^3 + (6A + 3B + 2C + D)x^2$$
$$+ (6B + 3C + 2D + E)x + (9A + 6C + 2E) \tag{14}$$

Equating corresponding coefficients in (14) yields the following system of five linear equations in five unknowns:

$$\begin{aligned} A + B &= 3 \\ 2B + C &= 4 \\ 6A + 3B + 2C + D &= 16 \\ 6B + 3C + 2D + E &= 20 \\ 9A + 6C + 2E &= 9 \end{aligned} \tag{15}$$

Efficient methods for solving systems of linear equations such as this are studied in a branch of mathematics called *linear algebra*; those methods are outside the scope of this text. However, as a practical matter most linear systems of any size are solved by computer, and most computer algebra systems have commands that in many cases can solve linear systems exactly. In this particular case we can simplify the work by first substituting $x = -2$

in (13), which yields $A = 1$. Substituting this known value of A in (15) yields the simpler system

$$\begin{aligned} B &= 2 \\ 2B + C &= 4 \\ 3B + 2C + D &= 10 \\ 6B + 3C + 2D + E &= 20 \\ 6C + 2E &= 0 \end{aligned} \tag{16}$$

This system can be solved by starting at the top and working down, first substituting $B = 2$ in the second equation to get $C = 0$, then substituting the known values of B and C in the third equation to get $D = 4$, and so forth. This yields

$$A = 1, \quad B = 2, \quad C = 0, \quad D = 4, \quad E = 0$$

Thus, (12) becomes

$$\frac{3x^4 + 4x^3 + 16x^2 + 20x + 9}{(x+2)(x^2+3)^2} = \frac{1}{x+2} + \frac{2x}{x^2+3} + \frac{4x}{(x^2+3)^2}$$

and so

$$\int \frac{3x^4 + 4x^3 + 16x^2 + 20x + 9}{(x+2)(x^2+3)^2}\,dx$$

$$= \int \frac{dx}{x+2} + \int \frac{2x}{x^2+3}\,dx + 4\int \frac{x}{(x^2+3)^2}\,dx$$

$$= \ln|x+2| + \ln(x^2+3) - \frac{2}{x^2+3} + C \blacktriangleleft$$

■ INTEGRATING IMPROPER RATIONAL FUNCTIONS

Although the method of partial fractions only applies to proper rational functions, an improper rational function can be integrated by performing a long division and expressing the function as the quotient plus the remainder over the divisor. The remainder over the divisor will be a proper rational function, which can then be decomposed into partial fractions. This idea is illustrated in the following example.

▶ **Example 5**   Evaluate $\displaystyle\int \frac{3x^4 + 3x^3 - 5x^2 + x - 1}{x^2 + x - 2}\,dx$.

Solution.   The integrand is an improper rational function since the numerator has degree 4 and the denominator has degree 2. Thus, we first perform the long division

$$\begin{array}{r} 3x^2 + 1 \\ x^2 + x - 2 \overline{\smash{\big)}\ 3x^4 + 3x^3 - 5x^2 + x - 1} \\ \underline{3x^4 + 3x^3 - 6x^2 } \\ x^2 + x - 1 \\ \underline{x^2 + x - 2} \\ 1 \end{array}$$

It follows that the integrand can be expressed as

$$\frac{3x^4 + 3x^3 - 5x^2 + x - 1}{x^2 + x - 2} = (3x^2 + 1) + \frac{1}{x^2 + x - 2}$$

and hence

$$\int \frac{3x^4 + 3x^3 - 5x^2 + x - 1}{x^2 + x - 2}\,dx = \int (3x^2 + 1)\,dx + \int \frac{dx}{x^2 + x - 2}$$

The second integral on the right now involves a proper rational function and can thus be evaluated by a partial fraction decomposition. Using the result of Example 1 we obtain

$$\int \frac{3x^4 + 3x^3 - 5x^2 + x - 1}{x^2 + x - 2} \, dx = x^3 + x + \frac{1}{3} \ln \left| \frac{x-1}{x+2} \right| + C \blacktriangleleft$$

■ **CONCLUDING REMARKS**

There are some cases in which the method of partial fractions is inappropriate. For example, it would be inefficient to use partial fractions to perform the integration

$$\int \frac{3x^2 + 2}{x^3 + 2x - 8} \, dx = \ln |x^3 + 2x - 8| + C$$

since the substitution $u = x^3 + 2x - 8$ is more direct. Similarly, the integration

$$\int \frac{2x-1}{x^2+1} \, dx = \int \frac{2x}{x^2+1} \, dx - \int \frac{dx}{x^2+1} = \ln(x^2+1) - \tan^{-1} x + C$$

requires only a little algebra since the integrand is already in partial fraction form.

✔ **QUICK CHECK EXERCISES 7.5**   *(See page 523 for answers.)*

1. A partial fraction is a rational function of the form _____ or of the form _____.

2. (a) What is a proper rational function?
   (b) What condition must the degree of the numerator and the degree of the denominator of a rational function satisfy for the method of partial fractions to be applicable directly?
   (c) If the condition in part (b) is not satisfied, what must you do if you want to use partial fractions?

3. Suppose that the function $f(x) = P(x)/Q(x)$ is a proper rational function.
   (a) For each factor of $Q(x)$ of the form $(ax + b)^m$, the partial fraction decomposition of f contains the following sum of m partial fractions: _____

(b) For each factor of $Q(x)$ of the form $(ax^2 + bx + c)^m$, where $ax^2 + bx + c$ is an irreducible quadratic, the partial fraction decomposition of f contains the following sum of m partial fractions: _____

4. Complete the partial fraction decomposition.
   (a) $\dfrac{-3}{(x+1)(2x-1)} = \dfrac{A}{x+1} - \dfrac{2}{2x-1}$
   (b) $\dfrac{2x^2 - 3x}{(x^2+1)(3x+2)} = \dfrac{B}{3x+2} - \dfrac{1}{x^2+1}$

5. Evaluate the integral.
   (a) $\displaystyle\int \frac{3}{(x+1)(1-2x)} \, dx$ (b) $\displaystyle\int \frac{2x^2 - 3x}{(x^2+1)(3x+2)} \, dx$

EXERCISE SET 7.5   [C] CAS

1–8 Write out the form of the partial fraction decomposition. (Do not find the numerical values of the coefficients.) ■

1. $\dfrac{3x - 1}{(x-3)(x+4)}$

2. $\dfrac{5}{x(x^2 - 4)}$

3. $\dfrac{2x - 3}{x^3 - x^2}$

4. $\dfrac{x^2}{(x+2)^3}$

5. $\dfrac{1 - x^2}{x^3(x^2 + 2)}$

6. $\dfrac{3x}{(x-1)(x^2+6)}$

7. $\dfrac{4x^3 - x}{(x^2 + 5)^2}$

8. $\dfrac{1 - 3x^4}{(x-2)(x^2+1)^2}$

9–34 Evaluate the integral. ■

9. $\displaystyle\int \frac{dx}{x^2 - 3x - 4}$

10. $\displaystyle\int \frac{dx}{x^2 - 6x - 7}$

11. $\displaystyle\int \frac{11x + 17}{2x^2 + 7x - 4} \, dx$

12. $\displaystyle\int \frac{5x - 5}{3x^2 - 8x - 3} \, dx$

13. $\displaystyle\int \frac{2x^2 - 9x - 9}{x^3 - 9x} \, dx$

14. $\displaystyle\int \frac{dx}{x(x^2 - 1)}$

15. $\displaystyle\int \frac{x^2 - 8}{x + 3} \, dx$

16. $\displaystyle\int \frac{x^2 + 1}{x - 1} \, dx$

17. $\displaystyle\int \frac{3x^2 - 10}{x^2 - 4x + 4} \, dx$

18. $\displaystyle\int \frac{x^2}{x^2 - 3x + 2} \, dx$

19. $\displaystyle\int \frac{2x - 3}{x^2 - 3x - 10} \, dx$

20. $\displaystyle\int \frac{3x + 1}{3x^2 + 2x - 1} \, dx$

21. $\displaystyle\int \frac{x^5 + x^2 + 2}{x^3 - x} \, dx$

22. $\displaystyle\int \frac{x^5 - 4x^3 + 1}{x^3 - 4x} \, dx$

23. $\displaystyle\int \frac{2x^2 + 3}{x(x - 1)^2} \, dx$

24. $\displaystyle\int \frac{3x^2 - x + 1}{x^3 - x^2} \, dx$

25. $\displaystyle\int \frac{2x^2 - 10x + 4}{(x + 1)(x - 3)^2}\,dx$ **26.** $\displaystyle\int \frac{2x^2 - 2x - 1}{x^3 - x^2}\,dx$

27. $\displaystyle\int \frac{x^2}{(x + 1)^3}\,dx$ **28.** $\displaystyle\int \frac{2x^2 + 3x + 3}{(x + 1)^3}\,dx$

29. $\displaystyle\int \frac{2x^2 - 1}{(4x - 1)(x^2 + 1)}\,dx$ **30.** $\displaystyle\int \frac{dx}{x^3 + 2x}$

31. $\displaystyle\int \frac{x^3 + 3x^2 + x + 9}{(x^2 + 1)(x^2 + 3)}\,dx$ **32.** $\displaystyle\int \frac{x^3 + x^2 + x + 2}{(x^2 + 1)(x^2 + 2)}\,dx$

33. $\displaystyle\int \frac{x^3 - 2x^2 + 2x - 2}{x^2 + 1}\,dx$

34. $\displaystyle\int \frac{x^4 + 6x^3 + 10x^2 + x}{x^2 + 6x + 10}\,dx$

35–38 True–False Determine whether the statement is true or false. Explain your answer. ■

35. The technique of partial fractions is used for integrals whose integrands are ratios of polynomials.

36. The integrand in
$$\int \frac{3x^4 + 5}{(x^2 + 1)^2}\,dx$$
is a proper rational function.

37. The partial fraction decomposition of
$$\frac{2x + 3}{x^2} \quad \text{is} \quad \frac{2}{x} + \frac{3}{x^2}.$$

38. If $f(x) = P(x)/(x + 5)^3$ is a proper rational function, then the partial fraction decomposition of $f(x)$ has terms with constant numerators and denominators $(x + 5)$, $(x + 5)^2$, and $(x + 5)^3$.

39–42 Evaluate the integral by making a substitution that converts the integrand to a rational function. ■

39. $\displaystyle\int \frac{\cos\theta}{\sin^2\theta + 4\sin\theta - 5}\,d\theta$ **40.** $\displaystyle\int \frac{e^t}{e^{2t} - 4}\,dt$

41. $\displaystyle\int \frac{e^{3x}}{e^{2x} + 4}\,dx$ **42.** $\displaystyle\int \frac{5 + 2\ln x}{x(1 + \ln x)^2}\,dx$

43. Find the volume of the solid generated when the region enclosed by $y = x^2/(9 - x^2)$, $y = 0$, $x = 0$, and $x = 2$ is revolved about the x-axis.

44. Find the area of the region under the curve $y = 1/(1 + e^x)$, over the interval $[-\ln 5, \ln 5]$. [*Hint:* Make a substitution that converts the integrand to a rational function.]

C **45–46** Use a CAS to evaluate the integral in two ways: (i) integrate directly; (ii) use the CAS to find the partial fraction decomposition and integrate the decomposition. Integrate by hand to check the results. ■

45. $\displaystyle\int \frac{x^2 + 1}{(x^2 + 2x + 3)^2}\,dx$

46. $\displaystyle\int \frac{x^5 + x^4 + 4x^3 + 4x^2 + 4x + 4}{(x^2 + 2)^3}\,dx$

C **47–48** Integrate by hand and check your answers using a CAS. ■

47. $\displaystyle\int \frac{dx}{x^4 - 3x^3 - 7x^2 + 27x - 18}$

48. $\displaystyle\int \frac{dx}{16x^3 - 4x^2 + 4x - 1}$

FOCUS ON CONCEPTS

49. Show that
$$\int_0^1 \frac{x}{x^4 + 1}\,dx = \frac{\pi}{8}$$

50. Use partial fractions to derive the integration formula
$$\int \frac{1}{a^2 - x^2}\,dx = \frac{1}{2a}\ln\left|\frac{a + x}{a - x}\right| + C$$

51. Suppose that $ax^2 + bx + c$ is a quadratic polynomial and that the integration
$$\int \frac{1}{ax^2 + bx + c}\,dx$$
produces a function with no inverse tangent terms. What does this tell you about the roots of the polynomial?

52. Suppose that $ax^2 + bx + c$ is a quadratic polynomial and that the integration
$$\int \frac{1}{ax^2 + bx + c}\,dx$$
produces a function with neither logarithmic nor inverse tangent terms. What does this tell you about the roots of the polynomial?

53. Does there exist a quadratic polynomial $ax^2 + bx + c$ such that the integration
$$\int \frac{x}{ax^2 + bx + c}\,dx$$
produces a function with no logarithmic terms? If so, give an example; if not, explain why no such polynomial can exist.

54. Writing Suppose that $P(x)$ is a cubic polynomial. State the general form of the partial fraction decomposition for
$$f(x) = \frac{P(x)}{(x + 5)^4}$$
and state the implications of this decomposition for evaluating the integral $\int f(x)\,dx$.

55. Writing Consider the functions
$$f(x) = \frac{1}{x^2 - 4} \quad \text{and} \quad g(x) = \frac{x}{x^2 - 4}$$
Each of the integrals $\int f(x)\,dx$ and $\int g(x)\,dx$ can be evaluated using partial fractions and using at least one other integration technique. Demonstrate two different techniques for evaluating each of these integrals, and then discuss the considerations that would determine which technique you would use.

✔ **QUICK CHECK ANSWERS 7.5**

1. $\dfrac{A}{(ax+b)^k}$; $\dfrac{Ax+B}{(ax^2+bx+c)^k}$ 2. (a) A proper rational function is a rational function in which the degree of the numerator is less than the degree of the denominator. (b) The degree of the numerator must be less than the degree of the denominator. (c) Divide the denominator into the numerator, which results in the sum of a polynomial and a proper rational function.

3. (a) $\dfrac{A_1}{ax+b}+\dfrac{A_2}{(ax+b)^2}+\cdots+\dfrac{A_m}{(ax+b)^m}$ (b) $\dfrac{A_1x+B_1}{ax^2+bx+c}+\dfrac{A_2x+B_2}{(ax^2+bx+c)^2}+\cdots+\dfrac{A_mx+B_m}{(ax^2+bx+c)^m}$

4. (a) $A=1$ (b) $B=2$

5. (a) $\displaystyle\int \dfrac{3}{(x+1)(1-2x)}\,dx = \ln\left|\dfrac{x+1}{1-2x}\right|+C$ (b) $\displaystyle\int \dfrac{2x^2-3x}{(x^2+1)(3x+2)}\,dx = \dfrac{2}{3}\ln|3x+2|-\tan^{-1}x+C$

7.6 USING COMPUTER ALGEBRA SYSTEMS AND TABLES OF INTEGRALS

In this section we will discuss how to integrate using tables, and we will see some special substitutions to try when an integral doesn't match any of the forms in an integral table. In particular, we will discuss a method for integrating rational functions of $\sin x$ and $\cos x$. We will also address some of the issues that relate to using computer algebra systems for integration. Readers who are not using computer algebra systems can skip that material.

■ INTEGRAL TABLES

Tables of integrals are useful for eliminating tedious hand computation. The endpapers of this text contain a relatively brief table of integrals that we will refer to as the ***Endpaper Integral Table***; more comprehensive tables are published in standard reference books such as the *CRC Standard Mathematical Tables and Formulae*, CRC Press, Inc., 2002.

All integral tables have their own scheme for classifying integrals according to the form of the integrand. For example, the Endpaper Integral Table classifies the integrals into 15 categories; *Basic Functions*, *Reciprocals of Basic Functions*, *Powers of Trigonometric Functions*, *Products of Trigonometric Functions*, and so forth. The first step in working with tables is to read through the classifications so that you understand the classification scheme and know where to look in the table for integrals of different types.

■ PERFECT MATCHES

If you are lucky, the integral you are attempting to evaluate will match up perfectly with one of the forms in the table. However, when looking for matches you may have to make an adjustment for the variable of integration. For example, the integral

$$\int x^2 \sin x \, dx$$

is a perfect match with Formula (46) in the Endpaper Integral Table, except for the letter used for the variable of integration. Thus, to apply Formula (46) to the given integral we need to change the variable of integration in the formula from u to x. With that minor modification we obtain

$$\int x^2 \sin x \, dx = 2x \sin x + (2-x^2)\cos x + C$$

Here are some more examples of perfect matches.

▶ **Example 1**   Use the Endpaper Integral Table to evaluate

$$\text{(a)} \int \sin 7x \cos 2x \, dx \qquad \text{(b)} \int x^2 \sqrt{7 + 3x} \, dx$$

$$\text{(c)} \int \frac{\sqrt{2 - x^2}}{x} \, dx \qquad \text{(d)} \int (x^3 + 7x + 1) \sin \pi x \, dx$$

Solution (a).   The integrand can be classified as a product of trigonometric functions. Thus, from Formula (40) with $m = 7$ and $n = 2$ we obtain

$$\int \sin 7x \cos 2x \, dx = -\frac{\cos 9x}{18} - \frac{\cos 5x}{10} + C$$

Solution (b).   The integrand can be classified as a power of x multiplying $\sqrt{a + bx}$. Thus, from Formula (103) with $a = 7$ and $b = 3$ we obtain

$$\int x^2 \sqrt{7 + 3x} \, dx = \frac{2}{2835} (135x^2 - 252x + 392)(7 + 3x)^{3/2} + C$$

Solution (c).   The integrand can be classified as a power of x dividing $\sqrt{a^2 - x^2}$. Thus, from Formula (79) with $a = \sqrt{2}$ we obtain

$$\int \frac{\sqrt{2 - x^2}}{x} \, dx = \sqrt{2 - x^2} - \sqrt{2} \ln \left| \frac{\sqrt{2} + \sqrt{2 - x^2}}{x} \right| + C$$

Solution (d).   The integrand can be classified as a polynomial multiplying a trigonometric function. Thus, we apply Formula (58) with $p(x) = x^3 + 7x + 1$ and $a = \pi$. The successive nonzero derivatives of $p(x)$ are

$$p'(x) = 3x^2 + 7, \quad p''(x) = 6x, \quad p'''(x) = 6$$

and so

$$\int (x^3 + 7x + 1) \sin \pi x \, dx$$

$$= -\frac{x^3 + 7x + 1}{\pi} \cos \pi x + \frac{3x^2 + 7}{\pi^2} \sin \pi x + \frac{6x}{\pi^3} \cos \pi x - \frac{6}{\pi^4} \sin \pi x + C \blacktriangleleft$$

■ **MATCHES REQUIRING SUBSTITUTIONS**

Sometimes an integral that does not match any table entry can be made to match by making an appropriate substitution.

▶ **Example 2**   Use the Endpaper Integral Table to evaluate

$$\text{(a)} \int e^{\pi x} \sin^{-1}(e^{\pi x}) \, dx \qquad \text{(b)} \int x \sqrt{x^2 - 4x + 5} \, dx$$

Solution (a).   The integrand does not even come close to matching any of the forms in the table. However, a little thought suggests the substitution

$$u = e^{\pi x}, \quad du = \pi e^{\pi x} \, dx$$

from which we obtain

$$\int e^{\pi x} \sin^{-1}(e^{\pi x}) \, dx = \frac{1}{\pi} \int \sin^{-1} u \, du$$

The integrand is now a basic function, and Formula (7) yields

$$\int e^{\pi x} \sin^{-1}(e^{\pi x}) \, dx = \frac{1}{\pi} \left[u \sin^{-1} u + \sqrt{1 - u^2} \right] + C$$

$$= \frac{1}{\pi} \left[e^{\pi x} \sin^{-1}(e^{\pi x}) + \sqrt{1 - e^{2\pi x}} \right] + C$$

Solution (b). Again, the integrand does not closely match any of the forms in the table. However, a little thought suggests that it may be possible to bring the integrand closer to the form $x\sqrt{x^2 + a^2}$ by completing the square to eliminate the term involving x inside the radical. Doing this yields

$$\int x\sqrt{x^2 - 4x + 5} \, dx = \int x\sqrt{(x^2 - 4x + 4) + 1} \, dx = \int x\sqrt{(x - 2)^2 + 1} \, dx \quad (1)$$

At this point we are closer to the form $x\sqrt{x^2 + a^2}$, but we are not quite there because of the $(x - 2)^2$ rather than x^2 inside the radical. However, we can resolve that problem with the substitution

$$u = x - 2, \quad du = dx$$

With this substitution we have $x = u + 2$, so (1) can be expressed in terms of u as

$$\int x\sqrt{x^2 - 4x + 5} \, dx = \int (u + 2)\sqrt{u^2 + 1} \, du = \int u\sqrt{u^2 + 1} \, du + 2 \int \sqrt{u^2 + 1} \, du$$

The first integral on the right is now a perfect match with Formula (84) with $a = 1$, and the second is a perfect match with Formula (72) with $a = 1$. Thus, applying these formulas we obtain

$$\int x\sqrt{x^2 - 4x + 5} \, dx = \frac{1}{3}(u^2 + 1)^{3/2} + 2 \left[\frac{1}{2} u\sqrt{u^2 + 1} + \frac{1}{2} \ln(u + \sqrt{u^2 + 1}) \right] + C$$

If we now replace u by $x - 2$ (in which case $u^2 + 1 = x^2 - 4x + 5$), we obtain

$$\int x\sqrt{x^2 - 4x + 5} \, dx = \frac{1}{3}(x^2 - 4x + 5)^{3/2} + (x - 2)\sqrt{x^2 - 4x + 5}$$

$$+ \ln\left(x - 2 + \sqrt{x^2 - 4x + 5}\right) + C$$

Although correct, this form of the answer has an unnecessary mixture of radicals and fractional exponents. If desired, we can "clean up" the answer by writing

$$(x^2 - 4x + 5)^{3/2} = (x^2 - 4x + 5)\sqrt{x^2 - 4x + 5}$$

from which it follows that (verify)

$$\int x\sqrt{x^2 - 4x + 5} \, dx = \frac{1}{3}(x^2 - x - 1)\sqrt{x^2 - 4x + 5}$$

$$+ \ln\left(x - 2 + \sqrt{x^2 - 4x + 5}\right) + C \blacktriangleleft$$

■ **MATCHES REQUIRING REDUCTION FORMULAS**
In cases where the entry in an integral table is a reduction formula, that formula will have to be applied first to reduce the given integral to a form in which it can be evaluated.

▶ **Example 3**   Use the Endpaper Integral Table to evaluate $\displaystyle\int \frac{x^3}{\sqrt{1+x}}\,dx$.

Solution.   The integrand can be classified as a power of x multiplying the reciprocal of $\sqrt{a+bx}$. Thus, from Formula (107) with $a=1$, $b=1$, and $n=3$, followed by Formula (106), we obtain

$$\int \frac{x^3}{\sqrt{1+x}}\,dx = \frac{2x^3\sqrt{1+x}}{7} - \frac{6}{7}\int \frac{x^2}{\sqrt{1+x}}\,dx$$

$$= \frac{2x^3\sqrt{1+x}}{7} - \frac{6}{7}\left[\frac{2}{15}(3x^2 - 4x + 8)\sqrt{1+x}\right] + C$$

$$= \left(\frac{2x^3}{7} - \frac{12x^2}{35} + \frac{16x}{35} - \frac{32}{35}\right)\sqrt{1+x} + C \blacktriangleleft$$

■ **SPECIAL SUBSTITUTIONS**

The Endpaper Integral Table has numerous entries involving an exponent of 3/2 or involving square roots (exponent 1/2), but it has no entries with other fractional exponents. However, integrals involving fractional powers of x can often be simplified by making the substitution $u = x^{1/n}$ in which n is the least common multiple of the denominators of the exponents. The resulting integral will then involve integer powers of u.

▶ **Example 4**   Evaluate

$$\text{(a) } \int \frac{\sqrt{x}}{1 + \sqrt[3]{x}}\,dx \qquad \text{(b) } \int \sqrt{1 + e^x}\,dx$$

Solution (a).   The integrand contains $x^{1/2}$ and $x^{1/3}$, so we make the substitution $u = x^{1/6}$, from which we obtain

$$x = u^6, \quad dx = 6u^5\,du$$

Thus,

$$\int \frac{\sqrt{x}}{1 + \sqrt[3]{x}}\,dx = \int \frac{(u^6)^{1/2}}{1 + (u^6)^{1/3}}(6u^5)\,du = 6\int \frac{u^8}{1 + u^2}\,du$$

By long division

$$\frac{u^8}{1 + u^2} = u^6 - u^4 + u^2 - 1 + \frac{1}{1 + u^2}$$

from which it follows that

$$\int \frac{\sqrt{x}}{1 + \sqrt[3]{x}}\,dx = 6\int \left(u^6 - u^4 + u^2 - 1 + \frac{1}{1 + u^2}\right)du$$

$$= \tfrac{6}{7}u^7 - \tfrac{6}{5}u^5 + 2u^3 - 6u + 6\tan^{-1}u + C$$

$$= \tfrac{6}{7}x^{7/6} - \tfrac{6}{5}x^{5/6} + 2x^{1/2} - 6x^{1/6} + 6\tan^{-1}(x^{1/6}) + C$$

Solution (b).   The integral does not match any of the forms in the Endpaper Integral Table. However, the table does include several integrals containing $\sqrt{a+bu}$. This suggests the substitution $u = e^x$, from which we obtain

$$x = \ln u, \quad dx = \frac{1}{u}\,du$$

Thus, from Formula (110) with $a = 1$ and $b = 1$, followed by Formula (108), we obtain

$$\int \sqrt{1 + e^x} \, dx = \int \frac{\sqrt{1 + u}}{u} \, du$$

$$= 2\sqrt{1 + u} + \int \frac{du}{u\sqrt{1 + u}}$$

$$= 2\sqrt{1 + u} + \ln \left| \frac{\sqrt{1 + u} - 1}{\sqrt{1 + u} + 1} \right| + C$$

$$= 2\sqrt{1 + e^x} + \ln \left[\frac{\sqrt{1 + e^x} - 1}{\sqrt{1 + e^x} + 1} \right] + C \qquad \boxed{\text{Absolute value not needed}} \blacktriangleleft$$

Try finding the antiderivative in Example 4(b) using the substitution

$$u = \sqrt{1 + e^x}$$

Functions that consist of finitely many sums, differences, quotients, and products of $\sin x$ and $\cos x$ are called ***rational functions of sin x and cos x***. Some examples are

$$\frac{\sin x + 3 \cos^2 x}{\cos x + 4 \sin x}, \qquad \frac{\sin x}{1 + \cos x - \cos^2 x}, \qquad \frac{3 \sin^5 x}{1 + 4 \sin x}$$

The Endpaper Integral Table gives a few formulas for integrating rational functions of $\sin x$ and $\cos x$ under the heading *Reciprocals of Basic Functions*. For example, it follows from Formula (18) that

$$\int \frac{1}{1 + \sin x} \, dx = \tan x - \sec x + C \tag{2}$$

However, since the integrand is a rational function of $\sin x$, it may be desirable in a particular application to express the value of the integral in terms of $\sin x$ and $\cos x$ and rewrite (2) as

$$\int \frac{1}{1 + \sin x} \, dx = \frac{\sin x - 1}{\cos x} + C$$

Many rational functions of $\sin x$ and $\cos x$ can be evaluated by an ingenious method that was discovered by the mathematician Karl Weierstrass (see p. 102 for biography). The idea is to make the substitution

$$u = \tan(x/2), \qquad -\pi/2 < x/2 < \pi/2$$

from which it follows that

$$x = 2 \tan^{-1} u, \qquad dx = \frac{2}{1 + u^2} \, du$$

To implement this substitution we need to express $\sin x$ and $\cos x$ in terms of u. For this purpose we will use the identities

$$\sin x = 2 \sin(x/2) \cos(x/2) \tag{3}$$
$$\cos x = \cos^2(x/2) - \sin^2(x/2) \tag{4}$$

and the following relationships suggested by Figure 7.6.1:

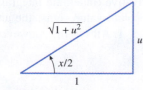

$$\sin(x/2) = \frac{u}{\sqrt{1 + u^2}} \quad \text{and} \quad \cos(x/2) = \frac{1}{\sqrt{1 + u^2}}$$

Substituting these expressions in (3) and (4) yields

$$\sin x = 2 \left(\frac{u}{\sqrt{1 + u^2}} \right) \left(\frac{1}{\sqrt{1 + u^2}} \right) = \frac{2u}{1 + u^2}$$

$$\cos x = \left(\frac{1}{\sqrt{1 + u^2}} \right)^2 - \left(\frac{u}{\sqrt{1 + u^2}} \right)^2 = \frac{1 - u^2}{1 + u^2}$$

▲ **Figure 7.6.1**

In summary, we have shown that the substitution $u = \tan(x/2)$ can be implemented in a rational function of $\sin x$ and $\cos x$ by letting

$$\sin x = \frac{2u}{1+u^2}, \quad \cos x = \frac{1-u^2}{1+u^2}, \quad dx = \frac{2}{1+u^2}\,du \tag{5}$$

▶ **Example 5** Evaluate $\displaystyle\int \frac{dx}{1 - \sin x + \cos x}$.

Solution. The integrand is a rational function of $\sin x$ and $\cos x$ that does not match any of the formulas in the Endpaper Integral Table, so we make the substitution $u = \tan(x/2)$. Thus, from (5) we obtain

$$\int \frac{dx}{1 - \sin x + \cos x} = \int \frac{\dfrac{2\,du}{1+u^2}}{1 - \left(\dfrac{2u}{1+u^2}\right) + \left(\dfrac{1-u^2}{1+u^2}\right)}$$

$$= \int \frac{2\,du}{(1+u^2) - 2u + (1-u^2)}$$

$$= \int \frac{du}{1-u} = -\ln|1-u| + C = -\ln|1 - \tan(x/2)| + C \quad \blacktriangleleft$$

> The substitution $u = \tan(x/2)$ will convert any rational function of $\sin x$ and $\cos x$ to an ordinary rational function of u. However, the method can lead to cumbersome partial fraction decompositions, so it may be worthwhile to consider other methods as well when hand computations are being used.

■ **INTEGRATING WITH COMPUTER ALGEBRA SYSTEMS**

Integration tables are rapidly giving way to computerized integration using computer algebra systems. However, as with many powerful tools, a knowledgeable operator is an important component of the system.

Sometimes computer algebra systems do not produce the most general form of the indefinite integral. For example, the integral formula

$$\int \frac{dx}{x-1} = \ln|x-1| + C$$

which can be obtained by inspection or by using the substitution $u = x - 1$, is valid for $x > 1$ or for $x < 1$. However, not all computer algebra systems produce this form of the answer. Some typical answers produced by various implementations of *Mathematica*, *Maple*, and the CAS on a handheld calculator are

$$\ln(-1+x), \quad \ln(x-1), \quad \ln(|x-1|)$$

Observe that none of the systems include the constant of integration—the answer produced is a particular antiderivative and not the most general antiderivative (indefinite integral). Observe also that only one of these answers includes the absolute value signs; the antiderivatives produced by the other systems are valid only for $x > 1$. All systems, however, are able to calculate the definite integral

$$\int_0^{1/2} \frac{dx}{x-1} = -\ln 2$$

correctly. Now let us examine how these systems handle the integral

$$\int x\sqrt{x^2 - 4x + 5}\,dx = \tfrac{1}{3}(x^2 - x - 1)\sqrt{x^2 - 4x + 5}$$
$$+ \ln(x - 2 + \sqrt{x^2 - 4x + 5}) \tag{6}$$

which we obtained in Example 2(b) (with the constant of integration included). Some CAS implementations produce this result in slightly different algebraic forms, but a version of *Maple* produces the result

$$\int x\sqrt{x^2 - 4x + 5}\,dx = \tfrac{1}{3}(x^2 - 4x + 5)^{3/2} + \tfrac{1}{2}(2x - 4)\sqrt{x^2 - 4x + 5} + \sinh^{-1}(x - 2)$$

This can be rewritten as (6) by expressing the fractional exponent in radical form and expressing $\sinh^{-1}(x - 2)$ in logarithmic form using Theorem 6.9.4 (verify). A version of *Mathematica* produces the result

$$\int x\sqrt{x^2 - 4x + 5}\,dx = \tfrac{1}{3}(x^2 - x - 1)\sqrt{x^2 - 4x + 5} - \sinh^{-1}(2 - x)$$

which can be rewritten in form (6) by using Theorem 6.9.4 together with the identity $\sinh^{-1}(-x) = -\sinh^{-1}x$ (verify).

Computer algebra systems can sometimes produce inconvenient or unnatural answers to integration problems. For example, various computer algebra systems produced the following results when asked to integrate $(x + 1)^7$:

> Expanding the expression
>
> $$\frac{(x + 1)^8}{8}$$
>
> produces a constant term of $\tfrac{1}{8}$, whereas the second expression in (7) has no constant term. What is the explanation?

$$\frac{(x + 1)^8}{8}, \qquad \frac{1}{8}x^8 + x^7 + \frac{7}{2}x^6 + 7x^5 + \frac{35}{4}x^4 + 7x^3 + \frac{7}{2}x^2 + x \tag{7}$$

The first form is in keeping with the hand computation

$$\int (x + 1)^7\,dx = \frac{(x + 1)^8}{8} + C$$

that uses the substitution $u = x + 1$, whereas the second form is based on expanding $(x + 1)^7$ and integrating term by term.

In Example 2(a) of Section 7.3 we showed that

$$\int \sin^4 x \cos^5 x\,dx = \tfrac{1}{5}\sin^5 x - \tfrac{2}{7}\sin^7 x + \tfrac{1}{9}\sin^9 x + C$$

However, a version of *Mathematica* integrates this as

$$\tfrac{3}{128}\sin x - \tfrac{1}{192}\sin 3x - \tfrac{1}{320}\sin 5x + \tfrac{1}{1792}\sin 7x + \tfrac{1}{2304}\sin 9x$$

whereas other computer algebra systems essentially integrate it as

$$-\tfrac{1}{9}\sin^3 x \cos^6 x - \tfrac{1}{21}\sin x \cos^6 x + \tfrac{1}{105}\cos^4 x \sin x + \tfrac{4}{315}\cos^2 x \sin x + \tfrac{8}{315}\sin x$$

Although these three results look quite different, they can be obtained from one another using appropriate trigonometric identities.

■ COMPUTER ALGEBRA SYSTEMS HAVE LIMITATIONS

A computer algebra system combines a set of integration rules (such as substitution) with a library of functions that it can use to construct antiderivatives. Such libraries contain elementary functions, such as polynomials, rational functions, trigonometric functions, as well as various nonelementary functions that arise in engineering, physics, and other applied fields. Just as our Endpaper Integral Table has only 121 indefinite integrals, these libraries are not exhaustive of all possible integrands. If the system cannot manipulate the integrand to a form matching one in its library, the program will give some indication that it cannot evaluate the integral. For example, when asked to evaluate the integral

> **TECHNOLOGY MASTERY**
>
> Sometimes integrals that cannot be evaluated by a CAS in their given form can be evaluated by first rewriting them in a different form or by making a substitution. If you have a CAS, make a u-substitution in (8) that will enable you to evaluate the integral with your CAS. Then evaluate the integral.

$$\int (1 + \ln x)\sqrt{1 + (x \ln x)^2}\,dx \tag{8}$$

all of the systems mentioned above respond by displaying some form of the unevaluated integral as an answer, indicating that they could not perform the integration.

Sometimes computer algebra systems respond by expressing an integral in terms of another integral. For example, if you try to integrate e^{x^2} using *Mathematica*, *Maple*, or

Sage, you will obtain an expression involving erf (which stands for *error function*). The function erf(x) is defined as

$$\text{erf}(x) = \frac{2}{\sqrt{\pi}} \int_0^x e^{-t^2}\, dt$$

so all three programs essentially rewrite the given integral in terms of a closely related integral. From one point of view this is what we did in integrating $1/x$, since the natural logarithm function is (formally) defined as

$$\ln x = \int_1^x \frac{1}{t}\, dt$$

(see Section 5.10).

▶ **Example 6**   A particle moves along an x-axis in such a way that its velocity $v(t)$ at time t is

$$v(t) = 30 \cos^7 t \sin^4 t \quad (t \ge 0)$$

Graph the position versus time curve for the particle, given that the particle is at $x = 1$ when $t = 0$.

Solution.   Since $dx/dt = v(t)$ and $x = 1$ when $t = 0$, the position function $x(t)$ is given by

$$x(t) = 1 + \int_0^t v(s)\, ds$$

Some computer algebra systems will allow this expression to be entered directly into a command for plotting functions, but it is often more efficient to perform the integration first. The authors' integration utility yields

$$x = \int 30 \cos^7 t \sin^4 t\, dt$$
$$= -\tfrac{30}{11} \sin^{11} t + 10 \sin^9 t - \tfrac{90}{7} \sin^7 t + 6 \sin^5 t + C$$

where we have added the required constant of integration. Using the initial condition $x(0) = 1$, we substitute the values $x = 1$ and $t = 0$ into this equation to find that $C = 1$, so

$$x(t) = -\tfrac{30}{11} \sin^{11} t + 10 \sin^9 t - \tfrac{90}{7} \sin^7 t + 6 \sin^5 t + 1 \quad (t \ge 0)$$

The graph of x versus t is shown in Figure 7.6.2. ◀

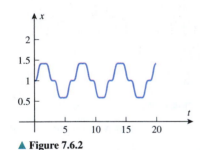

▲ **Figure 7.6.2**

✔ **QUICK CHECK EXERCISES 7.6**   *(See page 533 for answers.)*

1. Find an integral formula in the Endpaper Integral Table that can be used to evaluate the integral. Do not evaluate the integral.

 (a) $\displaystyle\int \frac{2x}{3x+4}\, dx$ _____

 (b) $\displaystyle\int \frac{1}{x\sqrt{5x-4}}\, dx$ _____

 (c) $\displaystyle\int x\sqrt{3x+2}\, dx$ _____

 (d) $\displaystyle\int x^2 \ln x\, dx$ _____

2. In each part, make the indicated u-substitution, and then find an integral formula in the Endpaper Integral Table that

can be used to evaluate the integral. Do not evaluate the integral.

 (a) $\displaystyle\int \frac{x}{1+e^{x^2}}\, dx; \ u = x^2$ _____

 (b) $\displaystyle\int e^{\sqrt{x}}\, dx; \ u = \sqrt{x}$ _____

 (c) $\displaystyle\int \frac{e^x}{1+\sin(e^x)}\, dx; \ u = e^x$ _____

 (d) $\displaystyle\int \frac{1}{(1-4x^2)^{3/2}}\, dx; \ u = 2x$ _____

3. In each part, use the Endpaper Integral Table to evaluate the integral. (If necessary, first make an appropriate substitution or complete the square.)   *(cont.)*

(a) $\displaystyle\int \frac{1}{4-x^2}\,dx = \underline{\hspace{2cm}}$

(b) $\displaystyle\int \cos 2x \cos x \, dx = \underline{\hspace{2cm}}$

(c) $\displaystyle\int \frac{e^{3x}}{\sqrt{1-e^{2x}}}\,dx = \underline{\hspace{2cm}}$

(d) $\displaystyle\int \frac{x}{x^2-4x+8}\,dx = \underline{\hspace{2cm}}$

EXERCISE SET 7.6   [C] CAS

[C] **1–24** (a) Use the Endpaper Integral Table to evaluate the given integral. (b) If you have a CAS, use it to evaluate the integral, and then confirm that the result is equivalent to the one that you found in part (a). ■

1. $\displaystyle\int \frac{4x}{3x-1}\,dx$

2. $\displaystyle\int \frac{x}{(4-5x)^2}\,dx$

3. $\displaystyle\int \frac{1}{x(2x+5)}\,dx$

4. $\displaystyle\int \frac{1}{x^2(1-5x)}\,dx$

5. $\displaystyle\int x\sqrt{2x+3}\,dx$

6. $\displaystyle\int \frac{x}{\sqrt{2-x}}\,dx$

7. $\displaystyle\int \frac{1}{x\sqrt{4-3x}}\,dx$

8. $\displaystyle\int \frac{1}{x\sqrt{3x-4}}\,dx$

9. $\displaystyle\int \frac{1}{16-x^2}\,dx$

10. $\displaystyle\int \frac{1}{x^2-9}\,dx$

11. $\displaystyle\int \sqrt{x^2-3}\,dx$

12. $\displaystyle\int \frac{\sqrt{x^2-5}}{x^2}\,dx$

13. $\displaystyle\int \frac{x^2}{\sqrt{x^2+4}}\,dx$

14. $\displaystyle\int \frac{1}{x^2\sqrt{x^2-2}}\,dx$

15. $\displaystyle\int \sqrt{9-x^2}\,dx$

16. $\displaystyle\int \frac{\sqrt{4-x^2}}{x^2}\,dx$

17. $\displaystyle\int \frac{\sqrt{4-x^2}}{x}\,dx$

18. $\displaystyle\int \frac{1}{x\sqrt{6x-x^2}}\,dx$

19. $\displaystyle\int \sin 3x \sin 4x\,dx$

20. $\displaystyle\int \sin 2x \cos 5x\,dx$

21. $\displaystyle\int x^3 \ln x\,dx$

22. $\displaystyle\int \frac{\ln x}{\sqrt{x^3}}\,dx$

23. $\displaystyle\int e^{-2x} \sin 3x\,dx$

24. $\displaystyle\int e^x \cos 2x\,dx$

[C] **25–36** (a) Make the indicated u-substitution, and then use the Endpaper Integral Table to evaluate the integral. (b) If you have a CAS, use it to evaluate the integral, and then confirm that the result is equivalent to the one that you found in part (a). ■

25. $\displaystyle\int \frac{e^{4x}}{(4-3e^{2x})^2}\,dx,\ u=e^{2x}$

26. $\displaystyle\int \frac{\sin 2x}{(\cos 2x)(3-\cos 2x)}\,dx,\ u=\cos 2x$

27. $\displaystyle\int \frac{1}{\sqrt{x}(9x+4)}\,dx,\ u=3\sqrt{x}$

28. $\displaystyle\int \frac{\cos 4x}{9+\sin^2 4x}\,dx,\ u=\sin 4x$

29. $\displaystyle\int \frac{1}{\sqrt{4x^2-9}}\,dx,\ u=2x$

30. $\displaystyle\int x\sqrt{2x^4+3}\,dx,\ u=\sqrt{2}x^2$

31. $\displaystyle\int \frac{4x^5}{\sqrt{2-4x^4}}\,dx,\ u=2x^2$

32. $\displaystyle\int \frac{1}{x^2\sqrt{3-4x^2}}\,dx,\ u=2x$

33. $\displaystyle\int \frac{\sin^2(\ln x)}{x}\,dx,\ u=\ln x$

34. $\displaystyle\int e^{-2x}\cos^2(e^{-2x})\,dx,\ u=e^{-2x}$

35. $\displaystyle\int xe^{-2x}\,dx,\ u=-2x$

36. $\displaystyle\int \ln(3x+1)\,dx,\ u=3x+1$

[C] **37–48** (a) Make an appropriate u-substitution, and then use the Endpaper Integral Table to evaluate the integral. (b) If you have a CAS, use it to evaluate the integral (no substitution), and then confirm that the result is equivalent to that in part (a). ■

37. $\displaystyle\int \frac{\cos 3x}{(\sin 3x)(\sin 3x+1)^2}\,dx$

38. $\displaystyle\int \frac{\ln x}{x\sqrt{4\ln x-1}}\,dx$

39. $\displaystyle\int \frac{x}{16x^4-1}\,dx$

40. $\displaystyle\int \frac{e^x}{3-4e^{2x}}\,dx$

41. $\displaystyle\int e^x\sqrt{3-4e^{2x}}\,dx$

42. $\displaystyle\int \frac{\sqrt{4-9x^2}}{x^2}\,dx$

43. $\displaystyle\int \sqrt{5x-9x^2}\,dx$

44. $\displaystyle\int \frac{1}{x\sqrt{x-5x^2}}\,dx$

45. $\displaystyle\int 4x\sin 2x\,dx$

46. $\displaystyle\int \cos\sqrt{x}\,dx$

47. $\displaystyle\int e^{-\sqrt{x}}\,dx$

48. $\displaystyle\int x\ln(2+x^2)\,dx$

[C] **49–52** (a) Complete the square, make an appropriate u-substitution, and then use the Endpaper Integral Table to evaluate the integral. (b) If you have a CAS, use it to evaluate the integral (no substitution or square completion), and then confirm that the result is equivalent to that in part (a). ■

49. $\displaystyle\int \frac{1}{x^2+6x-7}\,dx$

50. $\displaystyle\int \sqrt{3-2x-x^2}\,dx$

51. $\displaystyle\int \frac{x}{\sqrt{5 + 4x - x^2}}\, dx$ **52.** $\displaystyle\int \frac{x}{x^2 + 6x + 13}\, dx$

[c] **53–64** (a) Make an appropriate u-substitution of the form $u = x^{1/n}$ or $u = (x + a)^{1/n}$, and then evaluate the integral. (b) If you have a CAS, use it to evaluate the integral, and then confirm that the result is equivalent to the one that you found in part (a). ■

53. $\displaystyle\int x\sqrt{x - 2}\, dx$ **54.** $\displaystyle\int \frac{x}{\sqrt{x + 1}}\, dx$

55. $\displaystyle\int x^5 \sqrt{x^3 + 1}\, dx$ **56.** $\displaystyle\int \frac{1}{x\sqrt{x^3 - 1}}\, dx$

57. $\displaystyle\int \frac{dx}{x - \sqrt[3]{x}}$ **58.** $\displaystyle\int \frac{dx}{\sqrt{x} + \sqrt[3]{x}}$

59. $\displaystyle\int \frac{dx}{x(1 - x^{1/4})}$ **60.** $\displaystyle\int \frac{\sqrt{x}}{x + 1}\, dx$

61. $\displaystyle\int \frac{dx}{x^{1/2} - x^{1/3}}$ **62.** $\displaystyle\int \frac{1 + \sqrt{x}}{1 - \sqrt{x}}\, dx$

63. $\displaystyle\int \frac{x^3}{\sqrt{1 + x^2}}\, dx$ **64.** $\displaystyle\int \frac{x}{(x + 3)^{1/5}}\, dx$

[c] **65–70** (a) Make u-substitution (5) to convert the integrand to a rational function of u, and then evaluate the integral. (b) If you have a CAS, use it to evaluate the integral (no substitution), and then confirm that the result is equivalent to that in part (a). ■

65. $\displaystyle\int \frac{dx}{1 + \sin x + \cos x}$ **66.** $\displaystyle\int \frac{dx}{2 + \sin x}$

67. $\displaystyle\int \frac{d\theta}{1 - \cos \theta}$ **68.** $\displaystyle\int \frac{dx}{4 \sin x - 3 \cos x}$

69. $\displaystyle\int \frac{dx}{\sin x + \tan x}$ **70.** $\displaystyle\int \frac{\sin x}{\sin x + \tan x}\, dx$

71–72 Use any method to solve for x. ■

71. $\displaystyle\int_2^x \frac{1}{t(4 - t)}\, dt = 0.5,\ 2 < x < 4$

72. $\displaystyle\int_1^x \frac{1}{t\sqrt{2t - 1}}\, dt = 1,\ x > \tfrac{1}{2}$

73–76 Use any method to find the area of the region enclosed by the curves. ■

73. $y = \sqrt{25 - x^2},\ y = 0,\ x = 0,\ x = 4$

74. $y = \sqrt{9x^2 - 4},\ y = 0,\ x = 2$

75. $y = \dfrac{1}{25 - 16x^2},\ y = 0,\ x = 0,\ x = 1$

76. $y = \sqrt{x} \ln x,\ y = 0,\ x = 4$

77–80 Use any method to find the volume of the solid generated when the region enclosed by the curves is revolved about the y-axis. ■

77. $y = \cos x,\ y = 0,\ x = 0,\ x = \pi/2$

78. $y = \sqrt{x - 4},\ y = 0,\ x = 8$

79. $y = e^{-x},\ y = 0,\ x = 0,\ x = 3$

80. $y = \ln x,\ y = 0,\ x = 5$

81–82 Use any method to find the arc length of the curve. ■

81. $y = 2x^2,\ 0 \le x \le 2$ **82.** $y = 3 \ln x,\ 1 \le x \le 3$

83–84 Use any method to find the area of the surface generated by revolving the curve about the x-axis. ■

83. $y = \sin x,\ 0 \le x \le \pi$ **84.** $y = 1/x,\ 1 \le x \le 4$

[c] **85–86** Information is given about the motion of a particle moving along a coordinate line.
(a) Use a CAS to find the position function of the particle for $t \ge 0$.
(b) Graph the position versus time curve. ■

85. $v(t) = 20 \cos^6 t \sin^3 t,\ s(0) = 2$

86. $a(t) = e^{-t} \sin 2t \sin 4t,\ v(0) = 0,\ s(0) = 10$

FOCUS ON CONCEPTS

87. (a) Use the substitution $u = \tan(x/2)$ to show that
$$\int \sec x\, dx = \ln \left| \frac{1 + \tan(x/2)}{1 - \tan(x/2)} \right| + C$$
and confirm that this is consistent with Formula (22) of Section 7.3.
(b) Use the result in part (a) to show that
$$\int \sec x\, dx = \ln \left| \tan \left(\frac{\pi}{4} + \frac{x}{2} \right) \right| + C$$

88. Use the substitution $u = \tan(x/2)$ to show that
$$\int \csc x\, dx = \frac{1}{2} \ln \left[\frac{1 - \cos x}{1 + \cos x} \right] + C$$
and confirm that this is consistent with the result in Exercise 65(a) of Section 7.3.

89. Find a substitution that can be used to integrate rational functions of $\sinh x$ and $\cosh x$ and use your substitution to evaluate
$$\int \frac{dx}{2 \cosh x + \sinh x}$$
without expressing the integrand in terms of e^x and e^{-x}.

[c] **90–93** Some integrals that can be evaluated by hand cannot be evaluated by all computer algebra systems. Evaluate the integral by hand, and determine if it can be evaluated on your CAS. ■

90. $\displaystyle\int \frac{x^3}{\sqrt{1 - x^8}}\, dx$

91. $\displaystyle\int (\cos^{32} x \sin^{30} x - \cos^{30} x \sin^{32} x)\, dx$

92. $\displaystyle\int \sqrt{x - \sqrt{x^2 - 4}}\, dx$ [Hint: $\tfrac{1}{2}(\sqrt{x + 2} - \sqrt{x - 2})^2 = ?$]

93. $\int \dfrac{1}{x^{10} + x} \, dx$

[*Hint:* Rewrite the denominator as $x^{10}(1 + x^{-9})$.]

C 94. Let

$$f(x) = \frac{-2x^5 + 26x^4 + 15x^3 + 6x^2 + 20x + 43}{x^6 - x^5 - 18x^4 - 2x^3 - 39x^2 - x - 20}$$

(a) Use a CAS to factor the denominator, and then write down the form of the partial fraction decomposition. You need not find the values of the constants.

(b) Check your answer in part (a) by using the CAS to find the partial fraction decomposition of f.

(c) Integrate f by hand, and then check your answer by integrating with the CAS.

✔ **QUICK CHECK ANSWERS 7.6**

1. (a) Formula (60) (b) Formula (108) (c) Formula (102) (d) Formula (50)   **2.** (a) Formula (25) (b) Formula (51)

(c) Formula (18) (d) Formula (97)   **3.** (a) $\dfrac{1}{4} \ln \left| \dfrac{x + 2}{x - 2} \right| + C$ (b) $\dfrac{1}{6} \sin 3x + \dfrac{1}{2} \sin x + C$ (c) $-\dfrac{e^x}{2} \sqrt{1 - e^{2x}} + \dfrac{1}{2} \sin^{-1} e^x + C$

(d) $\dfrac{1}{2} \ln \left(x^2 - 4x + 8 \right) + \tan^{-1} \dfrac{x - 2}{2} + C$

7.7 NUMERICAL INTEGRATION; SIMPSON'S RULE

If it is necessary to evaluate a definite integral of a function for which an antiderivative cannot be found, then one must settle for some kind of numerical approximation of the integral. In Section 5.4 we considered three such approximations in the context of areas—left endpoint approximation, right endpoint approximation, and midpoint approximation. In this section we will extend those methods to general definite integrals, and we will develop some new methods that often provide more accuracy with less computation. We will also discuss the errors that arise in integral approximations.

■ **A REVIEW OF RIEMANN SUM APPROXIMATIONS**

Recall from Section 5.5 that the definite integral of a continuous function f over an interval $[a, b]$ may be computed as

$$\int_a^b f(x) \, dx = \lim_{\max \Delta x_k \to 0} \sum_{k=1}^n f(x_k^*) \Delta x_k$$

where the sum that appears on the right side is called a Riemann sum. In this formula, Δx_k is the width of the kth subinterval of a partition $a = x_0 < x_1 < x_2 < \cdots < x_n = b$ of $[a, b]$ into n subintervals, and x_k^* denotes an arbitrary point in the kth subinterval. If we take all subintervals of the same width, so that $\Delta x_k = (b - a)/n$, then as n increases the Riemann sum will eventually be a good approximation to the definite integral. We denote this by writing

$$\int_a^b f(x) \, dx \approx \left(\frac{b - a}{n} \right) [f(x_1^*) + f(x_2^*) + \cdots + f(x_n^*)] \tag{1}$$

If we denote the values of f at the endpoints of the subintervals by

$$y_0 = f(a), \quad y_1 = f(x_1), \quad y_2 = f(x_2), \ldots, y_{n-1} = f(x_{n-1}), \quad y_n = f(x_n)$$

and the values of f at the midpoints of the subintervals by

$$y_{m_1}, y_{m_2}, \ldots, y_{m_n}$$

then it follows from (1) that the left endpoint, right endpoint, and midpoint approximations discussed in Section 5.4 can be expressed as shown in Table 7.7.1. Although we originally

Table 7.7.1

LEFT ENDPOINT APPROXIMATION	RIGHT ENDPOINT APPROXIMATION	MIDPOINT APPROXIMATION
$\displaystyle\int_a^b f(x)\,dx \approx \left(\frac{b-a}{n}\right)[y_0 + y_1 + \cdots + y_{n-1}]$	$\displaystyle\int_a^b f(x)\,dx \approx \left(\frac{b-a}{n}\right)[y_1 + y_2 + \cdots + y_n]$	$\displaystyle\int_a^b f(x)\,dx \approx \left(\frac{b-a}{n}\right)[y_{m_1} + y_{m_2} + \cdots + y_{m_n}]$

obtained these results for nonnegative functions in the context of approximating areas, they are applicable to any function that is continuous on $[a, b]$.

■ TRAPEZOIDAL APPROXIMATION

It will be convenient in this section to denote the left endpoint, right endpoint, and midpoint approximations with n subintervals by L_n, R_n, and M_n, respectively. Of the three approximations, the midpoint approximation is most widely used in applications. If we take the average of L_n and R_n, then we obtain another important approximation denoted by

$$T_n = \tfrac{1}{2}(L_n + R_n)$$

called the **trapezoidal approximation**:

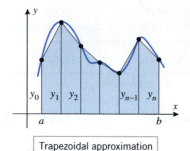

Trapezoidal approximation

▲ **Figure 7.7.1**

> **Trapezoidal Approximation**
>
> $$\int_a^b f(x)\,dx \approx T_n = \left(\frac{b-a}{2n}\right)[y_0 + 2y_1 + \cdots + 2y_{n-1} + y_n] \qquad (2)$$

The name "trapezoidal approximation" results from the fact that in the case where f is nonnegative on the interval of integration, the approximation T_n is the sum of the trapezoidal areas shown in Figure 7.7.1 (see Exercise 51).

▶ **Example 1** In Table 7.7.2 we have approximated

$$\ln 2 = \int_1^2 \frac{1}{x}\,dx$$

using the midpoint approximation and the trapezoidal approximation.[*] In each case we used $n = 10$ subdivisions of the interval $[1, 2]$, so that

$$\underbrace{\frac{b-a}{n} = \frac{2-1}{10} = 0.1}_{\text{Midpoint}} \quad \text{and} \quad \underbrace{\frac{b-a}{2n} = \frac{2-1}{20} = 0.05}_{\text{Trapezoidal}} \blacktriangleleft$$

---

[*]Throughout this section we will show numerical values to nine places to the right of the decimal point. If your calculating utility does not show this many places, then you will need to make the appropriate adjustments. What is important here is that you understand the principles being discussed.

Table 7.7.2

MIDPOINT AND TRAPEZOIDAL APPROXIMATIONS FOR $\int_1^2 \frac{1}{x}\,dx$

MIDPOINT APPROXIMATION			TRAPEZOIDAL APPROXIMATION				
	MIDPOINT			ENDPOINT		MULTIPLIER	
i	m_i	$y_{m_i} = f(m_i) = 1/m_i$	i	x_i	$y_i = f(x_i) = 1/x_i$	w_i	$w_i y_i$
1	1.05	0.952380952	0	1.0	1.000000000	1	1.000000000
2	1.15	0.869565217	1	1.1	0.909090909	2	1.818181818
3	1.25	0.800000000	2	1.2	0.833333333	2	1.666666667
4	1.35	0.740740741	3	1.3	0.769230769	2	1.538461538
5	1.45	0.689655172	4	1.4	0.714285714	2	1.428571429
6	1.55	0.645161290	5	1.5	0.666666667	2	1.333333333
7	1.65	0.606060606	6	1.6	0.625000000	2	1.250000000
8	1.75	0.571428571	7	1.7	0.588235294	2	1.176470588
9	1.85	0.540540541	8	1.8	0.555555556	2	1.111111111
10	1.95	0.512820513	9	1.9	0.526315789	2	1.052631579
		6.928353603	10	2.0	0.500000000	1	0.500000000
							13.875428063

$$\int_1^2 \frac{1}{x}\,dx \approx (0.1)(6.928353603) \approx 0.692835360$$

$$\int_1^2 \frac{1}{x}\,dx \approx (0.05)(13.875428063) \approx 0.693771403$$

By rewriting (3) and (4) in the form

$$\int_a^b f(x)\,dx = \text{approximation} + \text{error}$$

we see that positive values of E_M and E_T correspond to underestimates and negative values to overestimates.

■ **COMPARISON OF THE MIDPOINT AND TRAPEZOIDAL APPROXIMATIONS**

We define the *errors* in the midpoint and trapezoidal approximations to be

$$E_M = \int_a^b f(x)\,dx - M_n \quad \text{and} \quad E_T = \int_a^b f(x)\,dx - T_n \tag{3–4}$$

respectively, and we define $|E_M|$ and $|E_T|$ to be the *absolute errors* in these approximations. The absolute errors are nonnegative and do not distinguish between underestimates and overestimates.

▶ **Example 2**   The value of ln 2 to nine decimal places is

$$\ln 2 = \int_1^2 \frac{1}{x}\,dx \approx 0.693147181 \tag{5}$$

so we see from Tables 7.7.2 and 7.7.3 that the absolute errors in approximating ln 2 by M_{10} and T_{10} are

$$|E_M| = |\ln 2 - M_{10}| \approx 0.000311821$$

$$|E_T| = |\ln 2 - T_{10}| \approx 0.000624222$$

Thus, the midpoint approximation is more accurate than the trapezoidal approximation in this case. ◀

Table 7.7.3

ln 2 (NINE DECIMAL PLACES)	APPROXIMATION	ERROR
0.693147181	$M_{10} \approx 0.692835360$	$E_M = \ln 2 - M_{10} \approx 0.000311821$
0.693147181	$T_{10} \approx 0.693771403$	$E_T = \ln 2 - T_{10} \approx -0.000624222$

It is not accidental in Example 2 that the midpoint approximation of ln 2 was more accurate than the trapezoidal approximation. To see why this is so, we first need to look at the midpoint approximation from another point of view. To simplify our explanation, we will assume that f is nonnegative on $[a, b]$, though the conclusions we reach will be true without this assumption.

If f is a differentiable function, then the midpoint approximation is sometimes called the ***tangent line approximation*** because for each subinterval of $[a, b]$ the area of the rectangle used in the midpoint approximation is equal to the area of the trapezoid whose upper boundary is the tangent line to $y = f(x)$ at the midpoint of the subinterval (Figure 7.7.2). The equality of these areas follows from the fact that the shaded areas in the figure are congruent. We will now show how this point of view about midpoint approximations can be used to establish useful criteria for determining which of M_n or T_n produces the better approximation of a given integral.

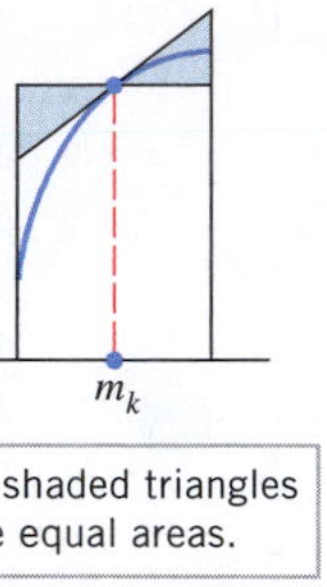

In Figure 7.7.3a we have isolated a subinterval of $[a, b]$ on which the graph of a function f is concave down, and we have shaded the areas that represent the errors in the midpoint and trapezoidal approximations over the subinterval. In Figure 7.7.3b we show a succession of four illustrations which make it evident that the error from the midpoint approximation is less than that from the trapezoidal approximation. If the graph of f were concave up, analogous figures would lead to the same conclusion. (This argument, due to Frank Buck, appeared in *The College Mathematics Journal*, Vol. 16, No. 1, 1985.)

The shaded triangles have equal areas.

▲ **Figure 7.7.2**

Justify the conclusions in each step of Figure 7.7.3b.

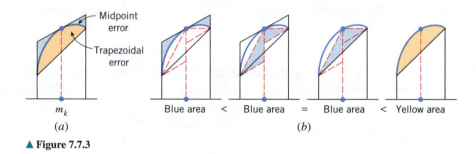

▲ **Figure 7.7.3**

Figure 7.7.3a also suggests that on a subinterval where the graph is concave down, the midpoint approximation is larger than the value of the integral and the trapezoidal approximation is smaller. On an interval where the graph is concave up it is the other way around. In summary, we have the following result, which we state without formal proof:

7.7.1   THEOREM   *Let f be continuous on $[a, b]$, and let $|E_M|$ and $|E_T|$ be the absolute errors that result from the midpoint and trapezoidal approximations of $\int_a^b f(x)\,dx$ using n subintervals.*

(a)   *If the graph of f is either concave up or concave down on (a, b), then $|E_M| < |E_T|$, that is, the absolute error from the midpoint approximation is less than that from the trapezoidal approximation.*

(b)   *If the graph of f is concave down on (a, b), then*

$$T_n < \int_a^b f(x)\,dx < M_n$$

(c)   *If the graph of f is concave up on (a, b), then*

$$M_n < \int_a^b f(x)\,dx < T_n$$

▶ **Example 3** Since the graph of $f(x) = 1/x$ is continuous on the interval $[1, 2]$ and concave up on the interval $(1, 2)$, it follows from part (a) of Theorem 7.7.1 that M_n will always provide a better approximation than T_n for

$$\int_1^2 \frac{1}{x}\, dx = \ln 2$$

Moreover, if follows from part (c) of Theorem 7.7.1 that $M_n < \ln 2 < T_n$ for every positive integer n. Note that this is consistent with our computations in Example 2. ◀

WARNING

Do not conclude that the midpoint approximation is always better than the trapezoidal approximation; the trapezoidal approximation may be better if the function changes concavity on the interval of integration.

▶ **Example 4** The midpoint and trapezoidal approximations can be used to approximate sin 1 by using the integral

$$\sin 1 = \int_0^1 \cos x\, dx$$

Since $f(x) = \cos x$ is continuous on $[0, 1]$ and concave down on $(0, 1)$, it follows from parts (a) and (b) of Theorem 7.7.1 that the absolute error in M_n will be less than that in T_n, and that $T_n < \sin 1 < M_n$ for every positive integer n. This is consistent with the results in Table 7.7.4 for $n = 5$ (intermediate computations are omitted). ◀

Table 7.7.4

sin 1 (NINE DECIMAL PLACES)	APPROXIMATION	ERROR
0.841470985	$M_5 \approx 0.842875074$	$E_M = \sin 1 - M_5 \approx -0.001404089$
0.841470985	$T_5 \approx 0.838664210$	$E_T = \sin 1 - T_5 \approx\ \ 0.002806775$

▶ **Example 5** Table 7.7.5 shows approximations for $\sin 3 = \int_0^3 \cos x\, dx$ using the midpoint and trapezoidal approximations with $n = 10$ subdivisions of the interval $[0, 3]$. Note that $|E_M| < |E_T|$ and $T_{10} < \sin 3 < M_{10}$, although these results are not guaranteed by Theorem 7.7.1 since $f(x) = \cos x$ changes concavity on the interval $[0, 3]$. ◀

Table 7.7.5

sin 3 (NINE DECIMAL PLACES)	APPROXIMATION	ERROR
0.141120008	$M_{10} \approx 0.141650601$	$E_M = \sin 3 - M_{10} \approx -0.000530592$
0.141120008	$T_{10} \approx 0.140060017$	$E_T = \sin 3 - T_{10} \approx\ \ 0.001059991$

■ **SIMPSON'S RULE**

When the left and right endpoint approximations are averaged to produce the trapezoidal approximation, a better approximation often results. We now see how a weighted average of the midpoint and trapezoidal approximations can yield an even better approximation.

The numerical evidence in Tables 7.7.3, 7.7.4, and 7.7.5 reveals that $E_T \approx -2E_M$, so that $2E_M + E_T \approx 0$ in these instances. This suggests that

$$3 \int_a^b f(x)\, dx = 2 \int_a^b f(x)\, dx + \int_a^b f(x)\, dx$$

$$= 2(M_k + E_M) + (T_k + E_T)$$

$$= (2M_k + T_k) + (2E_M + E_T)$$

$$\approx 2M_k + T_k$$

This gives

$$\int_a^b f(x)\,dx \approx \tfrac{1}{3}(2M_k + T_k) \tag{6}$$

The midpoint approximation M_k in (6) requires the evaluation of f at k points in the interval $[a, b]$, and the trapezoidal approximation T_k in (6) requires the evaluation of f at $k + 1$ points in $[a, b]$. Thus, $\tfrac{1}{3}(2M_k + T_k)$ uses $2k + 1$ values of f, taken at equally spaced points in the interval $[a, b]$. These points are obtained by partitioning $[a, b]$ into $2k$ equal subintervals indicated by the left endpoints, right endpoints, and midpoints used in T_k and M_k, respectively. Setting $n = 2k$, we use S_n to denote the weighted average of M_k and T_k in (6). That is,

$$S_n = S_{2k} = \tfrac{1}{3}(2M_k + T_k) \quad \text{or} \quad S_n = \tfrac{1}{3}(2M_{n/2} + T_{n/2}) \tag{7}$$

Table 7.7.6 displays the approximations S_n corresponding to the data in Tables 7.7.3 to 7.7.5.

Table 7.7.6

FUNCTION VALUE (NINE DECIMAL PLACES)	APPROXIMATION	ERROR
$\ln 2 \approx 0.693147181$	$\int_1^2 (1/x)\,dx \approx S_{20} = \tfrac{1}{3}(2M_{10} + T_{10}) \approx 0.693147375$	-0.000000194
$\sin 1 \approx 0.841470985$	$\int_0^1 \cos x\,dx \approx S_{10} = \tfrac{1}{3}(2M_5 + T_5) \approx 0.841471453$	-0.000000468
$\sin 3 \approx 0.141120008$	$\int_0^3 \cos x\,dx \approx S_{20} = \tfrac{1}{3}(2M_{10} + T_{10}) \approx 0.141120406$	-0.000000398

Using the midpoint approximation formula in Table 7.7.1 and Formula (2) for the trapezoidal approximation, we can derive a similar formula for S_n. We start by partitioning the interval $[a, b]$ into an *even* number of equal subintervals. If n is the number of subintervals, then each subinterval has length $(b - a)/n$. Label the endpoints of these subintervals successively by $a = x_0, x_1, x_2, \ldots, x_n = b$. Then $x_0, x_2, x_4, \ldots, x_n$ define a partition of $[a, b]$ into $n/2$ equal intervals, each of length $2(b - a)/n$, and the midpoints of these subintervals are $x_1, x_3, x_5, \ldots, x_{n-1}$, respectively, as illustrated in Figure 7.7.4. Using $y_i = f(x_i)$, we have

$$2M_{n/2} = 2\left(\frac{2(b - a)}{n}\right)[y_1 + y_3 + \cdots + y_{n-1}]$$

$$= \left(\frac{b - a}{n}\right)[4y_1 + 4y_3 + \cdots + 4y_{n-1}]$$

Noting that $(b - a)/[2(n/2)] = (b - a)/n$, we can express $T_{n/2}$ as

$$T_{n/2} = \left(\frac{b - a}{n}\right)[y_0 + 2y_2 + 2y_4 + \cdots + 2y_{n-2} + y_n]$$

Thus, $S_n = \tfrac{1}{3}(2M_{n/2} + T_{n/2})$ can be expressed as

$$S_n = \frac{1}{3}\left(\frac{b - a}{n}\right)[y_0 + 4y_1 + 2y_2 + 4y_3 + 2y_4 + \cdots + 2y_{n-2} + 4y_{n-1} + y_n] \tag{8}$$

The approximation

$$\int_a^b f(x)\,dx \approx S_n \tag{9}$$

with S_n as given in (8) is known as **Simpson's rule**. We denote the error in this approximation by

$$E_S = \int_a^b f(x)\,dx - S_n \tag{10}$$

As before, the absolute error in the approximation (9) is given by $|E_S|$.

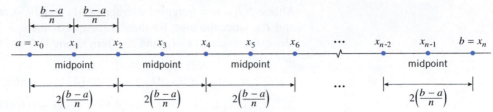

▶ Figure 7.7.4

▶ **Example 6** In Table 7.7.7 we have used Simpson's rule with $n = 10$ subintervals to obtain the approximation

$$\ln 2 = \int_1^2 \frac{1}{x}\,dx \approx S_{10} = 0.693150231$$

For this approximation,

$$\frac{1}{3}\left(\frac{b-a}{n}\right) = \frac{1}{3}\left(\frac{2-1}{10}\right) = \frac{1}{30}$$

Table 7.7.7

AN APPROXIMATION TO $\ln 2$ USING SIMPSON'S RULE

i	ENDPOINT x_i	$y_i = f(x_i) = 1/x_i$	MULTIPLIER w_i	$w_i y_i$
0	1.0	1.000000000	1	1.000000000
1	1.1	0.909090909	4	3.636363636
2	1.2	0.833333333	2	1.666666667
3	1.3	0.769230769	4	3.076923077
4	1.4	0.714285714	2	1.428571429
5	1.5	0.666666667	4	2.666666667
6	1.6	0.625000000	2	1.250000000
7	1.7	0.588235294	4	2.352941176
8	1.8	0.555555556	2	1.111111111
9	1.9	0.526315789	4	2.105263158
10	2.0	0.500000000	1	0.500000000
				20.794506921

$$\int_1^2 \frac{1}{x}\,dx \approx \left(\tfrac{1}{30}\right)(20.794506921) \approx 0.693150231$$

Thomas Simpson (1710–1761) English mathematician. Simpson was the son of a weaver. He was trained to follow in his father's footsteps and had little formal education in his early life. His interest in science and mathematics was aroused in 1724, when he witnessed an eclipse of the Sun and received two books from a peddler, one on astrology and the other on arithmetic. Simpson quickly absorbed their contents and soon became a successful local fortune teller. His improved financial situation enabled him to give up weaving and marry his landlady. Then in 1733 some mysterious "unfortunate incident" forced him to move. He settled in Derby, where he taught in an evening school and worked at weaving during the day. In 1736 he moved to London and published his first mathematical work in a periodical called the *Ladies' Diary* (of which he later became the editor). In 1737 he published a successful calculus textbook that enabled him to give up weaving completely and concentrate on textbook writing and teaching. His fortunes improved further in 1740 when one Robert Heath accused him of plagiarism. The publicity was marvelous, and Simpson proceeded to dash off a succession of best-selling textbooks: *Algebra* (ten editions plus translations), *Geometry* (twelve editions plus translations), *Trigonometry* (five editions plus translations), and numerous others. It is interesting to note that Simpson did not discover the rule that bears his name—it was a well-known result by Simpson's time.

Although S_{10} is a weighted average of M_5 and T_5, it makes sense to compare S_{10} to M_{10} and T_{10}, since the sums for these three approximations involve the same number of terms. Using the values for M_{10} and T_{10} from Example 2 and the value for S_{10} in Table 7.7.7, we have

$$|E_M| = |\ln 2 - M_{10}| \approx |0.693147181 - 0.692835360| = 0.000311821$$

$$|E_T| = |\ln 2 - T_{10}| \approx |0.693147181 - 0.693771403| = 0.000624222$$

$$|E_S| = |\ln 2 - S_{10}| \approx |0.693147181 - 0.693150231| = 0.000003050$$

Comparing these absolute errors, it is clear that S_{10} is a much more accurate approximation of $\ln 2$ than either M_{10} or T_{10}. ◄

■ GEOMETRIC INTERPRETATION OF SIMPSON'S RULE

The midpoint (or tangent line) approximation and the trapezoidal approximation for a definite integral are based on approximating a segment of the curve $y = f(x)$ by line segments. Intuition suggests that we might improve on these approximations using parabolic arcs rather than line segments, thereby accounting for concavity of the curve $y = f(x)$ more closely.

At the heart of this idea is a formula, sometimes called the ***one-third rule***. The one-third rule expresses a definite integral of a quadratic function $g(x) = Ax^2 + Bx + C$ in terms of the values Y_0, Y_1, and Y_2 of g at the left endpoint, midpoint, and right endpoint, respectively, of the interval of integration $[m - \Delta x, m + \Delta x]$ (see Figure 7.7.5):

$$\int_{m-\Delta x}^{m+\Delta x} (Ax^2 + Bx + C)\, dx = \frac{\Delta x}{3}[Y_0 + 4Y_1 + Y_2] \tag{11}$$

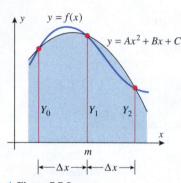

▲ **Figure 7.7.5**

Verification of the one-third rule is left for the reader (Exercise 53). By applying the one-third rule to subintervals $[x_{2k-2}, x_{2k}]$, $k = 1, \ldots, n/2$, one arrives at Formula (8) for Simpson's rule (Exercise 54). Thus, Simpson's rule corresponds to the integral of a piecewise-quadratic approximation to $f(x)$.

■ ERROR BOUNDS

With all the methods studied in this section, there are two sources of error: the *intrinsic* or *truncation error* due to the approximation formula, and the *roundoff error* introduced in the calculations. In general, increasing n reduces the truncation error but increases the roundoff error, since more computations are required for larger n. In practical applications, it is important to know how large n must be taken to ensure that a specified degree of accuracy is obtained. The analysis of roundoff error is complicated and will not be considered here. However, the following theorems, which are proved in books on numerical analysis, provide upper bounds on the truncation errors in the midpoint, trapezoidal, and Simpson's rule approximations.

7.7.2   THEOREM (*Midpoint and Trapezoidal Error Bounds*)   *If f'' is continuous on $[a, b]$ and if K_2 is the maximum value of $|f''(x)|$ on $[a, b]$, then*

$$(a)\quad |E_M| = \left| \int_a^b f(x)\, dx - M_n \right| \le \frac{(b-a)^3 K_2}{24n^2} \tag{12}$$

$$(b)\quad |E_T| = \left| \int_a^b f(x)\, dx - T_n \right| \le \frac{(b-a)^3 K_2}{12n^2} \tag{13}$$

7.7.3 THEOREM (*Simpson Error Bound*) *If $f^{(4)}$ is continuous on $[a, b]$ and if K_4 is the maximum value of $|f^{(4)}(x)|$ on $[a, b]$, then*

$$|E_S| = \left| \int_a^b f(x)\,dx - S_n \right| \le \frac{(b-a)^5 K_4}{180 n^4} \qquad (14)$$

▶ **Example 7** Find an upper bound on the absolute error that results from approximating

$$\ln 2 = \int_1^2 \frac{1}{x}\,dx$$

using (a) the midpoint approximation M_{10}, (b) the trapezoidal approximation T_{10}, and (c) Simpson's rule S_{10}, each with $n = 10$ subintervals.

Solution. We will apply Formulas (12), (13), and (14) with

$$f(x) = \frac{1}{x}, \quad a = 1, \quad b = 2, \quad \text{and} \quad n = 10$$

We have

$$f'(x) = -\frac{1}{x^2}, \quad f''(x) = \frac{2}{x^3}, \quad f'''(x) = -\frac{6}{x^4}, \quad f^{(4)}(x) = \frac{24}{x^5}$$

Thus,

$$|f''(x)| = \left| \frac{2}{x^3} \right| = \frac{2}{x^3}, \quad |f^{(4)}(x)| = \left| \frac{24}{x^5} \right| = \frac{24}{x^5}$$

Note that the upper bounds calculated in Example 7 are consistent with the values $|E_M|, |E_T|$, and $|E_S|$ calculated in Example 6 but are considerably greater than those values. It is quite common that the upper bounds on the absolute errors given in Theorems 7.7.2 and 7.7.3 substantially exceed the actual absolute errors. However, that does not diminish the utility of these bounds.

where we have dropped the absolute values because $f''(x)$ and $f^{(4)}(x)$ have positive values for $1 \le x \le 2$. Since $|f''(x)|$ and $|f^4(x)|$ are continuous and decreasing on $[1, 2]$, both functions have their maximum values at $x = 1$; for $|f''(x)|$ this maximum value is 2 and for $|f^4(x)|$ the maximum value is 24. Thus we can take $K_2 = 2$ in (12) and (13), and $K_4 = 24$ in (14). This yields

$$|E_M| \le \frac{(b-a)^3 K_2}{24 n^2} = \frac{1^3 \cdot 2}{24 \cdot 10^2} \approx 0.000833333$$

$$|E_T| \le \frac{(b-a)^3 K_2}{12 n^2} = \frac{1^3 \cdot 2}{12 \cdot 10^2} \approx 0.001666667$$

$$|E_S| \le \frac{(b-a)^5 K_4}{180 n^4} = \frac{1^5 \cdot 24}{180 \cdot 10^4} \approx 0.000013333 \;\blacktriangleleft$$

▶ **Example 8** How many subintervals should be used in approximating

$$\ln 2 = \int_1^2 \frac{1}{x}\,dx$$

by Simpson's rule for five decimal-place accuracy?

Solution. To obtain five decimal-place accuracy, we must choose the number of subintervals so that

$$|E_S| \le 0.000005 = 5 \times 10^{-6}$$

From (14), this can be achieved by taking n in Simpson's rule to satisfy

$$\frac{(b-a)^5 K_4}{180 n^4} \le 5 \times 10^{-6}$$

Taking $a = 1$, $b = 2$, and $K_4 = 24$ (found in Example 7) in this inequality yields

$$\frac{1^5 \cdot 24}{180 \cdot n^4} \leq 5 \times 10^{-6}$$

which, on taking reciprocals, can be rewritten as

$$n^4 \geq \frac{2 \times 10^6}{75} = \frac{8 \times 10^4}{3}$$

Thus,

$$n \geq \frac{20}{\sqrt[4]{6}} \approx 12.779$$

Since n must be an even integer, the smallest value of n that satisfies this requirement is $n = 14$. Thus, the approximation S_{14} using 14 subintervals will produce five decimal-place accuracy. ◄

REMARK In cases where it is difficult to find the values of K_2 and K_4 in Formulas (12), (13), and (14), these constants may be replaced by any larger constants. For example, suppose that a constant K can be easily found with the certainty that $|f''(x)| < K$ on the interval. Then $K_2 \leq K$ and

$$|E_T| \leq \frac{(b-a)^3 K_2}{12n^2} \leq \frac{(b-a)^3 K}{12n^2} \tag{15}$$

so the right side of (15) is also an upper bound on the value of $|E_T|$. Using K, however, will likely increase the computed value of n needed for a given error tolerance. Many applications involve the resolution of competing practical issues, illustrated here through the trade-off between the convenience of finding a crude bound for $|f''(x)|$ versus the efficiency of using the smallest possible n for a desired accuracy.

▶ **Example 9** How many subintervals should be used in approximating

$$\int_0^1 \cos(x^2)\,dx$$

by the midpoint approximation for three decimal-place accuracy?

Solution. To obtain three decimal-place accuracy, we must choose n so that

$$|E_M| \leq 0.0005 = 5 \times 10^{-4} \tag{16}$$

From (12) with $f(x) = \cos(x^2)$, $a = 0$, and $b = 1$, an upper bound on $|E_M|$ is given by

$$|E_M| \leq \frac{K_2}{24n^2} \tag{17}$$

where $|K_2|$ is the maximum value of $|f''(x)|$ on the interval $[0, 1]$. However,

$$f'(x) = -2x\sin(x^2)$$
$$f''(x) = -4x^2\cos(x^2) - 2\sin(x^2) = -[4x^2\cos(x^2) + 2\sin(x^2)]$$

so that

$$|f''(x)| = |4x^2\cos(x^2) + 2\sin(x^2)| \tag{18}$$

It would be tedious to look for the maximum value of this function on the interval $[0, 1]$. For x in $[0, 1]$, it is easy to see that each of the expressions x^2, $\cos(x^2)$, and $\sin(x^2)$ is bounded in absolute value by 1, so $|4x^2\cos(x^2) + 2\sin(x^2)| \leq 4 + 2 = 6$ on $[0, 1]$. We can improve on this by using a graphing utility to sketch $|f''(x)|$, as shown in Figure 7.7.6. It is evident from the graph that

$$|f''(x)| < 4 \quad \text{for} \quad 0 \leq x \leq 1$$

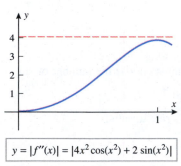

$$y = |f''(x)| = |4x^2\cos(x^2) + 2\sin(x^2)|$$

▲ **Figure 7.7.6**

Thus, it follows from (17) that

$$|E_M| \leq \frac{K_2}{24n^2} < \frac{4}{24n^2} = \frac{1}{6n^2}$$

and hence we can satisfy (16) by choosing n so that

$$\frac{1}{6n^2} < 5 \times 10^{-4}$$

which, on taking reciprocals, can be written as

$$n^2 > \frac{10^4}{30} \quad \text{or} \quad n > \frac{10^2}{\sqrt{30}} \approx 18.257$$

The smallest integer value of n satisfying this inequality is $n = 19$. Thus, the midpoint approximation M_{19} using 19 subintervals will produce three decimal-place accuracy. ◄

■ A COMPARISON OF THE THREE METHODS

Of the three methods studied in this section, Simpson's rule generally produces more accurate results than the midpoint or trapezoidal approximations for the same amount of work. To make this plausible, let us express (12), (13), and (14) in terms of the subinterval width

$$\Delta x = \frac{b - a}{n}$$

We obtain

$$|E_M| \leq \frac{1}{24} K_2 (b - a)(\Delta x)^2 \tag{19}$$

$$|E_T| \leq \frac{1}{12} K_2 (b - a)(\Delta x)^2 \tag{20}$$

$$|E_S| \leq \frac{1}{180} K_4 (b - a)(\Delta x)^4 \tag{21}$$

(verify). For Simpson's rule, the upper bound on the absolute error is proportional to $(\Delta x)^4$, whereas the upper bound on the absolute error for the midpoint and trapezoidal approximations is proportional to $(\Delta x)^2$. Thus, reducing the interval width by a factor of 10, for example, reduces the error bound by a factor of 100 for the midpoint and trapezoidal approximations but reduces the error bound by a factor of 10,000 for Simpson's rule. This suggests that, as n increases, the accuracy of Simpson's rule improves much more rapidly than that of the other approximations.

As a final note, observe that if $f(x)$ is a polynomial of degree 3 or less, then we have $f^{(4)}(x) = 0$ for all x, so $K_4 = 0$ in (14) and consequently $|E_S| = 0$. Thus, Simpson's rule gives exact results for polynomials of degree 3 or less. Similarly, the midpoint and trapezoidal approximations give exact results for polynomials of degree 1 or less. (You should also be able to see that this is so geometrically.)

✔ QUICK CHECK EXERCISES 7.7   (See page 547 for answers.)

1. Let T_n be the trapezoidal approximation for the definite integral of $f(x)$ over an interval $[a, b]$ using n subintervals.
   (a) Expressed in terms of L_n and R_n (the left and right endpoint approximations), $T_n = $ _____.
   (b) Expressed in terms of the function values $y_0, y_1, \ldots, y_n$ at the endpoints of the subintervals, $T_n = $ _____.

2. Let I denote the definite integral of f over an interval $[a, b]$ with T_n and M_n the respective trapezoidal and midpoint approximations of I for a given n. Assume that the graph of f is concave up on the interval $[a, b]$ and order the quantities T_n, M_n, and I from smallest to largest:
   _____ < _____ < _____.

3. Let S_6 be the Simpson's rule approximation for $\int_a^b f(x)\,dx$ using $n = 6$ subintervals.

(a) Expressed in terms of M_3 and T_3 (the midpoint and trapezoidal approximations), $S_6 = $ _____.

(b) Using the function values $y_0, y_1, y_2, \ldots, y_6$ at the endpoints of the subintervals, $S_6 = $ _____.

4. Assume that $f^{(4)}$ is continuous on $[0, 1]$ and that $f^{(k)}(x)$ satisfies $|f^{(k)}(x)| \le 1$ on $[0, 1]$, $k = 1, 2, 3, 4$. Find an upper bound on the absolute error that results from approximating the integral of f over $[0, 1]$ using (a) the midpoint approximation M_{10}; (b) the trapezoidal approximation T_{10}; and (c) Simpson's rule S_{10}.

5. Approximate $\int_1^3 \frac{1}{x^2}\,dx$ using the indicated method.

(a) $M_1 = $ _____
(b) $T_1 = $ _____
(c) $S_2 = $ _____

EXERCISE SET 7.7   ⊂ CAS

1–6 Approximate the integral using (a) the midpoint approximation M_{10}, (b) the trapezoidal approximation T_{10}, and (c) Simpson's rule approximation S_{20} using Formula (7). In each case, find the exact value of the integral and approximate the absolute error. Express your answers to at least four decimal places. ■

1. $\int_0^3 \sqrt{x+1}\,dx$ **2.** $\int_4^9 \frac{1}{\sqrt{x}}\,dx$ **3.** $\int_0^{\pi/2} \cos x\,dx$

4. $\int_0^2 \sin x\,dx$ **5.** $\int_1^3 e^{-2x}\,dx$ **6.** $\int_0^3 \frac{1}{3x+1}\,dx$

7–12 Use inequalities (12), (13), and (14) to find upper bounds on the errors in parts (a), (b), or (c) of the indicated exercise. ■

7. Exercise 1   **8.** Exercise 2   **9.** Exercise 3

10. Exercise 4   **11.** Exercise 5   **12.** Exercise 6

13–18 Use inequalities (12), (13), and (14) to find a number n of subintervals for (a) the midpoint approximation M_n, (b) the trapezoidal approximation T_n, and (c) Simpson's rule approximation S_n to ensure that the absolute error will be less than the given value. ■

13. Exercise 1; 5×10^{-4} **14.** Exercise 2; 5×10^{-4}

15. Exercise 3; 10^{-3} **16.** Exercise 4; 10^{-3}

17. Exercise 5; 10^{-4} **18.** Exercise 6; 10^{-4}

19–22 True–False Determine whether the statement is true or false. Explain your answer. ■

19. The midpoint approximation, M_n, is the average of the left and right endpoint approximations, L_n and R_n, respectively.

20. If $f(x)$ is concave down on the interval (a, b), then the trapezoidal approximation T_n underestimates $\int_a^b f(x)\,dx$.

21. The Simpson's rule approximation S_{50} for $\int_a^b f(x)\,dx$ is a weighted average of the approximations M_{50} and T_{50}, where M_{50} is given twice the weight of T_{50} in the average.

22. Simpson's rule approximation S_{50} for $\int_a^b f(x)\,dx$ corresponds to $\int_a^b q(x)\,dx$, where the graph of q is composed of 25 parabolic segments joined at points on the graph of f.

23–24 Find a function $g(x)$ of the form

$$g(x) = Ax^2 + Bx + C$$

whose graph contains the points $(m - \Delta x, f(m - \Delta x))$, $(m, f(m))$, and $(m + \Delta x, f(m + \Delta x))$, for the given function $f(x)$ and the given values of m and Δx. Then verify Formula (11):

$$\int_{m-\Delta x}^{m+\Delta x} g(x)\,dx = \frac{\Delta x}{3}[Y_0 + 4Y_1 + Y_2]$$

where $Y_0 = f(m - \Delta x)$, $Y_1 = f(m)$, and $Y_2 = f(m + \Delta x)$. ■

23. $f(x) = \frac{1}{x}$; $m = 3$, $\Delta x = 1$

24. $f(x) = \sin^2(\pi x)$; $m = \frac{1}{6}$, $\Delta x = \frac{1}{6}$

25–30 Approximate the integral using Simpson's rule S_{10} and compare your answer to that produced by a calculating utility with a numerical integration capability. Express your answers to at least four decimal places. ■

25. $\int_{-1}^1 e^{-x^2}\,dx$ **26.** $\int_0^3 \frac{x}{\sqrt{2x^3+1}}\,dx$

27. $\int_{-1}^2 x\sqrt{2+x^3}\,dx$ **28.** $\int_0^\pi \frac{x}{2+\sin x}\,dx$

29. $\int_0^1 \cos(x^2)\,dx$ **30.** $\int_1^2 (\ln x)^{3/2}\,dx$

31–32 The exact value of the given integral is π (verify). Approximate the integral using (a) the midpoint approximation M_{10}, (b) the trapezoidal approximation T_{10}, and (c) Simpson's rule approximation S_{20} using Formula (7). Approximate the absolute error and express your answers to at least four decimal places. ■

31. $\int_0^2 \frac{8}{x^2+4}\,dx$ **32.** $\int_0^3 \frac{4}{9}\sqrt{9-x^2}\,dx$

33. In Example 8 we showed that taking $n = 14$ subdivisions ensures that the approximation of

$$\ln 2 = \int_1^2 \frac{1}{x}\,dx$$

by Simpson's rule is accurate to five decimal places. Confirm this by comparing the approximation of $\ln 2$ produced by Simpson's rule with $n = 14$ to the value produced directly by your calculating utility.

34. In each part, determine whether a trapezoidal approximation would be an underestimate or an overestimate for the definite integral.

(a) $\int_0^1 \cos(x^2)\,dx$ (b) $\int_{3/2}^2 \cos(x^2)\,dx$

35–36 Find a value of n to ensure that the absolute error in approximating the integral by the midpoint approximation will be less than 10^{-4}. Estimate the absolute error, and express your answers to at least four decimal places. ■

35. $\int_0^2 x \sin x\,dx$ **36.** $\int_0^1 e^{\cos x}\,dx$

37–38 Show that the inequalities (12) and (13) are of no value in finding an upper bound on the absolute error that results from approximating the integral using either the midpoint approximation or the trapezoidal approximation. ■

37. $\int_0^1 x\sqrt{x}\,dx$ **38.** $\int_0^1 \sin\sqrt{x}\,dx$

39–40 Use Simpson's rule approximation S_{10} to approximate the length of the curve over the stated interval. Express your answers to at least four decimal places. ■

39. $y = \sin x$ from $x = 0$ to $x = \pi$

40. $y = x^{-2}$ from $x = 1$ to $x = 2$

FOCUS ON CONCEPTS

41. A graph of the speed v versus time t curve for a test run of a Mitsubishi Galant ES is shown in the accompanying figure. Estimate the speeds at times $t = 0$, 2.5, 5, 7.5, 10, 12.5, 15 s from the graph, convert to ft/s using 1 mi/h $= \frac{22}{15}$ ft/s, and use these speeds and Simpson's rule to approximate the number of feet traveled during the first 15 s. Round your answer to the nearest foot. [*Hint:* Distance traveled $= \int_0^{15} v(t)\,dt$.]

Source: Data from *Car and Driver*, November 2003.

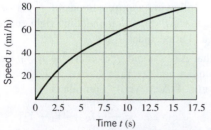

Time t (s)   ◀ **Figure Ex-41**

42. A graph of the acceleration a versus time t for an object moving on a straight line is shown in the accompanying figure. Estimate the accelerations at $t = 0, 1, 2, \ldots, 8$ seconds (s) from the graph and use Simpson's rule to approximate the change in velocity from $t = 0$ to $t = 8$ s. Round your answer to the nearest tenth cm/s. [*Hint:* Change in velocity $= \int_0^8 a(t)\,dt$.]

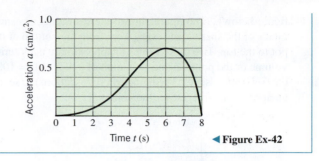

Time t (s)   ◀ **Figure Ex-42**

43–46 Numerical integration methods can be used in problems where only measured or experimentally determined values of the integrand are available. Use Simpson's rule to estimate the value of the relevant integral in these exercises. ■

43. The accompanying table gives the speeds, in miles per second, at various times for a test rocket that was fired upward from the surface of the Earth. Use these values to approximate the number of miles traveled during the first 180 s. Round your answer to the nearest tenth of a mile. [*Hint:* Distance traveled $= \int_0^{180} v(t)\,dt$.]

TIME t (s)	SPEED v (mi/s)
0	0.00
30	0.03
60	0.08
90	0.16
120	0.27
150	0.42
180	0.65

◀ **Table Ex-43**

44. The accompanying table gives the speeds of a bullet at various distances from the muzzle of a rifle. Use these values to approximate the number of seconds for the bullet to travel 1800 ft. Express your answer to the nearest hundredth of a second. [*Hint:* If v is the speed of the bullet and x is the distance traveled, then $v = dx/dt$ so that $dt/dx = 1/v$ and $t = \int_0^{1800} (1/v)\,dx$.]

DISTANCE x (ft)	SPEED v (ft/s)
0	3100
300	2908
600	2725
900	2549
1200	2379
1500	2216
1800	2059

◀ **Table Ex-44**

45. Measurements of a pottery shard recovered from an archaeological dig reveal that the shard came from a pot with a flat bottom and circular cross sections (see the accompanying figure below). The figure shows interior radius measurements of the shard made every 4 cm from the bottom of the pot to the top. Use those values to approximate the interior volume of the pot to the nearest tenth of a liter (1 L = 1000 cm^3). [*Hint:* Use 6.2.3 (volume by cross sections) to set up an appropriate integral for the volume.]

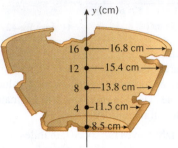

◀ **Figure Ex-45**

46. Engineers want to construct a straight and level road 600 ft long and 75 ft wide by making a vertical cut through an intervening hill (see the accompanying figure). Heights of the hill above the centerline of the proposed road, as obtained at various points from a contour map of the region, are shown in the accompanying figure. To estimate the construction costs, the engineers need to know the volume of earth that must be removed. Approximate this volume, rounded to the nearest cubic foot. [*Hint:* First set up an integral for the cross-sectional area of the cut along the centerline of the road, then assume that the height of the hill does not vary between the centerline and edges of the road.]

HORIZONTAL DISTANCE x (ft)	HEIGHT h (ft)
0	0
100	7
200	16
300	24
400	25
500	16
600	0

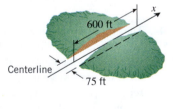

▲ **Figure Ex-46**

47. Let $f(x) = \cos(x^2)$.
(a) Use a CAS to approximate the maximum value of $|f''(x)|$ on the interval [0, 1].
(b) How large must n be in the midpoint approximation of $\int_0^1 f(x)\,dx$ to ensure that the absolute error is less than 5×10^{-4}? Compare your result with that obtained in Example 9.
(c) Estimate the integral using the midpoint approximation with the value of n obtained in part (b).

48. Let $f(x) = \sqrt{1 + x^3}$.
(a) Use a CAS to approximate the maximum value of $|f''(x)|$ on the interval [0, 1].

(b) How large must n be in the trapezoidal approximation of $\int_0^1 f(x)\,dx$ to ensure that the absolute error is less than 10^{-3}?
(c) Estimate the integral using the trapezoidal approximation with the value of n obtained in part (b).

49. Let $f(x) = \cos(x - x^2)$.
(a) Use a CAS to approximate the maximum value of $|f^{(4)}(x)|$ on the interval [0, 1].
(b) How large must the value of n be in the approximation S_n of $\int_0^1 f(x)\,dx$ by Simpson's rule to ensure that the absolute error is less than 10^{-4}?
(c) Estimate the integral using Simpson's rule approximation S_n with the value of n obtained in part (b).

50. Let $f(x) = \sqrt{2 + x^3}$.
(a) Use a CAS to approximate the maximum value of $|f^{(4)}(x)|$ on the interval [0, 1].
(b) How large must the value of n be in the approximation S_n of $\int_0^1 f(x)\,dx$ by Simpson's rule to ensure that the absolute error is less than 10^{-6}?
(c) Estimate the integral using Simpson's rule approximation S_n with the value of n obtained in part (b).

FOCUS ON CONCEPTS

51. (a) Verify that the average of the left and right endpoint approximations as given in Table 7.7.1 gives Formula (2) for the trapezoidal approximation.
(b) Suppose that f is a continuous nonnegative function on the interval $[a, b]$ and partition $[a, b]$ with equally spaced points, $a = x_0 < x_1 < \cdots < x_n = b$. Find the area of the trapezoid under the line segment joining points $(x_k, f(x_k))$ and $(x_{k+1}, f(x_{k+1}))$ and above the interval $[x_k, x_{k+1}]$. Show that the right side of Formula (2) is the sum of these trapezoidal areas (Figure 7.7.1).

52. Let f be a function that is positive, continuous, decreasing, and concave down on the interval $[a, b]$. Assuming that $[a, b]$ is subdivided into n equal subintervals, arrange the following approximations of $\int_a^b f(x)\,dx$ in order of increasing value: left endpoint, right endpoint, midpoint, and trapezoidal.

53. Suppose that $\Delta x > 0$ and $g(x) = Ax^2 + Bx + C$. Let m be a number and set $Y_0 = g(m - \Delta x)$, $Y_1 = g(m)$, and $Y_2 = g(m + \Delta x)$. Verify Formula (11):
$$\int_{m-\Delta x}^{m+\Delta x} g(x)\,dx = \frac{\Delta x}{3}[Y_0 + 4Y_1 + Y_2]$$

54. Suppose that f is a continuous nonnegative function on the interval $[a, b]$, n is even, and $[a, b]$ is partitioned using $n + 1$ equally spaced points, $a = x_0 < x_1 < \cdots < x_n = b$. Set $y_0 = f(x_0)$, $y_1 = f(x_1)$, ..., $y_n = f(x_n)$. Let $g_1, g_2, \ldots, g_{n/2}$ be the quadratic functions of the form $g_i(x) = Ax^2 + Bx + C$ so that *(cont.)*

- the graph of g_1 passes through the points (x_0, y_0), (x_1, y_1), and (x_2, y_2);
- the graph of g_2 passes through the points (x_2, y_2), (x_3, y_3), and (x_4, y_4);
- ...
- the graph of $g_{n/2}$ passes through the points (x_{n-2}, y_{n-2}), (x_{n-1}, y_{n-1}), and (x_n, y_n).

Verify that Formula (8) computes the area under a piecewise quadratic function by showing that

$$\sum_{j=1}^{n/2} \left(\int_{x_{2j-2}}^{x_{2j}} g_j(x)\, dx \right)$$

$$= \frac{1}{3}\left(\frac{b-a}{n}\right)[y_0 + 4y_1 + 2y_2 + 4y_3 + 2y_4 + \cdots$$

$$+ 2y_{n-2} + 4y_{n-1} + y_n]$$

55. Writing Discuss two different circumstances under which numerical integration is necessary.

56. Writing For the numerical integration methods of this section, better accuracy of an approximation was obtained by increasing the number of subdivisions of the interval. Another strategy is to use the same number of subintervals, but to select subintervals of differing lengths. Discuss a scheme for doing this to approximate $\int_0^4 \sqrt{x}\, dx$ using a trapezoidal approximation with 4 subintervals. Comment on the advantages and disadvantages of your scheme.

✔**QUICK CHECK ANSWERS 7.7**

1. (a) $\frac{1}{2}(L_n + R_n)$ (b) $\left(\dfrac{b-a}{2n}\right)[y_0 + 2y_1 + \cdots + 2y_{n-1} + y_n]$ **2.** $M_n < I < T_n$ **3.** (a) $\frac{2}{3}M_3 + \frac{1}{3}T_3$

(b) $\left(\dfrac{b-a}{18}\right)(y_0 + 4y_1 + 2y_2 + 4y_3 + 2y_4 + 4y_5 + y_6)$ **4.** (a) $\dfrac{1}{2400}$ (b) $\dfrac{1}{1200}$ (c) $\dfrac{1}{1,800,000}$

5. (a) $M_1 = \frac{1}{2}$ (b) $T_1 = \frac{10}{9}$ (c) $S_2 = \frac{19}{27}$

7.8 IMPROPER INTEGRALS

Up to now we have focused on definite integrals with continuous integrands and finite intervals of integration. In this section we will extend the concept of a definite integral to include infinite intervals of integration and integrands that become infinite within the interval of integration.

■ **IMPROPER INTEGRALS**

It is assumed in the definition of the definite integral

$$\int_a^b f(x)\, dx$$

that $[a, b]$ is a finite interval and that the limit that defines the integral exists; that is, the function f is integrable. We observed in Theorems 5.5.2 and 5.5.8 that continuous functions are integrable, as are bounded functions with finitely many points of discontinuity. We also observed in Theorem 5.5.8 that functions that are not bounded on the interval of integration are not integrable. Thus, for example, a function with a vertical asymptote within the interval of integration would not be integrable.

Our main objective in this section is to extend the concept of a definite integral to allow for infinite intervals of integration and integrands with vertical asymptotes within the interval of integration. We will call the vertical asymptotes *infinite discontinuities*, and we will call integrals with infinite intervals of integration or infinite discontinuities within the interval of integration ***improper integrals***. Here are some examples:

- Improper integrals with infinite intervals of integration:

$$\int_1^{+\infty} \frac{dx}{x^2}, \quad \int_{-\infty}^0 e^x \, dx, \quad \int_{-\infty}^{+\infty} \frac{dx}{1+x^2}$$

- Improper integrals with infinite discontinuities in the interval of integration:

$$\int_{-3}^3 \frac{dx}{x^2}, \quad \int_1^2 \frac{dx}{x-1}, \quad \int_0^{\pi} \tan x \, dx$$

- Improper integrals with infinite discontinuities and infinite intervals of integration:

$$\int_0^{+\infty} \frac{dx}{\sqrt{x}}, \quad \int_{-\infty}^{+\infty} \frac{dx}{x^2-9}, \quad \int_1^{+\infty} \sec x \, dx$$

■ INTEGRALS OVER INFINITE INTERVALS

To motivate a reasonable definition for improper integrals of the form

$$\int_a^{+\infty} f(x) \, dx$$

let us begin with the case where f is continuous and nonnegative on $[a, +\infty)$, so we can think of the integral as the area under the curve $y = f(x)$ over the interval $[a, +\infty)$ (Figure 7.8.1). At first, you might be inclined to argue that this area is infinite because the region has infinite extent. However, such an argument would be based on vague intuition rather than precise mathematical logic, since the concept of area has only been defined over intervals of *finite extent*. Thus, before we can make any reasonable statements about the area of the region in Figure 7.8.1, we need to begin by defining what we mean by the area of this region. For that purpose, it will help to focus on a specific example.

Suppose we are interested in the area A of the region that lies below the curve $y = 1/x^2$ and above the interval $[1, +\infty)$ on the x-axis. Instead of trying to find the entire area at once, let us begin by calculating the portion of the area that lies above a finite interval $[1, b]$, where $b > 1$ is arbitrary. That area is

$$\int_1^b \frac{dx}{x^2} = -\frac{1}{x}\bigg]_1^b = 1 - \frac{1}{b}$$

(Figure 7.8.2). If we now allow b to increase so that $b \to +\infty$, then the portion of the area over the interval $[1, b]$ will begin to fill out the area over the entire interval $[1, +\infty)$ (Figure 7.8.3), and hence we can reasonably define the area A under $y = 1/x^2$ over the interval $[1, +\infty)$ to be

$$A = \int_1^{+\infty} \frac{dx}{x^2} = \lim_{b \to +\infty} \int_1^b \frac{dx}{x^2} = \lim_{b \to +\infty} \left(1 - \frac{1}{b}\right) = 1 \tag{1}$$

Thus, the area has a finite value of 1 and is not infinite as we first conjectured.

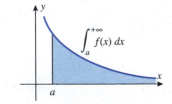

▲ **Figure 7.8.1**

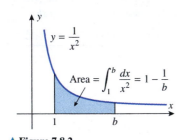

▲ **Figure 7.8.2**

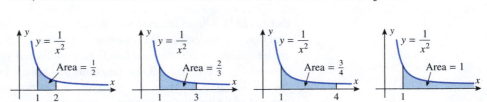

▶ **Figure 7.8.3**

With the preceding discussion as our guide, we make the following definition (which is applicable to functions with both positive and negative values).

If f is nonnegative over the interval $[a, +\infty)$, then the improper integral in Definition 7.8.1 can be interpreted to be the area under the graph of f over the interval $[a, +\infty)$. If the integral converges, then the area is finite and equal to the value of the integral, and if the integral diverges, then the area is regarded to be infinite.

7.8.1 DEFINITION The *improper integral of f over the interval* $[a, +\infty)$ is defined to be

$$\int_a^{+\infty} f(x)\, dx = \lim_{b \to +\infty} \int_a^b f(x)\, dx$$

In the case where the limit exists, the improper integral is said to **converge**, and the limit is defined to be the value of the integral. In the case where the limit does not exist, the improper integral is said to **diverge**, and it is not assigned a value.

▶ **Example 1** Evaluate

$$\text{(a)} \int_1^{+\infty} \frac{dx}{x^3} \qquad \text{(b)} \int_1^{+\infty} \frac{dx}{x}$$

Solution (a). Following the definition, we replace the infinite upper limit by a finite upper limit b, and then take the limit of the resulting integral. This yields

$$\int_1^{+\infty} \frac{dx}{x^3} = \lim_{b \to +\infty} \int_1^b \frac{dx}{x^3} = \lim_{b \to +\infty} \left[-\frac{1}{2x^2} \right]_1^b = \lim_{b \to +\infty} \left(\frac{1}{2} - \frac{1}{2b^2} \right) = \frac{1}{2}$$

Since the limit is finite, the integral converges and its value is $1/2$.

Solution (b).

$$\int_1^{+\infty} \frac{dx}{x} = \lim_{b \to +\infty} \int_1^b \frac{dx}{x} = \lim_{b \to +\infty} \left[\ln x \right]_1^b = \lim_{b \to +\infty} \ln b = +\infty$$

In this case the integral diverges and hence has no value. ◀

Because the functions $1/x^3$, $1/x^2$, and $1/x$ are nonnegative over the interval $[1, +\infty)$, it follows from (1) and the last example that over this interval the area under $y = 1/x^3$ is $\frac{1}{2}$, the area under $y = 1/x^2$ is 1, and the area under $y = 1/x$ is infinite. However, on the surface the graphs of the three functions seem very much alike (Figure 7.8.4), and there is nothing to suggest why one of the areas should be infinite and the other two finite. One explanation is that $1/x^3$ and $1/x^2$ approach zero more rapidly than $1/x$ as $x \to +\infty$, so that the area over the interval $[1, b]$ accumulates less rapidly under the curves $y = 1/x^3$ and $y = 1/x^2$ than under $y = 1/x$ as $b \to +\infty$, and the difference is just enough that the first two areas are finite and the third is infinite.

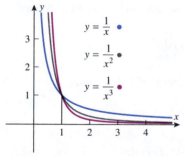

▲ **Figure 7.8.4**

▶ **Example 2** For what values of p does the integral $\int_1^{+\infty} \frac{dx}{x^p}$ converge?

Solution. We know from the preceding example that the integral diverges if $p = 1$, so let us assume that $p \neq 1$. In this case we have

$$\int_1^{+\infty} \frac{dx}{x^p} = \lim_{b \to +\infty} \int_1^b x^{-p}\, dx = \lim_{b \to +\infty} \frac{x^{1-p}}{1-p} \Bigg]_1^b = \lim_{b \to +\infty} \left[\frac{b^{1-p}}{1-p} - \frac{1}{1-p} \right]$$

If $p > 1$, then the exponent $1 - p$ is negative and $b^{1-p} \to 0$ as $b \to +\infty$; and if $p < 1$, then the exponent $1 - p$ is positive and $b^{1-p} \to +\infty$ as $b \to +\infty$. Thus, the integral converges if $p > 1$ and diverges otherwise. In the convergent case the value of the integral is

$$\int_1^{+\infty} \frac{dx}{x^p} = \left[0 - \frac{1}{1-p} \right] = \frac{1}{p-1} \quad (p > 1) \quad ◀$$

The following theorem summarizes this result.

7.8.2   THEOREM

$$\int_{1}^{+\infty} \frac{dx}{x^p} = \begin{cases} \dfrac{1}{p-1} & \text{if } p > 1 \\ \text{diverges} & \text{if } p \le 1 \end{cases}$$

▶ **Example 3**   Evaluate $\displaystyle\int_{0}^{+\infty} (1-x)e^{-x}\, dx$.

Solution.   We begin by evaluating the indefinite integral using integration by parts. Setting $u = 1 - x$ and $dv = e^{-x}\, dx$ yields

$$\int (1-x)e^{-x}\, dx = -e^{-x}(1-x) - \int e^{-x}\, dx = -e^{-x} + xe^{-x} + e^{-x} + C = xe^{-x} + C$$

Thus,

$$\int_{0}^{+\infty} (1-x)e^{-x}\, dx = \lim_{b \to +\infty} \int_{0}^{b} (1-x)e^{-x}\, dx = \lim_{b \to +\infty} \Big[xe^{-x} \Big]_{0}^{b} = \lim_{b \to +\infty} \frac{b}{e^{b}}$$

The limit is an indeterminate form of type ∞/∞, so we will apply L'Hôpital's rule by differentiating the numerator and denominator with respect to b. This yields

$$\int_{0}^{+\infty} (1-x)e^{-x}\, dx = \lim_{b \to +\infty} \frac{1}{e^{b}} = 0$$

We can interpret this to mean that the net signed area between the graph of $y = (1-x)e^{-x}$ and the interval $[0, +\infty)$ is 0 (Figure 7.8.5). ◀

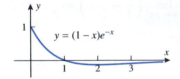

$y = (1-x)e^{-x}$

The net signed area between the graph and the interval $[0, +\infty)$ is zero.

▲ **Figure 7.8.5**

If f is nonnegative over the interval $(-\infty, +\infty)$, then the improper integral

$$\int_{-\infty}^{+\infty} f(x)\, dx$$

can be interpreted to be the area under the graph of f over the interval $(-\infty, +\infty)$. The area is finite and equal to the value of the integral if the integral converges and is infinite if it diverges.

7.8.3   DEFINITION   The *improper integral of f over the interval* $(-\infty, b]$ is defined to be

$$\int_{-\infty}^{b} f(x)\, dx = \lim_{a \to -\infty} \int_{a}^{b} f(x)\, dx \tag{2}$$

The integral is said to **converge** if the limit exists and **diverge** if it does not.

The *improper integral of f over the interval* $(-\infty, +\infty)$ is defined as

$$\int_{-\infty}^{+\infty} f(x)\, dx = \int_{-\infty}^{c} f(x)\, dx + \int_{c}^{+\infty} f(x)\, dx \tag{3}$$

where c is any real number. The improper integral is said to **converge** if *both* terms converge and **diverge** if *either* term diverges.

▶ **Example 4**   Evaluate $\displaystyle\int_{-\infty}^{+\infty} \frac{dx}{1+x^2}$.

Solution.   We will evaluate the integral by choosing $c = 0$ in (3). With this value for c we obtain

$$\int_{0}^{+\infty} \frac{dx}{1+x^2} = \lim_{b \to +\infty} \int_{0}^{b} \frac{dx}{1+x^2} = \lim_{b \to +\infty} \Big[\tan^{-1} x \Big]_{0}^{b} = \lim_{b \to +\infty} (\tan^{-1} b) = \frac{\pi}{2}$$

$$\int_{-\infty}^{0} \frac{dx}{1+x^2} = \lim_{a \to -\infty} \int_{a}^{0} \frac{dx}{1+x^2} = \lim_{a \to -\infty} \Big[\tan^{-1} x \Big]_{a}^{0} = \lim_{a \to -\infty} (-\tan^{-1} a) = \frac{\pi}{2}$$

Although we usually choose $c = 0$ in (3), the choice does not matter because it can be proved that neither the convergence nor the value of the integral is affected by the choice of c.

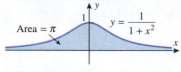

Area = π

$y = \dfrac{1}{1 + x^2}$

▲ **Figure 7.8.6**

Thus, the integral converges and its value is

$$\int_{-\infty}^{+\infty} \frac{dx}{1 + x^2} = \int_{-\infty}^{0} \frac{dx}{1 + x^2} + \int_{0}^{+\infty} \frac{dx}{1 + x^2} = \frac{\pi}{2} + \frac{\pi}{2} = \pi$$

Since the integrand is nonnegative on the interval $(-\infty, +\infty)$, the integral represents the area of the region shown in Figure 7.8.6. ◄

■ INTEGRALS WHOSE INTEGRANDS HAVE INFINITE DISCONTINUITIES

Next we will consider improper integrals whose integrands have infinite discontinuities. We will start with the case where the interval of integration is a finite interval $[a, b]$ and the infinite discontinuity occurs at the right-hand endpoint.

To motivate an appropriate definition for such an integral let us consider the case where f is nonnegative on $[a, b]$, so we can interpret the improper integral $\int_a^b f(x)\,dx$ as the area of the region in Figure 7.8.7a. The problem of finding the area of this region is complicated by the fact that it extends indefinitely in the positive y-direction. However, instead of trying to find the entire area at once, we can proceed indirectly by calculating the portion of the area over the interval $[a, k]$, where $a \le k < b$, and then letting k approach b to fill out the area of the entire region (Figure 7.8.7b). Motivated by this idea, we make the following definition.

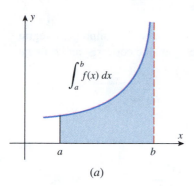

(a)

$\displaystyle\int_a^b f(x)\,dx$

(b)

$\displaystyle\int_a^k f(x)\,dx$

▲ **Figure 7.8.7**

> **7.8.4   DEFINITION**   If f is continuous on the interval $[a, b]$, except for an infinite discontinuity at b, then the ***improper integral of f over the interval $[a, b]$*** is defined as
>
> $$\int_a^b f(x)\,dx = \lim_{k \to b^-} \int_a^k f(x)\,dx \qquad (4)$$
>
> In the case where the indicated limit exists, the improper integral is said to ***converge***, and the limit is defined to be the value of the integral. In the case where the limit does not exist, the improper integral is said to ***diverge***, and it is not assigned a value.

▶ **Example 5**   Evaluate $\displaystyle\int_0^1 \frac{dx}{\sqrt{1 - x}}$.

Solution.   The integral is improper because the integrand approaches $+\infty$ as x approaches the upper limit 1 from the left (Figure 7.8.8). From (4),

$$\int_0^1 \frac{dx}{\sqrt{1 - x}} = \lim_{k \to 1^-} \int_0^k \frac{dx}{\sqrt{1 - x}} = \lim_{k \to 1^-} \left[-2\sqrt{1 - x} \right]_0^k$$

$$= \lim_{k \to 1^-} \left[-2\sqrt{1 - k} + 2 \right] = 2 \quad ◄$$

$y = \dfrac{1}{\sqrt{1 - x}}$

▲ **Figure 7.8.8**

Improper integrals with an infinite discontinuity at the left-hand endpoint or inside the interval of integration are defined as follows.

7.8.5 DEFINITION If f is continuous on the interval $[a, b]$, except for an infinite discontinuity at a, then the **improper integral of f over the interval $[a, b]$** is defined as

$$\int_a^b f(x)\,dx = \lim_{k \to a^+} \int_k^b f(x)\,dx \tag{5}$$

The integral is said to **converge** if the indicated limit exists and **diverge** if it does not.

If f is continuous on the interval $[a, b]$, except for an infinite discontinuity at a point c in (a, b), then the **improper integral of f over the interval $[a, b]$** is defined as

$$\int_a^b f(x)\,dx = \int_a^c f(x)\,dx + \int_c^b f(x)\,dx \tag{6}$$

where the two integrals on the right side are themselves improper. The improper integral on the left side is said to **converge** if *both* terms on the right side converge and **diverge** if *either* term on the right side diverges (Figure 7.8.9).

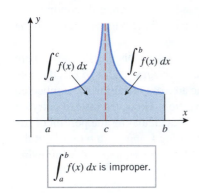

$\int_a^b f(x)\,dx$ is improper.

▲ **Figure 7.8.9**

▶ **Example 6** Evaluate

$$\text{(a)} \int_1^2 \frac{dx}{1-x} \qquad \text{(b)} \int_1^4 \frac{dx}{(x-2)^{2/3}}$$

Solution (a). The integral is improper because the integrand approaches $-\infty$ as x approaches the lower limit 1 from the right (Figure 7.8.10). From Definition 7.8.5 we obtain

$$\int_1^2 \frac{dx}{1-x} = \lim_{k \to 1^+} \int_k^2 \frac{dx}{1-x} = \lim_{k \to 1^+} \left[-\ln|1-x| \right]_k^2$$

$$= \lim_{k \to 1^+} \left[-\ln|-1| + \ln|1-k| \right] = \lim_{k \to 1^+} \ln|1-k| = -\infty$$

so the integral diverges.

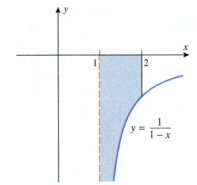

$y = \dfrac{1}{1-x}$

▲ **Figure 7.8.10**

Solution (b). The integral is improper because the integrand approaches $+\infty$ at $x = 2$, which is inside the interval of integration. From Definition 7.8.5 we obtain

$$\int_1^4 \frac{dx}{(x-2)^{2/3}} = \int_1^2 \frac{dx}{(x-2)^{2/3}} + \int_2^4 \frac{dx}{(x-2)^{2/3}} \tag{7}$$

and we must investigate the convergence of both improper integrals on the right. Since

$$\int_1^2 \frac{dx}{(x-2)^{2/3}} = \lim_{k \to 2^-} \int_1^k \frac{dx}{(x-2)^{2/3}} = \lim_{k \to 2^-} \left[3(k-2)^{1/3} - 3(1-2)^{1/3} \right] = 3$$

$$\int_2^4 \frac{dx}{(x-2)^{2/3}} = \lim_{k \to 2^+} \int_k^4 \frac{dx}{(x-2)^{2/3}} = \lim_{k \to 2^+} \left[3(4-2)^{1/3} - 3(k-2)^{1/3} \right] = 3\sqrt[3]{2}$$

we have from (7) that

$$\int_1^4 \frac{dx}{(x-2)^{2/3}} = 3 + 3\sqrt[3]{2} \quad \blacktriangleleft$$

WARNING It is sometimes tempting to apply the Fundamental Theorem of Calculus directly to an improper integral without taking the appropriate limits. To illustrate what can go wrong with this procedure, suppose we fail to recognize that the integral

$$\int_0^2 \frac{dx}{(x-1)^2} \tag{8}$$

is improper and mistakenly evaluate this integral as

$$-\frac{1}{x-1}\Big]_0^2 = -1 - (1) = -2$$

This result is clearly incorrect because the integrand is never negative and hence the integral cannot be negative! To evaluate (8) correctly we should first write

$$\int_0^2 \frac{dx}{(x-1)^2} = \int_0^1 \frac{dx}{(x-1)^2} + \int_1^2 \frac{dx}{(x-1)^2}$$

and then treat each term as an improper integral. For the first term,

$$\int_0^1 \frac{dx}{(x-1)^2} = \lim_{k \to 1^-} \int_0^k \frac{dx}{(x-1)^2} = \lim_{k \to 1^-}\left[-\frac{1}{k-1}-1\right] = +\infty$$

so (8) diverges.

■ ARC LENGTH AND SURFACE AREA USING IMPROPER INTEGRALS

In Definitions 6.4.2 and 6.5.2 for arc length and surface area we required the function f to be smooth (continuous first derivative) to ensure the integrability in the resulting formula. However, smoothness is overly restrictive since some of the most basic formulas in geometry involve functions that are not smooth but lead to convergent improper integrals. Accordingly, let us agree to extend the definitions of arc length and surface area to allow functions that are not smooth, but for which the resulting integral in the formula converges.

▶ **Example 7** Derive the formula for the circumference of a circle of radius r.

Solution. For convenience, let us assume that the circle is centered at the origin, in which case its equation is $x^2 + y^2 = r^2$. We will find the arc length of the portion of the circle that lies in the first quadrant and then multiply by 4 to obtain the total circumference (Figure 7.8.11).

Since the equation of the upper semicircle is $y = \sqrt{r^2 - x^2}$, it follows from Formula (4) of Section 6.4 that the circumference C is

$$C = 4\int_0^r \sqrt{1 + (dy/dx)^2}\, dx = 4\int_0^r \sqrt{1 + \left(-\frac{x}{\sqrt{r^2 - x^2}}\right)^2}\, dx$$

$$= 4r\int_0^r \frac{dx}{\sqrt{r^2 - x^2}}$$

This integral is improper because of the infinite discontinuity at $x = r$, and hence we evaluate it by writing

$$C = 4r\lim_{k \to r^-} \int_0^k \frac{dx}{\sqrt{r^2 - x^2}}$$

$$= 4r\lim_{k \to r^-}\left[\sin^{-1}\left(\frac{x}{r}\right)\right]_0^k \qquad \boxed{\begin{array}{l}\text{Formula (77) in the}\\\text{Endpaper Integral Table}\end{array}}$$

$$= 4r\lim_{k \to r^-}\left[\sin^{-1}\left(\frac{k}{r}\right) - \sin^{-1}0\right]$$

$$= 4r[\sin^{-1}1 - \sin^{-1}0] = 4r\left(\frac{\pi}{2} - 0\right) = 2\pi r \quad ◀$$

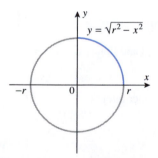

$y = \sqrt{r^2 - x^2}$

▲ **Figure 7.8.11**

✔ QUICK CHECK EXERCISES 7.8   (See page 557 for answers.)

1. In each part, determine whether the integral is improper, and if so, explain why. Do not evaluate the integrals.

(a) $\displaystyle\int_{\pi/4}^{3\pi/4} \cot x \, dx$

(b) $\displaystyle\int_{\pi/4}^{\pi} \cot x \, dx$

(c) $\displaystyle\int_{0}^{+\infty} \frac{1}{x^2+1} \, dx$

(d) $\displaystyle\int_{1}^{+\infty} \frac{1}{x^2-1} \, dx$

2. Express each improper integral in Quick Check Exercise 1 in terms of one or more appropriate limits. Do not evaluate the limits.

3. The improper integral

$$\int_{1}^{+\infty} x^{-p} \, dx$$

converges to _____ provided _____.

4. Evaluate the integrals that converge.

(a) $\displaystyle\int_{0}^{+\infty} e^{-x} \, dx$

(b) $\displaystyle\int_{0}^{+\infty} e^{x} \, dx$

(c) $\displaystyle\int_{0}^{1} \frac{1}{x^3} \, dx$

(d) $\displaystyle\int_{0}^{1} \frac{1}{\sqrt[3]{x^2}} \, dx$

EXERCISE SET 7.8    Graphing Utility   C CAS

1. In each part, determine whether the integral is improper, and if so, explain why.

(a) $\displaystyle\int_{1}^{5} \frac{dx}{x-3}$

(b) $\displaystyle\int_{1}^{5} \frac{dx}{x+3}$

(c) $\displaystyle\int_{0}^{1} \ln x \, dx$

(d) $\displaystyle\int_{1}^{+\infty} e^{-x} \, dx$

(e) $\displaystyle\int_{-\infty}^{+\infty} \frac{dx}{\sqrt[3]{x-1}}$

(f) $\displaystyle\int_{0}^{\pi/4} \tan x \, dx$

2. In each part, determine all values of p for which the integral is improper.

(a) $\displaystyle\int_{0}^{1} \frac{dx}{x^p}$

(b) $\displaystyle\int_{1}^{2} \frac{dx}{x-p}$

(c) $\displaystyle\int_{0}^{1} e^{-px} \, dx$

3–32 Evaluate the integrals that converge. ■

3. $\displaystyle\int_{0}^{+\infty} e^{-2x} \, dx$

4. $\displaystyle\int_{-1}^{+\infty} \frac{x}{1+x^2} \, dx$

5. $\displaystyle\int_{3}^{+\infty} \frac{2}{x^2-1} \, dx$

6. $\displaystyle\int_{0}^{+\infty} xe^{-x^2} \, dx$

7. $\displaystyle\int_{e}^{+\infty} \frac{1}{x \ln^3 x} \, dx$

8. $\displaystyle\int_{2}^{+\infty} \frac{1}{x\sqrt{\ln x}} \, dx$

9. $\displaystyle\int_{-\infty}^{0} \frac{dx}{(2x-1)^3}$

10. $\displaystyle\int_{-\infty}^{3} \frac{dx}{x^2+9}$

11. $\displaystyle\int_{-\infty}^{0} e^{3x} \, dx$

12. $\displaystyle\int_{-\infty}^{0} \frac{e^x \, dx}{3-2e^x}$

13. $\displaystyle\int_{-\infty}^{+\infty} x \, dx$

14. $\displaystyle\int_{-\infty}^{+\infty} \frac{x}{\sqrt{x^2+2}} \, dx$

15. $\displaystyle\int_{-\infty}^{+\infty} \frac{x}{(x^2+3)^2} \, dx$

16. $\displaystyle\int_{-\infty}^{+\infty} \frac{e^{-t}}{1+e^{-2t}} \, dt$

17. $\displaystyle\int_{0}^{4} \frac{dx}{(x-4)^2}$

18. $\displaystyle\int_{0}^{8} \frac{dx}{\sqrt[3]{x}}$

19. $\displaystyle\int_{0}^{\pi/2} \tan x \, dx$

20. $\displaystyle\int_{0}^{4} \frac{dx}{\sqrt{4-x}}$

21. $\displaystyle\int_{0}^{1} \frac{dx}{\sqrt{1-x^2}}$

22. $\displaystyle\int_{-3}^{1} \frac{x \, dx}{\sqrt{9-x^2}}$

23. $\displaystyle\int_{\pi/3}^{\pi/2} \frac{\sin x}{\sqrt{1-2\cos x}} \, dx$

24. $\displaystyle\int_{0}^{\pi/4} \frac{\sec^2 x}{1-\tan x} \, dx$

25. $\displaystyle\int_{0}^{3} \frac{dx}{x-2}$

26. $\displaystyle\int_{-2}^{2} \frac{dx}{x^2}$

27. $\displaystyle\int_{-1}^{8} x^{-1/3} \, dx$

28. $\displaystyle\int_{0}^{1} \frac{dx}{(x-1)^{2/3}}$

29. $\displaystyle\int_{0}^{+\infty} \frac{1}{x^2} \, dx$

30. $\displaystyle\int_{1}^{+\infty} \frac{dx}{x\sqrt{x^2-1}}$

31. $\displaystyle\int_{0}^{1} \frac{dx}{\sqrt{x}(x+1)}$

32. $\displaystyle\int_{0}^{+\infty} \frac{dx}{\sqrt{x}(x+1)}$

33–36 True–False Determine whether the statement is true or false. Explain your answer. ■

33. $\displaystyle\int_{1}^{+\infty} x^{-4/3} \, dx$ converges to 3.

34. If f is continuous on $[a, +\infty]$ and $\lim_{x \to +\infty} f(x) = 1$, then $\int_{a}^{+\infty} f(x) \, dx$ converges.

35. $\displaystyle\int_{1}^{2} \frac{1}{x(x-3)} \, dx$ is an improper integral.

36. $\displaystyle\int_{-1}^{1} \frac{1}{x^3} \, dx = 0$

37–40 Make the u-substitution and evaluate the resulting definite integral. ■

37. $\displaystyle\int_{0}^{+\infty} \frac{e^{-\sqrt{x}}}{\sqrt{x}} \, dx; \ u = \sqrt{x}$ [Note: $u \to +\infty$ as $x \to +\infty$.]

38. $\displaystyle\int_{12}^{+\infty} \frac{dx}{\sqrt{x}(x+4)}; \ u = \sqrt{x}$ [Note: $u \to +\infty$ as $x \to +\infty$.]

39. $\displaystyle\int_{0}^{+\infty} \frac{e^{-x}}{\sqrt{1-e^{-x}}} \, dx; \ u = 1 - e^{-x}$

[Note: $u \to 1$ as $x \to +\infty$.]

40. $\displaystyle\int_0^{+\infty} \frac{e^{-x}}{\sqrt{1-e^{-2x}}}\,dx;\ u=e^{-x}$

41–42 Express the improper integral as a limit, and then evaluate that limit with a CAS. Confirm the answer by evaluating the integral directly with the CAS. ■

41. $\displaystyle\int_0^{+\infty} e^{-x}\cos x\,dx$ **42.** $\displaystyle\int_0^{+\infty} xe^{-3x}\,dx$

43. In each part, try to evaluate the integral exactly with a CAS. If your result is not a simple numerical answer, then use the CAS to find a numerical approximation of the integral.

(a) $\displaystyle\int_{-\infty}^{+\infty} \frac{1}{x^8+x+1}\,dx$ (b) $\displaystyle\int_0^{+\infty} \frac{1}{\sqrt{1+x^3}}\,dx$

(c) $\displaystyle\int_1^{+\infty} \frac{\ln x}{e^x}\,dx$ (d) $\displaystyle\int_1^{+\infty} \frac{\sin x}{x^2}\,dx$

44. In each part, confirm the result with a CAS.

(a) $\displaystyle\int_0^{+\infty} \frac{\sin x}{\sqrt{x}}\,dx = \sqrt{\frac{\pi}{2}}$ (b) $\displaystyle\int_{-\infty}^{+\infty} e^{-x^2}\,dx = \sqrt{\pi}$

(c) $\displaystyle\int_0^1 \frac{\ln x}{1+x}\,dx = -\frac{\pi^2}{12}$

45. Find the length of the curve $y=(4-x^{2/3})^{3/2}$ over the interval $[0, 8]$.

46. Find the length of the curve $y=\sqrt{4-x^2}$ over the interval $[0, 2]$.

47–48 Use L'Hôpital's rule to help evaluate the improper integral. ■

47. $\displaystyle\int_0^1 \ln x\,dx$ **48.** $\displaystyle\int_1^{+\infty} \frac{\ln x}{x^2}\,dx$

49. Find the area of the region between the x-axis and the curve $y=e^{-3x}$ for $x \ge 0$.

50. Find the area of the region between the x-axis and the curve $y=8/(x^2-4)$ for $x \ge 4$.

51. Suppose that the region between the x-axis and the curve $y=e^{-x}$ for $x \ge 0$ is revolved about the x-axis.
(a) Find the volume of the solid that is generated.
(b) Find the surface area of the solid.

FOCUS ON CONCEPTS

52. Suppose that f and g are continuous functions and that

$$0 \le f(x) \le g(x)$$

if $x \ge a$. Give a reasonable informal argument using areas to explain why the following results are true.
(a) If $\int_a^{+\infty} f(x)\,dx$ diverges, then $\int_a^{+\infty} g(x)\,dx$ diverges.
(b) If $\int_a^{+\infty} g(x)\,dx$ converges, then $\int_a^{+\infty} f(x)\,dx$ converges and $\int_a^{+\infty} f(x)\,dx \le \int_a^{+\infty} g(x)\,dx$.
[*Note:* The results in this exercise are sometimes called *comparison tests* for improper integrals.]

53–56 Use the results in Exercise 52. ■

53. (a) Confirm graphically and algebraically that

$$e^{-x^2} \le e^{-x} \quad (x \ge 1)$$

(b) Evaluate the integral

$$\int_1^{+\infty} e^{-x}\,dx$$

(c) What does the result obtained in part (b) tell you about the integral

$$\int_1^{+\infty} e^{-x^2}\,dx?$$

54. (a) Confirm graphically and algebraically that

$$\frac{1}{2x+1} \le \frac{e^x}{2x+1} \quad (x \ge 0)$$

(b) Evaluate the integral

$$\int_0^{+\infty} \frac{dx}{2x+1}$$

(c) What does the result obtained in part (b) tell you about the integral

$$\int_0^{+\infty} \frac{e^x}{2x+1}\,dx?$$

55. Let R be the region to the right of $x=1$ that is bounded by the x-axis and the curve $y=1/x$. When this region is revolved about the x-axis it generates a solid whose surface is known as *Gabriel's Horn* (for reasons that should be clear from the accompanying figure). Show that the solid has a finite volume but its surface has an infinite area. [*Note:* It has been suggested that if one could saturate the interior of the solid with paint and allow it to seep through to the surface, then one could paint an infinite surface with a finite amount of paint! What do you think?]

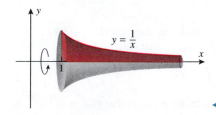

◀ **Figure Ex-55**

56. In each part, use Exercise 52 to determine whether the integral converges or diverges. If it converges, then use part (b) of that exercise to find an upper bound on the value of the integral.

(a) $\displaystyle\int_2^{+\infty} \frac{\sqrt{x^3+1}}{x}\,dx$ (b) $\displaystyle\int_2^{+\infty} \frac{x}{x^5+1}\,dx$

(c) $\displaystyle\int_0^{+\infty} \frac{xe^x}{2x+1}\,dx$

FOCUS ON CONCEPTS

57. Sketch the region whose area is
$$\int_0^{+\infty} \frac{dx}{1+x^2}$$
and use your sketch to show that
$$\int_0^{+\infty} \frac{dx}{1+x^2} = \int_0^1 \sqrt{\frac{1-y}{y}}\, dy$$

58. (a) Give a reasonable informal argument, based on areas, that explains why the integrals
$$\int_0^{+\infty} \sin x\, dx \quad \text{and} \quad \int_0^{+\infty} \cos x\, dx$$
diverge.

(b) Show that $\displaystyle\int_0^{+\infty} \frac{\cos\sqrt{x}}{\sqrt{x}}\, dx$ diverges.

59. In electromagnetic theory, the magnetic potential at a point on the axis of a circular coil is given by
$$u = \frac{2\pi N I r}{k} \int_a^{+\infty} \frac{dx}{(r^2+x^2)^{3/2}}$$
where N, I, r, k, and a are constants. Find u.

C **60.** The *average speed*, $\bar{v}$, of the molecules of an ideal gas is given by
$$\bar{v} = \frac{4}{\sqrt{\pi}}\left(\frac{M}{2RT}\right)^{3/2} \int_0^{+\infty} v^3 e^{-Mv^2/(2RT)}\, dv$$
and the *root-mean-square speed*, v_{rms}, by
$$v_{\text{rms}}^2 = \frac{4}{\sqrt{\pi}}\left(\frac{M}{2RT}\right)^{3/2} \int_0^{+\infty} v^4 e^{-Mv^2/(2RT)}\, dv$$
where v is the molecular speed, T is the gas temperature, M is the molecular weight of the gas, and R is the gas constant.
(a) Use a CAS to show that
$$\int_0^{+\infty} x^3 e^{-a^2 x^2}\, dx = \frac{1}{2a^4}, \quad a > 0$$
and use this result to show that $\bar{v} = \sqrt{8RT/(\pi M)}$.
(b) Use a CAS to show that
$$\int_0^{+\infty} x^4 e^{-a^2 x^2}\, dx = \frac{3\sqrt{\pi}}{8a^5}, \quad a > 0$$
and use this result to show that $v_{\text{rms}} = \sqrt{3RT/M}$.

61. In Exercise 25 of Section 6.6, we determined the work required to lift a 6000 lb satellite to an orbital position that is 1000 mi above the Earth's surface. The ideas discussed in that exercise will be needed here.
(a) Find a definite integral that represents the work required to lift a 6000 lb satellite to a position b miles above the Earth's surface.
(b) Find a definite integral that represents the work required to lift a 6000 lb satellite an "infinite distance" above the Earth's surface. Evaluate the integral. [*Note:* The result obtained here is sometimes called the work required to "escape" the Earth's gravity.]

62–63 A *transform* is a formula that converts or "transforms" one function into another. Transforms are used in applications to convert a difficult problem into an easier problem whose solution can then be used to solve the original difficult problem. The *Laplace transform* of a function $f(t)$, which plays an important role in the study of differential equations, is denoted by $\mathcal{L}\{f(t)\}$ and is defined by
$$\mathcal{L}\{f(t)\} = \int_0^{+\infty} e^{-st} f(t)\, dt$$
In this formula s is treated as a constant in the integration process; thus, the Laplace transform has the effect of transforming $f(t)$ into a function of s. Use this formula in these exercises. ■

62. Show that
(a) $\mathcal{L}\{1\} = \dfrac{1}{s}, \ s > 0$ \qquad **(b)** $\mathcal{L}\{e^{2t}\} = \dfrac{1}{s-2}, \ s > 2$
(c) $\mathcal{L}\{\sin t\} = \dfrac{1}{s^2+1}, \ s > 0$
(d) $\mathcal{L}\{\cos t\} = \dfrac{s}{s^2+1}, \ s > 0.$

63. In each part, find the Laplace transform.
(a) $f(t) = t, \ s > 0$ \qquad **(b)** $f(t) = t^2, \ s > 0$
(c) $f(t) = \begin{cases} 0, & t < 3 \\ 1, & t \ge 3 \end{cases}, \ s > 0$

C **64.** Later in the text, we will show that
$$\int_0^{+\infty} e^{-x^2}\, dx = \tfrac{1}{2}\sqrt{\pi}$$
Confirm that this is reasonable by using a CAS or a calculator with a numerical integration capability.

65. Use the result in Exercise 64 to show that
(a) $\displaystyle\int_{-\infty}^{+\infty} e^{-ax^2}\, dx = \sqrt{\frac{\pi}{a}}, \ a > 0$
(b) $\displaystyle\frac{1}{\sqrt{2\pi}\sigma} \int_{-\infty}^{+\infty} e^{-x^2/2\sigma^2}\, dx = 1, \ \sigma > 0.$

66–67 A convergent improper integral over an infinite interval can be approximated by first replacing the infinite limit(s) of integration by finite limit(s), then using a numerical integration technique, such as Simpson's rule, to approximate the integral with finite limit(s). This technique is illustrated in these exercises. ■

66. Suppose that the integral in Exercise 64 is approximated by first writing it as
$$\int_0^{+\infty} e^{-x^2}\, dx = \int_0^K e^{-x^2}\, dx + \int_K^{+\infty} e^{-x^2}\, dx$$
then dropping the second term, and then applying Simpson's rule to the integral
$$\int_0^K e^{-x^2}\, dx$$
The resulting approximation has two sources of error: the error from Simpson's rule and the error
$$E = \int_K^{+\infty} e^{-x^2}\, dx \qquad \text{(cont.)}$$

that results from discarding the second term. We call E the *truncation error*.

(a) Approximate the integral in Exercise 64 by applying Simpson's rule with $n = 10$ subdivisions to the integral

$$\int_0^3 e^{-x^2}\, dx$$

Round your answer to four decimal places and compare it to $\frac{1}{2}\sqrt{\pi}$ rounded to four decimal places.

(b) Use the result that you obtained in Exercise 52 and the fact that $e^{-x^2} \le \frac{1}{3}xe^{-x^2}$ for $x \ge 3$ to show that the truncation error for the approximation in part (a) satisfies $0 < E < 2.1 \times 10^{-5}$.

67. (a) It can be shown that

$$\int_0^{+\infty} \frac{1}{x^6 + 1}\, dx = \frac{\pi}{3}$$

Approximate this integral by applying Simpson's rule with $n = 20$ subdivisions to the integral

$$\int_0^4 \frac{1}{x^6 + 1}\, dx$$

Round your answer to three decimal places and compare it to $\pi/3$ rounded to three decimal places.

(b) Use the result that you obtained in Exercise 52 and the fact that $1/(x^6 + 1) < 1/x^6$ for $x \ge 4$ to show that the truncation error for the approximation in part (a) satisfies $0 < E < 2 \times 10^{-4}$.

68. For what values of p does $\displaystyle\int_0^{+\infty} e^{px}\, dx$ converge?

69. Show that $\displaystyle\int_0^1 dx/x^p$ converges if $p < 1$ and diverges if $p \ge 1$.

C **70.** It is sometimes possible to convert an improper integral into a "proper" integral having the same value by making an appropriate substitution. Evaluate the following integral by making the indicated substitution, and investigate what happens if you evaluate the integral directly using a CAS.

$$\int_0^1 \sqrt{\frac{1+x}{1-x}}\, dx; \quad u = \sqrt{1-x}$$

71–72 Transform the given improper integral into a proper integral by making the stated u-substitution; then approximate the proper integral by Simpson's rule with $n = 10$ subdivisions. Round your answer to three decimal places. ∎

71. $\displaystyle\int_0^1 \frac{\cos x}{\sqrt{x}}\, dx; \quad u = \sqrt{x}$

72. $\displaystyle\int_0^1 \frac{\sin x}{\sqrt{1-x}}\, dx; \quad u = \sqrt{1-x}$

73. Writing What is "improper" about an integral over an infinite interval? Explain why Definition 5.5.1 for $\int_a^b f(x)\, dx$ fails for $\int_a^{+\infty} f(x)\, dx$. Discuss a strategy for assigning a value to $\int_a^{+\infty} f(x)\, dx$.

74. Writing What is "improper" about a definite integral over an interval on which the integrand has an infinite discontinuity? Explain why Definition 5.5.1 for $\int_a^b f(x)\, dx$ fails if the graph of f has a vertical asymptote at $x = a$. Discuss a strategy for assigning a value to $\int_a^b f(x)\, dx$ in this circumstance.

✔ QUICK CHECK ANSWERS 7.8

1. (a) proper (b) improper, since $\cot x$ has an infinite discontinuity at $x = \pi$ (c) improper, since there is an infinite interval of integration (d) improper, since there is an infinite interval of integration and the integrand has an infinite discontinuity at $x = 1$

2. (b) $\displaystyle\lim_{b \to \pi^-} \int_{\pi/4}^b \cot x\, dx$ (c) $\displaystyle\lim_{b \to +\infty} \int_0^b \frac{1}{x^2 + 1}\, dx$ (d) $\displaystyle\lim_{a \to 1^+} \int_a^2 \frac{1}{x^2 - 1}\, dx + \lim_{b \to +\infty} \int_2^b \frac{1}{x^2 - 1}\, dx$ **3.** $\dfrac{1}{p - 1}$; $p > 1$

4. (a) 1 (b) diverges (c) diverges (d) 3

CHAPTER 7 REVIEW EXERCISES

1–6 Evaluate the given integral with the aid of an appropriate u-substitution. ∎

1. $\displaystyle\int \sqrt{4 + 9x}\, dx$

2. $\displaystyle\int \frac{1}{\sec \pi x}\, dx$

3. $\displaystyle\int \sqrt{\cos x}\, \sin x\, dx$

4. $\displaystyle\int \frac{dx}{x \ln x}$

5. $\displaystyle\int x \tan^2(x^2) \sec^2(x^2)\, dx$

6. $\displaystyle\int_0^9 \frac{\sqrt{x}}{x + 9}\, dx$

7. (a) Evaluate the integral

$$\int \frac{1}{\sqrt{2x - x^2}}\, dx$$

three ways: using the substitution $u = \sqrt{x}$, using the substitution $u = \sqrt{2 - x}$, and completing the square.

(b) Show that the answers in part (a) are equivalent.

8. Evaluate the integral $\displaystyle\int_0^1 \frac{x^3}{\sqrt{x^2+1}}\,dx$

   (a) using integration by parts

   (b) using the substitution $u = \sqrt{x^2+1}$.

9–12 Use integration by parts to evaluate the integral. ■

9. $\displaystyle\int xe^{-x}\,dx$ **10.** $\displaystyle\int x\sin 2x\,dx$

11. $\displaystyle\int \ln(2x+3)\,dx$ **12.** $\displaystyle\int_0^{1/2} \tan^{-1}(2x)\,dx$

13. Evaluate $\int 8x^4 \cos 2x\,dx$ using tabular integration by parts.

14. A particle moving along the x-axis has velocity function $v(t) = t^2 e^{-t}$. How far does the particle travel from time $t = 0$ to $t = 5$?

15–20 Evaluate the integral. ■

15. $\displaystyle\int \sin^2 5\theta\,d\theta$ **16.** $\displaystyle\int \sin^3 2x \cos^2 2x\,dx$

17. $\displaystyle\int \sin x \cos 2x\,dx$ **18.** $\displaystyle\int_0^{\pi/6} \sin 2x \cos 4x\,dx$

19. $\displaystyle\int \sin^4 2x\,dx$ **20.** $\displaystyle\int x\cos^5(x^2)\,dx$

21–26 Evaluate the integral by making an appropriate trigonometric substitution. ■

21. $\displaystyle\int \frac{x^2}{\sqrt{9-x^2}}\,dx$ **22.** $\displaystyle\int \frac{dx}{x^2\sqrt{16-x^2}}$

23. $\displaystyle\int \frac{dx}{\sqrt{x^2-1}}$ **24.** $\displaystyle\int \frac{x^2}{\sqrt{x^2-25}}\,dx$

25. $\displaystyle\int \frac{x^2}{\sqrt{9+x^2}}\,dx$ **26.** $\displaystyle\int \frac{\sqrt{1+4x^2}}{x}\,dx$

27–32 Evaluate the integral using the method of partial fractions. ■

27. $\displaystyle\int \frac{dx}{x^2+3x-4}$ **28.** $\displaystyle\int \frac{dx}{x^2+8x+7}$

29. $\displaystyle\int \frac{x^2+2}{x+2}\,dx$ **30.** $\displaystyle\int \frac{x^2+x-16}{(x-1)(x-3)^2}\,dx$

31. $\displaystyle\int \frac{x^2}{(x+2)^3}\,dx$ **32.** $\displaystyle\int \frac{dx}{x^3+x}$

33. Consider the integral $\displaystyle\int \frac{1}{x^3-x}\,dx$.

   (a) Evaluate the integral using the substitution $x = \sec\theta$. For what values of x is your result valid?

   (b) Evaluate the integral using the substitution $x = \sin\theta$. For what values of x is your result valid?

   (c) Evaluate the integral using the method of partial fractions. For what values of x is your result valid?

34. Find the area of the region that is enclosed by the curves $y = (x-3)/(x^3+x^2)$, $y = 0$, $x = 1$, and $x = 2$.

35–40 Use the Endpaper Integral Table to evaluate the integral. ■

35. $\displaystyle\int \sin 7x \cos 9x\,dx$ **36.** $\displaystyle\int (x^3-x^2)e^{-x}\,dx$

37. $\displaystyle\int x\sqrt{x-x^2}\,dx$ **38.** $\displaystyle\int \frac{dx}{x\sqrt{4x+3}}$

39. $\displaystyle\int \tan^2 2x\,dx$ **40.** $\displaystyle\int \frac{3x-1}{2+x^2}\,dx$

41–42 Approximate the integral using (a) the midpoint approximation M_{10}, (b) the trapezoidal approximation T_{10}, and (c) Simpson's rule approximation S_{20}. In each case, find the exact value of the integral and approximate the absolute error. Express your answers to at least four decimal places. ■

41. $\displaystyle\int_1^3 \frac{1}{\sqrt{x+1}}\,dx$ **42.** $\displaystyle\int_{-1}^1 \frac{1}{1+x^2}\,dx$

43–44 Use inequalities (12), (13), and (14) of Section 7.7 to find upper bounds on the errors in parts (a), (b), or (c) of the indicated exercise. ■

43. Exercise 41   **44.** Exercise 42

45–46 Use inequalities (12), (13), and (14) of Section 7.7 to find a number n of subintervals for (a) the midpoint approximation M_n, (b) the trapezoidal approximation T_n, and (c) Simpson's rule approximation S_n to ensure the absolute error will be less than 10^{-4}. ■

45. Exercise 41   **46.** Exercise 42

47–50 Evaluate the integral if it converges. ■

47. $\displaystyle\int_0^{+\infty} e^{-x}\,dx$ **48.** $\displaystyle\int_{-\infty}^2 \frac{dx}{x^2+4}$

49. $\displaystyle\int_0^9 \frac{dx}{\sqrt{9-x}}$ **50.** $\displaystyle\int_0^1 \frac{1}{2x-1}\,dx$

51. Find the area that is enclosed between the x-axis and the curve $y = (\ln x - 1)/x^2$ for $x \geq e$.

52. Find the volume of the solid that is generated when the region between the x-axis and the curve $y = e^{-x}$ for $x \geq 0$ is revolved about the y-axis.

53. Find a positive value of a that satisfies the equation

$$\int_0^{+\infty} \frac{1}{x^2+a^2}\,dx = 1$$

54. Consider the following methods for evaluating integrals: u-substitution, integration by parts, partial fractions, reduction formulas, and trigonometric substitutions. In each part, state the approach that you would try first to evaluate the integral. If none of them seems appropriate, then say so. You need not evaluate the integral.

   (a) $\displaystyle\int x\sin x\,dx$ (b) $\displaystyle\int \cos x \sin x\,dx$

(cont.)

(c) $\int \tan^7 x \, dx$

(d) $\int \tan^7 x \sec^2 x \, dx$

(e) $\int \dfrac{3x^2}{x^3 + 1} \, dx$

(f) $\int \dfrac{3x^2}{(x + 1)^3} \, dx$

(g) $\int \tan^{-1} x \, dx$

(h) $\int \sqrt{4 - x^2} \, dx$

(i) $\int x\sqrt{4 - x^2} \, dx$

55–74 Evaluate the integral.

55. $\int \dfrac{dx}{(3 + x^2)^{3/2}}$

56. $\int x \cos 3x \, dx$

57. $\int_0^{\pi/4} \tan^7 \theta \, d\theta$

58. $\int \dfrac{\cos \theta}{\sin^2 \theta - 6 \sin \theta + 12} \, d\theta$

59. $\int \sin^2 2x \cos^3 2x \, dx$

60. $\int_0^4 \dfrac{1}{(x - 3)^2} \, dx$

61. $\int e^{2x} \cos 3x \, dx$

62. $\int_{-1/\sqrt{2}}^{1/\sqrt{2}} (1 - 2x^2)^{3/2} \, dx$

63. $\int \dfrac{dx}{(x - 1)(x + 2)(x - 3)}$

64. $\int_0^{1/3} \dfrac{dx}{(4 - 9x^2)^2}$

65. $\int_4^8 \dfrac{\sqrt{x - 4}}{x} \, dx$

66. $\int_0^{\ln 2} \sqrt{e^x - 1} \, dx$

67. $\int \dfrac{1}{\sqrt{e^x + 1}} \, dx$

68. $\int \dfrac{dx}{x(x^2 + x + 1)}$

69. $\int_0^{1/2} \sin^{-1} x \, dx$

70. $\int \tan^5 4x \sec^4 4x \, dx$

71. $\int \dfrac{x + 3}{\sqrt{x^2 + 2x + 2}} \, dx$

72. $\int \dfrac{\sec^2 \theta}{\tan^3 \theta - \tan^2 \theta} \, d\theta$

73. $\int_a^{+\infty} \dfrac{x}{(x^2 + 1)^2} \, dx$

74. $\int_0^{+\infty} \dfrac{dx}{a^2 + b^2 x^2}, \quad a, b > 0$

CHAPTER 7 MAKING CONNECTIONS   [C] CAS

1. Recall from Theorem 3.3.1 and the discussion preceding it that if $f'(x) > 0$, then the function f is increasing and has an inverse function. Parts (a), (b), and (c) of this problem show that if this condition is satisfied and if f' is continuous, then a definite integral of f^{-1} can be expressed in terms of a definite integral of f.

(a) Use integration by parts to show that

$$\int_a^b f(x) \, dx = bf(b) - af(a) - \int_a^b x f'(x) \, dx$$

(b) Use the result in part (a) to show that if $y = f(x)$, then

$$\int_a^b f(x) \, dx = bf(b) - af(a) - \int_{f(a)}^{f(b)} f^{-1}(y) \, dy$$

(c) Show that if we let $\alpha = f(a)$ and $\beta = f(b)$, then the result in part (b) can be written as

$$\int_\alpha^\beta f^{-1}(x) \, dx = \beta f^{-1}(\beta) - \alpha f^{-1}(\alpha) - \int_{f^{-1}(\alpha)}^{f^{-1}(\beta)} f(x) \, dx$$

2. In each part, use the result in Exercise 1 to obtain the equation, and then confirm that the equation is correct by performing the integrations.

(a) $\int_0^{1/2} \sin^{-1} x \, dx = \tfrac{1}{2} \sin^{-1} \left(\tfrac{1}{2} \right) - \int_0^{\pi/6} \sin x \, dx$

(b) $\int_e^{e^2} \ln x \, dx = (2e^2 - e) - \int_1^2 e^x \, dx$

3. The **Gamma function**, $\Gamma(x)$, is defined as

$$\Gamma(x) = \int_0^{+\infty} t^{x-1} e^{-t} \, dt$$

It can be shown that this improper integral converges if and only if $x > 0$.

(a) Find $\Gamma(1)$.

(b) Prove: $\Gamma(x + 1) = x\Gamma(x)$ for all $x > 0$. [*Hint:* Use integration by parts.]

(c) Use the results in parts (a) and (b) to find $\Gamma(2)$, $\Gamma(3)$, and $\Gamma(4)$; and then make a conjecture about $\Gamma(n)$ for positive integer values of n.

(d) Show that $\Gamma \left(\tfrac{1}{2} \right) = \sqrt{\pi}$. [*Hint:* See Exercise 64 of Section 7.8.]

(e) Use the results obtained in parts (b) and (d) to show that $\Gamma \left(\tfrac{3}{2} \right) = \tfrac{1}{2} \sqrt{\pi}$ and $\Gamma \left(\tfrac{5}{2} \right) = \tfrac{3}{4} \sqrt{\pi}$.

4. Refer to the Gamma function defined in Exercise 3 to show that

(a) $\int_0^1 (\ln x)^n \, dx = (-1)^n \Gamma(n + 1), \quad n > 0$

[*Hint:* Let $t = -\ln x$.]

(b) $\int_0^{+\infty} e^{-x^n} \, dx = \Gamma \left(\dfrac{n + 1}{n} \right), \quad n > 0.$

[*Hint:* Let $t = x^n$. Use the result in Exercise 3(b).]

[C] **5.** A **simple pendulum** consists of a mass that swings in a vertical plane at the end of a massless rod of length L, as shown in the accompanying figure. Suppose that a simple pendulum is displaced through an angle θ_0 and released from rest. It can be shown that in the absence of friction, the time T required for the pendulum to make one complete back-and-forth swing, called the **period**, is given by

$$T = \sqrt{\frac{8L}{g}} \int_0^{\theta_0} \frac{1}{\sqrt{\cos\theta - \cos\theta_0}}\, d\theta \qquad (1)$$

where $\theta = \theta(t)$ is the angle the pendulum makes with the vertical at time t. The improper integral in (1) is difficult to evaluate numerically. By a substitution outlined below it can be shown that the period can be expressed as

$$T = 4\sqrt{\frac{L}{g}} \int_0^{\pi/2} \frac{1}{\sqrt{1 - k^2\sin^2\phi}}\, d\phi \qquad (2)$$

where $k = \sin(\theta_0/2)$. The integral in (2) is called a **complete elliptic integral of the first kind** and is more easily evaluated by numerical methods.

(a) Obtain (2) from (1) by substituting

$$\cos\theta = 1 - 2\sin^2(\theta/2)$$

$$\cos\theta_0 = 1 - 2\sin^2(\theta_0/2)$$

$$k = \sin(\theta_0/2)$$

and then making the change of variable

$$\sin\phi = \frac{\sin(\theta/2)}{\sin(\theta_0/2)} = \frac{\sin(\theta/2)}{k}$$

(b) Use (2) and the numerical integration capability of your CAS to estimate the period of a simple pendulum for which $L = 1.5$ ft, $\theta_0 = 20°$, and $g = 32$ ft/s^2.

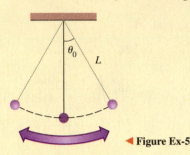

◀ **Figure Ex-5**

EXPANDING THE CALCULUS HORIZON

To learn how numerical integration can be applied to the cost analysis of an engineering project, see the module entitled **Railroad Design** at:

www.wiley.com/college/anton

MATHEMATICAL MODELING WITH DIFFERENTIAL EQUATIONS

Photo by Milton Bell, Texas Archeological Research Laboratory, The University of Texas at Austin.

In the 1920's, excavation of an archeological site in Folsom, N.M. uncovered a collection of prehistoric stone spearheads now known as "Folsom points." In 1950, carbon dating of charred bison bones found nearby confirmed that human hunters lived in the area between 9000 B.C. and 8000 B.C. We will study carbon dating in this chapter.

Many of the principles in science and engineering concern relationships between changing quantities. Since rates of change are represented mathematically by derivatives, it should not be surprising that such principles are often expressed in terms of differential equations. We introduced the concept of a differential equation in Section 5.2, but in this chapter we will go into more detail. We will discuss some important mathematical models that involve differential equations, and we will discuss some methods for solving and approximating solutions of some of the basic types of differential equations. However, we will only be able to touch the surface of this topic, leaving many important topics in differential equations to courses that are devoted completely to the subject.

8.1 MODELING WITH DIFFERENTIAL EQUATIONS

In this section we will introduce some basic terminology and concepts concerning differential equations. We will also discuss the general idea of modeling with differential equations, and we will encounter important models that can be applied to demography, medicine, ecology, and physics. In later sections of this chapter we will investigate methods that may be used to solve these differential equations.

Table 8.1.1

DIFFERENTIAL EQUATION	ORDER
$\dfrac{dy}{dx} = 3y$	1
$\dfrac{d^2y}{dx^2} - 6\dfrac{dy}{dx} + 8y = 0$	2
$\dfrac{d^3y}{dt^3} - t\dfrac{dy}{dt} + (t^2 - 1)y = e^t$	3
$y' - y = e^{2x}$	1
$y'' + y' = \cos t$	2

■ TERMINOLOGY

Recall from Section 5.2 that a ***differential equation*** is an equation involving one or more derivatives of an unknown function. In this section we will denote the unknown function by $y = y(x)$ unless the differential equation arises from an applied problem involving time, in which case we will denote it by $y = y(t)$. The ***order*** of a differential equation is the order of the highest derivative that it contains. Some examples are given in Table 8.1.1. The last two equations in that table are expressed in "prime" notation, which does not specify the independent variable explicitly. However, you will usually be able to tell from the equation itself or from the context in which it arises whether to interpret y' as dy/dx or dy/dt.

■ SOLUTIONS OF DIFFERENTIAL EQUATIONS

A function $y = y(x)$ is a ***solution*** of a differential equation on an open interval if the equation is satisfied identically on the interval when y and its derivatives are substituted into the equation. For example, $y = e^{2x}$ is a solution of the differential equation

$$\frac{dy}{dx} - y = e^{2x} \tag{1}$$

on the interval $(-\infty, +\infty)$, since substituting y and its derivative into the left side of this equation yields

$$\frac{dy}{dx} - y = \frac{d}{dx}[e^{2x}] - e^{2x} = 2e^{2x} - e^{2x} = e^{2x}$$

for all real values of x. However, this is not the only solution on $(-\infty, +\infty)$; for example, the function

$$y = e^{2x} + Ce^{x} \tag{2}$$

is also a solution for every real value of the constant C, since

$$\frac{dy}{dx} - y = \frac{d}{dx}[e^{2x} + Ce^{x}] - (e^{2x} + Ce^{x}) = (2e^{2x} + Ce^{x}) - (e^{2x} + Ce^{x}) = e^{2x}$$

After developing some techniques for solving equations such as (1), we will be able to show that *all* solutions of (1) on $(-\infty, +\infty)$ can be obtained by substituting values for the constant C in (2). On a given interval, a solution of a differential equation from which all solutions on that interval can be derived by substituting values for arbitrary constants is called a **general solution** of the equation on the interval. Thus (2) is a general solution of (1) on the interval $(-\infty, +\infty)$.

The graph of a solution of a differential equation is called an **integral curve** for the equation, so the general solution of a differential equation produces a family of integral curves corresponding to the different possible choices for the arbitrary constants. For example, Figure 8.1.1 shows some integral curves for (1), which were obtained by assigning values to the arbitrary constant in (2).

The first-order equation (1) has a single arbitrary constant in its general solution (2). Usually, the general solution of an nth-order differential equation will contain n arbitrary constants. This is plausible, since n integrations are needed to recover a function from its nth derivative.

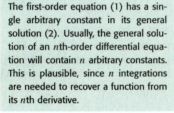

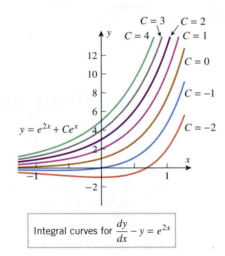

Integral curves for $\dfrac{dy}{dx} - y = e^{2x}$

▶ **Figure 8.1.1**

■ INITIAL-VALUE PROBLEMS

When an applied problem leads to a differential equation, there are usually conditions in the problem that determine specific values for the arbitrary constants. As a rule of thumb, it requires n conditions to determine values for all n arbitrary constants in the general solution of an nth-order differential equation (one condition for each constant). For a first-order equation, the single arbitrary constant can be determined by specifying the value of the unknown function $y(x)$ at an arbitrary x-value x_0, say $y(x_0) = y_0$. This is called an **initial condition**, and the problem of solving a first-order equation subject to an initial condition is called a **first-order initial-value problem**. Geometrically, the initial condition $y(x_0) = y_0$ has the effect of isolating the integral curve that passes through the point (x_0, y_0) from the complete family of integral curves.

▶ **Example 1**   The solution of the initial-value problem

$$\frac{dy}{dx} - y = e^{2x}, \quad y(0) = 3$$

can be obtained by substituting the initial condition $x = 0$, $y = 3$ in the general solution (2) to find C. We obtain

$$3 = e^0 + Ce^0 = 1 + C$$

Thus, $C = 2$, and the solution of the initial-value problem, which is obtained by substituting this value of C in (2), is

$$y = e^{2x} + 2e^x$$

Geometrically, this solution is realized as the integral curve in Figure 8.1.1 that passes through the point $(0, 3)$. ◀

Since many of the fundamental laws of the physical and social sciences involve rates of change, it should not be surprising that such laws are modeled by differential equations. Here are some examples of the modeling process.

■ UNINHIBITED POPULATION GROWTH

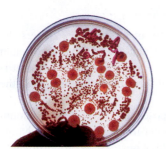

Hank Morgan–Rainbow/Getty Images
When the number of bacteria is small, an uninhibited population growth model can be used to model the growth of bacteria in a petri dish.

One of the simplest models of population growth is based on the observation that when populations (people, plants, bacteria, and fruit flies, for example) are not constrained by environmental limitations, they tend to grow at a rate that is proportional to the size of the population—the larger the population, the more rapidly it grows.

To translate this principle into a mathematical model, suppose that $y = y(t)$ denotes the population at time t. At each point in time, the rate of increase of the population with respect to time is dy/dt, so the assumption that the rate of growth is proportional to the population is described by the differential equation

$$\frac{dy}{dt} = ky \tag{3}$$

where k is a positive constant of proportionality that can usually be determined experimentally. Thus, if the population is known at some point in time, say $y = y_0$ at time $t = 0$, then a formula for the population $y(t)$ can be obtained by solving the initial-value problem

$$\frac{dy}{dt} = ky, \quad y(0) = y_0$$

■ INHIBITED POPULATION GROWTH; LOGISTIC MODELS

The uninhibited population growth model was predicated on the assumption that the population $y = y(t)$ was not constrained by the environment. While this assumption is reasonable as long as the size of the population is relatively small, environmental effects become increasingly important as the population grows. In general, populations grow within ecological systems that can only support a certain number of individuals; the number L of such individuals is called the ***carrying capacity*** of the system. When $y > L$, the population exceeds the capacity of the ecological system and tends to decrease toward L; when $y < L$, the population is below the capacity of the ecological system and tends to increase toward L; when $y = L$, the population is in balance with the capacity of the ecological system and tends to remain stable.

To translate this into a mathematical model, we must look for a differential equation in which $y > 0$, $L > 0$, and

$$\frac{dy}{dt} < 0 \quad \text{if} \quad \frac{y}{L} > 1, \qquad \frac{dy}{dt} > 0 \quad \text{if} \quad \frac{y}{L} < 1, \qquad \frac{dy}{dt} = 0 \quad \text{if} \quad \frac{y}{L} = 1$$

Moreover, when the population is far below the carrying capacity (i.e., $y/L \approx 0$), then the environmental constraints should have little effect, and the growth rate should behave like the uninhibited population model. Thus, we want

$$\frac{dy}{dt} \approx ky \quad \text{if} \quad \frac{y}{L} \approx 0$$

A simple differential equation that meets all of these requirements is

$$\frac{dy}{dt} = k\left(1 - \frac{y}{L}\right)y$$

where k is a positive constant of proportionality. Thus if k and L can be determined experimentally, and if the population is known at some point, say $y(0) = y_0$, then a formula for the population $y(t)$ can be determined by solving the initial-value problem

$$\frac{dy}{dt} = k\left(1 - \frac{y}{L}\right)y, \quad y(0) = y_0 \tag{4}$$

This theory of population growth is due to the Belgian mathematician P. F. Verhulst (1804–1849), who introduced it in 1838 and described it as "logistic growth."[*] Thus, the differential equation in (4) is called the *logistic differential equation*, and the growth model described by (4) is called the *logistic model*.

■ PHARMACOLOGY

When a drug (say, penicillin or aspirin) is administered to an individual, it enters the bloodstream and then is absorbed by the body over time. Medical research has shown that the amount of a drug that is present in the bloodstream tends to decrease at a rate that is proportional to the amount of the drug present—the more of the drug that is present in the bloodstream, the more rapidly it is absorbed by the body.

To translate this principle into a mathematical model, suppose that $y = y(t)$ is the amount of the drug present in the bloodstream at time t. At each point in time, the rate of change in y with respect to t is dy/dt, so the assumption that the rate of decrease is proportional to the amount y in the bloodstream translates into the differential equation

$$\frac{dy}{dt} = -ky \tag{5}$$

where k is a positive constant of proportionality that depends on the drug and can be determined experimentally. The negative sign is required because y decreases with time. Thus, if the initial dosage of the drug is known, say $y = y_0$ at time $t = 0$, then a formula for $y(t)$ can be obtained by solving the initial-value problem

$$\frac{dy}{dt} = -ky, \quad y(0) = y_0$$

■ SPREAD OF DISEASE

Suppose that a disease begins to spread in a population of L individuals. Logic suggests that at each point in time the rate at which the disease spreads will depend on how many individuals are already affected and how many are not—as more individuals are affected, the opportunity to spread the disease tends to increase, but at the same time there are fewer individuals who are not affected, so the opportunity to spread the disease tends to decrease. Thus, there are two conflicting influences on the rate at which the disease spreads.

---

[*] Verhulst's model fell into obscurity for nearly a hundred years because he did not have sufficient census data to test its validity. However, interest in the model was revived during the 1930s when biologists used it successfully to describe the growth of fruit fly and flour beetle populations. Verhulst himself used the model to predict that an upper limit of Belgium's population would be approximately 9,400,000. In 2006 the population was about 10,379,000.

To translate this into a mathematical model, suppose that $y = y(t)$ is the number of individuals who have the disease at time t, so of necessity the number of individuals who do not have the disease at time t is $L - y$. As the value of y increases, the value of $L - y$ decreases, so the conflicting influences of the two factors on the rate of spread dy/dt are taken into account by the differential equation

$$\frac{dy}{dt} = ky(L - y)$$

where k is a positive constant of proportionality that depends on the nature of the disease and the behavior patterns of the individuals and can be determined experimentally. Thus, if the number of affected individuals is known at some point in time, say $y = y_0$ at time $t = 0$, then a formula for $y(t)$ can be obtained by solving the initial-value problem

$$\frac{dy}{dt} = ky(L - y), \quad y(0) = y_0 \tag{6}$$

> Show that the model for the spread of disease can be viewed as a logistic model with constant of proportionality kL by rewriting (6) appropriately.

■ NEWTON'S LAW OF COOLING

If a hot object is placed into a cool environment, the object will cool at a rate proportional to the difference in temperature between the object and the environment. Similarly, if a cold object is placed into a warm environment, the object will warm at a rate that is again proportional to the difference in temperature between the object and the environment. Together, these observations comprise a result known as *Newton's Law of Cooling*. (Newton's Law of Cooling appeared previously in the exercises of Section 2.2 and was mentioned briefly in Section 5.8.) To translate this into a mathematical model, suppose that $T = T(t)$ is the temperature of the object at time t and that T_e is the temperature of the environment, which is assumed to be constant. Since the rate of change dT/dt is proportional to $T - T_e$, we have

$$\frac{dT}{dt} = k(T - T_e)$$

where k is a constant of proportionality. Moreover, since dT/dt is positive when $T < T_e$, and is negative when $T > T_e$, the sign of k must be *negative*. Thus if the temperature of the object is known at some time, say $T = T_0$ at time $t = 0$, then a formula for the temperature $T(t)$ can be obtained by solving the initial-value problem

$$\frac{dT}{dt} = k(T - T_e), \quad T(0) = T_0 \tag{7}$$

■ VIBRATIONS OF SPRINGS

We conclude this section with an engineering model that leads to a second-order differential equation.

As shown in Figure 8.1.2, consider a block of mass m attached to the end of a horizontal spring. Assume that the block is then set into vibratory motion by pulling the spring beyond its natural position and releasing it at time $t = 0$. We will be interested in finding a mathematical model that describes the vibratory motion of the block over time.

To translate this problem into mathematical form, we introduce a horizontal x-axis whose positive direction is to the right and whose origin is at the right end of the spring when the spring is in its natural position (Figure 8.1.3). Our goal is to find a model for the coordinate $x = x(t)$ of the point of attachment of the block to the spring as a function of time. In developing this model, we will assume that the only force on the mass m is the restoring force of the spring, and we will ignore the influence of other forces such as friction, air resistance, and so forth. Recall from Hooke's Law (Section 6.6) that when the connection point has coordinate $x(t)$, the restoring force is $-kx(t)$, where k is the spring constant. [The negative sign is due to the fact that the restoring force is to the left when $x(t)$ is positive, and the restoring force is to the right when $x(t)$ is negative.] It follows from Newton's

Natural position

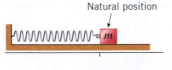

Stretched

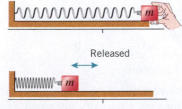

Released

▲ Figure 8.1.2

Natural position

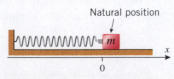

▲ Figure 8.1.3

Second Law of Motion [Equation (5) of Section 6.6] that this restoring force is equal to the product of the mass m and the acceleration d^2x/dt^2 of the mass. In other words, we have

$$m\frac{d^2x}{dt^2} = -kx$$

which is a second-order differential equation for x. If at time $t = 0$ the mass is released from rest at position $x(0) = x_0$, then a formula for $x(t)$ can be found by solving the initial-value problem

$$m\frac{d^2x}{dt^2} = -kx, \quad x(0) = x_0, \quad x'(0) = 0 \tag{8}$$

[If at time $t = 0$ the mass is given an initial velocity $v_0 \neq 0$, then the condition $x'(0) = 0$ must be replaced by $x'(0) = v_0$.]

✔ QUICK CHECK EXERCISES 8.1   (See page 568 for answers.)

1. Match each differential equation with its family of solutions.

(a) $x\dfrac{dy}{dx} = y$ _____

(b) $y'' = 4y$ _____

(c) $\dfrac{dy}{dx} = 2x$ _____

(d) $\dfrac{d^2y}{dx^2} = -4y$ _____

(i) $y = x^2 + C$

(ii) $y = C_1 \sin 2x + C_2 \cos 2x$

(iii) $y = C_1 e^{2x} + C_2 e^{-2x}$

(iv) $y = Cx$

2. If $y = C_1 e^{2x} + C_2 x e^{2x}$ is the general solution of a differential equation, then the order of the equation is _____, and a solution to the differential equation that satisfies the initial conditions $y(0) = 1$, $y'(0) = 4$ is given by $y = $ _____.

3. The graph of a differentiable function $y = y(x)$ passes through the point $(0, 1)$ and at every point $P(x, y)$ on the graph the tangent line is perpendicular to the line through P and the origin. Find an initial-value problem whose solution is $y(x)$.

4. A glass of ice water with a temperature of $36°\text{F}$ is placed in a room with a constant temperature of $68°\text{F}$. Assuming that Newton's Law of Cooling applies, find an initial-value problem whose solution is the temperature of water t minutes after it is placed in the room. [*Note:* The differential equation will involve a constant of proportionality.]

EXERCISE SET 8.1

1. Confirm that $y = 3e^{x^3}$ is a solution of the initial-value problem $y' = 3x^2 y$, $y(0) = 3$.

2. Confirm that $y = \frac{1}{4}x^4 + 2\cos x + 1$ is a solution of the initial-value problem $y' = x^3 - 2\sin x$, $y(0) = 3$.

3–4 State the order of the differential equation, and confirm that the functions in the given family are solutions. ■

3. (a) $(1 + x)\dfrac{dy}{dx} = y$; $\ y = c(1 + x)$

   (b) $y'' + y = 0$; $\ y = c_1 \sin t + c_2 \cos t$

4. (a) $2\dfrac{dy}{dx} + y = x - 1$; $\ y = ce^{-x/2} + x - 3$

   (b) $y'' - y = 0$; $\ y = c_1 e^t + c_2 e^{-t}$

5–8 True–False Determine whether the statement is true or false. Explain your answer. ■

5. The equation

$$\left(\frac{dy}{dx}\right)^2 = \frac{dy}{dx} + 2y$$

is an example of a second-order differential equation.

6. The differential equation

$$\frac{dy}{dx} = 2y + 1$$

has a solution that is constant.

7. We expect the general solution of the differential equation

$$\frac{d^3y}{dx^3} + 3\frac{d^2y}{dx^2} - \frac{dy}{dx} + 4y = 0$$

to involve three arbitrary constants.

8. If every solution to a differential equation can be expressed in the form $y = Ae^{x+b}$ for some choice of constants A and b, then the differential equation must be of second order.

9–14 In each part, verify that the functions are solutions of the differential equation by substituting the functions into the equation. ■

9. $y'' + y' - 2y = 0$

   (a) e^{-2x} and e^x

   (b) $c_1 e^{-2x} + c_2 e^x$ (c_1, c_2 constants)

10. $y'' - y' - 6y = 0$
(a) e^{-2x} and e^{3x}
(b) $c_1 e^{-2x} + c_2 e^{3x}$ (c_1, c_2 constants)

11. $y'' - 4y' + 4y = 0$
(a) e^{2x} and xe^{2x}
(b) $c_1 e^{2x} + c_2 xe^{2x}$ (c_1, c_2 constants)

12. $y'' - 8y' + 16y = 0$
(a) e^{4x} and xe^{4x}
(b) $c_1 e^{4x} + c_2 xe^{4x}$ (c_1, c_2 constants)

13. $y'' + 4y = 0$
(a) $\sin 2x$ and $\cos 2x$
(b) $c_1 \sin 2x + c_2 \cos 2x$ (c_1, c_2 constants)

14. $y'' + 4y' + 13y = 0$
(a) $e^{-2x} \sin 3x$ and $e^{-2x} \cos 3x$
(b) $e^{-2x}(c_1 \sin 3x + c_2 \cos 3x)$ (c_1, c_2 constants)

15–20 Use the results of Exercises 9–14 to find a solution to the initial-value problem. ■

15. $y'' + y' - 2y = 0$, $y(0) = -1$, $y'(0) = -4$

16. $y'' - y' - 6y = 0$, $y(0) = 1$, $y'(0) = 8$

17. $y'' - 4y' + 4y = 0$, $y(0) = 2$, $y'(0) = 2$

18. $y'' - 8y' + 16y = 0$, $y(0) = 1$, $y'(0) = 1$

19. $y'' + 4y = 0$, $y(0) = 1$, $y'(0) = 2$

20. $y'' + 4y' + 13y = 0$, $y(0) = -1$, $y'(0) = -1$

21–26 Find a solution to the initial-value problem. ■

21. $y' + 4x = 2$, $y(0) = 3$

22. $y'' + 6x = 0$, $y(0) = 1$, $y'(0) = 2$

23. $y' - y^2 = 0$, $y(1) = 2$ [*Hint:* Assume the solution has an inverse function $x = x(y)$. Find, and solve, a differential equation that involves $x'(y)$.]

24. $y' = 1 + y^2$, $y(0) = 0$ (See Exercise 23.)

25. $x^2 y' + 2xy = 0$, $y(1) = 2$ [*Hint:* Interpret the left-hand side of the equation as the derivative of a product of two functions.]

26. $xy' + y = e^x$, $y(1) = 1 + e$ (See Exercise 25.)

FOCUS ON CONCEPTS

27. (a) Suppose that a quantity $y = y(t)$ increases at a rate that is proportional to the square of the amount present, and suppose that at time $t = 0$, the amount present is y_0. Find an initial-value problem whose solution is $y(t)$.
(b) Suppose that a quantity $y = y(t)$ decreases at a rate that is proportional to the square of the amount present, and suppose that at a time $t = 0$, the amount present is y_0. Find an initial-value problem whose solution is $y(t)$.

28. (a) Suppose that a quantity $y = y(t)$ changes in such a way that $dy/dt = k\sqrt{y}$, where $k > 0$. Describe how y changes in words.

(b) Suppose that a quantity $y = y(t)$ changes in such a way that $dy/dt = -ky^3$, where $k > 0$. Describe how y changes in words.

29. (a) Suppose that a particle moves along an s-axis in such a way that its velocity $v(t)$ is always half of $s(t)$. Find a differential equation whose solution is $s(t)$.
(b) Suppose that an object moves along an s-axis in such a way that its acceleration $a(t)$ is always twice the velocity. Find a differential equation whose solution is $s(t)$.

30. Suppose that a body moves along an s-axis through a resistive medium in such a way that the velocity $v = v(t)$ decreases at a rate that is twice the square of the velocity.
(a) Find a differential equation whose solution is the velocity $v(t)$.
(b) Find a differential equation whose solution is the position $s(t)$.

31. Consider a solution $y = y(t)$ to the uninhibited population growth model.
(a) Use Equation (3) to explain why y will be an increasing function of t.
(b) Use Equation (3) to explain why the graph $y = y(t)$ will be concave up.

32. Consider the logistic model for population growth.
(a) Explain why there are two constant solutions to this model.
(b) For what size of the population will the population be growing most rapidly?

33. Consider the model for the spread of disease.
(a) Explain why there are two constant solutions to this model.
(b) For what size of the infected population is the disease spreading most rapidly?

34. Explain why there is exactly one constant solution to the Newton's Law of Cooling model.

35. Show that if c_1 and c_2 are any constants, the function

$$x = x(t) = c_1 \cos\left(\sqrt{\frac{k}{m}}\, t\right) + c_2 \sin\left(\sqrt{\frac{k}{m}}\, t\right)$$

is a solution to the differential equation for the vibrating spring. (The corresponding motion of the spring is referred to as *simple harmonic motion*.)

36. (a) Use the result of Exercise 35 to solve the initial-value problem in (8).
(b) Find the amplitude, period, and frequency of your answer to part (a), and interpret each of these in terms of the motion of the spring.

37. **Writing** Select one of the models in this section and write a paragraph that discusses conditions under which the model would not be appropriate. How might you modify the model to take those conditions into account?

8.2 SEPARATION OF VARIABLES

In this section we will discuss a method, called "separation of variables," that can be used to solve a large class of first-order differential equations of a particular form. We will use this method to investigate mathematical models for exponential growth and decay, including population models and carbon dating.

▪ FIRST-ORDER SEPARABLE EQUATIONS

We will now consider a method of solution that can often be applied to first-order equations that are expressible in the form

$$h(y)\frac{dy}{dx} = g(x) \tag{1}$$

Some writers define a separable equation to be one that can be written in the form $dy/dx = G(x)H(y)$. Explain why this is equivalent to our definition.

Such first-order equations are said to be *separable*. Some examples of separable equations are given in Table 8.2.1. The name "separable" arises from the fact that Equation (1) can be rewritten in the differential form

$$h(y)\,dy = g(x)\,dx \tag{2}$$

in which the expressions involving x and y appear on opposite sides. The process of rewriting (1) in form (2) is called *separating variables*.

Table 8.2.1

EQUATION	FORM (1)	$h(y)$	$g(x)$
$\dfrac{dy}{dx} = \dfrac{x}{y}$	$y\dfrac{dy}{dx} = x$	y	x
$\dfrac{dy}{dx} = x^2y^3$	$\dfrac{1}{y^3}\dfrac{dy}{dx} = x^2$	$\dfrac{1}{y^3}$	x^2
$\dfrac{dy}{dx} = y$	$\dfrac{1}{y}\dfrac{dy}{dx} = 1$	$\dfrac{1}{y}$	1
$\dfrac{dy}{dx} = y - \dfrac{y}{x}$	$\dfrac{1}{y}\dfrac{dy}{dx} = 1 - \dfrac{1}{x}$	$\dfrac{1}{y}$	$1 - \dfrac{1}{x}$

To motivate a method for solving separable equations, assume that $h(y)$ and $g(x)$ are continuous functions of their respective variables, and let $H(y)$ and $G(x)$ denote antiderivatives of $h(y)$ and $g(x)$, respectively. Consider the equation that results if we integrate both sides of (2), the left side with respect to y and the right side with respect to x. We then have

$$\int h(y)\,dy = \int g(x)\,dx \tag{3}$$

or, equivalently,

$$H(y) = G(x) + C \tag{4}$$

where C denotes a constant. We claim that a differentiable function $y = y(x)$ is a solution to (1) if and only if y satisfies Equation (4) for some choice of the constant C.

Suppose that $y = y(x)$ is a solution to (1). It then follows from the chain rule that

$$\frac{d}{dx}[H(y)] = \frac{dH}{dy}\frac{dy}{dx} = h(y)\frac{dy}{dx} = g(x) = \frac{dG}{dx} \tag{5}$$

Since the functions $H(y)$ and $G(x)$ have the same derivative with respect to x, they must differ by a constant (Theorem 4.8.3). It then follows that y satisfies (4) for an appropriate choice of C. Conversely, if $y = y(x)$ is defined implicitly by Equation (4), then implicit differentiation shows that (5) is satisfied, and thus $y(x)$ is a solution to (1) (Exercise 67). Because of this, it is common practice to refer to Equation (4) as the "solution" to (1).

In summary, we have the following procedure for solving (1), called *separation of variables*:

Separation of Variables

Step 1. Separate the variables in (1) by rewriting the equation in the differential form

$$h(y)\,dy = g(x)\,dx$$

Step 2. Integrate both sides of the equation in Step 1 (the left side with respect to y and the right side with respect to x):

$$\int h(y)\,dy = \int g(x)\,dx$$

Step 3. If $H(y)$ is any antiderivative of $h(y)$ and $G(x)$ is any antiderivative of $g(x)$, then the equation

$$H(y) = G(x) + C$$

will generally define a family of solutions implicitly. In some cases it may be possible to solve this equation explicitly for y.

▶ **Example 1** Solve the differential equation

$$\frac{dy}{dx} = -4xy^2$$

and then solve the initial-value problem

$$\frac{dy}{dx} = -4xy^2, \quad y(0) = 1$$

For an initial-value problem in which the differential equation is separable, you can either use the initial condition to solve for C, as in Example 1, or replace the indefinite integrals in Step 2 by definite integrals (Exercise 68).

Solution. For $y \neq 0$ we can write the differential equation in form (1) as

$$\frac{1}{y^2}\frac{dy}{dx} = -4x$$

Separating variables and integrating yields

$$\frac{1}{y^2}\,dy = -4x\,dx$$

$$\int \frac{1}{y^2}\,dy = \int -4x\,dx$$

or

$$-\frac{1}{y} = -2x^2 + C$$

Solving for y as a function of x, we obtain

$$y = \frac{1}{2x^2 - C}$$

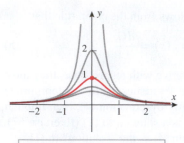

Integral curves for $\dfrac{dy}{dx} = -4xy^2$

▲ **Figure 8.2.1**

The solution of an initial-value problem in x and y can sometimes be expressed explicitly as a function of x [as in Formula (6) of Example 1], or explicitly as a function of y [as in Formula (8) of Example 2]. However, sometimes the solution cannot be expressed in either such form, so the only option is to express it implicitly as an equation in x and y.

The initial condition $y(0) = 1$ requires that $y = 1$ when $x = 0$. Substituting these values into our solution yields $C = -1$ (verify). Thus, a solution to the initial-value problem is

$$y = \frac{1}{2x^2 + 1} \tag{6}$$

Some integral curves and our solution of the initial-value problem are graphed in Figure 8.2.1. ◄

One aspect of our solution to Example 1 deserves special comment. Had the initial condition been $y(0) = 0$ instead of $y(0) = 1$, the method we used would have failed to yield a solution to the resulting initial-value problem (Exercise 25). This is due to the fact that we assumed $y \neq 0$ in order to rewrite the equation $dy/dx = -4xy^2$ in the form

$$\frac{1}{y^2}\frac{dy}{dx} = -4x$$

It is important to be aware of such assumptions when manipulating a differential equation algebraically.

▶ **Example 2** Solve the initial-value problem

$$(4y - \cos y)\frac{dy}{dx} - 3x^2 = 0, \quad y(0) = 0$$

Solution. We can write the differential equation in form (1) as

$$(4y - \cos y)\frac{dy}{dx} = 3x^2$$

Separating variables and integrating yields

$$(4y - \cos y)\,dy = 3x^2\,dx$$

$$\int (4y - \cos y)\,dy = \int 3x^2\,dx$$

or

$$2y^2 - \sin y = x^3 + C \tag{7}$$

For the initial-value problem, the initial condition $y(0) = 0$ requires that $y = 0$ if $x = 0$. Substituting these values into (7) to determine the constant of integration yields $C = 0$ (verify). Thus, the solution of the initial-value problem is

$$2y^2 - \sin y = x^3$$

or

$$x = \sqrt[3]{2y^2 - \sin y} \quad ◄ \tag{8}$$

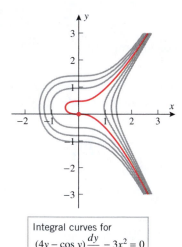

Integral curves for

$(4y - \cos y)\dfrac{dy}{dx} - 3x^2 = 0$

▲ **Figure 8.2.2**

Some integral curves and the solution of the initial-value problem in Example 2 are graphed in Figure 8.2.2.

Initial-value problems often result from geometrical questions, as in the following example.

▶ **Example 3** Find a curve in the xy-plane that passes through $(0, 3)$ and whose tangent line at a point (x, y) has slope $2x/y^2$.

TECHNOLOGY MASTERY

Some computer algebra systems can graph implicit equations. Figure 8.2.2 shows the graphs of (7) for $C = 0, \pm 1, \pm 2$, and ± 3. If you have a CAS that can graph implicit equations, try to duplicate this figure.

Solution. Since the slope of the tangent line is dy/dx, we have

$$\frac{dy}{dx} = \frac{2x}{y^2} \tag{9}$$

and, since the curve passes through $(0, 3)$, we have the initial condition

$$y(0) = 3$$

Equation (9) is separable and can be written as

$$y^2 \, dy = 2x \, dx$$

so

$$\int y^2 \, dy = \int 2x \, dx \quad \text{or} \quad \tfrac{1}{3} y^3 = x^2 + C$$

It follows from the initial condition that $y = 3$ if $x = 0$. Substituting these values into the last equation yields $C = 9$ (verify), so the equation of the desired curve is

$$\tfrac{1}{3} y^3 = x^2 + 9 \quad \text{or} \quad y = (3x^2 + 27)^{1/3} \blacktriangleleft$$

■ EXPONENTIAL GROWTH AND DECAY MODELS

The population growth and pharmacology models developed in Section 8.1 are examples of a general class of models called *exponential models*. In general, exponential models arise in situations where a quantity increases or decreases at a rate that is proportional to the amount of the quantity present. More precisely, we make the following definition.

8.2.1 DEFINITION A quantity $y = y(t)$ is said to have an *exponential growth model* if it increases at a rate that is proportional to the amount of the quantity present, and it is said to have an *exponential decay model* if it decreases at a rate that is proportional to the amount of the quantity present. Thus, for an exponential growth model, the quantity $y(t)$ satisfies an equation of the form

$$\frac{dy}{dt} = ky \quad (k > 0) \tag{10}$$

and for an exponential decay model, the quantity $y(t)$ satisfies an equation of the form

$$\frac{dy}{dt} = -ky \quad (k > 0) \tag{11}$$

The constant k is called the *growth constant* or the *decay constant*, as appropriate.

Equations (10) and (11) are separable since they have the form of (1), but with t rather than x as the independent variable. To illustrate how these equations can be solved, suppose that a positive quantity $y = y(t)$ has an exponential growth model and that we know the amount of the quantity at some point in time, say $y = y_0$ when $t = 0$. Thus, a formula for $y(t)$ can be obtained by solving the initial-value problem

$$\frac{dy}{dt} = ky, \quad y(0) = y_0$$

Separating variables and integrating yields

$$\int \frac{1}{y} \, dy = \int k \, dt$$

or (since $y > 0$)

$$\ln y = kt + C \tag{12}$$

The initial condition implies that $y = y_0$ when $t = 0$. Substituting these values in (12) yields $C = \ln y_0$ (verify). Thus,

$$\ln y = kt + \ln y_0$$

from which it follows that

$$y = e^{\ln y} = e^{kt + \ln y_0}$$

or, equivalently,

$$y = y_0 e^{kt} \tag{13}$$

We leave it for you to show that if $y = y(t)$ has an exponential decay model, and if $y(0) = y_0$, then

$$y = y_0 e^{-kt} \tag{14}$$

■ INTERPRETING THE GROWTH AND DECAY CONSTANTS

The significance of the constant k in Formulas (13) and (14) can be understood by reexamining the differential equations that gave rise to these formulas. For example, in the case of the exponential growth model, Equation (10) can be rewritten as

$$k = \frac{dy/dt}{y} \tag{15}$$

It is standard practice in applications to call (15) the *growth rate*, even though it is misleading (the growth rate is dy/dt). However, the practice is so common that we will follow it here.

which states that the growth rate as a fraction of the entire population remains constant over time, and this constant is k. For this reason, k is called the ***relative growth rate*** of the population. It is usual to express the relative growth rate as a percentage. Thus, a relative growth rate of 3% per unit of time in an exponential growth model means that $k = 0.03$. Similarly, the constant k in an exponential decay model is called the ***relative decay rate***.

▶ **Example 4** According to United Nations data, the world population in 1998 was approximately 5.9 billion and growing at a rate of about 1.33% per year. Assuming an exponential growth model, estimate the world population at the beginning of the year 2023.

Solution. We assume that the population at the beginning of 1998 was 5.9 billion and let

$t = $ time elapsed from the beginning of 1998 (in years)

$y = $ world population (in billions)

Since the beginning of 1998 corresponds to $t = 0$, it follows from the given data that

$$y_0 = y(0) = 5.9 \text{ (billion)}$$

Since the growth rate is 1.33% ($k = 0.0133$), it follows from (13) that the world population at time t will be

$$y(t) = y_0 e^{kt} = 5.9 e^{0.0133t} \tag{16}$$

In Example 4 the growth rate was given, so there was no need to calculate it. If the growth rate or decay rate is unknown, then it can be calculated using the initial condition and the value of y at another point in time (Exercise 44).

Since the beginning of the year 2023 corresponds to an elapsed time of $t = 25$ years ($2023 - 1998 = 25$), it follows from (16) that the world population by the year 2023 will be

$$y(25) = 5.9 e^{0.0133(25)} \approx 8.2$$

which is a population of approximately 8.2 billion. ◀

■ DOUBLING TIME AND HALF-LIFE

If a quantity y has an exponential growth model, then the time required for the original size to double is called the ***doubling time***, and if y has an exponential decay model, then the time required for the original size to reduce by half is called the ***half-life***. As it turns out, doubling time and half-life depend only on the growth or decay rate and not on the amount present initially. To see why this is so, suppose that $y = y(t)$ has an exponential growth model

$$y = y_0 e^{kt} \tag{17}$$

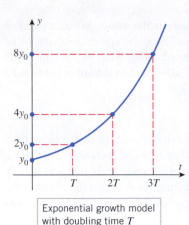

Exponential growth model
with doubling time T

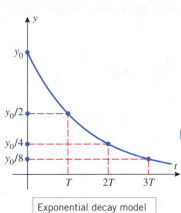

Exponential decay model
with half-life T

▲ Figure 8.2.3

and let T denote the amount of time required for y to double in size. Thus, at time $t = T$ the value of y will be $2y_0$, and hence from (17)

$$2y_0 = y_0 e^{kT} \quad \text{or} \quad e^{kT} = 2$$

Taking the natural logarithm of both sides yields $kT = \ln 2$, which implies that the doubling time is

$$T = \frac{1}{k} \ln 2 \tag{18}$$

We leave it as an exercise to show that Formula (18) also gives the half-life of an exponential decay model. Observe that this formula does not involve the initial amount y_0, so that in an exponential growth or decay model, the quantity y doubles (or reduces by half) every T units (Figure 8.2.3).

▶ **Example 5** It follows from (18) that with a continued growth rate of 1.33% per year, the doubling time for the world population will be

$$T = \frac{1}{0.0133} \ln 2 \approx 52.116$$

or approximately 52 years. Thus, with a continued 1.33% annual growth rate the population of 5.9 billion in 1998 will double to 11.8 billion by the year 2050 and will double again to 23.6 billion by 2102. ◀

■ RADIOACTIVE DECAY

It is a fact of physics that radioactive elements disintegrate spontaneously in a process called *radioactive decay*. Experimentation has shown that the rate of disintegration is proportional to the amount of the element present, which implies that the amount $y = y(t)$ of a radioactive element present as a function of time has an exponential decay model.

Every radioactive element has a specific half-life; for example, the half-life of radioactive carbon-14 is about 5730 years. Thus, from (18), the decay constant for this element is

$$k = \frac{1}{T} \ln 2 = \frac{\ln 2}{5730} \approx 0.000121$$

and this implies that if there are y_0 units of carbon-14 present at time $t = 0$, then the number of units present after t years will be approximately

$$y(t) = y_0 e^{-0.000121t} \tag{19}$$

▶ **Example 6** If 100 grams of radioactive carbon-14 are stored in a cave for 1000 years, how many grams will be left at that time?

Solution. From (19) with $y_0 = 100$ and $t = 1000$, we obtain

$$y(1000) = 100e^{-0.000121(1000)} = 100e^{-0.121} \approx 88.6$$

Thus, about 88.6 grams will be left. ◀

■ CARBON DATING

When the nitrogen in the Earth's upper atmosphere is bombarded by cosmic radiation, the radioactive element carbon-14 is produced. This carbon-14 combines with oxygen to form carbon dioxide, which is ingested by plants, which in turn are eaten by animals. In this way all living plants and animals absorb quantities of radioactive carbon-14. In 1947 the American nuclear scientist W. F. Libby[*] proposed the theory that the percentage of

[*]W. F. Libby, "Radiocarbon Dating," *American Scientist*, Vol. 44, 1956, pp. 98–112.

carbon-14 in the atmosphere and in living tissues of plants is the same. When a plant or animal dies, the carbon-14 in the tissue begins to decay. Thus, the age of an artifact that contains plant or animal material can be estimated by determining what percentage of its original carbon-14 content remains. Various procedures, called **carbon dating** or **carbon-14 dating**, have been developed for measuring this percentage.

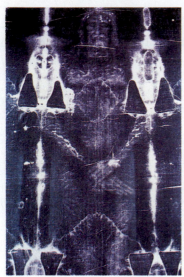

Patrick Mesner/Liaison Agency, Inc./Getty Images
The Shroud of Turin

▶ **Example 7**   In 1988 the Vatican authorized the British Museum to date a cloth relic known as the Shroud of Turin, possibly the burial shroud of Jesus of Nazareth. This cloth, which first surfaced in 1356, contains the negative image of a human body that was widely believed to be that of Jesus. The report of the British Museum showed that the fibers in the cloth contained between 92% and 93% of their original carbon-14. Use this information to estimate the age of the shroud.

Solution.   From (19), the fraction of the original carbon-14 that remains after t years is

$$\frac{y(t)}{y_0} = e^{-0.000121t}$$

Taking the natural logarithm of both sides and solving for t, we obtain

$$t = -\frac{1}{0.000121} \ln\left(\frac{y(t)}{y_0}\right)$$

Thus, taking $y(t)/y_0$ to be 0.93 and 0.92, we obtain

$$t = -\frac{1}{0.000121} \ln(0.93) \approx 600$$

$$t = -\frac{1}{0.000121} \ln(0.92) \approx 689$$

This means that when the test was done in 1988, the shroud was between 600 and 689 years old, thereby placing its origin between 1299 A.D. and 1388 A.D. Thus, if one accepts the validity of carbon-14 dating, the Shroud of Turin cannot be the burial shroud of Jesus of Nazareth. ◀

✔**QUICK CHECK EXERCISES 8.2**   *(See page 579 for answers.)*

1. Solve the first-order separable equation

$$h(y)\frac{dy}{dx} = g(x)$$

by completing the following steps:

Step 1. Separate the variables by writing the equation in the differential form _____.

Step 2. Integrate both sides of the equation in Step 1: _____.

Step 3. If $H(y)$ is any antiderivative of $h(y)$, $G(x)$ is any antiderivative of $g(x)$, and C is an unspecified constant, then, as suggested by Step 2, the equation _____ will generally define a family of solutions to $h(y)\,dy/dx = g(x)$ implicitly.

2. Suppose that a quantity $y = y(t)$ has an exponential growth model with growth constant $k > 0$.
   (a) $y(t)$ satisfies a first-order differential equation of the form $dy/dt = $ _____.
   (b) In terms of k, the doubling time of the quantity is _____.
   (c) If $y_0 = y(0)$ is the initial amount of the quantity, then an explicit formula for $y(t)$ is given by $y(t) = $ _____.

3. Suppose that a quantity $y = y(t)$ has an exponential decay model with decay constant $k > 0$.
   (a) $y(t)$ satisfies a first-order differential equation of the form $dy/dt = $ _____.
   (b) In terms of k, the half-life of the quantity is _____.
   (c) If $y_0 = y(0)$ is the initial amount of the quantity, then an explicit formula for $y(t)$ is given by $y(t) = $ _____.

4. The initial-value problem

$$\frac{dy}{dx} = -\frac{x}{y}, \quad y(0) = 1$$

has solution $y(x) = \underline{\qquad}$.

EXERCISE SET 8.2   Graphing Utility [c] CAS

1–10 Solve the differential equation by separation of variables. Where reasonable, express the family of solutions as explicit functions of x. ■

1. $\dfrac{dy}{dx} = \dfrac{y}{x}$

2. $\dfrac{dy}{dx} = 2(1 + y^2)x$

3. $\dfrac{\sqrt{1+x^2}}{1+y}\dfrac{dy}{dx} = -x$

4. $(1+x^4)\dfrac{dy}{dx} = \dfrac{x^3}{y}$

5. $(2 + 2y^2)y' = e^x y$

6. $y' = -xy$

7. $e^{-y}\sin x - y'\cos^2 x = 0$

8. $y' - (1+x)(1+y^2) = 0$

9. $\dfrac{dy}{dx} - \dfrac{y^2 - y}{\sin x} = 0$

10. $y - \dfrac{dy}{dx}\sec x = 0$

11–14 Solve the initial-value problem by separation of variables. ■

11. $y' = \dfrac{3x^2}{2y + \cos y}, \quad y(0) = \pi$

12. $y' - xe^y = 2e^y, \quad y(0) = 0$

13. $\dfrac{dy}{dt} = \dfrac{2t + 1}{2y - 2}, \quad y(0) = -1$

14. $y'\cosh^2 x - y\cosh 2x = 0, \quad y(0) = 3$

15. (a) Sketch some typical integral curves of the differential equation $y' = y/2x$.
(b) Find an equation for the integral curve that passes through the point $(2, 1)$.

16. (a) Sketch some typical integral curves of the differential equation $y' = -x/y$.
(b) Find an equation for the integral curve that passes through the point $(3, 4)$.

17–18 Solve the differential equation and then use a graphing utility to generate five integral curves for the equation. ■

17. $(x^2 + 4)\dfrac{dy}{dx} + xy = 0$

18. $(\cos y)y' = \cos x$

19–20 Solve the differential equation. If you have a CAS with implicit plotting capability, use the CAS to generate five integral curves for the equation. ■

19. $y' = \dfrac{x^2}{1 - y^2}$

20. $y' = \dfrac{y}{1 + y^2}$

21–24 True–False Determine whether the statement is true or false. Explain your answer. ■

21. Every differential equation of the form $y' = f(y)$ is separable.

22. A differential equation of the form

$$h(x)\frac{dy}{dx} = g(y)$$

is not separable.

23. If a radioactive element has a half-life of 1 minute, and if a container holds 32 g of the element at 1:00 P.M., then the amount remaining at 1:05 P.M. will be 1 g.

24. If a population is growing exponentially, then the time it takes the population to quadruple is independent of the size of the population.

25. Suppose that the initial condition in Example 1 had been $y(0) = 0$. Show that none of the solutions generated in Example 1 satisfy this initial condition, and then solve the initial-value problem

$$\frac{dy}{dx} = -4xy^2, \quad y(0) = 0$$

Why does the method of Example 1 fail to produce this particular solution?

26. Find all ordered pairs (x_0, y_0) such that if the initial condition in Example 1 is replaced by $y(x_0) = y_0$, the solution of the resulting initial-value problem is defined for all real numbers.

27. Find an equation of a curve with x-intercept 2 whose tangent line at any point (x, y) has slope xe^{-y}.

28. Use a graphing utility to generate a curve that passes through the point $(1, 1)$ and whose tangent line at (x, y) is perpendicular to the line through (x, y) with slope $-2y/(3x^2)$.

29. Suppose that an initial population of 10,000 bacteria grows exponentially at a rate of 2% per hour and that $y = y(t)$ is the number of bacteria present t hours later.
(a) Find an initial-value problem whose solution is $y(t)$.
(b) Find a formula for $y(t)$.
(c) How long does it take for the initial population of bacteria to double?
(d) How long does it take for the population of bacteria to reach 45,000?

30. A cell of the bacterium *E. coli* divides into two cells every 20 minutes when placed in a nutrient culture. Let $y = y(t)$ be the number of cells that are present t minutes after a single cell is placed in the culture. Assume that the growth of the bacteria is approximated by an exponential growth model.
(a) Find an initial-value problem whose solution is $y(t)$.
(b) Find a formula for $y(t)$. *(cont.)*

(c) How many cells are present after 2 hours?

(d) How long does it take for the number of cells to reach 1,000,000?

31. Radon-222 is a radioactive gas with a half-life of 3.83 days. This gas is a health hazard because it tends to get trapped in the basements of houses, and many health officials suggest that homeowners seal their basements to prevent entry of the gas. Assume that 5.0×10^7 radon atoms are trapped in a basement at the time it is sealed and that $y(t)$ is the number of atoms present t days later.

(a) Find an initial-value problem whose solution is $y(t)$.

(b) Find a formula for $y(t)$.

(c) How many atoms will be present after 30 days?

(d) How long will it take for 90% of the original quantity of gas to decay?

32. Polonium-210 is a radioactive element with a half-life of 140 days. Assume that 10 milligrams of the element are placed in a lead container and that $y(t)$ is the number of milligrams present t days later.

(a) Find an initial-value problem whose solution is $y(t)$.

(b) Find a formula for $y(t)$.

(c) How many milligrams will be present after 10 weeks?

(d) How long will it take for 70% of the original sample to decay?

33. Suppose that 100 fruit flies are placed in a breeding container that can support at most 10,000 flies. Assuming that the population grows exponentially at a rate of 2% per day, how long will it take for the container to reach capacity?

34. Suppose that the town of Grayrock had a population of 10,000 in 1998 and a population of 12,000 in 2003. Assuming an exponential growth model, in what year will the population reach 20,000?

35. A scientist wants to determine the half-life of a certain radioactive substance. She determines that in exactly 5 days a 10.0-milligram sample of the substance decays to 3.5 milligrams. Based on these data, what is the half-life?

36. Suppose that 30% of a certain radioactive substance decays in 5 years.

(a) What is the half-life of the substance in years?

(b) Suppose that a certain quantity of this substance is stored in a cave. What percentage of it will remain after t years?

FOCUS ON CONCEPTS

37. (a) Make a conjecture about the effect on the graphs of $y = y_0 e^{kt}$ and $y = y_0 e^{-kt}$ of varying k and keeping y_0 fixed. Confirm your conjecture with a graphing utility.

(b) Make a conjecture about the effect on the graphs of $y = y_0 e^{kt}$ and $y = y_0 e^{-kt}$ of varying y_0 and keeping k fixed. Confirm your conjecture with a graphing utility.

38. (a) What effect does increasing y_0 and keeping k fixed have on the doubling time or half-life of an exponential model? Justify your answer.

(b) What effect does increasing k and keeping y_0 fixed have on the doubling time and half-life of an exponential model? Justify your answer.

39. (a) There is a trick, called the **Rule of 70**, that can be used to get a quick estimate of the doubling time or half-life of an exponential model. According to this rule, the doubling time or half-life is roughly 70 divided by the percentage growth or decay rate. For example, we showed in Example 5 that with a continued growth rate of 1.33% per year the world population would double every 52 years. This result agrees with the Rule of 70, since $70/1.33 \approx 52.6$. Explain why this rule works.

(b) Use the Rule of 70 to estimate the doubling time of a population that grows exponentially at a rate of 1% per year.

(c) Use the Rule of 70 to estimate the half-life of a population that decreases exponentially at a rate of 3.5% per hour.

(d) Use the Rule of 70 to estimate the growth rate that would be required for a population growing exponentially to double every 10 years.

40. Find a formula for the tripling time of an exponential growth model.

41. In 1950, a research team digging near Folsom, New Mexico, found charred bison bones along with some leaf-shaped projectile points (called the "Folsom points") that had been made by a Paleo-Indian hunting culture. It was clear from the evidence that the bison had been cooked and eaten by the makers of the points, so that carbon-14 dating of the bones made it possible for the researchers to determine when the hunters roamed North America. Tests showed that the bones contained between 27% and 30% of their original carbon-14. Use this information to show that the hunters lived roughly between 9000 B.C. and 8000 B.C.

42. (a) Use a graphing utility to make a graph of p_{rem} versus t, where p_{rem} is the percentage of carbon-14 that remains in an artifact after t years.

(b) Use the graph to estimate the percentage of carbon-14 that would have to have been present in the 1988 test of the Shroud of Turin for it to have been the burial shroud of Jesus of Nazareth (see Example 7).

43. (a) It is currently accepted that the half-life of carbon-14 might vary ± 40 years from its nominal value of 5730 years. Does this variation make it possible that the Shroud of Turin dates to the time of Jesus of Nazareth (see Example 7)?

(b) Review the subsection of Section 3.5 entitled Error Propagation, and then estimate the percentage error that results in the computed age of an artifact from an $r\%$ error in the half-life of carbon-14.

44. Suppose that a quantity y has an exponential growth model $y = y_0 e^{kt}$ or an exponential decay model $y = y_0 e^{-kt}$, and it is known that $y = y_1$ if $t = t_1$. In each case find a formula for k in terms of y_0, y_1, and t_1, assuming that $t_1 \neq 0$.

45. (a) Show that if a quantity $y = y(t)$ has an exponential model, and if $y(t_1) = y_1$ and $y(t_2) = y_2$, then the doubling time or the half-life T is

$$T = \left| \frac{(t_2 - t_1) \ln 2}{\ln(y_2/y_1)} \right|$$

(b) In a certain 1-hour period the number of bacteria in a colony increases by 25%. Assuming an exponential growth model, what is the doubling time for the colony?

46. Suppose that P dollars is invested at an annual interest rate of $r \times 100\%$. If the accumulated interest is credited to the account at the end of the year, then the interest is said to be *compounded annually*; if it is credited at the end of each 6-month period, then it is said to be *compounded semiannually*; and if it is credited at the end of each 3-month period, then it is said to be *compounded quarterly*. The more frequently the interest is compounded, the better it is for the investor since more of the interest is itself earning interest.

(a) Show that if interest is compounded n times a year at equally spaced intervals, then the value A of the investment after t years is

$$A = P \left(1 + \frac{r}{n}\right)^{nt}$$

(b) One can imagine interest to be compounded each day, each hour, each minute, and so forth. Carried to the limit one can conceive of interest compounded at each instant of time; this is called *continuous compounding*. Thus, from part (a), the value A of P dollars after t years when invested at an annual rate of $r \times 100\%$, compounded continuously, is

$$A = \lim_{n \to +\infty} P \left(1 + \frac{r}{n}\right)^{nt}$$

Use the fact that $\lim_{x \to 0} (1 + x)^{1/x} = e$ to prove that $A = Pe^{rt}$.

(c) Use the result in part (b) to show that money invested at continuous compound interest increases at a rate proportional to the amount present.

47. (a) If $1000 is invested at 8% per year compounded continuously (Exercise 46), what will the investment be worth after 5 years?

(b) If it is desired that an investment at 8% per year compounded continuously should have a value of $10,000 after 10 years, how much should be invested now?

(c) How long does it take for an investment at 8% per year compounded continuously to double in value?

48. What is the effective annual interest rate for an interest rate of $r\%$ per year compounded continuously?

49. Assume that $y = y(t)$ satisfies the logistic equation with $y_0 = y(0)$ the initial value of y.
(a) Use separation of variables to derive the solution

$$y = \frac{y_0 L}{y_0 + (L - y_0)e^{-kt}}$$

(b) Use part (a) to show that $\lim_{t \to +\infty} y(t) = L$.

50. Use your answer to Exercise 49 to derive a solution to the model for the spread of disease [Equation (6) of Section 8.1].

51. The graph of a solution to the logistic equation is known as a *logistic curve*, and if $y_0 > 0$, it has one of four general shapes, depending on the relationship between y_0 and L. In each part, assume that $k = 1$ and use a graphing utility to plot a logistic curve satisfying the given condition.
(a) $y_0 > L$ **(b)** $y_0 = L$
(c) $L/2 \leq y_0 < L$ **(d)** $0 < y_0 < L/2$

52–53 The graph of a logistic model

$$y = \frac{y_0 L}{y_0 + (L - y_0)e^{-kt}}$$

is shown. Estimate y_0, L, and k.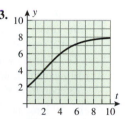

52.

53.

54. Plot a solution to the initial-value problem

$$\frac{dy}{dt} = 0.98 \left(1 - \frac{y}{5}\right) y, \quad y_0 = 1$$

55. Suppose that the growth of a population $y = y(t)$ is given by the logistic equation

$$y = \frac{60}{5 + 7e^{-t}}$$

(a) What is the population at time $t = 0$?
(b) What is the carrying capacity L?
(c) What is the constant k?
(d) When does the population reach half of the carrying capacity?
(e) Find an initial-value problem whose solution is $y(t)$.

56. Suppose that the growth of a population $y = y(t)$ is given by the logistic equation

$$y = \frac{1000}{1 + 999e^{-0.9t}}$$

(a) What is the population at time $t = 0$?
(b) What is the carrying capacity L?
(c) What is the constant k?

(cont.)

(d) When does the population reach 75% of the carrying capacity?

(e) Find an initial-value problem whose solution is $y(t)$.

57. Suppose that a university residence hall houses 1000 students. Following the semester break, 20 students in the hall return with the flu, and 5 days later 35 students have the flu.

(a) Use the result of Exercise 50 to find the number of students who will have the flu t days after returning to school.

(b) Make a table that illustrates how the flu spreads day to day over a 2-week period.

(c) Use a graphing utility to generate a graph that illustrates how the flu spreads over a 2-week period.

58. Suppose that at time $t = 0$ an object with temperature T_0 is placed in a room with constant temperature T_a. If $T_0 < T_a$, then the temperature of the object will increase, and if $T_0 > T_a$, then the temperature will decrease. Assuming that Newton's Law of Cooling applies, show that in both cases the temperature $T(t)$ at time t is given by

$$T(t) = T_a + (T_0 - T_a)e^{-kt}$$

where k is a positive constant.

59. A cup of water with a temperature of $95°C$ is placed in a room with a constant temperature of $21°C$.

(a) Assuming that Newton's Law of Cooling applies, use the result of Exercise 58 to find the temperature of the water t minutes after it is placed in the room. [*Note:* The solution will involve a constant of proportionality.]

(b) How many minutes will it take for the water to reach a temperature of $51°C$ if it cools to $85°C$ in 1 minute?

60. A glass of lemonade with a temperature of $40°F$ is placed in a room with a constant temperature of $70°F$, and 1 hour later its temperature is $52°F$. Show that t hours after the lemonade is placed in the room its temperature is approximated by $T = 70 - 30e^{-0.5t}$.

61. A rocket, fired upward from rest at time $t = 0$, has an initial mass of m_0 (including its fuel). Assuming that the fuel is consumed at a constant rate k, the mass m of the rocket, while fuel is being burned, will be given by $m = m_0 - kt$. It can be shown that if air resistance is neglected and the fuel gases are expelled at a constant speed c relative to the rocket, then the velocity v of the rocket will satisfy the equation

$$m \frac{dv}{dt} = ck - mg$$

where g is the acceleration due to gravity.

(a) Find $v(t)$ keeping in mind that the mass m is a function of t.

(b) Suppose that the fuel accounts for 80% of the initial mass of the rocket and that all of the fuel is consumed in 100 s. Find the velocity of the rocket in meters per second at the instant the fuel is exhausted. [*Note:* Take $g = 9.8 \text{ m/s}^2$ and $c = 2500 \text{ m/s}$.]

62. A bullet of mass m, fired straight up with an initial velocity of v_0, is slowed by the force of gravity and a drag force of air resistance kv^2, where k is a positive constant. As the bullet moves upward, its velocity v satisfies the equation

$$m \frac{dv}{dt} = -(kv^2 + mg)$$

where g is the constant acceleration due to gravity.

(a) Show that if $x = x(t)$ is the height of the bullet above the barrel opening at time t, then

$$mv \frac{dv}{dx} = -(kv^2 + mg)$$

(b) Express x in terms of v given that $x = 0$ when $v = v_0$.

(c) Assuming that

$$v_0 = 988 \text{ m/s}, \quad g = 9.8 \text{ m/s}^2$$
$$m = 3.56 \times 10^{-3} \text{ kg}, \quad k = 7.3 \times 10^{-6} \text{ kg/m}$$

use the result in part (b) to find out how high the bullet rises. [*Hint:* Find the velocity of the bullet at its highest point.]

63–64 Suppose that a tank containing a liquid is vented to the air at the top and has an outlet at the bottom through which the liquid can drain. It follows from *Torricelli's law* in physics that if the outlet is opened at time $t = 0$, then at each instant the depth of the liquid $h(t)$ and the area $A(h)$ of the liquid's surface are related by

$$A(h) \frac{dh}{dt} = -k\sqrt{h}$$

where k is a positive constant that depends on such factors as the viscosity of the liquid and the cross-sectional area of the outlet. Use this result in these exercises, assuming that h is in feet, $A(h)$ is in square feet, and t is in seconds. ■

63. Suppose that the cylindrical tank in the accompanying figure is filled to a depth of 4 feet at time $t = 0$ and that the constant in Torricelli's law is $k = 0.025$.

(a) Find $h(t)$.

(b) How many minutes will it take for the tank to drain completely?

64. Follow the directions of Exercise 63 for the cylindrical tank in the accompanying figure, assuming that the tank is filled to a depth of 4 feet at time $t = 0$ and that the constant in Torricelli's law is $k = 0.025$.

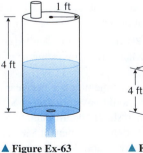

▲ Figure Ex-63

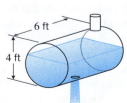

▲ Figure Ex-64

65. Suppose that a particle moving along the x-axis encounters a resisting force that results in an acceleration of $a = dv/dt = -\frac{1}{32}v^2$. If $x = 0$ cm and $v = 128$ cm/s at time $t = 0$, find the velocity v and position x as a function of t for $t \geq 0$.

66. Suppose that a particle moving along the x-axis encounters a resisting force that results in an acceleration of $a = dv/dt = -0.02\sqrt{v}$. Given that $x = 0$ cm and $v = 9$ cm/s at time $t = 0$, find the velocity v and position x as a function of t for $t \geq 0$.

FOCUS ON CONCEPTS

67. Use implicit differentiation to prove that any differentiable function defined implicitly by Equation (4) will be a solution to (1).

68. Prove that a solution to the initial-value problem

$$h(y)\frac{dy}{dx} = g(x), \quad y(x_0) = y_0$$

is defined implicitly by the equation

$$\int_{y_0}^{y} h(r)\,dr = \int_{x_0}^{x} g(s)\,ds$$

69. Let L denote a tangent line at (x, y) to a solution of Equation (1), and let (x_1, y_1), (x_2, y_2) denote any two points on L. Prove that Equation (2) is satisfied by $dy = \Delta y = y_2 - y_1$ and $dx = \Delta x = x_2 - x_1$.

70. Writing A student objects to the method of separation of variables because it often produces an equation in x and y instead of an explicit function $y = f(x)$. Discuss the pros and cons of this student's position.

71. Writing A student objects to Step 2 in the method of separation of variables because one side of the equation is integrated with respect to x while the other side is integrated with respect to y. Answer this student's objection. [*Hint*: Recall the method of integration by substitution.]

✔**QUICK CHECK ANSWERS 8.2**

1. Step 1: $h(y)\,dy = g(x)\,dx$; Step 2: $\displaystyle\int h(y)\,dy = \int g(x)\,dx$; Step 3: $H(y) = G(x) + C$ **2.** (a) ky (b) $\dfrac{\ln 2}{k}$ (c) $y_0 e^{kt}$

3. (a) $-ky$ (b) $\dfrac{\ln 2}{k}$ (c) $y_0 e^{-kt}$ **4.** $y = \sqrt{1 - x^2}$

8.3 **SLOPE FIELDS; EULER'S METHOD**

In this section we will reexamine the concept of a slope field and we will discuss a method for approximating solutions of first-order equations numerically. Numerical approximations are important in cases where the differential equation cannot be solved exactly.

■ FUNCTIONS OF TWO VARIABLES

We will be concerned here with first-order equations that are expressed with the derivative by itself on one side of the equation. For example,

$$y' = x^3 \quad \text{and} \quad y' = \sin(xy)$$

The first of these equations involves only x on the right side, so it has the form $y' = f(x)$. However, the second equation involves both x and y on the right side, so it has the form $y' = f(x, y)$, where the symbol $f(x, y)$ stands for a function of the two variables x and y. Later in the text we will study functions of two variables in more depth, but for now it will suffice to think of $f(x, y)$ as a formula that produces a unique output when values of x and y are given as inputs. For example, if

$$f(x, y) = x^2 + 3y$$

In applied problems involving time, it is usual to use t as the independent variable, in which case one would be concerned with equations of the form $y' = f(t, y)$, where $y' = dy/dt$.

and if the inputs are $x = 2$ and $y = -4$, then the output is

$$f(2, -4) = 2^2 + 3(-4) = 4 - 12 = -8$$

SLOPE FIELDS

In Section 5.2 we introduced the concept of a slope field in the context of differential equations of the form $y' = f(x)$; the same principles apply to differential equations of the form

$$y' = f(x, y)$$

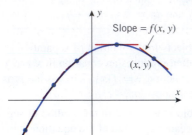

At each point (x, y) on an integral curve of $y' = f(x, y)$, the tangent line has slope $f(x, y)$.

▲ **Figure 8.3.1**

To see why this is so, let us review the basic idea. If we interpret y' as the slope of a tangent line, then the differential equation states that at each point (x, y) on an integral curve, the slope of the tangent line is equal to the value of f at that point (Figure 8.3.1). For example, suppose that $f(x, y) = y - x$, in which case we have the differential equation

$$y' = y - x \tag{1}$$

A geometric description of the set of integral curves can be obtained by choosing a rectangular grid of points in the xy-plane, calculating the slopes of the tangent lines to the integral curves at the gridpoints, and drawing small segments of the tangent lines through those points. The resulting picture is called a **slope field** or a **direction field** for the differential equation because it shows the "slope" or "direction" of the integral curves at the gridpoints. The more gridpoints that are used, the better the description of the integral curves. For example, Figure 8.3.2 shows two slope fields for (1)—the first was obtained by hand calculation using the 49 gridpoints shown in the accompanying table, and the second, which gives a clearer picture of the integral curves, was obtained using 625 gridpoints and a CAS.

VALUES OF $f(x, y) = y - x$

	$y = -3$	$y = -2$	$y = -1$	$y = 0$	$y = 1$	$y = 2$	$y = 3$
$x = -3$	0	1	2	3	4	5	6
$x = -2$	-1	0	1	2	3	4	5
$x = -1$	-2	-1	0	1	2	3	4
$x = 0$	-3	-2	-1	0	1	2	3
$x = 1$	-4	-3	-2	-1	0	1	2
$x = 2$	-5	-4	-3	-2	-1	0	1
$x = 3$	-6	-5	-4	-3	-2	-1	0

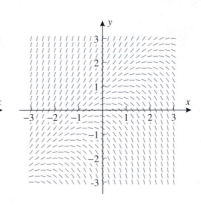

▲ **Figure 8.3.2**

It so happens that Equation (1) can be solved exactly using a method we will introduce in Section 8.4. We leave it for you to confirm that the general solution of this equation is

$$y = x + 1 + Ce^x \tag{2}$$

Confirm that the first slope field in Figure 8.3.2 is consistent with the accompanying table in that figure.

Figure 8.3.3 shows some of the integral curves superimposed on the slope field. Note that it was not necessary to have the general solution to construct the slope field. Indeed, slope fields are important precisely because they can be constructed in cases where the differential equation cannot be solved exactly.

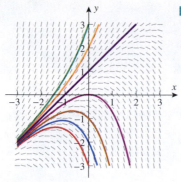

▲ Figure 8.3.3

■ EULER'S METHOD

Consider an initial-value problem of the form

$$y' = f(x, y), \quad y(x_0) = y_0$$

The slope field for the differential equation $y' = f(x, y)$ gives us a way to visualize the solution of the initial-value problem, since the graph of the solution is the integral curve that passes through the point (x_0, y_0). The slope field will also help us to develop a method for approximating the solution to the initial-value problem numerically.

We will not attempt to approximate $y(x)$ for all values of x; rather, we will choose some small increment Δx and focus on approximating the values of $y(x)$ at a succession of x-values spaced Δx units apart, starting from x_0. We will denote these x-values by

$$x_1 = x_0 + \Delta x, \quad x_2 = x_1 + \Delta x, \quad x_3 = x_2 + \Delta x, \quad x_4 = x_3 + \Delta x, \ldots$$

and we will denote the approximations of $y(x)$ at these points by

$$y_1 \approx y(x_1), \quad y_2 \approx y(x_2), \quad y_3 \approx y(x_3), \quad y_4 \approx y(x_4), \ldots$$

The technique that we will describe for obtaining these approximations is called ***Euler's Method***. Although there are better approximation methods available, many of them use Euler's Method as a starting point, so the underlying concepts are important to understand.

The basic idea behind Euler's Method is to start at the known initial point (x_0, y_0) and draw a line segment in the direction determined by the slope field until we reach the point (x_1, y_1) with x-coordinate $x_1 = x_0 + \Delta x$ (Figure 8.3.4). If Δx is small, then it is reasonable to expect that this line segment will not deviate much from the integral curve $y = y(x)$, and thus y_1 should closely approximate $y(x_1)$. To obtain the subsequent approximations, we repeat the process using the slope field as a guide at each step. Starting at the endpoint (x_1, y_1), we draw a line segment determined by the slope field until we reach the point (x_2, y_2) with x-coordinate $x_2 = x_1 + \Delta x$, and from that point we draw a line segment determined by the slope field to the point (x_3, y_3) with x-coordinate $x_3 = x_2 + \Delta x$, and so forth. As indicated in Figure 8.3.4, this procedure produces a polygonal path that tends to follow the integral curve closely, so it is reasonable to expect that the y-values $y_2, y_3, y_4, \ldots$ will closely approximate $y(x_2), y(x_3), y(x_4), \ldots$.

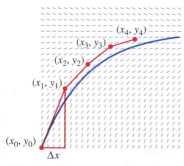

▲ Figure 8.3.4

To explain how the approximations $y_1, y_2, y_3, \ldots$ can be computed, let us focus on a typical line segment. As indicated in Figure 8.3.5, assume that we have found the point (x_n, y_n), and we are trying to determine the next point (x_{n+1}, y_{n+1}), where $x_{n+1} = x_n + \Delta x$. Since the slope of the line segment joining the points is determined by the slope field at the starting point, the slope is $f(x_n, y_n)$, and hence

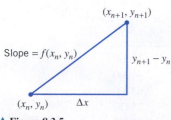

▲ Figure 8.3.5

$$\frac{y_{n+1} - y_n}{x_{n+1} - x_n} = \frac{y_{n+1} - y_n}{\Delta x} = f(x_n, y_n)$$

which we can rewrite as

$$y_{n+1} = y_n + f(x_n, y_n)\Delta x$$

This formula, which is the heart of Euler's Method, tells us how to use each approximation to compute the next approximation.

Euler's Method

To approximate the solution of the initial-value problem

$$y' = f(x, y), \quad y(x_0) = y_0$$

proceed as follows:

Step 1. Choose a nonzero number Δx to serve as an *increment* or *step size* along the x-axis, and let

$$x_1 = x_0 + \Delta x, \quad x_2 = x_1 + \Delta x, \quad x_3 = x_2 + \Delta x, \dots$$

Step 2. Compute successively

$$\begin{aligned} y_1 &= y_0 + f(x_0, y_0)\Delta x \\ y_2 &= y_1 + f(x_1, y_1)\Delta x \\ y_3 &= y_2 + f(x_2, y_2)\Delta x \end{aligned}$$

$$\vdots$$

$$y_{n+1} = y_n + f(x_n, y_n)\Delta x$$

The numbers $y_1, y_2, y_3, \dots$ in these equations are the approximations of $y(x_1)$, $y(x_2)$, $y(x_3)$, ...

▶ **Example 1**   Use Euler's Method with a step size of 0.1 to make a table of approximate values of the solution of the initial-value problem

$$y' = y - x, \quad y(0) = 2 \tag{3}$$

over the interval $0 \le x \le 1$.

Solution.   In this problem we have $f(x, y) = y - x$, $x_0 = 0$, and $y_0 = 2$. Moreover, since the step size is 0.1, the x-values at which the approximate values will be obtained are

$$x_1 = 0.1, \quad x_2 = 0.2, \quad x_3 = 0.3, \dots, \quad x_9 = 0.9, \quad x_{10} = 1$$

The first three approximations are

$$\begin{aligned} y_1 &= y_0 + f(x_0, y_0)\Delta x = 2 + (2 - 0)(0.1) = 2.2 \\ y_2 &= y_1 + f(x_1, y_1)\Delta x = 2.2 + (2.2 - 0.1)(0.1) = 2.41 \\ y_3 &= y_2 + f(x_2, y_2)\Delta x = 2.41 + (2.41 - 0.2)(0.1) = 2.631 \end{aligned}$$

Here is a way of organizing all 10 approximations rounded to five decimal places:

EULER'S METHOD FOR $y' = y - x$, $y(0) = 2$ WITH $\Delta x = 0.1$

n	x_n	y_n	$f(x_n, y_n)\Delta x$	$y_{n+1} = y_n + f(x_n, y_n)\Delta x$
0	0	2.00000	0.20000	2.20000
1	0.1	2.20000	0.21000	2.41000
2	0.2	2.41000	0.22100	2.63100
3	0.3	2.63100	0.23310	2.86410
4	0.4	2.86410	0.24641	3.11051
5	0.5	3.11051	0.26105	3.37156
6	0.6	3.37156	0.27716	3.64872
7	0.7	3.64872	0.29487	3.94359
8	0.8	3.94359	0.31436	4.25795
9	0.9	4.25795	0.33579	4.59374
10	1.0	4.59374	—	—

Observe that each entry in the last column becomes the next entry in the third column. This is reminiscent of Newton's Method in which each successive approximation is used to find the next. ◄

■ ACCURACY OF EULER'S METHOD

It follows from (3) and the initial condition $y(0) = 2$ that the exact solution of the initial-value problem in Example 1 is

$$y = x + 1 + e^x$$

Thus, in this case we can compare the approximate values of $y(x)$ produced by Euler's Method with decimal approximations of the exact values (Table 8.3.1). In Table 8.3.1 the *absolute error* is calculated as

$$|\text{exact value} - \text{approximation}|$$

and the *percentage error* as

$$\frac{|\text{exact value} - \text{approximation}|}{|\text{exact value}|} \times 100\%$$

As a rule of thumb, the absolute error in an approximation produced by Euler's Method is proportional to the step size. Thus, reducing the step size by half reduces the absolute and percentage errors by roughly half. However, reducing the step size increases the amount of computation, thereby increasing the potential for more roundoff error. Such matters are discussed in courses on differential equations or numerical analysis.

Notice that the absolute error tends to increase as x moves away from x_0.

Table 8.3.1

x	EXACT SOLUTION	EULER APPROXIMATION	ABSOLUTE ERROR	PERCENTAGE ERROR
0	2.00000	2.00000	0.00000	0.00
0.1	2.20517	2.20000	0.00517	0.23
0.2	2.42140	2.41000	0.01140	0.47
0.3	2.64986	2.63100	0.01886	0.71
0.4	2.89182	2.86410	0.02772	0.96
0.5	3.14872	3.11051	0.03821	1.21
0.6	3.42212	3.37156	0.05056	1.48
0.7	3.71375	3.64872	0.06503	1.75
0.8	4.02554	3.94359	0.08195	2.04
0.9	4.35960	4.25795	0.10165	2.33
1.0	4.71828	4.59374	0.12454	2.64

✔ QUICK CHECK EXERCISES 8.3   (See page 586 for answers.)

1. Match each differential equation with its slope field.
   (a) $y' = 2xy^2$ _____   (b) $y' = e^{-y}$ _____
   (c) $y' = y$ _____   (d) $y' = 2xy$ _____

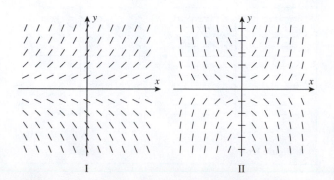

I   II

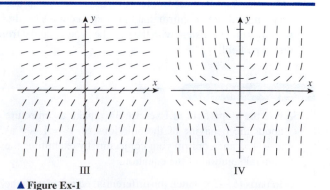

III   IV

▲ Figure Ex-1

2. The slope field for $y' = y/x$ at the 16 gridpoints (x, y), where $x = -2, -1, 1, 2$ and $y = -2, -1, 1, 2$ is shown in

the accompanying figure. Use this slope field and geometric reasoning to find the integral curve that passes through the point (1, 2).

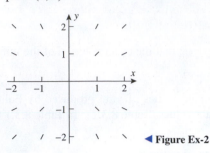

◀ **Figure Ex-2**

3. When using Euler's Method on the initial-value problem $y' = f(x, y)$, $y(x_0) = y_0$, we obtain y_{n+1} from y_n, x_n, and Δx by means of the formula $y_{n+1} =$ _____.

4. Consider the initial-value problem $y' = y$, $y(0) = 1$.
(a) Use Euler's Method with two steps to approximate $y(1)$.
(b) What is the exact value of $y(1)$?

EXERCISE SET 8.3   ◪ Graphing Utility

1. Sketch the slope field for $y' = xy/4$ at the 25 gridpoints (x, y), where $x = -2, -1, \ldots, 2$ and $y = -2, -1, \ldots, 2$.

2. Sketch the slope field for $y' + y = 2$ at the 25 gridpoints (x, y), where $x = 0, 1, \ldots, 4$ and $y = 0, 1, \ldots, 4$.

3. A slope field for the differential equation $y' = 1 - y$ is shown in the accompanying figure. In each part, sketch the graph of the solution that satisfies the initial condition.
(a) $y(0) = -1$ (b) $y(0) = 1$ (c) $y(0) = 2$

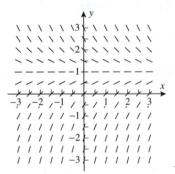

◀ **Figure Ex-3**

◪ **4.** Solve the initial-value problems in Exercise 3, and use a graphing utility to confirm that the integral curves for these solutions are consistent with the sketches you obtained from the slope field.

FOCUS ON CONCEPTS

5. Use the slope field in Exercise 3 to make a conjecture about the behavior of the solutions of $y' = 1 - y$ as $x \to +\infty$, and confirm your conjecture by examining the general solution of the equation.

6. In parts (a)–(f), match the differential equation with the slope field, and explain your reasoning.
(a) $y' = 1/x$ (b) $y' = 1/y$
(c) $y' = e^{-x^2}$ (d) $y' = y^2 - 1$

(e) $y' = \dfrac{x + y}{x - y}$ (f) $y' = (\sin x)(\sin y)$

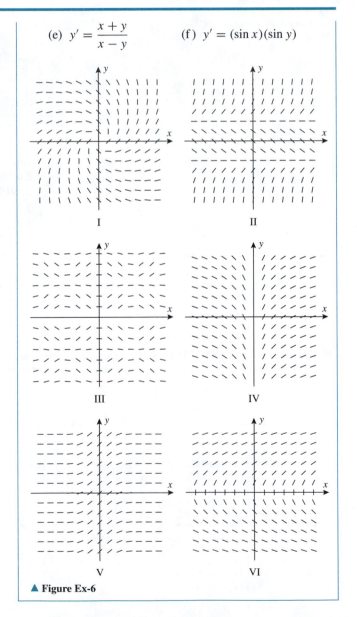

▲ **Figure Ex-6**

7–10 Use Euler's Method with the given step size Δx or Δt to approximate the solution of the initial-value problem over the stated interval. Present your answer as a table and as a graph. ■

7. $dy/dx = \sqrt[3]{y}$, $y(0) = 1$, $0 \le x \le 4$, $\Delta x = 0.5$

8. $dy/dx = x - y^2$, $y(0) = 1$, $0 \le x \le 2$, $\Delta x = 0.25$

9. $dy/dt = \cos y$, $y(0) = 1$, $0 \le t \le 2$, $\Delta t = 0.5$

10. $dy/dt = e^{-y}$, $y(0) = 0$, $0 \le t \le 1$, $\Delta t = 0.1$

11. Consider the initial-value problem
$$y' = \sin \pi t, \quad y(0) = 0$$
Use Euler's Method with five steps to approximate $y(1)$.

12–15 True–False Determine whether the statement is true or false. Explain your answer. ■

12. If the graph of $y = f(x)$ is an integral curve for a slope field, then so is any vertical translation of this graph.

13. Every integral curve for the slope field $dy/dx = e^{xy}$ is the graph of an increasing function of x.

14. Every integral curve for the slope field $dy/dx = e^y$ is concave up.

15. If $p(y)$ is a cubic polynomial in y, then the slope field $dy/dx = p(y)$ has an integral curve that is a horizontal line.

FOCUS ON CONCEPTS

16. (a) Show that the solution of the initial-value problem $y' = e^{-x^2}$, $y(0) = 0$ is
$$y(x) = \int_0^x e^{-t^2}\, dt$$

(b) Use Euler's Method with $\Delta x = 0.05$ to approximate the value of
$$y(1) = \int_0^1 e^{-t^2}\, dt$$
and compare the answer to that produced by a calculating utility with a numerical integration capability.

17. The accompanying figure shows a slope field for the differential equation $y' = -x/y$.
(a) Use the slope field to estimate $y\left(\frac{1}{2}\right)$ for the solution that satisfies the given initial condition $y(0) = 1$.
(b) Compare your estimate to the exact value of $y\left(\frac{1}{2}\right)$.

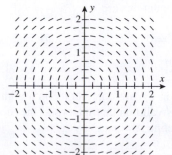

◀ **Figure Ex-17**

18. Refer to slope field II in Quick Check Exercise 1.
(a) Does the slope field appear to have a horizontal line as an integral curve?
(b) Use the differential equation for the slope field to verify your answer to part (a).

19. Refer to the slope field in Exercise 3 and consider the integral curve through $(0, -1)$.
(a) Use the slope field to estimate where the integral curve intersects the x-axis.
(b) Compare your estimate in part (a) with the exact value of the x-intercept for the integral curve.

20. Consider the initial-value problem
$$\frac{dy}{dx} = \frac{\sqrt{y}}{2}, \quad y(0) = 1$$
(a) Use Euler's Method with step sizes of $\Delta x = 0.2$, 0.1, and 0.05 to obtain three approximations of $y(1)$.
(b) Find $y(1)$ exactly.

21. A slope field of the form $y' = f(y)$ is said to be **au-tonomous**.
(a) Explain why the tangent segments along any horizontal line will be parallel for an autonomous slope field.
(b) The word *autonomous* means "independent." In what sense is an autonomous slope field independent?
(c) Suppose that $G(y)$ is an antiderivative of $1/[f(y)]$ and that C is a constant. Explain why any differentiable function defined implicitly by $G(y) - x = C$ will be a solution to the equation $y' = f(y)$.

22. (a) Solve the equation $y' = \sqrt{y}$ and show that every nonconstant solution has a graph that is everywhere concave up.
(b) Explain how the conclusion in part (a) may be obtained directly from the equation $y' = \sqrt{y}$ without solving.

23. (a) Find a slope field whose integral curve through $(1, 1)$ satisfies $xy^3 - x^2y = 0$ by differentiating this equation implicitly.
(b) Prove that if $y(x)$ is any integral curve of the slope field in part (a), then $x[y(x)]^3 - x^2y(x)$ will be a constant function.
(c) Find an equation that implicitly defines the integral curve through $(-1, -1)$ of the slope field in part (a).

24. (a) Find a slope field whose integral curve through $(0, 0)$ satisfies $xe^y + ye^x = 0$ by differentiating this equation implicitly.
(b) Prove that if $y(x)$ is any integral curve of the slope field in part (a), then $xe^{y(x)} + y(x)e^x$ will be a constant function.
(c) Find an equation that implicitly defines the integral curve through $(1, 1)$ of the slope field in part (a).

25. Consider the initial-value problem $y' = y$, $y(0) = 1$, and let y_n denote the approximation of $y(1)$ using Euler's Method with n steps.
(a) What would you conjecture is the exact value of $\lim_{n \to +\infty} y_n$? Explain your reasoning.
(b) Find an explicit formula for y_n and use it to verify your conjecture in part (a).

26. Writing Explain the connection between Euler's Method and the local linear approximation discussed in Section 3.5.

27. Writing Given a slope field, what features of an integral curve might be discussed from the slope field? Apply your ideas to the slope field in Exercise 3.

✔ **QUICK CHECK ANSWERS 8.3**

1. (a) IV (b) III (c) I (d) II   **2.** $y = 2x$, $x > 0$ **3.** $y_n + f(x_n, y_n)\Delta x$ **4.** (a) 2.25 (b) e

---

8.4 **FIRST-ORDER DIFFERENTIAL EQUATIONS AND APPLICATIONS**

In this section we will discuss a general method that can be used to solve a large class of first-order differential equations. We will use this method to solve differential equations related to the problems of mixing liquids and free fall retarded by air resistance.

■ **FIRST-ORDER LINEAR EQUATIONS**

The simplest first-order equations are those that can be written in the form

$$\frac{dy}{dx} = q(x) \tag{1}$$

Such equations can often be solved by integration. For example, if

$$\frac{dy}{dx} = x^3 \tag{2}$$

then

$$y = \int x^3 \, dx = \frac{x^4}{4} + C$$

is the general solution of (2) on the interval $(-\infty, +\infty)$. More generally, a first-order differential equation is called *linear* if it is expressible in the form

$$\frac{dy}{dx} + p(x)y = q(x) \tag{3}$$

Equation (1) is the special case of (3) that results when the function $p(x)$ is identically 0. Some other examples of first-order linear differential equations are

$$\frac{dy}{dx} + x^2 y = e^x, \qquad \frac{dy}{dx} + (\sin x)y + x^3 = 0, \qquad \frac{dy}{dx} + 5y = 2$$

$$\boxed{p(x) = x^2, q(x) = e^x} \qquad \boxed{p(x) = \sin x, q(x) = -x^3} \qquad \boxed{p(x) = 5, q(x) = 2}$$

We will assume that the functions $p(x)$ and $q(x)$ in (3) are continuous on a common interval, and we will look for a general solution that is valid on that interval. One method for doing this is based on the observation that if we define $\mu = \mu(x)$ by

$$\mu = e^{\int p(x)\,dx} \tag{4}$$

then

$$\frac{d\mu}{dx} = e^{\int p(x)\,dx} \cdot \frac{d}{dx} \int p(x)\,dx = \mu p(x)$$

Thus,

$$\frac{d}{dx}(\mu y) = \mu\frac{dy}{dx} + \frac{d\mu}{dx}y = \mu\frac{dy}{dx} + \mu p(x)y \tag{5}$$

If (3) is multiplied through by μ, it becomes

$$\mu\frac{dy}{dx} + \mu p(x)y = \mu q(x)$$

Combining this with (5) we have

$$\frac{d}{dx}(\mu y) = \mu q(x) \tag{6}$$

This equation can be solved for y by integrating both sides with respect to x and then dividing through by μ to obtain

$$y = \frac{1}{\mu}\int \mu q(x)\,dx \tag{7}$$

which is a general solution of (3) on the interval. The function μ in (4) is called an *integrating factor* for (3), and this method for finding a general solution of (3) is called the *method of integrating factors*. Although one could simply memorize Formula (7), we recommend solving first-order linear equations by actually carrying out the steps used to derive this formula:

The Method of Integrating Factors

Step 1. Calculate the integrating factor

$$\mu = e^{\int p(x)\,dx}$$

Since any μ will suffice, we can take the constant of integration to be zero in this step.

Step 2. Multiply both sides of (3) by μ and express the result as

$$\frac{d}{dx}(\mu y) = \mu q(x)$$

Step 3. Integrate both sides of the equation obtained in Step 2 and then solve for y. Be sure to include a constant of integration in this step.

▶ **Example 1**   Solve the differential equation

$$\frac{dy}{dx} - y = e^{2x}$$

Solution.   Comparing the given equation to (3), we see that we have a first-order linear equation with $p(x) = -1$ and $q(x) = e^{2x}$. These coefficients are continuous on the interval $(-\infty, +\infty)$, so the method of integrating factors will produce a general solution on this interval. The first step is to compute the integrating factor. This yields

$$\mu = e^{\int p(x)\,dx} = e^{\int (-1)\,dx} = e^{-x}$$

Next we multiply both sides of the given equation by μ to obtain

$$e^{-x}\frac{dy}{dx} - e^{-x}y = e^{-x}e^{2x}$$

which we can rewrite as

$$\frac{d}{dx}[e^{-x}y] = e^x$$

Integrating both sides of this equation with respect to x we obtain

$$e^{-x}y = e^x + C$$

Finally, solving for y yields the general solution

$$y = e^{2x} + Ce^x \; \blacktriangleleft$$

Confirm that the solution obtained in Example 1 agrees with that obtained by substituting the integrating factor into Formula (7).

A differential equation of the form

$$P(x)\frac{dy}{dx} + Q(x)y = R(x)$$

can be solved by dividing through by $P(x)$ to put the equation in the form of (3) and then applying the method of integrating factors. However, the resulting solution will only be valid on intervals where $p(x) = Q(x)/P(x)$ and $q(x) = R(x)/P(x)$ are both continuous.

▶ **Example 2** Solve the initial-value problem

$$x\frac{dy}{dx} - y = x, \quad y(1) = 2$$

Solution. This differential equation can be written in the form of (3) by dividing through by x. This yields

$$\frac{dy}{dx} - \frac{1}{x}y = 1 \tag{8}$$

where $q(x) = 1$ is continuous on $(-\infty, +\infty)$ and $p(x) = -1/x$ is continuous on $(-\infty, 0)$ and $(0, +\infty)$. Since we need $p(x)$ and $q(x)$ to be continuous on a common interval, and since our initial condition requires a solution for $x = 1$, we will find a general solution of (8) on the interval $(0, +\infty)$. On this interval we have $|x| = x$, so that

$$\int p(x)\,dx = -\int \frac{1}{x}\,dx = -\ln|x| = -\ln x \qquad \boxed{\begin{array}{l}\text{Taking the constant of}\\\text{integration to be 0}\end{array}}$$

Thus, an integrating factor that will produce a general solution on the interval $(0, +\infty)$ is

$$\mu = e^{\int p(x)\,dx} = e^{-\ln x} = e^{\ln(1/x)} = \frac{1}{x}$$

Multiplying both sides of Equation (8) by this integrating factor yields

$$\frac{1}{x}\frac{dy}{dx} - \frac{1}{x^2}y = \frac{1}{x}$$

or

$$\frac{d}{dx}\left[\frac{1}{x}y\right] = \frac{1}{x}$$

It is not accidental that the initial-value problem in Example 2 has a unique solution. If the coefficients of (3) are continuous on an open interval that contains the point x_0, then for any y_0 there will be a unique solution of (3) on that interval that satisfies the initial condition $y(x_0) = y_0$ [Exercise 29(b)].

Therefore, on the interval $(0, +\infty)$,

$$\frac{1}{x} y = \int \frac{1}{x}\, dx = \ln x + C$$

from which it follows that

$$y = x \ln x + Cx \qquad (9)$$

The initial condition $y(1) = 2$ requires that $y = 2$ if $x = 1$. Substituting these values into (9) and solving for C yields $C = 2$ (verify), so the solution of the initial-value problem is

$$y = x \ln x + 2x \blacktriangleleft$$

We conclude this section with some applications of first-order differential equations.

■ MIXING PROBLEMS

In a typical mixing problem, a tank is filled to a specified level with a solution that contains a known amount of some soluble substance (say salt). The thoroughly stirred solution is allowed to drain from the tank at a known rate, and at the same time a solution with a known concentration of the soluble substance is added to the tank at a known rate that may or may not differ from the draining rate. As time progresses, the amount of the soluble substance in the tank will generally change, and the usual mixing problem seeks to determine the amount of the substance in the tank at a specified time. This type of problem serves as a model for many kinds of problems: discharge and filtration of pollutants in a river, injection and absorption of medication in the bloodstream, and migrations of species into and out of an ecological system, for example.

5 gal/min

100 gal

5 gal/min

▲ **Figure 8.4.1**

▶ **Example 3** At time $t = 0$, a tank contains 4 lb of salt dissolved in 100 gal of water. Suppose that brine containing 2 lb of salt per gallon of brine is allowed to enter the tank at a rate of 5 gal/min and that the mixed solution is drained from the tank at the same rate (Figure 8.4.1). Find the amount of salt in the tank after 10 minutes.

Solution. Let $y(t)$ be the amount of salt (in pounds) after t minutes. We are given that $y(0) = 4$, and we want to find $y(10)$. We will begin by finding a differential equation that is satisfied by $y(t)$. To do this, observe that dy/dt, which is the rate at which the amount of salt in the tank changes with time, can be expressed as

$$\frac{dy}{dt} = \text{rate in} - \text{rate out} \qquad (10)$$

where *rate in* is the rate at which salt enters the tank and *rate out* is the rate at which salt leaves the tank. But the rate at which salt enters the tank is

$$\text{rate in} = (2\ \text{lb/gal}) \cdot (5\ \text{gal/min}) = 10\ \text{lb/min}$$

Since brine enters and drains from the tank at the same rate, the volume of brine in the tank stays constant at 100 gal. Thus, after t minutes have elapsed, the tank contains $y(t)$ lb of salt per 100 gal of brine, and hence the rate at which salt leaves the tank at that instant is

$$\text{rate out} = \left(\frac{y(t)}{100}\ \text{lb/gal} \right) \cdot (5\ \text{gal/min}) = \frac{y(t)}{20}\ \text{lb/min}$$

Therefore, (10) can be written as

$$\frac{dy}{dt} = 10 - \frac{y}{20} \qquad \text{or} \qquad \frac{dy}{dt} + \frac{y}{20} = 10$$

which is a first-order linear differential equation satisfied by $y(t)$. Since we are given that $y(0) = 4$, the function $y(t)$ can be obtained by solving the initial-value problem

$$\frac{dy}{dt} + \frac{y}{20} = 10, \quad y(0) = 4$$

The integrating factor for the differential equation is

$$\mu = e^{\int (1/20)\,dt} = e^{t/20}$$

If we multiply the differential equation through by μ, then we obtain

$$\frac{d}{dt}(e^{t/20}y) = 10e^{t/20}$$

$$e^{t/20}y = \int 10e^{t/20}\,dt = 200e^{t/20} + C$$

$$y(t) = 200 + Ce^{-t/20} \tag{11}$$

The initial condition states that $y = 4$ when $t = 0$. Substituting these values into (11) and solving for C yields $C = -196$ (verify), so

$$y(t) = 200 - 196e^{-t/20} \tag{12}$$

The graph of (12) is shown in Figure 8.4.2. At time $t = 10$ the amount of salt in the tank is

$$y(10) = 200 - 196e^{-0.5} \approx 81.1 \text{ lb} \blacktriangleleft$$

Notice that it follows from (11) that

$$\lim_{t \to +\infty} y(t) = 200$$

for all values of C, so regardless of the amount of salt that is present in the tank initially, the amount of salt in the tank will eventually stabilize at 200 lb. This can also be seen geometrically from the slope field for the differential equation shown in Figure 8.4.3. This slope field suggests the following: If the amount of salt present in the tank is greater than 200 lb initially, then the amount of salt will decrease steadily over time toward a limiting value of 200 lb; and if the amount of salt is less than 200 lb initially, then it will increase steadily toward a limiting value of 200 lb. The slope field also suggests that if the amount present initially is exactly 200 lb, then the amount of salt in the tank will stay constant at 200 lb. This can also be seen from (11), since $C = 0$ in this case (verify).

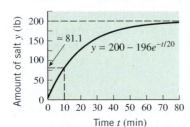

▲ Figure 8.4.2

The graph shown in Figure 8.4.2 suggests that $y(t) \to 200$ as $t \to +\infty$. This means that over an extended period of time the amount of salt in the tank tends toward 200 lb. Give an informal physical argument to explain why this result is to be expected.

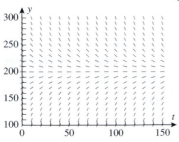

▲ Figure 8.4.3

■ A MODEL OF FREE-FALL MOTION RETARDED BY AIR RESISTANCE

In Section 5.7 we considered the free-fall model of an object moving along a vertical axis near the surface of the Earth. It was assumed in that model that there is no air resistance and that the only force acting on the object is the Earth's gravity. Our goal here is to find a model that takes air resistance into account. For this purpose we make the following assumptions:

- The object moves along a vertical s-axis whose origin is at the surface of the Earth and whose positive direction is up (Figure 5.7.7).
- At time $t = 0$ the height of the object is s_0 and the velocity is v_0.
- The only forces on the object are the force $F_G = -mg$ of the Earth's gravity acting down and the force F_R of air resistance acting opposite to the direction of motion. The force F_R is called the **drag force**.

In the case of free-fall motion retarded by air resistance, the net force acting on the object is

$$F_G + F_R = -mg + F_R$$

and the acceleration is d^2s/dt^2, so Newton's Second Law of Motion [Equation (5) of Section 6.6] implies that

$$-mg + F_R = m\frac{d^2s}{dt^2} \tag{13}$$

Experimentation has shown that the force F_R of air resistance depends on the shape of the object and its speed—the greater the speed, the greater the drag force. There are many possible models for air resistance, but one of the most basic assumes that the drag force F_R is proportional to the velocity of the object, that is,

$$F_R = -cv$$

where c is a positive constant that depends on the object's shape and properties of the air.[*] (The minus sign ensures that the drag force is opposite to the direction of motion.) Substituting this in (13) and writing d^2s/dt^2 as dv/dt, we obtain

$$-mg - cv = m\frac{dv}{dt}$$

Dividing by m and rearranging we obtain

$$\frac{dv}{dt} + \frac{c}{m}v = -g$$

which is a first-order linear differential equation in the unknown function $v = v(t)$ with $p(t) = c/m$ and $q(t) = -g$ [see (3)]. For a specific object, the coefficient c can be determined experimentally, so we will assume that m, g, and c are known constants. Thus, the velocity function $v = v(t)$ can be obtained by solving the initial-value problem

$$\frac{dv}{dt} + \frac{c}{m}v = -g, \quad v(0) = v_0 \tag{14}$$

Once the velocity function is found, the position function $s = s(t)$ can be obtained by solving the initial-value problem

$$\frac{ds}{dt} = v(t), \quad s(0) = s_0 \tag{15}$$

In Exercise 25 we will ask you to solve (14) and show that

$$v(t) = e^{-ct/m}\left(v_0 + \frac{mg}{c}\right) - \frac{mg}{c} \tag{16}$$

Note that

$$\lim_{t \to +\infty} v(t) = -\frac{mg}{c} \tag{17}$$

(verify). Thus, the speed $|v(t)|$ does not increase indefinitely, as in free fall; rather, because of the air resistance, it approaches a finite limiting speed v_τ given by

$$v_\tau = \left|-\frac{mg}{c}\right| = \frac{mg}{c} \tag{18}$$

This is called the **terminal speed** of the object, and (17) is called its **terminal velocity**.

REMARK | Intuition suggests that near the limiting velocity, the velocity $v(t)$ changes very slowly; that is, $dv/dt \approx 0$. Thus, it should not be surprising that the limiting velocity can be obtained informally from (14) by setting $dv/dt = 0$ in the differential equation and solving for v. This yields

$$v = -\frac{mg}{c}$$

which agrees with (17).

---

[*]Other common models assume that $F_R = -cv^2$ or, more generally, $F_R = -cv^p$ for some value of p.

✔ **QUICK CHECK EXERCISES 8.4**   *(See page 594 for answers.)*

1. Solve the first-order linear differential equation

$$\frac{dy}{dx} + p(x)y = q(x)$$

by completing the following steps:

Step 1. Calculate the integrating factor $\mu = $ _____.

Step 2. Multiply both sides of the equation by the integrating factor and express the result as

$$\frac{d}{dx}[\underline{\qquad}] = \underline{\qquad}$$

Step 3. Integrate both sides of the equation obtained in Step 2 and solve for $y = $ _____.

2. An integrating factor for

$$\frac{dy}{dx} + \frac{y}{x} = q(x)$$

is _____.

3. At time $t = 0$, a tank contains 30 oz of salt dissolved in 60 gal of water. Then brine containing 5 oz of salt per gallon of brine is allowed to enter the tank at a rate of 3 gal/min and the mixed solution is drained from the tank at the same rate. Give an initial-value problem satisfied by the amount of salt $y(t)$ in the tank at time t. Do not solve the problem.

EXERCISE SET 8.4   ∿ Graphing Utility

1–6 Solve the differential equation by the method of integrating factors.

1. $\dfrac{dy}{dx} + 4y = e^{-3x}$

2. $\dfrac{dy}{dx} + 2xy = x$

3. $y' + y = \cos(e^x)$

4. $2\dfrac{dy}{dx} + 4y = 1$

5. $(x^2 + 1)\dfrac{dy}{dx} + xy = 0$

6. $\dfrac{dy}{dx} + y + \dfrac{1}{1 - e^x} = 0$

7–10 Solve the initial-value problem. ■

7. $x\dfrac{dy}{dx} + y = x, \quad y(1) = 2$

8. $x\dfrac{dy}{dx} - y = x^2, \quad y(1) = -1$

9. $\dfrac{dy}{dx} - 2xy = 2x, \quad y(0) = 3$

10. $\dfrac{dy}{dt} + y = 2, \quad y(0) = 1$

11–14 True–False Determine whether the statement is true or false. Explain your answer. ■

11. If y_1 and y_2 are two solutions to a first-order linear differential equation, then $y = y_1 + y_2$ is also a solution.

12. If the first-order linear differential equation

$$\frac{dy}{dx} + p(x)y = q(x)$$

has a solution that is a constant function, then $q(x)$ is a constant multiple of $p(x)$.

13. In a mixing problem, we expect the concentration of the dissolved substance within the tank to approach a finite limit over time.

14. In our model for free-fall motion retarded by air resistance, the terminal velocity is proportional to the weight of the falling object.

15. A slope field for the differential equation $y' = 2y - x$ is shown in the accompanying figure. In each part, sketch the graph of the solution that satisfies the initial condition.
(a) $y(1) = 1$ (b) $y(0) = -1$ (c) $y(-1) = 0$

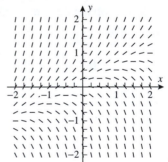

◀ **Figure Ex-15**

∿ **16.** Solve the initial-value problems in Exercise 15, and use a graphing utility to confirm that the integral curves for these solutions are consistent with the sketches you obtained from the slope field.

FOCUS ON CONCEPTS

17. Use the slope field in Exercise 15 to make a conjecture about the effect of y_0 on the behavior of the solution of the initial-value problem $y' = 2y - x$, $y(0) = y_0$ as $x \to +\infty$, and check your conjecture by examining the solution of the initial-value problem.

18. Consider the slope field in Exercise 15.
(a) Use Euler's Method with $\Delta x = 0.1$ to estimate $y\left(\frac{1}{2}\right)$ for the solution that satisfies the initial condition $y(0) = 1$.

(b) Would you conjecture your answer in part (a) to be greater than or less than the actual value of $y\left(\frac{1}{2}\right)$? Explain.

(c) Check your conjecture in part (b) by finding the exact value of $y\left(\frac{1}{2}\right)$.

---

19. (a) Use Euler's Method with a step size of $\Delta x = 0.2$ to approximate the solution of the initial-value problem

$$y' = x + y, \quad y(0) = 1$$

over the interval $0 \le x \le 1$.

(b) Solve the initial-value problem exactly, and calculate the error and the percentage error in each of the approximations in part (a).

(c) Sketch the exact solution and the approximate solution together.

20. It was stated at the end of Section 8.3 that reducing the step size in Euler's Method by half reduces the error in each approximation by about half. Confirm that the error in $y(1)$ is reduced by about half if a step size of $\Delta x = 0.1$ is used in Exercise 19.

21. At time $t = 0$, a tank contains 25 oz of salt dissolved in 50 gal of water. Then brine containing 4 oz of salt per gallon of brine is allowed to enter the tank at a rate of 2 gal/min and the mixed solution is drained from the tank at the same rate.
(a) How much salt is in the tank at an arbitrary time t?
(b) How much salt is in the tank after 25 min?

22. A tank initially contains 200 gal of pure water. Then at time $t = 0$ brine containing 5 lb of salt per gallon of brine is allowed to enter the tank at a rate of 20 gal/min and the mixed solution is drained from the tank at the same rate.
(a) How much salt is in the tank at an arbitrary time t?
(b) How much salt is in the tank after 30 min?

23. A tank with a 1000 gal capacity initially contains 500 gal of water that is polluted with 50 lb of particulate matter. At time $t = 0$, pure water is added at a rate of 20 gal/min and the mixed solution is drained off at a rate of 10 gal/min. How much particulate matter is in the tank when it reaches the point of overflowing?

24. The water in a polluted lake initially contains 1 lb of mercury salts per 100,000 gal of water. The lake is circular with diameter 30 m and uniform depth 3 m. Polluted water is pumped from the lake at a rate of 1000 gal/h and is replaced with fresh water at the same rate. Construct a table that shows the amount of mercury in the lake (in lb) at the end of each hour over a 12-hour period. Discuss any assumptions you made. [*Note:* Use 1 m^3 = 264 gal.]

25. (a) Use the method of integrating factors to derive solution (16) to the initial-value problem (14). [*Note:* Keep in mind that c, m, and g are constants.]
(b) Show that (16) can be expressed in terms of the terminal speed (18) as

$$v(t) = e^{-gt/v_\tau}(v_0 + v_\tau) - v_\tau$$

(c) Show that if $s(0) = s_0$, then the position function of the object can be expressed as

$$s(t) = s_0 - v_\tau t + \frac{v_\tau}{g}(v_0 + v_\tau)(1 - e^{-gt/v_\tau})$$

26. Suppose a fully equipped skydiver weighing 240 lb has a terminal speed of 120 ft/s with a closed parachute and 24 ft/s with an open parachute. Suppose further that this skydiver is dropped from an airplane at an altitude of 10,000 ft, falls for 25 s with a closed parachute, and then falls the rest of the way with an open parachute.
(a) Assuming that the skydiver's initial vertical velocity is zero, use Exercise 25 to find the skydiver's vertical velocity and height at the time the parachute opens. [*Note:* Take $g = 32$ ft/s^2.]
(b) Use a calculating utility to find a numerical solution for the total time that the skydiver is in the air.

27. The accompanying figure is a schematic diagram of a basic *RL* series electrical circuit that contains a power source with a time-dependent voltage of $V(t)$ volts (V), a resistor with a constant resistance of R ohms (Ω), and an inductor with a constant inductance of L henrys (H). If you don't know anything about electrical circuits, don't worry; all you need to know is that electrical theory states that a current of $I(t)$ amperes (A) flows through the circuit where $I(t)$ satisfies the differential equation

$$L\frac{dI}{dt} + RI = V(t)$$

(a) Find $I(t)$ if $R = 10\,\Omega$, $L = 5$ H, V is a constant 20 V, and $I(0) = 0$ A.
(b) What happens to the current over a long period of time?

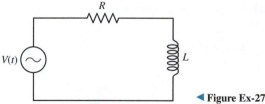

◀ **Figure Ex-27**

28. Find $I(t)$ for the electrical circuit in Exercise 27 if $R = 6\,\Omega$, $L = 3$ H, $V(t) = 3\sin t$ V, and $I(0) = 15$ A.

FOCUS ON CONCEPTS

29. (a) Prove that any function $y = y(x)$ defined by Equation (7) will be a solution to (3).
(b) Consider the initial-value problem

$$\frac{dy}{dx} + p(x)y = q(x), \quad y(x_0) = y_0$$

where the functions $p(x)$ and $q(x)$ are both continuous on some open interval. Using the general solution for a first-order linear equation, prove that this initial-value problem has a unique solution on the interval.

30. (a) Prove that solutions need not be unique for nonlinear initial-value problems by finding two solutions to

$$y\frac{dy}{dx} = x, \quad y(0) = 0$$

(b) Prove that solutions need not exist for nonlinear initial-value problems by showing that there is no solution for

$$y\frac{dy}{dx} = -x, \quad y(0) = 0$$

31. Writing Explain why the quantity μ in the *Method of Integrating Factors* is called an "integrating factor" and explain its role in this method.

32. Writing Suppose that a given first-order differential equation can be solved both by the method of integrating factors and by separation of variables. Discuss the advantages and disadvantages of each method.

✔ QUICK CHECK ANSWERS 8.4

1. Step 1: $e^{\int p(x)\,dx}$;   Step 2: μy, $\mu q(x)$;   Step 3: $\dfrac{1}{\mu}\displaystyle\int \mu q(x)\,dx$ **2.** x **3.** $\dfrac{dy}{dt} + \dfrac{y}{20} = 15$, $y(0) = 30$

CHAPTER 8 REVIEW EXERCISES   [C] CAS

1. Give an informal explanation of why one might expect the number of arbitrary constants in the general solution of a differential equation to be equal to the order of the equation.

2. Which of the given differential equations are separable?

(a) $\dfrac{dy}{dx} = f(x)g(y)$ **(b)** $\dfrac{dy}{dx} = \dfrac{f(x)}{g(y)}$

(c) $\dfrac{dy}{dx} = f(x) + g(y)$ **(d)** $\dfrac{dy}{dx} = \sqrt{f(x)g(y)}$

3–5 Solve the differential equation by the method of separation of variables.

3. $\dfrac{dy}{dx} = (1 + y^2)x^2$ **4.** $3\tan y - \dfrac{dy}{dx}\sec x = 0$

5. $(1 + y^2)y' = e^x y$

6–8 Solve the initial-value problem by the method of separation of variables. ■

6. $y' = 1 + y^2$, $y(0) = 1$ **7.** $y' = \dfrac{y^5}{x(1 + y^4)}$, $y(1) = 1$

8. $y' = 4y^2 \sec^2 2x$, $y(\pi/8) = 1$

9. Sketch the integral curve of $y' = -2xy^2$ that passes through the point $(0, 1)$.

10. Sketch the integral curve of $2yy' = 1$ that passes through the point $(0, 1)$ and the integral curve that passes through the point $(0, -1)$.

11. Sketch the slope field for $y' = xy/8$ at the 25 gridpoints (x, y), where $x = 0, 1, \ldots, 4$ and $y = 0, 1, \ldots, 4$.

12. Solve the differential equation $y' = xy/8$, and find a family of integral curves for the slope field in Exercise 11.

13–14 Use Euler's Method with the given step size Δx to approximate the solution of the initial-value problem over the stated interval. Present your answer as a table and as a graph. ■

13. $dy/dx = \sqrt{y}$, $y(0) = 1$, $0 \le x \le 4$, $\Delta x = 0.5$

14. $dy/dx = \sin y$, $y(0) = 1$, $0 \le x \le 2$, $\Delta x = 0.5$

15. Consider the initial-value problem

$$y' = \cos 2\pi t, \quad y(0) = 1$$

Use Euler's Method with five steps to approximate $y(1)$.

16. Use Euler's Method with a step size of $\Delta t = 0.1$ to approximate the solution of the initial-value problem

$$y' = 1 + 5t - y, \quad y(1) = 5$$

over the interval $[1, 2]$.

17. Cloth found in an Egyptian pyramid contains 78.5% of its original carbon-14. Estimate the age of the cloth.

18. Suppose that an initial population of 5000 bacteria grows exponentially at a rate of 1% per hour and that $y = y(t)$ is the number of bacteria present after t hours.

(a) Find an initial-value problem whose solution is $y(t)$.
(b) Find a formula for $y(t)$.
(c) What is the doubling time for the population?
(d) How long does it take for the population of bacteria to reach 30,000?

19–20 Solve the differential equation by the method of integrating factors. ■

19. $\dfrac{dy}{dx} + 3y = e^{-2x}$ **20.** $\dfrac{dy}{dx} + y - \dfrac{1}{1 + e^x} = 0$

21–23 Solve the initial-value problem by the method of integrating factors. ■

21. $y' - xy = x, \quad y(0) = 3$

22. $xy' + 2y = 4x^2, \quad y(1) = 2$

23. $y' \cosh x + y \sinh x = \cosh^2 x, \quad y(0) = 2$

c 24. (a) Solve the initial-value problem

$$y' - y = x \sin 3x, \quad y(0) = 1$$

by the method of integrating factors, using a CAS to perform any difficult integrations.

(b) Use the CAS to solve the initial-value problem directly, and confirm that the answer is consistent with that obtained in part (a).

(c) Graph the solution.

25. Classify the following first-order differential equations as separable, linear, both, or neither.

(a) $\dfrac{dy}{dx} - 3y = \sin x$ (b) $\dfrac{dy}{dx} + xy = x$

(c) $y\dfrac{dy}{dx} - x = 1$ (d) $\dfrac{dy}{dx} + xy^2 = \sin(xy)$

26. Determine whether the methods of integrating factors and separation of variables produce the same solutions of the differential equation

$$\frac{dy}{dx} - 4xy = x$$

27. A tank contains 1000 gal of fresh water. At time $t = 0$ min, brine containing 5 oz of salt per gallon of brine is poured into the tank at a rate of 10 gal/min, and the mixed solution is drained from the tank at the same rate. After 15 min that process is stopped and fresh water is poured into the tank at the rate of 5 gal/min, and the mixed solution is drained from the tank at the same rate. Find the amount of salt in the tank at time $t = 30$ min.

28. Suppose that a room containing 1200 ft³ of air is free of carbon monoxide. At time $t = 0$ cigarette smoke containing 4% carbon monoxide is introduced at the rate of 0.1 ft³/min, and the well-circulated mixture is vented from the room at the same rate.

(a) Find a formula for the percentage of carbon monoxide in the room at time t.

(b) Extended exposure to air containing 0.012% carbon monoxide is considered dangerous. How long will it take to reach this level?

Source: This is based on a problem from William E. Boyce and Richard C. DiPrima, *Elementary Differential Equations*, 7th ed., John Wiley & Sons, New York, 2001.

CHAPTER 8 MAKING CONNECTIONS

1. Consider the first-order differential equation

$$\frac{dy}{dx} + py = q$$

where p and q are constants. If $y = y(x)$ is a solution to this equation, define $u = u(x) = q - py(x)$.

(a) Without solving the differential equation, show that u grows exponentially as a function of x if $p < 0$, and decays exponentially as a function of x if $0 < p$.

(b) Use the result of part (a) and Equations (13–14) of Section 8.2 to solve the initial-value problem

$$\frac{dy}{dx} + 2y = 4, \quad y(0) = -1$$

2. Consider a differential equation of the form

$$\frac{dy}{dx} = f(ax + by + c)$$

where f is a function of a single variable. If $y = y(x)$ is a solution to this equation, define $u = u(x) = ax + by(x) + c$.

(a) Find a separable differential equation that is satisfied by the function u.

(b) Use your answer to part (a) to solve

$$\frac{dy}{dx} = \frac{1}{x + y}$$

3. A first-order differential equation is ***homogeneous*** if it can be written in the form

$$\frac{dy}{dx} = f\left(\frac{y}{x}\right) \quad \text{for } x \neq 0$$

where f is a function of a single variable. If $y = y(x)$ is a solution to a first-order homogeneous differential equation, define $u = u(x) = y(x)/x$.

(a) Find a separable differential equation that is satisfied by the function u.

(b) Use your answer to part (a) to solve

$$\frac{dy}{dx} = \frac{x - y}{x + y}$$

4. A first-order differential equation is called a ***Bernoulli equation*** if it can be written in the form

$$\frac{dy}{dx} + p(x)y = q(x)y^n \quad \text{for } n \neq 0, 1$$

If $y = y(x)$ is a solution to a Bernoulli equation, define $u = u(x) = [y(x)]^{1-n}$.

(a) Find a first-order linear differential equation that is satisfied by u.

(b) Use your answer to part (a) to solve the initial-value problem

$$x\frac{dy}{dx} - y = -2xy^2, \quad y(1) = \frac{1}{2}$$

9
INFINITE SERIES

iStockphoto

Perspective creates the illusion that the sequence of railroad ties continues indefinitely but converges toward a single point infinitely far away.

In this chapter we will be concerned with infinite series, which are sums that involve infinitely many terms. Infinite series play a fundamental role in both mathematics and science—they are used, for example, to approximate trigonometric functions and logarithms, to solve differential equations, to evaluate difficult integrals, to create new functions, and to construct mathematical models of physical laws. Since it is impossible to add up infinitely many numbers directly, one goal will be to define exactly what we mean by the sum of an infinite series. However, unlike finite sums, it turns out that not all infinite series actually have a sum, so we will need to develop tools for determining which infinite series have sums and which do not. Once the basic ideas have been developed we will begin to apply our work; we will show how infinite series are used to evaluate such quantities as ln 2, e, sin 3°, and π, how they are used to create functions, and finally, how they are used to model physical laws.

9.1 SEQUENCES

In everyday language, the term "sequence" means a succession of things in a definite order—chronological order, size order, or logical order, for example. In mathematics, the term "sequence" is commonly used to denote a succession of numbers whose order is determined by a rule or a function. In this section, we will develop some of the basic ideas concerning sequences of numbers.

■ DEFINITION OF A SEQUENCE

Stated informally, an *infinite sequence*, or more simply a *sequence*, is an unending succession of numbers, called *terms*. It is understood that the terms have a definite order; that is, there is a first term a_1, a second term a_2, a third term a_3, a fourth term a_4, and so forth. Such a sequence would typically be written as

$$a_1, a_2, a_3, a_4, \ldots$$

where the dots are used to indicate that the sequence continues indefinitely. Some specific examples are

$$1, 2, 3, 4, \ldots, \qquad 1, \tfrac{1}{2}, \tfrac{1}{3}, \tfrac{1}{4}, \ldots,$$
$$2, 4, 6, 8, \ldots, \qquad 1, -1, 1, -1, \ldots$$

Each of these sequences has a definite pattern that makes it easy to generate additional terms if we assume that those terms follow the same pattern as the displayed terms. However, such patterns can be deceiving, so it is better to have a rule or formula for generating the terms. One way of doing this is to look for a function that relates each term in the sequence to its term number. For example, in the sequence

$$2, 4, 6, 8, \ldots$$

each term is twice the term number; that is, the nth term in the sequence is given by the formula $2n$. We denote this by writing the sequence as

$$2, 4, 6, 8, \ldots, 2n, \ldots$$

We call the function $f(n) = 2n$ the *general term* of this sequence. Now, if we want to know a specific term in the sequence, we need only substitute its term number in the formula for the general term. For example, the 37th term in the sequence is $2 \cdot 37 = 74$.

▶ **Example 1**   In each part, find the general term of the sequence.

(a) $\frac{1}{2}, \frac{2}{3}, \frac{3}{4}, \frac{4}{5}, \ldots$ (b) $\frac{1}{2}, \frac{1}{4}, \frac{1}{8}, \frac{1}{16}, \ldots$

(c) $\frac{1}{2}, -\frac{2}{3}, \frac{3}{4}, -\frac{4}{5}, \ldots$ (d) $1, 3, 5, 7, \ldots$

Table 9.1.1

TERM NUMBER	1	2	3	4	$\cdots$	n	$\cdots$
TERM	$\frac{1}{2}$	$\frac{2}{3}$	$\frac{3}{4}$	$\frac{4}{5}$	$\cdots$	$\frac{n}{n+1}$	$\cdots$

Solution (a).   In Table 9.1.1, the four known terms have been placed below their term numbers, from which we see that the numerator is the same as the term number and the denominator is one greater than the term number. This suggests that the nth term has numerator n and denominator $n + 1$, as indicated in the table. Thus, the sequence can be expressed as

$$\frac{1}{2}, \frac{2}{3}, \frac{3}{4}, \frac{4}{5}, \ldots, \frac{n}{n+1}, \ldots$$

Table 9.1.2

TERM NUMBER	1	2	3	4	$\cdots$	n	$\cdots$
TERM	$\frac{1}{2}$	$\frac{1}{2^2}$	$\frac{1}{2^3}$	$\frac{1}{2^4}$	$\cdots$	$\frac{1}{2^n}$	$\cdots$

Solution (b).   In Table 9.1.2, the denominators of the four known terms have been expressed as powers of 2 and the first four terms have been placed below their term numbers, from which we see that the exponent in the denominator is the same as the term number. This suggests that the denominator of the nth term is 2^n, as indicated in the table. Thus, the sequence can be expressed as

$$\frac{1}{2}, \frac{1}{4}, \frac{1}{8}, \frac{1}{16}, \ldots, \frac{1}{2^n}, \ldots$$

Solution (c).   This sequence is identical to that in part (a), except for the alternating signs. Thus, the nth term in the sequence can be obtained by multiplying the nth term in part (a) by $(-1)^{n+1}$. This factor produces the correct alternating signs, since its successive values, starting with $n = 1$, are $1, -1, 1, -1, \ldots$. Thus, the sequence can be written as

$$\frac{1}{2}, -\frac{2}{3}, \frac{3}{4}, -\frac{4}{5}, \ldots, (-1)^{n+1}\frac{n}{n+1}, \ldots$$

Table 9.1.3

TERM NUMBER	1	2	3	4	$\cdots$	n	$\cdots$
TERM	1	3	5	7	$\cdots$	$2n-1$	$\cdots$

Solution (d).   In Table 9.1.3, the four known terms have been placed below their term numbers, from which we see that each term is one less than twice its term number. This suggests that the nth term in the sequence is $2n - 1$, as indicated in the table. Thus, the sequence can be expressed as

$$1, 3, 5, 7, \ldots, 2n - 1, \ldots \blacktriangleleft$$

When the general term of a sequence

$$a_1, a_2, a_3, \ldots, a_n, \ldots \qquad (1)$$

is known, there is no need to write out the initial terms, and it is common to write only the general term enclosed in braces. Thus, (1) might be written as

$$\{a_n\}_{n=1}^{+\infty} \quad \text{or as} \quad \{a_n\}_{n=1}^{\infty}$$

For example, here are the four sequences in Example 1 expressed in brace notation.

A sequence cannot be uniquely determined from a few initial terms. For example, the sequence whose general term is

$$f(n) = \tfrac{1}{3}(3 - 5n + 6n^2 - n^3)$$

has 1, 3, and 5 as its first three terms, but its fourth term is also 5.

SEQUENCE	BRACE NOTATION
$\dfrac{1}{2}, \dfrac{2}{3}, \dfrac{3}{4}, \dfrac{4}{5}, \dots, \dfrac{n}{n+1}, \dots$	$\left\{\dfrac{n}{n+1}\right\}_{n=1}^{+\infty}$
$\dfrac{1}{2}, \dfrac{1}{4}, \dfrac{1}{8}, \dfrac{1}{16}, \dots, \dfrac{1}{2^n}, \dots$	$\left\{\dfrac{1}{2^n}\right\}_{n=1}^{+\infty}$
$\dfrac{1}{2}, -\dfrac{2}{3}, \dfrac{3}{4}, -\dfrac{4}{5}, \dots, (-1)^{n+1}\dfrac{n}{n+1}, \dots$	$\left\{(-1)^{n+1}\dfrac{n}{n+1}\right\}_{n=1}^{+\infty}$
$1, 3, 5, 7, \dots, 2n-1, \dots$	$\{2n-1\}_{n=1}^{+\infty}$

The letter n in (1) is called the **index** for the sequence. It is not essential to use n for the index; any letter not reserved for another purpose can be used. For example, we might view the general term of the sequence $a_1, a_2, a_3, \dots$ to be the kth term, in which case we would denote this sequence as $\{a_k\}_{k=1}^{+\infty}$. Moreover, it is not essential to start the index at 1; sometimes it is more convenient to start it at 0 (or some other integer). For example, consider the sequence

$$1, \frac{1}{2}, \frac{1}{2^2}, \frac{1}{2^3}, \dots$$

One way to write this sequence is

$$\left\{\frac{1}{2^{n-1}}\right\}_{n=1}^{+\infty}$$

However, the general term will be simpler if we think of the initial term in the sequence as the zeroth term, in which case we can write the sequence as

$$\left\{\frac{1}{2^n}\right\}_{n=0}^{+\infty}$$

We began this section by describing a sequence as an unending succession of numbers. Although this conveys the general idea, it is not a satisfactory mathematical definition because it relies on the term "succession," which is itself an undefined term. To motivate a precise definition, consider the sequence

$$2, 4, 6, 8, \dots, 2n, \dots$$

If we denote the general term by $f(n) = 2n$, then we can write this sequence as

$$f(1), f(2), f(3), \dots, f(n), \dots$$

which is a "list" of values of the function

$$f(n) = 2n, \quad n = 1, 2, 3, \dots$$

whose domain is the set of positive integers. This suggests the following definition.

9.1.1 **DEFINITION** A *sequence* is a function whose domain is a set of integers.

Typically, the domain of a sequence is the set of positive integers or the set of nonnegative integers. We will regard the expression $\{a_n\}_{n=1}^{+\infty}$ to be an alternative notation for the function $f(n) = a_n, n = 1, 2, 3, \dots$, and we will regard $\{a_n\}_{n=0}^{+\infty}$ to be an alternative notation for the function $f(n) = a_n, n = 0, 1, 2, 3, \dots$.

GRAPHS OF SEQUENCES

Since sequences are functions, it makes sense to talk about the graph of a sequence. For example, the graph of the sequence $\{1/n\}_{n=1}^{+\infty}$ is the graph of the equation

$$y = \frac{1}{n}, \quad n = 1, 2, 3, \ldots$$

When the starting value for the index of a sequence is not relevant to the discussion, it is common to use a notation such as $\{a_n\}$ in which there is no reference to the starting value of n. We can distinguish between different sequences by using different letters for their general terms; thus, $\{a_n\}$, $\{b_n\}$, and $\{c_n\}$ denote three different sequences.

Because the right side of this equation is defined only for positive integer values of n, the graph consists of a succession of isolated points (Figure 9.1.1a). This is different from the graph of

$$y = \frac{1}{x}, \quad x \geq 1$$

which is a continuous curve (Figure 9.1.1b).

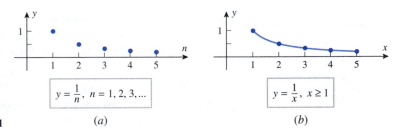

| $y = \frac{1}{n}$, $n = 1, 2, 3, \ldots$ | $y = \frac{1}{x}$, $x \geq 1$ |

▶ **Figure 9.1.1**   (a)   (b)

LIMIT OF A SEQUENCE

Since sequences are functions, we can inquire about their limits. However, because a sequence $\{a_n\}$ is only defined for integer values of n, the only limit that makes sense is the limit of a_n as $n \to +\infty$. In Figure 9.1.2 we have shown the graphs of four sequences, each of which behaves differently as $n \to +\infty$:

- The terms in the sequence $\{n + 1\}$ increase without bound.
- The terms in the sequence $\{(-1)^{n+1}\}$ oscillate between -1 and 1.
- The terms in the sequence $\{n/(n + 1)\}$ increase toward a "limiting value" of 1.
- The terms in the sequence $\left\{1 + \left(-\frac{1}{2}\right)^n\right\}$ also tend toward a "limiting value" of 1, but do so in an oscillatory fashion.

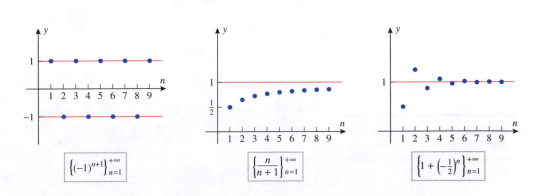

▲ **Figure 9.1.2**

Informally speaking, the limit of a sequence $\{a_n\}$ is intended to describe how a_n behaves as $n \to +\infty$. To be more specific, we will say that *a sequence $\{a_n\}$ approaches a limit L if the terms in the sequence eventually become arbitrarily close to L.* Geometrically, this

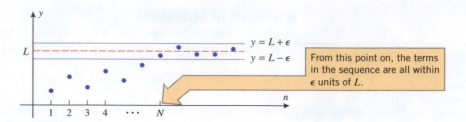

▶ **Figure 9.1.3**

means that for any positive number ϵ there is a point in the sequence after which all terms lie between the lines $y = L - \epsilon$ and $y = L + \epsilon$ (Figure 9.1.3).

The following definition makes these ideas precise.

How would you define these limits?

$$\lim_{n \to +\infty} a_n = +\infty$$

$$\lim_{n \to +\infty} a_n = -\infty$$

9.1.2   DEFINITION A sequence $\{a_n\}$ is said to *converge* to the *limit* L if given any $\epsilon > 0$, there is a positive integer N such that $|a_n - L| < \epsilon$ for $n \geq N$. In this case we write

$$\lim_{n \to +\infty} a_n = L$$

A sequence that does not converge to some finite limit is said to *diverge*.

▶ **Example 2**   The first two sequences in Figure 9.1.2 diverge, and the second two converge to 1; that is,

$$\lim_{n \to +\infty} \frac{n}{n + 1} = 1 \quad \text{and} \quad \lim_{n \to +\infty} \left[1 + \left(-\tfrac{1}{2} \right)^n \right] = 1 \quad \blacktriangleleft$$

The following theorem, which we state without proof, shows that the familiar properties of limits apply to sequences. This theorem ensures that the algebraic techniques used to find limits of the form $\lim_{x \to +\infty}$ can also be used for limits of the form $\lim_{n \to +\infty}$.

9.1.3   THEOREM   *Suppose that the sequences $\{a_n\}$ and $\{b_n\}$ converge to limits L_1 and L_2, respectively, and c is a constant. Then:*

(a) $\displaystyle \lim_{n \to +\infty} c = c$

(b) $\displaystyle \lim_{n \to +\infty} ca_n = c \lim_{n \to +\infty} a_n = cL_1$

(c) $\displaystyle \lim_{n \to +\infty} (a_n + b_n) = \lim_{n \to +\infty} a_n + \lim_{n \to +\infty} b_n = L_1 + L_2$

(d) $\displaystyle \lim_{n \to +\infty} (a_n - b_n) = \lim_{n \to +\infty} a_n - \lim_{n \to +\infty} b_n = L_1 - L_2$

(e) $\displaystyle \lim_{n \to +\infty} (a_n b_n) = \lim_{n \to +\infty} a_n \cdot \lim_{n \to +\infty} b_n = L_1 L_2$

(f) $\displaystyle \lim_{n \to +\infty} \left(\frac{a_n}{b_n} \right) = \frac{\displaystyle \lim_{n \to +\infty} a_n}{\displaystyle \lim_{n \to +\infty} b_n} = \frac{L_1}{L_2} \quad (\text{if } L_2 \neq 0)$

Additional limit properties follow from those in Theorem 9.1.3. For example, use part (e) to show that if $a_n \to L$ and m is a positive integer, then

$$\lim_{n \to +\infty} (a_n)^m = L^m$$

If the general term of a sequence is $f(n)$, where $f(x)$ is a function defined on the entire interval $[1, +\infty)$, then the values of $f(n)$ can be viewed as "sample values" of $f(x)$ taken

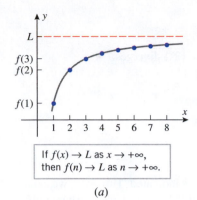

If $f(x) \to L$ as $x \to +\infty$,
then $f(n) \to L$ as $n \to +\infty$.

(a)

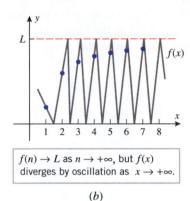

$f(n) \to L$ as $n \to +\infty$, but $f(x)$
diverges by oscillation as $x \to +\infty$.

(b)

▲ **Figure 9.1.4**

at the positive integers. Thus,

$$\text{if } f(x) \to L \text{ as } x \to +\infty, \quad \text{then} \quad f(n) \to L \text{ as } n \to +\infty$$

(Figure 9.1.4*a*). However, the converse is not true; that is, one cannot infer that $f(x) \to L$ as $x \to +\infty$ from the fact that $f(n) \to L$ as $n \to +\infty$ (Figure 9.1.4*b*).

▶ **Example 3** In each part, determine whether the sequence converges or diverges by examining the limit as $n \to +\infty$.

(a) $\left\{ \dfrac{n}{2n+1} \right\}_{n=1}^{+\infty}$

(b) $\left\{ (-1)^{n+1} \dfrac{n}{2n+1} \right\}_{n=1}^{+\infty}$

(c) $\left\{ (-1)^{n+1} \dfrac{1}{n} \right\}_{n=1}^{+\infty}$

(d) $\{8 - 2n\}_{n=1}^{+\infty}$

Solution (a). Dividing numerator and denominator by n and using Theorem 9.1.3 yields

$$\lim_{n \to +\infty} \frac{n}{2n+1} = \lim_{n \to +\infty} \frac{1}{2 + 1/n} = \frac{\displaystyle\lim_{n \to +\infty} 1}{\displaystyle\lim_{n \to +\infty} (2 + 1/n)} = \frac{\displaystyle\lim_{n \to +\infty} 1}{\displaystyle\lim_{n \to +\infty} 2 + \lim_{n \to +\infty} 1/n}$$

$$= \frac{1}{2+0} = \frac{1}{2}$$

Thus, the sequence converges to $\frac{1}{2}$.

Solution (b). This sequence is the same as that in part (a), except for the factor of $(-1)^{n+1}$, which oscillates between $+1$ and -1. Thus, the terms in this sequence oscillate between positive and negative values, with the odd-numbered terms being identical to those in part (a) and the even-numbered terms being the negatives of those in part (a). Since the sequence in part (a) has a limit of $\frac{1}{2}$, it follows that the odd-numbered terms in this sequence approach $\frac{1}{2}$, and the even-numbered terms approach $-\frac{1}{2}$. Therefore, this sequence has no limit—it diverges.

Solution (c). Since $1/n \to 0$, the product $(-1)^{n+1}(1/n)$ oscillates between positive and negative values, with the odd-numbered terms approaching 0 through positive values and the even-numbered terms approaching 0 through negative values. Thus,

$$\lim_{n \to +\infty} (-1)^{n+1} \frac{1}{n} = 0$$

so the sequence converges to 0.

Solution (d). $\displaystyle\lim_{n \to +\infty} (8 - 2n) = -\infty$, so the sequence $\{8 - 2n\}_{n=1}^{+\infty}$ diverges. ◀

▶ **Example 4** In each part, determine whether the sequence converges, and if so, find its limit.

(a) $1, \dfrac{1}{2}, \dfrac{1}{2^2}, \dfrac{1}{2^3}, \ldots, \dfrac{1}{2^n}, \ldots$

(b) $1, 2, 2^2, 2^3, \ldots, 2^n, \ldots$

Solution. Replacing n by x in the first sequence produces the power function $(1/2)^x$, and replacing n by x in the second sequence produces the power function 2^x. Now recall that if $0 < b < 1$, then $b^x \to 0$ as $x \to +\infty$, and if $b > 1$, then $b^x \to +\infty$ as $x \to +\infty$ (Figure 0.5.1).

Thus,

$$\lim_{n \to +\infty} \frac{1}{2^n} = 0 \quad \text{and} \quad \lim_{n \to +\infty} 2^n = +\infty$$

So, the sequence $\{1/2^n\}$ converges to 0, but the sequence $\{2^n\}$ diverges. ◄

▶ **Example 5** Find the limit of the sequence $\left\{ \dfrac{n}{e^n} \right\}_{n=1}^{+\infty}$.

Solution. The expression

$$\lim_{n \to +\infty} \frac{n}{e^n}$$

is an indeterminate form of type ∞/∞, so L'Hôpital's rule is indicated. However, we cannot apply this rule directly to n/e^n because the functions n and e^n have been defined here only at the positive integers, and hence are not differentiable functions. To circumvent this problem we extend the domains of these functions to all real numbers, here implied by replacing n by x, and apply L'Hôpital's rule to the limit of the quotient x/e^x. This yields

$$\lim_{x \to +\infty} \frac{x}{e^x} = \lim_{x \to +\infty} \frac{1}{e^x} = 0$$

from which we can conclude that

$$\lim_{n \to +\infty} \frac{n}{e^n} = 0 \quad ◄$$

▶ **Example 6** Show that $\lim\limits_{n \to +\infty} \sqrt[n]{n} = 1$.

Solution.

$$\lim_{n \to +\infty} \sqrt[n]{n} = \lim_{n \to +\infty} n^{1/n} = \lim_{n \to +\infty} e^{(1/n) \ln n} = e^0 = 1 \qquad \boxed{\begin{array}{l} \text{By L'Hôpital's rule} \\ \text{applied to } (1/x) \ln x \end{array}} \quad ◄$$

Sometimes the even-numbered and odd-numbered terms of a sequence behave sufficiently differently that it is desirable to investigate their convergence separately. The following theorem, whose proof is omitted, is helpful for that purpose.

> **9.1.4 THEOREM** *A sequence converges to a limit L if and only if the sequences of even-numbered terms and odd-numbered terms both converge to L.*

▶ **Example 7** The sequence

$$\frac{1}{2}, \frac{1}{3}, \frac{1}{2^2}, \frac{1}{3^2}, \frac{1}{2^3}, \frac{1}{3^3}, \dots$$

converges to 0, since the even-numbered terms and the odd-numbered terms both converge to 0, and the sequence

$$1, \tfrac{1}{2}, 1, \tfrac{1}{3}, 1, \tfrac{1}{4}, \dots$$

diverges, since the odd-numbered terms converge to 1 and the even-numbered terms converge to 0. ◄

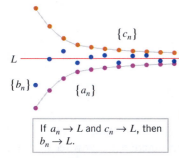

If $a_n \to L$ and $c_n \to L$, then $b_n \to L$.

▲ **Figure 9.1.5**

■ **THE SQUEEZING THEOREM FOR SEQUENCES**

The following theorem, illustrated in Figure 9.1.5, is an adaptation of the Squeezing Theorem (1.6.4) to sequences. This theorem will be useful for finding limits of sequences that cannot be obtained directly. The proof is omitted.

9.1.5 **THEOREM** (*The Squeezing Theorem for Sequences*) Let $\{a_n\}$, $\{b_n\}$, and $\{c_n\}$ be sequences such that

$$a_n \leq b_n \leq c_n \quad (\text{for all values of } n \text{ beyond some index } N)$$

If the sequences $\{a_n\}$ and $\{c_n\}$ have a common limit L as $n \to +\infty$, then $\{b_n\}$ also has the limit L as $n \to +\infty$.

Recall that if n is a positive integer, then $n!$ (read "n factorial") is the product of the first n positive integers. In addition, it is convenient to define $0! = 1$.

▶ **Example 8** Use numerical evidence to make a conjecture about the limit of the sequence

$$\left\{ \frac{n!}{n^n} \right\}_{n=1}^{+\infty}$$

and then confirm that your conjecture is correct.

Solution. Table 9.1.4, which was obtained with a calculating utility, suggests that the limit of the sequence may be 0. To confirm this we need to examine the limit of

$$a_n = \frac{n!}{n^n}$$

Table 9.1.4

n	$\dfrac{n!}{n^n}$
1	1.0000000000
2	0.5000000000
3	0.2222222222
4	0.0937500000
5	0.0384000000
6	0.0154320988
7	0.0061198990
8	0.0024032593
9	0.0009366567
10	0.0003628800
11	0.0001399059
12	0.0000537232

as $n \to +\infty$. Although this is an indeterminate form of type ∞/∞, L'Hôpital's rule is not helpful because we have no definition of $x!$ for values of x that are not integers. However, let us write out some of the initial terms and the general term in the sequence:

$$a_1 = 1, \quad a_2 = \frac{1 \cdot 2}{2 \cdot 2} = \frac{1}{2}, \quad a_3 = \frac{1 \cdot 2 \cdot 3}{3 \cdot 3 \cdot 3} = \frac{2}{9} < \frac{1}{3}, \quad a_4 = \frac{1 \cdot 2 \cdot 3 \cdot 4}{4 \cdot 4 \cdot 4 \cdot 4} = \frac{3}{32} < \frac{1}{4}, \cdots$$

If $n > 1$, the general term of the sequence can be rewritten as

$$a_n = \frac{1 \cdot 2 \cdot 3 \cdots n}{n \cdot n \cdot n \cdots n} = \frac{1}{n} \left(\frac{2 \cdot 3 \cdots n}{n \cdot n \cdots n} \right)$$

from which it follows that $a_n \leq 1/n$ (why?). It is now evident that

$$0 \leq a_n \leq \frac{1}{n}$$

However, the two outside expressions have a limit of 0 as $n \to +\infty$; thus, the Squeezing Theorem for Sequences implies that $a_n \to 0$ as $n \to +\infty$, which confirms our conjecture. ◀

The following theorem is often useful for finding the limit of a sequence with both positive and negative terms—it states that if the sequence $\{|a_n|\}$ that is obtained by taking the absolute value of each term in the sequence $\{a_n\}$ converges to 0, then $\{a_n\}$ also converges to 0.

9.1.6 **THEOREM** If $\lim\limits_{n \to +\infty} |a_n| = 0$, then $\lim\limits_{n \to +\infty} a_n = 0$.

PROOF Depending on the sign of a_n, either $a_n = |a_n|$ or $a_n = -|a_n|$. Thus, in all cases we have

$$-|a_n| \leq a_n \leq |a_n|$$

However, the limit of the two outside terms is 0, and hence the limit of a_n is 0 by the Squeezing Theorem for Sequences. ∎

▶ **Example 9**   Consider the sequence

$$1, -\frac{1}{2}, \frac{1}{2^2}, -\frac{1}{2^3}, \ldots, (-1)^n \frac{1}{2^n}, \ldots$$

If we take the absolute value of each term, we obtain the sequence

$$1, \frac{1}{2}, \frac{1}{2^2}, \frac{1}{2^3}, \ldots, \frac{1}{2^n}, \ldots$$

which, as shown in Example 4, converges to 0. Thus, from Theorem 9.1.6 we have

$$\lim_{n \to +\infty} \left[(-1)^n \frac{1}{2^n} \right] = 0 \quad \blacktriangleleft$$

■ **SEQUENCES DEFINED RECURSIVELY**

Some sequences do not arise from a formula for the general term, but rather from a formula or set of formulas that specify how to generate each term in the sequence from terms that precede it; such sequences are said to be defined ***recursively***, and the defining formulas are called ***recursion formulas***. A good example is the mechanic's rule for approximating square roots. In Exercise 25 of Section 4.7 you were asked to show that

$$x_1 = 1, \quad x_{n+1} = \frac{1}{2} \left(x_n + \frac{a}{x_n} \right) \tag{2}$$

describes the sequence produced by Newton's Method to approximate $\sqrt{a}$ as a zero of the function $f(x) = x^2 - a$. Table 9.1.5 shows the first five terms in an application of the mechanic's rule to approximate $\sqrt{2}$.

Table 9.1.5

n	$x_1 = 1, \quad x_{n+1} = \frac{1}{2}\left(x_n + \frac{2}{x_n}\right)$	DECIMAL APPROXIMATION
	$x_1 = 1$ (Starting value)	1.00000000000
1	$x_2 = \frac{1}{2}\left[1 + \frac{2}{1}\right] = \frac{3}{2}$	1.50000000000
2	$x_3 = \frac{1}{2}\left[\frac{3}{2} + \frac{2}{3/2}\right] = \frac{17}{12}$	1.41666666667
3	$x_4 = \frac{1}{2}\left[\frac{17}{12} + \frac{2}{17/12}\right] = \frac{577}{408}$	1.41421568627
4	$x_5 = \frac{1}{2}\left[\frac{577}{408} + \frac{2}{577/408}\right] = \frac{665,857}{470,832}$	1.41421356237
5	$x_6 = \frac{1}{2}\left[\frac{665,857}{470,832} + \frac{2}{665,857/470,832}\right] = \frac{886,731,088,897}{627,013,566,048}$	1.41421356237

It would take us too far afield to investigate the convergence of sequences defined recursively, but we will conclude this section with a useful technique that can sometimes be used to compute limits of such sequences.

▶ **Example 10**   Assuming that the sequence in Table 9.1.5 converges, show that the limit is $\sqrt{2}$.

Solution. Assume that $x_n \to L$, where L is to be determined. Since $n + 1 \to +\infty$ as $n \to +\infty$, it is also true that $x_{n+1} \to L$ as $n \to +\infty$. Thus, if we take the limit of the expression

$$x_{n+1} = \frac{1}{2}\left(x_n + \frac{2}{x_n}\right)$$

as $n \to +\infty$, we obtain

$$L = \frac{1}{2}\left(L + \frac{2}{L}\right)$$

which can be rewritten as $L^2 = 2$. The negative solution of this equation is extraneous because $x_n > 0$ for all n, so $L = \sqrt{2}$. ◄

✔ QUICK CHECK EXERCISES 9.1   (See page 607 for answers.)

1. Consider the sequence $4, 6, 8, 10, 12, \ldots$.
   (a) If $\{a_n\}_{n=1}^{+\infty}$ denotes this sequence, then $a_1 = $ _____, $a_4 = $ _____, and $a_7 = $ _____. The general term is $a_n = $ _____.
   (b) If $\{b_n\}_{n=0}^{+\infty}$ denotes this sequence, then $b_0 = $ _____, $b_4 = $ _____, and $b_8 = $ _____. The general term is $b_n = $ _____.

2. What does it mean to say that a sequence $\{a_n\}$ converges?

3. Consider sequences $\{a_n\}$ and $\{b_n\}$, where $a_n \to 2$ as $n \to +\infty$ and $b_n = (-1)^n$. Determine which of the following se-quences converge and which diverge. If a sequence con-verges, indicate its limit.
   (a) $\{b_n\}$ (b) $\{3a_n - 1\}$ (c) $\{b_n^2\}$
   (d) $\{a_n + b_n\}$ (e) $\left\{\dfrac{1}{a_n^2 + 3}\right\}$ (f) $\left\{\dfrac{b_n}{1000}\right\}$

4. Suppose that $\{a_n\}$, $\{b_n\}$, and $\{c_n\}$ are sequences such that $a_n \leq b_n \leq c_n$ for all $n \geq 10$, and that $\{a_n\}$ and $\{c_n\}$ both converge to 12. Then the _____ Theorem for Sequences implies that $\{b_n\}$ converges to _____.

EXERCISE SET 9.1   ~ Graphing Utility

1. In each part, find a formula for the general term of the se-quence, starting with $n = 1$.
   (a) $1, \dfrac{1}{3}, \dfrac{1}{9}, \dfrac{1}{27}, \ldots$ (b) $1, -\dfrac{1}{3}, \dfrac{1}{9}, -\dfrac{1}{27}, \ldots$
   (c) $\dfrac{1}{2}, \dfrac{3}{4}, \dfrac{5}{6}, \dfrac{7}{8}, \ldots$ (d) $\dfrac{1}{\sqrt{\pi}}, \dfrac{4}{\sqrt[3]{\pi}}, \dfrac{9}{\sqrt[4]{\pi}}, \dfrac{16}{\sqrt[5]{\pi}}, \ldots$

2. In each part, find two formulas for the general term of the sequence, one starting with $n = 1$ and the other with $n = 0$.
   (a) $1, -r, r^2, -r^3, \ldots$ (b) $r, -r^2, r^3, -r^4, \ldots$

3. (a) Write out the first four terms of the sequence $\{1 + (-1)^n\}$, starting with $n = 0$.
   (b) Write out the first four terms of the sequence $\{\cos n\pi\}$, starting with $n = 0$.
   (c) Use the results in parts (a) and (b) to express the gen-eral term of the sequence $4, 0, 4, 0, \ldots$ in two different ways, starting with $n = 0$.

4. In each part, find a formula for the general term using fac-torials and starting with $n = 1$.
   (a) $1 \cdot 2, 1 \cdot 2 \cdot 3 \cdot 4, 1 \cdot 2 \cdot 3 \cdot 4 \cdot 5 \cdot 6,$ $1 \cdot 2 \cdot 3 \cdot 4 \cdot 5 \cdot 6 \cdot 7 \cdot 8, \ldots$
   (b) $1, 1 \cdot 2 \cdot 3, 1 \cdot 2 \cdot 3 \cdot 4 \cdot 5, 1 \cdot 2 \cdot 3 \cdot 4 \cdot 5 \cdot 6 \cdot 7, \ldots$

5–6 Let f be the function $f(x) = \cos\left(\dfrac{\pi}{2}x\right)$ and define se-quences $\{a_n\}$ and $\{b_n\}$ by $a_n = f(2n)$ and $b_n = f(2n + 1)$. ■

5. (a) Does $\lim_{x \to +\infty} f(x)$ exist? Explain.
   (b) Evaluate a_1, a_2, a_3, a_4, and a_5.
   (c) Does $\{a_n\}$ converge? If so, find its limit.

6. (a) Evaluate b_1, b_2, b_3, b_4, and b_5.
   (b) Does $\{b_n\}$ converge? If so, find its limit.
   (c) Does $\{f(n)\}$ converge? If so, find its limit.

7–22 Write out the first five terms of the sequence, determine whether the sequence converges, and if so find its limit. ■

7. $\left\{\dfrac{n}{n+2}\right\}_{n=1}^{+\infty}$ 8. $\left\{\dfrac{n^2}{2n+1}\right\}_{n=1}^{+\infty}$ 9. $\{2\}_{n=1}^{+\infty}$

10. $\left\{\ln\left(\dfrac{1}{n}\right)\right\}_{n=1}^{+\infty}$ 11. $\left\{\dfrac{\ln n}{n}\right\}_{n=1}^{+\infty}$ 12. $\left\{n\sin\dfrac{\pi}{n}\right\}_{n=1}^{+\infty}$

13. $\{1 + (-1)^n\}_{n=1}^{+\infty}$ 14. $\left\{\dfrac{(-1)^{n+1}}{n^2}\right\}_{n=1}^{+\infty}$

15. $\left\{(-1)^n\dfrac{2n^3}{n^3+1}\right\}_{n=1}^{+\infty}$ 16. $\left\{\dfrac{n}{2^n}\right\}_{n=1}^{+\infty}$

17. $\left\{\dfrac{(n+1)(n+2)}{2n^2}\right\}_{n=1}^{+\infty}$ 18. $\left\{\dfrac{\pi^n}{4^n}\right\}_{n=1}^{+\infty}$

19. $\{n^2 e^{-n}\}_{n=1}^{+\infty}$ 20. $\{\sqrt{n^2 + 3n} - n\}_{n=1}^{+\infty}$

21. $\left\{\left(\dfrac{n+3}{n+1}\right)^{n}\right\}_{n=1}^{+\infty}$ **22.** $\left\{\left(1-\dfrac{2}{n}\right)^{n}\right\}_{n=1}^{+\infty}$

23–30 Find the general term of the sequence, starting with $n=1$, determine whether the sequence converges, and if so find its limit. ■

23. $\dfrac{1}{2}, \dfrac{3}{4}, \dfrac{5}{6}, \dfrac{7}{8}, \dots$ **24.** $0, \dfrac{1}{2^2}, \dfrac{2}{3^2}, \dfrac{3}{4^2}, \dots$

25. $\dfrac{1}{3}, -\dfrac{1}{9}, \dfrac{1}{27}, -\dfrac{1}{81}, \dots$ **26.** $-1, 2, -3, 4, -5, \dots$

27. $\left(1-\dfrac{1}{2}\right), \left(\dfrac{1}{3}-\dfrac{1}{2}\right), \left(\dfrac{1}{3}-\dfrac{1}{4}\right), \left(\dfrac{1}{5}-\dfrac{1}{4}\right), \dots$

28. $3, \dfrac{3}{2}, \dfrac{3}{2^2}, \dfrac{3}{2^3}, \dots$

29. $(\sqrt{2}-\sqrt{3}), (\sqrt{3}-\sqrt{4}), (\sqrt{4}-\sqrt{5}), \dots$

30. $\dfrac{1}{3^5}, -\dfrac{1}{3^6}, \dfrac{1}{3^7}, -\dfrac{1}{3^8}, \dots$

31–34 True–False Determine whether the statement is true or false. Explain your answer. ■

31. Sequences are functions.

32. If $\{a_n\}$ and $\{b_n\}$ are sequences such that $\{a_n + b_n\}$ converges, then $\{a_n\}$ and $\{b_n\}$ converge.

33. If $\{a_n\}$ diverges, then $a_n \to +\infty$ or $a_n \to -\infty$.

34. If the graph of $y = f(x)$ has a horizontal asymptote as $x \to +\infty$, then the sequence $\{f(n)\}$ converges.

35–36 Use numerical evidence to make a conjecture about the limit of the sequence, and then use the Squeezing Theorem for Sequences (Theorem 9.1.5) to confirm that your conjecture is correct. ■

35. $\displaystyle\lim_{n \to +\infty} \dfrac{\sin^2 n}{n}$ **36.** $\displaystyle\lim_{n \to +\infty} \left(\dfrac{1+n}{2n}\right)^n$

FOCUS ON CONCEPTS

37. Give two examples of sequences, all of whose terms are between -10 and 10, that do not converge. Use graphs of your sequences to explain their properties.

38. (a) Suppose that f satisfies $\lim_{x \to 0^+} f(x) = +\infty$. Is it possible that the sequence $\{f(1/n)\}$ converges? Explain.
    (b) Find a function f such that $\lim_{x \to 0^+} f(x)$ does not exist but the sequence $\{f(1/n)\}$ converges.

39. (a) Starting with $n = 1$, write out the first six terms of the sequence $\{a_n\}$, where
$$a_n = \begin{cases} 1, & \text{if } n \text{ is odd} \\ n, & \text{if } n \text{ is even} \end{cases}$$
    (b) Starting with $n = 1$, and considering the even and odd terms separately, find a formula for the general term of the sequence
$$1, \dfrac{1}{2^2}, 3, \dfrac{1}{2^4}, 5, \dfrac{1}{2^6}, \dots$$

(c) Starting with $n = 1$, and considering the even and odd terms separately, find a formula for the general term of the sequence
$$1, \dfrac{1}{3}, \dfrac{1}{3}, \dfrac{1}{5}, \dfrac{1}{5}, \dfrac{1}{7}, \dfrac{1}{7}, \dfrac{1}{9}, \dfrac{1}{9}, \dots$$

(d) Determine whether the sequences in parts (a), (b), and (c) converge. For those that do, find the limit.

40. For what positive values of b does the sequence $b, 0, b^2, 0, b^3, 0, b^4, \dots$ converge? Justify your answer.

41. Assuming that the sequence given in Formula (2) of this section converges, use the method of Example 10 to show that the limit of this sequence is $\sqrt{a}$.

42. Consider the sequence
$$a_1 = \sqrt{6}$$
$$a_2 = \sqrt{6 + \sqrt{6}}$$
$$a_3 = \sqrt{6 + \sqrt{6 + \sqrt{6}}}$$
$$a_4 = \sqrt{6 + \sqrt{6 + \sqrt{6 + \sqrt{6}}}}$$
$$\vdots$$
(a) Find a recursion formula for a_{n+1}.
(b) Assuming that the sequence converges, use the method of Example 10 to find the limit.

43. (a) A bored student enters the number 0.5 in a calculator display and then repeatedly computes the square of the number in the display. Taking $a_0 = 0.5$, find a formula for the general term of the sequence $\{a_n\}$ of numbers that appear in the display.
(b) Try this with a calculator and make a conjecture about the limit of a_n.
(c) Confirm your conjecture by finding the limit of a_n.
(d) For what values of a_0 will this procedure produce a convergent sequence?

44. Let
$$f(x) = \begin{cases} 2x, & 0 \le x < 0.5 \\ 2x - 1, & 0.5 \le x < 1 \end{cases}$$
Does the sequence $f(0.2), f(f(0.2)), f(f(f(0.2))), \dots$ converge? Justify your reasoning.

45. (a) Use a graphing utility to generate the graph of the equation $y = (2^x + 3^x)^{1/x}$, and then use the graph to make a conjecture about the limit of the sequence
$$\{(2^n + 3^n)^{1/n}\}_{n=1}^{+\infty}$$
(b) Confirm your conjecture by calculating the limit.

46. Consider the sequence $\{a_n\}_{n=1}^{+\infty}$ whose nth term is
$$a_n = \dfrac{1}{n} \sum_{k=1}^{n} \dfrac{1}{1 + (k/n)}$$
Show that $\lim_{n \to +\infty} a_n = \ln 2$ by interpreting a_n as the Riemann sum of a definite integral.

47. The sequence whose terms are $1, 1, 2, 3, 5, 8, 13, 21, \ldots$ is called the **Fibonacci sequence** in honor of the Italian mathematician Leonardo ("Fibonacci") da Pisa (c. 1170–1250). This sequence has the property that after starting with two 1's, each term is the sum of the preceding two.

(a) Denoting the sequence by $\{a_n\}$ and starting with $a_1 = 1$ and $a_2 = 1$, show that

$$\frac{a_{n+2}}{a_{n+1}} = 1 + \frac{a_n}{a_{n+1}} \quad \text{if } n \geq 1$$

(b) Give a reasonable informal argument to show that if the sequence $\{a_{n+1}/a_n\}$ converges to some limit L, then the sequence $\{a_{n+2}/a_{n+1}\}$ must also converge to L.

(c) Assuming that the sequence $\{a_{n+1}/a_n\}$ converges, show that its limit is $(1 + \sqrt{5})/2$.

48. If we accept the fact that the sequence $\{1/n\}_{n=1}^{+\infty}$ converges to the limit $L = 0$, then according to Definition 9.1.2, for every $\epsilon > 0$, there exists a positive integer N such that $|a_n - L| = |(1/n) - 0| < \epsilon$ when $n \geq N$. In each part, find the smallest possible value of N for the given value of ϵ.

(a) $\epsilon = 0.5$ (b) $\epsilon = 0.1$ (c) $\epsilon = 0.001$

49. If we accept the fact that the sequence

$$\left\{ \frac{n}{n+1} \right\}_{n=1}^{+\infty}$$

converges to the limit $L = 1$, then according to Definition 9.1.2, for every $\epsilon > 0$ there exists an integer N such that

$$|a_n - L| = \left| \frac{n}{n+1} - 1 \right| < \epsilon$$

when $n \geq N$. In each part, find the smallest value of N for the given value of ϵ.

(a) $\epsilon = 0.25$ (b) $\epsilon = 0.1$ (c) $\epsilon = 0.001$

50. Use Definition 9.1.2 to prove that

(a) the sequence $\{1/n\}_{n=1}^{+\infty}$ converges to 0

(b) the sequence $\left\{ \frac{n}{n+1} \right\}_{n=1}^{+\infty}$ converges to 1.

51. **Writing** Discuss, with examples, various ways that a sequence could diverge.

52. **Writing** Discuss the convergence of the sequence $\{r^n\}$ considering the cases $|r| < 1$, $|r| > 1$, $r = 1$, and $r = -1$ separately.

✔**QUICK CHECK ANSWERS 9.1**

1. (a) $4; 10; 16; 2n + 2$ (b) $4; 12; 20; 2n + 4$ **2.** $\lim\limits_{n \to +\infty} a_n$ exists   **3.** (a) diverges (b) converges to 5 (c) converges to 1

(d) diverges (e) converges to $\frac{1}{7}$ (f) diverges   **4.** Squeezing; 12

9.2 MONOTONE SEQUENCES

There are many situations in which it is important to know whether a sequence converges, but the value of the limit is not relevant to the problem at hand. In this section we will study several techniques that can be used to determine whether a sequence converges.

■ TERMINOLOGY

We begin with some terminology.

9.2.1 **DEFINITION** A sequence $\{a_n\}_{n=1}^{+\infty}$ is called

strictly increasing if	$a_1 < a_2 < a_3 < \cdots < a_n < \cdots$
increasing if	$a_1 \leq a_2 \leq a_3 \leq \cdots \leq a_n \leq \cdots$
strictly decreasing if	$a_1 > a_2 > a_3 > \cdots > a_n > \cdots$
decreasing if	$a_1 \geq a_2 \geq a_3 \geq \cdots \geq a_n \geq \cdots$

A sequence that is either increasing or decreasing is said to be ***monotone***, and a sequence that is either strictly increasing or strictly decreasing is said to be ***strictly monotone***.

Note that an increasing sequence need not be strictly increasing, and a decreasing sequence need not be strictly decreasing.

Some examples are given in Table 9.2.1 and their corresponding graphs are shown in Figure 9.2.1. The first and second sequences in Table 9.2.1 are strictly monotone; the third and fourth sequences are monotone but not strictly monotone; and the fifth sequence is neither strictly monotone nor monotone.

Table 9.2.1

SEQUENCE	DESCRIPTION
$\dfrac{1}{2}, \dfrac{2}{3}, \dfrac{3}{4}, \ldots, \dfrac{n}{n+1}, \ldots$	Strictly increasing
$1, \dfrac{1}{2}, \dfrac{1}{3}, \ldots, \dfrac{1}{n}, \ldots$	Strictly decreasing
$1, 1, 2, 2, 3, 3, \ldots$	Increasing; not strictly increasing
$1, 1, \dfrac{1}{2}, \dfrac{1}{2}, \dfrac{1}{3}, \dfrac{1}{3}, \ldots$	Decreasing; not strictly decreasing
$1, -\dfrac{1}{2}, \dfrac{1}{3}, -\dfrac{1}{4}, \ldots, (-1)^{n+1}\dfrac{1}{n}, \ldots$	Neither increasing nor decreasing

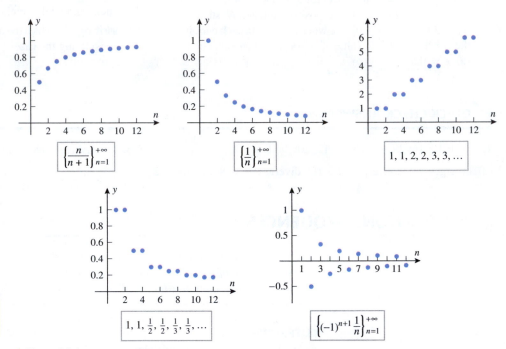

> **Can a sequence be both increasing and decreasing? Explain.**

▲ **Figure 9.2.1**

■ TESTING FOR MONOTONICITY

Frequently, one can *guess* whether a sequence is monotone or strictly monotone by writing out some of the initial terms. However, to be certain that the guess is correct, one must give a precise mathematical argument. Table 9.2.2 provides two ways of doing this, one based

Table 9.2.2

DIFFERENCE BETWEEN SUCCESSIVE TERMS	RATIO OF SUCCESSIVE TERMS	CONCLUSION
$a_{n+1} - a_n > 0$	$a_{n+1}/a_n > 1$	Strictly increasing
$a_{n+1} - a_n < 0$	$a_{n+1}/a_n < 1$	Strictly decreasing
$a_{n+1} - a_n \geq 0$	$a_{n+1}/a_n \geq 1$	Increasing
$a_{n+1} - a_n \leq 0$	$a_{n+1}/a_n \leq 1$	Decreasing

on differences of successive terms and the other on ratios of successive terms. It is assumed in the latter case that the terms are positive. One must show that the specified conditions hold for *all* pairs of successive terms.

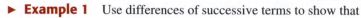

▶ **Example 1**　Use differences of successive terms to show that

$$\frac{1}{2}, \frac{2}{3}, \frac{3}{4}, \ldots, \frac{n}{n+1}, \ldots$$

(Figure 9.2.2) is a strictly increasing sequence.

Solution.　The pattern of the initial terms suggests that the sequence is strictly increasing. To prove that this is so, let

$$a_n = \frac{n}{n+1}$$

We can obtain a_{n+1} by replacing n by $n+1$ in this formula. This yields

$$a_{n+1} = \frac{n+1}{(n+1)+1} = \frac{n+1}{n+2}$$

Thus, for $n \geq 1$

$$a_{n+1} - a_n = \frac{n+1}{n+2} - \frac{n}{n+1} = \frac{n^2 + 2n + 1 - n^2 - 2n}{(n+1)(n+2)} = \frac{1}{(n+1)(n+2)} > 0$$

which proves that the sequence is strictly increasing. ◀

▶ **Example 2**　Use ratios of successive terms to show that the sequence in Example 1 is strictly increasing.

Solution.　As shown in the solution of Example 1,

$$a_n = \frac{n}{n+1} \quad \text{and} \quad a_{n+1} = \frac{n+1}{n+2}$$

Forming the ratio of successive terms we obtain

$$\frac{a_{n+1}}{a_n} = \frac{(n+1)/(n+2)}{n/(n+1)} = \frac{n+1}{n+2} \cdot \frac{n+1}{n} = \frac{n^2 + 2n + 1}{n^2 + 2n} \tag{1}$$

from which we see that $a_{n+1}/a_n > 1$ for $n \geq 1$. This proves that the sequence is strictly increasing. ◀

The following example illustrates still a third technique for determining whether a sequence is strictly monotone.

▶ **Example 3**　In Examples 1 and 2 we proved that the sequence

$$\frac{1}{2}, \frac{2}{3}, \frac{3}{4}, \ldots, \frac{n}{n+1}, \ldots$$

is strictly increasing by considering the difference and ratio of successive terms. Alternatively, we can proceed as follows. Let

$$f(x) = \frac{x}{x+1}$$

so that the nth term in the given sequence is $a_n = f(n)$. The function f is increasing for $x \geq 1$ since

$$f'(x) = \frac{(x+1)(1) - x(1)}{(x+1)^2} = \frac{1}{(x+1)^2} > 0$$

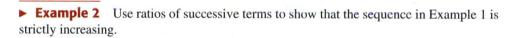

▲ **Figure 9.2.2**

Table 9.2.3

DERIVATIVE OF f FOR $x \geq 1$	CONCLUSION FOR THE SEQUENCE WITH $a_n = f(n)$
$f'(x) > 0$	Strictly increasing
$f'(x) < 0$	Strictly decreasing
$f'(x) \geq 0$	Increasing
$f'(x) \leq 0$	Decreasing

Thus,

$$a_n = f(n) < f(n+1) = a_{n+1}$$

which proves that the given sequence is strictly increasing. ◄

In general, if $f(n) = a_n$ is the nth term of a sequence, and if f is differentiable for $x \geq 1$, then the results in Table 9.2.3 can be used to investigate the monotonicity of the sequence.

■ PROPERTIES THAT HOLD EVENTUALLY

Sometimes a sequence will behave erratically at first and then settle down into a definite pattern. For example, the sequence

$$9, -8, -17, 12, 1, 2, 3, 4, \ldots \tag{2}$$

is strictly increasing from the fifth term on, but the sequence as a whole cannot be classified as strictly increasing because of the erratic behavior of the first four terms. To describe such sequences, we introduce the following terminology.

> **9.2.2   DEFINITION**   If discarding finitely many terms from the beginning of a sequence produces a sequence with a certain property, then the original sequence is said to have that property *eventually*.

For example, although we cannot say that sequence (2) is strictly increasing, we can say that it is eventually strictly increasing.

▶ **Example 4**   Show that the sequence $\left\{ \dfrac{10^n}{n!} \right\}_{n=1}^{+\infty}$ is eventually strictly decreasing.

Solution.   We have

$$a_n = \frac{10^n}{n!} \quad \text{and} \quad a_{n+1} = \frac{10^{n+1}}{(n+1)!}$$

so

$$\frac{a_{n+1}}{a_n} = \frac{10^{n+1}/(n+1)!}{10^n/n!} = \frac{10^{n+1}n!}{10^n(n+1)!} = 10\frac{n!}{(n+1)n!} = \frac{10}{n+1} \tag{3}$$

From (3), $a_{n+1}/a_n < 1$ for all $n \geq 10$, so the sequence is eventually strictly decreasing, as confirmed by the graph in Figure 9.2.3. ◄

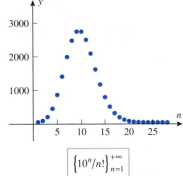

$\left\{ 10^n/n! \right\}_{n=1}^{+\infty}$

▲ **Figure 9.2.3**

■ AN INTUITIVE VIEW OF CONVERGENCE

Informally stated, the convergence or divergence of a sequence does not depend on the behavior of its *initial terms*, but rather on how the terms behave *eventually*. For example, the sequence

$$3, -9, -13, 17, 1, \frac{1}{2}, \frac{1}{3}, \frac{1}{4}, \ldots$$

eventually behaves like the sequence

$$1, \frac{1}{2}, \frac{1}{3}, \ldots, \frac{1}{n}, \ldots$$

and hence has a limit of 0.

■ CONVERGENCE OF MONOTONE SEQUENCES

The following two theorems, whose proofs are discussed at the end of this section, show that a monotone sequence either converges or becomes infinite—divergence by oscillation cannot occur.

> **9.2.3** **THEOREM** *If a sequence $\{a_n\}$ is eventually increasing, then there are two possibilities:*
>
> (a) *There is a constant M, called an **upper bound** for the sequence, such that $a_n \leq M$ for all n, in which case the sequence converges to a limit L satisfying $L \leq M$.*
>
> (b) *No upper bound exists, in which case $\lim\limits_{n \to +\infty} a_n = +\infty$.*

> **9.2.4** **THEOREM** *If a sequence $\{a_n\}$ is eventually decreasing, then there are two possibilities:*
>
> (a) *There is a constant M, called a **lower bound** for the sequence, such that $a_n \geq M$ for all n, in which case the sequence converges to a limit L satisfying $L \geq M$.*
>
> (b) *No lower bound exists, in which case $\lim\limits_{n \to +\infty} a_n = -\infty$.*

Theorems 9.2.3 and 9.2.4 are examples of *existence theorems*; they tell us whether a limit exists, but they do not provide a method for finding it.

▶ **Example 5** Show that the sequence $\left\{ \dfrac{10^n}{n!} \right\}_{n=1}^{+\infty}$ converges and find its limit.

Solution. We showed in Example 4 that the sequence is eventually strictly decreasing. Since all terms in the sequence are positive, it is bounded below by $M = 0$, and hence Theorem 9.2.4 guarantees that it converges to a nonnegative limit L. However, the limit is not evident directly from the formula $10^n/n!$ for the nth term, so we will need some ingenuity to obtain it.

It follows from Formula (3) of Example 4 that successive terms in the given sequence are related by the recursion formula

$$a_{n+1} = \frac{10}{n+1} a_n \tag{4}$$

where $a_n = 10^n/n!$. We will take the limit as $n \to +\infty$ of both sides of (4) and use the fact that

$$\lim_{n \to +\infty} a_{n+1} = \lim_{n \to +\infty} a_n = L$$

We obtain

$$L = \lim_{n \to +\infty} a_{n+1} = \lim_{n \to +\infty} \left(\frac{10}{n+1} a_n \right) = \lim_{n \to +\infty} \frac{10}{n+1} \lim_{n \to +\infty} a_n = 0 \cdot L = 0$$

so that

$$L = \lim_{n \to +\infty} \frac{10^n}{n!} = 0 \blacktriangleleft$$

In the exercises we will show that the technique illustrated in the last example can be adapted to obtain

$$\lim_{n \to +\infty} \frac{x^n}{n!} = 0 \tag{5}$$

for any real value of x (Exercise 29). This result will be useful in our later work.

■ THE COMPLETENESS AXIOM

In this text we have accepted the familiar properties of real numbers without proof, and indeed, we have not even attempted to define the term *real number*. Although this is sufficient for many purposes, it was recognized by the late nineteenth century that the study of limits and functions in calculus requires a precise axiomatic formulation of the real numbers analogous to the axiomatic development of Euclidean geometry. Although we will not attempt to pursue this development, we will need to discuss one of the axioms about real numbers in order to prove Theorems 9.2.3 and 9.2.4. But first we will introduce some terminology.

If S is a nonempty set of real numbers, then we call u an ***upper bound*** for S if u is greater than or equal to every number in S, and we call l a ***lower bound*** for S if l is smaller than or equal to every number in S. For example, if S is the set of numbers in the interval $(1, 3)$, then $u = 10, 4, 3.2$, and 3 are upper bounds for S and $l = -10, 0, 0.5$, and 1 are lower bounds for S. Observe also that $u = 3$ is the smallest of all upper bounds and $l = 1$ is the largest of all lower bounds. The existence of a smallest upper bound and a largest lower bound for S is not accidental; it is a consequence of the following axiom.

9.2.5 **AXIOM** (***The Completeness Axiom***) *If a nonempty set S of real numbers has an upper bound, then it has a smallest upper bound (called the **least upper bound**), and if a nonempty set S of real numbers has a lower bound, then it has a largest lower bound (called the **greatest lower bound**).*

PROOF OF THEOREM 9.2.3

(a) We will prove the result for increasing sequences, and leave it for the reader to adapt the argument to sequences that are eventually increasing. Assume there exists a number M such that $a_n \leq M$ for $n = 1, 2, \ldots$. Then M is an upper bound for the set of terms in the sequence. By the Completeness Axiom there is a least upper bound for the terms; call it L. Now let ϵ be any positive number. Since L is the least upper bound for the terms, $L - \epsilon$ is not an upper bound for the terms, which means that there is at least one term a_N such that

$$a_N > L - \epsilon$$

Moreover, since $\{a_n\}$ is an increasing sequence, we must have

$$a_n \geq a_N > L - \epsilon \tag{6}$$

when $n \geq N$. But a_n cannot exceed L since L is an upper bound for the terms. This observation together with (6) tells us that $L \geq a_n > L - \epsilon$ for $n \geq N$, so all terms from the Nth on are within ϵ units of L. This is exactly the requirement to have

$$\lim_{n \to +\infty} a_n = L$$

Finally, $L \leq M$ since M is an upper bound for the terms and L is the least upper bound. This proves part (a).

(b) If there is no number M such that $a_n \leq M$ for $n = 1, 2, \ldots$, then no matter how large we choose M, there is a term a_N such that

$$a_N > M$$

and, since the sequence is increasing,

$$a_n \geq a_N > M$$

when $n \geq N$. Thus, the terms in the sequence become arbitrarily large as n increases. That is,

$$\lim_{n \to +\infty} a_n = +\infty \quad \blacksquare$$

We omit the proof of Theorem 9.2.4 since it is similar to that of 9.2.3.

✔ **QUICK CHECK EXERCISES 9.2**   *(See page 614 for answers.)*

1. Classify each sequence as (I) increasing, (D) decreasing, or (N) neither increasing nor decreasing.

   ———— $\{2n\}$ ———— $\{2^{-n}\}$

   ———— $\left\{\dfrac{5-n}{n^2}\right\}$ ———— $\left\{\dfrac{-1}{n^2}\right\}$

   ———— $\left\{\dfrac{(-1)^n}{n^2}\right\}$

2. Classify each sequence as (M) monotonic, (S) strictly monotonic, or (N) not monotonic.

   ———— $\{n + (-1)^n\}$ ———— $\{2n + (-1)^n\}$

   ———— $\{3n + (-1)^n\}$

3. Since
   $$\frac{n/[2(n+1)]}{(n-1)/(2n)} = \frac{n^2}{n^2 - 1} > \text{————}$$
   the sequence $\{(n-1)/(2n)\}$ is strictly ————.

4. Since
   $$\frac{d}{dx}[(x-8)^2] > 0 \text{ for } x > \text{————}$$
   the sequence $\{(n-8)^2\}$ is ———— strictly ————.

EXERCISE SET 9.2

1–6 Use the difference $a_{n+1} - a_n$ to show that the given sequence $\{a_n\}$ is strictly increasing or strictly decreasing. ■

1. $\left\{\dfrac{1}{n}\right\}_{n=1}^{+\infty}$ 2. $\left\{1 - \dfrac{1}{n}\right\}_{n=1}^{+\infty}$ 3. $\left\{\dfrac{n}{2n+1}\right\}_{n=1}^{+\infty}$

4. $\left\{\dfrac{n}{4n-1}\right\}_{n=1}^{+\infty}$ 5. $\{n - 2^n\}_{n=1}^{+\infty}$ 6. $\{n - n^2\}_{n=1}^{+\infty}$

7–12 Use the ratio a_{n+1}/a_n to show that the given sequence $\{a_n\}$ is strictly increasing or strictly decreasing. ■

7. $\left\{\dfrac{n}{2n+1}\right\}_{n=1}^{+\infty}$ 8. $\left\{\dfrac{2^n}{1+2^n}\right\}_{n=1}^{+\infty}$ 9. $\{ne^{-n}\}_{n=1}^{+\infty}$

10. $\left\{\dfrac{10^n}{(2n)!}\right\}_{n=1}^{+\infty}$ 11. $\left\{\dfrac{n^n}{n!}\right\}_{n=1}^{+\infty}$ 12. $\left\{\dfrac{5^n}{2^{(n^2)}}\right\}_{n=1}^{+\infty}$

13–16 True–False Determine whether the statement is true or false. Explain your answer. ■

13. If $a_{n+1} - a_n > 0$ for all $n \geq 1$, then the sequence $\{a_n\}$ is strictly increasing.

14. A sequence $\{a_n\}$ is monotone if $a_{n+1} - a_n \neq 0$ for all $n \geq 1$.

15. Any bounded sequence converges.

16. If $\{a_n\}$ is eventually increasing, then $a_{100} < a_{200}$.

17–20 Use differentiation to show that the given sequence is strictly increasing or strictly decreasing. ■

17. $\left\{\dfrac{n}{2n+1}\right\}_{n=1}^{+\infty}$ 18. $\left\{\dfrac{\ln(n+2)}{n+2}\right\}_{n=1}^{+\infty}$

19. $\{\tan^{-1} n\}_{n=1}^{+\infty}$ 20. $\{ne^{-2n}\}_{n=1}^{+\infty}$

21–24 Show that the given sequence is eventually strictly increasing or eventually strictly decreasing. ■

21. $\{2n^2 - 7n\}_{n=1}^{+\infty}$ 22. $\{n^3 - 4n^2\}_{n=1}^{+\infty}$

23. $\left\{\dfrac{n!}{3^n}\right\}_{n=1}^{+\infty}$ 24. $\{n^5 e^{-n}\}_{n=1}^{+\infty}$

FOCUS ON CONCEPTS

25. Suppose that $\{a_n\}$ is a monotone sequence such that $1 \leq a_n \leq 2$ for all n. Must the sequence converge? If so, what can you say about the limit?

26. Suppose that $\{a_n\}$ is a monotone sequence such that $a_n \leq 2$ for all n. Must the sequence converge? If so, what can you say about the limit?

27. Let $\{a_n\}$ be the sequence defined recursively by $a_1 = \sqrt{2}$ and $a_{n+1} = \sqrt{2 + a_n}$ for $n \geq 1$.
    (a) List the first three terms of the sequence.
    (b) Show that $a_n < 2$ for $n \geq 1$.
    (c) Show that $a_{n+1}^2 - a_n^2 = (2 - a_n)(1 + a_n)$ for $n \geq 1$.
    (d) Use the results in parts (b) and (c) to show that $\{a_n\}$ is a strictly increasing sequence. [*Hint:* If x and y are positive real numbers such that $x^2 - y^2 > 0$, then it follows by factoring that $x - y > 0$.]
    (e) Show that $\{a_n\}$ converges and find its limit L.

28. Let $\{a_n\}$ be the sequence defined recursively by $a_1 = 1$ and $a_{n+1} = \frac{1}{2}[a_n + (3/a_n)]$ for $n \geq 1$.
    (a) Show that $a_n \geq \sqrt{3}$ for $n \geq 2$. [*Hint:* What is the minimum value of $\frac{1}{2}[x + (3/x)]$ for $x > 0$?]
    (b) Show that $\{a_n\}$ is eventually decreasing. [*Hint:* Examine $a_{n+1} - a_n$ or a_{n+1}/a_n and use the result in part (a).]
    (c) Show that $\{a_n\}$ converges and find its limit L.

29. The goal of this exercise is to establish Formula (5), namely,
    $$\lim_{n \to +\infty} \frac{x^n}{n!} = 0$$
    Let $a_n = |x|^n/n!$ and observe that the case where $x = 0$ is obvious, so we will focus on the case where $x \neq 0$.
    (a) Show that
    $$a_{n+1} = \frac{|x|}{n+1} a_n$$
    (b) Show that the sequence $\{a_n\}$ is eventually strictly decreasing.

(cont.)

(c) Show that the sequence $\{a_n\}$ converges.

30. (a) Compare appropriate areas in the accompanying figure to deduce the following inequalities for $n \geq 2$:

$$\int_1^n \ln x \, dx < \ln n! < \int_1^{n+1} \ln x \, dx$$

(b) Use the result in part (a) to show that

$$\frac{n^n}{e^{n-1}} < n! < \frac{(n+1)^{n+1}}{e^n}, \quad n > 1$$

(c) Use the Squeezing Theorem for Sequences (Theorem 9.1.5) and the result in part (b) to show that

$$\lim_{n \to +\infty} \frac{\sqrt[n]{n!}}{n} = \frac{1}{e}$$

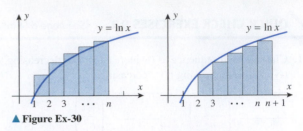

▲ **Figure Ex-30**

31. Use the left inequality in Exercise 30(b) to show that

$$\lim_{n \to +\infty} \sqrt[n]{n!} = +\infty$$

32. Writing Give an example of an increasing sequence that is not eventually strictly increasing. What can you conclude about the terms of any such sequence? Explain.

33. Writing Discuss the appropriate use of "eventually" for various properties of sequences. For example, which is a useful expression: "eventually bounded" or "eventually monotone"?

9.3 INFINITE SERIES

The purpose of this section is to discuss sums that contain infinitely many terms. The most familiar examples of such sums occur in the decimal representations of real numbers. For example, when we write $\frac{1}{3}$ in the decimal form $\frac{1}{3} = 0.3333\ldots$, we mean

$$\frac{1}{3} = 0.3 + 0.03 + 0.003 + 0.0003 + \cdots$$

which suggests that the decimal representation of $\frac{1}{3}$ can be viewed as a sum of infinitely many real numbers.

■ **SUMS OF INFINITE SERIES**

Our first objective is to define what is meant by the "sum" of infinitely many real numbers. We begin with some terminology.

9.3.1 DEFINITION An *infinite series* is an expression that can be written in the form

$$\sum_{k=1}^{\infty} u_k = u_1 + u_2 + u_3 + \cdots + u_k + \cdots$$

The numbers $u_1, u_2, u_3, \ldots$ are called the *terms* of the series.

Since it is impossible to add infinitely many numbers together directly, sums of infinite series are defined and computed by an indirect limiting process. To motivate the basic idea, consider the decimal

$$0.3333\ldots \tag{1}$$

This can be viewed as the infinite series

$$0.3 + 0.03 + 0.003 + 0.0003 + \cdots$$

or, equivalently,

$$\frac{3}{10} + \frac{3}{10^2} + \frac{3}{10^3} + \frac{3}{10^4} + \cdots \tag{2}$$

Since (1) is the decimal expansion of $\frac{1}{3}$, any reasonable definition for the sum of an infinite series should yield $\frac{1}{3}$ for the sum of (2). To obtain such a definition, consider the following sequence of (finite) sums:

$$s_1 = \frac{3}{10} = 0.3$$

$$s_2 = \frac{3}{10} + \frac{3}{10^2} = 0.33$$

$$s_3 = \frac{3}{10} + \frac{3}{10^2} + \frac{3}{10^3} = 0.333$$

$$s_4 = \frac{3}{10} + \frac{3}{10^2} + \frac{3}{10^3} + \frac{3}{10^4} = 0.3333$$

$$\vdots$$

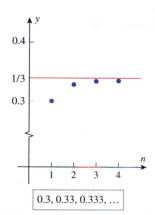

0.3, 0.33, 0.333, …

▲ **Figure 9.3.1**

The sequence of numbers s_1, s_2, s_3, s_4, … (Figure 9.3.1) can be viewed as a succession of approximations to the "sum" of the infinite series, which we want to be $\frac{1}{3}$. As we progress through the sequence, more and more terms of the infinite series are used, and the approximations get better and better, suggesting that the desired sum of $\frac{1}{3}$ might be the *limit* of this sequence of approximations. To see that this is so, we must calculate the limit of the general term in the sequence of approximations, namely,

$$s_n = \frac{3}{10} + \frac{3}{10^2} + \cdots + \frac{3}{10^n} \tag{3}$$

The problem of calculating

$$\lim_{n \to +\infty} s_n = \lim_{n \to +\infty} \left(\frac{3}{10} + \frac{3}{10^2} + \cdots + \frac{3}{10^n} \right)$$

is complicated by the fact that both the last term and the number of terms in the sum change with n. It is best to rewrite such limits in a closed form in which the number of terms does not vary, if possible. (See the discussion of closed form and open form following Example 2 in Section 5.4.) To do this, we multiply both sides of (3) by $\frac{1}{10}$ to obtain

$$\frac{1}{10} s_n = \frac{3}{10^2} + \frac{3}{10^3} + \cdots + \frac{3}{10^n} + \frac{3}{10^{n+1}} \tag{4}$$

and then subtract (4) from (3) to obtain

$$s_n - \frac{1}{10} s_n = \frac{3}{10} - \frac{3}{10^{n+1}}$$

$$\frac{9}{10} s_n = \frac{3}{10} \left(1 - \frac{1}{10^n} \right)$$

$$s_n = \frac{1}{3} \left(1 - \frac{1}{10^n} \right)$$

Since $1/10^n \to 0$ as $n \to +\infty$, it follows that

$$\lim_{n \to +\infty} s_n = \lim_{n \to +\infty} \frac{1}{3} \left(1 - \frac{1}{10^n} \right) = \frac{1}{3}$$

which we denote by writing

$$\frac{1}{3} = \frac{3}{10} + \frac{3}{10^2} + \frac{3}{10^3} + \cdots + \frac{3}{10^n} + \cdots$$

Motivated by the preceding example, we are now ready to define the general concept of the "sum" of an infinite series

$$u_1 + u_2 + u_3 + \cdots + u_k + \cdots$$

We begin with some terminology: Let s_n denote the sum of the initial terms of the series, up to and including the term with index n. Thus,

$$s_1 = u_1$$
$$s_2 = u_1 + u_2$$
$$s_3 = u_1 + u_2 + u_3$$
$$\vdots$$
$$s_n = u_1 + u_2 + u_3 + \cdots + u_n = \sum_{k=1}^{n} u_k$$

The number s_n is called the ***nth partial sum*** of the series and the sequence $\{s_n\}_{n=1}^{+\infty}$ is called the ***sequence of partial sums***.

As n increases, the partial sum $s_n = u_1 + u_2 + \cdots + u_n$ includes more and more terms of the series. Thus, if s_n tends toward a limit as $n \to +\infty$, it is reasonable to view this limit as the sum of *all* the terms in the series. This suggests the following definition.

9.3.2   DEFINITION   Let $\{s_n\}$ be the sequence of partial sums of the series

$$u_1 + u_2 + u_3 + \cdots + u_k + \cdots$$

If the sequence $\{s_n\}$ converges to a limit S, then the series is said to ***converge*** to S, and S is called the ***sum*** of the series. We denote this by writing

$$S = \sum_{k=1}^{\infty} u_k$$

If the sequence of partial sums diverges, then the series is said to ***diverge***. A divergent series has no sum.

▶ **Example 1**   Determine whether the series

$$1 - 1 + 1 - 1 + 1 - 1 + \cdots$$

converges or diverges. If it converges, find the sum.

Solution.   It is tempting to conclude that the sum of the series is zero by arguing that the positive and negative terms cancel one another. However, this is *not correct*; the problem is that algebraic operations that hold for finite sums do not carry over to infinite series in all cases. Later, we will discuss conditions under which familiar algebraic operations can be applied to infinite series, but for this example we turn directly to Definition 9.3.2. The partial sums are

$$s_1 = 1$$
$$s_2 = 1 - 1 = 0$$
$$s_3 = 1 - 1 + 1 = 1$$
$$s_4 = 1 - 1 + 1 - 1 = 0$$

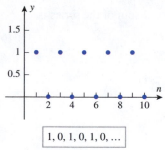

1, 0, 1, 0, 1, 0, ...

▲ Figure 9.3.2

and so forth. Thus, the sequence of partial sums is

$$1, 0, 1, 0, 1, 0, \ldots$$

(Figure 9.3.2). Since this is a divergent sequence, the given series diverges and consequently has no sum. ◄

■ GEOMETRIC SERIES

In many important series, each term is obtained by multiplying the preceding term by some fixed constant. Thus, if the initial term of the series is a and each term is obtained by multiplying the preceding term by r, then the series has the form

$$\sum_{k=0}^{\infty} ar^k = a + ar + ar^2 + ar^3 + \cdots + ar^k + \cdots \quad (a \neq 0) \tag{5}$$

Such series are called **geometric series**, and the number r is called the **ratio** for the series. Here are some examples:

$$1 + 2 + 4 + 8 + \cdots + 2^k + \cdots \qquad \boxed{a = 1, r = 2}$$

$$\frac{3}{10} + \frac{3}{10^2} + \frac{3}{10^3} + \cdots + \frac{3}{10^k} + \cdots \qquad \boxed{a = \tfrac{3}{10}, r = \tfrac{1}{10}}$$

$$\frac{1}{2} - \frac{1}{4} + \frac{1}{8} - \frac{1}{16} + \cdots + (-1)^{k+1}\frac{1}{2^k} + \cdots \qquad \boxed{a = \tfrac{1}{2}, r = -\tfrac{1}{2}}$$

$$1 + 1 + 1 + \cdots + 1 + \cdots \qquad \boxed{a = 1, r = 1}$$

$$1 - 1 + 1 - 1 + \cdots + (-1)^{k+1} + \cdots \qquad \boxed{a = 1, r = -1}$$

$$1 + x + x^2 + x^3 + \cdots + x^k + \cdots \qquad \boxed{a = 1, r = x}$$

The following theorem is the fundamental result on convergence of geometric series.

Sometimes it is desirable to start the index of summation of an infinite series at $k = 0$ rather than $k = 1$, in which case we would call u_0 the *zeroth term* and $s_0 = u_0$ the *zeroth partial sum.* One can prove that changing the starting value for the index of summation of an infinite series has no effect on the convergence, the divergence, or the sum. If we had started the index at $k = 1$ in (5), then the series would be expressed as

$$\sum_{k=1}^{\infty} ar^{k-1}$$

Since this expression is more complicated than (5), we started the index at $k = 0$.

**9.3.3   THEOREM   ** *A geometric series*

$$\sum_{k=0}^{\infty} ar^k = a + ar + ar^2 + \cdots + ar^k + \cdots \quad (a \neq 0)$$

converges if $|r| < 1$ *and diverges if* $|r| \geq 1$. *If the series converges, then the sum is*

$$\sum_{k=0}^{\infty} ar^k = \frac{a}{1-r}$$

PROOF   Let us treat the case $|r| = 1$ first. If $r = 1$, then the series is

$$a + a + a + a + \cdots$$

so the nth partial sum is $s_n = (n+1)a$ and

$$\lim_{n \to +\infty} s_n = \lim_{n \to +\infty} (n+1)a = \pm\infty$$

(the sign depending on whether a is positive or negative). This proves divergence. If $r = -1$, the series is

$$a - a + a - a + \cdots$$

so the sequence of partial sums is

$$a, 0, a, 0, a, 0, \ldots$$

which diverges.

Now let us consider the case where $|r| \neq 1$. The nth partial sum of the series is

$$s_n = a + ar + ar^2 + \cdots + ar^n \tag{6}$$

Multiplying both sides of (6) by r yields

$$rs_n = ar + ar^2 + \cdots + ar^n + ar^{n+1} \tag{7}$$

and subtracting (7) from (6) gives

$$s_n - rs_n = a - ar^{n+1}$$

or

$$(1 - r)s_n = a - ar^{n+1} \tag{8}$$

Since $r \neq 1$ in the case we are considering, this can be rewritten as

$$s_n = \frac{a - ar^{n+1}}{1 - r} = \frac{a}{1 - r}(1 - r^{n+1}) \tag{9}$$

If $|r| < 1$, then r^{n+1} goes to 0 as $n \to +\infty$ (can you see why?), so $\{s_n\}$ converges. From (9)

$$\lim_{n \to +\infty} s_n = \frac{a}{1 - r}$$

If $|r| > 1$, then either $r > 1$ or $r < -1$. In the case $r > 1$, r^{n+1} increases without bound as $n \to +\infty$, and in the case $r < -1$, r^{n+1} oscillates between positive and negative values that grow in magnitude, so $\{s_n\}$ diverges in both cases. ■

> Note that (6) is an open form for s_n, while (9) is a closed form for s_n. In general, one needs a closed form to calculate the limit.

▶ **Example 2**   In each part, determine whether the series converges, and if so find its sum.

$$\text{(a) } \sum_{k=0}^{\infty} \frac{5}{4^k} \qquad \text{(b) } \sum_{k=1}^{\infty} 3^{2k} 5^{1-k}$$

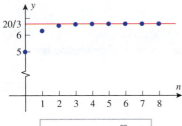

Partial sums for $\sum_{k=0}^{\infty} \frac{5}{4^k}$

▲ **Figure 9.3.3**

Solution (a).   This is a geometric series with $a = 5$ and $r = \frac{1}{4}$. Since $|r| = \frac{1}{4} < 1$, the series converges and the sum is

$$\frac{a}{1 - r} = \frac{5}{1 - \frac{1}{4}} = \frac{20}{3}$$

(Figure 9.3.3).

Solution (b).   This is a geometric series in concealed form, since we can rewrite it as

$$\sum_{k=1}^{\infty} 3^{2k} 5^{1-k} = \sum_{k=1}^{\infty} \frac{9^k}{5^{k-1}} = \sum_{k=1}^{\infty} 9\left(\frac{9}{5}\right)^{k-1}$$

Since $r = \frac{9}{5} > 1$, the series diverges. ◀

▶ **Example 3**   Find the rational number represented by the repeating decimal

$$0.784784784\ldots$$

Solution.   We can write

$$0.784784784\ldots = 0.784 + 0.000784 + 0.000000784 + \cdots$$

so the given decimal is the sum of a geometric series with $a = 0.784$ and $r = 0.001$. Thus,

$$0.784784784\ldots = \frac{a}{1 - r} = \frac{0.784}{1 - 0.001} = \frac{0.784}{0.999} = \frac{784}{999} \quad ◀$$

TECHNOLOGY MASTERY

Computer algebra systems have commands for finding sums of convergent series. If you have a CAS, use it to compute the sums in Examples 2 and 3.

▶ **Example 4** In each part, find all values of x for which the series converges, and find the sum of the series for those values of x.

(a) $\displaystyle\sum_{k=0}^{\infty} x^k$ (b) $3 - \dfrac{3x}{2} + \dfrac{3x^2}{4} - \dfrac{3x^3}{8} + \cdots + \dfrac{3(-1)^k}{2^k}x^k + \cdots$

Solution (a). The expanded form of the series is

$$\sum_{k=0}^{\infty} x^k = 1 + x + x^2 + \cdots + x^k + \cdots$$

The series is a geometric series with $a = 1$ and $r = x$, so it converges if $|x| < 1$ and diverges otherwise. When the series converges its sum is

$$\sum_{k=0}^{\infty} x^k = \frac{1}{1 - x}$$

Solution (b). This is a geometric series with $a = 3$ and $r = -x/2$. It converges if $|-x/2| < 1$, or equivalently, when $|x| < 2$. When the series converges its sum is

$$\sum_{k=0}^{\infty} 3\left(-\frac{x}{2}\right)^k = \frac{3}{1 - \left(-\dfrac{x}{2}\right)} = \frac{6}{2 + x} \quad \blacktriangleleft$$

■ TELESCOPING SUMS

▶ **Example 5** Determine whether the series

$$\sum_{k=1}^{\infty} \frac{1}{k(k+1)} = \frac{1}{1 \cdot 2} + \frac{1}{2 \cdot 3} + \frac{1}{3 \cdot 4} + \frac{1}{4 \cdot 5} + \cdots$$

converges or diverges. If it converges, find the sum.

Solution. The nth partial sum of the series is

$$s_n = \sum_{k=1}^{n} \frac{1}{k(k+1)} = \frac{1}{1 \cdot 2} + \frac{1}{2 \cdot 3} + \frac{1}{3 \cdot 4} + \cdots + \frac{1}{n(n+1)}$$

We will begin by rewriting s_n in closed form. This can be accomplished by using the method of partial fractions to obtain (verify)

$$\frac{1}{k(k+1)} = \frac{1}{k} - \frac{1}{k+1}$$

from which we obtain the sum

$$s_n = \sum_{k=1}^{n} \left(\frac{1}{k} - \frac{1}{k+1}\right)$$

$$= \left(1 - \frac{1}{2}\right) + \left(\frac{1}{2} - \frac{1}{3}\right) + \left(\frac{1}{3} - \frac{1}{4}\right) + \cdots + \left(\frac{1}{n} - \frac{1}{n+1}\right)$$

$$= 1 + \left(-\frac{1}{2} + \frac{1}{2}\right) + \left(-\frac{1}{3} + \frac{1}{3}\right) + \cdots + \left(-\frac{1}{n} + \frac{1}{n}\right) - \frac{1}{n+1}$$

$$= 1 - \frac{1}{n+1} \tag{10}$$

The sum in (10) is an example of a *telescoping sum*. The name is derived from the fact that in simplifying the sum, one term in each parenthetical expression cancels one term in the next parenthetical expression, until the entire sum collapses (like a folding telescope) into just two terms.

Thus,

$$\sum_{k=1}^{\infty} \frac{1}{k(k+1)} = \lim_{n \to +\infty} s_n = \lim_{n \to +\infty} \left(1 - \frac{1}{n+1}\right) = 1 \quad \blacktriangleleft$$

■ HARMONIC SERIES

One of the most important of all diverging series is the *harmonic series*,

$$\sum_{k=1}^{\infty} \frac{1}{k} = 1 + \frac{1}{2} + \frac{1}{3} + \frac{1}{4} + \frac{1}{5} + \cdots$$

which arises in connection with the overtones produced by a vibrating musical string. It is not immediately evident that this series diverges. However, the divergence will become apparent when we examine the partial sums in detail. Because the terms in the series are all positive, the partial sums

$$s_1 = 1, \quad s_2 = 1 + \frac{1}{2}, \quad s_3 = 1 + \frac{1}{2} + \frac{1}{3}, \quad s_4 = 1 + \frac{1}{2} + \frac{1}{3} + \frac{1}{4}, \cdots$$

form a strictly increasing sequence

$$s_1 < s_2 < s_3 < \cdots < s_n < \cdots$$

(Figure 9.3.4a). Thus, by Theorem 9.2.3 we can prove divergence by demonstrating that there is no constant M that is greater than or equal to *every* partial sum. To this end, we will consider some selected partial sums, namely, $s_2, s_4, s_8, s_{16}, s_{32}, \ldots$. Note that the subscripts are successive powers of 2, so that these are the partial sums of the form s_{2^n} (Figure 9.3.4b). These partial sums satisfy the inequalities

$$s_2 = 1 + \tfrac{1}{2} > \tfrac{1}{2} + \tfrac{1}{2} = \tfrac{2}{2}$$

$$s_4 = s_2 + \tfrac{1}{3} + \tfrac{1}{4} > s_2 + \left(\tfrac{1}{4} + \tfrac{1}{4}\right) = s_2 + \tfrac{1}{2} > \tfrac{3}{2}$$

$$s_8 = s_4 + \tfrac{1}{5} + \tfrac{1}{6} + \tfrac{1}{7} + \tfrac{1}{8} > s_4 + \left(\tfrac{1}{8} + \tfrac{1}{8} + \tfrac{1}{8} + \tfrac{1}{8}\right) = s_4 + \tfrac{1}{2} > \tfrac{4}{2}$$

$$s_{16} = s_8 + \tfrac{1}{9} + \tfrac{1}{10} + \tfrac{1}{11} + \tfrac{1}{12} + \tfrac{1}{13} + \tfrac{1}{14} + \tfrac{1}{15} + \tfrac{1}{16}$$

$$> s_8 + \left(\tfrac{1}{16} + \tfrac{1}{16} + \tfrac{1}{16} + \tfrac{1}{16} + \tfrac{1}{16} + \tfrac{1}{16} + \tfrac{1}{16} + \tfrac{1}{16}\right) = s_8 + \tfrac{1}{2} > \tfrac{5}{2}$$

$$\vdots$$

$$s_{2^n} > \frac{n+1}{2}$$

If M is any constant, we can find a positive integer n such that $(n+1)/2 > M$. But for this n

$$s_{2^n} > \frac{n+1}{2} > M$$

so that no constant M is greater than or equal to *every* partial sum of the harmonic series. This proves divergence.

This divergence proof, which predates the discovery of calculus, is due to a French bishop and teacher, Nicole Oresme (1323–1382). This series eventually attracted the interest of Johann and Jakob Bernoulli (p. 700) and led them to begin thinking about the general concept of convergence, which was a new idea at that time.

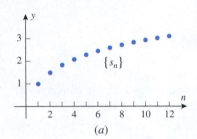

(a)

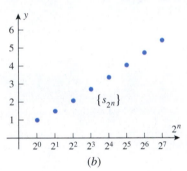

(b)

Partial sums for the harmonic series

▲ **Figure 9.3.4**

Courtesy Lilly Library, Indiana University

This is a proof of the divergence of the harmonic series, as it appeared in an appendix of Jakob Bernoulli's posthumous publication, Ars Conjectandi, which appeared in 1713.

✔ **QUICK CHECK EXERCISES 9.3**   *(See page 623 for answers.)*

1. In mathematics, the terms "sequence" and "series" have different meanings: a _____ is a succession, whereas a _____ is a sum.

2. Consider the series

$$\sum_{k=1}^{\infty} \frac{1}{2^k}$$

If $\{s_n\}$ is the sequence of partial sums for this series, then $s_1 = $ _____, $s_2 = $ _____, $s_3 = $ _____, $s_4 = $ _____, and $s_n = $ _____.

3. What does it mean to say that a series $\sum u_k$ converges?

4. A geometric series is a series of the form

$$\sum_{k=0}^{\infty} \underline{\hspace{1cm}}$$

This series converges to _____ if _____. This series diverges if _____.

5. The harmonic series has the form

$$\sum_{k=1}^{\infty} \underline{\hspace{1cm}}$$

Does the harmonic series converge or diverge?

EXERCISE SET 9.3   ☐c CAS

1–2 In each part, find exact values for the first four partial sums, find a closed form for the nth partial sum, and determine whether the series converges by calculating the limit of the nth partial sum. If the series converges, then state its sum. ■

1. (a) $2 + \dfrac{2}{5} + \dfrac{2}{5^2} + \cdots + \dfrac{2}{5^{k-1}} + \cdots$

   (b) $\dfrac{1}{4} + \dfrac{2}{4} + \dfrac{2^2}{4} + \cdots + \dfrac{2^{k-1}}{4} + \cdots$

   (c) $\dfrac{1}{2 \cdot 3} + \dfrac{1}{3 \cdot 4} + \dfrac{1}{4 \cdot 5} + \cdots + \dfrac{1}{(k+1)(k+2)} + \cdots$

2. (a) $\displaystyle\sum_{k=1}^{\infty} \left(\frac{1}{4}\right)^k$ (b) $\displaystyle\sum_{k=1}^{\infty} 4^{k-1}$ (c) $\displaystyle\sum_{k=1}^{\infty} \left(\frac{1}{k+3} - \frac{1}{k+4}\right)$

3–14 Determine whether the series converges, and if so find its sum. ■

3. $\displaystyle\sum_{k=1}^{\infty} \left(-\frac{3}{4}\right)^{k-1}$

4. $\displaystyle\sum_{k=1}^{\infty} \left(\frac{2}{3}\right)^{k+2}$

5. $\displaystyle\sum_{k=1}^{\infty} (-1)^{k-1} \frac{7}{6^{k-1}}$

6. $\displaystyle\sum_{k=1}^{\infty} \left(-\frac{3}{2}\right)^{k+1}$

7. $\displaystyle\sum_{k=1}^{\infty} \frac{1}{(k+2)(k+3)}$

8. $\displaystyle\sum_{k=1}^{\infty} \left(\frac{1}{2^k} - \frac{1}{2^{k+1}}\right)$

9. $\displaystyle\sum_{k=1}^{\infty} \frac{1}{9k^2 + 3k - 2}$

10. $\displaystyle\sum_{k=2}^{\infty} \frac{1}{k^2 - 1}$

11. $\displaystyle\sum_{k=3}^{\infty} \frac{1}{k-2}$

12. $\displaystyle\sum_{k=5}^{\infty} \left(\frac{e}{\pi}\right)^{k-1}$

13. $\displaystyle\sum_{k=1}^{\infty} \frac{4^{k+2}}{7^{k-1}}$

14. $\displaystyle\sum_{k=1}^{\infty} 5^{3k} 7^{1-k}$

15. Match a series from one of Exercises 3, 5, 7, or 9 with the graph of its sequence of partial sums.

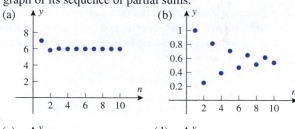

16. Match a series from one of Exercises 4, 6, 8, or 10 with the graph of its sequence of partial sums.

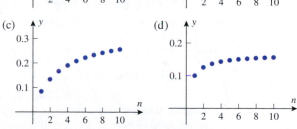
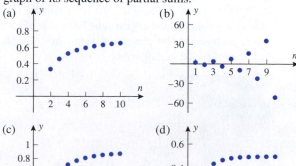
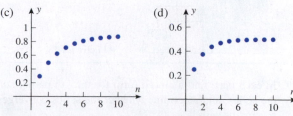

17–20 True–False Determine whether the statement is true or false. Explain your answer. ■

17. An infinite series converges if its sequence of terms converges.

18. The geometric series $a + ar + ar^2 + \cdots + ar^n + \cdots$ converges provided $|r| < 1$.

19. The harmonic series diverges.

20. An infinite series converges if its sequence of partial sums is bounded and monotone.

21–24 Express the repeating decimal as a fraction. ■

21. $0.9999\ldots$ **22.** $0.4444\ldots$

23. $5.373737\ldots$ **24.** $0.451141414\ldots$

25. Recall that a *terminating decimal* is a decimal whose digits are all 0 from some point on ($0.5 = 0.50000\ldots$, for example). Show that a decimal of the form $0.a_1 a_2 \ldots a_n 9999\ldots$, where $a_n \neq 9$, can be expressed as a terminating decimal.

FOCUS ON CONCEPTS

26. The great Swiss mathematician Leonhard Euler (biography on p. 3) sometimes reached incorrect conclusions in his pioneering work on infinite series. For example, Euler deduced that

$$\tfrac{1}{2} = 1 - 1 + 1 - 1 + \cdots$$

and

$$-1 = 1 + 2 + 4 + 8 + \cdots$$

by substituting $x = -1$ and $x = 2$ in the formula

$$\frac{1}{1-x} = 1 + x + x^2 + x^3 + \cdots$$

What was the problem with his reasoning?

27. A ball is dropped from a height of 10 m. Each time it strikes the ground it bounces vertically to a height that is $\frac{3}{4}$ of the preceding height. Find the total distance the ball will travel if it is assumed to bounce infinitely often.

28. The accompanying figure shows an "infinite staircase" constructed from cubes. Find the total volume of the staircase, given that the largest cube has a side of length 1 and each successive cube has a side whose length is half that of the preceding cube.

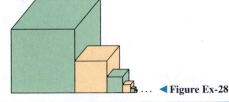

◀ **Figure Ex-28**

29. In each part, find a closed form for the nth partial sum of the series, and determine whether the series converges. If so, find its sum.

(a) $\ln \dfrac{1}{2} + \ln \dfrac{2}{3} + \ln \dfrac{3}{4} + \cdots + \ln \dfrac{k}{k+1} + \cdots$

(b) $\ln \left(1 - \dfrac{1}{4}\right) + \ln \left(1 - \dfrac{1}{9}\right) + \ln \left(1 - \dfrac{1}{16}\right) + \cdots$

$$+ \ln \left(1 - \dfrac{1}{(k+1)^2}\right) + \cdots$$

30. Use geometric series to show that

(a) $\displaystyle\sum_{k=0}^{\infty} (-1)^k x^k = \dfrac{1}{1+x}$ if $-1 < x < 1$

(b) $\displaystyle\sum_{k=0}^{\infty} (x-3)^k = \dfrac{1}{4-x}$ if $2 < x < 4$

(c) $\displaystyle\sum_{k=0}^{\infty} (-1)^k x^{2k} = \dfrac{1}{1+x^2}$ if $-1 < x < 1$.

31. In each part, find all values of x for which the series converges, and find the sum of the series for those values of x.

(a) $x - x^3 + x^5 - x^7 + x^9 - \cdots$

(b) $\dfrac{1}{x^2} + \dfrac{2}{x^3} + \dfrac{4}{x^4} + \dfrac{8}{x^5} + \dfrac{16}{x^6} + \cdots$

(c) $e^{-x} + e^{-2x} + e^{-3x} + e^{-4x} + e^{-5x} + \cdots$

32. Show that for all real values of x

$$\sin x - \frac{1}{2}\sin^2 x + \frac{1}{4}\sin^3 x - \frac{1}{8}\sin^4 x + \cdots = \frac{2\sin x}{2 + \sin x}$$

33. Let a_1 be any real number, and let $\{a_n\}$ be the sequence defined recursively by

$$a_{n+1} = \tfrac{1}{2}(a_n + 1)$$

Make a conjecture about the limit of the sequence, and confirm your conjecture by expressing a_n in terms of a_1 and taking the limit.

34. Show: $\displaystyle\sum_{k=1}^{\infty} \frac{\sqrt{k+1} - \sqrt{k}}{\sqrt{k^2 + k}} = 1$.

35. Show: $\displaystyle\sum_{k=1}^{\infty} \left(\frac{1}{k} - \frac{1}{k+2}\right) = \frac{3}{2}$.

36. Show: $\dfrac{1}{1 \cdot 3} + \dfrac{1}{2 \cdot 4} + \dfrac{1}{3 \cdot 5} + \cdots = \dfrac{3}{4}$.

37. Show: $\dfrac{1}{1 \cdot 3} + \dfrac{1}{3 \cdot 5} + \dfrac{1}{5 \cdot 7} + \cdots = \dfrac{1}{2}$.

38. In his *Treatise on the Configurations of Qualities and Motions* (written in the 1350s), the French Bishop of Lisieux, Nicole Oresme, used a geometric method to find the sum of the series

$$\sum_{k=1}^{\infty} \frac{k}{2^k} = \frac{1}{2} + \frac{2}{4} + \frac{3}{8} + \frac{4}{16} + \cdots$$

In part (*a*) of the accompanying figure, each term in the series is represented by the area of a rectangle, and in part (b) the configuration in part (a) has been divided into rectangles with areas $A_1, A_2, A_3, \dots$. Find the sum $A_1 + A_2 + A_3 + \cdots$.

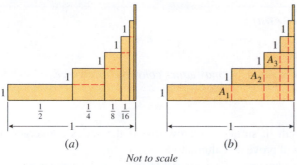

(a)   (b)

Not to scale

▲ **Figure Ex-38**

39. As shown in the accompanying figure, suppose that an angle θ is bisected using a straightedge and compass to produce ray R_1, then the angle between R_1 and the initial side is bisected to produce ray R_2. Thereafter, rays $R_3, R_4, R_5, \dots$ are constructed in succession by bisecting the angle between the preceding two rays. Show that the sequence of angles that these rays make with the initial side has a limit of $\theta/3$.

Source: This problem is based on "Trisection of an Angle in an Infinite Number of Steps" by Eric Kincannon, which appeared in *The College Mathematics Journal*, Vol. 21, No. 5, November 1990.

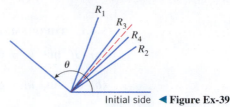

Initial side   ◀ **Figure Ex-39**

c **40.** In each part, use a CAS to find the sum of the series if it converges, and then confirm the result by hand calculation.

(a) $\displaystyle\sum_{k=1}^{\infty} (-1)^{k+1} 2^k 3^{2-k}$ (b) $\displaystyle\sum_{k=1}^{\infty} \frac{3^{3k}}{5^{k-1}}$ (c) $\displaystyle\sum_{k=1}^{\infty} \frac{1}{4k^2 - 1}$

41. **Writing** Discuss the similarities and differences between what it means for a sequence to converge and what it means for a series to converge.

42. **Writing** Read about Zeno's dichotomy paradox in an appropriate reference work and relate the paradox in a setting that is familiar to you. Discuss a connection between the paradox and geometric series.

✔ **QUICK CHECK ANSWERS 9.3**

1. sequence; series   **2.** $\dfrac{1}{2}; \dfrac{3}{4}; \dfrac{7}{8}; \dfrac{15}{16}; 1 - \dfrac{1}{2^n}$ **3.** The sequence of partial sums converges.

4. $ar^k \ (a \neq 0); \dfrac{a}{1-r}; |r| < 1; |r| \geq 1$ **5.** $\dfrac{1}{k}$; diverge

9.4 CONVERGENCE TESTS

In the last section we showed how to find the sum of a series by finding a closed form for the nth partial sum and taking its limit. However, it is relatively rare that one can find a closed form for the nth partial sum of a series, so alternative methods are needed for finding the sum of a series. One possibility is to prove that the series converges, and then to approximate the sum by a partial sum with sufficiently many terms to achieve the desired degree of accuracy. In this section we will develop various tests that can be used to determine whether a given series converges or diverges.

■ **THE DIVERGENCE TEST**

In stating general results about convergence or divergence of series, it is convenient to use the notation $\sum u_k$ as a generic notation for a series, thus avoiding the issue of whether the sum begins with $k = 0$ or $k = 1$ or some other value. Indeed, we will see shortly that the starting index value is irrelevant to the issue of convergence. The kth term in an infinite series $\sum u_k$ is called the **general term** of the series. The following theorem establishes a relationship between the limit of the general term and the convergence properties of a series.

9.4.1 **THEOREM** (*The Divergence Test*)

(*a*) *If* $\lim\limits_{k \to +\infty} u_k \neq 0$, *then the series* $\sum u_k$ *diverges.*

(*b*) *If* $\lim\limits_{k \to +\infty} u_k = 0$, *then the series* $\sum u_k$ *may either converge or diverge.*

PROOF (*a*) To prove this result, it suffices to show that if the series converges, then $\lim_{k \to +\infty} u_k = 0$ (why?). We will prove this alternative form of (*a*).

Let us assume that the series converges. The general term u_k can be written as

$$u_k = s_k - s_{k-1} \tag{1}$$

where s_k is the sum of the terms through u_k and s_{k-1} is the sum of the terms through u_{k-1}. If S denotes the sum of the series, then $\lim_{k \to +\infty} s_k = S$, and since $(k - 1) \to +\infty$ as $k \to +\infty$, we also have $\lim_{k \to +\infty} s_{k-1} = S$. Thus, from (1)

$$\lim_{k \to +\infty} u_k = \lim_{k \to +\infty} (s_k - s_{k-1}) = S - S = 0$$

PROOF (*b*) To prove this result, it suffices to produce both a convergent series and a divergent series for which $\lim_{k \to +\infty} u_k = 0$. The following series both have this property:

$$\frac{1}{2} + \frac{1}{2^2} + \cdots + \frac{1}{2^k} + \cdots \quad \text{and} \quad 1 + \frac{1}{2} + \frac{1}{3} + \cdots + \frac{1}{k} + \cdots$$

The first is a convergent geometric series and the second is the divergent harmonic series. ■

WARNING

The converse of Theorem 9.4.2 is false; i.e., showing that

$$\lim_{k \to +\infty} u_k = 0$$

does not prove that $\sum u_k$ converges, since this property may hold for divergent as well as convergent series. This is illustrated in the proof of part (*b*) of Theorem 9.4.1.

The alternative form of part (*a*) given in the preceding proof is sufficiently important that we state it separately for future reference.

9.4.2 **THEOREM** *If the series* $\sum u_k$ *converges, then* $\lim\limits_{k \to +\infty} u_k = 0$.

▶ **Example 1** The series

$$\sum_{k=1}^{\infty} \frac{k}{k+1} = \frac{1}{2} + \frac{2}{3} + \frac{3}{4} + \cdots + \frac{k}{k+1} + \cdots$$

diverges since

$$\lim_{k \to +\infty} \frac{k}{k+1} = \lim_{k \to +\infty} \frac{1}{1 + 1/k} = 1 \neq 0 \blacktriangleleft$$

■ **ALGEBRAIC PROPERTIES OF INFINITE SERIES**

For brevity, the proof of the following result is omitted.

9.4.3 **THEOREM**

See Exercises 27 and 28 for an exploration of what happens when $\sum u_k$ or $\sum v_k$ diverge.

(a) *If $\sum u_k$ and $\sum v_k$ are convergent series, then $\sum(u_k + v_k)$ and $\sum(u_k - v_k)$ are convergent series and the sums of these series are related by*

$$\sum_{k=1}^{\infty}(u_k + v_k) = \sum_{k=1}^{\infty} u_k + \sum_{k=1}^{\infty} v_k$$

$$\sum_{k=1}^{\infty}(u_k - v_k) = \sum_{k=1}^{\infty} u_k - \sum_{k=1}^{\infty} v_k$$

(b) *If c is a nonzero constant, then the series $\sum u_k$ and $\sum cu_k$ both converge or both diverge. In the case of convergence, the sums are related by*

$$\sum_{k=1}^{\infty} cu_k = c\sum_{k=1}^{\infty} u_k$$

(c) *Convergence or divergence is unaffected by deleting a finite number of terms from a series; in particular, for any positive integer K, the series*

$$\sum_{k=1}^{\infty} u_k = u_1 + u_2 + u_3 + \cdots$$

$$\sum_{k=K}^{\infty} u_k = u_K + u_{K+1} + u_{K+2} + \cdots$$

both converge or both diverge.

WARNING

Do not read too much into part (c) of Theorem 9.4.3. Although convergence is not affected when finitely many terms are deleted from the beginning of a convergent series, the *sum* of the series is changed by the removal of those terms.

▶ **Example 2** Find the sum of the series

$$\sum_{k=1}^{\infty}\left(\frac{3}{4^k} - \frac{2}{5^{k-1}}\right)$$

Solution. The series

$$\sum_{k=1}^{\infty}\frac{3}{4^k} = \frac{3}{4} + \frac{3}{4^2} + \frac{3}{4^3} + \cdots$$

is a convergent geometric series $\left(a = \frac{3}{4}, r = \frac{1}{4}\right)$, and the series

$$\sum_{k=1}^{\infty}\frac{2}{5^{k-1}} = 2 + \frac{2}{5} + \frac{2}{5^2} + \frac{2}{5^3} + \cdots$$

is also a convergent geometric series $\left(a = 2, r = \frac{1}{5}\right)$. Thus, from Theorems 9.4.3(a) and 9.3.3 the given series converges and

$$\sum_{k=1}^{\infty}\left(\frac{3}{4^k} - \frac{2}{5^{k-1}}\right) = \sum_{k=1}^{\infty}\frac{3}{4^k} - \sum_{k=1}^{\infty}\frac{2}{5^{k-1}}$$

$$= \frac{\frac{3}{4}}{1 - \frac{1}{4}} - \frac{2}{1 - \frac{1}{5}} = -\frac{3}{2} \quad ◀$$

▶ **Example 3** Determine whether the following series converge or diverge.

$$\text{(a)} \sum_{k=1}^{\infty} \frac{5}{k} = 5 + \frac{5}{2} + \frac{5}{3} + \cdots + \frac{5}{k} + \cdots \qquad \text{(b)} \sum_{k=10}^{\infty} \frac{1}{k} = \frac{1}{10} + \frac{1}{11} + \frac{1}{12} + \cdots$$

Solution. The first series is a constant times the divergent harmonic series, and hence diverges by part (b) of Theorem 9.4.3. The second series results by deleting the first nine terms from the divergent harmonic series, and hence diverges by part (c) of Theorem 9.4.3. ◀

▪ THE INTEGRAL TEST

The expressions

$$\sum_{k=1}^{\infty} \frac{1}{k^2} \quad \text{and} \quad \int_{1}^{+\infty} \frac{1}{x^2}\, dx$$

are related in that the integrand in the improper integral results when the index k in the general term of the series is replaced by x and the limits of summation in the series are replaced by the corresponding limits of integration. The following theorem shows that there is a relationship between the convergence of the series and the integral.

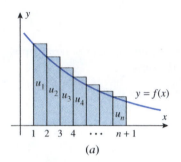

(a)

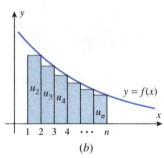

(b)

▲ **Figure 9.4.1**

9.4.4 **THEOREM** (*The Integral Test*) *Let* $\sum u_k$ *be a series with positive terms. If* f *is a function that is decreasing and continuous on an interval* $[a, +\infty)$ *and such that* $u_k = f(k)$ *for all* $k \geq a$, *then*

$$\sum_{k=1}^{\infty} u_k \quad \text{and} \quad \int_{a}^{+\infty} f(x)\, dx$$

both converge or both diverge.

The proof of the integral test is deferred to the end of this section. However, the gist of the proof is captured in Figure 9.4.1: if the integral diverges, then so does the series (Figure 9.4.1a), and if the integral converges, then so does the series (Figure 9.4.1b).

▶ **Example 4** Show that the integral test applies, and use the integral test to determine whether the following series converge or diverge.

$$\text{(a)} \sum_{k=1}^{\infty} \frac{1}{k} \qquad \text{(b)} \sum_{k=1}^{\infty} \frac{1}{k^2}$$

Solution (a). We already know that this is the divergent harmonic series, so the integral test will simply illustrate another way of establishing the divergence.

Note first that the series has positive terms, so the integral test is applicable. If we replace k by x in the general term $1/k$, we obtain the function $f(x) = 1/x$, which is decreasing and continuous for $x \geq 1$ (as required to apply the integral test with $a = 1$). Since

$$\int_{1}^{+\infty} \frac{1}{x}\, dx = \lim_{b \to +\infty} \int_{1}^{b} \frac{1}{x}\, dx = \lim_{b \to +\infty} [\ln b - \ln 1] = +\infty$$

the integral diverges and consequently so does the series.

Solution (b). Note first that the series has positive terms, so the integral test is applicable. If we replace k by x in the general term $1/k^2$, we obtain the function $f(x) = 1/x^2$, which is decreasing and continuous for $x \geq 1$. Since

$$\int_1^{+\infty} \frac{1}{x^2}\,dx = \lim_{b\to+\infty}\int_1^b \frac{dx}{x^2} = \lim_{b\to+\infty}\left[-\frac{1}{x}\right]_1^b = \lim_{b\to+\infty}\left[1 - \frac{1}{b}\right] = 1$$

the integral converges and consequently the series converges by the integral test with $a = 1$. ◄

■ ***p*-SERIES**

The series in Example 4 are special cases of a class of series called ***p*-series** or ***hyperharmonic series***. A p-series is an infinite series of the form

$$\sum_{k=1}^{\infty}\frac{1}{k^p} = 1 + \frac{1}{2^p} + \frac{1}{3^p} + \cdots + \frac{1}{k^p} + \cdots$$

where $p > 0$. Examples of p-series are

$$\sum_{k=1}^{\infty}\frac{1}{k} = 1 + \frac{1}{2} + \frac{1}{3} + \cdots + \frac{1}{k} + \cdots \qquad \boxed{p = 1}$$

$$\sum_{k=1}^{\infty}\frac{1}{k^2} = 1 + \frac{1}{2^2} + \frac{1}{3^2} + \cdots + \frac{1}{k^2} + \cdots \qquad \boxed{p = 2}$$

$$\sum_{k=1}^{\infty}\frac{1}{\sqrt{k}} = 1 + \frac{1}{\sqrt{2}} + \frac{1}{\sqrt{3}} + \cdots + \frac{1}{\sqrt{k}} + \cdots \qquad \boxed{p = \tfrac{1}{2}}$$

The following theorem tells when a p-series converges.

9.4.5 THEOREM (*Convergence of p-Series*)

$$\sum_{k=1}^{\infty}\frac{1}{k^p} = 1 + \frac{1}{2^p} + \frac{1}{3^p} + \cdots + \frac{1}{k^p} + \cdots$$

converges if $p > 1$ and diverges if $0 < p \leq 1$.

PROOF To establish this result when $p \neq 1$, we will use the integral test.

$$\int_1^{+\infty}\frac{1}{x^p}\,dx = \lim_{b\to+\infty}\int_1^b x^{-p}\,dx = \lim_{b\to+\infty}\frac{x^{1-p}}{1-p}\Big]_1^b = \lim_{b\to+\infty}\left[\frac{b^{1-p}}{1-p} - \frac{1}{1-p}\right]$$

Assume first that $p > 1$. Then $1 - p < 0$, so $b^{1-p} \to 0$ as $b \to +\infty$. Thus, the integral converges [its value is $-1/(1-p)$] and consequently the series also converges.

Now assume that $0 < p < 1$. It follows that $1 - p > 0$ and $b^{1-p} \to +\infty$ as $b \to +\infty$, so the integral and the series diverge. The case $p = 1$ is the harmonic series, which was previously shown to diverge. ■

▶ **Example 5**

$$1 + \frac{1}{\sqrt[3]{2}} + \frac{1}{\sqrt[3]{3}} + \cdots + \frac{1}{\sqrt[3]{k}} + \cdots$$

diverges since it is a p-series with $p = \tfrac{1}{3} < 1$. ◄

■ PROOF OF THE INTEGRAL TEST

Before we can prove the integral test, we need a basic result about convergence of series with *nonnegative* terms. If $u_1 + u_2 + u_3 + \cdots + u_k + \cdots$ is such a series, then its sequence of partial sums is increasing, that is,

$$s_1 \leq s_2 \leq s_3 \leq \cdots \leq s_n \leq \cdots$$

Thus, from Theorem 9.2.3 the sequence of partial sums converges to a limit S if and only if it has some upper bound M, in which case $S \leq M$. If no upper bound exists, then the sequence of partial sums diverges. Since convergence of the sequence of partial sums corresponds to convergence of the series, we have the following theorem.

9.4.6 THEOREM *If $\sum u_k$ is a series with nonnegative terms, and if there is a constant M such that*

$$s_n = u_1 + u_2 + \cdots + u_n \leq M$$

for every n, then the series converges and the sum S satisfies $S \leq M$. If no such M exists, then the series diverges.

In words, this theorem implies that *a series with nonnegative terms converges if and only if its sequence of partial sums is bounded above.*

PROOF OF THEOREM 9.4.4 We need only show that the series converges when the integral converges and that the series diverges when the integral diverges. For simplicity, we will limit the proof to the case where $a = 1$. Assume that $f(x)$ satisfies the hypotheses of the theorem for $x \geq 1$. Since

$$f(1) = u_1, \ f(2) = u_2, \ldots, \ f(n) = u_n, \ldots$$

the values of $u_1, u_2, \ldots, u_n, \ldots$ can be interpreted as the areas of the rectangles shown in Figure 9.4.2.

The following inequalities result by comparing the areas under the curve $y = f(x)$ to the areas of the rectangles in Figure 9.4.2 for $n > 1$:

$$\int_1^{n+1} f(x)\, dx < u_1 + u_2 + \cdots + u_n = s_n \qquad \boxed{\text{Figure } 9.4.2a}$$

$$s_n - u_1 = u_2 + u_3 + \cdots + u_n < \int_1^n f(x)\, dx \qquad \boxed{\text{Figure } 9.4.2b}$$

These inequalities can be combined as

$$\int_1^{n+1} f(x)\, dx < s_n < u_1 + \int_1^n f(x)\, dx \qquad (2)$$

If the integral $\int_1^{+\infty} f(x)\, dx$ converges to a finite value L, then from the right-hand inequality in (2)

$$s_n < u_1 + \int_1^n f(x)\, dx < u_1 + \int_1^{+\infty} f(x)\, dx = u_1 + L$$

Thus, each partial sum is less than the finite constant $u_1 + L$, and the series converges by Theorem 9.4.6. On the other hand, if the integral $\int_1^{+\infty} f(x)\, dx$ diverges, then

$$\lim_{n \to +\infty} \int_1^{n+1} f(x)\, dx = +\infty$$

so that from the left-hand inequality in (2), $s_n \to +\infty$ as $n \to +\infty$. This implies that the series also diverges. ■

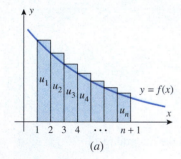

(a)

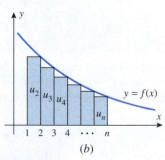

(b)

▲ **Figure 9.4.2**

✔ **QUICK CHECK EXERCISES 9.4**   *(See page 631 for answers.)*

1. The divergence test says that if _____ $\neq 0$, then the series $\sum u_k$ diverges.

2. Given that

$$a_1 = 3, \quad \sum_{k=1}^{\infty} a_k = 1, \quad \text{and} \quad \sum_{k=1}^{\infty} b_k = 5$$

it follows that

$$\sum_{k=2}^{\infty} a_k = \underline{\hspace{1cm}} \quad \text{and} \quad \sum_{k=1}^{\infty} (2a_k + b_k) = \underline{\hspace{1cm}}$$

3. Since $\int_{1}^{+\infty} (1/\sqrt{x})\,dx = +\infty$, the _____ test applied to the series $\sum_{k=1}^{\infty}$ _____ shows that this series _____.

4. A p-series is a series of the form

$$\sum_{k=1}^{\infty} \underline{\hspace{1cm}}$$

This series converges if _____. This series diverges if _____.

EXERCISE SET 9.4   ~ Graphing Utility   [c] CAS

1. Use Theorem 9.4.3 to find the sum of each series.

(a) $\left(\dfrac{1}{2} + \dfrac{1}{4}\right) + \left(\dfrac{1}{2^2} + \dfrac{1}{4^2}\right) + \cdots + \left(\dfrac{1}{2^k} + \dfrac{1}{4^k}\right) + \cdots$

(b) $\displaystyle\sum_{k=1}^{\infty} \left(\dfrac{1}{5^k} - \dfrac{1}{k(k+1)}\right)$

2. Use Theorem 9.4.3 to find the sum of each series.

(a) $\displaystyle\sum_{k=2}^{\infty} \left[\dfrac{1}{k^2 - 1} - \dfrac{7}{10^{k-1}}\right]$ (b) $\displaystyle\sum_{k=1}^{\infty} \left[7^{-k}3^{k+1} - \dfrac{2^{k+1}}{5^k}\right]$

3–4 For each given p-series, identify p and determine whether the series converges.

3. (a) $\displaystyle\sum_{k=1}^{\infty} \dfrac{1}{k^3}$ (b) $\displaystyle\sum_{k=1}^{\infty} \dfrac{1}{\sqrt{k}}$ (c) $\displaystyle\sum_{k=1}^{\infty} k^{-1}$ (d) $\displaystyle\sum_{k=1}^{\infty} k^{-2/3}$

4. (a) $\displaystyle\sum_{k=1}^{\infty} k^{-4/3}$ (b) $\displaystyle\sum_{k=1}^{\infty} \dfrac{1}{\sqrt[4]{k}}$ (c) $\displaystyle\sum_{k=1}^{\infty} \dfrac{1}{\sqrt[3]{k^5}}$ (d) $\displaystyle\sum_{k=1}^{\infty} \dfrac{1}{k^\pi}$

5–6 Apply the divergence test and state what it tells you about the series.

5. (a) $\displaystyle\sum_{k=1}^{\infty} \dfrac{k^2 + k + 3}{2k^2 + 1}$ (b) $\displaystyle\sum_{k=1}^{\infty} \left(1 + \dfrac{1}{k}\right)^k$

(c) $\displaystyle\sum_{k=1}^{\infty} \cos k\pi$ (d) $\displaystyle\sum_{k=1}^{\infty} \dfrac{1}{k!}$

6. (a) $\displaystyle\sum_{k=1}^{\infty} \dfrac{k}{e^k}$ (b) $\displaystyle\sum_{k=1}^{\infty} \ln k$

(c) $\displaystyle\sum_{k=1}^{\infty} \dfrac{1}{\sqrt{k}}$ (d) $\displaystyle\sum_{k=1}^{\infty} \dfrac{\sqrt{k}}{\sqrt{k} + 3}$

7–8 Confirm that the integral test is applicable and use it to determine whether the series converges. ■

7. (a) $\displaystyle\sum_{k=1}^{\infty} \dfrac{1}{5k + 2}$ (b) $\displaystyle\sum_{k=1}^{\infty} \dfrac{1}{1 + 9k^2}$

8. (a) $\displaystyle\sum_{k=1}^{\infty} \dfrac{k}{1 + k^2}$ (b) $\displaystyle\sum_{k=1}^{\infty} \dfrac{1}{(4 + 2k)^{3/2}}$

9–24 Determine whether the series converges. ■

9. $\displaystyle\sum_{k=1}^{\infty} \dfrac{1}{k + 6}$ **10.** $\displaystyle\sum_{k=1}^{\infty} \dfrac{3}{5k}$ **11.** $\displaystyle\sum_{k=1}^{\infty} \dfrac{1}{\sqrt{k + 5}}$

12. $\displaystyle\sum_{k=1}^{\infty} \dfrac{1}{\sqrt[k]{e}}$ **13.** $\displaystyle\sum_{k=1}^{\infty} \dfrac{1}{\sqrt[3]{2k - 1}}$ **14.** $\displaystyle\sum_{k=3}^{\infty} \dfrac{\ln k}{k}$

15. $\displaystyle\sum_{k=1}^{\infty} \dfrac{k}{\ln(k + 1)}$ **16.** $\displaystyle\sum_{k=1}^{\infty} ke^{-k^2}$ **17.** $\displaystyle\sum_{k=1}^{\infty} \left(1 + \dfrac{1}{k}\right)^{-k}$

18. $\displaystyle\sum_{k=1}^{\infty} \dfrac{k^2 + 1}{k^2 + 3}$ **19.** $\displaystyle\sum_{k=1}^{\infty} \dfrac{\tan^{-1} k}{1 + k^2}$ **20.** $\displaystyle\sum_{k=1}^{\infty} \dfrac{1}{\sqrt{k^2 + 1}}$

21. $\displaystyle\sum_{k=1}^{\infty} k^2 \sin^2\left(\dfrac{1}{k}\right)$ **22.** $\displaystyle\sum_{k=1}^{\infty} k^2 e^{-k^3}$

23. $\displaystyle\sum_{k=5}^{\infty} 7k^{-1.01}$ **24.** $\displaystyle\sum_{k=1}^{\infty} \operatorname{sech}^2 k$

25–26 Use the integral test to investigate the relationship between the value of p and the convergence of the series. ■

25. $\displaystyle\sum_{k=2}^{\infty} \dfrac{1}{k(\ln k)^p}$ **26.** $\displaystyle\sum_{k=3}^{\infty} \dfrac{1}{k(\ln k)[\ln(\ln k)]^p}$

FOCUS ON CONCEPTS

27. Suppose that the series $\sum u_k$ converges and the series $\sum v_k$ diverges. Show that the series $\sum(u_k + v_k)$ and $\sum(u_k - v_k)$ both diverge. [*Hint:* Assume that $\sum(u_k + v_k)$ converges and use Theorem 9.4.3 to obtain a contradiction.]

28. Find examples to show that if the series $\sum u_k$ and $\sum v_k$ both diverge, then the series $\sum(u_k + v_k)$ and $\sum(u_k - v_k)$ may either converge or diverge.

29–30 Use the results of Exercises 27 and 28, if needed, to determine whether each series converges or diverges. ■

29. (a) $\displaystyle\sum_{k=1}^{\infty}\left[\left(\frac{2}{3}\right)^{k-1}+\frac{1}{k}\right]$ (b) $\displaystyle\sum_{k=1}^{\infty}\left[\frac{1}{3k+2}-\frac{1}{k^{3/2}}\right]$

30. (a) $\displaystyle\sum_{k=2}^{\infty}\left[\frac{1}{k(\ln k)^2}-\frac{1}{k^2}\right]$ (b) $\displaystyle\sum_{k=2}^{\infty}\left[ke^{-k^2}+\frac{1}{k\ln k}\right]$

31–34 True–False Determine whether the statement is true or false. Explain your answer. ■

31. If $\sum u_k$ converges to L, then $\sum(1/u_k)$ converges to $1/L$.

32. If $\sum cu_k$ diverges for some constant c, then $\sum u_k$ must diverge.

33. The integral test can be used to prove that a series diverges.

34. The series $\displaystyle\sum_{k=1}^{\infty}\frac{1}{p^k}$ is a p-series.

[c] **35.** Use a CAS to confirm that

$$\sum_{k=1}^{\infty}\frac{1}{k^2}=\frac{\pi^2}{6}\quad\text{and}\quad\sum_{k=1}^{\infty}\frac{1}{k^4}=\frac{\pi^4}{90}$$

and then use these results in each part to find the sum of the series.

(a) $\displaystyle\sum_{k=1}^{\infty}\frac{3k^2-1}{k^4}$ (b) $\displaystyle\sum_{k=3}^{\infty}\frac{1}{k^2}$ (c) $\displaystyle\sum_{k=2}^{\infty}\frac{1}{(k-1)^4}$

36–40 Exercise 36 will show how a partial sum can be used to obtain upper and lower bounds on the sum of a series when the hypotheses of the integral test are satisfied. This result will be needed in Exercises 37–40. ■

36. (a) Let $\sum_{k=1}^{\infty}u_k$ be a convergent series with positive terms, and let f be a function that is decreasing and continuous on $[n, +\infty)$ and such that $u_k = f(k)$ for $k \geq n$. Use an area argument and the accompanying figure to show that

$$\int_{n+1}^{+\infty}f(x)\,dx<\sum_{k=n+1}^{\infty}u_k<\int_{n}^{+\infty}f(x)\,dx$$

(b) Show that if S is the sum of the series $\sum_{k=1}^{\infty}u_k$ and s_n is the nth partial sum, then

$$s_n+\int_{n+1}^{+\infty}f(x)\,dx<S<s_n+\int_{n}^{+\infty}f(x)\,dx$$

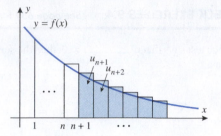

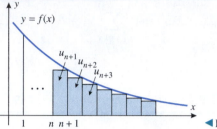

◀ **Figure Ex-36**

37. (a) It was stated in Exercise 35 that

$$\sum_{k=1}^{\infty}\frac{1}{k^2}=\frac{\pi^2}{6}$$

Show that if s_n is the nth partial sum of this series, then

$$s_n+\frac{1}{n+1}<\frac{\pi^2}{6}<s_n+\frac{1}{n}$$

(b) Calculate s_3 exactly, and then use the result in part (a) to show that

$$\frac{29}{18}<\frac{\pi^2}{6}<\frac{61}{36}$$

(c) Use a calculating utility to confirm that the inequalities in part (b) are correct.

(d) Find upper and lower bounds on the error that results if the sum of the series is approximated by the 10th partial sum.

38. In each part, find upper and lower bounds on the error that results if the sum of the series is approximated by the 10th partial sum.

(a) $\displaystyle\sum_{k=1}^{\infty}\frac{1}{(2k+1)^2}$ (b) $\displaystyle\sum_{k=1}^{\infty}\frac{1}{k^2+1}$ (c) $\displaystyle\sum_{k=1}^{\infty}\frac{k}{e^k}$

39. It was stated in Exercise 35 that

$$\sum_{k=1}^{\infty}\frac{1}{k^4}=\frac{\pi^4}{90}$$

(a) Let s_n be the nth partial sum of the series above. Show that

$$s_n+\frac{1}{3(n+1)^3}<\frac{\pi^4}{90}<s_n+\frac{1}{3n^3}$$

(b) We can use a partial sum of the series to approximate $\pi^4/90$ to three decimal-place accuracy by capturing the sum of the series in an interval of length 0.001 (or less). Find the smallest value of n such that the interval containing $\pi^4/90$ in part (a) has a length of 0.001 or less.

(c) Approximate $\pi^4/90$ to three decimal places using the midpoint of an interval of width at most 0.001 that contains the sum of the series. Use a calculating utility to confirm that your answer is within 0.0005 of $\pi^4/90$.

40. We showed in Section 9.3 that the harmonic series $\sum_{k=1}^{\infty} 1/k$ diverges. Our objective in this problem is to demonstrate that although the partial sums of this series approach $+\infty$, they increase extremely slowly.

(a) Use inequality (2) to show that for $n \geq 2$

$$\ln(n+1) < s_n < 1 + \ln n$$

(b) Use the inequalities in part (a) to find upper and lower bounds on the sum of the first million terms in the series.

(c) Show that the sum of the first billion terms in the series is less than 22.

(d) Find a value of n so that the sum of the first n terms is greater than 100.

41. Use a graphing utility to confirm that the integral test applies to the series $\sum_{k=1}^{\infty} k^2 e^{-k}$, and then determine whether the series converges.

42. (a) Show that the hypotheses of the integral test are satisfied by the series $\sum_{k=1}^{\infty} 1/(k^3 + 1)$.

(b) Use a CAS and the integral test to confirm that the series converges.

(c) Construct a table of partial sums for $n = 10, 20, 30, \ldots, 100$, showing at least six decimal places.

(d) Based on your table, make a conjecture about the sum of the series to three decimal-place accuracy.

(e) Use part (b) of Exercise 36 to check your conjecture.

✔ **QUICK CHECK ANSWERS 9.4**

1. $\lim\limits_{k \to +\infty} u_k$ **2.** $-2; 7$ **3.** integral; $\dfrac{1}{\sqrt{k}}$; diverges   **4.** $\dfrac{1}{k^p}; p > 1; 0 < p \leq 1$

9.5 THE COMPARISON, RATIO, AND ROOT TESTS

In this section we will develop some more basic convergence tests for series with nonnegative terms. Later, we will use some of these tests to study the convergence of Taylor series.

■ THE COMPARISON TEST

We will begin with a test that is useful in its own right and is also the building block for other important convergence tests. The underlying idea of this test is to use the known convergence or divergence of a series to deduce the convergence or divergence of another series.

9.5.1   THEOREM (*The Comparison Test*)   *Let $\sum_{k=1}^{\infty} a_k$ and $\sum_{k=1}^{\infty} b_k$ be series with nonnegative terms and suppose that*

$$a_1 \leq b_1, \ a_2 \leq b_2, \ a_3 \leq b_3, \ldots, a_k \leq b_k, \ldots$$

(a) If the "bigger series" Σb_k converges, then the "smaller series" Σa_k also converges.

(b) If the "smaller series" Σa_k diverges, then the "bigger series" Σb_k also diverges.

It is not essential in Theorem 9.5.1 that the condition $a_k \leq b_k$ hold for all k, as stated; the conclusions of the theorem remain true if this condition is eventually true.

We have left the proof of this theorem for the exercises; however, it is easy to visualize why the theorem is true by interpreting the terms in the series as areas of rectangles

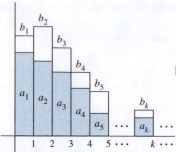

For each rectangle, a_k denotes the area of the blue portion and b_k denotes the combined area of the white and blue portions.

▲ **Figure 9.5.1**

(Figure 9.5.1). The comparison test states that if the total area $\sum b_k$ is finite, then the total area $\sum a_k$ must also be finite; and if the total area $\sum a_k$ is infinite, then the total area $\sum b_k$ must also be infinite.

■ **USING THE COMPARISON TEST**

There are two steps required for using the comparison test to determine whether a series $\sum u_k$ with positive terms converges:

Step 1. Guess at whether the series $\sum u_k$ converges or diverges.

Step 2. Find a series that proves the guess to be correct. That is, if we guess that $\sum u_k$ diverges, we must find a divergent series whose terms are "smaller" than the corresponding terms of $\sum u_k$, and if we guess that $\sum u_k$ converges, we must find a convergent series whose terms are "bigger" than the corresponding terms of $\sum u_k$.

In most cases, the series $\sum u_k$ being considered will have its general term u_k expressed as a fraction. To help with the guessing process in the first step, we have formulated two principles that are based on the form of the denominator for u_k. These principles sometimes *suggest* whether a series is likely to converge or diverge. We have called these "informal principles" because they are not intended as formal theorems. In fact, we will not guarantee that they *always* work. However, they work often enough to be useful.

9.5.2   INFORMAL PRINCIPLE   *Constant terms in the denominator of u_k can usually be deleted without affecting the convergence or divergence of the series.*

9.5.3   INFORMAL PRINCIPLE   *If a polynomial in k appears as a factor in the numerator or denominator of u_k, all but the leading term in the polynomial can usually be discarded without affecting the convergence or divergence of the series.*

▶ **Example 1**   Use the comparison test to determine whether the following series converge or diverge.

$$\text{(a) } \sum_{k=1}^{\infty} \frac{1}{\sqrt{k} - \frac{1}{2}} \qquad \text{(b) } \sum_{k=1}^{\infty} \frac{1}{2k^2 + k}$$

Solution (a).   According to Principle 9.5.2, we should be able to drop the constant in the denominator without affecting the convergence or divergence. Thus, the given series is likely to behave like

$$\sum_{k=1}^{\infty} \frac{1}{\sqrt{k}} \qquad (1)$$

which is a divergent p-series $\left(p = \frac{1}{2}\right)$. Thus, we will guess that the given series diverges and try to prove this by finding a divergent series that is "smaller" than the given series. However, series (1) does the trick since

$$\frac{1}{\sqrt{k} - \frac{1}{2}} > \frac{1}{\sqrt{k}} \quad \text{for } k = 1, 2, \ldots$$

Thus, we have proved that the given series diverges.

Solution (b).   According to Principle 9.5.3, we should be able to discard all but the leading term in the polynomial without affecting the convergence or divergence. Thus, the given series is likely to behave like

$$\sum_{k=1}^{\infty} \frac{1}{2k^2} = \frac{1}{2} \sum_{k=1}^{\infty} \frac{1}{k^2} \tag{2}$$

which converges since it is a constant times a convergent p-series ($p = 2$). Thus, we will guess that the given series converges and try to prove this by finding a convergent series that is "bigger" than the given series. However, series (2) does the trick since

$$\frac{1}{2k^2 + k} < \frac{1}{2k^2} \quad \text{for } k = 1, 2, \ldots$$

Thus, we have proved that the given series converges. ◄

■ THE LIMIT COMPARISON TEST

In the last example, Principles 9.5.2 and 9.5.3 provided the guess about convergence or divergence as well as the series needed to apply the comparison test. Unfortunately, it is not always so straightforward to find the series required for comparison, so we will now consider an alternative to the comparison test that is usually easier to apply. The proof is given in Appendix J.

9.5.4   THEOREM (*The Limit Comparison Test*)   *Let $\sum a_k$ and $\sum b_k$ be series with positive terms and suppose that*

$$\rho = \lim_{k \to +\infty} \frac{a_k}{b_k}$$

If ρ is finite and $\rho > 0$, then the series both converge or both diverge.

The cases where $\rho = 0$ or $\rho = +\infty$ are discussed in the exercises (Exercise 54).

To use the limit comparison test we must again first guess at the convergence or divergence of $\sum a_k$ and then find a series $\sum b_k$ that supports our guess. The following example illustrates this principle.

► **Example 2**   Use the limit comparison test to determine whether the following series converge or diverge.

(a) $\displaystyle\sum_{k=1}^{\infty} \frac{1}{\sqrt{k}+1}$ (b) $\displaystyle\sum_{k=1}^{\infty} \frac{1}{2k^2 + k}$ (c) $\displaystyle\sum_{k=1}^{\infty} \frac{3k^3 - 2k^2 + 4}{k^7 - k^3 + 2}$

Solution (a).   As in Example 1, Principle 9.5.2 suggests that the series is likely to behave like the divergent p-series (1). To prove that the given series diverges, we will apply the limit comparison test with

$$a_k = \frac{1}{\sqrt{k}+1} \quad \text{and} \quad b_k = \frac{1}{\sqrt{k}}$$

We obtain

$$\rho = \lim_{k \to +\infty} \frac{a_k}{b_k} = \lim_{k \to +\infty} \frac{\sqrt{k}}{\sqrt{k}+1} = \lim_{k \to +\infty} \frac{1}{1 + \dfrac{1}{\sqrt{k}}} = 1$$

Since ρ is finite and positive, it follows from Theorem 9.5.4 that the given series diverges.

Solution (b). As in Example 1, Principle 9.5.3 suggests that the series is likely to behave like the convergent series (2). To prove that the given series converges, we will apply the limit comparison test with

$$a_k = \frac{1}{2k^2 + k} \quad \text{and} \quad b_k = \frac{1}{2k^2}$$

We obtain

$$\rho = \lim_{k \to +\infty} \frac{a_k}{b_k} = \lim_{k \to +\infty} \frac{2k^2}{2k^2 + k} = \lim_{k \to +\infty} \frac{2}{2 + \dfrac{1}{k}} = 1$$

Since ρ is finite and positive, it follows from Theorem 9.5.4 that the given series converges, which agrees with the conclusion reached in Example 1 using the comparison test.

Solution (c). From Principle 9.5.3, the series is likely to behave like

$$\sum_{k=1}^{\infty} \frac{3k^3}{k^7} = \sum_{k=1}^{\infty} \frac{3}{k^4} \tag{3}$$

which converges since it is a constant times a convergent p-series. Thus, the given series is likely to converge. To prove this, we will apply the limit comparison test to series (3) and the given series. We obtain

$$\rho = \lim_{k \to +\infty} \frac{\dfrac{3k^3 - 2k^2 + 4}{k^7 - k^3 + 2}}{\dfrac{3}{k^4}} = \lim_{k \to +\infty} \frac{3k^7 - 2k^6 + 4k^4}{3k^7 - 3k^3 + 6} = 1$$

Since ρ is finite and nonzero, it follows from Theorem 9.5.4 that the given series converges, since (3) converges. ◄

■ THE RATIO TEST

The comparison test and the limit comparison test hinge on first making a guess about convergence and then finding an appropriate series for comparison, both of which can be difficult tasks in cases where Principles 9.5.2 and 9.5.3 cannot be applied. In such cases the next test can often be used, since it works exclusively with the terms of the given series—it requires neither an initial guess about convergence nor the discovery of a series for comparison. Its proof is given in Appendix J.

9.5.5 THEOREM (*The Ratio Test*) *Let $\sum u_k$ be a series with positive terms and suppose that*

$$\rho = \lim_{k \to +\infty} \frac{u_{k+1}}{u_k}$$

(a) *If $\rho < 1$, the series converges.*

(b) *If $\rho > 1$ or $\rho = +\infty$, the series diverges.*

(c) *If $\rho = 1$, the series may converge or diverge, so that another test must be tried.*

► **Example 3** Each of the following series has positive terms, so the ratio test applies. In each part, use the ratio test to determine whether the following series converge or diverge.

(a) $\displaystyle\sum_{k=1}^{\infty} \frac{1}{k!}$ (b) $\displaystyle\sum_{k=1}^{\infty} \frac{k}{2^k}$ (c) $\displaystyle\sum_{k=1}^{\infty} \frac{k^k}{k!}$ (d) $\displaystyle\sum_{k=3}^{\infty} \frac{(2k)!}{4^k}$ (e) $\displaystyle\sum_{k=1}^{\infty} \frac{1}{2k - 1}$

Solution (a). The series converges, since

$$\rho = \lim_{k \to +\infty} \frac{u_{k+1}}{u_k} = \lim_{k \to +\infty} \frac{1/(k+1)!}{1/k!} = \lim_{k \to +\infty} \frac{k!}{(k+1)!} = \lim_{k \to +\infty} \frac{1}{k+1} = 0 < 1$$

Solution (b). The series converges, since

$$\rho = \lim_{k \to +\infty} \frac{u_{k+1}}{u_k} = \lim_{k \to +\infty} \frac{k+1}{2^{k+1}} \cdot \frac{2^k}{k} = \frac{1}{2} \lim_{k \to +\infty} \frac{k+1}{k} = \frac{1}{2} < 1$$

Solution (c). The series diverges, since

$$\rho = \lim_{k \to +\infty} \frac{u_{k+1}}{u_k} = \lim_{k \to +\infty} \frac{(k+1)^{k+1}}{(k+1)!} \cdot \frac{k!}{k^k} = \lim_{k \to +\infty} \frac{(k+1)^k}{k^k} = \lim_{k \to +\infty} \left(1 + \frac{1}{k}\right)^k = e > 1$$

> See Formula (7) of Section 1.3

Solution (d). The series diverges, since

$$\rho = \lim_{k \to +\infty} \frac{u_{k+1}}{u_k} = \lim_{k \to +\infty} \frac{[2(k+1)]!}{4^{k+1}} \cdot \frac{4^k}{(2k)!} = \lim_{k \to +\infty} \left(\frac{(2k+2)!}{(2k)!} \cdot \frac{1}{4}\right)$$

$$= \lim_{k \to +\infty} \left(\frac{(2k+2)(2k+1)(2k)!}{(2k)!} \cdot \frac{1}{4}\right) = \frac{1}{4} \lim_{k \to +\infty} (2k+2)(2k+1) = +\infty$$

Solution (e). The ratio test is of no help since

$$\rho = \lim_{k \to +\infty} \frac{u_{k+1}}{u_k} = \lim_{k \to +\infty} \frac{1}{2(k+1)-1} \cdot \frac{2k-1}{1} = \lim_{k \to +\infty} \frac{2k-1}{2k+1} = 1$$

However, the integral test proves that the series diverges since

$$\int_1^{+\infty} \frac{dx}{2x-1} = \lim_{b \to +\infty} \int_1^b \frac{dx}{2x-1} = \lim_{b \to +\infty} \frac{1}{2} \ln(2x-1) \Big]_1^b = +\infty$$

Both the comparison test and the limit comparison test would also have worked here (verify). ◄

■ THE ROOT TEST

In cases where it is difficult or inconvenient to find the limit required for the ratio test, the next test is sometimes useful. Since its proof is similar to the proof of the ratio test, we will omit it.

> **9.5.6 THEOREM (*The Root Test*)** Let $\sum u_k$ be a series with positive terms and suppose that
> $$\rho = \lim_{k \to +\infty} \sqrt[k]{u_k} = \lim_{k \to +\infty} (u_k)^{1/k}$$
>
> (a) If $\rho < 1$, the series converges.
>
> (b) If $\rho > 1$ or $\rho = +\infty$, the series diverges.
>
> (c) If $\rho = 1$, the series may converge or diverge, so that another test must be tried.

► **Example 4** Use the root test to determine whether the following series converge or diverge.

$$\text{(a)} \sum_{k=2}^{\infty} \left(\frac{4k-5}{2k+1}\right)^k \qquad \text{(b)} \sum_{k=1}^{\infty} \frac{1}{(\ln(k+1))^k}$$

Solution (a). The series diverges, since

$$\rho = \lim_{k \to +\infty} (u_k)^{1/k} = \lim_{k \to +\infty} \frac{4k-5}{2k+1} = 2 > 1$$

Solution (b). The series converges, since

$$\rho = \lim_{k \to +\infty} (u_k)^{1/k} = \lim_{k \to +\infty} \frac{1}{\ln(k+1)} = 0 < 1 \ \blacktriangleleft$$

✔ QUICK CHECK EXERCISES 9.5   (See page 637 for answers.)

1–4 Select between *converges* or *diverges* to fill the first blank. ■

1. The series

$$\sum_{k=1}^{\infty} \frac{2k^2+1}{2k^{8/3}-1}$$

_____ by comparison with the *p*-series $\sum_{k=1}^{\infty}$ _____.

2. Since

$$\lim_{k \to +\infty} \frac{(k+1)^3/3^{k+1}}{k^3/3^k} = \lim_{k \to +\infty} \frac{\left(1+\frac{1}{k}\right)^3}{3} = \frac{1}{3}$$

the series $\sum_{k=1}^{\infty} k^3/3^k$ _____ by the _____ test.

3. Since

$$\lim_{k \to +\infty} \frac{(k+1)!/3^{k+1}}{k!/3^k} = \lim_{k \to +\infty} \frac{k+1}{3} = +\infty$$

the series $\sum_{k=1}^{\infty} k!/3^k$ _____ by the _____ test.

4. Since

$$\lim_{k \to +\infty} \left(\frac{1}{k^{k/2}}\right)^{1/k} = \lim_{k \to +\infty} \frac{1}{k^{1/2}} = 0$$

the series $\sum_{k=1}^{\infty} 1/k^{k/2}$ _____ by the _____ test.

EXERCISE SET 9.5

1–2 Make a guess about the convergence or divergence of the series, and confirm your guess using the comparison test. ■

1. (a) $\sum_{k=1}^{\infty} \frac{1}{5k^2-k}$ (b) $\sum_{k=1}^{\infty} \frac{3}{k-\frac{1}{4}}$

2. (a) $\sum_{k=2}^{\infty} \frac{k+1}{k^2-k}$ (b) $\sum_{k=1}^{\infty} \frac{2}{k^4+k}$

3. In each part, use the comparison test to show that the series converges.

(a) $\sum_{k=1}^{\infty} \frac{1}{3^k+5}$ (b) $\sum_{k=1}^{\infty} \frac{5\sin^2 k}{k!}$

4. In each part, use the comparison test to show that the series diverges.

(a) $\sum_{k=1}^{\infty} \frac{\ln k}{k}$ (b) $\sum_{k=1}^{\infty} \frac{k}{k^{3/2}-\frac{1}{2}}$

5–10 Use the limit comparison test to determine whether the series converges. ■

5. $\sum_{k=1}^{\infty} \frac{4k^2-2k+6}{8k^7+k-8}$ **6.** $\sum_{k=1}^{\infty} \frac{1}{9k+6}$

7. $\sum_{k=1}^{\infty} \frac{5}{3^k+1}$ **8.** $\sum_{k=1}^{\infty} \frac{k(k+3)}{(k+1)(k+2)(k+5)}$

9. $\sum_{k=1}^{\infty} \frac{1}{\sqrt[3]{8k^2-3k}}$ **10.** $\sum_{k=1}^{\infty} \frac{1}{(2k+3)^{17}}$

11–16 Use the ratio test to determine whether the series converges. If the test is inconclusive, then say so. ■

11. $\sum_{k=1}^{\infty} \frac{3^k}{k!}$ **12.** $\sum_{k=1}^{\infty} \frac{4^k}{k^2}$ **13.** $\sum_{k=1}^{\infty} \frac{1}{5k}$

14. $\sum_{k=1}^{\infty} k\left(\frac{1}{2}\right)^k$ **15.** $\sum_{k=1}^{\infty} \frac{k!}{k^3}$ **16.** $\sum_{k=1}^{\infty} \frac{k}{k^2+1}$

17–20 Use the root test to determine whether the series converges. If the test is inconclusive, then say so. ■

17. $\sum_{k=1}^{\infty} \left(\frac{3k+2}{2k-1}\right)^k$ **18.** $\sum_{k=1}^{\infty} \left(\frac{k}{100}\right)^k$

19. $\sum_{k=1}^{\infty} \frac{k}{5^k}$ **20.** $\sum_{k=1}^{\infty} (1-e^{-k})^k$

21–24 True–False Determine whether the statement is true or false. Explain your answer. ■

21. The limit comparison test decides convergence based on a limit of the quotient of consecutive terms in a series.

22. If $\lim_{k \to +\infty}(u_{k+1}/u_k) = 5$, then $\sum u_k$ diverges.

23. If $\lim_{k \to +\infty}(k^2 u_k) = 5$, then $\sum u_k$ converges.

24. The root test decides convergence based on a limit of kth roots of terms in the sequence of partial sums for a series.

25–47 Use any method to determine whether the series converges. ∎

25. $\sum_{k=0}^{\infty} \dfrac{7^k}{k!}$ **26.** $\sum_{k=1}^{\infty} \dfrac{1}{2k+1}$ **27.** $\sum_{k=1}^{\infty} \dfrac{k^2}{5^k}$

28. $\sum_{k=1}^{\infty} \dfrac{k!10^k}{3^k}$ **29.** $\sum_{k=1}^{\infty} k^{50}e^{-k}$ **30.** $\sum_{k=1}^{\infty} \dfrac{k^2}{k^3+1}$

31. $\sum_{k=1}^{\infty} \dfrac{\sqrt{k}}{k^3+1}$ **32.** $\sum_{k=1}^{\infty} \dfrac{4}{2+3^k k}$

33. $\sum_{k=1}^{\infty} \dfrac{1}{\sqrt{k(k+1)}}$ **34.** $\sum_{k=1}^{\infty} \dfrac{2+(-1)^k}{5^k}$

35. $\sum_{k=1}^{\infty} \dfrac{1}{1+\sqrt{k}}$ **36.** $\sum_{k=1}^{\infty} \dfrac{k!}{k^k}$ **37.** $\sum_{k=1}^{\infty} \dfrac{\ln k}{e^k}$

38. $\sum_{k=1}^{\infty} \dfrac{k!}{e^{k^2}}$ **39.** $\sum_{k=0}^{\infty} \dfrac{(k+4)!}{4!k!4^k}$ **40.** $\sum_{k=1}^{\infty} \left(\dfrac{k}{k+1}\right)^{k^2}$

41. $\sum_{k=1}^{\infty} \dfrac{1}{4+2^{-k}}$ **42.** $\sum_{k=1}^{\infty} \dfrac{\sqrt{k}\ln k}{k^3+1}$ **43.** $\sum_{k=1}^{\infty} \dfrac{\tan^{-1} k}{k^2}$

44. $\sum_{k=1}^{\infty} \dfrac{5^k+k}{k!+3}$ **45.** $\sum_{k=0}^{\infty} \dfrac{(k!)^2}{(2k)!}$ **46.** $\sum_{k=1}^{\infty} \dfrac{[\pi(k+1)]^k}{k^{k+1}}$

47. $\sum_{k=1}^{\infty} \dfrac{\ln k}{3^k}$

48. For what positive values of α does the series $\sum_{k=1}^{\infty}(\alpha^k/k^\alpha)$ converge?

49–50 Find the general term of the series and use the ratio test to show that the series converges. ∎

49. $1 + \dfrac{1 \cdot 2}{1 \cdot 3} + \dfrac{1 \cdot 2 \cdot 3}{1 \cdot 3 \cdot 5} + \dfrac{1 \cdot 2 \cdot 3 \cdot 4}{1 \cdot 3 \cdot 5 \cdot 7} + \cdots$

50. $1 + \dfrac{1 \cdot 3}{3!} + \dfrac{1 \cdot 3 \cdot 5}{5!} + \dfrac{1 \cdot 3 \cdot 5 \cdot 7}{7!} + \cdots$

51. Show that $\ln x < \sqrt{x}$ if $x > 0$, and use this result to investigate the convergence of

(a) $\sum_{k=1}^{\infty} \dfrac{\ln k}{k^2}$ (b) $\sum_{k=2}^{\infty} \dfrac{1}{(\ln k)^2}$

FOCUS ON CONCEPTS

52. (a) Make a conjecture about the convergence of the series $\sum_{k=1}^{\infty} \sin(\pi/k)$ by considering the local linear approximation of $\sin x$ at $x = 0$.
(b) Try to confirm your conjecture using the limit comparison test.

53. (a) We will see later that the polynomial $1 - x^2/2$ is the "local quadratic" approximation for $\cos x$ at $x = 0$. Make a conjecture about the convergence of the series

$$\sum_{k=1}^{\infty}\left[1 - \cos\left(\dfrac{1}{k}\right)\right]$$

by considering this approximation.
(b) Try to confirm your conjecture using the limit comparison test.

54. Let $\sum a_k$ and $\sum b_k$ be series with positive terms. Prove:
(a) If $\lim_{k \to +\infty}(a_k/b_k) = 0$ and $\sum b_k$ converges, then $\sum a_k$ converges.
(b) If $\lim_{k \to +\infty}(a_k/b_k) = +\infty$ and $\sum b_k$ diverges, then $\sum a_k$ diverges.

55. Use Theorem 9.4.6 to prove the comparison test (Theorem 9.5.1).

56. Writing What does the ratio test tell you about the convergence of a geometric series? Discuss similarities between geometric series and series to which the ratio test applies.

57. Writing Given an infinite series, discuss a strategy for deciding what convergence test to use.

✔**QUICK CHECK ANSWERS 9.5**

1. diverges; $1/k^{2/3}$ **2.** converges; ratio **3.** diverges; ratio **4.** converges; root

9.6 ALTERNATING SERIES; ABSOLUTE AND CONDITIONAL CONVERGENCE

Up to now we have focused exclusively on series with nonnegative terms. In this section we will discuss series that contain both positive and negative terms.

■ ALTERNATING SERIES

Series whose terms alternate between positive and negative, called *alternating series*, are of special importance. Some examples are

$$\sum_{k=1}^{\infty}(-1)^{k+1}\frac{1}{k} = 1 - \frac{1}{2} + \frac{1}{3} - \frac{1}{4} + \frac{1}{5} - \cdots$$

$$\sum_{k=1}^{\infty}(-1)^{k}\frac{1}{k} = -1 + \frac{1}{2} - \frac{1}{3} + \frac{1}{4} - \frac{1}{5} + \cdots$$

In general, an alternating series has one of the following two forms:

$$\sum_{k=1}^{\infty}(-1)^{k+1}a_k = a_1 - a_2 + a_3 - a_4 + \cdots \tag{1}$$

$$\sum_{k=1}^{\infty}(-1)^{k}a_k = -a_1 + a_2 - a_3 + a_4 - \cdots \tag{2}$$

where the a_k's are assumed to be positive in both cases.

The following theorem is the key result on convergence of alternating series.

9.6.1   THEOREM (*Alternating Series Test*)   *An alternating series of either form* (1) *or form* (2) *converges if the following two conditions are satisfied:*

(a) $a_1 \geq a_2 \geq a_3 \geq \cdots \geq a_k \geq \cdots$

(b) $\displaystyle\lim_{k \to +\infty} a_k = 0$

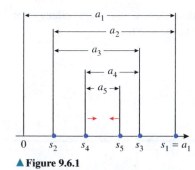

▲ **Figure 9.6.1**

It is not essential for condition (*a*) in Theorem 9.6.1 to hold for all terms; an alternating series will converge if condition (*b*) is true and condition (*a*) holds eventually.

PROOF   We will consider only alternating series of form (1). The idea of the proof is to show that if conditions (*a*) and (*b*) hold, then the sequences of even-numbered and odd-numbered partial sums converge to a common limit S. It will then follow from Theorem 9.1.4 that the entire sequence of partial sums converges to S.

Figure 9.6.1 shows how successive partial sums satisfying conditions (*a*) and (*b*) appear when plotted on a horizontal axis. The even-numbered partial sums

$$s_2, s_4, s_6, s_8, \ldots, s_{2n}, \ldots$$

form an increasing sequence bounded above by a_1, and the odd-numbered partial sums

$$s_1, s_3, s_5, \ldots, s_{2n-1}, \ldots$$

form a decreasing sequence bounded below by 0. Thus, by Theorems 9.2.3 and 9.2.4, the even-numbered partial sums converge to some limit S_E and the odd-numbered partial sums converge to some limit S_O. To complete the proof we must show that $S_E = S_O$. But the

$(2n)$-th term in the series is $-a_{2n}$, so that $s_{2n} - s_{2n-1} = -a_{2n}$, which can be written as

$$s_{2n-1} = s_{2n} + a_{2n}$$

However, $2n \to +\infty$ and $2n - 1 \to +\infty$ as $n \to +\infty$, so that

$$S_O = \lim_{n \to +\infty} s_{2n-1} = \lim_{n \to +\infty} (s_{2n} + a_{2n}) = S_E + 0 = S_E$$

which completes the proof. ■

<table>
<tr><td>

If an alternating series violates condition (*b*) of the alternating series test, then the series must diverge by the divergence test (Theorem 9.4.1).

</td></tr>
</table>

▶ **Example 1**   Use the alternating series test to show that the following series converge.

$$\text{(a)} \sum_{k=1}^{\infty} (-1)^{k+1} \frac{1}{k} \qquad \text{(b)} \sum_{k=1}^{\infty} (-1)^{k+1} \frac{k+3}{k(k+1)}$$

Solution (a).   The two conditions in the alternating series test are satisfied since

<table>
<tr><td>

The series in part (a) of Example 1 is called the **alternating harmonic series**. Note that this series converges, whereas the harmonic series diverges.

</td></tr>
</table>

$$a_k = \frac{1}{k} > \frac{1}{k+1} = a_{k+1} \quad \text{and} \quad \lim_{k \to +\infty} a_k = \lim_{k \to +\infty} \frac{1}{k} = 0$$

Solution (b).   The two conditions in the alternating series test are satisfied since

$$\frac{a_{k+1}}{a_k} = \frac{k+4}{(k+1)(k+2)} \cdot \frac{k(k+1)}{k+3} = \frac{k^2 + 4k}{k^2 + 5k + 6} = \frac{k^2 + 4k}{(k^2 + 4k) + (k+6)} < 1$$

so

$$a_k > a_{k+1}$$

and

$$\lim_{k \to +\infty} a_k = \lim_{k \to +\infty} \frac{k+3}{k(k+1)} = \lim_{k \to +\infty} \frac{\dfrac{1}{k} + \dfrac{3}{k^2}}{1 + \dfrac{1}{k}} = 0 \blacktriangleleft$$

■ **APPROXIMATING SUMS OF ALTERNATING SERIES**

The following theorem is concerned with the error that results when the sum of an alternating series is approximated by a partial sum.

9.6.2   THEOREM   *If an alternating series satisfies the hypotheses of the alternating series test, and if S is the sum of the series, then:*

(*a*)   *S lies between any two successive partial sums; that is, either*

$$s_n \le S \le s_{n+1} \quad or \quad s_{n+1} \le S \le s_n \tag{3}$$

depending on which partial sum is larger.

(*b*)   *If S is approximated by s_n, then the absolute error $|S - s_n|$ satisfies*

$$|S - s_n| \le a_{n+1} \tag{4}$$

Moreover, the sign of the error $S - s_n$ is the same as that of the coefficient of a_{n+1}.

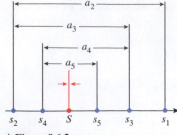

▲ **Figure 9.6.2**

PROOF   We will prove the theorem for series of form (1). Referring to Figure 9.6.2 and keeping in mind our observation in the proof of Theorem 9.6.1 that the odd-numbered partial sums form a decreasing sequence converging to S and the even-numbered partial sums form an increasing sequence converging to S, we see that successive partial sums oscillate from one side of S to the other in smaller and smaller steps with the odd-numbered partial sums being larger than S and the even-numbered partial sums being smaller than S. Thus, depending on whether n is even or odd, we have

$$s_n \leq S \leq s_{n+1} \quad \text{or} \quad s_{n+1} \leq S \leq s_n$$

which proves (3). Moreover, in either case we have

$$|S - s_n| \leq |s_{n+1} - s_n| \tag{5}$$

But $s_{n+1} - s_n = \pm a_{n+1}$ (the sign depending on whether n is even or odd). Thus, it follows from (5) that $|S - s_n| \leq a_{n+1}$, which proves (4). Finally, since the odd-numbered partial sums are larger than S and the even-numbered partial sums are smaller than S, it follows that $S - s_n$ has the same sign as the coefficient of a_{n+1} (verify). ∎

REMARK   In words, inequality (4) states that for a series satisfying the hypotheses of the alternating series test, the magnitude of the error that results from approximating S by s_n is at most that of the first term that is *not* included in the partial sum. Also, note that if $a_1 > a_2 > \cdots > a_k > \cdots$, then inequality (4) can be strengthened to $|S - s_n| < a_{n+1}$.

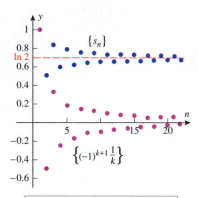

Graph of the sequences of terms and nth partial sums for the alternating harmonic series

▲ **Figure 9.6.3**

▶ **Example 2**   Later in this chapter we will show that the sum of the alternating harmonic series is

$$\ln 2 = 1 - \frac{1}{2} + \frac{1}{3} - \frac{1}{4} + \cdots + (-1)^{k+1}\frac{1}{k} + \cdots$$

This is illustrated in Figure 9.6.3.

(a) Accepting this to be so, find an upper bound on the magnitude of the error that results if $\ln 2$ is approximated by the sum of the first eight terms in the series.

(b) Find a partial sum that approximates $\ln 2$ to one decimal-place accuracy (the nearest tenth).

Solution (a).   It follows from the strengthened form of (4) that

$$|\ln 2 - s_8| < a_9 = \frac{1}{9} < 0.12 \tag{6}$$

As a check, let us compute s_8 exactly. We obtain

$$s_8 = 1 - \frac{1}{2} + \frac{1}{3} - \frac{1}{4} + \frac{1}{5} - \frac{1}{6} + \frac{1}{7} - \frac{1}{8} = \frac{533}{840}$$

Thus, with the help of a calculator

$$|\ln 2 - s_8| = \left|\ln 2 - \frac{533}{840}\right| \approx 0.059$$

This shows that the error is well under the estimate provided by upper bound (6).

Solution (b).   For one decimal-place accuracy, we must choose a value of n for which $|\ln 2 - s_n| \leq 0.05$. However, it follows from the strengthened form of (4) that

$$|\ln 2 - s_n| < a_{n+1}$$

so it suffices to choose n so that $a_{n+1} \leq 0.05$.

One way to find n is to use a calculating utility to obtain numerical values for a_1, a_2, a_3, ... until you encounter the first value that is less than or equal to 0.05. If you do this, you will find that it is $a_{20} = 0.05$; this tells us that partial sum s_{19} will provide the desired accuracy. Another way to find n is to solve the inequality

$$\frac{1}{n+1} \le 0.05$$

algebraically. We can do this by taking reciprocals, reversing the sense of the inequality, and then simplifying to obtain $n \ge 19$. Thus, s_{19} will provide the required accuracy, which is consistent with the previous result.

With the help of a calculating utility, the value of s_{19} is approximately $s_{19} \approx 0.7$ and the value of $\ln 2$ obtained directly is approximately $\ln 2 \approx 0.69$, which agrees with s_{19} when rounded to one decimal place. ◄

> As Example 2 illustrates, the alternating harmonic series does not provide an efficient way to approximate ln 2, since too many terms and hence too much computation is required to achieve reasonable accuracy. Later, we will develop better ways to approximate logarithms.

■ ABSOLUTE CONVERGENCE

The series

$$1 - \frac{1}{2} - \frac{1}{2^2} + \frac{1}{2^3} + \frac{1}{2^4} - \frac{1}{2^5} - \frac{1}{2^6} + \cdots$$

does not fit in any of the categories studied so far—it has mixed signs but is not alternating. We will now develop some convergence tests that can be applied to such series.

9.6.3 DEFINITION   A series

$$\sum_{k=1}^{\infty} u_k = u_1 + u_2 + \cdots + u_k + \cdots$$

is said to **converge absolutely** if the series of absolute values

$$\sum_{k=1}^{\infty} |u_k| = |u_1| + |u_2| + \cdots + |u_k| + \cdots$$

converges and is said to **diverge absolutely** if the series of absolute values diverges.

▶ **Example 3**   Determine whether the following series converge absolutely.

(a) $1 - \dfrac{1}{2} - \dfrac{1}{2^2} + \dfrac{1}{2^3} + \dfrac{1}{2^4} - \dfrac{1}{2^5} - \cdots$ (b) $1 - \dfrac{1}{2} + \dfrac{1}{3} - \dfrac{1}{4} + \dfrac{1}{5} - \cdots$

Solution (a).   The series of absolute values is the convergent geometric series

$$1 + \frac{1}{2} + \frac{1}{2^2} + \frac{1}{2^3} + \frac{1}{2^4} + \frac{1}{2^5} + \cdots$$

so the given series converges absolutely.

Solution (b).   The series of absolute values is the divergent harmonic series

$$1 + \frac{1}{2} + \frac{1}{3} + \frac{1}{4} + \frac{1}{5} + \cdots$$

so the given series diverges absolutely. ◄

It is important to distinguish between the notions of convergence and absolute convergence. For example, the series in part (b) of Example 3 converges, since it is the alternating harmonic series, yet we demonstrated that it does not converge absolutely. However, the following theorem shows that *if a series converges absolutely, then it converges.*

Theorem 9.6.4 provides a way of inferring convergence of a series with positive and negative terms from a related series with nonnegative terms (the series of absolute values). This is important because most of the convergence tests that we have developed apply only to series with nonnegative terms.

9.6.4   THEOREM   *If the series*

$$\sum_{k=1}^{\infty} |u_k| = |u_1| + |u_2| + \cdots + |u_k| + \cdots$$

converges, then so does the series

$$\sum_{k=1}^{\infty} u_k = u_1 + u_2 + \cdots + u_k + \cdots$$

PROOF   We will write the series $\sum u_k$ as

$$\sum_{k=1}^{\infty} u_k = \sum_{k=1}^{\infty} [(u_k + |u_k|) - |u_k|]   \tag{7}$$

We are assuming that $\sum |u_k|$ converges, so that if we can show that $\sum (u_k + |u_k|)$ converges, then it will follow from (7) and Theorem 9.4.3(*a*) that $\sum u_k$ converges. However, the value of $u_k + |u_k|$ is either 0 or $2|u_k|$, depending on the sign of u_k. Thus, in all cases it is true that

$$0 \le u_k + |u_k| \le 2|u_k|$$

But $\sum 2|u_k|$ converges, since it is a constant times the convergent series $\sum |u_k|$; hence $\sum (u_k + |u_k|)$ converges by the comparison test. ∎

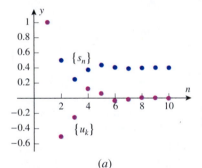

(a)

▶ **Example 4**   Show that the following series converge.

$$\text{(a) } 1 - \frac{1}{2} - \frac{1}{2^2} + \frac{1}{2^3} + \frac{1}{2^4} - \frac{1}{2^5} - \frac{1}{2^6} + \cdots \qquad \text{(b) } \sum_{k=1}^{\infty} \frac{\cos k}{k^2}$$

Solution (a).   Observe that this is not an alternating series because the signs alternate in pairs after the first term. Thus, we have no convergence test that can be applied directly. However, we showed in Example 3(a) that the series converges absolutely, so Theorem 9.6.4 implies that it converges (Figure 9.6.4*a*).

Solution (b).   With the help of a calculating utility, you will be able to verify that the signs of the terms in this series vary irregularly. Thus, we will test for absolute convergence. The series of absolute values is

$$\sum_{k=1}^{\infty} \left| \frac{\cos k}{k^2} \right|$$

However,

$$\left| \frac{\cos k}{k^2} \right| \le \frac{1}{k^2}$$

(b)

Graphs of the sequences of terms and nth partial sums for the series in Example 4

▲ **Figure 9.6.4**

But $\sum 1/k^2$ is a convergent p-series ($p = 2$), so the series of absolute values converges by the comparison test. Thus, the given series converges absolutely and hence converges (Figure 9.6.4b). ◄

■ CONDITIONAL CONVERGENCE

Although Theorem 9.6.4 is a useful tool for series that converge absolutely, it provides no information about the convergence or divergence of a series that diverges absolutely. For example, consider the two series

$$1 - \frac{1}{2} + \frac{1}{3} - \frac{1}{4} + \cdots + (-1)^{k+1}\frac{1}{k} + \cdots \tag{8}$$

$$-1 - \frac{1}{2} - \frac{1}{3} - \frac{1}{4} - \cdots - \frac{1}{k} - \cdots \tag{9}$$

Both of these series diverge absolutely, since in each case the series of absolute values is the divergent harmonic series

$$1 + \frac{1}{2} + \frac{1}{3} + \cdots + \frac{1}{k} + \cdots$$

However, series (8) converges, since it is the alternating harmonic series, and series (9) diverges, since it is a constant times the divergent harmonic series. As a matter of terminology, a series that converges but diverges absolutely is said to *converge conditionally* (or to be *conditionally convergent*). Thus, (8) is a conditionally convergent series.

▶ **Example 5**   In Example 1(b) we used the alternating series test to show that the series

$$\sum_{k=1}^{\infty}(-1)^{k+1}\frac{k+3}{k(k+1)}$$

converges. Determine whether this series converges absolutely or converges conditionally.

Solution.   We test the series for absolute convergence by examining the series of absolute values:

$$\sum_{k=1}^{\infty}\left|(-1)^{k+1}\frac{k+3}{k(k+1)}\right| = \sum_{k=1}^{\infty}\frac{k+3}{k(k+1)}$$

Principle 9.5.3 suggests that the series of absolute values should behave like the divergent p-series with $p = 1$. To prove that the series of absolute values diverges, we will apply the limit comparison test with

$$a_k = \frac{k+3}{k(k+1)} \quad \text{and} \quad b_k = \frac{1}{k}$$

We obtain

$$\rho = \lim_{k \to +\infty}\frac{a_k}{b_k} = \lim_{k \to +\infty}\frac{k(k+3)}{k(k+1)} = \lim_{k \to +\infty}\frac{k+3}{k+1} = 1$$

Since ρ is finite and positive, it follows from the limit comparison test that the series of absolute values diverges. Thus, the original series converges and also diverges absolutely, and so converges conditionally. ◄

■ THE RATIO TEST FOR ABSOLUTE CONVERGENCE

Although one cannot generally infer convergence or divergence of a series from absolute divergence, the following variation of the ratio test provides a way of deducing divergence from absolute divergence in certain situations. We omit the proof.

9.6.5   **THEOREM** (*Ratio Test for Absolute Convergence*)   *Let $\sum u_k$ be a series with nonzero terms and suppose that*

$$\rho = \lim_{k \to +\infty} \frac{|u_{k+1}|}{|u_k|}$$

(a)   *If $\rho < 1$, then the series $\sum u_k$ converges absolutely and therefore converges.*

(b)   *If $\rho > 1$ or if $\rho = +\infty$, then the series $\sum u_k$ diverges.*

(c)   *If $\rho = 1$, no conclusion about convergence or absolute convergence can be drawn from this test.*

▶ **Example 6**   Use the ratio test for absolute convergence to determine whether the series converges.

$$\text{(a) } \sum_{k=1}^{\infty} (-1)^k \frac{2^k}{k!} \qquad \text{(b) } \sum_{k=1}^{\infty} (-1)^k \frac{(2k-1)!}{3^k}$$

Solution (a).   Taking the absolute value of the general term u_k we obtain

$$|u_k| = \left| (-1)^k \frac{2^k}{k!} \right| = \frac{2^k}{k!}$$

Thus,

$$\rho = \lim_{k \to +\infty} \frac{|u_{k+1}|}{|u_k|} = \lim_{k \to +\infty} \frac{2^{k+1}}{(k+1)!} \cdot \frac{k!}{2^k} = \lim_{k \to +\infty} \frac{2}{k+1} = 0 < 1$$

which implies that the series converges absolutely and therefore converges.

Solution (b).   Taking the absolute value of the general term u_k we obtain

$$|u_k| = \left| (-1)^k \frac{(2k-1)!}{3^k} \right| = \frac{(2k-1)!}{3^k}$$

Thus,

$$\rho = \lim_{k \to +\infty} \frac{|u_{k+1}|}{|u_k|} = \lim_{k \to +\infty} \frac{[2(k+1)-1]!}{3^{k+1}} \cdot \frac{3^k}{(2k-1)!}$$

$$= \lim_{k \to +\infty} \frac{1}{3} \cdot \frac{(2k+1)!}{(2k-1)!} = \frac{1}{3} \lim_{k \to +\infty} (2k)(2k+1) = +\infty$$

which implies that the series diverges.   ◀

■ **SUMMARY OF CONVERGENCE TESTS**

We conclude this section with a summary of convergence tests that can be used for reference. The skill of selecting a good test is developed through lots of practice. In some instances a test may be inconclusive, so another test must be tried.

Summary of Convergence Tests

NAME	STATEMENT	COMMENTS
Divergence Test (9.4.1)	If $\lim\limits_{k \to +\infty} u_k \neq 0$, then $\sum u_k$ diverges.	If $\lim\limits_{k \to +\infty} u_k = 0$, then $\sum u_k$ may or may not converge.
Integral Test (9.4.4)	Let $\sum u_k$ be a series with positive terms. If f is a function that is decreasing and continuous on an interval $[a, +\infty)$ and such that $u_k = f(k)$ for all $k \geq a$, then $$\sum_{k=1}^{\infty} u_k \quad \text{and} \quad \int_a^{+\infty} f(x)\, dx$$ both converge or both diverge.	This test only applies to series that have positive terms. Try this test when $f(x)$ is easy to integrate.
Comparison Test (9.5.1)	Let $\sum_{k=1}^{\infty} a_k$ and $\sum_{k=1}^{\infty} b_k$ be series with nonnegative terms such that $$a_1 \leq b_1,\ a_2 \leq b_2,\ \ldots,\ a_k \leq b_k,\ \ldots$$ If $\sum b_k$ converges, then $\sum a_k$ converges, and if $\sum a_k$ diverges, then $\sum b_k$ diverges.	This test only applies to series with nonnegative terms. Try this test as a last resort; other tests are often easier to apply.
Limit Comparison Test (9.5.4)	Let $\sum a_k$ and $\sum b_k$ be series with positive terms and let $$\rho = \lim_{k \to +\infty} \frac{a_k}{b_k}$$ If $0 < \rho < +\infty$, then both series converge or both diverge.	This is easier to apply than the comparison test, but still requires some skill in choosing the series $\sum b_k$ for comparison.
Ratio Test (9.5.5)	Let $\sum u_k$ be a series with positive terms and suppose that $$\rho = \lim_{k \to +\infty} \frac{u_{k+1}}{u_k}$$ (a) Series converges if $\rho < 1$. (b) Series diverges if $\rho > 1$ or $\rho = +\infty$. (c) The test is inconclusive if $\rho = 1$.	Try this test when u_k involves factorials or kth powers.
Root Test (9.5.6)	Let $\sum u_k$ be a series with positive terms and suppose that $$\rho = \lim_{k \to +\infty} \sqrt[k]{u_k}$$ (a) The series converges if $\rho < 1$. (b) The series diverges if $\rho > 1$ or $\rho = +\infty$. (c) The test is inconclusive if $\rho = 1$.	Try this test when u_k involves kth powers.
Alternating Series Test (9.6.1)	If $a_k > 0$ for $k = 1, 2, 3, \ldots$, then the series $$a_1 - a_2 + a_3 - a_4 + \cdots$$ $$-a_1 + a_2 - a_3 + a_4 - \cdots$$ converge if the following conditions hold: (a) $a_1 \geq a_2 \geq a_3 \geq \cdots$ (b) $\lim\limits_{k \to +\infty} a_k = 0$	This test applies only to alternating series.
Ratio Test for Absolute Convergence (9.6.5)	Let $\sum u_k$ be a series with nonzero terms and suppose that $$\rho = \lim_{k \to +\infty} \frac{|u_{k+1}|}{|u_k|}$$ (a) The series converges absolutely if $\rho < 1$. (b) The series diverges if $\rho > 1$ or $\rho = +\infty$. (c) The test is inconclusive if $\rho = 1$.	The series need not have positive terms and need not be alternating to use this test.

✔**QUICK CHECK EXERCISES 9.6** (*See page 648 for answers.*)

1. What characterizes an *alternating* series?

2. (a) The series

$$\sum_{k=1}^{\infty} \frac{(-1)^{k+1}}{k^2}$$

converges by the alternating series test since _____ and _____.

(b) If

$$S = \sum_{k=1}^{\infty} \frac{(-1)^{k+1}}{k^2} \quad \text{and} \quad s_9 = \sum_{k=1}^{9} \frac{(-1)^{k+1}}{k^2}$$

then $|S - s_9| <$ _____.

3. Classify each sequence as conditionally convergent, absolutely convergent, or divergent.

(a) $\displaystyle\sum_{k=1}^{\infty} (-1)^{k+1} \frac{1}{k}$: _____

(b) $\displaystyle\sum_{k=1}^{\infty} (-1)^{k} \frac{3k-1}{9k+15}$: _____

(c) $\displaystyle\sum_{k=1}^{\infty} (-1)^{k} \frac{1}{k(k+2)}$: _____

(d) $\displaystyle\sum_{k=1}^{\infty} (-1)^{k+1} \frac{1}{\sqrt[4]{k^3}}$: _____

4. Given that

$$\lim_{k \to +\infty} \frac{(k+1)^4/4^{k+1}}{k^4/4^k} = \lim_{k \to +\infty} \frac{\left(1 + \dfrac{1}{k}\right)^4}{4} = \frac{1}{4}$$

is the series $\sum_{k=1}^{\infty} (-1)^k k^4/4^k$ conditionally convergent, absolutely convergent, or divergent?

EXERCISE SET 9.6 □C CAS

1–2 Show that the series converges by confirming that it satisfies the hypotheses of the alternating series test (Theorem 9.6.1). ■

1. $\displaystyle\sum_{k=1}^{\infty} \frac{(-1)^{k+1}}{2k+1}$

2. $\displaystyle\sum_{k=1}^{\infty} (-1)^{k+1} \frac{k}{3^k}$

3–6 Determine whether the alternating series converges; justify your answer. ■

3. $\displaystyle\sum_{k=1}^{\infty} (-1)^{k+1} \frac{k+1}{3k+1}$

4. $\displaystyle\sum_{k=1}^{\infty} (-1)^{k+1} \frac{k+1}{\sqrt{k}+1}$

5. $\displaystyle\sum_{k=1}^{\infty} (-1)^{k+1} e^{-k}$

6. $\displaystyle\sum_{k=3}^{\infty} (-1)^{k} \frac{\ln k}{k}$

7–12 Use the ratio test for absolute convergence (Theorem 9.6.5) to determine whether the series converges or diverges. If the test is inconclusive, say so. ■

7. $\displaystyle\sum_{k=1}^{\infty} \left(-\frac{3}{5}\right)^k$

8. $\displaystyle\sum_{k=1}^{\infty} (-1)^{k+1} \frac{2^k}{k!}$

9. $\displaystyle\sum_{k=1}^{\infty} (-1)^{k+1} \frac{3^k}{k^2}$

10. $\displaystyle\sum_{k=1}^{\infty} (-1)^{k} \frac{k}{5^k}$

11. $\displaystyle\sum_{k=1}^{\infty} (-1)^{k} \frac{k^3}{e^k}$

12. $\displaystyle\sum_{k=1}^{\infty} (-1)^{k+1} \frac{k^k}{k!}$

13–28 Classify each series as absolutely convergent, conditionally convergent, or divergent. ■

13. $\displaystyle\sum_{k=1}^{\infty} \frac{(-1)^{k+1}}{3k}$

14. $\displaystyle\sum_{k=1}^{\infty} \frac{(-1)^{k+1}}{k^{4/3}}$

15. $\displaystyle\sum_{k=1}^{\infty} \frac{(-4)^{k}}{k^2}$

16. $\displaystyle\sum_{k=1}^{\infty} \frac{(-1)^{k+1}}{k!}$

17. $\displaystyle\sum_{k=1}^{\infty} \frac{\cos k\pi}{k}$

18. $\displaystyle\sum_{k=3}^{\infty} \frac{(-1)^k \ln k}{k}$

19. $\displaystyle\sum_{k=1}^{\infty} (-1)^{k+1} \frac{k+2}{k(k+3)}$

20. $\displaystyle\sum_{k=1}^{\infty} \frac{(-1)^{k+1} k^2}{k^3+1}$

21. $\displaystyle\sum_{k=1}^{\infty} \sin \frac{k\pi}{2}$

22. $\displaystyle\sum_{k=1}^{\infty} \frac{\sin k}{k^3}$

23. $\displaystyle\sum_{k=2}^{\infty} \frac{(-1)^k}{k \ln k}$

24. $\displaystyle\sum_{k=1}^{\infty} \frac{(-1)^k}{\sqrt{k(k+1)}}$

25. $\displaystyle\sum_{k=2}^{\infty} \left(-\frac{1}{\ln k}\right)^k$

26. $\displaystyle\sum_{k=1}^{\infty} \frac{k \cos k\pi}{k^2+1}$

27. $\displaystyle\sum_{k=1}^{\infty} \frac{(-1)^{k+1} k!}{(2k-1)!}$

28. $\displaystyle\sum_{k=1}^{\infty} (-1)^{k+1} \frac{3^{2k-1}}{k^2+1}$

29–32 True–False Determine whether the statement is true or false. Explain your answer. ■

29. An alternating series is one whose terms alternate between even and odd.

30. If a series satisfies the hypothesis of the alternating series test, then the sequence of partial sums of the series oscillates between overestimates and underestimates for the sum of the series.

31. If a series converges, then either it converges absolutely or it converges conditionally.

32. If $\sum (u_k)^2$ converges, then $\sum u_k$ converges absolutely.

33–36 Each series satisfies the hypotheses of the alternating series test. For the stated value of n, find an upper bound on the absolute error that results if the sum of the series is approximated by the nth partial sum. ■

33. $\displaystyle\sum_{k=1}^{\infty} \frac{(-1)^{k+1}}{k}$; $n = 7$ **34.** $\displaystyle\sum_{k=1}^{\infty} \frac{(-1)^{k+1}}{k!}$; $n = 5$

35. $\displaystyle\sum_{k=1}^{\infty} \frac{(-1)^{k+1}}{\sqrt{k}}$; $n = 99$

36. $\displaystyle\sum_{k=1}^{\infty} \frac{(-1)^{k+1}}{(k+1)\ln(k+1)}$; $n = 3$

37–40 Each series satisfies the hypotheses of the alternating series test. Find a value of n for which the nth partial sum is ensured to approximate the sum of the series to the stated accuracy. ■

37. $\displaystyle\sum_{k=1}^{\infty} \frac{(-1)^{k+1}}{k}$; $|\text{error}| < 0.0001$

38. $\displaystyle\sum_{k=1}^{\infty} \frac{(-1)^{k+1}}{k!}$; $|\text{error}| < 0.00001$

39. $\displaystyle\sum_{k=1}^{\infty} \frac{(-1)^{k+1}}{\sqrt{k}}$;   two decimal places

40. $\displaystyle\sum_{k=1}^{\infty} \frac{(-1)^{k+1}}{(k+1)\ln(k+1)}$;   one decimal place

41–42 Find an upper bound on the absolute error that results if s_{10} is used to approximate the sum of the given *geometric* series. Compute s_{10} rounded to four decimal places and compare this value with the exact sum of the series. ■

41. $\dfrac{3}{4} - \dfrac{3}{8} + \dfrac{3}{16} - \dfrac{3}{32} + \cdots$ **42.** $1 - \dfrac{2}{3} + \dfrac{4}{9} - \dfrac{8}{27} + \cdots$

43–46 Each series satisfies the hypotheses of the alternating series test. Approximate the sum of the series to two decimal-place accuracy. ■

43. $1 - \dfrac{1}{3!} + \dfrac{1}{5!} - \dfrac{1}{7!} + \cdots$ **44.** $1 - \dfrac{1}{2!} + \dfrac{1}{4!} - \dfrac{1}{6!} + \cdots$

45. $\dfrac{1}{1\cdot 2} - \dfrac{1}{2\cdot 2^2} + \dfrac{1}{3\cdot 2^3} - \dfrac{1}{4\cdot 2^4} + \cdots$

46. $\dfrac{1}{1^5 + 4\cdot 1} - \dfrac{1}{3^5 + 4\cdot 3} + \dfrac{1}{5^5 + 4\cdot 5} - \dfrac{1}{7^5 + 4\cdot 7} + \cdots$

FOCUS ON CONCEPTS

C 47. The purpose of this exercise is to show that the error bound in part (b) of Theorem 9.6.2 can be overly conservative in certain cases.
   (a) Use a CAS to confirm that
   $$\frac{\pi}{4} = 1 - \frac{1}{3} + \frac{1}{5} - \frac{1}{7} + \cdots$$
   (b) Use the CAS to show that $|(\pi/4) - s_{25}| < 10^{-2}$.

   (c) According to the error bound in part (b) of Theorem 9.6.2, what value of n is required to ensure that $|(\pi/4) - s_n| < 10^{-2}$?

48. Prove: If a series $\sum a_k$ converges absolutely, then the series $\sum a_k^2$ converges.

49. (a) Find examples to show that if $\sum a_k$ converges, then $\sum a_k^2$ may diverge or converge.
   (b) Find examples to show that if $\sum a_k^2$ converges, then $\sum a_k$ may diverge or converge.

50. Let $\sum u_k$ be a series and define series $\sum p_k$ and $\sum q_k$ so that
$$p_k = \begin{cases} u_k, & u_k > 0 \\ 0, & u_k \leq 0 \end{cases} \quad \text{and} \quad q_k = \begin{cases} 0, & u_k \geq 0 \\ -u_k, & u_k < 0 \end{cases}$$

   (a) Show that $\sum u_k$ converges absolutely if and only if $\sum p_k$ and $\sum q_k$ both converge.
   (b) Show that if one of $\sum p_k$ or $\sum q_k$ converges and the other diverges, then $\sum u_k$ diverges.
   (c) Show that if $\sum u_k$ converges conditionally, then both $\sum p_k$ and $\sum q_k$ diverge.

51. It can be proved that the terms of any conditionally convergent series can be rearranged to give either a divergent series or a conditionally convergent series whose sum is any given number S. For example, we stated in Example 2 that
$$\ln 2 = 1 - \frac{1}{2} + \frac{1}{3} - \frac{1}{4} + \frac{1}{5} - \frac{1}{6} + \cdots$$
Show that we can rearrange this series so that its sum is $\frac{1}{2}\ln 2$ by rewriting it as
$$\left(1 - \frac{1}{2} - \frac{1}{4}\right) + \left(\frac{1}{3} - \frac{1}{6} - \frac{1}{8}\right) + \left(\frac{1}{5} - \frac{1}{10} - \frac{1}{12}\right) + \cdots$$
[*Hint:* Add the first two terms in each grouping.]

52–54 Exercise 51 illustrates that one of the nuances of "conditional" convergence is that the sum of a series that converges conditionally depends on the order that the terms of the series are summed. Absolutely convergent series are more dependable, however. It can be proved that any series that is constructed from an absolutely convergent series by rearranging the terms will also be absolutely convergent and has the same sum as the original series. Use this fact together with parts (a) and (b) of Theorem 9.4.3 in these exercises. ■

52. It was stated in Exercise 35 of Section 9.4 that
$$\frac{\pi^2}{6} = 1 + \frac{1}{2^2} + \frac{1}{3^2} + \frac{1}{4^2} + \cdots$$
Use this to show that
$$\frac{\pi^2}{8} = 1 + \frac{1}{3^2} + \frac{1}{5^2} + \frac{1}{7^2} + \cdots$$

53. Use the series for $\pi^2/6$ given in the preceding exercise to show that
$$\frac{\pi^2}{12} = 1 - \frac{1}{2^2} + \frac{1}{3^2} - \frac{1}{4^2} + \cdots$$

54. It was stated in Exercise 35 of Section 9.4 that

$$\frac{\pi^4}{90} = 1 + \frac{1}{2^4} + \frac{1}{3^4} + \frac{1}{4^4} + \cdots$$

Use this to show that

$$\frac{\pi^4}{96} = 1 + \frac{1}{3^4} + \frac{1}{5^4} + \frac{1}{7^4} + \cdots$$

55. Writing Consider the series

$$1 - \frac{1}{2} + \frac{2}{3} - \frac{1}{3} + \frac{2}{4} - \frac{1}{4} + \frac{2}{5} - \frac{1}{5} + \cdots$$

Determine whether this series converges and use this series as an example in a discussion of the importance of hypotheses (*a*) and (*b*) of the alternating series test (Theorem 9.6.1).

56. Writing Discuss the ways that conditional convergence is "conditional." In particular, describe how one could rearrange the terms of a conditionally convergent series $\sum u_k$ so that the resulting series diverges, either to $+\infty$ or to $-\infty$. [*Hint:* See Exercise 50.]

✔ **QUICK CHECK ANSWERS 9.6**

1. Terms alternate between positive and negative.   **2.** (a) $1 \geq \frac{1}{4} \geq \frac{1}{9} \geq \cdots \geq \frac{1}{k^2} \geq \frac{1}{(k+1)^2} \geq \cdots;$ $\lim\limits_{k \to +\infty} \frac{1}{k^2} = 0$ (b) $\frac{1}{100}$
3. (a) conditionally convergent (b) divergent (c) absolutely convergent (d) conditionally convergent   **4.** absolutely convergent

---

9.7 MACLAURIN AND TAYLOR POLYNOMIALS

In a local linear approximation the tangent line to the graph of a function is used to obtain a linear approximation of the function near the point of tangency. In this section we will consider how one might improve on the accuracy of local linear approximations by using higher-order polynomials as approximating functions. We will also investigate the error associated with such approximations.

■ LOCAL QUADRATIC APPROXIMATIONS

Recall from Formula (1) in Section 3.5 that the local linear approximation of a function f at x_0 is

$$f(x) \approx f(x_0) + f'(x_0)(x - x_0) \tag{1}$$

In this formula, the approximating function

$$p(x) = f(x_0) + f'(x_0)(x - x_0)$$

is a first-degree polynomial satisfying $p(x_0) = f(x_0)$ and $p'(x_0) = f'(x_0)$ (verify). Thus, the local linear approximation of f at x_0 has the property that its value and the value of its first derivative match those of f at x_0.

If the graph of a function f has a pronounced "bend" at x_0, then we can expect that the accuracy of the local linear approximation of f at x_0 will decrease rapidly as we progress away from x_0 (Figure 9.7.1). One way to deal with this problem is to approximate the function f at x_0 by a polynomial p of degree 2 with the property that the value of p and the values of its first two derivatives match those of f at x_0. This ensures that the graphs of f and p not only have the same tangent line at x_0, but they also bend in the same direction at x_0 (both concave up or concave down). As a result, we can expect that the graph of p will remain close to the graph of f over a larger interval around x_0 than the graph of the local linear approximation. The polynomial p is called the *local quadratic approximation of f at* $x = x_0$.

To illustrate this idea, let us try to find a formula for the local quadratic approximation of a function f at $x = 0$. This approximation has the form

$$f(x) \approx c_0 + c_1 x + c_2 x^2 \tag{2}$$

where c_0, c_1, and c_2 must be chosen so that the values of

$$p(x) = c_0 + c_1 x + c_2 x^2$$

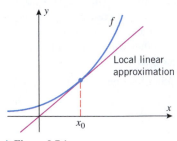

▲ **Figure 9.7.1**

Local linear approximation

and its first two derivatives match those of f at 0. Thus, we want

$$p(0) = f(0), \quad p'(0) = f'(0), \quad p''(0) = f''(0) \tag{3}$$

But the values of $p(0)$, $p'(0)$, and $p''(0)$ are as follows:

$$
\begin{array}{ll}
p(x) = c_0 + c_1 x + c_2 x^2 & p(0) = c_0 \\
p'(x) = c_1 + 2c_2 x & p'(0) = c_1 \\
p''(x) = 2c_2 & p''(0) = 2c_2
\end{array}
$$

Thus, it follows from (3) that

$$c_0 = f(0), \quad c_1 = f'(0), \quad c_2 = \frac{f''(0)}{2}$$

and substituting these in (2) yields the following formula for the local quadratic approximation of f at $x = 0$:

$$f(x) \approx f(0) + f'(0)x + \frac{f''(0)}{2}x^2 \tag{4}$$

▶ **Example 1**   Find the local linear and quadratic approximations of e^x at $x = 0$, and graph e^x and the two approximations together.

Solution.   If we let $f(x) = e^x$, then $f'(x) = f''(x) = e^x$; and hence

$$f(0) = f'(0) = f''(0) = e^0 = 1$$

Thus, from (4) the local quadratic approximation of e^x at $x = 0$ is

$$e^x \approx 1 + x + \frac{x^2}{2}$$

and the local linear approximation (which is the linear part of the local quadratic approximation) is

$$e^x \approx 1 + x$$

The graphs of e^x and the two approximations are shown in Figure 9.7.2. As expected, the local quadratic approximation is more accurate than the local linear approximation near $x = 0$. ◀

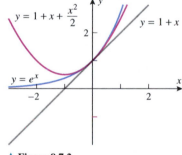

▲ **Figure 9.7.2**

MACLAURIN POLYNOMIALS

It is natural to ask whether one can improve on the accuracy of a local quadratic approximation by using a polynomial of degree 3. Specifically, one might look for a polynomial of degree 3 with the property that its value and the values of its first three derivatives match

Colin Maclaurin (1698–1746) Scottish mathematician. Maclaurin's father, a minister, died when the boy was only six months old, and his mother when he was nine years old. He was then raised by an uncle who was also a minister. Maclaurin entered Glasgow University as a divinity student but switched to mathematics after one year. He received his Master's degree at age 17 and, in spite of his youth, began teaching at Marischal College in Aberdeen, Scotland. He met Isaac Newton during a visit to London in 1719 and from that time on became Newton's disciple. During that era, some of Newton's analytic methods were bitterly attacked by major mathematicians and much of Maclaurin's important mathematical work resulted from his efforts to defend Newton's ideas geometrically. Maclaurin's work, *A Treatise of Fluxions* (1742), was the first systematic formulation of Newton's methods. The treatise was so carefully done that it was a standard of mathematical rigor in calculus until the work of Cauchy in 1821. Maclaurin was also an outstanding experimentalist; he devised numerous ingenious mechanical devices, made important astronomical observations, performed actuarial computations for insurance societies, and helped to improve maps of the islands around Scotland.

those of f at a point; and if this provides an improvement in accuracy, why not go on to polynomials of even higher degree? Thus, we are led to consider the following general problem.

9.7.1 PROBLEM   Given a function f that can be differentiated n times at $x = x_0$, find a polynomial p of degree n with the property that the value of p and the values of its first n derivatives match those of f at x_0.

We will begin by solving this problem in the case where $x_0 = 0$. Thus, we want a polynomial

$$p(x) = c_0 + c_1 x + c_2 x^2 + c_3 x^3 + \cdots + c_n x^n \tag{5}$$

such that

$$f(0) = p(0), \quad f'(0) = p'(0), \quad f''(0) = p''(0), \dots, \quad f^{(n)}(0) = p^{(n)}(0) \tag{6}$$

But

$$
\begin{aligned}
p(x) &= c_0 + c_1 x + c_2 x^2 + c_3 x^3 + \cdots + c_n x^n \\
p'(x) &= c_1 + 2c_2 x + 3c_3 x^2 + \cdots + nc_n x^{n-1} \\
p''(x) &= 2c_2 + 3 \cdot 2c_3 x + \cdots + n(n-1)c_n x^{n-2} \\
p'''(x) &= 3 \cdot 2c_3 + \cdots + n(n-1)(n-2)c_n x^{n-3} \\
&\vdots \\
p^{(n)}(x) &= n(n-1)(n-2) \cdots (1)c_n
\end{aligned}
$$

Thus, to satisfy (6) we must have

$$
\begin{aligned}
f(0) &= p(0) &&= c_0 \\
f'(0) &= p'(0) &&= c_1 \\
f''(0) &= p''(0) &&= 2c_2 = 2!c_2 \\
f'''(0) &= p'''(0) &&= 3 \cdot 2c_3 = 3!c_3 \\
&\vdots \\
f^{(n)}(0) &= p^{(n)}(0) &&= n(n-1)(n-2) \cdots (1)c_n = n!c_n
\end{aligned}
$$

which yields the following values for the coefficients of $p(x)$:

$$c_0 = f(0), \quad c_1 = f'(0), \quad c_2 = \frac{f''(0)}{2!}, \quad c_3 = \frac{f'''(0)}{3!}, \dots, \quad c_n = \frac{f^{(n)}(0)}{n!}$$

The polynomial that results by using these coefficients in (5) is called the *nth Maclaurin polynomial for f*.

Local linear approximations and local quadratic approximations at $x = 0$ of a function f are special cases of the MacLaurin polynomials for f. Verify that $f(x) \approx p_1(x)$ is the local linear approximation of f at $x = 0$, and $f(x) \approx p_2(x)$ is the local quadratic approximation at $x = 0$.

9.7.2 DEFINITION   If f can be differentiated n times at 0, then we define the **nth Maclaurin polynomial for f** to be

$$p_n(x) = f(0) + f'(0)x + \frac{f''(0)}{2!}x^2 + \frac{f'''(0)}{3!}x^3 + \cdots + \frac{f^{(n)}(0)}{n!}x^n \tag{7}$$

Note that the polynomial in (7) has the property that its value and the values of its first n derivatives match the values of f and its first n derivatives at $x = 0$.

▶ **Example 2**   Find the Maclaurin polynomials p_0, p_1, p_2, p_3, and p_n for e^x.

Solution.   Let $f(x) = e^x$. Thus,

$$f'(x) = f''(x) = f'''(x) = \cdots = f^{(n)}(x) = e^x$$

and

$$f(0) = f'(0) = f''(0) = f'''(0) = \cdots = f^{(n)}(0) = e^0 = 1$$

Therefore,

$$p_0(x) = f(0) = 1$$

$$p_1(x) = f(0) + f'(0)x = 1 + x$$

$$p_2(x) = f(0) + f'(0)x + \frac{f''(0)}{2!}x^2 = 1 + x + \frac{x^2}{2!} = 1 + x + \frac{1}{2}x^2$$

$$p_3(x) = f(0) + f'(0)x + \frac{f''(0)}{2!}x^2 + \frac{f'''(0)}{3!}x^3$$

$$= 1 + x + \frac{x^2}{2!} + \frac{x^3}{3!} = 1 + x + \frac{1}{2}x^2 + \frac{1}{6}x^3$$

$$p_n(x) = f(0) + f'(0)x + \frac{f''(0)}{2!}x^2 + \cdots + \frac{f^{(n)}(0)}{n!}x^n$$

$$= 1 + x + \frac{x^2}{2!} + \cdots + \frac{x^n}{n!} \quad ◀$$

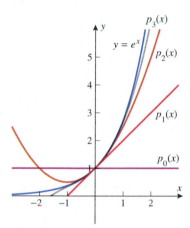

▲ **Figure 9.7.3**

Figure 9.7.3 shows the graph of e^x (in blue) and the graph of the first four Maclaurin polynomials. Note that the graphs of $p_1(x)$, $p_2(x)$, and $p_3(x)$ are virtually indistinguishable from the graph of e^x near $x = 0$, so these polynomials are good approximations of e^x for x near 0. However, the farther x is from 0, the poorer these approximations become. This is typical of the Maclaurin polynomials for a function $f(x)$; they provide good approximations of $f(x)$ near 0, but the accuracy diminishes as x progresses away from 0. It is usually the case that the higher the degree of the polynomial, the larger the interval on which it provides a specified accuracy. Accuracy issues will be investigated later.

Augustin Louis Cauchy (**1789–1857**) French mathematician. Cauchy's early education was acquired from his father, a barrister and master of the classics. Cauchy entered L'Ecole Polytechnique in 1805 to study engineering, but because of poor health, was advised to concentrate on mathematics. His major mathematical work began in 1811 with a series of brilliant solutions to some difficult outstanding problems. In 1814 he wrote a treatise on integrals that was to become the basis for modern complex variable theory; in 1816 there followed a classic paper on wave propagation in liquids that won a prize from the French Academy; and in 1822 he wrote a paper that formed the basis of modern elasticity theory. Cauchy's mathematical contributions for the next 35 years were brilliant and staggering in quantity, over 700 papers filling 26 modern volumes. Cauchy's work initiated the era of modern analysis. He brought to mathematics standards of precision and rigor undreamed of by Leibniz and Newton.

Cauchy's life was inextricably tied to the political upheavals of the time. A strong partisan of the Bourbons, he left his wife and children in 1830 to follow the Bourbon king Charles X into exile. For his loyalty he was made a baron by the ex-king. Cauchy eventually returned to France, but refused to accept a university position until the government waived its requirement that he take a loyalty oath.

It is difficult to get a clear picture of the man. Devoutly Catholic, he sponsored charitable work for unwed mothers, criminals, and relief for Ireland. Yet other aspects of his life cast him in an unfavorable light. The Norwegian mathematician Abel described him as, "mad, infinitely Catholic, and bigoted." Some writers praise his teaching, yet others say he rambled incoherently and, according to a report of the day, he once devoted an entire lecture to extracting the square root of seventeen to ten decimal places by a method well known to his students. In any event, Cauchy is undeniably one of the greatest minds in the history of science.

▶ **Example 3** Find the nth Maclaurin polynomials for

(a) $\sin x$ (b) $\cos x$

Solution (a). In the Maclaurin polynomials for $\sin x$, only the odd powers of x appear explicitly. To see this, let $f(x) = \sin x$; thus,

$$
\begin{aligned}
f(x) &= \sin x & f(0) &= 0 \\
f'(x) &= \cos x & f'(0) &= 1 \\
f''(x) &= -\sin x & f''(0) &= 0 \\
f'''(x) &= -\cos x & f'''(0) &= -1
\end{aligned}
$$

Since $f^{(4)}(x) = \sin x = f(x)$, the pattern $0, 1, 0, -1$ will repeat as we evaluate successive derivatives at 0. Therefore, the successive Maclaurin polynomials for $\sin x$ are

$$p_0(x) = 0$$

$$p_1(x) = 0 + x$$

$$p_2(x) = 0 + x + 0$$

$$p_3(x) = 0 + x + 0 - \frac{x^3}{3!}$$

$$p_4(x) = 0 + x + 0 - \frac{x^3}{3!} + 0$$

$$p_5(x) = 0 + x + 0 - \frac{x^3}{3!} + 0 + \frac{x^5}{5!}$$

$$p_6(x) = 0 + x + 0 - \frac{x^3}{3!} + 0 + \frac{x^5}{5!} + 0$$

$$p_7(x) = 0 + x + 0 - \frac{x^3}{3!} + 0 + \frac{x^5}{5!} + 0 - \frac{x^7}{7!}$$

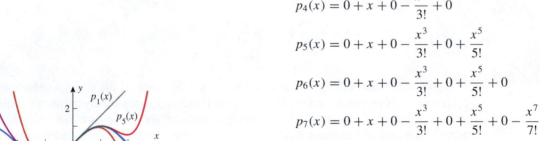

Because of the zero terms, each even-order Maclaurin polynomial [after $p_0(x)$] is the same as the preceding odd-order Maclaurin polynomial. That is,

$$p_{2k+1}(x) = p_{2k+2}(x) = x - \frac{x^3}{3!} + \frac{x^5}{5!} - \frac{x^7}{7!} + \cdots + (-1)^k \frac{x^{2k+1}}{(2k+1)!} \quad (k = 0, 1, 2, \ldots)$$

The graphs of $\sin x$, $p_1(x)$, $p_3(x)$, $p_5(x)$, and $p_7(x)$ are shown in Figure 9.7.4.

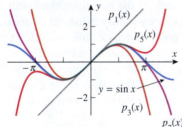

▲ **Figure 9.7.4**

Solution (b). In the Maclaurin polynomials for $\cos x$, only the even powers of x appear explicitly; the computations are similar to those in part (a). The reader should be able to show that

$$p_0(x) = p_1(x) = 1$$

$$p_2(x) = p_3(x) = 1 - \frac{x^2}{2!}$$

$$p_4(x) = p_5(x) = 1 - \frac{x^2}{2!} + \frac{x^4}{4!}$$

$$p_6(x) = p_7(x) = 1 - \frac{x^2}{2!} + \frac{x^4}{4!} - \frac{x^6}{6!}$$

In general, the Maclaurin polynomials for $\cos x$ are given by

$$p_{2k}(x) = p_{2k+1}(x) = 1 - \frac{x^2}{2!} + \frac{x^4}{4!} - \frac{x^6}{6!} + \cdots + (-1)^k \frac{x^{2k}}{(2k)!} \quad (k = 0, 1, 2, \ldots)$$

The graphs of $\cos x$, $p_0(x)$, $p_2(x)$, $p_4(x)$, and $p_6(x)$ are shown in Figure 9.7.5. ◀

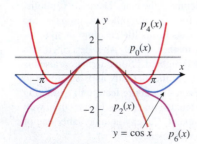

▲ **Figure 9.7.5**

■ TAYLOR POLYNOMIALS

Up to now we have focused on approximating a function f in the vicinity of $x = 0$. Now we will consider the more general case of approximating f in the vicinity of an arbitrary domain value x_0. The basic idea is the same as before; we want to find an nth-degree polynomial p with the property that its value and the values of its first n derivatives match those of f at x_0. However, rather than expressing $p(x)$ in powers of x, it will simplify the computations if we express it in powers of $x - x_0$; that is,

$$p(x) = c_0 + c_1(x - x_0) + c_2(x - x_0)^2 + \cdots + c_n(x - x_0)^n \tag{8}$$

We will leave it as an exercise for you to imitate the computations used in the case where $x_0 = 0$ to show that

$$c_0 = f(x_0), \quad c_1 = f'(x_0), \quad c_2 = \frac{f''(x_0)}{2!}, \quad c_3 = \frac{f'''(x_0)}{3!}, \dots, \quad c_n = \frac{f^{(n)}(x_0)}{n!}$$

Substituting these values in (8) we obtain a polynomial called the *nth Taylor polynomial about $x = x_0$ for f*.

Local linear approximations and local quadratic approximations at $x = x_0$ of a function f are special cases of the Taylor polynomials for f. Verify that $f(x) \approx p_1(x)$ is the local linear approximation of f at $x = x_0$, and $f(x) \approx p_2(x)$ is the local quadratic approximation at $x = x_0$.

9.7.3 DEFINITION If f can be differentiated n times at x_0, then we define the *nth Taylor polynomial for f about $x = x_0$* to be

$$p_n(x) = f(x_0) + f'(x_0)(x - x_0) + \frac{f''(x_0)}{2!}(x - x_0)^2$$
$$+ \frac{f'''(x_0)}{3!}(x - x_0)^3 + \cdots + \frac{f^{(n)}(x_0)}{n!}(x - x_0)^n \tag{9}$$

The Maclaurin polynomials are the special cases of the Taylor polynomials in which $x_0 = 0$. Thus, theorems about Taylor polynomials also apply to Maclaurin polynomials.

▶ **Example 4** Find the first four Taylor polynomials for $\ln x$ about $x = 2$.

Solution. Let $f(x) = \ln x$. Thus,

$$
\begin{array}{ll}
f(x) = \ln x & f(2) = \ln 2 \\
f'(x) = 1/x & f'(2) = 1/2 \\
f''(x) = -1/x^2 & f''(2) = -1/4 \\
f'''(x) = 2/x^3 & f'''(2) = 1/4
\end{array}
$$

Brook Taylor (1685–1731) English mathematician. Taylor was born of well-to-do parents. Musicians and artists were entertained frequently in the Taylor home, which undoubtedly had a lasting influence on him. In later years, Taylor published a definitive work on the mathematical theory of perspective and obtained major mathematical results about the vibrations of strings. There also exists an unpublished work, *On Musick*, that was intended to be part of a joint paper with Isaac Newton. Taylor's life was scarred with unhappiness, illness, and tragedy. Because his first wife was not rich enough to suit his father, the two men argued bitterly and parted ways. Subsequently, his wife died in childbirth. Then, after he remarried, his second wife also died in childbirth, though his daughter survived. Taylor's most productive period was from 1714 to 1719, during which time he wrote on a wide range of subjects—magnetism, capillary action, thermometers, perspective, and calculus. In his final years, Taylor devoted his writing efforts to religion and philosophy. According to Taylor, the results that bear his name were motivated by coffeehouse conversations about works of Newton on planetary motion and works of Halley ("Halley's comet") on roots of polynomials. Unfortunately, Taylor's writing style was so terse and hard to understand that he never received credit for many of his innovations.

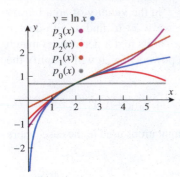

▲ **Figure 9.7.6**

Substituting in (9) with $x_0 = 2$ yields

$$p_0(x) = f(2) = \ln 2$$

$$p_1(x) = f(2) + f'(2)(x - 2) = \ln 2 + \tfrac{1}{2}(x - 2)$$

$$p_2(x) = f(2) + f'(2)(x - 2) + \frac{f''(2)}{2!}(x - 2)^2 = \ln 2 + \tfrac{1}{2}(x - 2) - \tfrac{1}{8}(x - 2)^2$$

$$p_3(x) = f(2) + f'(2)(x - 2) + \frac{f''(2)}{2!}(x - 2)^2 + \frac{f'''(2)}{3!}(x - 2)^3$$

$$= \ln 2 + \tfrac{1}{2}(x - 2) - \tfrac{1}{8}(x - 2)^2 + \tfrac{1}{24}(x - 2)^3$$

The graph of $\ln x$ (in blue) and its first four Taylor polynomials about $x = 2$ are shown in Figure 9.7.6. As expected, these polynomials produce their best approximations of $\ln x$ near 2. ◄

■ SIGMA NOTATION FOR TAYLOR AND MACLAURIN POLYNOMIALS

Frequently, we will want to express Formula (9) in sigma notation. To do this, we use the notation $f^{(k)}(x_0)$ to denote the kth derivative of f at $x = x_0$, and we make the convention that $f^{(0)}(x_0)$ denotes $f(x_0)$. This enables us to write

$$\sum_{k=0}^{n} \frac{f^{(k)}(x_0)}{k!}(x - x_0)^k = f(x_0) + f'(x_0)(x - x_0)$$

$$+ \frac{f''(x_0)}{2!}(x - x_0)^2 + \cdots + \frac{f^{(n)}(x_0)}{n!}(x - x_0)^n \quad (10)$$

In particular, we can write the nth Maclaurin polynomial for $f(x)$ as

$$\sum_{k=0}^{n} \frac{f^{(k)}(0)}{k!}x^k = f(0) + f'(0)x + \frac{f''(0)}{2!}x^2 + \cdots + \frac{f^{(n)}(0)}{n!}x^n \quad (11)$$

► **Example 5** Find the nth Maclaurin polynomial for

$$\frac{1}{1 - x}$$

and express it in sigma notation.

Solution. Let $f(x) = 1/(1 - x)$. The values of f and its first k derivatives at $x = 0$ are as follows:

$$f(x) = \frac{1}{1 - x} \qquad\qquad f(0) = 1 = 0!$$

$$f'(x) = \frac{1}{(1 - x)^2} \qquad\qquad f'(0) = 1 = 1!$$

$$f''(x) = \frac{2}{(1 - x)^3} \qquad\qquad f''(0) = 2 = 2!$$

$$f'''(x) = \frac{3 \cdot 2}{(1 - x)^4} \qquad\qquad f'''(0) = 3!$$

$$f^{(4)}(x) = \frac{4 \cdot 3 \cdot 2}{(1 - x)^5} \qquad\qquad f^{(4)}(0) = 4!$$

$$\vdots \qquad\qquad\qquad \vdots$$

$$f^{(k)}(x) = \frac{k!}{(1 - x)^{k+1}} \qquad f^{(k)}(0) = k!$$

TECHNOLOGY MASTERY

Computer algebra systems have commands for generating Taylor polynomials of any specified degree. If you have a CAS, use it to find some of the Maclaurin and Taylor polynomials in Examples 3, 4, and 5.

Thus, substituting $f^{(k)}(0) = k!$ into Formula (11) yields the nth Maclaurin polynomial for $1/(1 - x)$:

$$p_n(x) = \sum_{k=0}^{n} x^k = 1 + x + x^2 + \cdots + x^n \quad (n = 0, 1, 2, \ldots) \blacktriangleleft$$

▶ **Example 6** Find the nth Taylor polynomial for $1/x$ about $x = 1$ and express it in sigma notation.

Solution. Let $f(x) = 1/x$. The computations are similar to those in Example 5. We leave it for you to show that

$$f(1) = 1, \quad f'(1) = -1, \quad f''(1) = 2!, \quad f'''(1) = -3!,$$
$$f^{(4)}(1) = 4!, \ldots, \quad f^{(k)}(1) = (-1)^k k!$$

Thus, substituting $f^{(k)}(1) = (-1)^k k!$ into Formula (10) with $x_0 = 1$ yields the nth Taylor polynomial for $1/x$:

$$\sum_{k=0}^{n} (-1)^k (x - 1)^k = 1 - (x - 1) + (x - 1)^2 - (x - 1)^3 + \cdots + (-1)^n (x - 1)^n \blacktriangleleft$$

■ **THE nTH REMAINDER**

It will be convenient to have a notation for the error in the approximation $f(x) \approx p_n(x)$. Accordingly, we will let $R_n(x)$ denote the difference between $f(x)$ and its nth Taylor polynomial; that is,

$$R_n(x) = f(x) - p_n(x) = f(x) - \sum_{k=0}^{n} \frac{f^{(k)}(x_0)}{k!}(x - x_0)^k \tag{12}$$

This can also be written as

$$f(x) = p_n(x) + R_n(x) = \sum_{k=0}^{n} \frac{f^{(k)}(x_0)}{k!}(x - x_0)^k + R_n(x) \tag{13}$$

The function $R_n(x)$ is called the **nth remainder** for the Taylor series of f, and Formula (13) is called **Taylor's formula with remainder**.

Finding a bound for $R_n(x)$ gives an indication of the accuracy of the approximation $p_n(x) \approx f(x)$. The following theorem, which is proved in Appendix J, provides such a bound.

The bound for $|R_n(x)|$ in (14) is called the **Lagrange error bound**.

9.7.4 THEOREM (*The Remainder Estimation Theorem*) *If the function f can be differentiated $n + 1$ times on an interval containing the number x_0, and if M is an upper bound for $|f^{(n+1)}(x)|$ on the interval, that is, $|f^{(n+1)}(x)| \leq M$ for all x in the interval, then*

$$|R_n(x)| \leq \frac{M}{(n + 1)!}|x - x_0|^{n+1} \tag{14}$$

for all x in the interval.

▶ **Example 7** Use an nth Maclaurin polynomial for e^x to approximate e to five decimal-place accuracy.

Solution. We note first that the exponential function e^x has derivatives of all orders for every real number x. From Example 2, the nth Maclaurin polynomial for e^x is

$$\sum_{k=0}^{n} \frac{x^k}{k!} = 1 + x + \frac{x^2}{2!} + \cdots + \frac{x^n}{n!}$$

from which we have

$$e = e^1 \approx \sum_{k=0}^{n} \frac{1^k}{k!} = 1 + 1 + \frac{1}{2!} + \cdots + \frac{1}{n!}$$

Thus, our problem is to determine how many terms to include in a Maclaurin polynomial for e^x to achieve five decimal-place accuracy; that is, we want to choose n so that the absolute value of the nth remainder at $x = 1$ satisfies

$$|R_n(1)| \le 0.000005$$

To determine n we use the Remainder Estimation Theorem with $f(x) = e^x$, $x = 1$, $x_0 = 0$, and the interval $[0, 1]$. In this case it follows from (14) that

$$|R_n(1)| \le \frac{M}{(n+1)!} \cdot |1 - 0|^{n+1} = \frac{M}{(n+1)!} \tag{15}$$

where M is an upper bound on the value of $f^{(n+1)}(x) = e^x$ for x in the interval $[0, 1]$. However, e^x is an increasing function, so its maximum value on the interval $[0, 1]$ occurs at $x = 1$; that is, $e^x \le e$ on this interval. Thus, we can take $M = e$ in (15) to obtain

$$|R_n(1)| \le \frac{e}{(n+1)!} \tag{16}$$

Unfortunately, this inequality is not very useful because it involves e, which is the very quantity we are trying to approximate. However, if we accept that $e < 3$, then we can replace (16) with the following less precise, but more easily applied, inequality:

$$|R_n(1)| \le \frac{3}{(n+1)!}$$

Thus, we can achieve five decimal-place accuracy by choosing n so that

$$\frac{3}{(n+1)!} \le 0.000005 \quad \text{or} \quad (n+1)! \ge 600{,}000$$

Since $9! = 362{,}880$ and $10! = 3{,}628{,}800$, the smallest value of n that meets this criterion is $n = 9$. Thus, to five decimal-place accuracy

$$e \approx 1 + 1 + \frac{1}{2!} + \frac{1}{3!} + \frac{1}{4!} + \frac{1}{5!} + \frac{1}{6!} + \frac{1}{7!} + \frac{1}{8!} + \frac{1}{9!} \approx 2.71828$$

As a check, a calculator's 12-digit representation of e is $e \approx 2.71828182846$, which agrees with the preceding approximation when rounded to five decimal places. ◀

▶ **Example 8** Use the Remainder Estimation Theorem to find an interval containing $x = 0$ throughout which $f(x) = \cos x$ can be approximated by $p(x) = 1 - (x^2/2!)$ to three decimal-place accuracy.

Solution. We note first that $f(x) = \cos x$ has derivatives of all orders for every real number x, so the first hypothesis of the Remainder Estimation Theorem is satisfied over any interval that we choose. The given polynomial $p(x)$ is both the second and the third

Maclaurin polynomial for $\cos x$; we will choose the degree n of the polynomial to be as large as possible, so we will take $n = 3$. Our problem is to determine an interval on which the absolute value of the third remainder at x satisfies

$$|R_3(x)| \le 0.0005$$

We will use the Remainder Estimation Theorem with $f(x) = \cos x$, $n = 3$, and $x_0 = 0$. It follows from (14) that

$$|R_3(x)| \le \frac{M}{(3+1)!}|x - 0|^{3+1} = \frac{M|x|^4}{24} \qquad (17)$$

where M is an upper bound for $|f^{(4)}(x)| = |\cos x|$. Since $|\cos x| \le 1$ for every real number x, we can take $M = 1$ in (17) to obtain

$$|R_3(x)| \le \frac{|x|^4}{24} \qquad (18)$$

Thus we can achieve three decimal-place accuracy by choosing values of x for which

$$\frac{|x|^4}{24} \le 0.0005 \quad \text{or} \quad |x| \le 0.3309$$

so the interval $[-0.3309, 0.3309]$ is one option. We can check this answer by graphing $|f(x) - p(x)|$ over the interval $[-0.3309, 0.3309]$ (Figure 9.7.7). ◄

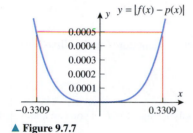

▲ **Figure 9.7.7**

QUICK CHECK EXERCISES 9.7 *(See page 659 for answers.)*

1. If f can be differentiated three times at 0, then the third Maclaurin polynomial for f is $p_3(x) =$ _____.

2. The third Maclaurin polynomial for $f(x) = e^{2x}$ is
$$p_3(x) = \underline{\qquad} + \underline{\qquad} x$$
$$+ \underline{\qquad} x^2 + \underline{\qquad} x^3$$

3. If $f(2) = 3$, $f'(2) = -4$, and $f''(2) = 10$, then the second Taylor polynomial for f about $x = 2$ is $p_2(x) =$ _____.

4. The third Taylor polynomial for $f(x) = x^5$ about $x = -1$ is
$$p_3(x) = \underline{\qquad} + \underline{\qquad} (x + 1)$$
$$+ \underline{\qquad} (x + 1)^2 + \underline{\qquad} (x + 1)^3$$

5. (a) If a function f has nth Taylor polynomial $p_n(x)$ about $x = x_0$, then the nth remainder $R_n(x)$ is defined by $R_n(x) =$ _____.

   (b) Suppose that a function f can be differentiated five times on an interval containing $x_0 = 2$ and that $|f^{(5)}(x)| \le 20$ for all x in the interval. Then the fourth remainder satisfies $|R_4(x)| \le$ _____ for all x in the interval.

EXERCISE SET 9.7 ⬚ Graphing Utility

⬚ **1–2** In each part, find the local quadratic approximation of f at $x = x_0$, and use that approximation to find the local linear approximation of f at x_0. Use a graphing utility to graph f and the two approximations on the same screen. ■

1. (a) $f(x) = e^{-x}$; $x_0 = 0$ (b) $f(x) = \cos x$; $x_0 = 0$

2. (a) $f(x) = \sin x$; $x_0 = \pi/2$ (b) $f(x) = \sqrt{x}$; $x_0 = 1$

3. (a) Find the local quadratic approximation of $\sqrt{x}$ at $x_0 = 1$.
   (b) Use the result obtained in part (a) to approximate $\sqrt{1.1}$, and compare your approximation to that produced directly by your calculating utility. [*Note:* See Example 1 of Section 3.5.]

4. (a) Find the local quadratic approximation of $\cos x$ at $x_0 = 0$.
   (b) Use the result obtained in part (a) to approximate $\cos 2°$, and compare the approximation to that produced directly by your calculating utility.

5. Use an appropriate local quadratic approximation to approximate $\tan 61°$, and compare the result to that produced directly by your calculating utility.

6. Use an appropriate local quadratic approximation to approximate $\sqrt{36.03}$, and compare the result to that produced directly by your calculating utility.

7–16 Find the Maclaurin polynomials of orders $n = 0, 1, 2, 3$, and 4, and then find the nth Maclaurin polynomials for the function in sigma notation. ■

7. e^{-x} **8.** e^{ax} **9.** $\cos \pi x$

10. $\sin \pi x$ **11.** $\ln(1 + x)$ **12.** $\dfrac{1}{1 + x}$

13. $\cosh x$ **14.** $\sinh x$ **15.** $x \sin x$

16. xe^x

17–24 Find the Taylor polynomials of orders $n = 0, 1, 2, 3$, and 4 about $x = x_0$, and then find the nth Taylor polynomial for the function in sigma notation. ■

17. e^x; $x_0 = 1$ **18.** e^{-x}; $x_0 = \ln 2$

19. $\dfrac{1}{x}$; $x_0 = -1$ **20.** $\dfrac{1}{x + 2}$; $x_0 = 3$

21. $\sin \pi x$; $x_0 = \dfrac{1}{2}$ **22.** $\cos x$; $x_0 = \dfrac{\pi}{2}$

23. $\ln x$; $x_0 = 1$ **24.** $\ln x$; $x_0 = e$

25. (a) Find the third Maclaurin polynomial for
$$f(x) = 1 + 2x - x^2 + x^3$$
(b) Find the third Taylor polynomial about $x = 1$ for
$$f(x) = 1 + 2(x - 1) - (x - 1)^2 + (x - 1)^3$$

26. (a) Find the nth Maclaurin polynomial for
$$f(x) = c_0 + c_1 x + c_2 x^2 + \cdots + c_n x^n$$
(b) Find the nth Taylor polynomial about $x = 1$ for
$$f(x) = c_0 + c_1(x - 1) + c_2(x - 1)^2 + \cdots + c_n(x - 1)^n$$

27–30 Find the first four distinct Taylor polynomials about $x = x_0$, and use a graphing utility to graph the given function and the Taylor polynomials on the same screen. ■

27. $f(x) = e^{-2x}$; $x_0 = 0$ **28.** $f(x) = \sin x$; $x_0 = \pi/2$

29. $f(x) = \cos x$; $x_0 = \pi$ **30.** $\ln(x + 1)$; $x_0 = 0$

31–34 True–False Determine whether the statement is true or false. Explain your answer. ■

31. The equation of a tangent line to a differentiable function is a first-degree Taylor polynomial for that function.

32. The graph of a function f and the graph of its Maclaurin polynomial have a common y-intercept.

33. If $p_6(x)$ is the sixth-degree Taylor polynomial for a function f about $x = x_0$, then $p_6^{(4)}(x_0) = 4! f^{(4)}(x_0)$.

34. If $p_4(x)$ is the fourth-degree Maclaurin polynomial for e^x, then
$$|e^2 - p_4(2)| \le \frac{9}{5!}$$

35–36 Use the method of Example 7 to approximate the given expression to the specified accuracy. Check your answer to that produced directly by your calculating utility. ■

35. $\sqrt{e}$; four decimal-place accuracy

36. $1/e$; three decimal-place accuracy

FOCUS ON CONCEPTS

37. Which of the functions graphed in the following figure is most likely to have $p(x) = 1 - x + 2x^2$ as its second-order Maclaurin polynomial? Explain your reasoning.

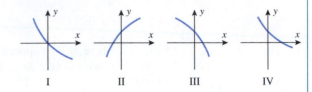

I   II   III   IV

38. Suppose that the values of a function f and its first three derivatives at $x = 1$ are
$$f(1) = 2, \quad f'(1) = -3, \quad f''(1) = 0, \quad f'''(1) = 6$$
Find as many Taylor polynomials for f as you can about $x = 1$.

39. Let $p_1(x)$ and $p_2(x)$ be the local linear and local quadratic approximations of $f(x) = e^{\sin x}$ at $x = 0$.
(a) Use a graphing utility to generate the graphs of $f(x)$, $p_1(x)$, and $p_2(x)$ on the same screen for $-1 \le x \le 1$.
(b) Construct a table of values of $f(x)$, $p_1(x)$, and $p_2(x)$ for $x = -1.00$, -0.75, -0.50, -0.25, 0, 0.25, 0.50, 0.75, 1.00. Round the values to three decimal places.
(c) Generate the graph of $|f(x) - p_1(x)|$, and use the graph to determine an interval on which $p_1(x)$ approximates $f(x)$ with an error of at most ± 0.01. [*Suggestion:* Review the discussion relating to Figure 3.5.4.]
(d) Generate the graph of $|f(x) - p_2(x)|$, and use the graph to determine an interval on which $p_2(x)$ approximates $f(x)$ with an error of at most ± 0.01.

40. (a) The accompanying figure shows a sector of radius r and central angle 2α. Assuming that the angle α is small, use the local quadratic approximation of $\cos \alpha$ at $\alpha = 0$ to show that $x \approx r\alpha^2/2$.
(b) Assuming that the Earth is a sphere of radius 4000 mi, use the result in part (a) to approximate the maximum amount by which a 100 mi arc along the equator will diverge from its chord.

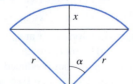

◀ **Figure Ex-40**

41. (a) Find an interval $[0, b]$ over which e^x can be approximated by $1 + x + (x^2/2!)$ to three decimal-place accuracy throughout the interval.

(b) Check your answer in part (a) by graphing

$$\left| e^x - \left(1 + x + \frac{x^2}{2!} \right) \right|$$

over the interval you obtained.

42. Show that the nth Taylor polynomial for $\sinh x$ about $x = \ln 4$ is

$$\sum_{k=0}^{n} \frac{16 - (-1)^k}{8k!} (x - \ln 4)^k$$

43–46 Use the Remainder Estimation Theorem to find an interval containing $x = 0$ over which $f(x)$ can be approximated by $p(x)$ to three decimal-place accuracy throughout the interval. Check your answer by graphing $|f(x) - p(x)|$ over the interval you obtained. ■

43. $f(x) = \sin x; \ p(x) = x - \dfrac{x^3}{3!}$

44. $f(x) = \cos x; \ p(x) = 1 - \dfrac{x^2}{2!} + \dfrac{x^4}{4!}$

45. $f(x) = \dfrac{1}{1 + x^2}; \ p(x) = 1 - x^2 + x^4$

46. $f(x) = \ln(1 + x); \ p(x) = x - \dfrac{x^2}{2} + \dfrac{x^3}{3}$

✔ **QUICK CHECK ANSWERS 9.7**

1. $f(0) + f'(0)x + \dfrac{f''(0)}{2!}x^2 + \dfrac{f'''(0)}{3!}x^3$ **2.** $1; 2; 2; \frac{4}{3}$ **3.** $3 - 4(x - 2) + 5(x - 2)^2$ **4.** $-1; 5; -10; 10$

5. (a) $f(x) - p_n(x)$ (b) $\frac{1}{6}|x - 2|^5$

9.8 MACLAURIN AND TAYLOR SERIES; POWER SERIES

Recall from the last section that the nth Taylor polynomial $p_n(x)$ at $x = x_0$ for a function f was defined so its value and the values of its first n derivatives match those of f at x_0. This being the case, it is reasonable to expect that for values of x near x_0 the values of $p_n(x)$ will become better and better approximations of $f(x)$ as n increases, and may possibly converge to $f(x)$ as $n \to +\infty$. We will explore this idea in this section.

■ MACLAURIN AND TAYLOR SERIES

In Section 9.7 we defined the nth Maclaurin polynomial for a function f as

$$\sum_{k=0}^{n} \frac{f^{(k)}(0)}{k!} x^k = f(0) + f'(0)x + \frac{f''(0)}{2!}x^2 + \cdots + \frac{f^{(n)}(0)}{n!}x^n$$

and the nth Taylor polynomial for f about $x = x_0$ as

$$\sum_{k=0}^{n} \frac{f^{(k)}(x_0)}{k!} (x - x_0)^k = f(x_0) + f'(x_0)(x - x_0)$$

$$+ \frac{f''(x_0)}{2!}(x - x_0)^2 + \cdots + \frac{f^{(n)}(x_0)}{n!}(x - x_0)^n$$

It is not a big step to extend the notions of Maclaurin and Taylor polynomials to series by not stopping the summation index at n. Thus, we have the following definition.

9.8.1   **DEFINITION**   If f has derivatives of all orders at x_0, then we call the series

$$\sum_{k=0}^{\infty} \frac{f^{(k)}(x_0)}{k!}(x - x_0)^k = f(x_0) + f'(x_0)(x - x_0) + \frac{f''(x_0)}{2!}(x - x_0)^2$$

$$+ \cdots + \frac{f^{(k)}(x_0)}{k!}(x - x_0)^k + \cdots \qquad (1)$$

the *Taylor series for f about x = x₀*. In the special case where $x_0 = 0$, this series becomes

$$\sum_{k=0}^{\infty} \frac{f^{(k)}(0)}{k!}x^k = f(0) + f'(0)x + \frac{f''(0)}{2!}x^2 + \cdots + \frac{f^{(k)}(0)}{k!}x^k + \cdots \qquad (2)$$

in which case we call it the *Maclaurin series for f*.

Note that the nth Maclaurin and Taylor polynomials are the nth partial sums for the corresponding Maclaurin and Taylor series.

▶ **Example 1**   Find the Maclaurin series for

(a) e^x (b) $\sin x$ (c) $\cos x$ (d) $\dfrac{1}{1 - x}$

Solution (a).   In Example 2 of Section 9.7 we found that the nth Maclaurin polynomial for e^x is

$$p_n(x) = \sum_{k=0}^{n} \frac{x^k}{k!} = 1 + x + \frac{x^2}{2!} + \cdots + \frac{x^n}{n!}$$

Thus, the Maclaurin series for e^x is

$$\sum_{k=0}^{\infty} \frac{x^k}{k!} = 1 + x + \frac{x^2}{2!} + \cdots + \frac{x^k}{k!} + \cdots$$

Solution (b).   In Example 3(a) of Section 9.7 we found that the Maclaurin polynomials for $\sin x$ are given by

$$p_{2k+1}(x) = p_{2k+2}(x) = x - \frac{x^3}{3!} + \frac{x^5}{5!} - \frac{x^7}{7!} + \cdots + (-1)^k \frac{x^{2k+1}}{(2k+1)!} \qquad (k = 0, 1, 2, \ldots)$$

Thus, the Maclaurin series for $\sin x$ is

$$\sum_{k=0}^{\infty} (-1)^k \frac{x^{2k+1}}{(2k+1)!} = x - \frac{x^3}{3!} + \frac{x^5}{5!} - \frac{x^7}{7!} + \cdots + (-1)^k \frac{x^{2k+1}}{(2k+1)!} + \cdots$$

Solution (c).   In Example 3(b) of Section 9.7 we found that the Maclaurin polynomials for $\cos x$ are given by

$$p_{2k}(x) = p_{2k+1}(x) = 1 - \frac{x^2}{2!} + \frac{x^4}{4!} - \frac{x^6}{6!} + \cdots + (-1)^k \frac{x^{2k}}{(2k)!} \qquad (k = 0, 1, 2, \ldots)$$

Thus, the Maclaurin series for $\cos x$ is

$$\sum_{k=0}^{\infty} (-1)^k \frac{x^{2k}}{(2k)!} = 1 - \frac{x^2}{2!} + \frac{x^4}{4!} - \frac{x^6}{6!} + \cdots + (-1)^k \frac{x^{2k}}{(2k)!} + \cdots$$

Solution (d). In Example 5 of Section 9.7 we found that the nth Maclaurin polynomial for $1/(1-x)$ is

$$p_n(x) = \sum_{k=0}^{n} x^k = 1 + x + x^2 + \cdots + x^n \quad (n = 0, 1, 2, \ldots)$$

Thus, the Maclaurin series for $1/(1-x)$ is

$$\sum_{k=0}^{\infty} x^k = 1 + x + x^2 + \cdots + x^k + \cdots \quad \blacktriangleleft$$

▶ **Example 2** Find the Taylor series for $1/x$ about $x = 1$.

Solution. In Example 6 of Section 9.7 we found that the nth Taylor polynomial for $1/x$ about $x = 1$ is

$$\sum_{k=0}^{n} (-1)^k (x-1)^k = 1 - (x-1) + (x-1)^2 - (x-1)^3 + \cdots + (-1)^n (x-1)^n$$

Thus, the Taylor series for $1/x$ about $x = 1$ is

$$\sum_{k=0}^{\infty} (-1)^k (x-1)^k = 1 - (x-1) + (x-1)^2 - (x-1)^3 + \cdots + (-1)^k (x-1)^k + \cdots \quad \blacktriangleleft$$

■ **POWER SERIES IN *x***

Maclaurin and Taylor series differ from the series that we have considered in Sections 9.3 to 9.6 in that their terms are not merely constants, but instead involve a variable. These are examples of *power series*, which we now define.

If $c_0, c_1, c_2, \ldots$ are constants and x is a variable, then a series of the form

$$\sum_{k=0}^{\infty} c_k x^k = c_0 + c_1 x + c_2 x^2 + \cdots + c_k x^k + \cdots \tag{3}$$

is called a *power series in **x***. Some examples are

$$\sum_{k=0}^{\infty} x^k = 1 + x + x^2 + x^3 + \cdots$$

$$\sum_{k=0}^{\infty} \frac{x^k}{k!} = 1 + x + \frac{x^2}{2!} + \frac{x^3}{3!} + \cdots$$

$$\sum_{k=0}^{\infty} (-1)^k \frac{x^{2k}}{(2k)!} = 1 - \frac{x^2}{2!} + \frac{x^4}{4!} - \frac{x^6}{6!} + \cdots$$

From Example 1, these are the Maclaurin series for the functions $1/(1-x)$, e^x, and $\cos x$, respectively. Indeed, every Maclaurin series

$$\sum_{k=0}^{\infty} \frac{f^{(k)}(0)}{k!} x^k = f(0) + f'(0)x + \frac{f''(0)}{2!} x^2 + \cdots + \frac{f^{(k)}(0)}{k!} x^k + \cdots$$

is a power series in x.

■ RADIUS AND INTERVAL OF CONVERGENCE

If a numerical value is substituted for x in a power series $\sum c_k x^k$, then the resulting series of numbers may either converge or diverge. This leads to the problem of determining the set of x-values for which a given power series converges; this is called its **convergence set**.

Observe that every power series in x converges at $x = 0$, since substituting this value in (3) produces the series

$$c_0 + 0 + 0 + 0 + \cdots + 0 + \cdots$$

whose sum is c_0. In some cases $x = 0$ may be the only number in the convergence set; in other cases the convergence set is some finite or infinite interval containing $x = 0$. This is the content of the following theorem, whose proof will be omitted.

9.8.2 THEOREM *For any power series in x, exactly one of the following is true:*

(a) *The series converges only for $x = 0$.*

(b) *The series converges absolutely (and hence converges) for all real values of x.*

(c) *The series converges absolutely (and hence converges) for all x in some finite open interval $(-R, R)$ and diverges if $x < -R$ or $x > R$. At either of the values $x = R$ or $x = -R$, the series may converge absolutely, converge conditionally, or diverge, depending on the particular series.*

This theorem states that the convergence set for a power series in x is always an interval centered at $x = 0$ (possibly just the value $x = 0$ itself or possibly infinite). For this reason, the convergence set of a power series in x is called the **interval of convergence**. In the case where the convergence set is the single value $x = 0$ we say that the series has **radius of convergence 0**, in the case where the convergence set is $(-\infty, +\infty)$ we say that the series has **radius of convergence $+\infty$**, and in the case where the convergence set extends between $-R$ and R we say that the series has **radius of convergence R** (Figure 9.8.1).

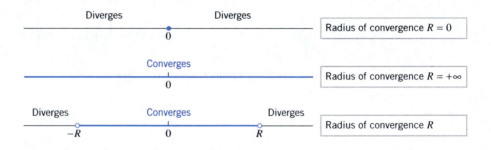

▶ **Figure 9.8.1**

■ FINDING THE INTERVAL OF CONVERGENCE

The usual procedure for finding the interval of convergence of a power series is to apply the ratio test for absolute convergence (Theorem 9.6.5). The following example illustrates how this works.

▶ **Example 3** Find the interval of convergence and radius of convergence of the following power series.

$$\text{(a) } \sum_{k=0}^{\infty} x^k \qquad \text{(b) } \sum_{k=0}^{\infty} \frac{x^k}{k!} \qquad \text{(c) } \sum_{k=0}^{\infty} k! x^k \qquad \text{(d) } \sum_{k=0}^{\infty} \frac{(-1)^k x^k}{3^k (k+1)}$$

Solution (a).   Applying the ratio test for absolute convergence to the given series, we obtain

$$\rho = \lim_{k \to +\infty} \left| \frac{u_{k+1}}{u_k} \right| = \lim_{k \to +\infty} \left| \frac{x^{k+1}}{x^k} \right| = \lim_{k \to +\infty} |x| = |x|$$

so the series converges absolutely if $\rho = |x| < 1$ and diverges if $\rho = |x| > 1$. The test is inconclusive if $|x| = 1$ (i.e., if $x = 1$ or $x = -1$), which means that we will have to investigate convergence at these values separately. At these values the series becomes

$$\sum_{k=0}^{\infty} 1^k = 1 + 1 + 1 + 1 + \cdots \qquad \boxed{x = 1}$$

$$\sum_{k=0}^{\infty} (-1)^k = 1 - 1 + 1 - 1 + \cdots \qquad \boxed{x = -1}$$

both of which diverge; thus, the interval of convergence for the given power series is $(-1, 1)$, and the radius of convergence is $R = 1$.

Solution (b).   Applying the ratio test for absolute convergence to the given series, we obtain

$$\rho = \lim_{k \to +\infty} \left| \frac{u_{k+1}}{u_k} \right| = \lim_{k \to +\infty} \left| \frac{x^{k+1}}{(k+1)!} \cdot \frac{k!}{x^k} \right| = \lim_{k \to +\infty} \left| \frac{x}{k+1} \right| = 0$$

Since $\rho < 1$ for all x, the series converges absolutely for all x. Thus, the interval of convergence is $(-\infty, +\infty)$ and the radius of convergence is $R = +\infty$.

Solution (c).   If $x \neq 0$, then the ratio test for absolute convergence yields

$$\rho = \lim_{k \to +\infty} \left| \frac{u_{k+1}}{u_k} \right| = \lim_{k \to +\infty} \left| \frac{(k+1)! x^{k+1}}{k! x^k} \right| = \lim_{k \to +\infty} |(k+1)x| = +\infty$$

Therefore, the series diverges for all nonzero values of x. Thus, the interval of convergence is the single value $x = 0$ and the radius of convergence is $R = 0$.

Solution (d).   Since $|(-1)^k| = |(-1)^{k+1}| = 1$, we obtain

$$\rho = \lim_{k \to +\infty} \left| \frac{u_{k+1}}{u_k} \right| = \lim_{k \to +\infty} \left| \frac{x^{k+1}}{3^{k+1}(k+2)} \cdot \frac{3^k(k+1)}{x^k} \right|$$

$$= \lim_{k \to +\infty} \left[\frac{|x|}{3} \cdot \left(\frac{k+1}{k+2} \right) \right]$$

$$= \frac{|x|}{3} \lim_{k \to +\infty} \left(\frac{1 + (1/k)}{1 + (2/k)} \right) = \frac{|x|}{3}$$

The ratio test for absolute convergence implies that the series converges absolutely if $|x| < 3$ and diverges if $|x| > 3$. The ratio test fails to provide any information when $|x| = 3$, so the cases $x = -3$ and $x = 3$ need separate analyses. Substituting $x = -3$ in the given series yields

$$\sum_{k=0}^{\infty} \frac{(-1)^k(-3)^k}{3^k(k+1)} = \sum_{k=0}^{\infty} \frac{(-1)^k(-1)^k 3^k}{3^k(k+1)} = \sum_{k=0}^{\infty} \frac{1}{k+1}$$

which is the divergent harmonic series $1 + \frac{1}{2} + \frac{1}{3} + \frac{1}{4} + \cdots$. Substituting $x = 3$ in the given series yields

$$\sum_{k=0}^{\infty} \frac{(-1)^k 3^k}{3^k(k+1)} = \sum_{k=0}^{\infty} \frac{(-1)^k}{k+1} = 1 - \frac{1}{2} + \frac{1}{3} - \frac{1}{4} + \cdots$$

which is the conditionally convergent alternating harmonic series. Thus, the interval of convergence for the given series is $(-3, 3]$ and the radius of convergence is $R = 3$. ◄

■ POWER SERIES IN $x - x_0$

If x_0 is a constant, and if x is replaced by $x - x_0$ in (3), then the resulting series has the form

$$\sum_{k=0}^{\infty} c_k(x - x_0)^k = c_0 + c_1(x - x_0) + c_2(x - x_0)^2 + \cdots + c_k(x - x_0)^k + \cdots$$

This is called a ***power series in*** $x - x_0$. Some examples are

$$\sum_{k=0}^{\infty} \frac{(x - 1)^k}{k + 1} = 1 + \frac{(x - 1)}{2} + \frac{(x - 1)^2}{3} + \frac{(x - 1)^3}{4} + \cdots \qquad \boxed{x_0 = 1}$$

$$\sum_{k=0}^{\infty} \frac{(-1)^k(x + 3)^k}{k!} = 1 - (x + 3) + \frac{(x + 3)^2}{2!} - \frac{(x + 3)^3}{3!} + \cdots \qquad \boxed{x_0 = -3}$$

The first of these is a power series in $x - 1$ and the second is a power series in $x + 3$. Note that a power series in x is a power series in $x - x_0$ in which $x_0 = 0$. More generally, the Taylor series

$$\sum_{k=0}^{\infty} \frac{f^{(k)}(x_0)}{k!}(x - x_0)^k$$

is a power series in $x - x_0$.

The main result on convergence of a power series in $x - x_0$ can be obtained by substituting $x - x_0$ for x in Theorem 9.8.2. This leads to the following theorem.

9.8.3   THEOREM   *For a power series* $\sum c_k(x - x_0)^k$, *exactly one of the following statements is true:*

(a)   The series converges only for $x = x_0$.

(b)   The series converges absolutely (and hence converges) for all real values of x.

(c)   The series converges absolutely (and hence converges) for all x *in some finite open interval* $(x_0 - R, x_0 + R)$ *and diverges if* $x < x_0 - R$ *or* $x > x_0 + R$. *At either of the values* $x = x_0 - R$ *or* $x = x_0 + R$, *the series may converge absolutely, converge conditionally, or diverge, depending on the particular series.*

It follows from this theorem that the set of values for which a power series in $x - x_0$ converges is always an interval centered at $x = x_0$; we call this the ***interval of convergence*** (Figure 9.8.2). In part (*a*) of Theorem 9.8.3 the interval of convergence reduces to the single value $x = x_0$, in which case we say that the series has ***radius of convergence*** $R = 0$; in part

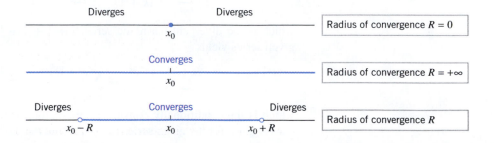

▶ **Figure 9.8.2**

(b) the interval of convergence is infinite (the entire real line), in which case we say that the series has **radius of convergence $R = +\infty$**; and in part (c) the interval extends between $x_0 - R$ and $x_0 + R$, in which case we say that the series has **radius of convergence R**.

▶ **Example 4** Find the interval of convergence and radius of convergence of the series

$$\sum_{k=1}^{\infty} \frac{(x-5)^k}{k^2}$$

Solution. We apply the ratio test for absolute convergence.

$$\rho = \lim_{k \to +\infty} \left| \frac{u_{k+1}}{u_k} \right| = \lim_{k \to +\infty} \left| \frac{(x-5)^{k+1}}{(k+1)^2} \cdot \frac{k^2}{(x-5)^k} \right|$$

$$= \lim_{k \to +\infty} \left[|x-5| \left(\frac{k}{k+1} \right)^2 \right]$$

$$= |x-5| \lim_{k \to +\infty} \left(\frac{1}{1 + (1/k)} \right)^2 = |x-5|$$

Thus, the series converges absolutely if $|x-5| < 1$, or $-1 < x - 5 < 1$, or $4 < x < 6$. The series diverges if $x < 4$ or $x > 6$.

To determine the convergence behavior at the endpoints $x = 4$ and $x = 6$, we substitute these values in the given series. If $x = 6$, the series becomes

$$\sum_{k=1}^{\infty} \frac{1^k}{k^2} = \sum_{k=1}^{\infty} \frac{1}{k^2} = 1 + \frac{1}{2^2} + \frac{1}{3^2} + \frac{1}{4^2} + \cdots$$

which is a convergent p-series ($p = 2$). If $x = 4$, the series becomes

$$\sum_{k=1}^{\infty} \frac{(-1)^k}{k^2} = -1 + \frac{1}{2^2} - \frac{1}{3^2} + \frac{1}{4^2} - \cdots$$

Since this series converges absolutely, the interval of convergence for the given series is $[4, 6]$. The radius of convergence is $R = 1$ (Figure 9.8.3). ◀

> It will always be a waste of time to test for convergence at the endpoints of the interval of convergence using the ratio test, since ρ will always be 1 at those points if
>
> $$\lim_{k \to +\infty} \left| \frac{u_{k+1}}{u_k} \right|$$
>
> exists. Explain why this must be so.

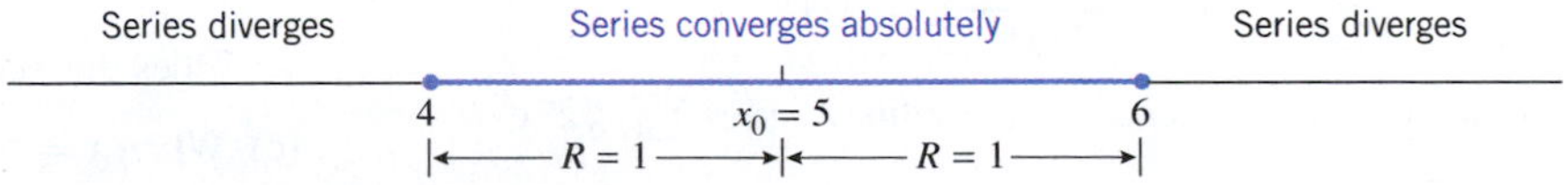

▶ **Figure 9.8.3**

Series diverges	Series converges absolutely	Series diverges

$$\overset{4}{\longleftarrow R = 1 \longrightarrow} \overset{x_0 = 5}{\longleftarrow R = 1 \longrightarrow} \overset{6}{|}$$

■ **FUNCTIONS DEFINED BY POWER SERIES**

If a function f is expressed as a power series on some interval, then we say that the power series ***represents*** f on that interval. For example, we saw in Example 4(a) of Section 9.3 that

$$\frac{1}{1-x} = \sum_{k=0}^{\infty} x^k$$

if $|x| < 1$, so this power series represents the function $1/(1-x)$ on the interval $-1 < x < 1$.

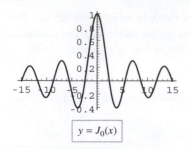

$y = J_0(x)$

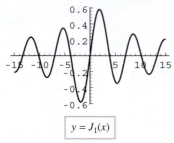

$y = J_1(x)$

Generated by Mathematica

▲ **Figure 9.8.4**

TECHNOLOGY MASTERY

Many computer algebra systems have the Bessel functions as part of their libraries. If you have a CAS with Bessel functions, use it to generate the graphs in Figure 9.8.4.

Sometimes new functions actually originate as power series, and the properties of the functions are developed by working with their power series representations. For example, the functions

$$J_0(x) = \sum_{k=0}^{\infty} \frac{(-1)^k x^{2k}}{2^{2k}(k!)^2} = 1 - \frac{x^2}{2^2(1!)^2} + \frac{x^4}{2^4(2!)^2} - \frac{x^6}{2^6(3!)^2} + \cdots \tag{4}$$

and

$$J_1(x) = \sum_{k=0}^{\infty} \frac{(-1)^k x^{2k+1}}{2^{2k+1}(k!)(k+1)!} = \frac{x}{2} - \frac{x^3}{2^3(1!)(2!)} + \frac{x^5}{2^5(2!)(3!)} - \cdots \tag{5}$$

which are called **Bessel functions** in honor of the German mathematician and astronomer Friedrich Wilhelm Bessel (1784–1846), arise naturally in the study of planetary motion and in various problems that involve heat flow.

To find the domains of these functions, we must determine where their defining power series converge. For example, in the case of $J_0(x)$ we have

$$\rho = \lim_{k \to +\infty} \left| \frac{u_{k+1}}{u_k} \right| = \lim_{k \to +\infty} \left| \frac{x^{2(k+1)}}{2^{2(k+1)}[(k+1)!]^2} \cdot \frac{2^{2k}(k!)^2}{x^{2k}} \right|$$

$$= \lim_{k \to +\infty} \left| \frac{x^2}{4(k+1)^2} \right| = 0 < 1$$

so the series converges for all x; that is, the domain of $J_0(x)$ is $(-\infty, +\infty)$. We leave it as an exercise (Exercise 59) to show that the power series for $J_1(x)$ also converges for all x. Computer-generated graphs of $J_0(x)$ and $J_1(x)$ are shown in Figure 9.8.4.

✔**QUICK CHECK EXERCISES 9.8**   *(See page 668 for answers.)*

1. If f has derivatives of all orders at x_0, then the Taylor series for f about $x = x_0$ is defined to be

$$\sum_{k=0}^{\infty} \underline{\qquad}$$

2. Since

$$\lim_{k \to +\infty} \left| \frac{2^{k+1} x^{k+1}}{2^k x^k} \right| = 2|x|$$

the radius of convergence for the infinite series $\sum_{k=0}^{\infty} 2^k x^k$ is _____.

3. Since

$$\lim_{k \to +\infty} \left| \frac{(3^{k+1} x^{k+1})/(k+1)!}{(3^k x^k)/k!} \right| = \lim_{k \to +\infty} \left| \frac{3x}{k+1} \right| = 0$$

the interval of convergence for the series $\sum_{k=0}^{\infty} (3^k/k!) x^k$ is

_____.

4. (a) Since

$$\lim_{k \to +\infty} \left| \frac{(x-4)^{k+1}/\sqrt{k+1}}{(x-4)^k/\sqrt{k}} \right| = \lim_{k \to +\infty} \left| \sqrt{\frac{k}{k+1}} (x-4) \right|$$

$$= |x-4|$$

the radius of convergence for the infinite series $\sum_{k=1}^{\infty} (1/\sqrt{k})(x-4)^k$ is _____.

(b) When $x = 3$,

$$\sum_{k=1}^{\infty} \frac{1}{\sqrt{k}} (x-4)^k = \sum_{k=1}^{\infty} \frac{1}{\sqrt{k}} (-1)^k$$

Does this series converge or diverge?

(c) When $x = 5$,

$$\sum_{k=1}^{\infty} \frac{1}{\sqrt{k}} (x-4)^k = \sum_{k=1}^{\infty} \frac{1}{\sqrt{k}}$$

Does this series converge or diverge?

(d) The interval of convergence for the infinite series $\sum_{k=1}^{\infty} (1/\sqrt{k})(x-4)^k$ is _____.

EXERCISE SET 9.8    Graphing Utility   [c] CAS

1–10 Use sigma notation to write the Maclaurin series for the function. ■

1. e^{-x} **2.** e^{ax} **3.** $\cos \pi x$ **4.** $\sin \pi x$

5. $\ln(1+x)$ **6.** $\dfrac{1}{1+x}$ **7.** $\cosh x$

8. $\sinh x$ **9.** $x \sin x$ **10.** xe^x

11–18 Use sigma notation to write the Taylor series about $x = x_0$ for the function. ■

11. e^x; $x_0 = 1$ **12.** e^{-x}; $x_0 = \ln 2$

13. $\dfrac{1}{x}$; $x_0 = -1$ **14.** $\dfrac{1}{x+2}$; $x_0 = 3$

15. $\sin \pi x$; $x_0 = \dfrac{1}{2}$ **16.** $\cos x$; $x_0 = \dfrac{\pi}{2}$

17. $\ln x$; $x_0 = 1$ **18.** $\ln x$; $x_0 = e$

19–22 Find the interval of convergence of the power series, and find a familiar function that is represented by the power series on that interval. ■

19. $1 - x + x^2 - x^3 + \cdots + (-1)^k x^k + \cdots$

20. $1 + x^2 + x^4 + \cdots + x^{2k} + \cdots$

21. $1 + (x-2) + (x-2)^2 + \cdots + (x-2)^k + \cdots$

22. $1 - (x+3) + (x+3)^2 - (x+3)^3$
$$+ \cdots + (-1)^k (x+3)^k | \cdots$$

23. Suppose that the function f is represented by the power series
$$f(x) = 1 - \frac{x}{2} + \frac{x^2}{4} - \frac{x^3}{8} + \cdots + (-1)^k \frac{x^k}{2^k} + \cdots$$
(a) Find the domain of f.   (b) Find $f(0)$ and $f(1)$.

24. Suppose that the function f is represented by the power series
$$f(x) = 1 - \frac{x-5}{3} + \frac{(x-5)^2}{3^2} - \frac{(x-5)^3}{3^3} + \cdots$$
(a) Find the domain of f.   (b) Find $f(3)$ and $f(6)$.

25–28 True–False Determine whether the statement is true or false. Explain your answer. ■

25. If a power series in x converges conditionally at $x = 3$, then the series converges if $|x| < 3$ and diverges if $|x| > 3$.

26. The ratio test is often useful to determine convergence at the endpoints of the interval of convergence of a power series.

27. The Maclaurin series for a polynomial function has radius of convergence $+\infty$.

28. The series $\displaystyle\sum_{k=0}^{\infty} \frac{x^k}{k!}$ converges if $|x| < 1$.

29–48 Find the radius of convergence and the interval of convergence. ■

29. $\displaystyle\sum_{k=0}^{\infty} \frac{x^k}{k+1}$ **30.** $\displaystyle\sum_{k=0}^{\infty} 3^k x^k$ **31.** $\displaystyle\sum_{k=0}^{\infty} \frac{(-1)^k x^k}{k!}$

32. $\displaystyle\sum_{k=0}^{\infty} \frac{k!}{2^k} x^k$ **33.** $\displaystyle\sum_{k=1}^{\infty} \frac{5^k}{k^2} x^k$ **34.** $\displaystyle\sum_{k=2}^{\infty} \frac{x^k}{\ln k}$

35. $\displaystyle\sum_{k=1}^{\infty} \frac{x^k}{k(k+1)}$ **36.** $\displaystyle\sum_{k=0}^{\infty} \frac{(-2)^k x^{k+1}}{k+1}$

37. $\displaystyle\sum_{k=1}^{\infty} (-1)^{k-1} \frac{x^k}{\sqrt{k}}$ **38.** $\displaystyle\sum_{k=0}^{\infty} \frac{(-1)^k x^{2k}}{(2k)!}$

39. $\displaystyle\sum_{k=0}^{\infty} \frac{3^k}{k!} x^k$ **40.** $\displaystyle\sum_{k=2}^{\infty} (-1)^{k+1} \frac{x^k}{k(\ln k)^2}$

41. $\displaystyle\sum_{k=0}^{\infty} \frac{x^k}{1+k^2}$ **42.** $\displaystyle\sum_{k=0}^{\infty} \frac{(x-3)^k}{2^k}$

43. $\displaystyle\sum_{k=1}^{\infty} (-1)^{k+1} \frac{(x+1)^k}{k}$ **44.** $\displaystyle\sum_{k=0}^{\infty} (-1)^k \frac{(x-4)^k}{(k+1)^2}$

45. $\displaystyle\sum_{k=0}^{\infty} \left(\frac{3}{4}\right)^k (x+5)^k$ **46.** $\displaystyle\sum_{k=1}^{\infty} \frac{(2k+1)!}{k^3} (x-2)^k$

47. $\displaystyle\sum_{k=0}^{\infty} \frac{\pi^k (x-1)^{2k}}{(2k+1)!}$ **48.** $\displaystyle\sum_{k=0}^{\infty} \frac{(2x-3)^k}{4^{2k}}$

49. Use the root test to find the interval of convergence of
$$\sum_{k=2}^{\infty} \frac{x^k}{(\ln k)^k}$$

50. Find the domain of the function
$$f(x) = \sum_{k=1}^{\infty} \frac{1 \cdot 3 \cdot 5 \cdots (2k-1)}{(2k-2)!} x^k$$

51. Show that the series
$$1 - \frac{x}{2!} + \frac{x^2}{4!} - \frac{x^3}{6!} + \cdots$$
is the Maclaurin series for the function
$$f(x) = \begin{cases} \cos\sqrt{x}, & x \geq 0 \\ \cosh\sqrt{-x}, & x < 0 \end{cases}$$

[*Hint:* Use the Maclaurin series for $\cos x$ and $\cosh x$ to obtain series for $\cos\sqrt{x}$, where $x \geq 0$, and $\cosh\sqrt{-x}$, where $x \leq 0$.]

FOCUS ON CONCEPTS

~ **52.** If a function f is represented by a power series on an interval, then the graphs of the partial sums can be used as approximations to the graph of f.
(a) Use a graphing utility to generate the graph of $1/(1-x)$ together with the graphs of the first four partial sums of its Maclaurin series over the interval $(-1, 1)$.
(b) In general terms, where are the graphs of the partial sums the most accurate?

53. Prove:

(a) If f is an even function, then all odd powers of x in its Maclaurin series have coefficient 0.

(b) If f is an odd function, then all even powers of x in its Maclaurin series have coefficient 0.

54. Suppose that the power series $\sum c_k(x - x_0)^k$ has radius of convergence R and p is a nonzero constant. What can you say about the radius of convergence of the power series $\sum pc_k(x - x_0)^k$? Explain your reasoning. [*Hint:* See Theorem 9.4.3.]

55. Suppose that the power series $\sum c_k(x - x_0)^k$ has a finite radius of convergence R, and the power series $\sum d_k(x - x_0)^k$ has a radius of convergence of $+\infty$. What can you say about the radius of convergence of $\sum (c_k + d_k)(x - x_0)^k$? Explain your reasoning.

56. Suppose that the power series $\sum c_k(x - x_0)^k$ has a finite radius of convergence R_1 and the power series $\sum d_k(x - x_0)^k$ has a finite radius of convergence R_2. What can you say about the radius of convergence of $\sum (c_k + d_k)(x - x_0)^k$? Explain your reasoning. [*Hint:* The case $R_1 = R_2$ requires special attention.]

57. Show that if p is a positive integer, then the power series

$$\sum_{k=0}^{\infty} \frac{(pk)!}{(k!)^p} x^k$$

has a radius of convergence of $1/p^p$.

58. Show that if p and q are positive integers, then the power series

$$\sum_{k=0}^{\infty} \frac{(k + p)!}{k!(k + q)!} x^k$$

has a radius of convergence of $+\infty$.

59. Show that the power series representation of the Bessel function $J_1(x)$ converges for all x [Formula (5)].

60. Approximate the values of the Bessel functions $J_0(x)$ and $J_1(x)$ at $x = 1$, each to four decimal-place accuracy.

c **61.** If the constant p in the general p-series is replaced by a variable x for $x > 1$, then the resulting function is called the ***Riemann zeta function*** and is denoted by

$$\zeta(x) = \sum_{k=1}^{\infty} \frac{1}{k^x}$$

(a) Let s_n be the nth partial sum of the series for $\zeta(3.7)$. Find n such that s_n approximates $\zeta(3.7)$ to two decimal-place accuracy, and calculate s_n using this value of n. [*Hint:* Use the right inequality in Exercise 36(b) of Section 9.4 with $f(x) = 1/x^{3.7}$.]

(b) Determine whether your CAS can evaluate the Riemann zeta function directly. If so, compare the value produced by the CAS to the value of s_n obtained in part (a).

62. Prove: If $\lim_{k \to +\infty} |c_k|^{1/k} = L$, where $L \neq 0$, then $1/L$ is the radius of convergence of the power series $\sum_{k=0}^{\infty} c_k x^k$.

63. Prove: If the power series $\sum_{k=0}^{\infty} c_k x^k$ has radius of convergence R, then the series $\sum_{k=0}^{\infty} c_k x^{2k}$ has radius of convergence $\sqrt{R}$.

64. Prove: If the interval of convergence of the series $\sum_{k=0}^{\infty} c_k(x - x_0)^k$ is $(x_0 - R, x_0 + R]$, then the series converges conditionally at $x_0 + R$.

65. **Writing** The sine function can be defined geometrically from the unit circle or analytically from its Maclaurin series. Discuss the advantages of each representation with regard to providing information about the sine function.

✔**QUICK CHECK ANSWERS 9.8**

1. $\dfrac{f^{(k)}(x_0)}{k!}(x - x_0)^k$ **2.** $\dfrac{1}{2}$ **3.** $(-\infty, +\infty)$ **4.** (a) 1 (b) converges (c) diverges (d) $[3, 5)$

9.9 CONVERGENCE OF TAYLOR SERIES

In this section we will investigate when a Taylor series for a function converges to that function on some interval, and we will consider how Taylor series can be used to approximate values of trigonometric, exponential, and logarithmic functions.

■ THE CONVERGENCE PROBLEM FOR TAYLOR SERIES

Recall that the nth Taylor polynomial for a function f about $x = x_0$ has the property that its value and the values of its first n derivatives match those of f at x_0. As n increases, more and more derivatives match up, so it is reasonable to hope that for values of x near x_0 the values of the Taylor polynomials might converge to the value of $f(x)$; that is,

$$f(x) = \lim_{n \to +\infty} \sum_{k=0}^{n} \frac{f^{(k)}(x_0)}{k!}(x - x_0)^k \tag{1}$$

However, the nth Taylor polynomial for f is the nth partial sum of the Taylor series for f, so (1) is equivalent to stating that the Taylor series for f converges at x, and its sum is $f(x)$. Thus, we are led to consider the following problem.

Problem 9.9.1 is concerned not only with whether the Taylor series of a function f converges, but also whether it converges to the function f itself. Indeed, it is possible for a Taylor series of a function f to converge to values different from $f(x)$ for certain values of x (Exercise 14).

9.9.1 PROBLEM Given a function f that has derivatives of all orders at $x = x_0$, determine whether there is an open interval containing x_0 such that $f(x)$ is the sum of its Taylor series about $x = x_0$ at each point in the interval; that is,

$$f(x) = \sum_{k=0}^{\infty} \frac{f^{(k)}(x_0)}{k!}(x - x_0)^k \tag{2}$$

for all values of x in the interval.

One way to show that (1) holds is to show that

$$\lim_{n \to +\infty} \left[f(x) - \sum_{k=0}^{n} \frac{f^{(k)}(x_0)}{k!}(x - x_0)^k \right] = 0$$

However, the difference appearing on the left side of this equation is the nth remainder for the Taylor series [Formula (12) of Section 9.7]. Thus, we have the following result.

9.9.2 THEOREM *The equality*

$$f(x) = \sum_{k=0}^{\infty} \frac{f^{(k)}(x_0)}{k!}(x - x_0)^k$$

holds at a point x if and only if $\lim_{n \to +\infty} R_n(x) = 0$.

■ **ESTIMATING THE nTH REMAINDER**

It is relatively rare that one can prove directly that $R_n(x) \to 0$ as $n \to +\infty$. Usually, this is proved indirectly by finding appropriate bounds on $|R_n(x)|$ and applying the Squeezing Theorem for Sequences. The Remainder Estimation Theorem (Theorem 9.7.4) provides a useful bound for this purpose. Recall that this theorem asserts that if M is an upper bound for $|f^{(n+1)}(x)|$ on an interval containing x_0, then

$$|R_n(x)| \leq \frac{M}{(n+1)!}|x - x_0|^{n+1} \tag{3}$$

for all x in that interval.

The following example illustrates how the Remainder Estimation Theorem is applied.

▶ **Example 1** Show that the Maclaurin series for $\cos x$ converges to $\cos x$ for all x; that is,

$$\cos x = \sum_{k=0}^{\infty} (-1)^k \frac{x^{2k}}{(2k)!} = 1 - \frac{x^2}{2!} + \frac{x^4}{4!} - \frac{x^6}{6!} + \cdots \qquad (-\infty < x < +\infty)$$

Solution. From Theorem 9.9.2 we must show that $R_n(x) \to 0$ for all x as $n \to +\infty$. For this purpose let $f(x) = \cos x$, so that for all x we have

$$f^{(n+1)}(x) = \pm \cos x \quad \text{or} \quad f^{(n+1)}(x) = \pm \sin x$$

In all cases we have $|f^{(n+1)}(x)| \leq 1$, so we can apply (3) with $M = 1$ and $x_0 = 0$ to conclude that

$$0 \leq |R_n(x)| \leq \frac{|x|^{n+1}}{(n+1)!} \tag{4}$$

However, it follows from Formula (5) of Section 9.2 with $n + 1$ in place of n and $|x|$ in place of x that

$$\lim_{n \to +\infty} \frac{|x|^{n+1}}{(n+1)!} = 0 \tag{5}$$

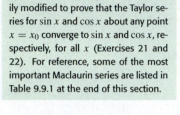

The method of Example 1 can be easily modified to prove that the Taylor series for $\sin x$ and $\cos x$ about any point $x = x_0$ converge to $\sin x$ and $\cos x$, respectively, for all x (Exercises 21 and 22). For reference, some of the most important Maclaurin series are listed in Table 9.9.1 at the end of this section.

Using this result and the Squeezing Theorem for Sequences (Theorem 9.1.5), it follows from (4) that $|R_n(x)| \to 0$ and hence that $R_n(x) \to 0$ as $n \to +\infty$ (Theorem 9.1.6). Since this is true for all x, we have proved that the Maclaurin series for $\cos x$ converges to $\cos x$ for all x. This is illustrated in Figure 9.9.1, where we can see how successive partial sums approximate the cosine curve more and more closely. ◀

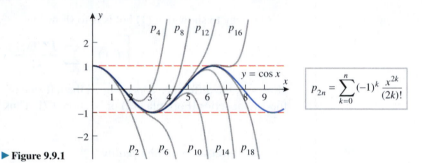

▶ **Figure 9.9.1**

$$P_{2n} = \sum_{k=0}^{n} (-1)^k \frac{x^{2k}}{(2k)!}$$

■ APPROXIMATING TRIGONOMETRIC FUNCTIONS

In general, to approximate the value of a function f at a point x using a Taylor series, there are two basic questions that must be answered:

- About what point x_0 should the Taylor series be expanded?
- How many terms in the series should be used to achieve the desired accuracy?

In response to the first question, x_0 needs to be a point at which the derivatives of f can be evaluated easily, since these values are needed for the coefficients in the Taylor series. Furthermore, if the function f is being evaluated at x, then x_0 should be chosen as close as possible to x, since Taylor series tend to converge more rapidly near x_0. For example, to approximate $\sin 3°$ ($= \pi/60$ radians), it would be reasonable to take $x_0 = 0$, since $\pi/60$ is close to 0 and the derivatives of $\sin x$ are easy to evaluate at 0. On the other hand, to approximate $\sin 85°$ ($= 17\pi/36$ radians), it would be more natural to take $x_0 = \pi/2$, since $17\pi/36$ is close to $\pi/2$ and the derivatives of $\sin x$ are easy to evaluate at $\pi/2$.

In response to the second question posed above, the number of terms required to achieve a specific accuracy needs to be determined on a problem-by-problem basis. The next example gives two methods for doing this.

▶ **Example 2** Use the Maclaurin series for $\sin x$ to approximate $\sin 3°$ to five decimal-place accuracy.

Solution. In the Maclaurin series

$$\sin x = \sum_{k=0}^{\infty} (-1)^k \frac{x^{2k+1}}{(2k+1)!} = x - \frac{x^3}{3!} + \frac{x^5}{5!} - \frac{x^7}{7!} + \cdots \qquad (6)$$

the angle x is assumed to be in radians (because the differentiation formulas for the trigonometric functions were derived with this assumption). Since $3° = \pi/60$ radians, it follows from (6) that

$$\sin 3° = \sin \frac{\pi}{60} = \left(\frac{\pi}{60}\right) - \frac{(\pi/60)^3}{3!} + \frac{(\pi/60)^5}{5!} - \frac{(\pi/60)^7}{7!} + \cdots \qquad (7)$$

We must now determine how many terms in the series are required to achieve five decimal-place accuracy. We will consider two possible approaches, one using the Remainder Estimation Theorem (Theorem 9.7.4) and the other using the fact that (7) satisfies the hypotheses of the alternating series test (Theorem 9.6.1).

Method 1. (*The Remainder Estimation Theorem*)

Since we want to achieve five decimal-place accuracy, our goal is to choose n so that the absolute value of the nth remainder at $x = \pi/60$ does not exceed $0.000005 = 5 \times 10^{-6}$; that is,

$$\left| R_n \left(\frac{\pi}{60} \right) \right| \leq 0.000005 \qquad (8)$$

However, if we let $f(x) = \sin x$, then $f^{(n+1)}(x)$ is either $\pm \sin x$ or $\pm \cos x$, and in either case $|f^{(n+1)}(x)| \leq 1$ for all x. Thus, it follows from the Remainder Estimation Theorem with $M = 1$, $x_0 = 0$, and $x = \pi/60$ that

$$\left| R_n \left(\frac{\pi}{60} \right) \right| \leq \frac{(\pi/60)^{n+1}}{(n+1)!}$$

Thus, we can satisfy (8) by choosing n so that

$$\frac{(\pi/60)^{n+1}}{(n+1)!} \leq 0.000005$$

With the help of a calculating utility you can verify that the smallest value of n that meets this criterion is $n = 3$. Thus, to achieve five decimal-place accuracy we need only keep terms up to the third power in (7). This yields

$$\sin 3° \approx \left(\frac{\pi}{60} \right) - \frac{(\pi/60)^3}{3!} \approx 0.05234 \qquad (9)$$

(verify). As a check, a calculator gives $\sin 3° \approx 0.05233595624$, which agrees with (9) when rounded to five decimal places.

Method 2. (*The Alternating Series Test*)

We leave it for you to check that (7) satisfies the hypotheses of the alternating series test (Theorem 9.6.1).

Let s_n denote the sum of the terms in (7) up to and including the nth power of $\pi/60$. Since the exponents in the series are odd integers, the integer n must be odd, and the exponent of the first term *not* included in the sum s_n must be $n + 2$. Thus, it follows from part (*b*) of Theorem 9.6.2 that

$$|\sin 3° - s_n| < \frac{(\pi/60)^{n+2}}{(n+2)!}$$

This means that for five decimal-place accuracy we must look for the first positive odd integer n such that

$$\frac{(\pi/60)^{n+2}}{(n+2)!} \leq 0.000005$$

With the help of a calculating utility you can verify that the smallest value of n that meets this criterion is $n = 3$. This agrees with the result obtained above using the Remainder Estimation Theorem and hence leads to approximation (9) as before. ◄

■ ROUNDOFF AND TRUNCATION ERROR

There are two types of errors that occur when computing with series. The first, called *truncation error*, is the error that results when a series is approximated by a partial sum; and the second, called *roundoff error*, is the error that arises from approximations in numerical computations. For example, in our derivation of (9) we took $n = 3$ to keep the truncation error below 0.000005. However, to evaluate the partial sum we had to approximate π, thereby introducing roundoff error. Had we not exercised some care in choosing this approximation, the roundoff error could easily have degraded the final result.

Methods for estimating and controlling roundoff error are studied in a branch of mathematics called *numerical analysis*. However, as a rule of thumb, to achieve n decimal-place accuracy in a final result, all intermediate calculations must be accurate to at least $n + 1$ decimal places. Thus, in (9) at least six decimal-place accuracy in π is required to achieve the five decimal-place accuracy in the final numerical result. As a practical matter, a good working procedure is to perform all intermediate computations with the maximum number of digits that your calculating utility can handle and then round at the end.

■ APPROXIMATING EXPONENTIAL FUNCTIONS

► **Example 3** Show that the Maclaurin series for e^x converges to e^x for all x; that is,

$$e^x = \sum_{k=0}^{\infty} \frac{x^k}{k!} = 1 + x + \frac{x^2}{2!} + \frac{x^3}{3!} + \cdots + \frac{x^k}{k!} + \cdots \qquad (-\infty < x < +\infty)$$

Solution. Let $f(x) = e^x$, so that

$$f^{(n+1)}(x) = e^x$$

We want to show that $R_n(x) \to 0$ as $n \to +\infty$ for all x in the interval $-\infty < x < +\infty$. However, it will be helpful here to consider the cases $x \le 0$ and $x > 0$ separately. If $x \le 0$, then we will take the interval in the Remainder Estimation Theorem (Theorem 9.7.4) to be $[x, 0]$, and if $x > 0$, then we will take it to be $[0, x]$. Since $f^{(n+1)}(x) = e^x$ is an increasing function, it follows that if c is in the interval $[x, 0]$, then

$$|f^{(n+1)}(c)| \le |f^{(n+1)}(0)| = e^0 = 1$$

and if c is in the interval $[0, x]$, then

$$|f^{(n+1)}(c)| \le |f^{(n+1)}(x)| = e^x$$

Thus, we can apply Theorem 9.7.4 with $M = 1$ in the case where $x \le 0$ and with $M = e^x$ in the case where $x > 0$. This yields

$$0 \le |R_n(x)| \le \frac{|x|^{n+1}}{(n+1)!} \qquad \text{if } x \le 0$$

$$0 \le |R_n(x)| \le e^x \frac{|x|^{n+1}}{(n+1)!} \qquad \text{if } x > 0$$

Thus, in both cases it follows from (5) and the Squeezing Theorem for Sequences that $|R_n(x)| \to 0$ as $n \to +\infty$, which in turn implies that $R_n(x) \to 0$ as $n \to +\infty$. Since this is true for all x, we have proved that the Maclaurin series for e^x converges to e^x for all x. ◄

Since the Maclaurin series for e^x converges to e^x for all x, we can use partial sums of the Maclaurin series to approximate powers of e to arbitrary precision. Recall that in Example 7 of Section 9.7 we were able to use the Remainder Estimation Theorem to determine that evaluating the ninth Maclaurin polynomial for e^x at $x = 1$ yields an approximation for e with five decimal-place accuracy:

$$e \approx 1 + 1 + \frac{1}{2!} + \frac{1}{3!} + \frac{1}{4!} + \frac{1}{5!} + \frac{1}{6!} + \frac{1}{7!} + \frac{1}{8!} + \frac{1}{9!} \approx 2.71828$$

■ APPROXIMATING LOGARITHMS

The Maclaurin series

$$\ln(1 + x) = x - \frac{x^2}{2} + \frac{x^3}{3} - \frac{x^4}{4} + \cdots \qquad (-1 < x \le 1) \qquad (10)$$

is the starting point for the approximation of natural logarithms. Unfortunately, the usefulness of this series is limited because of its slow convergence and the restriction $-1 < x \le 1$. However, if we replace x by $-x$ in this series, we obtain

$$\ln(1 - x) = -x - \frac{x^2}{2} - \frac{x^3}{3} - \frac{x^4}{4} - \cdots \qquad (-1 \le x < 1) \qquad (11)$$

and on subtracting (11) from (10) we obtain

$$\ln\left(\frac{1+x}{1-x}\right) = 2\left(x + \frac{x^3}{3} + \frac{x^5}{5} + \frac{x^7}{7} + \cdots\right) \qquad (-1 < x < 1) \qquad (12)$$

Series (12), first obtained by James Gregory in 1668, can be used to compute the natural logarithm of any positive number y by letting

$$y = \frac{1+x}{1-x}$$

or, equivalently,

$$x = \frac{y-1}{y+1} \qquad (13)$$

and noting that $-1 < x < 1$. For example, to compute $\ln 2$ we let $y = 2$ in (13), which yields $x = \frac{1}{3}$. Substituting this value in (12) gives

$$\ln 2 = 2\left[\frac{1}{3} + \frac{\left(\frac{1}{3}\right)^3}{3} + \frac{\left(\frac{1}{3}\right)^5}{5} + \frac{\left(\frac{1}{3}\right)^7}{7} + \cdots\right] \qquad (14)$$

In Exercise 19 we will ask you to show that five decimal-place accuracy can be achieved using the partial sum with terms up to and including the 13th power of $\frac{1}{3}$. Thus, to five decimal-place accuracy

$$\ln 2 \approx 2\left[\frac{1}{3} + \frac{\left(\frac{1}{3}\right)^3}{3} + \frac{\left(\frac{1}{3}\right)^5}{5} + \frac{\left(\frac{1}{3}\right)^7}{7} + \cdots + \frac{\left(\frac{1}{3}\right)^{13}}{13}\right] \approx 0.69315$$

(verify). As a check, a calculator gives $\ln 2 \approx 0.69314718056$, which agrees with the preceding approximation when rounded to five decimal places.

■ APPROXIMATING π

In the next section we will show that

$$\tan^{-1} x = x - \frac{x^3}{3} + \frac{x^5}{5} - \frac{x^7}{7} + \cdots \qquad (-1 \le x \le 1) \qquad (15)$$

Letting $x = 1$, we obtain

$$\frac{\pi}{4} = \tan^{-1} 1 = 1 - \frac{1}{3} + \frac{1}{5} - \frac{1}{7} + \cdots$$

In Example 2 of Section 9.6, we stated without proof that

$$\ln 2 = 1 - \frac{1}{2} + \frac{1}{3} - \frac{1}{4} + \frac{1}{5} - \cdots$$

This result can be obtained by letting $x = 1$ in (10), but as indicated in the text discussion, this series converges too slowly to be of practical use.

James Gregory
(1638–1675) Scottish mathematician and astronomer. Gregory, the son of a minister, was famous in his time as the inventor of the Gregorian reflecting telescope, so named in his honor. Although he is not generally ranked with the great mathematicians, much of his work relating to calculus was studied by Leibniz and Newton and undoubtedly influenced some of their discoveries. There is a manuscript, discovered posthumously, which shows that Gregory had anticipated Taylor series well before Taylor.

or

$$\pi = 4\left[1 - \frac{1}{3} + \frac{1}{5} - \frac{1}{7} + \cdots\right]$$

This famous series, obtained by Leibniz in 1674, converges too slowly to be of computational value. A more practical procedure for approximating π uses the identity

$$\frac{\pi}{4} = \tan^{-1}\frac{1}{2} + \tan^{-1}\frac{1}{3} \tag{16}$$

which was derived in Exercise 58 of Section 0.4. By using this identity and series (15) to approximate $\tan^{-1}\frac{1}{2}$ and $\tan^{-1}\frac{1}{3}$, the value of π can be approximated efficiently to any degree of accuracy.

■ BINOMIAL SERIES

If m is a real number, then the Maclaurin series for $(1 + x)^m$ is called the **binomial series**; it is given by

$$1 + mx + \frac{m(m - 1)}{2!}x^2 + \frac{m(m - 1)(m - 2)}{3!}x^3 + \cdots + \frac{m(m - 1)\cdots(m - k + 1)}{k!}x^k + \cdots$$

In the case where m is a nonnegative integer, the function $f(x) = (1 + x)^m$ is a polynomial of degree m, so

$$f^{(m+1)}(0) = f^{(m+2)}(0) = f^{(m+3)}(0) = \cdots = 0$$

Let $f(x) = (1 + x)^m$. Verify that

$f(0) = 1$

$f'(0) = m$

$f''(0) = m(m - 1)$

$f'''(0) = m(m - 1)(m - 2)$

$\vdots$

$f^{(k)}(0) = m(m - 1)\cdots(m - k + 1)$

and the binomial series reduces to the familiar binomial expansion

$$(1 + x)^m = 1 + mx + \frac{m(m - 1)}{2!}x^2 + \frac{m(m - 1)(m - 2)}{3!}x^3 + \cdots + x^m$$

which is valid for $-\infty < x < +\infty$.

It can be proved that if m is not a nonnegative integer, then the binomial series converges to $(1 + x)^m$ if $|x| < 1$. Thus, for such values of x

$$(1 + x)^m = 1 + mx + \frac{m(m - 1)}{2!}x^2 + \cdots + \frac{m(m - 1)\cdots(m - k + 1)}{k!}x^k + \cdots \tag{17}$$

or in sigma notation,

$$(1 + x)^m = 1 + \sum_{k=1}^{\infty} \frac{m(m - 1)\cdots(m - k + 1)}{k!}x^k \quad \text{if } |x| < 1 \tag{18}$$

▶ **Example 4** Find binomial series for

$$\text{(a)} \ \frac{1}{(1 + x)^2} \qquad \text{(b)} \ \frac{1}{\sqrt{1 + x}}$$

Solution (a). Since the general term of the binomial series is complicated, you may find it helpful to write out some of the beginning terms of the series, as in Formula (17), to see developing patterns. Substituting $m = -2$ in this formula yields

$$\frac{1}{(1 + x)^2} = (1 + x)^{-2} = 1 + (-2)x + \frac{(-2)(-3)}{2!}x^2$$

$$+ \frac{(-2)(-3)(-4)}{3!}x^3 + \frac{(-2)(-3)(-4)(-5)}{4!}x^4 + \cdots$$

$$= 1 - 2x + \frac{3!}{2!}x^2 - \frac{4!}{3!}x^3 + \frac{5!}{4!}x^4 - \cdots$$

$$= 1 - 2x + 3x^2 - 4x^3 + 5x^4 - \cdots$$

$$= \sum_{k=0}^{\infty} (-1)^k(k + 1)x^k$$

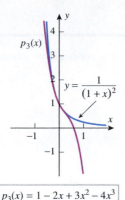

$p_3(x) = 1 - 2x + 3x^2 - 4x^3$

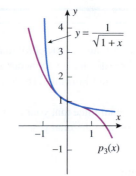

$p_3(x) = 1 - \frac{1}{2}x + \frac{3}{8}x^2 - \frac{5}{16}x^3$

▲ **Figure 9.9.2**

Solution (b).   Substituting $m = -\frac{1}{2}$ in (17) yields

$$\frac{1}{\sqrt{1+x}} = 1 - \frac{1}{2}x + \frac{\left(-\frac{1}{2}\right)\left(-\frac{1}{2}-1\right)}{2!}x^2 + \frac{\left(-\frac{1}{2}\right)\left(-\frac{1}{2}-1\right)\left(-\frac{1}{2}-2\right)}{3!}x^3 + \cdots$$

$$= 1 - \frac{1}{2}x + \frac{1\cdot 3}{2^2\cdot 2!}x^2 - \frac{1\cdot 3\cdot 5}{2^3\cdot 3!}x^3 + \cdots$$

$$= 1 + \sum_{k=1}^{\infty}(-1)^k\frac{1\cdot 3\cdot 5\cdots(2k-1)}{2^k k!}x^k \quad \blacktriangleleft$$

Figure 9.9.2 shows the graphs of the functions in Example 4 compared to their third-degree Maclaurin polynomials.

■ **SOME IMPORTANT MACLAURIN SERIES**

For reference, Table 9.9.1 lists the Maclaurin series for some of the most important functions, together with a specification of the intervals over which the Maclaurin series converge to those functions. Some of these results are derived in the exercises and others will be derived in the next section using some special techniques that we will develop.

Table 9.9.1

SOME IMPORTANT MACLAURIN SERIES

MACLAURIN SERIES	INTERVAL OF CONVERGENCE
$\dfrac{1}{1-x} = \displaystyle\sum_{k=0}^{\infty} x^k = 1 + x + x^2 + x^3 + \cdots$	$-1 < x < 1$
$\dfrac{1}{1+x^2} = \displaystyle\sum_{k=0}^{\infty} (-1)^k x^{2k} = 1 - x^2 + x^4 - x^6 + \cdots$	$-1 < x < 1$
$e^x = \displaystyle\sum_{k=0}^{\infty} \dfrac{x^k}{k!} = 1 + x + \dfrac{x^2}{2!} + \dfrac{x^3}{3!} + \dfrac{x^4}{4!} + \cdots$	$-\infty < x < +\infty$
$\sin x = \displaystyle\sum_{k=0}^{\infty} (-1)^k \dfrac{x^{2k+1}}{(2k+1)!} = x - \dfrac{x^3}{3!} + \dfrac{x^5}{5!} - \dfrac{x^7}{7!} + \cdots$	$-\infty < x < +\infty$
$\cos x = \displaystyle\sum_{k=0}^{\infty} (-1)^k \dfrac{x^{2k}}{(2k)!} = 1 - \dfrac{x^2}{2!} + \dfrac{x^4}{4!} - \dfrac{x^6}{6!} + \cdots$	$-\infty < x < +\infty$
$\ln(1+x) = \displaystyle\sum_{k=1}^{\infty} (-1)^{k+1} \dfrac{x^k}{k} = x - \dfrac{x^2}{2} + \dfrac{x^3}{3} - \dfrac{x^4}{4} + \cdots$	$-1 < x \le 1$
$\tan^{-1} x = \displaystyle\sum_{k=0}^{\infty} (-1)^k \dfrac{x^{2k+1}}{2k+1} = x - \dfrac{x^3}{3} + \dfrac{x^5}{5} - \dfrac{x^7}{7} + \cdots$	$-1 \le x \le 1$
$\sinh x = \displaystyle\sum_{k=0}^{\infty} \dfrac{x^{2k+1}}{(2k+1)!} = x + \dfrac{x^3}{3!} + \dfrac{x^5}{5!} + \dfrac{x^7}{7!} + \cdots$	$-\infty < x < +\infty$
$\cosh x = \displaystyle\sum_{k=0}^{\infty} \dfrac{x^{2k}}{(2k)!} = 1 + \dfrac{x^2}{2!} + \dfrac{x^4}{4!} + \dfrac{x^6}{6!} + \cdots$	$-\infty < x < +\infty$
$(1+x)^m = 1 + \displaystyle\sum_{k=1}^{\infty} \dfrac{m(m-1)\cdots(m-k+1)}{k!}x^k$	$-1 < x < 1^*$ ($m \ne 0, 1, 2, \ldots$)

*The behavior at the endpoints depends on m: For $m > 0$ the series converges absolutely at both endpoints; for $m \le -1$ the series diverges at both endpoints; and for $-1 < m < 0$ the series converges conditionally at $x = 1$ and diverges at $x = -1$.

✔ **QUICK CHECK EXERCISES 9.9**   *(See page 677 for answers.)*

1. $\cos x = \sum\limits_{k=0}^{\infty}$ _____

2. $e^x = \sum\limits_{k=0}^{\infty}$ _____

3. $\ln(1+x) = \sum\limits_{k=1}^{\infty}$ _____ for x in the interval _____.

4. If m is a real number but not a nonnegative integer, the *binomial series*

$$1 + \sum\limits_{k=1}^{\infty} \text{_____}$$

converges to $(1+x)^m$ if $|x| <$ _____.

EXERCISE SET 9.9   Graphing Utility   CAS

1. Use the Remainder Estimation Theorem and the method of Example 1 to prove that the Taylor series for $\sin x$ about $x = \pi/4$ converges to $\sin x$ for all x.

2. Use the Remainder Estimation Theorem and the method of Example 3 to prove that the Taylor series for e^x about $x = 1$ converges to e^x for all x.

3–10 Approximate the specified function value as indicated and check your work by comparing your answer to the function value produced directly by your calculating utility. ■

3. Approximate $\sin 4°$ to five decimal-place accuracy using both of the methods given in Example 2.

4. Approximate $\cos 3°$ to three decimal-place accuracy using both of the methods given in Example 2.

5. Approximate $\cos 0.1$ to five decimal-place accuracy using the Maclaurin series for $\cos x$.

6. Approximate $\tan^{-1} 0.1$ to three decimal-place accuracy using the Maclaurin series for $\tan^{-1} x$.

7. Approximate $\sin 85°$ to four decimal-place accuracy using an appropriate Taylor series.

8. Approximate $\cos(-175°)$ to four decimal-place accuracy using a Taylor series.

9. Approximate $\sinh 0.5$ to three decimal-place accuracy using the Maclaurin series for $\sinh x$.

10. Approximate $\cosh 0.1$ to three decimal-place accuracy using the Maclaurin series for $\cosh x$.

11. (a) Use Formula (12) in the text to find a series that converges to $\ln 1.25$.
(b) Approximate $\ln 1.25$ using the first two terms of the series. Round your answer to three decimal places, and compare the result to that produced directly by your calculating utility.

12. (a) Use Formula (12) to find a series that converges to $\ln 3$.
(b) Approximate $\ln 3$ using the first two terms of the series. Round your answer to three decimal places, and compare the result to that produced directly by your calculating utility.

FOCUS ON CONCEPTS

13. (a) Use the Maclaurin series for $\tan^{-1} x$ to approximate $\tan^{-1} \frac{1}{2}$ and $\tan^{-1} \frac{1}{3}$ to three decimal-place accuracy.
(b) Use the results in part (a) and Formula (16) to approximate π.
(c) Would you be willing to guarantee that your answer in part (b) is accurate to three decimal places? Explain your reasoning.
(d) Compare your answer in part (b) to that produced by your calculating utility.

14. The purpose of this exercise is to show that the Taylor series of a function f may possibly converge to a value different from $f(x)$ for certain values of x. Let

$$f(x) = \begin{cases} e^{-1/x^2}, & x \neq 0 \\ 0, & x = 0 \end{cases}$$

(a) Use the definition of a derivative to show that $f'(0) = 0$.
(b) With some difficulty it can be shown that if $n \geq 2$ then $f^{(n)}(0) = 0$. Accepting this fact, show that the Maclaurin series of f converges for all x, but converges to $f(x)$ only at $x = 0$.

 15. (a) Find an upper bound on the error that can result if $\cos x$ is approximated by $1 - (x^2/2!) + (x^4/4!)$ over the interval $[-0.2, 0.2]$.
(b) Check your answer in part (a) by graphing

$$\left| \cos x - \left(1 - \frac{x^2}{2!} + \frac{x^4}{4!}\right) \right|$$

over the interval.

 16. (a) Find an upper bound on the error that can result if $\ln(1+x)$ is approximated by x over the interval $[-0.01, 0.01]$.
(b) Check your answer in part (a) by graphing

$$|\ln(1+x) - x|$$

over the interval.

17. Use Formula (17) for the binomial series to obtain the Maclaurin series for

   (a) $\dfrac{1}{1+x}$ (b) $\sqrt[3]{1+x}$ (c) $\dfrac{1}{(1+x)^3}$.

18. If m is any real number, and k is a nonnegative integer, then we define the **binomial coefficient** $\dbinom{m}{k}$ by the formulas $\dbinom{m}{0} = 1$ and

$$\binom{m}{k} = \frac{m(m-1)(m-2)\cdots(m-k+1)}{k!}$$

   for $k \geq 1$. Express Formula (17) in the text in terms of binomial coefficients.

19. In this exercise we will use the Remainder Estimation Theorem to determine the number of terms that are required in Formula (14) to approximate $\ln 2$ to five decimal-place accuracy. For this purpose let

$$f(x) = \ln\frac{1+x}{1-x} = \ln(1+x) - \ln(1-x) \quad (-1 < x < 1)$$

   (a) Show that

$$f^{(n+1)}(x) = n!\left[\frac{(-1)^n}{(1+x)^{n+1}} + \frac{1}{(1-x)^{n+1}}\right]$$

   (b) Use the triangle inequality [Theorem 0.1.4(d)] to show that

$$|f^{(n+1)}(x)| \leq n!\left[\frac{1}{(1+x)^{n+1}} + \frac{1}{(1-x)^{n+1}}\right]$$

   (c) Since we want to achieve five decimal-place accuracy, our goal is to choose n so that the absolute value of the nth remainder at $x = \frac{1}{3}$ does not exceed the value $0.000005 = 0.5 \times 10^{-5}$; that is, $\left|R_n\left(\frac{1}{3}\right)\right| \leq 0.000005$. Use the Remainder Estimation Theorem to show that this condition will be satisfied if n is chosen so that

$$\frac{M}{(n+1)!}\left(\frac{1}{3}\right)^{n+1} \leq 0.000005$$

   where $|f^{(n+1)}(x)| \leq M$ on the interval $[0, \frac{1}{3}]$.

   (d) Use the result in part (b) to show that M can be taken as

$$M = n!\left[1 + \frac{1}{\left(\frac{2}{3}\right)^{n+1}}\right]$$

   (e) Use the results in parts (c) and (d) to show that five decimal-place accuracy will be achieved if n satisfies

$$\frac{1}{n+1}\left[\left(\frac{1}{3}\right)^{n+1} + \left(\frac{1}{2}\right)^{n+1}\right] \leq 0.000005$$

   and then show that the smallest value of n that satisfies this condition is $n = 13$.

20. Use Formula (12) and the method of Exercise 19 to approximate $\ln\left(\frac{5}{3}\right)$ to five decimal-place accuracy. Then check your work by comparing your answer to that produced directly by your calculating utility.

21. Prove: The Taylor series for $\cos x$ about any value $x = x_0$ converges to $\cos x$ for all x.

22. Prove: The Taylor series for $\sin x$ about any value $x = x_0$ converges to $\sin x$ for all x.

23. Research has shown that the proportion p of the population with IQs (intelligence quotients) between α and β is approximately

$$p = \frac{1}{16\sqrt{2\pi}}\int_\alpha^\beta e^{-\frac{1}{2}\left(\frac{x-100}{16}\right)^2}\,dx$$

   Use the first three terms of an appropriate Maclaurin series to estimate the proportion of the population that has IQs between 100 and 110.

[c] 24. (a) In 1706 the British astronomer and mathematician John Machin discovered the following formula for $\pi/4$, called **Machin's formula**:

$$\frac{\pi}{4} = 4\tan^{-1}\frac{1}{5} - \tan^{-1}\frac{1}{239}$$

   Use a CAS to approximate $\pi/4$ using Machin's formula to 25 decimal places.

   (b) In 1914 the brilliant Indian mathematician Srinivasa Ramanujan (1887–1920) showed that

$$\frac{1}{\pi} = \frac{\sqrt{8}}{9801}\sum_{k=0}^\infty \frac{(4k)!(1103 + 26{,}390k)}{(k!)^4 396^{4k}}$$

   Use a CAS to compute the first four partial sums in **Ramanujan's formula**.

✔ **QUICK CHECK ANSWERS 9.9**

1. $(-1)^k\dfrac{x^{2k}}{(2k)!}$ 2. $\dfrac{x^k}{k!}$ 3. $(-1)^{k+1}\dfrac{x^k}{k}$; $(-1, 1]$ 4. $\dfrac{m(m-1)\cdots(m-k+1)}{k!}x^k$; 1

9.10   **DIFFERENTIATING AND INTEGRATING POWER SERIES; MODELING WITH TAYLOR SERIES**

In this section we will discuss methods for finding power series for derivatives and integrals of functions, and we will discuss some practical methods for finding Taylor series that can be used in situations where it is difficult or impossible to find the series directly.

■ **DIFFERENTIATING POWER SERIES**

We begin by considering the following problem.

9.10.1   PROBLEM   Suppose that a function f is represented by a power series on an open interval. How can we use the power series to find the derivative of f on that interval?

The solution to this problem can be motivated by considering the Maclaurin series for $\sin x$:

$$\sin x = x - \frac{x^3}{3!} + \frac{x^5}{5!} - \frac{x^7}{7!} + \cdots \qquad (-\infty < x < +\infty)$$

Of course, we already know that the derivative of $\sin x$ is $\cos x$; however, we are concerned here with using the Maclaurin series to deduce this. The solution is easy—all we need to do is differentiate the Maclaurin series term by term and observe that the resulting series is the Maclaurin series for $\cos x$:

$$\frac{d}{dx}\left[x - \frac{x^3}{3!} + \frac{x^5}{5!} - \frac{x^7}{7!} + \cdots \right] = 1 - 3\frac{x^2}{3!} + 5\frac{x^4}{5!} - 7\frac{x^6}{7!} + \cdots$$

$$= 1 - \frac{x^2}{2!} + \frac{x^4}{4!} - \frac{x^6}{6!} + \cdots = \cos x$$

Here is another example.

$$\frac{d}{dx}[e^x] = \frac{d}{dx}\left[1 + x + \frac{x^2}{2!} + \frac{x^3}{3!} + \frac{x^4}{4!} + \cdots \right]$$

$$= 1 + 2\frac{x}{2!} + 3\frac{x^2}{3!} + 4\frac{x^3}{4!} + \cdots = 1 + x + \frac{x^2}{2!} + \frac{x^3}{3!} + \cdots = e^x$$

The preceding computations suggest that if a function f is represented by a power series on an open interval, then a power series representation of f' on that interval can be obtained by differentiating the power series for f term by term. This is stated more precisely in the following theorem, which we give without proof.

9.10.2   THEOREM (*Differentiation of Power Series*)   *Suppose that a function f is represented by a power series in $x - x_0$ that has a nonzero radius of convergence R; that is,*

$$f(x) = \sum_{k=0}^{\infty} c_k (x - x_0)^k \qquad (x_0 - R < x < x_0 + R)$$

Then:

(a)   *The function f is differentiable on the interval $(x_0 - R, x_0 + R)$.*

(b)   *If the power series representation for f is differentiated term by term, then the resulting series has radius of convergence R and converges to f' on the interval $(x_0 - R, x_0 + R)$; that is,*

$$f'(x) = \sum_{k=0}^{\infty} \frac{d}{dx}[c_k (x - x_0)^k] \qquad (x_0 - R < x < x_0 + R)$$

This theorem has an important implication about the differentiability of functions that are represented by power series. According to the theorem, the power series for f' has the same radius of convergence as the power series for f, and this means that the theorem can be applied to f' as well as f. However, if we do this, then we conclude that f' is differentiable on the interval $(x_0 - R, x_0 + R)$, and the power series for f'' has the same radius of convergence as the power series for f and f'. We can now repeat this process ad infinitum, applying the theorem successively to f'', f''', ..., $f^{(n)}$, ... to conclude that f has derivatives of all orders on the interval $(x_0 - R, x_0 + R)$. Thus, we have established the following result.

9.10.3 **THEOREM** *If a function f can be represented by a power series in $x - x_0$ with a nonzero radius of convergence R, then f has derivatives of all orders on the interval $(x_0 - R, x_0 + R)$.*

In short, it is only the most "well-behaved" functions that can be represented by power series; that is, if a function f does not possess derivatives of all orders on an interval $(x_0 - R, x_0 + R)$, then it cannot be represented by a power series in $x - x_0$ on that interval.

▶ **Example 1** In Section 9.8, we showed that the Bessel function $J_0(x)$, represented by the power series

$$J_0(x) = \sum_{k=0}^{\infty} \frac{(-1)^k x^{2k}}{2^{2k}(k!)^2} \tag{1}$$

has radius of convergence $+\infty$ [see Formula (7) of that section and the related discussion]. Thus, $J_0(x)$ has derivatives of all orders on the interval $(-\infty, +\infty)$, and these can be obtained by differentiating the series term by term. For example, if we write (1) as

$$J_0(x) = 1 + \sum_{k=1}^{\infty} \frac{(-1)^k x^{2k}}{2^{2k}(k!)^2}$$

and differentiate term by term, we obtain

$$J_0'(x) = \sum_{k=1}^{\infty} \frac{(-1)^k (2k) x^{2k-1}}{2^{2k}(k!)^2} = \sum_{k=1}^{\infty} \frac{(-1)^k x^{2k-1}}{2^{2k-1} k!(k-1)!} \; ◀$$

See Exercise 45 for a relationship between $J_0'(x)$ and $J_1(x)$.

REMARK | The computations in this example use some techniques that are worth noting. First, when a power series is expressed in sigma notation, the formula for the general term of the series will often not be of a form that can be used for differentiating the constant term. Thus, if the series has a nonzero constant term, as here, it is usually a good idea to split it off from the summation before differentiating. Second, observe how we simplified the final formula by canceling the factor k from one of the factorials in the denominator. This is a standard simplification technique.

■ **INTEGRATING POWER SERIES**

Since the derivative of a function that is represented by a power series can be obtained by differentiating the series term by term, it should not be surprising that an antiderivative of a function represented by a power series can be obtained by integrating the series term by term. For example, we know that $\sin x$ is an antiderivative of $\cos x$. Here is how this result can be obtained by integrating the Maclaurin series for $\cos x$ term by term:

$$\int \cos x\, dx = \int \left[1 - \frac{x^2}{2!} + \frac{x^4}{4!} - \frac{x^6}{6!} + \cdots\right] dx$$

$$= \left[x - \frac{x^3}{3(2!)} + \frac{x^5}{5(4!)} - \frac{x^7}{7(6!)} + \cdots\right] + C$$

$$= \left[x - \frac{x^3}{3!} + \frac{x^5}{5!} - \frac{x^7}{7!} + \cdots\right] + C = \sin x + C$$

The same idea applies to definite integrals. For example, by direct integration we have

$$\int_0^1 \frac{dx}{1 + x^2} = \tan^{-1} x \bigg]_0^1 = \tan^{-1} 1 - \tan 0 = \frac{\pi}{4} - 0 = \frac{\pi}{4}$$

and we will show later in this section that

$$\frac{\pi}{4} = 1 - \frac{1}{3} + \frac{1}{5} - \frac{1}{7} + \cdots \qquad (2)$$

Thus,

$$\int_0^1 \frac{dx}{1 + x^2} = 1 - \frac{1}{3} + \frac{1}{5} - \frac{1}{7} + \cdots$$

Here is how this result can be obtained by integrating the Maclaurin series for $1/(1 + x^2)$ term by term (see Table 9.9.1):

$$\int_0^1 \frac{dx}{1 + x^2} = \int_0^1 [1 - x^2 + x^4 - x^6 + \cdots]\, dx$$

$$= x - \frac{x^3}{3} + \frac{x^5}{5} - \frac{x^7}{7} + \cdots \bigg]_0^1 = 1 - \frac{1}{3} + \frac{1}{5} - \frac{1}{7} + \cdots$$

The preceding computations are justified by the following theorem, which we give without proof.

Theorems 9.10.2 and 9.10.4 tell us how to use a power series representation of a function f to produce power series representations of $f'(x)$ and $\int f(x)\, dx$ that have the same radius of convergence as f. However, the *intervals* of convergence for these series may not be the same because their convergence behavior may differ at the endpoints of the interval. (See Exercises 25 and 26.)

9.10.4   THEOREM (*Integration of Power Series*)   *Suppose that a function f is represented by a power series in $x - x_0$ that has a nonzero radius of convergence R; that is,*

$$f(x) = \sum_{k=0}^{\infty} c_k (x - x_0)^k \qquad (x_0 - R < x < x_0 + R)$$

(a)   *If the power series representation of f is integrated term by term, then the resulting series has radius of convergence R and converges to an antiderivative for $f(x)$ on the interval $(x_0 - R, x_0 + R)$; that is,*

$$\int f(x)\, dx = \sum_{k=0}^{\infty} \left[\frac{c_k}{k + 1}(x - x_0)^{k+1}\right] + C \qquad (x_0 - R < x < x_0 + R)$$

(b)   *If α and β are points in the interval $(x_0 - R, x_0 + R)$, and if the power series representation of f is integrated term by term from α to β, then the resulting series converges absolutely on the interval $(x_0 - R, x_0 + R)$ and*

$$\int_{\alpha}^{\beta} f(x)\, dx = \sum_{k=0}^{\infty} \left[\int_{\alpha}^{\beta} c_k (x - x_0)^k\, dx\right]$$

■ POWER SERIES REPRESENTATIONS MUST BE TAYLOR SERIES

For many functions it is difficult or impossible to find the derivatives that are required to obtain a Taylor series. For example, to find the Maclaurin series for $1/(1+x^2)$ directly would require some tedious derivative computations (try it). A more practical approach is to substitute $-x^2$ for x in the geometric series

$$\frac{1}{1-x} = 1 + x + x^2 + x^3 + x^4 + \cdots \qquad (-1 < x < 1)$$

to obtain

$$\frac{1}{1+x^2} = 1 - x^2 + x^4 - x^6 + x^8 - \cdots$$

However, there are two questions of concern with this procedure:

- Where does the power series that we obtained for $1/(1+x^2)$ actually converge to $1/(1+x^2)$?

- How do we know that the power series we have obtained is actually the Maclaurin series for $1/(1+x^2)$?

The first question is easy to resolve. Since the geometric series converges to $1/(1-x)$ if $|x| < 1$, the second series will converge to $1/(1+x^2)$ if $|-x^2| < 1$ or $|x^2| < 1$. However, this is true if and only if $|x| < 1$, so the power series we obtained for the function $1/(1+x^2)$ converges to this function if $-1 < x < 1$.

The second question is more difficult to answer and leads us to the following general problem.

> **9.10.5 PROBLEM** Suppose that a function f is represented by a power series in $x - x_0$ that has a nonzero radius of convergence. What relationship exists between the given power series and the Taylor series for f about $x = x_0$?

The answer is that they are the same; and here is the theorem that proves it.

Theorem 9.10.6 tells us that no matter how we arrive at a power series representation of a function f, be it by substitution, by differentiation, by integration, or by some algebraic process, that series will be the Taylor series for f about $x = x_0$, provided the series converges to f on some open interval containing x_0.

> **9.10.6 THEOREM** *If a function f is represented by a power series in $x - x_0$ on some open interval containing x_0, then that power series is the Taylor series for f about $x = x_0$.*

PROOF Suppose that

$$f(x) = c_0 + c_1(x - x_0) + c_2(x - x_0)^2 + \cdots + c_k(x - x_0)^k + \cdots$$

for all x in some open interval containing x_0. To prove that this is the Taylor series for f about $x = x_0$, we must show that

$$c_k = \frac{f^{(k)}(x_0)}{k!} \quad \text{for} \quad k = 0, 1, 2, 3, \ldots$$

However, the assumption that the series converges to $f(x)$ on an open interval containing x_0 ensures that it has a nonzero radius of convergence R; hence we can differentiate term by term in accordance with Theorem 9.10.2. Thus,

$$f(x) = c_0 + c_1(x - x_0) + c_2(x - x_0)^2 + c_3(x - x_0)^3 + c_4(x - x_0)^4 + \cdots$$

$$f'(x) = c_1 + 2c_2(x - x_0) + 3c_3(x - x_0)^2 + 4c_4(x - x_0)^3 + \cdots$$

$$f''(x) = 2!c_2 + (3 \cdot 2)c_3(x - x_0) + (4 \cdot 3)c_4(x - x_0)^2 + \cdots$$

$$f'''(x) = 3!c_3 + (4 \cdot 3 \cdot 2)c_4(x - x_0) + \cdots$$

$$\vdots$$

On substituting $x = x_0$, all the powers of $x - x_0$ drop out, leaving

$$f(x_0) = c_0, \quad f'(x_0) = c_1, \quad f''(x_0) = 2!c_2, \quad f'''(x_0) = 3!c_3, \ldots$$

from which we obtain

$$c_0 = f(x_0), \quad c_1 = f'(x_0), \quad c_2 = \frac{f''(x_0)}{2!}, \quad c_3 = \frac{f'''(x_0)}{3!}, \ldots$$

which shows that the coefficients $c_0, c_1, c_2, c_3, \ldots$ are precisely the coefficients in the Taylor series about x_0 for $f(x)$. ■

SOME PRACTICAL WAYS TO FIND TAYLOR SERIES

▶ **Example 2** Find Taylor series for the given functions about the given x_0.

$$\text{(a) } e^{-x^2}, \quad x_0 = 0 \qquad \text{(b) } \ln x, \quad x_0 = 1 \qquad \text{(c) } \frac{1}{x}, \quad x_0 = 1$$

Solution (a). The simplest way to find the Maclaurin series for e^{-x^2} is to substitute $-x^2$ for x in the Maclaurin series

$$e^x = 1 + x + \frac{x^2}{2!} + \frac{x^3}{3!} + \frac{x^4}{4!} + \cdots \tag{3}$$

to obtain

$$e^{-x^2} = 1 - x^2 + \frac{x^4}{2!} - \frac{x^6}{3!} + \frac{x^8}{4!} - \cdots$$

Since (3) converges for all values of x, so will the series for e^{-x^2}.

Solution (b). We begin with the Maclaurin series for $\ln(1 + x)$, which can be found in Table 9.9.1:

$$\ln(1 + x) = x - \frac{x^2}{2} + \frac{x^3}{3} - \frac{x^4}{4} + \cdots \qquad (-1 < x \le 1)$$

Substituting $x - 1$ for x in this series gives

$$\ln(1 + [x - 1]) = \ln x = (x - 1) - \frac{(x - 1)^2}{2} + \frac{(x - 1)^3}{3} - \frac{(x - 1)^4}{4} + \cdots \tag{4}$$

Since the original series converges when $-1 < x \le 1$, the interval of convergence for (4) will be $-1 < x - 1 \le 1$ or, equivalently, $0 < x \le 2$.

Solution (c). Since $1/x$ is the derivative of $\ln x$, we can differentiate the series for $\ln x$ found in (b) to obtain

$$\frac{1}{x} = 1 - \frac{2(x - 1)}{2} + \frac{3(x - 1)^2}{3} - \frac{4(x - 1)^3}{4} + \cdots$$

$$= 1 - (x - 1) + (x - 1)^2 - (x - 1)^3 + \cdots \tag{5}$$

By Theorem 9.10.2, we know that the radius of convergence for (5) is the same as that for (4), which is $R = 1$. Thus the interval of convergence for (5) must be at least $0 < x < 2$. Since the behaviors of (4) and (5) may differ at the endpoints $x = 0$ and $x = 2$, those must be checked separately. When $x = 0$, (5) becomes

$$1 - (-1) + (-1)^2 - (-1)^3 + \cdots = 1 + 1 + 1 + 1 + \cdots$$

which diverges by the divergence test. Similarly, when $x = 2$, (5) becomes

$$1 - 1 + 1^2 - 1^3 + \cdots = 1 - 1 + 1 - 1 + \cdots$$

which also diverges by the divergence test. Thus the interval of convergence for (5) is $0 < x < 2$. ◄

▶ **Example 3** Find the Maclaurin series for $\tan^{-1} x$.

Solution. It would be tedious to find the Maclaurin series directly. A better approach is to start with the formula

$$\int \frac{1}{1 + x^2}\, dx = \tan^{-1} x + C$$

and integrate the Maclaurin series

$$\frac{1}{1 + x^2} = 1 - x^2 + x^4 - x^6 + x^8 - \cdots \qquad (-1 < x < 1)$$

term by term. This yields

$$\tan^{-1} x + C = \int \frac{1}{1 + x^2}\, dx = \int [1 - x^2 + x^4 - x^6 + x^8 - \cdots]\, dx$$

or

$$\tan^{-1} x = \left[x - \frac{x^3}{3} + \frac{x^5}{5} - \frac{x^7}{7} + \frac{x^9}{9} - \cdots \right] - C$$

The constant of integration can be evaluated by substituting $x = 0$ and using the condition $\tan^{-1} 0 = 0$. This gives $C = 0$, so that

$$\tan^{-1} x = x - \frac{x^3}{3} + \frac{x^5}{5} - \frac{x^7}{7} + \frac{x^9}{9} - \cdots \qquad (-1 < x < 1) \qquad (6)$$

◄

REMARK Observe that neither Theorem 9.10.2 nor Theorem 9.10.3 addresses what happens at the endpoints of the interval of convergence. However, it can be proved that if the Taylor series for f about $x = x_0$ converges to $f(x)$ for all x in the interval $(x_0 - R, x_0 + R)$, and if the Taylor series converges at the right endpoint $x_0 + R$, then the value that it converges to at that point is the limit of $f(x)$ as $x \to x_0 + R$ from the left; and if the Taylor series converges at the left endpoint $x_0 - R$, then the value that it converges to at that point is the limit of $f(x)$ as $x \to x_0 - R$ from the right.

For example, the Maclaurin series for $\tan^{-1} x$ given in (6) converges at both $x = -1$ and $x = 1$, since the hypotheses of the alternating series test (Theorem 9.6.1) are satisfied at those points. Thus, the continuity of $\tan^{-1} x$ on the interval $[-1, 1]$ implies that at $x = 1$ the Maclaurin series converges to

$$\lim_{x \to 1^-} \tan^{-1} x = \tan^{-1} 1 = \frac{\pi}{4}$$

and at $x = -1$ it converges to

$$\lim_{x \to -1^+} \tan^{-1} x = \tan^{-1}(-1) = -\frac{\pi}{4}$$

This shows that the Maclaurin series for $\tan^{-1} x$ actually converges to $\tan^{-1} x$ on the closed interval $-1 \leq x \leq 1$. Moreover, the convergence at $x = 1$ establishes Formula (2).

■ **APPROXIMATING DEFINITE INTEGRALS USING TAYLOR SERIES**

Taylor series provide an alternative to Simpson's rule and other numerical methods for approximating definite integrals.

▶ **Example 4**   Approximate the integral

$$\int_0^1 e^{-x^2}\, dx$$

to three decimal-place accuracy by expanding the integrand in a Maclaurin series and integrating term by term.

Solution.   We found in Example 2(a) that the Maclaurin series for e^{-x^2} is

$$e^{-x^2} = 1 - x^2 + \frac{x^4}{2!} - \frac{x^6}{3!} + \frac{x^8}{4!} - \cdots$$

Therefore,

$$\int_0^1 e^{-x^2}\, dx = \int_0^1 \left[1 - x^2 + \frac{x^4}{2!} - \frac{x^6}{3!} + \frac{x^8}{4!} - \cdots\right] dx$$

$$= \left[x - \frac{x^3}{3} + \frac{x^5}{5(2!)} - \frac{x^7}{7(3!)} + \frac{x^9}{9(4!)} - \cdots\right]_0^1$$

$$= 1 - \frac{1}{3} + \frac{1}{5\cdot 2!} - \frac{1}{7\cdot 3!} + \frac{1}{9\cdot 4!} - \cdots$$

$$= \sum_{k=0}^{\infty} \frac{(-1)^k}{(2k+1)k!}$$

Since this series clearly satisfies the hypotheses of the alternating series test (Theorem 9.6.1), it follows from Theorem 9.6.2 that if we approximate the integral by s_n (the nth partial sum of the series), then

$$\left|\int_0^1 e^{-x^2}\, dx - s_n\right| < \frac{1}{[2(n+1)+1](n+1)!} = \frac{1}{(2n+3)(n+1)!}$$

Thus, for three decimal-place accuracy we must choose n such that

$$\frac{1}{(2n+3)(n+1)!} \leq 0.0005 = 5 \times 10^{-4}$$

With the help of a calculating utility you can show that the smallest value of n that satisfies this condition is $n = 5$. Thus, the value of the integral to three decimal-place accuracy is

$$\int_0^1 e^{-x^2}\, dx \approx 1 - \frac{1}{3} + \frac{1}{5\cdot 2!} - \frac{1}{7\cdot 3!} + \frac{1}{9\cdot 4!} - \frac{1}{11\cdot 5!} \approx 0.747$$

What advantages does the method of Example 4 have over Simpson's rule? What are its disadvantages?

As a check, a calculator with a built-in numerical integration capability produced the approximation 0.746824, which agrees with our result when rounded to three decimal places. ◀

■ **FINDING TAYLOR SERIES BY MULTIPLICATION AND DIVISION**

The following examples illustrate some algebraic techniques that are sometimes useful for finding Taylor series.

$$1 - x^2 + \frac{x^4}{2} - \cdots$$
$$\times$$
$$x - \frac{x^3}{3} + \frac{x^5}{5} - \cdots$$
$$\overline{x - x^3 + \frac{x^5}{2} - \cdots}$$
$$-\frac{x^3}{3} + \frac{x^5}{3} - \frac{x^7}{6} + \cdots$$
$$\frac{x^5}{5} - \frac{x^7}{5} + \cdots$$
$$\overline{x - \frac{4}{3}x^3 + \frac{31}{30}x^5 - \cdots}$$

► **Example 5**   Find the first three nonzero terms in the Maclaurin series for the function $f(x) = e^{-x^2} \tan^{-1} x$.

Solution.   Using the series for e^{-x^2} and $\tan^{-1} x$ obtained in Examples 2 and 3 gives

$$e^{-x^2} \tan^{-1} x = \left(1 - x^2 + \frac{x^4}{2} - \cdots\right)\left(x - \frac{x^3}{3} + \frac{x^5}{5} - \cdots\right)$$

Multiplying, as shown in the margin, we obtain

$$e^{-x^2} \tan^{-1} x = x - \frac{4}{3}x^3 + \frac{31}{30}x^5 - \cdots$$

More terms in the series can be obtained by including more terms in the factors. Moreover, one can prove that a series obtained by this method converges at each point in the intersection of the intervals of convergence of the factors (and possibly on a larger interval). Thus, we can be certain that the series we have obtained converges for all x in the interval $-1 \leq x \leq 1$ (why?). ◄

$$x + \frac{x^3}{3} + \frac{2x^5}{15} + \cdots$$
$$1 - \frac{x^2}{2} + \frac{x^4}{24} - \cdots \overline{\left) x - \frac{x^3}{6} + \frac{x^5}{120} - \cdots\right.}$$
$$x - \frac{x^3}{2} + \frac{x^5}{24} - \cdots$$
$$\overline{\frac{x^3}{3} - \frac{x^5}{30} + \cdots}$$
$$\frac{x^3}{3} - \frac{x^5}{6} + \cdots$$
$$\overline{\frac{2x^5}{15} + \cdots}$$

► **Example 6**   Find the first three nonzero terms in the Maclaurin series for $\tan x$.

Solution.   Using the first three terms in the Maclaurin series for $\sin x$ and $\cos x$, we can express $\tan x$ as

$$\tan x = \frac{\sin x}{\cos x} = \frac{x - \dfrac{x^3}{3!} + \dfrac{x^5}{5!} - \cdots}{1 - \dfrac{x^2}{2!} + \dfrac{x^4}{4!} - \cdots}$$

Dividing, as shown in the margin, we obtain

$$\tan x = x + \frac{x^3}{3} + \frac{2x^5}{15} + \cdots \quad ◄$$

TECHNOLOGY MASTERY

If you have a CAS, use its capability for multiplying and dividing polynomials to perform the computations in Examples 5 and 6.

■ MODELING PHYSICAL LAWS WITH TAYLOR SERIES

Taylor series provide an important way of modeling physical laws. To illustrate the idea we will consider the problem of modeling the period of a simple pendulum (Figure 9.10.1). As explained in Chapter 7 Making Connections Exercise 5, the period T of such a pendulum is given by

$$T = 4\sqrt{\frac{L}{g}} \int_0^{\pi/2} \frac{1}{\sqrt{1 - k^2 \sin^2 \phi}} \, d\phi \tag{7}$$

where

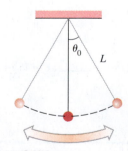

▲ **Figure 9.10.1**

L = length of the supporting rod

g = acceleration due to gravity

$k = \sin(\theta_0/2)$, where θ_0 is the initial angle of displacement from the vertical

The integral, which is called a ***complete elliptic integral of the first kind***, cannot be expressed in terms of elementary functions and is often approximated by numerical methods. Unfortunately, numerical values are so specific that they often give little insight into general physical principles. However, if we expand the integrand of (7) in a series and integrate term by term, then we can generate an infinite series that can be used to construct various mathematical models for the period T that give a deeper understanding of the behavior of the pendulum.

© ACE STOCK LIMITED/Alamy

Understanding the motion of a pendulum played a critical role in the advance of accurate time-keeping with the development of the pendulum clock in the 17th century.

To obtain a series for the integrand, we will substitute $-k^2 \sin^2 \phi$ for x in the binomial series for $1/\sqrt{1+x}$ that we derived in Example 4(b) of Section 9.9. If we do this, then we can rewrite (7) as

$$T = 4\sqrt{\frac{L}{g}} \int_0^{\pi/2} \left[1 + \frac{1}{2}k^2 \sin^2 \phi + \frac{1 \cdot 3}{2^2 2!}k^4 \sin^4 \phi + \frac{1 \cdot 3 \cdot 5}{2^3 3!}k^6 \sin^6 \phi + \cdots \right] d\phi \quad (8)$$

If we integrate term by term, then we can produce a series that converges to the period T. However, one of the most important cases of pendulum motion occurs when the initial displacement is small, in which case all subsequent displacements are small, and we can assume that $k = \sin(\theta_0/2) \approx 0$. In this case we expect the convergence of the series for T to be rapid, and we can approximate the sum of the series by dropping all but the constant term in (8). This yields

$$T = 2\pi\sqrt{\frac{L}{g}} \quad (9)$$

which is called the *first-order model* of T or the model for *small vibrations*. This model can be improved on by using more terms in the series. For example, if we use the first two terms in the series, we obtain the *second-order model*

$$T = 2\pi\sqrt{\frac{L}{g}} \left(1 + \frac{k^2}{4} \right) \quad (10)$$

(verify).

✔ **QUICK CHECK EXERCISES 9.10**   (*See page 689 for answers.*)

1. The Maclaurin series for e^{-x^2} obtained by substituting $-x^2$ for x in the series

$$e^x = \sum_{k=0}^{\infty} \frac{x^k}{k!}$$

is $e^{-x^2} = \sum_{k=0}^{\infty} \underline{\qquad}$.

2. $\dfrac{d}{dx}\left[\sum_{k=1}^{\infty} (-1)^{k+1} \dfrac{x^k}{k} \right] = \underline{\qquad} + \underline{\qquad} x$

$+ \underline{\qquad} x^2 + \underline{\qquad} x^3 + \cdots$

$= \sum_{k=0}^{\infty} \underline{\qquad}$

3. $\left(\sum_{k=0}^{\infty} \dfrac{x^k}{k!} \right)\left(\sum_{k=0}^{\infty} \dfrac{x^k}{k+1} \right)$

$= \left(1 + x + \dfrac{x^2}{2!} + \cdots \right)\left(1 + \dfrac{x}{2} + \dfrac{x^2}{3} + \cdots \right)$

$= \underline{\qquad} + \underline{\qquad} x + \underline{\qquad} x^2 + \cdots$

4. Suppose that $f(1) = 4$ and $f'(x) = \displaystyle\sum_{k=0}^{\infty} \dfrac{(-1)^k}{(k+1)!}(x-1)^k$

(a) $f''(1) = \underline{\qquad}$

(b) $f(x) = \underline{\qquad} + \underline{\qquad}(x-1)$

$+ \underline{\qquad}(x-1)^2 + \underline{\qquad}(x-1)^3 + \cdots$

$= \underline{\qquad} + \displaystyle\sum_{k=1}^{\infty} \underline{\qquad}$

EXERCISE SET 9.10   [c] CAS

1. In each part, obtain the Maclaurin series for the function by making an appropriate substitution in the Maclaurin series for $1/(1-x)$. Include the general term in your answer, and state the radius of convergence of the series.

(a) $\dfrac{1}{1+x}$ (b) $\dfrac{1}{1-x^2}$ (c) $\dfrac{1}{1-2x}$ (d) $\dfrac{1}{2-x}$

2. In each part, obtain the Maclaurin series for the function by making an appropriate substitution in the Maclaurin series for $\ln(1+x)$. Include the general term in your answer, and

state the radius of convergence of the series.

(a) $\ln(1-x)$ (b) $\ln(1+x^2)$

(c) $\ln(1+2x)$ (d) $\ln(2+x)$

3. In each part, obtain the first four nonzero terms of the Maclaurin series for the function by making an appropriate substitution in one of the binomial series obtained in Example 4 of Section 9.9.

(a) $(2+x)^{-1/2}$ (b) $(1-x^2)^{-2}$

4. (a) Use the Maclaurin series for $1/(1 - x)$ to find the Maclaurin series for $1/(a - x)$, where $a \neq 0$, and state the radius of convergence of the series.

(b) Use the binomial series for $1/(1 + x)^2$ obtained in Example 4 of Section 9.9 to find the first four nonzero terms in the Maclaurin series for $1/(a + x)^2$, where $a \neq 0$, and state the radius of convergence of the series.

5–8 Find the first four nonzero terms of the Maclaurin series for the function by making an appropriate substitution in a known Maclaurin series and performing any algebraic operations that are required. State the radius of convergence of the series. ■

5. (a) $\sin 2x$ (b) e^{-2x} (c) e^{x^2} (d) $x^2 \cos \pi x$

6. (a) $\cos 2x$ (b) $x^2 e^x$ (c) $x e^{-x}$ (d) $\sin(x^2)$

7. (a) $\dfrac{x^2}{1 + 3x}$ (b) $x \sinh 2x$ (c) $x(1 - x^2)^{3/2}$

8. (a) $\dfrac{x}{x - 1}$ (b) $3 \cosh(x^2)$ (c) $\dfrac{x}{(1 + 2x)^3}$

9–10 Find the first four nonzero terms of the Maclaurin series for the function by using an appropriate trigonometric identity or property of logarithms and then substituting in a known Maclaurin series. ■

9. (a) $\sin^2 x$ (b) $\ln[(1 + x^3)^{12}]$

10. (a) $\cos^2 x$ (b) $\ln\left(\dfrac{1 - x}{1 + x}\right)$

11. (a) Use a known Maclaurin series to find the Taylor series of $1/x$ about $x = 1$ by expressing this function as

$$\frac{1}{x} = \frac{1}{1 - (1 - x)}$$

(b) Find the interval of convergence of the Taylor series.

12. Use the method of Exercise 11 to find the Taylor series of $1/x$ about $x = x_0$, and state the interval of convergence of the Taylor series.

13–14 Find the first four nonzero terms of the Maclaurin series for the function by multiplying the Maclaurin series of the factors. ■

13. (a) $e^x \sin x$ (b) $\sqrt{1 + x} \ln(1 + x)$

14. (a) $e^{-x^2} \cos x$ (b) $(1 + x^2)^{4/3}(1 + x)^{1/3}$

15–16 Find the first four nonzero terms of the Maclaurin series for the function by dividing appropriate Maclaurin series. ■

15. (a) $\sec x \;\left(= \dfrac{1}{\cos x}\right)$ (b) $\dfrac{\sin x}{e^x}$

16. (a) $\dfrac{\tan^{-1} x}{1 + x}$ (b) $\dfrac{\ln(1 + x)}{1 - x}$

17. Use the Maclaurin series for e^x and e^{-x} to derive the Maclaurin series for $\sinh x$ and $\cosh x$. Include the general terms in your answers and state the radius of convergence of each series.

18. Use the Maclaurin series for $\sinh x$ and $\cosh x$ to obtain the first four nonzero terms in the Maclaurin series for $\tanh x$.

19–20 Find the first five nonzero terms of the Maclaurin series for the function by using partial fractions and a known Maclaurin series. ■

19. $\dfrac{4x - 2}{x^2 - 1}$ **20.** $\dfrac{x^3 + x^2 + 2x - 2}{x^2 - 1}$

21–22 Confirm the derivative formula by differentiating the appropriate Maclaurin series term by term. ■

21. (a) $\dfrac{d}{dx}[\cos x] = -\sin x$ (b) $\dfrac{d}{dx}[\ln(1 + x)] = \dfrac{1}{1 + x}$

22. (a) $\dfrac{d}{dx}[\sinh x] = \cosh x$ (b) $\dfrac{d}{dx}[\tan^{-1} x] = \dfrac{1}{1 + x^2}$

23–24 Confirm the integration formula by integrating the appropriate Maclaurin series term by term. ■

23. (a) $\displaystyle\int e^x \, dx = e^x + C$

(b) $\displaystyle\int \sinh x \, dx = \cosh x + C$

24. (a) $\displaystyle\int \sin x \, dx = -\cos x + C$

(b) $\displaystyle\int \dfrac{1}{1 + x} \, dx = \ln(1 + x) + C$

25. Consider the series

$$\sum_{k=0}^{\infty} \frac{x^{k+1}}{(k + 1)(k + 2)}$$

Determine the intervals of convergence for this series and for the series obtained by differentiating this series term by term.

26. Consider the series

$$\sum_{k=1}^{\infty} \frac{(-3)^k}{k} x^k$$

Determine the intervals of convergence for this series and for the series obtained by integrating this series term by term.

27. (a) Use the Maclaurin series for $1/(1 - x)$ to find the Maclaurin series for

$$f(x) = \frac{x}{1 - x^2}$$

(b) Use the Maclaurin series obtained in part (a) to find $f^{(5)}(0)$ and $f^{(6)}(0)$.

(c) What can you say about the value of $f^{(n)}(0)$?

28. Let $f(x) = x^2 \cos 2x$. Use the method of Exercise 27 to find $f^{(99)}(0)$.

29–30 The limit of an indeterminate form as $x \to x_0$ can sometimes be found by expanding the functions involved in Taylor series about $x = x_0$ and taking the limit of the series term by term. Use this method to find the limits in these exercises. ■

29. (a) $\displaystyle\lim_{x \to 0} \frac{\sin x}{x}$ (b) $\displaystyle\lim_{x \to 0} \frac{\tan^{-1} x - x}{x^3}$

30. (a) $\displaystyle\lim_{x \to 0} \frac{1 - \cos x}{\sin x}$ (b) $\displaystyle\lim_{x \to 0} \frac{\ln \sqrt{1 + x} - \sin 2x}{x}$

31–34 Use Maclaurin series to approximate the integral to three decimal-place accuracy. ◼

31. $\displaystyle\int_0^1 \sin(x^2)\,dx$

32. $\displaystyle\int_0^{1/2} \tan^{-1}(2x^2)\,dx$

33. $\displaystyle\int_0^{0.2} \sqrt[3]{1+x^4}\,dx$

34. $\displaystyle\int_0^{1/2} \frac{dx}{\sqrt[4]{x^2+1}}$

FOCUS ON CONCEPTS

35. (a) Find the Maclaurin series for e^{x^4}. What is the radius of convergence?

(b) Explain two different ways to use the Maclaurin series for e^{x^4} to find a series for $x^3 e^{x^4}$. Confirm that both methods produce the same series.

36. (a) Differentiate the Maclaurin series for $1/(1-x)$, and use the result to show that

$$\sum_{k=1}^{\infty} kx^k = \frac{x}{(1-x)^2} \quad \text{for } -1 < x < 1$$

(b) Integrate the Maclaurin series for $1/(1-x)$, and use the result to show that

$$\sum_{k=1}^{\infty} \frac{x^k}{k} = -\ln(1-x) \quad \text{for } -1 < x < 1$$

(c) Use the result in part (b) to show that

$$\sum_{k=1}^{\infty} (-1)^{k+1}\frac{x^k}{k} = \ln(1+x) \quad \text{for } -1 < x < 1$$

(d) Show that the series in part (c) converges if $x = 1$.

(e) Use the remark following Example 3 to show that

$$\sum_{k=1}^{\infty} (-1)^{k+1}\frac{x^k}{k} = \ln(1+x) \quad \text{for } -1 < x \leq 1$$

---

37. Use the results in Exercise 36 to find the sum of the series.

(a) $\displaystyle\sum_{k=1}^{\infty} \frac{k}{3^k} = \frac{1}{3} + \frac{2}{3^2} + \frac{3}{3^3} + \frac{4}{3^4} + \cdots$

(b) $\displaystyle\sum_{k=1}^{\infty} \frac{1}{k(4^k)} = \frac{1}{4} + \frac{1}{2(4^2)} + \frac{1}{3(4^3)} + \frac{1}{4(4^4)} + \cdots$

38. Use the results in Exercise 36 to find the sum of each series.

(a) $\displaystyle\sum_{k=1}^{\infty} (-1)^{k+1}\frac{1}{k} = 1 - \frac{1}{2} + \frac{1}{3} - \frac{1}{4} + \cdots$

(b) $\displaystyle\sum_{k=1}^{\infty} \frac{(e-1)^k}{ke^k} = \frac{e-1}{e} + \frac{(e-1)^2}{2(e^2)} - \frac{(e-1)^3}{3(e^3)} + \cdots$

39. (a) Use the relationship

$$\int \frac{1}{\sqrt{1+x^2}}\,dx = \sinh^{-1}x + C$$

to find the first four nonzero terms in the Maclaurin series for $\sinh^{-1}x$.

(b) Express the series in sigma notation.

(c) What is the radius of convergence?

---

40. (a) Use the relationship

$$\int \frac{1}{\sqrt{1-x^2}}\,dx = \sin^{-1}x + C$$

to find the first four nonzero terms in the Maclaurin series for $\sin^{-1}x$.

(b) Express the series in sigma notation.

(c) What is the radius of convergence?

41. We showed by Formula (19) of Section 8.2 that if there are y_0 units of radioactive carbon-14 present at time $t = 0$, then the number of units present t years later is

$$y(t) = y_0 e^{-0.000121t}$$

(a) Express $y(t)$ as a Maclaurin series.

(b) Use the first two terms in the series to show that the number of units present after 1 year is approximately $(0.999879)y_0$.

(c) Compare this to the value produced by the formula for $y(t)$.

C **42.** Suppose that a simple pendulum with a length of $L = 1$ meter is given an initial displacement of $\theta_0 = 5°$ from the vertical.

(a) Approximate the period T of the pendulum using Formula (9) for the first-order model of T. [*Note:* Take $g = 9.8 \text{ m/s}^2$.]

(b) Approximate the period of the pendulum using Formula (10) for the second-order model.

(c) Use the numerical integration capability of a CAS to approximate the period of the pendulum from Formula (7), and compare it to the values obtained in parts (a) and (b).

43. Use the first three nonzero terms in Formula (8) and the Wallis sine formula in the Endpaper Integral Table (Formula 122) to obtain a model for the period of a simple pendulum.

44. Recall that the gravitational force exerted by the Earth on an object is called the object's *weight* (or more precisely, its *Earth weight*). If an object of mass m is on the surface of the Earth (mean sea level), then the magnitude of its weight is mg, where g is the acceleration due to gravity at the Earth's surface. A more general formula for the magnitude of the gravitational force that the Earth exerts on an object of mass m is

$$F = \frac{mgR^2}{(R+h)^2}$$

where R is the radius of the Earth and h is the height of the object above the Earth's surface.

(a) Use the binomial series for $1/(1+x)^2$ obtained in Example 4 of Section 9.9 to express F as a Maclaurin series in powers of h/R.

(b) Show that if $h = 0$, then $F = mg$.

(c) Show that if $h/R \approx 0$, then $F \approx mg - (2mgh/R)$. [*Note:* The quantity $2mgh/R$ can be thought of as a "correction term" for the weight that takes the object's height above the Earth's surface into account.]

(d) If we assume that the Earth is a sphere of radius $R = 4000$ mi at mean sea level, by approximately what percentage does a person's weight change in going from mean sea level to the top of Mt. Everest (29,028 ft)?

45. (a) Show that the Bessel function $J_0(x)$ given by Formula (4) of Section 9.8 satisfies the differential equation $xy'' + y' + xy = 0$. (This is called the **Bessel equation of order zero**.)

(b) Show that the Bessel function $J_1(x)$ given by Formula (5) of Section 9.8 satisfies the differential equation $x^2 y'' + xy' + (x^2 - 1)y = 0$. (This is called the **Bessel equation of order one**.)

(c) Show that $J_0'(x) = -J_1(x)$.

46. Prove: If the power series $\sum_{k=0}^{\infty} a_k x^k$ and $\sum_{k=0}^{\infty} b_k x^k$ have the same sum on an interval $(-r, r)$, then $a_k = b_k$ for all values of k.

47. **Writing** Evaluate the limit

$$\lim_{x \to 0} \frac{x - \sin x}{x^3}$$

in two ways: using L'Hôpital's rule and by replacing $\sin x$ by its Maclaurin series. Discuss how the use of a series can give qualitative information about how the value of an indeterminate limit is approached.

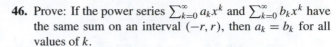

QUICK CHECK ANSWERS 9.10

1. $(-1)^k \dfrac{x^{2k}}{k!}$ **2.** $1; -1; 1; -1; (-1)^k x^k$ **3.** $1; \dfrac{3}{2}; \dfrac{4}{3}$ **4.** (a) $-\dfrac{1}{2}$ (b) $4; 1; -\dfrac{1}{4}; \dfrac{1}{18}; 4; (-1)^{k+1} \dfrac{(x-1)^k}{k \cdot (k!)}$

CHAPTER 9 REVIEW EXERCISES

1. What is the difference between an infinite sequence and an infinite series?

2. What is meant by the sum of an infinite series?

3. (a) What is a geometric series? Give some examples of convergent and divergent geometric series.

(b) What is a p-series? Give some examples of convergent and divergent p-series.

4. State conditions under which an alternating series is guaranteed to converge.

5. (a) What does it mean to say that an infinite series converges absolutely?

(b) What relationship exists between convergence and absolute convergence of an infinite series?

6. State the Remainder Estimation Theorem, and describe some of its uses.

7. If a power series in $x - x_0$ has radius of convergence R, what can you say about the set of x-values at which the series converges?

8. (a) Write down the formula for the Maclaurin series for f in sigma notation.

(b) Write down the formula for the Taylor series for f about $x = x_0$ in sigma notation.

9. Are the following statements true or false? If true, state a theorem to justify your conclusion; if false, then give a counterexample.

(a) If $\sum u_k$ converges, then $u_k \to 0$ as $k \to +\infty$.

(b) If $u_k \to 0$ as $k \to +\infty$, then $\sum u_k$ converges.

(c) If $f(n) = a_n$ for $n = 1, 2, 3, \ldots$, and if $a_n \to L$ as $n \to +\infty$, then $f(x) \to L$ as $x \to +\infty$.

(d) If $f(n) = a_n$ for $n = 1, 2, 3, \ldots$, and if $f(x) \to L$ as $x \to +\infty$, then $a_n \to L$ as $n \to +\infty$.

(e) If $0 < a_n < 1$, then $\{a_n\}$ converges.

(f) If $0 < u_k < 1$, then $\sum u_k$ converges.

(g) If $\sum u_k$ and $\sum v_k$ converge, then $\sum (u_k + v_k)$ diverges.

(h) If $\sum u_k$ and $\sum v_k$ diverge, then $\sum (u_k - v_k)$ converges.

(i) If $0 \le u_k \le v_k$ and $\sum v_k$ converges, then $\sum u_k$ converges.

(j) If $0 \le u_k \le v_k$ and $\sum u_k$ diverges, then $\sum v_k$ diverges.

(k) If an infinite series converges, then it converges absolutely.

(l) If an infinite series diverges absolutely, then it diverges.

10. State whether each of the following is true or false. Justify your answers.

(a) The function $f(x) = x^{1/3}$ has a Maclaurin series.

(b) $1 + \frac{1}{2} - \frac{1}{3} + \frac{1}{3} - \frac{1}{3} + \frac{1}{4} - \frac{1}{4} + \cdots = 1$

(c) $1 + \frac{1}{2} - \frac{1}{2} + \frac{1}{2} - \frac{1}{2} + \frac{1}{2} - \frac{1}{2} + \cdots = 1$

11. Find the general term of the sequence, starting with $n = 1$, determine whether the sequence converges, and if so find its limit.

(a) $\dfrac{3}{2^2 - 1^2}, \dfrac{4}{3^2 - 2^2}, \dfrac{5}{4^2 - 3^2}, \ldots$

(b) $\dfrac{1}{3}, -\dfrac{2}{5}, \dfrac{3}{7}, -\dfrac{4}{9}, \ldots$

12. Suppose that the sequence $\{a_k\}$ is defined recursively by

$$a_0 = c, \quad a_{k+1} = \sqrt{a_k}$$

Assuming that the sequence converges, find its limit if

(a) $c = \frac{1}{2}$ (b) $c = \frac{3}{2}$.

13. Show that the sequence is eventually strictly monotone.

(a) $\{(n - 10)^4\}_{n=0}^{+\infty}$ (b) $\left\{ \dfrac{100^n}{(2n)!(n!)} \right\}_{n=1}^{+\infty}$

14. (a) Give an example of a bounded sequence that diverges.

(b) Give an example of a monotonic sequence that diverges.

15–20 Use any method to determine whether the series converge. ■

15. (a) $\displaystyle\sum_{k=1}^{\infty} \frac{1}{5^k}$ (b) $\displaystyle\sum_{k=1}^{\infty} \frac{1}{5^k+1}$

16. (a) $\displaystyle\sum_{k=1}^{\infty} (-1)^k \frac{k+4}{k^2+k}$ (b) $\displaystyle\sum_{k=1}^{\infty} (-1)^{k+1} \left(\frac{k+2}{3k-1}\right)^k$

17. (a) $\displaystyle\sum_{k=1}^{\infty} \frac{1}{k^3+2k+1}$ (b) $\displaystyle\sum_{k=1}^{\infty} \frac{1}{(3+k)^{2/5}}$

18. (a) $\displaystyle\sum_{k=1}^{\infty} \frac{\ln k}{k\sqrt{k}}$ (b) $\displaystyle\sum_{k=1}^{\infty} \frac{k^{4/3}}{8k^2+5k+1}$

19. (a) $\displaystyle\sum_{k=1}^{\infty} \frac{9}{\sqrt{k}+1}$ (b) $\displaystyle\sum_{k=1}^{\infty} \frac{\cos(1/k)}{k^2}$

20. (a) $\displaystyle\sum_{k=1}^{\infty} \frac{k^{-1/2}}{2+\sin^2 k}$ (b) $\displaystyle\sum_{k=1}^{\infty} \frac{(-1)^{k+1}}{k^2+1}$

21. Find a formula for the exact error that results when the sum of the geometric series $\sum_{k=0}^{\infty}(1/5)^k$ is approximated by the sum of the first 100 terms in the series.

22. Suppose that $\displaystyle\sum_{k=1}^{n} u_k = 2 - \frac{1}{n}$. Find

(a) u_{100} (b) $\displaystyle\lim_{k \to +\infty} u_k$ (c) $\displaystyle\sum_{k=1}^{\infty} u_k$.

23. In each part, determine whether the series converges; if so, find its sum.

(a) $\displaystyle\sum_{k=1}^{\infty} \left(\frac{3}{2^k} - \frac{2}{3^k}\right)$ (b) $\displaystyle\sum_{k=1}^{\infty} [\ln(k+1) - \ln k]$

(c) $\displaystyle\sum_{k=1}^{\infty} \frac{1}{k(k+2)}$ (d) $\displaystyle\sum_{k=1}^{\infty} [\tan^{-1}(k+1) - \tan^{-1} k]$

24. It can be proved that

$$\lim_{n \to +\infty} \sqrt[n]{n!} = +\infty \quad \text{and} \quad \lim_{n \to +\infty} \frac{\sqrt[n]{n!}}{n} = \frac{1}{e}$$

In each part, use these limits and the root test to determine whether the series converges.

(a) $\displaystyle\sum_{k=0}^{\infty} \frac{2^k}{k!}$ (b) $\displaystyle\sum_{k=0}^{\infty} \frac{k^k}{k!}$

25. Let a, b, and p be positive constants. For which values of p does the series $\displaystyle\sum_{k=1}^{\infty} \frac{1}{(a+bk)^p}$ converge?

26. Find the interval of convergence of

$$\sum_{k=0}^{\infty} \frac{(x-x_0)^k}{b^k} \quad (b > 0)$$

27. (a) Show that $k^k \geq k!$.

(b) Use the comparison test to show that $\displaystyle\sum_{k=1}^{\infty} k^{-k}$ converges.

(c) Use the root test to show that the series converges.

28. Does the series $1 - \frac{2}{3} + \frac{3}{5} - \frac{4}{7} + \frac{5}{9} + \cdots$ converge? Justify your answer.

29. (a) Find the first five Maclaurin polynomials of the function $p(x) = 1 - 7x + 5x^2 + 4x^3$.

(b) Make a general statement about the Maclaurin polynomials of a polynomial of degree n.

30. Show that the approximation

$$\sin x \approx x - \frac{x^3}{3!} + \frac{x^5}{5!}$$

is accurate to four decimal places if $0 \leq x \leq \pi/4$.

31. Use a Maclaurin series and properties of alternating series to show that $|\ln(1+x) - x| \leq x^2/2$ if $0 < x < 1$.

32. Use Maclaurin series to approximate the integral

$$\int_0^1 \frac{1 - \cos x}{x} \, dx$$

to three decimal-place accuracy.

33. In parts (a)–(d), find the sum of the series by associating it with some Maclaurin series.

(a) $2 + \dfrac{4}{2!} + \dfrac{8}{3!} + \dfrac{16}{4!} + \cdots$

(b) $\pi - \dfrac{\pi^3}{3!} + \dfrac{\pi^5}{5!} - \dfrac{\pi^7}{7!} + \cdots$

(c) $1 - \dfrac{e^2}{2!} + \dfrac{e^4}{4!} - \dfrac{e^6}{6!} + \cdots$

(d) $1 - \ln 3 + \dfrac{(\ln 3)^2}{2!} - \dfrac{(\ln 3)^3}{3!} + \cdots$

34. In each part, write out the first four terms of the series, and then find the radius of convergence.

(a) $\displaystyle\sum_{k=1}^{\infty} \frac{1 \cdot 2 \cdot 3 \cdots k}{1 \cdot 4 \cdot 7 \cdots (3k-2)} x^k$

(b) $\displaystyle\sum_{k=1}^{\infty} (-1)^k \frac{1 \cdot 2 \cdot 3 \cdots k}{1 \cdot 3 \cdot 5 \cdots (2k-1)} x^{2k+1}$

35. Use an appropriate Taylor series for $\sqrt[3]{x}$ to approximate $\sqrt[3]{28}$ to three decimal-place accuracy, and check your answer by comparing it to that produced directly by your calculating utility.

36. Differentiate the Maclaurin series for xe^x and use the result to show that

$$\sum_{k=0}^{\infty} \frac{k+1}{k!} = 2e$$

37. Use the supplied Maclaurin series for $\sin x$ and $\cos x$ to find the first four nonzero terms of the Maclaurin series for the given functions.

$$\sin x = \sum_{k=0}^{\infty} (-1)^k \frac{x^{2k+1}}{(2k+1)!}$$

$$\cos x = \sum_{k=0}^{\infty} (-1)^k \frac{x^{2k}}{(2k)!}$$

(a) $\sin x \cos x$ (b) $\frac{1}{2}\sin 2x$

CHAPTER 9 MAKING CONNECTIONS

1. As shown in the accompanying figure, suppose that lines L_1 and L_2 form an angle θ, $0 < \theta < \pi/2$, at their point of intersection P. A point P_0 is chosen that is on L_1 and a units from P. Starting from P_0 a zig-zag path is constructed by successively going back and forth between L_1 and L_2 along a perpendicular from one line to the other. Find the following sums in terms of θ and a.
   (a) $P_0P_1 + P_1P_2 + P_2P_3 + \cdots$
   (b) $P_0P_1 + P_2P_3 + P_4P_5 + \cdots$
   (c) $P_1P_2 + P_3P_4 + P_5P_6 + \cdots$

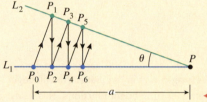

◀ **Figure Ex-1**

2. (a) Find A and B such that
$$\frac{6^k}{(3^{k+1} - 2^{k+1})(3^k - 2^k)} = \frac{2^k A}{3^k - 2^k} + \frac{2^k B}{3^{k+1} - 2^{k+1}}$$

   (b) Use the result in part (a) to find a closed form for the nth partial sum of the series
$$\sum_{k=1}^{\infty} \frac{6^k}{(3^{k+1} - 2^{k+1})(3^k - 2^k)}$$

   and then find the sum of the series.

   Source: This exercise is adapted from a problem that appeared in the Forty-Fifth Annual William Lowell Putnam Competition.

3. Show that the alternating p-series
$$1 - \frac{1}{2^p} + \frac{1}{3^p} - \frac{1}{4^p} + \cdots + (-1)^{k+1}\frac{1}{k^p} + \cdots$$

   converges absolutely if $p > 1$, converges conditionally if $0 < p \leq 1$, and diverges if $p \leq 0$.

4. As illustrated in the accompanying figure, a bug, starting at point A on a 180 cm wire, walks the length of the wire, stops and walks in the opposite direction for half the length of the wire, stops again and walks in the opposite direction for one-third the length of the wire, stops again and walks in the opposite direction for one-fourth the length of the wire, and so forth until it stops for the 1000th time.

   (a) Give upper and lower bounds on the distance between the bug and point A when it finally stops. [*Hint:* As stated in Example 2 of Section 9.6, assume that the sum of the alternating harmonic series is ln 2.]
   (b) Give upper and lower bounds on the total distance that the bug has traveled when it finally stops. [*Hint:* Use inequality (2) of Section 9.4.]

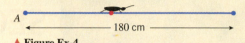

A ◀————— 180 cm —————▶

▲ **Figure Ex-4**

5. In Section 6.6 we defined the kinetic energy K of a particle with mass m and velocity v to be $K = \frac{1}{2}mv^2$ [see Formula (7) of that section]. In this formula the mass m is assumed to be constant, and K is called the ***Newtonian kinetic energy***. However, in Albert Einstein's relativity theory the mass m increases with the velocity and the kinetic energy K is given by the formula
$$K = m_0c^2\left[\frac{1}{\sqrt{1 - (v/c)^2}} - 1\right]$$

   in which m_0 is the mass of the particle when its velocity is zero, and c is the speed of light. This is called the ***relativistic kinetic energy***. Use an appropriate binomial series to show that if the velocity is small compared to the speed of light (i.e., $v/c \approx 0$), then the Newtonian and relativistic kinetic energies are in close agreement.

6. In Section 8.4 we studied the motion of a falling object that has mass m and is retarded by air resistance. We showed that if the initial velocity is v_0 and the drag force F_R is proportional to the velocity, that is, $F_R = -cv$, then the velocity of the object at time t is
$$v(t) = e^{-ct/m}\left(v_0 + \frac{mg}{c}\right) - \frac{mg}{c}$$

   where g is the acceleration due to gravity [see Formula (16) of Section 8.4].

   (a) Use a Maclaurin series to show that if $ct/m \approx 0$, then the velocity can be approximated as
$$v(t) \approx v_0 - \left(\frac{cv_0}{m} + g\right)t$$

   (b) Improve on the approximation in part (a).

EXPANDING THE CALCULUS HORIZON

To learn how ecologists use mathematical models based on the process of iteration to study the growth and decline of animal populations, see the module entitled **Iteration and Dynamical Systems** at:

www.wiley.com/college/anton

Dwight R. Kuhn

10

PARAMETRIC AND POLAR CURVES; CONIC SECTIONS

Mathematical curves, such as the spirals in the center of a sunflower, can be described conveniently using ideas developed in this chapter.

In this chapter we will study alternative ways of expressing curves in the plane. We will begin by studying parametric curves: curves described in terms of component functions. This study will include methods for finding tangent lines to parametric curves. We will then introduce polar coordinate systems and discuss methods for finding tangent lines to polar curves, arc length of polar curves, and areas enclosed by polar curves. Our attention will then turn to a review of the basic properties of conic sections: parabolas, ellipses, and hyperbolas. Finally, we will consider conic sections in the context of polar coordinates and discuss some applications in astronomy.

10.1 PARAMETRIC EQUATIONS; TANGENT LINES AND ARC LENGTH FOR PARAMETRIC CURVES

Graphs of functions must pass the vertical line test, a limitation that excludes curves with self-intersections or even such basic curves as circles. In this section we will study an alternative method for describing curves algebraically that is not subject to the severe restriction of the vertical line test. We will then derive formulas required to find slopes, tangent lines, and arc lengths of these parametric curves. We will conclude with an investigation of a classic parametric curve known as the cycloid.

■ PARAMETRIC EQUATIONS

Suppose that a particle moves along a curve C in the xy-plane in such a way that its x- and y-coordinates, as functions of time, are

$$x = f(t), \quad y = g(t)$$

We call these the **parametric equations** of motion for the particle and refer to C as the **trajectory** of the particle or the **graph** of the equations (Figure 10.1.1). The variable t is called the **parameter** for the equations.

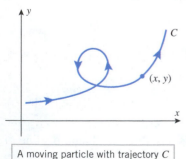

A moving particle with trajectory C

▲ **Figure 10.1.1**

▶ **Example 1** Sketch the trajectory over the time interval $0 \le t \le 10$ of the particle whose parametric equations of motion are

$$x = t - 3\sin t, \quad y = 4 - 3\cos t \tag{1}$$

Solution. One way to sketch the trajectory is to choose a representative succession of times, plot the (x, y) coordinates of points on the trajectory at those times, and connect the points with a smooth curve. The trajectory in Figure 10.1.2 was obtained in this way from the data in Table 10.1.1 in which the approximate coordinates of the particle are given at time increments of 1 unit. Observe that there is no t-axis in the picture; the values of t appear only as labels on the plotted points, and even these are usually omitted unless it is important to emphasize the locations of the particle at specific times. ◄

Table 10.1.1

t	x	y
0	0.0	1.0
1	−1.5	2.4
2	−0.7	5.2
3	2.6	7.0
4	6.3	6.0
5	7.9	3.1
6	6.8	1.1
7	5.0	1.7
8	5.0	4.4
9	7.8	6.7
10	11.6	6.5

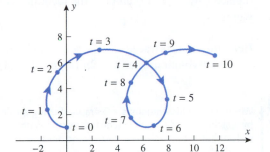

▲ **Figure 10.1.2**

Although parametric equations commonly arise in problems of motion with time as the parameter, they arise in other contexts as well. Thus, unless the problem dictates that the parameter t in the equations

$$x = f(t), \quad y = g(t)$$

represents time, it should be viewed simply as an independent variable that varies over some interval of real numbers. (In fact, there is no need to use the letter t for the parameter; any letter not reserved for another purpose can be used.) If no restrictions on the parameter are stated explicitly or implied by the equations, then it is understood that it varies from $-\infty$ to $+\infty$. To indicate that a parameter t is restricted to an interval $[a, b]$, we will write

$$x = f(t), \quad y = g(t) \qquad (a \leq t \leq b)$$

► **Example 2** Find the graph of the parametric equations

$$x = \cos t, \quad y = \sin t \qquad (0 \leq t \leq 2\pi) \tag{2}$$

Solution. One way to find the graph is to eliminate the parameter t by noting that

$$x^2 + y^2 = \sin^2 t + \cos^2 t = 1$$

Thus, the graph is contained in the unit circle $x^2 + y^2 = 1$. Geometrically, the parameter t can be interpreted as the angle swept out by the radial line from the origin to the point $(x, y) = (\cos t, \sin t)$ on the unit circle (Figure 10.1.3). As t increases from 0 to 2π, the point traces the circle counterclockwise, starting at $(1, 0)$ when $t = 0$ and completing one full revolution when $t = 2\pi$. One can obtain different portions of the circle by varying the interval over which the parameter varies. For example,

$$x = \cos t, \quad y = \sin t \qquad (0 \leq t \leq \pi) \tag{3}$$

represents just the upper semicircle in Figure 10.1.3. ◄

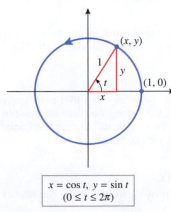

$x = \cos t, y = \sin t$
$(0 \leq t \leq 2\pi)$

▲ **Figure 10.1.3**

■ ORIENTATION

The direction in which the graph of a pair of parametric equations is traced as the parameter increases is called the *direction of increasing parameter* or sometimes the *orientation* imposed on the curve by the equations. Thus, we make a distinction between a *curve*, which is a set of points, and a *parametric curve*, which is a curve with an orientation imposed on it by a set of parametric equations. For example, we saw in Example 2 that the circle represented parametrically by (2) is traced counterclockwise as t increases and hence has *counterclockwise orientation*. As shown in Figures 10.1.2 and 10.1.3, the orientation of a parametric curve can be indicated by arrowheads.

To obtain parametric equations for the unit circle with *clockwise orientation*, we can replace t by $-t$ in (2) and use the identities $\cos(-t) = \cos t$ and $\sin(-t) = -\sin t$. This yields

$$x = \cos t, \quad y = -\sin t \qquad (0 \le t \le 2\pi)$$

Here, the circle is traced clockwise by a point that starts at $(1, 0)$ when $t = 0$ and completes one full revolution when $t = 2\pi$ (Figure 10.1.4).

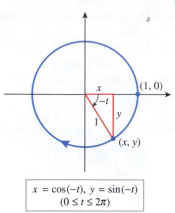

$x = \cos(-t),\ y = \sin(-t)$
$(0 \le t \le 2\pi)$

▲ **Figure 10.1.4**

TECHNOLOGY MASTERY When parametric equations are graphed using a calculator, the orientation can often be determined by watching the direction in which the graph is traced on the screen. However, many computers graph so fast that it is often hard to discern the orientation. See if you can use your graphing utility to confirm that (3) has a counterclockwise orientation.

▶ **Example 3**   Graph the parametric curve

$$x = 2t - 3, \quad y = 6t - 7$$

by eliminating the parameter, and indicate the orientation on the graph.

Solution.   To eliminate the parameter we will solve the first equation for t as a function of x, and then substitute this expression for t into the second equation:

$$t = \left(\tfrac{1}{2}\right)(x + 3)$$

$$y = 6\left(\tfrac{1}{2}\right)(x + 3) - 7$$

$$y = 3x + 2$$

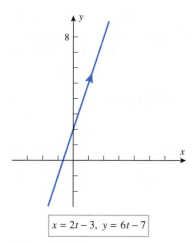

$x = 2t - 3,\ y = 6t - 7$

▲ **Figure 10.1.5**

Thus, the graph is a line of slope 3 and y-intercept 2. To find the orientation we must look to the original equations; the direction of increasing t can be deduced by observing that x increases as t increases *or* by observing that y increases as t increases. Either piece of information tells us that the line is traced left to right as shown in Figure 10.1.5. ◀

REMARK   Not all parametric equations produce curves with definite orientations; if the equations are badly behaved, then the point tracing the curve may leap around sporadically or move back and forth, failing to determine a definite direction. For example, if

$$x = \sin t, \quad y = \sin^2 t$$

then the point (x, y) moves along the parabola $y = x^2$. However, the value of x varies periodically between -1 and 1, so the point (x, y) moves periodically back and forth along the parabola between the points $(-1, 1)$ and $(1, 1)$ (as shown in Figure 10.1.6). Later in the text we will discuss restrictions that eliminate such erratic behavior, but for now we will just avoid such complications.

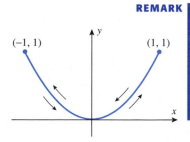

▲ **Figure 10.1.6**

■ EXPRESSING ORDINARY FUNCTIONS PARAMETRICALLY

An equation $y = f(x)$ can be expressed in parametric form by introducing the parameter $t = x$; this yields the parametric equations

$$x = t, \quad y = f(t)$$

For example, the portion of the curve $y = \cos x$ over the interval $[-2\pi, 2\pi]$ can be expressed parametrically as

$$x = t, \quad y = \cos t \qquad (-2\pi \leq t \leq 2\pi)$$

(Figure 10.1.7).

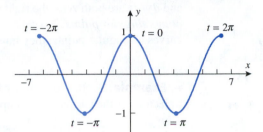

▶ **Figure 10.1.7**

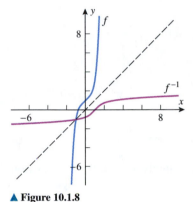

▲ **Figure 10.1.8**

If a function f is one-to-one, then it has an inverse function f^{-1}. In this case the equation $y = f^{-1}(x)$ is equivalent to $x = f(y)$. We can express the graph of f^{-1} in parametric form by introducing the parameter $y = t$; this yields the parametric equations

$$x = f(t), \quad y = t$$

For example, Figure 10.1.8 shows the graph of $f(x) = x^5 + x + 1$ and its inverse. The graph of f can be repesented parametrically as

$$x = t, \quad y = t^5 + t + 1$$

and the graph of f^{-1} can be represented parametrically as

$$x = t^5 + t + 1, \quad y = t$$

TANGENT LINES TO PARAMETRIC CURVES

We will be concerned with curves that are given by parametric equations

$$x = f(t), \quad y = g(t)$$

in which $f(t)$ and $g(t)$ have continuous first derivatives with respect to t. It can be proved that if $dx/dt \neq 0$, then y is a differentiable function of x, in which case the chain rule implies that

$$\frac{dy}{dx} = \frac{dy/dt}{dx/dt} \tag{4}$$

This formula makes it possible to find dy/dx directly from the parametric equations without eliminating the parameter.

▶ **Example 4** Find the slope of the tangent line to the unit circle

$$x = \cos t, \quad y = \sin t \qquad (0 \leq t \leq 2\pi)$$

at the point where $t = \pi/6$ (Figure 10.1.9).

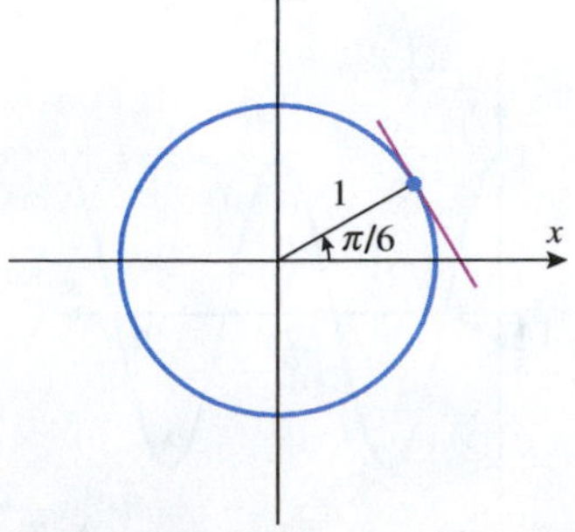

Solution. From (4), the slope at a general point on the circle is

$$\frac{dy}{dx} = \frac{dy/dt}{dx/dt} = \frac{\cos t}{-\sin t} = -\cot t \tag{5}$$

Thus, the slope at $t = \pi/6$ is

$$\left.\frac{dy}{dx}\right|_{t=\pi/6} = -\cot\frac{\pi}{6} = -\sqrt{3} \quad ◀$$

▲ **Figure 10.1.9**

Note that Formula (5) makes sense geometrically because the radius from the origin to the point $P(\cos t, \sin t)$ has slope $m = \tan t$. Thus the tangent line at P, being perpendicular to the radius, has slope

$$-\frac{1}{m} = -\frac{1}{\tan t} = -\cot t$$

(Figure 10.1.10).

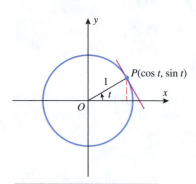

Radius OP has slope $m = \tan t$.

▲ **Figure 10.1.10**

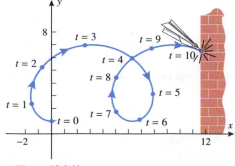

Stanislovas Kairys/iStockphoto

The complicated motion of a paper airplane is best described mathematically using parametric equations.

It follows from Formula (4) that the tangent line to a parametric curve will be horizontal at those points where $dy/dt = 0$ and $dx/dt \neq 0$, since $dy/dx = 0$ at such points. Two different situations occur when $dx/dt = 0$. At points where $dx/dt = 0$ and $dy/dt \neq 0$, the right side of (4) has a nonzero numerator and a zero denominator; we will agree that the curve has **infinite slope** and a **vertical tangent line** at such points. At points where dx/dt and dy/dt are both zero, the right side of (4) becomes an indeterminate form; we call such points **singular points**. No general statement can be made about the behavior of parametric curves at singular points; they must be analyzed case by case.

▶ **Example 5**   In a disastrous first flight, an experimental paper airplane follows the trajectory of the particle in Example 1:

$$x = t - 3\sin t, \quad y = 4 - 3\cos t \qquad (t \geq 0)$$

but crashes into a wall at time $t = 10$ (Figure 10.1.11).

(a) At what times was the airplane flying horizontally?

(b) At what times was it flying vertically?

Solution (a).   The airplane was flying horizontally at those times when $dy/dt = 0$ and $dx/dt \neq 0$. From the given trajectory we have

$$\frac{dy}{dt} = 3\sin t \quad \text{and} \quad \frac{dx}{dt} = 1 - 3\cos t \tag{6}$$

Setting $dy/dt = 0$ yields the equation $3\sin t = 0$, or, more simply, $\sin t = 0$. This equation has four solutions in the time interval $0 \leq t \leq 10$:

$$t = 0, \quad t = \pi, \quad t = 2\pi, \quad t = 3\pi$$

Since $dx/dt = 1 - 3\cos t \neq 0$ for these values of t (verify), the airplane was flying horizontally at times

$$t = 0, \quad t = \pi \approx 3.14, \quad t = 2\pi \approx 6.28, \quad \text{and} \quad t = 3\pi \approx 9.42$$

which is consistent with Figure 10.1.11.

Solution (b).   The airplane was flying vertically at those times when $dx/dt = 0$ and $dy/dt \neq 0$. Setting $dx/dt = 0$ in (6) yields the equation

$$1 - 3\cos t = 0 \quad \text{or} \quad \cos t = \tfrac{1}{3}$$

This equation has three solutions in the time interval $0 \leq t \leq 10$ (Figure 10.1.12):

$$t = \cos^{-1}\tfrac{1}{3}, \quad t = 2\pi - \cos^{-1}\tfrac{1}{3}, \quad t = 2\pi + \cos^{-1}\tfrac{1}{3}$$

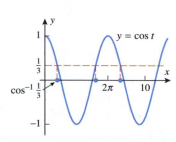

▲ **Figure 10.1.11**

▲ **Figure 10.1.12**

Since $dy/dt = 3\sin t$ is not zero at these points (why?), it follows that the airplane was flying vertically at times

$$t = \cos^{-1}\tfrac{1}{3} \approx 1.23, \quad t \approx 2\pi - 1.23 \approx 5.05, \quad t \approx 2\pi + 1.23 \approx 7.51$$

which again is consistent with Figure 10.1.11. ◄

▶ **Example 6** The curve represented by the parametric equations

$$x = t^2, \quad y = t^3 \qquad (-\infty < t < +\infty)$$

is called a **semicubical parabola**. The parameter t can be eliminated by cubing x and squaring y, from which it follows that $y^2 = x^3$. The graph of this equation, shown in Figure 10.1.13, consists of two branches: an upper branch obtained by graphing $y = x^{3/2}$ and a lower branch obtained by graphing $y = -x^{3/2}$. The two branches meet at the origin, which corresponds to $t = 0$ in the parametric equations. This is a singular point because the derivatives $dx/dt = 2t$ and $dy/dt = 3t^2$ are both zero there. ◄

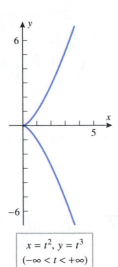

$$x = t^2, \, y = t^3$$
$$(-\infty < t < +\infty)$$

▲ **Figure 10.1.13**

▶ **Example 7** Without eliminating the parameter, find dy/dx and d^2y/dx^2 at $(1, 1)$ and $(1, -1)$ on the semicubical parabola given by the parametric equations in Example 6.

Solution. From (4) we have

$$\frac{dy}{dx} = \frac{dy/dt}{dx/dt} = \frac{3t^2}{2t} = \frac{3}{2}t \qquad (t \neq 0) \tag{7}$$

and from (4) applied to $y' = dy/dx$ we have

$$\frac{d^2y}{dx^2} = \frac{dy'}{dx} = \frac{dy'/dt}{dx/dt} = \frac{3/2}{2t} = \frac{3}{4t} \tag{8}$$

Since the point $(1, 1)$ on the curve corresponds to $t = 1$ in the parametric equations, it follows from (7) and (8) that

$$\left.\frac{dy}{dx}\right|_{t=1} = \frac{3}{2} \quad \text{and} \quad \left.\frac{d^2y}{dx^2}\right|_{t=1} = \frac{3}{4}$$

Similarly, the point $(1, -1)$ corresponds to $t = -1$ in the parametric equations, so applying (7) and (8) again yields

$$\left.\frac{dy}{dx}\right|_{t=-1} = -\frac{3}{2} \quad \text{and} \quad \left.\frac{d^2y}{dx^2}\right|_{t=-1} = -\frac{3}{4}$$

Note that the values we obtained for the first and second derivatives are consistent with the graph in Figure 10.1.13, since at $(1, 1)$ on the upper branch the tangent line has positive slope and the curve is concave up, and at $(1, -1)$ on the lower branch the tangent line has negative slope and the curve is concave down.

Finally, observe that we were able to apply Formulas (7) and (8) for both $t = 1$ and $t = -1$, even though the points $(1, 1)$ and $(1, -1)$ lie on different branches. In contrast, had we chosen to perform the same computations by eliminating the parameter, we would have had to obtain separate derivative formulas for $y = x^{3/2}$ and $y = -x^{3/2}$. ◄

WARNING

Although it is true that

$$\frac{dy}{dx} = \frac{dy/dt}{dx/dt}$$

you cannot conclude that d^2y/dx^2 is the quotient of d^2y/dt^2 and d^2x/dt^2. To illustrate that this conclusion is erroneous, show that for the parametric curve in Example 7,

$$\left.\frac{d^2y}{dx^2}\right|_{t=1} \neq \left.\frac{d^2y/dt^2}{d^2x/dt^2}\right|_{t=1}$$

■ **ARC LENGTH OF PARAMETRIC CURVES**

The following result provides a formula for finding the arc length of a curve from parametric equations for the curve. Its derivation is similar to that of Formula (3) in Section 6.4 and will be omitted.

Formulas (4) and (5) in Section 6.4 can be viewed as special cases of (9). For example, Formula (4) in Section 6.4 can be obtained from (9) by writing $y = f(x)$ parametrically as

$$x = t, \quad y = f(t)$$

and Formula (5) in Section 6.4 can be obtained by writing $x = g(y)$ parametrically as

$$x = g(t), \quad y = t$$

10.1.1   **ARC LENGTH FORMULA FOR PARAMETRIC CURVES**   If no segment of the curve represented by the parametric equations

$$x = x(t), \quad y = y(t) \qquad (a \le t \le b)$$

is traced more than once as t increases from a to b, and if dx/dt and dy/dt are continuous functions for $a \le t \le b$, then the arc length L of the curve is given by

$$L = \int_a^b \sqrt{\left(\frac{dx}{dt}\right)^2 + \left(\frac{dy}{dt}\right)^2} \, dt \qquad (9)$$

▶ **Example 8**   Use (9) to find the circumference of a circle of radius a from the parametric equations

$$x = a \cos t, \quad y = a \sin t \qquad (0 \le t \le 2\pi)$$

Solution.

$$L = \int_0^{2\pi} \sqrt{\left(\frac{dx}{dt}\right)^2 + \left(\frac{dy}{dt}\right)^2} \, dt = \int_0^{2\pi} \sqrt{(-a \sin t)^2 + (a \cos t)^2} \, dt$$

$$= \int_0^{2\pi} a \, dt = at \Big]_0^{2\pi} = 2\pi a \quad ◀$$

■ **THE CYCLOID (THE APPLE OF DISCORD)**

The results of this section can be used to investigate a curve known as a *cycloid*. This curve, which is one of the most significant in the history of mathematics, can be generated by a point on a circle that rolls along a straight line (Figure 10.1.14). This curve has a fascinating history, which we will discuss shortly; but first we will show how to obtain parametric equations for it. For this purpose, let us assume that the circle has radius a and rolls along the positive x-axis of a rectangular coordinate system. Let $P(x, y)$ be the point on the circle that traces the cycloid, and assume that P is initially at the origin. We will take as our parameter the angle θ that is swept out by the radial line to P as the circle rolls (Figure 10.1.14). It is standard here to regard θ as positive, even though it is generated by a clockwise rotation.

The motion of P is a combination of the movement of the circle's center parallel to the x-axis and the rotation of P about the center. As the radial line sweeps out an angle θ, the point P traverses an arc of length $a\theta$, and the circle moves a distance $a\theta$ along the x-axis. Thus, as suggested by Figure 10.1.15, the center moves to the point $(a\theta, a)$, and the coordinates of P are

$$x = a\theta - a \sin \theta, \quad y = a - a \cos \theta \qquad (10)$$

These are the equations of the cycloid in terms of the parameter θ.

One of the reasons the cycloid is important in the history of mathematics is that the study of its properties helped to spur the development of early versions of differentiation and integration. Work on the cycloid was carried out by some of the most famous names in seventeenth century mathematics, including Johann and Jakob Bernoulli, Descartes, L'Hôpital, Newton, and Leibniz. The curve was named the "cycloid" by the Italian mathematician and astronomer, Galileo, who spent over 40 years investigating its properties. An early problem of interest was that of constructing tangent lines to the cycloid. This problem was first solved by Descartes, and then by Fermat, whom Descartes had challenged with the question. A modern solution to this problem follows directly from the parametric equations (10) and Formula (4). For example, using Formula (4), it is straightforward to show that the x-intercepts of the cycloid are cusps and that there is a horizontal tangent line to the cycloid halfway between adjacent x-intercepts (Exercise 60).

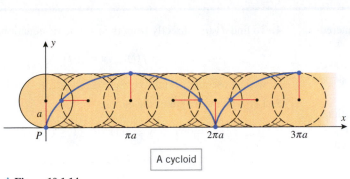

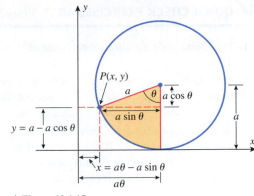

A cycloid

▲ **Figure 10.1.14**

▲ **Figure 10.1.15**

▲ **Figure 10.1.16**

Another early problem was determining the arc length of an arch of the cycloid. This was solved in 1658 by the famous British architect and mathematician, Sir Christopher Wren. He showed that the arc length of one arch of the cycloid is exactly eight times the radius of the generating circle. [For a solution to this problem using Formula (9), see Exercise 71.]

The cycloid is also important historically because it provides the solution to two famous mathematical problems—the *brachistochrone problem* (from Greek words meaning "shortest time") and the *tautochrone problem* (from Greek words meaning "equal time"). The brachistochrone problem is to determine the shape of a wire along which a bead might slide from a point P to another point Q, not directly below, in the *shortest time*. The tautochrone problem is to find the shape of a wire from P to Q such that two beads started at any points on the wire between P and Q reach Q in the *same amount of time*. The solution to both problems turns out to be an inverted cycloid (Figure 10.1.16).

In June of 1696, Johann Bernoulli posed the brachistochrone problem in the form of a challenge to other mathematicians. At first, one might conjecture that the wire should form a straight line, since that shape results in the shortest distance from P to Q. However, the inverted cycloid allows the bead to fall more rapidly at first, building up sufficient speed to reach Q in the shortest time, even though it travels a longer distance. The problem was solved by Newton, Leibniz, and L'Hôpital, as well as by Johann Bernoulli and his older brother Jakob; it was formulated and solved *incorrectly* years earlier by Galileo, who thought the answer was a circular arc. In fact, Johann was so impressed with his brother Jakob's solution that he claimed it to be his own. (This was just one of many disputes about the cycloid that eventually led to the curve being known as the "apple of discord.") One solution of the brachistochrone problem leads to the differential equation

$$\left(1 + \left(\frac{dy}{dx}\right)^2\right) y = 2a \tag{11}$$

where a is a positive constant. We leave it as an exercise (Exercise 72) to show that the cycloid provides a solution to this differential equation.

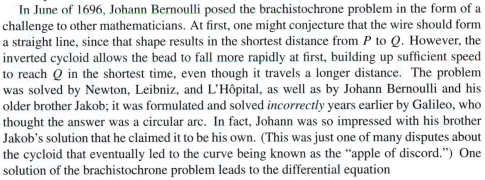

Newton's solution of the brachistochrone problem in his own handwriting

✔ QUICK CHECK EXERCISES 10.1   *(See page 705 for answers.)*

1. Find parametric equations for a circle of radius 2, centered at $(3, 5)$.

2. The graph of the curve described by the parametric equations $x = 4t - 1$, $y = 3t + 2$ is a straight line with slope _____ and y-intercept _____.

3. Suppose that a parametric curve C is given by the equations $x = f(t)$, $y = g(t)$ for $0 \leq t \leq 1$. Find parametric equations for C that reverse the direction the curve is traced as the parameter increases from 0 to 1.

4. To find dy/dx directly from the parametric equations

$$x = f(t), \quad y = g(t)$$

we can use the formula $dy/dx = $ _____.

5. Let L be the length of the curve

$$x = \ln t, \quad y = \sin t \quad (1 \leq t \leq \pi)$$

An integral expression for L is _____.

EXERCISE SET 10.1    Graphing Utility   CAS

1. **(a)** By eliminating the parameter, sketch the trajectory over the time interval $0 \leq t \leq 5$ of the particle whose parametric equations of motion are

$$x = t - 1, \quad y = t + 1$$

(b) Indicate the direction of motion on your sketch.

(c) Make a table of x- and y-coordinates of the particle at times $t = 0, 1, 2, 3, 4, 5$.

(d) Mark the position of the particle on the curve at the times in part (c), and label those positions with the values of t.

Johann (left) **and Jakob** (right) **Bernoulli** Members of an amazing Swiss family that included several generations of outstanding mathematicians and scientists. Nikolaus Bernoulli (1623–1708), a druggist, fled from Antwerp to escape religious persecution and ultimately settled in Basel, Switzerland. There he had three sons, Jakob I (also called Jacques or James), Nikolaus, and Johann I (also called Jean or John). The Roman numerals are used to distinguish family members with identical names (see the family tree below). Following Newton and Leibniz, the Bernoulli brothers, Jakob I and Johann I, are considered by some to be the two most important founders of calculus. Jakob I was self-taught in mathematics. His father wanted him to study for the ministry, but he turned to mathematics and in 1686 became a professor at the University of Basel. When he started working in mathematics, he knew nothing of Newton's and Leibniz' work. He eventually became familiar with Newton's results, but because so little of Leibniz' work was published, Jakob duplicated many of Leibniz' results.

Jakob's younger brother Johann I was urged to enter into business by his father. Instead, he turned to medicine and studied mathematics under the guidance of his older brother. He eventually became a mathematics professor at Gröningen in Holland, and then, when Jakob died in 1705, Johann succeeded him as mathematics professor at Basel. Throughout their lives, Jakob I and Johann I had a mutual passion for criticizing each other's work, which frequently erupted into ugly confrontations. Leibniz tried to mediate the disputes, but Jakob, who resented Leibniz' superior intellect, accused him of siding with Johann, and thus Leibniz became entangled in the arguments. The brothers often worked on common problems that they posed as challenges to one another. Johann, interested in gaining fame, often used unscrupulous means to make himself appear the originator of his brother's results; Jakob occasionally retaliated. Thus, it is often difficult to determine who deserves credit for many results. However, both men made major contributions to the development of calculus. In addition to his work on calculus, Jakob helped establish fundamental principles in probability, including the Law of Large Numbers, which is a cornerstone of modern probability theory.

Among the other members of the Bernoulli family, Daniel, son of Johann I, is the most famous. He was a professor of mathematics at St. Petersburg Academy in Russia and subsequently a professor of anatomy and then physics at Basel. He did work in calculus and probability, but is best known for his work in physics. A basic law of fluid flow, called Bernoulli's principle, is named in his honor. He won the annual prize of the French Academy 10 times for work on vibrating strings, tides of the sea, and kinetic theory of gases.

Johann II succeeded his father as professor of mathematics at Basel. His research was on the theory of heat and sound. Nikolaus I was a mathematician and law scholar who worked on probability and series. On the recommendation of Leibniz, he was appointed professor of mathematics at Padua and then went to Basel as a professor of logic and then law. Nikolaus II was professor of jurisprudence in Switzerland and then professor of mathematics at St. Petersburg Academy. Johann III was a professor of mathematics and astronomy in Berlin and Jakob II succeeded his uncle Daniel as professor of mathematics at St. Petersburg Academy in Russia. Truly an incredible family!

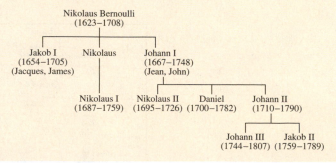

2. (a) By eliminating the parameter, sketch the trajectory over the time interval $0 \le t \le 1$ of the particle whose parametric equations of motion are

$$x = \cos(\pi t), \quad y = \sin(\pi t)$$

(b) Indicate the direction of motion on your sketch.
(c) Make a table of x- and y-coordinates of the particle at times $t = 0, 0.25, 0.5, 0.75, 1$.
(d) Mark the position of the particle on the curve at the times in part (c), and label those positions with the values of t.

3–12 Sketch the curve by eliminating the parameter, and indicate the direction of increasing t. ■

3. $x = 3t - 4, \ y = 6t + 2$

4. $x = t - 3, \ y = 3t - 7 \quad (0 \le t \le 3)$

5. $x = 2\cos t, \ y = 5\sin t \quad (0 \le t \le 2\pi)$

6. $x = \sqrt{t}, \ y = 2t + 4$

7. $x = 3 + 2\cos t, \ y = 2 + 4\sin t \quad (0 \le t \le 2\pi)$

8. $x = \sec t, \ y = \tan t \quad (\pi \le t < 3\pi/2)$

9. $x = \cos 2t, \ y = \sin t \quad (-\pi/2 \le t \le \pi/2)$

10. $x = 4t + 3, \ y = 16t^2 - 9$

11. $x = 2\sin^2 t, \ y = 3\cos^2 t \quad (0 \le t \le \pi/2)$

12. $x = \sec^2 t, \ y = \tan^2 t \quad (0 \le t < \pi/2)$

13–18 Find parametric equations for the curve, and check your work by generating the curve with a graphing utility. ■

13. A circle of radius 5, centered at the origin, oriented clockwise.

14. The portion of the circle $x^2 + y^2 = 1$ that lies in the third quadrant, oriented counterclockwise.

15. A vertical line intersecting the x-axis at $x = 2$, oriented upward.

16. The ellipse $x^2/4 + y^2/9 = 1$, oriented counterclockwise.

17. The portion of the parabola $x = y^2$ joining $(1, -1)$ and $(1, 1)$, oriented down to up.

18. The circle of radius 4, centered at $(1, -3)$, oriented counterclockwise.

19. (a) Use a graphing utility to generate the trajectory of a particle whose equations of motion over the time interval $0 \le t \le 5$ are

$$x = 6t - \tfrac{1}{2}t^3, \quad y = 1 + \tfrac{1}{2}t^2$$

(b) Make a table of x- and y-coordinates of the particle at times $t = 0, 1, 2, 3, 4, 5$.
(c) At what times is the particle on the y-axis?
(d) During what time interval is $y < 5$?
(e) At what time does the x-coordinate of the particle reach a maximum?

20. (a) Use a graphing utility to generate the trajectory of a paper airplane whose equations of motion for $t \ge 0$ are

$$x = t - 2\sin t, \quad y = 3 - 2\cos t$$

(b) Assuming that the plane flies in a room in which the floor is at $y = 0$, explain why the plane will not crash into the floor. [For simplicity, ignore the physical size of the plane by treating it as a particle.]
(c) How high must the ceiling be to ensure that the plane does not touch or crash into it?

21–22 Graph the equation using a graphing utility. ■

21. (a) $x = y^2 + 2y + 1$
(b) $x = \sin y, \ -2\pi \le y \le 2\pi$

22. (a) $x = y + 2y^3 - y^5$
(b) $x = \tan y, \ -\pi/2 < y < \pi/2$

FOCUS ON CONCEPTS

23. In each part, match the parametric equation with one of the curves labeled (I)–(VI), and explain your reasoning.
(a) $x = \sqrt{t}, \ y = \sin 3t$ (b) $x = 2\cos t, \ y = 3\sin t$
(c) $x = t\cos t, \ y = t\sin t$
(d) $x = \dfrac{3t}{1+t^3}, \ y = \dfrac{3t^2}{1+t^3}$
(e) $x = \dfrac{t^3}{1+t^2}, \ y = \dfrac{2t^2}{1+t^2}$
(f) $x = \tfrac{1}{2}\cos t, \ y = \sin 2t$

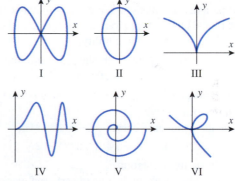

▲ Figure Ex-23

24. (a) Identify the orientation of the curves in Exercise 23.
(b) Explain why the parametric curve

$$x = t^2, \quad y = t^4 \quad (-1 \le t \le 1)$$

does not have a definite orientation.

25. (a) Suppose that the line segment from the point $P(x_0, y_0)$ to $Q(x_1, y_1)$ is represented parametrically by

$$x = x_0 + (x_1 - x_0)t,$$
$$y = y_0 + (y_1 - y_0)t \quad (0 \le t \le 1)$$

and that $R(x, y)$ is the point on the line segment corresponding to a specified value of t (see the accompanying figure on the next page). Show that $t = r/q$, where r is the distance from P to R and q is the distance from P to Q.

(cont.)

(b) What value of t produces the midpoint between points P and Q?

(c) What value of t produces the point that is three-fourths of the way from P to Q?

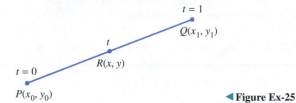

◀ **Figure Ex-25**

26. Find parametric equations for the line segment joining $P(2, -1)$ and $Q(3, 1)$, and use the result in Exercise 25 to find

(a) the midpoint between P and Q

(b) the point that is one-fourth of the way from P to Q

(c) the point that is three-fourths of the way from P to Q.

27. (a) Show that the line segment joining the points (x_0, y_0) and (x_1, y_1) can be represented parametrically as

$$x = x_0 + (x_1 - x_0)\frac{t - t_0}{t_1 - t_0},$$
$$(t_0 \le t \le t_1)$$
$$y = y_0 + (y_1 - y_0)\frac{t - t_0}{t_1 - t_0}$$

(b) Which way is the line segment oriented?

(c) Find parametric equations for the line segment traced from $(3, -1)$ to $(1, 4)$ as t varies from 1 to 2, and check your result with a graphing utility.

28. (a) By eliminating the parameter, show that if a and c are not both zero, then the graph of the parametric equations

$$x = at + b, \quad y = ct + d \quad (t_0 \le t \le t_1)$$

is a line segment.

(b) Sketch the parametric curve

$$x = 2t - 1, \quad y = t + 1 \quad (1 \le t \le 2)$$

and indicate its orientation.

(c) What can you say about the line in part (a) if a or c (but not both) is zero?

(d) What do the equations represent if a and c are both zero?

⌁ **29–32** Use a graphing utility and parametric equations to display the graphs of f and f^{-1} on the same screen. ∎

29. $f(x) = x^3 + 0.2x - 1, \quad -1 \le x \le 2$

30. $f(x) = \sqrt{x^2 + 2} + x, \quad -5 \le x \le 5$

31. $f(x) = \cos(\cos 0.5x), \quad 0 \le x \le 3$

32. $f(x) = x + \sin x, \quad 0 \le x \le 6$

33–36 True–False Determine whether the statement is true or false. Explain your answer. ∎

33. The equation $y = 1 - x^2$ can be described parametrically by $x = \sin t$, $y = \cos^2 t$.

34. The graph of the parametric equations $x = f(t)$, $y = t$ is the reflection of the graph of $y = f(x)$ about the x-axis.

35. For the parametric curve $x = x(t)$, $y = 3t^4 - 2t^3$, the derivative of y with respect to x is computed by

$$\frac{dy}{dx} = \frac{12t^3 - 6t^2}{x'(t)}$$

36. The curve represented by the parametric equations

$$x = t^3, \quad y = t + t^6 \quad (-\infty < t < +\infty)$$

is concave down for $t < 0$.

37. Parametric curves can be defined piecewise by using different formulas for different values of the parameter. Sketch the curve that is represented piecewise by the parametric equations

$$\begin{cases} x = 2t, & y = 4t^2 & (0 \le t \le \frac{1}{2}) \\ x = 2 - 2t, & y = 2t & (\frac{1}{2} \le t \le 1) \end{cases}$$

38. Find parametric equations for the rectangle in the accompanying figure, assuming that the rectangle is traced counterclockwise as t varies from 0 to 1, starting at $(\frac{1}{2}, \frac{1}{2})$ when $t = 0$. [*Hint:* Represent the rectangle piecewise, letting t vary from 0 to $\frac{1}{4}$ for the first edge, from $\frac{1}{4}$ to $\frac{1}{2}$ for the second edge, and so forth.]

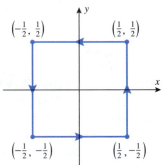

◀ **Figure Ex-38**

⌁ **39.** (a) Find parametric equations for the ellipse that is centered at the origin and has intercepts $(4, 0)$, $(-4, 0)$, $(0, 3)$, and $(0, -3)$.

(b) Find parametric equations for the ellipse that results by translating the ellipse in part (a) so that its center is at $(-1, 2)$.

(c) Confirm your results in parts (a) and (b) using a graphing utility.

⌁ **40.** We will show later in the text that if a projectile is fired from ground level with an initial speed of v_0 meters per second at an angle α with the horizontal, and if air resistance is neglected, then its position after t seconds, relative to the coordinate system in the accompanying figure on the next page is

$$x = (v_0 \cos \alpha)t, \quad y = (v_0 \sin \alpha)t - \frac{1}{2}gt^2$$

where $g \approx 9.8 \text{ m/s}^2$.

(a) By eliminating the parameter, show that the trajectory lies on the graph of a quadratic polynomial.

(b) Use a graphing utility to sketch the trajectory if $\alpha = 30°$ and $v_0 = 1000$ m/s.

(c) Using the trajectory in part (b), how high does the shell rise?

(cont.)

(d) Using the trajectory in part (b), how far does the shell travel horizontally?

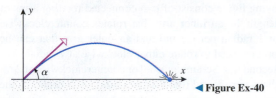

◀ **Figure Ex-40**

FOCUS ON CONCEPTS

41. (a) Find the slope of the tangent line to the parametric curve $x = t/2$, $y = t^2 + 1$ at $t = -1$ and at $t = 1$ without eliminating the parameter.
(b) Check your answers in part (a) by eliminating the parameter and differentiating an appropriate function of x.

42. (a) Find the slope of the tangent line to the parametric curve $x = 3\cos t$, $y = 4\sin t$ at $t = \pi/4$ and at $t = 7\pi/4$ without eliminating the parameter.
(b) Check your answers in part (a) by eliminating the parameter and differentiating an appropriate function of x.

43. For the parametric curve in Exercise 41, make a conjecture about the sign of d^2y/dx^2 at $t = -1$ and at $t = 1$, and confirm your conjecture without eliminating the parameter.

44. For the parametric curve in Exercise 42, make a conjecture about the sign of d^2y/dx^2 at $t = \pi/4$ and at $t = 7\pi/4$, and confirm your conjecture without eliminating the parameter.

45–50 Find dy/dx and d^2y/dx^2 at the given point without eliminating the parameter. ■

45. $x = \sqrt{t}$, $y = 2t + 4$; $t = 1$

46. $x = \frac{1}{2}t^2 + 1$, $y = \frac{1}{3}t^3 - t$; $t = 2$

47. $x = \sec t$, $y = \tan t$; $t = \pi/3$

48. $x = \sinh t$, $y = \cosh t$; $t = 0$

49. $x = \theta + \cos\theta$, $y = 1 + \sin\theta$; $\theta = \pi/6$

50. $x = \cos\phi$, $y = 3\sin\phi$; $\phi = 5\pi/6$

51. (a) Find the equation of the tangent line to the curve
$$x = e^t, \quad y = e^{-t}$$
at $t = 1$ without eliminating the parameter.
(b) Find the equation of the tangent line in part (a) by eliminating the parameter.

52. (a) Find the equation of the tangent line to the curve
$$x = 2t + 4, \quad y = 8t^2 - 2t + 4$$
at $t = 1$ without eliminating the parameter.
(b) Find the equation of the tangent line in part (a) by eliminating the parameter.

53–54 Find all values of t at which the parametric curve has (a) a horizontal tangent line and (b) a vertical tangent line. ■

53. $x = 2\sin t$, $y = 4\cos t$ $(0 \leq t \leq 2\pi)$

54. $x = 2t^3 - 15t^2 + 24t + 7$, $y = t^2 + t + 1$

55. In the mid-1850s the French physicist Jules Antoine Lissajous (1822–1880) became interested in parametric equations of the form
$$x = \sin at, \quad y = \sin bt$$
in the course of studying vibrations that combine two perpendicular sinusoidal motions. If a/b is a rational number, then the combined effect of the oscillations is a periodic motion along a path called a *Lissajous curve*.
(a) Use a graphing utility to generate the complete graph of the Lissajous curves corresponding to $a = 1$, $b = 2$; $a = 2$, $b = 3$; $a = 3$, $b = 4$; and $a = 4$, $b = 5$.
(b) The Lissajous curve
$$x = \sin t, \quad y = \sin 2t \quad (0 \leq t \leq 2\pi)$$
crosses itself at the origin (see Figure Ex-55). Find equations for the two tangent lines at the origin.

56. The *prolate cycloid*
$$x = 2 - \pi\cos t, \quad y = 2t - \pi\sin t \quad (-\pi \leq t \leq \pi)$$
crosses itself at a point on the x-axis (see the accompanying figure). Find equations for the two tangent lines at that point.

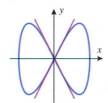

▲ **Figure Ex-55** ▲ **Figure Ex-56**

57. Show that the curve $x = t^2$, $y = t^3 - 4t$ intersects itself at the point $(4, 0)$, and find equations for the two tangent lines to the curve at the point of intersection.

58. Show that the curve with parametric equations
$$x = t^2 - 3t + 5, \quad y = t^3 + t^2 - 10t + 9$$
intersects itself at the point $(3, 1)$, and find equations for the two tangent lines to the curve at the point of intersection.

59. (a) Use a graphing utility to generate the graph of the parametric curve
$$x = \cos^3 t, \quad y = \sin^3 t \quad (0 \leq t \leq 2\pi)$$
and make a conjecture about the values of t at which singular points occur.
(b) Confirm your conjecture in part (a) by calculating appropriate derivatives.

60. Verify that the cycloid described by Formula (10) has cusps at its x-intercepts and horizontal tangent lines at midpoints between adjacent x-intercepts (see Figure 10.1.14).

61. (a) What is the slope of the tangent line at time t to the trajectory of the paper airplane in Example 5?

(b) What was the airplane's approximate angle of inclination when it crashed into the wall?

62. Suppose that a bee follows the trajectory

$$x = t - 2\cos t, \quad y = 2 - 2\sin t \quad (0 \le t \le 10)$$

(a) At what times was the bee flying horizontally?

(b) At what times was the bee flying vertically?

63. Consider the family of curves described by the parametric equations

$$x = a\cos t + h, \quad y = b\sin t + k \quad (0 \le t < 2\pi)$$

where $a \ne 0$ and $b \ne 0$. Describe the curves in this family if

(a) h and k are fixed but a and b can vary

(b) a and b are fixed but h and k can vary

(c) $a = 1$ and $b = 1$, but h and k vary so that $h = k + 1$.

64. (a) Use a graphing utility to study how the curves in the family

$$x = 2a\cos^2 t, \quad y = 2a\cos t \sin t \quad (-2\pi < t < 2\pi)$$

change as a varies from 0 to 5.

(b) Confirm your conclusion algebraically.

(c) Write a brief paragraph that describes your findings.

65–70 Find the exact arc length of the curve over the stated interval.

65. $x = t^2, \quad y = \frac{1}{3}t^3 \quad (0 \le t \le 1)$

66. $x = \sqrt{t} - 2, \quad y = 2t^{3/4} \quad (1 \le t \le 16)$

67. $x = \cos 3t, \quad y = \sin 3t \quad (0 \le t \le \pi)$

68. $x = \sin t + \cos t, \quad y = \sin t - \cos t \quad (0 \le t \le \pi)$

69. $x = e^{2t}(\sin t + \cos t), \quad y = e^{2t}(\sin t - \cos t) \quad (-1 \le t \le 1)$

70. $x = 2\sin^{-1} t, \quad y = \ln(1 - t^2) \quad (0 \le t \le \frac{1}{2})$

C **71.** (a) Use Formula (9) to show that the length L of one arch of a cycloid is given by

$$L = a\int_0^{2\pi} \sqrt{2(1 - \cos\theta)}\, d\theta$$

(b) Use a CAS to show that L is eight times the radius of the wheel that generates the cycloid (see the accompanying figure).

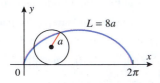

$L = 8a$

◀ **Figure Ex-71**

72. Use the parametric equations in Formula (10) to verify that the cycloid provides one solution to the differential equation

$$\left(1 + \left(\frac{dy}{dx}\right)^2\right) y = 2a$$

where a is a positive constant.

73. The amusement park rides illustrated in the accompanying figure consist of two connected rotating arms of length 1—an inner arm that rotates counterclockwise at 1 radian per second and an outer arm that can be programmed to rotate either clockwise at 2 radians per second (the Scrambler ride) or counterclockwise at 2 radians per second (the Calypso ride). The center of the rider cage is at the end of the outer arm.

(a) Show that in the Scrambler ride the center of the cage has parametric equations

$$x = \cos t + \cos 2t, \quad y = \sin t - \sin 2t$$

(b) Find parametric equations for the center of the cage in the Calypso ride, and use a graphing utility to confirm that the center traces the curve shown in the accompanying figure.

(c) Do you think that a rider travels the same distance in one revolution of the Scrambler ride as in one revolution of the Calypso ride? Justify your conclusion.

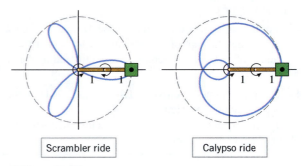

| Scrambler ride | Calypso ride |

▲ **Figure Ex-73**

74. (a) If a thread is unwound from a fixed circle while being held taut (i.e., tangent to the circle), then the end of the thread traces a curve called an ***involute of a circle***. Show that if the circle is centered at the origin, has radius a, and the end of the thread is initially at the point $(a, 0)$, then the involute can be expressed parametrically as

$$x = a(\cos\theta + \theta\sin\theta), \quad y = a(\sin\theta - \theta\cos\theta)$$

where θ is the angle shown in part (a) of the accompanying figure on the next page.

(b) Assuming that the dog in part (b) of the accompanying figure on the next page unwinds its leash while keeping it taut, for what values of θ in the interval $0 \le \theta \le 2\pi$ will the dog be walking North? South? East? West?

(c) Use a graphing utility to generate the curve traced by the dog, and show that it is consistent with your answer in part (b).

(a)   (b)

▲ **Figure Ex-74**

75–80 If $f'(t)$ and $g'(t)$ are continuous functions, and if no segment of the curve

$$x = f(t), \quad y = g(t) \quad (a \le t \le b)$$

is traced more than once, then it can be shown that the area of the surface generated by revolving this curve about the x-axis is

$$S = \int_a^b 2\pi y \sqrt{\left(\frac{dx}{dt}\right)^2 + \left(\frac{dy}{dt}\right)^2}\, dt$$

and the area of the surface generated by revolving the curve about the y-axis is

$$S = \int_a^b 2\pi x \sqrt{\left(\frac{dx}{dt}\right)^2 + \left(\frac{dy}{dt}\right)^2}\, dt$$

[The derivations are similar to those used to obtain Formulas (4) and (5) in Section 6.5.] Use the formulas above in these exercises. ■

75. Find the area of the surface generated by revolving $x = t^2$, $y = 3t$ $(0 \le t \le 2)$ about the x-axis.

76. Find the area of the surface generated by revolving the curve $x = e^t \cos t$, $y = e^t \sin t$ $(0 \le t \le \pi/2)$ about the x-axis.

77. Find the area of the surface generated by revolving the curve $x = \cos^2 t$, $y = \sin^2 t$ $(0 \le t \le \pi/2)$ about the y-axis.

78. Find the area of the surface generated by revolving $x = 6t$, $y = 4t^2$ $(0 \le t \le 1)$ about the y-axis.

79. By revolving the semicircle

$$x = r \cos t, \quad y = r \sin t \quad (0 \le t \le \pi)$$

about the x-axis, show that the surface area of a sphere of radius r is $4\pi r^2$.

80. The equations

$$x = a\phi - a \sin\phi, \quad y = a - a\cos\phi \quad (0 \le \phi \le 2\pi)$$

represent one arch of a cycloid. Show that the surface area generated by revolving this curve about the x-axis is given by $S = 64\pi a^2/3$.

81. Writing Consult appropriate reference works and write an essay on American mathematician Nathaniel Bowditch (1773–1838) and his investigation of *Bowditch curves* (better known as Lissajous curves; see Exercise 55).

82. Writing What are some of the advantages of expressing a curve parametrically rather than in the form $y = f(x)$?

✔ **QUICK CHECK ANSWERS 10.1**

1. $x = 3 + 2\cos t$, $y = 5 + 2\sin t$ $(0 \le t \le 2\pi)$ **2.** $\frac{3}{4}$; 2.75 **3.** $x = f(1 - t)$, $y = g(1 - t)$ **4.** $\dfrac{dy/dt}{dx/dt} = \dfrac{g'(t)}{f'(t)}$

5. $\displaystyle\int_1^\pi \sqrt{(1/t)^2 + \cos^2 t}\, dt$

10.2 POLAR COORDINATES

Up to now we have specified the location of a point in the plane by means of coordinates relative to two perpendicular coordinate axes. However, sometimes a moving point has a special affinity for some fixed point, such as a planet moving in an orbit under the central attraction of the Sun. In such cases, the path of the particle is best described by its angular direction and its distance from the fixed point. In this section we will discuss a new kind of coordinate system that is based on this idea.

■ **POLAR COORDINATE SYSTEMS**

A *polar coordinate system* in a plane consists of a fixed point O, called the *pole* (or *origin*), and a ray emanating from the pole, called the *polar axis*. In such a coordinate system

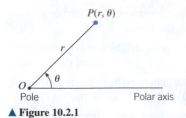

▲ Figure 10.2.1

we can associate with each point P in the plane a pair of ***polar coordinates*** (r, θ), where r is the distance from P to the pole and θ is an angle from the polar axis to the ray OP (Figure 10.2.1). The number r is called the ***radial coordinate*** of P and the number θ the ***angular coordinate*** (or ***polar angle***) of P. In Figure 10.2.2, the points $(6, \pi/4)$, $(5, 2\pi/3)$, $(3, 5\pi/4)$, and $(4, 11\pi/6)$ are plotted in polar coordinate systems. If P is the pole, then $r = 0$, but there is no clearly defined polar angle. We will agree that an arbitrary angle can be used in this case; that is, $(0, \theta)$ are polar coordinates of the pole for all choices of θ.

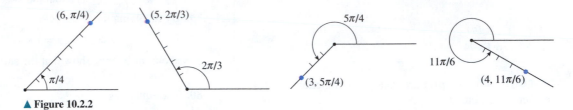

▲ Figure 10.2.2

The polar coordinates of a point are not unique. For example, the polar coordinates

$$(1, 7\pi/4), \quad (1, -\pi/4), \quad \text{and} \quad (1, 15\pi/4)$$

all represent the same point (Figure 10.2.3).

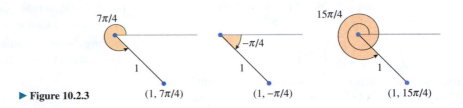

▶ Figure 10.2.3

In general, if a point P has polar coordinates (r, θ), then

$$(r, \theta + 2n\pi) \quad \text{and} \quad (r, \theta - 2n\pi)$$

are also polar coordinates of P for any nonnegative integer n. Thus, every point has infinitely many pairs of polar coordinates.

As defined above, the radial coordinate r of a point P is nonnegative, since it represents the distance from P to the pole. However, it will be convenient to allow for negative values of r as well. To motivate an appropriate definition, consider the point P with polar coordinates $(3, 5\pi/4)$. As shown in Figure 10.2.4, we can reach this point by rotating the polar axis through an angle of $5\pi/4$ and then moving 3 units from the pole along the terminal side of the angle, or we can reach the point P by rotating the polar axis through an angle of $\pi/4$ and then moving 3 units from the pole along the extension of the terminal side. This suggests that the point $(3, 5\pi/4)$ might also be denoted by $(-3, \pi/4)$, with the minus sign serving to indicate that the point is on the *extension* of the angle's terminal side rather than on the terminal side itself.

In general, the terminal side of the angle $\theta + \pi$ is the extension of the terminal side of θ, so we define negative radial coordinates by agreeing that

$$(-r, \theta) \quad \text{and} \quad (r, \theta + \pi)$$

are polar coordinates of the same point.

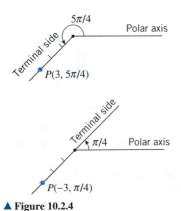

▲ Figure 10.2.4

■ RELATIONSHIP BETWEEN POLAR AND RECTANGULAR COORDINATES

Frequently, it will be useful to superimpose a rectangular xy-coordinate system on top of a polar coordinate system, making the positive x-axis coincide with the polar axis. If this is done, then every point P will have both rectangular coordinates (x, y) and polar coordinates

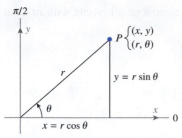

▲ Figure 10.2.5

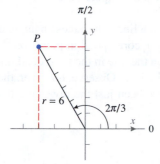

▲ Figure 10.2.6

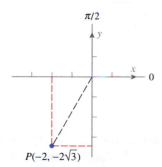

▲ Figure 10.2.7

(r, θ). As suggested by Figure 10.2.5, these coordinates are related by the equations

$$x = r \cos \theta, \quad y = r \sin \theta \tag{1}$$

These equations are well suited for finding x and y when r and θ are known. However, to find r and θ when x and y are known, it is preferable to use the identities $\sin^2 \theta + \cos^2 \theta = 1$ and $\tan \theta = \sin \theta / \cos \theta$ to rewrite (1) as

$$r^2 = x^2 + y^2, \quad \tan \theta = \frac{y}{x} \tag{2}$$

▶ **Example 1** Find the rectangular coordinates of the point P whose polar coordinates are $(r, \theta) = (6, 2\pi/3)$ (Figure 10.2.6).

Solution. Substituting the polar coordinates $r = 6$ and $\theta = 2\pi/3$ in (1) yields

$$x = 6 \cos \frac{2\pi}{3} = 6 \left(-\frac{1}{2} \right) = -3$$

$$y = 6 \sin \frac{2\pi}{3} = 6 \left(\frac{\sqrt{3}}{2} \right) = 3\sqrt{3}$$

Thus, the rectangular coordinates of P are $(x, y) = (-3, 3\sqrt{3})$. ◀

▶ **Example 2** Find polar coordinates of the point P whose rectangular coordinates are $(-2, -2\sqrt{3})$ (Figure 10.2.7).

Solution. We will find the polar coordinates (r, θ) of P that satisfy the conditions $r > 0$ and $0 \leq \theta < 2\pi$. From the first equation in (2),

$$r^2 = x^2 + y^2 = (-2)^2 + (-2\sqrt{3})^2 = 4 + 12 = 16$$

so $r = 4$. From the second equation in (2),

$$\tan \theta = \frac{y}{x} = \frac{-2\sqrt{3}}{-2} = \sqrt{3}$$

From this and the fact that $(-2, -2\sqrt{3})$ lies in the third quadrant, it follows that the angle satisfying the requirement $0 \leq \theta < 2\pi$ is $\theta = 4\pi/3$. Thus, $(r, \theta) = (4, 4\pi/3)$ are polar coordinates of P. All other polar coordinates of P are expressible in the form

$$\left(4, \frac{4\pi}{3} + 2n\pi \right) \quad \text{or} \quad \left(-4, \frac{\pi}{3} + 2n\pi \right)$$

where n is an integer. ◀

■ GRAPHS IN POLAR COORDINATES

We will now consider the problem of graphing equations in r and θ, where θ is assumed to be measured in radians. Some examples of such equations are

$$r = 1, \quad \theta = \pi/4, \quad r = \theta, \quad r = \sin \theta, \quad r = \cos 2\theta$$

In a rectangular coordinate system the graph of an equation in x and y consists of all points whose coordinates (x, y) satisfy the equation. However, in a polar coordinate system, points have infinitely many different pairs of polar coordinates, so that a given point may have some polar coordinates that satisfy an equation and others that do not. Given an equation in r and θ, we define its *graph in polar coordinates* to consist of all points with *at least one* pair of coordinates (r, θ) that satisfy the equation.

▶ **Example 3**   Sketch the graphs of

$$\text{(a) } r = 1 \qquad \text{(b) } \theta = \frac{\pi}{4}$$

in polar coordinates.

Solution (a).   For all values of θ, the point $(1, \theta)$ is 1 unit away from the pole. Since θ is arbitrary, the graph is the circle of radius 1 centered at the pole (Figure 10.2.8a).

Solution (b).   For all values of r, the point $(r, \pi/4)$ lies on a line that makes an angle of $\pi/4$ with the polar axis (Figure 10.2.8b). Positive values of r correspond to points on the line in the first quadrant and negative values of r to points on the line in the third quadrant. Thus, in absence of any restriction on r, the graph is the entire line. Observe, however, that had we imposed the restriction $r \geq 0$, the graph would have been just the ray in the first quadrant. ◄

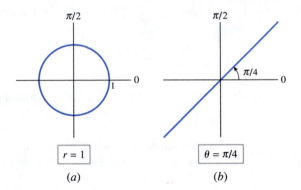

▶ **Figure 10.2.8** (a) (b)

Equations $r = f(\theta)$ that express r as a function of θ are especially important. One way to graph such an equation is to choose some typical values of θ, calculate the corresponding values of r, and then plot the resulting pairs (r, θ) in a polar coordinate system. The next two examples illustrate this process.

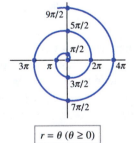

▲ **Figure 10.2.9**

Graph the spiral $r = \theta$ $(\theta \leq 0)$. Compare your graph to that in Figure 10.2.9.

▶ **Example 4**   Sketch the graph of $r = \theta$ $(\theta \geq 0)$ in polar coordinates by plotting points.

Solution.   Observe that as θ increases, so does r; thus, the graph is a curve that spirals out from the pole as θ increases. A reasonably accurate sketch of the spiral can be obtained by plotting the points that correspond to values of θ that are integer multiples of $\pi/2$, keeping in mind that the value of r is always equal to the value of θ (Figure 10.2.9). ◄

▶ **Example 5**   Sketch the graph of the equation $r = \sin \theta$ in polar coordinates by plotting points.

Solution.   Table 10.2.1 shows the coordinates of points on the graph at increments of $\pi/6$.

These points are plotted in Figure 10.2.10. Note, however, that there are 13 points listed in the table but only 6 distinct plotted points. This is because the pairs from $\theta = \pi$ on yield duplicates of the preceding points. For example, $(-1/2, 7\pi/6)$ and $(1/2, \pi/6)$ represent the same point. ◄

Table 10.2.1

θ (RADIANS)	0	$\frac{\pi}{6}$	$\frac{\pi}{3}$	$\frac{\pi}{2}$	$\frac{2\pi}{3}$	$\frac{5\pi}{6}$	π	$\frac{7\pi}{6}$	$\frac{4\pi}{3}$	$\frac{3\pi}{2}$	$\frac{5\pi}{3}$	$\frac{11\pi}{6}$	2π
$r = \sin\theta$	0	$\frac{1}{2}$	$\frac{\sqrt{3}}{2}$	1	$\frac{\sqrt{3}}{2}$	$\frac{1}{2}$	0	$-\frac{1}{2}$	$-\frac{\sqrt{3}}{2}$	-1	$-\frac{\sqrt{3}}{2}$	$-\frac{1}{2}$	0
(r, θ)	$(0,0)$	$\left(\frac{1}{2},\frac{\pi}{6}\right)$	$\left(\frac{\sqrt{3}}{2},\frac{\pi}{3}\right)$	$\left(1,\frac{\pi}{2}\right)$	$\left(\frac{\sqrt{3}}{2},\frac{2\pi}{3}\right)$	$\left(\frac{1}{2},\frac{5\pi}{6}\right)$	$(0,\pi)$	$\left(-\frac{1}{2},\frac{7\pi}{6}\right)$	$\left(-\frac{\sqrt{3}}{2},\frac{4\pi}{3}\right)$	$\left(-1,\frac{3\pi}{2}\right)$	$\left(-\frac{\sqrt{3}}{2},\frac{5\pi}{3}\right)$	$\left(-\frac{1}{2},\frac{11\pi}{6}\right)$	$(0,2\pi)$

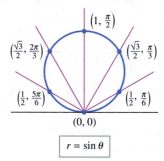

▲ **Figure 10.2.10**

Observe that the points in Figure 10.2.10 appear to lie on a circle. We can confirm that this is so by expressing the polar equation $r = \sin\theta$ in terms of x and y. To do this, we multiply the equation through by r to obtain

$$r^2 = r\sin\theta$$

which now allows us to apply Formulas (1) and (2) to rewrite the equation as

$$x^2 + y^2 = y$$

Rewriting this equation as $x^2 + y^2 - y = 0$ and then completing the square yields

$$x^2 + \left(y - \tfrac{1}{2}\right)^2 = \tfrac{1}{4}$$

which is a circle of radius $\frac{1}{2}$ centered at the point $\left(0, \frac{1}{2}\right)$ in the xy-plane.

It is often useful to view the equation $r = f(\theta)$ as an equation in rectangular coordinates (rather than polar coordinates) and graphed in a rectangular θr-coordinate system. For example, Figure 10.2.11 shows the graph of $r = \sin\theta$ displayed using rectangular θr-coordinates. This graph can actually help to visualize how the polar graph in Figure 10.2.10 is generated:

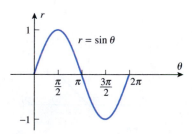

▲ **Figure 10.2.11**

- At $\theta = 0$ we have $r = 0$, which corresponds to the pole $(0, 0)$ on the polar graph.
- As θ varies from 0 to $\pi/2$, the value of r increases from 0 to 1, so the point (r, θ) moves along the circle from the pole to the high point at $(1, \pi/2)$.
- As θ varies from $\pi/2$ to π, the value of r decreases from 1 back to 0, so the point (r, θ) moves along the circle from the high point back to the pole.
- As θ varies from π to $3\pi/2$, the values of r are negative, varying from 0 to -1. Thus, the point (r, θ) moves along the circle from the pole to the high point at $(1, \pi/2)$, which is the same as the point $(-1, 3\pi/2)$. This duplicates the motion that occurred for $0 \leq \theta \leq \pi/2$.
- As θ varies from $3\pi/2$ to 2π, the value of r varies from -1 to 0. Thus, the point (r, θ) moves along the circle from the high point back to the pole, duplicating the motion that occurred for $\pi/2 \leq \theta \leq \pi$.

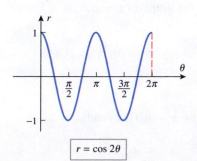

▲ **Figure 10.2.12**

▶ **Example 6**   Sketch the graph of $r = \cos 2\theta$ in polar coordinates.

Solution.   Instead of plotting points, we will use the graph of $r = \cos 2\theta$ in rectangular coordinates (Figure 10.2.12) to visualize how the polar graph of this equation is generated. The analysis and the resulting polar graph are shown in Figure 10.2.13. This curve is called a *four-petal rose*. ◄

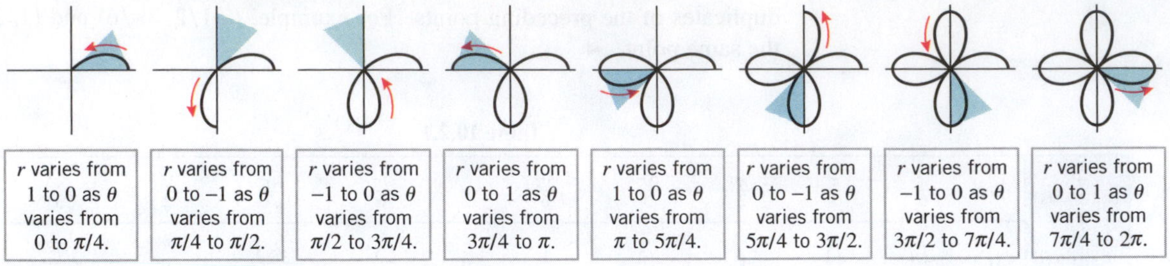

r varies from 1 to 0 as θ varies from 0 to π/4.
r varies from 0 to −1 as θ varies from π/4 to π/2.
r varies from −1 to 0 as θ varies from π/2 to 3π/4.
r varies from 0 to 1 as θ varies from 3π/4 to π.
r varies from 1 to 0 as θ varies from π to 5π/4.
r varies from 0 to −1 as θ varies from 5π/4 to 3π/2.
r varies from −1 to 0 as θ varies from 3π/2 to 7π/4.
r varies from 0 to 1 as θ varies from 7π/4 to 2π.

▲ **Figure 10.2.13**

SYMMETRY TESTS

Observe that the polar graph of $r = \cos 2\theta$ in Figure 10.2.13 is symmetric about the x-axis and the y-axis. This symmetry could have been predicted from the following theorem, which is suggested by Figure 10.2.14 (we omit the proof).

10.2.1 THEOREM (*Symmetry Tests*)

(*a*) *A curve in polar coordinates is symmetric about the x-axis if replacing θ by $-\theta$ in its equation produces an equivalent equation (Figure 10.2.14a).*

(*b*) *A curve in polar coordinates is symmetric about the y-axis if replacing θ by $\pi - \theta$ in its equation produces an equivalent equation (Figure 10.2.14b).*

(*c*) *A curve in polar coordinates is symmetric about the origin if replacing θ by $\theta + \pi$, or replacing r by $-r$ in its equation produces an equivalent equation (Figure 10.2.14c).*

The converse of each part of Theorem 10.2.1 is false. See Exercise 79.

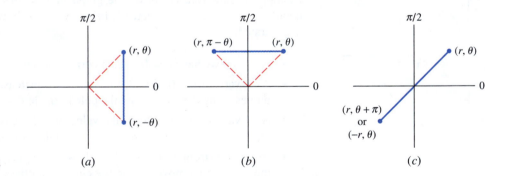

π/2
(r, θ)
0
(r, −θ)
(a)
π/2
(r, π − θ)
(r, θ)
0
(b)
π/2
(r, θ)
0
(r, θ + π)
or
(−r, θ)
(c)

► **Figure 10.2.14**

► **Example 7** Use Theorem 10.2.1 to confirm that the graph of $r = \cos 2\theta$ in Figure 10.2.13 is symmetric about the x-axis and y-axis.

Solution. To test for symmetry about the x-axis, we replace θ by $-\theta$. This yields

$$r = \cos(-2\theta) = \cos 2\theta$$

Thus, replacing θ by $-\theta$ does not alter the equation.

To test for symmetry about the y-axis, we replace θ by $\pi - \theta$. This yields

$$r = \cos 2(\pi - \theta) = \cos(2\pi - 2\theta) = \cos(-2\theta) = \cos 2\theta$$

Thus, replacing θ by $\pi - \theta$ does not alter the equation. ◄

A graph that is symmetric about both the x-axis and the y-axis is also symmetric about the origin. Use Theorem 10.2.1(c) to verify that the curve in Example 7 is symmetric about the origin.

▶ **Example 8** Sketch the graph of $r = a(1 - \cos\theta)$ in polar coordinates, assuming a to be a positive constant.

Solution. Observe first that replacing θ by $-\theta$ does not alter the equation, so we know in advance that the graph is symmetric about the polar axis. Thus, if we graph the upper half of the curve, then we can obtain the lower half by reflection about the polar axis.

As in our previous examples, we will first graph the equation in rectangular θr-coordinates. This graph, which is shown in Figure 10.2.15a, can be obtained by rewriting the given equation as $r = a - a\cos\theta$, from which we see that the graph in rectangular θr-coordinates can be obtained by first reflecting the graph of $r = a\cos\theta$ about the x-axis to obtain the graph of $r = -a\cos\theta$, and then translating that graph up a units to obtain the graph of $r = a - a\cos\theta$. Now we can see the following:

- As θ varies from 0 to $\pi/3$, r increases from 0 to $a/2$.
- As θ varies from $\pi/3$ to $\pi/2$, r increases from $a/2$ to a.
- As θ varies from $\pi/2$ to $2\pi/3$, r increases from a to $3a/2$.
- As θ varies from $2\pi/3$ to π, r increases from $3a/2$ to $2a$.

This produces the polar curve shown in Figure 10.2.15b. The rest of the curve can be obtained by continuing the preceding analysis from π to 2π or, as noted above, by reflecting the portion already graphed about the x-axis (Figure 10.2.15c). This heart-shaped curve is called a *cardioid* (from the Greek word *kardia* meaning "heart"). ◀

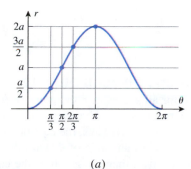

(a)

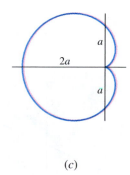

$$r = a(1 - \cos\theta)$$

(b)

(c)

▲ **Figure 10.2.15**

▶ **Example 9** Sketch the graph of $r^2 = 4\cos 2\theta$ in polar coordinates.

Solution. This equation does not express r as a function of θ, since solving for r in terms of θ yields two functions:

$$r = 2\sqrt{\cos 2\theta} \quad \text{and} \quad r = -2\sqrt{\cos 2\theta}$$

Thus, to graph the equation $r^2 = 4\cos 2\theta$ we will have to graph the two functions separately and then combine those graphs.

We will start with the graph of $r = 2\sqrt{\cos 2\theta}$. Observe first that this equation is not changed if we replace θ by $-\theta$ or if we replace θ by $\pi - \theta$. Thus, the graph is symmetric about the x-axis and the y-axis. This means that the entire graph can be obtained by graphing the portion in the first quadrant, reflecting that portion about the y-axis to obtain the portion in the second quadrant, and then reflecting those two portions about the x-axis to obtain the portions in the third and fourth quadrants.

To begin the analysis, we will graph the equation $r = 2\sqrt{\cos 2\theta}$ in rectangular θr-coordinates (see Figure 10.2.16a). Note that there are gaps in that graph over the intervals $\pi/4 < \theta < 3\pi/4$ and $5\pi/4 < \theta < 7\pi/4$ because $\cos 2\theta$ is negative for those values of θ. From this graph we can see the following:

- As θ varies from 0 to $\pi/4$, r decreases from 2 to 0.
- As θ varies from $\pi/4$ to $\pi/2$, no points are generated on the polar graph.

This produces the portion of the graph shown in Figure 10.2.16b. As noted above, we can complete the graph by a reflection about the y-axis followed by a reflection about the x-axis (Figure 10.2.16c). The resulting propeller-shaped graph is called a **lemniscate** (from the Greek word *lemniscos* for a looped ribbon resembling the number 8). We leave it for you to verify that the equation $r = 2\sqrt{\cos 2\theta}$ has the same graph as $r = -2\sqrt{\cos 2\theta}$, but traced in a diagonally opposite manner. Thus, the graph of the equation $r^2 = 4\cos 2\theta$ consists of two identical superimposed lemniscates. ◄

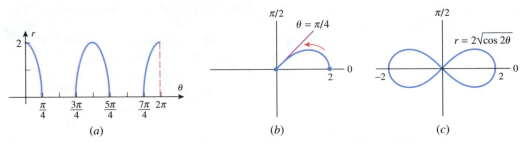

(a)   (b)   (c)

▲ **Figure 10.2.16**

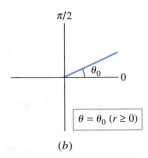

(a)

(b)

▲ **Figure 10.2.17**

■ FAMILIES OF LINES AND RAYS THROUGH THE POLE

If θ_0 is a fixed angle, then for all values of r the point (r, θ_0) lies on the line that makes an angle of $\theta = \theta_0$ with the polar axis; and, conversely, every point on this line has a pair of polar coordinates of the form (r, θ_0). Thus, the equation $\theta = \theta_0$ represents the line that passes through the pole and makes an angle of θ_0 with the polar axis (Figure 10.2.17a). If r is restricted to be nonnegative, then the graph of the equation $\theta = \theta_0$ is the ray that emanates from the pole and makes an angle of θ_0 with the polar axis (Figure 10.2.17b). Thus, as θ_0 varies, the equation $\theta = \theta_0$ produces either a family of lines through the pole or a family of rays through the pole, depending on the restrictions on r.

■ FAMILIES OF CIRCLES

We will consider three families of circles in which a is assumed to be a positive constant:

$$r = a \qquad r = 2a\cos\theta \qquad r = 2a\sin\theta \qquad (3\text{–}5)$$

The equation $r = a$ represents a circle of radius a centered at the pole (Figure 10.2.18a). Thus, as a varies, this equation produces a family of circles centered at the pole. For families (4) and (5), recall from plane geometry that a triangle that is inscribed in a circle with a diameter of the circle for a side must be a right triangle. Thus, as indicated in Figures 10.2.18b and 10.2.18c, the equation $r = 2a\cos\theta$ represents a circle of radius a, centered on the x-axis and tangent to the y-axis at the origin; similarly, the equation $r = 2a\sin\theta$ represents a circle of radius a, centered on the y-axis and tangent to the x-axis at the origin. Thus, as a varies, Equations (4) and (5) produce the families illustrated in Figures 10.2.18d and 10.2.18e.

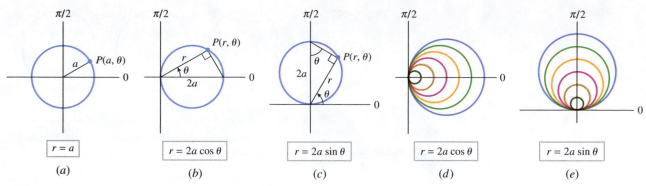

▲ Figure 10.2.18

■ FAMILIES OF ROSE CURVES

In polar coordinates, equations of the form

$$r = a\sin n\theta \qquad r = a\cos n\theta \tag{6--7}$$

in which $a > 0$ and n is a positive integer represent families of flower-shaped curves called **roses** (Figure 10.2.19). The rose consists of n equally spaced petals of radius a if n is odd and $2n$ equally spaced petals of radius a if n is even. It can be shown that a rose with an even number of petals is traced out exactly once as θ varies over the interval $0 \le \theta < 2\pi$ and a rose with an odd number of petals is traced out exactly once as θ varies over the interval $0 \le \theta < \pi$ (Exercise 78). A four-petal rose of radius 1 was graphed in Example 6.

ROSE CURVES

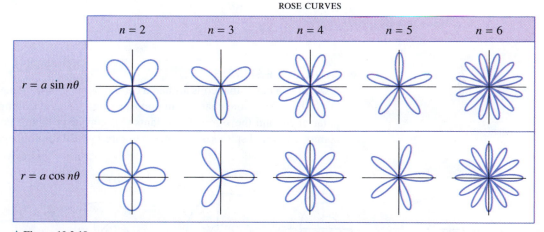

▲ Figure 10.2.19

■ FAMILIES OF CARDIOIDS AND LIMAÇONS

Equations with any of the four forms

$$r = a \pm b\sin\theta \qquad r = a \pm b\cos\theta \tag{8--9}$$

in which $a > 0$ and $b > 0$ represent polar curves called **limaçons** (from the Latin word *limax* for a snail-like creature that is commonly called a "slug"). There are four possible shapes for a limaçon that are determined by the ratio a/b (Figure 10.2.20). If $a = b$ (the case $a/b = 1$), then the limaçon is called a **cardioid** because of its heart-shaped appearance, as noted in Example 8.

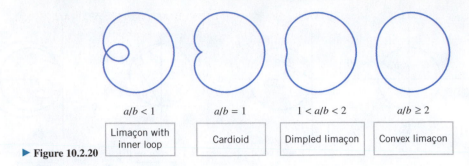

$a/b < 1$	$a/b = 1$	$1 < a/b < 2$	$a/b \geq 2$
Limaçon with inner loop	Cardioid	Dimpled limaçon	Convex limaçon

▶ **Figure 10.2.20**

▶ **Example 10**   Figure 10.2.21 shows the family of limaçons $r = a + \cos\theta$ with the constant a varying from 0.25 to 2.50 in steps of 0.25. In keeping with Figure 10.2.20, the limaçons evolve from the loop type to the convex type. As a increases from the starting value of 0.25, the loops get smaller and smaller until the cardioid is reached at $a = 1$. As a increases further, the limaçons evolve through the dimpled type into the convex type. ◀

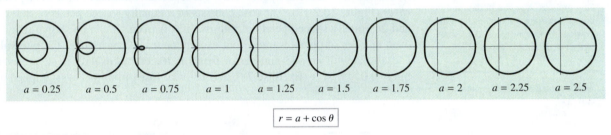

| $a = 0.25$ | $a = 0.5$ | $a = 0.75$ | $a = 1$ | $a = 1.25$ | $a = 1.5$ | $a = 1.75$ | $a = 2$ | $a = 2.25$ | $a = 2.5$ |

$$r = a + \cos\theta$$

▲ **Figure 10.2.21**

■ FAMILIES OF SPIRALS

A *spiral* is a curve that coils around a central point. Spirals generally have "left-hand" and "right-hand" versions that coil in opposite directions, depending on the restrictions on the polar angle and the signs of constants that appear in their equations. Some of the more common types of spirals are shown in Figure 10.2.22 for nonnegative values of θ, a, and b.

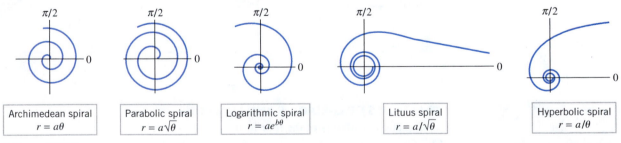

Archimedean spiral	Parabolic spiral	Logarithmic spiral	Lituus spiral	Hyperbolic spiral
$r = a\theta$	$r = a\sqrt{\theta}$	$r = ae^{b\theta}$	$r = a/\sqrt{\theta}$	$r = a/\theta$

▲ **Figure 10.2.22**

■ SPIRALS IN NATURE

Spirals of many kinds occur in nature. For example, the shell of the chambered nautilus (below) forms a logarithmic spiral, and a coiled sailor's rope forms an Archimedean spiral. Spirals also occur in flowers, the tusks of certain animals, and in the shapes of galaxies.

Thomas Taylor/Photo Researchers

The shell of the chambered nautilus reveals a logarithmic spiral. The animal lives in the outermost chamber.

Rex Ziak/Stone/Getty Images

A sailor's coiled rope forms an Archimedean spiral.

Courtesy NASA & The Hubble Heritage Team

A spiral galaxy.

■ GENERATING POLAR CURVES WITH GRAPHING UTILITIES

For polar curves that are too complicated for hand computation, graphing utilities can be used. Although many graphing utilities are capable of graphing polar curves directly, some are not. However, if a graphing utility is capable of graphing parametric equations, then it can be used to graph a polar curve $r = f(\theta)$ by converting this equation to parametric form. This can be done by substituting $f(\theta)$ for r in (1). This yields

$$x = f(\theta)\cos\theta, \quad y = f(\theta)\sin\theta \tag{10}$$

which is a pair of parametric equations for the polar curve in terms of the parameter θ.

▶ **Example 11** Express the polar equation

$$r = 2 + \cos\frac{5\theta}{2}$$

parametrically, and generate the polar graph from the parametric equations using a graphing utility.

Solution. Substituting the given expression for r in $x = r\cos\theta$ and $y = r\sin\theta$ yields the parametric equations

$$x = \left[2 + \cos\frac{5\theta}{2}\right]\cos\theta, \quad y = \left[2 + \cos\frac{5\theta}{2}\right]\sin\theta$$

Next, we need to find an interval over which to vary θ to produce the entire graph. To find such an interval, we will look for the smallest number of complete revolutions that must occur until the value of r begins to repeat. Algebraically, this amounts to finding the smallest positive integer n such that

$$2 + \cos\left(\frac{5(\theta + 2n\pi)}{2}\right) = 2 + \cos\frac{5\theta}{2}$$

or

$$\cos\left(\frac{5\theta}{2} + 5n\pi\right) = \cos\frac{5\theta}{2}$$

TECHNOLOGY MASTERY

Use a graphing utility to duplicate the curve in Figure 10.2.23. If your graphing utility requires that t be used as the parameter, then you will have to replace θ by t in (10) to generate the graph.

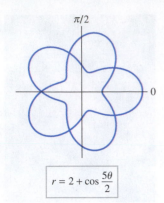

$$r = 2 + \cos\frac{5\theta}{2}$$

▲ **Figure 10.2.23**

For this equality to hold, the quantity $5n\pi$ must be an even multiple of π; the smallest n for which this occurs is $n = 2$. Thus, the entire graph will be traced in two revolutions, which means it can be generated from the parametric equations

$$x = \left[2 + \cos\frac{5\theta}{2}\right]\cos\theta, \quad y = \left[2 + \cos\frac{5\theta}{2}\right]\sin\theta \quad (0 \le \theta \le 4\pi)$$

This yields the graph in Figure 10.2.23. ◀

✔ QUICK CHECK EXERCISES 10.2   (See page 719 for answers.)

1. (a) Rectangular coordinates of a point (x, y) may be recovered from its polar coordinates (r, θ) by means of the equations $x = $ _____ and $y = $ _____.

(b) Polar coordinates (r, θ) may be recovered from rectangular coordinates (x, y) by means of the equations $r^2 = $ _____ and $\tan\theta = $ _____.

2. Find the rectangular coordinates of the points whose polar coordinates are given.

(a) $(4, \pi/3)$ (b) $(2, -\pi/6)$

(c) $(6, -2\pi/3)$ (d) $(4, 5\pi/4)$

3. In each part, find polar coordinates satisfying the stated conditions for the point whose rectangular coordinates are $(1, \sqrt{3})$.

(a) $r \ge 0$ and $0 \le \theta < 2\pi$
(b) $r \le 0$ and $0 \le \theta < 2\pi$

4. In each part, state the name that describes the polar curve most precisely: a rose, a line, a circle, a limaçon, a cardioid, a spiral, a lemniscate, or none of these.

(a) $r = 1 - \theta$ (b) $r = 1 + 2\sin\theta$
(c) $r = \sin 2\theta$ (d) $r = \cos^2\theta$
(e) $r = \csc\theta$ (f) $r = 2 + 2\cos\theta$
(g) $r = -2\sin\theta$

EXERCISE SET 10.2   ⁓ Graphing Utility

1–2 Plot the points in polar coordinates.

1. (a) $(3, \pi/4)$ (b) $(5, 2\pi/3)$ (c) $(1, \pi/2)$
(d) $(4, 7\pi/6)$ (e) $(-6, -\pi)$ (f) $(-1, 9\pi/4)$

2. (a) $(2, -\pi/3)$ (b) $(3/2, -7\pi/4)$ (c) $(-3, 3\pi/2)$
(d) $(-5, -\pi/6)$ (e) $(2, 4\pi/3)$ (f) $(0, \pi)$

3–4 Find the rectangular coordinates of the points whose polar coordinates are given. ■

3. (a) $(6, \pi/6)$ (b) $(7, 2\pi/3)$ (c) $(-6, -5\pi/6)$
(d) $(0, -\pi)$ (e) $(7, 17\pi/6)$ (f) $(-5, 0)$

4. (a) $(-2, \pi/4)$ (b) $(6, -\pi/4)$ (c) $(4, 9\pi/4)$
(d) $(3, 0)$ (e) $(-4, -3\pi/2)$ (f) $(0, 3\pi)$

5. In each part, a point is given in rectangular coordinates. Find two pairs of polar coordinates for the point, one pair satisfying $r \ge 0$ and $0 \le \theta < 2\pi$, and the second pair satisfying $r \ge 0$ and $-2\pi < \theta \le 0$.

(a) $(-5, 0)$ (b) $(2\sqrt{3}, -2)$ (c) $(0, -2)$
(d) $(-8, -8)$ (e) $(-3, 3\sqrt{3})$ (f) $(1, 1)$

6. In each part, find polar coordinates satisfying the stated conditions for the point whose rectangular coordinates are $(-\sqrt{3}, 1)$.

(a) $r \ge 0$ and $0 \le \theta < 2\pi$
(b) $r \le 0$ and $0 \le \theta < 2\pi$
(c) $r \ge 0$ and $-2\pi < \theta \le 0$
(d) $r \le 0$ and $-\pi < \theta \le \pi$

7–8 Use a calculating utility, where needed, to approximate the polar coordinates of the points whose rectangular coordinates are given. ■

7. (a) $(3, 4)$ (b) $(6, -8)$ (c) $(-1, \tan^{-1} 1)$

8. (a) $(-3, 4)$ (b) $(-3, 1.7)$ (c) $\left(2, \sin^{-1}\frac{1}{2}\right)$

9–10 Identify the curve by transforming the given polar equation to rectangular coordinates. ■

9. (a) $r = 2$ (b) $r\sin\theta = 4$
(c) $r = 3\cos\theta$ (d) $r = \dfrac{6}{3\cos\theta + 2\sin\theta}$

10. (a) $r = 5\sec\theta$ (b) $r = 2\sin\theta$
(c) $r = 4\cos\theta + 4\sin\theta$ (d) $r = \sec\theta\tan\theta$

11–12 Express the given equations in polar coordinates. ■

11. (a) $x = 3$ (b) $x^2 + y^2 = 7$
(c) $x^2 + y^2 + 6y = 0$ (d) $9xy = 4$

12. (a) $y = -3$ (b) $x^2 + y^2 = 5$
(c) $x^2 + y^2 + 4x = 0$ (d) $x^2(x^2 + y^2) = y^2$

13–16 A graph is given in a rectangular θr-coordinate system. Sketch the corresponding graph in polar coordinates. ■

13. 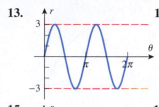   **14.**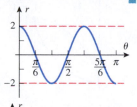

15.   **16.**

17–20 Find an equation for the given polar graph. [*Note:* Numeric labels on these graphs represent distances to the origin.] ■

17. (a)   (b)   (c)

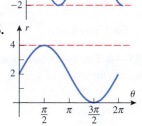

Circle   Circle   Cardioid

18. (a)   (b)   (c)

Limaçon   Circle   Three-petal rose

19. (a)   (b)   (c)

Four-petal rose   Limaçon   Lemniscate

20. (a)   (b)   (c)

Cardioid   Five-petal rose   Circle

21–46 Sketch the curve in polar coordinates. ■

21. $\theta = \dfrac{\pi}{3}$ **22.** $\theta = -\dfrac{3\pi}{4}$ **23.** $r = 3$

24. $r = 4\cos\theta$ **25.** $r = 6\sin\theta$ **26.** $r - 2 = 2\cos\theta$

27. $r = 3(1 + \sin\theta)$ **28.** $r = 5 - 5\sin\theta$

29. $r = 4 - 4\cos\theta$ **30.** $r = 1 + 2\sin\theta$

31. $r = -1 - \cos\theta$ **32.** $r = 4 + 3\cos\theta$

33. $r = 3 - \sin\theta$ **34.** $r = 3 + 4\cos\theta$

35. $r - 5 = 3\sin\theta$ **36.** $r = 5 - 2\cos\theta$

37. $r = -3 - 4\sin\theta$ **38.** $r^2 = \cos 2\theta$

39. $r^2 = 16\sin 2\theta$ **40.** $r = 4\theta \quad (\theta \geq 0)$

41. $r = 4\theta \quad (\theta \leq 0)$ **42.** $r = 4\theta$

43. $r = -2\cos 2\theta$ **44.** $r = 3\sin 2\theta$

45. $r = 9\sin 4\theta$ **46.** $r = 2\cos 3\theta$

47–50 True–False Determine whether the statement is true or false. Explain your answer. ■

47. The polar coordinate pairs $(-1, \pi/3)$ and $(1, -2\pi/3)$ describe the same point.

48. If the graph of $r = f(\theta)$ drawn in rectangular θr-coordinates is symmetric about the r-axis, then the graph of $r = f(\theta)$ drawn in polar coordinates is symmetric about the x-axis.

49. The portion of the polar graph of $r = \sin 2\theta$ for values of θ between $\pi/2$ and π is contained in the second quadrant.

50. The graph of a dimpled limaçon passes through the polar origin.

51–55 Determine a shortest parameter interval on which a complete graph of the polar equation can be generated, and then use a graphing utility to generate the polar graph. ■

51. $r = \cos\dfrac{\theta}{2}$ **52.** $r = \sin\dfrac{\theta}{2}$

53. $r = 1 - 2\sin\dfrac{\theta}{4}$ **54.** $r = 0.5 + \cos\dfrac{\theta}{3}$

55. $r = \cos\dfrac{\theta}{5}$

56. The accompanying figure shows the graph of the "butterfly curve"

$$r = e^{\cos\theta} - 2\cos 4\theta + \sin^3\dfrac{\theta}{4}$$

Determine a shortest parameter interval on which the complete butterfly can be generated, and then check your answer using a graphing utility.

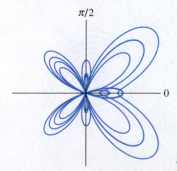

◀ **Figure Ex-56**

57. The accompanying figure shows the Archimedean spiral $r = \theta/2$ produced with a graphing calculator.

(a) What interval of values for θ do you think was used to generate the graph?

(b) Duplicate the graph with your own graphing utility.

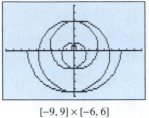

$$[-9, 9] \times [-6, 6]$$
$$x\text{Scl} = 1, y\text{Scl} = 1$$

◀ **Figure Ex-57**

58. Find equations for the two families of circles in the accompanying figure.

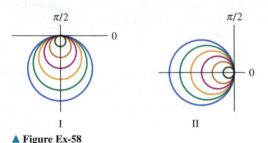

I   II

▲ **Figure Ex-58**

59. (a) Show that if a varies, then the polar equation

$$r = a \sec \theta \quad (-\pi/2 < \theta < \pi/2)$$

describes a family of lines perpendicular to the polar axis.

(b) Show that if b varies, then the polar equation

$$r = b \csc \theta \quad (0 < \theta < \pi)$$

describes a family of lines parallel to the polar axis.

FOCUS ON CONCEPTS

60. The accompanying figure shows graphs of the Archimedean spiral $r = \theta$ and the parabolic spiral $r = \sqrt{\theta}$. Which is which? Explain your reasoning.

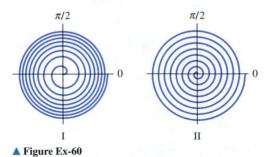

I   II

▲ **Figure Ex-60**

61–62 A polar graph of $r = f(\theta)$ is given over the stated interval. Sketch the graph of

(a) $r = f(-\theta)$

(b) $r = f\left(\theta - \dfrac{\pi}{2}\right)$

(c) $r = f\left(\theta + \dfrac{\pi}{2}\right)$

(d) $r = -f(\theta)$. ∎

61. $0 \leq \theta \leq \pi/2$

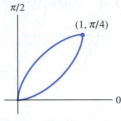

▲ **Figure Ex-61**

62. $\pi/2 \leq \theta \leq \pi$

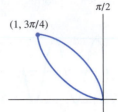

▲ **Figure Ex-62**

63–64 Use the polar graph from the indicated exercise to sketch the graph of

(a) $r = f(\theta) + 1$

(b) $r = 2f(\theta) - 1$. ∎

63. Exercise 61   **64.** Exercise 62

65. Show that if the polar graph of $r = f(\theta)$ is rotated counterclockwise around the origin through an angle α, then $r = f(\theta - \alpha)$ is an equation for the rotated curve. [*Hint:* If (r_0, θ_0) is any point on the original graph, then $(r_0, \theta_0 + \alpha)$ is a point on the rotated graph.]

66. Use the result in Exercise 65 to find an equation for the lemniscate that results when the lemniscate in Example 9 is rotated counterclockwise through an angle of $\pi/2$.

67. Use the result in Exercise 65 to find an equation for the cardioid $r = 1 + \cos \theta$ after it has been rotated through the given angle, and check your answer with a graphing utility.

(a) $\dfrac{\pi}{4}$ (b) $\dfrac{\pi}{2}$ (c) π (d) $\dfrac{5\pi}{4}$

68. (a) Show that if A and B are not both zero, then the graph of the polar equation

$$r = A \sin \theta + B \cos \theta$$

is a circle. Find its radius.

(b) Derive Formulas (4) and (5) from the formula given in part (a).

69. Find the highest point on the cardioid $r = 1 + \cos \theta$.

70. Find the leftmost point on the upper half of the cardioid $r = 1 + \cos \theta$.

71. Show that in a polar coordinate system the distance d between the points (r_1, θ_1) and (r_2, θ_2) is

$$d = \sqrt{r_1^2 + r_2^2 - 2r_1 r_2 \cos(\theta_1 - \theta_2)}$$

72–74 Use the formula obtained in Exercise 71 to find the distance between the two points indicated in polar coordinates. ∎

72. $(3, \pi/6)$ and $(2, \pi/3)$

73. Successive tips of the four-petal rose $r = \cos 2\theta$. Check your answer using geometry.

74. Successive tips of the three-petal rose $r = \sin 3\theta$. Check your answer using trigonometry.

75. In the late seventeenth century the Italian astronomer Giovanni Domenico Cassini (1625–1712) introduced the family of curves

$$(x^2 + y^2 + a^2)^2 - b^4 - 4a^2x^2 = 0 \quad (a > 0, b > 0)$$

in his studies of the relative motions of the Earth and the Sun. These curves, which are called *Cassini ovals*, have one of the three basic shapes shown in the accompanying figure.

(a) Show that if $a = b$, then the polar equation of the Cassini oval is $r^2 = 2a^2 \cos 2\theta$, which is a lemniscate.

(b) Use the formula in Exercise 71 to show that the lemniscate in part (a) is the curve traced by a point that moves in such a way that the product of its distances from the polar points $(a, 0)$ and (a, π) is a^2.

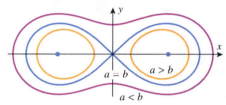

▲ **Figure Ex-75**

76–77 Vertical and horizontal asymptotes of polar curves can sometimes be detected by investigating the behavior of $x = r \cos \theta$ and $y = r \sin \theta$ as θ varies. This idea is used in these exercises. ■

76. Show that the *hyperbolic spiral* $r = 1/\theta$ ($\theta > 0$) has a horizontal asymptote at $y = 1$ by showing that $y \to 1$ and $x \to +\infty$ as $\theta \to 0^+$. Confirm this result by generating the spiral with a graphing utility.

77. Show that the spiral $r = 1/\theta^2$ does not have any horizontal asymptotes.

78. Prove that a rose with an even number of petals is traced out exactly once as θ varies over the interval $0 \le \theta < 2\pi$ and a rose with an odd number of petals is traced out exactly once as θ varies over the interval $0 \le \theta < \pi$.

79. (a) Use a graphing utility to confirm that the graph of $r = 2 - \sin(\theta/2)$ ($0 \le \theta \le 4\pi$) is symmetric about the x-axis.

(b) Show that replacing θ by $-\theta$ in the polar equation $r = 2 - \sin(\theta/2)$ does not produce an equivalent equation. Why does this not contradict the symmetry demonstrated in part (a)?

80. Writing Use a graphing utility to investigate how the family of polar curves $r = 1 + a \cos n\theta$ is affected by changing the values of a and n, where a is a positive real number and n is a positive integer. Write a brief paragraph to explain your conclusions.

81. Writing Why do you think the adjective "polar" was chosen in the name "polar coordinates"?

✔ **QUICK CHECK ANSWERS 10.2**

1. (a) $r \cos \theta$; $r \sin \theta$ (b) $x^2 + y^2$; y/x **2.** (a) $(2, 2\sqrt{3})$ (b) $(\sqrt{3}, -1)$ (c) $(-3, -3\sqrt{3})$ (d) $(-2\sqrt{2}, -2\sqrt{2})$
3. (a) $(2, \pi/3)$ (b) $(-2, 4\pi/3)$ **4.** (a) spiral (b) limaçon (c) rose (d) none of these (e) line (f) cardioid (g) circle

---

10.3 TANGENT LINES, ARC LENGTH, AND AREA FOR POLAR CURVES

In this section we will derive the formulas required to find slopes, tangent lines, and arc lengths of polar curves. We will then show how to find areas of regions that are bounded by polar curves.

■ TANGENT LINES TO POLAR CURVES

Our first objective in this section is to find a method for obtaining slopes of tangent lines to polar curves of the form $r = f(\theta)$ in which r is a differentiable function of θ. We showed in the last section that a curve of this form can be expressed parametrically in terms of the parameter θ by substituting $f(\theta)$ for r in the equations $x = r \cos \theta$ and $y = r \sin \theta$. This yields

$$x = f(\theta) \cos \theta, \quad y = f(\theta) \sin \theta$$

from which we obtain

$$\frac{dx}{d\theta} = -f(\theta)\sin\theta + f'(\theta)\cos\theta = -r\sin\theta + \frac{dr}{d\theta}\cos\theta$$

$$\frac{dy}{d\theta} = f(\theta)\cos\theta + f'(\theta)\sin\theta = r\cos\theta + \frac{dr}{d\theta}\sin\theta$$

$$(1)$$

Thus, if $dx/d\theta$ and $dy/d\theta$ are continuous and if $dx/d\theta \neq 0$, then y is a differentiable function of x, and Formula (4) in Section 10.1 with θ in place of t yields

$$\frac{dy}{dx} = \frac{dy/d\theta}{dx/d\theta} = \frac{r\cos\theta + \sin\theta\,\dfrac{dr}{d\theta}}{-r\sin\theta + \cos\theta\,\dfrac{dr}{d\theta}} \qquad (2)$$

▶ **Example 1**   Find the slope of the tangent line to the circle $r = 4\cos\theta$ at the point where $\theta = \pi/4$.

Solution.   From (2) with $r = 4\cos\theta$, so that $dr/d\theta = -4\sin\theta$, we obtain

$$\frac{dy}{dx} = \frac{4\cos^2\theta - 4\sin^2\theta}{-8\sin\theta\cos\theta} = -\frac{\cos^2\theta - \sin^2\theta}{2\sin\theta\cos\theta}$$

Using the double-angle formulas for sine and cosine,

$$\frac{dy}{dx} = -\frac{\cos 2\theta}{\sin 2\theta} = -\cot 2\theta$$

Thus, at the point where $\theta = \pi/4$ the slope of the tangent line is

$$m = \frac{dy}{dx}\bigg|_{\theta=\pi/4} = -\cot\frac{\pi}{2} = 0$$

which implies that the circle has a horizontal tangent line at the point where $\theta = \pi/4$ (Figure 10.3.1). ◀

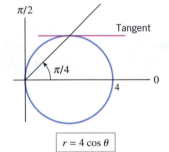

$r = 4\cos\theta$

▲ **Figure 10.3.1**

▶ **Example 2**   Find the points on the cardioid $r = 1 - \cos\theta$ at which there is a horizontal tangent line, a vertical tangent line, or a singular point.

Solution.   A horizontal tangent line will occur where $dy/d\theta = 0$ and $dx/d\theta \neq 0$, a vertical tangent line where $dy/d\theta \neq 0$ and $dx/d\theta = 0$, and a singular point where $dy/d\theta = 0$ and $dx/d\theta = 0$. We could find these derivatives from the formulas in (1). However, an alternative approach is to go back to basic principles and express the cardioid parametrically by substituting $r = 1 - \cos\theta$ in the conversion formulas $x = r\cos\theta$ and $y = r\sin\theta$. This yields

$$x = (1 - \cos\theta)\cos\theta, \quad y = (1 - \cos\theta)\sin\theta \qquad (0 \leq \theta \leq 2\pi)$$

Differentiating these equations with respect to θ and then simplifying yields (verify)

$$\frac{dx}{d\theta} = \sin\theta(2\cos\theta - 1), \quad \frac{dy}{d\theta} = (1 - \cos\theta)(1 + 2\cos\theta)$$

Thus, $dx/d\theta = 0$ if $\sin\theta = 0$ or $\cos\theta = \frac{1}{2}$, and $dy/d\theta = 0$ if $\cos\theta = 1$ or $\cos\theta = -\frac{1}{2}$. We leave it for you to solve these equations and show that the solutions of $dx/d\theta = 0$ on the interval $0 \leq \theta \leq 2\pi$ are

$$\frac{dx}{d\theta} = 0: \quad \theta = 0, \ \frac{\pi}{3}, \ \pi, \ \frac{5\pi}{3}, \ 2\pi$$

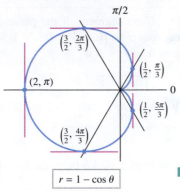

$r = 1 - \cos\theta$

▲ Figure 10.3.2

and the solutions of $dy/d\theta = 0$ on the interval $0 \leq \theta \leq 2\pi$ are

$$\frac{dy}{d\theta} = 0: \quad \theta = 0, \quad \frac{2\pi}{3}, \quad \frac{4\pi}{3}, \quad 2\pi$$

Thus, horizontal tangent lines occur at $\theta = 2\pi/3$ and $\theta = 4\pi/3$; vertical tangent lines occur at $\theta = \pi/3$, π, and $5\pi/3$; and singular points occur at $\theta = 0$ and $\theta = 2\pi$ (Figure 10.3.2). Note, however, that $r = 0$ at both singular points, so there is really only one singular point on the cardioid—the pole. ◄

■ TANGENT LINES TO POLAR CURVES AT THE ORIGIN

Formula (2) reveals some useful information about the behavior of a polar curve $r = f(\theta)$ that passes through the origin. If we assume that $r = 0$ and $dr/d\theta \neq 0$ when $\theta = \theta_0$, then it follows from Formula (2) that the slope of the tangent line to the curve at $\theta = \theta_0$ is

$$\frac{dy}{dx} = \frac{0 + \sin\theta_0 \dfrac{dr}{d\theta}}{0 + \cos\theta_0 \dfrac{dr}{d\theta}} = \frac{\sin\theta_0}{\cos\theta_0} = \tan\theta_0$$

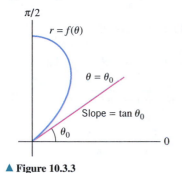

▲ Figure 10.3.3

(Figure 10.3.3). However, $\tan\theta_0$ is also the slope of the line $\theta = \theta_0$, so we can conclude that this line is tangent to the curve at the origin. Thus, we have established the following result.

10.3.1   THEOREM   *If the polar curve $r = f(\theta)$ passes through the origin at $\theta = \theta_0$, and if $dr/d\theta \neq 0$ at $\theta = \theta_0$, then the line $\theta = \theta_0$ is tangent to the curve at the origin.*

This theorem tells us that equations of the tangent lines at the origin to the curve $r = f(\theta)$ can be obtained by solving the equation $f(\theta) = 0$. It is important to keep in mind, however, that $r = f(\theta)$ may be zero for more than one value of θ, so there may be more than one tangent line at the origin. This is illustrated in the next example.

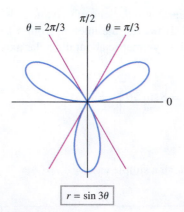

$r = \sin 3\theta$

▲ Figure 10.3.4

▶ **Example 3**   The three-petal rose $r = \sin 3\theta$ in Figure 10.3.4 has three tangent lines at the origin, which can be found by solving the equation

$$\sin 3\theta = 0$$

It was shown in Exercise 78 of Section 10.2 that the complete rose is traced once as θ varies over the interval $0 \leq \theta < \pi$, so we need only look for solutions in this interval. We leave it for you to confirm that these solutions are

$$\theta = 0, \quad \theta = \frac{\pi}{3}, \quad \text{and} \quad \theta = \frac{2\pi}{3}$$

Since $dr/d\theta = 3\cos 3\theta \neq 0$ for these values of θ, these three lines are tangent to the rose at the origin, which is consistent with the figure. ◄

■ ARC LENGTH OF A POLAR CURVE

A formula for the arc length of a polar curve $r = f(\theta)$ can be derived by expressing the curve in parametric form and applying Formula (9) of Section 10.1 for the arc length of a parametric curve. We leave it as an exercise to show the following.

10.3.2 **ARC LENGTH FORMULA FOR POLAR CURVES** If no segment of the polar curve $r = f(\theta)$ is traced more than once as θ increases from α to β, and if $dr/d\theta$ is continuous for $\alpha \leq \theta \leq \beta$, then the arc length L from $\theta = \alpha$ to $\theta = \beta$ is

$$L = \int_\alpha^\beta \sqrt{[f(\theta)]^2 + [f'(\theta)]^2}\, d\theta = \int_\alpha^\beta \sqrt{r^2 + \left(\frac{dr}{d\theta}\right)^2}\, d\theta \qquad (3)$$

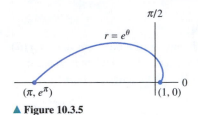

$r = e^\theta$

(π, e^π) $(1, 0)$

▲ **Figure 10.3.5**

▶ **Example 4** Find the arc length of the spiral $r = e^\theta$ in Figure 10.3.5 between $\theta = 0$ and $\theta = \pi$.

Solution.

$$L = \int_\alpha^\beta \sqrt{r^2 + \left(\frac{dr}{d\theta}\right)^2}\, d\theta = \int_0^\pi \sqrt{(e^\theta)^2 + (e^\theta)^2}\, d\theta$$

$$= \int_0^\pi \sqrt{2}\, e^\theta\, d\theta = \sqrt{2}\, e^\theta \Big]_0^\pi = \sqrt{2}(e^\pi - 1) \approx 31.3 \blacktriangleleft$$

▶ **Example 5** Find the total arc length of the cardioid $r = 1 + \cos\theta$.

Solution. The cardioid is traced out once as θ varies from $\theta = 0$ to $\theta = 2\pi$. Thus,

$$L = \int_\alpha^\beta \sqrt{r^2 + \left(\frac{dr}{d\theta}\right)^2}\, d\theta = \int_0^{2\pi} \sqrt{(1 + \cos\theta)^2 + (-\sin\theta)^2}\, d\theta$$

$$= \sqrt{2} \int_0^{2\pi} \sqrt{1 + \cos\theta}\, d\theta$$

$$= 2 \int_0^{2\pi} \sqrt{\cos^2 \tfrac{1}{2}\theta}\, d\theta \qquad \boxed{\text{Identity (45) of Appendix B}}$$

$$= 2 \int_0^{2\pi} \left|\cos \tfrac{1}{2}\theta\right| d\theta$$

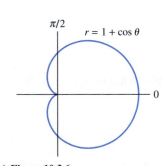

$\pi/2$

$r = 1 + \cos\theta$

0

▲ **Figure 10.3.6**

Since $\cos \tfrac{1}{2}\theta$ changes sign at π, we must split the last integral into the sum of two integrals: the integral from 0 to π plus the integral from π to 2π. However, the integral from π to 2π is equal to the integral from 0 to π, since the cardioid is symmetric about the polar axis (Figure 10.3.6). Thus,

$$L = 2 \int_0^{2\pi} \left|\cos \tfrac{1}{2}\theta\right| d\theta = 4 \int_0^\pi \cos \tfrac{1}{2}\theta\, d\theta = 8 \sin \tfrac{1}{2}\theta \Big]_0^\pi = 8 \blacktriangleleft$$

■ **AREA IN POLAR COORDINATES**

We begin our investigation of area in polar coordinates with a simple case.

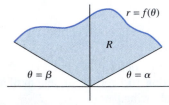

$r = f(\theta)$

R

$\theta = \beta$ $\theta = \alpha$

▲ **Figure 10.3.7**

10.3.3 **AREA PROBLEM IN POLAR COORDINATES** Suppose that α and β are angles that satisfy the condition

$$\alpha < \beta \leq \alpha + 2\pi$$

and suppose that $f(\theta)$ is continuous and nonnegative for $\alpha \leq \theta \leq \beta$. Find the area of the region R enclosed by the polar curve $r = f(\theta)$ and the rays $\theta = \alpha$ and $\theta = \beta$ (Figure 10.3.7).

In rectangular coordinates we obtained areas under curves by dividing the region into an increasing number of vertical strips, approximating the strips by rectangles, and taking a limit. In polar coordinates rectangles are clumsy to work with, and it is better to partition the region into *wedges* by using rays

$$\theta = \theta_1, \ \theta = \theta_2, \ \ldots, \ \theta = \theta_{n-1}$$

such that

$$\alpha < \theta_1 < \theta_2 < \cdots < \theta_{n-1} < \beta$$

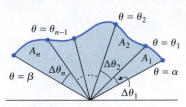

▲ **Figure 10.3.8**

(Figure 10.3.8). As shown in that figure, the rays divide the region R into n wedges with areas $A_1, A_2, \ldots, A_n$ and central angles $\Delta\theta_1, \Delta\theta_2, \ldots, \Delta\theta_n$. The area of the entire region can be written as

$$A = A_1 + A_2 + \cdots + A_n = \sum_{k=1}^{n} A_k \tag{4}$$

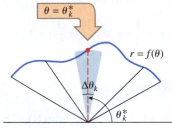

▲ **Figure 10.3.9**

If $\Delta\theta_k$ is small, then we can approximate the area A_k of the kth wedge by the area of a sector with central angle $\Delta\theta_k$ and radius $f(\theta_k^*)$, where $\theta = \theta_k^*$ is any ray that lies in the kth wedge (Figure 10.3.9). Thus, from (4) and Formula (5) of Appendix B for the area of a sector, we obtain

$$A = \sum_{k=1}^{n} A_k \approx \sum_{k=1}^{n} \tfrac{1}{2}[f(\theta_k^*)]^2 \Delta\theta_k \tag{5}$$

If we now increase n in such a way that $\max \Delta\theta_k \to 0$, then the sectors will become better and better approximations of the wedges and it is reasonable to expect that (5) will approach the exact value of the area A (Figure 10.3.10); that is,

$$A = \lim_{\max \Delta\theta_k \to 0} \sum_{k=1}^{n} \tfrac{1}{2}[f(\theta_k^*)]^2 \Delta\theta_k = \int_{\alpha}^{\beta} \tfrac{1}{2}[f(\theta)]^2 \, d\theta$$

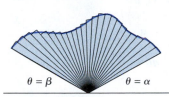

▲ **Figure 10.3.10**

Note that the discussion above can easily be adapted to the case where $f(\theta)$ is nonpositive for $\alpha \leq \theta \leq \beta$. We summarize this result below.

10.3.4   **AREA IN POLAR COORDINATES**   If α and β are angles that satisfy the condition

$$\alpha < \beta \leq \alpha + 2\pi$$

and if $f(\theta)$ is continuous and either nonnegative or nonpositive for $\alpha \leq \theta \leq \beta$, then the area A of the region R enclosed by the polar curve $r = f(\theta)$ ($\alpha \leq \theta \leq \beta$) and the lines $\theta = \alpha$ and $\theta = \beta$ is

$$A = \int_{\alpha}^{\beta} \tfrac{1}{2}[f(\theta)]^2 \, d\theta = \int_{\alpha}^{\beta} \tfrac{1}{2} r^2 \, d\theta \tag{6}$$

The hardest part of applying (6) is determining the limits of integration. This can be done as follows:

Area in Polar Coordinates: Limits of Integration

Step 1. Sketch the region R whose area is to be determined.

Step 2. Draw an arbitrary "radial line" from the pole to the boundary curve $r = f(\theta)$.

Step 3. Ask, "Over what interval of values must θ vary in order for the radial line to sweep out the region R?"

Step 4. Your answer in Step 3 will determine the lower and upper limits of integration.

▶ **Example 6**   Find the area of the region in the first quadrant that is within the cardioid $r = 1 - \cos\theta$.

Solution.   The region and a typical radial line are shown in Figure 10.3.11. For the radial line to sweep out the region, θ must vary from 0 to $\pi/2$. Thus, from (6) with $\alpha = 0$ and $\beta = \pi/2$, we obtain

$$A = \int_0^{\pi/2} \frac{1}{2} r^2 \, d\theta = \frac{1}{2} \int_0^{\pi/2} (1 - \cos\theta)^2 \, d\theta = \frac{1}{2} \int_0^{\pi/2} (1 - 2\cos\theta + \cos^2\theta) \, d\theta$$

With the help of the identity $\cos^2\theta = \frac{1}{2}(1 + \cos 2\theta)$, this can be rewritten as

$$A = \frac{1}{2} \int_0^{\pi/2} \left(\frac{3}{2} - 2\cos\theta + \frac{1}{2}\cos 2\theta \right) d\theta = \frac{1}{2} \left[\frac{3}{2}\theta - 2\sin\theta + \frac{1}{4}\sin 2\theta \right]_0^{\pi/2} = \frac{3}{8}\pi - 1 \blacktriangleleft$$

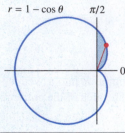

$r = 1 - \cos\theta$ $\pi/2$

0

The shaded region is swept out by the radial line as θ varies from 0 to $\pi/2$.

▲ **Figure 10.3.11**

▶ **Example 7**   Find the entire area within the cardioid of Example 6.

Solution.   For the radial line to sweep out the entire cardioid, θ must vary from 0 to 2π. Thus, from (6) with $\alpha = 0$ and $\beta = 2\pi$,

$$A = \int_0^{2\pi} \frac{1}{2} r^2 \, d\theta = \frac{1}{2} \int_0^{2\pi} (1 - \cos\theta)^2 \, d\theta$$

If we proceed as in Example 6, this reduces to

$$A = \frac{1}{2} \int_0^{2\pi} \left(\frac{3}{2} - 2\cos\theta + \frac{1}{2}\cos 2\theta \right) d\theta = \frac{3\pi}{2}$$

Alternative Solution.   Since the cardioid is symmetric about the x-axis, we can calculate the portion of the area above the x-axis and double the result. In the portion of the cardioid above the x-axis, θ ranges from 0 to π, so that

$$A = 2 \int_0^{\pi} \frac{1}{2} r^2 \, d\theta = \int_0^{\pi} (1 - \cos\theta)^2 \, d\theta = \frac{3\pi}{2} \blacktriangleleft$$

■ **USING SYMMETRY**

Although Formula (6) is applicable if $r = f(\theta)$ is negative, area computations can some-times be simplified by using symmetry to restrict the limits of integration to intervals where $r \geq 0$. This is illustrated in the next example.

▶ **Example 8**   Find the area of the region enclosed by the rose curve $r = \cos 2\theta$.

Solution.   Referring to Figure 10.2.13 and using symmetry, the area in the first quadrant that is swept out for $0 \leq \theta \leq \pi/4$ is one-eighth of the total area inside the rose. Thus, from Formula (6)

$$A = 8 \int_0^{\pi/4} \frac{1}{2} r^2 \, d\theta = 4 \int_0^{\pi/4} \cos^2 2\theta \, d\theta$$

$$= 4 \int_0^{\pi/4} \frac{1}{2}(1 + \cos 4\theta) \, d\theta = 2 \int_0^{\pi/4} (1 + \cos 4\theta) \, d\theta$$

$$= 2\theta + \frac{1}{2}\sin 4\theta \Big]_0^{\pi/4} = \frac{\pi}{2} \blacktriangleleft$$

Sometimes the most natural way to satisfy the restriction $\alpha < \beta \leq \alpha + 2\pi$ required by Formula (6) is to use a negative value for α. For example, suppose that we are interested in finding the area of the shaded region in Figure 10.3.12a. The first step would be to determine the intersections of the cardioid $r = 4 + 4\cos\theta$ and the circle $r = 6$, since this information is needed for the limits of integration. To find the points of intersection, we can equate the two expressions for r. This yields

$$4 + 4\cos\theta = 6 \quad \text{or} \quad \cos\theta = \frac{1}{2}$$

which is satisfied by the positive angles

$$\theta = \frac{\pi}{3} \quad \text{and} \quad \theta = \frac{5\pi}{3}$$

However, there is a problem here because the radial lines to the circle and cardioid do not sweep through the shaded region shown in Figure 10.3.12b as θ varies over the interval $\pi/3 \leq \theta \leq 5\pi/3$. There are two ways to circumvent this problem—one is to take advantage of the symmetry by integrating over the interval $0 \leq \theta \leq \pi/3$ and doubling the result, and the second is to use a negative lower limit of integration and integrate over the interval $-\pi/3 \leq \theta \leq \pi/3$ (Figure 10.3.12c). The two methods are illustrated in the next example.

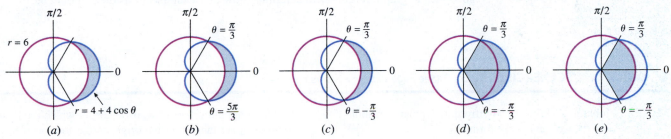

▲ **Figure 10.3.12**

▶ **Example 9**   Find the area of the region that is inside of the cardioid $r = 4 + 4\cos\theta$ and outside of the circle $r = 6$.

Solution Using a Negative Angle.   The area of the region can be obtained by subtracting the areas in Figures 10.3.12d and 10.3.12e:

$$A = \int_{-\pi/3}^{\pi/3} \frac{1}{2}(4 + 4\cos\theta)^2\, d\theta - \int_{-\pi/3}^{\pi/3} \frac{1}{2}(6)^2\, d\theta \qquad \boxed{\begin{array}{l}\text{Area inside cardioid}\\ \text{minus area inside circle.}\end{array}}$$

$$= \int_{-\pi/3}^{\pi/3} \frac{1}{2}[(4 + 4\cos\theta)^2 - 36]\, d\theta = \int_{-\pi/3}^{\pi/3} (16\cos\theta + 8\cos^2\theta - 10)\, d\theta$$

$$= \left[16\sin\theta + (4\theta + 2\sin 2\theta) - 10\theta\right]_{-\pi/3}^{\pi/3} = 18\sqrt{3} - 4\pi$$

Solution Using Symmetry.   Using symmetry, we can calculate the area above the polar axis and double it. This yields (verify)

$$A = 2\int_{0}^{\pi/3} \frac{1}{2}[(4 + 4\cos\theta)^2 - 36]\, d\theta = 2(9\sqrt{3} - 2\pi) = 18\sqrt{3} - 4\pi$$

which agrees with the preceding result. ◀

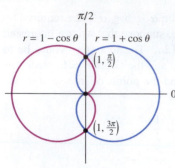

▲ **Figure 10.3.13**

The orbits intersect, but the satellites do not collide.

▲ **Figure 10.3.14**

■ INTERSECTIONS OF POLAR GRAPHS

In the last example we found the intersections of the cardioid and circle by equating their expressions for r and solving for θ. However, because a point can be represented in different ways in polar coordinates, this procedure will not always produce all of the intersections. For example, the cardioids

$$r = 1 - \cos\theta \quad \text{and} \quad r = 1 + \cos\theta \tag{7}$$

intersect at three points: the pole, the point $(1, \pi/2)$, and the point $(1, 3\pi/2)$ (Figure 10.3.13). Equating the right-hand sides of the equations in (7) yields $1 - \cos\theta = 1 + \cos\theta$ or $\cos\theta = 0$, so

$$\theta = \frac{\pi}{2} + k\pi, \quad k = 0, \pm 1, \pm 2, \ldots$$

Substituting any of these values in (7) yields $r = 1$, so that we have found only two distinct points of intersection, $(1, \pi/2)$ and $(1, 3\pi/2)$; the pole has been missed. This problem occurs because the two cardioids pass through the pole at different values of θ—the cardioid $r = 1 - \cos\theta$ passes through the pole at $\theta = 0$, and the cardioid $r = 1 + \cos\theta$ passes through the pole at $\theta = \pi$.

The situation with the cardioids is analogous to two satellites circling the Earth in intersecting orbits (Figure 10.3.14). The satellites will not collide unless they reach the same point at the same time. In general, when looking for intersections of polar curves, it is a good idea to graph the curves to determine how many intersections there should be.

✔ QUICK CHECK EXERCISES 10.3   *(See page 729 for answers.)*

1. (a) To obtain dy/dx directly from the polar equation $r = f(\theta)$, we can use the formula

$$\frac{dy}{dx} = \frac{dy/d\theta}{dx/d\theta} = \underline{\hspace{2cm}}$$

(b) Use the formula in part (a) to find dy/dx directly from the polar equation $r = \csc\theta$.

2. (a) What conditions on $f(\theta_0)$ and $f'(\theta_0)$ guarantee that the line $\theta = \theta_0$ is tangent to the polar curve $r = f(\theta)$ at the origin?

(b) What are the values of θ_0 in $[0, 2\pi]$ at which the lines $\theta = \theta_0$ are tangent at the origin to the four-petal rose $r = \cos 2\theta$?

3. (a) To find the arc length L of the polar curve $r = f(\theta)$ ($\alpha \le \theta \le \beta$), we can use the formula $L = \underline{\hspace{2cm}}$.

(b) The polar curve $r = \sec\theta$ ($0 \le \theta \le \pi/4$) has arc length $L = \underline{\hspace{2cm}}$.

4. The area of the region enclosed by a nonnegative polar curve $r = f(\theta)$ ($\alpha \le \theta \le \beta$) and the lines $\theta = \alpha$ and $\theta = \beta$ is given by the definite integral $\underline{\hspace{2cm}}$.

5. Find the area of the circle $r = a$ by integration.

EXERCISE SET 10.3   ⌇ Graphing Utility   [c] CAS

1–6 Find the slope of the tangent line to the polar curve for the given value of θ. ■

1. $r = 2\sin\theta$; $\theta = \pi/6$

2. $r = 1 + \cos\theta$; $\theta = \pi/2$

3. $r = 1/\theta$; $\theta = 2$

4. $r = a\sec 2\theta$; $\theta = \pi/6$

5. $r = \sin 3\theta$; $\theta = \pi/4$

6. $r = 4 - 3\sin\theta$; $\theta = \pi$

7–8 Calculate the slopes of the tangent lines indicated in the accompanying figures. ■

7. $r = 2 + 2\sin\theta$

8. $r = 1 - 2\sin\theta$

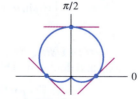

▲ **Figure Ex-7**

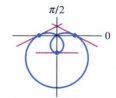

▲ **Figure Ex-8**

9–10 Find polar coordinates of all points at which the polar curve has a horizontal or a vertical tangent line. ■

9. $r = a(1 + \cos\theta)$

10. $r = a\sin\theta$

11–12 Use a graphing utility to make a conjecture about the number of points on the polar curve at which there is a horizontal tangent line, and confirm your conjecture by finding appropriate derivatives. ■

11. $r = \sin\theta \cos^2\theta$ **12.** $r = 1 - 2\sin\theta$

13–18 Sketch the polar curve and find polar equations of the tangent lines to the curve at the pole. ■

13. $r = 2\cos 3\theta$ **14.** $r = 4\sin\theta$ **15.** $r = 4\sqrt{\cos 2\theta}$

16. $r = \sin 2\theta$ **17.** $r = 1 - 2\cos\theta$ **18.** $r = 2\theta$

19–22 Use Formula (3) to calculate the arc length of the polar curve. ■

19. The entire circle $r = a$

20. The entire circle $r = 2a\cos\theta$

21. The entire cardioid $r = a(1 - \cos\theta)$

22. $r = e^{3\theta}$ from $\theta = 0$ to $\theta = 2$

23. (a) Show that the arc length of one petal of the rose $r = \cos n\theta$ is given by

$$2\int_0^{\pi/(2n)} \sqrt{1 + (n^2 - 1)\sin^2 n\theta}\; d\theta$$

   (b) Use the numerical integration capability of a calculating utility to approximate the arc length of one petal of the four-petal rose $r = \cos 2\theta$.
   (c) Use the numerical integration capability of a calculating utility to approximate the arc length of one petal of the n-petal rose $r = \cos n\theta$ for $n = 2, 3, 4, \ldots, 20$; then make a conjecture about the limit of these arc lengths as $n \to +\infty$.

24. (a) Sketch the spiral $r = e^{-\theta/8}$ $(0 \le \theta < +\infty)$.
   (b) Find an improper integral for the total arc length of the spiral.
   (c) Show that the integral converges and find the total arc length of the spiral.

25. Write down, but do not evaluate, an integral for the area of each shaded region.

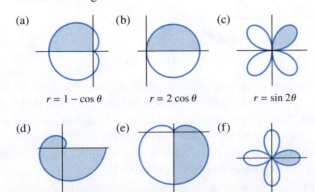

(a) $r = 1 - \cos\theta$ (b) $r = 2\cos\theta$ (c) $r = \sin 2\theta$

(d) $r = \theta$ (e) $r = 1 - \sin\theta$ (f) $r = \cos 2\theta$

26. Find the area of the shaded region in Exercise 25(d).

27. In each part, find the area of the circle by integration.
   (a) $r = 2a\sin\theta$ (b) $r = 2a\cos\theta$

28. (a) Show that $r = 2\sin\theta + 2\cos\theta$ is a circle.
   (b) Find the area of the circle using a geometric formula and then by integration.

29–34 Find the area of the region described. ■

29. The region that is enclosed by the cardioid $r = 2 + 2\sin\theta$.

30. The region in the first quadrant within the cardioid $r = 1 + \cos\theta$.

31. The region enclosed by the rose $r = 4\cos 3\theta$.

32. The region enclosed by the rose $r = 2\sin 2\theta$.

33. The region enclosed by the inner loop of the limaçon $r = 1 + 2\cos\theta$. [*Hint:* $r \le 0$ over the interval of integration.]

34. The region swept out by a radial line from the pole to the curve $r = 2/\theta$ as θ varies over the interval $1 \le \theta \le 3$.

35–38 Find the area of the shaded region. ■

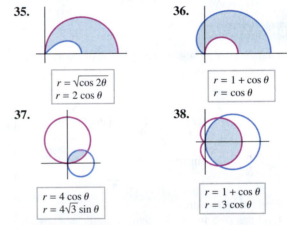

35. $r = \sqrt{\cos 2\theta}$
   $r = 2\cos\theta$

36. $r = 1 + \cos\theta$
   $r = \cos\theta$

37. $r = 4\cos\theta$
   $r = 4\sqrt{3}\sin\theta$

38. $r = 1 + \cos\theta$
   $r = 3\cos\theta$

39–46 Find the area of the region described. ■

39. The region inside the circle $r = 3\sin\theta$ and outside the cardioid $r = 1 + \sin\theta$.

40. The region outside the cardioid $r = 2 - 2\cos\theta$ and inside the circle $r = 4$.

41. The region inside the cardioid $r = 2 + 2\cos\theta$ and outside the circle $r = 3$.

42. The region that is common to the circles $r = 2\cos\theta$ and $r = 2\sin\theta$.

43. The region between the loops of the limaçon $r = \frac{1}{2} + \cos\theta$.

44. The region inside the cardioid $r = 2 + 2\cos\theta$ and to the right of the line $r\cos\theta = \frac{3}{2}$.

45. The region inside the circle $r = 2$ and to the right of the line $r = \sqrt{2}\sec\theta$.

46. The region inside the rose $r = 2a\cos 2\theta$ and outside the circle $r = a\sqrt{2}$.

47–50 True–False Determine whether the statement is true or false. Explain your answer. ■

47. The x-axis is tangent to the polar curve $r = \cos(\theta/2)$ at $\theta = 3\pi$.

48. The arc length of the polar curve $r = \sqrt{\theta}$ for $0 \leq \theta \leq \pi/2$ is given by

$$L = \int_0^{\pi/2} \sqrt{1 + \frac{1}{4\theta}}\, d\theta$$

49. The area of a sector with central angle θ taken from a circle of radius r is θr^2.

50. The expression

$$\frac{1}{2} \int_{-\pi/4}^{\pi/4} (1 - \sqrt{2}\cos\theta)^2\, d\theta$$

computes the area enclosed by the inner loop of the limaçon $r = 1 - \sqrt{2}\cos\theta$.

FOCUS ON CONCEPTS

51. (a) Find the error: The area that is inside the lemniscate $r^2 = a^2\cos 2\theta$ is

$$A = \int_0^{2\pi} \tfrac{1}{2} r^2\, d\theta = \int_0^{2\pi} \tfrac{1}{2} a^2 \cos 2\theta\, d\theta$$

$$= \tfrac{1}{4} a^2 \sin 2\theta \Big]_0^{2\pi} = 0$$

(b) Find the correct area.

(c) Find the area inside the lemniscate $r^2 = 4\cos 2\theta$ and outside the circle $r = \sqrt{2}$.

52. Find the area inside the curve $r^2 = \sin 2\theta$.

53. A radial line is drawn from the origin to the spiral $r = a\theta$ ($a > 0$ and $\theta \geq 0$). Find the area swept out during the second revolution of the radial line that was not swept out during the first revolution.

54. As illustrated in the accompanying figure, suppose that a rod with one end fixed at the pole of a polar coordinate system rotates counterclockwise at the constant rate of 1 rad/s. At time $t = 0$ a bug on the rod is 10 mm from the pole and is moving outward along the rod at the constant speed of 2 mm/s.

(a) Find an equation of the form $r = f(\theta)$ for the path of motion of the bug, assuming that $\theta = 0$ when $t = 0$.

(b) Find the distance the bug travels along the path in part (a) during the first 5 s. Round your answer to the nearest tenth of a millimeter.

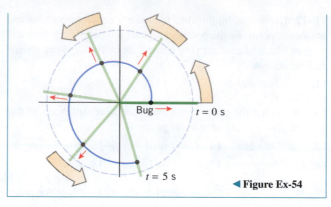

Bug → $t = 0$ s

$t = 5$ s   ◀ **Figure Ex-54**

C 55. (a) Show that the Folium of Descartes $x^3 - 3xy + y^3 = 0$ can be expressed in polar coordinates as

$$r = \frac{3\sin\theta\cos\theta}{\cos^3\theta + \sin^3\theta}$$

(b) Use a CAS to show that the area inside of the loop is $\frac{3}{2}$ (Figure 3.1.3a).

C 56. (a) What is the area that is enclosed by one petal of the rose $r = a\cos n\theta$ if n is an even integer?

(b) What is the area that is enclosed by one petal of the rose $r = a\cos n\theta$ if n is an odd integer?

(c) Use a CAS to show that the total area enclosed by the rose $r = a\cos n\theta$ is $\pi a^2/2$ if the number of petals is even. [*Hint:* See Exercise 78 of Section 10.2.]

(d) Use a CAS to show that the total area enclosed by the rose $r = a\cos n\theta$ is $\pi a^2/4$ if the number of petals is odd.

57. One of the most famous problems in Greek antiquity was "squaring the circle," that is, using a straightedge and compass to construct a square whose area is equal to that of a given circle. It was proved in the nineteenth century that no such construction is possible. However, show that the shaded areas in the accompanying figure are equal, thereby "squaring the crescent."

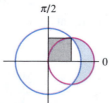

$\pi/2$

0

◀ **Figure Ex-57**

58. Use a graphing utility to generate the polar graph of the equation $r = \cos 3\theta + 2$, and find the area that it encloses.

59. Use a graphing utility to generate the graph of the ***bifolium*** $r = 2\cos\theta \sin^2\theta$, and find the area of the upper loop.

60. Use Formula (9) of Section 10.1 to derive the arc length formula for polar curves, Formula (3).

61. As illustrated in the accompanying figure, let $P(r, \theta)$ be a point on the polar curve $r = f(\theta)$, let ψ be the smallest counterclockwise angle from the extended radius OP to the tangent line at P, and let ϕ be the angle of inclination of the tangent line. Derive the formula

$$\tan \psi = \frac{r}{dr/d\theta}$$

by substituting $\tan \phi$ for dy/dx in Formula (2) and applying the trigonometric identity

$$\tan(\phi - \theta) = \frac{\tan \phi - \tan \theta}{1 + \tan \phi \tan \theta}$$

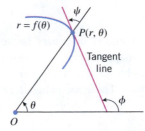

◀ **Figure Ex-61**

62–63 Use the formula for ψ obtained in Exercise 61. ∎

62. (a) Use the trigonometric identity

$$\tan \frac{\theta}{2} = \frac{1 - \cos \theta}{\sin \theta}$$

to show that if (r, θ) is a point on the cardioid

$$r = 1 - \cos \theta \qquad (0 \leq \theta < 2\pi)$$

then $\psi = \theta/2$.

(b) Sketch the cardioid and show the angle ψ at the points where the cardioid crosses the y-axis.

(c) Find the angle ψ at the points where the cardioid crosses the y-axis.

63. Show that for a logarithmic spiral $r = ae^{b\theta}$, the angle from the radial line to the tangent line is constant along the spiral (see the accompanying figure). [*Note:* For this reason, logarithmic spirals are sometimes called **equiangular spirals**.]

◀ **Figure Ex-63**

64. (a) In the discussion associated with Exercises 75–80 of Section 10.1, formulas were given for the area of the

surface of revolution that is generated by revolving a parametric curve about the x-axis or y-axis. Use those formulas to derive the following formulas for the areas of the surfaces of revolution that are generated by revolving the portion of the polar curve $r = f(\theta)$ from $\theta = \alpha$ to $\theta = \beta$ about the polar axis and about the line $\theta = \pi/2$:

$$S = \int_\alpha^\beta 2\pi r \sin \theta \sqrt{r^2 + \left(\frac{dr}{d\theta}\right)^2}\, d\theta \qquad \boxed{\text{About } \theta = 0}$$

$$S = \int_\alpha^\beta 2\pi r \cos \theta \sqrt{r^2 + \left(\frac{dr}{d\theta}\right)^2}\, d\theta \qquad \boxed{\text{About } \theta = \pi/2}$$

(b) State conditions under which these formulas hold.

65–68 Sketch the surface, and use the formulas in Exercise 64 to find the surface area. ∎

65. The surface generated by revolving the circle $r = \cos \theta$ about the line $\theta = \pi/2$.

66. The surface generated by revolving the spiral $r = e^\theta$ $(0 \leq \theta \leq \pi/2)$ about the line $\theta = \pi/2$.

67. The "apple" generated by revolving the upper half of the cardioid $r = 1 - \cos \theta$ $(0 \leq \theta \leq \pi)$ about the polar axis.

68. The sphere of radius a generated by revolving the semicircle $r = a$ in the upper half-plane about the polar axis.

69. Writing
(a) Show that if $0 \leq \theta_1 < \theta_2 \leq \pi$ and if r_1 and r_2 are positive, then the area A of a triangle with vertices $(0, 0)$, (r_1, θ_1), and (r_2, θ_2) is

$$A = \tfrac{1}{2} r_1 r_2 \sin(\theta_2 - \theta_1)$$

(b) Use the formula obtained in part (a) to describe an approach to answer Area Problem 10.3.3 that uses an approximation of the region R by triangles instead of circular wedges. Reconcile your approach with Formula (6).

70. Writing In order to find the area of a region bounded by two polar curves it is often necessary to determine their points of intersection. Give an example to illustrate that the points of intersection of curves $r = f(\theta)$ and $r = g(\theta)$ may not coincide with solutions to $f(\theta) = g(\theta)$. Discuss some strategies for determining intersection points of polar curves and provide examples to illustrate your strategies.

✔ **QUICK CHECK ANSWERS 10.3**

1. (a) $\dfrac{r \cos \theta + \sin \theta \dfrac{dr}{d\theta}}{-r \sin \theta + \cos \theta \dfrac{dr}{d\theta}}$ (b) $\dfrac{dy}{dx} = 0$ **2.** (a) $f(\theta_0) = 0,\ f'(\theta_0) \neq 0$ (b) $\theta_0 = \dfrac{\pi}{4}, \dfrac{3\pi}{4}, \dfrac{5\pi}{4}, \dfrac{7\pi}{4}$

3. (a) $\displaystyle\int_\alpha^\beta \sqrt{r^2 + \left(\frac{dr}{d\theta}\right)^2}\, d\theta$ (b) 1 **4.** $\displaystyle\int_\alpha^\beta \tfrac{1}{2}[f(\theta)]^2\, d\theta = \int_\alpha^\beta \tfrac{1}{2} r^2\, d\theta$ **5.** $\displaystyle\int_0^{2\pi} \tfrac{1}{2} a^2\, d\theta = \pi a^2$

10.4   CONIC SECTIONS

In this section* we will discuss some of the basic geometric properties of parabolas, ellipses, and hyperbolas. These curves play an important role in calculus and also arise naturally in a broad range of applications in such fields as planetary motion, design of telescopes and antennas, geodetic positioning, and medicine, to name a few.

■ CONIC SECTIONS

Circles, ellipses, parabolas, and hyperbolas are called *conic sections* or *conics* because they can be obtained as intersections of a plane with a double-napped circular cone (Figure 10.4.1). If the plane passes through the vertex of the double-napped cone, then the intersection is a point, a pair of intersecting lines, or a single line. These are called *degenerate conic sections*.

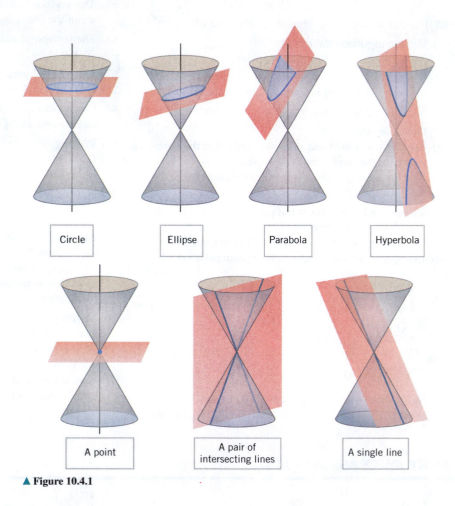

| Circle | Ellipse | Parabola | Hyperbola |

| A point | A pair of intersecting lines | A single line |

▲ Figure 10.4.1

---

*Some students may already be familiar with the material in this section, in which case it can be treated as a review. Instructors who want to spend some additional time on precalculus review may want to allocate more than one lecture on this material.

■ DEFINITIONS OF THE CONIC SECTIONS

Although we could derive properties of parabolas, ellipses, and hyperbolas by defining them as intersections with a double-napped cone, it will be better suited to calculus if we begin with equivalent definitions that are based on their geometric properties.

> **10.4.1 DEFINITION** A *parabola* is the set of all points in the plane that are equidistant from a fixed line and a fixed point not on the line.

The line is called the *directrix* of the parabola, and the point is called the *focus* (Figure 10.4.2). A parabola is symmetric about the line that passes through the focus at right angles to the directrix. This line, called the *axis* or the *axis of symmetry* of the parabola, intersects the parabola at a point called the *vertex*.

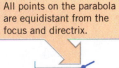

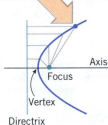

All points on the parabola are equidistant from the focus and directrix.

Axis

Focus

Vertex

Directrix

▲ **Figure 10.4.2**

> **10.4.2 DEFINITION** An *ellipse* is the set of all points in the plane, the sum of whose distances from two fixed points is a given positive constant that is greater than the distance between the fixed points.

The two fixed points are called the *foci* (plural of "focus") of the ellipse, and the midpoint of the line segment joining the foci is called the *center* (Figure 10.4.3a). To help visualize Definition 10.4.2, imagine that two ends of a string are tacked to the foci and a pencil traces a curve as it is held tight against the string (Figure 10.4.3b). The resulting curve will be an ellipse since the sum of the distances to the foci is a constant, namely, the total length of the string. Note that if the foci coincide, the ellipse reduces to a circle. For ellipses other than circles, the line segment through the foci and across the ellipse is called the *major axis* (Figure 10.4.3c), and the line segment across the ellipse, through the center, and perpendicular to the major axis is called the *minor axis*. The endpoints of the major axis are called *vertices*.

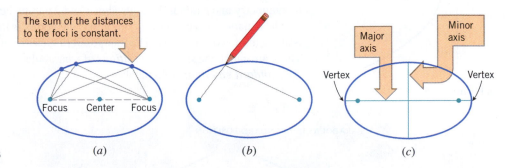

The sum of the distances to the foci is constant.

Focus Center Focus

Major axis

Minor axis

Vertex

Vertex

(a)   *(b)*   *(c)*

▶ **Figure 10.4.3**

> **10.4.3 DEFINITION** A *hyperbola* is the set of all points in the plane, the difference of whose distances from two fixed distinct points is a given positive constant that is less than the distance between the fixed points.

The two fixed points are called the *foci* of the hyperbola, and the term "difference" that is used in the definition is understood to mean the distance to the farther focus minus the distance to the closer focus. As a result, the points on the hyperbola form two *branches*, each "wrapping around" the closer focus (Figure 10.4.4a). The midpoint of the line segment joining the foci is called the **center** of the hyperbola, the line through the foci is called the **focal axis**, and the line through the center that is perpendicular to the focal axis is called the **conjugate axis**. The hyperbola intersects the focal axis at two points called the **vertices**.

Associated with every hyperbola is a pair of lines, called the **asymptotes** of the hyperbola. These lines intersect at the center of the hyperbola and have the property that as a point P moves along the hyperbola away from the center, the vertical distance between P and one of the asymptotes approaches zero (Figure 10.4.4b).

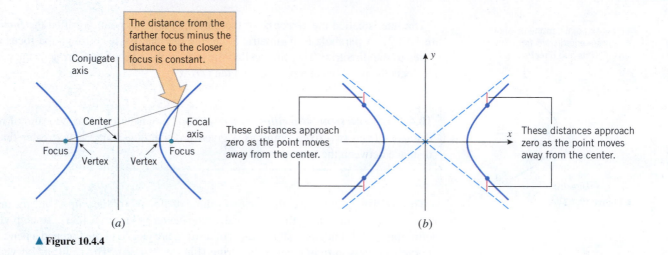

The distance from the farther focus minus the distance to the closer focus is constant.

(a)

(b)

These distances approach zero as the point moves away from the center.

These distances approach zero as the point moves away from the center.

▲ **Figure 10.4.4**

■ EQUATIONS OF PARABOLAS IN STANDARD POSITION

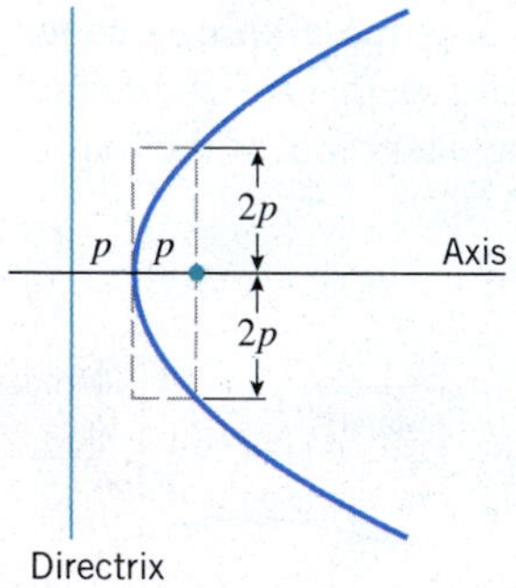

It is traditional in the study of parabolas to denote the distance between the focus and the vertex by p. The vertex is equidistant from the focus and the directrix, so the distance between the vertex and the directrix is also p; consequently, the distance between the focus and the directrix is $2p$ (Figure 10.4.5). As illustrated in that figure, the parabola passes through two of the corners of a box that extends from the vertex to the focus along the axis of symmetry and extends $2p$ units above and $2p$ units below the axis of symmetry.

The equation of a parabola is simplest if the vertex is the origin and the axis of symmetry is along the x-axis or y-axis. The four possible such orientations are shown in Figure 10.4.6. These are called the **standard positions** of a parabola, and the resulting equations are called the **standard equations** of a parabola.

Directrix

▲ **Figure 10.4.5**

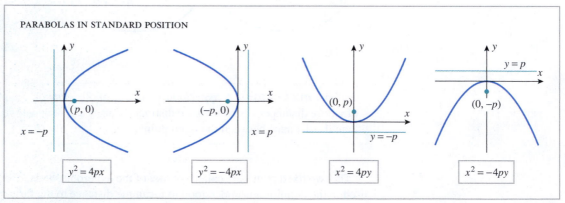

PARABOLAS IN STANDARD POSITION

$y^2 = 4px$ $y^2 = -4px$ $x^2 = 4py$ $x^2 = -4py$

▲ **Figure 10.4.6**

To illustrate how the equations in Figure 10.4.6 are obtained, we will derive the equation for the parabola with focus $(p, 0)$ and directrix $x = -p$. Let $P(x, y)$ be any point on the parabola. Since P is equidistant from the focus and directrix, the distances PF and PD in Figure 10.4.7 are equal; that is,

$$PF = PD \tag{1}$$

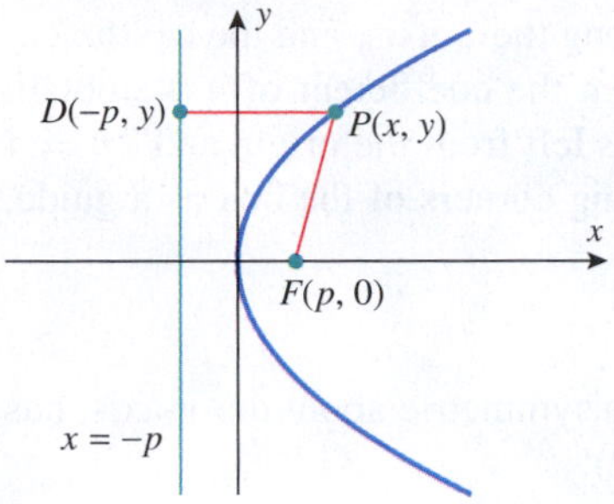

where $D(-p, y)$ is the foot of the perpendicular from P to the directrix. From the distance formula, the distances PF and PD are

$$PF = \sqrt{(x - p)^2 + y^2} \quad \text{and} \quad PD = \sqrt{(x + p)^2} \tag{2}$$

Substituting in (1) and squaring yields

$$(x - p)^2 + y^2 = (x + p)^2 \tag{3}$$

and after simplifying

$$y^2 = 4px \tag{4}$$

The derivations of the other equations in Figure 10.4.6 are similar.

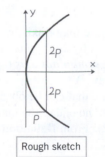

▲ Figure 10.4.7

■ A TECHNIQUE FOR SKETCHING PARABOLAS

Parabolas can be sketched from their *standard equations* using four basic steps:

Sketching a Parabola from Its Standard Equation

Step 1. Determine whether the axis of symmetry is along the x-axis or the y-axis. Referring to Figure 10.4.6, the axis of symmetry is along the x-axis if the equation has a y^2-term, and it is along the y-axis if it has an x^2-term.

Step 2. Determine which way the parabola opens. If the axis of symmetry is along the x-axis, then the parabola opens to the right if the coefficient of x is positive, and it opens to the left if the coefficient is negative. If the axis of symmetry is along the y-axis, then the parabola opens up if the coefficient of y is positive, and it opens down if the coefficient is negative.

Step 3. Determine the value of p and draw a box extending p units from the origin along the axis of symmetry in the direction in which the parabola opens and extending $2p$ units on each side of the axis of symmetry.

Step 4. Using the box as a guide, sketch the parabola so that its vertex is at the origin and it passes through the corners of the box (Figure 10.4.8).

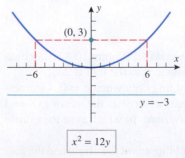

Rough sketch

▲ Figure 10.4.8

▶ **Example 1** Sketch the graphs of the parabolas

(a) $x^2 = 12y$ (b) $y^2 + 8x = 0$

and show the focus and directrix of each.

Solution (a). This equation involves x^2, so the axis of symmetry is along the y-axis, and the coefficient of y is positive, so the parabola opens upward. From the coefficient of y, we obtain $4p = 12$ or $p = 3$. Drawing a box extending $p = 3$ units up from the origin and $2p = 6$ units to the left and $2p = 6$ units to the right of the y-axis, then using corners of the box as a guide, yields the graph in Figure 10.4.9.

The focus is $p = 3$ units from the vertex along the axis of symmetry in the direction in which the parabola opens, so its coordinates are $(0, 3)$. The directrix is perpendicular to the axis of symmetry at a distance of $p = 3$ units from the vertex on the opposite side from the focus, so its equation is $y = -3$.

$x^2 = 12y$

▲ Figure 10.4.9

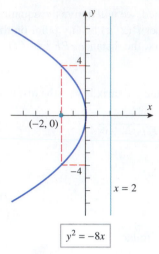

▲ Figure 10.4.10

Solution (b).   We first rewrite the equation in the standard form

$$y^2 = -8x$$

This equation involves y^2, so the axis of symmetry is along the x-axis, and the coefficient of x is negative, so the parabola opens to the left. From the coefficient of x we obtain $4p = 8$, so $p = 2$. Drawing a box extending $p = 2$ units left from the origin and $2p = 4$ units above and $2p = 4$ units below the x-axis, then using corners of the box as a guide, yields the graph in Figure 10.4.10.   ◄

▶ **Example 2**   Find an equation of the parabola that is symmetric about the y-axis, has its vertex at the origin, and passes through the point $(5, 2)$.

Solution.   Since the parabola is symmetric about the y-axis and has its vertex at the origin, the equation is of the form

$$x^2 = 4py \quad \text{or} \quad x^2 = -4py$$

where the sign depends on whether the parabola opens up or down. But the parabola must open up since it passes through the point $(5, 2)$, which lies in the first quadrant. Thus, the equation is of the form

$$x^2 = 4py \tag{5}$$

Since the parabola passes through $(5, 2)$, we must have $5^2 = 4p \cdot 2$ or $4p = \frac{25}{2}$. Therefore, (5) becomes

$$x^2 = \tfrac{25}{2}y \quad ◄$$

■ EQUATIONS OF ELLIPSES IN STANDARD POSITION

It is traditional in the study of ellipses to denote the length of the major axis by $2a$, the length of the minor axis by $2b$, and the distance between the foci by $2c$ (Figure 10.4.11). The number a is called the **semimajor axis** and the number b the **semiminor axis** (standard but odd terminology, since a and b are numbers, not geometric axes).

There is a basic relationship between the numbers a, b, and c that can be obtained by examining the sum of the distances to the foci from a point P at the end of the major axis and from a point Q at the end of the minor axis (Figure 10.4.12). From Definition 10.4.2, these sums must be equal, so we obtain

$$2\sqrt{b^2 + c^2} = (a - c) + (a + c)$$

from which it follows that

$$a = \sqrt{b^2 + c^2} \tag{6}$$

or, equivalently,

$$c = \sqrt{a^2 - b^2} \tag{7}$$

From (6), the distance from a focus to an end of the minor axis is a (Figure 10.4.13), which implies that for *all* points on the ellipse the sum of the distances to the foci is $2a$.

It also follows from (6) that $a \geq b$ with the equality holding only when $c = 0$. Geometrically, this means that the major axis of an ellipse is at least as large as the minor axis and that the two axes have equal length only when the foci coincide, in which case the ellipse is a circle.

The equation of an ellipse is simplest if the center of the ellipse is at the origin and the foci are on the x-axis or y-axis. The two possible such orientations are shown in Figure 10.4.14.

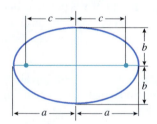

▲ Figure 10.4.11

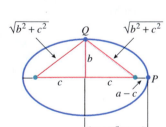

▲ Figure 10.4.12

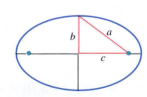

▲ Figure 10.4.13

These are called the *standard positions* of an ellipse, and the resulting equations are called the *standard equations* of an ellipse.

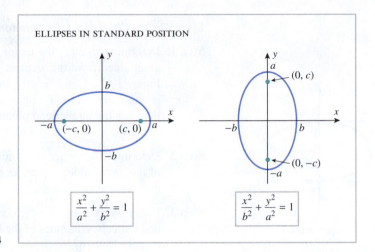

ELLIPSES IN STANDARD POSITION

$$\frac{x^2}{a^2} + \frac{y^2}{b^2} = 1$$

$$\frac{x^2}{b^2} + \frac{y^2}{a^2} = 1$$

▶ **Figure 10.4.14**

To illustrate how the equations in Figure 10.4.14 are obtained, we will derive the equation for the ellipse with foci on the x-axis. Let $P(x, y)$ be any point on that ellipse. Since the sum of the distances from P to the foci is $2a$, it follows (Figure 10.4.15) that

$$PF' + PF = 2a$$

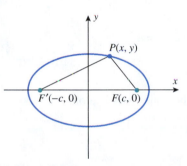

▲ **Figure 10.4.15**

so

$$\sqrt{(x + c)^2 + y^2} + \sqrt{(x - c)^2 + y^2} = 2a$$

Transposing the second radical to the right side of the equation and squaring yields

$$(x + c)^2 + y^2 = 4a^2 - 4a\sqrt{(x - c)^2 + y^2} + (x - c)^2 + y^2$$

and, on simplifying,

$$\sqrt{(x - c)^2 + y^2} = a - \frac{c}{a}x \tag{8}$$

Squaring again and simplifying yields

$$\frac{x^2}{a^2} + \frac{y^2}{a^2 - c^2} = 1$$

which, by virtue of (6), can be written as

$$\frac{x^2}{a^2} + \frac{y^2}{b^2} = 1 \tag{9}$$

Conversely, it can be shown that any point whose coordinates satisfy (9) has $2a$ as the sum of its distances from the foci, so that such a point is on the ellipse.

■ **A TECHNIQUE FOR SKETCHING ELLIPSES**

Ellipses can be sketched from their *standard equations* using three basic steps:

> ### *Sketching an Ellipse from Its Standard Equation*
>
> **Step 1.** Determine whether the major axis is on the x-axis or the y-axis. This can be ascertained from the sizes of the denominators in the equation. Referring to Figure 10.4.14, and keeping in mind that $a^2 > b^2$ (since $a > b$), the major axis is along the x-axis if x^2 has the larger denominator, and it is along the y-axis if y^2 has the larger denominator. If the denominators are equal, the ellipse is a circle.
>
> **Step 2.** Determine the values of a and b and draw a box extending a units on each side of the center along the major axis and b units on each side of the center along the minor axis.
>
> **Step 3.** Using the box as a guide, sketch the ellipse so that its center is at the origin and it touches the sides of the box where the sides intersect the coordinate axes (Figure 10.4.16).

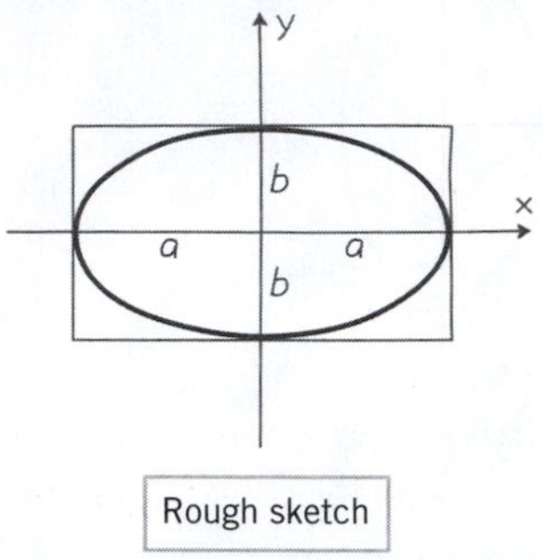

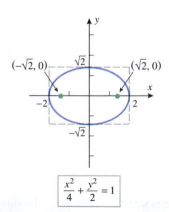

Rough sketch

▲ **Figure 10.4.16**

▶ **Example 3**   Sketch the graphs of the ellipses

$$\text{(a) } \frac{x^2}{9} + \frac{y^2}{16} = 1 \qquad \text{(b) } x^2 + 2y^2 = 4$$

showing the foci of each.

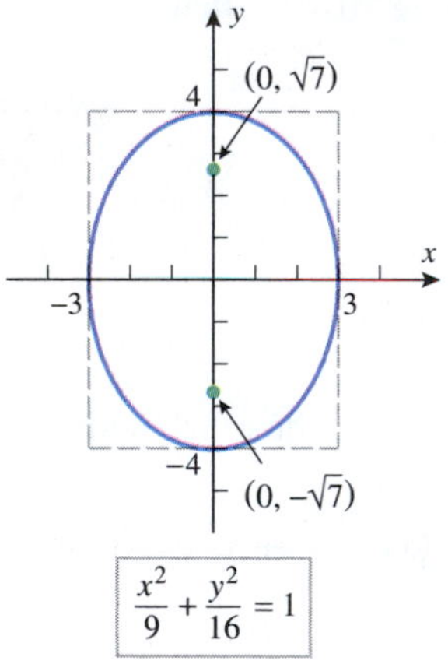

Solution (a).   Since y^2 has the larger denominator, the major axis is along the y-axis. Moreover, since $a^2 > b^2$, we must have $a^2 = 16$ and $b^2 = 9$, so

$$a = 4 \quad \text{and} \quad b = 3$$

Drawing a box extending 4 units on each side of the origin along the y-axis and 3 units on each side of the origin along the x-axis as a guide yields the graph in Figure 10.4.17.

The foci lie c units on each side of the center along the major axis, where c is given by (7). From the values of a^2 and b^2 above, we obtain

$$c = \sqrt{a^2 - b^2} = \sqrt{16 - 9} = \sqrt{7} \approx 2.6$$

Thus, the coordinates of the foci are $(0, \sqrt{7})$ and $(0, -\sqrt{7})$, since they lie on the y-axis.

Solution (b).   We first rewrite the equation in the standard form

$$\frac{x^2}{4} + \frac{y^2}{2} = 1$$

Since x^2 has the larger denominator, the major axis lies along the x-axis, and we have $a^2 = 4$ and $b^2 = 2$. Drawing a box extending $a = 2$ units on each side of the origin along the x-axis and extending $b = \sqrt{2} \approx 1.4$ units on each side of the origin along the y-axis as a guide yields the graph in Figure 10.4.18.

From (7), we obtain

$$c = \sqrt{a^2 - b^2} = \sqrt{2} \approx 1.4$$

Thus, the coordinates of the foci are $(\sqrt{2}, 0)$ and $(-\sqrt{2}, 0)$, since they lie on the x-axis. ◀

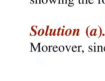

$(0, \sqrt{7})$

$(0, -\sqrt{7})$

$$\frac{x^2}{9} + \frac{y^2}{16} = 1$$

▲ **Figure 10.4.17**

$(-\sqrt{2}, 0)$ $(\sqrt{2}, 0)$

$$\frac{x^2}{4} + \frac{y^2}{2} = 1$$

▲ **Figure 10.4.18**

▶ **Example 4**   Find an equation for the ellipse with foci $(0, \pm 2)$ and major axis with endpoints $(0, \pm 4)$.

Solution.   From Figure 10.4.14, the equation has the form

$$\frac{x^2}{b^2} + \frac{y^2}{a^2} = 1$$

and from the given information, $a = 4$ and $c = 2$. It follows from (6) that

$$b^2 = a^2 - c^2 = 16 - 4 = 12$$

so the equation of the ellipse is

$$\frac{x^2}{12} + \frac{y^2}{16} = 1 \quad ◀$$

■ EQUATIONS OF HYPERBOLAS IN STANDARD POSITION

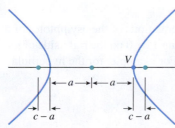

▲ **Figure 10.4.19**

It is traditional in the study of hyperbolas to denote the distance between the vertices by $2a$, the distance between the foci by $2c$ (Figure 10.4.19), and to define the quantity b as

$$b = \sqrt{c^2 - a^2} \tag{10}$$

This relationship, which can also be expressed as

$$c = \sqrt{a^2 + b^2} \tag{11}$$

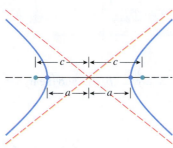

▲ **Figure 10.4.20**

is pictured geometrically in Figure 10.4.20. As illustrated in that figure, and as we will show later in this section, the asymptotes pass through the corners of a box extending b units on each side of the center along the conjugate axis and a units on each side of the center along the focal axis. The number a is called the ***semifocal axis*** of the hyperbola and the number b the ***semiconjugate axis***. (As with the semimajor and semiminor axes of an ellipse, these are numbers, not geometric axes.)

If V is one vertex of a hyperbola, then, as illustrated in Figure 10.4.21, the distance from V to the farther focus minus the distance from V to the closer focus is

$$[(c - a) + 2a] - (c - a) = 2a$$

Thus, for *all* points on a hyperbola, the distance to the farther focus minus the distance to the closer focus is $2a$.

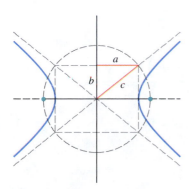

▲ **Figure 10.4.21**

The equation of a hyperbola has an especially convenient form if the center of the hyperbola is at the origin and the foci are on the x-axis or y-axis. The two possible such orientations are shown in Figure 10.4.22. These are called the ***standard positions*** of a hyperbola, and the resulting equations are called the ***standard equations*** of a hyperbola.

The derivations of these equations are similar to those already given for parabolas and ellipses, so we will leave them as exercises. However, to illustrate how the equations of the asymptotes are derived, we will derive those equations for the hyperbola

$$\frac{x^2}{a^2} - \frac{y^2}{b^2} = 1$$

We can rewrite this equation as

$$y^2 = \frac{b^2}{a^2}(x^2 - a^2)$$

which is equivalent to the pair of equations

$$y = \frac{b}{a}\sqrt{x^2 - a^2} \quad \text{and} \quad y = -\frac{b}{a}\sqrt{x^2 - a^2}$$

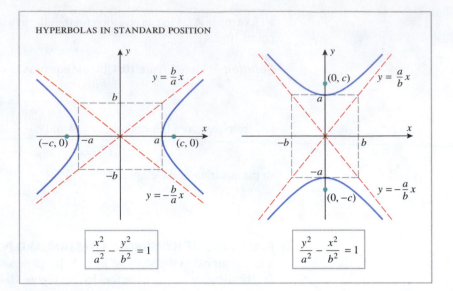

HYPERBOLAS IN STANDARD POSITION

▶ **Figure 10.4.22**

Thus, in the first quadrant, the vertical distance between the line $y = (b/a)x$ and the hyperbola can be written as

$$\frac{b}{a}x - \frac{b}{a}\sqrt{x^2 - a^2}$$

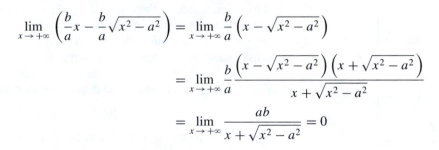

▲ **Figure 10.4.23**

(Figure 10.4.23). But this distance tends to zero as $x \to +\infty$ since

$$\lim_{x \to +\infty}\left(\frac{b}{a}x - \frac{b}{a}\sqrt{x^2 - a^2}\right) = \lim_{x \to +\infty}\frac{b}{a}\left(x - \sqrt{x^2 - a^2}\right)$$

$$= \lim_{x \to +\infty}\frac{b}{a}\frac{\left(x - \sqrt{x^2 - a^2}\right)\left(x + \sqrt{x^2 - a^2}\right)}{x + \sqrt{x^2 - a^2}}$$

$$= \lim_{x \to +\infty}\frac{ab}{x + \sqrt{x^2 - a^2}} = 0$$

The analysis in the remaining quadrants is similar.

■ **A QUICK WAY TO FIND ASYMPTOTES**

There is a trick that can be used to avoid memorizing the equations of the asymptotes of a hyperbola. They can be obtained, when needed, by replacing 1 by 0 on the right side of the hyperbola equation, and then solving for y in terms of x. For example, for the hyperbola

$$\frac{x^2}{a^2} - \frac{y^2}{b^2} = 1$$

we would write

$$\frac{x^2}{a^2} - \frac{y^2}{b^2} = 0 \quad \text{or} \quad y^2 = \frac{b^2}{a^2}x^2 \quad \text{or} \quad y = \pm\frac{b}{a}x$$

which are the equations for the asymptotes.

■ A TECHNIQUE FOR SKETCHING HYPERBOLAS

Hyperbolas can be sketched from their *standard equations* using four basic steps:

Sketching a Hyperbola from Its Standard Equation

Step 1. Determine whether the focal axis is on the x-axis or the y-axis. This can be ascertained from the location of the minus sign in the equation. Referring to Figure 10.4.22, the focal axis is along the x-axis when the minus sign precedes the y^2-term, and it is along the y-axis when the minus sign precedes the x^2-term.

Step 2. Determine the values of a and b and draw a box extending a units on either side of the center along the focal axis and b units on either side of the center along the conjugate axis. (The squares of a and b can be read directly from the equation.)

Step 3. Draw the asymptotes along the diagonals of the box.

Step 4. Using the box and the asymptotes as a guide, sketch the graph of the hyperbola (Figure 10.4.24).

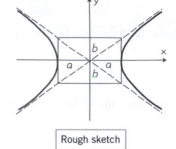

Rough sketch

▲ **Figure 10.4.24**

▶ **Example 5**   Sketch the graphs of the hyperbolas

$$\text{(a)} \quad \frac{x^2}{4} - \frac{y^2}{9} = 1 \qquad \text{(b)} \quad y^2 - x^2 = 1$$

showing their vertices, foci, and asymptotes.

Solution (a).   The minus sign precedes the y^2-term, so the focal axis is along the x-axis. From the denominators in the equation we obtain

$$a^2 = 4 \quad \text{and} \quad b^2 = 9$$

Since a and b are positive, we must have $a = 2$ and $b = 3$. Recalling that the vertices lie a units on each side of the center on the focal axis, it follows that their coordinates in this case are $(2, 0)$ and $(-2, 0)$. Drawing a box extending $a = 2$ units along the x-axis on each side of the origin and $b = 3$ units on each side of the origin along the y-axis, then drawing the asymptotes along the diagonals of the box as a guide, yields the graph in Figure 10.4.25.

To obtain equations for the asymptotes, we replace 1 by 0 in the given equation; this yields

$$\frac{x^2}{4} - \frac{y^2}{9} = 0 \quad \text{or} \quad y = \pm\frac{3}{2}x$$

The foci lie c units on each side of the center along the focal axis, where c is given by (11). From the values of a^2 and b^2 above we obtain

$$c = \sqrt{a^2 + b^2} = \sqrt{4 + 9} = \sqrt{13} \approx 3.6$$

Since the foci lie on the x-axis in this case, their coordinates are $(\sqrt{13}, 0)$ and $(-\sqrt{13}, 0)$.

Solution (b).   The minus sign precedes the x^2-term, so the focal axis is along the y-axis. From the denominators in the equation we obtain $a^2 = 1$ and $b^2 = 1$, from which it follows that

$$a = 1 \quad \text{and} \quad b = 1$$

Thus, the vertices are at $(0, -1)$ and $(0, 1)$. Drawing a box extending $a = 1$ unit on either side of the origin along the y-axis and $b = 1$ unit on either side of the origin along the x-axis, then drawing the asymptotes, yields the graph in Figure 10.4.26. Since the box is actually

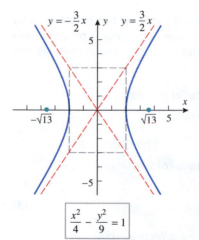

$$\frac{x^2}{4} - \frac{y^2}{9} = 1$$

▲ **Figure 10.4.25**

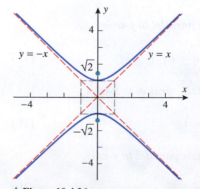

▲ **Figure 10.4.26**

A hyperbola in which $a = b$, as in part (b) of Example 5, is called an *equilateral hyperbola*. Such hyperbolas always have perpendicular asymptotes.

a square, the asymptotes are perpendicular and have equations $y = \pm x$. This can also be seen by replacing 1 by 0 in the given equation, which yields $y^2 - x^2 = 0$ or $y = \pm x$. Also,

$$c = \sqrt{a^2 + b^2} = \sqrt{1 + 1} = \sqrt{2}$$

so the foci, which lie on the y-axis, are $(0, -\sqrt{2})$ and $(0, \sqrt{2})$. ◄

▶ **Example 6**   Find the equation of the hyperbola with vertices $(0, \pm 8)$ and asymptotes $y = \pm \frac{4}{3}x$.

Solution.   Since the vertices are on the y-axis, the equation of the hyperbola has the form $(y^2/a^2) - (x^2/b^2) = 1$ and the asymptotes are

$$y = \pm \frac{a}{b}x$$

From the locations of the vertices we have $a = 8$, so the given equations of the asymptotes yield

$$y = \pm \frac{a}{b}x = \pm \frac{8}{b}x = \pm \frac{4}{3}x$$

from which it follows that $b = 6$. Thus, the hyperbola has the equation

$$\frac{y^2}{64} - \frac{x^2}{36} = 1 \quad ◄$$

■ TRANSLATED CONICS

Equations of conics that are translated from their standard positions can be obtained by replacing x by $x - h$ and y by $y - k$ in their standard equations. For a parabola, this translates the vertex from the origin to the point (h, k); and for ellipses and hyperbolas, this translates the center from the origin to the point (h, k).

Parabolas with vertex (h, k) and axis parallel to x-axis

$$(y - k)^2 = 4p(x - h) \quad \text{[Opens right]} \tag{12}$$

$$(y - k)^2 = -4p(x - h) \quad \text{[Opens left]} \tag{13}$$

Parabolas with vertex (h, k) and axis parallel to y-axis

$$(x - h)^2 = 4p(y - k) \quad \text{[Opens up]} \tag{14}$$

$$(x - h)^2 = -4p(y - k) \quad \text{[Opens down]} \tag{15}$$

Ellipse with center (h, k) and major axis parallel to x-axis

$$\frac{(x - h)^2}{a^2} + \frac{(y - k)^2}{b^2} = 1 \quad [b < a] \tag{16}$$

Ellipse with center (h, k) and major axis parallel to y-axis

$$\frac{(x - h)^2}{b^2} + \frac{(y - k)^2}{a^2} = 1 \quad [b < a] \tag{17}$$

Hyperbola with center (h, k) and focal axis parallel to x-axis

$$\frac{(x - h)^2}{a^2} - \frac{(y - k)^2}{b^2} = 1 \tag{18}$$

Hyperbola with center (h, k) and focal axis parallel to y-axis

$$\frac{(y - k)^2}{a^2} - \frac{(x - h)^2}{b^2} = 1 \tag{19}$$

▶ **Example 7** Find an equation for the parabola that has its vertex at $(1, 2)$ and its focus at $(4, 2)$.

Solution. Since the focus and vertex are on a horizontal line, and since the focus is to the right of the vertex, the parabola opens to the right and its equation has the form

$$(y - k)^2 = 4p(x - h)$$

Since the vertex and focus are 3 units apart, we have $p = 3$, and since the vertex is at $(h, k) = (1, 2)$, we obtain

$$(y - 2)^2 = 12(x - 1) \blacktriangleleft$$

Sometimes the equations of translated conics occur in expanded form, in which case we are faced with the problem of identifying the graph of a quadratic equation in x and y:

$$Ax^2 + Cy^2 + Dx + Ey + F = 0 \tag{20}$$

The basic procedure for determining the nature of such a graph is to complete the squares of the quadratic terms and then try to match up the resulting equation with one of the forms of a translated conic.

▶ **Example 8** Describe the graph of the equation

$$y^2 - 8x - 6y - 23 = 0$$

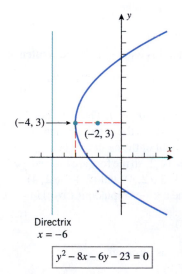

Directrix
$x = -6$

$\boxed{y^2 - 8x - 6y - 23 = 0}$

▲ **Figure 10.4.27**

Solution. The equation involves quadratic terms in y but none in x, so we first take all of the y-terms to one side:

$$y^2 - 6y = 8x + 23$$

Next, we complete the square on the y-terms by adding 9 to both sides:

$$(y - 3)^2 = 8x + 32$$

Finally, we factor out the coefficient of the x-term to obtain

$$(y - 3)^2 = 8(x + 4)$$

This equation is of form (12) with $h = -4$, $k = 3$, and $p = 2$, so the graph is a parabola with vertex $(-4, 3)$ opening to the right. Since $p = 2$, the focus is 2 units to the right of the vertex, which places it at the point $(-2, 3)$; and the directrix is 2 units to the left of the vertex, which means that its equation is $x = -6$. The parabola is shown in Figure 10.4.27. ◀

▶ **Example 9** Describe the graph of the equation

$$16x^2 + 9y^2 - 64x - 54y + 1 = 0$$

Solution. This equation involves quadratic terms in both x and y, so we will group the x-terms and the y-terms on one side and put the constant on the other:

$$(16x^2 - 64x) + (9y^2 - 54y) = -1$$

Next, factor out the coefficients of x^2 and y^2 and complete the squares:

$$16(x^2 - 4x + 4) + 9(y^2 - 6y + 9) = -1 + 64 + 81$$

or

$$16(x - 2)^2 + 9(y - 3)^2 = 144$$

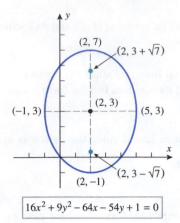

$$16x^2 + 9y^2 - 64x - 54y + 1 = 0$$

▲ **Figure 10.4.28**

Finally, divide through by 144 to introduce a 1 on the right side:

$$\frac{(x-2)^2}{9} + \frac{(y-3)^2}{16} = 1$$

This is an equation of form (17), with $h = 2$, $k = 3$, $a^2 = 16$, and $b^2 = 9$. Thus, the graph of the equation is an ellipse with center $(2, 3)$ and major axis parallel to the y-axis. Since $a = 4$, the major axis extends 4 units above and 4 units below the center, so its endpoints are $(2, 7)$ and $(2, -1)$ (Figure 10.4.28). Since $b = 3$, the minor axis extends 3 units to the left and 3 units to the right of the center, so its endpoints are $(-1, 3)$ and $(5, 3)$. Since

$$c = \sqrt{a^2 - b^2} = \sqrt{16 - 9} = \sqrt{7}$$

the foci lie $\sqrt{7}$ units above and below the center, placing them at the points $(2, 3 + \sqrt{7})$ and $(2, 3 - \sqrt{7})$. ◄

▶ **Example 10** Describe the graph of the equation

$$x^2 - y^2 - 4x + 8y - 21 = 0$$

Solution. This equation involves quadratic terms in both x and y, so we will group the x-terms and the y-terms on one side and put the constant on the other:

$$(x^2 - 4x) - (y^2 - 8y) = 21$$

We leave it for you to verify by completing the squares that this equation can be written as

$$\frac{(x-2)^2}{9} - \frac{(y-4)^2}{9} = 1 \qquad (21)$$

This is an equation of form (18) with $h = 2$, $k = 4$, $a^2 = 9$, and $b^2 = 9$. Thus, the equation represents a hyperbola with center $(2, 4)$ and focal axis parallel to the x-axis. Since $a = 3$, the vertices are located 3 units to the left and 3 units to the right of the center, or at the points $(-1, 4)$ and $(5, 4)$. From (11), $c = \sqrt{a^2 + b^2} = \sqrt{9 + 9} = 3\sqrt{2}$, so the foci are located $3\sqrt{2}$ units to the left and right of the center, or at the points $(2 - 3\sqrt{2}, 4)$ and $(2 + 3\sqrt{2}, 4)$.

The equations of the asymptotes may be found using the trick of replacing 1 by 0 in (21) to obtain

$$\frac{(x-2)^2}{9} - \frac{(y-4)^2}{9} = 0$$

This can be written as $y - 4 = \pm(x - 2)$, which yields the asymptotes

$$y = x + 2 \quad \text{and} \quad y = -x + 6$$

With the aid of a box extending $a = 3$ units left and right of the center and $b = 3$ units above and below the center, we obtain the sketch in Figure 10.4.29. ◄

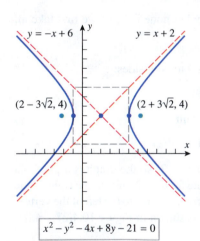

$$x^2 - y^2 - 4x + 8y - 21 = 0$$

▲ **Figure 10.4.29**

■ **REFLECTION PROPERTIES OF THE CONIC SECTIONS**

Parabolas, ellipses, and hyperbolas have certain reflection properties that make them extremely valuable in various applications. In the exercises we will ask you to prove the following results.

10.4.4 **THEOREM** (*Reflection Property of Parabolas*) *The tangent line at a point P on a parabola makes equal angles with the line through P parallel to the axis of symmetry and the line through P and the focus (Figure 10.4.30a).*

10.4.5   THEOREM (*Reflection Property of Ellipses*)   *A line tangent to an ellipse at a point P makes equal angles with the lines joining P to the foci (Figure 10.4.30b).*

10.4.6   THEOREM (*Reflection Property of Hyperbolas*)   *A line tangent to a hyperbola at a point P makes equal angles with the lines joining P to the foci (Figure 10.4.30c).*

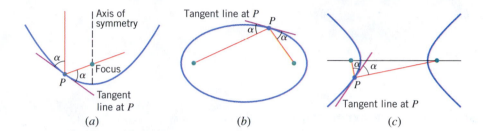

▶ **Figure 10.4.30**   (a)   (b)   (c)

■ APPLICATIONS OF THE CONIC SECTIONS

John Mead/Science Photo Library/Photo Researchers

Incoming signals are reflected by the parabolic antenna to the receiver at the focus.

Fermat's principle in optics implies that light reflects off of a surface at an angle equal to its angle of incidence. (See Exercise 62 in Section 4.5.) In particular, if a reflecting surface is generated by revolving a parabola about its axis of symmetry, it follows from Theorem 10.4.4 that all light rays entering parallel to the axis will be reflected to the focus (Figure 10.4.31a); conversely, if a light source is located at the focus, then the reflected rays will all be parallel to the axis (Figure 10.4.31b). This principle is used in certain telescopes to reflect the approximately parallel rays of light from the stars and planets off of a parabolic mirror to an eyepiece at the focus; and the parabolic reflectors in flashlights and automobile headlights utilize this principle to form a parallel beam of light rays from a bulb placed at the focus. The same optical principles apply to radar signals and sound waves, which explains the parabolic shape of many antennas.

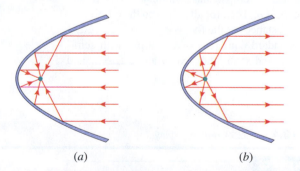

▶ **Figure 10.4.31**   (a)   (b)

Visitors to various rooms in the United States Capitol Building and in St. Paul's Cathedral in London are often astonished by the "whispering gallery" effect in which two people at opposite ends of the room can hear one another's whispers very clearly. Such rooms have ceilings with elliptical cross sections and common foci. Thus, when the two people stand at the foci, their whispers are reflected directly to one another off of the elliptical ceiling.

Hyperbolic navigation systems, which were developed in World War II as navigational aids to ships, are based on the definition of a hyperbola. With these systems the ship receives

synchronized radio signals from two widely spaced transmitters with known positions. The ship's electronic receiver measures the difference in reception times between the signals and then uses that difference to compute the difference $2a$ between its distances from the two transmitters. This information places the ship somewhere on the hyperbola whose foci are at the transmitters and whose points have $2a$ as the difference in their distances from the foci. By repeating the process with a second set of transmitters, the position of the ship can be approximated as the intersection of two hyperbolas (Figure 10.4.32). (The modern global positioning system (GPS) is based on the same principle.)

▲ **Figure 10.4.32**

✔ QUICK CHECK EXERCISES 10.4   *(See page 748 for answers.)*

1. Identify the conic.
  (a) The set of points in the plane, the sum of whose distances to two fixed points is a positive constant greater than the distance between the fixed points is _____.
  (b) The set of points in the plane, the difference of whose distances to two fixed points is a positive constant less than the distance between the fixed points is _____.
  (c) The set of points in the plane that are equidistant from a fixed line and a fixed point not on the line is _____.

2. (a) The equation of the parabola with focus $(p, 0)$ and directrix $x = -p$ is _____.
  (b) The equation of the parabola with focus $(0, p)$ and directrix $y = -p$ is _____.

3. (a) Suppose that an ellipse has semimajor axis a and semiminor axis b. Then for all points on the ellipse, the sum of the distances to the foci is equal to _____.
  (b) The two standard equations of an ellipse with semimajor axis a and semiminor axis b are _____ and _____.

  (c) Suppose that an ellipse has semimajor axis a, semiminor axis b, and foci $(\pm c, 0)$. Then c may be obtained from a and b by the equation $c =$ _____.

4. (a) Suppose that a hyperbola has semifocal axis a and semiconjugate axis b. Then for all points on the hyperbola, the difference of the distance to the farther focus minus the distance to the closer focus is equal to _____.
  (b) The two standard equations of a hyperbola with semifocal axis a and semiconjugate axis b are _____ and _____.
  (c) Suppose that a hyperbola in standard position has semifocal axis a, semiconjugate axis b, and foci $(\pm c, 0)$. Then c may be obtained from a and b by the equation $c =$ _____. The equations of the asymptotes of this hyperbola are $y = \pm$ _____.

EXERCISE SET 10.4   Graphing Utility

FOCUS ON CONCEPTS

1. In parts (a)–(f), find the equation of the conic.

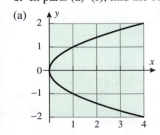

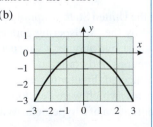

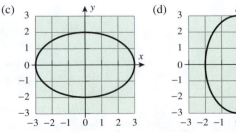

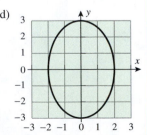

(e)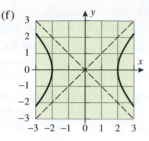

(f)

2. (a) Find the focus and directrix for each parabola in Exercise 1.

(b) Find the foci of the ellipses in Exercise 1.

(c) Find the foci and the equations of the asymptotes of the hyperbolas in Exercise 1.

3–6 Sketch the parabola, and label the focus, vertex, and directrix. ■

3. (a) $y^2 = 4x$ (b) $x^2 = -8y$

4. (a) $y^2 = -10x$ (b) $x^2 = 4y$

5. (a) $(y-1)^2 = -12(x+4)$ (b) $(x-1)^2 = 2\left(y - \frac{1}{2}\right)$

6. (a) $y^2 - 6y - 2x + 1 = 0$ (b) $y = 4x^2 + 8x + 5$

7–10 Sketch the ellipse, and label the foci, vertices, and ends of the minor axis. ■

7. (a) $\dfrac{x^2}{16} + \dfrac{y^2}{9} = 1$ (b) $9x^2 + y^2 = 9$

8. (a) $\dfrac{x^2}{25} + \dfrac{y^2}{4} = 1$ (b) $4x^2 + y^2 = 36$

9. (a) $(x+3)^2 + 4(y-5)^2 = 16$

(b) $\frac{1}{4}x^2 + \frac{1}{9}(y+2)^2 - 1 = 0$

10. (a) $9x^2 + 4y^2 - 18x + 24y + 9 = 0$

(b) $5x^2 + 9y^2 + 20x - 54y = -56$

11–14 Sketch the hyperbola, and label the vertices, foci, and asymptotes. ■

11. (a) $\dfrac{x^2}{16} - \dfrac{y^2}{9} = 1$ (b) $9y^2 - x^2 = 36$

12. (a) $\dfrac{y^2}{9} - \dfrac{x^2}{25} = 1$ (b) $16x^2 - 25y^2 = 400$

13. (a) $\dfrac{(y+4)^2}{3} - \dfrac{(x-2)^2}{5} = 1$

(b) $16(x+1)^2 - 8(y-3)^2 = 16$

14. (a) $x^2 - 4y^2 + 2x + 8y - 7 = 0$

(b) $16x^2 - y^2 - 32x - 6y = 57$

15–18 Find an equation for the parabola that satisfies the given conditions. ■

15. (a) Vertex $(0, 0)$; focus $(3, 0)$.

(b) Vertex $(0, 0)$; directrix $y = \frac{1}{4}$.

16. (a) Focus $(6, 0)$; directrix $x = -6$.

(b) Focus $(1, 1)$; directrix $y = -2$.

17. Axis $y = 0$; passes through $(3, 2)$ and $(2, -\sqrt{2})$.

18. Vertex $(5, -3)$; axis parallel to the y-axis; passes through $(9, 5)$.

19–22 Find an equation for the ellipse that satisfies the given conditions. ■

19. (a) Ends of major axis $(\pm 3, 0)$; ends of minor axis $(0, \pm 2)$.

(b) Length of minor axis 8; foci $(0, \pm 3)$.

20. (a) Foci $(\pm 1, 0)$; $b = \sqrt{2}$.

(b) $c = 2\sqrt{3}$; $a = 4$; center at the origin; foci on a coordinate axis (two answers).

21. (a) Ends of major axis $(0, \pm 6)$; passes through $(-3, 2)$.

(b) Foci $(-1, 1)$ and $(-1, 3)$; minor axis of length 4.

22. (a) Center at $(0, 0)$; major and minor axes along the coordinate axes; passes through $(3, 2)$ and $(1, 6)$.

(b) Foci $(2, 1)$ and $(2, -3)$; major axis of length 6.

23–26 Find an equation for a hyperbola that satisfies the given conditions. [*Note:* In some cases there may be more than one hyperbola.] ■

23. (a) Vertices $(\pm 2, 0)$; foci $(\pm 3, 0)$.

(b) Vertices $(0, \pm 2)$; asymptotes $y = \pm \frac{2}{3}x$.

24. (a) Asymptotes $y = \pm \frac{3}{2}x$; $b = 4$.

(b) Foci $(0, \pm 5)$; asymptotes $y = \pm 2x$.

25. (a) Asymptotes $y = \pm \frac{3}{4}x$; $c = 5$.

(b) Foci $(\pm 3, 0)$; asymptotes $y = \pm 2x$.

26. (a) Vertices $(0, 6)$ and $(6, 6)$; foci 10 units apart.

(b) Asymptotes $y = x - 2$ and $y = -x + 4$; passes through the origin.

27–30 True–False Determine whether the statement is true or false. Explain your answer. ■

27. A hyperbola is the set of all points in the plane that are equidistant from a fixed line and a fixed point not on the line.

28. If an ellipse is not a circle, then the foci of an ellipse lie on the major axis of the ellipse.

29. If a parabola has equation $y^2 = 4px$, where p is a positive constant, then the perpendicular distance from the parabola's focus to its directrix is p.

30. The hyperbola $(y^2/a^2) - x^2 = 1$ has asymptotes the lines $y = \pm x/a$.

31. (a) As illustrated in the accompanying figure, a parabolic arch spans a road 40 ft wide. How high is the arch if a center section of the road 20 ft wide has a minimum clearance of 12 ft?

(b) How high would the center be if the arch were the upper half of an ellipse?

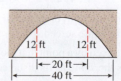

◄ **Figure Ex-31**

32. (a) Find an equation for the parabolic arch with base b and height h, shown in the accompanying figure.

(b) Find the area under the arch.

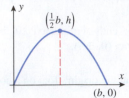

$\blacktriangleleft$ **Figure Ex-32**

33. Show that the vertex is the closest point on a parabola to the focus. [*Suggestion:* Introduce a convenient coordinate system and use Definition 10.4.1.]

34. As illustrated in the accompanying figure, suppose that a comet moves in a parabolic orbit with the Sun at its focus and that the line from the Sun to the comet makes an angle of 60° with the axis of the parabola when the comet is 40 million miles from the center of the Sun. Use the result in Exercise 33 to determine how close the comet will come to the center of the Sun.

35. For the parabolic reflector in the accompanying figure, how far from the vertex should the light source be placed to produce a beam of parallel rays?

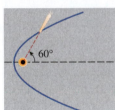

$\blacktriangle$ **Figure Ex-34**

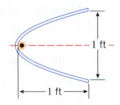

$\blacktriangle$ **Figure Ex-35**

36. (a) Show that the right and left branches of the hyperbola

$$\frac{x^2}{a^2} - \frac{y^2}{b^2} = 1$$

can be represented parametrically as

$$x = a\cosh t, \quad y = b\sinh t \quad (-\infty < t < +\infty)$$
$$x = -a\cosh t, \quad y = b\sinh t \quad (-\infty < t < +\infty)$$

(b) Use a graphing utility to generate both branches of the hyperbola $x^2 - y^2 = 1$ on the same screen.

37. (a) Show that the right and left branches of the hyperbola

$$\frac{x^2}{a^2} - \frac{y^2}{b^2} = 1$$

can be represented parametrically as

$$x = a\sec t, \quad y = b\tan t \quad (-\pi/2 < t < \pi/2)$$
$$x = -a\sec t, \quad y = b\tan t \quad (-\pi/2 < t < \pi/2)$$

(b) Use a graphing utility to generate both branches of the hyperbola $x^2 - y^2 = 1$ on the same screen.

38. Find an equation of the parabola traced by a point that moves so that its distance from $(2, 4)$ is the same as its distance to the x-axis.

39. Find an equation of the ellipse traced by a point that moves so that the sum of its distances to $(4, 1)$ and $(4, 5)$ is 12.

40. Find the equation of the hyperbola traced by a point that moves so that the difference between its distances to $(0, 0)$ and $(1, 1)$ is 1.

41. Show that an ellipse with semimajor axis a and semiminor axis b has area $A = \pi ab$.

FOCUS ON CONCEPTS

42. Show that if a plane is not parallel to the axis of a right circular cylinder, then the intersection of the plane and cylinder is an ellipse (possibly a circle). [*Hint:* Let θ be the angle shown in the accompanying figure, introduce coordinate axes as shown, and express x' and y' in terms of x and y.]

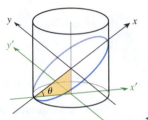

$\blacktriangleleft$ **Figure Ex-42**

43. As illustrated in the accompanying figure, a carpenter needs to cut an elliptical hole in a sloped roof through which a circular vent pipe of diameter D is to be inserted vertically. The carpenter wants to draw the outline of the hole on the roof using a pencil, two tacks, and a piece of string (as in Figure 10.4.3b). The center point of the ellipse is known, and common sense suggests that its major axis must be perpendicular to the drip line of the roof. The carpenter needs to determine the length L of the string and the distance T between a tack and the center point. The architect's plans show that the pitch of the roof is p (pitch = rise over run; see the accompanying figure). Find T and L in terms of D and p.

Source: This exercise is based on an article by William H. Enos, which appeared in the *Mathematics Teacher*, Feb. 1991, p. 148.

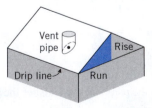

$\blacktriangleleft$ **Figure Ex-43**

44. As illustrated in the accompanying figure on the next page, suppose that two observers are stationed at the points $F_1(c, 0)$ and $F_2(-c, 0)$ in an xy-coordinate system. Suppose also that the sound of an explosion in the xy-plane is heard by the F_1 observer t seconds before it is heard by the F_2 observer. Assuming that the speed of sound is a constant v, show that the explosion occurred somewhere on the hyperbola

$$\frac{x^2}{v^2t^2/4} - \frac{y^2}{c^2 - (v^2t^2/4)} = 1$$

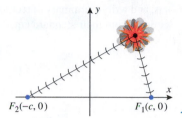

◀ **Figure Ex-44**

45. As illustrated in the accompanying figure, suppose that two transmitting stations are positioned 100 km apart at points $F_1(50, 0)$ and $F_2(-50, 0)$ on a straight shoreline in an xy-coordinate system. Suppose also that a ship is traveling parallel to the shoreline but 200 km at sea. Find the coordinates of the ship if the stations transmit a pulse simultaneously, but the pulse from station F_1 is received by the ship 100 microseconds sooner than the pulse from station F_2. [*Hint:* Use the formula obtained in Exercise 44, assuming that the pulses travel at the speed of light (299,792,458 m/s).]

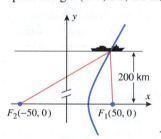

◀ **Figure Ex-45**

46. A nuclear cooling tower is to have a height of h feet and the shape of the solid that is generated by revolving the region R enclosed by the right branch of the hyperbola $1521x^2 - 225y^2 = 342,225$ and the lines $x = 0$, $y = -h/2$, and $y = h/2$ about the y-axis.
(a) Find the volume of the tower.
(b) Find the lateral surface area of the tower.

47. Let R be the region that is above the x-axis and enclosed between the curve $b^2x^2 - a^2y^2 = a^2b^2$ and the line $x = \sqrt{a^2 + b^2}$.
(a) Sketch the solid generated by revolving R about the x-axis, and find its volume.
(b) Sketch the solid generated by revolving R about the y-axis, and find its volume.

48. Prove: The line tangent to the parabola $x^2 = 4py$ at the point (x_0, y_0) is $x_0x = 2p(y + y_0)$.

49. Prove: The line tangent to the ellipse

$$\frac{x^2}{a^2} + \frac{y^2}{b^2} = 1$$

at the point (x_0, y_0) has the equation

$$\frac{xx_0}{a^2} + \frac{yy_0}{b^2} = 1$$

50. Prove: The line tangent to the hyperbola

$$\frac{x^2}{a^2} - \frac{y^2}{b^2} = 1$$

at the point (x_0, y_0) has the equation

$$\frac{xx_0}{a^2} - \frac{yy_0}{b^2} = 1$$

51. Use the results in Exercises 49 and 50 to show that if an ellipse and a hyperbola have the same foci, then at each point of intersection their tangent lines are perpendicular.

52. Consider the second-degree equation

$$Ax^2 + Cy^2 + Dx + Ey + F = 0$$

where A and C are not both 0. Show by completing the square:
(a) If $AC > 0$, then the equation represents an ellipse, a circle, a point, or has no graph.
(b) If $AC < 0$, then the equation represents a hyperbola or a pair of intersecting lines.
(c) If $AC = 0$, then the equation represents a parabola, a pair of parallel lines, or has no graph.

53. In each part, use the result in Exercise 52 to make a statement about the graph of the equation, and then check your conclusion by completing the square and identifying the graph.
(a) $x^2 - 5y^2 - 2x - 10y - 9 = 0$
(b) $x^2 - 3y^2 - 6y - 3 = 0$
(c) $4x^2 + 8y^2 + 16x + 16y + 20 = 0$
(d) $3x^2 + y^2 + 12x + 2y + 13 = 0$
(e) $x^2 + 8x + 2y + 14 = 0$
(f) $5x^2 + 40x + 2y + 94 = 0$

54. Derive the equation $x^2 = 4py$ in Figure 10.4.6.

55. Derive the equation $(x^2/b^2) + (y^2/a^2) = 1$ given in Figure 10.4.14.

56. Derive the equation $(x^2/a^2) - (y^2/b^2) = 1$ given in Figure 10.4.22.

57. Prove Theorem 10.4.4. [*Hint:* Choose coordinate axes so that the parabola has the equation $x^2 = 4py$. Show that the tangent line at $P(x_0, y_0)$ intersects the y-axis at $Q(0, -y_0)$ and that the triangle whose three vertices are at P, Q, and the focus is isosceles.]

58. Given two intersecting lines, let L_2 be the line with the larger angle of inclination ϕ_2, and let L_1 be the line with the smaller angle of inclination ϕ_1. We define the **angle** θ **between** L_1 **and** L_2 by $\theta = \phi_2 - \phi_1$. (See the accompanying figure on the next page.)
(a) Prove: If L_1 and L_2 are not perpendicular, then

$$\tan \theta = \frac{m_2 - m_1}{1 + m_1m_2}$$

where L_1 and L_2 have slopes m_1 and m_2.
(b) Prove Theorem 10.4.5. [*Hint:* Introduce coordinates so that the equation $(x^2/a^2) + (y^2/b^2) = 1$ describes the ellipse, and use part (a).]

(cont.)

(c) Prove Theorem 10.4.6. [*Hint:* Introduce coordinates so that the equation $(x^2/a^2) - (y^2/b^2) = 1$ describes the hyperbola, and use part (a).]

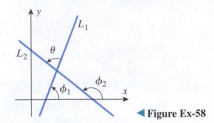

◀ **Figure Ex-58**

59. Writing Suppose that you want to draw an ellipse that has given values for the lengths of the major and minor axes by using the method shown in Figure 10.4.3*b*. Assuming that the axes are drawn, explain how a compass can be used to locate the positions for the tacks.

60. Writing List the forms for standard equations of parabolas, ellipses, and hyperbolas, and write a summary of techniques for sketching conic sections from their standard equations.

✔ **QUICK CHECK ANSWERS 10.4**

1. (a) an ellipse (b) a hyperbola (c) a parabola   **2.** (a) $y^2 = 4px$ (b) $x^2 = 4py$
3. (a) $2a$ (b) $\dfrac{x^2}{a^2} + \dfrac{y^2}{b^2} = 1$; $\dfrac{x^2}{b^2} + \dfrac{y^2}{a^2} = 1$ (c) $\sqrt{a^2 - b^2}$ **4.** (a) $2a$ (b) $\dfrac{x^2}{a^2} - \dfrac{y^2}{b^2} = 1$; $\dfrac{y^2}{a^2} - \dfrac{x^2}{b^2} = 1$ (c) $\sqrt{a^2 + b^2}$; $\dfrac{b}{a}x$

10.5   ROTATION OF AXES; SECOND-DEGREE EQUATIONS

In the preceding section we obtained equations of conic sections with axes parallel to the coordinate axes. In this section we will study the equations of conics that are "tilted" relative to the coordinate axes. This will lead us to investigate rotations of coordinate axes.

■ QUADRATIC EQUATIONS IN x AND y

We saw in Examples 8 to 10 of the preceding section that equations of the form

$$Ax^2 + Cy^2 + Dx + Ey + F = 0 \tag{1}$$

can represent conic sections. Equation (1) is a special case of the more general equation

$$Ax^2 + Bxy + Cy^2 + Dx + Ey + F = 0 \tag{2}$$

which, if A, B, and C are not all zero, is called a ***quadratic equation*** in x and y. It is usually the case that the graph of any second-degree equation is a conic section. If $B = 0$, then (2) reduces to (1) and the conic section has its axis or axes parallel to the coordinate axes. However, if $B \neq 0$, then (2) contains a ***cross-product term*** Bxy, and the graph of the conic section represented by the equation has its axis or axes "tilted" relative to the coordinate axes. As an illustration, consider the ellipse with foci $F_1(1, 2)$ and $F_2(-1, -2)$ and such that the sum of the distances from each point $P(x, y)$ on the ellipse to the foci is 6 units. Expressing this condition as an equation, we obtain (Figure 10.5.1)

$$\sqrt{(x-1)^2 + (y-2)^2} + \sqrt{(x+1)^2 + (y+2)^2} = 6$$

Squaring both sides, then isolating the remaining radical, then squaring again ultimately yields

$$8x^2 - 4xy + 5y^2 = 36$$

as the equation of the ellipse. This is of form (2) with $A = 8$, $B = -4$, $C = 5$, $D = 0$, $E = 0$, and $F = -36$.

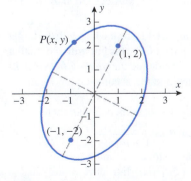

▲ **Figure 10.5.1**

ROTATION OF AXES

To study conics that are tilted relative to the coordinate axes it is frequently helpful to rotate the coordinate axes, so that the rotated coordinate axes are parallel to the axes of the conic. Before we can discuss the details, we need to develop some ideas about rotation of coordinate axes.

In Figure 10.5.2a the axes of an xy-coordinate system have been rotated about the origin through an angle θ to produce a new $x'y'$-coordinate system. As shown in the figure, each point P in the plane has coordinates (x', y') as well as coordinates (x, y). To see how the two are related, let r be the distance from the common origin to the point P, and let α be the angle shown in Figure 10.5.2b. It follows that

$$x = r\cos(\theta + \alpha), \quad y = r\sin(\theta + \alpha) \tag{3}$$

and

$$x' = r\cos\alpha, \quad y' = r\sin\alpha \tag{4}$$

Using familiar trigonometric identities, the relationships in (3) can be written as

$$x = r\cos\theta\cos\alpha - r\sin\theta\sin\alpha$$
$$y = r\sin\theta\cos\alpha + r\cos\theta\sin\alpha$$

and on substituting (4) in these equations we obtain the following relationships called the *rotation equations*:

$$\begin{aligned} x &= x'\cos\theta - y'\sin\theta \\ y &= x'\sin\theta + y'\cos\theta \end{aligned} \tag{5}$$

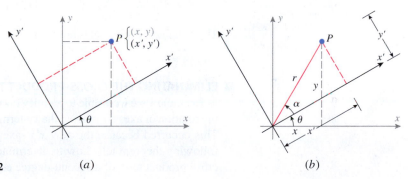

▶ **Figure 10.5.2**   (a)   (b)

▶ **Example 1**   Suppose that the axes of an xy-coordinate system are rotated through an angle of $\theta = 45°$ to obtain an $x'y'$-coordinate system. Find the equation of the curve

$$x^2 - xy + y^2 - 6 = 0$$

in $x'y'$-coordinates.

Solution.   Substituting $\sin\theta = \sin 45° = 1/\sqrt{2}$ and $\cos\theta = \cos 45° = 1/\sqrt{2}$ in (5) yields the rotation equations

$$x = \frac{x'}{\sqrt{2}} - \frac{y'}{\sqrt{2}} \quad \text{and} \quad y = \frac{x'}{\sqrt{2}} + \frac{y'}{\sqrt{2}}$$

Substituting these into the given equation yields

$$\left(\frac{x'}{\sqrt{2}} - \frac{y'}{\sqrt{2}}\right)^2 - \left(\frac{x'}{\sqrt{2}} - \frac{y'}{\sqrt{2}}\right)\left(\frac{x'}{\sqrt{2}} + \frac{y'}{\sqrt{2}}\right) + \left(\frac{x'}{\sqrt{2}} + \frac{y'}{\sqrt{2}}\right)^2 - 6 = 0$$

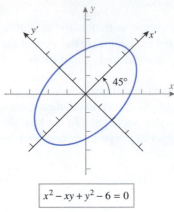

$$x^2 - xy + y^2 - 6 = 0$$

▲ **Figure 10.5.3**

or

$$\frac{x'^2 - 2x'y' + y'^2 - x'^2 + y'^2 + x'^2 + 2x'y' + y'^2}{2} = 6$$

or

$$\frac{x'^2}{12} + \frac{y'^2}{4} = 1$$

which is the equation of an ellipse (Figure 10.5.3). ◀

If the rotation equations (5) are solved for x' and y' in terms of x and y, one obtains (Exercise 16):

$$\begin{aligned} x' &= x \cos\theta + y \sin\theta \\ y' &= -x \sin\theta + y \cos\theta \end{aligned} \tag{6}$$

▶ **Example 2**   Find the new coordinates of the point $(2, 4)$ if the coordinate axes are rotated through an angle of $\theta = 30°$.

Solution.   Using the rotation equations in (6) with $x = 2$, $y = 4$, $\cos\theta = \cos 30° = \sqrt{3}/2$, and $\sin\theta = \sin 30° = 1/2$, we obtain

$$\begin{aligned} x' &= 2(\sqrt{3}/2) + 4(1/2) = \sqrt{3} + 2 \\ y' &= -2(1/2) + 4(\sqrt{3}/2) = -1 + 2\sqrt{3} \end{aligned}$$

Thus, the new coordinates are $(\sqrt{3} + 2, -1 + 2\sqrt{3})$. ◀

■ ELIMINATING THE CROSS-PRODUCT TERM

In Example 1 we were able to identify the curve $x^2 - xy + y^2 - 6 = 0$ as an ellipse because the rotation of axes eliminated the xy-term, thereby reducing the equation to a familiar form. This occurred because the new $x'y'$-axes were aligned with the axes of the ellipse. The following theorem tells how to determine an appropriate rotation of axes to eliminate the cross-product term of a second-degree equation in x and y.

10.5.1   THEOREM   *If the equation*

$$Ax^2 + Bxy + Cy^2 + Dx + Ey + F = 0 \tag{7}$$

is such that $B \neq 0$, and if an $x'y'$-coordinate system is obtained by rotating the xy-axes through an angle θ satisfying

$$\cot 2\theta = \frac{A - C}{B} \tag{8}$$

then, in $x'y'$-coordinates, Equation (7) will have the form

$$A'x'^2 + C'y'^2 + D'x' + E'y' + F' = 0$$

It is always possible to satisfy (8) with an angle θ in the interval

$$0 < \theta < \pi/2$$

We will always choose θ in this way.

PROOF   Substituting (5) into (7) and simplifying yields

$$A'x'^2 + B'x'y' + C'y'^2 + D'x' + E'y' + F' = 0$$

where

$$A' = A \cos^2 \theta + B \cos \theta \sin \theta + C \sin^2 \theta$$

$$B' = B(\cos^2 \theta - \sin^2 \theta) + 2(C - A) \sin \theta \cos \theta$$

$$C' = A \sin^2 \theta - B \sin \theta \cos \theta + C \cos^2 \theta$$

$$D' = D \cos \theta + E \sin \theta \qquad (9)$$

$$E' = -D \sin \theta + E \cos \theta$$

$$F' = F$$

(Verify.) To complete the proof we must show that $B' = 0$ if

$$\cot 2\theta = \frac{A - C}{B}$$

or, equivalently,

$$\frac{\cos 2\theta}{\sin 2\theta} = \frac{A - C}{B} \qquad (10)$$

However, by using the trigonometric double-angle formulas, we can rewrite B' in the form

$$B' = B \cos 2\theta - (A - C) \sin 2\theta$$

Thus, $B' = 0$ if θ satisfies (10). ∎

▶ **Example 3**   Identify and sketch the curve $xy = 1$.

Solution.   As a first step, we will rotate the coordinate axes to eliminate the cross-product term. Comparing the given equation to (7), we have

$$A = 0, \quad B = 1, \quad C = 0$$

Thus, the desired angle of rotation must satisfy

$$\cot 2\theta = \frac{A - C}{B} = \frac{0 - 0}{1} = 0$$

This condition can be met by taking $2\theta = \pi/2$ or $\theta = \pi/4 = 45°$. Making the substitutions $\cos \theta = \cos 45° = 1/\sqrt{2}$ and $\sin \theta = \sin 45° = 1/\sqrt{2}$ in (5) yields

$$x = \frac{x'}{\sqrt{2}} - \frac{y'}{\sqrt{2}} \quad \text{and} \quad y = \frac{x'}{\sqrt{2}} + \frac{y'}{\sqrt{2}}$$

Substituting these in the equation $xy = 1$ yields

$$\left(\frac{x'}{\sqrt{2}} - \frac{y'}{\sqrt{2}} \right) \left(\frac{x'}{\sqrt{2}} + \frac{y'}{\sqrt{2}} \right) = 1 \quad \text{or} \quad \frac{x'^2}{2} - \frac{y'^2}{2} = 1$$

which is the equation in the $x'y'$-coordinate system of an equilateral hyperbola with vertices at $(\sqrt{2}, 0)$ and $(-\sqrt{2}, 0)$ in that coordinate system (Figure 10.5.4). ◀

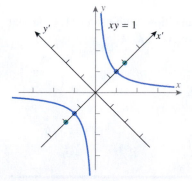

▲ **Figure 10.5.4**

In problems where it is inconvenient to solve

$$\cot 2\theta = \frac{A - C}{B}$$

for θ, the values of $\sin \theta$ and $\cos \theta$ needed for the rotation equations can be obtained by first calculating $\cos 2\theta$ and then computing $\sin \theta$ and $\cos \theta$ from the identities

$$\sin \theta = \sqrt{\frac{1 - \cos 2\theta}{2}} \quad \text{and} \quad \cos \theta = \sqrt{\frac{1 + \cos 2\theta}{2}}$$

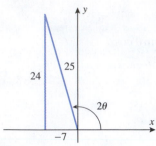

▲ Figure 10.5.5

▶ **Example 4**   Identify and sketch the curve

$$153x^2 - 192xy + 97y^2 - 30x - 40y - 200 = 0$$

Solution.   We have $A = 153$, $B = -192$, and $C = 97$, so

$$\cot 2\theta = \frac{A - C}{B} = -\frac{56}{192} = -\frac{7}{24}$$

Since θ is to be chosen in the range $0 < \theta < \pi/2$, this relationship is represented by the triangle in Figure 10.5.5. From that triangle we obtain $\cos 2\theta = -\frac{7}{25}$, which implies that

$$\cos \theta = \sqrt{\frac{1 + \cos 2\theta}{2}} = \sqrt{\frac{1 - \frac{7}{25}}{2}} = \frac{3}{5}$$

$$\sin \theta = \sqrt{\frac{1 - \cos 2\theta}{2}} = \sqrt{\frac{1 + \frac{7}{25}}{2}} = \frac{4}{5}$$

Substituting these values in (5) yields the rotation equations

$$x = \tfrac{3}{5}x' - \tfrac{4}{5}y' \quad \text{and} \quad y = \tfrac{4}{5}x' + \tfrac{3}{5}y'$$

and substituting these in turn in the given equation yields

$$\tfrac{153}{25}(3x' - 4y')^2 - \tfrac{192}{25}(3x' - 4y')(4x' + 3y') + \tfrac{97}{25}(4x' + 3y')^2$$

$$- \tfrac{30}{5}(3x' - 4y') - \tfrac{40}{5}(4x' + 3y') - 200 = 0$$

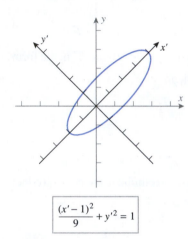

$$\frac{(x' - 1)^2}{9} + y'^2 = 1$$

▲ Figure 10.5.6

which simplifies to

$$25x'^2 + 225y'^2 - 50x' - 200 = 0$$

or

$$x'^2 + 9y'^2 - 2x' - 8 = 0$$

Completing the square yields

$$\frac{(x' - 1)^2}{9} + y'^2 = 1$$

which is the equation in the $x'y'$-coordinate system of an ellipse with center $(1, 0)$ in that coordinate system and semiaxes $a = 3$ and $b = 1$ (Figure 10.5.6). ◀

There is a method for deducing the kind of curve represented by a second-degree equation directly from the equation itself without rotating coordinate axes. For a discussion of this topic, see the section on the ***discriminant*** that appears in Web Appendix K.

✔ **QUICK CHECK EXERCISES 10.5**   *(See page 754 for answers.)*

1. Suppose that an xy-coordinate system is rotated θ radians to produce a new $x'y'$-coordinate system.
   (a) x and y may be obtained from x', y', and θ using the rotation equations $x = $ _____ and $y = $ _____.
   (b) x' and y' may be obtained from x, y, and θ using the equations $x' = $ _____ and $y' = $ _____.

2. If the equation

   $$Ax^2 + Bxy + Cy^2 + Dx + Ey + F = 0$$

   is such that $B \neq 0$, then the xy-term in this equation can be

eliminated by a rotation of axes through an angle θ satisfying $\cot 2\theta = $ _____.

3. In each part, determine a rotation angle θ that will eliminate the xy-term.
   (a) $2x^2 + xy + 2y^2 + x - y = 0$
   (b) $x^2 + 2\sqrt{3}xy + 3y^2 - 2x + y = 1$
   (c) $3x^2 + \sqrt{3}xy + 2y^2 + y = 0$

4. Express $2x^2 + xy + 2y^2 = 1$ in the $x'y'$-coordinate system obtained by rotating the xy-coordinate system through the angle $\theta = \pi/4$.

EXERCISE SET 10.5

1. Let an $x'y'$-coordinate system be obtained by rotating an xy-coordinate system through an angle of $\theta = 60°$.
   (a) Find the $x'y'$-coordinates of the point whose xy-coordinates are $(-2, 6)$.
   (b) Find an equation of the curve $\sqrt{3}xy + y^2 = 6$ in $x'y'$-coordinates.
   (c) Sketch the curve in part (b), showing both xy-axes and $x'y'$-axes.

2. Let an $x'y'$-coordinate system be obtained by rotating an xy-coordinate system through an angle of $\theta = 30°$.
   (a) Find the $x'y'$-coordinates of the point whose xy-coordinates are $(1, -\sqrt{3})$.
   (b) Find an equation of the curve $2x^2 + 2\sqrt{3}xy = 3$ in $x'y'$-coordinates.
   (c) Sketch the curve in part (b), showing both xy-axes and $x'y'$-axes.

3–12 Rotate the coordinate axes to remove the xy-term. Then identify the type of conic and sketch its graph. ■

3. $xy = -9$ 4. $x^2 - xy + y^2 - 2 = 0$

5. $x^2 + 4xy - 2y^2 - 6 = 0$

6. $31x^2 + 10\sqrt{3}xy + 21y^2 - 144 = 0$

7. $x^2 + 2\sqrt{3}xy + 3y^2 + 2\sqrt{3}x - 2y = 0$

8. $34x^2 - 24xy + 41y^2 - 25 = 0$

9. $9x^2 - 24xy + 16y^2 - 80x - 60y + 100 = 0$

10. $5x^2 - 6xy + 5y^2 - 8\sqrt{2}x + 8\sqrt{2}y = 8$

11. $52x^2 - 72xy + 73y^2 + 40x + 30y - 75 = 0$

12. $6x^2 + 24xy - y^2 - 12x + 26y + 11 = 0$

13. Let an $x'y'$-coordinate system be obtained by rotating an xy-coordinate system through an angle of $45°$. Use (6) to find an equation of the curve $3x'^2 + y'^2 = 6$ in xy-coordinates.

14. Let an $x'y'$-coordinate system be obtained by rotating an xy-coordinate system through an angle of $30°$. Use (5) to find an equation in $x'y'$-coordinates of the curve $y = x^2$.

15. Let an $x'y'$-coordinate system be obtained by rotating an xy-coordinate system through an angle θ. Prove: For every value of θ, the equation $x^2 + y^2 = r^2$ becomes the equation $x'^2 + y'^2 = r^2$. Give a geometric explanation.

16. Derive (6) by solving the rotation equations in (5) for x' and y' in terms of x and y.

17. Let an $x'y'$-coordinate system be obtained by rotating an xy-coordinate system through an angle θ. Explain how to find the xy-coordinates of a point whose $x'y'$-coordinates are known.

18. Let an $x'y'$-coordinate system be obtained by rotating an xy-coordinate system through an angle θ. Explain how to find the xy-equation of a line whose $x'y'$-equation is known.

19–22 Show that the graph of the given equation is a parabola. Find its vertex, focus, and directrix. ■

19. $x^2 + 2xy + y^2 + 4\sqrt{2}x - 4\sqrt{2}y = 0$

20. $x^2 - 2\sqrt{3}xy + 3y^2 - 8\sqrt{3}x - 8y = 0$

21. $9x^2 - 24xy + 16y^2 - 80x - 60y + 100 = 0$

22. $x^2 + 2\sqrt{3}xy + 3y^2 + 16\sqrt{3}x - 16y - 96 = 0$

23–26 Show that the graph of the given equation is an ellipse. Find its foci, vertices, and the ends of its minor axis. ■

23. $288x^2 - 168xy + 337y^2 - 3600 = 0$

24. $25x^2 - 14xy + 25y^2 - 288 = 0$

25. $31x^2 + 10\sqrt{3}xy + 21y^2 - 32x + 32\sqrt{3}y - 80 = 0$

26. $43x^2 - 14\sqrt{3}xy + 57y^2 - 36\sqrt{3}x - 36y - 540 = 0$

27–30 Show that the graph of the given equation is a hyperbola. Find its foci, vertices, and asymptotes. ■

27. $x^2 - 10\sqrt{3}xy + 11y^2 + 64 = 0$

28. $17x^2 - 312xy + 108y^2 - 900 = 0$

29. $32y^2 - 52xy - 7x^2 + 72\sqrt{5}x - 144\sqrt{5}y + 900 = 0$

30. $2\sqrt{2}y^2 + 5\sqrt{2}xy + 2\sqrt{2}x^2 + 18x + 18y + 36\sqrt{2} = 0$

31. Show that the graph of the equation

$$\sqrt{x} + \sqrt{y} = 1$$

is a portion of a parabola. [*Hint:* First rationalize the equation and then perform a rotation of axes.]

32. Derive the expression for B' in (9).

33. Use (9) to prove that $B^2 - 4AC = B'^2 - 4A'C'$ for all values of θ.

34. Use (9) to prove that $A + C = A' + C'$ for all values of θ.

35. Prove: If $A = C$ in (7), then the cross-product term can be eliminated by rotating through $45°$.

36. Prove: If $B \neq 0$, then the graph of $x^2 + Bxy + F = 0$ is a hyperbola if $F \neq 0$ and two intersecting lines if $F = 0$.

✔ QUICK CHECK ANSWERS 10.5

1. (a) $x' \cos\theta - y' \sin\theta$; $x' \sin\theta + y' \cos\theta$ (b) $x \cos\theta + y \sin\theta$; $-x \sin\theta + y \cos\theta$ 2. $\dfrac{A-C}{B}$ 3. (a) $\dfrac{\pi}{4}$ (b) $\dfrac{\pi}{3}$ (c) $\dfrac{\pi}{6}$
4. $5x'^2 + 3y'^2 = 2$

10.6 CONIC SECTIONS IN POLAR COORDINATES

It will be shown later in the text that if an object moves in a gravitational field that is directed toward a fixed point (such as the center of the Sun), then the path of that object must be a conic section with the fixed point at a focus. For example, planets in our solar system move along elliptical paths with the Sun at a focus, and the comets move along parabolic, elliptical, or hyperbolic paths with the Sun at a focus, depending on the conditions under which they were born. For applications of this type it is usually desirable to express the equations of the conic sections in polar coordinates with the pole at a focus. In this section we will show how to do this.

■ THE FOCUS–DIRECTRIX CHARACTERIZATION OF CONICS
To obtain polar equations for the conic sections we will need the following theorem.

> It is an unfortunate historical accident that the letter e is used for the base of the natural logarithm as well as for the eccentricity of conic sections. However, as a practical matter the appropriate interpretation will usually be clear from the context in which the letter is used.

10.6.1 THEOREM (*Focus–Directrix Property of Conics*) *Suppose that a point P moves in the plane determined by a fixed point (called the **focus**) and a fixed line (called the **directrix**), where the focus does not lie on the directrix. If the point moves in such a way that its distance to the focus divided by its distance to the directrix is some constant e (called the **eccentricity**), then the curve traced by the point is a conic section. Moreover, the conic is*

(a) a parabola if $e = 1$ *(b) an ellipse if $0 < e < 1$*   *(c) a hyperbola if $e > 1$.*

We will not give a formal proof of this theorem; rather, we will use the specific cases in Figure 10.6.1 to illustrate the basic ideas. For the parabola, we will take the directrix to be $x = -p$, as usual; and for the ellipse and the hyperbola we will take the directrix to be $x = a^2/c$. We want to show in all three cases that if P is a point on the graph, F is the focus, and D is the directrix, then the ratio PF/PD is some constant e, where $e = 1$ for the parabola, $0 < e < 1$ for the ellipse, and $e > 1$ for the hyperbola. We will give the arguments for the parabola and ellipse and leave the argument for the hyperbola as an exercise.

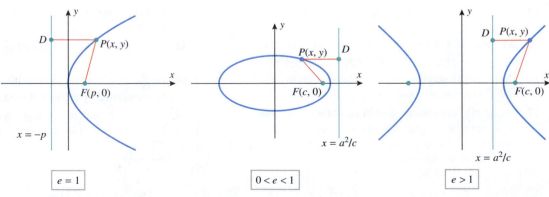

▲ Figure 10.6.1

For the parabola, the distance PF to the focus is equal to the distance PD to the directrix, so that $PF/PD = 1$, which is what we wanted to show. For the ellipse, we rewrite Equation (8) of Section 10.4 as

$$\sqrt{(x-c)^2 + y^2} = a - \frac{c}{a}x = \frac{c}{a}\left(\frac{a^2}{c} - x\right)$$

But the expression on the left side is the distance PF, and the expression in the parentheses on the right side is the distance PD, so we have shown that

$$PF = \frac{c}{a}PD$$

Thus, PF/PD is constant, and the eccentricity is

$$e = \frac{c}{a} \qquad (1)$$

If we rule out the degenerate case where $a = 0$ or $c = 0$, then it follows from Formula (7) of Section 10.4 that $0 < c < a$, so $0 < e < 1$, which is what we wanted to show.

We will leave it as an exercise to show that the eccentricity of the hyperbola in Figure 10.6.1 is also given by Formula (1), but in this case it follows from Formula (11) of Section 10.4 that $c > a$, so $e > 1$.

■ ECCENTRICITY OF AN ELLIPSE AS A MEASURE OF FLATNESS

The eccentricity of an ellipse can be viewed as a measure of its flatness—as e approaches 0 the ellipses become more and more circular, and as e approaches 1 they become more and more flat (Figure 10.6.2). Table 10.6.1 shows the orbital eccentricities of various celestial objects. Note that most of the planets actually have fairly circular orbits.

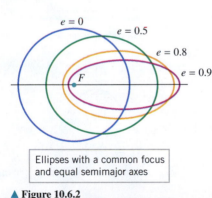

▲ Figure 10.6.2

Table 10.6.1

CELESTIAL BODY	ECCENTRICITY
Mercury	0.206
Venus	0.007
Earth	0.017
Mars	0.093
Jupiter	0.048
Saturn	0.056
Uranus	0.046
Neptune	0.010
Pluto	0.249
Halley's comet	0.970

■ POLAR EQUATIONS OF CONICS

Our next objective is to derive polar equations for the conic sections from their focus–directrix characterizations. We will assume that the focus is at the pole and the directrix is either parallel or perpendicular to the polar axis. If the directrix is parallel to the polar axis, then it can be above or below the pole; and if the directrix is perpendicular to the polar axis, then it can be to the left or right of the pole. Thus, there are four cases to consider. We will derive the formulas for the case in which the directrix is perpendicular to the polar axis and to the right of the pole.

As illustrated in Figure 10.6.3, let us assume that the directrix is perpendicular to the polar axis and d units to the right of the pole, where the constant d is known. If P is a point

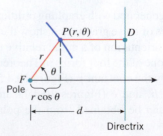

▲ Figure 10.6.3

on the conic and if the eccentricity of the conic is e, then it follows from Theorem 10.6.1 that $PF/PD = e$ or, equivalently, that

$$PF = ePD \qquad (2)$$

However, it is evident from Figure 10.6.3 that $PF = r$ and $PD = d - r\cos\theta$. Thus, (2) can be written as

$$r = e(d - r\cos\theta)$$

which can be solved for r and expressed as

$$r = \frac{ed}{1 + e\cos\theta}$$

(verify). Observe that this single polar equation can represent a parabola, an ellipse, or a hyperbola, depending on the value of e. In contrast, the rectangular equations for these conics all have different forms. The derivations in the other three cases are similar.

10.6.2   THEOREM   *If a conic section with eccentricity e is positioned in a polar coordinate system so that its focus is at the pole and the corresponding directrix is d units from the pole and is either parallel or perpendicular to the polar axis, then the equation of the conic has one of four possible forms, depending on its orientation:*

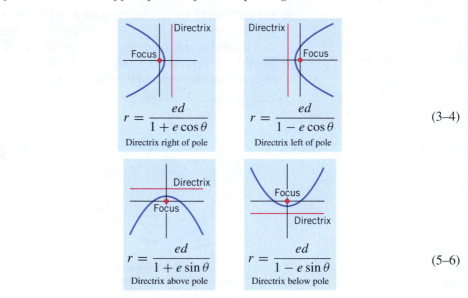

$$r = \frac{ed}{1 + e\cos\theta} \qquad\qquad r = \frac{ed}{1 - e\cos\theta} \qquad\qquad (3\text{–}4)$$
Directrix right of pole   Directrix left of pole

$$r = \frac{ed}{1 + e\sin\theta} \qquad\qquad r = \frac{ed}{1 - e\sin\theta} \qquad\qquad (5\text{–}6)$$
Directrix above pole   Directrix below pole

■ SKETCHING CONICS IN POLAR COORDINATES

Precise graphs of conic sections in polar coordinates can be generated with graphing utilities. However, it is often useful to be able to make quick sketches of these graphs that show their orientations and give some sense of their dimensions. The orientation of a conic relative to the polar axis can be deduced by matching its equation with one of the four forms in Theorem 10.6.2. The key dimensions of a parabola are determined by the constant p (Figure 10.4.5) and those of ellipses and hyperbolas by the constants a, b, and c (Figures 10.4.11 and 10.4.20). Thus, we need to show how these constants can be obtained from the polar equations.

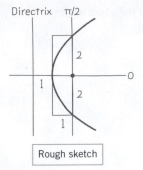

Directrix π/2

Rough sketch

▲ **Figure 10.6.4**

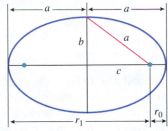

▲ **Figure 10.6.5**

In words, Formula (8) states that a is the **arithmetic average** (also called the **arithmetic mean**) of r_0 and r_1, and Formula (10) states that b is the **geometric mean** of r_0 and r_1.

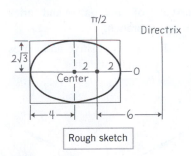

π/2

Directrix

Center

Rough sketch

▲ **Figure 10.6.6**

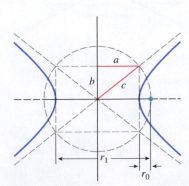

▲ **Figure 10.6.7**

▶ **Example 1** Sketch the graph of $r = \dfrac{2}{1 - \cos\theta}$ in polar coordinates.

Solution. The equation is an exact match to (4) with $d = 2$ and $e = 1$. Thus, the graph is a parabola with the focus at the pole and the directrix 2 units to the left of the pole. This tells us that the parabola opens to the right along the polar axis and $p = 1$. Thus, the parabola looks roughly like that sketched in Figure 10.6.4. ◀

All of the important geometric information about an ellipse can be obtained from the values of a, b, and c in Figure 10.6.5. One way to find these values from the polar equation of an ellipse is based on finding the distances from the focus to the vertices. As shown in the figure, let r_0 be the distance from the focus to the closest vertex and r_1 the distance to the farthest vertex. Thus,

$$r_0 = a - c \quad \text{and} \quad r_1 = a + c \tag{7}$$

from which it follows that

$$a = \tfrac{1}{2}(r_1 + r_0) \qquad c = \tfrac{1}{2}(r_1 - r_0) \tag{8–9}$$

Moreover, it also follows from (7) that

$$r_0 r_1 = a^2 - c^2 = b^2$$

Thus,

$$b = \sqrt{r_0 r_1} \tag{10}$$

▶ **Example 2** Find the constants a, b, and c for the ellipse $r = \dfrac{6}{2 + \cos\theta}$.

Solution. This equation does not match any of the forms in Theorem 10.6.2 because they all require a constant term of 1 in the denominator. However, we can put the equation into one of these forms by dividing the numerator and denominator by 2 to obtain

$$r = \dfrac{3}{1 + \tfrac{1}{2}\cos\theta}$$

This is an exact match to (3) with $d = 6$ and $e = \tfrac{1}{2}$, so the graph is an ellipse with the directrix 6 units to the right of the pole. The distance r_0 from the focus to the closest vertex can be obtained by setting $\theta = 0$ in this equation, and the distance r_1 to the farthest vertex can be obtained by setting $\theta = \pi$. This yields

$$r_0 = \dfrac{3}{1 + \tfrac{1}{2}\cos 0} = \dfrac{3}{\tfrac{3}{2}} = 2, \quad r_1 = \dfrac{3}{1 + \tfrac{1}{2}\cos\pi} = \dfrac{3}{\tfrac{1}{2}} = 6$$

Thus, from Formulas (8), (10), and (9), respectively, we obtain

$$a = \tfrac{1}{2}(r_1 + r_0) = 4, \quad b = \sqrt{r_0 r_1} = 2\sqrt{3}, \quad c = \tfrac{1}{2}(r_1 - r_0) = 2$$

The ellipse looks roughly like that sketched in Figure 10.6.6. ◀

All of the important information about a hyperbola can be obtained from the values of a, b, and c in Figure 10.6.7. As with the ellipse, one way to find these values from the polar equation of a hyperbola is based on finding the distances from the focus to the vertices. As shown in the figure, let r_0 be the distance from the focus to the closest vertex and r_1 the distance to the farthest vertex. Thus,

$$r_0 = c - a \quad \text{and} \quad r_1 = c + a \tag{11}$$

from which it follows that

$$a = \tfrac{1}{2}(r_1 - r_0) \qquad c = \tfrac{1}{2}(r_1 + r_0) \tag{12–13}$$

Moreover, it also follows from (11) that

> In words, Formula (13) states that c is the **arithmetic mean** of r_0 and r_1, and Formula (14) states that b is the **geometric mean** of r_0 and r_1.

$$r_0 r_1 = c^2 - a^2 = b^2$$

from which it follows that

$$b = \sqrt{r_0 r_1} \tag{14}$$

▶ **Example 3**   Sketch the graph of $r = \dfrac{2}{1 + 2\sin\theta}$ in polar coordinates.

Solution.   This equation is an exact match to (5) with $d = 1$ and $e = 2$. Thus, the graph is a hyperbola with its directrix 1 unit above the pole. However, it is not so straightforward to compute the values of r_0 and r_1, since hyperbolas in polar coordinates are generated in a strange way as θ varies from 0 to 2π. This can be seen from Figure 10.6.8a, which is the graph of the given equation in rectangular θr-coordinates. It follows from this graph that the corresponding polar graph is generated in pieces (see Figure 10.6.8b):

- As θ varies over the interval $0 \le \theta < 7\pi/6$, the value of r is positive and varies from 2 down to 2/3 and then to $+\infty$, which generates part of the lower branch.

- As θ varies over the interval $7\pi/6 < \theta \le 3\pi/2$, the value of r is negative and varies from $-\infty$ to -2, which generates the right part of the upper branch.

- As θ varies over the interval $3\pi/2 \le \theta < 11\pi/6$, the value of r is negative and varies from -2 to $-\infty$, which generates the left part of the upper branch.

- As θ varies over the interval $11\pi/6 < \theta \le 2\pi$, the value of r is positive and varies from $+\infty$ to 2, which fills in the missing piece of the lower right branch.

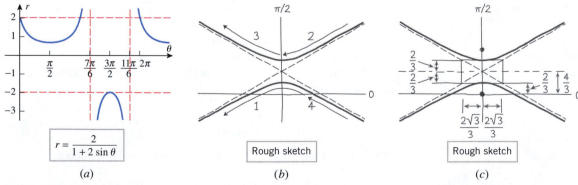

$$r = \frac{2}{1 + 2\sin\theta}$$

Rough sketch

(a) (b) (c)

▲ **Figure 10.6.8**

> To obtain a rough sketch of a hyperbola, it is generally sufficient to locate the center, the asymptotes, and the points where $\theta = 0$, $\theta = \pi/2$, $\theta = \pi$, and $\theta = 3\pi/2$.

It is now clear that we can obtain r_0 by setting $\theta = \pi/2$ and r_1 by setting $\theta = 3\pi/2$. Keeping in mind that r_0 and r_1 are positive, this yields

$$r_0 = \frac{2}{1 + 2\sin(\pi/2)} = \frac{2}{3}, \quad r_1 = \left| \frac{2}{1 + 2\sin(3\pi/2)} \right| = \left| \frac{2}{-1} \right| = 2$$

Thus, from Formulas (12), (14), and (13), respectively, we obtain

$$a = \frac{1}{2}(r_1 - r_0) = \frac{2}{3}, \quad b = \sqrt{r_0 r_1} = \frac{2\sqrt{3}}{3}, \quad c = \frac{1}{2}(r_1 + r_0) = \frac{4}{3}$$

Thus, the hyperbola looks roughly like that sketched in Figure 10.6.8c. ◄

■ APPLICATIONS IN ASTRONOMY

In 1609 Johannes Kepler published a book known as *Astronomia Nova* (or sometimes *Commentaries on the Motions of Mars*) in which he succeeded in distilling thousands of years of observational astronomy into three beautiful laws of planetary motion (Figure 10.6.9).

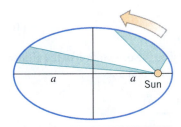

Equal areas are swept out in equal times, and the square of the period T is proportional to a^3.

▲ **Figure 10.6.9**

10.6.3 KEPLER'S LAWS

- First law (*Law of Orbits*). Each planet moves in an elliptical orbit with the Sun at a focus.
- Second law (*Law of Areas*). The radial line from the center of the Sun to the center of a planet sweeps out equal areas in equal times.
- Third law (*Law of Periods*). The square of a planet's period (the time it takes the planet to complete one orbit about the Sun) is proportional to the cube of the semimajor axis of its orbit.

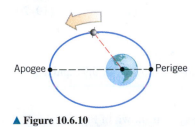

▲ **Figure 10.6.10**

Kepler's laws, although stated for planetary motion around the Sun, apply to all orbiting celestial bodies that are subjected to a *single* central gravitational force—artificial satellites subjected only to the central force of Earth's gravity and moons subjected only to the central gravitational force of a planet, for example. Later in the text we will derive Kepler's laws from basic principles, but for now we will show how they can be used in basic astronomical computations.

In an elliptical orbit, the closest point to the focus is called the *perigee* and the farthest point the *apogee* (Figure 10.6.10). The distances from the focus to the perigee and apogee

---

Johannes Kepler (1571–1630) German astronomer and physicist. Kepler, whose work provided our contemporary view of planetary motion, led a fascinating but ill-starred life. His alcoholic father made him work in a family-owned tavern as a child, later withdrawing him from elementary school and hiring him out as a field laborer, where the boy contracted smallpox, permanently crippling his hands and impairing his eyesight. In later years, Kepler's first wife and several children died, his mother was accused of witchcraft, and being a Protestant he was often subjected to persecution by Catholic authorities. He was often impoverished, eking out a living as an astrologer and prognosticator. Looking back on his unhappy childhood, Kepler described his father as "criminally inclined" and "quarrelsome" and his mother as "garrulous" and "bad-tempered." However, it was his mother who left an indelible mark on the six-year-old Kepler by showing him the comet of 1577; and in later life he personally prepared her defense against the witchcraft charges. Kepler became acquainted with the work of Copernicus as a student at the University of Tübingen, where he received his master's degree in 1591. He continued on as a theological student, but at the urging of the university officials he abandoned his clerical studies and accepted a position as a mathematician and teacher in Graz, Austria. However, he was expelled from the city when it came under Catholic control, and in 1600 he finally moved on to Prague, where he became an assistant at the observatory of the famous Danish astronomer Tycho Brahe. Brahe was a brilliant and meticulous astronomical observer who amassed the most accurate astronomical data known at that time; and when Brahe died in 1601 Kepler inherited the treasure-trove of data. After eight years of intense labor, Kepler deciphered the underlying principles buried in the data and in 1609 published his monumental work, *Astronomia Nova*, in which he stated his first two laws of planetary motion. Commenting on his discovery of elliptical orbits, Kepler wrote, "I was almost driven to madness in considering and calculating this matter. I could not find out why the planet would rather go on an elliptical orbit (rather than a circle). Oh ridiculous me!" It ultimately remained for Isaac Newton to discover the laws of gravitation that explained the reason for elliptical orbits.

are called the **perigee distance** and **apogee distance**, respectively. For orbits around the Sun, it is more common to use the terms **perihelion** and **aphelion**, rather than perigee and apogee, and to measure time in Earth years and distances in astronomical units (AU), where 1 AU is the semimajor axis a of the Earth's orbit (approximately 150×10^6 km or 92.9×10^6 mi). With this choice of units, the constant of proportionality in Kepler's third law is 1, since $a = 1$ AU produces a period of $T = 1$ Earth year. In this case Kepler's third law can be expressed as

$$T = a^{3/2} \tag{15}$$

Shapes of elliptical orbits are often specified by giving the eccentricity e and the semimajor axis a, so it is useful to express the polar equations of an ellipse in terms of these constants. Figure 10.6.11, which can be obtained from the ellipse in Figure 10.6.1 and the relationship $c = ea$, implies that the distance d between the focus and the directrix is

$$d = \frac{a}{e} - c = \frac{a}{e} - ea = \frac{a(1 - e^2)}{e} \tag{16}$$

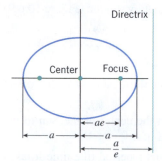

▲ Figure 10.6.11

from which it follows that $ed = a(1 - e^2)$. Thus, depending on the orientation of the ellipse, the formulas in Theorem 10.6.2 can be expressed in terms of a and e as

$$r = \frac{a(1 - e^2)}{1 \pm e \cos \theta} \qquad r = \frac{a(1 - e^2)}{1 \pm e \sin \theta} \tag{17–18}$$

| $+$: Directrix right of pole | $+$: Directrix above pole |
| $-$: Directrix left of pole | $-$: Directrix below pole |

Moreover, it is evident from Figure 10.6.11 that the distances from the focus to the closest and farthest vertices can be expressed in terms of a and e as

$$r_0 = a - ea = a(1 - e) \quad \text{and} \quad r_1 = a + ea = a(1 + e) \tag{19–20}$$

▲ Figure 10.6.12

▶ **Example 4** Halley's comet (last seen in 1986) has an eccentricity of 0.97 and a semimajor axis of $a = 18.1$ AU.

(a) Find the equation of its orbit in the polar coordinate system shown in Figure 10.6.12.

(b) Find the period of its orbit.

(c) Find its perihelion and aphelion distances.

Solution (a). From (17), the polar equation of the orbit has the form

$$r = \frac{a(1 - e^2)}{1 + e \cos \theta}$$

But $a(1 - e^2) = 18.1[1 - (0.97)^2] \approx 1.07$. Thus, the equation of the orbit is

$$r = \frac{1.07}{1 + 0.97 \cos \theta}$$

Solution (b). From (15), with $a = 18.1$, the period of the orbit is

$$T = (18.1)^{3/2} \approx 77 \text{ years}$$

Solution (c). Since the perihelion and aphelion distances are the distances to the closest and farthest vertices, respectively, it follows from (19) and (20) that

$$r_0 = a - ea = a(1 - e) = 18.1(1 - 0.97) \approx 0.543 \text{ AU}$$
$$r_1 = a + ea = a(1 + e) = 18.1(1 + 0.97) \approx 35.7 \text{ AU}$$

Science Photo Library/Photo Researchers
Halley's comet photographed April 21, 1910 in Peru.

or since $1 \text{ AU} \approx 150 \times 10^6$ km, the perihelion and aphelion distances in kilometers are

$$r_0 = 18.1(1 - 0.97)(150 \times 10^6) \approx 81,500,000 \text{ km}$$
$$r_1 = 18.1(1 + 0.97)(150 \times 10^6) \approx 5,350,000,000 \text{ km} \blacktriangleleft$$

Minimum
distance

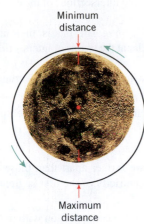

Maximum
distance

▲ **Figure 10.6.13**

▶ **Example 5** An Apollo lunar lander orbits the Moon in an elliptic orbit with eccentricity $e = 0.12$ and semimajor axis $a = 2015$ km. Assuming the Moon to be a sphere of radius 1740 km, find the minimum and maximum heights of the lander above the lunar surface (Figure 10.6.13).

Solution. If we let r_0 and r_1 denote the minimum and maximum distances from the center of the Moon, then the minimum and maximum distances from the surface of the Moon will be

$$d_{\min} = r_0 - 1740$$
$$d_{\max} = r_1 - 1740$$

or from Formulas (19) and (20)

$$d_{\min} = r_0 - 1740 = a(1 - e) - 1740 = 2015(0.88) - 1740 = 33.2 \text{ km}$$
$$d_{\max} = r_1 - 1740 = a(1 + e) - 1740 = 2015(1.12) - 1740 = 516.8 \text{ km} \blacktriangleleft$$

✔ **QUICK CHECK EXERCISES 10.6**   *(See page 763 for answers.)*

1. In each part, name the conic section described.
   (a) The set of points whose distance to the point $(2, 3)$ is half the distance to the line $x + y = 1$ is _____.
   (b) The set of points whose distance to the point $(2, 3)$ is equal to the distance to the line $x + y = 1$ is _____.
   (c) The set of points whose distance to the point $(2, 3)$ is twice the distance to the line $x + y = 1$ is _____.

2. In each part: (i) Identify the polar graph as a parabola, an ellipse, or a hyperbola; (ii) state whether the directrix is above, below, to the left, or to the right of the pole; and (iii) find the distance from the pole to the directrix.
   (a) $r = \dfrac{1}{4 + \cos \theta}$ (b) $r = \dfrac{1}{1 - 4 \cos \theta}$

   (c) $r = \dfrac{1}{4 + 4 \sin \theta}$ (d) $r = \dfrac{4}{1 - \sin \theta}$

3. If the distance from a vertex of an ellipse to the nearest focus is r_0, and if the distance from that vertex to the farthest focus is r_1, then the semimajor axis is $a =$ _____ and the semiminor axis is $b =$ _____.

4. If the distance from a vertex of a hyperbola to the nearest focus is r_0, and if the distance from that vertex to the farthest focus is r_1, then the semifocal axis is $a =$ _____ and the semiconjugate axis is $b =$ _____.

EXERCISE SET 10.6   Graphing Utility

1–2 Find the eccentricity and the distance from the pole to the directrix, and sketch the graph in polar coordinates. ■

1. (a) $r = \dfrac{3}{2 - 2 \cos \theta}$ (b) $r = \dfrac{3}{2 + \sin \theta}$

2. (a) $r = \dfrac{4}{2 + 3 \cos \theta}$ (b) $r = \dfrac{5}{3 + 3 \sin \theta}$

 3–4 Use Formulas (3)–(6) to identify the type of conic and its orientation. Check your answer by generating the graph with a graphing utility. ■

3. (a) $r = \dfrac{8}{1 - \sin \theta}$ (b) $r = \dfrac{16}{4 + 3 \sin \theta}$

4. (a) $r = \dfrac{4}{2 - 3 \sin \theta}$ (b) $r = \dfrac{12}{4 + \cos \theta}$

5–6 Find a polar equation for the conic that has its focus at the pole and satisfies the stated conditions. Points are in polar coordinates and directrices in rectangular coordinates for simplicity. (In some cases there may be more than one conic that satisfies the conditions.) ■

5. (a) Ellipse; $e = \frac{3}{4}$; directrix $x = 2$.
   (b) Parabola; directrix $x = 1$.
   (c) Hyperbola; $e = \frac{4}{3}$; directrix $y = 3$.

6. (a) Ellipse; ends of major axis $(2, \pi/2)$ and $(6, 3\pi/2)$.
   (b) Parabola; vertex $(2, \pi)$.
   (c) Hyperbola; $e = \sqrt{2}$; vertex $(2, 0)$.

7–8 Find the distances from the pole to the vertices, and then apply Formulas (8)–(10) to find the equation of the ellipse in rectangular coordinates. ■

7. (a) $r = \dfrac{6}{2 + \sin\theta}$ (b) $r = \dfrac{1}{2 - \cos\theta}$

8. (a) $r = \dfrac{6}{5 + 2\cos\theta}$ (b) $r = \dfrac{8}{4 - 3\sin\theta}$

9–10 Find the distances from the pole to the vertices, and then apply Formulas (12)–(14) to find the equation of the hyperbola in rectangular coordinates. ■

9. (a) $r = \dfrac{3}{1 + 2\sin\theta}$ (b) $r = \dfrac{5}{2 - 3\cos\theta}$

10. (a) $r = \dfrac{4}{1 - 2\sin\theta}$ (b) $r = \dfrac{15}{2 + 8\cos\theta}$

11–12 Find a polar equation for the ellipse that has its focus at the pole and satisfies the stated conditions. ■

11. (a) Directrix to the right of the pole; $a = 8$; $e = \frac{1}{2}$.
   (b) Directrix below the pole; $a = 4$; $e = \frac{3}{5}$.

12. (a) Directrix to the left of the pole; $b = 4$; $e = \frac{3}{5}$.
   (b) Directrix above the pole; $c = 5$; $e = \frac{1}{5}$.

13. Find the polar equation of an equilateral hyperbola with a focus at the pole and vertex $(5, 0)$.

FOCUS ON CONCEPTS

14. Prove that a hyperbola is an equilateral hyperbola if and only if $e = \sqrt{2}$.

15. (a) Show that the coordinates of the point P on the hyperbola in Figure 10.6.1 satisfy the equation
$$\sqrt{(x - c)^2 + y^2} = \frac{c}{a}x - a$$
   (b) Use the result obtained in part (a) to show that $PF/PD = c/a$.

16. (a) Show that the eccentricity of an ellipse can be expressed in terms of r_0 and r_1 as
$$e = \frac{r_1 - r_0}{r_1 + r_0}$$
   (b) Show that
$$\frac{r_1}{r_0} = \frac{1 + e}{1 - e}$$

17. (a) Show that the eccentricity of a hyperbola can be expressed in terms of r_0 and r_1 as
$$e = \frac{r_1 + r_0}{r_1 - r_0}$$
   (b) Show that
$$\frac{r_1}{r_0} = \frac{e + 1}{e - 1}$$

---

18. (a) Sketch the curves
$$r = \frac{1}{1 + \cos\theta} \quad \text{and} \quad r = \frac{1}{1 - \cos\theta}$$
   (b) Find polar coordinates of the intersections of the curves in part (a).
   (c) Show that the curves are *orthogonal*, that is, their tangent lines are perpendicular at the points of intersection.

19–22 True–False Determine whether the statement is true or false. Explain your answer. ■

19. If an ellipse is not a circle, then the eccentricity of the ellipse is less than one.

20. A parabola has eccentricity greater than one.

21. If one ellipse has foci that are farther apart than those of a second ellipse, then the eccentricity of the first is greater than that of the second.

22. If d is a positive constant, then the conic section with polar equation
$$r = \frac{d}{1 + \cos\theta}$$
is a parabola.

23–28 Use the following values, where needed:

radius of the Earth $= 4000$ mi $= 6440$ km
1 year (Earth year) $= 365$ days (Earth days)
1 AU $= 92.9 \times 10^6$ mi $= 150 \times 10^6$ km ■

23. The dwarf planet Pluto has eccentricity $e = 0.249$ and semimajor axis $a = 39.5$ AU.
   (a) Find the period T in years.
   (b) Find the perihelion and aphelion distances.
   (c) Choose a polar coordinate system with the center of the Sun at the pole, and find a polar equation of Pluto's orbit in that coordinate system.
   (d) Make a sketch of the orbit with reasonably accurate proportions.

24. (a) Let a be the semimajor axis of a planet's orbit around the Sun, and let T be its period. Show that if T is measured in days and a is measured in kilometers, then $T = (365 \times 10^{-9})(a/150)^{3/2}$.
   (b) Use the result in part (a) to find the period of the planet Mercury in days, given that its semimajor axis is $a = 57.95 \times 10^6$ km.
   (c) Choose a polar coordinate system with the Sun at the pole, and find an equation for the orbit of Mercury in that coordinate system given that the eccentricity of the orbit is $e = 0.206$.
   (d) Use a graphing utility to generate the orbit of Mercury from the equation obtained in part (c).

25. The Hale–Bopp comet, discovered independently on July 23, 1995 by Alan Hale and Thomas Bopp, has an orbital eccentricity of $e = 0.9951$ and a period of 2380 years.
   (a) Find its semimajor axis in astronomical units (AU).
   (b) Find its perihelion and aphelion distances. *(cont.)*

(c) Choose a polar coordinate system with the center of the Sun at the pole, and find an equation for the Hale–Bopp orbit in that coordinate system.

(d) Make a sketch of the Hale–Bopp orbit with reasonably accurate proportions.

26. Mars has a perihelion distance of 204,520,000 km and an aphelion distance of 246,280,000 km.

(a) Use these data to calculate the eccentricity, and compare your answer to the value given in Table 10.6.1.

(b) Find the period of Mars.

(c) Choose a polar coordinate system with the center of the Sun at the pole, and find an equation for the orbit of Mars in that coordinate system.

(d) Use a graphing utility to generate the orbit of Mars from the equation obtained in part (c).

27. *Vanguard 1* was launched in March 1958 into an orbit around the Earth with eccentricity $e = 0.21$ and semimajor axis 8864.5 km. Find the minimum and maximum heights of *Vanguard 1* above the surface of the Earth.

28. The planet Jupiter is believed to have a rocky core of radius 10,000 km surrounded by two layers of hydrogen— a 40,000 km thick layer of compressed metallic-like hydrogen and a 20,000 km thick layer of ordinary molecular hydrogen. The visible features, such as the Great Red Spot, are at the outer surface of the molecular hydrogen layer.

On November 6, 1997 the spacecraft *Galileo* was placed in a Jovian orbit to study the moon Europa. The orbit had eccentricity 0.814580 and semimajor axis 3,514,918.9 km. Find *Galileo*'s minimum and maximum heights above the molecular hydrogen layer (see the accompanying figure).

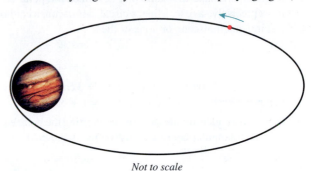

Not to scale

▲ **Figure Ex-28**

29. **Writing** Discuss how a hyperbola's eccentricity e affects the shape of the hyperbola. How is the shape affected as e approaches 1? As e approaches $+\infty$? Draw some pictures to illustrate your conclusions.

30. **Writing** Discuss the relationship between the eccentricity e of an ellipse and the distance z between the directrix and center of the ellipse. For example, if the foci remain fixed, what happens to z as e approaches 0?

✔ QUICK CHECK ANSWERS 10.6

1. (a) an ellipse (b) a parabola (c) a hyperbola 2. (a) (i) ellipse (ii) to the right of the pole (iii) distance $= 1$
(b) (i) hyperbola (ii) to the left of the pole (iii) distance $= \frac{1}{4}$ (c) (i) parabola (ii) above the pole (iii) distance $= \frac{1}{4}$
(d) (i) parabola (ii) below the pole (iii) distance $= 4$ 3. $\frac{1}{2}(r_1 + r_0)$; $\sqrt{r_0 r_1}$ 4. $\frac{1}{2}(r_1 - r_0)$; $\sqrt{r_0 r_1}$

CHAPTER 10 REVIEW EXERCISES    Graphing Utility

1. Find parametric equations for the portion of the circle $x^2 + y^2 = 2$ that lies outside the first quadrant, oriented clockwise. Check your work by generating the curve with a graphing utility.

2. (a) Suppose that the equations $x = f(t)$, $y = g(t)$ describe a curve C as t increases from 0 to 1. Find parametric equations that describe the same curve C but traced in the opposite direction as t increases from 0 to 1.

(b) Check your work using the parametric graphing feature of a graphing utility by generating the line segment between $(1, 2)$ and $(4, 0)$ in both possible directions as t increases from 0 to 1.

3. (a) Find the slope of the tangent line to the parametric curve $x = t^2 + 1$, $y = t/2$ at $t = -1$ and $t = 1$ without eliminating the parameter.

(b) Check your answers in part (a) by eliminating the parameter and differentiating a function of x.

4. Find dy/dx and $d^2 y/dx^2$ at $t = 2$ for the parametric curve $x = \frac{1}{2}t^2$, $y = \frac{1}{3}t^3$.

5. Find all values of t at which a tangent line to the parametric curve $x = 2\cos t$, $y = 4\sin t$ is

(a) horizontal   (b) vertical.

6. Find the exact arc length of the curve

$$x = 1 - 5t^4, \quad y = 4t^5 - 1 \quad (0 \le t \le 1)$$

7. In each part, find the rectangular coordinates of the point whose polar coordinates are given.

(a) $(-8, \pi/4)$ (b) $(7, -\pi/4)$ (c) $(8, 9\pi/4)$
(d) $(5, 0)$ (e) $(-2, -3\pi/2)$ (f) $(0, \pi)$

8. Express the point whose xy-coordinates are $(-1, 1)$ in polar coordinates with

(a) $r > 0$, $0 \le \theta < 2\pi$ (b) $r < 0$, $0 \le \theta < 2\pi$
(c) $r > 0$, $-\pi < \theta \le \pi$ (d) $r < 0$, $-\pi < \theta \le \pi$.

9. In each part, use a calculating utility to approximate the polar coordinates of the point whose rectangular coordinates are given.

(a) $(4, 3)$ (b) $(2, -5)$ (c) $(1, \tan^{-1} 1)$

10. In each part, state the name that describes the polar curve most precisely: a rose, a line, a circle, a limaçon, a cardioid, a spiral, a lemniscate, or none of these.

(a) $r = 3\cos\theta$ (b) $r = \cos 3\theta$

(c) $r = \dfrac{3}{\cos\theta}$ (d) $r = 3 - \cos\theta$

(e) $r = 1 - 3\cos\theta$ (f) $r^2 = 3\cos\theta$

(g) $r = (3\cos\theta)^2$ (h) $r = 1 + 3\theta$

11. In each part, identify the curve by converting the polar equation to rectangular coordinates. Assume that $a > 0$.

(a) $r = a\sec^2\dfrac{\theta}{2}$ (b) $r^2\cos 2\theta = a^2$

(c) $r = 4\csc\left(\theta - \dfrac{\pi}{4}\right)$ (d) $r = 4\cos\theta + 8\sin\theta$

12. In each part, express the given equation in polar coordinates.

(a) $x = 7$ (b) $x^2 + y^2 = 9$

(c) $x^2 + y^2 - 6y = 0$ (d) $4xy = 9$

13–17 Sketch the curve in polar coordinates. ■

13. $\theta = \dfrac{\pi}{6}$ **14.** $r = 6\cos\theta$

15. $r = 3(1 - \sin\theta)$ **16.** $r^2 = \sin 2\theta$

17. $r = 3 - \cos\theta$

18. (a) Show that the maximum value of the y-coordinate of points on the curve $r = 1/\sqrt{\theta}$ for θ in the interval $(0, \pi]$ occurs when $\tan\theta = 2\theta$.

(b) Use a calculating utility to solve the equation in part (a) to at least four decimal-place accuracy.

(c) Use the result of part (b) to approximate the maximum value of y for $0 < \theta \le \pi$.

19. (a) Find the minimum and maximum x-coordinates of points on the cardioid $r = 1 - \cos\theta$.

(b) Find the minimum and maximum y-coordinates of points on the cardioid in part (a).

20. Determine the slope of the tangent line to the polar curve $r = 1 + \sin\theta$ at $\theta = \pi/4$.

21. A parametric curve of the form

$$x = a\cot t + b\cos t, \quad y = a + b\sin t \quad (0 < t < 2\pi)$$

is called a ***conchoid of Nicomedes*** (see the accompanying figure for the case $0 < a < b$).

(a) Describe how the conchoid

$$x = \cot t + 4\cos t, \quad y = 1 + 4\sin t$$

is generated as t varies over the interval $0 < t < 2\pi$.

(b) Find the horizontal asymptote of the conchoid given in part (a).

(c) For what values of t does the conchoid in part (a) have a horizontal tangent line? A vertical tangent line?

(d) Find a polar equation $r = f(\theta)$ for the conchoid in part (a), and then find polar equations for the tangent lines to the conchoid at the pole.

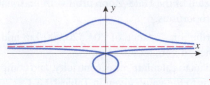

◄ **Figure Ex-21**

22. (a) Find the arc length of the polar curve $r = 1/\theta$ for $\pi/4 \le \theta \le \pi/2$.

(b) What can you say about the arc length of the portion of the curve that lies inside the circle $r = 1$?

23. Find the area of the region that is enclosed by the cardioid $r = 2 + 2\cos\theta$.

24. Find the area of the region in the first quadrant within the cardioid $r = 1 + \sin\theta$.

25. Find the area of the region that is common to the circles $r = 1$, $r = 2\cos\theta$, and $r = 2\sin\theta$.

26. Find the area of the region that is inside the cardioid $r = a(1 + \sin\theta)$ and outside the circle $r = a\sin\theta$.

27–30 Sketch the parabola, and label the focus, vertex, and directrix. ■

27. $y^2 = 6x$ **28.** $x^2 = -9y$

29. $(y + 1)^2 = -7(x - 4)$ **30.** $\left(x - \frac{1}{2}\right)^2 = 2(y - 1)$

31–34 Sketch the ellipse, and label the foci, the vertices, and the ends of the minor axis. ■

31. $\dfrac{x^2}{4} + \dfrac{y^2}{25} = 1$ **32.** $4x^2 + 9y^2 = 36$

33. $9(x - 1)^2 + 16(y - 3)^2 = 144$

34. $3(x + 2)^2 + 4(y + 1)^2 = 12$

35–37 Sketch the hyperbola, and label the vertices, foci, and asymptotes. ■

35. $\dfrac{x^2}{16} - \dfrac{y^2}{4} = 1$ **36.** $9y^2 - 4x^2 = 36$

37. $\dfrac{(x - 2)^2}{9} - \dfrac{(y - 4)^2}{4} = 1$

38. In each part, sketch the graph of the conic section with reasonably accurate proportions.

(a) $x^2 - 4x + 8y + 36 = 0$

(b) $3x^2 + 4y^2 - 30x - 8y + 67 = 0$

(c) $4x^2 - 5y^2 - 8x - 30y - 21 = 0$

39–41 Find an equation for the conic described. ■

39. A parabola with vertex $(0, 0)$ and focus $(0, -4)$.

40. An ellipse with the ends of the major axis $(0, \pm\sqrt{5})$ and the ends of the minor axis $(\pm 1, 0)$.

41. A hyperbola with vertices $(0, \pm 3)$ and asymptotes $y = \pm x$.

42. It can be shown in the accompanying figure that hanging cables form parabolic arcs rather than catenaries if they are subjected to uniformly distributed downward forces along their length. For example, if the weight of the roadway in a suspension bridge is assumed to be uniformly distributed along the supporting cables, then the cables can be modeled by parabolas.

(a) Assuming a parabolic model, find an equation for the cable in the accompanying figure, taking the y-axis to be vertical and the origin at the low point of the cable.

(b) Find the length of the cable between the supports.

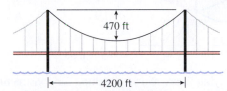

470 ft

|← 4200 ft →|

▲ **Figure Ex-42**

43. It will be shown later in this text that if a projectile is launched with speed v_0 at an angle α with the horizontal and at a height y_0 above ground level, then the resulting trajectory relative to the coordinate system in the accompanying figure will have parametric equations

$$x = (v_0 \cos\alpha)t, \quad y = y_0 + (v_0 \sin\alpha)t - \tfrac{1}{2}gt^2$$

where g is the acceleration due to gravity.

(a) Show that the trajectory is a parabola.

(b) Find the coordinates of the vertex.

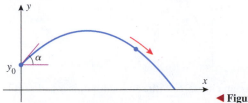

◀ **Figure Ex-43**

44. Mickey Mantle is recognized as baseball's unofficial king of long home runs. On April 17, 1953 Mantle blasted a pitch by Chuck Stobbs of the hapless Washington Senators out of Griffith Stadium, just clearing the 50 ft wall at the 391 ft marker in left center. Assuming that the ball left the bat at a height of 3 ft above the ground and at an angle of $45°$, use the parametric equations in Exercise 43 with $g = 32$ ft/s^2 to find

(a) the speed of the ball as it left the bat

(b) the maximum height of the ball

(c) the distance along the ground from home plate to where the ball struck the ground.

45–47 Rotate the coordinate axes to remove the xy-term, and then name the conic. ■

45. $x^2 + y^2 - 3xy - 3 = 0$

46. $7x^2 + 2\sqrt{3}xy + 5y^2 - 4 = 0$

47. $4\sqrt{5}x^2 + 4\sqrt{5}xy + \sqrt{5}y^2 + 5x - 10y = 0$

48. Rotate the coordinate axes to show that the graph of

$$17x^2 - 312xy + 108y^2 + 1080x - 1440y + 4500 = 0$$

is a hyperbola. Then find its vertices, foci, and asymptotes.

49. In each part: (i) Identify the polar graph as a parabola, an ellipse, or a hyperbola; (ii) state whether the directrix is above, below, to the left, or to the right of the pole; and (iii) find the distance from the pole to the directrix.

(a) $r = \dfrac{1}{3 + \cos\theta}$ (b) $r = \dfrac{1}{1 - 3\cos\theta}$

(c) $r = \dfrac{1}{3(1 + \sin\theta)}$ (d) $r = \dfrac{3}{1 - \sin\theta}$

50–51 Find an equation in xy-coordinates for the conic section that satisfies the given conditions. ■

50. (a) Ellipse with eccentricity $e = \tfrac{2}{7}$ and ends of the minor axis at the points $(0, \pm 3)$.

(b) Parabola with vertex at the origin, focus on the y-axis, and directrix passing through the point $(7, 4)$.

(c) Hyperbola that has the same foci as the ellipse $3x^2 + 16y^2 = 48$ and asymptotes $y = \pm 2x/3$.

51. (a) Ellipse with center $(-3, 2)$, vertex $(2, 2)$, and eccentricity $e = \tfrac{4}{5}$.

(b) Parabola with focus $(-2, -2)$ and vertex $(-2, 0)$.

(c) Hyperbola with vertex $(-1, 7)$ and asymptotes $y - 5 = \pm 8(x + 1)$.

52. Use the parametric equations $x = a \cos t$, $y = b \sin t$ to show that the circumference C of an ellipse with semimajor axis a and eccentricity e is

$$C = 4a \int_0^{\pi/2} \sqrt{1 - e^2 \sin^2 u}\, du$$

53. Use Simpson's rule or the numerical integration capability of a graphing utility to approximate the circumference of the ellipse $4x^2 + 9y^2 = 36$ from the integral obtained in Exercise 52.

54. (a) Calculate the eccentricity of the Earth's orbit, given that the ratio of the distance between the center of the Earth and the center of the Sun at perihelion to the distance between the centers at aphelion is $\tfrac{59}{61}$.

(b) Find the distance between the center of the Earth and the center of the Sun at perihelion, given that the average value of the perihelion and aphelion distances between the centers is 93 million miles.

(c) Use the result in Exercise 52 and Simpson's rule or the numerical integration capability of a graphing utility to approximate the distance that the Earth travels in 1 year (one revolution around the Sun).

CHAPTER 10 MAKING CONNECTIONS   C CAS

C **1.** Recall from Section 5.10 that the Fresnel sine and cosine functions are defined as

$$S(x) = \int_0^x \sin\left(\frac{\pi t^2}{2}\right) dt \quad \text{and} \quad C(x) = \int_0^x \cos\left(\frac{\pi t^2}{2}\right) dt$$

The following parametric curve, which is used to study amplitudes of light waves in optics, is called a **clothoid** or **Cornu spiral** in honor of the French scientist Marie Alfred Cornu (1841–1902):

$$x = C(t) = \int_0^t \cos\left(\frac{\pi u^2}{2}\right) du$$
$$y = S(t) = \int_0^t \sin\left(\frac{\pi u^2}{2}\right) du \qquad (-\infty < t < +\infty)$$

(a) Use a CAS to graph the Cornu spiral.
(b) Describe the behavior of the spiral as $t \to +\infty$ and as $t \to -\infty$.
(c) Find the arc length of the spiral for $-1 \le t \le 1$.

2. (a) The accompanying figure shows an ellipse with semimajor axis a and semiminor axis b. Express the coordinates of the points P, Q, and R in terms of t.
(b) How does the geometric interpretation of the parameter t differ between a circle

$$x = a\cos t, \quad y = a\sin t$$

and an ellipse

$$x = a\cos t, \quad y = b\sin t?$$

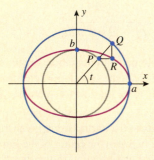

◀ **Figure Ex-2**

3. The accompanying figure shows Kepler's method for constructing a parabola. A piece of string the length of the left edge of the drafting triangle is tacked to the vertex Q of the

triangle and the other end to a fixed point F. A pencil holds the string taut against the base of the triangle as the edge opposite Q slides along a horizontal line L below F. Show that the pencil traces an arc of a parabola with focus F and directrix L.

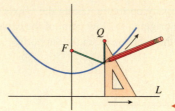

◀ **Figure Ex-3**

4. The accompanying figure shows a method for constructing a hyperbola. A corner of a ruler is pinned to a fixed point F_1 and the ruler is free to rotate about that point. A piece of string whose length is less than that of the ruler is tacked to a point F_2 and to the free corner Q of the ruler on the same edge as F_1. A pencil holds the string taut against the top edge of the ruler as the ruler rotates about the point F_1. Show that the pencil traces an arc of a hyperbola with foci F_1 and F_2.

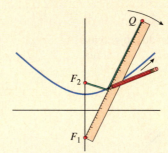

◀ **Figure Ex-4**

5. Consider an ellipse E with semimajor axis a and semiminor axis b, and set $c = \sqrt{a^2 - b^2}$.
(a) Show that the ellipsoid that results when E is revolved about its major axis has volume $V = \frac{4}{3}\pi ab^2$ and surface area

$$S = 2\pi ab\left(\frac{b}{a} + \frac{a}{c}\sin^{-1}\frac{c}{a}\right)$$

(b) Show that the ellipsoid that results when E is revolved about its minor axis has volume $V = \frac{4}{3}\pi a^2 b$ and surface area

$$S = 2\pi ab\left(\frac{a}{b} + \frac{b}{c}\ln\frac{a+c}{b}\right)$$

EXPANDING THE CALCULUS HORIZON

To learn how polar coordinates and conic sections can be used to analyze the possibility of a collision between a comet and Earth, see the module entitled **Comet Collision** at:

www.wiley.com/college/anton

GRAPHING FUNCTIONS USING CALCULATORS AND COMPUTER ALGEBRA SYSTEMS

■ GRAPHING CALCULATORS AND COMPUTER ALGEBRA SYSTEMS

The development of new technology has significantly changed how and where mathematicians, engineers, and scientists perform their work, as well as their approach to problem solving. Among the most significant of these developments are programs called *Computer Algebra Systems* (abbreviated CAS), the most common being *Mathematica* and *Maple*.[*] Computer algebra systems not only have graphing capabilities, but, as their name suggests, they can perform many of the symbolic computations that occur in algebra, calculus, and branches of higher mathematics. For example, it is a trivial task for a CAS to perform the factorization

$$x^6 + 23x^5 + 147x^4 - 139x^3 - 3464x^2 - 2112x + 23040 = (x+5)(x-3)^2(x+8)^3$$

or the exact numerical computation

$$\left(\frac{63456}{3177295} - \frac{43907}{22854377}\right)^3 = \frac{225191245716420829125932023012866923}{382895955819369204449565945369203764688375}$$

Technology has also made it possible to generate graphs of equations and functions in seconds that in the past might have taken hours. Figure A.1 shows the graphs of the function $f(x) = x^4 - x^3 - 2x^2$ produced with various graphing utilities; the first two were generated with the CAS programs, *Mathematica* and *Maple*, and the third with a graphing calculator. Graphing calculators produce coarser graphs than most computer programs but have the advantage of being compact and portable.

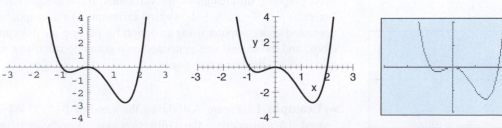

Generated by Mathematica   *Generated by Maple*   *Generated by a graphing calculator*

▲ **Figure A.1**

---

[*]*Mathematica* is a product of Wolfram Research, Inc.; *Maple* is a product of Waterloo Maple Software, Inc.

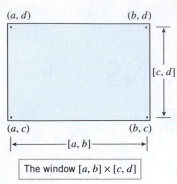

▲ Figure A.2

The window $[a, b] \times [c, d]$

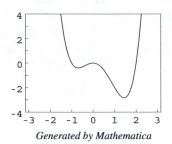

Generated by Mathematica

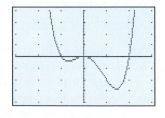

Generated by a graphing calculator

▲ Figure A.3

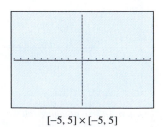

$[-5, 5] \times [-5, 5]$
$x\mathrm{Scl} = 0.5, \; y\mathrm{Scl} = 10$

▲ Figure A.4

■ **VIEWING WINDOWS**

Graphing utilities can only show a portion of the xy-plane in the viewing screen, so the first step in graphing an equation is to determine which rectangular portion of the xy-plane you want to display. This region is called the ***viewing window*** (or ***viewing rectangle***). For example, in Figure A.1 the viewing window extends over the interval $[-3, 3]$ in the x-direction and over the interval $[-4, 4]$ in the y-direction, so we denote the viewing window by $[-3, 3] \times [-4, 4]$ (read "$[-3, 3]$ by $[-4, 4]$"). In general, if the viewing window is $[a, b] \times [c, d]$, then the window extends between $x = a$ and $x = b$ in the x-direction and between $y = c$ and $y = d$ in the y-direction. We will call $[a, b]$ the ***x-interval*** for the window and $[c, d]$ the ***y-interval*** for the window (Figure A.2).

Different graphing utilities designate viewing windows in different ways. For example, the first two graphs in Figure A.1 were produced by the commands

```
Plot[x^4 - x^3 -2*x^2, {x, -3, 3}, PlotRange->{-4, 4}]
```
(*Mathematica*)
```
plot(x^4 - x^3 -2*x^2, x = -3..3, y = -4..4);
```
(*Maple*)

and the last graph was produced on a graphing calculator by pressing the GRAPH button after setting the values for the variables that determine the x-interval and y-interval to be

$$x\mathrm{Min} = -3, \quad x\mathrm{Max} = 3, \quad y\mathrm{Min} = -4, \quad y\mathrm{Max} = 4$$

■ **TICK MARKS AND GRID LINES**

To help locate points in a viewing window, graphing utilities provide methods for drawing ***tick marks*** (also called ***scale marks***). With computer programs such as *Mathematica* and *Maple*, there are specific commands for designating the spacing between tick marks, but if the user does not specify the spacing, then the programs make certain *default* choices. For example, in the first two parts of Figure A.1, the tick marks shown were the default choices.

On some graphing calculators the spacing between tick marks is determined by two ***scale variables*** (also called ***scale factors***), which we will denote by

$$x\mathrm{Scl} \quad \text{and} \quad y\mathrm{Scl}$$

(The notation varies among calculators.) These variables specify the spacing between the tick marks in the x- and y-directions, respectively. For example, in the third part of Figure A.1 the window and tick marks were designated by the settings

$$x\mathrm{Min} = -3 \qquad x\mathrm{Max} = 3$$
$$y\mathrm{Min} = -4 \qquad y\mathrm{Max} = 4$$
$$x\mathrm{Scl} = 1 \qquad y\mathrm{Scl} = 1$$

Most graphing utilities allow for variations in the design and positioning of tick marks. For example, Figure A.3 shows two variations of the graphs in Figure A.1; the first was generated on a computer using an option for placing the ticks and numbers on the edges of a box, and the second was generated on a graphing calculator using an option for drawing grid lines to simulate graph paper.

▶ **Example 1**   Figure A.4 shows the window $[-5, 5] \times [-5, 5]$ with the tick marks spaced 0.5 unit apart in the x-direction and 10 units apart in the y-direction. No tick marks are actually visible in the y-direction because the tick mark at the origin is covered by the x-axis, and all other tick marks in that direction fall outside of the viewing window.   ◀

▶ **Example 2**   Figure A.5 shows the window $[-10, 10] \times [-10, 10]$ with the tick marks spaced 0.1 unit apart in the x- and y-directions. In this case the tick marks are so close together that they create thick lines on the coordinate axes. When this occurs you will usually want to increase the scale factors to reduce the number of tick marks to make them legible. ◀

Graphing calculators provide a way of clearing all settings and returning them to *default values*. For example, on one calculator the default window is $[-10, 10] \times [-10, 10]$ and the default scale factors are xScl $= 1$ and yScl $= 1$. Read your documentation to determine the default values for your calculator and how to restore the default settings. If you are using a CAS, read your documentation to determine the commands for specifying the spacing between tick marks.

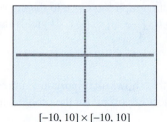

$[-10, 10] \times [-10, 10]$
xScl $= 0.1$, yScl $= 0.1$

▲ **Figure A.5**

■ CHOOSING A VIEWING WINDOW

When the graph of a function extends indefinitely in some direction, no single viewing window can show it all. In such cases the choice of the viewing window can affect one's perception of how the graph looks. For example, Figure A.6 shows a computer-generated graph of $y = 9 - x^2$, and Figure A.7 shows four views of this graph generated on a calculator.

- In part (*a*) the graph falls completely outside of the window, so the window is blank (except for the ticks and axes).

- In part (*b*) the graph is broken into two pieces because it passes in and out of the window.

- In part (*c*) the graph appears to be a straight line because we have zoomed in on a very small segment of the curve.

- In part (*d*) we have a more revealing picture of the graph shape because the window encompasses the high point on the graph and the intersections with the x-axis.

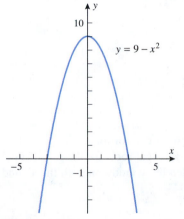

▲ **Figure A.6**

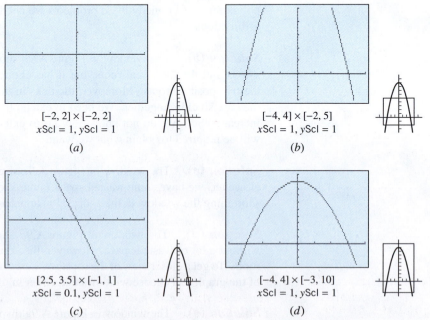

$[-2, 2] \times [-2, 2]$
xScl $= 1$, yScl $= 1$
(*a*)

$[-4, 4] \times [-2, 5]$
xScl $= 1$, yScl $= 1$
(*b*)

$[2.5, 3.5] \times [-1, 1]$
xScl $= 0.1$, yScl $= 1$
(*c*)

$[-4, 4] \times [-3, 10]$
xScl $= 1$, yScl $= 1$
(*d*)

▲ **Figure A.7** Four views of $y = 9 - x^2$

The following example illustrates how the domain and range of a function can be used to find a good viewing window when the graph of the function does not extend indefinitely in both the x- and y-directions.

▶ **Example 3**   Use the domain and range of the function $f(x) = \sqrt{12 - 3x^2}$ to determine a viewing window that contains the entire graph.

Solution.   The natural domain of f is $[-2, 2]$ and the range is $[0, \sqrt{12}]$ (verify), so the entire graph will be contained in the viewing window $[-2, 2] \times [0, \sqrt{12}]$. For clarity, it is desirable to use a slightly larger window to avoid having the graph too close to the edges of the screen. For example, taking the viewing window to be $[-3, 3] \times [-1, 4]$ yields the graph in Figure A.8. ◀

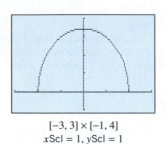

$[-3, 3] \times [-1, 4]$
$x\text{Scl} = 1, y\text{Scl} = 1$

▲ **Figure A.8**

Sometimes it will be impossible to find a single window that shows all important features of a graph, in which case you will need to decide what is most important for the problem at hand and choose the window appropriately.

▶ **Example 4**   Graph the equation $y = x^3 - 12x^2 + 18$ in the following windows and discuss the advantages and disadvantages of each window.

(a) $[-10, 10] \times [-10, 10]$ with $x\text{Scl} = 1, y\text{Scl} = 1$

(b) $[-20, 20] \times [-20, 20]$ with $x\text{Scl} = 1, y\text{Scl} = 1$

(c) $[-20, 20] \times [-300, 20]$ with $x\text{Scl} = 1, y\text{Scl} = 20$

(d) $[-5, 15] \times [-300, 20]$ with $x\text{Scl} = 1, y\text{Scl} = 20$

(e) $[1, 2] \times [-1, 1]$ with $x\text{Scl} = 0.1, y\text{Scl} = 0.1$

Solution (a).   The window in Figure A.9a has chopped off the portion of the graph that intersects the y-axis, and it shows only two of three possible real roots for the given cubic polynomial. To remedy these problems we need to widen the window in both the x- and y-directions.

Solution (b).   The window in Figure A.9b shows the intersection of the graph with the y-axis and the three real roots, but it has chopped off the portion of the graph between the two positive roots. Moreover, the ticks in the y-direction are nearly illegible because they are so close together. We need to extend the window in the negative y-direction and increase $y\text{Scl}$. We do not know how far to extend the window, so some experimentation will be required to obtain what we want.

Solution (c).   The window in Figure A.9c shows all of the main features of the graph. However, we have some wasted space in the x-direction. We can improve the picture by shortening the window in the x-direction appropriately.

Solution (d).   The window in Figure A.9d shows all of the main features of the graph without a lot of wasted space. However, the window does not provide a clear view of the roots. To get a closer view of the roots we must forget about showing all of the main features of the graph and choose windows that zoom in on the roots themselves.

Solution (e).   The window in Figure A.9e displays very little of the graph, but it clearly shows that the root in the interval $[1, 2]$ is approximately 1.3. ◀

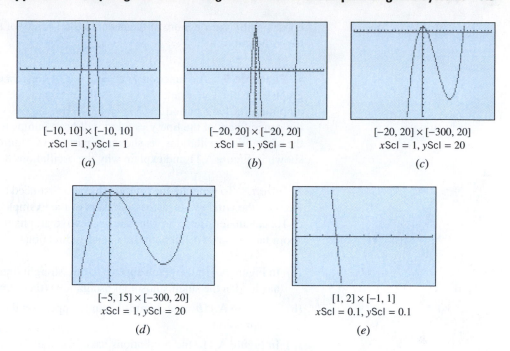

$[-10, 10] \times [-10, 10]$
xScl $= 1$, yScl $= 1$

(a)

$[-20, 20] \times [-20, 20]$
xScl $= 1$, yScl $= 1$

(b)

$[-20, 20] \times [-300, 20]$
xScl $= 1$, yScl $= 20$

(c)

$[-5, 15] \times [-300, 20]$
xScl $= 1$, yScl $= 20$

(d)

$[1, 2] \times [-1, 1]$
xScl $= 0.1$, yScl $= 0.1$

(e)

▶ **Figure A.9**

TECHNOLOGY MASTERY   Sometimes you will want to determine the viewing window by choosing the x-interval and allowing the graphing utility to determine a y-interval that encompasses the maximum and minimum values of the function over the x-interval. Most graphing utilities provide some method for doing this, so read your documentation to determine how to use this feature. Allowing the graphing utility to determine the y-interval of the window takes some of the guesswork out of problems like that in part (b) of the preceding example.

■ ZOOMING

The process of enlarging or reducing the size of a viewing window is called *zooming*. If you reduce the size of the window, you see *less* of the graph as a whole, but more detail of the part shown; this is called *zooming in*. In contrast, if you enlarge the size of the window, you see *more* of the graph as a whole, but less detail of the part shown; this is called *zooming out*. Most graphing calculators provide menu items for zooming in or zooming out by fixed factors. For example, on one calculator the amount of enlargement or reduction is controlled by setting values for two *zoom factors*, denoted by xFact and yFact. If

$$x\text{Fact} = 10 \quad \text{and} \quad y\text{Fact} = 5$$

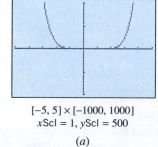

$[-5, 5] \times [-1000, 1000]$
xScl $= 1$, yScl $= 500$

(a)

$[-5, 5] \times [-10, 10]$
xScl $= 1$, yScl $= 1$

(b)

▲ **Figure A.10**

then each time a zoom command is executed the viewing window is enlarged or reduced by a factor of 10 in the x-direction and a factor of 5 in the y-direction. With computer programs such as *Mathematica* and *Maple*, zooming is controlled by adjusting the x-interval and y-interval directly; however, there are ways to automate this by programming.

■ COMPRESSION

Enlarging the viewing window for a graph has the geometric effect of compressing the graph, since more of the graph is packed into the calculator screen. If the compression is sufficiently great, then some of the detail in the graph may be lost. Thus, the choice of the viewing window frequently depends on whether you want to see more of the graph or more of the detail. Figure A.10 shows two views of the equation

$$y = x^5(x - 2)$$

In part (a) of the figure the y-interval is very large, resulting in a vertical compression that obscures the detail in the vicinity of the x-axis. In part (b) the y-interval is smaller, and consequently we see more of the detail in the vicinity of the x-axis but less of the graph in the y-direction.

▶ **Example 5** The function $f(x) = x + 0.01 \sin(50\pi x)$ is the sum of $f_1(x) = x$, whose graph is the line $y = x$, and $f_2(x) = 0.01 \sin(50\pi x)$, whose graph is a sinusoidal curve with amplitude 0.01 and period $2\pi/50\pi = 0.04$. This suggests that the graph of $f(x)$ will follow the general path of the line $y = x$ but will have bumps resulting from the contributions of the sinusoidal oscillations, as shown in part (c) of Figure A.11. Generate the four graphs shown in Figure A.11 and explain why the oscillations are visible only in part (c).

Solution. To generate the four graphs, you first need to put your utility in radian mode.[*] Because the windows in successive parts of the example are decreasing in size by a factor of 10, calculator users can generate successive graphs by using the zoom feature with the zoom factors set to 10 in both the x- and y-directions.

(a) In Figure A.11a the graph appears to be a straight line because the vertical compression has hidden the small sinusoidal oscillations (their amplitude is only 0.01).

(b) In Figure A.11b small bumps begin to appear on the line because there is less vertical compression.

(c) In Figure A.11c the oscillations have become clear because the vertical scale is more in keeping with the amplitude of the oscillations.

(d) In Figure A.11d the graph appears to be a straight line because we have zoomed in on such a small portion of the curve. ◀

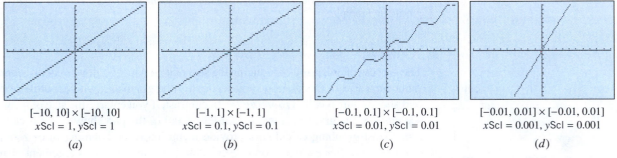

$[-10, 10] \times [-10, 10]$	$[-1, 1] \times [-1, 1]$	$[-0.1, 0.1] \times [-0.1, 0.1]$	$[-0.01, 0.01] \times [-0.01, 0.01]$
xScl $= 1$, yScl $= 1$	xScl $= 0.1$, yScl $= 0.1$	xScl $= 0.01$, yScl $= 0.01$	xScl $= 0.001$, yScl $= 0.001$
(a)	(b)	(c)	(d)

▲ **Figure A.11**

■ ASPECT RATIO DISTORTION

Figure A.12a shows a circle of radius 5 and two perpendicular lines graphed in the window $[-10, 10] \times [-10, 10]$ with xScl $= 1$ and yScl $= 1$. However, the circle is distorted and the lines do not appear perpendicular because the calculator has not used the same length for 1 unit on the x-axis and 1 unit on the y-axis. (Compare the spacing between the ticks on the axes.) This is called ***aspect ratio distortion***. Many calculators provide a menu item for automatically correcting the distortion by adjusting the viewing window appropriately. For example, one calculator makes this correction to the viewing window $[-10, 10] \times [-10, 10]$ by changing it to

$$[-16.9970674487, 16.9970674487] \times [-10, 10]$$

(Figure A.12b). With computer programs such as *Mathematica* and *Maple*, aspect ratio distortion is controlled with adjustments to the physical dimensions of the viewing window on the computer screen, rather than altering the x- and y-intervals of the viewing window.

---

[*]In this text we follow the convention that angles are measured in radians unless degree measure is specified.

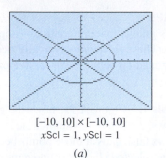 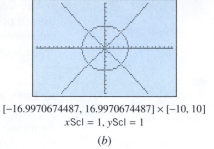

$[-10, 10] \times [-10, 10]$
$x\text{Scl} = 1, y\text{Scl} = 1$

$[-16.9970674487, 16.9970674487] \times [-10, 10]$
$x\text{Scl} = 1, y\text{Scl} = 1$

▶ Figure A.12   (a)   (b)

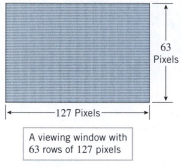

63
Pixels

──── 127 Pixels ────

A viewing window with
63 rows of 127 pixels

▲ Figure A.13

TECHNOLOGY MASTERY

If you are using a graphing calculator,
read the documentation to determine
its resolution.

■ SAMPLING ERROR

The viewing window of a graphing utility is composed of a rectangular grid of small rectangular blocks called *pixels*. For black-and-white displays each pixel has two states—an activated (or dark) state and a deactivated (or light state). A graph is formed by activating appropriate pixels to produce the curve shape. In one popular calculator the grid of pixels consists of 63 rows of 127 pixels each (Figure A.13), in which case we say that the screen has a *resolution* of 127×63 (pixels in each row × number of rows). A typical resolution for a computer screen is 1280×1024. The greater the resolution, the smoother the graphs tend to appear on the screen.

The procedure that a graphing utility uses to generate a graph is similar to plotting points by hand: When an equation is entered and a window is chosen, the utility *selects* the x-coordinates of certain pixels (the choice of which depends on the window being used) and *computes* the corresponding y-coordinates. It then activates the pixels whose coordinates most closely match those of the calculated points and uses a built-in algorithm to activate additional intermediate pixels to create the curve shape. This process is not perfect, and it is possible that a particular window will produce a false impression about the graph shape because important characteristics of the graph occur between the computed points. This is called *sampling error*. For example, Figure A.14 shows the graph of $y = \cos(10\pi x)$ produced by a popular calculator in four different windows. (Your calculator may produce different results.) The graph in part (a) has the correct shape, but the other three do not because of sampling error:

- In part (b) the plotted pixels happened to fall at the peaks of the cosine curve, giving the false impression that the graph is a horizontal line.

- In part (c) the plotted pixels fell at successively higher points along the graph.

- In part (d) the plotted points fell in some regular pattern that created yet another misleading impression of the graph shape.

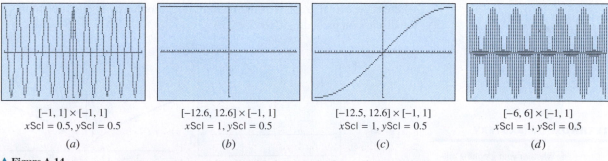

$[-1, 1] \times [-1, 1]$
$x\text{Scl} = 0.5, y\text{Scl} = 0.5$
(a)

$[-12.6, 12.6] \times [-1, 1]$
$x\text{Scl} = 1, y\text{Scl} = 0.5$
(b)

$[-12.5, 12.6] \times [-1, 1]$
$x\text{Scl} = 1, y\text{Scl} = 0.5$
(c)

$[-6, 6] \times [-1, 1]$
$x\text{Scl} = 1, y\text{Scl} = 0.5$
(d)

▲ Figure A.14

REMARK For trigonometric graphs with rapid oscillations, Figure A.14 suggests that restricting the x-interval to a few periods is likely to produce a more accurate representation about the graph shape.

FALSE GAPS

Sometimes graphs that are continuous appear to have gaps when they are generated on a calculator. These *false gaps* typically occur where the graph rises so rapidly that vertical space is opened up between successive pixels.

▶ **Example 6**   Figure A.15 shows the graph of the semicircle $y = \sqrt{9 - x^2}$ in two viewing windows. Although this semicircle has x-intercepts at the points $x = \pm 3$, part (a) of the figure shows false gaps at those points because there are no pixels with x-coordinates ± 3 in the window selected. In part (b) no gaps occur because there are pixels with x-coordinates $x = \pm 3$ in the window being used.   ◀

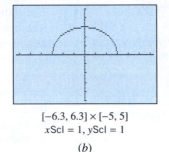

$[-5, 5] \times [-5, 5]$
xScl = 1, yScl = 1

$[-6.3, 6.3] \times [-5, 5]$
xScl = 1, yScl = 1

(a)   (b)

▶ **Figure A.15**

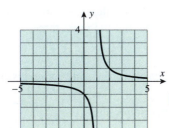

$[-10, 10] \times [-10, 10]$
xScl = 1, yScl = 1

$y = 1/(x - 1)$ with false line segments

(a)

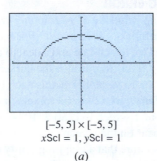

Actual curve shape of $y = 1/(x - 1)$

(b)

▲ **Figure A.16**

FALSE LINE SEGMENTS

In addition to creating false gaps in continuous graphs, calculators can err in the opposite direction by placing *false line segments* in the gaps of discontinuous curves.

▶ **Example 7**   Figure A.16a shows the graph of $y = 1/(x - 1)$ in the default window on a calculator. Although the graph appears to contain vertical line segments near $x = 1$, they should not be there. There is actually a gap in the curve at $x = 1$, since a division by zero occurs at that point (Figure A.16b).   ◀

ERRORS OF OMISSION

Most graphing utilities use logarithms to evaluate functions with fractional exponents such as $f(x) = x^{2/3} = \sqrt[3]{x^2}$. However, because logarithms are only defined for positive numbers, many (but not all) graphing utilities will omit portions of the graphs of functions with fractional exponents. For example, one calculator graphs $y = x^{2/3}$ as in Figure A.17a, whereas the actual graph is as in Figure A.17b. (For a way to circumvent this problem, see the discussion preceding Exercise 23.)

TECHNOLOGY MASTERY

Determine whether your graphing utility produces the graph of the equation $y = x^{2/3}$ for both positive and negative values of x.

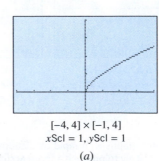

$[-4, 4] \times [-1, 4]$
xScl = 1, yScl = 1

(a)

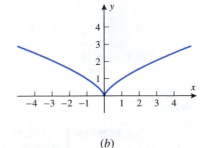

(b)

▶ **Figure A.17**

■ WHAT IS THE TRUE SHAPE OF A GRAPH?

Although graphing utilities are powerful tools for generating graphs quickly, they can produce misleading graphs as a result of compression, sampling error, false gaps, and false line segments. In short, *graphing utilities can suggest graph shapes*, *but they cannot establish them with certainty*. Thus, the more you know about the functions you are graphing, the easier it will be to choose good viewing windows, and the better you will be able to judge the reasonableness of the results produced by your graphing utility.

■ MORE INFORMATION ON GRAPHING AND CALCULATING UTILITIES

The main source of information about your graphing utility is its own documentation, and from time to time in this book we suggest that you refer to that documentation to learn some particular technique.

■ GENERATING PARAMETRIC CURVES WITH GRAPHING UTILITIES

Many graphing utilities allow you to graph equations of the form $y = f(x)$ but not equations of the form $x = g(y)$. Sometimes you will be able to rewrite $x = g(y)$ in the form $y = f(x)$; however, if this is inconvenient or impossible, then you can graph $x = g(y)$ by introducing a parameter $t = y$ and expressing the equation in the parametric form $x = g(t)$, $y = t$. (You may have to experiment with various intervals for t to produce a complete graph.)

▶ **Example 8**   Use a graphing utility to graph the equation $x = 3y^5 - 5y^3 + 1$.

Solution.   If we let $t = y$ be the parameter, then the equation can be written in parametric form as

$$x = 3t^5 - 5t^3 + 1, \quad y = t$$

Figure A.18 shows the graph of these equations for $-1.5 \leq t \leq 1.5$. ◀

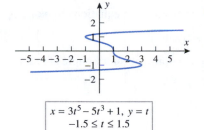

$x = 3t^5 - 5t^3 + 1, \ y = t$
$-1.5 \leq t \leq 1.5$

▲ **Figure A.18**

Some parametric curves are so complex that it is virtually impossible to visualize them without using some kind of graphing utility. Figure A.19 shows three such curves.

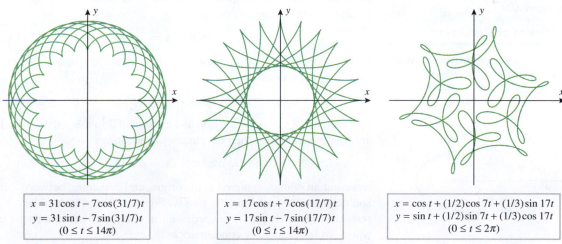

$x = 31\cos t - 7\cos(31/7)t$
$y = 31\sin t - 7\sin(31/7)t$
$(0 \leq t \leq 14\pi)$

$x = 17\cos t + 7\cos(17/7)t$
$y = 17\sin t - 7\sin(17/7)t$
$(0 \leq t \leq 14\pi)$

$x = \cos t + (1/2)\cos 7t + (1/3)\sin 17t$
$y = \sin t + (1/2)\sin 7t + (1/3)\cos 17t$
$(0 \leq t \leq 2\pi)$

▲ **Figure A.19**

■ GRAPHING INVERSE FUNCTIONS WITH GRAPHING UTILITIES

Most graphing utilities cannot graph inverse functions directly. However, there is a way of graphing inverse functions by expressing the graphs parametrically. To see how this can be done, suppose that we are interested in graphing the inverse of a one-to-one function f.

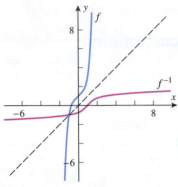

▲ **Figure A.20**

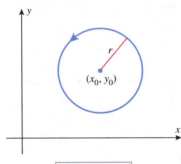

$$x = x_0 + r \cos t$$
$$y = y_0 + r \sin t$$
$$(0 \le t \le 2\pi)$$

▲ **Figure A.21**

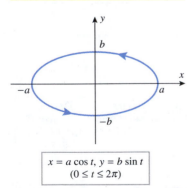

$$x = a \cos t, \ y = b \sin t$$
$$(0 \le t \le 2\pi)$$

▲ **Figure A.22**

We know that the equation $y = f(x)$ can be expressed parametrically as

$$x = t, \quad y = f(t) \tag{1}$$

and we know that the graph of f^{-1} can be obtained by interchanging x and y, since this reflects the graph of f about the line $y = x$. Thus, from (1) the graph of f^{-1} can be represented parametrically as

$$x = f(t), \quad y = t \tag{2}$$

For example, Figure A.20 shows the graph of $f(x) = x^5 + x + 1$ and its inverse generated with a graphing utility. The graph of f was generated from the parametric equations

$$x = t, \quad y = t^5 + t + 1$$

and the graph of f^{-1} was generated from the parametric equations

$$x = t^5 + t + 1, \quad y = t$$

■ TRANSLATION

If a parametric curve C is given by the equations $x = f(t)$, $y = g(t)$, then adding a constant to $f(t)$ translates the curve C in the x-direction, and adding a constant to $g(t)$ translates it in the y-direction. Thus, a circle of radius r, centered at (x_0, y_0) can be represented parametrically as

$$x = x_0 + r \cos t, \quad y = y_0 + r \sin t \quad (0 \le t \le 2\pi)$$

(Figure A.21). If desired, we can eliminate the parameter from these equations by noting that

$$(x - x_0)^2 + (y - y_0)^2 = (r \cos t)^2 + (r \sin t)^2 = r^2$$

Thus, we have obtained the familiar equation in rectangular coordinates for a circle of radius r, centered at (x_0, y_0):

$$(x - x_0)^2 + (y - y_0)^2 = r^2$$

■ SCALING

If a parametric curve C is given by the equations $x = f(t)$, $y = g(t)$, then multiplying $f(t)$ by a constant stretches or compresses C in the x-direction, and multiplying $g(t)$ by a constant stretches or compresses C in the y-direction. For example, we would expect the parametric equations

$$x = 3 \cos t, \quad y = 2 \sin t \quad (0 \le t \le 2\pi)$$

to represent an ellipse, centered at the origin, since the graph of these equations results from stretching the unit circle

$$x = \cos t, \quad y = \sin t \quad (0 \le t \le 2\pi)$$

by a factor of 3 in the x-direction and a factor of 2 in the y-direction. In general, if a and b are positive constants, then the parametric equations

$$x = a \cos t, \quad y = b \sin t \quad (0 \le t \le 2\pi) \tag{3}$$

represent an ellipse, centered at the origin, and extending between $-a$ and a on the x-axis and between $-b$ and b on the y-axis (Figure A.22). The numbers a and b are called the **semiaxes** of the ellipse. If desired, we can eliminate the parameter t in (3) and rewrite the equations in rectangular coordinates as

$$\frac{x^2}{a^2} + \frac{y^2}{b^2} = 1 \tag{4}$$

EXERCISE SET A   ∿ Graphing Utility

∿ **1–4** Use a graphing utility to generate the graph of f in the given viewing windows, and specify the window that you think gives the best view of the graph. ■

1. $f(x) = x^4 - x^2$
   (a) $[-50, 50] \times [-50, 50]$ (b) $[-5, 5] \times [-5, 5]$
   (c) $[-2, 2] \times [-2, 2]$ (d) $[-2, 2] \times [-1, 1]$
   (e) $[-1.5, 1.5] \times [-0.5, 0.5]$

2. $f(x) = x^5 - x^3$
   (a) $[-50, 50] \times [-50, 50]$ (b) $[-5, 5] \times [-5, 5]$
   (c) $[-2, 2] \times [-2, 2]$ (d) $[-2, 2] \times [-1, 1]$
   (e) $[-1.5, 1.5] \times [-0.5, 0.5]$

3. $f(x) = x^2 + 12$
   (a) $[-1, 1] \times [13, 15]$ (b) $[-2, 2] \times [11, 15]$
   (c) $[-4, 4] \times [10, 28]$ (d) A window of your choice

4. $f(x) = -12 - x^2$
   (a) $[-1, 1] \times [-15, -13]$ (b) $[-2, 2] \times [-15, -11]$
   (c) $[-4, 4] \times [-28, -10]$ (d) A window of your choice

∿ **5–6** Use the domain and range of f to determine a viewing window that contains the entire graph, and generate the graph in that window. ■

5. $f(x) = \sqrt{16 - 2x^2}$ 6. $f(x) = \sqrt{3 - 2x - x^2}$

∿ **7–14** Generate the graph of f in a viewing window that you think is appropriate. ■

7. $f(x) = x^2 - 9x - 36$ 8. $f(x) = \dfrac{x + 7}{x - 9}$

9. $f(x) = 2\cos(80x)$ 10. $f(x) = 12\sin(x/80)$

11. $f(x) = 300 - 10x^2 + 0.01x^3$

12. $f(x) = x(30 - 2x)(25 - 2x)$

13. $f(x) = x^2 + \dfrac{1}{x}$ 14. $f(x) = \sqrt{11x - 18}$

∿ **15–16** Generate the graph of f and determine whether your graphs contain false line segments. Sketch the actual graph and see if you can make the false line segments disappear by changing the viewing window. ■

15. $f(x) = \dfrac{x}{x^2 - 1}$ 16. $f(x) = \dfrac{x^2}{4 - x^2}$

∿ **17.** The graph of the equation $x^2 + y^2 = 16$ is a circle of radius 4 centered at the origin.
   (a) Find a function whose graph is the upper semicircle and graph it.
   (b) Find a function whose graph is the lower semicircle and graph it.
   (c) Graph the upper and lower semicircles together. If the combined graphs do not appear circular, see if you can adjust the viewing window to eliminate the aspect ratio distortion.
   (d) Graph the portion of the circle in the first quadrant.
   (e) Is there a function whose graph is the right half of the circle? Explain.

18. In each part, graph the equation by solving for y in terms of x and graphing the resulting functions together.
   (a) $x^2/4 + y^2/9 = 1$ (b) $y^2 - x^2 = 1$

∿ **19.** Read the documentation for your graphing utility to determine how to graph functions involving absolute values, and graph the given equation.
   (a) $y = |x|$ (b) $y = |x - 1|$
   (c) $y = |x| - 1$ (d) $y = |\sin x|$
   (e) $y = \sin|x|$ (f) $y = |x| - |x + 1|$

∿ **20.** Based on your knowledge of the absolute value function, sketch the graph of $f(x) = |x|/x$. Check your result using a graphing utility.

21–22 Most graphing utilities provide some way of graphing functions that are defined piecewise; read the documentation for your graphing utility to find out how to do this. However, if your goal is just to find the general shape of the graph, you can graph each portion of the function separately and combine the pieces with a hand-drawn sketch. Use this method in these exercises. ■

21. Draw the graph of
$$f(x) = \begin{cases} \sqrt[3]{x - 2}, & x \le 2 \\ x^3 - 2x - 4, & x > 2 \end{cases}$$

22. Draw the graph of
$$f(x) = \begin{cases} x^3 - x^2, & x \le 1 \\ \dfrac{1}{1 - x}, & 1 < x < 4 \\ x^2 \cos\sqrt{x}, & 4 \le x \end{cases}$$

∿ **23–24** We noted in the text that for functions involving fractional exponents (or radicals), graphing utilities sometimes omit portions of the graph. If $f(x) = x^{p/q}$, where p/q is a positive fraction in *lowest terms*, then you can circumvent this problem as follows:
   • If p is even and q is odd, then graph $g(x) = |x|^{p/q}$ instead of $f(x)$.
   • If p is odd and q is odd, then graph $g(x) = (|x|/x)|x|^{p/q}$ instead of $f(x)$. ■

23. (a) Generate the graphs of $f(x) = x^{2/5}$ and $g(x) = |x|^{2/5}$, and determine whether your graphing utility missed part of the graph of f.
   (b) Generate the graphs of the functions $f(x) = x^{1/5}$ and $g(x) = (|x|/x)|x|^{1/5}$, and determine whether your graphing utility missed part of the graph of f.
   (c) Generate a graph of the function $f(x) = (x - 1)^{4/5}$ that shows all of its important features.
   (d) Generate a graph of the function $f(x) = (x + 1)^{3/4}$ that shows all of its important features.

24. The graphs of $y = (x^2 - 4)^{2/3}$ and $y = [(x^2 - 4)^2]^{1/3}$ should be the same. Does your graphing utility produce

the same graph for both equations? If not, what do you think is happening?

25. In each part, graph the function for various values of c, and write a paragraph or two that describes how changes in c affect the graph in each case.

(a) $y = cx^2$ (b) $y = x^2 + cx$ (c) $y = x^2 + x + c$

26. The graph of an equation of the form $y^2 = x(x-a)(x-b)$ (where $0 < a < b$) is called a *bipartite cubic*. The accompanying figure shows a typical graph of this type.

(a) Graph the bipartite cubic $y^2 = x(x-1)(x-2)$ by solving for y in terms of x and graphing the two resulting functions.

(b) Find the x-intercepts of the bipartite cubic

$$y^2 = x(x-a)(x-b)$$

and make a conjecture about how changes in the values of a and b would affect the graph. Test your conjecture by graphing the bipartite cubic for various values of a and b.

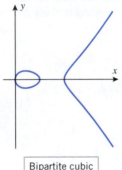

Bipartite cubic

◀ **Figure Ex-26**

27. Based on your knowledge of the graphs of $y = x$ and $y = \sin x$, make a sketch of the graph of $y = x \sin x$. Check your conclusion using a graphing utility.

28. What do you think the graph of $y = \sin(1/x)$ looks like? Test your conclusion using a graphing utility. [*Suggestion:* Examine the graph on a succession of smaller and smaller intervals centered at $x = 0$.]

29–30 Graph the equation using a graphing utility. ◼

29. (a) $x = y^2 + 2y + 1$

(b) $x = \sin y, \quad -2\pi \le y \le 2\pi$

30. (a) $x = y + 2y^3 - y^5$

(b) $x = \tan y, \quad -\pi/2 < y < \pi/2$

31–34 Use a graphing utility and parametric equations to display the graphs of f and f^{-1} on the same screen. ◼

31. $f(x) = x^3 + 0.2x - 1, \quad -1 \le x \le 2$

32. $f(x) = \sqrt{x^2 + 2} + x, \quad -5 \le x \le 5$

33. $f(x) = \cos(\cos 0.5x), \quad 0 \le x \le 3$

34. $f(x) = x + \sin x, \quad 0 \le x \le 6$

35. (a) Find parametric equations for the ellipse that is centered at the origin and has intercepts $(4, 0)$, $(-4, 0)$, $(0, 3)$, and $(0, -3)$.

(b) Find parametric equations for the ellipse that results by translating the ellipse in part (a) so that its center is at $(-1, 2)$.

(c) Confirm your results in parts (a) and (b) using a graphing utility.

B

TRIGONOMETRY REVIEW

◼ ANGLES

Angles in the plane can be generated by rotating a ray about its endpoint. The starting position of the ray is called the **initial side** of the angle, the final position is called the **terminal side** of the angle, and the point at which the initial and terminal sides meet is called the **vertex** of the angle. We allow for the possibility that the ray may make more than one complete revolution. Angles are considered to be **positive** if generated counterclockwise and **negative** if generated clockwise (Figure B.1).

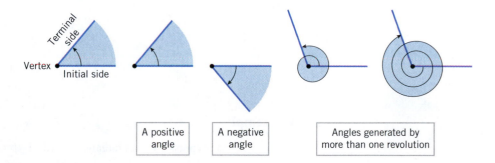

▶ **Figure B.1**

There are two standard measurement systems for describing the size of an angle: **degree measure** and **radian measure**. In degree measure, one degree (written $1°$) is the measure of an angle generated by $1/360$ of one revolution. Thus, there are $360°$ in an angle of one revolution, $180°$ in an angle of one-half revolution, $90°$ in an angle of one-quarter revolution (a *right angle*), and so forth. Degrees are divided into sixty equal parts, called **minutes**, and minutes are divided into sixty equal parts, called **seconds**. Thus, one minute (written $1'$) is $1/60$ of a degree, and one second (written $1''$) is $1/60$ of a minute. Smaller subdivisions of a degree are expressed as fractions of a second.

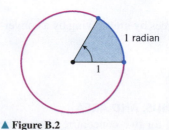

▲ **Figure B.2**

In radian measure, angles are measured by the length of the arc that the angle subtends on a circle of radius 1 when the vertex is at the center. One unit of arc on a circle of radius 1 is called one **radian** (written 1 radian or 1 rad) (Figure B.2), and hence the entire circumference of a circle of radius 1 is 2π radians. It follows that an angle of $360°$ subtends an arc of 2π radians, an angle of $180°$ subtends an arc of π radians, an angle of $90°$ subtends an arc of $\pi/2$ radians, and so forth. Figure B.3 and Table B.1 show the relationship between degree measure and radian measure for some important positive angles.

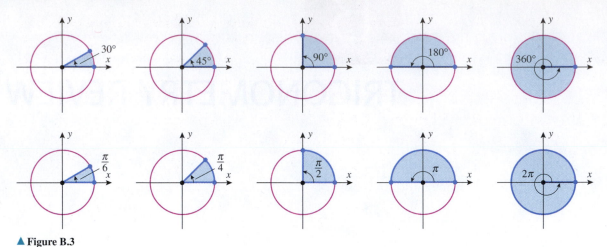

▲ **Figure B.3**

Observe that in Table B.1, angles in degrees are designated by the degree symbol, but angles in radians have no units specified. This is standard practice—when no units are specified for an angle, it is understood that the units are radians.

Table B.1

DEGREES	30°	45°	60°	90°	120°	135°	150°	180°	270°	360°
RADIANS	$\dfrac{\pi}{6}$	$\dfrac{\pi}{4}$	$\dfrac{\pi}{3}$	$\dfrac{\pi}{2}$	$\dfrac{2\pi}{3}$	$\dfrac{3\pi}{4}$	$\dfrac{5\pi}{6}$	π	$\dfrac{3\pi}{2}$	2π

From the fact that π radians corresponds to $180°$, we obtain the following formulas, which are useful for converting from degrees to radians and conversely.

$$1° = \frac{\pi}{180}\,\text{rad} \approx 0.01745\,\text{rad} \tag{1}$$

$$1\,\text{rad} = \left(\frac{180}{\pi}\right)^{\!\circ} \approx 57°\,17'\,44.8'' \tag{2}$$

▶ **Example 1**

(a) Express $146°$ in radians.   (b) Express 3 radians in degrees.

Solution (a). From (1), degrees can be converted to radians by multiplying by a conversion factor of $\pi/180$. Thus,

$$146° = \left(\frac{\pi}{180} \cdot 146\right)\text{rad} = \frac{73\pi}{90}\,\text{rad} \approx 2.5482\,\text{rad}$$

Solution (b). From (2), radians can be converted to degrees by multiplying by a conversion factor of $180/\pi$. Thus,

$$3\,\text{rad} = \left(3 \cdot \frac{180}{\pi}\right)^{\!\circ} = \left(\frac{540}{\pi}\right)^{\!\circ} \approx 171.9° \quad \blacktriangleleft$$

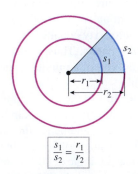

$$\frac{s_1}{s_2} = \frac{r_1}{r_2}$$

▲ **Figure B.4**

■ **RELATIONSHIPS BETWEEN ARC LENGTH, ANGLE, RADIUS, AND AREA**

There is a theorem from plane geometry which states that for two concentric circles, the ratio of the arc lengths subtended by a central angle is equal to the ratio of the corresponding radii (Figure B.4). In particular, if s is the arc length subtended on a circle of radius r by a central angle of θ radians, then by comparison with the arc length subtended by that angle on a circle of radius 1 we obtain

$$\frac{s}{\theta} = \frac{r}{1}$$

from which we obtain the following relationships between the central angle θ, the radius r, and the subtended arc length s when θ is in radians (Figure B.5):

$$\theta = s/r \qquad \text{and} \qquad s = r\theta \qquad\qquad (3\text{–}4)$$

The shaded region in Figure B.5 is called a *sector*. It is a theorem from plane geometry that the ratio of the area A of this sector to the area of the entire circle is the same as the ratio of the central angle of the sector to the central angle of the entire circle; thus, if the angles are in radians, we have

$$\frac{A}{\pi r^2} = \frac{\theta}{2\pi}$$

Solving for A yields the following formula for the area of a sector in terms of the radius r and the angle θ in radians:

$$A = \tfrac{1}{2} r^2 \theta \qquad\qquad (5)$$

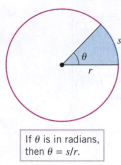

If θ is in radians, then $\theta = s/r$.

▲ **Figure B.5**

■ TRIGONOMETRIC FUNCTIONS FOR RIGHT TRIANGLES

The *sine*, *cosine*, *tangent*, *cosecant*, *secant*, and *cotangent* of a positive acute angle θ can be defined as ratios of the sides of a right triangle. Using the notation from Figure B.6, these definitions take the following form:

$$\sin\theta = \frac{\text{side opposite } \theta}{\text{hypotenuse}} = \frac{y}{r}, \qquad \csc\theta = \frac{\text{hypotenuse}}{\text{side opposite } \theta} = \frac{r}{y}$$

$$\cos\theta = \frac{\text{side adjacent to } \theta}{\text{hypotenuse}} = \frac{x}{r}, \qquad \sec\theta = \frac{\text{hypotenuse}}{\text{side adjacent to } \theta} = \frac{r}{x} \qquad (6)$$

$$\tan\theta = \frac{\text{side opposite } \theta}{\text{side adjacent to } \theta} = \frac{y}{x}, \qquad \cot\theta = \frac{\text{side adjacent to } \theta}{\text{side opposite } \theta} = \frac{x}{y}$$

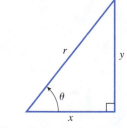

▲ **Figure B.6**

We will call sin, cos, tan, csc, sec, and cot the *trigonometric functions*. Because similar triangles have proportional sides, the values of the trigonometric functions depend only on the size of θ and not on the particular right triangle used to compute the ratios. Moreover, in these definitions it does not matter whether θ is measured in degrees or radians.

▶ **Example 2** Recall from geometry that the two legs of a $45°{-}45°{-}90°$ triangle are of equal size and that the hypotenuse of a $30°{-}60°{-}90°$ triangle is twice the shorter leg, where the shorter leg is opposite the $30°$ angle. These facts and the Theorem of Pythagoras yield Figure B.7. From that figure we obtain the results in Table B.2. ◀

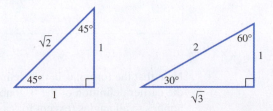

▶ **Figure B.7**

Table B.2

$\sin 45° = 1/\sqrt{2},$	$\cos 45° = 1/\sqrt{2},$	$\tan 45° = 1$
$\csc 45° = \sqrt{2},$	$\sec 45° = \sqrt{2},$	$\cot 45° = 1$
$\sin 30° = 1/2,$	$\cos 30° = \sqrt{3}/2,$	$\tan 30° = 1/\sqrt{3}$
$\csc 30° = 2,$	$\sec 30° = 2/\sqrt{3},$	$\cot 30° = \sqrt{3}$
$\sin 60° = \sqrt{3}/2,$	$\cos 60° = 1/2,$	$\tan 60° = \sqrt{3}$
$\csc 60° = 2/\sqrt{3},$	$\sec 60° = 2,$	$\cot 60° = 1/\sqrt{3}$

■ ANGLES IN RECTANGULAR COORDINATE SYSTEMS

Because the angles of a right triangle are between $0°$ and $90°$, the formulas in (6) are not directly applicable to negative angles or to angles greater than $90°$. To extend the trigonometric functions to include these cases, it will be convenient to consider angles in rectangular coordinate systems. An angle is said to be in *standard position* in an xy-coordinate system if its vertex is at the origin and its initial side is on the positive x-axis (Figure B.8).

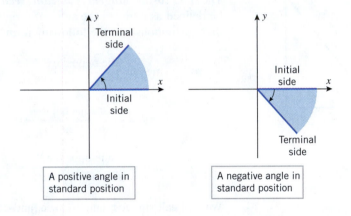

A positive angle in standard position

A negative angle in standard position

▶ **Figure B.8**

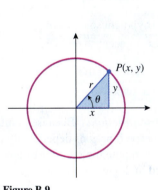

▲ **Figure B.9**

To define the trigonometric functions of an angle θ in standard position, construct a circle of radius r, centered at the origin, and let $P(x, y)$ be the intersection of the terminal side of θ with this circle (Figure B.9). We make the following definition.

B.1 **DEFINITION**

$$\sin \theta = \frac{y}{r}, \quad \cos \theta = \frac{x}{r}, \quad \tan \theta = \frac{y}{x}$$

$$\csc \theta = \frac{r}{y}, \quad \sec \theta = \frac{r}{x}, \quad \cot \theta = \frac{x}{y}$$

Note that the formulas in this definition agree with those in (6), so there is no conflict with the earlier definition of the trigonometric functions for triangles. However, this definition applies to all angles (except for cases where a zero denominator occurs).

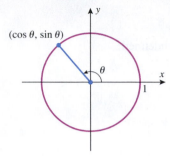

▲ Figure B.10

In the special case where $r = 1$, we have $\sin\theta = y$ and $\cos\theta = x$, so the terminal side of the angle θ intersects the unit circle at the point $(\cos\theta, \sin\theta)$ (Figure B.10). It follows from Definition B.1 that the remaining trigonometric functions of θ are expressible as (verify)

$$\tan\theta = \frac{\sin\theta}{\cos\theta}, \quad \cot\theta = \frac{\cos\theta}{\sin\theta} = \frac{1}{\tan\theta}, \quad \sec\theta = \frac{1}{\cos\theta}, \quad \csc\theta = \frac{1}{\sin\theta} \quad (7\text{–}10)$$

These observations suggest the following procedure for evaluating the trigonometric functions of common angles:

- Construct the angle θ in standard position in an xy-coordinate system.

- Find the coordinates of the intersection of the terminal side of the angle and the unit circle; the x- and y-coordinates of this intersection are the values of $\cos\theta$ and $\sin\theta$, respectively.

- Use Formulas (7) through (10) to find the values of the remaining trigonometric functions from the values of $\cos\theta$ and $\sin\theta$.

▶ **Example 3** Evaluate the trigonometric functions of $\theta = 150°$.

Solution. Construct a unit circle and place the angle $\theta = 150°$ in standard position (Figure B.11). Since $\angle AOP$ is $30°$ and $\triangle OAP$ is a $30°$–$60°$–$90°$ triangle, the leg AP has length $\frac{1}{2}$ (half the hypotenuse) and the leg OA has length $\sqrt{3}/2$ by the Theorem of Pythagoras. Thus, the coordinates of P are $(-\sqrt{3}/2, 1/2)$, from which we obtain

$$\sin 150° = \frac{1}{2}, \quad \cos 150° = -\frac{\sqrt{3}}{2}, \quad \tan 150° = \frac{\sin 150°}{\cos 150°} = \frac{1/2}{-\sqrt{3}/2} = -\frac{1}{\sqrt{3}}$$

$$\csc 150° = \frac{1}{\sin 150°} = 2, \quad \sec 150° = \frac{1}{\cos 150°} = -\frac{2}{\sqrt{3}}$$

$$\cot 150° = \frac{1}{\tan 150°} = -\sqrt{3} \blacktriangleleft$$

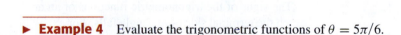

▲ Figure B.11

▶ **Example 4** Evaluate the trigonometric functions of $\theta = 5\pi/6$.

Solution. Since $5\pi/6 = 150°$, this problem is equivalent to that of Example 3. From that example we obtain

$$\sin\frac{5\pi}{6} = \frac{1}{2}, \quad \cos\frac{5\pi}{6} = -\frac{\sqrt{3}}{2}, \quad \tan\frac{5\pi}{6} = -\frac{1}{\sqrt{3}}$$

$$\csc\frac{5\pi}{6} = 2, \quad \sec\frac{5\pi}{6} = -\frac{2}{\sqrt{3}}, \quad \cot\frac{5\pi}{6} = -\sqrt{3} \blacktriangleleft$$

▶ **Example 5** Evaluate the trigonometric functions of $\theta = -\pi/2$.

Solution. As shown in Figure B.12, the terminal side of $\theta = -\pi/2$ intersects the unit circle at the point $(0, -1)$, so

$$\sin(-\pi/2) = -1, \quad \cos(-\pi/2) = 0$$

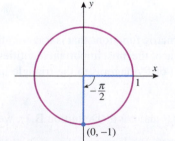

▲ Figure B.12

and from Formulas (7) through (10),

$$\tan(-\pi/2) = \frac{\sin(-\pi/2)}{\cos(-\pi/2)} = \frac{-1}{0} \quad \text{(undefined)}$$

$$\cot(-\pi/2) = \frac{\cos(-\pi/2)}{\sin(-\pi/2)} = \frac{0}{-1} = 0$$

$$\sec(-\pi/2) = \frac{1}{\cos(-\pi/2)} = \frac{1}{0} \quad \text{(undefined)}$$

$$\csc(-\pi/2) = \frac{1}{\sin(-\pi/2)} = \frac{1}{-1} = -1 \quad \blacktriangleleft$$

The reader should be able to obtain all of the results in Table B.3 by the methods illustrated in the last three examples. The dashes indicate quantities that are undefined.

Table B.3

	$\theta = 0$ (0°)	$\pi/6$ (30°)	$\pi/4$ (45°)	$\pi/3$ (60°)	$\pi/2$ (90°)	$2\pi/3$ (120°)	$3\pi/4$ (135°)	$5\pi/6$ (150°)	π (180°)	$3\pi/2$ (270°)	2π (360°)
$\sin\theta$	0	1/2	$1/\sqrt{2}$	$\sqrt{3}/2$	1	$\sqrt{3}/2$	$1/\sqrt{2}$	1/2	0	-1	0
$\cos\theta$	1	$\sqrt{3}/2$	$1/\sqrt{2}$	1/2	0	$-1/2$	$-1/\sqrt{2}$	$-\sqrt{3}/2$	-1	0	1
$\tan\theta$	0	$1/\sqrt{3}$	1	$\sqrt{3}$	—	$-\sqrt{3}$	-1	$-1/\sqrt{3}$	0	—	0
$\csc\theta$	—	2	$\sqrt{2}$	$2/\sqrt{3}$	1	$2/\sqrt{3}$	$\sqrt{2}$	2	—	-1	—
$\sec\theta$	1	$2/\sqrt{3}$	$\sqrt{2}$	2	—	-2	$-\sqrt{2}$	$-2/\sqrt{3}$	-1	—	1
$\cot\theta$	—	$\sqrt{3}$	1	$1/\sqrt{3}$	0	$-1/\sqrt{3}$	-1	$-\sqrt{3}$	—	0	—

REMARK | It is only in special cases that exact values for trigonometric functions can be obtained; usually, a calculating utility or a computer program will be required.

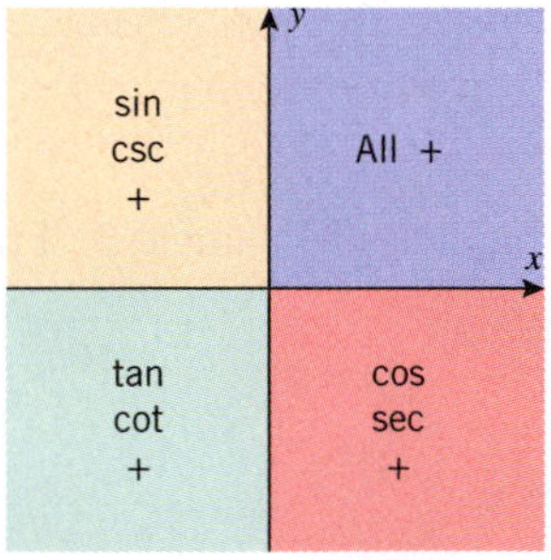

▲ **Figure B.13**

The signs of the trigonometric functions of an angle are determined by the quadrant in which the terminal side of the angle falls. For example, if the terminal side falls in the first quadrant, then x and y are positive in Definition B.1, so all of the trigonometric functions have positive values. If the terminal side falls in the second quadrant, then x is negative and y is positive, so sin and csc are positive, but all other trigonometric functions are negative. The diagram in Figure B.13 shows which trigonometric functions are positive in the various quadrants. The reader will find it instructive to check that the results in Table B.3 are consistent with Figure B.13.

■ TRIGONOMETRIC IDENTITIES

A *trigonometric identity* is an equation involving trigonometric functions that is true for all angles for which both sides of the equation are defined. One of the most important identities in trigonometry can be derived by applying the Theorem of Pythagoras to the triangle in Figure B.9 to obtain

$$x^2 + y^2 = r^2$$

Dividing both sides by r^2 and using the definitions of $\sin\theta$ and $\cos\theta$ (Definition B.1), we obtain the following fundamental result:

$$\sin^2\theta + \cos^2\theta = 1 \tag{11}$$

The following identities can be obtained from (11) by dividing through by $\cos^2\theta$ and $\sin^2\theta$, respectively, then applying Formulas (7) through (10):

$$\tan^2\theta + 1 = \sec^2\theta \tag{12}$$

$$1 + \cot^2\theta = \csc^2\theta \tag{13}$$

If (x, y) is a point on the unit circle, then the points $(-x, y)$, $(-x, -y)$, and $(x, -y)$ also lie on the unit circle (why?), and the four points form corners of a rectangle with sides parallel to the coordinate axes (Figure B.14a). The x- and y-coordinates of each corner represent the cosine and sine of an angle in standard position whose terminal side passes through the corner; hence we obtain the identities in parts (b), (c), and (d) of Figure B.14 for sine and cosine. Dividing those identities leads to identities for the tangent. In summary:

$$\sin(\pi - \theta) = \sin\theta, \qquad \sin(\pi + \theta) = -\sin\theta, \qquad \sin(-\theta) = -\sin\theta \tag{14--16}$$

$$\cos(\pi - \theta) = -\cos\theta, \qquad \cos(\pi + \theta) = -\cos\theta, \qquad \cos(-\theta) = \cos\theta \tag{17--19}$$

$$\tan(\pi - \theta) = -\tan\theta, \qquad \tan(\pi + \theta) = \tan\theta, \qquad \tan(-\theta) = -\tan\theta \tag{20--22}$$

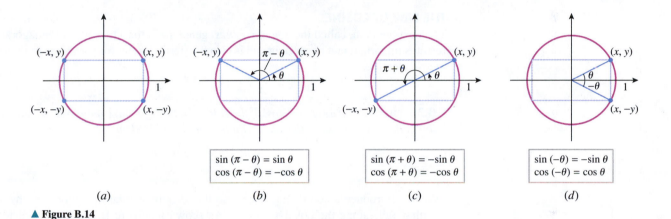

(a) (b) (c) (d)

▲ **Figure B.14**

Two angles in standard position that have the same terminal side must have the same values for their trigonometric functions since their terminal sides intersect the unit circle at the same point. In particular, two angles whose radian measures differ by a multiple of 2π have the same terminal side and hence have the same values for their trigonometric functions. This yields the identities

$$\sin\theta = \sin(\theta + 2\pi) = \sin(\theta - 2\pi) \tag{23}$$

$$\cos\theta = \cos(\theta + 2\pi) = \cos(\theta - 2\pi) \tag{24}$$

and more generally,

$$\sin\theta = \sin(\theta \pm 2n\pi), \quad n = 0, 1, 2, \ldots \tag{25}$$

$$\cos\theta = \cos(\theta \pm 2n\pi), \quad n = 0, 1, 2, \ldots \tag{26}$$

Identity (21) implies that

$$\tan\theta = \tan(\theta + \pi) \qquad \text{and} \qquad \tan\theta = \tan(\theta - \pi) \tag{27--28}$$

Identity (27) is just (21) with the terms in the sum reversed, and identity (28) follows from (21) by substituting $\theta - \pi$ for θ. These two identities state that adding or subtracting π

from an angle does not affect the value of the tangent of the angle. It follows that the same is true for any multiple of π; thus,

$$\tan\theta = \tan(\theta \pm n\pi), \quad n = 0, 1, 2, \dots \tag{29}$$

Figure B.15 shows complementary angles θ and $(\pi/2) - \theta$ of a right triangle. It follows from (6) that

$$\sin\theta = \frac{\text{side opposite } \theta}{\text{hypotenuse}} = \frac{\text{side adjacent to } (\pi/2) - \theta}{\text{hypotenuse}} = \cos\left(\frac{\pi}{2} - \theta\right)$$

$$\cos\theta = \frac{\text{side adjacent to } \theta}{\text{hypotenuse}} = \frac{\text{side opposite } (\pi/2) - \theta}{\text{hypotenuse}} = \sin\left(\frac{\pi}{2} - \theta\right)$$

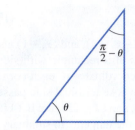

▲ **Figure B.15**

which yields the identities

$$\sin\left(\frac{\pi}{2} - \theta\right) = \cos\theta, \quad \cos\left(\frac{\pi}{2} - \theta\right) = \sin\theta, \quad \tan\left(\frac{\pi}{2} - \theta\right) = \cot\theta \tag{30–32}$$

where the third identity results from dividing the first two. These identities are also valid for angles that are not acute and for negative angles as well.

■ **THE LAW OF COSINES**

The next theorem, called the **law of cosines**, generalizes the Theorem of Pythagoras. This result is important in its own right and is also the starting point for some important trigonometric identities.

> **B.2   THEOREM** (*Law of Cosines*)   *If the sides of a triangle have lengths a, b, and c, and if θ is the angle between the sides with lengths a and b, then*
>
> $$c^2 = a^2 + b^2 - 2ab\cos\theta$$

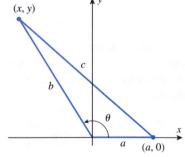

▲ **Figure B.16**

PROOF   Introduce a coordinate system so that θ is in standard position and the side of length a falls along the positive x-axis. As shown in Figure B.16, the side of length a extends from the origin to $(a, 0)$ and the side of length b extends from the origin to some point (x, y). From the definition of $\sin\theta$ and $\cos\theta$ we have $\sin\theta = y/b$ and $\cos\theta = x/b$, so

$$y = b\sin\theta, \quad x = b\cos\theta \tag{33}$$

From the distance formula in Theorem H.1 of Appendix H, we obtain

$$c^2 = (x - a)^2 + (y - 0)^2$$

so that, from (33),

$$c^2 = (b\cos\theta - a)^2 + b^2\sin^2\theta$$

$$= a^2 + b^2(\cos^2\theta + \sin^2\theta) - 2ab\cos\theta$$

$$= a^2 + b^2 - 2ab\cos\theta$$

which completes the proof.   ■

We will now show how the law of cosines can be used to obtain the following identities, called the **addition formulas** for sine and cosine:

$$\sin(\alpha + \beta) = \sin\alpha\cos\beta + \cos\alpha\sin\beta \tag{34}$$

$$\cos(\alpha + \beta) = \cos\alpha\cos\beta - \sin\alpha\sin\beta \tag{35}$$

$$\sin(\alpha - \beta) = \sin\alpha\cos\beta - \cos\alpha\sin\beta \qquad (36)$$

$$\cos(\alpha - \beta) = \cos\alpha\cos\beta + \sin\alpha\sin\beta \qquad (37)$$

We will derive (37) first. In our derivation we will assume that $0 \le \beta < \alpha < 2\pi$ (Figure B.17). As shown in the figure, the terminal sides of α and β intersect the unit circle at the points $P_1(\cos\alpha, \sin\alpha)$ and $P_2(\cos\beta, \sin\beta)$. If we denote the lengths of the sides of triangle OP_1P_2 by OP_1, P_1P_2, and OP_2, then $OP_1 = OP_2 = 1$ and, from the distance formula in Theorem H.1 of Appendix H,

$$
\begin{aligned}
(P_1P_2)^2 &= (\cos\beta - \cos\alpha)^2 + (\sin\beta - \sin\alpha)^2 \\
&= (\sin^2\alpha + \cos^2\alpha) + (\sin^2\beta + \cos^2\beta) - 2(\cos\alpha\cos\beta + \sin\alpha\sin\beta) \\
&= 2 - 2(\cos\alpha\cos\beta + \sin\alpha\sin\beta)
\end{aligned}
$$

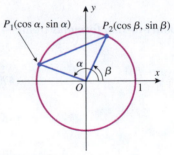

$P_1(\cos\alpha, \sin\alpha)$ $P_2(\cos\beta, \sin\beta)$

▲ **Figure B.17**

But angle $P_2OP_1 = \alpha - \beta$, so that the law of cosines yields

$$
\begin{aligned}
(P_1P_2)^2 &= (OP_1)^2 + (OP_2)^2 - 2(OP_1)(OP_2)\cos(\alpha - \beta) \\
&= 2 - 2\cos(\alpha - \beta)
\end{aligned}
$$

Equating the two expressions for $(P_1P_2)^2$ and simplifying, we obtain

$$\cos(\alpha - \beta) = \cos\alpha\cos\beta + \sin\alpha\sin\beta$$

which completes the derivation of (37).

We can use (31) and (37) to derive (36) as follows:

$$
\begin{aligned}
\sin(\alpha - \beta) &= \cos\left[\frac{\pi}{2} - (\alpha - \beta)\right] = \cos\left[\left(\frac{\pi}{2} - \alpha\right) - (-\beta)\right] \\
&= \cos\left(\frac{\pi}{2} - \alpha\right)\cos(-\beta) + \sin\left(\frac{\pi}{2} - \alpha\right)\sin(-\beta) \\
&= \cos\left(\frac{\pi}{2} - \alpha\right)\cos\beta - \sin\left(\frac{\pi}{2} - \alpha\right)\sin\beta \\
&= \sin\alpha\cos\beta - \cos\alpha\sin\beta
\end{aligned}
$$

Identities (34) and (35) can be obtained from (36) and (37) by substituting $-\beta$ for β and using the identities

$$\sin(-\beta) = -\sin\beta, \qquad \cos(-\beta) = \cos\beta$$

We leave it for the reader to derive the identities

$$\tan(\alpha + \beta) = \frac{\tan\alpha + \tan\beta}{1 - \tan\alpha\tan\beta} \qquad \tan(\alpha - \beta) = \frac{\tan\alpha - \tan\beta}{1 + \tan\alpha\tan\beta} \qquad (38\text{--}39)$$

Identity (38) can be obtained by dividing (34) by (35) and then simplifying. Identity (39) can be obtained from (38) by substituting $-\beta$ for β and simplifying.

In the special case where $\alpha = \beta$, identities (34), (35), and (38) yield the **double-angle formulas**

$$\sin 2\alpha = 2\sin\alpha\cos\alpha \qquad (40)$$

$$\cos 2\alpha = \cos^2\alpha - \sin^2\alpha \qquad (41)$$

$$\tan 2\alpha = \frac{2\tan\alpha}{1 - \tan^2\alpha} \qquad (42)$$

By using the identity $\sin^2\alpha + \cos^2\alpha = 1$, (41) can be rewritten in the alternative forms

$$\cos 2\alpha = 2\cos^2\alpha - 1 \qquad \text{and} \qquad \cos 2\alpha = 1 - 2\sin^2\alpha \qquad (43\text{--}44)$$

If we replace α by $\alpha/2$ in (43) and (44) and use some algebra, we obtain the *half-angle formulas*

$$\cos^2\frac{\alpha}{2} = \frac{1 + \cos\alpha}{2} \qquad \text{and} \qquad \sin^2\frac{\alpha}{2} = \frac{1 - \cos\alpha}{2} \qquad (45\text{--}46)$$

We leave it for the exercises to derive the following *product-to-sum formulas* from (34) through (37):

$$\sin\alpha\cos\beta = \frac{1}{2}[\sin(\alpha - \beta) + \sin(\alpha + \beta)] \qquad (47)$$

$$\sin\alpha\sin\beta = \frac{1}{2}[\cos(\alpha - \beta) - \cos(\alpha + \beta)] \qquad (48)$$

$$\cos\alpha\cos\beta = \frac{1}{2}[\cos(\alpha - \beta) + \cos(\alpha + \beta)] \qquad (49)$$

We also leave it for the exercises to derive the following *sum-to-product formulas*:

$$\sin\alpha + \sin\beta = 2\sin\frac{\alpha + \beta}{2}\cos\frac{\alpha - \beta}{2} \qquad (50)$$

$$\sin\alpha - \sin\beta = 2\cos\frac{\alpha + \beta}{2}\sin\frac{\alpha - \beta}{2} \qquad (51)$$

$$\cos\alpha + \cos\beta = 2\cos\frac{\alpha + \beta}{2}\cos\frac{\alpha - \beta}{2} \qquad (52)$$

$$\cos\alpha - \cos\beta = -2\sin\frac{\alpha + \beta}{2}\sin\frac{\alpha - \beta}{2} \qquad (53)$$

■ FINDING AN ANGLE FROM THE VALUE OF ITS TRIGONOMETRIC FUNCTIONS

There are numerous situations in which it is necessary to find an unknown angle from a known value of one of its trigonometric functions. The following example illustrates a method for doing this.

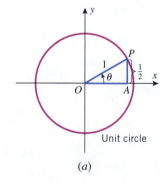

(a)

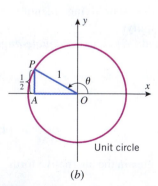

Unit circle

(b)

▲ **Figure B.18**

▶ **Example 6**   Find θ if $\sin\theta = \frac{1}{2}$.

Solution.   We begin by looking for positive angles that satisfy the equation. Because $\sin\theta$ is positive, the angle θ must terminate in the first or second quadrant. If it terminates in the first quadrant, then the hypotenuse of $\triangle OAP$ in Figure B.18a is double the leg AP, so

$$\theta = 30° = \frac{\pi}{6} \text{ radians}$$

If θ terminates in the second quadrant (Figure B.18b), then the hypotenuse of $\triangle OAP$ is double the leg AP, so $\angle AOP = 30°$, which implies that

$$\theta = 180° - 30° = 150° = \frac{5\pi}{6} \text{ radians}$$

Now that we have found these two solutions, all other solutions are obtained by adding or subtracting multiples of $360°$ (2π radians) to or from them. Thus, the entire set of solutions is given by the formulas

$$\theta = 30° \pm n \cdot 360°, \quad n = 0, 1, 2, \ldots$$

and

$$\theta = 150° \pm n \cdot 360°, \quad n = 0, 1, 2, \ldots$$

or in radian measure,

$$\theta = \frac{\pi}{6} \pm n \cdot 2\pi, \quad n = 0, 1, 2, \ldots$$

and

$$\theta = \frac{5\pi}{6} \pm n \cdot 2\pi, \quad n = 0, 1, 2, \ldots \blacktriangleleft$$

■ ANGLE OF INCLINATION

The slope of a nonvertical line L is related to the angle that L makes with the positive x-axis. If ϕ is the smallest positive angle measured counterclockwise from the x-axis to L, then the slope of the line can be expressed as

$$m = \tan \phi \tag{54}$$

(Figure B.19a). The angle ϕ, which is called the *angle of inclination* of the line, satisfies $0° \leq \phi < 180°$ in degree measure (or, equivalently, $0 \leq \phi < \pi$ in radian measure). If ϕ is an acute angle, then $m = \tan \phi$ is positive and the line slopes up to the right, and if ϕ is an obtuse angle, then $m = \tan \phi$ is negative and the line slopes down to the right. For example, a line whose angle of inclination is $45°$ has slope $m = \tan 45° = 1$, and a line whose angle of inclination is $135°$ has a slope of $m = \tan 135° = -1$ (Figure B.19b). Figure B.20 shows a convenient way of using the line $x = 1$ as a "ruler" for visualizing the relationship between lines of various slopes.

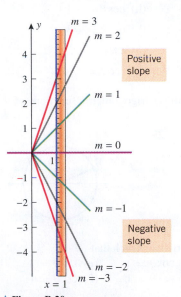

▲ Figure B.20

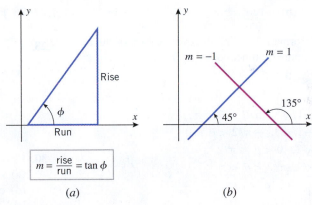

▲ Figure B.19

EXERCISE SET B

1–2 Express the angles in radians. ■

1. (a) $75°$ (b) $390°$ (c) $20°$ (d) $138°$

2. (a) $420°$ (b) $15°$ (c) $225°$ (d) $165°$

3–4 Express the angles in degrees. ■

3. (a) $\pi/15$ (b) 1.5 (c) $8\pi/5$ (d) 3π

4. (a) $\pi/10$ (b) 2 (c) $2\pi/5$ (d) $7\pi/6$

5–6 Find the exact values of all six trigonometric functions of θ. ■

5. (a)    (b)   (c)

6. (a) 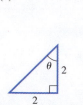   (b)    (c)

7–12 The angle θ is an acute angle of a right triangle. Solve the problems by drawing an appropriate right triangle. Do *not* use a calculator. ■

7. Find $\sin\theta$ and $\cos\theta$ given that $\tan\theta = 3$.

8. Find $\sin\theta$ and $\tan\theta$ given that $\cos\theta = \frac{2}{3}$.

9. Find $\tan\theta$ and $\csc\theta$ given that $\sec\theta = \frac{5}{2}$.

10. Find $\cot\theta$ and $\sec\theta$ given that $\csc\theta = 4$.

11. Find the length of the side adjacent to θ given that the hypotenuse has length 6 and $\cos\theta = 0.3$.

12. Find the length of the hypotenuse given that the side opposite θ has length 2.4 and $\sin\theta = 0.8$.

13–14 The value of an angle θ is given. Find the values of all six trigonometric functions of θ without using a calculator. ■

13. (a) $225°$ (b) $-210°$ (c) $5\pi/3$ (d) $-3\pi/2$

14. (a) $330°$ (b) $-120°$ (c) $9\pi/4$ (d) -3π

15–16 Use the information to find the exact values of the remaining five trigonometric functions of θ. ■

15. (a) $\cos\theta = \frac{3}{5}, \quad 0 < \theta < \pi/2$

(b) $\cos\theta = \frac{3}{5}, \quad -\pi/2 < \theta < 0$

(c) $\tan\theta = -1/\sqrt{3}, \quad \pi/2 < \theta < \pi$

(d) $\tan\theta = -1/\sqrt{3}, \quad -\pi/2 < \theta < 0$

(e) $\csc\theta = \sqrt{2}, \quad 0 < \theta < \pi/2$

(f) $\csc\theta = \sqrt{2}, \quad \pi/2 < \theta < \pi$

16. (a) $\sin\theta = \frac{1}{4}, \quad 0 < \theta < \pi/2$

(b) $\sin\theta = \frac{1}{4}, \quad \pi/2 < \theta < \pi$

(c) $\cot\theta = \frac{1}{3}, \quad 0 < \theta < \pi/2$

(d) $\cot\theta = \frac{1}{3}, \quad \pi < \theta < 3\pi/2$

(e) $\sec\theta = -\frac{5}{2}, \quad \pi/2 < \theta < \pi$

(f) $\sec\theta = -\frac{5}{2}, \quad \pi < \theta < 3\pi/2$

17–18 Use a calculating utility to find x to four decimal places. ■

17. (a) 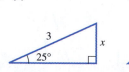   (b)

18. (a) 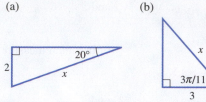   (b)

19. In each part, let θ be an acute angle of a right triangle. Express the remaining five trigonometric functions in terms of a.

(a) $\sin\theta = a/3$ (b) $\tan\theta = a/5$ (c) $\sec\theta = a$

20–27 Find all values of θ (in radians) that satisfy the given equation. Do not use a calculator. ■

20. (a) $\cos\theta = -1/\sqrt{2}$ (b) $\sin\theta = -1/\sqrt{2}$

21. (a) $\tan\theta = -1$ (b) $\cos\theta = \frac{1}{2}$

22. (a) $\sin\theta = -\frac{1}{2}$ (b) $\tan\theta = \sqrt{3}$

23. (a) $\tan\theta = 1/\sqrt{3}$ (b) $\sin\theta = -\sqrt{3}/2$

24. (a) $\sin\theta = -1$ (b) $\cos\theta = -1$

25. (a) $\cot\theta = -1$ (b) $\cot\theta = \sqrt{3}$

26. (a) $\sec\theta = -2$ (b) $\csc\theta = -2$

27. (a) $\csc\theta = 2/\sqrt{3}$ (b) $\sec\theta = 2/\sqrt{3}$

28–29 Find the values of all six trigonometric functions of θ. ■

28. 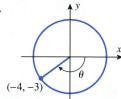   **29.**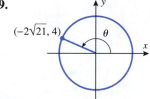

30. Find all values of θ (in radians) such that

(a) $\sin\theta = 1$ (b) $\cos\theta = 1$ (c) $\tan\theta = 1$

(d) $\csc\theta = 1$ (e) $\sec\theta = 1$ (f) $\cot\theta = 1$.

31. Find all values of θ (in radians) such that

(a) $\sin\theta = 0$ (b) $\cos\theta = 0$ (c) $\tan\theta = 0$

(d) $\csc\theta$ is undefined   (e) $\sec\theta$ is undefined

(f) $\cot\theta$ is undefined.

32. How could you use a ruler and protractor to approximate $\sin 17°$ and $\cos 17°$?

33. Find the length of the circular arc on a circle of radius 4 cm subtended by an angle of

(a) $\pi/6$ (b) $150°$.

34. Find the radius of a circular sector that has an angle of $\pi/3$ and a circular arc length of 7 units.

35. A point P moving counterclockwise on a circle of radius 5 cm traverses an arc length of 2 cm. What is the angle swept out by a radius from the center to P?

36. Find a formula for the area A of a circular sector in terms of its radius r and arc length s.

37. As shown in the accompanying figure, a right circular cone is made from a circular piece of paper of radius R by cutting out a sector of angle θ radians and gluing the cut edges of the remaining piece together. Find
(a) the radius r of the base of the cone in terms of R and θ.
(b) the height h of the cone in terms of R and θ.

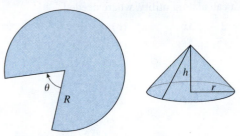

▲ **Figure Ex-37**

38. As shown in the accompanying figure, let r and L be the radius of the base and the slant height of a right circular cone. Show that the lateral surface area, S, of the cone is $S = \pi r L$. [*Hint:* As shown in the figure in Exercise 37, the lateral surface of the cone becomes a circular sector when cut along a line from the vertex to the base and flattened.]

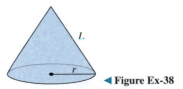

◀ **Figure Ex-38**

39. Two sides of a triangle have lengths of 3 cm and 7 cm and meet at an angle of $60°$. Find the area of the triangle.

40. Let ABC be a triangle whose angles at A and B are $30°$ and $45°$. If the side opposite the angle B has length 9, find the lengths of the remaining sides and the size of the angle C.

41. A 10-foot ladder leans against a house and makes an angle of $67°$ with level ground. How far is the top of the ladder above the ground? Express your answer to the nearest tenth of a foot.

42. From a point 120 feet on level ground from a building, the angle of elevation to the top of the building is $76°$. Find the height of the building. Express your answer to the nearest foot.

43. An observer on level ground is at a distance d from a building. The angles of elevation to the bottoms of the windows on the second and third floors are α and β, respectively. Find the distance h between the bottoms of the windows in terms of α, β, and d.

44. From a point on level ground, the angle of elevation to the top of a tower is α. From a point that is d units closer to the tower, the angle of elevation is β. Find the height h of the tower in terms of α, β, and d.

45–46 Do *not* use a calculator in these exercises. ▪

45. If $\cos \theta = \frac{2}{3}$ and $0 < \theta < \pi/2$, find
(a) $\sin 2\theta$
(b) $\cos 2\theta$.

46. If $\tan \alpha = \frac{3}{4}$ and $\tan \beta = 2$, where $0 < \alpha < \pi/2$ and $0 < \beta < \pi/2$, find
(a) $\sin(\alpha - \beta)$
(b) $\cos(\alpha + \beta)$.

47. Express $\sin 3\theta$ and $\cos 3\theta$ in terms of $\sin \theta$ and $\cos \theta$.

48–58 Derive the given identities. ▪

48. $\dfrac{\cos \theta \sec \theta}{1 + \tan^2 \theta} = \cos^2 \theta$

49. $\dfrac{\cos \theta \tan \theta + \sin \theta}{\tan \theta} = 2 \cos \theta$

50. $2 \csc 2\theta = \sec \theta \csc \theta$ **51.** $\tan \theta + \cot \theta = 2 \csc 2\theta$

52. $\dfrac{\sin 2\theta}{\sin \theta} - \dfrac{\cos 2\theta}{\cos \theta} = \sec \theta$

53. $\dfrac{\sin \theta + \cos 2\theta - 1}{\cos \theta - \sin 2\theta} = \tan \theta$

54. $\sin 3\theta + \sin \theta = 2 \sin 2\theta \cos \theta$

55. $\sin 3\theta - \sin \theta = 2 \cos 2\theta \sin \theta$

56. $\tan \dfrac{\theta}{2} = \dfrac{1 - \cos \theta}{\sin \theta}$ **57.** $\tan \dfrac{\theta}{2} = \dfrac{\sin \theta}{1 + \cos \theta}$

58. $\cos \left(\dfrac{\pi}{3} + \theta \right) + \cos \left(\dfrac{\pi}{3} - \theta \right) = \cos \theta$

59–60 In these exercises, refer to an arbitrary triangle ABC in which the side of length a is opposite angle A, the side of length b is opposite angle B, and the side of length c is opposite angle C. ▪

59. Prove: The area of a triangle ABC can be written as
$$\text{area} = \tfrac{1}{2} bc \sin A$$
Find two other similar formulas for the area.

60. Prove the **law of sines**: In any triangle, the ratios of the sides to the sines of the opposite angles are equal; that is,
$$\frac{a}{\sin A} = \frac{b}{\sin B} = \frac{c}{\sin C}$$

61. Use identities (34) through (37) to express each of the following in terms of $\sin \theta$ or $\cos \theta$.
(a) $\sin \left(\dfrac{\pi}{2} + \theta \right)$
(b) $\cos \left(\dfrac{\pi}{2} + \theta \right)$
(c) $\sin \left(\dfrac{3\pi}{2} - \theta \right)$
(d) $\cos \left(\dfrac{3\pi}{2} + \theta \right)$

62. Derive identities (38) and (39).

63. Derive identity
(a) (47)
(b) (48)
(c) (49).

64. If $A = \alpha + \beta$ and $B = \alpha - \beta$, then $\alpha = \frac{1}{2}(A + B)$ and $\beta = \frac{1}{2}(A - B)$ (verify). Use this result and identities (47) through (49) to derive identity
(a) (50)
(b) (52)
(c) (53).

65. Substitute $-\beta$ for β in identity (50) to derive identity (51).

66. (a) Express $3 \sin \alpha + 5 \cos \alpha$ in the form
$$C \sin(\alpha + \phi)$$

(b) Show that a sum of the form
$$A \sin \alpha + B \cos \alpha$$
can be rewritten in the form $C \sin(\alpha + \phi)$.

67. Show that the length of the diagonal of the parallelogram in the accompanying figure is
$$d = \sqrt{a^2 + b^2 + 2ab \cos \theta}$$

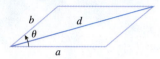

◀ **Figure Ex-67**

68–69 Find the angle of inclination of the line with slope m to the nearest degree. Use a calculating utility, where needed. ▪

68. (a) $m = \frac{1}{2}$ (b) $m = -1$
    (c) $m = 2$ (d) $m = -57$

69. (a) $m = -\frac{1}{2}$ (b) $m = 1$
    (c) $m = -2$ (d) $m = 57$

70–71 Find the angle of inclination of the line to the nearest degree. Use a calculating utility, where needed. ▪

70. (a) $3y = 2 - \sqrt{3}x$ (b) $y - 4x + 7 = 0$

71. (a) $y = \sqrt{3}x + 2$ (b) $y + 2x + 5 = 0$

SOLVING POLYNOMIAL EQUATIONS

We will assume in this appendix that you know how to divide polynomials using long division and synthetic division. If you need to review those techniques, refer to an algebra book.

◼ A BRIEF REVIEW OF POLYNOMIALS

Recall that if n is a nonnegative integer, then a **polynomial of degree n** is a function that can be written in the following forms, depending on whether you want the powers of x in ascending or descending order:

$$c_0 + c_1x + c_2x^2 + \cdots + c_nx^n \quad (c_n \neq 0)$$
$$c_nx^n + c_{n-1}x^{n-1} + \cdots + c_1x + c_0 \quad (c_n \neq 0)$$

The numbers $c_0, c_1, \ldots, c_n$ are called the **coefficients** of the polynomial. The coefficient c_n (which multiplies the highest power of x) is called the **leading coefficient**, the term c_nx^n is called the **leading term**, and the coefficient c_0 is called the **constant term**. Polynomials of degree 1, 2, 3, 4, and 5 are called **linear**, **quadratic**, **cubic**, **quartic**, and **quintic**, respectively. For simplicity, general polynomials of low degree are often written without subscripts on the coefficients:

$$p(x) = a \qquad \text{Constant polynomial}$$

$$p(x) = ax + b \quad (a \neq 0) \qquad \text{Linear polynomial}$$

$$p(x) = ax^2 + bx + c \quad (a \neq 0) \qquad \text{Quadratic polynomial}$$

$$p(x) = ax^3 + bx^2 + cx + d \quad (a \neq 0) \qquad \text{Cubic polynomial}$$

When you attempt to factor a polynomial completely, one of three things can happen:

- You may be able to decompose the polynomial into distinct linear factors using only real numbers; for example,

$$x^3 + x^2 - 2x = x(x^2 + x - 2) = x(x - 1)(x + 2)$$

- You may be able to decompose the polynomial into linear factors using only real numbers, but some of the factors may be repeated; for example,

$$x^6 - 3x^4 + 2x^3 = x^3(x^3 - 3x + 2) = x^3(x - 1)^2(x + 2) \tag{1}$$

- You may be able to decompose the polynomial into linear and quadratic factors using only real numbers, but you may not be able to decompose the quadratic factors into linear factors using only real numbers (such quadratic factors are said to be **irreducible** over the real numbers); for example,

$$x^4 - 1 = (x^2 - 1)(x^2 + 1) = (x - 1)(x + 1)(x^2 + 1)$$
$$= (x - 1)(x + 1)(x - i)(x + i)$$

Here, the factor $x^2 + 1$ is irreducible over the real numbers.

In general, if $p(x)$ is a polynomial of degree n with leading coefficient a, and if complex numbers are allowed, then $p(x)$ can be factored as

$$p(x) = a(x - r_1)(x - r_2) \cdots (x - r_n) \tag{2}$$

where $r_1, r_2, \ldots, r_n$ are called the **zeros** of $p(x)$ or the **roots** of the equation $p(x) = 0$, and (2) is called the **complete linear factorization** of $p(x)$. If some of the factors in (2) are repeated, then they can be combined; for example, if the first k factors are distinct and the rest are repetitions of the first k, then (2) can be expressed in the form

$$p(x) = a(x - r_1)^{m_1}(x - r_2)^{m_2} \cdots (x - r_k)^{m_k} \tag{3}$$

where $r_1, r_2, \ldots, r_k$ are the *distinct* roots of $p(x) = 0$. The exponents $m_1, m_2, \ldots, m_k$ tell us how many times the various factors occur in the complete linear factorization; for example, in (3) the factor $(x - r_1)$ occurs m_1 times, the factor $(x - r_2)$ occurs m_2 times, and so forth. Some techniques for factoring polynomials are discussed later in this appendix. In general, if a factor $(x - r)$ occurs m times in the complete linear factorization of a polynomial, then we say that r is a root or zero of **multiplicity m**, and if $(x - r)$ has no repetitions (i.e., r has multiplicity 1), then we say that r is a **simple** root or zero. For example, it follows from (1) that the equation $x^6 - 3x^4 + 2x^3 = 0$ can be expressed as

$$x^3(x - 1)^2(x + 2) = 0 \tag{4}$$

so this equation has three distinct roots—a root $x = 0$ of multiplicity 3, a root $x = 1$ of multiplicity 2, and a simple root $x = -2$.

Note that in (3) the multiplicities of the roots must add up to n, since $p(x)$ has degree n; that is,

$$m_1 + m_2 + \cdots + m_k = n$$

For example, in (4) the multiplicities add up to 6, which is the same as the degree of the polynomial.

It follows from (2) that a polynomial of degree n can have at most n distinct roots; if all of the roots are simple, then there will be *exactly n*, but if some are repeated, then there will be fewer than n. However, when counting the roots of a polynomial, it is standard practice to count multiplicities, since that convention allows us to say that a polynomial of degree n has n roots. For example, from (1) the six roots of the polynomial $p(x) = x^6 - 3x^4 + 2x^3$ are

$$r = 0, \ 0, \ 0, \ 1, \ 1, \ -2$$

In summary, we have the following important theorem.

C.1   THEOREM   *If complex roots are allowed, and if roots are counted according to their multiplicities, then a polynomial of degree n has exactly n roots.*

■ THE REMAINDER THEOREM

When two positive integers are divided, the numerator can be expressed as the quotient plus the remainder over the divisor, where the remainder is less than the divisor. For example,

$$\tfrac{17}{5} = 3 + \tfrac{2}{5}$$

If we multiply this equation through by 5, we obtain

$$17 = 5 \cdot 3 + 2$$

which states that the *numerator is the divisor times the quotient plus the remainder*.

The following theorem, which we state without proof, is an analogous result for division of polynomials.

C.2 **THEOREM** *If $p(x)$ and $s(x)$ are polynomials, and if $s(x)$ is not the zero polynomial, then $p(x)$ can be expressed as*

$$p(x) = s(x)q(x) + r(x)$$

where $q(x)$ and $r(x)$ are the quotient and remainder that result when $p(x)$ is divided by $s(x)$, and either $r(x)$ is the zero polynomial or the degree of $r(x)$ is less than the degree of $s(x)$.

In the special case where $p(x)$ is divided by a first-degree polynomial of the form $x - c$, the remainder must be some constant r, since it is either zero or has degree less than 1. Thus, Theorem C.2 implies that

$$p(x) = (x - c)q(x) + r$$

and this in turn implies that $p(c) = r$. In summary, we have the following theorem.

C.3 **THEOREM** (*Remainder Theorem*) *If a polynomial $p(x)$ is divided by $x - c$, then the remainder is $p(c)$.*

▶ **Example 1** According to the Remainder Theorem, the remainder on dividing

$$p(x) = 2x^3 + 3x^2 - 4x - 3$$

by $x + 4$ should be

$$p(-4) = 2(-4)^3 + 3(-4)^2 - 4(-4) - 3 = -67$$

Show that this is so.

Solution. By long division

$$
\require{enclose}
\begin{array}{r}
2x^2 - 5x + 16 \\
x + 4 \enclose{longdiv}{2x^3 + 3x^2 - 4x - 3} \\
\underline{2x^3 + 8x^2 } \\
-5x^2 - 4x \\
\underline{-5x^2 - 20x } \\
16x - 3 \\
\underline{16x + 64} \\
-67
\end{array}
$$

which shows that the remainder is -67.

Alternative Solution. Because we are dividing by an expression of the form $x - c$ (where $c = -4$), we can use synthetic division rather than long division. The computations are

$$
\begin{array}{r|rrrr}
-4 & 2 & 3 & -4 & -3 \\
   &   & -8 & 20 & -64 \\
\hline
   & 2 & -5 & 16 & -67
\end{array}
$$

which again shows that the remainder is -67. ◀

■ THE FACTOR THEOREM

To *factor* a polynomial $p(x)$ is to write it as a product of lower-degree polynomials, called *factors* of $p(x)$. For $s(x)$ to be a factor of $p(x)$ there must be no remainder when $p(x)$ is divided by $s(x)$. For example, if $p(x)$ can be factored as

$$p(x) = s(x)q(x) \tag{5}$$

then

$$\frac{p(x)}{s(x)} = q(x) \tag{6}$$

so dividing $p(x)$ by $s(x)$ produces a quotient $q(x)$ with no remainder. Conversely, (6) implies (5), so $s(x)$ is a factor of $p(x)$ if there is no remainder when $p(x)$ is divided by $s(x)$.

In the special case where $x - c$ is a factor of $p(x)$, the polynomial $p(x)$ can be expressed as

$$p(x) = (x - c)q(x)$$

which implies that $p(c) = 0$. Conversely, if $p(c) = 0$, then the Remainder Theorem implies that $x - c$ is a factor of $p(x)$, since the remainder is 0 when $p(x)$ is divided by $x - c$. These results are summarized in the following theorem.

C.4 **THEOREM** (*Factor Theorem*) *A polynomial $p(x)$ has a factor $x - c$ if and only if $p(c) = 0$.*

It follows from this theorem that the statements below say the same thing in different ways:

- $x - c$ is a factor of $p(x)$.
- $p(c) = 0$.
- c is a zero of $p(x)$.
- c is a root of the equation $p(x) = 0$.
- c is a solution of the equation $p(x) = 0$.
- c is an x-intercept of $y = p(x)$.

▶ **Example 2** Confirm that $x - 1$ is a factor of

$$p(x) = x^3 - 3x^2 - 13x + 15$$

by dividing $x - 1$ into $p(x)$ and checking that the remainder is zero.

Solution. By long division

$$
\begin{array}{r}
x^2 - 2x - 15 \\
x - 1 \overline{\smash{\big)}\ x^3 - 3x^2 - 13x + 15} \\
\underline{x^3 - x^2} \\
-2x^2 - 13x \\
\underline{-2x^2 + 2x} \\
-15x + 15 \\
\underline{-15x + 15} \\
0
\end{array}
$$

which shows that the remainder is zero.

Alternative Solution.   Because we are dividing by an expression of the form $x - c$, we can use synthetic division rather than long division. The computations are

$$\begin{array}{r|rrrr} 1 & 1 & -3 & -13 & 15 \\ & & 1 & -2 & -15 \\ \hline & 1 & -2 & -15 & 0 \end{array}$$

which again confirms that the remainder is zero. ◄

■ USING ONE FACTOR TO FIND OTHER FACTORS

If $x - c$ is a factor of $p(x)$, and if $q(x) = p(x)/(x - c)$, then

$$p(x) = (x - c)q(x) \tag{7}$$

so that additional linear factors of $p(x)$ can be obtained by factoring the quotient $q(x)$.

▶ **Example 3**   Factor

$$p(x) = x^3 - 3x^2 - 13x + 15 \tag{8}$$

completely into linear factors.

Solution.   We showed in Example 2 that $x - 1$ is a factor of $p(x)$ and we also showed that $p(x)/(x - 1) = x^2 - 2x - 15$. Thus,

$$x^3 - 3x^2 - 13x + 15 = (x - 1)(x^2 - 2x - 15)$$

Factoring $x^2 - 2x - 15$ by inspection yields

$$x^3 - 3x^2 - 13x + 15 = (x - 1)(x - 5)(x + 3)$$

which is the complete linear factorization of $p(x)$. ◄

■ METHODS FOR FINDING ROOTS

A general quadratic equation $ax^2 + bx + c = 0$ can be solved by using the quadratic formula to express the solutions of the equation in terms of the coefficients. Versions of this formula were known since Babylonian times, and by the seventeenth century formulas had been obtained for solving general cubic and quartic equations. However, attempts to find formulas for the solutions of general fifth-degree equations and higher proved fruitless. The reason for this became clear in 1829 when the French mathematician Evariste Galois (1811–1832) proved that it is impossible to express the solutions of a general fifth-degree equation or higher in terms of its coefficients using algebraic operations.

Today, we have powerful computer programs for finding the zeros of specific polynomials. For example, it takes only seconds for a computer algebra system, such as *Mathematica* or *Maple*, to show that the zeros of the polynomial

$$p(x) = 10x^4 - 23x^3 - 10x^2 + 29x + 6 \tag{9}$$

are

$$x = -1, \quad x = -\tfrac{1}{5}, \quad x = \tfrac{3}{2}, \quad \text{and} \quad x = 2 \tag{10}$$

The algorithms that these programs use to find the integer and rational zeros of a polynomial, if any, are based on the following theorem, which is proved in advanced algebra courses.

C.5 THEOREM *Suppose that*

$$p(x) = c_n x^n + c_{n-1} x^{n-1} + \cdots + c_1 x + c_0$$

is a polynomial with integer coefficients.

(a) *If r is an integer zero of $p(x)$, then r must be a divisor of the constant term c_0.*

(b) *If $r = a/b$ is a rational zero of $p(x)$ in which all common factors of a and b have been canceled, then a must be a divisor of the constant term c_0, and b must be a divisor of the leading coefficient c_n.*

For example, in (9) the constant term is 6 (which has divisors ± 1, ± 2, ± 3, and ± 6) and the leading coefficient is 10 (which has divisors ± 1, ± 2, ± 5, and ± 10). Thus, the only possible integer zeros of $p(x)$ are

$$\pm 1, \quad \pm 2, \quad \pm 3, \quad \pm 6$$

and the only possible noninteger rational zeros are

$$\pm\tfrac{1}{2}, \quad \pm\tfrac{1}{5}, \quad \pm\tfrac{1}{10}, \quad \pm\tfrac{2}{5}, \quad \pm\tfrac{3}{2}, \quad \pm\tfrac{3}{5}, \quad \pm\tfrac{3}{10}, \quad \pm\tfrac{6}{5}$$

Using a computer, it is a simple matter to evaluate $p(x)$ at each of the numbers in these lists to show that its only rational zeros are the numbers in (10).

▶ **Example 4** Solve the equation $x^3 + 3x^2 - 7x - 21 = 0$.

Solution. The solutions of the equation are the zeros of the polynomial

$$p(x) = x^3 + 3x^2 - 7x - 21$$

We will look for integer zeros first. All such zeros must divide the constant term, so the only possibilities are ± 1, ± 3, ± 7, and ± 21. Substituting these values into $p(x)$ (or using the method of Exercise 6) shows that $x = -3$ is an integer zero. This tells us that $x + 3$ is a factor of $p(x)$ and that $p(x)$ can be written as

$$x^3 + 3x^2 - 7x - 21 = (x + 3)q(x)$$

where $q(x)$ is the quotient that results when $x^3 + 3x^2 - 7x - 21$ is divided by $x + 3$. We leave it for you to perform the division and show that $q(x) = x^2 - 7$; hence,

$$x^3 + 3x^2 - 7x - 21 = (x + 3)(x^2 - 7) = (x + 3)(x + \sqrt{7})(x - \sqrt{7})$$

which tells us that the solutions of the given equation are $x = 3$, $x = \sqrt{7} \approx 2.65$, and $x = -\sqrt{7} \approx -2.65$. ◀

EXERCISE SET C   [C] CAS

1–2 Find the quotient $q(x)$ and the remainder $r(x)$ that result when $p(x)$ is divided by $s(x)$. ■

1. (a) $p(x) = x^4 + 3x^3 - 5x + 10$; $s(x) = x^2 - x + 2$
   (b) $p(x) = 6x^4 + 10x^2 + 5$; $s(x) = 3x^2 - 1$
   (c) $p(x) = x^5 + x^3 + 1$; $s(x) = x^2 + x$

2. (a) $p(x) = 2x^4 - 3x^3 + 5x^2 + 2x + 7$; $s(x) = x^2 - x + 1$
   (b) $p(x) = 2x^5 + 5x^4 - 4x^3 + 8x^2 + 1$; $s(x) = 2x^2 - x + 1$
   (c) $p(x) = 5x^6 + 4x^2 + 5$; $s(x) = x^3 + 1$

3–4 Use synthetic division to find the quotient $q(x)$ and the remainder $r(x)$ that result when $p(x)$ is divided by $s(x)$. ■

3. (a) $p(x) = 3x^3 - 4x - 1$; $s(x) = x - 2$
   (b) $p(x) = x^4 - 5x^2 + 4$; $s(x) = x + 5$
   (c) $p(x) = x^5 - 1$; $s(x) = x - 1$

4. (a) $p(x) = 2x^3 - x^2 - 2x + 1$; $s(x) = x - 1$
   (b) $p(x) = 2x^4 + 3x^3 - 17x^2 - 27x - 9$; $s(x) = x + 4$
   (c) $p(x) = x^7 + 1$; $s(x) = x - 1$

5. Let $p(x) = 2x^4 + x^3 - 3x^2 + x - 4$. Use synthetic division and the Remainder Theorem to find $p(0)$, $p(1)$, $p(-3)$, and $p(7)$.

6. Let $p(x)$ be the polynomial in Example 4. Use synthetic division and the Remainder Theorem to evaluate $p(x)$ at $x = \pm 1, \pm 3, \pm 7,$ and ± 21.

7. Let $p(x) = x^3 + 4x^2 + x - 6$. Find a polynomial $q(x)$ and a constant r such that
   (a) $p(x) = (x - 2)q(x) + r$
   (b) $p(x) = (x + 1)q(x) + r$.

8. Let $p(x) = x^5 - 1$. Find a polynomial $q(x)$ and a constant r such that
   (a) $p(x) = (x + 1)q(x) + r$
   (b) $p(x) = (x - 1)q(x) + r$.

9. In each part, make a list of all possible candidates for the rational zeros of $p(x)$.
   (a) $p(x) = x^7 + 3x^3 - x + 24$
   (b) $p(x) = 3x^4 - 2x^2 + 7x - 10$
   (c) $p(x) = x^{35} - 17$

10. Find all integer zeros of
$$p(x) = x^6 + 5x^5 - 16x^4 - 15x^3 - 12x^2 - 38x - 21$$

11–15 Factor the polynomials completely. ■

11. $p(x) = x^3 - 2x^2 - x + 2$

12. $p(x) = 3x^3 + x^2 - 12x - 4$

13. $p(x) = x^4 + 10x^3 + 36x^2 + 54x + 27$

14. $p(x) = 2x^4 + x^3 + 3x^2 + 3x - 9$

15. $p(x) = x^5 + 4x^4 - 4x^3 - 34x^2 - 45x - 18$

c 16. For each of the factorizations that you obtained in Exercises 11–15, check your answer using a CAS.

17–21 Find all real solutions of the equations. ■

17. $x^3 + 3x^2 + 4x + 12 = 0$

18. $2x^3 - 5x^2 - 10x + 3 = 0$

19. $3x^4 + 14x^3 + 14x^2 - 8x - 8 = 0$

20. $2x^4 - x^3 - 14x^2 - 5x + 6 = 0$

21. $x^5 - 2x^4 - 6x^3 + 5x^2 + 8x + 12 = 0$

c 22. For each of the equations you solved in Exercises 17–21, check your answer using a CAS.

23. Find all values of k for which $x - 1$ is a factor of the polynomial $p(x) = k^2x^3 - 7kx + 10$.

24. Is $x + 3$ a factor of $x^7 + 2187$? Justify your answer.

c 25. A 3 cm thick slice is cut from a cube, leaving a volume of 196 cm^3. Use a CAS to find the length of a side of the original cube.

26. (a) Show that there is no positive rational number that exceeds its cube by 1.
    (b) Does there exist a real number that exceeds its cube by 1? Justify your answer.

27. Use the Factor Theorem to show each of the following.
    (a) $x - y$ is a factor of $x^n - y^n$ for all positive integer values of n.
    (b) $x + y$ is a factor of $x^n - y^n$ for all positive even integer values of n.
    (c) $x + y$ is a factor of $x^n + y^n$ for all positive odd integer values of n.

SELECTED PROOFS

■ PROOFS OF BASIC LIMIT THEOREMS

An extensive excursion into proofs of limit theorems would be too time consuming to undertake, so we have selected a few proofs of results from Section 1.2 that illustrate some of the basic ideas.

D.1 THEOREM *Let a be any real number, let k be a constant, and suppose that* $\lim_{x \to a} f(x) = L_1$ *and that* $\lim_{x \to a} g(x) = L_2$. *Then:*

(a) $\displaystyle\lim_{x \to a} k = k$

(b) $\displaystyle\lim_{x \to a} [f(x) + g(x)] = \lim_{x \to a} f(x) + \lim_{x \to a} g(x) = L_1 + L_2$

(c) $\displaystyle\lim_{x \to a} [f(x)g(x)] = \left(\lim_{x \to a} f(x)\right)\left(\lim_{x \to a} g(x)\right) = L_1 L_2$

PROOF (a) We will apply Definition 1.4.1 with $f(x) = k$ and $L = k$. Thus, given $\epsilon > 0$, we must find a number $\delta > 0$ such that

$$|k - k| < \epsilon \quad \text{if} \quad 0 < |x - a| < \delta$$

or, equivalently,

$$0 < \epsilon \quad \text{if} \quad 0 < |x - a| < \delta$$

But the condition on the left side of this statement is *always* true, no matter how δ is chosen. Thus, any positive value for δ will suffice.

PROOF (b) We must show that given $\epsilon > 0$ we can find a number $\delta > 0$ such that

$$|(f(x) + g(x)) - (L_1 + L_2)| < \epsilon \quad \text{if} \quad 0 < |x - a| < \delta \tag{1}$$

However, from the limits of f and g in the hypothesis of the theorem we can find numbers δ_1 and δ_2 such that

$$|f(x) - L_1| < \epsilon/2 \quad \text{if} \quad 0 < |x - a| < \delta_1$$
$$|g(x) - L_2| < \epsilon/2 \quad \text{if} \quad 0 < |x - a| < \delta_2$$

Moreover, the inequalities on the left sides of these statements *both* hold if we replace δ_1 and δ_2 by any positive number δ that is less than both δ_1 and δ_2. Thus, for any such δ it follows that

$$|f(x) - L_1| + |g(x) - L_2| < \epsilon \quad \text{if} \quad 0 < |x - a| < \delta \tag{2}$$

However, it follows from the triangle inequality [Theorem F.5 of Appendix F] that
$$|(f(x) + g(x)) - (L_1 + L_2)| = |(f(x) - L_1) + (g(x) - L_2)|$$
$$\leq |f(x) - L_1| + |g(x) - L_2|$$
so that (1) follows from (2).

PROOF (c) We must show that given $\epsilon > 0$ we can find a number $\delta > 0$ such that
$$|f(x)g(x) - L_1 L_2| < \epsilon \quad \text{if} \quad 0 < |x - a| < \delta \qquad (3)$$
To find δ it will be helpful to express (3) in a different form. If we rewrite $f(x)$ and $g(x)$ as
$$f(x) = L_1 + (f(x) - L_1) \quad \text{and} \quad g(x) = L_2 + (g(x) - L_2)$$
then the inequality on the left side of (3) can be expressed as (verify)
$$|L_1(g(x) - L_2) + L_2(f(x) - L_1) + (f(x) - L_1)(g(x) - L_2)| < \epsilon \qquad (4)$$
Since
$$\lim_{x \to a} f(x) = L_1 \quad \text{and} \quad \lim_{x \to a} g(x) = L_2$$
we can find positive numbers $\delta_1, \delta_2, \delta_3$, and δ_4 such that

$$|f(x) - L_1| < \sqrt{\epsilon/3} \qquad \text{if} \quad 0 < |x - a| < \delta_1$$
$$|f(x) - L_1| < \frac{\epsilon}{3(1 + |L_2|)} \qquad \text{if} \quad 0 < |x - a| < \delta_2$$
$$|g(x) - L_2| < \sqrt{\epsilon/3} \qquad \text{if} \quad 0 < |x - a| < \delta_3 \qquad (5)$$
$$|g(x) - L_2| < \frac{\epsilon}{3(1 + |L_1|)} \qquad \text{if} \quad 0 < |x - a| < \delta_4$$

Moreover, the inequalities on the left sides of these four statements *all* hold if we replace $\delta_1, \delta_2, \delta_3$, and δ_4 by any positive number δ that is smaller than $\delta_1, \delta_2, \delta_3$, and δ_4. Thus, for any such δ it follows with the help of the triangle inequality that

$$|L_1(g(x) - L_2) + L_2(f(x) - L_1) + (f(x) - L_1)(g(x) - L_2)|$$
$$\leq |L_1(g(x) - L_2)| + |L_2(f(x) - L_1)| + |(f(x) - L_1)(g(x) - L_2)|$$
$$= |L_1||g(x) - L_2| + |L_2||f(x) - L_1| + |f(x) - L_1||g(x) - L_2|$$
$$< |L_1|\frac{\epsilon}{3(1 + |L_1|)} + |L_2|\frac{\epsilon}{3(1 + |L_2|)} + \sqrt{\epsilon/3}\sqrt{\epsilon/3} \qquad \boxed{\text{From (5)}}$$
$$= \frac{\epsilon}{3}\frac{|L_1|}{1 + |L_1|} + \frac{\epsilon}{3}\frac{|L_2|}{1 + |L_2|} + \frac{\epsilon}{3}$$
$$< \frac{\epsilon}{3} + \frac{\epsilon}{3} + \frac{\epsilon}{3} = \epsilon \qquad \boxed{\text{Since } \frac{|L_1|}{1 + |L_1|} < 1 \text{ and } \frac{|L_2|}{1 + |L_2|} < 1}$$

which shows that (4) holds for the δ selected. ∎

Do not be alarmed if the proof of part (c) seems difficult; it takes some experience with proofs of this type to develop a feel for choosing a valid δ. Your initial goal should be to understand the ideas and the computations.

PROOF OF A BASIC CONTINUITY PROPERTY
Next we will prove Theorem 1.5.5 for two-sided limits.

D.2 THEOREM (Theorem 1.5.5) *If $\lim_{x \to c} g(x) = L$ and if the function f is continuous at L, then $\lim_{x \to c} f(g(x)) = f(L)$. That is,*
$$\lim_{x \to c} f(g(x)) = f\left(\lim_{x \to c} g(x)\right)$$

PROOF   We must show that given $\epsilon > 0$, we can find a number $\delta > 0$ such that

$$|f(g(x)) - f(L)| < \epsilon \quad \text{if} \quad 0 < |x - c| < \delta \tag{6}$$

Since f is continuous at L, we have

$$\lim_{u \to L} f(u) = f(L)$$

and hence we can find a number $\delta_1 > 0$ such that

$$|f(u) - f(L)| < \epsilon \quad \text{if} \quad |u - L| < \delta_1$$

In particular, if $u = g(x)$, then

$$|f(g(x)) - f(L)| < \epsilon \quad \text{if} \quad |g(x) - L| < \delta_1 \tag{7}$$

But $\lim_{x \to c} g(x) = L$, and hence there is a number $\delta > 0$ such that

$$|g(x) - L| < \delta_1 \quad \text{if} \quad 0 < |x - c| < \delta \tag{8}$$

Thus, if x satisfies the condition on the right side of statement (8), then it follows that $g(x)$ satisfies the condition on the right side of statement (7), and this implies that the condition on the left side of statement (6) is satisfied, completing the proof. ∎

■ PROOF OF THE CHAIN RULE

Next we will prove the chain rule (Theorem 2.6.1), but first we need a preliminary result.

D.3   THEOREM   *If f is differentiable at x and if $y = f(x)$, then*

$$\Delta y = f'(x)\Delta x + \epsilon \Delta x$$

where $\epsilon \to 0$ as $\Delta x \to 0$ and $\epsilon = 0$ if $\Delta x = 0$.

PROOF   Define

$$\epsilon = \begin{cases} \dfrac{f(x + \Delta x) - f(x)}{\Delta x} - f'(x) & \text{if } \Delta x \neq 0 \\[2mm] 0 & \text{if } \Delta x = 0 \end{cases} \tag{9}$$

If $\Delta x \neq 0$, it follows from (9) that

$$\epsilon \Delta x = [f(x + \Delta x) - f(x)] - f'(x)\Delta x \tag{10}$$

But

$$\Delta y = f(x + \Delta x) - f(x) \tag{11}$$

so (10) can be written as

$$\epsilon \Delta x = \Delta y - f'(x)\Delta x$$

or

$$\Delta y = f'(x)\Delta x + \epsilon \Delta x \tag{12}$$

If $\Delta x = 0$, then (12) still holds (why?), so (12) is valid for all values of Δx. It remains to show that $\epsilon \to 0$ as $\Delta x \to 0$. But this follows from the assumption that f is differentiable at x, since

$$\lim_{\Delta x \to 0} \epsilon = \lim_{\Delta x \to 0} \left[\frac{f(x + \Delta x) - f(x)}{\Delta x} - f'(x) \right] = f'(x) - f'(x) = 0 \; ∎$$

We are now ready to prove the chain rule.

D.4 **THEOREM** *(Theorem 2.6.1)* *If g is differentiable at the point x and f is differentiable at the point $g(x)$, then the composition $f \circ g$ is differentiable at the point x. Moreover, if $y = f(g(x))$ and $u = g(x)$, then*

$$\frac{dy}{dx} = \frac{dy}{du} \cdot \frac{du}{dx}$$

PROOF Since g is differentiable at x and $u = g(x)$, it follows from Theorem D.3 that

$$\Delta u = g'(x)\Delta x + \epsilon_1 \Delta x \tag{13}$$

where $\epsilon_1 \to 0$ as $\Delta x \to 0$. And since $y = f(u)$ is differentiable at $u = g(x)$, it follows from Theorem D.3 that

$$\Delta y = f'(u)\Delta u + \epsilon_2 \Delta u \tag{14}$$

where $\epsilon_2 \to 0$ as $\Delta u \to 0$.

Factoring out the Δu in (14) and then substituting (13) yields

$$\Delta y = [f'(u) + \epsilon_2][g'(x)\Delta x + \epsilon_1 \Delta x]$$

or

$$\Delta y = [f'(u) + \epsilon_2][g'(x) + \epsilon_1]\Delta x$$

or if $\Delta x \neq 0$,

$$\frac{\Delta y}{\Delta x} = [f'(u) + \epsilon_2][g'(x) + \epsilon_1] \tag{15}$$

But (13) implies that $\Delta u \to 0$ as $\Delta x \to 0$, and hence $\epsilon_1 \to 0$ and $\epsilon_2 \to 0$ as $\Delta x \to 0$. Thus, from (15)

$$\lim_{\Delta x \to 0} \frac{\Delta y}{\Delta x} = f'(u)g'(x)$$

or

$$\frac{dy}{dx} = f'(u)g'(x) = \frac{dy}{du} \cdot \frac{du}{dx} \quad \blacksquare$$

■ PROOF THAT RELATIVE EXTREMA OCCUR AT CRITICAL POINTS

In this subsection we will prove Theorem 4.2.2, which states that the relative extrema of a function occur at critical points.

D.5 **THEOREM** *(Theorem 4.2.2)* *Suppose that f is a function defined on an open interval containing the point x_0. If f has a relative extremum at $x = x_0$, then $x = x_0$ is a critical point of f; that is, either $f'(x_0) = 0$ or f is not differentiable at x_0.*

PROOF Suppose that f has a relative maximum at x_0. There are two possibilities—either f is differentiable at x_0 or it is not. If it is not, then x_0 is a critical point for f and we are done. If f is differentiable at x_0, then we must show that $f'(x_0) = 0$. We will do this by showing that $f'(x_0) \geq 0$ and $f'(x_0) \leq 0$, from which it follows that $f'(x_0) = 0$. From the definition of a derivative we have

$$f'(x_0) = \lim_{h \to 0} \frac{f(x_0 + h) - f(x_0)}{h}$$

so that

$$f'(x_0) = \lim_{h \to 0^+} \frac{f(x_0 + h) - f(x_0)}{h} \tag{16}$$

and

$$f'(x_0) = \lim_{h \to 0^-} \frac{f(x_0 + h) - f(x_0)}{h} \tag{17}$$

Because f has a relative maximum at x_0, there is an open interval (a, b) containing x_0 in which $f(x) \leq f(x_0)$ for all x in (a, b).

Assume that h is sufficiently small so that $x_0 + h$ lies in the interval (a, b). Thus,

$$f(x_0 + h) \leq f(x_0) \quad \text{or equivalently} \quad f(x_0 + h) - f(x_0) \leq 0$$

Thus, if h is negative,

$$\frac{f(x_0 + h) - f(x_0)}{h} \geq 0 \tag{18}$$

and if h is positive,

$$\frac{f(x_0 + h) - f(x_0)}{h} \leq 0 \tag{19}$$

But an expression that never assumes negative values cannot approach a negative limit and an expression that never assumes positive values cannot approach a positive limit, so that

$$f'(x_0) = \lim_{h \to 0^-} \frac{f(x_0 + h) - f(x_0)}{h} \geq 0 \qquad \boxed{\text{From (17) and (18)}}$$

and

$$f'(x_0) = \lim_{h \to 0^+} \frac{f(x_0 + h) - f(x_0)}{h} \leq 0 \qquad \boxed{\text{From (16) and (19)}}$$

Since $f'(x_0) \geq 0$ and $f'(x_0) \leq 0$, it must be that $f'(x_0) = 0$.

A similar argument applies if f has a relative minimum at x_0. ∎

■ PROOFS OF TWO SUMMATION FORMULAS

We will prove parts (a) and (b) of Theorem 5.4.2. The proof of part (c) is similar to that of part (b) and is omitted.

D.6 **THEOREM** (*Theorem 5.4.2*)

(a) $\displaystyle\sum_{k=1}^{n} k = 1 + 2 + \cdots + n = \frac{n(n+1)}{2}$

(b) $\displaystyle\sum_{k=1}^{n} k^2 = 1^2 + 2^2 + \cdots + n^2 = \frac{n(n+1)(2n+1)}{6}$

(c) $\displaystyle\sum_{k=1}^{n} k^3 = 1^3 + 2^3 + \cdots + n^3 = \left[\frac{n(n+1)}{2}\right]^2$

PROOF (a)   Writing

$$\sum_{k=1}^{n} k$$

two ways, with summands in increasing order and in decreasing order, and then adding, we obtain

$$\sum_{k=1}^{n} k = \quad 1 \quad + \quad 2 \quad + \quad 3 \quad + \cdots + (n-2) + (n-1) + \quad n$$

$$\sum_{k=1}^{n} k = \quad n \quad + (n-1) + (n-2) + \cdots + \quad 3 \quad + \quad 2 \quad + \quad 1$$

$$2\sum_{k=1}^{n} k = (n+1) + (n+1) + (n+1) + \cdots + (n+1) + (n+1) + (n+1)$$

$$= n(n+1)$$

Thus,

$$\sum_{k=1}^{n} k = \frac{n(n+1)}{2}$$

PROOF (b)   Note that

$$(k+1)^3 - k^3 = k^3 + 3k^2 + 3k + 1 - k^3 = 3k^2 + 3k + 1$$

So,

$$\sum_{k=1}^{n}[(k+1)^3 - k^3] = \sum_{k=1}^{n}(3k^2 + 3k + 1) \tag{20}$$

Writing out the left side of (20) with the index running *down* from $k = n$ to $k = 1$, we have

$$\sum_{k=1}^{n}[(k+1)^3 - k^3] = [(n+1)^3 - n^3] + \cdots + [4^3 - 3^3] + [3^3 - 2^3] + [2^3 - 1^3]$$

$$= (n+1)^3 - 1 \tag{21}$$

Combining (21) and (20), and expanding the right side of (20) by using Theorem 5.4.1 and part (a) of this theorem yields

$$(n+1)^3 - 1 = 3\sum_{k=1}^{n}k^2 + 3\sum_{k=1}^{n}k + \sum_{k=1}^{n}1$$

$$= 3\sum_{k=1}^{n}k^2 + 3\frac{n(n+1)}{2} + n$$

So,

$$3\sum_{k=1}^{n}k^2 = [(n+1)^3 - 1] - 3\frac{n(n+1)}{2} - n$$

$$= (n+1)^3 - 3(n+1)\left(\frac{n}{2}\right) - (n+1)$$

$$= \frac{n+1}{2}[2(n+1)^2 - 3n - 2]$$

$$= \frac{n+1}{2}[2n^2 + n] = \frac{n(n+1)(2n+1)}{2}$$

Thus,

$$\sum_{k=1}^{n}k^2 = \frac{n(n+1)(2n+1)}{6} \quad \blacksquare$$

The sum in (21) is an example of a *telescoping sum*, since the cancellation of each of the two parts of an interior summand with parts of its neighboring summands allows the entire sum to collapse like a telescope.

■ PROOF OF THE LIMIT COMPARISON TEST

D.7   THEOREM *(Theorem 9.5.4)*   *Let $\sum a_k$ and $\sum b_k$ be series with positive terms and suppose that*

$$\rho = \lim_{k \to +\infty} \frac{a_k}{b_k}$$

If ρ is finite and $\rho > 0$, then the series both converge or both diverge.

PROOF We need only show that $\sum b_k$ converges when $\sum a_k$ converges and that $\sum b_k$ diverges when $\sum a_k$ diverges, since the remaining cases are logical implications of these (why?). The idea of the proof is to apply the comparison test to $\sum a_k$ and suitable multiples of $\sum b_k$. For this purpose let ϵ be any positive number. Since

$$\rho = \lim_{k \to +\infty} \frac{a_k}{b_k}$$

it follows that eventually the terms in the sequence $\{a_k/b_k\}$ must be within ϵ units of ρ; that is, there is a positive integer K such that for $k \geq K$ we have

$$\rho - \epsilon < \frac{a_k}{b_k} < \rho + \epsilon$$

In particular, if we take $\epsilon = \rho/2$, then for $k \geq K$ we have

$$\frac{1}{2}\rho < \frac{a_k}{b_k} < \frac{3}{2}\rho \quad \text{or} \quad \frac{1}{2}\rho b_k < a_k < \frac{3}{2}\rho b_k$$

Thus, by the comparison test we can conclude that

$$\sum_{k=K}^{\infty} \frac{1}{2}\rho b_k \quad \text{converges if} \quad \sum_{k=K}^{\infty} a_k \quad \text{converges} \tag{22}$$

$$\sum_{k=K}^{\infty} \frac{3}{2}\rho b_k \quad \text{diverges if} \quad \sum_{k=K}^{\infty} a_k \quad \text{diverges} \tag{23}$$

But the convergence or divergence of a series is not affected by deleting finitely many terms or by multiplying the general term by a nonzero constant, so (22) and (23) imply that

$$\sum_{k=1}^{\infty} b_k \quad \text{converges if} \quad \sum_{k=1}^{\infty} a_k \quad \text{converges}$$

$$\sum_{k=1}^{\infty} b_k \quad \text{diverges if} \quad \sum_{k=1}^{\infty} a_k \quad \text{diverges} \quad \blacksquare$$

■ PROOF OF THE RATIO TEST

D.8 **THEOREM** (*Theorem 9.5.5*) *Let $\sum u_k$ be a series with positive terms and suppose that*

$$\rho = \lim_{k \to +\infty} \frac{u_{k+1}}{u_k}$$

(a) *If $\rho < 1$, the series converges.*

(b) *If $\rho > 1$ or $\rho = +\infty$, the series diverges.*

(c) *If $\rho = 1$, the series may converge or diverge, so that another test must be tried.*

PROOF (a) The number ρ must be nonnegative since it is the limit of u_{k+1}/u_k, which is positive for all k. In this part of the proof we assume that $\rho < 1$, so that $0 \leq \rho < 1$.

We will prove convergence by showing that the terms of the given series are eventually less than the terms of a convergent geometric series. For this purpose, choose any real number r such that $0 < \rho < r < 1$. Since the limit of u_{k+1}/u_k is ρ, and $\rho < r$, the terms of the sequence $\{u_{k+1}/u_k\}$ must eventually be less than r. Thus, there is a positive integer K such that for $k \geq K$ we have

$$\frac{u_{k+1}}{u_k} < r \quad \text{or} \quad u_{k+1} < r u_k$$

This yields the inequalities

$$u_{K+1} < r u_K$$
$$u_{K+2} < r u_{K+1} < r^2 u_K$$
$$u_{K+3} < r u_{K+2} < r^3 u_K \qquad (24)$$
$$u_{K+4} < r u_{K+3} < r^4 u_K$$
$$\vdots$$

But $0 < r < 1$, so

$$r u_K + r^2 u_K + r^3 u_K + \cdots$$

is a convergent geometric series. From the inequalities in (24) and the comparison test it follows that

$$u_{K+1} + u_{K+2} + u_{K+3} + \cdots$$

must also be a convergent series. Thus, $u_1 + u_2 + u_3 + \cdots + u_k + \cdots$ converges by Theorem 9.4.3(c).

PROOF (b) In this part we will prove divergence by showing that the limit of the general term is not zero. Since the limit of u_{k+1}/u_k is ρ and $\rho > 1$, the terms in the sequence $\{u_{k+1}/u_k\}$ must eventually be greater than 1. Thus, there is a positive integer K such that for $k \geq K$ we have

$$\frac{u_{k+1}}{u_k} > 1 \quad \text{or} \quad u_{k+1} > u_k$$

This yields the inequalities

$$u_{K+1} > u_K$$
$$u_{K+2} > u_{K+1} > u_K$$
$$u_{K+3} > u_{K+2} > u_K \qquad (25)$$
$$u_{K+4} > u_{K+3} > u_K$$
$$\vdots$$

Since $u_K > 0$, it follows from the inequalities in (25) that $\lim_{k \to +\infty} u_k \neq 0$, and thus the series $u_1 + u_2 + \cdots + u_k + \cdots$ diverges by part (a) of Theorem 9.4.1. The proof in the case where $\rho = +\infty$ is omitted.

PROOF (c) The divergent harmonic series and the convergent p-series with $p = 2$ both have $\rho = 1$ (verify), so the ratio test does not distinguish between convergence and divergence when $\rho = 1$. ∎

■ PROOF OF THE REMAINDER ESTIMATION THEOREM

> **D.9 THEOREM (Theorem 9.7.4)** *If the function f can be differentiated $n + 1$ times on an interval containing the number x_0, and if M is an upper bound for $|f^{(n+1)}(x)|$ on the interval, that is, $|f^{(n+1)}(x)| \leq M$ for all x in the interval, then*
>
> $$|R_n(x)| \leq \frac{M}{(n+1)!} |x - x_0|^{n+1}$$
>
> *for all x in the interval.*

PROOF We are assuming that f can be differentiated $n + 1$ times on an interval containing the number x_0 and that

$$|f^{(n+1)}(x)| \leq M \qquad (26)$$

for all x in the interval. We want to show that

$$|R_n(x)| \leq \frac{M}{(n+1)!} |x - x_0|^{n+1} \qquad (27)$$

for all x in the interval, where

$$R_n(x) = f(x) - \sum_{k=0}^{n} \frac{f^{(k)}(x_0)}{k!}(x - x_0)^k \tag{28}$$

In our proof we will need the following two properties of $R_n(x)$:

$$R_n(x_0) = R_n'(x_0) = \cdots = R_n^{(n)}(x_0) = 0 \tag{29}$$

$$R_n^{(n+1)}(x) = f^{(n+1)}(x) \quad \text{for all } x \text{ in the interval} \tag{30}$$

These properties can be obtained by analyzing what happens if the expression for $R_n(x)$ in Formula (28) is differentiated j times and x_0 is then substituted in that derivative. If $j < n$, then the jth derivative of the summation in Formula (28) consists of a constant term $f^{(j)}(x_0)$ plus terms involving powers of $x - x_0$ (verify). Thus, $R_n^{(j)}(x_0) = 0$ for $j < n$, which proves all but the last equation in (29). For the last equation, observe that the nth derivative of the summation in (28) is the constant $f^{(n)}(x_0)$, so $R_n^{(n)}(x_0) = 0$. Formula (30) follows from the observation that the $(n + 1)$-st derivative of the summation in (28) is zero (why?).

Now to the main part of the proof. For simplicity we will give the proof for the case where $x \geq x_0$ and leave the case where $x < x_0$ for the reader. It follows from (26) and (30) that $|R_n^{(n+1)}(x)| \leq M$, and hence

$$-M \leq R_n^{(n+1)}(x) \leq M$$

Thus,

$$\int_{x_0}^{x} -M \, dt \leq \int_{x_0}^{x} R_n^{(n+1)}(t) \, dt \leq \int_{x_0}^{x} M \, dt \tag{31}$$

However, it follows from (29) that $R_n^{(n)}(x_0) = 0$, so

$$\int_{x_0}^{x} R_n^{(n+1)}(t) \, dt = R_n^{(n)}(t) \Big]_{x_0}^{x} = R_n^{(n)}(x)$$

Thus, performing the integrations in (31) we obtain the inequalities

$$-M(x - x_0) \leq R_n^{(n)}(x) \leq M(x - x_0)$$

Now we will integrate again. Replacing x by t in these inequalities, integrating from x_0 to x, and using $R_n^{(n-1)}(x_0) = 0$ yields

$$-\frac{M}{2}(x - x_0)^2 \leq R_n^{(n-1)}(x) \leq \frac{M}{2}(x - x_0)^2$$

If we keep repeating this process, then after $n + 1$ integrations we will obtain

$$-\frac{M}{(n + 1)!}(x - x_0)^{n+1} \leq R_n(x) \leq \frac{M}{(n + 1)!}(x - x_0)^{n+1}$$

which we can rewrite as

$$|R_n(x)| \leq \frac{M}{(n + 1)!}(x - x_0)^{n+1}$$

This completes the proof of (27), since the absolute value signs can be omitted in that formula when $x \geq x_0$ (which is the case we are considering). ∎

■ PROOF OF THE EQUALITY OF MIXED PARTIALS

D.10 **THEOREM** (*Theorem 13.3.2*) *Let f be a function of two variables. If f_{xy} and f_{yx} are continuous on some open disk, then $f_{xy} = f_{yx}$ on that disk.*

PROOF Suppose that f is a function of two variables with f_{xy} and f_{yx} both continuous on some open disk. Let (x, y) be a point in that disk and define the function

$$w(\Delta x, \Delta y) = f(x + \Delta x, y + \Delta y) - f(x + \Delta x, y) - f(x, y + \Delta y) + f(x, y)$$

Now fix y and Δy and let

$$g(x) = f(x, y + \Delta y) - f(x, y)$$

so that

$$w(\Delta x, \Delta y) = g(x + \Delta x) - g(x) \tag{32}$$

Since f is differentiable on an open disk containing (x, y), the function g will be differentiable on some interval containing x and $x + \Delta x$ for Δx small enough. The Mean-Value Theorem then applies to g on this interval, and thus there is a c between x and Δx with

$$g(x + \Delta x) - g(x) = g'(c)\Delta x$$

But

$$g'(c) = f_x(c, y + \Delta y) - f_x(c, y)$$

so from Equation (32)

$$w(\Delta x, \Delta y) = g(x + \Delta x) - g(x) = g'(c)\Delta x = (f_x(c, y + \Delta y) - f_x(c, y))\Delta x \tag{33}$$

Now let $h(y) = f_x(c, y)$. Since f_x is differentiable on an open disk containing (x, y), h will be differentiable on some interval containing y and $y + \Delta y$ for Δy small enough. Applying the Mean-Value Theorem to h on this interval gives a d between y and $y + \Delta y$ with

$$h(y + \Delta y) - h(y) = h'(d)\Delta y$$

But $h'(d) = f_{xy}(c, d)$, so by (33) and the definition of h we have

$$\begin{aligned} w(\Delta x, \Delta y) &= (f_x(c, y + \Delta y) - f_x(c, y))\Delta x \\ &= (h(y + \Delta y) - h(y))\Delta x = h'(d)\Delta y\Delta x \\ &= f_{xy}(c, d)\Delta y\Delta x \end{aligned}$$

and

$$f_{xy}(c, d) = \frac{w(\Delta x, \Delta y)}{\Delta y\Delta x} \tag{34}$$

Since c lies between x and Δx and d lies between y and Δy, (c, d) approaches (x, y) as $(\Delta x, \Delta y)$ approaches $(0, 0)$. It then follows from the continuity of f_{xy} and (34) that

$$f_{xy}(x, y) = \lim_{(\Delta x, \Delta y) \to (0,0)} f_{xy}(c, d) = \lim_{(\Delta x, \Delta y) \to (0,0)} \frac{w(\Delta x, \Delta y)}{\Delta y\Delta x}$$

In similar fashion to the above argument, it can be shown that

$$f_{yx}(x, y) = \lim_{(\Delta x, \Delta y) \to (0,0)} \frac{w(\Delta x, \Delta y)}{\Delta y\Delta x}$$

and the result follows. ■

■ PROOF OF THE TWO-VARIABLE CHAIN RULE FOR DERIVATIVES

D.11 **THEOREM** (*Theorem 13.5.1*)   *If $x = x(t)$ and $y = y(t)$ are differentiable at t, and if $z = f(x, y)$ is differentiable at the point $(x(t), y(t))$, then $z = f(x(t), y(t))$ is differentiable at t and*

$$\frac{dz}{dt} = \frac{\partial z}{\partial x}\frac{dx}{dt} + \frac{\partial z}{\partial y}\frac{dy}{dt}$$

PROOF Let Δx, Δy, and Δz denote the changes in x, y, and z, respectively, that correspond to a change of Δt in t. Then

$$\frac{dz}{dt} = \lim_{\Delta t \to 0} \frac{\Delta z}{\Delta t}, \quad \frac{dx}{dt} = \lim_{\Delta t \to 0} \frac{\Delta x}{\Delta t}, \quad \frac{dy}{dt} = \lim_{\Delta t \to 0} \frac{\Delta y}{\Delta t}$$

Since $f(x, y)$ is differentiable at $(x(t), y(t))$, it follows from (5) in Section 13.4 that

$$\Delta z = \frac{\partial z}{\partial x} \Delta x + \frac{\partial z}{\partial y} \Delta y + \epsilon(\Delta x, \Delta y)\sqrt{(\Delta x)^2 + (\Delta y)^2} \tag{35}$$

where the partial derivatives are evaluated at $(x(t), y(t))$ and where $\epsilon(\Delta x, \Delta y)$ satisfies $\epsilon(\Delta x, \Delta y) \to 0$ as $(\Delta x, \Delta y) \to (0, 0)$ and $\epsilon(0, 0) = 0$. Dividing both sides of (35) by Δt yields

$$\frac{\Delta z}{\Delta t} = \frac{\partial z}{\partial x}\frac{\Delta x}{\Delta t} + \frac{\partial z}{\partial y}\frac{\Delta y}{\Delta t} + \frac{\epsilon(\Delta x, \Delta y)\sqrt{(\Delta x)^2 + (\Delta y)^2}}{\Delta t} \tag{36}$$

Since

$$\lim_{\Delta t \to 0} \frac{\sqrt{(\Delta x)^2 + (\Delta y)^2}}{|\Delta t|} = \lim_{\Delta t \to 0} \sqrt{\left(\frac{\Delta x}{\Delta t}\right)^2 + \left(\frac{\Delta y}{\Delta t}\right)^2} = \sqrt{\left(\lim_{\Delta t \to 0} \frac{\Delta x}{\Delta t}\right)^2 + \left(\lim_{\Delta t \to 0} \frac{\Delta y}{\Delta t}\right)^2}$$

$$= \sqrt{\left(\frac{dx}{dt}\right)^2 + \left(\frac{dy}{dt}\right)^2}$$

we have

$$\lim_{\Delta t \to 0} \left| \frac{\epsilon(\Delta x, \Delta y)\sqrt{(\Delta x)^2 + (\Delta y)^2}}{\Delta t} \right| = \lim_{\Delta t \to 0} \frac{|\epsilon(\Delta x, \Delta y)|\sqrt{(\Delta x)^2 + (\Delta y)^2}}{|\Delta t|}$$

$$= \lim_{\Delta t \to 0} |\epsilon(\Delta x, \Delta y)| \cdot \lim_{\Delta t \to 0} \frac{\sqrt{(\Delta x)^2 + (\Delta y)^2}}{|\Delta t|}$$

$$= 0 \cdot \sqrt{\left(\frac{dx}{dt}\right)^2 + \left(\frac{dy}{dt}\right)^2} = 0$$

Therefore,

$$\lim_{\Delta t \to 0} \frac{\epsilon(\Delta x, \Delta y)\sqrt{(\Delta x)^2 + (\Delta y)^2}}{\Delta t} = 0$$

Taking the limit as $\Delta t \to 0$ of both sides of (36) then yields the equation

$$\frac{dz}{dt} = \frac{\partial z}{\partial x}\frac{dx}{dt} + \frac{\partial z}{\partial y}\frac{dy}{dt} \quad \blacksquare$$

ANSWERS TO ODD-NUMBERED EXERCISES

► **Exercise Set 0.1 (Page 12)**

1. (a) $-2.9, -2.0, 2.35, 2.9$ (b) none (c) $y = 0$ (d) $-1.75 \le x \le 2.15$ (e) $y_{max} = 2.8$ at $x = -2.6$; $y_{min} = -2.2$ at $x = 1.2$

3. (a) yes (b) yes (c) no (d) no

5. (a) 1999, about \$47,700 (b) 1993, \$41,600 (c) first year

7. (a) $-2; 10; 10; 25; 4; 27t^2 - 2$ (b) $0; 4; -4; 6; 2\sqrt{2}$; $f(3t) = 1/3t$ for $t > 1$ and $f(3t) = 6t$ for $t \le 1$

9. (a) domain: $x \ne 3$; range: $y \ne 0$ (b) domain: $x \ne 0$; range: $\{-1, 1\}$ (c) domain: $x \le -\sqrt{3}$ or $x \ge \sqrt{3}$; range: $y \ge 0$
(d) domain: $-\infty < x < +\infty$; range: $y \ge 2$
(e) domain: $x \ne \left(2n + \frac{1}{2}\right)\pi$, $n = 0, \pm 1, \pm 2, \dots$; range: $y \ge \frac{1}{2}$
(f) domain: $-2 \le x < 2$ or $x > 2$; range: $0 \le y < 2$ or $y > 2$

11. (a) no; births and deaths (b) decreases for 8 hours, takes a jump upward, and repeats

13.
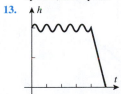

15. function; $y = \sqrt{25 - x^2}$

17. function; $y = \begin{cases} \sqrt{25 - x^2}, & -5 \le x \le 0 \\ -\sqrt{25 - x^2}, & 0 < x \le 5 \end{cases}$

19. False; for example, the graph of the function $f(x) = x^2 - 1$ crosses the x-axis at $x = \pm 1$.

21. False; the range also includes 0.

23. (a) $2, 4$ (b) none (c) $x \le 2$; $4 \le x$ (d) $y_{min} = -1$; no maximum

25. $h = L(1 - \cos\theta)$

27. (a) $f(x) = \begin{cases} 2x + 1, & x < 0 \\ 4x + 1, & x \ge 0 \end{cases}$ (b) $g(x) = \begin{cases} 1 - 2x, & x < 0 \\ 1, & 0 \le x < 1 \\ 2x - 1, & x \ge 1 \end{cases}$

29. (a) $V = (8 - 2x)(15 - 2x)x$
(b) $0 < x < 4$
(c) $0 < V < 90$, approximately
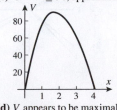
(d) V appears to be maximal for $x \approx 1.7$.

31. (a) $L = x + 2y$
(b) $L = x + 2000/x$
(c) $0 < x \le 100$
(d) $x \approx 45$ ft, $y \approx 22$ ft
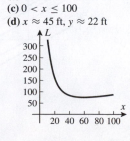

33. (a) $r \approx 3.4$, $h \approx 13.7$ (b) taller
(c) $r \approx 3.1$ cm, $h \approx 16.0$ cm, $C \approx 4.76$ cents

35. (i) $x = 1, -2$ (ii) $g(x) = x + 1$, all x

37. (a) $25°$F (b) $13°$F (c) $5°$F **39.** $15°$F

► **Exercise Set 0.2 (Page 24)**

1. (a)
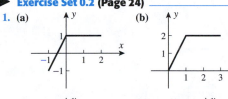
(b)

(c)
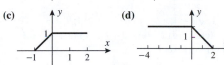
(d)

3. (a)
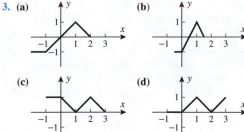
(b)

(c)
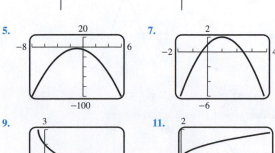
(d)

5.

7.

9.
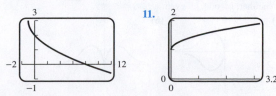

11.

A45

13.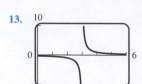

15.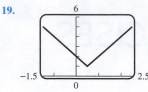

(c)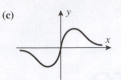

17.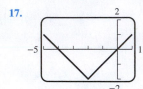

19.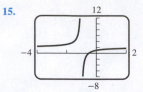

59. (a) even **(b)** odd **(c)** even **(d)** neither **(e)** odd **(f)** even

63. (a) y-axis
(b) origin
(c) x-axis, y-axis, origin

65.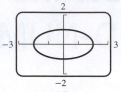

21.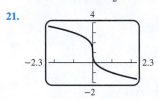

23.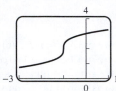

67.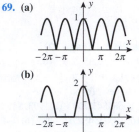

69. (a)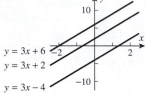

(b)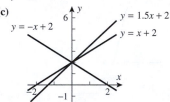

71. yes; $f(x) = x^k$, $g(x) = x^n$

25. (a)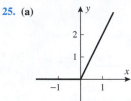

(b) $y = \begin{cases} 0, & x \le 0 \\ 2x, & x > 0 \end{cases}$

27. $3\sqrt{x-1}, x \ge 1$; $\sqrt{x-1}, x \ge 1$; $2x - 2, x \ge 1$; $2, x > 1$

29. (a) 3 **(b)** 9 **(c)** 2 **(d)** 2 **(e)** $\sqrt{2+h}$ **(f)** $(3+h)^3 + 1$

31. $1 - x, x \le 1$; $\sqrt{1 - x^2}, |x| \le 1$

33. $\dfrac{1}{1-2x}, x \ne \dfrac{1}{2}, 1$; $-\dfrac{1}{2x} - \dfrac{1}{2}, x \ne 0, 1$

35. (a) $g(x) = \sqrt{x}, h(x) = x + 2$ **(b)** $g(x) = |x|, h(x) = x^2 - 3x + 5$

37. (a) $g(x) = x^2, h(x) = \sin x$ **(b)** $g(x) = 3/x, h(x) = 5 + \cos x$

39. (a) $g(x) = x^3, h(x) = 1 + \sin(x^2)$
(b) $g(x) = \sqrt{x}, h(x) = 1 - \sqrt[3]{x}$

Responses to True–False questions may be abridged to save space.

41. True; see Definition 0.2.1.

43. True; see Theorem 0.2.3 and the definition of even function that follows.

▶ Exercise Set 0.3 (Page 35)

1. (a) $y = 3x + b$ **(c)**
(b) $y = 3x + 6$

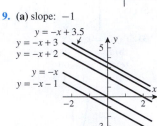

$y = 3x + 6$
$y = 3x + 2$
$y = 3x - 4$

3. (a) $y = mx + 2$ **(c)**
(b) $y = -x + 2$

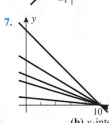

$y = -x + 2$
$y = 1.5x + 2$
$y = x + 2$

5. $y = \pm \dfrac{9 - x_0 x}{\sqrt{9 - x_0^2}}$

7.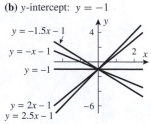

y-intercepts represent current value of item being depreciated.

9. (a) slope: -1

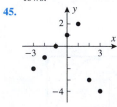

$y = -x + 3.5$
$y = -x + 3$
$y = -x + 2$
$y = -x$
$y = -x - 1$

(b) y-intercept: $y = -1$

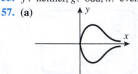

$y = -1.5x - 1$
$y = -x - 1$
$y = -1$
$y = 2x - 1$
$y = 2.5x - 1$

45.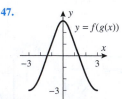

47.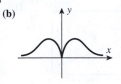

$y = f(g(x))$

49. $\pm 1.5, \pm 2$ **51.** $6x + 3h, 3w + 3x$ **53.** $-\dfrac{1}{x(x+h)}, -\dfrac{1}{xw}$

55. f: neither, g: odd, h: even

57. (a) **(b)**

(c) pass through $(-4, 2)$

$y = -1.5(x + 4) + 2$

$y = -(x + 4) + 2$

$y = 2$

$y = 2(x + 4) + 2$

$y = 2.5(x + 4) + 2$

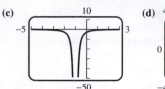

(d) x-intercept: $x = 1$

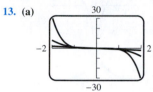

$y = 2(x - 1)$

$y = \frac{3}{2}(x - 1)$

$y = (x - 1)$

$y = -2(x - 1)$

$y = -3(x - 1)$

11. (a) VI
(b) IV
(c) III
(d) V
(e) I
(f) II

13. (a)

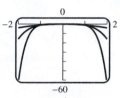 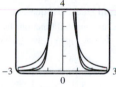

(b)

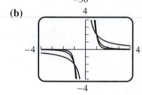 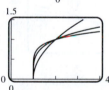

(c)

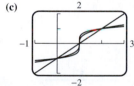 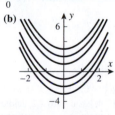

15. (a)

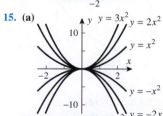

$y = 3x^2$ $y = 2x^2$

$y = x^2$

$y = -x^2$

$y = -3x^2$ $y = -2x^2$

(b)

(c)

17. (a)

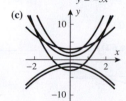

(b)

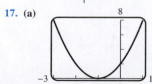

19.

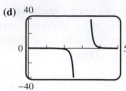 (c) (d)

21. (a) newton-meters (N·m) (b) 20 N·m

(c)

V (L)	0.25	0.5	1.0	1.5	2.0
P (N/m^2)	80×10^3	40×10^3	20×10^3	13.3×10^3	10×10^3

(d)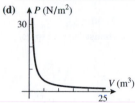

23. (a) $k = 0.000045$ N·m^2
(b) 0.000005 N
(d) The force becomes infinite; the force tends to zero.

(c)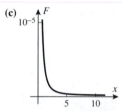

Responses to True–False questions may be abridged to save space.

25. True; see Figure 0.3.2(b).

27. False; the constant of proportionality is $2 \cdot 6 = 12$.

29. (a) II; $y = 1, x = -1, 2$ (b) I; $y = 0, x = -2, 3$ (c) IV; $y = 2$
(d) III; $y = 0, x = -2$

31. (a) $y = 3 \sin(x/2)$ (b) $y = 4 \cos 2x$ (c) $y = -5 \sin 4x$

33. (a) $y = \sin[x + (\pi/2)]$ (b) $y = 3 + 3 \sin(2x/9)$

(c) $y = 1 + 2 \sin\left(2x - \dfrac{\pi}{2}\right)$

35. (a) amplitude $= 3$, period $= \pi/2$ (b) amplitude $= 2$, period $= 2$

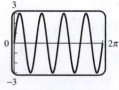 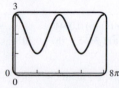

(c) amplitude $= 1$, period $= 4\pi$

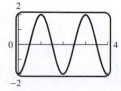

37. $x = \dfrac{5\sqrt{13}}{2} \sin(2\pi t + \tan^{-1} \frac{1}{2\sqrt{3}})$

▶ **Exercise Set 0.4 (Page 48)** _____

1. (a) yes (b) no (c) yes (d) no
3. (a) yes (b) yes (c) no (d) yes (e) no (f) no
5. (a) yes (b) no
7. (a) $8, -1, 0$
   (b) $[-2, 2], [-8, 8]$
   (c)

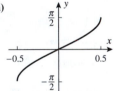

9. $\frac{1}{7}(x+6)$
11. $\sqrt[3]{(x+5)/3}$
13. $-\sqrt{3/x}$
15. $\begin{cases} (5/2) - x, & x > 1/2 \\ 1/x, & 0 < x \le 1/2 \end{cases}$
17. $x^{1/4} - 2$ for $x \ge 16$
19. $\frac{1}{2}(3 - x^2)$ for $x \le 0$

21. (a) $f^{-1}(x) = \dfrac{-b + \sqrt{b^2 - 4a(c - x)}}{2a}$

   (b) $f^{-1}(x) = \dfrac{-b - \sqrt{b^2 - 4a(c - x)}}{2a}$

23. (a) $y = (6.214 \times 10^{-4})x$ (b) $x = \dfrac{10^4}{6.214}y$
   (c) how many meters in y miles
25. (b) symmetric about the line $y = x$ **27.** 10

Responses to True–False questions may be abridged to save space.
31. False; $f^{-1}(2) = 2$
33. True; see Theorem 0.4.3.
35. $\frac{4}{5}, \frac{3}{5}, \frac{3}{4}, \frac{5}{3}, \frac{5}{4}$
37. (a) $0 \le x \le \pi$ (b) $-1 \le x \le 1$ (c) $-\pi/2 < x < \pi/2$
   (d) $-\infty < x < +\infty$ **39.** $\frac{24}{25}$
41. (a) $\dfrac{1}{\sqrt{1+x^2}}$ (b) $\dfrac{\sqrt{1-x^2}}{x}$ (c) $\dfrac{\sqrt{x^2-1}}{x}$ (d) $\dfrac{1}{\sqrt{x^2-1}}$
43. (a)   (b)

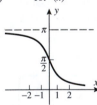

45. (a) 0.25545, error   (b) $|x| \le \sin 1$
47. (a) $\cot^{-1}(x)$ $\csc^{-1}(x)$

   (b) $\cot^{-1} x$: all x, $0 < y < \pi$
   $\csc^{-1} x$: $|x| \ge 1$, $0 < |y| < \pi/2$
49. (a) $55.0°$ (b) $33.6°$ (c) $25.8°$ **51.** (a) 21.1 hours   (b) 2.9 hours
53. $29°$

▶ **Exercise Set 0.5 (Page 61)** _____

1. (a) -4 (b) 4   (c) $\frac{1}{4}$ **3.** (a) 2.9690   (b) 0.0341
5. (a) 4   (b) -5 (c) 1   (d) $\frac{1}{2}$ **7.** (a) 1.3655   (b) -0.3011
9. (a) $2r + \dfrac{s}{2} + \dfrac{t}{2}$ (b) $s - 3r - t$
11. (a) $1 + \log x + \frac{1}{2}\log(x - 3)$ (b) $2 \ln |x| + 3 \ln \sin x - \frac{1}{2}\ln(x^2 + 1)$
13. $\log \frac{256}{3}$ **15.** $\ln \dfrac{\sqrt[3]{x}(x+1)^2}{\cos x}$ **17.** 0.01   **19.** e^2 **21.** 4

23. $\sqrt{3/2}$ **25.** $-\dfrac{\ln 3}{2 \ln 5}$ **27.** $\frac{1}{3}\ln \frac{7}{2}$ **29.** -2

31. (a) domain: $(-\infty, +\infty)$; range: $(-1, +\infty)$

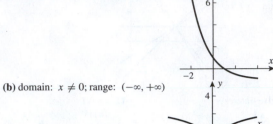

   (b) domain: $x \ne 0$; range: $(-\infty, +\infty)$

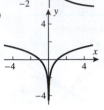

33. (a) domain: $x \ne 0$; range: $(-\infty, +\infty)$

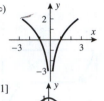

   (b) domain: $(-\infty, +\infty)$; range: $(0, 1]$

Responses to True–False questions may be abridged to save space.
35. False; exponential functions have constant base and variable exponent.
37. True; $\ln x = \log_e x$ **39.** $2.8777, -0.3174$
41.

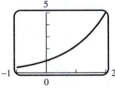

43. (a) no   (d) $y = (\sqrt{5})^x$
   (b) $y = 2^{x/4}$
   (c) $y = 2^{-x}$

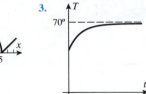

45. $\log \frac{1}{2} < 0$, so $3 \log \frac{1}{2} < 2 \log \frac{1}{2}$ **47.** 201 days
49. (a) 7.4, basic   (b) 4.2, acidic   (c) 6.4, acidic   (d) 5.9, acidic
51. (a) 140 dB, damage   (b) 120 dB, damage   (c) 80 dB, no damage
   (d) 75 dB, no damage
53. ≈ 200 **55.** (a) $\approx 5 \times 10^{16}$ J   (b) ≈ 0.67

▶ **Chapter 0 Review Exercises (Page 63)** _____
1.   **3.**

5. (a) $C = 5x^2 + (64/x)$ (b) $x > 0$ **9.**

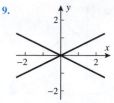

7. (a) $V = (6 - 2x)(5 - x)x$ ft^3
(b) $0 < x < 3$
(c) 3.57 ft $\times 3.79$ ft $\times 1.21$ ft

11.

x	-4	-3	-2	-1	0	1	2	3	4
$f(x)$	0	-1	2	1	3	-2	-3	4	-4
$g(x)$	3	2	1	-3	-1	-4	4	-2	0
$(f \circ g)(x)$	4	-3	-2	-1	1	0	-4	2	3
$(g \circ f)(x)$	-1	-3	4	-4	-2	1	2	0	3

13. $0, -2$ **15.** $1/(2 - x^2), x \neq \pm 1, \pm\sqrt{2}$
17. (a) odd   (b) even   (c) neither   (d) even
19. (a) circles of radius 1 centered on the parabola $y = x^2$
(b) parabolas congruent to $y = x^2$ that open up with vertices on the
line $y = x/2$
21. (a)

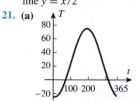

(b) January 11   (c) 122 days

23. $A: \left(-\frac{2}{3}\pi, 1 - \sqrt{3}\right)$; $B: \left(\frac{1}{3}\pi, 1 + \sqrt{3}\right)$; $C: \left(\frac{2}{3}\pi, 1 + \sqrt{3}\right)$;
$D: \left(\frac{5}{3}\pi, 1 - \sqrt{3}\right)$

27. (a) $\frac{1}{2}(x + 1)^{1/3}$ (b) none   (c) $\frac{1}{2}\ln(x - 1)$ (d) $\frac{x + 2}{x - 1}$
(e) $\dfrac{1}{2 + \sin^{-1} x}$ (f) $\tan\left(\dfrac{1}{3x} - \dfrac{1}{3}\right), x < -\dfrac{2}{3\pi - 2}$ or $x > \dfrac{2}{3\pi + 2}$
29. (a) $\frac{33}{65}$ (b) $\frac{56}{65}$ **31.** $\frac{10^{60}}{63360} \approx 1.6 \times 10^{55}$ miles   **33.** $15x + 2$
35. (a)

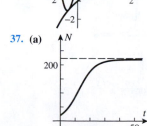

(b) $-\dfrac{\pi}{2}, 0, \dfrac{\pi}{2}, \pi, \dfrac{3\pi}{2}$; $-\dfrac{\pi}{4}, \dfrac{\pi}{4}, \dfrac{3\pi}{4}, \dfrac{5\pi}{4}$

37. (a)

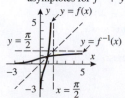

(b) about 10 years   (c) 220 sheep

39. (b) $3.654, 332105.108$
41. (a) f is increasing   (b) asymptotes for f: $x = 0$ and $x = \pi/2$;
asymptotes for f^{-1}: $y = 0$ (as $x \to -\infty$) and $y = \pi/2$ (as $x \to +\infty$)

► **Exercise Set 1.1 (Page 77)**
1. (a) 3   (b) 3   (c) 3   (d) 3
3. (a) -1 (b) 3   (c) does not exist   (d) 1
5. (a) 0   (b) 0   (c) 0   (d) 3   **7.** (a) $-\infty$ (b) $-\infty$ (c) $-\infty$ (d) 1
9. (a) 1   (b) $-\infty$ (c) does not exist   (d) -2
11.

x	-0.01	-0.001	-0.0001	0.0001	0.001	0.01
$f(x)$	0.99502	0.99950	0.99995	1.00005	1.00050	1.00502

The limit appears to be 1.
13. (a) $\frac{1}{3}$ (b) $+\infty$ (c) $-\infty$ **15.** (a) 3   (b) does not exist
Responses to True–False questions may be abridged to save space.
17. False; see Example 6.
19. False; the one-sided limits must also be equal.
21.

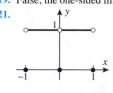

23.

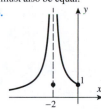

25.

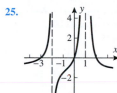

27. $y = -2x - 1$
29. $y = 4x - 3$

31. (a) rest length   (b) 0. As speed approaches c, length shrinks to zero.
33. The limit should be 1.

► **Exercise Set 1.2 (Page 87)**
1. (a) -6 (b) 13   (c) -8 (d) 16   (e) 2   (f) $-\frac{1}{2}$
3. 6   **5.** $\frac{3}{4}$ **7.** 4   **9.** $-\frac{4}{5}$ **11.** -3 **13.** $\frac{3}{2}$ **15.** $+\infty$
17. does not exist   **19.** $-\infty$ **21.** $+\infty$ **23.** does not exist   **25.** $+\infty$
27. $+\infty$ **29.** 6   **31.** (a) 2   (b) 2   (c) 2
Responses to True–False questions may be abridged to save space.
33. True; this is Theorem 1.2.2(a).   **35.** False; see Example 9.   **37.** $\frac{1}{4}$
39. (a) 3   (b)

41. (a) Theorem 1.2.2(a) does not apply.
(b) $\lim\limits_{x \to 0^+} \left(\dfrac{1}{x} - \dfrac{1}{x^2}\right) = \lim\limits_{x \to 0^+} \left(\dfrac{x - 1}{x^2}\right) = -\infty$ **43.** $a = 2$
45. The left and/or right limits could be $\pm\infty$; or the limit could exist and
equal any preassigned real number.

► **Exercise Set 1.3 (Page 96)**
1. (a) $-\infty$ (b) $+\infty$ **3.** (a) 0   (b) -1
5. (a) -12 (b) 21   (c) -15 (d) 25   (e) 2   (f) $-\frac{3}{5}$ (g) 0
(h) does not exist
7. (a)

x	0.1	0.01	0.001	0.0001	0.00001	0.000001
$f(x)$	1.471128	1.560797	1.569796	1.570696	1.570786	1.570795

The limit appears to be $\pi/2$.   (b) $\pi/2$

9. $-\infty$ **11.** $+\infty$ **13.** $\frac{3}{2}$ **15.** 0 **17.** 0 **19.** $-\infty$ **21.** $-\frac{1}{7}$

23. $\dfrac{-\sqrt[3]{5}}{2}$ **25.** $-\sqrt{5}$ **27.** $1/\sqrt{6}$ **29.** $\sqrt{3}$ **31.** 0 **33.** 1

35. 1 **37.** $-\infty$ **39.** e

Responses to True–False questions may be abridged to save space.

41. False; 1^∞ is an indeterminate form. The limit is e^2.

43. True; consider $f(x) = (\sin x)/x$.

45. $\displaystyle\lim_{t\to+\infty} n(t) = +\infty$; $\displaystyle\lim_{t\to+\infty} e(t) = c$ **47.** (a) $+\infty$ (b) -5

51. (a) no   (b) yes; $\tan x$ and $\sec x$ at $x = n\pi + \pi/2$, and $\cot x$ and $\csc x$ at $x = n\pi$, $n = 0, \pm 1, \pm 2, \ldots$

55. $+\infty$ **57.** $+\infty$ **59.** 1 **61.** e

65. (a)

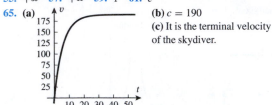

(b) $c = 190$
(c) It is the terminal velocity of the skydiver.

67. (a) e (c) e^a **69.** $x + 2$ **71.** $1 - x^2$ **73.** $\sin x$

▶ **Exercise Set 1.4 (Page 106)**

1. (a) $|x| < 0.1$ (b) $|x - 3| < 0.0025$ (c) $|x - 4| < 0.000125$

3. (a) $x_0 = 3.8025$, $x_1 = 4.2025$ (b) $\delta = 0.1975$

5. $\delta = 0.0442$ **7.** $\delta = 0.13$

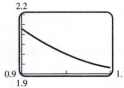

11. $|(3x - 5) - 1| = |3x - 6| = 3 \cdot |x - 2| < 3 \cdot \delta = \epsilon$

13. $\delta = 1$ **15.** $\delta = \frac{1}{3}\epsilon$ **17.** $\delta = \epsilon/2$ **19.** $\delta = \epsilon$ **21.** $\delta = \epsilon$

Responses to True–False questions may be abridged to save space.

23. True; $|f(x) - f(a)| = |m||x - a|$

25. True; constant functions

29. (b) 65 (c) $\epsilon/65$; 65; 65; $\epsilon/65$ **31.** $\delta = \min\left(1, \frac{1}{6}\epsilon\right)$

33. $\delta = \min(1, \epsilon/(1 + \epsilon))$ **35.** $\delta = 2\epsilon$

39. (a) $-\sqrt{\dfrac{1 - \epsilon}{\epsilon}}$; $\sqrt{\dfrac{1 - \epsilon}{\epsilon}}$ (b) $\sqrt{\dfrac{1 - \epsilon}{\epsilon}}$ (c) $-\sqrt{\dfrac{1 - \epsilon}{\epsilon}}$

41. 10 **43.** 999 **45.** -202 **47.** -57.5 **49.** $N = \dfrac{1}{\sqrt{\epsilon}}$

51. $N = -\dfrac{5}{2} - \dfrac{11}{2\epsilon}$ **53.** $N = (1 + 2/\epsilon)^2$

55. (a) $|x| < \frac{1}{10}$ (b) $|x - 1| < \frac{1}{1000}$
(c) $|x - 3| < \frac{1}{10\sqrt{10}}$ (d) $|x| < \frac{1}{10}$

57. $\delta = 1/\sqrt{M}$ **59.** $\delta = 1/M$ **61.** $\delta = 1/(-M)^{1/4}$ **63.** $\delta = \epsilon$

65. $\delta = \epsilon^2$ **67.** $\delta = \epsilon$ **69.** (a) $\delta = -1/M$ (b) $\delta = 1/M$

71. (a) $N = M - 1$ (b) $N = M - 1$

73. (a) 0.4 amps   (b) about 0.39474 to 0.40541 amps   (c) $3/(7.5 + \delta)$ to $3/(7.5 - \delta)$ (d) $\delta \approx 0.01870$ (e) current approaches $+\infty$

▶ **Exercise Set 1.5 (Page 118)**

1. (a) not continuous, $x = 2$ (b) not continuous, $x = 2$
(c) not continuous, $x = 2$ (d) continuous   (e) continuous
(f) continuous

3. (a) not continuous, $x = 1, 3$ (b) continuous
(c) not continuous, $x = 1$ (d) continuous
(e) not continuous, $x = 3$ (f) continuous

5. (a) no   (b) no   (c) no   (d) yes   (e) yes   (f) no   (g) yes

7. (a)

(b)

(c)

(d)

9. (a)

(b) One second could cost you one dollar.

11. none   **13.** none   **15.** $-1/2, 0$ **17.** $-1, 0, 1$ **19.** none

21. none

Responses to True–False questions may be abridged to save space.

23. True; the composition of continuous functions is continuous.

25. False; let f and g be the functions in Exercise 6.

27. True; $f(x) = \sqrt{f(x)} \cdot \sqrt{f(x)}$

29. (a) $k = 5$ (b) $k = \frac{4}{3}$ **31.** $k = 4, m = 5/3$

33. (a)

(b)

35. (a) $x = 0$, not removable   (b) $x = -3$, removable
(c) $x = 2$, removable; $x = -2$, not removable

37. (a) $x = \frac{1}{2}$, not removable;
at $x = -3$, removable
(b) $(2x - 1)(x + 3)$

45. $f(x) = 1$ for $0 \le x < 1$, $f(x) = -1$ for $1 \le x \le 2$

49. $x = -1.25, x = 0.75$ **51.** $x = 2.24$

▶ **Exercise Set 1.6 (Page 125)**

1. none   **3.** $x = n\pi, n = 0, \pm 1, \pm 2, \ldots$

5. $x = n\pi, n = 0, \pm 1, \pm 2, \ldots$

7. $2n\pi + (\pi/6), 2n\pi + (5\pi/6), n = 0, \pm 1, \pm 2, \ldots$

9. $\left[-\frac{1}{2}, \frac{1}{2}\right]$ **11.** $(0, 3)$ and $(3, +\infty)$ **13.** $(-\infty, -1]$ and $[1, +\infty)$

15. (a) $\sin x, x^3 + 7x + 1$ (b) $|x|, \sin x$ (c) $x^3, \cos x, x + 1$

17. 1 **19.** $-\pi/6$ **21.** 1 **23.** 3 **25.** $+\infty$ **27.** $\frac{7}{3}$

29. 0 **31.** 0 **33.** 1 **35.** 2 **37.** does not exist   **39.** 0

41. (a)

x	4	4.5	4.9	5.1	5.5	6
$f(x)$	0.093497	0.100932	0.100842	0.098845	0.091319	0.076497

The limit appears to be $\frac{1}{10}$. **(b)** $\frac{1}{10}$

Responses to True–False questions may be abridged to save space.

43. True; use the Squeezing Theorem.

45. False; consider $f(x) = \tan^{-1} x$.

47. (a) Using degrees instead of radians   **(b)** $\pi/180$

49. 1   **51.** $k = \frac{1}{2}$ **53. (a)** 1   **(b)** 0   **(c)** 1

55. $-\pi$ **57.** $-\sqrt{2}$ **59.** 1   **61.** 5

63. $\lim_{x \to 0} \sin(1/x)$ does not exist

65. The limit is 0.

67. (b)

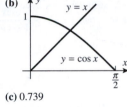

(c) 0.739

69. (a) Gravity is strongest at the poles and weakest at the equator.

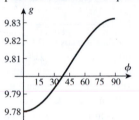

Chapter 1 Review Exercises (Page 128)

1. (a) 1   **(b)** does not exist   **(c)** does not exist   **(d)** 1   **(e)** 3   **(f)** 0
   (g) 0   **(h)** 2   **(i)** $\frac{1}{2}$

3. (a) 0.405   **5.** 1   **7.** $-3/2$ **9.** 32/3

11. (a) $y = 0$ **(b)** none   **(c)** $y = 2$ **13.** 1   **15.** $3 - k$ **17.** 0

19. e^{-3} **21.** $2001.60, 2009.66, 2013.62, 2013.75

23. (a) $2x/(x - 1)$ is one example.

25. (a) $\lim_{x \to 2} f(x) = 5$ **(b)** $\delta = 0.0045$

27. (a) $\delta = 0.0025$ **(b)** $\delta = 0.0025$ **(c)** $\delta = 1/9000$
   (Some larger values also work.)

31. (a) $-1, 1$ **(b)** none   **(c)** $-3, 0$ **33.** no; not continuous at $x = 2$

35. Consider $f(x) = x$ for $x \neq 0$, $f(0) = 1$, $a = -1$, $b = 1$, $k = 0$.

Chapter 1 Making Connections (Page 130)

Where correct answers to a Making Connections exercise may vary, no answer is listed. Sample answers for these questions are available on the Book Companion Site.

4. (a) The circle through the origin with center $\left(0, \frac{1}{8}\right)$
   (b) The circle through the origin with center $\left(0, \frac{1}{2}\right)$
   (c) The circle does not exist.
   (d) The circle through the origin with center $\left(0, \frac{1}{2}\right)$
   (e) The circle through $(0, 1)$ with center at the origin.
   (f) The circle through the origin with center $\left(0, \dfrac{1}{2g(0)}\right)$
   (g) The circle does not exist.

Exercise Set 2.1 (Page 140)

1. (a) 4 m/s   **(b)**

3. (a) 0 cm/s   **(b)** $t = 0, t = 2$, and $t = 4.2$ **(c)** maximum: $t = 1$;
   minimum: $t = 3$ **(d)** -7.5 cm/s

5. straight line with slope equal to the velocity

7. Answers may vary.   **9.** Answers may vary.

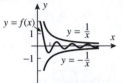

11. (a) 2   **(b)** 0   **(c)** $4x_0$ **13. (a)** $-\frac{1}{6}$ **(b)** $-\frac{1}{4}$ **(c)** $-1/x_0^2$
   (d)   **(d)**

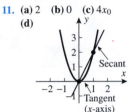   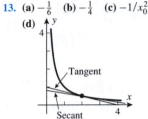

15. (a) $2x_0$ **(b)** -2 **17. (a)** $1 + \dfrac{1}{2\sqrt{x_0}}$ **(b)** $\frac{3}{2}$

Responses to True–False questions may be abridged to save space.

19. True; set $h = x - 1$, so $x = 1 + h$ and $h \to 0$ is equivalent to $x \to 1$.

21. False; velocity is a ratio of change in position to change in time.

23. (a) 72°F at about 4:30 P.M.   **(b)** 4°F/h   **(c)** -7°F/h at about 9 P.M.

25. (a) first year   **(d)**
   (b) 6 cm/year
   (c) 10 cm/year at about age 14

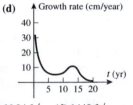

27. (a) 19,200 ft   **(b)** 480 ft/s   **(c)** 66.94 ft/s   **(d)** 1440 ft/s

Exercise Set 2.2 (Page 152)

1. $2, 0, -2, -1$ **5.**   **7.** $y = 5x - 16$.
3. (b) 3   **(c)** 3   **9.** $4x, y = 4x - 2$
   11. $3x^2; y = 0$

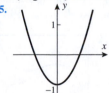

13. $\dfrac{1}{2\sqrt{x + 1}}; y = \frac{1}{6}x + \frac{5}{3}$ **15.** $-1/x^2$ **17.** $2x - 1$

19. $-1/(2x^{3/2})$ **21.** $8t + 1$

23. (a) D   **(b)** F   **(c)** B   **(d)** C   **(e)** A   **(f)** E

25. (a)   **(b)**

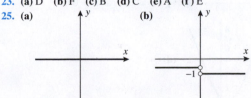

(c)

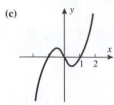

Responses to True–False questions may be abridged to save space.

27. False; $f'(a) = 0$ **29.** False; for example, $f(x) = |x|$

31. (a) $\sqrt{x}, 1$ (b) $x^2, 3$ **33.** -2

35. $y = -2x + 1$

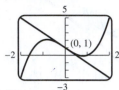

37. (b)

w	1.5	1.1	1.01	1.001	1.0001	1.00001
$[f(w) - f(1)]/(w - 1)$	1.6569	1.4355	1.3911	1.3868	1.3863	1.3863

39. (a) 0.04, 0.22, 0.88   (b) best: $\dfrac{f(2) - f(0)}{2 - 0}$; worst: $\dfrac{f(3) - f(1)}{3 - 1}$

41. (a) dollars per foot   (b) the price per additional foot   (c) positive
(d) $1000

43. (a) $F \approx 200$ lb, $dF/d\theta \approx 50$ lb/rad   (b) $\mu = 0.25$

45. (a) $T \approx 115°F$, $dT/dt \approx -3.35°F$/min   (b) $k = -0.084$

▶ **Exercise Set 2.3 (Page 161)** ———

1. $28x^6$ **3.** $24x^7 + 2$ **5.** 0   **7.** $-\frac{1}{3}(7x^6 + 2)$

9. $-3x^{-4} - 7x^{-8}$ **11.** $24x^{-9} + (1/\sqrt{x})$

13. $f'(x) = ex^{e-1} - \dfrac{\sqrt{10}}{x^{(1+\sqrt{10})}}$

15. $3ax^2 + 2bx + c$ **17.** 7   **19.** $2t - 1$ **21.** 15   **23.** -8 **25.** 0

27. 0   **29.** $32t$ **31.** $3\pi r^2$

Responses to True–False questions may be abridged to save space.

33. True; apply the difference and constant multiple rules.

35. False; $\dfrac{d}{dx}[4f(x) + x^3]\Big|_{x=2} = [4f'(x) + 3x^2]\Big|_{x=2} = 32$

37. (a) $4\pi r^2$ (b) 100π **39.** $y = 5x + 17$

41. (a) $42x - 10$ (b) 24   (c) $2/x^3$ (d) $700x^3 - 96x$

43. (a) $-210x^{-8} + 60x^2$ (b) $-6x^{-4}$ (c) $6a$

45. (a) 0
(b) 112
(c) 360

49. $(1, \frac{5}{6}), (2, \frac{2}{3})$

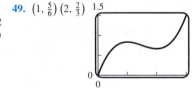

51. $y = 3x^2 - x - 2$ **53.** $x = \frac{1}{2}$

55. $(2 + \sqrt{3}, -6 - 4\sqrt{3}), (2 - \sqrt{3}, -6 + 4\sqrt{3})$

57. $-2x_0$ **61.** $-\dfrac{2GmM}{r^3}$ **63.** $f'(x) > 0$ for all $x \neq 0$

65. yes, 3

67. not differentiable at $x = 1$ **69.** (a) $x = \frac{2}{3}$ (b) $x = \pm 2$ **71.** (b) yes

73. (a) $n(n-1)(n-2)\cdots 1$ (b) 0   (c) $a_n n(n-1)(n-2)\cdots 1$

79. $-12/(2x + 1)^3$ **81.** $-2/(x + 1)^3$

▶ **Exercise Set 2.4 (Page 168)** ———

1. $4x + 1$ **3.** $4x^3$ **5.** $18x^2 - \frac{3}{2}x + 12$

7. $-15x^{-2} - 14x^{-3} + 48x^{-4} + 32x^{-5}$ **9.** $3x^2$

11. $\dfrac{-3x^2 - 8x + 3}{(x^2 + 1)^2}$ **13.** $\dfrac{3x^2 - 8x}{(3x - 4)^2}$ **15.** $\dfrac{x^{3/2} + 10x^{1/2} + 4 - 3x^{-1/2}}{(x + 3)^2}$

17. $2(1 + x^{-1})(x^{-3} + 7) + (2x + 1)(-x^{-2})(x^{-3} + 7) +$
$(2x + 1)(1 + x^{-1})(-3x^{-4})$ **19.** $3(7x^6 + 2)(x^7 + 2x - 3)^2$

21. -29 **23.** 0   **25.** (a) $-\frac{37}{4}$ (b) $-\frac{23}{16}$

27. (a) 10   (b) 19   (c) 9   (d) -1 **29.** $-2 \pm \sqrt{3}$ **31.** none   **33.** -2

37. $F''(x) = xf''(x) + 2f'(x)$

39. $R'(120) = 1800$; increasing the price by Δp dollars increases revenue by approximately $1800\Delta p$ dollars.

41. $f'(x) = -nx^{-n-1}$

▶ **Exercise Set 2.5 (Page 172)** ———

1. $-4\sin x + 2\cos x$ **3.** $4x^2 \sin x - 8x \cos x$

5. $(1 + 5\sin x - 5\cos x)/(5 + \sin x)^2$ **7.** $\sec x \tan x - \sqrt{2} \sec^2 x$

9. $-4\csc x \cot x + \csc^2 x$ **11.** $\sec^3 x + \sec x \tan^2 x$ **13.** $-\dfrac{\csc x}{1 + \csc x}$

15. 0   **17.** $\dfrac{1}{(1 + x\tan x)^2}$ **19.** $-x\cos x - 2\sin x$

21. $-x\sin x + 5\cos x$ **23.** $-4\sin x \cos x$

25. (a) $y = x$ (b) $y = 2x - (\pi/2) + 1$ (c) $y = 2x + (\pi/2) - 1$

29. (a) $x = \pm\pi/2, \pm 3\pi/2$ (b) $x = -3\pi/2, \pi/2$
(c) no horizontal tangent line   (d) $x = \pm 2\pi, \pm\pi, 0$

31. 0.087 ft/deg   **33.** 1.75 m/deg

Responses to True–False questions may be abridged to save space.

35. False; by the product rule, $g'(x) = f(x)\cos x + f'(x)\sin x$.

37. True; $f(x) = (\sin x)/(\cos x) = \tan x$, so $f'(x) = \sec^2 x$.

39. $-\cos x$ **41.** $3, 7, 11, \ldots$

43. (a) all x (b) all x (c) $x \neq (\pi/2) + n\pi, n = 0, \pm 1, \pm 2, \ldots$
(d) $x \neq n\pi, n = 0, \pm 1, \pm 2, \ldots$ (e) $x \neq (\pi/2) + n\pi, n = 0, \pm 1,$
$\pm 2, \ldots$ (f) $x \neq n\pi, n = 0, \pm 1, \pm 2, \ldots$ (g) $x \neq (2n + 1)\pi, n = 0,$
$\pm 1, \pm 2, \ldots$ (h) $x \neq n\pi/2, n = 0, \pm 1, \pm 2, \ldots$ (i) all x

▶ **Exercise Set 2.6 (Page 178)** ———

1. 6   **3.** (a) $(2x - 3)^5, 10(2x - 3)^4$ (b) $2x^5 - 3, 10x^4$

5. (a) -7 (b) -8 **7.** $37(x^3 + 2x)^{36}(3x^2 + 2)$

9. $-2\left(x^3 - \dfrac{7}{x}\right)^{-3}\left(3x^2 + \dfrac{7}{x^2}\right)$ **11.** $\dfrac{24(1 - 3x)}{(3x^2 - 2x + 1)^4}$

13. $\dfrac{3}{4\sqrt{x}\sqrt{4 + 3\sqrt{x}}}$ **15.** $-\dfrac{2}{x^3}\cos\left(\dfrac{1}{x^2}\right)$ **17.** $-20\cos^4 x \sin x$

19. $-\dfrac{3}{\sqrt{x}}\cos(3\sqrt{x})\sin(3\sqrt{x})$ **21.** $28x^6 \sec^2(x^7)\tan(x^7)$

23. $-\dfrac{5\sin(5x)}{2\sqrt{\cos(5x)}}$

25. $-3[x + \csc(x^3 + 3)]^{-4}[1 - 3x^2\csc(x^3 + 3)\cot(x^3 + 3)]$

27. $10x^3 \sin 5x \cos 5x + 3x^2 \sin^2 5x$

29. $-x^3 \sec\left(\dfrac{1}{x}\right)\tan\left(\dfrac{1}{x}\right) + 5x^4 \sec\left(\dfrac{1}{x}\right)$

31. $\sin(\cos x)\sin x$ **33.** $-6\cos^2(\sin 2x)\sin(\sin 2x)\cos 2x$

35. $35(5x + 8)^6(1 - \sqrt{x})^6 - \dfrac{3}{\sqrt{x}}(5x + 8)^7(1 - \sqrt{x})^5$

37. $\dfrac{33(x - 5)^2}{(2x + 1)^4}$ **39.** $-\dfrac{2(2x + 3)^2(52x^2 + 96x + 3)}{(4x^2 - 1)^9}$

41. $5[x\sin 2x + \tan^4(x^7)]^4[2x\cos 2x + \sin 2x + 28x^6 \tan^3(x^7)\sec^2(x^7)]$

43. $y = -x$ **45.** $y = -1$ **47.** $y = 8\sqrt{\pi}x - 8\pi$ **49.** $y = \frac{7}{2}x - \frac{3}{2}$

51. $-25x\cos(5x) - 10\sin(5x) - 2\cos(2x)$ **53.** $4(1 - x)^{-3}$

55. $3\cot^2\theta\csc^2\theta$ **57.** $\pi(b - a)\sin 2\pi\omega$

59. (a)

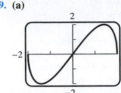

(c) $\dfrac{4 - 2x^2}{\sqrt{4 - x^2}}$

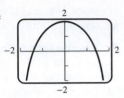

(d) $y - \sqrt{3} = \frac{2}{\sqrt{3}}(x - 1)$

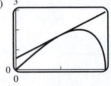

Responses to True–False questions may be abridged to save space.

61. False; by the chain rule, $\dfrac{d}{dx}[\sqrt{y}] = \dfrac{1}{2\sqrt{y}} \cdot \dfrac{dy}{dx} = \dfrac{f'(x)}{2\sqrt{f(x)}}$.

63. False; by the chain rule, $dy/dx = (-\sin[g(x)]) \cdot g'(x)$.

65. (c) $f = 1/T$ **(d)** amplitude $= 0.6$ cm, $T = 2\pi/15$ seconds per oscillation, $f = 15/(2\pi)$ oscillations per second

67. $\frac{7}{24}\sqrt{6}$ **69. (a)** $10\,\text{lb/in}^2, -2\,\text{lb/in}^2/\text{mi}$ **(b)** $-0.6\,\text{lb/in}^2/\text{s}$

71. $\begin{cases} \cos x, & 0 < x < \pi \\ -\cos x, & -\pi < x < 0 \end{cases}$

73. (c) $-\dfrac{1}{x}\cos\dfrac{1}{x} + \sin\dfrac{1}{x}$ **(d)** limit as x goes to 0 does not exist

75. (a) 21   **(b)** -36 **77.** $1/(2x)$ **79.** $\frac{2}{3}x$ **83.** $f'(g(h(x)))g'(h(x))h'(x)$

▶ Chapter 2 Review Exercises (Page 181)

3. (a) $2x$ **(b)** 4   **5.** 58.75 ft/s   **7. (a)** 13 mi/h   **(b)** 7 mi/h

9. (a) $-2/\sqrt{9 - 4x}$ **(b)** $1/(x + 1)^2$

11. (a) $x = -2, -1, 1, 3$ **(b)** $(-\infty, -2), (-1, 1), (3, +\infty)$

   (c) $(-2, -1), (1, 3)$ **(d)** 4

13. (a) 78 million people per year   **(b)** 1.3% per year

15. (a) $x^2\cos x + 2x\sin x$ **(c)** $4x\cos x + (2 - x^2)\sin x$

17. (a) $(6x^2 + 8x - 17)/(3x + 2)^2$ **(c)** $118/(3x + 2)^3$

19. (a) 2000 gal/min   **(b)** 2500 gal/min   **21. (a)** 3.6   **(b)** -0.777778

23. $f(1) = 0, f'(1) = 5$ **25.** $y = -16x, y = -145x/4$

29. (a) $8x^7 - \dfrac{3}{2\sqrt{x}} - 15x^{-4}$ **(b)** $(2x + 1)^{100}(1030x^2 + 10x - 1414)$

31. (a) $\dfrac{(x - 1)(15x + 1)}{2\sqrt{3x + 1}}$ **(b)** $-3(3x + 1)^2(3x + 2)/x^7$

33. $x = -\frac{7}{2}, -\frac{1}{2}, 2$ **35.** $y = \pm 2x$

37. $x = n\pi \pm (\pi/4), n = 0, \pm 1, \pm 2, \ldots$ **39.** $y = -3x + (1 + 9\pi/4)$

41. (a) $40\sqrt{3}$ **(b)** 7500

▶ Chapter 2 Making Connections (Page 184)

Where correct answers to a Making Connections exercise may vary, no answer is listed. Sample answers for these questions are available on the Book Companion Site.

2. (c) $k = 2$ **(d)** $h'(x) = 0$

3. (b) $f' \cdot g \cdot h \cdot k + f \cdot g' \cdot h \cdot k + f \cdot g \cdot h' \cdot k + f \cdot g \cdot h \cdot k'$

4. (c) $\dfrac{f' \cdot g \cdot h - f \cdot g' \cdot h + f \cdot g \cdot h'}{g^2}$

▶ Exercise Set 3.1 (Page 190)

1. (a) $(6x^2 - y - 1)/x$ **(b)** $4x - 2/x^2$ **3.** $-\dfrac{x}{y}$ **5.** $\dfrac{1 - 2xy - 3y^3}{x^2 + 9xy^2}$

7. $\dfrac{-y^{3/2}}{x^{3/2}}$ **9.** $\dfrac{1 - 2xy^2\cos(x^2y^2)}{2x^2y\cos(x^2y^2)}$

11. $\dfrac{1 - 3y^2\tan^2(xy^2 + y)\sec^2(xy^2 + y)}{3(2xy + 1)\tan^2(xy^2 + y)\sec^2(xy^2 + y)}$ **13.** $-\dfrac{8}{9y^3}$ **15.** $\dfrac{2y}{x^2}$

17. $\dfrac{\sin y}{(1 + \cos y)^3}$ **19.** $-1/\sqrt{3}, 1/\sqrt{3}$

Responses to True–False questions may be abridged to save space.

21. False; the graph of f need only coincide with a portion of the graph of the equation in x and y.

23. False; the equation is equivalent to $x^2 = y^2$ and $y = |x|$ satisfies this equation.

25. $-15^{-3/4} \approx -0.1312$ **27.** $-\frac{9}{13}$

31. (a)   **(c)** $x = -y^2$ or $x = y^2 + 1$

33. points $(2, 2), (-2, -2)$; $y' = -1$ at both points   **35.** $a = \frac{1}{4}, b = \frac{5}{4}$

39. (a)   **(c)** $2\sqrt[3]{2}/3$

▶ Exercise Set 3.2 (Page 195)

1. $1/x$ **3.** $1/(1 + x)$ **5.** $2x/(x^2 - 1)$ **7.** $\dfrac{1 - x^2}{x(1 + x^2)}$ **9.** $2/x$

11. $\dfrac{1}{2x\sqrt{\ln x}}$ **13.** $1 + \ln x$ **15.** $2x\log_2(3 - 2x) - \dfrac{2x^2}{(\ln 2)(3 - 2x)}$

17. $\dfrac{2x(1 + \log x) - x/(\ln 10)}{(1 + \log x)^2}$ **19.** $1/(x\ln x)$ **21.** $2\csc 2x$

23. $-\dfrac{1}{x}\sin(\ln x)$ **25.** $2\cot x/(\ln 10)$ **27.** $\dfrac{3}{x - 1} + \dfrac{8x}{x^2 + 1}$

29. $-\tan x + \dfrac{3x}{4 - 3x^2}$

Responses to True–False questions may be abridged to save space.

31. True; $\displaystyle\lim_{x \to 0^+}\dfrac{1}{x} = +\infty$ **33.** True; $1/x$ is an odd function.

35. $x\sqrt[3]{1 + x^2}\left[\dfrac{1}{x} + \dfrac{2x}{3(1 + x^2)}\right]$

37. $\dfrac{(x^2 - 8)^{1/3}\sqrt{x^3 + 1}}{x^6 - 7x + 5}\left[\dfrac{2x}{3(x^2 - 8)} + \dfrac{3x^2}{2(x^3 + 1)} - \dfrac{6x^5 - 7}{x^6 - 7x + 5}\right]$

39. (a) $-\dfrac{1}{x(\ln x)^2}$ **(b)** $-\dfrac{\ln 2}{x(\ln x)^2}$

41. $y = ex - 2$ **43.** $y = -x/e$ **45. (a)** $y = x/e$ **47.** $A(w) = w/2$

51. $f(x) = \ln(x + 1)$ **53. (a)** 3   **(b)** -5 **55. (a)** 0   **(b)** $\sqrt{2}$

▶ Exercise Set 3.3 (Page 201)

1. (b) $\frac{1}{9}$ **3.** $-2/x^2$ **5. (a)** no   **(b)** yes   **(c)** yes   **(d)** yes

7. $\dfrac{1}{15y^2 + 1}$ **9.** $\dfrac{1}{10y^4 + 3y^2}$ **13.** $f(x) + g(x), f(g(x))$

15. $7e^{7x}$ **17.** $x^2e^x(x + 3)$ **19.** $\dfrac{4}{(e^x + e^{-x})^2}$

21. $(x\sec^2 x + \tan x)e^{x\tan x}$ **23.** $(1 - 3e^{3x})e^{x - e^{3x}}$

25. $\dfrac{x - 1}{e^x - x}$ **27.** $2^x\ln 2$ **29.** $\pi^{\sin x}(\ln \pi)\cos x$

31. $(x^3 - 2x)^{\ln x}\left[\dfrac{3x^2 - 2}{x^3 - 2x}\ln x + \dfrac{1}{x}\ln(x^3 - 2x)\right]$

33. $(\ln x)^{\tan x}\left[\dfrac{\tan x}{x\ln x} + (\sec^2 x)\ln(\ln x)\right]$

35. $(\ln x)^{\ln x}\left[\dfrac{\ln(\ln x)}{x} + \dfrac{1}{x}\right]$ **37.** $3/\sqrt{1 - 9x^2}$

39. $-\dfrac{1}{|x|\sqrt{x^2-1}}$ **41.** $3x^2/(1+x^6)$ **43.** $-\dfrac{\sec^2 x}{\tan^2 x} = -\csc^2 x$

45. $\dfrac{e^x}{|x|\sqrt{x^2-1}} + e^x \sec^{-1} x$ **47.** 0   **49.** 0   **51.** $-\dfrac{1}{2\sqrt{x}(1+x)}$

Responses to True–False questions may be abridged to save space.

53. False; consider $y = Ae^x$.   **55.** True; use the chain rule.

59. $\dfrac{(3x^2 + \tan^{-1} y)(1+y^2)}{(1+y^2)e^y - x}$ **61. (b)** $1-(\sqrt{3}/3)$

63. (b) $y = (88x - 89)/7$ **69.** $r = 1, K = 12$

71. 3   **73.** $\ln 10$ **75.** 12π

▶ **Exercise Set 3.4 (Page 208)**

1. (a) 6   **(b)** $-\dfrac{1}{3}$ **3. (a)** -2 **(b)** $6\sqrt{5}$

5. (b) $A = x^2$ **(c)** $\dfrac{dA}{dt} = 2x\dfrac{dx}{dt}$ **(d)** $12\,\text{ft}^2/\text{min}$

7. (a) $\dfrac{dV}{dt} = \pi\left(r^2\dfrac{dh}{dt} + 2rh\dfrac{dr}{dt}\right)$ **(b)** $-20\pi\,\text{in}^3/\text{s}$; decreasing

9. (a) $\dfrac{d\theta}{dt} = \dfrac{\cos^2\theta}{x^2}\left(x\dfrac{dy}{dt} - y\dfrac{dx}{dt}\right)$ **(b)** $-\dfrac{5}{16}\,\text{rad/s}$; decreasing

11. $\dfrac{4\pi}{15}\,\text{in}^2/\text{min}$ **13.** $\dfrac{1}{\sqrt{\pi}}\,\text{mi/h}$ **15.** $4860\pi\,\text{cm}^3/\text{min}$ **17.** $\dfrac{5}{6}\,\text{ft/s}$

19. $\dfrac{125}{\sqrt{61}}\,\text{ft/s}$ **21.** $704\,\text{ft/s}$

23. (a) 500 mi, 1716 mi   **(b)** 1354 mi; 27.7 mi/min

25. $\dfrac{9}{20\pi}\,\text{ft/min}$ **27.** $125\pi\,\text{ft}^3/\text{min}$ **29.** $250\,\text{mi/h}$

31. $\dfrac{36\sqrt{69}}{25}\,\text{ft/min}$ **33.** $\dfrac{8\pi}{5}\,\text{km/s}$ **35.** $600\sqrt{7}\,\text{mi/h}$

37. (a) $-\dfrac{60}{7}$ units per second   **(b)** falling   **39.** -4 units per second

41. $x = \pm\sqrt{\dfrac{-5+\sqrt{33}}{2}}$ **43.** 4.5 cm/s; away   **47.** $\dfrac{20}{9\pi}\,\text{cm/s}$

▶ **Exercise Set 3.5 (Page 217)**

1. (a) $f(x) \approx 1 + 3(x-1)$ **(b)** $f(1+\Delta x) \approx 1 + 3\Delta x$ **(c)** 1.06

3. (a) $1 + \dfrac{1}{2}x$, 0.95, 1.05   **17.** $|x| < 1.692$

(b)

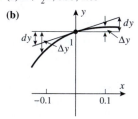

19. $|x| < 0.3158$ **21. (a)** 0.0174533   **(b)** $x_0 = 45°$ **(c)** 0.694765

23. 83.16   **25.** 8.0625   **27.** 8.9944   **29.** 0.1   **31.** 0.8573

33. 0.780398   **37. (a)** 4, 5   **(b)**

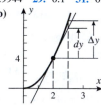

39. $3x^2\,dx$, $3x^2\Delta x + 3x(\Delta x)^2 + (\Delta x)^3$

41. $(2x - 2)\,dx$, $2x\Delta x + (\Delta x)^2 - 2\Delta x$

43. (a) $(12x^2 - 14x)\,dx$ **(b)** $(-x\sin x + \cos x)\,dx$

45. (a) $\dfrac{2-3x}{2\sqrt{1-x}}\,dx$ **(b)** $-17(1+x)^{-18}\,dx$

Responses to True–False questions may be abridged to save space.

47. False; $dy = (dy/dx)\,dx$ **49.** False; consider any linear function.

51. 0.0225   **53.** 0.0048   **55. (a)** $\pm 2\,\text{ft}^2$ **(b)** side: $\pm 1\%$; area: $\pm 2\%$

57. (a) opposite: ± 0.151 in; adjacent: ± 0.087 in

   (b) opposite: $\pm 3.0\%$; adjacent: $\pm 1.0\%$

59. $\pm 10\%$ **61.** $\pm 0.017\,\text{cm}^2$ **63.** $\pm 6\%$ **65.** $\pm 0.5\%$ **67.** $15\pi/2\,\text{cm}^3$

69. (a) $\alpha = 1.5 \times 10^{-5}/°\text{C}$ **(b)** 180.1 cm long

▶ **Exercise Set 3.6 (Page 226)**

1. (a) $\dfrac{2}{3}$ **(b)** $\dfrac{2}{3}$

Responses to True–False questions may be abridged to save space.

3. True; the expression $(\ln x)/x$ is undefined if $x \le 0$.

5. False; applying L'Hôpital's rule repeatedly shows that the limit is 0.

7. 1   **9.** 1   **11.** -1 **13.** 0   **15.** $-\infty$ **17.** 0   **19.** 2   **21.** 0

23. π **25.** $-\dfrac{5}{3}$ **27.** e^{-3} **29.** e^2 **31.** $e^{2/\pi}$ **33.** 0   **35.** $\dfrac{1}{2}$

37. $+\infty$ **39.** 1   **41.** 1   **43.** 1   **45.** 1   **47. (b)** 2

49. 0

51. e^3

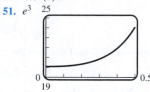

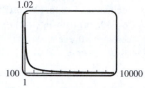

53. no horizontal asymptote   **55.** $y = 1$

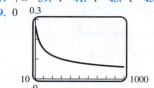

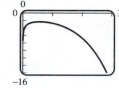

57. (a) 0   **(b)** $+\infty$ **(c)** 0   **(d)** $-\infty$ **(e)** $+\infty$ **(f)** $-\infty$ **59.** 1

61. does not exist   **63.** Vt/L **67. (a)** no   **(b)** Both limits equal 0.

69. does not exist

▶ **Chapter 3 Review Exercises (Page 228)**

1. (a) $\dfrac{2-3x^2-y}{x}$ **(b)** $-\dfrac{1}{x^2} - 2x$ **3.** $-\dfrac{y^2}{x^2}$

5. $\dfrac{y\sec(xy)\tan(xy)}{1 - x\sec(xy)\tan(xy)}$ **7.** $-\dfrac{21}{16y^3}$

9. $2/(2-\pi)$ **13.** $(\sqrt[3]{4}/3, \sqrt[3]{2}/3)$

15. $\dfrac{1}{x+1} + \dfrac{2}{x+2} - \dfrac{3}{x+3} - \dfrac{4}{x+4}$ **17.** $\dfrac{1}{x}$ **19.** $\dfrac{1}{3x(\ln x + 1)^{2/3}}$

21. $\dfrac{1}{(\ln 10)x \ln x}$ **23.** $\dfrac{3}{2x} + \dfrac{2x^3}{1+x^4}$ **25.** $2x$ **27.** $e^{\sqrt{x}}(2 + \sqrt{x})$

29. $\dfrac{2}{\pi(1+4x^2)}$ **31.** $e^x x^{(e^x)}\left(\ln x + \dfrac{1}{x}\right)$ **33.** $\dfrac{1}{|2x+1|\sqrt{x^2+x}}$

35. $\dfrac{x^3}{\sqrt{x^2+1}}\left(\dfrac{3}{x} - \dfrac{x}{x^2+1}\right)$

37. (b)

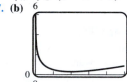

   (d) curve must have a horizontal tangent line between $x = 1$ and $x = e$

   (e) $x = 2$

39. e^2 **41.** $e^{1/e}$ **43.** No; for example, $f(x) = x^3$.   **45.** $\left(\dfrac{1}{3}, e\right)$

51. (a) 100

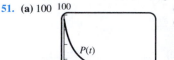

   (b) The population tends to 19.

   (c) The *rate* tends to zero.

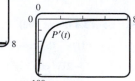

53. $+\infty, +\infty$: yes; $+\infty, -\infty$: no; $-\infty, +\infty$: no; $-\infty, -\infty$: yes

55. $+\infty$ **57.** $\dfrac{1}{9}$ **59.** $500\pi\,\text{m}^2/\text{min}$

61. (a) $-0.5, 1, 0.5$ (b) $\pi/4, 1, \pi/2$ (c) $3, -1.0$

63. (a) between 139.48 m and 144.55 m   (c) $|d\phi| \le 0.98°$

▶ **Chapter 3 Making Connections** (Page 230)

Answers are provided in the Student Solutions Manual.

▶ **Exercise Set 4.1** (Page 241)

1. (a) $f' > 0, f'' > 0$ (b) $f' > 0, f'' < 0$

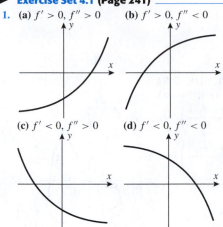

(c) $f' < 0, f'' > 0$ (d) $f' < 0, f'' < 0$

3. A: $dy/dx < 0, d^2y/dx^2 > 0$; B: $dy/dx > 0, d^2y/dx^2 < 0$;
C: $dy/dx < 0, d^2y/dx^2 < 0$ **5.** $x = -1, 0, 1, 2$

7. (a) $[4, 6]$ (b) $[1, 4], [6, 7]$ (c) $(1, 2), (3, 5)$ (d) $(2, 3), (5, 7)$
(e) $x = 2, 3, 5$

9. (a) $[1, 3]$ (b) $(-\infty, 1], [3, +\infty)$ (c) $(-\infty, 2), (4, +\infty)$ (d) $(2, 4)$
(e) $x = 2, 4$

Responses to True–False questions may be abridged to save space.

11. True; see definition of decreasing: $f(x_1) > f(x_2)$ whenever
$0 \le x_1 < x_2 \le 2$.

13. False; for example, $f(x) = (x - 1)^3$ is increasing on $[0, 2]$ and
$f'(1) = 0$.

15. (a) $[3/2, +\infty)$ (b) $(-\infty, 3/2]$ (c) $(-\infty, +\infty)$ (d) none   (e) none

17. (a) $(-\infty, +\infty)$ (b) none   (c) $(-1/2, +\infty)$ (d) $(-\infty, -1/2)$ (e) $-1/2$

19. (a) $[1, +\infty)$ (b) $(-\infty, 1]$ (c) $(-\infty, 0), \left(\frac{2}{3}, +\infty\right)$ (d) $\left(0, \frac{2}{3}\right)$ (e) $0, \frac{2}{3}$

21. (a) $\left[\frac{3 - \sqrt{5}}{2}, \frac{3 + \sqrt{5}}{2}\right]$ (b) $\left(-\infty, \frac{3 - \sqrt{5}}{2}\right], \left[\frac{3 + \sqrt{5}}{2}, +\infty\right)$

(c) $\left(0, \frac{4 - \sqrt{6}}{2}\right), \left(\frac{4 + \sqrt{6}}{2}, +\infty\right)$ (d) $(-\infty, 0), \left(\frac{4 - \sqrt{6}}{2}, \frac{4 + \sqrt{6}}{2}\right)$

(e) $0, \frac{4 \pm \sqrt{6}}{2}$

23. (a) $[-1/2, +\infty)$ (b) $(-\infty, -1/2]$ (c) $(-2, 1)$
(d) $(-\infty, -2), (1, +\infty)$ (e) $-2, 1$

25. (a) $[-1, 0], [1, +\infty)$ (b) $(-\infty, -1], [0, 1]$ (c) $(-\infty, 0), (0, +\infty)$
(d) none   (e) none

27. (a) $(-\infty, 0]$ (b) $[0, +\infty)$ (c) $(-\infty, -1), (1, +\infty)$
(d) $(-1, 1)$ (e) $-1, 1$

29. (a) $[0, +\infty)$ (b) $(-\infty, 0]$ (c) $(-2, 2)$
(d) $(-\infty, -2), (2, +\infty)$ (e) $-2, 2$

31. (a) $[0, +\infty)$ (b) $(-\infty, 0]$ (c) $\left(-\sqrt{\frac{1 + \sqrt{7}}{3}}, \sqrt{\frac{1 + \sqrt{7}}{3}}\right)$

(d) $\left(-\infty, -\sqrt{\frac{1 + \sqrt{7}}{3}}\right), \left(\sqrt{\frac{1 + \sqrt{7}}{3}}, +\infty\right)$ (e) $\pm\sqrt{\frac{1 + \sqrt{7}}{3}}$

33. increasing: $[-\pi/4, 3\pi/4]$; decreasing: $[-\pi, -\pi/4], [3\pi/4, \pi]$;
concave up: $(-3\pi/4, \pi/4)$; concave down: $(-\pi, -3\pi/4), (\pi/4, \pi)$;
inflection points: $-3\pi/4, \pi/4$

35. increasing: none; decreasing: $(-\pi, \pi)$; concave up: $(-\pi, 0)$;
concave down: $(0, \pi)$; inflection point: 0

37. increasing: $[-\pi, -3\pi/4], [-\pi/4, \pi/4], [3\pi/4, \pi]$; decreasing:
$[-3\pi/4, -\pi/4], [\pi/4, 3\pi/4]$; concave up: $(-\pi/2, 0), (\pi/2, \pi)$;
concave down: $(-\pi, -\pi/2), (0, \pi/2)$; inflection points: $0, \pm\pi/2$

39. (a)   (b)

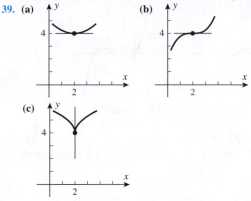

(c)

41. $1 + \frac{1}{3}x - \sqrt[3]{1 + x} \ge 0$ if $x > 0$ **43.** $x \ge \sin x$

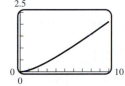

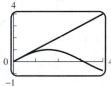

47.

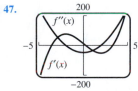

points of inflection at $x = -2, 2$;
concave up on $[-5, -2], [2, 5]$;
concave down on $[-2, 2]$;
increasing on $[-3.5829, 0.2513]$
and $[3.3316, 5]$;
decreasing on $[-5, -3.5829]$,
$[0.2513, 3.3316]$

49. $-2.464202, 0.662597, 2.701605$ **53.** (a) true   (b) false

57. (c) inflection point $(1, 0)$; concave up on $(1, +\infty)$;
concave down on $(-\infty, 1)$

63.

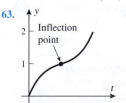

65.

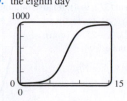

67. (a) $\dfrac{LAk}{(1 + A)^2}$ **69.** the eighth day

(c) $\dfrac{1}{k} \ln A$

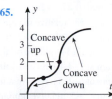

▶ **Exercise Set 4.2 (Page 252)**

1. (a) 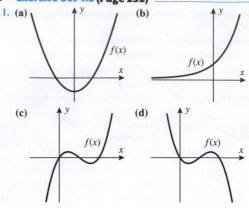 **(b)**

(c) **(d)**

5. (b) nothing **(c)** f has a relative minimum at $x = 1$, g has no relative extremum at $x = 1$.

7. critical: $0, \pm\sqrt{2}$; stationary: $0, \pm\sqrt{2}$

9. critical: $-3, 1$; stationary: $-3, 1$ **11.** critical: $0, \pm 5$; stationary: 0

13. critical: $n\pi/2$ for every integer n; stationary: $n\pi + \pi/2$ for every integer n

Responses to True–False questions may be abridged to save space.

15. False; for example, $f(x) = (x-1)^2(x-1.5)$ has a relative maximum at $x = 1$, but $f(2) = 0.5 > 0 = f(1)$.

17. False; to apply the second derivative test (Theorem 4.2.4) at $x = 1$, $f'(1)$ must equal 0.

19.   **21. (a)** none **(b)** $x = 1$ **(c)** none
(d)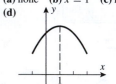

23. (a) 2 **(b)** 0 **(c)** 1, 3
(d)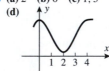

25. 0 (neither), $\sqrt[3]{5}$ (min) **27.** -2 (min), $2/3$ (max) **29.** 0 (min)

31. -1 (min), 1 (max) **33.** relative maximum at $(4/3, 19/3)$

35. relative maximum at $(\pi/4, 1)$; relative minimum at $(3\pi/4, -1)$

37. relative maximum at $(1, 1)$; relative minima at $(0, 0)$, $(2, 0)$

39. relative maximum at $(-1, 0)$; relative minimum at $(-3/5, -108/3125)$

41. relative maximum at $(-1, 1)$; relative minimum at $(0, 0)$

43. no relative extrema **45.** relative minimum at $(0, \ln 2)$

47. relative minimum at $(-\ln 2, -1/4)$

49. relative maximum at $(3/2, 9/4)$; relative minima at $(0, 0)$, $(3, 0)$

51. intercepts: $(0, -4)$, $(-1, 0)$, $(4, 0)$;
stationary point: $(3/2, -25/4)$ (min);
inflection points: none

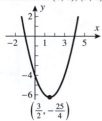

53. intercepts: $(0, 5)$, $\left(\dfrac{-7 \pm \sqrt{57}}{4}, 0\right)$, $(5, 0)$;
stationary points: $(-2, 49)$
(max), $(3, -76)$ (min);
inflection point: $(1/2, -27/2)$

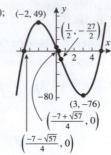

55. intercepts: $(-1, 0)$, $(0, 0)$, $(2, 0)$;
stationary points: $(-1, 0)$ (max),
$\left(\dfrac{1 - \sqrt{3}}{2}, \dfrac{9 - 6\sqrt{3}}{4}\right)$ (min),
$\left(\dfrac{1 + \sqrt{3}}{2}, \dfrac{9 + 6\sqrt{3}}{4}\right)$ (max);
inflection points: $\left(-\dfrac{1}{\sqrt{2}}, \dfrac{5}{4} - \sqrt{2}\right)$,
$\left(\dfrac{1}{\sqrt{2}}, \dfrac{5}{4} + \sqrt{2}\right)$,

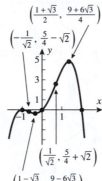

57. intercepts: $(0, -1)$, $(-1, 0)$, $(1, 0)$;
stationary points: $(-1/2, -27/16)$ (min),
$(1, 0)$ (neither);
inflection points: $(0, -1)$, $(1, 0)$

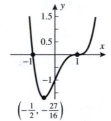

59. intercepts: $(-1, 0)$, $(0, 0)$, $(1, 0)$;
stationary points: $(-1, 0)$ (max),
$\left(-\dfrac{1}{\sqrt{5}}, -\dfrac{16}{25\sqrt{5}}\right)$ (min),
$\left(\dfrac{1}{\sqrt{5}}, \dfrac{16}{25\sqrt{5}}\right)$ (max), $(1, 0)$ (min);
inflection points: $\left(-\sqrt{\dfrac{3}{5}}, -\dfrac{4}{25}\sqrt{\dfrac{3}{5}}\right)$,
$(0, 0)$, $\left(\sqrt{\dfrac{3}{5}}, \dfrac{4}{25}\sqrt{\dfrac{3}{5}}\right)$

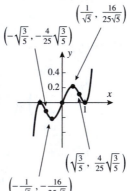

61. (a)

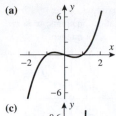

(b)

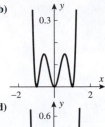

(c)

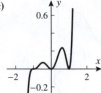

(d)

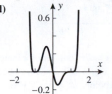

63. relative min of 0 at $x = \pi/2, \pi, 3\pi/2$;
relative max of 1 at $x = \pi/4, 3\pi/4, 5\pi/4, 7\pi/4$

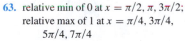

65. relative min of 0 at $x = \pi/2, 3\pi/2$;
relative max of 1 at $x = \pi$

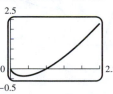

67. relative min of $-1/e$ at $x = 1/e$

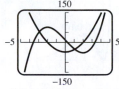

69. relative min of 0 at $x = 0$;
relative max of $1/e^2$ at $x = 1$

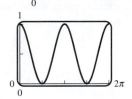

71. relative minima at $x = -3.58, 3.33$;
relative max at $x = 0.25$

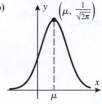

73. relative maximum at $x \approx -0.272$; relative minimum at $x \approx 0.224$
75. relative maximum at $x = 0$; relative minima at $x \approx \pm 0.618$
77. (a) 54 **(b)** 9
79. (b)

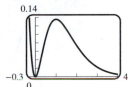

▶ **Exercise Set 4.3 (Page 264)**

1. stationary points: none;
inflection points: none;
asymptotes: $x = 4$, $y = -2$;
asymptote crossings: none

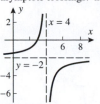

3. stationary points: none;
inflection point: $(0, 0)$;
asymptotes: $x = \pm 2$, $y = 0$;
asymptote crossings: $(0, 0)$

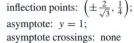

5. stationary point: $(0, 0)$;
inflection points: $\left(\pm \frac{2}{\sqrt{3}}, \frac{1}{4}\right)$;
asymptote: $y = 1$;
asymptote crossings: none

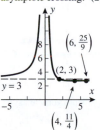

7. stationary point: $(0, -1)$;
inflection points: $(0, -1)$,
$\left(-\frac{1}{\sqrt[3]{2}}, -\frac{1}{3}\right)$;
asymptotes: $x = 1$, $y = 1$;
asymptote crossings: none

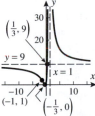

9. stationary point: $(4, 11/4)$;
inflection point: $(6, 25/9)$;
asymptotes: $x = 0$, $y = 3$;
asymptote crossing: $(2, 3)$

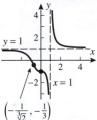

11. stationary point: $(-1/3, 0)$;
inflection point: $(-1, 1)$;
asymptotes: $x = 1$, $y = 9$;
asymptote crossing: $(1/3, 9)$

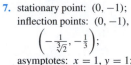

13. stationary points: none;
inflection points: none;
asymptotes: $x = 1$, $y = -1$;
asymptote crossings: none

15. (a)

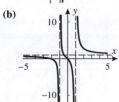

(b)

17.

19. stationary point: $\left(-\frac{1}{\sqrt[3]{2}}, \frac{3}{2}\sqrt[3]{2}\right)$; $\left(-\frac{1}{\sqrt[3]{2}}, \frac{3}{2}\sqrt[3]{2}\right)$

inflection point: $(1, 0)$;

asymptotes: $y = x^2, x = 0$;

asymptote crossings: none

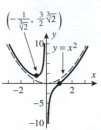

21. stationary points: $(-4, -27/2), (2, 0)$;

inflection point: $(2, 0)$;

asymptotes: $x = 0, y = x - 6$;

asymptote crossing: $(2/3, -16/3)$

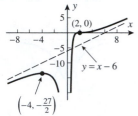

23. stationary points: $(-3, 23), (0, -4)$;

inflection point: $(0, -4)$;

asymptotes: $x = -2, y = x^2 - 2x$;

asymptote crossings: none

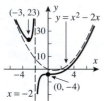

25. (a) VI **(b)** I **(c)** III **(d)** V **(e)** IV **(f)** II

Responses to True–False questions may be abridged to save space.

27. True; if deg $P >$ deg Q, then $f(x)$ is unbounded as $x \to \pm\infty$; if deg $P <$ deg Q, then $f(x) \to 0$ as $x \to \pm\infty$.

29. False; for example, $f(x) = (x - 1)^{1/3}$ is continuous (with vertical tangent line) at $x = 1$, but $f'(x) = \dfrac{1}{3(x-1)^{2/3}}$ has a vertical asymptote at $x = 1$.

31. critical points: $(\pm 1/2, 0)$; **33.** critical points: $(-1, 1), (0, 0)$;

inflection points: none inflection points: none

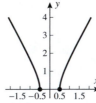
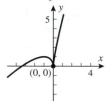

35. critical points: $(0, 0), (1, 3)$; inflection points: $(0, 0), (-2, -6\sqrt[3]{2})$.

It's hard to see all the important features in one graph, so two graphs are shown:

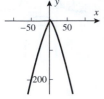

37. critical points: $(0, 4), (1, 3)$;

inflection points: $(0, 4), (8, 4)$

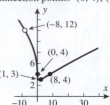

39. extrema: none;

inflection points: $x = \pi n$

for integers n

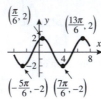

41. minima: $x = 7\pi/6 + 2\pi n$ for integers n;

maxima: $x = \pi/6 + 2\pi n$ for integers n;

inflection points: $x = 2\pi/3 + \pi n$

for integers n

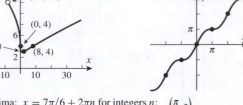

43. relative minima: 1 at $x = \pi$; -1 at $x = 0, 2\pi$;

relative maxima: $5/4$ at $x = -2\pi/3, 2\pi/3, 4\pi/3, 8\pi/3$;

inflection points where $\cos x = \dfrac{-1 \pm \sqrt{33}}{8}$: $(-2.57, 1.13)$,

$(-0.94, 0.06), (0.94, 0.06), (2.57, 1.13), (3.71, 1.13), (5.35, 0.06)$,

$(7.22, 0.06), (8.86, 1.13)$

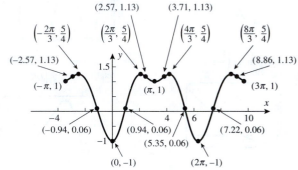

45. (a) $+\infty, 0$ **47. (a)** $0, +\infty$

(b)

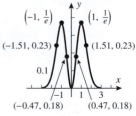

(b)

49. (a) $0, 0$

(b) relative max $= 1/e$ at $x = \pm 1$; relative min $= 0$ at $x = 0$;

inflection points where $x = \pm\sqrt{\dfrac{5 \pm \sqrt{17}}{4}}$:

about $(\pm 0.47, 0.18), (\pm 1.51, 0.23)$; asymptote: $y = 0$

51. **(a)** $-\infty, 0$
(b) relative max $= -e^2$
   at $x = 2$;
no relative min;
no inflection points;
asymptotes: $y = 0, x = 1$

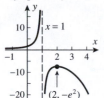
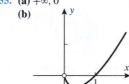

53. **(a)** $0, +\infty$
(b) critical points at $x = 0.2$;
relative min at $x = 0$,
relative max at $x = 2$;
points of inflection at $x = 2 \pm \sqrt{2}$;
horizontal asymptote $y = 0$
   as $x \to +\infty$
$$\lim_{x \to -\infty} f(x) = +\infty$$

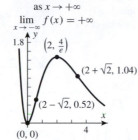
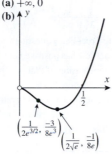

55. **(a)** $+\infty, 0$
(b)

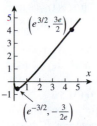

57. **(a)** $+\infty, 0$
(b)

59. **(a)** $+\infty, 0$
(b) no relative max; relative min $= -\dfrac{3}{2e}$ at $x = e^{-3/2}$;
inflection point: $(e^{3/2}, 3e/2)$;
no asymptotes. It's hard to see all the important features in one graph, so two graphs are shown:

61. **(a)**

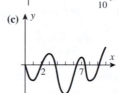

(b) relative max at $x = 1/b$;
inflection point at $x = 2/b$

63. **(a)** does not exist, 0
(b) $y = e^x$ and $y = e^x \cos x$ intersect for $x = 2\pi n$, and $y = -e^x$ and $y = e^x \cos x$ intersect for $x = 2\pi n + \pi$, for all integers n.

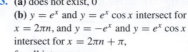
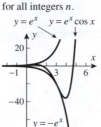

(c) Graphs for $a = 1$:

65. **(a)** $x = 1, 2.5, 3, 4$ **(b)** $(-\infty, 1], [2.5, 3]$
(c) relative max at $x = 1, 3$;
relative min at $x = 2.5$ **(d)** $x \approx 0.6, 1.9, 4$

67.

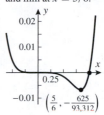

69. Graph misses zeros at $x = 0, 1$
and min at $x = 5/6$.

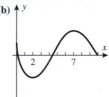

▶ **Exercise Set 4.4 (Page 272)**

1. relative maxima at $x = 2, 6$; absolute max at $x = 6$;
relative min at $x = 4$; absolute minima at $x = 0, 4$

3. **(a)**

(b)

(c)

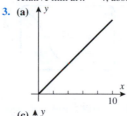

7. max $= 2$ at $x = 1, 2$;
min $= 1$ at $x = 3/2$
9. max $= 8$ at $x = 4$;
min $= -1$ at $x = 1$
11. maximum value $3/\sqrt{5}$ at $x = 1$;
minimum value $-3/\sqrt{5}$ at $x = -1$
13. max $= \sqrt{2} - \pi/4$ at $x = -\pi/4$;
min $= \pi/3 - \sqrt{3}$ at $x = \pi/3$
15. maximum value 17 at $x = -5$; minimum value 1 at $x = -3$.

Responses to True–False questions may be abridged to save space.

17. True; see the Extreme-Value Theorem (4.4.2).
19. True; see Theorem 4.4.3.
21. no maximum; min $= -9/4$ at $x = 1/2$
23. maximum value $f(1) = 1$; no minimum
25. no maximum or minimum
27. max $= -2 - 2\sqrt{2}$ at $x = -1 - \sqrt{2}$; no minimum
29. no maximum; min $= 0$ at $x = 0, 2$

31. maximum value 48 at $x = 8$;
minimum value 0 at $x = 0, 20$

33. no maximum or minimum

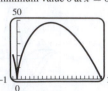

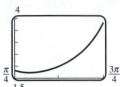

35. $\max = 2\sqrt{2} + 1$ at $x = 3\pi/4$; $\min = \sqrt{3}$ at $x = \pi/3$

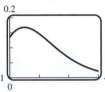

37. maximum value $\frac{27}{8}e^{-3}$ at $x = \frac{3}{2}$;
minimum value $64/e^8$ at $x = 4$

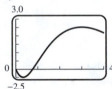

39. $\max = 5 \ln 10 - 9$ at $x = 3$;
$\min = 5 \ln(10/9) - 1$ at $x = 1/3$

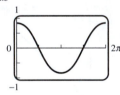

41. maximum value $\sin(1) \approx 0.84147$;
minimum value $-\sin(1) \approx -0.84147$

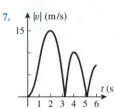

43. maximum value 2; minimum value $-\frac{1}{4}$

45. $\max = 3$ at $x = 2n\pi$; $\min = -3/2$ at $x = \pm 2\pi/3 + 2n\pi$
for any integer n **49.** 2, at $x = 1$

53. maximum $y = 4$ at $t = \pi, 3\pi$; minimum $y = 0$ at $t = 0, 2\pi$

▶ **Exercise Set 4.5 (Page 283)** ───────────

1. (a) 1 (b) $\frac{1}{2}$ **3.** 500 ft parallel to stream, 250 ft perpendicular
5. 500 ft ($3 fencing) × 750 ft ($2 fencing) **7.** 5 in × $\frac{12}{5}$ in
9. $10\sqrt{2} \times 10\sqrt{2}$ **11.** 80 ft ($1 fencing), 40 ft ($2 fencing)
15. maximum area is 108 when $x = 2$
17. maximum area is 144 when $x = 2$
19. 11,664 in^3 **21.** $\frac{200}{27}$ ft^3 **23.** base 10 cm square, height 20 cm
25. ends $\sqrt[3]{3V/4}$ units square, length $\frac{4}{3}\sqrt[3]{3V/4}$
27. height $= 2R/\sqrt{3}$, radius $= \sqrt{2/3}R$
31. height $=$ radius $= \sqrt[3]{500/\pi}$ cm **33.** $L/12$ by $L/12$ by $L/12$
35. height $= L/\sqrt{3}$, radius $= \sqrt{2/3}L$
37. height $= 2\sqrt[3]{75/\pi}$ cm, radius $= \sqrt{2}\sqrt[3]{75/\pi}$ cm
39. height $= 4R$, radius $= \sqrt{2}R$
41. $R(x) = 225x - 0.25x^2$; $R'(x) = 225 - 0.5x$; 450 tons
43. (a) 7000 units (b) yes (c) $15 **45.** 13,722 lb **47.** $3\sqrt{3}$
49. height $= r/\sqrt{2}$ **51.** $\left(\sqrt{2}, \frac{1}{2}\right)$ **53.** $\left(-1/\sqrt{3}, \frac{3}{4}\right)$
55. (a) π mi (b) $2\sin^{-1}(1/4)$ mi **57.** $4(1 + 2^{2/3})^{3/2}$ ft
59. 30 cm from the weaker source **61.** $\sqrt{24} = 2\sqrt{6}$ ft

▶ **Exercise Set 4.6 (Page 294)** ───────────

1. (a) positive, negative, slowing down
(b) positive, positive, speeding up
(c) negative, positive, slowing down
3. (a) left **5.**

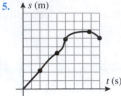

(b) negative
(c) speeding up
(d) slowing down

7.

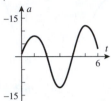

Responses to True–False questions may be abridged to save space.
9. False; a particle has positive velocity when its position versus time
graph is increasing; if that positive velocity is decreasing, the particle
would be slowing down.
11. False; acceleration is the derivative of velocity (with respect to time);
speed is the absolute value of velocity.
13. (a) 7.5 ft/s^2 (b) $t = 0$ s
15. (a)

t	s	v	a
1	0.71	0.56	-0.44
2	1	0	-0.62
3	0.71	-0.56	-0.44
4	0	-0.79	0
5	-0.71	-0.56	0.44

(b) stopped at $t = 2$;
moving right at $t = 1$;
moving left at $t = 3, 4, 5$
(c) speeding up at $t = 3$;
slowing down at $t = 1, 5$;
neither at $t = 2, 4$

17. (a) $v(t) = 3t^2 - 6t$, $a(t) = 6t - 6$
(b) $s(1) = -2$ ft, $v(1) = -3$ ft/s, $|v(1)| = 3$ ft/s, $a(1) = 0$ ft/s^2
(c) $t = 0, 2$ s (d) speeding up for $0 < t < 1$ and $2 < t$, slowing
down for $1 < t < 2$ (e) 58 ft
19. (a) $v(t) = 3\pi \sin(\pi t/3)$, $a(t) = \pi^2 \cos(\pi t/3)$ (b) $s(1) = 9/2$ ft,
$v(1) = $ speed $= 3\sqrt{3}\pi/2$ ft/s, $a(1) = \pi^2/2$ ft/s^2 (c) $t = 0$ s, 3 s
(d) speeding up: $0 < t < 1.5, 3 < t < 4.5$;
slowing down: $1.5 < t < 3, 4.5 < t < 5$ (e) 31.5 ft
21. (a) $v(t) = -\frac{1}{3}(t^2 - 6t + 8)e^{-t/3}$, $a(t) = \frac{1}{9}(t^2 - 12t + 26)e^{-t/3}$
(b) $s(1) = 9e^{-1/3}$ ft, $v(1) = -e^{-1/3}$ ft/s, speed $= e^{-1/3}$ ft/s, $a(1) = \frac{5}{3}e^{-1/3}$ ft/s^2 (c) $t = 2$ s, 4 s
(d) speeding up: $2 < t < 6 - \sqrt{10}, 4 < t < 6 + \sqrt{10}$; slowing
down: $0 < t < 2, 6 - \sqrt{10} < t < 4, 6 + \sqrt{10} < t$
(e) $8 - 24e^{-2/3} + 48e^{-4/3} - 33e^{-5/3}$
23. (a) $\sqrt{5}$ (b) $\sqrt{5}/10$

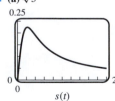

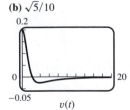

(c) speeding up for $\sqrt{5} < t < \sqrt{15}$;
slowing down for $0 < t < \sqrt{5}$ and $\sqrt{15} < t$

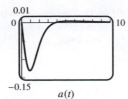

$a(t)$

25.

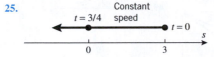

27.

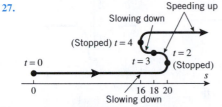

29.

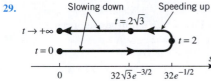

31.

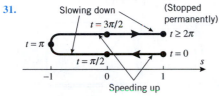

33. (a) 12 ft/s **(b)** $t = 2.2$ s, $s = -24.2$ ft
35. (a) $t = 2 \pm 1/\sqrt{3}$, $s = \ln 2$, $v = \pm\sqrt{3}$ **(b)** $t = 2, s = 0, a = 6$
37. (a)

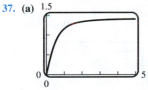

(b) $\sqrt{2}$

39. (b) $\frac{2}{3}$ unit **(c)** $0 \le t < 1$ and $t > 2$

▶ **Exercise Set 4.7 (Page 300)** _____
1. 1.414213562 **3.** 1.817120593 **5.** $x \approx 1.76929$
7. $x \approx 1.224439550$ **9.** $x \approx -1.24962$ **11.** $x \approx 1.02987$
13. $x \approx 4.493409458$ **15.** $x \approx 0.68233$

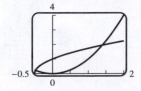

17. $-0.474626618, 1.395336994$ **19.** $x \approx 0.58853$ or 3.09636

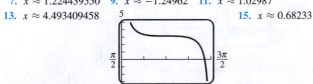

Responses to True–False questions may be abridged to save space.
21. True; $x = x_{n+1}$ is the x-intercept of the tangent line to $y = f(x)$ at
$x = x_n$.
23. False; for example, if $f(x) = x(x-3)^2$, Newton's Method fails (analogous to Figure 4.7.4) with $x_1 = 1$ and approximates the root $x = 3$
for $x_1 > 1$.
25. (b) 3.162277660 **27.** -4.098859132
29. $x = -1$ or $x \approx 0.17951$ **31.** (0.589754512, 0.347810385)
33. (b) $\theta \approx 2.99156$ rad or $171°$ **35.** $-1.220744085, 0.724491959$
37. $i = 0.053362$ or 5.33% **39. (a)** The values do not converge.

▶ **Exercise Set 4.8 (Page 308)** _____
1. $c = 4$ **3.** $c = \pi$ **5.** $c = 1$ **7.** $\frac{5}{4}$
9. (a) $[-2, 1]$
(b) $c \approx -1.29$
(c) -1.2885843

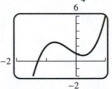

Responses to True–False questions may be abridged to save space.
11. False; Rolle's Theorem requires the additional hypothesis that f is differentiable on (a, b) and $f(a) = f(b) = 0$; see Example 2.
13. False; the Constant Difference Theorem applies to two functions with equal derivatives on an interval to conclude that the functions differ by a constant on the interval.
15. (b) $\tan x$ is not continuous on $[0, \pi]$. **25.** $f(x) = xe^x - e^x + 2$
35. (b) $f(x) = \sin x, g(x) = \cos x$
37. **41.** $a = 6, b = -3$

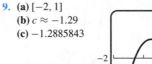

▶ **Chapter 4 Review Exercises (Page 310)** _____
1. (a) $f(x_1) < f(x_2); f(x_1) > f(x_2); f(x_1) = f(x_2)$
(b) $f' > 0; f' < 0; f' = 0$
3. (a) $\left[\frac{5}{2}, +\infty\right)$ **(b)** $\left(-\infty, \frac{5}{2}\right)$ **(c)** $(-\infty, +\infty)$ **(d)** none **(e)** none
5. (a) $[0, +\infty)$ **(b)** $(-\infty, 0]$ **(c)** $(-\sqrt{2/3}, \sqrt{2/3})$
(d) $(-\infty, -\sqrt{2/3}), (\sqrt{2/3}, +\infty)$ **(e)** $-\sqrt{2/3}, \sqrt{2/3}$
7. (a) $[-1, +\infty)$ **(b)** $(-\infty, -1]$ **(c)** $(-\infty, 0), (2, +\infty)$
(d) $(0, 2)$ **(e)** $0, 2$
9. (a) $(-\infty, 0]$ **(b)** $[0, +\infty)$ **(c)** $(-\infty, -1/\sqrt{2}), (1/\sqrt{2}, +\infty)$
(d) $(-1/\sqrt{2}, 1/\sqrt{2})$ **(e)** $\pm 1/\sqrt{2}$
11. increasing on $[\pi, 2\pi]$;
decreasing on $[0, \pi]$;
concave up on $(\pi/2, 3\pi/2)$;
concave down on $(0, \pi/2), (3\pi/2, 2\pi)$;
inflection points: $(\pi/2, 0), (3\pi/2, 0)$

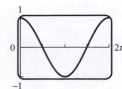

13. increasing on $[0, \pi/4], [3\pi/4, \pi]$;
decreasing on $[\pi/4, 3\pi/4]$;
concave up on $(\pi/2, \pi)$;
concave down on $(0, \pi/2)$;
inflection point: $(\pi/2, 0)$

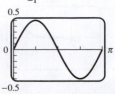

37.

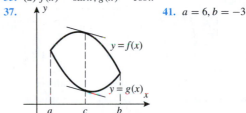

15. (a) **(b)** 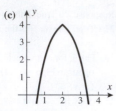 **(c)**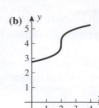

17. $-\dfrac{b}{2a} \le 0$ **19.** $x = -1$ **21. (a)** at an inflection point

25. (a) $x = \pm\sqrt{2}$ (stationary points)   **(b)** $x = 0$ (stationary point)

27. (a) relative max at $x = 1$, relative min at $x = 7$, neither at $x = 0$
   (b) relative max at $x = \pi/2, 3\pi/2$; relative min at $x = 7\pi/6, 11\pi/6$
   (c) relative max at $x = 5$

29. $\lim\limits_{x \to -\infty} f(x) = +\infty$, $\lim\limits_{x \to +\infty} f(x) = +\infty$;
   relative min at $x = 0$;
   points of inflection at $x = \frac{1}{2}, 1$;
   no asymptotes

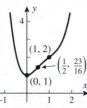

31. $\lim\limits_{x \to \pm\infty} f(x)$ does not exist; critical point at $x = 0$; relative min
   at $x = 0$; point of inflection when $1 + 4x^2 \tan(x^2 + 1) = 0$;
   vertical asymptotes at $x = \pm\sqrt{\pi\left(n + \frac{1}{2}\right) - 1}, n = 0, 1, 2, \ldots$

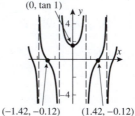

   $(-1.42, -0.12)$ $(1.42, -0.12)$

33. critical points at $x = -5, 0$; relative max at $x = -5$, relative min at
   $x = 0$; points of inflection at $x \approx -7.26, -1.44, 1.20$; horizontal
   asymptote $y = 1$ for $x \to \pm\infty$

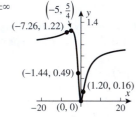

35. $\lim\limits_{x \to -\infty} f(x) = +\infty$, $\lim\limits_{x \to +\infty} f(x) = -\infty$;
   critical point at $x = 0$;
   no extrema;
      inflection point at $x = 0$
   (f changes concavity);
   no asymptotes

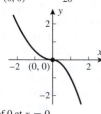

37. no relative extrema   **39.** relative min of 0 at $x = 0$

41. relative min of 0 at $x = 0$ **43.** relative min of 0 at $x = 0$

45. (a) 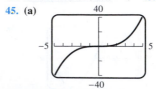   **(b)** relative max at $x = -\frac{1}{20}$;
   relative min at $x = \frac{1}{20}$

47. (a)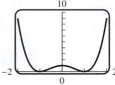

49. $f(x) = \dfrac{x^2 + x - 7}{3x^2 + x - 1}, x \ne \dfrac{1}{2}$

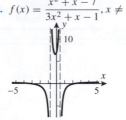

   horizontal asymptote $y = 1/3$
   vertical asymptotes at
   $x = (-1 \pm \sqrt{13})/6$

(c) The finer details can be seen
when graphing over a much smaller
x-window.

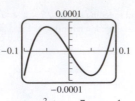

53. (a) true   **(b)** false

55. (a) no max; min $= -13/4$ at $x = 3/2$ **(b)** no max or min
   (c) no max; min $m = e^2/4$ at $x = 2$
   (d) no max; min $m = e^{-1/e}$ at $x = 1/e$

57. (a) minimum value 0 for $x = \pm 1$;   **(b)** max $= 1/2$ at $x = 1$;
   no maximum   min $= 0$ at $x = 0$

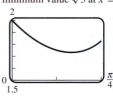   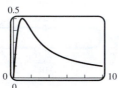

(c) maximum value 2 at $x = 0$;   **(d)** maximum value
   minimum value $\sqrt{3}$ at $x = \pi/6$ $f(-2 - \sqrt{3}) \approx 0.84$;
                                       minimum value
                                       $f(-2 + \sqrt{3}) \approx -0.06$

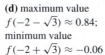   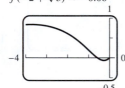

59. (a) 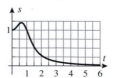   **(b)** minimum:
   $(-2.111985, -0.355116)$;
   maximum:
   $(0.372591, 2.012931)$

61. width $= 4\sqrt{2}$, height $= 3\sqrt{2}$ **63.** 2 in square

65. (a) yes   **(b)** yes

67. (a) $v = -2\dfrac{t(t^4 + 2t^2 - 1)}{(t^4 + 1)^2}, a = 2\dfrac{3t^8 + 10t^6 - 12t^4 - 6t^2 + 1}{(t^4 + 1)^3}$
   (b)

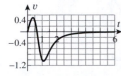   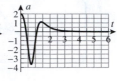

   (c) $t \approx 0.64, s \approx 1.2$ **(d)** $0 \le t \le 0.64$

(e) speeding up when $0 \le t < 0.36$ and $0.64 < t < 1.1$, otherwise slowing down   (f) maximum speed ≈ 1.05 when $t \approx 1.10$

69. $x \approx -2.11491, 0.25410, 1.86081$

71. $x \approx -1.165373043$ **73.** 249×10^6 km

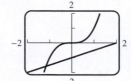

75. (a) yes, $c = 0$ (b) no
(c) yes, $c = \sqrt{\pi/2}$

77. use Rolle's Theorem

▶ Chapter 4 Making Connections (Page 314)

Where correct answers to a Making Connections exercise may vary, no answer is listed. Sample answers for these questions are available on the Book Companion Site.

1. (a) no zeros   (b) one   (c) $\lim\limits_{x \to +\infty} g'(x) = 0$

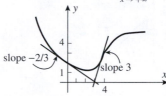

2. (a) $(-2.2, 4), (2, 1.2), (4.2, 3)$
(b) critical numbers at $x = -5.1, -2, 0.2, 2$;
local min at $x = -5.1, 2$; local max at $x = -2$;
no extrema at $x = 0.2$; $f''(1) \approx -1.2$

3. $x = -4, 5$ **4.** (d) $f(c) = 0$

6. (a) route (i): 10 s; route (iv): 10 s

(b) $2 \le x \le 5$; $\dfrac{4\sqrt{10}}{2.1} + \dfrac{5}{0.7} \approx 13.166$ s

(c) $0 \le x \le 2$; 10 s
(d) route (i) or (iv); 10 s

▶ Exercise Set 5.1 (Page 321)

1.

n	2	5	10	50	100
A_n	0.853553	0.749739	0.710509	0.676095	0.671463

3.

n	2	5	10	50	100
A_n	1.57080	1.93376	1.98352	1.99935	1.99984

5.

n	2	5	10	50	100
A_n	0.583333	0.645635	0.668771	0.688172	0.690653

7.

n	2	5	10	50	100
A_n	0.433013	0.659262	0.726130	0.774567	0.780106

9.

n	2	5	10	50	100
A_n	3.71828	2.85174	2.59327	2.39772	2.37398

11.

n	2	5	10	50	100
A_n	1.04720	0.75089	0.65781	0.58730	0.57894

13. $3(x-1)$ **15.** $x(x+2)$ **17.** $(x+3)(x-1)$

Responses to True–False questions may be abridged to save space.

19. False; the limit would be the area of the circle 4π.

21. True; this is the basis of the antiderivative method.

23. area $= A(6) - A(3)$ **27.** $f(x) = 2x; a = 2$

▶ Exercise Set 5.2 (Page 330)

1. (a) $\displaystyle\int \frac{x}{\sqrt{1+x^2}}\,dx = \sqrt{1+x^2} + C$

(b) $\displaystyle\int (x+1)e^x\,dx = xe^x + C$

5. $\dfrac{d}{dx}\left[\sqrt{x^3+5}\right] = \dfrac{3x^2}{2\sqrt{x^3+5}}$, so $\displaystyle\int \dfrac{3x^2}{2\sqrt{x^3+5}}\,dx = \sqrt{x^3+5} + C.$

7. $\dfrac{d}{dx}[\sin(2\sqrt{x})] = \dfrac{\cos(2\sqrt{x})}{\sqrt{x}}$, so $\displaystyle\int \dfrac{\cos(2\sqrt{x})}{\sqrt{x}}\,dx = \sin(2\sqrt{x}) + C.$

9. (a) $(x^9/9) + C$ (b) $\frac{7}{12}x^{12/7} + C$ (c) $\frac{2}{9}x^{9/2} + C$

11. $\dfrac{5}{2}x^2 - \dfrac{1}{6x^4} + C$ **13.** $-\dfrac{1}{2}x^{-2} - \dfrac{12}{5}x^{5/4} + \dfrac{8}{3}x^3 + C$

15. $(x^2/2) + (x^5/5) + C$ **17.** $3x^{4/3} - \frac{12}{7}x^{7/3} + \frac{3}{10}x^{10/3} + C$

19. $\dfrac{x^2}{2} - \dfrac{2}{x} + \dfrac{1}{3x^3} + C$ **21.** $2\ln|x| + 3e^x + C$

23. $-3\cos x - 2\tan x + C$ **25.** $\tan x + \sec x + C$

27. $\tan\theta + C$ **29.** $\sec x + C$ **31.** $\theta - \cos\theta + C$

33. $\frac{1}{2}\sin^{-1}x - 3\tan^{-1}x + C$ **35.** $\tan x - \sec x + C$

Responses to True–False questions may be abridged to save space.

37. True; this is Equations (1) and (2).

39. False; the initial condition is not satisfied since $y(0) = 2$.

41.

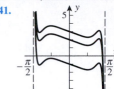

43. (a) $y(x) = \frac{3}{4}x^{4/3} + \frac{5}{4}$
(b) $y = -\cos t + t + 1 - \pi/3$
(c) $y(x) = \frac{2}{3}x^{3/2} + 2x^{1/2} - \frac{8}{3}$

45. (a) $y = 4e^x - 3$ (b) $y = \ln|t| + 5$

47. $s(t) = 16t^2 + 20$ **49.** $s(t) = 2t^{3/2} - 15$

51. $f(x) = \frac{4}{15}x^{5/2} + C_1 x + C_2$ **53.** $y = x^2 + x - 6$

55. $f(x) = \cos x + 1$ **57.** $y = x^3 - 6x + 7$

59. (a)

(b)

(c) $f(x) = \dfrac{x^2}{2} - 1$

61. (b)   **63.** (c)

67. (b) $\pi/2$ **69.** $\tan x - x + C$

71. (a) $\frac{1}{2}(x - \sin x) + C$ (b) $\frac{1}{2}(x + \sin x) + C$

73. $v = \dfrac{1087}{\sqrt{273}}T^{1/2}$ ft/s

▶ Exercise Set 5.3 (Page 338)

1. (a) $\dfrac{(x^2+1)^{24}}{24} + C$ (b) $-\dfrac{\cos^4 x}{4} + C$

3. (a) $\frac{1}{4}\tan(4x+1) + C$ (b) $\frac{1}{6}(1+2y^2)^{3/2} + C$

5. (a) $-\frac{1}{2}\cot^2 x + C$ (b) $\frac{1}{10}(1+\sin t)^{10} + C$

7. (a) $\frac{2}{7}(1+x)^{7/2} - \frac{4}{5}(1+x)^{5/2} + \frac{2}{3}(1+x)^{3/2} + C$
(b) $-\cot(\sin x) + C$

9. (a) $\ln|\ln x| + C$ (b) $-\frac{1}{5}e^{-5x} + C$

11. (a) $\frac{1}{3}\tan^{-1}(x^3) + C$ **(b)** $\sin^{-1}(\ln x) + C$

15. $\frac{1}{40}(4x-3)^{10} + C$ **17.** $-\frac{1}{7}\cos 7x + C$ **19.** $\frac{1}{4}\sec 4x + C$

21. $\frac{1}{2}e^{2x} + C$ **23.** $\frac{1}{2}\sin^{-1}(2x) + C$ **25.** $\frac{1}{21}(7t^2+12)^{3/2} + C$

27. $\dfrac{3}{2(1-2x)^2} + C$ **29.** $-\dfrac{1}{40(5x^4+2)^2} + C$ **31.** $e^{\sin x} + C$

33. $-\frac{1}{6}e^{-2x^3} + C$ **35.** $\tan^{-1}e^x + C$ **37.** $\frac{1}{5}\cos(5/x) + C$

39. $-\frac{1}{15}\cos^5 3t + C$ **41.** $\frac{1}{2}\tan(x^2) + C$ **43.** $-\frac{1}{6}(2-\sin 4\theta)^{3/2} + C$

45. $\sin^{-1}(\tan x) + C$ **47.** $\frac{1}{6}\sec^3 2x + C$ **49.** $-e^{-x} + C$

51. $-e^{-2\sqrt{x}} + C$ **53.** $\frac{1}{6}(2y+1)^{3/2} - \frac{1}{2}(2y+1)^{1/2} + C$

55. $-\frac{1}{2}\cos 2\theta + \frac{1}{6}\cos^3 2\theta + C$ **57.** $t + \ln|t| + C$

59. $\int [\ln(e^x) + \ln(e^{-x})]\,dx = C$

61. (a) $\sin^{-1}\left(\frac{1}{3}x\right) + C$ **(b)** $\frac{1}{\sqrt{5}}\tan^{-1}\left(\frac{x}{\sqrt{5}}\right) + C$

(c) $\dfrac{1}{\sqrt{\pi}}\sec^{-1}\left(\dfrac{x}{\sqrt{\pi}}\right) + C$

63. $\dfrac{1}{b}\dfrac{(a+bx)^{n+1}}{n+1} + C$ **65.** $\dfrac{1}{b(n+1)}\sin^{n+1}(a+bx) + C$

67. (a) $\frac{1}{2}\sin^2 x + C_1;\ -\frac{1}{2}\cos^2 x + C_2$ **(b)** They differ by a constant.

69. $\frac{2}{15}(5x+1)^{3/2} - \frac{158}{15}$ **71.** $y = -\frac{1}{2}e^{2t} + \frac{13}{2}$

73. (a) $\sqrt{x^2+1} + C$ **75.** $f(x) = \frac{2}{9}(3x+1)^{3/2} + \frac{7}{9}$

(b)
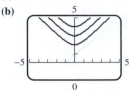

1. (a) 36 **(b)** 55 **(c)** 40 **(d)** 6 **(e)** 11 **(f)** 0 **3.** $\sum\limits_{k=1}^{10} k$ **5.** $\sum\limits_{k=1}^{10} 2k$

7. $\sum\limits_{k=1}^{6}(-1)^{k+1}(2k-1)$ **9. (a)** $\sum\limits_{k=1}^{50} 2k$ **(b)** $\sum\limits_{k=1}^{50}(2k-1)$ **11.** 5050

13. 2870 **15.** 214,365 **17.** $\frac{3}{2}(n+1)$ **19.** $\frac{1}{4}(n-1)^2$

Responses to True–False questions may be abridged to save space.

21. True; by parts (a) and (c) of Theorem 5.4.2.

23. False; consider $[a,b] = [-1,0]$.

25. (a) $\left(2+\dfrac{3}{n}\right)^4 \cdot \dfrac{3}{n}, \left(2+\dfrac{6}{n}\right)^4 \cdot \dfrac{3}{n}, \left(2+\dfrac{9}{n}\right)^4 \cdot \dfrac{3}{n},$

$\left(2+\dfrac{3(n-1)}{n}\right)^4 \cdot \dfrac{3}{n}, (2+3)^4 \cdot \dfrac{3}{n}$ **(b)** $\sum\limits_{k=0}^{n-1}\left(2+k\cdot\dfrac{3}{n}\right)^4 \dfrac{3}{n}$

27. (a) 46 **(b)** 52 **(c)** 58 **29. (a)** $\frac{\pi}{4}$ **(b)** 0 **(c)** $-\frac{\pi}{4}$

31. (a) 0.7188, 0.7058, 0.6982 **(b)** 0.6688, 0.6808, 0.6882

(c) 0.6928, 0.6931, 0.6931

33. (a) 4.8841, 5.1156, 5.2488 **(b)** 5.6841, 5.5156, 5.4088

(c) 5.3471, 5.3384, 5.3346

35. $\frac{15}{4}$ **37.** 18 **39.** 320 **41.** $\frac{15}{4}$ **43.** 18 **45.** 16 **47.** $\frac{1}{3}$ **49.** 0

51. $\frac{2}{3}$ **53. (b)** $\frac{1}{4}(b^4-a^4)$

55. $\dfrac{n^2+2n}{4}$ if n is even; $\dfrac{(n+1)^2}{4}$ if n is odd; **57.** $3^{17}-3^4$ **59.** $-\frac{399}{400}$

61. (b) $\frac{1}{2}$ **65. (a)** yes **(b)** yes

1. (a) $\frac{71}{6}$ **(b)** 2 **3. (a)** $-\frac{117}{16}$ **(b)** 3 **5.** $\displaystyle\int_{-1}^{2} x^2\,dx$

7. $\displaystyle\int_{-3}^{3} 4x(1-3x)\,dx$

9. (a) $\displaystyle\lim_{\max \Delta x_k \to 0} \sum_{k=1}^{n} 2x_k^* \Delta x_k;\, a=1, b=2$

(b) $\displaystyle\lim_{\max \Delta x_k \to 0} \sum_{k=1}^{n} \dfrac{x_k^*}{x_k^*+1}\Delta x_k;\, a=0, b=1$

13. (a) $A = \frac{9}{2}$ **(b)** $-A = -\frac{3}{2}$

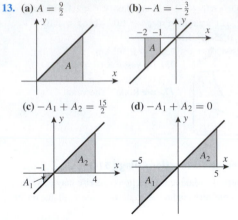

(c) $-A_1 + A_2 = \frac{15}{2}$ **(d)** $-A_1 + A_2 = 0$

15. (a) $A = 10$ **(b)** $A_1 - A_2 = 0$ by symmetry

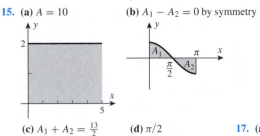

(c) $A_1 + A_2 = \frac{13}{2}$ **(d)** $\pi/2$

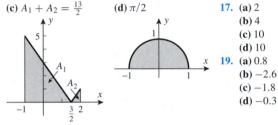

17. (a) 2
(b) 4
(c) 10
(d) 10
19. (a) 0.8
(b) -2.6
(c) -1.8
(d) -0.3

21. -1 **23.** 3 **25.** -4 **27.** $(1+\pi)/2$

Responses to True–False questions may be abridged to save space.

29. False; see Theorem 5.5.8(a).

31. False; consider $f(x) = x - 2$ on $[0,3]$.

33. (a) negative **(b)** positive **37.** $\frac{25}{2}\pi$ **39.** $\frac{5}{2}$

45. (a) integrable **(b)** integrable **(c)** not integrable **(d)** integrable

1. (a) $\int_0^2 (2-x)\,dx = 2$ **(b)** $\int_{-1}^{1} 2\,dx = 4$ **(c)** $\int_1^3 (x+1)\,dx = 6$

3. (a) $x^* = f(x^*) = 1$ **(b)** x^* is any point in $[-1,1], f(x^*) = 2$

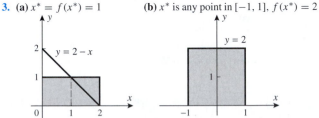

(c) $x^* = 2$, $f(x^*) = 3$

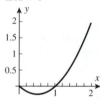

5. $\frac{65}{4}$ **7.** 14   **9.** $\frac{3}{2}$ **11. (a)** $\frac{4}{3}$ **(b)** -7 **13.** 48   **15.** 3   **17.** $\frac{845}{5}$
19. 0   **21.** $\sqrt{2}$ **23.** $5e^3 - 10$ **25.** $\pi/4$ **27.** $\pi/12$ **29.** -12
31. (a) $5/2$ **(b)** $2 - \frac{\sqrt{2}}{2}$ **33. (a)** $e + (1/e) - 2$ **(b)** 1

35. (a) $\frac{17}{6}$ **(b)** $F(x) = \begin{cases} \dfrac{x^2}{2}, & x \le 1 \\[2mm] \dfrac{x^3}{3} + \dfrac{1}{6}, & x > 1 \end{cases}$

Responses to True–False questions may be abridged to save space.
37. False; since $|x|$ is continuous, it has an antiderivative.
39. True; by the Fundamental Theorem of Calculus.
41. 0.6659; $\frac{2}{3}$ **43.** 3.1060; $\dfrac{1}{\tan 1}$ **45.** 12   **47.** $\frac{9}{2}$
49. area $= 1$ **51.** area $= e + e^{-1} - 2$

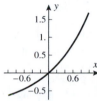

53. (b) degree mode, 0.93
55. (a) change in height from age 0 to age 10 years; inches
    (b) change in radius from time $t = 1$ s to time $t = 2$ s; centimeters
    (c) difference between speed of sound at $100°$ F and at $32°$ F; feet per
    second   **(d)** change in position from time t_1 to time t_2; centimeters
57. (a) $3x^2 - 3$ **59. (a)** $\sin(x^2)$ **(b)** $e^{\sqrt{x}}$ **61.** $-x \sec x$
63. (a) 0   **(b)** 5   **(c)** $\frac{4}{5}$
65. (a) $x = 3$ **(b)** increasing on $[3, +\infty)$, decreasing on $(-\infty, 3]$
    (c) concave up on $(-1, 7)$, concave down on $(-\infty, -1)$ and $(7, +\infty)$
67. (a) $(0, +\infty)$ **(b)** $x = 1$
69. (a) 120 gal   **(b)** 420 gal   **(c)** 2076.36 gal   **71.** 1

▶ **Exercise Set 5.7 (Page 382)**

1. (a) displacement $= 3$; distance $= 3$
    (b) displacement $= -3$; distance $= 3$
    (c) displacement $= -\frac{1}{2}$; distance $= \frac{3}{2}$
    (d) displacement $= \frac{3}{2}$; distance $= 2$
3. (a) 35.3 m/s   **(b)** 51.4 m/s   **5. (a)** $t^3 - t^2 + 1$ **(b)** $4t + 3 - \frac{1}{3}\sin 3t$
7. (a) $\frac{3}{2}t^2 + t - 4$ **(b)** $t + 1 - \ln t$
9. (a) displacement $= 1$ m; distance $= 1$ m
    (b) displacement $= -1$ m; distance $= 3$ m
11. (a) displacement $= \frac{9}{4}$ m; distance $= \frac{11}{4}$ m
    (b) displacement $= 2\sqrt{3} - 6$ m; distance $= 6 - 2\sqrt{3}$ m
13. 4, 13/3   **15.** 296/27, 296/27
17. (a) $s = 2/\pi$, $v = 1$, $|v| = 1$, $a = 0$
    (b) $s = \frac{1}{2}$, $v = -\frac{3}{2}$, $|v| = \frac{3}{2}$, $a = -3$ **19.** $t \approx 1.27$ s

21.

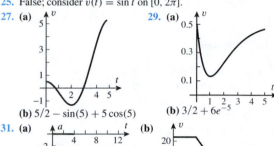

Responses to True–False questions may be abridged to save space.
23. True; if $a(t) = a_0$, then $v(t) = a_0 t + v_0$.
25. False; consider $v(t) = \sin t$ on $[0, 2\pi]$.
27. (a)   **29. (a)**

    (b) $5/2 - \sin(5) + 5\cos(5)$ **(b)** $3/2 + 6e^{-5}$
31. (a)   **(b)**

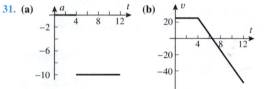

    (c) 120 cm, -20 cm   **(d)** 131.25 cm at $t = 6.5$ s
33. (a) $-\frac{121}{5}$ ft/s^2 **(b)** $\frac{70}{33}$ s   **(c)** $\frac{60}{11}$ s   **35.** 50 s, 5000 ft
37. (a) 16 ft/s, -48 ft/s   **(b)** 196 ft   **(c)** 112 ft/s
39. (a) 1 s   **(b)** $\frac{1}{2}$ s   **41. (a)** 6.122 s   **(b)** 183.7 m   **(c)** 6.122 s   **(d)** 60 m/s
43. (a) 5 s   **(b)** 272.5 m   **(c)** 10 s   **(d)** -49 m/s
    (e) 12.46 s   **(f)** 73.1 m/s   **45.** 113.42 ft/s

▶ **Exercise Set 5.8 (Page 388)**

1. (a) 4   **(c)**   **3.** 6
    (b) 2   **5.** $2/\pi$

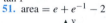

                               7. $\dfrac{1}{e-1}$

                               9. $\dfrac{\pi}{12(\sqrt{3} - 1)}$

11. $\dfrac{1 - e^{-8}}{8}$ **13. (a)** 5.28   **(b)** 4.305   **(c)** 4   **15. (a)** $-\frac{1}{6}$ **(b)** $\frac{1}{2}$
Responses to True–False questions may be abridged to save space.
19. False; let $g(x) = \cos x$; $f(x) = 0$ on $[0, 3\pi/2]$.
21. True; see Theorem 5.5.4(b).
23. (a) $\frac{263}{4}$ **(b)** 31   **25.** 1404π lb   **27.** 97 cars/min   **31.** 27

▶ **Exercise Set 5.9 (Page 394)**

1. (a) $\frac{1}{2}\int_1^5 u^3\, du$ **(b)** $\frac{3}{2}\int_9^{25} \sqrt{u}\, du$ **(c)** $\dfrac{1}{\pi}\displaystyle\int_{-\pi/2}^{\pi/2} \cos u\, du$
    (d) $\int_1^2 (u+1)u^5\, du$ **3. (a)** $\frac{1}{2}\int_{-1}^1 e^u\, du$ **(b)** $\int_1^2 u\, du$
5. 10   **7.** 0   **9.** $\frac{1192}{15}$ **11.** $8 - (4\sqrt{2})$ **13.** $-\frac{1}{48}$ **15.** $\ln \frac{21}{13}$
17. $\pi/6$ **19.** $25\pi/6$ **21.** $\pi/8$ **23.** $2/\pi$ m   **25.** 6   **27.** $\pi/18$
29. 2   **31.** $\frac{2}{3}(\sqrt{10} - 2\sqrt{2})$ **33.** $2(\sqrt{7} - \sqrt{3})$ **35.** 1   **37.** 0
39. $(\sqrt{3} - 1)/3$ **41.** $\frac{106}{405}$ **43.** $(\ln 3)/2$ **45.** $\pi/(6\sqrt{3})$ **47.** $\pi/9$
49. (a) $\frac{23}{4480}$ **51. (a)** $\frac{5}{3}$ **(b)** $\frac{5}{3}$ **(c)** $-\frac{1}{2}$ **55.** $\approx 48{,}233{,}500{,}000$
57. (a) 0.45   **(b)** 0.461   **59.** $(\ln 7)/2$ **61. (a)** $2/\pi$
65. (b) $\frac{3}{2}$ **(c)** $\pi/4$

▶ **Exercise Set 5.10 (Page 406)**

1. (a)

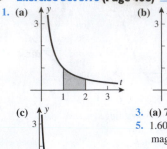

(b)

(c)

7. (a) $x^{-1}, x > 0$ **(b)** $x^2, x \neq 0$ **(c)** $-x^2, -\infty < x < +\infty$
(d) $-x, -\infty < x < +\infty$ **(e)** $x^3, x > 0$ **(f)** $\ln x + x, x > 0$
(g) $x - \sqrt[3]{x}, -\infty < x < +\infty$ **(h)** $e^x/x, x > 0$
9. (a) $e^{\pi \ln 3}$ **(b)** $e^{\sqrt{2} \ln 2}$ **11. (a)** $\sqrt{e}$ **(b)** e^2 **13.** $x^2 - x$
15. (a) $3/x$ **(b)** 1 **17. (a)** 0 **(b)** 0 **(c)** 1

Responses to True–False questions may be abridged to save space.

19. True; both equal $-\ln a$.
21. False; the integrand is unbounded on $[-1, e]$ and thus the integrand is undefined.
23. (a) $2x^3\sqrt{1+x^2}$ **(b)** $-\frac{2}{3}(x^2+1)^{3/2} + \frac{2}{5}(x^2+1)^{5/2} - \frac{4\sqrt{2}}{15}$
25. (a) $-\cos(x^3)$ **(b)** $-\tan^2 x$ **27.** $-3\dfrac{3x-1}{9x^2+1} + 2x\dfrac{x^2-1}{x^4+1}$
29. (a) $3x^2\sin^2(x^3) - 2x\sin^2(x^2)$ **(b)** $\dfrac{2}{1-x^2}$
31. (a) $F(0) = 0, F(3) = 0, F(5) = 6, F(7) = 6, F(10) = 3$
  (b) increasing on $\left[\frac{3}{2}, 6\right]$ and $\left[\frac{37}{4}, 10\right]$, decreasing on $\left[0, \frac{3}{2}\right]$ and $\left[6, \frac{37}{4}\right]$
  (c) maximum $\frac{15}{2}$ at $x = 6$, minimum $-\frac{9}{4}$ at $x = \frac{3}{2}$
  (d)

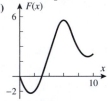

33. $F(x) = \begin{cases} (1-x^2)/2, & x < 0 \\ (1+x^2)/2, & x \geq 0 \end{cases}$ **35.** $y(x) = x^2 + \ln x + 1$
37. $y(x) = \tan x + \cos x - (\sqrt{2}/2)$
39. $P(x) = P_0 + \int_0^x r(t)\, dt$ individuals **41.** I is the derivative of II.
43. (a) $t = 3$ **(b)** $t = 1, 5$
  (c) $t = 5$ **(d)** $t = 3$
  (e) F is concave up on $\left(0, \frac{1}{2}\right)$ and $(2, 4)$,
  concave down on $\left(\frac{1}{2}, 2\right)$ and $(4, 5)$.

(f)

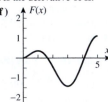

45. (a) relative maxima at $x = \pm\sqrt{4k+1}, k = 0, 1, \ldots$; relative minima
  at $x = \pm\sqrt{4k-1}, k = 1, 2, \ldots$
  (b) $x = \pm\sqrt{2k}, k = 1, 2, \ldots$, and at $x = 0$
47. $f(x) = 2e^{2x}, a = \ln 2$ **49.** 0.06

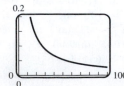

▶ **Chapter 5 Review Exercises (Page 408)**

1. $-\dfrac{1}{4x^2} + \dfrac{8}{3}x^{3/2} + C$ **3.** $-4\cos x + 2\sin x + C$
5. $3x^{1/3} - 5e^x + C$ **7.** $\tan^{-1} x + 2\sin^{-1} x + C$
9. (a) $y(x) = 2\sqrt{x} - \frac{2}{3}x^{3/2} - \frac{4}{3}$ **(b)** $y(x) = \sin x - 5e^x + 5$
  (c) $y(x) = \frac{5}{4} + \frac{3}{4}x^{4/3}$ **(d)** $y(x) = \frac{1}{2}e^{x^2} - \frac{1}{2}$
13. $\frac{1}{2}\sec^{-1}(x^2 - 1) + C$ **15.** $\frac{1}{3}\sqrt{5 + 2\sin 3x} + C$
17. $-\dfrac{1}{3a}\dfrac{1}{ax^3 + b} + C$
19. (a) $\displaystyle\sum_{k=0}^{14}(k+4)(k+1)$ **(b)** $\displaystyle\sum_{k=5}^{19}(k-1)(k-4)$
21. $\frac{32}{3}$ **23.** $0.35122, 0.42054, 0.38650$
27. (a) $\frac{3}{4}$ **(b)** $-\frac{3}{2}$ **(c)** $-\frac{35}{4}$ **(d)** -2 **(e)** not enough information
  (f) not enough information
29. (a) $2 + (\pi/2)$ **(b)** $\dfrac{1}{3}(10^{3/2} - 1) - \dfrac{9\pi}{4}$ **(c)** $\pi/8$
31. 48 **33.** $\frac{2}{3}$ **35.** $\frac{3}{2} - \sec 1$ **37.** $\frac{5}{2}$ **39.** $\frac{52}{3}$ **41.** $e^3 - e$
43. area $= \frac{1}{6}$ **45.** $\frac{22}{3}$
47. (a) $x^3 + 1$
49. e^{x^2}
51. $|x - 1|$
53. $\dfrac{\cos x}{1 + \sin^3 x}$

57. (b) $\dfrac{\pi}{2}$; $\tan^{-1} x + \tan^{-1}\left(\dfrac{1}{x}\right) = \dfrac{\pi}{2}$
59. (a) $F(x)$ is 0 if $x = 1$, positive if $x > 1$, and negative if $x < 1$.
  (b) $F(x)$ is 0 if $x = -1$, positive if $-1 < x \leq 2$,
  and negative if $-2 \leq x < -1$.
61. (a) $\frac{4}{3}$ **(b)** $e - 1$ **63.** $\frac{3}{10}$ **67.** $\frac{1}{4}t^4 - \frac{2}{3}t^3 + t + 1$
69. $t^2 - 3t + 7$ **71.** 12 m, 20 m **73.** $\frac{1}{3}$ m, $\frac{10}{3} - 2\sqrt{2}$ m
75. displacement $= -6$ m; distance $= \frac{13}{2}$ m
77. (a) 2.2 s **(b)** 387.2 ft **79.** $v_0/2$ ft/s **81.** $\frac{121}{5}$ **83.** $\frac{2}{3}$ **85.** 0
87. $2 - 2/\sqrt{e}$ **89. (a)** e^2 **(b)** $e^{1/3}$

▶ **Chapter 5 Making Connections (Page 412)**

Where correct answers to a Making Connections exercise may vary, no answer is listed. Sample answers for these questions are available on the Book Companion Site.

1. (b) $b^2 - a^2$ **2.** $16/3$ **3.** 12
4. (a) the sum for f is m times that for g
  (b) $m\displaystyle\int_0^1 g(x)\, dx = \int_0^m f(x)\, du$
5. (a) they are equal **(b)** $\displaystyle\int_2^3 g(x)\, dx = \int_4^9 f(u)\, du$

▶ **Exercise Set 6.1 (Page 419)**

1. $9/2$ **3.** 1 **5. (a)** $4/3$ **(b)** $4/3$ **7.** $49/192$ **9.** $1/2$ **11.** $\sqrt{2}$
13. $\frac{1}{2}$ **15.** $\pi - 1$ **17.** 24 **19.** $37/12$ **21.** $4\sqrt{2}$ **23.** $\frac{1}{2}$
25. $\ln 2 - \frac{1}{2}$

Responses to True–False questions may be abridged to save space.

27. True; use area Formula (1) with $f(x) = g(x) + c$.
29. True; the integrand must assume both positive and negative values.
  By the Intermediate-Value Theorem, the integrand must be equal to 0
  somewhere in $[a, b]$.
31. $k \approx 0.9973$ **33.** $9152/105$ **35.** $9/\sqrt[3]{4}$
37. (a) $4/3$ **(b)** $m = 2 - \sqrt[3]{4}$ **39.** 1.180898334
41. $0.4814, 2.3639, 1.1897$ **43.** 2.54270

45. racer 1's lead over racer 2 at time $t = 0$

47. **(a)** (area above graph of g and below graph of f) minus (area above graph of f and below graph of g)
   (b) area between graphs of f and g **49.** $a^2/6$

▶ **Exercise Set 6.2 (Page 428)**

1. 8π **3.** $13\pi/6$ **5.** $(1 - \sqrt{2}/2)\pi$ **7.** 8π **9.** $32/5$ **11.** $256\pi/3$
13. $2048\pi/15$ **15.** 4π **17.** $\pi^2/4$ **19.** $3/5$ **21.** 2π
23. $72\pi/5$ **25.** $\dfrac{\pi}{2}(e^2 - 1)$
Responses to True–False questions may be abridged to save space.
27. False; see the solids associated with Exercises 9 and 10.
29. False; see Example 2 where the cross-sectional area is a linear function of x.
31. $4\pi ab^2/3$ **33.** π **35.** $\int_a^b \pi[f(x) - k]^2\,dx$ **37.** **(b)** $40\pi/3$
39. $648\pi/5$ **41.** $\pi/2$ **43.** $\pi/15$ **45.** $40,000\pi$ ft^3 **47.** $1/30$
49. **(a)** $2\pi/3$ **(b)** $16/3$ **(c)** $4\sqrt{3}/3$ **51.** 0.710172176 **53.** π
57. **(b)** left ≈ 11.157; right ≈ 11.771; $V \approx$ average $= 11.464$ cm^3
59. $V = \begin{cases} 3\pi h^2, & 0 \le h < 2 \\ \frac{1}{3}\pi(12h^2 - h^3 - 4), & 2 \le h \le 4 \end{cases}$ **61.** $\frac{2}{3}r^3 \tan\theta$ **63.** $16r^3/3$

▶ **Exercise Set 6.3 (Page 436)**

1. $15\pi/2$ **3.** $\pi/3$ **5.** $2\pi/5$ **7.** 4π **9.** $20\pi/3$ **11.** $\pi \ln 2$ **13.** $\pi/2$
15. $\pi/5$
Responses to True–False questions may be abridged to save space.
17. True; this is a restatement of Formula (1).
19. True; see Formula (2).
21. $2\pi e^2$ **23.** 1.73680 **25.** **(a)** $7\pi/30$ **(b)** easier
27. **(a)** $\int_0^1 2\pi(1 - x)x\,dx$ **(b)** $\int_0^1 2\pi(1 + y)(1 - y)\,dy$
29. $7\pi/4$ **31.** $\pi r^2 h/3$ **33.** $V = \frac{4}{3}\pi(L/2)^3$ **35.** $b = 1$

▶ **Exercise Set 6.4 (Page 441)**

1. $L = \sqrt{5}$ **3.** $(85\sqrt{85} - 8)/243$ **5.** $\frac{1}{27}(80\sqrt{10} - 13\sqrt{13})$ **7.** $\frac{17}{6}$
Responses to True–False questions may be abridged to save space.
9. False; f' is undefined at the endpoints ± 1.
11. True; if $f(x) = mx + c$ over $[a, b]$, then $L = \sqrt{1 + m^2}(b - a)$, which is equal to the given sum.
13. $L = \ln(1 + \sqrt{2})$
15. **(a)**
   (b) dy/dx does not exist at $x = 0$.
   (c) $L = (13\sqrt{13} + 80\sqrt{10} - 16)/27$

17. **(a)** They are mirror images across the line $y = x$.
   (b) $\int_{1/2}^2 \sqrt{1 + 4x^2}\,dx$, $\int_{1/4}^4 \sqrt{1 + \dfrac{1}{4x}}\,dx$, $x = \sqrt{u}$ transforms the first integral into the second.
   (c) $\int_{1/4}^4 \sqrt{1 + \dfrac{1}{4y}}\,dy$, $\int_{1/2}^2 \sqrt{1 + 4y^2}\,dy$
   (d) $4.0724, 4.0716$
   (e) The first: Both are underestimates of the arc length, so the larger one is more accurate.
   (f) $4.0724, 4.0662$ **(g)** 4.0729

19. **(a)** They are mirror images across the line $y = x$.
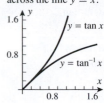
   (b) $\int_0^{\pi/3} \sqrt{1 + \sec^4 x}\,dx$, $\int_0^{\sqrt{3}} \sqrt{1 + \dfrac{1}{(1 + x^2)^2}}\,dx$, $x = \tan^{-1} u$ transforms the first integral into the second.
   (c) $\int_0^{\sqrt{3}} \sqrt{1 + \dfrac{1}{(1 + y^2)^2}}\,dy$, $\int_0^{\pi/3} \sqrt{1 + \sec^4 y}\,dy$
   (d) $2.0566, 2.0567$
   (e) The second: Both are underestimates of the arc length, so the larger one is more accurate.   **(f)** $2.0509, 2.0571$ **(g)** 2.0570
23. $k = 1.83$ **25.** 196.31 yards   **27.** $(2\sqrt{2} - 1)/3$
29. π **31.** $L = \sqrt{2}(e^{\pi/2} - 1)$ **33.** **(b)** 9.69 **(c)** 5.16 cm

▶ **Exercise Set 6.5 (Page 447)**

1. $35\pi\sqrt{2}$ **3.** 8π **5.** $40\pi\sqrt{82}$ **7.** 24π **9.** $16\pi/9$
11. $16,911\pi/1024$ **13.** $2\pi[\sqrt{2} + \ln(\sqrt{2} + 1)]$ **15.** $S \approx 22.94$
Responses to True–False questions may be abridged to save space.
17. True; use Formula (1) with $r_1 = 0, r_2 = r, l = \sqrt{r^2 + h^2}$.
19. True; the sum telescopes to the surface area of a cylinder.
21. 14.39 **23.** $S = \int_a^b 2\pi[f(x) + k]\sqrt{1 + [f'(x)]^2}\,dx$
33. $\dfrac{8}{3}\pi(17\sqrt{17} - 1)$ **35.** $\dfrac{\pi}{24}(17\sqrt{17} - 1)$

▶ **Exercise Set 6.6 (Page 456)**

1. 7.5 ft·lb   **3.** $d = 7/4$ **5.** 100 ft·lb   **7.** 160 J   **9.** 20 lb/ft
Responses to True–False questions may be abridged to save space.
11. False; the work done is the same.
13. True; joules   **15.** $47,385\pi$ ft·lb   **17.** $261,600$ J
19. **(a)** $926,640$ ft·lb   **(b)** hp of motor $= 0.468$ **21.** $75,000$ ft·lb
23. $120,000$ ft·tons   **25.** **(a)** $2,400,000,000/x^2$ lb
   (b) $(9.6 \times 10^{10})/(x + 4000)^2$ lb   **(c)** 2.5344×10^{10} ft·lb
27. $v_f = 100$ m/s
29. **(a)** decreases of 4.5×10^{14} J   **(b)** ≈ 0.107 **(c)** ≈ 8.24 bombs

▶ **Exercise Set 6.7 (Page 465)**

1. **(a)** positive: m_2 is at the fulcrum, so it can be ignored; masses m_1 and m_3 are equidistant from position 5, but $m_1 < m_3$, so the beam will rotate clockwise.   **(b)** The fulcrum should be placed $\frac{50}{7}$ units to the right of m_1.
3. $\left(\frac{1}{2}, \frac{1}{2}\right)$ **5.** $\left(1, \frac{1}{2}\right)$ **7.** $\left(\frac{2}{3}, \frac{1}{3}\right)$ **9.** $\left(\frac{5}{14}, \frac{38}{35}\right)$ **11.** $\left(\frac{2}{3}, \frac{1}{3}\right)$
13. $\left(-\frac{1}{2}, 4\right)$ **15.** $\left(\frac{1}{2}, \frac{8}{5}\right)$ **17.** $\left(\frac{9}{20}, \frac{9}{20}\right)$ **19.** $\left(\frac{49}{48}, \frac{7}{3} - \ln 2\right)$
23. $\frac{4}{3}$; $\left(\frac{3}{5}, \frac{3}{8}\right)$ **25.** 3; $\left(0, \frac{2}{3}\right)$ **27.** 8; $\left(\frac{\pi}{2}, \frac{\pi}{8}\right)$
29. $\ln 4 - 1$; $\left(\dfrac{4\ln 4 - 3}{4\ln 4 - 4}, \dfrac{(\ln 2)^2 + 1 - \ln 4}{\ln 4 - 1}\right)$
Responses to True–False questions may be abridged to save space.
31. True; use symmetry.   **33.** True; use symmetry.
35. $\left(\dfrac{2a}{3}, 0\right)$ **37.** $(\bar{x}, \bar{y}) = \left(0, \dfrac{(a + 2b)c}{3(a + b)}\right)$
41. $2\pi^2 abk$ **43.** $(a/3, b/3)$

▶ **Exercise Set 6.8 (Page 472)**

1. **(a)** $F = 31,200$ lb; $P = 312$ lb/ft^2
   (b) $F = 2,452,500$ N; $P = 98.1$ kPa
3. 499.2 lb   **5.** 8.175×10^5 N   **7.** $1,098,720$ N   **9.** yes
11. $\rho a^3/\sqrt{2}$ lb
Responses to True–False questions may be abridged to save space.
13. True; this is a consequence of inequalities (4).

15. False; by Equation (7) the force can be arbitrarily large for a fixed volume of water.

17. 63,648 lb   **19.** 9.81×10^9 N   **21.** (b) $80\rho_0$ lb/min

▶ **Exercise Set 6.9 (Page 482)**

1. (a) ≈ 10.0179 (b) ≈ 3.7622 (c) $15/17 \approx 0.8824$
   (d) ≈ -1.4436 (e) ≈ 1.7627 (f) ≈ 0.9730

3. (a) $\dfrac{4}{3}$ (b) $\dfrac{5}{4}$ (c) $\dfrac{312}{313}$ (d) $-\dfrac{63}{16}$

5.

	$\sinh x_0$	$\cosh x_0$	$\tanh x_0$	$\coth x_0$	$\operatorname{sech} x_0$	$\operatorname{csch} x_0$
(a)	2	$\sqrt{5}$	$2/\sqrt{5}$	$\sqrt{5}/2$	$1/\sqrt{5}$	$1/2$
(b)	3/4	5/4	3/5	5/3	4/5	4/3
(c)	4/3	5/3	4/5	5/4	3/5	3/4

9. $4\cosh(4x - 8)$ **11.** $-\dfrac{1}{x}\operatorname{csch}^2(\ln x)$

13. $\dfrac{1}{x^2}\operatorname{csch}\left(\dfrac{1}{x}\right)\coth\left(\dfrac{1}{x}\right)$ **15.** $\dfrac{2 + 5\cosh(5x)\sinh(5x)}{\sqrt{4x + \cosh^2(5x)}}$

17. $x^{5/2}\tanh(\sqrt{x})\operatorname{sech}^2(\sqrt{x}) + 3x^2\tanh^2(\sqrt{x})$

19. $\dfrac{1}{\sqrt{9 + x^2}}$ **21.** $\dfrac{1}{(\cosh^{-1}x)\sqrt{x^2 - 1}}$ **23.** $-\dfrac{(\tanh^{-1}x)^{-2}}{1 - x^2}$

25. $\dfrac{\sinh x}{|\sinh x|} = \begin{cases} 1, & x > 0 \\ -1, & x < 0 \end{cases}$ **27.** $-\dfrac{e^x}{2x\sqrt{1 - x}} + e^x\operatorname{sech}^{-1}x$

29. $\frac{1}{7}\sinh^7 x + C$ **31.** $\frac{2}{3}(\tanh x)^{3/2} + C$ **33.** $\ln(\cosh x) + C$

35. $37/375$ **37.** $\frac{1}{3}\sinh^{-1} 3x + C$ **39.** $-\operatorname{sech}^{-1}(e^x) + C$

41. $-\operatorname{csch}^{-1}|2x| + C$ **43.** $\frac{1}{2}\ln 3$

Responses to True–False questions may be abridged to save space.

45. True; see Figure 6.9.1   **47.** True; $f(x) = \sinh x$

49. $16/9$ **51.** 5π **53.** $\frac{3}{4}$

55. (a) $+\infty$ (b) $-\infty$ (c) 1   (d) -1 (e) $+\infty$ (f) $+\infty$

63. $|u| < 1$: $\tanh^{-1} u + C$; $|u| > 1$: $\tanh^{-1}(1/u) + C$

65. (a) $\ln 2$ (b) $1/2$ **71.** 405.9 ft

73. (a) 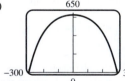   (b) 1480.2798 ft
   (c) ± 283.6249 ft
   (d) $82°$

75. (b) 14.44 m   (c) $15\ln 3 \approx 16.48$ m

▶ **Chapter 6 Review Exercises (Page 485)**

7. (a) $\int_a^b (f(x) - g(x))\,dx + \int_b^c (g(x) - f(x))\,dx + \int_c^d (f(x) - g(x))\,dx$ (b) $11/4$

9. $4352\pi/105$ **11.** $3/2 + \ln 4$ **13.** 9   **15.** $\dfrac{\pi}{6}\left(65^{3/2} - 37^{3/2}\right)$

17. (a) $W = \frac{1}{16}$ J   (b) 5 m   **19.** $\left(\frac{8}{5}, 0\right)$

21. (a) $F = \int_0^1 \rho x 3\,dx$ N   (b) $F = \int_1^4 \rho(1 + x)2x\,dx$ lb/ft^2

   (c) $F = \displaystyle\int_{-10}^0 9810|y|2\sqrt{\dfrac{125}{8}}(y + 10)\,dy$ N

▶ **Chapter 6 Making Connections (Page 487)**

Where correct answers to a Making Connections exercise may vary, no answer is listed. Sample answers for these questions are available on the Book Companion Site.

1. (a) πA_1 (b) $a = \dfrac{A_1}{2A_2}$ **2.** 1,010,807 ft·lb   **3.** $\displaystyle\int_0^a 2\pi r f(r)\,dr$

▶ **Exercise Set 7.1 (Page 490)**

1. $-2(x - 2)^4 + C$ **3.** $\frac{1}{2}\tan(x^2) + C$ **5.** $-\frac{1}{3}\ln(2 + \cos 3x) + C$

7. $\cosh(e^x) + C$ **9.** $e^{\tan x} + C$ **11.** $-\frac{1}{30}\cos^6 5x + C$

13. $\ln(e^x + \sqrt{e^{2x} + 4}) + C$ **15.** $2e^{\sqrt{x-1}} + C$ **17.** $2\sinh\sqrt{x} + C$

19. $-\dfrac{2}{\ln 3}3^{-\sqrt{x}} + C$ **21.** $\frac{1}{2}\coth\dfrac{2}{x} + C$ **23.** $-\frac{1}{4}\ln\left|\dfrac{2 + e^{-x}}{2 - e^{-x}}\right| + C$

25. $\sin^{-1}(e^x) + C$ **27.** $-\frac{1}{2}\cos(x^2) + C$ **29.** $-\dfrac{1}{\ln 16}4^{-x^2} + C$

31. (a) $\frac{1}{2}\sin^2 x + C$ (b) $-\frac{1}{4}\cos 2x + C$

33. (b) $\ln\left|\tan\dfrac{x}{2}\right| + C$ (c) $\ln\left|\cot\left(\dfrac{\pi}{4} - \dfrac{x}{2}\right)\right| + C$

▶ **Exercise Set 7.2 (Page 498)**

1. $-e^{-2x}\left(\dfrac{x}{2} + \dfrac{1}{4}\right) + C$ **3.** $x^2 e^x - 2xe^x + 2e^x + C$

5. $-\frac{1}{3}x\cos 3x + \frac{1}{9}\sin 3x + C$ **7.** $x^2\sin x + 2x\cos x - 2\sin x + C$

9. $\dfrac{x^2}{2}\ln x - \dfrac{x^2}{4} + C$ **11.** $x(\ln x)^2 - 2x\ln x + 2x + C$

13. $x\ln(3x - 2) - x - \frac{2}{3}\ln(3x - 2) + C$ **15.** $x\sin^{-1}x + \sqrt{1 - x^2} + C$

17. $x\tan^{-1}(3x) - \frac{1}{6}\ln(1 + 9x^2) + C$ **19.** $\frac{1}{2}e^x(\sin x - \cos x) + C$

21. $(x/2)[\sin(\ln x) - \cos(\ln x)] + C$ **23.** $x\tan x + \ln|\cos x| + C$

25. $\frac{1}{2}x^2 e^{x^2} - \frac{1}{2}e^{x^2} + C$ **27.** $\frac{1}{4}(3e^4 + 1)$ **29.** $(2e^3 + 1)/9$

31. $3\ln 3 - 2$ **33.** $\dfrac{5\pi}{6} - \sqrt{3} + 1$ **35.** $-\pi/2$

37. $\dfrac{1}{3}\left(2\sqrt{3}\pi - \dfrac{\pi}{2} - 2 + \ln 2\right)$

Responses to True–False questions may be abridged to save space.

39. True; see the subsection "Guidelines for Integration by Parts."

41. False; e^x isn't a factor of the integrand.

43. $2(\sqrt{x} - 1)e^{\sqrt{x}} + C$ **47.** $-(3x^2 + 5x + 7)e^{-x} + C$

49. $(4x^3 - 6x)\sin 2x - (2x^4 - 6x^2 + 3)\cos 2x + C$

51. $\dfrac{e^{ax}}{a^2 + b^2}(a\sin bx - b\cos bx) + C$ **53.** (a) $\frac{1}{2}\sin^2 x + C$

55. (a) $A = 1$ (b) $V = \pi(e - 2)$ **57.** $V = 2\pi^2$ **59.** $\pi^3 - 6\pi$

61. (a) $-\frac{1}{4}\sin^3 x\cos x - \frac{3}{8}\sin x\cos x + \frac{3}{8}x + C$ (b) $8/15$

65. (a) $\frac{1}{3}\tan^3 x - \tan x + x + C$ (b) $\frac{1}{3}\sec^2 x\tan x + \frac{2}{3}\tan x + C$
   (c) $x^3 e^x - 3x^2 e^x + 6xe^x - 6e^x + C$

69. $(x + 1)\ln(x + 1) - x + C$ **71.** $\frac{1}{2}(x^2 + 1)\tan^{-1}x - \frac{1}{2}x + C$

▶ **Exercise Set 7.3 (Page 506)**

1. $-\frac{1}{4}\cos^4 x + C$ **3.** $\dfrac{\theta}{2} - \dfrac{1}{20}\sin 10\theta + C$

5. $\dfrac{1}{3a}\cos^3 a\theta - \cos a\theta + C$ **7.** $\dfrac{1}{2a}\sin^2 ax + C$

9. $\frac{1}{3}\sin^3 t - \frac{1}{5}\sin^5 t + C$ **11.** $\frac{1}{8}x - \frac{1}{32}\sin 4x + C$

13. $-\frac{1}{10}\cos 5x + \frac{1}{2}\cos x + C$ **15.** $-\frac{1}{3}\cos(3x/2) - \cos(x/2) + C$

17. $2/3$ **19.** 0   **21.** $7/24$ **23.** $\frac{1}{2}\tan(2x - 1) + C$

25. $\ln|\cos(e^{-x})| + C$ **27.** $\frac{1}{4}\ln|\sec 4x + \tan 4x| + C$

29. $\frac{1}{3}\tan^3 x + C$ **31.** $\frac{1}{16}\sec^4 4x + C$ **33.** $\frac{1}{7}\sec^7 x - \frac{1}{5}\sec^5 x + C$

35. $\frac{1}{4}\sec^3 x\tan x - \frac{5}{8}\sec x\tan x + \frac{3}{8}\ln|\sec x + \tan x| + C$

37. $\frac{1}{3}\sec^3 t + C$ **39.** $\tan x + \frac{1}{3}\tan^3 x + C$

41. $\frac{1}{8}\tan^2 4x + \frac{1}{4}\ln|\cos 4x| + C$ **43.** $\frac{2}{3}\tan^{3/2}x + \frac{2}{7}\tan^{7/2}x + C$

45. $\dfrac{1}{2} - \dfrac{\pi}{8}$ **47.** $-\frac{1}{2} + \ln 2$ **49.** $-\frac{1}{5}\csc^5 x + \frac{1}{3}\csc^3 x + C$

51. $-\frac{1}{2}\csc^2 x - \ln|\sin x| + C$

Responses to True–False questions may be abridged to save space.

53. True; $\int \sin^5 x\cos^8 x\,dx = \int \sin x(1 - \cos^2 x)^2\cos^8 x\,dx = -\int(1 - u^2)^2 u^8\,du = -\int(u^8 - 2u^{10} + u^{12})\,du$

55. False; use this identity to help evaluate integrals of the form $\int \sin mx\cos nx\,dx$.

59. $L = \ln(\sqrt{2} + 1)$ **61.** $V = \pi/2$

67. $-\dfrac{1}{\sqrt{a^2+b^2}}\ln\left[\dfrac{\sqrt{a^2+b^2}+a\cos x - b\sin x}{a\sin x + b\cos x}\right]+C$

69. (a) $\frac{2}{3}$ (b) $3\pi/16$ (c) $\frac{8}{15}$ (d) $5\pi/32$

▶ **Exercise Set 7.4 (Page 513)**

1. $2\sin^{-1}(x/2)+\frac{1}{2}x\sqrt{4-x^2}+C$ **3.** $8\sin^{-1}\left(\frac{x}{4}\right)-\dfrac{x\sqrt{16-x^2}}{2}+C$

5. $\frac{1}{16}\tan^{-1}(x/2)+\dfrac{x}{8(4+x^2)}+C$ **7.** $\sqrt{x^2-9}-3\sec^{-1}(x/3)+C$

9. $-(x^2+2)\sqrt{1-x^2}+C$ **11.** $\dfrac{\sqrt{9x^2-4}}{4x}+C$ **13.** $\dfrac{x}{\sqrt{1-x^2}}+C$

15. $\ln|\sqrt{x^2-9}+x|+C$ **17.** $\dfrac{-x}{9\sqrt{4x^2-9}}+C$

19. $\frac{1}{2}\sin^{-1}(e^x)+\frac{1}{2}e^x\sqrt{1-e^{2x}}+C$ **21.** $2/3$ **23.** $(\sqrt{3}-\sqrt{2})/2$

25. $\dfrac{10\sqrt{3}+18}{243}$

Responses to True–False questions may be abridged to save space.

27. True; with the restriction $-\pi/2\le\theta\le\pi/2$, this substitution gives $\sqrt{a^2-x^2}=a\cos\theta$ and $dx=a\cos\theta\,d\theta$.

29. False; use the substitution $x=a\sec\theta$ with $0\le\theta<\pi/2\ (x\ge a)$ or $\pi/2\le\theta<\pi\ (x\le -a)$.

31. $\frac{1}{2}\ln(x^2+4)+C$

33. $L=\sqrt{5}-\sqrt{2}+\ln\dfrac{2+2\sqrt{2}}{1+\sqrt{5}}$ **35.** $S=\dfrac{\pi}{32}\left[18\sqrt{5}-\ln(2+\sqrt{5})\right]$

37. $\tan^{-1}(x-2)+C$ **39.** $\sin^{-1}\left(\dfrac{x-1}{2}\right)+C$

41. $\ln(x-3+\sqrt{(x-3)^2+1})+C$

43. $2\sin^{-1}\left(\dfrac{x+1}{2}\right)+\frac{1}{2}(x+1)\sqrt{3-2x-x^2}+C$

45. $\dfrac{1}{\sqrt{10}}\tan^{-1}\sqrt{\frac{2}{5}}(x+1)+C$ **47.** $\pi/6$

49. $u=\sin^2 x,\ \frac{1}{2}\displaystyle\int\sqrt{1-u^2}\,du$

$=\frac{1}{4}[\sin^2 x\sqrt{1-\sin^4 x}+\sin^{-1}(\sin^2 x)]+C$

51. (a) $\sinh^{-1}(x/3)+C$ (b) $\ln\left(\dfrac{\sqrt{x^2+9}}{3}+\dfrac{x}{3}\right)+C$

▶ **Exercise Set 7.5 (Page 521)**

1. $\dfrac{A}{x-3}+\dfrac{B}{x+4}$ **3.** $\dfrac{A}{x}+\dfrac{B}{x^2}+\dfrac{C}{x-1}$

5. $\dfrac{A}{x}+\dfrac{B}{x^2}+\dfrac{C}{x^3}+\dfrac{Dx+E}{x^2+2}$ **7.** $\dfrac{Ax+B}{x^2+5}+\dfrac{Cx+D}{(x^2+5)^2}$

9. $\frac{1}{5}\ln\left|\dfrac{x-4}{x+1}\right|+C$ **11.** $\frac{5}{2}\ln|2x-1|+3\ln|x+4|+C$

13. $\ln\left|\dfrac{x(x+3)^2}{x-3}\right|+C$ **15.** $\dfrac{x^2}{2}-3x+\ln|x+3|+C$

17. $3x+12\ln|x-2|-\dfrac{2}{x-2}+C$

19. $\ln|x^2-3x-10|+C$

21. $x+\dfrac{x^3}{3}+\ln\left|\dfrac{(x-1)^2(x+1)}{x^2}\right|+C$

23. $3\ln|x|-\ln|x-1|-\dfrac{5}{x-1}+C$

25. $\dfrac{2}{x-3}+\ln|x-3|+\ln|x+1|+C$

27. $\dfrac{2}{x+1}-\dfrac{1}{2(x+1)^2}+\ln|x+1|+C$

29. $-\frac{7}{34}\ln|4x-1|+\frac{6}{17}\ln(x^2+1)+\frac{3}{17}\tan^{-1}x+C$

31. $3\tan^{-1}x+\frac{1}{2}\ln(x^2+3)+C$

33. $\dfrac{x^2}{2}-2x+\frac{1}{2}\ln(x^2+1)+C$

Responses to True–False questions may be abridged to save space.

35. True; partial fractions rewrites proper rational functions $P(x)/Q(x)$ as a sum of terms of the form $\dfrac{A}{(Bx+C)^k}$ and/or $\dfrac{Dx+E}{(Fx^2+Gx+H)^k}$.

37. True; $\dfrac{2x+3}{x^2}=\dfrac{2x}{x^2}+\dfrac{3}{x^2}=\dfrac{2}{x}+\dfrac{3}{x^2}$.

39. $\frac{1}{6}\ln\left(\dfrac{1-\sin\theta}{5+\sin\theta}\right)+C$ **41.** $e^x-2\tan^{-1}\left(\frac{1}{2}e^x\right)+C$

43. $V=\pi\left(\frac{19}{5}-\frac{9}{4}\ln 5\right)$ **45.** $\dfrac{1}{\sqrt{2}}\tan^{-1}\left(\dfrac{x+1}{\sqrt{2}}\right)+\dfrac{1}{x^2+2x+3}+C$

47. $\frac{1}{8}\ln|x-1|-\frac{1}{5}\ln|x-2|+\frac{1}{12}\ln|x-3|-\frac{1}{120}\ln|x+3|+C$

▶ **Exercise Set 7.6 (Page 531)**

1. Formula (60): $\frac{4}{3}x+\frac{4}{9}\ln|3x-1|+C$

3. Formula (65): $\dfrac{1}{5}\ln\left|\dfrac{x}{5+2x}\right|+C$

5. Formula (102): $\frac{1}{5}(x-1)(2x+3)^{3/2}+C$

7. Formula (108): $\dfrac{1}{2}\ln\left|\dfrac{\sqrt{4-3x}-2}{\sqrt{4-3x}+2}\right|+C$

9. Formula (69): $\dfrac{1}{8}\ln\left|\dfrac{x+4}{x-4}\right|+C$

11. Formula (73): $\dfrac{x}{2}\sqrt{x^2-3}-\dfrac{3}{2}\ln|x+\sqrt{x^2-3}|+C$

13. Formula (95): $\dfrac{x}{2}\sqrt{x^2+4}-2\ln(x+\sqrt{x^2+4})+C$

15. Formula (74): $\dfrac{x}{2}\sqrt{9-x^2}+\dfrac{9}{2}\sin^{-1}\dfrac{x}{3}+C$

17. Formula (79): $\sqrt{4-x^2}-2\ln\left|\dfrac{2+\sqrt{4-x^2}}{x}\right|+C$

19. Formula (38): $-\dfrac{\sin 7x}{14}+\dfrac{1}{2}\sin x+C$

21. Formula (50): $\dfrac{x^4}{16}[4\ln x-1]+C$

23. Formula (42): $\dfrac{e^{-2x}}{13}[-2\sin(3x)-3\cos(3x)]+C$

25. Formula (62): $\dfrac{1}{2}\displaystyle\int\dfrac{u\,du}{(4-3u)^2}=\dfrac{1}{18}\left[\dfrac{4}{4-3e^{2x}}+\ln\left|4-3e^{2x}\right|\right]+C$

27. Formula (68): $\dfrac{2}{3}\displaystyle\int\dfrac{du}{u^2+4}=\dfrac{1}{3}\tan^{-1}\dfrac{3\sqrt{x}}{2}+C$

29. Formula (76): $\dfrac{1}{2}\displaystyle\int\dfrac{du}{\sqrt{u^2-9}}=\dfrac{1}{2}\ln|2x+\sqrt{4x^2-9}|+C$

31. Formula (81): $\dfrac{1}{4}\displaystyle\int\dfrac{u^2}{\sqrt{2-u^2}}\,du=-\dfrac{1}{4}x^2\sqrt{2-4x^4}$
$\qquad\qquad\qquad +\dfrac{1}{4}\sin^{-1}\left(\sqrt{2}x^2\right)+C$

33. Formula (26): $\displaystyle\int\sin^2 u\,du=\frac{1}{2}\ln x-\frac{1}{4}\sin(2\ln x)+C$

35. Formula (51): $\dfrac{1}{4}\displaystyle\int ue^u\,du=\dfrac{1}{4}(-2x-1)e^{-2x}+C$

37. $u=\sin 3x$, Formula (67): $\dfrac{1}{3}\displaystyle\int\dfrac{du}{u(u+1)^2}$
$\qquad =\dfrac{1}{3}\left(\dfrac{1}{\sin 3x+1}+\left|\dfrac{\sin 3x}{\sin 3x+1}\right|\right)+C$

39. $u=4x^2$, Formula (70): $\dfrac{1}{8}\displaystyle\int\dfrac{du}{u^2-1}=\dfrac{1}{16}\ln\left|\dfrac{4x^2-1}{4x^2+1}\right|+C$

41. $u=2e^x$, Formula (74): $\dfrac{1}{2}\displaystyle\int\sqrt{3-u^2}\,du=\dfrac{1}{2}e^x\sqrt{3-4e^{2x}}$
$\qquad\qquad\qquad +\dfrac{3}{4}\sin^{-1}\left(\dfrac{2e^x}{\sqrt{3}}\right)+C$

43. $u=3x$, Formula (112): $\dfrac{1}{3}\displaystyle\int\sqrt{\frac{5}{3}u-u^2}\,du=\dfrac{18x-5}{36}\sqrt{5x-9x^2}$
$\qquad\qquad +\dfrac{25}{216}\sin^{-1}\left(\dfrac{18x-5}{5}\right)+C$

45. $u = 2x$, Formula (44): $\int u \sin u \, du = \sin 2x - 2x \cos 2x + C$

47. $u = -\sqrt{x}$, Formula (51): $2 \int u e^u \, du = -2(\sqrt{x}+1)e^{-\sqrt{x}} + C$

49. $x^2 + 6x - 7 = (x+3)^2 - 16$, $u = x + 3$, Formula (70):
$$\int \frac{du}{u^2 - 16} = \frac{1}{8} \ln \left| \frac{x-1}{x+7} \right| + C$$

51. $x^2 - 4x - 5 = (x-2)^2 - 9$, $u = x - 2$, Formula (77):
$$\int \frac{u+2}{\sqrt{9-u^2}} \, du = -\sqrt{5 + 4x - x^2} + 2\sin^{-1}\left(\frac{x-2}{3}\right) + C$$

53. $u = \sqrt{x-2}$, $\frac{2}{5}(x-2)^{5/2} + \frac{4}{3}(x-2)^{3/2} + C$

55. $u = \sqrt{x^3 + 1}$,
$$\frac{2}{3} \int u^2(u^2 - 1) \, du = \frac{2}{15}(x^3+1)^{5/2} - \frac{2}{9}(x^3+1)^{3/2} + C$$

57. $u = x^{1/3}$, $\int \frac{3u^2}{u^3 - u} \, du = \frac{3}{2} \ln |x^{2/3} - 1| + C$

59. $u = x^{1/4}$, $4 \int \frac{1}{u(1-u)} \, du = 4 \ln \frac{x^{1/4}}{|1 - x^{1/4}|} + C$

61. $u = x^{1/6}$,
$$6 \int \frac{u^3}{u-1} \, du = 2x^{1/2} + 3x^{1/3} + 6x^{1/6} + 6 \ln |x^{1/6} - 1| + C$$

63. $u = \sqrt{1 + x^2}$, $\int (u^2 - 1) \, du = \frac{1}{3}(1 + x^2)^{3/2} - (1+x^2)^{1/2} + C$

65. $\int \dfrac{1}{1 + \dfrac{2u}{1+u^2} + \dfrac{1-u^2}{1+u^2}} \cdot \dfrac{2}{1+u^2} \, du = \int \dfrac{1}{u+1} \, du$
$$= \ln |\tan(x/2) + 1| + C$$

67. $\int \dfrac{d\theta}{1 - \cos\theta} = \int \dfrac{1}{u^2} \, du = -\cot(\theta/2) + C$

69. $\int \dfrac{1}{\dfrac{2u}{1+u^2} + \dfrac{2u}{1+u^2} \cdot \dfrac{1+u^2}{1-u^2}} \cdot \dfrac{2}{1+u^2} \, du = \int \dfrac{1-u^2}{2u} \, du$
$$= \frac{1}{2} \ln |\tan(x/2)| - \frac{1}{4} \tan^2(x/2) + C$$

71. $x = \dfrac{4e^2}{1+e^2}$ **73.** $A = 6 + \frac{25}{2} \sin^{-1} \frac{4}{5}$ **75.** $A = \frac{1}{40} \ln 9$

77. $V = \pi(\pi - 2)$ **79.** $V = 2\pi(1 - 4e^{-3})$

81. $L = \sqrt{65} + \frac{1}{8} \ln(8 + \sqrt{65})$ **83.** $S = 2\pi \left[\sqrt{2} + \ln(1 + \sqrt{2}) \right]$

85.

91. $\frac{1}{31} \cos^{31} x \sin^{31} x + C$

93. $-\frac{1}{9} \ln |1 + x^{-9}| + C$

Exercise Set 7.7 (Page 544)

1. $\int_0^3 \sqrt{x+1} \, dx = \frac{14}{3} \approx 4.66667$
(a) $M_{10} = 4.66760$; $|E_M| \approx 0.000933996$
(b) $T_{10} = 4.66480$; $|E_T| \approx 0.00187099$
(c) $S_{20} = 4.66667$; $|E_S| \approx 9.98365 \times 10^{-7}$

3. $\int_0^{\pi/2} \cos x \, dx = 1$
(a) $M_{10} = 1.00103$; $|E_M| \approx 0.00102882$
(b) $T_{10} = 0.997943$; $|E_T| \approx 0.00205701$
(c) $S_{20} = 1.00000$; $|E_S| \approx 2.11547 \times 10^{-7}$

5. $\int_1^3 e^{-2x} \, dx = \frac{-1 + e^4}{2e^6} \approx 0.0664283$
(a) $M_{10} = 0.0659875$; $|E_M| \approx 0.000440797$
(b) $T_{10} = 0.0673116$; $|E_T| \approx 0.000883357$
(c) $S_{20} = 0.0664289$; $|E_S| \approx 5.87673 \times 10^{-7}$

7. (a) $|E_M| \le \dfrac{9}{3200} = 0.0028125$
(b) $|E_T| \le \dfrac{9}{1600} = 0.005625$
(c) $|E_S| \le \dfrac{81}{10,240,000} \approx 7.91016 \times 10^{-6}$

9. (a) $|E_M| \le \dfrac{\pi^3}{19,200} \approx 0.00161491$
(b) $|E_T| \le \dfrac{\pi^3}{9600} \approx 0.00322982$
(c) $|E_S| \le \dfrac{\pi^5}{921,600,000} \approx 3.32053 \times 10^{-7}$

11. (a) $|E_M| \le \dfrac{1}{75e^2} \approx 0.00180447$
(b) $|E_T| \le \dfrac{2}{75e^2} \approx 0.00360894$
(c) $|E_S| \le \dfrac{1}{56,250e^2} \approx 2.40596 \times 10^{-6}$

13. (a) $n = 24$ (b) $n = 34$ (c) $n = 8$
15. (a) $n = 13$ (b) $n = 18$ (c) $n = 4$
17. (a) $n = 43$ (b) $n = 61$ (c) $n = 8$
Responses to True–False questions may be abridged to save space.
19. False; T_n is the average of L_n and R_n.
21. False; $S_{50} = \frac{2}{3} M_{25} + \frac{1}{3} T_{25}$
23. $g(x) = \frac{1}{24} x^2 - \frac{3}{8} x + \frac{13}{12}$
25. $S_{10} = 1.49367$; $\int_{-1}^1 e^{-x^2} \, dx \approx 1.49365$
27. $S_{10} = 3.80678$; $\int_{-1}^2 x\sqrt{1 + x^3} \, dx \approx 3.80554$
29. $S_{10} = 0.904524$; $\int_0^1 \cos x^2 \, dx \approx 0.904524$
31. (a) $M_{10} = 3.14243$; error $E_M \approx -0.000833331$
(b) $T_{10} = 3.13993$; error $E_T \approx 0.00166666$
(c) $S_{20} = 3.14159$; error $E_S \approx 6.20008 \times 10^{-10}$
33. $S_{14} = 0.693147984$, $|E_S| \approx 0.000000803 = 8.03 \times 10^{-7}$
35. $n = 116$ **39.** 3.82019 **41.** 1071 ft **43.** 37.9 mi **45.** 9.3 L
47. (a) max $|f''(x)| \approx 3.844880$ (b) $n = 18$ (c) 0.904741
49. (a) The maximum value of $|f^{(4)}(x)|$ is approximately 12.4282.
(b) $n = 6$ (c) $S_6 = 0.983347$

Exercise Set 7.8 (Page 554)

1. (a) improper; infinite discontinuity at $x = 3$ (b) not improper
(c) improper; infinite discontinuity at $x = 0$
(d) improper; infinite interval of integration
(e) improper; infinite interval of integration and infinite discontinuity
at $x = 1$ (f) not improper
3. $\frac{1}{2}$ **5.** $\ln 2$ **7.** $\frac{1}{2}$ **9.** $-\frac{1}{4}$ **11.** $\frac{1}{3}$ **13.** divergent **15.** 0
17. divergent **19.** divergent **21.** $\pi/2$ **23.** 1 **25.** divergent
27. $\frac{9}{2}$ **29.** divergent **31.** $\pi/2$
Responses to True–False questions may be abridged to save space.
33. True; see Theorem 7.8.2 with $p = \frac{4}{3} > 1$.
35. False; the integrand $\dfrac{1}{x(x-3)}$ is continuous on $[1, 2]$.
37. 2 **39.** 2 **41.** $\frac{1}{2}$
43. (a) 2.726585 (b) 2.804364 (c) 0.219384 (d) 0.504067 **45.** 12
47. -1 **49.** $\frac{1}{3}$ **51.** (a) $V = \pi/2$ (b) $S = \pi[\sqrt{2} + \ln(1 + \sqrt{2})]$
53. (b) $1/e$ (c) It is convergent. **55.** $V = \pi$
59. $\dfrac{2\pi N I}{kr} \left(1 - \dfrac{a}{\sqrt{r^2 + a^2}} \right)$
61. (b) 2.4×10^7 mi·lb **63.** (a) $\dfrac{1}{s^2}$ (b) $\dfrac{2}{s^3}$ (c) $\dfrac{e^{-3s}}{s}$
67. (a) 1.047 **71.** 1.809

▶ **Chapter 7 Review Exercises (Page 557)**

1. $\frac{2}{27}(4+9x)^{3/2}+C$ **3.** $-\frac{2}{3}\cos^{3/2}\theta+C$ **5.** $\frac{1}{6}\tan^3(x^2)+C$

7. (a) $2\sin^{-1}(\sqrt{x/2})+C$; $-2\sin^{-1}(\sqrt{2-x}/\sqrt{2})+C$; $\sin^{-1}(x-1)+C$

9. $-xe^{-x}-e^{-x}+C$ **11.** $x\ln(2x+3)-x+\frac{3}{2}\ln(2x+3)+C$

13. $(4x^4-12x^2+6)\sin(2x)+(8x^3-12x)\cos(2x)+C$

15. $\frac{1}{2}\theta-\frac{1}{20}\sin 10\theta+C$ **17.** $-\frac{1}{6}\cos 3x+\frac{1}{2}\cos x+C$

19. $-\frac{1}{8}\sin^3(2x)\cos 2x-\frac{3}{16}\cos 2x\sin 2x+\frac{3}{8}x+C$

21. $\frac{9}{2}\sin^{-1}(x/3)-\frac{1}{2}x\sqrt{9-x^2}+C$ **23.** $\ln|x+\sqrt{x^2-1}|+C$

25. $\dfrac{x\sqrt{x^2+9}}{2}-\dfrac{9\ln(|\sqrt{x^2+9}+x|)}{2}+C$ **27.** $\frac{1}{5}\ln\left|\dfrac{x-1}{x+4}\right|+C$

29. $\frac{1}{2}x^2-2x+6\ln|x+2|+C$

31. $\ln|x+2|+\dfrac{4}{x+2}-\dfrac{2}{(x+2)^2}+C$

35. Formula (40): $-\dfrac{\cos 16x}{32}+\dfrac{\cos 2x}{4}+C$

37. Formula (113): $\frac{1}{24}(8x^2-2x-3)\sqrt{x-x^2}+\frac{1}{16}\sin^{-1}(2x-1)+C$

39. Formula (28): $\frac{1}{2}\tan 2x-x+C$

41. $\displaystyle\int_1^3\dfrac{1}{\sqrt{x+1}}=4-2\sqrt{2}\approx 1.17157$
(a) $M_{10}=1.17138$; $|E_M|\approx 0.000190169$
(b) $T_{10}=1.17195$; $|E_T|\approx 0.000380588$
(c) $S_{20}=1.17157$; $|E_S|\approx 8.35151\times 10^{-8}$

43. (a) $|E_M|\le\dfrac{1}{1600\sqrt{2}}\approx 0.000441942$

(b) $|E_T|\le\dfrac{1}{800\sqrt{2}}\approx 0.000883883$

(c) $|E_S|\le\dfrac{7}{15{,}360{,}000\sqrt{2}}\approx 3.22249\times 10^{-7}$

45. (a) $n=22$ (b) $n=30$ (c) $n=6$ **47.** 1   **49.** 6

51. e^{-1} **53.** $a=\pi/2$ **55.** $\dfrac{x}{3\sqrt{3+x^2}}+C$ **57.** $\frac{5}{12}-\frac{1}{2}\ln 2$

59. $\frac{1}{6}\sin^3 2x-\frac{1}{10}\sin^5 2x+C$

61. $\frac{2}{13}e^{2x}\cos 3x+\frac{3}{13}e^{2x}\sin 3x+C$

63. $-\frac{1}{6}\ln|x-1|+\frac{1}{15}\ln|x+2|+\frac{1}{10}\ln|x-3|+C$

65. $4-\pi$ **67.** $\ln\dfrac{\sqrt{e^x+1}-1}{\sqrt{e^x+1}+1}+C$ **69.** $\dfrac{\pi}{12}+\dfrac{\sqrt{3}}{2}-1$

71. $\sqrt{x^2+2x+2}+2\ln(\sqrt{x^2+2x+2}+x+1)+C$ **73.** $\dfrac{1}{2(a^2+1)}$

▶ **Chapter 7 Making Connections (Page 559)**

Where correct answers to a Making Connections exercise may vary, no answer is listed. Sample answers for these questions are available on the Book Companion Site.

3. (a) $\Gamma(1)=1$ (c) $\Gamma(2)=1$, $\Gamma(3)=2$, $\Gamma(4)=6$

5. (b) 1.37078 seconds

▶ **Exercise Set 8.1 (Page 566)**

3. (a) first order   (b) second order

Responses to True–False questions may be abridged to save space.

5. False; only first-order derivatives appear.

7. True; it is third order.

15. $y(x)=e^{-2x}-2e^x$ **17.** $y(x)=2e^{2x}-2xe^{2x}$

19. $y(x)=\sin 2x+\cos 2x$ **21.** $y(x)=-2x^2+2x+3$

23. $y(x)=2/(3-2x)$ **25.** $y(x)=2/x^2$

27. (a) $\dfrac{dy}{dt}=ky^2$, $y(0)=y_0$ $(k>0)$

(b) $\dfrac{dy}{dt}=-ky^2$, $y(0)=y_0$ $(k>0)$

29. (a) $\dfrac{ds}{dt}=\frac{1}{2}s$ (b) $\dfrac{d^2s}{dt^2}=2\dfrac{ds}{dt}$ **33.** (b) $L/2$

▶ **Exercise Set 8.2 (Page 575)**

1. $y=Cx$ **3.** $y=Ce^{-\sqrt{1+x^2}}-1$, $C\ne 0$

5. $2\ln|y|+y^2=e^x+C$ **7.** $y=\ln(\sec x+C)$

9. $y=\dfrac{1}{1-C(\csc x-\cot x)}$, $y=0$

11. $y^2+\sin y=x^3+\pi^2$ **13.** $y^2-2y=t^2+t+3$

15. (a)

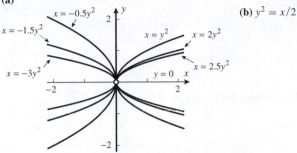

(b) $y^2=x/2$

17. $y=\dfrac{C}{\sqrt{x^2+4}}$

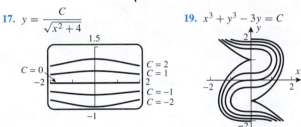

19. $x^3+y^3-3y=C$

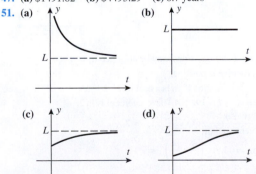

Responses to True–False questions may be abridged to save space.

21. True; since $\dfrac{1}{f(y)}\,dy=dx$.   **23.** True; since $\left(\frac{1}{2}\right)^5 32=1$.

27. $y=\ln\left(\dfrac{x^2}{2}-1\right)$

29. (a) $y'(t)=y(t)/50$, $y(0)=10{,}000$ (b) $y(t)=10{,}000e^{t/50}$
(c) $50\ln 2\approx 34.66$ hr   (d) $50\ln(4.5)\approx 75.20$ hr

31. (a) $\dfrac{dy}{dt}=-ky$, $k\approx 0.1810$ (b) $y=5.0\times 10^7e^{-0.181t}$
(c) $\approx 219{,}000$ atoms   (d) 12.72 days

33. $50\ln(100)\approx 230.26$ days   **35.** 3.30 days

39. (b) 70 years   (c) 20 years   (d) 7%

43. (a) no   (b) same, $r\%$ **45.** (b) $\ln(2)/\ln(5/4)\approx 3.106$ hr

47. (a) \$1491.82   (b) \$4493.29   (c) 8.7 years

51. (a)   (b)   (c)   (d)

53. $y_0\approx 2$, $L\approx 8$, $k\approx 0.5493$

55. (a) $y_0=5$ (b) $L=12$ (c) $k=1$ (d) $t=0.3365$
(e) $\dfrac{dy}{dt}=\frac{1}{2}y(12-y)$, $y(0)=5$

57. (a) $y=\dfrac{1000}{1+49e^{-0.115t}}$

(b)

t	0	1	2	3	4	5	6	7
$y(t)$	20	22	25	28	31	35	39	44

t	8	9	10	11	12	13	14
$y(t)$	49	54	61	67	75	83	93

(c)

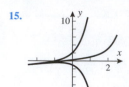

59. (a) $T = 21 + 74e^{-kt}$ (b) 6.22 min
61. (a) $v = c \ln \dfrac{m_0}{m_0 - kt} - gt$ (b) 3044 m/s
63. (a) $h \approx (2 - 0.003979t)^2$ (b) 8.4 min
65. (a) $v = 128/(4t + 1)$, $x = 32 \ln(4t + 1)$

▶ **Exercise Set 8.3 (Page 584)**

1.
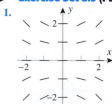

3.
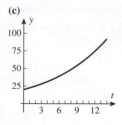

5. $y \to 1$ as $x \to +\infty$

7.

n	0	1	2	3	4	5	6	7	8
x_n	0	0.5	1	1.5	2	2.5	3	3.5	4
y_n	1.00	1.50	2.07	2.71	3.41	4.16	4.96	5.82	6.72

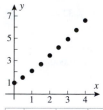

9.

n	0	1	2	3	4
t_n	0.00	0.50	1.00	1.50	2.00
y_n	1.00	1.27	1.42	1.49	1.53

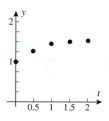

11. 0.62
Responses to True–False questions may be abridged to save space.
13. True; the derivative is positive.
15. True; $y = y_0$ is a solution if y_0 is a root of p.
17. (b) $y(1/2) = \sqrt{3}/2$ **19.** (b) The x-intercept is ln 2.
23. (a) $y' = \dfrac{2xy - y^3}{3xy^2 - x^2}$ (c) $xy^3 - x^2y = 2$
25. (b) $\displaystyle\lim_{n \to +\infty} y_n = \lim_{n \to +\infty}\left(\dfrac{n+1}{n}\right)^n = e$

▶ **Exercise Set 8.4 (Page 592)**

1. $y = e^{-3x} + Ce^{-4x}$ **3.** $y = e^{-x}\sin(e^x) + Ce^{-x}$ **5.** $y = \dfrac{C}{\sqrt{x^2 + 1}}$
7. $y = \dfrac{x}{2} + \dfrac{3}{2x}$ **9.** $y = 4e^{x^2} - 1$
Responses to True–False questions may be abridged to save space.
11. False; $y = x^2$ is a solution to $dy/dx = 2x$, but $y + y = 2x^2$ is not.
13. True; it will approach the concentration of the entering fluid.

15.

17. $\displaystyle\lim_{x \to +\infty} y = \begin{cases} +\infty & \text{if } y_0 \geq 1/4 \\ -\infty, & \text{if } y_0 < 1/4 \end{cases}$

19. (a)

n	0	1	2	3	4	5
x_n	0	0.2	0.4	0.6	0.8	1.0
y_n	1	1.20	1.48	1.86	2.35	2.98

(b) $y = -(x + 1) + 2e^x$

x_n	0	0.2	0.4	0.6	0.8	1.0
$y(x_n)$	1	1.24	1.58	2.04	2.65	3.44
Absolute error	0	0.04	0.10	0.19	0.30	0.46
Percentage error	0	3	6	9	11	13

21. (a) $200 - 175e^{-t/25}$ oz (b) 136 oz **23.** 25 lb
27. (a) $I(t) = 2 - 2e^{-2t}$ A (b) $I(t) \to 2$ A

▶ **Chapter 8 Review Exercises (Page 594)**

3. $y = \tan(x^3/3 + C)$ **5.** $\ln|y| + y^2/2 = e^x + C$ and $y = 0$
7. $y^{-4} + 4\ln(x/y) = 1$
9.
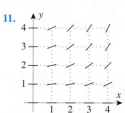

11.
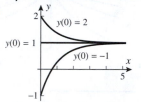

13.

n	0	1	2	3	4	5	6	7	8
x_n	0	0.5	1	1.5	2	2.5	3	3.5	4
y_n	1	1.50	2.11	2.84	3.68	4.64	5.72	6.91	8.23

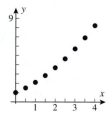

15. $y(1) \approx 1.00$

n	0	1	2	3	4	5
t_n	0	0.2	0.4	0.6	0.8	1.0
y_n	1.00	1.20	1.26	1.10	0.94	1.00

17. about 2000.6 years **19.** $y = e^{-2x} + Ce^{-3x}$
21. $y = -1 + 4e^{x^2/2}$ **23.** $y = 2\operatorname{sech} x + \frac{1}{2}(x\operatorname{sech} x + \sinh x)$
25. (a) linear (b) both (c) separable (d) neither **27.** about 646 oz

▶ **Chapter 8 Making Connections (Page 595)**

Where correct answers to a Making Connections exercise may vary, no
answer is listed. Sample answers for these questions are available on the
Book Companion Site.

1. (b) $y = 2 - 3e^{-2x}$
3. (a) $du/dx = (f(u) - u)/x$ (b) $x^2 - 2xy - y^2 = C$

▶ **Exercise Set 9.1 (Page 605)**

1. (a) $\dfrac{1}{3^{n-1}}$ (b) $\dfrac{(-1)^{n-1}}{3^{n-1}}$ (c) $\dfrac{2n-1}{2n}$ (d) $\dfrac{n^2}{\pi^{1/(n+1)}}$

3. (a) $2, 0, 2, 0$ (b) $1, -1, 1, -1$ (c) $2(1+(-1)^n)$; $2+2\cos n\pi$

5. (a) The limit doesn't exist due to repeated oscillation between -1 and 1.   (b) $-1; 1; -1; 1; -1$ (c) no

7. $\frac{1}{3}, \frac{2}{4}, \frac{3}{5}, \frac{4}{6}, \frac{5}{7}$; converges, $\displaystyle\lim_{n\to+\infty}\frac{n}{n+2}=1$

9. $2, 2, 2, 2, 2$; converges, $\displaystyle\lim_{n\to+\infty}2=2$

11. $\dfrac{\ln 1}{1}, \dfrac{\ln 2}{2}, \dfrac{\ln 3}{3}, \dfrac{\ln 4}{4}, \dfrac{\ln 5}{5}$; converges, $\displaystyle\lim_{n\to+\infty}\frac{\ln n}{n}=0$

13. $0, 2, 0, 2, 0$; diverges   15. $-1, \frac{16}{9}, -\frac{54}{28}, \frac{128}{65}, -\frac{250}{126}$; diverges

17. $\frac{6}{2}, \frac{12}{8}, \frac{20}{18}, \frac{30}{32}, \frac{42}{50}$; converges, $\displaystyle\lim_{n\to+\infty}\frac{1}{2}\left(1+\frac{1}{n}\right)\left(1+\frac{2}{n}\right)=\frac{1}{2}$

19. $e^{-1}, 4e^{-2}, 9e^{-3}, 16e^{-4}, 25e^{-5}$; converges, $\displaystyle\lim_{n\to+\infty}n^2e^{-n}=0$

21. $2, \left(\dfrac{5}{3}\right)^2, \left(\dfrac{6}{4}\right)^3, \left(\dfrac{7}{5}\right)^4, \left(\dfrac{8}{6}\right)^5$; converges, $\displaystyle\lim_{n\to+\infty}\left[\dfrac{n+3}{n+1}\right]^n=e^2$

23. $\left\{\dfrac{2n-1}{2n}\right\}_{n=1}^{+\infty}$; converges, $\displaystyle\lim_{n\to+\infty}\dfrac{2n-1}{2n}=1$

25. $\left\{(-1)^{n+1}\dfrac{1}{3^n}\right\}_{n=1}^{+\infty}$; converges, $\displaystyle\lim_{n\to+\infty}(-1)^{n+1}\dfrac{1}{3^n}=0$

27. $\left\{(-1)^{n+1}\left(\dfrac{1}{n}-\dfrac{1}{n+1}\right)\right\}_{n=1}^{+\infty}$;

    converges, $\displaystyle\lim_{n\to+\infty}(-1)^{n+1}\left(\dfrac{1}{n}-\dfrac{1}{n+1}\right)=0$

29. $\{\sqrt{n+1}-\sqrt{n+2}\}_{n=1}^{+\infty}$; converges, $\displaystyle\lim_{n\to+\infty}(\sqrt{n+1}-\sqrt{n+2})=0$

Responses to True–False questions may be abridged to save space.

31. True; a sequence is a function whose domain is a set of integers.

33. False; for example, $\{(-1)^{n+1}\}$ diverges with terms that oscillate between 1 and -1.   35. The limit is 0.

37. for example, $\{(-1)^n\}_{n=1}^{+\infty}$ and $\{\sin(\pi n/2)+1/n\}_{n=1}^{+\infty}$

39. (a) $1, 2, 1, 4, 1, 6$ (b) $a_n=\begin{cases}n, & n \text{ odd}\\ 1/2^n, & n \text{ even}\end{cases}$

    (c) $a_n=\begin{cases}1/n, & n \text{ odd}\\ 1/(n+1); & n \text{ even}\end{cases}$

    (d) (a) diverges; (b) diverges; (c) $\displaystyle\lim_{n\to+\infty}a_n=0$

43. (a) $(0.5)^{2n}$ (c) $\displaystyle\lim_{n\to+\infty}a_n=0$ (d) $-1\le a_0\le 1$

45. (a)   (b) $\displaystyle\lim_{n\to+\infty}(2^n+3^n)^{1/n}=3$

49. (a) $N=4$ (b) $N=10$ (c) $N=1000$

▶ **Exercise Set 9.2 (Page 613)**

1. strictly decreasing   3. strictly increasing   5. strictly decreasing

7. strictly increasing   9. strictly decreasing   11. strictly increasing

Responses to True–False questions may be abridged to save space.

13. True; $a_{n+1}-a_n>0$ for all n is equivalent to $a_1<a_2<a_3<\cdots<a_n<\cdots$.

15. False; for example, $\{(-1)^{n+1}\}=\{1,-1,1,-1,\ldots\}$ is bounded but diverges.

17. strictly increasing   19. strictly increasing

21. eventually strictly increasing   23. eventually strictly increasing

25. Yes; the limit lies in the interval $[1, 2]$.

27. (a) $\sqrt{2}, \sqrt{2+\sqrt{2}}, \sqrt{2+\sqrt{2+\sqrt{2}}}$ (e) $L=2$

▶ **Exercise Set 9.3 (Page 621)**

1. (a) $2, \frac{12}{5}, \frac{62}{25}, \frac{312}{125}$; $\frac{5}{2}\left(1-\left(\frac{1}{5}\right)^n\right)$; $\displaystyle\lim_{n\to+\infty}s_n=\frac{5}{2}$ (converges)

    (b) $\frac{1}{4}, \frac{3}{4}, \frac{7}{4}, \frac{15}{4}$; $-\frac{1}{4}(1-2^n)$; $\displaystyle\lim_{n\to+\infty}s_n=+\infty$ (diverges)

    (c) $\frac{1}{6}, \frac{1}{4}, \frac{3}{10}, \frac{1}{3}$; $\frac{1}{2}-\frac{1}{n+2}$; $\displaystyle\lim_{n\to+\infty}s_n=\frac{1}{2}$ (converges)

3. $\frac{4}{7}$ 5. 6 7. $\frac{1}{3}$ 9. $\frac{1}{6}$ 11. diverges   13. $\frac{448}{3}$

15. (a) Exercise 5   (b) Exercise 3   (c) Exercise 7   (d) Exercise 9

Responses to True–False questions may be abridged to save space.

17. False; an infinite series converges if its sequence of *partial sums* converges.

19. True; the sequence of partial sums $\{s_n\}$ for the harmonic series satisfies $s_{2^n}>\dfrac{n+1}{2}$, so this series diverges.

21. 1 23. $\frac{532}{99}$ 27. 70 m

29. (a) $S_n=-\ln(n+1)$; $\displaystyle\lim_{n\to+\infty}S_n=-\infty$ (diverges)

    (b) $S_n=\displaystyle\sum_{k=2}^{n+1}\left[\ln\dfrac{k-1}{k}-\ln\dfrac{k}{k+1}\right]$, $\displaystyle\lim_{n\to+\infty}S_n=-\ln 2$

31. (a) converges for $|x|<1$; $S=\dfrac{x}{1+x^2}$

    (b) converges for $|x|>2$; $S=\dfrac{1}{x^2-2x}$

    (c) converges for $x>0$; $S=\dfrac{1}{e^x-1}$

33. $a_n=\dfrac{1}{2^{n-1}}a_1+\dfrac{1}{2^{n-1}}+\dfrac{1}{2^{n-2}}+\cdots+\dfrac{1}{2}$, $\displaystyle\lim_{n\to+\infty}a_n=1$

▶ **Exercise Set 9.4 (Page 629)**

1. (a) $\frac{4}{3}$ (b) $-\frac{3}{4}$

3. (a) $p=3$, converges   (b) $p=\frac{1}{2}$, diverges

    (c) $p=1$, diverges   (d) $p=\frac{2}{3}$, diverges

5. (a) diverges   (b) diverges   (c) diverges   (d) no information

7. (a) diverges   (b) converges

9. diverges   11. diverges   13. diverges   15. diverges   17. diverges

19. converges   21. diverges   23. converges   25. converges for $p>1$

29. (a) diverges   (b) diverges

Responses to True–False questions may be abridged to save space.

31. False; for example, $\displaystyle\sum_{k=0}^{\infty}2^{-k}$ converges to 2, but $\displaystyle\sum\frac{1}{2^{-k}}=\sum 2^k$ diverges.

33. True; see Theorem 9.4.4.

35. (a) $(\pi^2/2)-(\pi^4/90)$ (b) $(\pi^2/6)-(5/4)$ (c) $\pi^4/90$

37. (d) $\frac{1}{11}<\frac{1}{6}\pi^2-s_{10}<\frac{1}{10}$

39. (a) $\displaystyle\int_n^{+\infty}\frac{1}{x^4}\,dx=\frac{1}{3n^3}$; apply Exercise 36(b)   (b) $n=6$

    (c) $\dfrac{\pi^4}{90}\approx 1.08238$

41. converges

▶ **Exercise Set 9.5 (Page 636)**

1. (a) converges   (b) diverges   5. converges   7. converges

9. diverges   11. converges   13. inconclusive   15. diverges

17. diverges   19. converges

Responses to True–False questions may be abridged to save space.

21. False; the limit comparison test uses a limit of the quotient of corresponding terms taken from two different sequences.

23. True; use the limit comparison test with the convergent series $\sum(1/k^2)$.

25. converges   27. converges   29. converges   31. converges

33. diverges   35. diverges   37. converges   39. converges

41. diverges   43. converges   45. converges   47. converges

49. $u_k = \dfrac{k!}{1 \cdot 3 \cdot 5 \cdots (2k-1)}$; $\rho = \lim\limits_{k \to +\infty} \dfrac{k+1}{2k+1} = \dfrac{1}{2}$; converges

51. **(a)** converges **(b)** diverges **53.** **(a)** converges

▶ **Exercise Set 9.6 (Page 646)**

3. diverges **5.** converges **7.** converges absolutely **9.** diverges

11. converges absolutely **13.** conditionally convergent **15.** divergent

17. conditionally convergent **19.** conditionally convergent

21. divergent **23.** conditionally convergent **25.** converges absolutely

27. converges absolutely

Responses to True–False questions may be abridged to save space.

29. False; an alternating series has terms that alternate between positive and negative.

31. True; if a series converges but diverges absolutely, then it converges conditionally.

33. |error| < 0.125 **35.** |error| < 0.1 **37.** $n = 9999$

39. $n = 39{,}999$ **41.** |error| < 0.00074; $s_{10} \approx 0.4995$; $S = 0.5$

43. 0.84 **45.** 0.41 **47.** **(c)** $n = 50$

49. **(a)** If $a_k = \dfrac{(-1)^k}{\sqrt{k}}$, then $\sum a_k$ converges and $\sum a_k^2$ diverges.

If $a_k = \dfrac{(-1)^k}{k}$, then $\sum a_k$ converges and $\sum a_k^2$ also converges.

(b) If $a_k = \dfrac{1}{k}$, then $\sum a_k^2$ converges and $\sum a_k$ diverges. If $a_k = \dfrac{1}{k^2}$, then $\sum a_k^2$ converges and $\sum a_k$ also converges.

▶ **Exercise Set 9.7 (Page 657)**

1. **(a)** $1 - x + \frac{1}{2}x^2$, $1 - x$ **(b)** $1 - \frac{1}{2}x^2$, 1

3. **(a)** $1 + \frac{1}{2}(x-1) - \frac{1}{8}(x-1)^2$ **(b)** 1.04875 **5.** 1.80397443

7. $p_0(x) = 1$, $p_1(x) = 1 - x$, $p_2(x) = 1 - x + \frac{1}{2}x^2$,

$p_3(x) = 1 - x + \frac{1}{2}x^2 - \frac{1}{3!}x^3$,

$p_4(x) = 1 - x + \frac{1}{2}x^2 - \frac{1}{3!}x^3 + \frac{1}{4!}x^4$; $\displaystyle\sum_{k=0}^{n} \frac{(-1)^k}{k!} x^k$

9. $p_0(x) = 1$, $p_1(x) = 1$, $p_2(x) = 1 - \dfrac{\pi^2}{2!}x^2$, $p_3(x) = 1 - \dfrac{\pi^2}{2!}x^2$,

$p_4(x) = 1 - \dfrac{\pi^2}{2!}x^2 + \dfrac{\pi^4}{4!}x^4$; $\displaystyle\sum_{k=0}^{\lfloor n/2 \rfloor} \frac{(-1)^k \pi^{2k}}{(2k)!} x^{2k}$ (See Exercise 70 of Section 0.2.)

11. $p_0(x) = 0$, $p_1(x) = x$, $p_2(x) = x - \frac{1}{2}x^2$, $p_3(x) = x - \frac{1}{2}x^2 + \frac{1}{3}x^3$,

$p_4(x) = x - \frac{1}{2}x^2 + \frac{1}{3}x^3 - \frac{1}{4}x^4$; $\displaystyle\sum_{k=1}^{n} \frac{(-1)^{k+1}}{k} x^k$

13. $p_0(x) = 1$, $p_1(x) = 1$, $p_2(x) = 1 + \dfrac{x^2}{2}$,

$p_3(x) = 1 + \dfrac{x^2}{2}$, $p_4(x) = 1 + \dfrac{x^2}{2} + \dfrac{x^4}{4!}$; $\displaystyle\sum_{k=0}^{\lfloor n/2 \rfloor} \frac{1}{(2k)!} x^{2k}$ (See Exercise 70 of Section 0.2.)

15. $p_0(x) = 0$, $p_1(x) = 0$, $p_2(x) = x^2$, $p_3(x) = x^2$,

$p_4(x) = x^2 - \frac{1}{6}x^4$; $\displaystyle\sum_{k=0}^{\lfloor n/2 \rfloor - 1} \frac{(-1)^k}{(2k+1)!} x^{2k+2}$ (See Exercise 70 of Section 0.2.)

17. $p_0(x) = e$, $p_1(x) = e + e(x-1)$,

$p_2(x) = e + e(x-1) + \dfrac{e}{2}(x-1)^2$,

$p_3(x) = e + e(x-1) + \dfrac{e}{2}(x-1)^2 + \dfrac{e}{3!}(x-1)^3$,

$p_4(x) = e + e(x-1) + \dfrac{e}{2}(x-1)^2 + \dfrac{e}{3!}(x-1)^3 + \dfrac{e}{4!}(x-1)^4$;

$\displaystyle\sum_{k=0}^{n} \frac{e}{k!}(x-1)^k$

19. $p_0(x) = -1$, $p_1(x) = -1 - (x+1)$,

$p_2(x) = -1 - (x+1) - (x+1)^2$,

$p_3(x) = -1 - (x+1) - (x+1)^2 - (x+1)^3$,

$p_4(x) = -1 - (x+1) - (x+1)^2 - (x+1)^3 - (x+1)^4$;

$\displaystyle\sum_{k=0}^{n} (-1)(x+1)^k$

21. $p_0(x) = p_1(x) = 1$, $p_2(x) = p_3(x) = 1 - \dfrac{\pi^2}{2}\left(x - \dfrac{1}{2}\right)^2$,

$p_4(x) = 1 - \dfrac{\pi^2}{2}\left(x - \dfrac{1}{2}\right)^2 + \dfrac{\pi^4}{4!}\left(x - \dfrac{1}{2}\right)^4$;

$\displaystyle\sum_{k=0}^{\lfloor n/2 \rfloor} \frac{(-1)^k \pi^{2k}}{(2k)!}\left(x - \dfrac{1}{2}\right)^{2k}$ (See Exercise 70 of Section 0.2.)

23. $p_0(x) = 0$, $p_1(x) = (x-1)$, $p_2(x) = (x-1) - \frac{1}{2}(x-1)^2$,

$p_3(x) = (x-1) - \frac{1}{2}(x-1)^2 + \frac{1}{3}(x-1)^3$,

$p_4(x) = (x-1) - \frac{1}{2}(x-1)^2 + \frac{1}{3}(x-1)^3 - \frac{1}{4}(x-1)^4$;

$\displaystyle\sum_{k=1}^{n} \frac{(-1)^{k-1}}{k}(x-1)^k$

25. **(a)** $p_3(x) = 1 + 2x - x^2 + x^3$

(b) $p_3(x) = 1 + 2(x-1) - (x-1)^2 + (x-1)^3$

27. $p_0(x) = 1$, $p_1(x) = 1 - 2x$,
$p_2(x) = 1 - 2x + 2x^2$,
$p_3(x) = 1 - 2x + 2x^2 - \frac{4}{3}x^3$

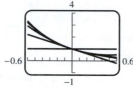

29. $p_0(x) = -1$, $p_2(x) = -1 + \frac{1}{2}(x - \pi)^2$,

$p_4(x) = -1 + \frac{1}{2}(x - \pi)^2 - \frac{1}{24}(x - \pi)^4$,

$p_6(x) = -1 + \frac{1}{2}(x - \pi)^2 - \frac{1}{24}(x - \pi)^4$
$+ \frac{1}{720}(x - \pi)^6$

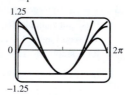

Responses to True–False questions may be abridged to save space.

31. True; $y = f(x_0) + f'(x_0)(x - x_0)$ is the first-degree Taylor polynomial for f about $x = x_0$.

33. False; $p_6^{(4)}(x_0) = f^{(4)}(x_0)$ **35.** 1.64870 **37.** IV

39. **(a)**

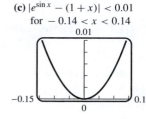

(b)

x	−1.000	−0.750	−0.500	−0.250	0.000	0.250	0.500	0.750	1.000
$f(x)$	0.431	0.506	0.619	0.781	1.000	1.281	1.615	1.977	2.320
$p_1(x)$	0.000	0.250	0.500	0.750	1.000	1.250	1.500	1.750	2.000
$p_2(x)$	0.500	0.531	0.625	0.781	1.000	1.281	1.625	2.031	2.500

(c) $|e^{\sin x} - (1 + x)| < 0.01$
for $-0.14 < x < 0.14$

(d) $\left| e^{\sin x} - \left(1 + x + \dfrac{x^2}{2}\right)\right| < 0.01$
for $-0.50 < x < 0.50$

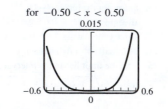

41. (a) $[0, 0.137]$ (b) 0.002

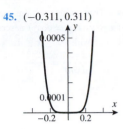

43. (a) $(-0.569, 0.569)$ **45.** $(-0.311, 0.311)$

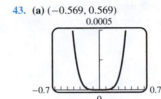

▶ **Exercise Set 9.8 (Page 667)**

1. $\displaystyle\sum_{k=0}^{\infty} \frac{(-1)^k}{k!} x^k$ **3.** $\displaystyle\sum_{k=0}^{\infty} \frac{(-1)^k \pi^{2k}}{(2k)!} x^{2k}$ **5.** $\displaystyle\sum_{k=1}^{\infty} \frac{(-1)^{k+1}}{k} x^k$

7. $\displaystyle\sum_{k=0}^{\infty} \frac{1}{(2k)!} x^{2k}$ **9.** $\displaystyle\sum_{k=0}^{\infty} \frac{(-1)^k}{(2k+1)!} x^{2k+2}$ **11.** $\displaystyle\sum_{k=0}^{\infty} \frac{e}{k!} (x-1)^k$

13. $\displaystyle\sum_{k=0}^{\infty} (-1)(x+1)^k$ **15.** $\displaystyle\sum_{k=0}^{\infty} \frac{(-1)^k \pi^{2k}}{(2k)!} \left(x - \frac{1}{2}\right)^{2k}$

17. $\displaystyle\sum_{k=1}^{\infty} \frac{(-1)^{k-1}}{k} (x-1)^k$ **19.** $-1 < x < 1;\ \dfrac{1}{1+x}$

21. $1 < x < 3;\ \dfrac{1}{3-x}$ **23.** (a) $-2 < x < 2$ (b) $f(0) = 1;\ f(1) = \frac{2}{3}$

Responses to True–False questions may be abridged to save space.

25. True; see Theorem 9.8.2(c).

27. True; the polynomial *is* the Maclaurin series and converges for all x.

29. $R = 1;\ [-1, 1)$ **31.** $R = +\infty;\ (-\infty, +\infty)$ **33.** $R = \frac{1}{5};\ [-\frac{1}{5}, \frac{1}{5}]$

35. $R = 1;\ [-1, 1]$ **37.** $R = 1;\ (-1, 1]$ **39.** $R = +\infty;\ (-\infty, +\infty)$

41. $R = 1;\ [-1, 1]$ **43.** $R = 1;\ (-2, 0]$ **45.** $R = \frac{4}{3};\ \left(-\frac{19}{3}, -\frac{11}{3}\right)$

47. $R = +\infty;\ (-\infty, +\infty)$ **49.** $(-\infty, +\infty)$ **55.** radius $= R$

61. (a) $n = 5;\ s_5 \approx 1.1026$ (b) $\zeta(3.7) \approx 1.10629$

▶ **Exercise Set 9.9 (Page 676)**

3. 0.069756 **5.** 0.99500 **7.** 0.99619 **9.** 0.5208

11. (a) $\displaystyle\sum_{k=1}^{\infty} 2\frac{(1/9)^{2k-1}}{2k-1}$ (b) 0.223

13. (a) $0.4635;\ 0.3218$ (b) 3.1412 (c) no

15. (a) error $\leq 9 \times 10^{-8}$ (b)

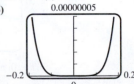

17. (a) $\displaystyle\sum_{k=0}^{\infty} (-1)^k x^k$ (b) $1 + \dfrac{x}{3} + \displaystyle\sum_{k=2}^{\infty} (-1)^{k-1} \frac{2 \cdot 5 \cdots (3k-4)}{3^k k!} x^k$

(c) $\displaystyle\sum_{k=0}^{\infty} (-1)^k \frac{(k+2)(k+1)}{2} x^k$ **23.** 23.406%

▶ **Exercise Set 9.10 (Page 686)**

1. (a) $1 - x + x^2 - \cdots + (-1)^k x^k + \cdots;\ R = 1$

(b) $1 + x^2 + x^4 + \cdots + x^{2k} + \cdots;\ R = 1$

(c) $1 + 2x + 4x^2 + \cdots + 2^k x^k + \cdots;\ R = \frac{1}{2}$

(d) $\dfrac{1}{2} + \dfrac{1}{2^2} x + \dfrac{1}{2^3} x^2 + \cdots + \dfrac{1}{2^{k+1}} x^k + \cdots;\ R = 2$

3. (a) $(2 + x)^{-1/2} = \dfrac{1}{2^{1/2}} - \dfrac{1}{2^{5/2}} x + \dfrac{1 \cdot 3}{2^{9/2} \cdot 2!} x^2 - \dfrac{1 \cdot 3 \cdot 5}{2^{13/2} \cdot 3!} x^3 + \cdots$

(b) $(1 - x^2)^{-2} = 1 + 2x^2 + 3x^4 + 4x^6 + \cdots$

5. (a) $2x - \dfrac{2^3}{3!} x^3 + \dfrac{2^5}{5!} x^5 - \dfrac{2^7}{7!} x^7 + \cdots;\ R = +\infty$

(b) $1 - 2x + 2x^2 - \dfrac{4}{3} x^3 + \cdots;\ R = +\infty$

(c) $1 + x^2 + \dfrac{1}{2!} x^4 + \dfrac{1}{3!} x^6 + \cdots;\ R = +\infty$

(d) $x^2 - \dfrac{\pi^2}{2} x^4 + \dfrac{\pi^4}{4!} x^6 - \dfrac{\pi^6}{6!} x^8 + \cdots;\ R = +\infty$

7. (a) $x^2 - 3x^3 + 9x^4 - 27x^5 + \cdots;\ R = \frac{1}{3}$

(b) $2x^2 + \dfrac{2^3}{3!} x^4 + \dfrac{2^5}{5!} x^6 + \dfrac{2^7}{7!} x^8 + \cdots;\ R = +\infty$

(c) $x - \dfrac{3}{2} x^3 + \dfrac{3}{8} x^5 + \dfrac{1}{16} x^7 + \cdots;\ R = 1$

9. (a) $x^2 - \dfrac{2^3}{4!} x^4 + \dfrac{2^5}{6!} x^6 - \dfrac{2^7}{8!} x^8 + \cdots$

(b) $12x^3 - 6x^6 + 4x^9 - 3x^{12} + \cdots$

11. (a) $1 - (x-1) + (x-1)^2 - \cdots + (-1)^k (x-1)^k + \cdots$ (b) $(0, 2)$

13. (a) $x + x^2 + \dfrac{x^3}{3} - \dfrac{x^5}{30} + \cdots$ (b) $x - \dfrac{x^3}{24} + \dfrac{x^4}{24} - \dfrac{71}{1920} x^5 + \cdots$

15. (a) $1 + \frac{1}{2} x^2 + \frac{5}{24} x^4 + \frac{61}{720} x^6 + \cdots$ (b) $x - x^2 + \frac{1}{3} x^3 - \frac{1}{30} x^5 + \cdots$

19. $2 - 4x + 2x^2 - 4x^3 + 2x^4 + \cdots$

25. $[-1, 1];\ [-1, 1)$ **27.** (a) $\displaystyle\sum_{k=0}^{\infty} x^{2k+1}$ (b) $f^{(5)}(0) = 5!,\ f^{(6)}(0) = 0$

(c) $f^{(n)}(0) = n! c_n = \begin{cases} n! & \text{if } n \text{ odd} \\ 0 & \text{if } n \text{ even} \end{cases}$

29. (a) 1 (b) $-\frac{1}{3}$ **31.** 0.3103 **33.** 0.200

35. (a) $\displaystyle\sum_{k=0}^{\infty} \frac{x^{4k}}{k!};\ R = +\infty$ **37.** (a) $3/4$ (b) $\ln(4/3)$

39. (a) $x - \dfrac{1}{6} x^3 + \dfrac{3}{40} x^5 - \dfrac{5}{112} x^7 + \cdots$

(b) $x + \displaystyle\sum_{k=1}^{\infty} (-1)^k \frac{1 \cdot 3 \cdot 5 \cdots (2k-1)}{2^k k! (2k+1)} x^{2k+1}$ (c) $R = 1$

41. (a) $y(t) = y_0 \displaystyle\sum_{k=0}^{\infty} \frac{(-1)^k (0.000121)^k t^k}{k!}$ (c) $0.9998790073 y_0$

43. $2\pi \sqrt{\dfrac{L}{g}} \left(1 + \dfrac{k^2}{4} + \dfrac{9k^4}{64}\right)$

▶ **Chapter 9 Review Exercises (Page 689)**

9. (a) true   (b) sometimes false   (c) sometimes false

(d) true   (e) sometimes false   (f) sometimes false

(g) false   (h) sometimes false   (i) true

(j) true   (k) sometimes false   (l) sometimes false

11. (a) $\left\{\dfrac{n+2}{(n+1)^2 - n^2}\right\}_{n=1}^{+\infty}$; converges, $\displaystyle\lim_{n \to +\infty} \frac{n+2}{(n+1)^2 - n^2} = \frac{1}{2}$

(b) $\left\{(-1)^{n+1} \dfrac{n}{2n+1}\right\}_{n=1}^{+\infty}$; diverges

15. (a) converges   (b) converges   **17.** (a) converges   (b) diverges

19. (a) diverges   (b) converges   **21.** $\dfrac{1}{4 \cdot 5^{99}}$

23. (a) 2 (b) diverges   (c) $3/4$ (d) $\pi/4$ **25.** $p > 1$

29. (a) $p_0(x) = 1,\ p_1(x) = 1 - 7x,\ p_2(x) = 1 - 7x + 5x^2$,

$p_3(x) = 1 - 7x + 5x^2 + 4x^3,\ p_4(x) = 1 - 7x + 5x^2 + 4x^3$

33. (a) $e^2 - 1$ (b) 0 (c) $\cos e$ (d) $\frac{1}{3}$

37. (a) $x - \frac{2}{3} x^3 + \frac{2}{15} x^5 - \frac{4}{315} x^7$ (b) $x - \frac{2}{3} x^3 + \frac{2}{15} x^5 - \frac{4}{315} x^7$

▶ **Chapter 9 Making Connections (Page 691)**

Where correct answers to a Making Connections exercise may vary, no answer is listed. Sample answers for these questions are available on the Book Companion Site.

1. (a) $\dfrac{a\sin\theta}{1-\cos\theta}$ **(b)** $a\csc\theta$ **(c)** $a\cot\theta$

2. (a) $A=1$, $B=-2$ **(b)** $s_n=2-\dfrac{2^{n+1}}{3^{n+1}-2^{n+1}}$; 2

4. (a) $124.58<d<124.77$ **(b)** $1243<s<1424$

6. (b) $v(t)\approx v_0-\left(\dfrac{cv_0}{m}+g\right)t+\dfrac{c^2}{2m^2}\left(v_0+\dfrac{mg}{c}\right)t^2$

▶ **Exercise Set 10.1 (Page 700)**

1. (a) $y=x+2\,(-1\le x\le 4)$ **3.**

(c)

t	0	1	2	3	4	5
x	-1	0	1	2	3	4
y	1	2	3	4	5	6

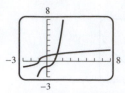

(d)

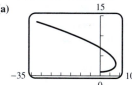

5.   **7.**

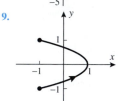   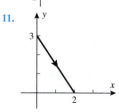

9.   **11.**

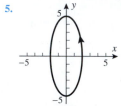   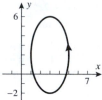

13. $x=5\cos t$, $y=-5\sin t\ (0\le t\le 2\pi)$ **15.** $x=2$, $y=t$

17. $x=t^2$, $y=t\ (-1\le t\le 1)$

19. (a)

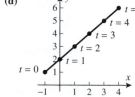

(b)

t	0	1	2	3	4	5
x	0	5.5	8	4.5	-8	-32.5
y	1	1.5	3	5.5	9	13.5

(c) $t=0,2\sqrt{3}$ **(d)** $0<t<2\sqrt{2}$ **(e)** $t=2$

21. (a)   **(b)**

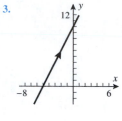

23. (a) IV   **(b)** II   **(c)** V   **(d)** VI   **(e)** III   **(f)** I   **25. (b)** $\frac{1}{2}$ **(c)** $\frac{3}{4}$

27. (b) from (x_0,y_0) to (x_1,y_1)

(c) $x=3-2(t-1)$, $y=-1+5(t-1)$

29.   **31.**

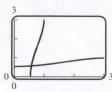

Responses to True–False questions may be abridged to save space.

33. False; $x=\sin t$, $y=\cos^2 t$ describe only the portion of the parabola $y=1-x^2$ with $-1\le x\le 1$.

35. True; $\dfrac{dy}{dx}=\dfrac{dy/dt}{dx/dt}=\dfrac{12t^3-6t^2}{x'(t)}$ **37.**

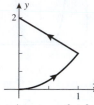

39. (a) $x=4\cos t$, $y=3\sin t$ **(b)** $x=-1+4\cos t$, $y=2+3\sin t$

41. $-4,4$ **43.** both are positive   **45.** $4,4$ **47.** $2/\sqrt{3},-1/(3\sqrt{3})$

49. $\sqrt{3},4$ **51.** $y=-e^{-2}x+2e^{-1}$

53. (a) $0,\pi,2\pi$ **(b)** $\pi/2,3\pi/2$

55. (a)   **(b)** $y=-2x$, $y=2x$

$a=1,b=2$ $a=2,b=3$

$a=3,b=4$ $a=4,b=5$

57. $y=2x-8$, $y=-2x+8$ **59.**

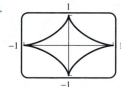

$t=0,\pi/2,\pi,3\pi/2,2\pi$

61. (a) $\dfrac{dy}{dx}=\dfrac{3\sin t}{1-3\cos t}$ **(b)** $\theta\approx -0.4345$

63. (a) ellipses with fixed center, varying axes of symmetry

(b) ellipses with varying center, fixed shape, size, and orientation

(c) circles of radius 1 with centers on line $y=x-1$

65. $\frac{1}{3}(5\sqrt{5}-8)$ **67.** 3π **69.** $\dfrac{\sqrt{10}}{2}(e^2-e^{-2})$

73. (b) $x=\cos t+\cos 2t$, $y=\sin t+\sin 2t$ **(c)** yes

75. $S=49\pi$ **77.** $S=\sqrt{2}\pi$

▶ **Exercise Set 10.2 (Page 716)** _____

1.

3. (a) $(3\sqrt{3}, 3)$
(b) $(-7/2, 7\sqrt{3}/2)$
(c) $(3\sqrt{3}, 3)$
(d) $(0, 0)$
(e) $(-7\sqrt{3}/2, 7/2)$
(f) $(-5, 0)$

5. (a) $(5, \pi), (5, -\pi)$ **(b)** $(4, 11\pi/6), (4, -\pi/6)$
(c) $(2, 3\pi/2), (2, -\pi/2)$ **(d)** $(8\sqrt{2}, 5\pi/4), (8\sqrt{2}, -3\pi/4)$
(e) $(6, 2\pi/3), (6, -4\pi/3)$ **(f)** $(\sqrt{2}, \pi/4), (\sqrt{2}, -7\pi/4)$

7. (a) $(5, 0.92730)$ **(b)** $(10, -0.92730)$ **(c)** $(1.27155, 2.47582)$

9. (a) circle   **(b)** line   **(c)** circle   **(d)** line

11. (a) $r = 3 \sec\theta$ **(b)** $r = \sqrt{7}$ **(c)** $r = -6 \sin\theta$
(d) $r^2 \cos\theta \sin\theta = 4/9$

13.

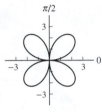

15.

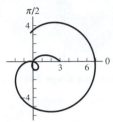

17. (a) $r = 5$ **(b)** $r = 6\cos\theta$ **(c)** $r = 1 - \cos\theta$
19. (a) $r = 3\sin 2\theta$ **(b)** $r = 3 + 2\sin\theta$ **(c)** $r^2 = 9\cos 2\theta$

21.

23.

25.

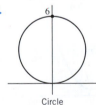

Circle

27.

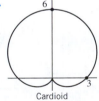

Cardioid

29.

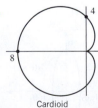

Cardioid

31.

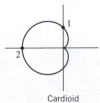

Cardioid

33.

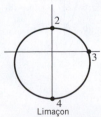

Limaçon

35.

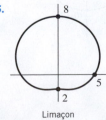

Limaçon

37.

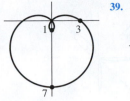

Limaçon

39.

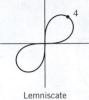

Lemniscate

41.

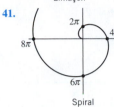

Spiral

43.

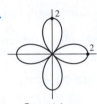

Four-petal rose

45.

Eight-petal rose

Responses to True–False questions may be abridged to save space.

47. True; $\left(-1, \dfrac{\pi}{3}\right)$ describes the same point as $\left(1, \dfrac{\pi}{3} + \pi\right)$, which describes the same point as $\left(1, \dfrac{\pi}{3} + \pi - 2\pi\right) = \left(1, -\dfrac{2\pi}{3}\right)$.

49. False; $-1 < \sin 2\theta < 0$ for $\pi/2 < \theta < \pi$, so this portion of the graph is in the fourth quadrant.

51. $0 \le \theta \le 4\pi$

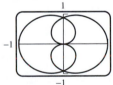

53. $0 \le \theta \le 8\pi$

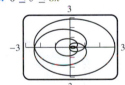

55. $0 \le \theta \le 5\pi$

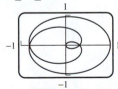

57. (a) $-4\pi < \theta < 4\pi$

61. (a)

(b)

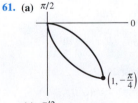

(c)

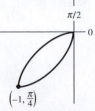

(d)

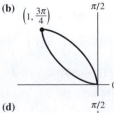

63. (a) $\pi/2$ **(b)**

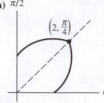

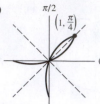

67. (a) $r = 1 + \dfrac{\sqrt{2}}{2}(\cos\theta + \sin\theta)$ **(b)** $r = 1 + \sin\theta$

   (c) $r = 1 - \cos\theta$ **(d)** $r = 1 - \dfrac{\sqrt{2}}{2}(\cos\theta + \sin\theta)$

69. $(3/2, \pi/3)$ **73.** $\sqrt{2}$

▶ **Exercise Set 10.3** (Page 726)

1. $\sqrt{3}$ **3.** $\dfrac{\tan 2 - 2}{2\tan 2 + 1}$ **5.** $1/2$ **7.** $1, 0, -1$

9. horizontal: $(3a/2, \pi/3), (0, \pi), (3a/2, 5\pi/3)$;
   vertical: $(2a, 0), (a/2, 2\pi/3), (a/2, 4\pi/3)$

11. $(0, 0), (\sqrt{2}/4, \pi/4), (\sqrt{2}/4, 3\pi/4)$

13.   **15.**

$\theta = \pi/2, \pm\pi/6$ $\theta = \pm\pi/4$

17.

$\theta = \pm\pi/3$

19. $L = 2\pi a$ **21.** $L = 8a$

23. (b) ≈ 2.42

(c)

n	2	3	4	5	6	7
L	2.42211	2.22748	2.14461	2.10100	2.07501	2.05816

n	8	9	10	11	12	13	14
L	2.04656	2.03821	2.03199	2.02721	2.02346	2.02046	2.01802

n	15	16	17	18	19	20
L	2.01600	2.01431	2.01288	2.01167	2.01062	2.00971

25. (a) $\displaystyle\int_{\pi/2}^{\pi} \tfrac{1}{2}(1 - \cos\theta)^2\, d\theta$ **(b)** $\displaystyle\int_{0}^{\pi/2} 2\cos^2\theta\, d\theta$

   (c) $\displaystyle\int_{0}^{\pi/2} \tfrac{1}{2}\sin^2 2\theta\, d\theta$ **(d)** $\displaystyle\int_{0}^{2\pi} \tfrac{1}{2}\theta^2\, d\theta$

   (e) $\displaystyle\int_{-\pi/2}^{\pi/2} \tfrac{1}{2}(1 - \sin\theta)^2\, d\theta$ **(f)** $\displaystyle\int_{0}^{\pi/4} \cos^2 2\theta\, d\theta$

27. (a) πa^2 **(b)** πa^2 **29.** 6π **31.** 4π **33.** $\pi - 3\sqrt{3}/2$

35. $\pi/2 - \tfrac{1}{4}$ **37.** $10\pi/3 - 4\sqrt{3}$ **39.** π **41.** $9\sqrt{3}/2 - \pi$

43. $(\pi + 3\sqrt{3})/4$ **45.** $\pi - 2$

Responses to True–False questions may be abridged to save space.

47. True; apply Theorem 10.3.1: $\cos\dfrac{\theta}{2}\Big|_{\theta=3\pi} = 0$ and $\dfrac{dr}{d\theta}\Big|_{\theta=3\pi} = \dfrac{1}{2} \neq 0$, so the line $\theta = 3\pi$ (the x-axis) is tangent to the curve at the origin.

49. False; the area of the sector is $\dfrac{\theta}{2\pi}\cdot\pi r^2 = \dfrac{1}{2}\theta r^2$.

51. (b) a^2 **(c)** $2\sqrt{3} - \dfrac{2\pi}{3}$ **53.** $8\pi^3 a^2$

59. $\pi/16$ 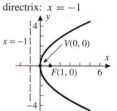   **65.** π^2

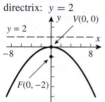

67. $32\pi/5$

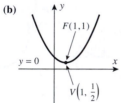

▶ **Exercise Set 10.4** (Page 744)

1. (a) $x = y^2$ **(b)** $-3y = x^2$ **(c)** $\dfrac{x^2}{9} + \dfrac{y^2}{4} = 1$ **(d)** $\dfrac{x^2}{4} + \dfrac{y^2}{9} = 1$

   (e) $y^2 - x^2 = 1$ **(f)** $\dfrac{x^2}{4} - \dfrac{y^2}{4} = 1$

3. (a) focus: $(1, 0)$;   **(b)** focus: $(0, -2)$;
   vertex: $(0, 0)$;   vertex: $(0, 0)$;
   directrix: $x = -1$ directrix: $y = 2$

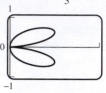

5. (a)   **(b)**

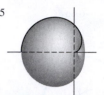

7. (a)   **(b)**

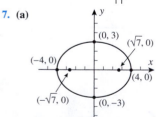

9. (a) $\dfrac{(x+3)^2}{16} + \dfrac{(y-5)^2}{4} = 1$

$c^2 = 16 - 4 = 12, c = 2\sqrt{3}$

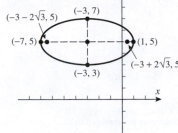

(b) $\dfrac{x^2}{4} + \dfrac{(y+2)^2}{9} = 1$

$c^2 = 9 - 4 = 5, c = 2\sqrt{5}$

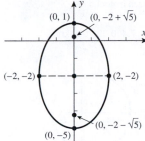

11. (a) vertices: $(\pm 4, 0)$;
foci: $(\pm 5, 0)$;
asymptotes: $y = \pm 3x/4$

(b) vertices: $(0, \pm 2)$;
foci: $(0, \pm 2\sqrt{10})$;
asymptotes: $y = \pm x/3$

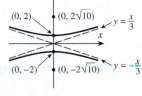

13. (a) $c^2 = 3 + 5 = 8, c = 2\sqrt{2}$

$y + 4 = \sqrt{\dfrac{3}{5}}(x-2)$

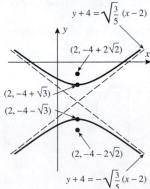

(b) $\dfrac{(x+1)^2}{1} - \dfrac{(y-3)^2}{2} = 1$

$c^2 = 1 + 2 = 3, c = \sqrt{3}$

$y - 3 = -\sqrt{2}(x+1)$

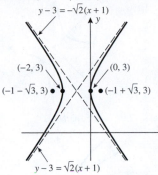

$y - 3 = \sqrt{2}(x+1)$

15. (a) $y^2 = 12x$ **(b)** $x^2 = -y$ **17.** $y^2 = 2(x-1)$

19. (a) $\dfrac{x^2}{9} + \dfrac{y^2}{4} = 1$ **(b)** $\dfrac{x^2}{16} + \dfrac{y^2}{25} = 1$

21. (a) $\dfrac{x^2}{81/8} + \dfrac{y^2}{36} = 1$ **(b)** $\dfrac{(x+1)^2}{4} + \dfrac{(y-2)^2}{5} = 1$

23. (a) $\dfrac{x^2}{4} - \dfrac{y^2}{5} = 1$ **(b)** $\dfrac{y^2}{4} - \dfrac{x^2}{9} = 1$

25. (a) $\dfrac{y^2}{9} - \dfrac{x^2}{16} = 1, \dfrac{x^2}{16} - \dfrac{y^2}{9} = 1$ **(b)** $\dfrac{x^2}{9/5} - \dfrac{y^2}{36/5} = 1$

Responses to True–False questions may be abridged to save space.

27. False; the description matches a parabola.

29. False; the distance from the parabola's focus to its directrix is $2p$; see Figure 10.4.6.

31. (a) 16 ft   **(b)** $8\sqrt{3}$ ft   **35.** $\frac{1}{16}$ ft

39. $\frac{1}{32}(x-4)^2 + \frac{1}{36}(y-3)^2 = 1$

43. $L = D\sqrt{1+p^2}, T = \frac{1}{2}pD$ **45.** $(64.612, 200)$

47. (a) $V = \dfrac{\pi b^2}{3a^2}(b^2 - 2a^2)\sqrt{a^2 + b^2} + \dfrac{2}{3}ab^2\pi$

(b) $V = \dfrac{2b^4}{3a}\pi$

53. (a) $(x-1)^2 - 5(y+1)^2 = 5$, hyperbola
(b) $x^2 - 3(y+1)^2 = 0, x = \pm\sqrt{3}(y+1)$, two lines
(c) $4(x+2)^2 + 8(y+1)^2 = 4$, ellipse
(d) $3(x+2)^2 + (y+1)^2 = 0$, the point $(-2, -1)$ (degenerate case)
(e) $(x+4)^2 + 2y = 2$, parabola
(f) $5(x+4)^2 + 2y = -14$, parabola

▶ **Exercise Set 10.5 (Page 753)**

1. (a) $x' = -1 + 3\sqrt{3}$, $y' = 3 + \sqrt{3}$ **3.** $y'^2 - x'^2 = 18$, hyperbola
(b) $3x'^2 - y'^2 = 12$
(c)

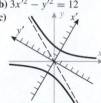
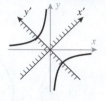

5. $\frac{1}{3}x'^2 - \frac{1}{2}y'^2 = 1$, hyperbola **7.** $y' = x'^2$, parabola

9. $y'^2 = 4(x' - 1)$, parabola **11.** $\frac{1}{4}(x' + 1)^2 + y'^2 = 1$, ellipse

13. $x^2 + xy + y^2 = 3$
19. vertex: $(0, 0)$; focus: $(-1/\sqrt{2}, 1/\sqrt{2})$; directrix: $y = x - \sqrt{2}$
21. vertex: $(4/5, 3/5)$; focus: $(8/5, 6/5)$; directrix: $4x + 3y = 0$
23. foci: $\pm(4\sqrt{7}/5, 3\sqrt{7}/5)$; vertices: $\pm(16/5, 12/5)$;
ends: $\pm(-9/5, 12/5)$
25. foci: $(1 - \sqrt{5}/2, -\sqrt{3} + \sqrt{15}/2)$, $(1 + \sqrt{5}/2, -\sqrt{3} - \sqrt{15}/2)$;
vertices: $(-1/2, \sqrt{3}/2)$, $(5/2, -5\sqrt{3}/2)$;
ends: $(1 + \sqrt{3}, 1 - \sqrt{3})$, $(1 - \sqrt{3}, -1 - \sqrt{3})$
27. foci: $\pm(\sqrt{15}, \sqrt{5})$; vertices: $\pm(2\sqrt{3}, 2)$;
asymptotes: $y = \dfrac{5\sqrt{3} \pm 8}{11}x$
29. foci: $\left(-\dfrac{4}{\sqrt{5}} \pm 2\sqrt{\dfrac{13}{5}}, \dfrac{8}{\sqrt{5}} \pm \sqrt{\dfrac{13}{5}}\right)$;
vertices: $(2/\sqrt{5}, 11/\sqrt{5})$, $(-2\sqrt{5}, \sqrt{5})$;
asymptotes: $y = 7x/4 + 3\sqrt{5}$, $y = -x/8 + 3\sqrt{5}/2$

▶ **Exercise Set 10.6 (Page 761)**

1. (a) $e = 1, d = \frac{3}{2}$ **(b)** $e = \frac{1}{2}, d = 3$

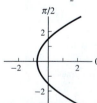
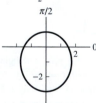

3. (a) parabola, opens up **(b)** ellipse, directrix above the pole

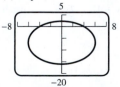

5. (a) $r = \dfrac{6}{4 + 3\cos\theta}$ **(b)** $r = \dfrac{1}{1 + \cos\theta}$ **(c)** $r = \dfrac{12}{3 + 4\sin\theta}$
7. (a) $r_0 = 2, r_1 = 6$; $\frac{1}{12}x^2 + \frac{1}{16}(y + 2)^2 = 1$
(b) $r_0 = \frac{1}{3}, r_1 = 1$; $\frac{9}{4}\left(x - \frac{1}{3}\right)^2 + 3y^2 = 1$
9. (a) $r_0 = 1, r_1 = 3$; $(y - 2)^2 - \dfrac{x^2}{3} = 1$
(b) $r_0 = 1, r_1 = 5$; $\dfrac{(x + 3)^2}{4} - \dfrac{y^2}{5} = 1$
11. (a) $r = \dfrac{12}{2 + \cos\theta}$ **(b)** $r = \dfrac{64}{25 - 15\sin\theta}$
13. $r = \dfrac{5\sqrt{2} + 5}{1 + \sqrt{2}\cos\theta}$ or $r = \dfrac{5\sqrt{2} - 5}{1 + \sqrt{2}\cos\theta}$
Responses to True–False questions may be abridged to save space.
19. True; the eccentricity e of an ellipse satisfies $0 < e < 1$ (Theorem 10.6.1).
21. False; eccentricity correlates to the "flatness" of an ellipse, which is independent of the distance between its foci.
23. (a) $T \approx 248$ yr
(b) $r_0 \approx 4{,}449{,}675{,}000$ km, $r_1 \approx 7{,}400{,}325{,}000$ km
(c) $r \approx \dfrac{37.05}{1 + 0.249\cos\theta}$ AU **(d)**

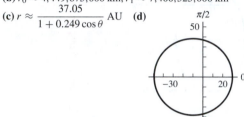

25. (a) $a \approx 178.26$ AU **(d)**
(b) $r_0 \approx 0.8735$ AU,
$r_1 \approx 355.64$ AU
(c) $r \approx \dfrac{1.74}{1 + 0.9951\cos\theta}$ AU

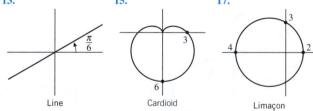

27. 563 km, 4286 km

▶ **Chapter 10 Review Exercises (Page 763)**

1. $x = \sqrt{2}\cos t$, $y = -\sqrt{2}\sin t$ $(0 \le t \le 3\pi/2)$ **3. (a)** $-1/4, 1/4$
5. (a) $t = \pi/2 + n\pi$ for $n = 0, \pm 1, \ldots$ **(b)** $t = n\pi$ for $n = 0, \pm 1, \ldots$
7. (a) $(-4\sqrt{2}, -4\sqrt{2})$ **(b)** $(7/\sqrt{2}, -7/\sqrt{2})$ **(c)** $(4\sqrt{2}, 4\sqrt{2})$
(d) $(5, 0)$ **(e)** $(0, -2)$ **(f)** $(0, 0)$
9. (a) $(5, 0.6435)$ **(b)** $(\sqrt{29}, 5.0929)$ **(c)** $(1.2716, 0.6658)$
11. (a) parabola **(b)** hyperbola **(c)** line **(d)** circle
13. **15.** **17.**

Line   Cardioid   Limaçon

19. (a) $-2, 1/4$ **(b)** $-3\sqrt{3}/4, 3\sqrt{3}/4$
21. (a) The top is traced from right to left as t goes from 0 to π. The bottom is traced from right to left as t goes from π to 2π, except for the loop, which is traced counterclockwise as t goes from $\pi + \sin^{-1}(1/4)$ to $2\pi - \sin^{-1}(1/4)$. **(b)** $y = 1$
(c) horizontal: $t = \pi/2, 3\pi/2$; vertical: $t = \pi + \sin^{-1}(1/\sqrt[3]{4})$, $2\pi - \sin^{-1}(1/\sqrt[3]{4})$
(d) $r = 4 + \csc\theta, \theta = \pi + \sin^{-1}(1/4), \theta = 2\pi - \sin^{-1}(1/4)$

23. $A = 6\pi$ **25.** $A = \dfrac{5\pi}{12} - \dfrac{\sqrt{3}}{2}$ **27.**

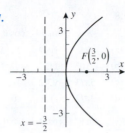

29.

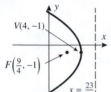

focus: $(9/4, -1)$;
vertex: $(4, -1)$;
directrix: $x = 23/4$

31.

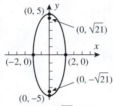

foci: $(0, \pm\sqrt{21})$;
vertices: $(0, \pm 5)$;
ends: $(\pm 2, 0)$

33.

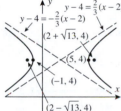

35.

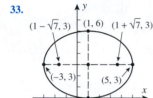

37.

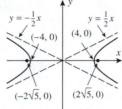

39. $x^2 = -16y$ **41.** $y^2 - x^2 = 9$

43. **(b)** $x = \dfrac{v_0^2}{g}\sin\alpha\cos\alpha$; $y = y_0 + \dfrac{v_0^2 \sin^2\alpha}{2g}$

45. $\theta = \pi/4$; $5(y')^2 - (x')^2 = 6$; hyperbola

47. $\theta = \tan^{-1}(1/2)$; $y' = (x')^2$; parabola

49. **(a)** (i) ellipse; (ii) right; (iii) 1 **(b)** (i) hyperbola (ii) left;
(iii) 1/3 **(c)** (i) parabola; (ii) above; (iii) 1/3 **(d)** (i) parabola;
(ii) below; (iii) 3

51. **(a)** $\dfrac{(x+3)^2}{25} + \dfrac{(y-2)^2}{9} = 1$ **(b)** $(x+2)^2 = -8y$

  (c) $\dfrac{(y-5)^2}{4} - 16(x+1)^2 = 1$

53. 15.86543959

▶ **Chapter 10 Making Connections** (Page 766)

Where correct answers to a Making Connections exercise may vary, no
answer is listed. Sample answers for these questions are available on the
Book Companion Site.

1. **(a)**

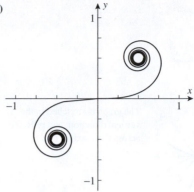

(c) $L = \displaystyle\int_{-1}^{1}\left[\cos^2\left(\dfrac{\pi t^2}{2}\right) + \sin^2\left(\dfrac{\pi t^2}{2}\right)\right] dt = 2$

2. **(a)** P: $(b\cos t, b\sin t)$;
  Q: $(a\cos t, a\sin t)$;
  R: $(a\cos t, b\sin t)$

▶ **Appendix A** (Page A1)

1. (e) **3.** (b), (c) **5.** $[-3, 3] \times [0, 5]$

7. $[-5, 14] \times [-60, 40]$ **9.** $[-0.1, 0.1] \times [-3, 3]$

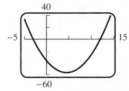

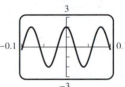

11. $[-400, 1050] \times [-1500000, 10000]$ **13.** $[-2, 2] \times [-20, 20]$

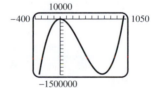

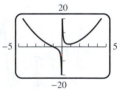

17. **(a)** $f(x) = \sqrt{16 - x^2}$ **(b)** $f(x) = -\sqrt{16 - x^2}$ **(e)** no

19. **(a)**

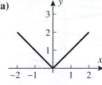

(b)

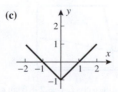

(c)

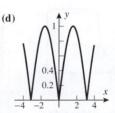

(d)

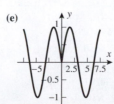

(e)

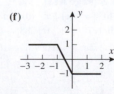

(f)

21.

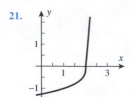

25. (a)

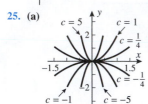

The graph is stretched in the vertical direction, and reflected across the x-axis if $c < 0$.

(b)

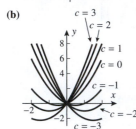

The graph is translated so its vertex is on the parabola $y = -x^2$.

(c)

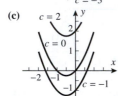

The graph is translated vertically.

27.

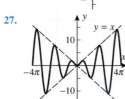

29. (a)

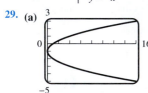

(b)

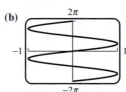

31.

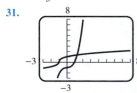

33.

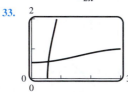

35. (a) $x = 4\cos t$, $y = 3\sin t$ **(b)** $x = -1 + 4\cos t$, $y = 2 + 3\sin t$

▶ **Appendix B (Page A13)**

1. (a) $\frac{5}{12}\pi$ **(b)** $\frac{13}{6}\pi$ **(c)** $\frac{1}{9}\pi$ **(d)** $\frac{23}{30}\pi$
3. (a) $12°$ **(b)** $(270/\pi)°$ **(c)** $288°$ **(d)** $540°$
5.

	$\sin\theta$	$\cos\theta$	$\tan\theta$	$\csc\theta$	$\sec\theta$	$\cot\theta$
(a)	$\sqrt{21}/5$	$2/5$	$\sqrt{21}/2$	$5/\sqrt{21}$	$5/2$	$2/\sqrt{21}$
(b)	$3/4$	$\sqrt{7}/4$	$3/\sqrt{7}$	$4/3$	$4/\sqrt{7}$	$\sqrt{7}/3$
(c)	$3/\sqrt{10}$	$1/\sqrt{10}$	3	$\sqrt{10}/3$	$\sqrt{10}$	$1/3$

7. $\sin\theta = 3/\sqrt{10}$, $\cos\theta = 1/\sqrt{10}$ **9.** $\tan\theta = \sqrt{21}/2$, $\csc\theta = 5/\sqrt{21}$

11. 1.8
13.

	θ	$\sin\theta$	$\cos\theta$	$\tan\theta$	$\csc\theta$	$\sec\theta$	$\cot\theta$
(a)	$225°$	$-1/\sqrt{2}$	$-1/\sqrt{2}$	1	$-\sqrt{2}$	$-\sqrt{2}$	1
(b)	$-210°$	$1/2$	$-\sqrt{3}/2$	$-1/\sqrt{3}$	2	$-2/\sqrt{3}$	$-\sqrt{3}$
(c)	$5\pi/3$	$-\sqrt{3}/2$	$1/2$	$-\sqrt{3}$	$-2/\sqrt{3}$	2	$-1/\sqrt{3}$
(d)	$-3\pi/2$	1	0	—	1	—	0

15.

	$\sin\theta$	$\cos\theta$	$\tan\theta$	$\csc\theta$	$\sec\theta$	$\cot\theta$
(a)	$4/5$	$3/5$	$4/3$	$5/4$	$5/3$	$3/4$
(b)	$-4/5$	$3/5$	$-4/3$	$-5/4$	$5/3$	$-3/4$
(c)	$1/2$	$-\sqrt{3}/2$	$-1/\sqrt{3}$	2	$-2/\sqrt{3}$	$-\sqrt{3}$
(d)	$-1/2$	$\sqrt{3}/2$	$-1/\sqrt{3}$	-2	$2/\sqrt{3}$	$-\sqrt{3}$
(e)	$1/\sqrt{2}$	$1/\sqrt{2}$	1	$\sqrt{2}$	$\sqrt{2}$	1
(f)	$1/\sqrt{2}$	$-1/\sqrt{2}$	-1	$\sqrt{2}$	$-\sqrt{2}$	-1

17. (a) 1.2679 **(b)** 3.5753
19.

	$\sin\theta$	$\cos\theta$	$\tan\theta$	$\csc\theta$	$\sec\theta$	$\cot\theta$
(a)	$a/3$	$\sqrt{9-a^2}/3$	$a/\sqrt{9-a^2}$	$3/a$	$3/\sqrt{9-a^2}$	$\sqrt{9-a^2}/a$
(b)	$a/\sqrt{a^2+25}$	$5/\sqrt{a^2+25}$	$a/5$	$\sqrt{a^2+25}/a$	$\sqrt{a^2+25}/5$	$5/a$
(c)	$\sqrt{a^2-1}/a$	$1/a$	$\sqrt{a^2-1}$	$a/\sqrt{a^2-1}$	a	$1/\sqrt{a^2-1}$

21. (a) $3\pi/4 \pm n\pi$, $n = 0, 1, 2, \ldots$
 (b) $\pi/3 \pm 2n\pi$ and $5\pi/3 \pm 2n\pi$, $n = 0, 1, 2, \ldots$
23. (a) $\pi/6 \pm n\pi$, $n = 0, 1, 2, \ldots$
 (b) $4\pi/3 \pm 2n\pi$ and $5\pi/3 \pm 2n\pi$, $n = 0, 1, 2, \ldots$
25. (a) $3\pi/4 \pm n\pi$, $n = 0, 1, 2, \ldots$
 (b) $\pi/6 \pm n\pi$, $n = 0, 1, 2, \ldots$
27. (a) $\pi/3 \pm 2n\pi$ and $2\pi/3 \pm 2n\pi$, $n = 0, 1, 2, \ldots$
 (b) $\pi/6 \pm 2n\pi$ and $11\pi/6 \pm 2n\pi$, $n = 0, 1, 2, \ldots$
29. $\sin\theta = 2/5$, $\cos\theta = -\sqrt{21}/5$, $\tan\theta = -2/\sqrt{21}$, $\csc\theta = 5/2$, $\sec\theta = -5/\sqrt{21}$, $\cot\theta = -\sqrt{21}/2$
31. (a) $\theta = \pm n\pi$, $n = 0, 1, 2, \ldots$ **(b)** $\theta = \pi/2 \pm n\pi$, $n = 0, 1, 2, \ldots$
 (c) $\theta = \pm n\pi$, $n = 0, 1, 2, \ldots$ **(d)** $\theta = \pm n\pi$, $n = 0, 1, 2, \ldots$
 (e) $\theta = \pi/2 \pm n\pi$, $n = 0, 1, 2, \ldots$ **(f)** $\theta = \pm n\pi$, $n = 0, 1, 2, \ldots$
33. (a) $2\pi/3$ cm **(b)** $10\pi/3$ cm **35.** $\frac{2}{5}$
37. (a) $\dfrac{2\pi - \theta}{2\pi} R$ **(b)** $\dfrac{\sqrt{4\pi\theta - \theta^2}}{2\pi} R$ **39.** $\frac{21}{4}\sqrt{3}$ **41.** 9.2 ft
43. $h = d(\tan\beta - \tan\alpha)$ **45. (a)** $4\sqrt{5}/9$ **(b)** $-\frac{1}{9}$
47. $\sin 3\theta = 3\sin\theta \cos^2\theta - \sin^3\theta$, $\cos 3\theta = \cos^3\theta - 3\sin^2\theta \cos\theta$
61. (a) $\cos\theta$ **(b)** $-\sin\theta$ **(c)** $-\cos\theta$ **(d)** $\sin\theta$
69. (a) $153°$ **(b)** $45°$ **(c)** $117°$ **(d)** $89°$ **71. (a)** $60°$ **(b)** $117°$

▶ **Appendix C (Page A27)**

1. (a) $q(x) = x^2 + 4x + 2$, $r(x) = -11x + 6$
 (b) $q(x) = 2x^2 + 4$, $r(x) = 9$
 (c) $q(x) = x^3 - x^2 + 2x - 2$, $r(x) = 2x + 1$
3. (a) $q(x) = 3x^2 + 6x + 8$, $r(x) = 15$
 (b) $q(x) = x^3 - 5x^2 + 20x - 100$, $r(x) = 504$
 (c) $q(x) = x^4 + x^3 + x^2 + x + 1$, $r(x) = 0$
5.

x	0	1	-3	7
$p(x)$	-4	-3	101	5001

7. (a) $q(x) = x^2 + 6x + 13$, $r = 20$ **(b)** $q(x) = x^2 + 3x - 2$, $r = -4$
9. (a) $\pm 1, \pm 2, \pm 3, \pm 4, \pm 6, \pm 8, \pm 12, \pm 24$
 (b) $\pm 1, \pm 2, \pm 5, \pm 10, \pm\frac{1}{3}, \pm\frac{2}{3}, \pm\frac{5}{3}, \pm\frac{10}{3}$ **(c)** $\pm 1, \pm 17$
11. $(x+1)(x-1)(x-2)$ **13.** $(x+3)^3(x+1)$
15. $(x+3)(x+2)(x+1)^2(x-3)$ **17.** -3 **19.** $-2, -\frac{2}{3}, -1 \pm \sqrt{3}$
21. $-2, 2, 3$ **23.** $2, 5$ **25.** 7 cm

INDEX

RATIONAL FUNCTIONS CONTAINING POWERS OF $a + bu$ IN THE DENOMINATOR

60. $\displaystyle\int \frac{u\,du}{a+bu} = \frac{1}{b^2}[bu - a\ln|a+bu|] + C$

64. $\displaystyle\int \frac{u\,du}{(a+bu)^3} = \frac{1}{b^2}\left[\frac{a}{2(a+bu)^2} - \frac{1}{a+bu}\right] + C$

61. $\displaystyle\int \frac{u^2\,du}{a+bu} = \frac{1}{b^3}\left[\frac{1}{2}(a+bu)^2 - 2a(a+bu) + a^2\ln|a+bu|\right] + C$

65. $\displaystyle\int \frac{du}{u(a+bu)} = \frac{1}{a}\ln\left|\frac{u}{a+bu}\right| + C$

62. $\displaystyle\int \frac{u\,du}{(a+bu)^2} = \frac{1}{b^2}\left[\frac{a}{a+bu} + \ln|a+bu|\right] + C$

66. $\displaystyle\int \frac{du}{u^2(a+bu)} = -\frac{1}{au} + \frac{b}{a^2}\ln\left|\frac{a+bu}{u}\right| + C$

63. $\displaystyle\int \frac{u^2\,du}{(a+bu)^2} = \frac{1}{b^3}\left[bu - \frac{a^2}{a+bu} - 2a\ln|a+bu|\right] + C$

67. $\displaystyle\int \frac{du}{u(a+bu)^2} = \frac{1}{a(a+bu)} + \frac{1}{a^2}\ln\left|\frac{u}{a+bu}\right| + C$

RATIONAL FUNCTIONS CONTAINING $a^2 \pm u^2$ IN THE DENOMINATOR ($a > 0$)

68. $\displaystyle\int \frac{du}{a^2+u^2} = \frac{1}{a}\tan^{-1}\frac{u}{a} + C$

70. $\displaystyle\int \frac{du}{u^2-a^2} = \frac{1}{2a}\ln\left|\frac{u-a}{u+a}\right| + C$

69. $\displaystyle\int \frac{du}{a^2-u^2} = \frac{1}{2a}\ln\left|\frac{u+a}{u-a}\right| + C$

71. $\displaystyle\int \frac{bu+c}{a^2+u^2}\,du = \frac{b}{2}\ln(a^2+u^2) + \frac{c}{a}\tan^{-1}\frac{u}{a} + C$

INTEGRALS OF $\sqrt{a^2+u^2}$, $\sqrt{a^2-u^2}$, $\sqrt{u^2-a^2}$ AND THEIR RECIPROCALS ($a > 0$)

72. $\displaystyle\int \sqrt{u^2+a^2}\,du = \frac{u}{2}\sqrt{u^2+a^2} + \frac{a^2}{2}\ln(u+\sqrt{u^2+a^2}) + C$

75. $\displaystyle\int \frac{du}{\sqrt{u^2+a^2}} = \ln(u+\sqrt{u^2+a^2}) + C$

73. $\displaystyle\int \sqrt{u^2-a^2}\,du = \frac{u}{2}\sqrt{u^2-a^2} - \frac{a^2}{2}\ln|u+\sqrt{u^2-a^2}| + C$

76. $\displaystyle\int \frac{du}{\sqrt{u^2-a^2}} = \ln|u+\sqrt{u^2-a^2}| + C$

74. $\displaystyle\int \sqrt{a^2-u^2}\,du = \frac{u}{2}\sqrt{a^2-u^2} + \frac{a^2}{2}\sin^{-1}\frac{u}{a} + C$

77. $\displaystyle\int \frac{du}{\sqrt{a^2-u^2}} = \sin^{-1}\frac{u}{a} + C$

POWERS OF u MULTIPLYING OR DIVIDING $\sqrt{a^2-u^2}$ OR ITS RECIPROCAL

78. $\displaystyle\int u^2\sqrt{a^2-u^2}\,du = \frac{u}{8}(2u^2-a^2)\sqrt{a^2-u^2} + \frac{a^4}{8}\sin^{-1}\frac{u}{a} + C$

81. $\displaystyle\int \frac{u^2\,du}{\sqrt{a^2-u^2}} = -\frac{u}{2}\sqrt{a^2-u^2} + \frac{a^2}{2}\sin^{-1}\frac{u}{a} + C$

79. $\displaystyle\int \frac{\sqrt{a^2-u^2}\,du}{u} = \sqrt{a^2-u^2} - a\ln\left|\frac{a+\sqrt{a^2-u^2}}{u}\right| + C$

82. $\displaystyle\int \frac{du}{u\sqrt{a^2-u^2}} = -\frac{1}{a}\ln\left|\frac{a+\sqrt{a^2-u^2}}{u}\right| + C$

80. $\displaystyle\int \frac{\sqrt{a^2-u^2}\,du}{u^2} = -\frac{\sqrt{a^2-u^2}}{u} - \sin^{-1}\frac{u}{a} + C$

83. $\displaystyle\int \frac{du}{u^2\sqrt{a^2-u^2}} = -\frac{\sqrt{a^2-u^2}}{a^2u} + C$

POWERS OF u MULTIPLYING OR DIVIDING $\sqrt{u^2\pm a^2}$ OR THEIR RECIPROCALS

84. $\displaystyle\int u\sqrt{u^2+a^2}\,du = \frac{1}{3}(u^2+a^2)^{3/2} + C$

90. $\displaystyle\int \frac{du}{u^2\sqrt{u^2\pm a^2}} = \mp\frac{\sqrt{u^2\pm a^2}}{a^2u} + C$

85. $\displaystyle\int u\sqrt{u^2-a^2}\,du = \frac{1}{3}(u^2-a^2)^{3/2} + C$

91. $\displaystyle\int u^2\sqrt{u^2+a^2}\,du = \frac{u}{8}(2u^2+a^2)\sqrt{u^2+a^2} - \frac{a^4}{8}\ln(u+\sqrt{u^2+a^2}) + C$

86. $\displaystyle\int \frac{du}{u\sqrt{u^2+a^2}} = -\frac{1}{a}\ln\left|\frac{a+\sqrt{u^2+a^2}}{u}\right| + C$

92. $\displaystyle\int u^2\sqrt{u^2-a^2}\,du = \frac{u}{8}(2u^2-a^2)\sqrt{u^2-a^2} - \frac{a^4}{8}\ln|u+\sqrt{u^2-a^2}| + C$

87. $\displaystyle\int \frac{du}{u\sqrt{u^2-a^2}} = \frac{1}{a}\sec^{-1}\left|\frac{u}{a}\right| + C$

93. $\displaystyle\int \frac{\sqrt{u^2+a^2}}{u^2}\,du = -\frac{\sqrt{u^2+a^2}}{u} + \ln(u+\sqrt{u^2+a^2}) + C$

94. $\displaystyle\int \frac{\sqrt{u^2-a^2}}{u^2}\,du = -\frac{\sqrt{u^2-a^2}}{u} + \ln|u+\sqrt{u^2-a^2}| + C$

88. $\displaystyle\int \frac{\sqrt{u^2-a^2}\,du}{u} = \sqrt{u^2-a^2} - a\sec^{-1}\left|\frac{u}{a}\right| + C$

95. $\displaystyle\int \frac{u^2}{\sqrt{u^2+a^2}}\,du = \frac{u}{2}\sqrt{u^2+a^2} - \frac{a^2}{2}\ln(u+\sqrt{u^2+a^2}) + C$

89. $\displaystyle\int \frac{\sqrt{u^2+a^2}\,du}{u} = \sqrt{u^2+a^2} - a\ln\left|\frac{a+\sqrt{u^2+a^2}}{u}\right| + C$

96. $\displaystyle\int \frac{u^2}{\sqrt{u^2-a^2}}\,du = \frac{u}{2}\sqrt{u^2-a^2} + \frac{a^2}{2}\ln|u+\sqrt{u^2-a^2}| + C$

INTEGRALS CONTAINING $(a^2+u^2)^{3/2}$, $(a^2-u^2)^{3/2}$, $(u^2-a^2)^{3/2}$ ($a > 0$)

97. $\displaystyle\int \frac{du}{(a^2-u^2)^{3/2}} = \frac{u}{a^2\sqrt{a^2-u^2}} + C$

100. $\displaystyle\int (u^2+a^2)^{3/2}\,du = \frac{u}{8}(2u^2+5a^2)\sqrt{u^2+a^2} + \frac{3a^4}{8}\ln(u+\sqrt{u^2+a^2}) + C$

98. $\displaystyle\int \frac{du}{(u^2\pm a^2)^{3/2}} = \pm\frac{u}{a^2\sqrt{u^2\pm a^2}} + C$

101. $\displaystyle\int (u^2-a^2)^{3/2}\,du = \frac{u}{8}(2u^2-5a^2)\sqrt{u^2-a^2} + \frac{3a^4}{8}\ln|u+\sqrt{u^2-a^2}| + C$

99. $\displaystyle\int (a^2-u^2)^{3/2}\,du = -\frac{u}{8}(2u^2-5a^2)\sqrt{a^2-u^2} + \frac{3a^4}{8}\sin^{-1}\frac{u}{a} + C$